ALTERNATIVE ENERGY AND SHALE GAS ENCYCLOPEDIA

ALTERNATIVE ENERGY AND SHALE GAS ENCYCLOPEDIA

Edited by

JAY H. LEHR
Editor-in-Chief

JACK KEELEY
Senior Editor

THOMAS B. KINGERY
Information Technology

WILEY SERIES ON ENERGY

WILEY

Published by John Wiley & Sons, Inc., Hoboken, New Jersey.
Published simultaneously in Canada.

For general information on our other products and services or for technical support, please contact our Customer Care Department within the United States at (800) 762-2974, outside the United States at (317) 572-3993 or fax (317) 572-4002.

Wiley also publishes its books in a variety of electronic formats. Some content that appears in print may not be available in electronic formats. For more information about Wiley products, visit our web site at www.wiley.com.

Library of Congress Cataloging-in-Publication Data:

Alternative energy and shale gas encyclopedia / edited by Jay H. Lehr, editor-in-chief ; Jack Keeley, senior editor ; Thomas B Kingery, Information Technology.
  pages cm
  Includes index.
  ISBN 978-0-470-89441-5 (cloth)
  1. Renewable energy sources–Encyclopedias.   2. Shale gas–Encyclopedias.   I. Lehr, Jay H., 1936- editor.
II. Keeley, J. W. (Jack W.), editor.   III. Kingery, Thomas B., editor.
  TJ807.4.E53 2015
  621.04203–dc23

                                                                                    2014049361

Printed in the United States of America.

10 9 8 7 6 5 4 3 2 1

CONTENTS

INTRODUCTION: ENERGY DRIVES *EVERYTHING*

Howard C. Hayden

INTRODUCTION

Everything we make, bend, heat, cool, cut, fasten, grow, harvest, move, or shape requires energy. That is, when we do *anything* to *anything*, we use energy. If we do it by hand, the energy source is the sun which produces the food we eat. The human labor part of the energy picture, however, is minuscule. Let me elaborate. The best coupling between man and machine is to put a person on a bicycle seat to use the strong leg muscles to push the pedals (which in turn might turn an electric generator). Whereas a good athlete might produce a few hundred watts—perhaps as much as a thousand watts—over a short period, it is a real chore to produce 100 W on a continuous basis for hours at a time. If we produce 100 W for a 10-hour period, the amount of electrical energy produced is 100 W × 10 hours, which is 1000 Wh, or 1 kilowatt-hour (kWh), for which the average price in the United States is about a dime. Not many people would be willing to work that hard, that long, for a mere 10 cents. For another comparison, a 2000-Calorie[1] daily diet is equivalent to about 100 W.

To expand that perspective just a bit, let us look at the amount of energy—over 100 exajoules[2]—used in the United States every year. Averaged over the 31.6 million seconds in the year, and over the roughly 315 million US citizens, our rate of energy consumption is about 11,000 W *per capita*, about 110 times as much power as the average human produces in the form of heat, or our athlete produces while on the bicycle seat. Alternatively, one may think of our energy consumption as being equivalent to having 110 servants tending

to our needs night and day. This is why we can accurately say that human labor is a minuscule part of the energy picture.

Over 90% of our energy comes from petroleum, natural gas, coal, and uranium. Of the 9% contribution from renewable energy sources, the venerable ones—hydro and biomass—provide over 80%.

The electricity that powers our appliances is not primary energy, but rather a *carrier* of energy. In a coal-fired power plant, for example, the fire boils water into high-pressure, high-temperature steam that turns a turbine that turns an electricity-producing generator. The generator energizes electrons, and our lights and machines in turn become energized by those electrons. The important detail is simply that the electrical power does not come from the socket, but rather from some primary source like coal or natural gas.

Scientists *began* to understand things pertaining to electricity at about the time of the American Revolution. It was not until 1882 that the first hydropower station produced commercial electrical power. So useful is electricity that, by now, a full 40% of our primary energy in the United States goes into the production of electricity.

SOME FUNDAMENTALS

Energy is somewhat hard to define. There is no instrument that directly measures energy. By and large, you cannot see it or touch it. *Work*, on the other hand, is fairly easy to define. Work is the equivalent of lifting a weight to some height above the starting point; indeed, it is the product of the weight and the vertical distance the weight moves.

Now we come to the definition: Energy is the *capacity* to do work. (That does not mean the *ability* to do work.) In other words, energy is numerically the same as the work it could do at an impossible 100% efficiency. In broadest terms, there

---

[1] One food Calorie (capitalized) is one kilocalorie.
[2] We will say more about this unit further on.

is energy due to position—called *potential energy*—such as the energy of an object that has been lifted. There is also energy due to motion—called *kinetic energy*—such as that of a falling ball. The two kinds of energy can appear together, such that of a ball dropped from a 20-story high building as it passes (say) the 12th floor, where it has potential energy with respect to the ground and also has kinetic energy due to its speed.

Chemical energy is potential energy. It is not due to gravity, but to the positions of the atoms and molecules with respect to one another. For example, when heptane molecules (the most common ones in gasoline) encounter oxygen molecules in air under conditions of high temperature (such as that created by a spark in a gasoline engine), the energy released a chemical bonds are broken and formed is thermal energy, often called heat.[3] Heat is a special kind of energy to which special rules apply. Energy (PE and KE) can be converted to heat with 100% efficiency. Heat can be converted to energy, but with an efficiency that is less than 100%.

Electromagnetic energy travels at the speed of light, but light is a minuscule part of the spectrum. The term *light* usually refers to visible light, but usually not to ultraviolet radiation, x-rays, gamma radiation, infrared, and radio waves, although the only meaningful distinction is that we can *see* light. The generic term is *electromagnetic radiation*.

ENERGY IS NOT POWER AND POWER IS NOT ENERGY

The distinction between *power* and *energy* is the same as the distinction between speed and location. Everybody understands that going 45 miles/h is not the same as being at 4th and Main. Similarly, everybody understands that 250 horsepower is not the same as the chemical energy in a liter of gasoline. But too many people fail to distinguish between *kilowatts* (a unit of power) and *kilowatt-hours* (a unit of energy).

Power is a term that has a well-defined technical meaning and several ill-defined meanings in common parlance (such as "political power," a "powerful idea," or "the power of the purse"). In common parlance, *power* means something related to the electrical socket. Indeed, power is provided by the grid, but electrical power is only one kind. For example, it takes power to run your car—the term *horsepower* is commonly used. But using *watts* for electrical power, horsepower for automobiles, and British Thermal Unit (BTU)/hour for a furnace is somewhat akin to measuring sugar in pounds, coal

in tons, and diamonds in carats. Why do that, when they are all weights? Similarly, power needs only one unit of measurement. By international agreement, that unit is the watt.

Let us begin with something simple. Suppose we have a cupful of gasoline. We can burn up that gasoline very slowly or we can burn it up in a hurry, possibly even in an explosion. In both cases, the same amount of energy is released, because that cup of gasoline has just so much energy, no more, no less. When the fuel is burned slowly, that is low power. When it is burned rapidly, that is high power.

For a very dramatic example, consider the electrical energy from a single large coal-fired power plant operating for one half of a day. The energy is used for heating, refrigeration, lighting, manufacturing, and many other useful ventures. But if that amount of energy were expended in a small area in about a microsecond, it would be an explosion like that of the atomic bombs that fell on Hiroshima and Nagasaki. That is extremely high *power*.

In all cases to which the term *power* applies, energy is being converted from one form to another. The faster the energy is converted, the higher the power.

Technically, then, power = energy converted ÷ the time interval.
Alternatively, energy converted = power × the time interval.

The electrical meter on your house measures power consumption, second by second, and keeps a running tally in units of kilowatt-hours.

UNITS OF MEASUREMENT

Energy has historically been measured in many units, including foot-pounds, ergs, joules, calories, kilocalories, and BTUs. There are also units that pertain to fuels: the gallon of gasoline, the barrel of oil, the therm, the cubic foot of natural gas, the heat content of a cord of white oak, and the ton of coal, to name a few.

Since power is energy per unit time, we need to consider common units for time: the second, the minute, the hour, the day, the summer, the winter, the year, and so forth. Using just the 12 listed units for energy and the 8 listed units of time, we can construct 96 units of power (such as kilocalories per month). A table of factors to convert from any one of them to any one of the others would have 96 by 96 entries, of 9216 entries. It is largely because of the vast profusion of units that most casual readers find the topic of energy confusing: how can you compare the things said in one book with those said in another?

The only sane thing to do is to insist on *one* unit for energy, *one* unit for time, and *one* unit for power. Then the attached numbers can tell their story. The units used *internationally*

---

[3]Thermodynamicists usually use the term *heat* to refer to thermal energy in transit, such as when a hot body is placed in contact with a cold one: heat flows from the hot one to the cold one. There are good reasons for doing so, but they need not concern us here.

(as well as *officially* by the US, though almost never by US agencies) are those in the *Le Système International d'unités* (abbreviated SI):

- Energy: the *joule*. Abbreviation: J, uppercase
- Time: the *second*. Abbreviation: s, lowercase
- Power: the *joule per second*, otherwise known as the *watt*. Abbreviation: W, uppercase

For many Americans, the term *joule* is unfamiliar. How big is it? For one, it is the energy expended by 1 W during 1 second: a *joule is a watt-second*. Think of the energy expended by a flashlight during 1 second. Alternatively, it is approximately the energy expended in lifting a 1-L bottle of soda 10 cm (about 4 inches, the "hand" known to horseman). Most importantly, even if the term *joule* is strange and unfamiliar, *it is a trivial matter to compare the amounts of energy if the same international unit is used for all applications.*

We pay the utility for the electrical energy we use, and the unit of measurement the utility uses is the kilowatt-hour (kWh). How much is that in joules? 1 kWh is 1000 W multiplied by the number of seconds in an hour ($60 \times 60 = 3600$), or 3,600,000 J.

No matter what units might be used, there would be an enormous range of numbers between (say) the energy consumed by walking up a flight of stairs and the energy consumed in the United States during an entire year. For example, in 2011, the annual energy consumption in the United States was 103,000,000,000,000,000,000 J. It is tedious work to keep track of all those zeroes, so we have shorthand methods. We write it as 1.03×10^{20} J (scientific notation), 103×10^{18} J (engineering notation, with exponents in multiples of 3), or 103 exajoules (abbreviated EJ).

The metric prefixes are given in Table I.1.

Many units have been defined for energy. Table I.2 presents conversion factors. Note that the BTU is actually defined in terms of the joule by a 12-digit conversion factor.

Often used informally as energy units are the heat contents of fuels. For example, we endlessly hear of nuclear weapon

TABLE I.1 Metric prefixes.

E (exa)	10^{18}
P (peta)	10^{15}
T (tera)	10^{12}
G (giga)	10^{9}
M (mega)	10^{6}
k (kilo)	10^{3}
m (milli)	10^{-3}
μ (micro)	10^{-6}
n (nano)	10^{-9}
p (pico)	10^{-12}
f (femto)	10^{-15}
a (atto)	10^{-18}

TABLE I.2 Energy conversion factors.

Multiply number of ↓	By ↓ to get joules
Watt-seconds	1
British Thermal Units (BTU)	1055.05585262
Quadrillion BTUs (quads)	1.055×10^{18}
Kilowatt-hours (kWh)	3.6×10^{6}
Horsepower-hours	2.69×10^{6}
Kilocalories (kcal)	4186.8
Calories (cal)	4.187
Therms ($=10^{5}$ BTU)	1.055×10^{8}
Foot-pounds	1.36
Ergs	1×10^{-7}
Watt-years	3.16×10^{7}

yields expressed in tons of TNT. Or, people will say that such-and-such project will displace a million tons of coal. Table I.3 gives heat contents of fuels by mass, and Table I.4 gives the heat content by volume. One kilogram of petroleum has a heat content of about 45 MJ/kg, and the same number holds approximately for all petroleum products. However, they have different densities. A gallon of propane weighs less than a gallon of gasoline, and therefore it has less heat content.

Combustion of fuels produces carbon dioxide and water vapor. Sometimes (for reasons of precision) it is important to distinguish between the higher heat value (HHV) and the lower heat value (LHV), the difference between them due to the heat of vaporization of water. Under almost all circumstances, either in heat engines or fuel cells, water *vapor* escapes, carrying with it the energy that converted the water to steam in the first place. The higher heat value (HHV) is the total energy per kilogram that is produced by the oxidation of the fuel; the lower heat value (LHV), is the HHV minus the energy of vaporization. The tables of the Energy Information Administration (EIA) give HHVs and some LHVs.

The most extreme case is that of hydrogen, for which the only by-product is water vapor. The HHV of hydrogen is 142 MJ/kg, but the LHV is 120 MJ/kg, some 15% less. From Table I.3, the difference between HHV and LHV is about 10% for both methane and methanol, and about 6% for petroleum fuels.

US Energy Usage

The best source of information about how we produce, consume, import, and export energy in the United States is the *Annual Energy Review* (*AER*), produced by the EIA of the Department of Energy (DOE). Through 2011, the EIA produced a PDF file of the *AER* that could be downloaded (http://www.eia.gov/totalenergy/data/annual). The EIA's website is now interactive and actually has more information than before; however, they no longer produce a single PDF of *AER*. A major failing of the *AER* is their use

TABLE I.3   Heat contents of fuels (by mass).

Multiply number of ↓	By ↓ to get joules (HHV)	By ↓ to get joules (LHV)
Kilogram of crude petroleum	$42.5–45.4 \times 10^6$	41.2×10^6
Kilogram of gasoline	47.5×10^6	44.5×10^6
Kilogram of diesel fuel	44.8×10^6	42.5×10^6
Kilogram of coal	15×10^6 to 30×10^6	
Kilogram of coal (US average)	24.1×10^6	
Kilogram of drymatter (biomass)	15×10^6	
Kilogram of hydrogen(Not a *source* of energy!)	141.9×10^6	119.9×10^6
Kilogram of methane (CH_4)	55.5×10^6	50.0×10^6
Kilogram of methanol (CH_3OH)	20.0×10^6	18.0×10^6
Kilogram of ethanol (EtOH)	30×10^6	
Kilogram of propane	50.4×10^6	45.6×10^6
Pound of coal (average)	10.9×10^6	
Ton of coal (average)	21.9×10^9	
Kilogram of TNT	2.1×10^6	

of arcane units: quadrillion BTU (a.k.a. quads) for primary energy, kilowatt-hours for electrical energy, BTU per kWh for reciprocal efficiency, barrels of oil, cubic feet of natural gas, tons of coal, *metric* tons of carbon, and so forth.

Because energy is eventually degraded to heat, there is some method to the EIA's madness. They refer to *primary* energy in heat units (quadrillion BTU), but to *electrical* energy in kilowatt-hours. This topic will be discussed further in our brief discussion of hydropower, but for now we simply point out that the EIA would be able to make the same distinction by using the subscripts "t" for *thermal*, and "e" for *electrical*. For example, many engineers now use the abbreviation W_t for thermal watts and W_e for electrical watts. In other words, the EIA should start using sensible units and get into the 20th century before the rest of us get out of the 21st.

As noted above, the United States used 103 EJ in 2011. But that factoid by itself does not tell us what our history of energy consumption is, how much energy is used for what purposes, or how much energy we get from what sources. Figure I.1 shows the history of US energy consumption from various sources. Notice that each horizontal line represents 10

TABLE I.4   Heat contents of fuels (by volume).

Multiply number of ↓	By ↓ to get joules HHV
Barrels of crude oil	6.12×10^9
Barrels of aviation gasoline	5.326×10^9
Barrels of motor gasoline	5.542×10^9
Barrels of propane	4.047×10^9
Barrels of kerosene	5.982×10^9
Gallon (US) of gasoline	0.131×10^9
Gallon of ethanol (EtOH)	0.095×10^9
Cubic feet of natural gas	0.001092×10^9
Ton of TNT	4.2×10^9
Cords of wood (white oak)	$31. \times 10^9$

times the value of the next-lower one. Roughly speaking, we use 100,000 times as much energy annually as our ancestors did in the middle 1600s. At that time, the energy source was almost exclusively firewood. Two centuries later, coal was making inroads, and petroleum was beginning to replace whale oil for lamps. Natural gas and hydropower made their debuts in the late 1800s, and nuclear power began producing electricity around 1960.

Figure I.1 shows a very dramatic increase in the US annual consumption of energy, but might leave the impression that we Americans are individually using a vast amount of energy compared to our forebears. Bear in mind, however, that the population in 1700 was fewer than 250,000; today it is about 1300 times as large at 315 million. It is interesting to calculate the energy consumption *per capita* in the US history. Figure I.2 shows that the *per-capita* energy consumption from

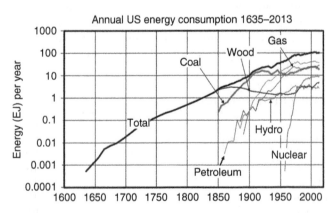

FIGURE I.1 Energy consumption in the United States since 1635, showing energy sources. The predominant source until about 1850 was firewood; since then its consumption has remained relatively constant. Notice that the vertical scale is logarithmic: each horizontal line represents a factor of 10 more than the line beneath it.

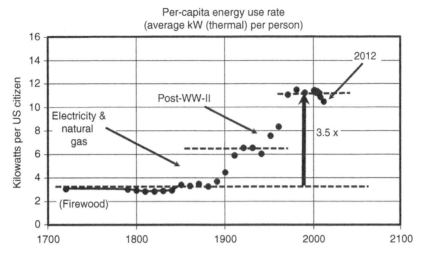

FIGURE I.2   Historical *per-capita* energy consumption in the United States. From the early 1700s, when firewood was the main energy source to the present, when we have automobiles, jet planes, satellites, GPS, elevators, electric lights, television, refrigeration, air conditioning, and fast internet, we have only tripled out *per-capita* energy consumption.

the early 1700s until the late 1800s was about 3.1 kW *per capita*, whereas it is now about 11 kW *per capita*. In other words, the overwhelming cause of our increase in energy consumption is due to increasing population; the *per-capita* growth during the last three centuries is a factor of 3.5. The reason for this low figure is that we have vastly improved our energy efficiency.

Overwhelmingly, the sources of our energy are the so-called "fossil fuels," coal, petroleum, and natural gas, as shown in Figure I.3. Almost all of our transportation is driven by petroleum, the main exception being electrically driven

commuter trains. Coal is used almost exclusively for producing electricity, as is nuclear energy. Natural gas is easier to use, so it is used not only for domestic heating, but also for producing electricity, as well as a negligible fraction of transportation. Contrary to the impression promoted by renewable energy enthusiasts, the great majority of our renewable energy is *not* wind and solar. Biomass (half being firewood, half being bio-liquids) and hydropower account for three-quarters of the renewable energy, as shown in Figure I.4. The bio-liquid component is primarily ethanol distilled from fermented corn and from (mostly Brazilian) sugar cane.

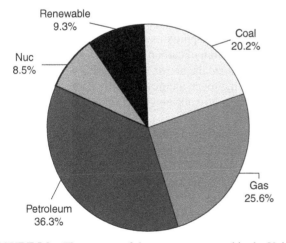

FIGURE I.3   The sources of the energy consumed in the United States in 2012. Renewable energy accounted for 9.3% of the energy, but the lion's share was the venerable ones: hydro and biomass (half of which was firewood).

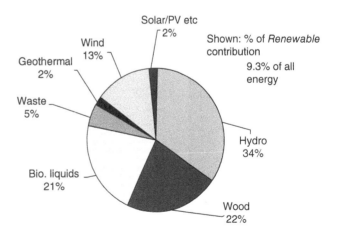

FIGURE I.4   The individual contributions of US renewable energy. All considered, they comprise 9.3% of all US energy consumption. The biological liquids are primarily corn ethanol added to gasoline.

US sources of electricity, 2011

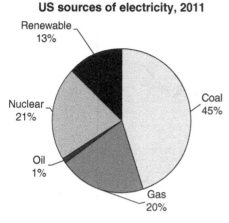

FIGURE I.5   Sources of US electricity, 2011. The coal fraction has diminished from 50% to 45% during the last few years, owing to the increasing availability of inexpensive natural gas and the EPA's strictures against coal.

As noted above, about 40% of our primary energy goes into the production of electricity. Figure I.5 shows how much each source contributes. Renewables account for 13% of our electricity, but again, the major fraction of the renewable contribution is from the venerable sources: firewood and hydro, as shown in Figure I.6. Owing to mandates (Renewable Portfolio Standards) and subsidies, wind has grown to produce 23% of our *renewable-source* electricity, or 3% of our electricity. Also notice that oil produces only about 1% of our electricity.

Physical Fundamentals

Fuels   The earth has vast reserves of coal, oil, and natural gas. They are usually called *fossil fuels*—implying that they are the remains of long-dead trees and animals. While it is true that methane (the main component of natural gas) is

US Renewable electricity generation, 2011

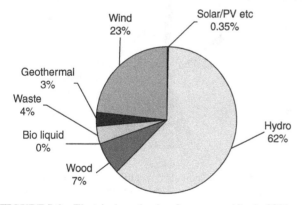

FIGURE I.6   Electrical production from renewables in 2011, as a percentage of total renewable kilowatt-hours.

produced in swamps and landfills, that explanation does not quite work for Titan, the largest moon of Saturn, where there are literally oceans of liquid methane. It seems likely that there may be some primordial methane trapped within the earth. It is controversial whether at least some of the earth's coal and oil are derived from that methane, but the question has relevance to the quantity of these fuels. That discussion, however, would take us too far afield.

Methane (CH_4) is chemically the simplest of the fuels and is a gas unless refrigerated to a very low temperature (approximately −162 C, or −260 F). As shown in Figure I.7, the methane molecule is a tetrahedral structure with a carbon atom in the center, and four hydrogen atoms at the vertices. When a methane molecule burns in air, the combustion produces one molecule of carbon dioxide (CO_2) and two molecules of water (H_2O).

Carbon has the property of forming long chains, side chains, hexagons, and many other complicated structures; without that property, petroleum—and for that matter DNA—would not exist. Figure I.8 shows the example of one petroleum molecule, pentane. It has 5 carbon atoms and 12 hydrogen atoms. As the number of carbon atoms increases, so does the number of hydrogen atoms. The ratio of their numbers—hydrogen to carbon—approaches 2. For example, nonane, the principal component of diesel fuel, has 9 carbon atoms and 20 hydrogens, so the ratio is 20:9, or 2.22. For this reason, we usually regard petroleum (which has both of these chemicals and *many* more) as being approximately CH_2.[4]

There are two natural forms of carbon: graphite and diamond. Coal is neither of these; it is not pure carbon. Nor is all coal the same. The broad classifications—lignite, sub-bituminous, bituminous, and anthracite—do not even tell the story, because coal from different mines will have different compositions, different quantities of water (yes, coal has lots of water), as well as different quantities of silicates, sulfur, mercury, and other contaminants. This topic is best left to experts, but for our considerations, the combustible part of coal is chemically approximately CH: one atom of hydrogen for every one atom of carbon.

To summarize the above, the conventional fuels differ in the ratio of hydrogen atoms to carbon atoms: exactly 4-to-1 for methane; approximately 2-to-1 for petroleum and one-to-one for coal.

Methane, with a 4:1 ratio has an HHV of 52.2 MJ/kg.

The 2:1 hydrogen/carbon ratio of all petroleum chemicals means that the HHV is closely the same for all of them, namely about 45 MJ/kg ± a few percent. The differences in the *volumetric* HHV values of the petroleum fuels are due to their differing densities. (The ethanol that is added to gasoline has only 72% as much energy on a per-liter or

---

[4]The long-chain molecules, known as *alkanes* have a generic formula C_nH_{2n+2}

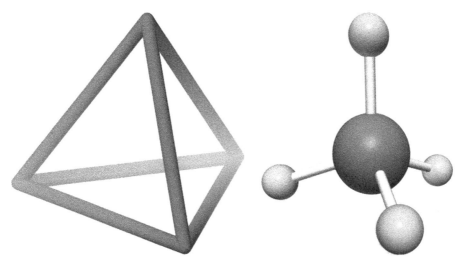

FIGURE I.7   Left, a regular tetrahedron. Place a carbon atom at the center and four hydrogen atoms at the vertices, and you get the structure of methane, the simplest of fuels.

per-gallon basis.) To get some perspective on the HHV, a car that gets 30 miles per gallon and is travelling at 60 miles per hour uses two gallons of gasoline per hour. Using the HHV for two gallons (262 MJ) and the 3600 seconds in an hour, we see that the fuel input power is 77 kW.

Anthracite coal has an HHV of 32 MJ/kg. Bituminous coal varies from 17 to 23 MJ/kg. The coal used in the United States in 2011 had an average HHV of 22.8 MJ/kg. Hydrogen, by comparison, has a much higher HHV, namely 141.8 MJ/kg. This high number does not mean that hydrogen has a special affinity for oxygen. Rather, the value is due to the fact that hydrogen atoms have a mass unit of 1, compared to 12 for carbon; the mass is in the denominator.

By comparison, nuclear fuel yields about 80 TJ/kg, which is 1.8 million times as high as the HHV for petroleum.

Uses of fuels 1: Heat-to work   All useful fuel-driven engines work by the expansion of a gas that is heated directly or indirectly by the fuel. In a steam engine, for example, water is boiled under high-pressure conditions, and the resulting steam expands in the engine (which may be a piston- or turbine-type). A nuclear reactor works similarly, in that its job is just to boil water. In a gasoline or diesel engine, the

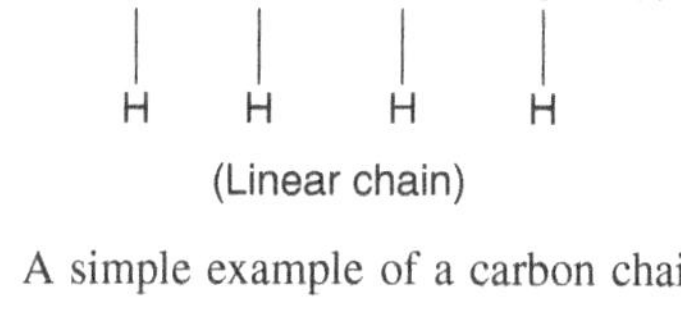

expanding gas is none other than the combustion products, mostly CO_2 and H_2O.

In the abstract, the engine is a system into which you inject energy, and there are two outputs: work and waste heat. The efficiency is the work accomplished divided by the heat energy input (alternatively, the power produced divided by the input power). Averaged over the entire electrical supply, the efficiency of electricity generation is about 35%. This fact has implications for how the EIA regards hydro and wind.

The EIA has tables of annual energy consumption in the United States. For hydro, the 2011 value is 3.171 quadrillion BTU. The amount of electricity produced by hydro in 2011 was 325.1 billion kWh. If we make a simple conversion of units, we find that 3.171 quadrillion BTU is equivalent to 929.3 billion kWh, which is 2.858 times the actual electrical energy produced. For wind, the figures are 1.168 quadrillion BTU and 119.7 billion kWh, resulting in a ratio of 2.859. No dishonesty is intended. The purpose of the EIA tables is to show how much energy would have been required if the electricity had been produced by heat engines operating at 35% efficiency, instead of hydro.

Why is the efficiency so low for producing electricity from fuels? It is not a matter of failing to understand the basic principle, which is that the higher the temperature of the expanding gas, the higher the efficiency of the engine. But efficiency has much to do with the properties of materials. For example, the higher the temperature, the more corrosive steam becomes. *Combined-cycle* power plants, consisting of two heat engines operating their own generators, have become very efficient. The first engine is a gas turbine with specially made blades (ceramic-coated, or single-crystal titanium) that can withstand very high temperatures. The second engine is a conventional steam-turbine system that boils water with the hot gases exhausted from the gas turbine. Between the

```
    H   H   H   H
    |   |   |   |
H — C — C — C — C — H
    |   |   |   |
    H   H   H   H
       (Linear chain)
```

FIGURE I.8   A simple example of a carbon chain molecule, in this case, pentane.

two, they manage to convert up to 60% of the fuel energy to electricity.

Uses of fuels 2: Fuel cells Electrolysis is a method for separating water into hydrogen and oxygen. Basically, DC electricity is run through water containing an electrolyte, such as NaCl. Hydrogen is liberated at the cathode, and oxygen is liberated at the anode. A fuel cell is the opposite. Hydrogen is injected at the anode and oxygen (often just air) is injected at the cathode. The hydrogen ions (a.k.a. protons) permeate through a barrier to combine with oxygen to form water. The system acts like a battery with many cells, each of about 0.7 V, wired in series. In other words, it converts chemical energy to electrical energy without the use of expanding gases. The efficiency is up to 60%. Fuel cells are particularly sensitive to contaminants in the fuel, but less so with high-temperature ceramics as the barrier.

Hydro Hydropower turbines and wind turbines have one important thing in common. They convert mechanical energy into electricity by simply spinning a generator. For over a century, it has been possible to convert the mechanical energy of a rotating shaft into electricity with over 95% efficiency.

In the case of hydropower, the source of energy is *potential* energy of water stored behind a dam. (There are some low-power, inefficient devices—"undershot wheels"—that convert kinetic energy of a moving stream.) In the case of wind power, the source of energy is the *kinetic* energy of moving air. An important difference is that hydro energy is stored (by accumulating water in a lake behind a dam), whereas wind power is available only when the wind blows.

The power output of a hydro plant is proportional to the rate of flow of water and to the elevation of the water surface above the point where water is discharged into the river. Assuming a reasonable efficiency of 85%, the power is given by

$$P_{out}(\text{watts}) = 8330 \times h(\text{meters})$$
$$\times \frac{\Delta V}{\Delta t}(\text{cubic meters per second}) \qquad (\text{I.1})$$

By and large, the flow rate $\Delta V/\Delta t$ is governed by the number of turbines being used. For example, Hoover Dam has 19 generators, each contributing about 5% of the 2.08 GW possible with all running. On yearly average, Hoover puts out about 480 MW. The annual capacity factor is therefore about 25% (480 MW ÷ 2080 MW). Hydro is rarely used for base-load power, but rather to handle the times during which the demand is high and/or fluctuating rapidly. Typically, hydropower plants have much more power available than the stream flow can handle on a long-term basis. High power is generated when needed, and that means high water flow, which reduces the water level. When demand drops and the water flow is shut off, the stream refills the lake.

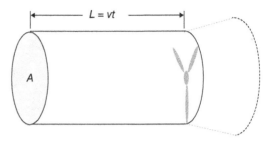

FIGURE I.9 A horizontal column of moving air reaches the wind turbine and expands as it slows down.

As hydropower is the easiest to use, it follows that the best dam sites are already in use. It produces about 8% of our electricity and is not likely to grow appreciably. No doubt, there are many rivers that can be dammed to produce (say) 50 MW, but just imagine having to write an Environmental Impact Statement — and having to fend off affected neighbors — for 20 of them, just to produce 1000 MW (enough for a city of 700,000).

Wind Imagine a horizontal circular column of moving air. Imagine a wind turbine of the same diameter facing into that column and extracting some of the kinetic energy (see Figure I.9). If the machine removed *all* of the energy, the column would stop, and no more energy would be available. In fact, the air has to keep moving, but at lower speed, because the wind turbine has slowed the air down. For the same amount of air to come in at high speed and leave at low speed, the slower air has to spread out to a larger diameter. This fact has implications for separating wind turbines from one-another. You do not want down-stream turbines to be in a low-speed area, and you do not want interference between side-by-side turbines. There is also the matter of turbulence, which can damage the machinery. Generally speaking, turbines are placed about 10 diameters apart in both directions.

The wind energy per unit time arriving at a wind turbine is proportional to the intercepted area (hence to R^2, the square of the radius). It is also proportional to the cube of the wind speed, because the kinetic energy is proportional to the square of the wind speed, and the rate at which it arrives is proportional to the speed v. It is also proportional to the density of the air (about 1.3 kg/m³), which varies with the elevation of the site. The wind turbine will convert a fraction (the *power coefficient*, for which we use a Greek η, eta) of that energy per unit time into useful power. If we represent the density by the Greek letter rho (ρ), the power output P_{out} is given by

$$P_{out} = (1/2)\eta\rho\pi R^2 v^3 \qquad (\text{I.2})$$

Before proceeding to a discussion of real turbines, we take note of two very important facts. Consider using larger turbines instead of smaller ones. If we double the diameter, a turbine will produce four times as much power. On the other hand, we must space the turbines twice as far apart—in each direction—so that the gross amount of land devoted to the array of turbines is four times as large. Therefore, the amount of power *per unit of land area* is independent of the size of the wind turbines. A good rule of thumb is that a large array of wind turbines *in an excellent site* will produce about 12.5 kW per hectare (10,000 square meters) (5 kW/acre) on a year-round average basis.

To explain this further, wind sites are rated in terms of the power-per-unit-area[5] carried in the wind, as shown in the following table from the Texas State Energy Conservation Office (NREL = National Renewable Energy Laboratory).

W/m²	NREL class
<100	1−
100–200	1+
200–300	2
300–400	3
400–500	4
500–600	5
600–800	6
>800	7

A wind turbine presents an interception area A, which is proportional to the square of the diameter of the turbine, and the wind turbines are spaced typically 10–15 diameters apart in both directions, so the gross land area must be proportional to the interception area. For example, for a 10-diameter (20-radius) separation, the ratio of turbine area to gross land area is $\pi R^2 / \pi (20R)^2 = 1/400$. A 500-W/m² site would be exposed to about 1.2 W/m² (12 kW/hectare) average power, but produce less, because the wind turbine has its own power coefficient.

The other important fact is the dramatic variation with wind speed. If the speed changes from 5 to 10 m/s, the power available becomes eight times as great. Alternatively, if the wind speed halves from 10 to 5 m/s, the power available drops by 87.5%. These rapid changes are unimportant when wind power is a small fraction of the power on the grid, but begin to be troublesome at about 10% penetration.

Capacity Factor for Wind Turbines   For a given site with its winds, capacity factor is a matter of engineering design. We can nail down two ends of the curve with bits of nonsense. What if we built a wind turbine using a child's pinwheel to

drive a 1-MW generator? We would get no energy at any time, and the power factor would be zero. What if we used a 50-m diameter turbine and had it drive a 1-W generator. Most likely, there would always be enough wind (or rotational energy in the rotor) to turn the generator for the full 8760 hours in a year. The capacity factor would be 100%.

In other words, the annual capacity factor (average power divided by nameplate power) depends on the size of the generator compared to the size of the turbine. For a couple of decades, wind turbines have been designed for an annual capacity factor of about 35%. In 2011, wind generated 120 billion kWh. At the end of 2010, the total wind nameplate power was 39.1 GW, and at the end of 2011 it was 45.2 GW. If we assume that the average nameplate power during 2011 was the average of the beginning and end capacities (42.15 GW), the wind turbines could have generated 369 billion constant high winds, so the capacity factor was 120 ÷ 369, or 32%.

Actual Wind Machines   At wind speeds below about 4 m/s, wind turbines produce no output at all. If they did, the amount of power would be trivially small in any case, owing to the v^3 dependence on wind speed (see Figure I.10). Then the power output rises dramatically until full power is achieved at about 16 m/s, after which the output is constant until 25 m/s when the machine shuts off to avoid damage.

The power coefficient η versus wind speed is shown in Figure I.11. It achieves a maximum value of about 40% at a wind speed of 8 m/s (29 km/h, 18 mph). The power coefficient is 25% or above for wind speeds from about 5 m/s (18 km/h,

[5]Technically speaking, it is not power, because no conversion of one type of energy to another is involved. Pedantically, the term should be energy per unit area per unit time crossing an area perpendicular to the wind.

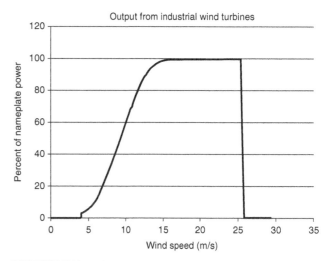

FIGURE I.10   The power output from a wind turbine, versus wind speed. Data from specification sheets of wind turbine manufacturers. The power rises abruptly, beginning at about 4 m/s until beginning to level off at around 13 m/s, reaching full power at 16 m/s. Above 25 m/s the machine must be shut down to avoid damage. Curves for major manufacturers are essentially identical.

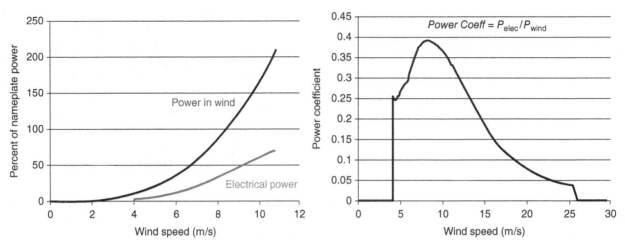

FIGURE I.11   Left: Wind turbine power and power in the wind, versus air speed. Right: The power coefficient, which is the ratio of generated power to wind power, versus wind speed. Data from specification sheets of wind turbine manufacturers.

11 mph) to about 13 m/s (47 km/h, 29 mph), which occurs about half the time.

The biggest machinery problem with industrial wind turbines has been the gear boxes. The larger the turbine diameter, the slower the rotation (but the tip speed remains at about seven times the wind speed, and can easily reach 80 m/s [200 miles per hour]). Typically they turn at 10–15 revolutions per minute, whereas the generator (depending on design) usually requires rotation rates in the high-hundreds to a few thousand RPM. The gear boxes are responsible for increasing the rotation rate, and they have very high stresses. It is possible, but expensive, to have direct-drive generators that use a large array of rare-earth permanent magnets around the periphery of the rotor, and coils appropriately placed on the stator.

The bases for wind turbines must be very substantial. Not only must the base have the strength to support the weight, but it must also deal with a tremendous torque ("twisting force"). For contemporary machines, it is equivalent to having a school bus out on the end of a plank the length of a football field. All in all, the generator, nacelle, support pipe, and base usually weigh over 1000 metric tons.

Geothermal Energy   The term *geothermal* has a historical meaning and a new one. The historical one is the obvious one, referring to heat from deep (or not so deep) in the earth. The new meaning refers to the use of the earth a few meters down as a heat reservoir.

Outcroppings of heat (such as at the Geysers in California) are sometimes hot enough to run a steam turbine. In 2011, geothermal heat generated 0.04% (4 pennies out of 100 dollars) of US electricity. The high-temperature environment is frequently corrosive, but putting that issue aside, there are three other problems. The higher the temperature, the higher the efficiency can be for a heat engine (such as a

steam engine). The hot outcroppings, however, are not usually nearly as hot as the steam in a conventional steam plant, so the efficiency is lower. In the second place, extraction of heat from the ground necessarily cools the region where the heat is extracted, and that lowers the efficiency of the engine. One must limit the extraction of heat to the rate at which heat is conducted to the site through the surrounding rock. Finally, the number of such outcroppings is limited. On the other hand, if technology develops for drilling very deep into the mantle, the heat supply is enormous.

Refrigerators and Heat Pumps   With regard to the new meaning of *geothermal*, it is necessary to take a brief look at refrigerators and heat pumps. A heat *engine* can be regarded in block form this way: It is a device into which you inject heat Q_{HIGH}, as shown schematically in Figure I.12. The device does some work W and rejects waste heat Q_{LOW} into the environment. Obviously, $Q_{HIGH} = Q_{LOW} + W$.

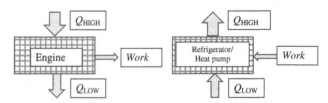

FIGURE I.12   Left: A schematic ideal engine. A quantity of heat Q_{HIGH} is fed into the engine from a high-temperature source. Work W is accomplished and waste heat Q_{LOW} is rejected to the low-temperature environment. The widths of the arrows represent amounts. Right: When work is put into the system (say by the serpentine belt of an automobile, or by an electric motor), the directions of the heat flow are reversed. Heat pumps and refrigerators are essentially identical. The only difference is the purpose. Refrigeration is concerned with the heat removed Q_{LOW} Heat pumps are concerned with the heat delivered Q_{HIGH}.

In a more realistic scenario, the machine is cyclic with some average input power and average useful power produced, and average heat power rejected into the environment. For example, we might inject 10 kW of heat power from steam into an engine. The machine produces (say) 3 kW of useful electrical power and discards 7 kW of heat into the environment. Simple enough. Generally speaking, the higher the source temperature, the higher the efficiency.

Now imagine turning that engine around, as shown on the right side of Figure I.12. Under ideal conditions, we could put in 3 kW of electricity and extract 7 kW of heat from the environment. These days, that does not take a lot of imagination, because that is exactly what refrigerators and air conditioners do. Yes, we do extract more heat energy from the refrigerated environment than we put in as electricity. Importantly, the 3 kW plus the 7 kW add up to 10 kW that is delivered back to the "hot" area, namely the room or the outdoors.

A heat pump is simply a refrigerator used for a different purpose. In this case, we are concerned with delivering heat from a cooler place to a hotter place. In our example, the 3-kW of electricity refrigerates one place to the tune of 7 kW and heats another place to the tune of 10 kW.

In actual practice, we can use an air conditioner in the summer to extract heat from the house and inject it into the comparatively cool ground which serves as a heat reservoir. That produces cooling more efficiently than rejecting heat into the hot summer air. By the same token, we can reverse the valves in the winter so as to refrigerate the relatively warm ground (compared to the outside air) and deliver the heat into the house. The use of the ground as a heat sink or source is what has led to the term *geothermal heat*, which many people would confuse with high-temperature heat in places like Yellowstone National Park.

The Seasonal Energy Efficiency Ratio (SEER) is the ratio of the seasonal heat extracted (for air conditioners) or delivered (for heat pumps) to the electrical energy input for commercial devices; however, there is a different unit in the numerator (BTUs) than in the denominator (Wh), and manufacturers usually do not bother telling customers that the units are BTU/Wh. To make a more understandable number, note that 1 BTU = 3.412 Wh. Commercial heat pumps and refrigerators have SEERs that run from about 10 BTU/Wh to as high as 18 BTU/Wh, corresponding to simple ratios of 2.93 (=10 BTU/Wh ÷ 3.412 BTU/Wh) to 5.86. In other words, one unit of energy from the electrical grid will deliver 2.93–5.86 units of heat. Engineers refer to these dimensionless ratios as the *coefficient of performance* (COP). A glance at Figure I.12 will reveal that the COP for a heat pump must be greater than that for the same device used as a refrigerator, because the heat delivered Q_{HIGH} must always exceed the heat Q_{LOW} removed from the cold reservoir. In fact, $COP_{heatpump} = COP_{A.C.} + 1$.

Solar Energy Solar energy has many forms, including biomass, solar heat, photovoltaics, wind energy, and hydropower. It is useful to distinguish between broad types. Biomass and solar devices (like heat collectors and photovoltaic cells) produce usable energy because of very local sunlight. In these cases, the solar energy is collected where, and only where, our devices capture sunlight.

By contrast, the winds blow in one place because of sunlight somewhere else, and hydropower produced at a dam is due to water that evaporates far away and rains out somewhere upstream. In these cases, the solar energy is captured *naturally*, carried elsewhere by air and water, and converted into a useful form by machinery. Hydro and wind are therefore discussed separately from other forms of solar energy.

Our magnificent sun is located 150 million kilometers away, is powered by hydrogen fusion, and has a surface temperature of about 5800 K. A one square-meter panel facing the sun at the location of the earth intercepts about 1400 W [thermal, W_t] of sunlight. Much sunlight is reflected by the atmosphere (clouds, water vapor . . .), and some is absorbed by the atmosphere, so that the sunlight reaching a square-meter panel at the surface and facing the sun intercepts about 950 W_t under clear-sky conditions at noon. The year-round average sunlight intensity ($\langle I_H \rangle$ known as *insolation*) on a *horizontal* surface in the United States, averaged over all of the lower-48 is 200 W/m², with ±20% place-to-place variations covering about 2/3 of the lower-48. For example, Hartford, CT, gets about 160 W/m², and Albuquerque, NM, gets about 240 W/m².

There is, therefore, a certain simplicity to energy obtained by direct sunlight. A given parcel of land of area A (unit: square meters (m²), using some apparatus, produces a certain quantity of useful energy E (S.I. unit: joules), during the year ($t = 31.6$ million seconds). The *useful* average power intensity $\langle I_U \rangle$ produced per unit area is therefore $\langle I_U \rangle = \frac{E/t}{A}$. Of course, the energy produced on the parcel of land is less than the solar energy that illuminated the parcel. We can define the overall efficiency η as the ratio of the two. We have

$$\langle I_U \rangle = \eta \langle I_H \rangle. \qquad (I.3)$$

The simplicity of equation (3) is that all we need to know to place solar energy in the broad context of energy production is the overall efficiency of the processes that convert solar energy into useful energy. This fact holds for photovoltaics, solar heat, growth of plants, and all other cases where sunlight is directly converted to energy.

Bio-fuels A relatively recent paper in *Chemical and Engineering News* "Chasing cheap feedstocks" (*C&EN*, Aug. 12, 2013, page 11) rated numerous energy fuels in terms of production, but in units that are not very illuminating. The *best* crops (well watered, well fertilized) produce about 10 tons

FIGURE I.13 Solar/thermal/electric generating systems in the Mojave Desert, California. Left: SEGS unit at Kramer Junction, California (2005); Right: Ivanpah installation (2013). Both systems use concentrating mirrors to produce super-heated steam to turn turbines.

of drymatter per acre per year. A little arithmetic with conversion factors (acre = 4047 m^2; ton = 907 kg; year = 3.16 × 10^7 s; heat content of drymatter ≈ 15 MJ/kg) brings the result that these fuels are produced at the rate of about 1.2 W/m^2 on year-round average basis. (Untended hardwood forests produce firewood at about 0.12 W/m^2.) Compared to sunlight at 200 W/m^2, this amounts to a solar efficiency of 0.06%. What we are discussing here is the heat energy that could be released if the drymatter were simply burned in place. Anything whatsoever that is done (such as mere collection and delivery) results in a lower efficiency yet.

There has been considerable debate about the amount of energy used in converting corn into ethanol, EtOH. The most optimistic estimates come from Shapouri, Duffield, and Wang.[6] Using data from a 9-state survey, they produced the following figures, which we will translate to SI: Annual corn yield is 125 bushels/acre; each bushel produces 2.64 gallons of ethanol; and the HHV of ethanol is 83,961 BTU/gallon. We add that there are 4047 m^2 in an acre, 1055 J in a BTU, and 31.6 million seconds in a year. The result is that the *gross* production of ethanol amounts to 0.228 W/m^2, averaged over the year. The purpose of the Shapouri group's paper was to show that the production of ethanol is energy net-positive, *i.e.,* there is more energy in the ethanol than it takes to produce it. Their figures are that the *net* amount of energy in a gallon of ethanol is 6732 BTU/gallon, a mere 8% of the HHV. Expressed in SI units, it amounts to a very pathetic 0.018 net watts per square meter.

The area of all of the contiguous 48 states (including water area and huge tracts of non-arable land) is 9.8 × 10^{12} m^2, and only 18% of the land (1.76 × 10^{12} m^2) is arable. The amount

of annual energy that could be produced by the best-tended best crops at 1.2 W/m^2 is about 67 EJ, only two-thirds of our consumption. The amount of *net* energy that could be produced using all arable land for production of ethanol (at 0.018 W/m^2) is 1.0 EJ, which is only 1% of our energy requirement.

Solar Heat Any box covered with glass or transparent plastic facing the sun collects solar heat. Its efficiency at collection depends upon the interior color (preferably black), the shape (large window area compared to area of the sides), insulation, and the properties of the transparent cover (in some cases, double glazing helps). Most importantly, the interior and exterior temperatures play a huge role. The higher the temperature difference between the interior and the exterior, the more the heat lost through the walls and the glazing. All commercial solar heat units have graphs showing the efficiency dropping off as the outside temperature gets lower.

Solar/Thermal Electricity One way to produce electricity from sunlight is to use mirrors to concentrate sunlight so as to create a high temperature, make superheated steam, and then run a conventional steam-turbine/generator set. Figure I.13 shows two ways of accomplishing that task. The shiny parabolic trough system on the left reflects sunlight to heat a pipe containing a high-temperature oil pumped to a heat exchanger to boil water. The troughs slowly rotate during the day as the sun moves across the sky. It is not necessary to make seasonal adjustments.

On the right of Figure I.13 is a new installation near Ivanpah Dry Lake. One-hundred-seventy-thousand (170,000) independent computer-controlled heliostats, each with two 7.5 m^2 mirrors, reflect sunlight to the dark regions at the tops of the towers to boil water. They are spread out over 1416 hectares (3500 acres), and the peak power is 392 MW. The

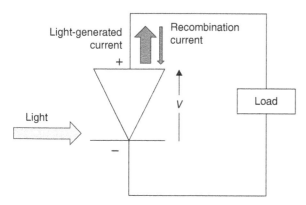

FIGURE I.14   Schematic photovoltaic cell. It is a diode (one-way valve for current), but the incident light makes the current go the opposite way from normal. As the voltage *V* rises, the recombination current rises dramatically, so it is necessary to draw sufficient current to keep the PV cell operating efficiently.

expected annual production is 1080 GWh, which amounts to an annual capacity factor of 31%. The system will generate year-round average power of 8.7 W/m^2 of land.

Photovoltaics (PV)   Photocells convert light directly into electricity, as shown in the schematic drawing in Figure I.14. All PV cells are diodes (rectifiers) that normally conduct electricity in only one direction. This one-way phenomenon occurs because dissimilar materials are in contact. Incident light, however, produces a reverse current.

The problem that bedevils PV cells is that to boost an electron to cross the barrier requires a given amount of energy.

This job is supposed to be done by sunlight, with a wide spectrum of wavelengths, hence a wide array of energies to do the boosting. Some light does not have enough energy, and some has excess energy that is simply wasted as heat. There are some multiple band-gap (energy-step size) devices with efficiencies well above 30%, but they are laboratory-size devices made with expensive, exotic materials.

There is a trade-off between efficiency and size, and it hinges on economics. Comparing two PV cells, let us assume that one has twice the efficiency of the other. Then, for a given amount of power, we can get by with half the physical area. That might seem like the cost would be cut in half, but the better cells might cost 100 (or 50) times as much. In that case, economics dictates the use of the less efficient cells. Alternatively, the land may be horrendously expensive, in which case economics favors the more efficient cells.

In any case, experience with large PV arrays shows that the annual capacity factor is usually less than 20%.

SUMMARY

Sunlight is free, and so are wind, geothermal heat, and water at high elevation. For that matter, coal, oil, and natural gas are also free. So is uranium. The problems lie in efficient, inexpensive delivery to users who need energy on demand. Winds are capricious, sunlight is not available at night, and the biggest and best hydropower sites are already in use. We have long since passed the point where biomass can make any significant increase to our energy supply. The following chapters, written by energy technologists, address the details.

LIST OF CONTRIBUTORS

Rafid Al-Khoury Faculty of Civil Engineering and Geosciences, Delft University of Technology, Delft, The Netherlands

Mathew C. Aneke Department of Chemical Engineering, Faculty of Science and Engineering, University of Hull, United Kingdom

Roberto D. Arce Physics Institute of Litoral (IFIS Litoral, UNL-CONICET), Santa Fe, Argentina; Facultad de Ingeniería Química, UNL, Santa Fe, Argentina

Amornchai Arpornwichanop Computational Process Engineering, Department of Chemical Engineering, Faculty of Engineering, Chulalongkorn University, Bangkok, Thailand

Suttichai Assabumrungrat Department of Chemical Engineering, Faculty of Engineering, Chulalongkorn University, Bangkok, Thailand

Suthida Authayanun Department of Chemical Engineering, Faculty of Engineering, Srinakharinwirot University, Nakhon Nayok, Thailand

Ömer Aydan Department of Civil Engineering and Architecture, University of the Ryukyus, Nishihara, Japan

M. Aram Azadpour Engineer consultant in digital logic functional modeling and digital logic design verification, Grapevine, TX, USA

R. Baile UMR CNRS 6134, Université de Corse, Corte, France

A. Barroso School of Engineering, University of Seville, Seville, Spain

Ali Bolatturk Department of Mechanical Engineering, Suleyman Demirel University, Isparta, Turkey

Richard S. Brown Pacific Northwest National Laboratory, Richland, WA, USA

Nicolás Budini Physics Institute of Litoral (IFIS Litoral, UNL-CONICET), Santa Fe, Argentina

Román H. Buitrago Physics Institute of Litoral (IFIS Litoral, UNL-CONICET), Santa Fe, Argentina; Facultad de Ingeniería Química, UNL, Santa Fe, Argentina

J. Cañas School of Engineering, University of Seville, Seville, Spain

Thomas J. Carlson Pacific Northwest National Laboratory, Richland, WA, USA

Roger H. Charlier Vrije Universiteit Brussel—VUB, Free University of Brussels, Brussels, Belgium; Florida Atlantic University, Boca Raton, FL, USA

Tin-Tin Chow Building Energy and Environmental Technology Research Unit, Division of Building Science and Technology, College of Science and Engineering, City University of Hong Kong, Hong Kong

Alison H. Colotelo Pacific Northwest National Laboratory, Richland, WA, USA

Angus C. W. Creech Institute of Energy Systems, University of Edinburgh, Edinburgh, Scotland

John P. Deane Energy Policy and Modelling Group, Environmental Research Institute, University College Cork, Ireland

Zhiqun (Daniel) Deng Pacific Northwest National Laboratory, Richland, WA, USA

Mark Diesendorf Interdisciplinary Environmental Studies, University of New South Wales, Sydney, Australia

Peter A. Dowd School of Civil, Environmental and Mining Engineering, University of Adelaide, Australia

Annette Evans Graduate School of the Environment, Macquarie University, Sydney, Australia

Tim J. Evans Graduate School of the Environment, Macquarie University, Sydney, Australia

Philip M. Fearnside National Institute for Research in Amazonia (INPA), Manaus, Brazil

Wolf-Gerrit Früh School of Engineering and Physical Sciences, Heriot-Watt University, Edinburgh, Scotland

Brian Ó Gallachóir Energy Policy and Modelling Group, Environmental Research Institute, University College Cork, Ireland

Winston Garcia-Gabin ABB Corporate Research Center, Västerås, Sweden

Alberto Gemelli Dipartimento di Ingegneria dell'Informazione, Università Politecnica delle Marche, Ancona, Italy

D. P. Georgiou Department of Mechanical Engineering and Aeronautics, Thermal Engines Lab, University of Patras, Rion-Patras, Greece

Daniel Gessler Alden Research Laboratory, Inc., Fort Collins, CO, USA

Fiona M. Gray School of Chemistry, University of St. Andrews, St. Andrews, UK

Zhiqiang Guan Queensland Geothermal Energy Centre of Excellence (QGECE), School of Mechanical and Mining Engineering, The University of Queensland, Brisbane, Australia

Hal Gurgenci Queensland Geothermal Energy Centre of Excellence (QGECE), School of Mechanical and Mining Engineering, The University of Queensland, Brisbane, Australia

Martin Hand Department of Earth Sciences, University of Adelaide, Australia

Joerg Hartmann Independent Sustainability Consultant, Estes Park, CO, USA

Howard C. Hayden University of Connecticut, Mansfield, CT, USA

Herbert Inhaber Risk Concepts, Las Vegas, NV, USA

Damon Honnery Department of Mechanical and Aerospace Engineering, Monash University—Clayton Campus, Melbourne, Australia

Kamel Hooman Queensland Geothermal Energy Centre of Excellence (QGECE), School of Mechanical and Mining Engineering, The University of Queensland, Brisbane, Australia

Mikel Iribas-Latour Wind Energy Department, Fundación CENER-CIEMAT, Sarriguren, Spain

Stefan Ivanell Uppsala University Campus Gotland, Section for Wind Energy, Visby, Sweden

Gary E. Johnson Pacific Northwest National Laboratory, Richland, WA, USA

Mehmet Kanoglu Department of Mechanical Engineering, University of Gaziantep, Gaziantep, Turkey

Hugo Abi Karam Department of Meteorology, Institute of Geosciences, Federal University of Rio de Janeiro, Brazil

Thomas B. Kingery Information Technology, Delaware, OH, USA

Urban Kjellén Norwegian University of Science and Technology, Trondheim, Norway; Statkraft, Oslo, Norway

Masami Kojima Energy and Extractives Global Practice, The World Bank, Washington, DC, USA

Jacob Ladenburg KORA, the Danish Institute for Local and Regional Government Research, Denmark

Ion-Doré Landau Automatic Control Department, GIPSA-LAB, Grenoble, France

Jay H. Lehr Editor-in-Chief, Science Director, Heartland Institute, Chicago, IL, USA

Jack Keeley Senior Editor, Former Chief of Ground Water Research, USEPA Kerr Water Research Laboratory, Ada, OK, USA

Pak Sing Leung Mechanical Engineering Division, Faculty of Engineering and Environment, Northumbria University, Newcastle upon Tyne, United Kingdom

Sauro Longhi Dipartimento di Ingegneria dell'-Informazione, Università Politecnica delle Marche, Ancona, Italy

Adriano Mancini Dipartimento di Ingegneria dell'-Informazione, Università Politecnica delle Marche, Ancona, Italy

J. C. Marín School of Engineering, University of Seville, Seville, Spain

Matthew C. Menkiti Department of Chemical Engineering, Nnamdi Azikiwe University Awka, Nigeria

Robert Mikkelsen Department of Wind Energy, DTU, Lyngby, Denmark

Rosemarie Mohais School of Civil, Environmental and Mining Engineering, University of Adelaide, Australia

Patrick Moriarty Department of Design, Monash University—Caulfield Campus, Melbourne, Australia

J. F. Muzy UMR CNRS 6134, Université de Corse, Corte, France

Greg F. Naterer Faculty of Engineering and Applied Science, Memorial University of Newfoundland, St. John's, Canada

Michael Negnevitsky Centre for Renewable Energy and Power Systems, University of Tasmania, Australia

Isaac Orr The Heartland Institute, Chicago, IL, USA

F. París School of Engineering, University of Seville, Seville, Spain

Antoine Peiffer Principle Power Inc., Berkeley, CA, USA

Gene R. Ploskey Pacific Northwest National Laboratory, Richland, WA, USA

Kevin Pope Faculty of Engineering and Applied Science, Memorial University of Newfoundland, St. John's, Canada

Panu Pratumnopharat Mechanical Engineering Department, Faculty of Engineering, Rajamangala University of Technology, Thanyaburi, Pathumtani, Thailand

Dominique Roddier Principle Power Inc., Berkeley, CA, USA

Marc A. Rosen Faculty of Engineering and Applied Science, University of Ontario Institute of Technology, Oshawa, Canada

Somnath Baidya Roy Department of Atmospheric Sciences, University of Illinois at Urbana-Champaign, Urbana, IL, USA

Dang Saebea Department of Chemical Engineering, Faculty of Engineering, Burapha University, Chonburi, Thailand

Petar Sarajcev Department of Power Systems, University of Split, FESB, Split, Croatia

Timothy C. Sassaman Alden Research Laboratory, Inc., Holden, MA, USA

Javier A. Schmidt Physics Institute of Litoral (IFIS Litoral, UNL-CONICET), Santa Fe, Argentina; Facultad de Ingeniería Química, UNL, Santa Fe, Argentina

Michael J. Smith Departamento de Química, Universidade do Minho, Braga, Portugal

Vladimir Strezov Graduate School of the Environment, Macquarie University, Sydney, Australia

Vorachatra Sukwattanajaroon Department of Chemical Engineering, Faculty of Engineering, Chulalongkorn University, Bangkok, Thailand

Bertrand F. Tchanche Ecole Supérieure d'Ingénieurs en Electronique et Electrotechnique (ESIEE), Amiens, France 14, Quai de la Somme, 80000 Amiens

N. G. Theodoropoulos Department of Mechanical Engineering and Aeronautics, Thermal Engines Lab, University of Patras, Rion-Patras, Greece

Alexandre C. Thys Florida Atlantic University, Boca Raton, FL, USA

Justin J. Traiteur Department of Atmospheric Sciences, University of Illinois at Urbana-Champaign, Urbana, IL, USA

Ka-Kui Tse Building Energy and Environmental Technology Research Unit, Division of Building Science and Technology, College of Science and Engineering, City University of Hong Kong, Hong Kong

Norman Tse Building Energy and Environmental Technology Research Unit, Division of Building Science and Technology, College of Science and Engineering, City University of Hong Kong, Hong Kong

Choashui Xu School of Civil, Environmental and Mining Engineering, University of Adelaide, Australia

Darine Zambrano Division of Systems and Control, Department of Information Technology, Uppsala University, Uppsala, Sweden

PART I

WIND

1

ACCEPTANCE OF WIND POWER: AN INTRODUCTION TO DRIVERS AND SOLUTIONS

Jacob Ladenburg

1.1 INTRODUCTION

Presently, on-land wind turbines located at windy locations are not far from producing electricity cost-effectively when compared to conventional brown power generation, such as coal, gas, and oil. However, wind turbines require relatively large areas in order to obtain significant generation capacities and each wind turbine is a large structure, which physically can be dominating in the open landscape. Accordingly, despite the improving costs of generation, the acceptance of wind turbines cannot be taken for granted, when new areas are to be developed. In many countries positive attitude toward wind power development is commonly found when people are asked to express their attitude toward wind power in general or, for example, specific wind power projects (Firestone et al., 2009; Haggett and Toke, 2006; Horbaty et al., 2012; Johansson and Laike, 2007; Todt et al., 2011). However, as the subsequent sections will point out, how the perception has evolved and how some people might have a negative attitude toward wind power, while others have a positive one, is a complex multidimensional matrix with nonzero off-diagonal elements.

First of all, it is important to recognize that negative attitudes are typically developed from the notion that wind turbines are not free from impacts and through their visibility in the landscape (Bishop and Miller, 2007; Manwell et al., 2002; Molnarova et al., 2012), the noise level (Chourpouliadis et al., 2012; Kaldellis et al., 2012; Pedersen et al. 2010) and potential impact on birds and bats (Aravena et al., 2006; Meyerhoff

et al., 2010) are perceived by some as an intrusion in the landscape. This is relative to not only other more natural landscape elements, but also other manmade structures (Álvarez-Farizo and Hanley, 2002; Strazzera et al., 2012). It is through these impacts and the influence the impacts have on the well-being of people and their perception of the urban landscape they live in that attitudes are formed. On the positive side, wind turbines generate renewable energy and some people might even perceive them as a natural and beautiful element in the landscape (Ladenburg and Lutzeyer, 2012). Accordingly, whether the individual person has a positive or negative attitude is a function of how the respondents perceive the costs and benefits of wind power. As it will be exemplified, how an individual perceive the costs and benefits is a function of the characteristics of the respondents, the type of experience they have with wind power and other energy sources, and particularly the type and level of wind power development. However, as it will also be addressed, a simple assessment of costs and benefits alone might be too simplistic, particularly if specific wind power projects are the subject for an attitude analysis. Grounded in specific planned projects, the process of how and why the wind turbines are going to be located at the specific site comes into play. More specifically, issues such as the trust in developers, fairness, and local involvement and participation influence acceptance strongly.

The aim of this chapter is to give the reader an introduction to some of the acceptance elements and a more detailed insight into the complexity of the drivers and solutions that can facilitate higher acceptance levels and vice versa.

Alternative Energy and Shale Gas Encyclopedia, First Edition. Edited by Jay H. Lehr and Jack Keeley.

1.2   THE TRADITIONAL DEMOGRAPHIC VARIABLES

A common approach in attitude studies is to use the socio-demographic characteristics of the respondents to explain variation in the attitude toward wind power, such as gender, education, income, and age. The attitude relations of the first three socio-demographic variables point in different directions; see, for example, Ek (2005), Firestone and Kempton (2007), Jacquet (2012), Johansson and Laike (2007), Klick and Smith (2010), Ladenburg and Dahlgaard (2012), Lilley et al. (2010), and Pedersen and Johansson (2012). However, the effect of age on acceptance seems to be more consistent. Except for a few surveys (Klick and Smith, 2010; Ladenburg, 2009a), the main headline seems to be that younger people are more positive toward wind power when compared to older people and this effect seems to be present in the case of both on-land and offshore wind power development (Bishop and Miller, 2007; Ek, 2005; Firestone and Kempton, 2007; Jacquet, 2012; Jones et al., 2011; Ladenburg, 2007; Ladenburg and Dahlgaard, 2012; Ladenburg and Möller, 2011; Lilley et al., 2010; Pedersen and Johansson, 2012).

The relevance of this apparent age effect is to be seen in a broader perspective. Ladenburg and Lutzeyer (2012) discuss the important question of whether these effects are persistent (a generation effect) or whether the effect is a function of the age of the person, who gradually becomes more negative toward wind power when aging (an aging effect) or whether potentially a combination of both. Which effect that drives acceptance is crucial for understanding the underlying perception of wind power when planning the future wind power landscape! If the effect is a pure aging effect we would expect acceptance of wind power to be a constant, weighted for the distribution of the relevant population naturally. Changes in acceptance that can be attributed to the age of the population will thus reflect changes in the distribution across age. If the distribution is unchanged, so would acceptance, all things equal. However, this becomes a completely different story if the effect is a generation effect. In that case, the studies point that acceptance might increase, the associated demand for wind power becomes larger, and the demand for impact mitigation is lowered. Accordingly, in 10–20 years acceptance toward wind power will be higher. This naturally requires that new generations are equally positive as the present younger generations are.

1.3   DYNAMIC EFFECTS FROM WIND POWER DEVELOPMENT

The above-mentioned socio-demographic effects related to the acceptance of wind power in principle assume no dependency on the wind power landscape people live in, the deployment of more wind power, and the experience people gain with wind power. However, with an increasing number of wind turbines in the landscape, the wind power footprints that drive acceptance become clearer and stronger. Consequently, it therefore becomes increasingly important to hold track on and have a feeling for the potential effects that wind power development and the associated experience with wind power can have on acceptance. Experiences can take many dimensions, ranging from having seen a turbine, seen a wind turbine occasionally, having a wind turbine in the viewshed, distance to the nearest wind turbine, and the number of wind turbines in the area where people live or have their daily or vocational traveling pattern. Independent of the type of experience, we would expect them to be more abundant in the relevant population as wind power capacity is increased and wind turbines in the landscape become more abundant. Accordingly, if these experience elements influence the acceptance of wind power development, this should ideally be accounted for in the larger wind energy planning perspective. By not accounting for the potentially dynamic negative effects, acceptance of wind power might suffer from high levels of opposition due to wind power development itself. In the following sections, some of the tested experiences variables will be elaborated on.

1.3.1   Viewshed

People who have a wind turbine in the viewshed have a completely different experience/relation to wind power compared to people who do not have a wind turbine in the viewshed. Being able to see a wind turbine whenever stepping outside and taking a look around might thus shape/tune the attitude in a particular manner. The viewshed effect might invoke a negative response and decrease acceptance/increase the demand for visual impact reductions seen in Ladenburg (2008), Ladenburg and Dubgaard (2007), and AMR-Interactive (2010). However, viewshed might not reduce acceptance (Johansson and Laike, 2007) and depending on the location of development it can even be positive (Ladenburg, 2010). Viewshed effects have typically been analyzed assuming that a "view is a view." But it must be acknowledged that a viewshed can vary between people. Some people might see a single turbine at a far distance in the horizon, while other people have one or several turbines at really close range. Though the literature seems to be weak on this point, the interaction between viewshed and spatial distribution of the wind turbines in the local areas is expected to influence acceptance (Ladenburg et al., 2013). Accordingly, it seems like the viewshed effects are heterogeneous.

1.3.2   Spatial Relations

Moving on from the viewshed effects to the spatial distribution of wind turbines in the local area, the proximity of the nearest wind turbine or wind farm seems to influence acceptance. However, as in the case of viewshed effects, the results point in different directions. Jacquet (2012) finds that distance to existing wind farms increases acceptance. Ladenburg and Möller (2011) find mixed effects, but a general decrease in acceptance of existing offshore wind farms as a function of the distance. Andersen et al. (1997) find that acceptance of a new turbine decreases with the distance to the nearest existing wind turbine. Warren et al. (2005) find negative effects from distance toward existing wind farms, while Johansson and Laike (2007) do not find an effect from distance in relation to the general attitude toward wind power. AMR-Interactive (2010) finds increased support as a function of the distance to existing and planned wind turbines toward a wind turbine 1–2 km from the residence. Swofford and Slattery (2010) find increased support toward wind power in general and a new wind farm as a function of the distance to existing wind turbines.

Accordingly, the variation in the distance effect suggests that it might not be distance alone that drives the attitude. Wind turbine characteristics, such as height, number of turbines, and the location of the turbines in the landscape, might also be important (Ek, 2006; Meyerhoff et al., 2010; Molnarova et al., 2012). For example, living 1 km from a single 500 kW turbine might have a relatively larger effect on attitude, compared to living 2 km from ten 2 MW turbines. An example of such an application is Ladenburg and Möller (2011), who find that both the distance to the nearest offshore wind farm and the number of wind turbines in the nearest offshore wind farm influence acceptance significantly. One explanation of the large variance in the spatial effects might also be found in the relative distribution and the number of turbines in the area. Thayer and Freeman (1987) estimate the frequency of wind turbine encounters on the perception of wind power and find that "familiar" respondents are more negative. Ladenburg (2010) finds some evidence that a respondent who sees more than 20 turbines daily is more negative toward existing offshore wind farms compared to respondents who see fewer turbines, though the number of turbines seen daily does not seem to influence the relative preferences for more onshore and offshore wind power development (Ladenburg, 2015). Ladenburg and Möller (2011) find that the number of turbines in the nearest offshore wind farm seems to influence acceptance of offshore wind farms. Krohn and Damborg (1999) find no effects from the number of turbines in the area on the attitude toward wind power in general. Jones et al. (2011) estimate the maximum ex ante perception of the number of turbines: more than 20% express that the region can sustain more than 100+ turbines. Ladenburg and Dahlgaard (2012) find that respondents seeing more than five turbines daily are more negative toward existing onshore wind turbines. However, the results also suggest that the negative cumulative effect is conditional on several parameters. Ladenburg et al. (2013) find that cumulative effects only seem to be triggered if people

have one or several onshore wind turbines in the viewshed from the residence or summerhouse. People who see many turbines daily but do not have turbines in the viewshed have an acceptance that matches the acceptance among people who see very few turbines. Accordingly, the study points that the cumulative effects from wind turbines that are seen through the daily traveling patterns are small and different from the cumulative effects from wind turbines at a closer range (viewshed).

In addition to these studies, Knapp and Ladenburg (2015) carry out a review focusing preferences for wind power location. They report multiple findings that people prefer the visual impacts from wind turbines to be mitigated. Accordingly, wind turbines should be located at some distance from residential areas (onshore) and far from the coastline (offshore).

1.4 OFFSHORE WIND FARMS A FUTURE SOLUTION?

The many test offshore wind power sites and especially the two large-scale wind farms at Nysted and Horns Rev in Denmark have given wind power developers and policy planners substitute wind power development opportunities. The industry and developers have grabbed the opportunity and pushed forward a rapid development particularly in the northern European waters (Bilgili et al., 2011). Offshore wind farms offer two particular differences compared to onland development. First, the wind regimes are better offshore. Second, by developing offshore, the wind turbines are typically located at a distance from the coast, which reduces the otherwise typical impacts associated with wind turbines (see section 1.1). As a natural consequence, people often prefer offshore development to onshore (Campbell et al., 2011; Ek, 2006; Ladenburg, 2008, Ladenburg, 2009b; Ek and Persson, 2014; Vecchiato 2014) though this might not always be the case (McCartney, 2006). However, the higher acceptance for offshore development comes with a cost, literally. Offshore wind farms are more costly to develop and operate and these costs are particularly large if locations far from shore and at deep waters are exploited (European Environment Agency, 2009). Naturally, these higher generation costs give strong economic incentives to find suitable developing sites, which are close to the shore or even just on the coast as done in, for example, some Danish wind farms. But locating offshore wind farms close to shore invokes an important opposition attribute of wind farms, namely the visual impacts. Several offshore studies thus strongly find evidence that near shore locations suffer from visual impacts, just as their onshore counterparts (Bishop and Miller, 2007; Krueger et al., 2011; Ladenburg and Dubgaard, 2007; Waldo, 2012; Westerberg et al., 2013; 2015)

Interestingly, the studies also give an intuitive solution to the problem; the wind farms should be located at larger distances from the shore. However, the optimal location is not necessarily at an "out of sight" range. Some levels of visual impacts seem to be acceptable from a consumer point of view. More specifically, the marginal benefits of locating an offshore wind farm an additional kilometer further from the coast are larger at near shore locations compared to locations further offshore; see Ladenburg and Lutzeyer (2012) for a review.

1.4.1 NIMBY, Trust, and Fairness Perceptions

Acceptance of specific wind power projects often meets local resistance, despite that people in general are positive toward wind power. Initially, the missing match between wind power acceptance on a general level and local wind power deployment had the Not-In-My-Back-Yard label attached (NIMBY; Devine-Wright, 2005; O'Hare, 1977), and NIMBY-induced attitudes have commonly been perceived as ignorant, selfish, and irrational uninformed (Aitken, 2010; Barry et al., 2008; Wolsink and Devilee, 2009). However, newer research has suggested that the NIMBY definition might only explain a smaller variance in the opposition toward local development (Devine-Wright, 2005; Jones and Eiser, 2009; Kempton et al., 2005; Wolsink, 2007). In his paper from 2007, Wolsink argues that the perception of equity and fairness seems to influence the presence of "backyard" motives and not selfishness. Devine-Wright and Howes (2010) argue that local opposition arises as a kind of place-protective action, which is triggered when emotional attachment of place-related identity is threatened. Eltham et al. (2008) illustrate that uncertainty and negative perceptions of the developers' motives, distrust of the developers, and disbelief in the planning system can have a negative influence on the outcome of wind farm projects; see also Devine-Wright (2005). In a review, Devine-Wright (2011) interestingly argues that breaking the cycle of NIMBYism will require new approaches how to think and practice public engagement in order to obtain a better connection between national policy making with the areas affected by specific projects. Finally, it is worth mentioning that Dewine-Wright jointly with a number of researchers focuses particularly on these issues in an edited book by Dewine-Wright (2011).

1.5 WIND POWER DEVELOPMENT AND THE DEMAND FOR OTHER RENEWABLE ENERGY SOURCES

Finally, the acceptance of wind power should be seen in the broader renewable energy perspective. In the pursuit of cleaner energy production and less dependency on brown energy sources, wind power is one among different Renewable Energy Sources (RES), mainly including solar energy (heat and photovoltage), biomass, and hydropower. In the majority of the studies, wind power and solar energy hold a higher level of demands compared to other RES (Cicia et al., 2012; Hanley and Nevin, 1999; Komarek et al., 2011; Scarpa and Willis, 2010). A common feature of the above-mentioned studies is that they do not control for the level of knowledge and experience the respondents have with RES, just as discussed in section 1.3 in the case of wind power. Using the estimated acceptance relation in a RES deployment scheme beyond a short time span would thus rest on the assumption that the acceptance that people have is independent of the experience people have with the different RES and thereby also the level of RES development. Ladenburg (2014) tests the effects of wind power experience in terms of having onshore and offshore wind turbines in the viewshed and/or living in areas with large offshore wind farms close to and far from shore. Having an onshore wind turbine in the viewshed reduces the demand for wind power. Interestingly, having an offshore wind farm in the viewshed increases the demand for wind power, except in the sample of respondents who have experience with large visual effects from offshore wind farms. In addition, the results also suggest that wind turbine experiences also influence the choice of solar energy in a negative way.

Fimereli et al. (2008) test these properties in a stated preference survey, using two categories of experience with RES. The first is whether or not the respondents have ever seen the specific type of RES or a coal/gas power station. The other variable is whether the respondent has lived near one of the RES or a coal/gas power station. Fimereli et al. (2008) find no effect from experience with wind power, biomass, or nuclear energy on the demand for the different energy sources. Accordingly, having lived near, for example, a wind turbine and having hands-on experience with the potential disamenities such as noise, flickering, or general visual impacts have not shaped the preferences/relative levels of acceptance compared to respondents who do not have these experiences. However, they find that respondents who have seen a gas or coal power station have a significantly higher demand for low carbon technologies relative to respondents who have not seen a gas or coal power station.

So both the studies point that the acceptance of wind power could potentially be triggered not only by the experience with other energy sources but also by the availability of other energy sources.

1.6 CONCLUSIONS

In this chapter some of the many elements in the matrix of wind power acceptance determinants have been presented. The aim was not only to give the reader an introduction to the topic but also shed light on the complexity of the formation of acceptance.

The complexity of the acceptance matrix makes it difficult to point out easy-to-apply solutions in order to maintain or increase acceptance, but individual topics hopefully give some insight into what drives acceptance and opposition. However, as put forward many of the presented elements are dependent on each other. Acceptance of wind power thus becomes a dynamic function with several layers of interactions. A good example is the spatial relations of wind power development. Acceptance is thus not a linear function of distances to wind turbines, having a wind turbine in the viewshed, or the number of turbines in the area. The studies presented here show that the elements intervene with each other and that the effect from one element might be dependent on the others. In addition, the aspects of local acceptances, trust, public involvement, socio-demographic relations, whether to develop onshore or offshore, and strategy to handle acceptance quickly become multifaceted. Hopefully the content and the many references in this chapter can be guidance and give the reader a constructive insight into the topic.

This chapter was written as part of project 0602-00205B under the Danish Agency for Science, Technology and Innovation and project 1305-00021B under the Danish Council for Strategic Research.

REFERENCES

Aitken, M. (2010). Why we still don't understand the social aspects of wind power: a critique of key assumptions within the literature. *Energy Policy*, 38:1834–1841.

Álvarez-Farizo, B., Hanley, N. (2002). Using conjoint analysis to quantify public preferences over the environmental impacts of wind farms. An example from Spain. *Energy Policy*, 30:107–116.

AMR-Interactive. (2010). Community attitudes to wind farms in NSW. Department of Environment, Climate Change and Water, New South Wales, Australia. Available at: http://www.environment.nsw.gov.au/resources/climatechange/10947WindFarms_Final.pdf

Andersen, K. H., Thomsen, M., Kruse, J. (1997). *Rapport om hvordan en dansk kommune blev selvforsynende med ren vindenergi og skabte ny indkomst til kommunens borgere*. Nordvestjysk Folkecenter for Vedvarende Energi (in Danish).

Aravena, C., Martinsson, P., Scarpa, R. (2006). The effect of a monetary attribute on the marginal rate of substitution in a choice experiment. Environmental and Resource Economics 3rd World Congress, Kyoto, Japan, July 3–7.

Barry, J., Ellis, G., Robinson, C. (2008). Cool rationalities and hot air: a rhetorical approach to understanding debates on renewable energy. *Global Environmental Politics*, 8:67–98.

Bilgili, M., Yasar, A., Simsek, E. (2011). Offshore wind power development in Europe and its comparison with onshore counterpart. *Renewable and Sustainable Energy Reviews*, 15:905–915.

Bishop, I. D., Miller, D. R. (2007). Visual assessment of off-shore wind turbines: the influence of distance, contrast, movement and social variables. *Renewable Energy*, 32:814–831.

Campbell, D., Aravena, C. D., Hutchinson, W. G. (2011). Cheap and expensive alternatives in stated choice experiments: are they equally considered by respondents? *Applied Economics Letters*, 18:743–747.

Chourpouliadis, C., Ioannou, E., Koras, A., Kalfas, A. I. (2012). Comparative study of the power production and noise emissions impact from two wind farms. *Energy Conversion and Management*, 60:233–242.

Cicia, G., Cembalo, L., Del Giudice, T., Palladino, A. (2012). Fossil energy versus nuclear, wind, solar and agricultural biomass: insights from an Italian national survey. *Energy Policy*, 42:59–66.

Devine-Wright, P. (2005). Beyond NIMBYism: towards an integrated framework for understanding public perceptions of wind energy. *Wind Energy*, 8:125–139.

Devine-Wright, P. (2011). Public engagement with large-scale renewable energy technologies: breaking the cycle of NIMBYism. *Wiley Interdisciplinary Reviews: Climate Change*, 2(1):19–26.

Devine-Wright, P. (editor) (2011). *Renewable Energy and the Public: From NIMBY to Participation*. Earth Scan.

Devine-Wright, P., Howes, Y. (2010). Disruption to place attachment and the protection of restorative environments: a wind energy case study. *Journal of Environmental Psychology*, 30(3):271–280.

Ek, K., Persson, L. (2014). Wind farms—Where and how to place them? A choice experiment approach to measure consumer preferences for characteristics of wind farm establishments in Sweden. *Ecological Economics*, 105:193–203.

Ek, C. (2006). Quantifying the environmental impacts of renewable energy. In: *Environmental Valuation in Developed Countries: Case Studies*, Pearce, D. W. (editor). Edward Elgar Publishing, Northampton, MA, pp. 181–200.

Ek, K. (2005). Public and private attitudes towards "green" electricity: the case of Swedish wind power. *Energy Policy*, 33:1677–1689.

Eltham, D. C., Harrison, G. P., Allen, S. J. (2008). Change in public attitudes towards a Cornish wind farm: implications for planning. *Energy Policy*, 36:23–33

European Environment Agency. (2009). Europe's onshore and offshore wind energy potential – an assessment of environmental and economic constraints. EEA Technical report, No. 6/2009. European Environment Agency.

Fimereli, E., Mourato, S., Pearson, P. (2008). Measuring preferences for low-carbon energy technologies in SE England: the case of electricity generation. European Association of Environmental and Resource Economists 16th Annual Conference, Gothenburg, Sweden.

Firestone, J., Kempton, W. (2007). Public opinion about large off-shore wind power: underlying factors. *Energy Policy*, 35:1584–1598.

Firestone, J., Kempton, W., Krueger, A. (2009). Public acceptance of offshore wind power projects in the USA. *Wind Energy*, 12:183–202.

Haggett, C., Toke, D. (2006). Crossing the great divide – using multi-method analysis to understand opposition to windfarms. *Public Administration*, 84:103–120.

Hanley, N., Nevin, C. (1999). Appraising renewable energy developments in remote communities: the case of the North Assynt Estate, Scotland. *Energy Policy*, 27:527–547.

Horbaty, R., Huber, S., Ellis, G. (2012). Large-scale wind deployment, social acceptance. *Wiley Interdisciplinary Reviews: Energy and Environment*, 1:194–205.

Jacquet, J. B. (2012). Landowner attitudes toward natural gas and wind farm development in northern Pennsylvania. *Energy Policy*, 50:677–688.

Johansson, M., Laike, T. (2007). Intention to respond to local wind turbines: the role of attitudes and visual perception. *Wind Energy* 10:435–451.

Jones, C. R., Eiser, J. R. (2009). Identifying predictors of attitudes towards local onshore wind development with reference to an English case study. *Energy Policy*, 37:4604–4614.

Jones, C. R., Orr, B. J., Eiser, J. R. (2011). When is enough, enough? Identifying predictors of capacity estimates for onshore wind-power development in a region of the UK. *Energy Policy*, 39:4563–4577.

Kaldellis, J. K., Garakis, K., Kapsali, M. (2012). Noise impact assessment on the basis of onsite acoustic noise immission measurements for a representative wind farm. *Renewable Energy*, 41:306–314.

Kempton, W., Firestone, J., Lilley, J., Rouleau, T., Whitaker, P. (2005). The offshore wind power debate: views from cape cod. *Coastal Management*, 33:119–149.

Klick, H., Smith, E. R. A. N. (2010). Public understanding of and support for wind power in the United States. *Renewable Energy*, 35:1585–1591.

Komarek, T. M., Lupi, F., Kaplowitz, M. D. (2011). Valuing energy policy attributes for environmental management: choice experiment evidence from a research institution. *Energy Policy*, 39:5105–5115.

Knapp, L., Ladenburg, J. (2015). Spatial relationships and economic preferences for wind power—a review. *Energies*, 8(6):6177–6201.

Krohn, S., Damborg, S. (1999). On public attitudes towards wind power. *Renewable Energy*, 16:954–960.

Krueger, A. D., Parsons, G. R., Firestone, J. (2011). Valuing the visual disamenity of offshore wind power projects at varying distances from the shore: an application on the Delaware shoreline. *Land Economics*, 87:268–283.

Ladenburg, J. (2007). *Attitudes and Preferences for the Future Wind Power Development in Denmark and Testing the Validity of Choice Experiments*. Institute of Food and Resource Economics, University of Copenhagen, Copenhagen, p. 136.

Ladenburg, J. (2008). Attitudes towards on-land and offshore wind power development in Denmark; choice of development strategy. *Renewable Energy*, 33:111–118.

Ladenburg, J. (2009a). Visual impact assessment of offshore wind farms and prior experience. *Applied Energy*, 86:380–387.

Ladenburg, J. (2009b). Stated public preferences for on-land and offshore wind power generation—a review. *Wind Energy*, 12:171–181.

Ladenburg, J. (2010). Attitudes towards offshore wind farms – the role of beach visits on attitude and demographic and attitude relations. *Energy Policy*, 38:1297–1304.

Ladenburg, J. (2014). Dynamic properties of the preferences for renewable energy sources—a wind power experience-based approach. *Energy*, 76(11): 542–551.

Ladenburg, J. (2015). Does more wind energy influence the choice of location for wind power development? Assessing the cumulative effects of daily wind turbine encounters in Denmark. *Energy Research & Social Science*, 10:26–30.

Ladenburg, J., Dahlgaard, J.-O. (2012). Attitudes, threshold levels and cumulative effects of the daily wind-turbine encounters. *Applied Energy*, 98:40–46.

Ladenburg, J., Dubgaard, A. (2007). Willingness to pay for reduced visual disamenities from offshore wind farms in Denmark. *Energy Policy*, 35:4059–4071.

Ladenburg, J., Lutzeyer, S. (2012). The economics of visual disamenity reductions of offshore wind farms - review and suggestions from an emerging field. *Renewable and Sustainable Energy Reviews*, 16:6793–6802.

Ladenburg, J., Möller, B. (2011). Attitude and acceptance of offshore wind farms—the influence of travel time and wind farm attributes. *Renewable and Sustainable Energy Reviews*, 15:4223–4235.

Ladenburg, J., Termansen, M., Hasler, B. (2013). Assessing acceptability of two onshore wind power development schemes: A test of viewshed effects and the cumulative effects of wind turbines. *Energy*, 54:45–54.

Lilley, M. B., Firestone, J., Kempton, W. (2010). The effect of wind power installations on coastal tourism. *Energies* 3:1–22.

Manwell, J. F., McGowan, J. G., Rogers, A. L. (2002). *Wind Energy Systems: Environmental Aspects and Impacts, Wind Energy Explained*. John Wiley & Sons, Ltd, pp. 469–510.

McCartney, A. (2006). The social value of seascapes in the Jurien Bay Marine Park: an assessment of positive and negative preferences for change. *Journal of Agricultural Economics*, 57:577–594.

Meyerhoff, J., Ohl, C., Hartje, V. (2010). Landscape externalities from onshore wind power. *Energy Policy*, 38:82–92.

Molnarova, K., Sklenicka, P., Stiborek, J., Svobodova, K., Salek, M., Brabec, E. (2012). Visual preferences for wind turbines: location, numbers and respondent characteristics. *Applied Energy*, 92:269–278.

O'Hare, M. (1977). Not on MY block you don't: facility siting and the strategic importance of compensation. *Public Policy*, 25:407–458.

Pedersen, E., Johansson, M. (2012). Wind power or uranium mine: appraisal of two energy-related environmental changes in a local context. *Energy Policy*, 44:312–319.

Pedersen, E., van den Berg, F., Bakker, R., Bouma, J. (2010). Can road traffic mask sound from wind turbines? response to wind turbine sound at different levels of road traffic sound. *Energy Policy*, 38:2520–2527.

Scarpa, R., Willis, K. (2010). Willingness-to-pay for renewable energy: primary and discretionary choice of British households' for micro-generation technologies. *Energy Economics*, 32:129–136.

Strazzera, E., Mura, M., Contu, D. (2012). Combining choice experiments with psychometric scales to assess the social acceptability of wind energy projects: a latent class approach. *Energy Policy*, 48:334–347.

Swofford, J., Slattery, M. (2010). Public attitudes of wind energy in Texas: local communities in close proximity to wind farms and their effect on decision-making. *Energy Policy*, 38:2508–2519.

Thayer, R. L., Freeman, C. M. (1987). Altamont: public perceptions of a wind energy landscape. *Landscape and Urban Planning*, 14:379–398.

Todt, O., González, M. I., Estévez, B. (2011). Conflict in the Sea of Trafalgar: offshore wind energy and its context. *Wind Energy*, 14:699–706.

Vecchiato, D. (2014). How do you like wind farms? Understanding prefences about new energy landscapes with choice experiments. *Aestimum*, p. 15–37.

Waldo, Å. (2012). Offshore wind power in Sweden—a qualitative analysis of attitudes with particular focus on opponents. *Energy Policy*, 41:692–702.

Warren, C., Lumsden, C., O'Dowd, S., Birnie, R. (2005). 'Green on green': public perceptions of wind power in Scotland and Ireland. *Journal of Environmental Planning and Management*, 48:853–875.

Westerberg, V., Jacobsen, J. B., Lifran, R. (2013). The case for offshore wind farms, artificial reefs and sustainable tourism in the French Mediterranean. *Tourism Management*, 34:172–183.

Westerberg, V., Jacobsen, J. B., Lifran, R. (2015). Offshore wind farms in Southern Europe—Determining tourist preference and social acceptance. *Energy Research & Social Science*, 10:165–179.

Wolsink, M. (2007). Wind power implementation: the nature of public attitudes: equity and fairness instead of 'backyard motives'. *Renewable and Sustainable Energy Reviews*, 11:1188–1207.

Wolsink, M., Devilee, J. (2009). The motives for accepting or rejecting waste infrastructure facilities. Shifting the focus from the planners' perspective to fairness and community commitment. *Journal of Environmental Planning and Management*, 52:217–236.

2

WIND POWER FORECASTING TECHNIQUES

MICHAEL NEGNEVITSKY

2.1 INTRODUCTION

Wind power is currently the fastest growing power generation sector in the world—worldwide growth in wind power generation has been at 40% a year for the last 10 years [1]. With this rapid growth of generation, new challenges are being introduced to power markets. Wind power is an intermittent source. For example, at a wind site in Tasmania, Australia, the average change in mean wind speed vectors over a 2.5-minute period is about 44.4% and the average change in maximum wind speeds over the same period is 55.0%. This demonstrates that the wind speed and its direction can change rapidly by a large amount over a very short period of time. Great variability in wind farm generation has both technical and commercial implications for the efficient planning and operation of power systems.

Large wind farms tend to use large wind turbines. Large turbines have a significant rotational inertia when considering a short time frame. As a result, wind patterns shorter than 10 seconds usually have a negligible effect on the turbine output, and it is a standard practice in wind generation to neglect short-term gusts. However, trends over several minutes are of vital importance.

In order to accommodate the increase in wind power generation, we need to change the way that wind generation is scheduled. Both utilities and electricity market operators will be required to forecast the wind generation to improve system performance. Wind generation will have to be bid into the market like any other generation. In order to handle these requirements, power system operators will need to have better wind forecasting models and tools.

Forecasting is a vital part of business planning in today's competitive environment, regardless of the field in which you work. To be ready for a situation before it occurs gives a distinct advantage over the competition. Wind prediction is complex due to the wind's high degree of volatility and deviation. This means that standard time length definitions for power engineering do not strictly apply. The phrase "short-term forecasting," in the case of wind forecasting, normally means periods from minutes out to half an hour while long-term forecasting means periods from several hours out to a few days.

This difference between the two forecasting time periods is important when it comes to creating a prediction system. Three main classes of different techniques have been used for wind power forecasting. These are numerical weather prediction (NWP) methods, statistical methods, and methods based on applications of artificial neural networks (ANN) and adaptive neuro-fuzzy inference systems (ANFISs). The NWP methods are dominant for forecasts over 10 hours ahead. However, statistical and ANN-based methods appear to be more accurate over shorter periods (minutes to a few hours) [2–4]. They are also much simpler than the NWP methods.

2.2 NUMERICAL WEATHER PREDICTION MODELS

Until recently long-term forecasting (several hours to days ahead) has been the major focus of the research on wind speed prediction. This research has been dominated by NWP models. NWP models operate by solving conservation equations (mass, momentum, heat, water, etc.) numerically at given locations on a spatial grid. These NWP models are very complex and it takes several hours to obtain a solution on a supercomputer. Future behavior of the atmosphere is

Alternative Energy and Shale Gas Encyclopedia, First Edition. Edited by Jay H. Lehr and Jack Keeley.
© 2016 John Wiley & Sons, Inc. Published 2016 by John Wiley & Sons, Inc.

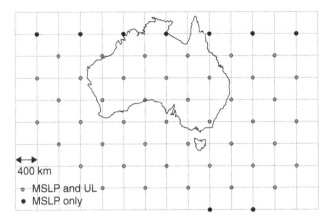

FIGURE 2.1   NWP grid for synoptic prediction.

predicted by utilizing both current conditions and past trends. Also, the NWP method is not standard and different NWP models are run by Bureaus of Meteorology in Europe and the United States—the two major developers of wind generation technology. These two NWP models are then used as a basis for several different wind speed prediction systems.

A major factor that influences the accuracy of NWP models is the resolution of the grid. Grid spacings finer than those used for general meteorological forecasting are required to produce sufficient results to be useful for wind speed forecasting at wind farm sites. Figures 2.1 and 2.2 give clear examples of the difference between meteorologically useful grid spacing and the grid spacing useful for wind power. Furthermore, the required resolution for a particular site depends on the complexity of the surrounding terrain and the phenomena that affect the wind patterns.

In terms of weather forecasting, there were two main forecast scales: planetary and synoptic. The planetary scale is

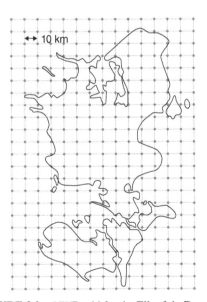

FIGURE 2.2   NWP grid for the Elkraft in Denmark.

used for extremely large weather patterns that affect the entire world. The synoptic scale is smaller, but still focuses on large weather patterns, of a scale of hundreds of kilometers across. This was acceptable for coarse weather prediction of patterns such as fronts, highs, and lows. However, as more precise forecasts were sought after, a new scale for forecasting was developed. This scale was mesoscale weather forecasting [5].

The mesoscale predictions usually incorporate some level of expertise. This is most often in the form of heuristics, or "if–then" rules. It should also be noted that most NWP models are developed by meteorologists and have a purpose other than forecasting wind power generation. To this end, another convention for time periods has been defined: mesoscale alpha (2 days to 6 hours), beta (6 hours to 30 minutes), and gamma (30 minutes to 3 minutes). Often beta and gamma are considered together, which directly parallels the power forecasting terminology, short-term wind prediction.

The main focus of NWP models is on the synoptic or planetary scales, and these models are very successful in this area. With the need for shorter term forecasts, NWP methods have been supplemented to increase the accuracy over the smaller time steps.

Short-term NWP models operate by solving conservation equations (mass, momentum, heat, water, etc.) numerically at given locations on a spatial grid that is three dimensional: latitude, longitude, and elevation. This is considered across the fourth dimension of time as well. The finer the spatial grid, the finer the ability to forecast. NWP models of synoptic scale (or greater) must be assumed hydrostatic (equilibrium of the vertical pressure gradient in the atmosphere) and therefore cannot model thermally driven winds such as sea breeze. Mesoscale models, however, can be non-hydrostatic and can predict smaller scale wind patterns such as land/sea breeze and venturi effect (mountain winds). In fact, the shorter term predictions can be dominated by terrain and non-hydrostatic phenomena. Thus, mesoscale NWP models need to include accurate digital elevation models (DEMs) to represent the topography over which the weather prediction occurs.

Predictions from an NWP model have four limiting factors: data variability; DEM resolution; grid spacing; and computation time. An NWP would also suffer from inaccurate input data, as is true for any forecast model.

The variability of the data cannot be avoided and will always limit the accuracy of an NWP model. As forecasts become more precise, so too must the DEM. This becomes a limiting factor once the resolution gets small enough that forests and towns become significant factors as these features can change rapidly (causing need for remodeling the environment).

However, even if these two obstacles could be overcome, the spacing of the monitoring stations can be an issue. For any numerical model, five grids points are required to resolve a wave's structure. Thus, for a 5-km (tight) grid separation, the smallest weather pattern that can be accurately modeled

TABLE 2.1 Advantages and disadvantages of Numerical Weather Prediction (NWP) models.

Advantages	Disadvantages
Can be accurate for forecasts of several days	Accuracy is totally dependent upon the accuracy of the model developed by the Bureau of Meteorology
Can be used for several wind farm sites	Complex to construct and operate
Well-established procedures/software	Inaccurate for short forecast times
Typically provide forecasts for every hour out to the maximum forecast lead time	Data at the model grid points must be translated to "realistic" wind speeds

would be 20 km across. This is insufficient for very short term wind forecasting.

The last limiting factor is that NWP models are heavily mathematical and are usually run on super computers, even then taking over an hour to obtain a result. This by itself may limit the usefulness of NWP methods for online applications in power systems. Table 2.1 provides some pros and cons of NWP models.

2.3 PERSISTENCE MODELS

Wind can be represented as a time series, which is defined as a set of observations of a parameter, or set of parameters, taken at a number of discrete time intervals. These intervals are usually (although not always) of a regular length. Most time series-based forecasting systems rely on evenly spaced data, and if the time step between data points is not consistent, or there is data missing, this may cause problems. A generalized example of a time series can be presented as follows:

$$X = \{\cdots, x(t-3), x(t-2), x(t-1), x(t)\} \qquad (2.1)$$

A real-world time series can exhibit chaotic behavior, making them difficult to predict. A wind speed time series, such as shown in Figure 2.3, clearly demonstrates this type of behavior.

One reason for NWP method's failure on the short-term time scale is that wind has a high variation over brief periods. An example of the variation in wind data is shown in Table 2.2. These values show that for the wind vector, the standard deviation is high when compared to the mean value, but more significantly, the standard deviation of the *change* in wind speed vector is even higher compared to its mean. This level of variation in a data set causes problems when trying to achieve accurate predictions.

The concept of persistence, also known as *naïve predictor*, was developed by meteorologists as a simple short-term forecasting approximation to augment NWP methods. Persistence relies on the high correlation between the current value of a wind parameter and the value in the immediate future and simply equates the next forecast value to the current value:

$$x_{\text{prediction}} = x(t) = x(t+1) \qquad (2.2)$$

This method was originally developed as a comparison tool to supplement the NWP models [6, 7]. Historically, the accuracy of short-term forecasting was not very important, and therefore the results obtained from persistence forecasts were sufficient [8]. The main advantage of persistence models is that they can be implemented easily. As a result, persistence models were almost universally accepted by the power industry for short-term wind prediction. Also persistence often provides a benchmark for short-term wind forecasting against which to measure the accuracy of alternative techniques [9, 10].

However, development of large-scale wind farms and their rapid integration into power grids will inevitably require a much greater accuracy of wind power generation forecasts over shorter periods of time (minutes rather than hours).

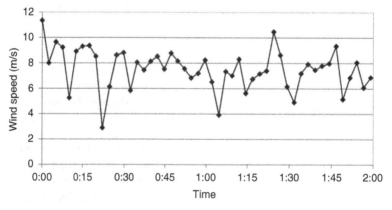

FIGURE 2.3 Typical time-series plot for a wind site in Tasmania, Australia.

TABLE 2.2   Wind variation for a site in Tasmania, Australia.

Measured variables	Mean ($\bar{x}$)	Standard deviation (σ)	Ratio ($\sigma/\bar{x}$)
Wind vector (m/s)	4.9	3.0	0.61
Change in wind vector over 2.5 minutes (m/s)	1.8	3.0	1.67

Persistence models will not be able to satisfy the industry needs any longer—new tools will be needed for very short term wind power forecasting.

2.4   CHOOSING FORECAST PARAMETERS

When considering short and very short term wind power forecasting there are a number of choices in relation to which parameters to forecast and what level of aggregation to consider. The first choice relates to whether to forecast wind parameters and infer or calculate wind power from those predictions or alternatively to forecast wind power directly using a suitable model.

Forecasting wind parameters (e.g., wind speed or wind vector magnitude) is a more fundamental approach and the intermediate results of predicted wind parameter values will be independent of turbine type and performance. However, where the goal is to forecast wind power output, conversion of forecast wind parameters into power output is required.

The general relationship between wind speed and the output power of a wind turbine is given by

$$P = \frac{1}{2} \cdot C_\text{p} \cdot \rho \cdot A \cdot V^3 \qquad (2.3)$$

where P is the output power (W); C_p is the rotor coefficient of performance; ρ is the air mass density (kg/m^3); A is the swept blade area (m^2); v is the speed of the wind perpendicular to the turbine (m/s).

For a particular turbine, the coefficient of performance (C_p) varies nonlinearly with wind speed and the overall turbine output performance is usually represented using a rating curve of power output versus wind speed provided by the manufacturer. However, a single rating curve does not adequately represent the dynamic response of a turbine output to varying wind speed and direction in the short-term time frame. Consequently if wind parameters for a turbine are accurately forecast using a particular model, there is no simple relationship to obtain accurate power output from these parameters and a second nonlinear model will be required. This suggests that where short-term wind power forecasting is required, a single model providing direct prediction of power output is the best approach.

Another forecasting parameter choice relates to the level of aggregation to be considered. The most flexible approach is to consider short-term wind power forecasting on a per turbine basis. By producing a forecasting model for each turbine, the output for a wind farm can be easily scaled and adjusted depending on turbine availability and wind farm configuration changes over time. In addition, inputs which are local and specific to each turbine such as wind speed at the hub and current turbine orientation in relation to the wind direction (yaw angle) could be considered in the model.

Alternatively, aggregated short-term wind power forecasting for a group of wind turbines may be easier to manage and may provide some inherent error cancellation. Aggregation could be considered for a sub-circuit of a wind farm or for an entire wind farm. While aggregation of a number of wind farms in a region could also be considered, this is expected to be less useful for short-term forecasting where local effects will dominate for the time frames being considered.

2.5   STATISTICAL AND NEURAL NETWORK METHODS

In wind power prediction, statistical methods and neural networks are generally aimed at very short term forecasting [11, 12]. Most statistical models for wind forecasting use auto-recursive algorithms. This means they use the difference between the predicted and actual wind speeds in the immediate past to tune the model parameters [7]. The accuracy of these methods degrades rapidly with increasing prediction lead time.

While NWP models take into account meteorological information as well as local weather conditions, statistical models predict wind power by using only measured values (both historic and current) of wind power. NWP models perform well for longer term wind forecasting (e.g., usually more than 6 hours ahead), while statistical models are preferable for short-term predictions [7, 13]. However, most commonly used statistical models are linear and rely on the traditional autoregressive moving average technique [14]. Such models are often not portable and need an expert to create an individualized model for each wind farm. Due to the lack of portability, changes in weather conditions often require significant changes in the model itself. A change in the model requires the attention of an expert again for the retuning of the statistical models. Finally, the lack of ability to predict direction in most forecasting systems causes difficulties in taking advantage of short gusts of wind [15].

Prediction research is a growth area and, increasingly often, this research involves the use of artificial intelligence techniques such as ANNs and ANFISs [16]. Neural network-based models are nonlinear. They use large time-series data

TABLE 2.3 Advantages and disadvantages of statistical models.

Advantages	Disadvantages
Accurate for short forecast times	Inaccurate for long forecast periods
Adapt to the current situation using auto-recursive mathematical algorithms	May require a testing and tuning process to produce good results
Can provide average or peak wind speeds over short periods	Require assistance of domain expert for each specific site

TABLE 2.4 Advantages and disadvantages of ANN-based models.

Advantages	Disadvantages
Accurate for shorter forecast times	Inaccurate for longer forecast periods
Learn historic patterns to predict future patterns	May require over a year of training data to learn seasonal patterns
Require little understanding of the weather dynamics, except to choose sensible inputs	Special networks must be used for online adaptation
Can provide average or peak wind speeds over short periods	Require individual networks for each forecast lead time unless special networks are used

sets to learn the relationship between the input data and output wind speeds [17–19]. Nonlinear models can accurately capture the effects of nonlinearity of wind characteristics and, as a result, they usually outperform persistence and statistical models. Tables 2.3 and 2.4 show advantages and disadvantages of statistical and neural network-based models.

2.6 ADAPTIVE NEURO-FUZZY INFERENCE SYSTEMS

Fuzzy systems and neural networks are natural complementary tools in building intelligent systems. While neural networks are low-level computational structures that perform well when dealing with raw data, fuzzy logic deals with reasoning on a higher level. However, fuzzy systems lack the ability to learn and cannot adjust themselves. The merger of a neural network with a fuzzy system into one integrated system therefore offers a promising approach to building very short term wind prediction models.

A neuro-fuzzy system is, in fact, a neural network that is functionally equivalent to a fuzzy inference model. For example, an ANFIS proposed by Roger Jang [20] is a six-layer feedforward neural network. Figure 2.4 shows the ANFIS architecture. For simplicity, we assume that the ANFIS has two inputs—$x1$ and $x2$—and one output—y. For additional simplicity, each input is represented by only two fuzzy sets, although three or more are not uncommon. Extra membership functions (MFs) will increase the accuracy of the results but will take longer to train. For this example, the two fuzzy sets are converted to the output by a first-order polynomial. The ANFIS in Figure 2.4 implements four rules, shown in Figure 2.5, where $x1$ and $x2$ are input variables; $A1$ and $A2$ are fuzzy sets on the universe of discourse $X1$; $B1$ and $B2$ are fuzzy sets on the universe of discourse $X2$; and $\{k_{i0}, k_{i1}, k_{i2}\}$ is a set of parameters specified for rule i.

Layer 1 is the input layer. Neurons in this layer simply pass external crisp signals to Layer 2.

Layer 2 is the fuzzification layer. Neurons in this layer perform fuzzification. In Jang's model, fuzzification neurons normally use a bell activation function.

Layer 3 is the rule layer. Each neuron in this layer corresponds to a single fuzzy rule. A rule neuron receives

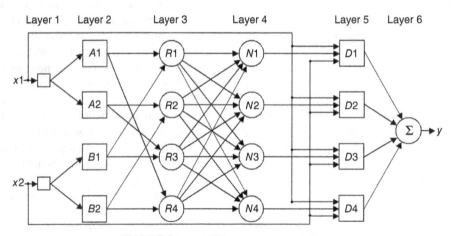

FIGURE 2.4 ANFIS model representation.

Rule 1		Rule 2	
IF	$x1$ is $A1$	IF	$x1$ is $A2$
AND	$x2$ is $B1$	AND	$x2$ is $B1$
	$y_1 = k_{10} + k_{11}x1$		$y_2 = k_{20} + k_{21}x1$
THEN	$+ k_{12}x2$	THEN	$+ k_{22}x2$
Rule 3:		Rule 4:	
IF	$x1$ is $A1$	IF	$x1$ is $A2$
AND	$x2$ is $B2$	AND	$x2$ is $B2$
	$y_3 = k_{30} + k_{31}x1$		$y_4 = k_{40} + k_{41}x1$
THEN	$+ k_{32}x2$	THEN	$+ k_{42}x2$

FIGURE 2.4   An example of ANFIS rules.

inputs from the respective fuzzification neurons and calculates the firing strength of the rule it represents.

Layer 4 is the normalization layer. Each neuron in this layer receives inputs from all neurons in the rule layer and calculates the normalized firing strength of a given rule—the ratio of the firing strength of a given rule to the sum of firing strengths of all rules. It represents the contribution of a given rule to the final result.

Layer 5 is the defuzzification layer. Each neuron in this layer is connected to the respective normalization neuron and also receives the initial inputs, $x1$ and $x2$. A defuzzification neuron calculates the weighted consequent value of a given rule.

Layer 6 is represented by a single summation neuron. This neuron calculates the sum of outputs of all defuzzification neurons and produces the overall ANFIS output, y.

An ANFIS uses a hybrid learning algorithm that combines the least-squares estimator and the gradient descent method [20]. First, initial activation functions are assigned to each membership neuron. The function centers of the neurons connected to input x_i are set so that the domain of x_i is divided equally, and the widths and slopes are set to allow sufficient overlapping of the respective functions. In the ANFIS training algorithm, each training epoch is composed from a forward pass and a backward pass. In the forward pass, a training set of input patterns (an input vector) is presented to the ANFIS, neuron outputs are calculated on the layer-by-layer basis, and rule consequent parameters are identified.

2.7   CASE STUDY

2.7.1   Defining the Problem

Currently, there is no very short term wind prediction package commercially available. Some research focuses on a time scale of an hour to 2 hours ahead [21]. There has also been attempts of using ANNs to forecast intervals as low as 10 minutes ahead [18, 19]. However, this research does not consider one of the major intervals for wind power management— the need for 2- to 3-minute predictions for governing power

output due to gusts [22]. This chapter addresses this time scale and provides 2.5-minute-ahead forecasts of wind vectors [23].

2.7.2   Developing the Model

The ANFIS model design is flexible and capable of handling rapidly fluctuating data patterns. This meant that it covered the criterion necessary for very short term wind prediction. This model has two major goals. The first one is to increase prediction accuracy (and at least outperform the simple-to-implement industry standard of persistence). The second one, which is no less important, is to make a system that can be easily installed at a variety of different sites.

To use the ANFIS model, the user first must define a number of variables:

the size of the training set;

the number of training epochs (iterations);

the type of the fuzzy MFs; and

the number of MFs associated with each input.

The size and diversity of the training set is important. If the set has insufficient variation to properly model the characteristics of the data, the training of the ANFIS will fail.

The standard MF for an ANFIS is a "bell function," such as the Gaussian distribution function [20].

The number of MFs depends upon the complexity of the system and the size of the training set. As a general rule, more MFs will better represent a complex system but will take longer to train—especially for a large training set. In this case study, two MFs were used for each input.

Table 2.5 demonstrates the relationship between training times and the number of MFs. It should be noted that this test was done on Tasmanian electricity demand data as they are far easier to predict and so smaller training sets could be used. Table 2.5 also includes a variable number of inputs: six or four. More inputs to the ANFIS model will result in a wider variety of data in each pass. This will result in more selective training, producing better predictions. However, as with MFs, a greater number of inputs will have the consequence of a longer training time.

TABLE 2.5   Comparison of single epoch training times for Tasmanian electricity demand data.

Number of inputs	Number of MFs	Training times (s)	RMS error
6	3	54,436	0.0296
6	2	264	0.0296
4	3	220	0.0345
4	2	8	0.0362

2.7.3   Choosing the Inputs

In order to predict a time series with an ANFIS model, it is important to select the right inputs for the system. Unfortunately, there are no rules available to expedite this process. Every data set has different deviations and rates of change. Thus, what might work for one data set will not necessarily be the best configuration for another.

If the system can be trained over a long period of time, the developer can afford to use larger data sets, more inputs, and more MFs; however, the developer must also consider the need for multiple training and testing runs. A system will need to be trained at least several times to show that it produces reliable results and many times before that in order to tune the parameters. Once the training set size is chosen, the number of MFs and the number of inputs are chosen to allow for training time and the available computing resources.

The most common method of choosing inputs for very short term time-series prediction is by considering a subset of the available data. Often these are known points from the same time series, chosen to best determine the desired prediction point. Equation (4) is a generalized example input to try to predict the point at $x(t + 1s)$. X is the input pattern, $x(t)$ is the value of the time series at the present time t, and s is the chosen time step:

$$X = [x(t - 3s)x(t - 2s)x(t - 1s)x(t)] \qquad (2.4)$$

Figure 2.6 provides a sample of wind data. The goal is to predict the point at 0:10:00. Using an input time step of 2.5 minutes, the equation would become

$$X = [x(0:00:00)x(0:02:30)x(0:05:00)(0:07:30)] \qquad (2.5)$$

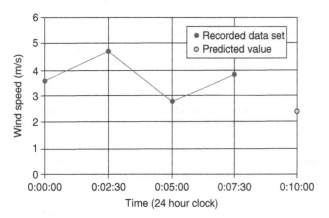

FIGURE 2.6   An example of predicting the wind speed value at 0:10:00.

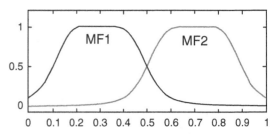

FIGURE 2.7   An example of bell membership functions.

This example uses four inputs and a single output. The number of data points in the input pattern can be adjusted as it is demonstrated below.

2.7.4   Choosing the Membership Functions

Initially, the ANFIS model was trained with a bell function, shown in Figure 2.7. This is an example of a soft-edged MF as the transitions have no abrupt corners. Note that the boundaries of these curves are extremely wide. When tested on less variable data, such as electricity demand, this model gave accurate results (less than 1% mean absolute error over a half hour interval). However, the results were not satisfactory for highly stochastic (or "spiky") test data. Large events in the time series would result in large error spikes. In order to improve robustness, the MF type was altered. It was found that hard-edged MFs were more robust and remove large overshoots.

The final design resulted in using a bell MF and overcoming the overshoot/undershoot issue by putting a hard-limiter in the system to truncate predictions outside a defined range—selected by simple maximum/minimum statistical data analysis. The use of data analysis was chosen instead of expert input so that the procedure could be easily replicated at other wind sites without the need for external input.

2.7.5   Predicting Difference to Aid Transportability

Wind forecasting models are site specific. Each recording site produces unique data. A comparison of recordings from the same wind farm (over very short term) showed that even sites in close proximity produce significantly different data. To test this, two adjacent wind turbine towers on the same wind farm were compared. The results are shown in Table 2.6.

TABLE 2.6   Comparison of wind turbine towers from the same wind farm.

Wind speed ([m/s) Mean for Tower A	Mean for Tower B	Mean absolute difference
0.80	0.77	0.19

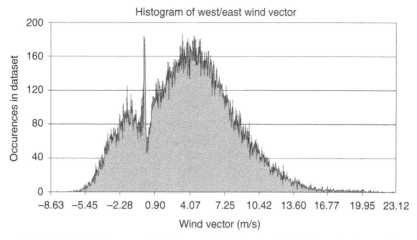

FIGURE 2.8 Histogram of Spring wind data from a site in Tasmania, Australia.

Although quantitative proof of time-series (data-based) prediction is difficult, and no system can be guaranteed as being portable to another site, there are steps that can be taken to improve portability. It was determined that while different sites may have varying wind patterns, most sites will behave similarly when considered on a very short term basis. To further increase the portability, this idea of similar very short term behavior is able to be exaggerated through using the difference between data points as the training data and as the output. The forecast difference will then be added to the present data point to determine the effective offset.

For the wind site used in this case study, some histograms were developed. Figure 2.8 shows a wind vector histogram for Spring. It shows that westerlies are the predominant wind pattern but also have a large reading at zero. Figure 2.9 shows the differenced wind vector's histogram. It resembles a Laplacian distribution, which shows a marked likelihood

for little or no alteration. This resulted in much better predictions. Using histograms such as these, it is possible to create a probabilistic model to develop additional data should it be required for training at a later stage.

2.7.6 Results of the Case Study

The case study used a wind site in Tasmania as the data set, providing a 21-month time series in steps of 2.5 minutes. This was to be the forecast period as well. The data were converted from speed and direction into $u - v$ vectors. For this case study, only one vector was considered at one tower height. This allows the training of more systems (within finite computing potential) resulting in a better comparison of the various techniques.

In order to adequately test the performance of the proposed ANFIS system, multiple architectures were evaluated

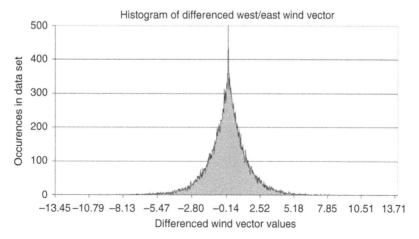

FIGURE 2.9 Histogram of the differenced values of Spring wind data from a site in Tasmania, Australia.

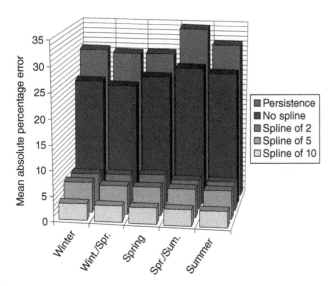

FIGURE 2.10 Chart of prediction errors for various systems, tested over a period of 8 months on a wind site in Tasmania, Australia.

architecture used is identical to the system used without a spline except that the data used as inputs (and outputs) were interpolated using a spline of 2, 5, and 10 points. The differences between these results are small, but the difference between these three results and the results using no spline or a persistence model is significant. It is thus concluded that the results were improved by using some degree of spline. It is important to demonstrate that the persistence model would not gain from such a method. Persistence by nature does not change between known points, so each intermediate prediction would be identical in value.

Further tests were developed to study the effectiveness of the model on a variety of training data. The results of these additional trials are summarized in Figure 2.11. The goal was to check whether additional data from a nearby data source could be used to improve prediction. In three out of the four cases, the change was negligible. The exceptional case was a consequence of the addition of data (from the same site) but recording the maximum value over the interval instead of the mean value. This proved to increase the mean absolute percentage error.

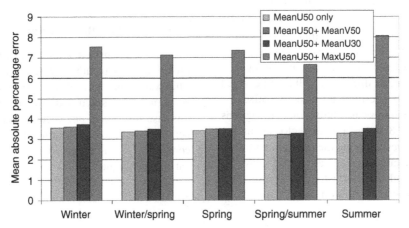

FIGURE 2.11 Chart of errors comparing various instances of ANFIS models using data from a wind site in Tasmania, Australia.

on the same data. A persistence model was also developed for comparison. Persistence is presently an industry benchmark for very short term wind forecasting and so is the most indicative assessment.

The ANFIS model was developed in several different formats. This was to highlight the usefulness of intermediary splines through data for very short term forecasting. The results are shown in Figure 2.10. The chart also includes the results from the persistence model. A useful comparison is available through considering the persistence results and the ANFIS model with no spline. The ANFIS model shows some improvement, in the order of 5%.

The tests using splined data as the input to the ANFIS model resulted in an improvement in prediction. The system

REFERENCES

[1] Available at: http://www.thewindpower.net (accessed April, 2012).

[2] Chen, P., Pedersen, T., Jensen-Bak, B., Chen, Z. (2010). ARIMA-based time series model of stochastic wind power generation. *IEEE Transaction on Power Systems*, 25(2):667–675.

[3] Pritchard, G. (2011). Short-term variations in wind power: some quantile-type models for probabilistic forecasting. *Wind Energy*, 14(2):255–269.

[4] Bessa, R. J., Miranda, V., Botterud, A., Wang, J. (2011). 'Good' or 'bad' wind power forecasts: a relative concept. *Wind Energy*, 14(5):625–636.

[5] Curry, K. (2001). *Definition of the Mesoscale.* University Corporation for Atmospheric Research.

[6] Landberg, L. (1994). Short-term prediction of local wind conditions. PhD Thesis, Risø-R-702(EN), Risø National Laboratory, Denmark, ISBN 87-550-1916-1.

[7] Nielsen, T., Madsen, H. A., Tøfting, J. (1999). Experiences with statistical methods for wind power prediction. *Proceedings of the European Wind Energy Conference*, Nice, France, pp. 1066–1069.

[8] Nielsen, T. S., Madsen, H., Tofting, J. (2000). WPPT, a tool for on-line wind power prediction. *Proceedings of the International Energy Agency Expert Meeting on Wind Forecasting Techniques*, Boulder, USA.

[9] Landberg, L., Giebel, G., Nielsen, H. A., Nielsen, T., Madsen, H. (2003). Short-term prediction: an overview, *Wind Energy*, 6:273–280.

[10] Potter, C., Negnevitsky, M., Jacka, K. (2005). Evaluating the use of power data prediction against wind data prediction for short-term windpower operation. *World Wind Energy Conference (WWEC)*, vol. 4. Available at: www.wwec2005.com/secure/finalpapers.shtml.

[11] Bathurst, G. N., Weatherill, J., Strbac, G. (2002). Trading wind generation in short term energy markets. *IEEE Transactions on Power Systems*, 17(3):782–789.

[12] Landberg, L., Giebel, G., Nielsen, H. A., Nielsen, T., Madsen, H. (2003). Short-term prediction: an overview. *Wind Energy*, 6:273–280.

[13] Sideratos G., Hatziargyriou, N. D. (2007). An advanced statistical method for wind power forecasting. *IEEE Transactions on Power Systems*, 22(1):258–265.

[14] Rajagopalan, S., Santoso, S. (2009). Wind power forecasting and error analysis using the autoregressive moving average modeling. *Proceedings of the IEEE/PES General Meeting*, Calgary, Alberta, Canada, July 26–30.

[15] Negnevitsky, M., Potter, C. (2006). Innovative short-term wind generation prediction techniques (Panel Paper). *Proceedings of the IEEE/PES General Meeting*, Montreal, Canada, pp. 18–22.

[16] Negnevitsky, M. (2011). *Artificial Intelligence: A Guide to Intelligent Systems*, 3rd edition. Addison-Wesley, Reading, MA.

[17] Mohandes, M. A., Rehman, S., Halawani, T. O. (1998). A neural networks approach for wind speed prediction. *Renewable Energy*, 13:345–354.

[18] Li, S., et al. (2001). Using neural networks to estimate wind turbine power generation. *IEEE Transactions on Energy Conversion*, 16(3):276–282.

[19] Catalao, J. P. S., Mendes, V. M. F. (2009). An artificial neural network approach for short-term wind power forecasting in Portugal. *Proceedings of the 15th International Conference on Intelligent System Applications to Power System*.

[20] Jang, J.-S. R. (1993). ANFIS: adaptive network-based fuzzy inference systems. *IEEE Transactions on Systems, Man, and Cybernetics*, 23(3):665–685.

[21] Zeng, J., Qiao, W. (2012). Short-term wind power prediction using a wavelet. Support vector machine. *IEEE Transactions on Sustainable Energy*, 3(2):255–664.

[22] Barton, J. P., Ineld, D. G. (2004). Energy storage and its use with intermittent renewable energy. *IEEE Transactions on Energy Conversion*, 19(2):441–448.

[23] Potter, C., Negnevitsky, M. (2006). Very short-term wind forecasting for Tasmanian power generation. *IEEE Transactions on Power Systems*, 21(2):965–972.

3

MAXIMIZING THE LOADING IN WIND TURBINE PLANTS: (A) THE BETZ LIMIT, (B) DUCTING THE TURBINE

D. P. GEORGIOU AND N. G. THEODOROPOULOS

3.1 THE WIND TURBINE EFFICIENCY

Any turbine generates power $(\dot{W})$ proportional to the volumetric flux through it and the corresponding total pressure difference. In wind turbine applications this improvement is correlated against

(i) the wind volumetric flux intercepted by the turbine rotor and
(ii) the dynamic head of the wind blowing far upstream.

The effectiveness of a wind turbine concept is usually expressed through the "power factor" (C_P), where

$$C_P = \frac{\dot{W}}{(\rho A V_R)\left(\frac{1}{2}V_R^2\right)} = \frac{\dot{W}}{\left(\frac{1}{2}\rho A_R V_0^2\right)} \qquad (3.1)$$

Here $\dot{W}$, ρ, A_R, and V_0 represent the actual power generated by the rotor, the air density, the rotor cross-sectional area, and wind far upstream velocity, respectively. In addition,

$$\dot{W} = \eta_R (A_R V_R)(\Delta P_{0R}) \qquad (3.2)$$

where η_R, V_R, and ΔP_{0R} represent the rotor efficiency, the wind velocity, and the corresponding total pressure rise across the rotor, respectively. For an incompressible flow, the velocity remains (nearby) constant across the turbine, so that $\Delta P_{0R} = \Delta P_R$, where ΔP_R is the static pressure drop. The

Betz analysis assumes that $\eta_R \approx 1.0$. Equation (3.1) may take the form

$$C_P = C_T Y \qquad (3.3)$$

where

$$C_T = \frac{\Delta P_R}{\frac{1}{2}\rho V_0^2} = \frac{T}{\frac{1}{2}\rho A_R V_0^2} \qquad (3.4)$$

is called the "loading factor (CT)" and represents the thrust applied on the rotor by the blowing wind, while

$$Y = V_R / V_0 \qquad (3.5)$$

represents the wind "acceleration factor" from far upstream up to the rotor plane.

3.2 THE BETZ LIMIT

Although the German Albert Betz has been credited with the first attempt to develop a quick estimate of the loading and power coefficients for horizontal axis wind turbines, Bergey [1] and van Kuik [2] have shown that, at about the same time, similar proposals were made by Lanchester in England and Joukowsky in Russia. The analysis proposed by all these people was based on the actuator disc concept proposed by Froude [3] and Rankine [4] for the analysis of the ship propeller. The method employs the application of the mass, momentum, and mechanical energy conservation

Alternative Energy and Shale Gas Encyclopedia, First Edition. Edited by Jay H. Lehr and Jack Keeley.
© 2016 John Wiley & Sons, Inc. Published 2016 by John Wiley & Sons, Inc.

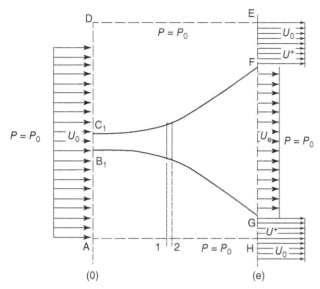

FIGURE 3.1   The control volume employed by Betz in the development of his model.

in a control volume defined by the streamtube enclosing the wind flux passing through the rotor of the turbine. This corresponds to the area $C_1FGB_1C_1$ in Figure 3.1 and was based on the assumption that the static pressures along the sides C_1F and GB_1 are constant, equal to the surrounding atmospheric pressure (P_0). Velocities in the inlet plane (B_1C_1) U_0 and the exit plane (FG) U_e are of uniform distribution. Similarly, the static pressure in these two planes is equal to the atmospheric pressure (P_0). Today, the analysis is somewhat modified, by extending the sidewalls to the "faraway" limit (where the static pressure is indeed equal to P_0), with the added assumption that the wind velocity outside the control volume $C_1FGB_1C_1$ remains undisturbed. Otherwise the analysis follows the same steps.

In addition,

1. the flow is strictly steady, one-dimensional, isentropic, and irrotational;
2. the rotor (actuator disk) is infinitesimally thin with a projection to the flow area A_R and uniform velocity and thrust loading over the entire cross section.

The power generated by the actuator disk is due to the static pressure drop of the fluid, as it passes through the disk, that is,

$$\dot{W} = \Delta P_R A_R V_R \qquad (3.6)$$

If the station numbering employed here refers to upstream (0), in front of the rotor (1), behind the rotor (2), and far downstream (e) conditions, the application of the Bernoulli

equation (i.e., the energy conservation equation for incompressible flows) in front and behind the disk implies that

$$P_0 + \frac{1}{2}\rho V_0^2 = P_1 + \frac{1}{2}\rho V_1^2 \qquad (3.7a)$$

$$P_2 + \frac{1}{2}\rho V_2^2 = P_e + \frac{1}{2}\rho V_e^2 \qquad (3.7b)$$

Taking into consideration the facts that $V_1 = V_2 = V_R$ and $P_0 = P_e$, it can easily be deduced that

$$C_T = 1 - X^2 \qquad (3.8)$$

where

$$X = \frac{V_e}{V_0} = \text{The exit deceleration} \qquad (3.9)$$

The parallel application of the momentum conservation equation between the rotor and the inlet/exit planes of the control volume $C_1FGB_1C_1$ leads to

$$T = \dot{m}(V_0 - V_e) = \rho A_R V_R (V_0 - V_e) \qquad (3.10)$$

where $\dot{m}$ = fluid mass flux inside the streamtube = $\rho A_R V_R$. Hence

$$C_T = \frac{T}{\frac{1}{2}\rho V_0^2 A_R} = 2\left(\frac{V_R}{V_0}\right)\left(1 - \frac{V_e}{V_0}\right) = 2Y(1-X)$$

$$(3.11)$$

Combining equations (3.8) and (3.9), it is deduced that

$$Y = 0.5(1 + X) \qquad (3.12)$$

Actually, instead of the parameter Y, Betz employed a flow perturbation parameter, the "induction factor (a)," where

$$Y = 1 - \alpha \qquad (3.13)$$

Equation (3.12) implies that

$$X = 1 - 2\alpha \qquad (3.14)$$

and

$$C_P = \frac{\Delta P_R A_R V_R}{\frac{1}{2}\rho V_0^3 A_R} = C_T Y = 4\alpha(1 - \alpha)^2 \qquad (3.15)$$

while

$$C_T = 4\alpha(1 - \alpha) \qquad (3.16)$$

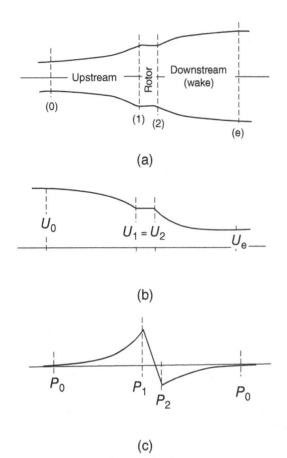

(a)

(b)

(c)

FIGURE 3.2 The evolution of (a) the cross-sectional area of the streamtube, (b) the streamwise velocity (U), and (c) the static pressure of the stream.

Equation (3.14) implies that

$$0 \leq \alpha \leq 0.5 \qquad (3.17)$$

The corresponding optimum value for the C_P coefficient appears when $\partial C_P / \partial \alpha = 0$. This leads to the conclusion that $\alpha = 1/3$ and $C_{P,max} = 16/27 = 0.592$.

The last conclusion is known as the "Betz limit." In other words, an ideal wind turbine may absorb no more than 59.2% of the kinetic energy of the wind intercepted by the turbine rotor, while only 70% of this flux actually passes through the rotor. Figure 3.2 illustrates the evolution of the cross-sectional area, the streamwise velocity (U), and the static pressure (p) inside the streamtube $C_1FGB_1C_1$, as defined in Figure 3.1.

The $C_P(a)$ and $C_T(a)$ relationships predicted by the above analysis are illustrated in Figure 3.3.

Only the prediction that $C_{P,max} = 0.592$ when $a = 0.3$ has been used extensively. No actual wind turbine plant built so far has exceeded this limit. The $C_P(a)$ and $C_T(a)$ relationships, however, have not been verified experimentally.

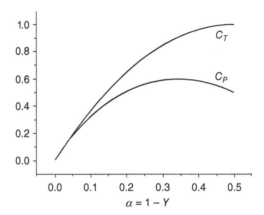

FIGURE 3.3 The $C_P(a)$ and $C_T(a)$ relationships as predicted by Betz.

Actual design starts with the blade element methods [5]. The $C_T(a)$ relationship has been measured on propellers when they operate as brakes [6].

These data agree fairly well with the predictions made by the theory proposed by Betz and the other two as long as $a < 0.5$. For values $a > 0.5$, the data diverge.

Since its publication, the Betz model prediction has been employed as the reference against which the actual performance of a given wind turbine must be compared. The drawbacks, however, of the model were recognized quite some time ago. During the 1970s, various studies attempted to improve it, mainly by introducing "corrections" involving the rotation in the rotor wake [7]. Inglis [6] discusses some of these studies.

3.2.1 The Georgiou–Theodoropoulos Improvement

A model explaining the strong deviation between the predictions of the well-known Betz model and the actual results for the performance of highly loaded wind turbines was provided by Georgiou and Theodoropoulos [8]. The new model accounts for the inner and outer stream interactions by reformulating the relevant one-dimensional flow equations.

This model assumes that the outer stream is squeezed by the inner streamtube, as the latter expands (Figure 3.4a), as indicated by experimental data produced by Jung et al. [9] (Figure 3.4b). Georgiou and Theodoropoulos assumed that the inner streamtube sustains an overpressure δP_0 on its streamwise periphery, as illustrated in Figure 3.5.

As a result, the equation transforms into

$$P_0 A_0 + (P_0 + \bar{\delta P_0})(A_e - A_0) + \dot{m} U_0 \\ = P_0 A_e + \Delta P_R \cdot A_R + \dot{m} U_e \qquad (3.18)$$

Here $\bar{\delta P_0}$ represents the "average" value of the overpressure and $\dot{m}$ the mass flux through the inner streamtube. The

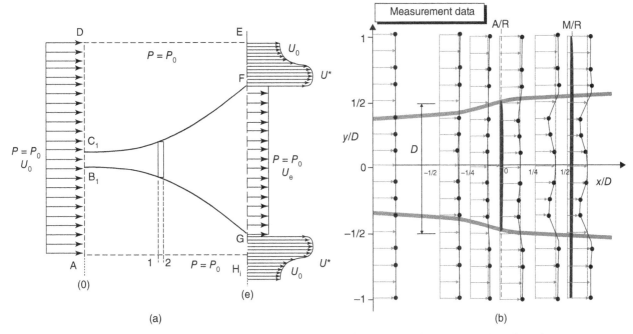

FIGURE 3.4 The control volume proposed by Georgiou and Theodoropoulos [8] based on the experimental results of Jung et al. [9].

corresponding form of the energy equation is

$$P_0 \left(\frac{\dot{m}}{\rho} \right) + \dot{m} \left(\frac{V_0^2}{2} \right) + \dot{W}_S = P_0 \left(\frac{\dot{m}}{\rho} \right) + \dot{m} \left(\frac{V_e^2}{2} \right) + \dot{W}_R$$

$$(3.19)$$

where $\dot{W}_S$ and $\dot{W}_R$ represent the work transfers due to $\overline{\delta P_0}$ and the rotor action, respectively. Of course

$$\dot{W}_R = \Delta P_{12} \left(\frac{\dot{m}}{\rho} \right) \qquad (3.20)$$

and

$$\dot{W}_S = \delta \bar{P}_0 \left(\dot{m} / \rho \right) = \delta \bar{P}_0 A_R V_R \qquad (3.21)$$

After some algebra, equations (3.18) and (3.19) may be transformed into the following two equations for the

thrust coefficient (C_T):

$$C_T = 2Y(1 - X) \left(1 + \frac{C_{PSD}}{2X} \right) \text{ (from equation (3.18))}$$

$$(3.22a)$$

$$C_T = (1 - X^2) + \frac{\dot{W}_S}{\frac{1}{2} \dot{m} V_0^2} = (1 - X^2)$$

$$+ C_{PSD} \text{ (from equation (3.19))} \qquad (3.22b)$$

that result from the energy equation.

Here

$$C_{PSD} = \frac{\delta P_0}{\frac{1}{2} \rho V_0^2} \qquad (3.23)$$

Equations (3.22a) and (3.22b) may be solved for the Y parameter so that

$$Y = \frac{1}{2}(1 + X) \frac{1 + \dfrac{C_{PSD}}{(1 - X)}}{1 + \dfrac{C_{PSD}}{2X}} \qquad (3.24)$$

The coefficient C_{PSD} cannot be deduced from experiments. However, it is a fact that (1) as $X \to 1$ the coefficient $C_{PSD} \to 0$ (i.e., no interaction between the two streams) and (2) as $X \to 0$, the coefficient $C_{PSD} \to 1$ (the entire outer stream is deflected by 90° as the exhaust velocity drops to

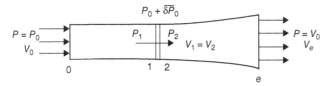

FIGURE 3.5 The overpressure on the streamwise periphery due to the squeezing effect of the outer stream flow.

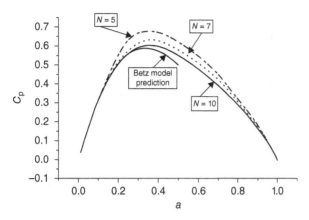

FIGURE 3.6   The predicted $C_P(a)$ relationship in the new and Betz's models.

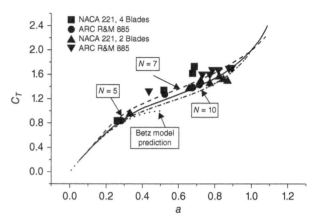

FIGURE 3.7   The predicted $C_T(a)$ relationship in the new and Betz's models.

zero). One may assume that C_{PSD} could be expressed as a polynomial series for the parameter $(1 - X)$:

$$C_{PSD} = \sum a_n \cdot (1 - X)^n \qquad (3.25)$$

The simplest expression is

$$C_{PSD} = (1 - X)^N \qquad (3.26)$$

which gives

$$C_T = 1 - X^2 + (1 - X)^N \qquad (3.27)$$

Given N, one may plot the $C_P(a)$ (Figure 3.6) and $C_T(a)$ (Figure 3.7) relationships. Figure 3.7 compares the $C_T(a)$ relationship for the standard Betz model, the present model

for various N values, and test data from "windmilling propeller" tests as given by Glauert [10]. The figures suggest that $N = 7$ will provide quite accurate results.

The new model suggests that the relationship $Y = 0.5(1+2X)$ produced by the Betz model is not valid for $a > 0.3$. Actually $Y \to 0$ as $X \to 1$, that is, high loading tends to stall the rotor turbine.

3.3   THE DUCTED WIND TURBINE

Starting with Betz himself [5], his limit on the turbine loading kindled interest on the "ducted wind turbine" concept. This concept allows for a significant contraction on the wind stream passing through the rotor, which leads to smaller

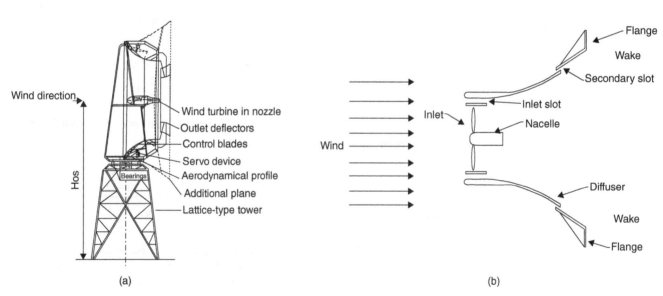

FIGURE 3.8   Some ducted wind turbine concepts: (a) the Frankovic–Vralovic concept; (b) a schematic of the Vortec 7 plant.

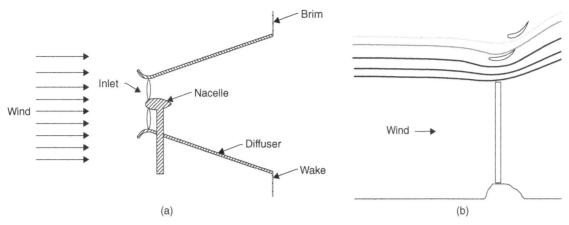

FIGURE 3.9 More modern ducted wind turbine concepts: (a) the flanged concept; (b) the Grassmann et al. concept [22].

diameter rotor, as long as the stream is capable of exhausting at the exit of the plant. All relevant studies have come to the conclusion that the "ducted" wind turbine efficiency may exceed the Betz limit [11]. For the period up to 1980, the relevant work has been reviewed by Igra [12, 13]. Foreman et al. [14] were among the first to develop an estimate of the "optimum" enhancement ratio (r), that is, the ratio of the actual C_P of the plant to the Betz limit value ($C_P = 0.592$). They concluded that this was mainly a function of the diffusion achieved behind the turbine. Taking into account typical component efficiencies, they estimated that $r < 2$. Georgalas et al. [15] and Politis and Koras [16] showed that $r > 2$ was possible. Frankovic and Vralovic [17] went even higher, to $r > 3$. Their concept is illustrated in Figure 3.8a.

An extensive analytical and experimental study was undertaken by the "Vortec" group in New Zealand who built a full-scale prototype, as illustrated in Figure 3.8b [18], with rather poor economic results. The actual data by the Vortec 7 plant produced C_P values above the Betz limit, although not by much, $C_{P,max} \cong 0.65$.

Over the last 10 years, some more complex "ducted" wind turbine concepts (and their "sea current" counterparts) were studied. They incorporate a diffuser exhausting into the subatmospheric pressure of the wake region generated by a "bluff body" (usually the diffuser itself). The most typical of these concepts is the "flanged" or "brimmed" diffuser (Figure 3.9a), under development in Japan [19–21]. The forced flow separation behind the flange leads to a lower pressure. In Scotland, another team is developing the "building integrated ducted wind turbine" [22], which exploits the flow blockage created by the building (which leads to an accelerated wind) and the corresponding roof under-pressure, where the ducted turbine exhausts. Finally, the "partially static" turbine was developed and tested by Grassmann et al. [22–24] as illustrated in Figure 3.9b. The experimental data for the Grassmann concept are presented in Figure 3.10 for both ducted and un-ducted form. The ducted concept generates a C_P value nearly double that of the un-ducted.

Van Bussel [25] has attempted to reevaluate the merits of the ducted wind turbine concept. He proposes to define a new "power coefficient" (i.e., plant efficiency) based on the wind kinetic power intercepted by a cross-sectional area equal to that of the diffuser exit (instead of the conventional one based on the turbine rotor area). He showed that now all plants exhibited maximum efficiency levels below the Betz limit.

Jamieson [26] has provided a generalized model of the optimum wind turbine power coefficient. He concluded that $C_{P,max} = (8/9)(1 - a_m)$, where a_m is the value of the induction factor at optimum (i.e., $C_{P,max}$) operating conditions.

Based on the station numbering illustrated in Figure 3.11, Georgiou and Theodoropoulos [27] predicted that

$$C_P = \eta_T \left(1 + C_{PW} - \left(C_{LS} + \frac{1}{T} \right) Y^2 \right) Y \quad (3.28)$$

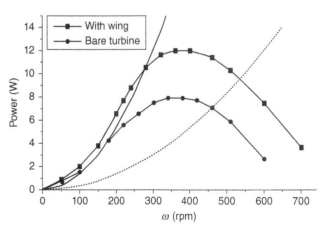

FIGURE 3.10 Data for the Grassmann concept [22], for ducted and un-ducted form, for different speeds of rotation, ω, at a wind speed of 5 m/s.

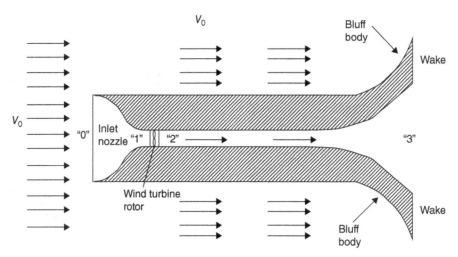

FIGURE 3.11 Station numbering in the ducted wind turbine.

where the parameter

$$\Delta C_{\text{PW}} = \frac{P_0 - P_3}{\frac{1}{2}\rho V_0^2} \qquad (3.29)$$

represents the nondimensional under-pressure at the wake of the exhausting duct, while C_{LS} represents the combined total pressure loss coefficient for the entire duct.

Figure 3.12 illustrates their prediction for ($C_{\text{P,max}}$) under representative ideal ($C_{\text{P,ID}}$) flow conditions in terms of the rotor efficiency (η_T) and the parameters C_{LS} and F_1, where

$$F_1 = 1 - X^2 + \Delta C_{\text{PW}} \qquad (3.30)$$

when $\left(\dfrac{\partial C_{\text{P}}}{\partial Y}\right)_{X=\text{ct}} = 0.$

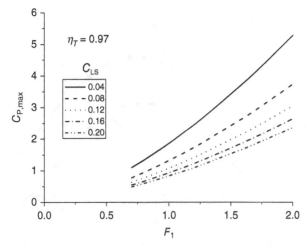

FIGURE 3.12 The power coefficient of the ducted wind turbine plants.

REFERENCES

[1] Bergey, K. H. (2007). The Lanchester–Betz limit. *Journal Energy*, 3:382–384.

[2] van Kuik, G. A. M. (2007). The Lanchester-Betz-Joukowsky limit. *Wind Energy*, 10:289–291.

[3] Froude, W. (1889). On the part played in propulsion by differences in fluid pressures. *Transactions of the Institute of Naval Architects*, 30:390–405.

[4] Rankine, W. J. M. (1865). On the mechanical principles of the action of propellers. *Transactions of the Institute of Naval Architects*, 6:13–30.

[5] Betz, A. (1920). Das Maximum der Theoretisch moglichen ausuntzung des windes durch windmotoren. *Zeitschrift fur das Gesamte Turbinewesen*, 26:307–309.

[6] Inglis, D. R. (1979). A windmill's theoretical maximum extraction of power from the wind. *American Journal of Physics*, 47:416–420.

[7] de Paor, A. M. (1982). Aerodynamic design of optimum wind turbines. *Applied Energy*, 12:221–228.

[8] Georgiou, D. P., Theodoropoulos, N. G. (2011). A momentum explanation for the unsatisfactory Betz model prediction in highly loaded wind turbines. *Wind Energy*, 14(5):653–660.

[9] Jung, S. N., No, T. S., Ryu, K. W. (2004). Aerodynamic performance prediction of a 30 kW counter-rotating wind turbine system. *Renewable Energy*, 30:631–644.

[10] Glauert, H. (1926). The analysis of experimental results in the windmilling brake and vortex ring states of an airscrew. Report 1026, Aeronautical Research Committee Reports and Memoranda, London, Her Majesty's Stationery Office.

[11] Hansen, M. O. L., Sorensen, N. N., Flay, R. G. J. (2000). Effect of placing a diffuser around a wind turbine. *Wind Energy*, 3:207–213.

[12] Igra, O. (1981). Research and development for shrouded wind turbines. *Energy Conversion and Management*, 21:13–48.

[13] Igra, O. (1977). The shrouded aero-generator. *Energy*, 2:429–439.

[14] Foreman, K. M., Gilbert, B., Oman, R. A. (1978). Diffuser augmentation of the wind turbines. *Solar Energy*, 20:305–311.

[15] Georgalas, C. G., Koras, A. D., Raptis, S. N. (1991). Parameterization of the power enhancement calculated for the ducted rotors with large tip clearance. *Wind Engineering*, 15:128–136.

[16] Politis, G. K., Koras, A. D. (1995). A performance prediction method for ducted medium loaded horizontal axis wind turbines. *Wind Engineering*, 19:273–288.

[17] Frankovic, B., Vralovic, I. (2001). New high profitable wind turbines. *Renewable Energy*, 24:491–499.

[18] Phillips, D. G., Flay, R. G. J., Nash, T. A. (1999). Aerodynamic analysis and monitoring of the Vortec 7 diffuser-augmented wind turbine. *IPENZ Transactions*, 26(1):13–19.

[19] Abe, K., Nishida, M., Sakurai, A., Ohya, Y., Kihara, H., Wada, E., Sato, K. (2005). Experimental and numerical investigations of flow fields behind a small wind turbine with a flanged diffuser. *Journal of Wind Engineering & Industrial Aerodynamics*, 93:951–970.

[20] Abe, K., Ohya, Y. (2004). Investigation of flow field around flanged diffusers using CFD. *Journal of Wind Engineering & Industrial Aerodynamics*, 92:315–330.

[21] Ohya, Y., Karasudani, T., Sakurai, A. (2002). Development of high-performance windturbine with brimmed diffuser. *Journal of the Japan Society for Aeronautical and Space Science*, 50:477–482.

[22] Grassmann, H., Bet, F., Cabras, G., Ceschia, M., Cobai, D., DelPapa, C. (2003). A partially static turbine—first experimental results. *Renewable Energy*, 28:1779–1785.

[23] Dannecker, R. K. W., Grant, A. D. (2003). Investigations of a building integrated ducted wind turbine module. *Wind Energy*, 5:53–71.

[24] Grassmann, H., Bet, F., Ceschia, M., Ganis, M. L. (2003). On the physics of partially static turbines. *Renewable Energy*, 29:491–499.

[25] van Bussel, G. J. W. (1999). An assessment of the performance of diffuser augmented wind turbines (DAWTs). Proceedings of the ASME/JSME Joint Engineering Conference, San Francisco.

[26] Jamieson, P. (2008). Generalized limits for energy extraction in a linear constant velocity flow field. *Wind Energy*, 11:445–457.

[27] Georgiou, D. P., Theodoropoulos, N. G. (2010). Grounding and the influence of the total pressure losses in ducted wind turbines. *Wind Energy*, 13(6):517–527.

4

MODELING WIND TURBINE WAKES FOR WIND FARMS

Angus C. W. Creech and Wolf-Gerrit Früh

4.1 INTRODUCTION

This section sets the scene for the importance of wake modeling by first placing it in the context of the other wind farm design tasks, observations of the effect of a wind turbine's wake on other turbines in a farm, and consequence on the overall performance of the wind farm. To conclude this section we will return to the behavior of an isolated turbine wake, to illustrate the main processes affecting turbine wakes which need to be adequately covered by wake models.

4.1.1 Wind Farm Design

Designing a large wind farm involves a number of tasks, including estimating the available wind resource from local and regional data, environmental assessments (such as visual intrusion, noise, and impacts on wild life), the local wind distribution at the proposed locations of the turbine, and many more. It is therefore important to provide tools for these tasks which are at the same time accurate, reliable, and cost-effective. With the development in computing power, it is now common practice to carry out flow modeling of the air flow through the proposed wind farm, and there are a number of commercial products and services to provide this.

However, a full flow model including just the most important factors would require representation of the topography at a fairly high resolution and allow for small-scale convection through heat fluxes from the surface, turbulence generation, transport, and decay, as well the effect of the wind on wind turbines and vice versa. All these processes would then have to be applied to a variety of meteorological conditions in terms of wind speeds, wind direction, and atmospheric stability as is needed to obtain a fair representation of the chosen wind farm site. Furthermore, in the design stage these modeling activities only become really useful if they can be used to optimize the layout of the turbines, which implies repeating all the meteorological conditions for a range of turbine locations.

Considering this range of demands, it is immediately obvious that such a task is only possible by representing many of these processes through highly simplified parameterizations. For example, the major effect of heat fluxes on the flow structures relevant to large turbine rotors is the vertical shear. Instead of modeling the actual heat fluxes and resulting buoyancy forces in the models, they are frequently only represented by imposing a typical wind shear profile as boundary conditions. The aerodynamics of the turbine blades would in direct simulations require a full description of the rotating solid blades with boundary conditions requiring a very high resolution; these are usually replaced by a body force determined from the performance of the rotor in the form of what is known as an actuator disk, or through the lift and drag coefficients from boundary element momentum theory, or similar approaches. Similarly, the wake behind a turbine and its gradual decay is usually represented through a much simplified empirical or semiempirical parameterization.

4.1.2 Observed Wake Effects

As wind farms become progressively larger, where many new projects have hundreds of megawatt-rated turbines, it is becoming very important to know what effect a turbine has on other turbines in the farm and what the combined effect of a large wind farm could be on the environment. These effects could be electrical, for example, on power quality issues from connecting the farm to the grid. On the aerodynamic

Alternative Energy and Shale Gas Encyclopedia, First Edition. Edited by Jay H. Lehr and Jack Keeley.
© 2016 John Wiley & Sons, Inc. Published 2016 by John Wiley & Sons, Inc.

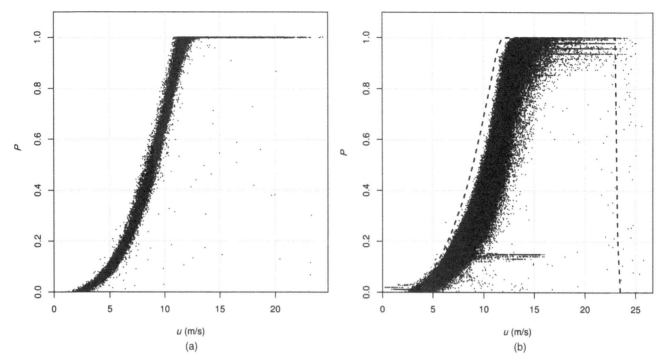

FIGURE 4.1   Turbine and farm performance curves. *Data source*: Vattenfall.

side, a wind turbine affects the flow and causes in a wake downstream of the turbine rotor in which the mean velocity is decreased, a rotation of the air induced, and substantial turbulence generated: another turbine placed in such a wake will experience very different air flow conditions than an isolated turbine. The consequences are that a downstream turbine may produce less power due to the reduced mean flow and that turbine blades may experience stronger fluctuations in their loadings and stresses due to the induced turbulence.

From the perspective of a wind farm developer, it is important to maximize the net income from electricity production from a given available area. Hence, it is important to optimize the spacing of turbines to install as many as possible while spacing them with sufficient distance to limit the wake effects.

It is best to illustrate the magnitude from the results of the Lillgrund offshore wind farm which consists of a relatively regular array of 48 fairly closely spaced turbines [1]. Figure 4.1 compares the power curve for one of the turbines of that wind farm with the power curve for the entire wind farm, the former scaled against the rated power of the turbine, the latter against the installed capacity of the wind farm. Figure 4.1a shows the power output for one of the turbines in the front row relative to the prevailing wind direction; the cases were included only when the wind came from a 120° sector surrounding that direction. The wind speed is that measured from the nacelle anemometer and the power output shows clearly the cut-in wind speed, the increase to the rated power, and the range where the turbine operates at the rated power,

and the scatter is relatively tight. Contrasted with this, Figure 4.1b shows the total wind farm output against nacelle wind speed of the most exposed wind for all wind directions. Not only does the wind farm performance show a much larger scatter but the entire performance curve is shifted to the right, to higher wind speeds. To better show the shift of the performance curve, the averaged performance from the individual turbine in Figure 4.1a is superimposed as the dashed line. The additional horizontal lines correspond to individual turbines being switched off rather than external wind conditions. However, the consistent shift of the curve to higher wind speeds can only be explained by shading of turbines situated in the wake of upstream turbines, and the much larger scatter can be explained by the fact that it depended very much on the wind direction as to how much turbines in the second and further rows are affected by upstream wakes.

The sensitivity of turbines in the second and further rows to the wind direction is shown in Figure 4.2, in which the power output of turbines compared to the front turbine is shown against the wind direction, where the wind direction of 230° is the case where the turbine row is fully aligned with the upstream wind. Due to the nonlinear nature of the turbine's power curve, the analysis centers on the wind speed range where the power output changes with speed, that is, between the cut-in wind speed and the rated wind speed. In this case, the turbines only produce between 30% and 40% of the front turbine. The wind directions of 105° and 250° are those where the wakes from the front row miss the turbines in the second row but those in the third row are now

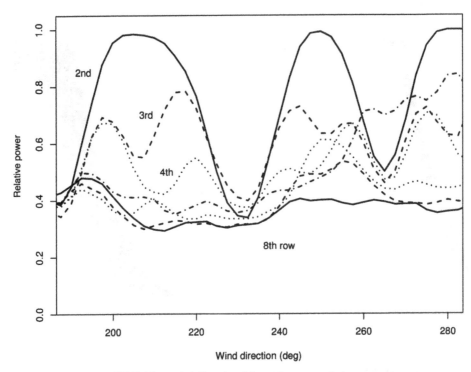

FIGURE 4.2 Wind directional dependence on relative power.

directly downwind of the front row turbine. These observations clearly show that the reduced velocity in the wind turbine wake has a strong effect on the productivity of a turbine in the wake. Some have suggested that the overall wind farm might actually generate more power if the front turbines are operated somewhat below their optimum conditions which allows the other turbines to produce more [2, 3].

Another feature evident in Figure 4.2, very much in common with observations from other wind farms [4, 5] and numerical models such as Churchfield et al. [6], is that once the available power has been reduced by a second turbine, it does not get reduced much further for a third turbine. In fact, in some cases, the third turbine performs better than the second, as is the case here for full alignment: the turbines in third to fifth rows perform noticeably better than the second turbine. This phenomenon is currently explained such that the turbulence induced by turbines—especially those operating in the turbulent inflow conditions of a wake—helps to disperse the wake much better than a wake with a lower turbulence level. However, it has to be borne in mind that other studies present a continued decrease of the average performance for turbines deeper into the wind farm array, a phenomenon usually referred to as the "deep array effect." To interpret these plots one has to bear in mind that they have to be obtained through averaging of wind speeds, weather conditions, and, most importantly, over a finite wind direction sector. For example, Figure 4.2 was obtained by averaging over a ±2.5° window, whereas the deep array effect becomes visible when averaging over a larger sector, such as ±10° or

even more. Considering that the actual wind is never fixed in speed and direction throughout the array over the averaging period of an individual observation (typically a few minutes) and that the nominal wind direction is determined either from a single point or as an average across the array, it is not clear how large the averaging sector should be to give the most appropriate representation of effect.

4.1.3 Individual Turbine Wake Recovery

Again, from field measurements, wind tunnel tests, and numerical models, it is well known that upstream turbulence strongly affects the decay of the wake [7], where a higher level of turbulence helps to erode the wake faster. This is illustrated in Figure 4.3, in which the decay of the wake is plotted for a number of different situations, including a direct comparison between a measured wake behind an 950 kW turbine and a computational model of it *in situ*. The exact nature and processes by which the upstream turbulence affects the wake recovery is not yet well understood and requires detailed experiments and numerical models; while the induced wake turbulence helps with the wake recovery and thereby power output, at the same time, it increases fatigue loads on the turbine by up to 80% [8], which would be expected to shorten the life span of turbine blades substantially.

Finally, some have observed that the relative yaw position of the turbines also affects the performance of the downstream turbine [9, 10].

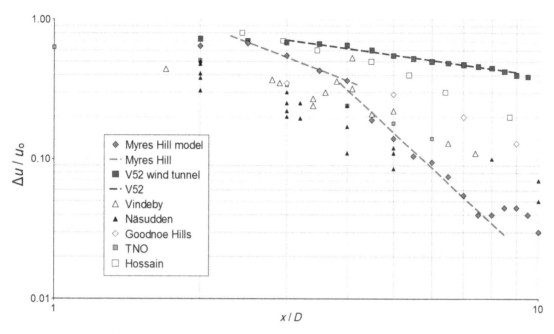

FIGURE 4.3   Environmental effects in single turbine wake recovery. From Creech et al. [7].

4.1.4   Outline

This short overview serves to highlight the fact that different fluid dynamical processes associated with turbine wakes have several very strong effects on other turbines in a wind farm, with implications for the lifetime of turbine blades, the optimum design of wind farms, and the optimum operation of wind farms. It is therefore very important to build up a good understanding of these processes and to develop a range of models, some of which would have to be fairly simple to be feasibly incorporated into the planning or operational tools and some of which would have to try to explore the full dynamics at the expense of a high complexity of these models. A key earlier review of wind turbine and wind farm wake modeling was that by Crespo, Hernández, and Frandsen in 1999 [11]. Since then, computing power of workstations and clusters has increased tremendously, opening up the possibilities of resolving more and more detail in direct computational fluid dynamics (CFD) simulations. But it is still impossible to configure a CFD model to include a number of wind turbines represented as rotating solid boundaries in a computational domain representing the surrounding topography. At the same time, the number of large wind farms has increased and is continuing to increase rapidly with a need to be able to represent wake effects in a relatively simple way, both for wakes from individual turbines on other turbines in the wind farm and for the effect the wind farm as a whole has on its environment, including possible wind farms downstream of other wind farms. Reflecting this, there were a number of activities noteworthy at the time of this entry, including the EUROMECH colloquium 508 on *Wind*

Turbine Wakes in 2009 [12] and the IEA WAKEBENCH initiative inaugurated in 2010 [13].

In line with the variety of methods used, the remainder of this chapter will introduce a hierarchy of models, starting with simple empirical models in Section 4.3, and gradually building up to models of the full time-varying three-dimensional fluid dynamics of the flow around a single wind turbine or a wind farm. After a section introducing the principles of CFD in Section 4.4, Section 4.5 introduces the methods to model the effect of wind turbines on the fluid in a CFD model. Section 4.6 then describes CFD-based wind turbine and wind farm models which are finally compared in Section 4.7.

4.2   EMPIRICAL METHODS TO ESTIMATE WAKE RECOVERY

For many years, the wake decay has been parameterized in flow models of proposed wind farms through a small number of models or assumptions, most notably using the Park model [14, 15] or through eddy viscosity [16]. A systematic survey has compared a range of these approaches with SODAR measurement from the Vindeby offshore wind farm [17] which highlighted the spread of results and the need to both interpret their results accordingly and develop more accurate methods.

Jensen's Park model is a simple method based on momentum conservation and assuming entrainment of freestream air surrounding the wake into the wake. In the simple models, a constant wake entrainment factor α is used and the wake expands linearly with distance, x, behind the rotor of radius R_T, given by $r_w = R_T (1 + \alpha x)$. The velocity in the wake as a function of downstream distance, x, from the rotor is given in Ref. [14] as

$$u(x) = u_0 \left[1 - 2a \left(1 + \alpha \left(\frac{x}{r_w} \right) \right)^{-2} \right] \quad (4.1)$$

where a is the axial flow induction factor as used in the actuator disc theory and related to the thrust coefficient of the turbine as $C_T = 4a(1 - a)$. The wake entrainment factor depends on the freestream turbulence level and velocity shear. For example, from Mosetti et al. [18] it can be empirically calculated from the hub height, z_{hub}, and the surface roughness, z_0, as

$$\alpha = \frac{0.5}{\ln \left(z_{hub}/z_0 \right)} \quad (4.2)$$

In practice, this approach is implemented on the basis of the thrust coefficients for the wind turbine as supplied by the manufacturers and using recommended entrainment or wake decay coefficients; for example, WAsP [17] uses

$$u(x) = u_0 \left[1 - \left(1 - \sqrt{1 - C_T} \right) \left(\frac{D}{D + 2k_w x} \right)^2 \right] \quad (4.3)$$

where the D is the rotor diameter and k_w the wake decay coefficient, with suggested values of $k_w = 0.05$ for offshore and 0.075 for onshore cases.

A variant of the Jensen model is the Frandsen model [19] which does not assume a linearly expanding wake but specifies a wake diameter based on the thrust coefficient of the turbine and a wake entrainment factor as

$$D_W = \left(\beta^{k/2} + \alpha \frac{x}{D} \right)^{-1/k} \quad (4.4)$$

with $\beta = \frac{1 + \sqrt{1 - C_T}}{2\sqrt{1 - C_T}}$, the constant α typically recommended to be $\alpha = 0.05$ and k between 2 and 3, usually taken as $k = 3$ (see [19]), and a wake velocity of

$$u(x) = u_0 \left[1 - \frac{1}{2} C_T \left(\frac{D}{D_W} \right)^2 \right] \quad (4.5)$$

In the application of this method for an array of turbines, the wake behind each individual turbine is allowed to expand until it either reaches the ground or meets another wake. Once this happens, the wakes are merged while the total momentum deficit is conserved.

The eddy viscosity approach originally developed by Ainslie [16] uses the momentum equations with an eddy viscosity term replacing the molecular diffusion for axisymmetric wake flow. As Crespo et al. [11] have pointed out, one weakness of the axisymmetric wake flow is that it cannot resolve wind shear caused by surface roughness or atmospheric stability. The UPMWAKE model [20, 21] was developed to allow for a turbine to be situated in a nonuniform flow. This model assumes an initial analytical velocity profile which represents the effect of atmospheric stability, as given by the Monin–Obukhov length, and the surface roughness, which is then perturbed by the wind turbine using Reynolds-averaged Navier–Stokes (RANS) CFD modeling with a k–ε turbulence model. A recent comparison showed that both the Park models and the eddy viscosity models consistently overestimate the electricity production from large wind farms, where most of the errors arose from turbines in the fourth row relative to the wind and further [22].

One aspect which all methods struggle with is that even the freestream wind is never constant, and the fluctuations in wind speed and direction continually affect not only the generation of the wake at the turbine but also the advection of the developing wake downstream. As a result, the wake is never a smoothly expanding wake along a straight trajectory but it meanders in the lateral direction. One simple approach to capture the effects of wake meandering is to superimpose a simple wake model on a wind which is fluctuating in direction, for example, using a stochastic model [23].

4.3 COMPUTATIONAL FLUID DYNAMICS

As CFD techniques represent a very active area of much of recent wake modeling research, this section is devoted to outlining the most common methods used, including finite element, finite volume, finite difference, and vorticity methods; turbulence modeling is also discussed in brief. Both vector and Einstein notations are used below to provide maximum clarity.

4.3.1 Basic Equations

At its simplest level, CFD problems for wind energy problems must solve the equations for incompressible Newtonian fluids, that is, the momentum and continuity equations, which can be written in Einstein notation, respectively, as

$$\rho \left(\frac{\partial u_i}{\partial t} + u_j \frac{\partial u_i}{\partial x_j} \right) = -\frac{\partial p}{\partial x_i} + \mu \frac{\partial}{\partial x_j} \left(\frac{\partial u_i}{\partial x_j} \right) + F_i \quad (4.6)$$

and

$$\frac{\partial u_i}{\partial x_i} = 0 \quad (4.7)$$

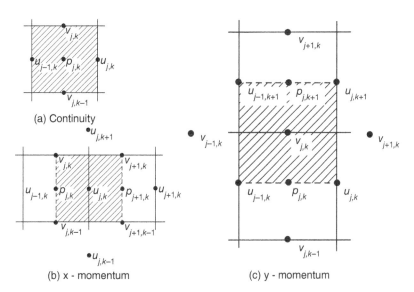

FIGURE 4.4   A two-dimensional finite difference grid. Courtesy of Fletcher [24].

where u_i represents the velocity field, p the pressure field, ρ the density of air, μ the molecular viscosity of the air, and F_i any body forces acting on the fluid, such as gravity.

In addition to these, boundary conditions may be applied, which can take the form of Dirichlet conditions for a given field ψ:

$$\psi = a \qquad (4.8)$$

Von Neuman, for coordinate components x_i:

$$\frac{\partial \psi}{\partial x_i} = a \qquad (4.9)$$

or Robin conditions, a compound of the two:

$$\frac{\partial \psi}{\partial x_i} + b\psi = a \qquad (4.10)$$

where a and b are given constants; ψ can represent either the velocity components u_i or the pressure p.

The momentum equation is a second-order, partial differential equation, coupled to the continuity equation. For many realistic scenarios, solving both these simultaneously and explicitly is impossible. In recent decades, there have been efforts to solve them numerically; we will cover the most common methods of discretizing and solving these equations below.

4.3.1.1 Finite Difference. Finite difference techniques have long been used in CFD, being one of the least complex to implement. Here the simulation domain is represented by a regular grid, with the velocity field defined at regular intervals; often a staggered grid approach is implemented, so that

the pressure terms are defined in the center of each grid cell, as shown in Figure 4.4.

The finite difference approach has the benefit of being straightforward to assemble into matrix form, given the spacing between each grid-point which does not change with time. However, this also means that problems with complex geometries are poorly represented and that grid resolution cannot be arbitrarily concentrated to better resolve local flow structures.

Coupling Pressure and Velocity. As pressure and velocity are coupled via the momentum equation (6) and mass continuity equation (7), at each time-step these must both be solved for while ensuring continuity. A common technique for doing so was first introduced by Patankar and Spalding [25] in their SIMPLE algorithm (semi-implicit method for pressure linked equations), whereby pressure and continuity terms are linked semi-implicitly. Before the first time-step, the pressure values on the grid are initialized to some value, usually to zero, then the values for the velocity and pressure are iteratively updated, until both $\underline{u}$ and p converge toward stable values. The steps of the iteration are as follows (from Fletcher [24], p. 362–365):

1. Calculate the guess for the correct value for the velocity field $\underline{u}^*$.
2. Obtain the correction to the pressure δp.
3. Calculate of the velocity correction $\underline{u}^c$.
4. Update the pressure via $p^{n+1} = p^n + \alpha_\mathrm{p}\delta p$, where α_p is the pressure relaxation parameter.
5. Set $\underline{u}^{n+1} = \underline{u}^* + \underline{u}^c$ ready for the next iteration.
6. If the velocity and pressure fields have not converged, go to step 1.

SIMPLE methods are typically slow to converge, and so two alternative schemes were developed. These are SIM-PLEC [26], which approximates terms in the pressure correction equations previously neglected in SIMPLE, and SIM-PLER [27], which solves the pressure correction term first so that it can be used to correct the velocity field to satisfy continuity. For further reading on linking pressure–velocity, see Ferziger et al. [28, section 7] and Fletcher [24, section 17.2.3]. Note, these methods are not exclusive to finite difference techniques and are also used in finite volume and finite element formulations.

4.3.1.2   Finite Volume.
In finite volume methods, the simulation volume is divided into small, finite control volumes (CVs), which represent cells on a mesh or grid. The nodes are defined in the center of these CVs. The differential equations governing the flow, for example, the momentum and continuity of mass equations, are integrated over each CV. The advantage of this approach is that by design it satisfies the conservation of mass and momentum in every CV, even on relatively coarse grids. It is for this reason that it is a popular technique with commercial CFD software packages such as ANSYS CFX [29]. The starting point is to take the integral, rather than differential, form of the momentum and conservation equations, which for an incompressible fluid are

$$\rho \frac{\partial}{\partial t} \int_V u_i \, dV$$

$$= \int_V \left[-\frac{\partial p}{\partial x_i} - \left(\frac{\partial u_i}{\partial t} + u_j \frac{\partial u_i}{\partial x_j} \right) + \mu \frac{\partial}{\partial x_j} \left(\frac{\partial u_i}{\partial x_j} \right) + F_i \right] dV \tag{4.11}$$

and

$$\int_S u_i n_i \, dS = 0 \tag{4.12}$$

where V represents the volume of a particular finite volume, S its surface, and n_i a vector normal to that surface. This equation guarantees that for each finite volume, momentum is conserved. For more on finite volume techniques an excellent introduction can be found in Ferziger [28].

4.3.1.3   Finite Element.
Finite element methods (FEM) have come to the fore recently particularly in oceanographic modeling [30–32], due to their ability to handle unstructured meshes, an ability which allows them to adapt to models with complex boundaries, such as irregular terrain or bathymetry. An excellent example of this would be the tidal simulation of Venetian Lagoon by Canu et al. [33], shown in Figure 4.5, which shows the unstructured triangular mesh fitted to the

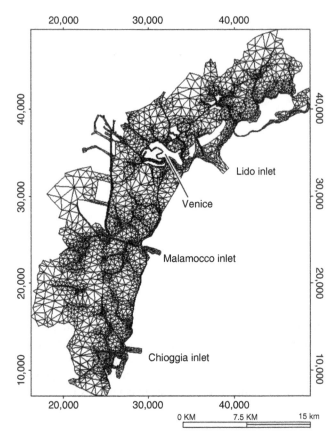

FIGURE 4.5   Finite element simulation of the Venetian Lagoon. Courtesy of Canu et al. [33].

coastline and its resolution concentrated in particular areas of interest.

In FEM, the simulation domain is discretized into elements which in two dimensions can be triangles, squares, or hexahedra, or in three dimensions tetrahedral or cuboids. Triangular and tetrahedral meshes are common, due to being simple to implement in unstructured meshes. Each element is defined by the nodes, for example, the values of momentum and pressure at the corner of each element; nodes are shared with neighboring elements. Each field is defined within an element using basis functions, so the velocity and pressure fields can be written as

$$\underline{u} = \sum_{n=1}^{N} \Phi_n \underline{u}_n \tag{4.13}$$

$$p = \sum_{n=1}^{N} \Psi_n p_n \tag{4.14}$$

where N is the number of mesh nodes, with Φ_n and Ψ_n the basis functions for velocity and pressure, respectively. The above equations are the global representation of velocity and

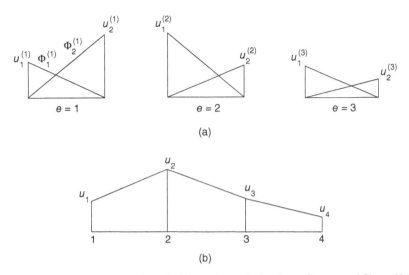

FIGURE 4.6   (a) Local and (b) global linear interpolation for u. Courtesy of Chung [34].

pressure; in finite element analysis, we start with a local representation, with interpolation of each occurring within each element. These can be of any order but a common choice is linear for both, often referred to as a P1P1 element pair (first-order pressure, first-order velocity). Figure 4.6 illustrates this with the velocity u on a one-dimensional line discretized into three elements, with four nodes in total. It shows how the velocity field between nodes 1 and 2 (in b) is the sum of the two interpolation functions, $\Phi_1^{(1)}$ and $\Phi_1^{(2)}$, within element $e = 1$ (in a). As one of the most straightforward implementations, it is less computationally expensive than higher order formulations. Through the Galerkin method, the local matrices are assembled into a global form of the momentum and continuity equations to be solved. A thorough description of the Galerkin method and CFD using finite element analysis in general can be found in many fluid dynamics books, such as Chung [34].

As stated previously, the key advantage of finite element analysis methods is their ability to handle irregular terrain, and so their rise in physical oceanographic modeling. This very same virtue allows them to be adapted to modeling wind farms in hilly areas and has been used by Creech et al. [7] for this purpose. However, their main disadvantage is that they do not naturally preserve the continuity of mass, as finite volume techniques do. To do so requires further controls, as has been implemented by Cotter [35].

4.3.1.4 *Vorticity Equation.* Since the lift F_{L} generated by an ideal aerofoil is given by the circulation, Γ, induced in the flow by the aerofoil such that $F_{\mathrm{L}} = -\rho u_0 \Gamma$, it is intuitive to model the evolution of the wake based on the conservation equations for vorticity rather than momentum. This can be implemented in an Eulerian approach by solving the vorticity equation on a domain-filling mesh or grid, as found in Hansen et al. [36], which in the most basic form with only molecular

diffusion is

$$\frac{D\underline{\omega}}{Dt} = \frac{\partial \underline{\omega}}{\partial t} + \left(\underline{u} \cdot \nabla\right)\underline{\omega} = \left(\underline{\omega} \cdot \nabla\right)\underline{u} - \underline{\omega}\left(\nabla \cdot \underline{u}\right) - v\nabla^2\underline{\omega} \tag{4.15}$$

or, allowing for a turbulent eddy viscosity, v_{t}, and body forces $\underline{f}$ (Vermeer et al. [37]),

$$\frac{\partial \underline{\omega}}{\partial t} + \nabla \times \left(\underline{\omega} \times \underline{u}\right) - v\nabla^2 \left[\left(1 + \frac{v_t}{v}\right)\underline{\omega}\right] + \nabla \times \underline{f} + \underline{Q}_\omega \tag{4.16}$$

on a domain-filling mesh or grid, where $\underline{\omega} = \nabla \times \underline{u}$ is the vorticity vector and v the kinematic viscosity; continuity completes this through $\nabla \cdot \underline{u} = 0$. These two conditions can be combined into a single set of Poisson equations as $\nabla^2 \underline{u} = -\nabla \times \underline{\omega}$.

While it appears a very attractive approach by eliminating the pressure term and implicitly satisfying continuity, it is not well suited for high Reynolds number cases or complex geometries [38].

4.3.1.5 *Vortex Particle Methods.* Alternatively, a Lagrangian method such as the vortex particle method can be adopted [37, 39], where the motion and evolution of a large number of individual vortices are computed based on Biot–Savart's law of the velocities induced by a vorticity distribution as

$$\underline{u}(\underline{x}') = \frac{1}{4\pi} \iiint \frac{\underline{\omega} \times (\underline{x} - \underline{x}')}{|\underline{x} - \underline{x}'|^3} \mathrm{d}\underline{x} \tag{4.17}$$

where the integration has to be carried out over the entire part of the domain which has nonzero vorticity.

Using this, the global flow field is then determined as the superposition of the external flow and the flow induced by the vorticity generated by the turbine blades. This superposition usually implies that the external flow has zero vorticity and simply acts as a background flow to advect the vortices. A number of studies have used this approach for a hierarchy of complexity but always requiring a prescribed vorticity generated by the turbine rotor. Once the vorticity is specified, the transport of the vorticity can be calculated either in a prescribed wake or a free wake.

In the simplest vortex models the rotor is represented by a hub vortex and a set of vortex rings, as if the rotor had an infinite number of blades, such as is found in Miller [40] and Øye [41], the former with a fixed wake and the latter including a wake expansion. Despite this, or maybe because of their simplicity, these models were able to simulate the wake reasonably well.

To move from an idealized rotor disc to a rotor with a finite number of specific blades, it is necessary to prescribe the vortex lines generated from each blade section rather than the azimuthally averaged vorticity ring. The most common approach is to use the vorticity generated by the bound circulation as specified through the Kutta–Joukowsky theorem. As this assumes two-dimensional and steady flow over an aerofoil section, this circulation has to be corrected to account for transient effects, such as dynamic stall [42], and three-dimensional effects, such as rotational augmentation [43]. Further developments of this approach include the so-called boundary integral equation method (BIEM) [44], in which the actual rotor blade geometry is represented through vorticity sources at the surface in addition to the vorticity doublets, while the vorticity away from the surfaces is described by doublets alone.

The vorticity generated by the turbine is then used in a model which either prescribes the wake structure [45, 46] or has a free wake [47]. The prescribed vortex method, where the position of each vortex is prescribed, has the advantage of being computationally very cheap and robust but is only valid if its position is adequately prescribed, which limits this to fairly simple and well-known situations. On the other hand, the free vortex method is not restricted to cases where the wake is known but includes the calculation of the wake velocities from the basic equations; an example of this is shown in Figure 4.7a and compared to a commercial RANS CFD package in Figure 4.7b. The cost of this is that the integration of the Biot–Savart equation must be calculated for each point in the domain. Furthermore, the calculations are prone to computational instability from the singularities which arise when two vortices are approaching each other.

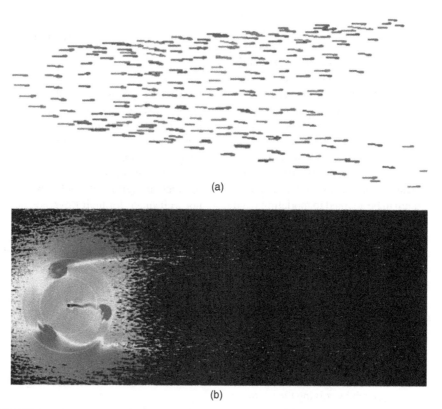

(a)

(b)

FIGURE 4.7   Comparison of the wake velocities calculated using (a) a discrete vortex method and (b) a commercial RANS CFD package (*Source*: Li and Çalişal [48]).

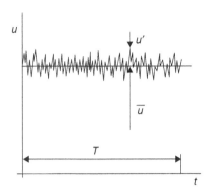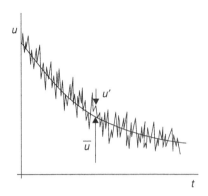

FIGURE 4.8 Time-averaged velocity for statistically steady flow (left) and unsteady flow (right) with ensemble averaging. Courtesy of Ferziger and Perić [28].

4.3.2 Turbulence

Most of the flows considered in wind turbine modeling are unsteady, turbulent flows, and so cannot be considered laminar. As such, turbulence must be taken into account for accurate wake simulation. For practical reasons, it is often treated with a degree of implicitness—the most common formulations being the fully implicit RANS or the semi-explicit large eddy simulation method.

4.3.2.1 Reynolds-Averaged Navier–Stokes. RANS takes the approach that unsteadiness in the flow is considered as fluctuations about a steady state. This means that there is no distinction between above grid or sub-grid turbulence and that both are implicitly modeled. While this approach cannot be expected to predict all types of turbulence flow accurately, RANS is a relatively inexpensive technique and so a popular choice for modeling fluid dynamics problems.

The velocity and pressure fields can be separated into steady and fluctuating components, as shown in Figure 4.8:

$$u = \bar{u} + u' \tag{4.18}$$
$$p = \bar{p} + p' \tag{4.19}$$

where $\bar{u}$ and $\bar{p}$ represent the time- and ensemble-averaged values, with u' and p' the fluctuations. The momentum and continuity equations for an incompressible Newtonian fluid then become

$$\rho\frac{\partial \bar{u}_i}{\partial t} + \frac{\partial}{\partial x_j}\left(\rho\bar{u}_i\bar{u}_j + \rho\overline{u'_i u'_j}\right) = -\frac{\partial \bar{p}}{\partial x_i} + \frac{\partial \bar{\tau}_{ij}}{\partial x_j} \tag{4.20}$$

$$\frac{\partial \bar{u}_i}{\partial x_i} = 0 \tag{4.21}$$

where $\bar{\tau}_{ij}$ represents the mean viscous stress tensor, that is,

$$\bar{\tau}_{ij} = \mu\left(\frac{\partial \bar{u}_i}{\partial x_j} + \frac{\partial \bar{u}_j}{\partial x_i}\right) \tag{4.22}$$

From the above equations, of particular interest is the term $\rho\overline{u'_i u'_j}$ in equation (20), which represents the Reynolds stress, that is, the fluid stress due to the fluctuations of the flow from the steady state. This must be solved implicitly, as it cannot be represented explicitly in terms of the mean values. Boussinesq [49] postulated that the Reynolds stress could be formulated as

$$-\overline{\rho u'_i u'_j} = \mu_t\left(\frac{\partial \bar{u}_i}{\partial x_j} + \frac{\partial \bar{u}_j}{\partial x_i}\right) - \frac{2}{3}\rho\delta_{ij}k \tag{4.23}$$

where μ_t is the eddy viscosity, δ_{ij} is the Kronecker delta, and k is the turbulent kinetic energy (TKE). μ_t represents the additional viscosity due to fluctuations (i.e., eddies) about the mean velocity values and k the additional kinetic energy that arises because of them. Solving for μ_t and k closes the turbulence, and common approaches to do this are the $k - \varepsilon$ and $k - \omega$ models [50, 51].

4.3.2.2 Large Eddy Simulation. Large eddy simulation (LES) takes the approach of resolving turbulence in a partially statistical, partially explicit manner. The turbulence is separated into large and small scales by applying a filter to the velocity field, so that in numerical simulations flow features larger than the grid resolution are resolved explicitly, while features smaller than the grid resolution are resolved implicitly. We define each component of the filtered velocity field as

$$\bar{u}_i = \int G(x, x')u_i(x, x')\,dx' \tag{4.24}$$

where $G(x, x')$ is the filter kernel. This filter is a localized function and has an associated length scale, Δ, which represents the division between small-scale and large scale turbulence. Examples of possible filter kernels are a simple local average or a Gaussian. For an incompressible fluid, the

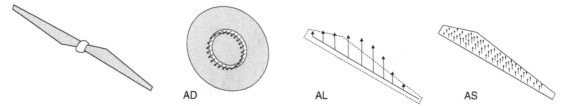

FIGURE 4.9 Indirect representations of turbine blades: actuator disc (AD), actuator line (AL), and surface (AS) (Sanderse et al. [52]).

Navier–Stokes equations become

$$\rho\left(\frac{\partial \bar{u}_i}{\partial t} + \bar{u}_j\frac{\partial \bar{u}_i}{\partial x_j}\right) = -\frac{\partial p}{\partial x_i} + \mu\frac{\partial}{\partial x_j}\left(\frac{\partial \bar{u}_i}{\partial x_j}\right) - \frac{\partial \tau_{ij}^{\Delta}}{\partial x_j} \quad (4.25)$$

$$\frac{\partial \bar{u}_i}{\partial x_i} = 0 \quad (4.26)$$

The additional term $\frac{\partial \tau_{ij}^{\Delta}}{\partial x_j}$ represents the additional stress due to small eddies that are not resolved explicitly, and the sub-grid stress tensor is defined as

$$\tau_{ij}^{\Delta} = \overline{u_i u_j} - \bar{u}_i \bar{u}_j \quad (4.27)$$

τ_{ij}^{Δ} is needed to close equation (4.25), and so a SGS model is required. A common choice is the Smagorinsky model, which defines the SGS stress as

$$\tau_{ij}^{\Delta} = 2\mu_t \bar{S}_{ij} \quad (4.28)$$

where μ_t is the sub-grid eddy viscosity and $\bar{S}_{ij}$ is the strain rate of the resolved velocity field. The eddy viscosity term is then defined as

$$\mu_t = C_S^2 \rho \Delta^2 \left|\bar{S}\right| \quad (4.29)$$

where C_S is a parameter to be defined and $\left|\bar{S}\right| = \sqrt{(\bar{S}_{ij}\bar{S}_{ij})}$. Estimates for isotropic turbulence usually give $C_S \approx 0.2$; however, the Smagorinsky constant, as it is often known, is often *not* constant but varies with the type of flow and can be a function of the Reynolds number as well as other flow parameters.

4.4 ROTOR MODELING TECHNIQUES

Several approaches exist to modeling with the turbines within simulation, most of which use indirect methods to represent the turbine blades. In principle, it is possible to simulate the full fluid dynamics by specifying the entire turbine through solid boundaries in the CFD domain and expect the boundary conditions and shear forces to lead the flow to respond to the turbine blade forces. However, this requires considerable computing resources; in contrast, indirect methods afford turbine models a degree of simplicity, so making them cheap from a computational perspective. Within these models, there are increasing degrees of complexity, from the simpler generalized actuator discs, to actuator lines and surfaces (see Figure 4.9); these come with an associated rise in processing requirements.

4.4.1 Actuator Discs

4.4.1.1 Uniform Thrust. The simplest representation of the turbine is by a disc of uniformly distributed thrust, which has a surface area A. The total force on the fluid, opposing the direction of flow, is specified as

$$f_t = -\frac{1}{2}C_T \rho u_0^2 \pi R^2 \quad (4.30)$$

where u_0 is a reference wind speed at hub height some distance upwind and C_T is the thrust coefficient for turbine, usually a function of u_0 and R the rotor radius. The calculation of u_0 can be problematic, especially in simulations with complex flow, from upwind turbines or irregular terrain. Finding u_0 is not straightforward due to the effect of the thrust on the flow, and so the relation $u_0 = u_{local}/(1 - a)$ is used, where a is the axial induction factor and u_{local} the local flow field. To determine a, Prospathopoulos et al. [53] presented at iterative procedure whereby an initial guess at u_0 is made to determine C_T. Taking u_{local} from the flow field round the disc, we can find a, and so a new value for u_0. This is repeated until convergence is achieved. A slightly different method was used by Calaf [54] and Meyers [55], replacing u_{local} with $\bar{u}_{disc}$, an average calculated across the area of the disc. This was used to directly calculate the backthrust, that is,

$$f_t = -\frac{1}{2}C_T' \rho \bar{u}_{disc}^2 \pi R^2 \quad (4.31)$$

where the modified thrust coefficient $C_T' = C_T/(1 - a)^2$. Instead of using an iterative solution, typical values for C_T and a were taken from the Wind Energy Handbook [56], giving $C_T' = 4/3$.

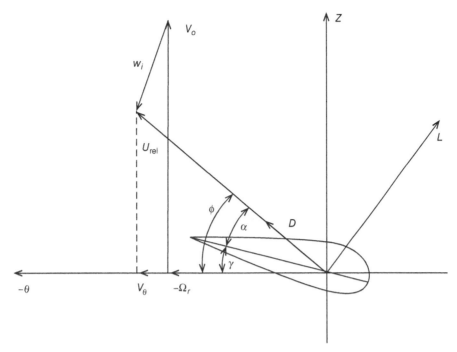

FIGURE 4.10 Cross-section of an aerofoil, showing the direction of the lift and drag forces, the direction of u_{rel}, the angle of attack α, and the local pitch angle γ. From Sorensen et al. [59].

4.4.1.2 Blade Element Momentum Theory. A further refinement to the actuator disc is to consider annular variations across the disc, rather than uniform, by incorporating parts of blade element momentum (BEM) theory to calculate lift and drag forces. Although disc representation means the blades cannot be discrete, the influence of the blades is spread azimuthally in infinitely thin rings centered on the hub. The lift and drag forces per span wise unit length on a blade section, shown in Figure 4.10, are commonly written as

$$\underline{f}_{2D}(\underline{x}) = \frac{1}{2}\rho u_{rel}^2 c\left[\underline{e}_L C_L + \underline{e}_D C_D\right] \qquad (4.32)$$

where C_L and C_D are the lift and drag coefficients, which vary with the angle of attack α and the Reynolds number Re; $\underline{e}_L$ and $\underline{e}_D$ are unit vectors in the direction of lift and drag, respectively; u_{rel} is the local relative speed of the air to the blade surface; and c is the chord length of the blade at distance r from the hub. Note, the inclusion of c allows blade geometry variation.

The determination of C_L and C_D can be made from lift/drag coefficient graphs, plotting the coefficient graphs as a function of α for a variety of Reynolds numbers. Graphs for aerofoils commonly used in wind turbines, such as the NACA 63- and 64- series, are readily available [57]. As such graphs do not incorporate the unsteady effects typical in turbulent flow, Masson et al. [58] incorporated a dynamic stall model to give accurate aerofoil performance with fluctuating values of attack.

The angle of attack α can be determined in several ways. Shen and Hansen look at the flow field upwind of the rotor [60]; another approach is to derive α from $\bar{u}_{disc}$, the angular velocity of the blades, and local blade twist [7]:

$$u_{rel} = \sqrt{(r\omega)^2 + \bar{u}_{disc}^2} \qquad (4.33)$$

This can present problems under high rotor loading where $u_0 \gg \bar{u}_{disc}$; this is similar to an issue addressed in actuator discs by others [53–55].

Recently, reactive turbine models have coupled lift and drag forces to turbine models, to simulate the unsteady behavior of rotors and mimic power generation through a generator (see Creech et al. [7]). As the force on the fluid at each point on the actuator disc is known at each time-step, the total torque acting on the fluid, τ_{fluid}, is also known. This must by necessity be balanced by the torque accelerating the blades, and the resistive torque of the generator, such that

$$\tau_{fluid} = -(\tau_{blades} + \tau_{gen}) \qquad (4.34)$$

If we choose a simple approximation of a generator and assume that it is a direct function of angular velocity [61], then we can write

$$\tau_{gen} = k\omega^2 \qquad (4.35)$$

where k is a constant defined from the performance specifications of the turbine, and we can also calculate the power generated:

$$P = \tau_{\text{gen}} \omega^2 \qquad (4.36)$$

Now that both the generator and fluid torque are known, the torque accelerating the blades can be determined, and if the momentum of inertia of the blades can be calculated, so can the angular acceleration of the rotor. In this way, turbine performance and blade loading can be entirely dynamic, which is useful for large eddy simulation turbulence models. Note that this or similar approaches can also be applied to the other actuator methods outlined below and have indeed been by Churchfield et al. [6], who applied a torque controller based upon a reference model at NREL for a 5 MW turbine [62].

4.4.1.3 Numerical Discretization. As simulations will be either grid or mesh based, the disc thrust will likely be spread over several cells. This necessitates the use of a kernel to spread the forces axially across a volume, that is,

$$\underline{f}_{\text{mesh}} = \underline{f}_{\text{disk}} \otimes \eta_\varepsilon \qquad (4.37)$$

Sørensen et al. [63] proposed a Gaussian kernel, η_ε, where the parameter ε controls the standard deviation and so the "smearing" of the thrust; larger values of ε give greater stability, smaller values greater accuracy. Others simply spread the thrust evenly in the axial direction across a cylindrical volume [7] making the kernel even and $\underline{f}_{\text{mesh}}$ dependent on the thickness of the cylinder.

4.4.2 Actuator Lines and Surfaces

Further accuracy can be achieved through using the actuator line method, first described by Sørensen and Shen [59]. Here, the forces on the fluid are concentrated around lines which represent the blades. In cylindrical coordinates (θ, r, z) these are

$$f_\theta = \frac{\rho c u_{\text{rel}}^2}{2r\,\mathrm{d}\theta\,\mathrm{d}z} \left(C_{\text{L}} \sin \phi - C_{\text{D}} \cos \phi \right) \qquad (4.38)$$

$$f_z = \frac{\rho c u_{\text{rel}}^2}{2r\,\mathrm{d}\theta\,\mathrm{d}z} \left(C_{\text{L}} \sin \phi + C_{\text{D}} \cos \phi \right) \qquad (4.39)$$

A regularization kernel is applied as before in equation (37); however, the kernel is now a Gaussian function of the distance from the force points on the rotor blades, that is,

$$\eta_\varepsilon(r) = \frac{1}{\varepsilon^3 \pi^{3/2}} \exp \left(-\frac{r}{\varepsilon} \right) \qquad (4.40)$$

FIGURE 4.11 Vorticity field from an actuator line simulation at freestream wind speed of 10 m/s (Sørensen et al. [59]).

where r is the distance from the force points. The total force on three-bladed turbine is then written as

$$\underline{f}_e(\underline{x}) = \sum_{i=1}^{3} \int_0^R \underline{f}_{2\text{D}}(r) \eta_\varepsilon(|\underline{x} - r\underline{e}_i|)\,\mathrm{d}r \qquad (4.41)$$

where $\underline{e}_i$ is a unit vector in the direction of blade i.

Sørensen et al. [59] ran axisymmetric simulations of a Nordtank turbine in a wind tunnel and found good agreement with measured power output at varied wind speeds. They also found that tip-shed vortices were being produced, which quickly diffused into continuous vortex sheets (see Figure 4.11). Despite being numerically intensive compared to other techniques, more recently actuator line methods have been applied by others to single and multiple wind turbine models, for example, Troldberg et al. [64–66] and in particular the wind farm simulations of Churchfield et al. [6]; these are covered in Section 4.6.2 below.

Lastly, there is actuator surface method, an extension of the actuator line concept by Shen et al. for horizontal axis turbines [67] and vertical axis turbines [68]. Rather than representing the aerofoil cross-section as one point on a line, the actuator surface method spreads the force over a notional surface representing the blade aerofoil. Thus, the body force exerted on the aerofoil cross-section becomes

$$\underline{f}_{2\text{D}}(\underline{x}) = \frac{1}{2} \rho u_{\text{rel}}^2 c \left(C_{\text{L}} \underline{e}_{\text{L}} + C_{\text{D}} \underline{e}_{\text{D}} \right) F_{\text{dist}}(\underline{x}) \qquad (4.42)$$

where $F_{\text{dist}}(\underline{x})$ is the distribution of pressure force along the blade chord, which is calculated from derived empirical

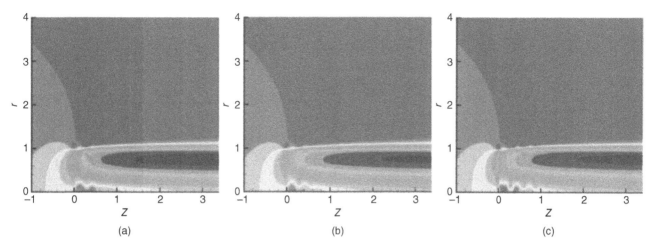

FIGURE 4.12 Comparison from RANS simulations in Shen et al. [67]: (a) the actuator line method, (b) the actuator surface method, and (c) the actuator surface method with fine mesh.

formulae. When compared with actuator line techniques [68], RANS simulations found improvements in the recreation of tip vortices, and the velocity field was more realistic than that produced by actuator line techniques (see Figure 4.12).

4.4.3 Direct Representation

Direct, or discrete, representation of single wind turbines for full wake simulation is rare, due to the prohibitive computational cost, and so unheard of for wind farms. However, given the dramatic increase in processing power in recent years, this may not always be the case. Wußow et al. [69] used LES to model the turbulent flow behind full-scale turbine, with the

rotation rate of the blades set by the incoming freestream wind speed. A three-dimensional domain of 4 million finite volume cells extended from one rotor diameter upwind to 3 diameters downwind, as shown in Figure 4.13.

Significant wake meandering was a key feature of the wakes produced over the 5 minutes of simulation time, and reasonable agreement was found with measured hub-height velocity data from 2 diameters downwind; however, the authors noted that the fixed pitch and rotor speed of the model was not a realistic representation of the behavior of a turbine in turbulent wind conditions—previously, the aerodynamics calculations necessary for more dynamic behavior were ruled out due to their prohibitive computational complexity.

FIGURE 4.13 Velocity magnitude in the wake of a discrete turbine simulation (Wußow et al. [69]).

4.5   WIND TURBINE SIMULATIONS

This section describes wind turbine and wind farm models using the two main approaches of modeling turbulent flow, first addressing the RANS models and then large eddy simulation (LES).

4.5.1   RANS Models

RANS' computationally inexpensive approach is an attractive prospect for far-wake simulation, and consequently modeling of wind farms. However, reviews by Sanderse [52] and Sumner [70] have found that the $k - \varepsilon$ and $k - \omega$ turbulence models tend to create excess diffusion in the wind turbine wakes, due to the limited applicability of the Boussinesq hypothesis.

In order to reduce the eddy viscosity in the near wake, the approach taken by El Kasmi and Masson [71] was to add a turbulent energy dissipation term to the momentum equation and apply it around the rotor to limit the TKE, and so turbulent eddy viscosity, in order to improve wake predictions. The validity of this approach was confirmed by Cabezón et al. [72], who compared RANS simulations of single turbine wakes using the standard $k - \varepsilon$ model with those using the El Kasmi and Masson additions; again the technique was shown to improve the near-wake turbulence and temper the excessive wake diffusion. Similar analysis using a $k - \omega$ turbulence closure scheme [53] has shown good agreement with Nibe wake data when incorporating the El Kasmi and Masson TKE limiting. A second technique based on realizability constraints on the Reynolds stress and Schwarz inequality [73] was also shown to perform well. A number of variants of the $k - \varepsilon$ model and a Reynolds stress model (RSM) were tested against measurements from a 330 kW wind turbine [74] with the result that the standard $k - \varepsilon$ model tended to underestimate the wake deficit, but the RSM model and specially tuned variants of the $k - \varepsilon$ model performed better. However, the performance of these depended strongly on the tuning of free parameters in the turbulence models. A whole wind farm in complex terrain was simulated with both analytical wake models and $k - \varepsilon$ RANS models [75], where it became apparent that neither method performed better than others but that all showed considerable variation of prediction error for the different turbines in the farm.

However, as these explicit alterations to the TKE are nonphysical and exist primarily to combat the overestimation of μ_t, it has been questioned whether El Kasmi and Masson's method is applicable to multiple turbine wakes [76]. In their turbine array simulations, Ammara et al. [77] did not include explicit turbulence source terms at the turbines, but nonetheless showed reasonable agreement with experimental data from the MOD-2 test wind turbine array at Goldendale.

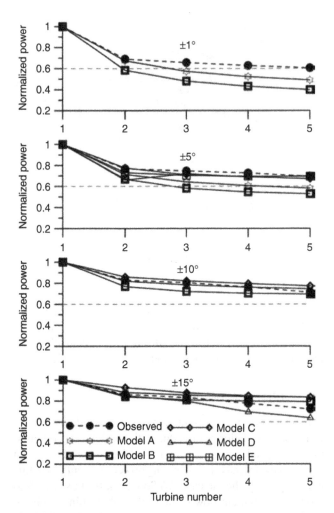

FIGURE 4.14   Comparison of models and measurements from Horns Rev for wind direction 270°, showing the power deficits on downwind turbines due to wake effects, at a hub-height freestream wind speed of approximately 8 m/s. The width of the wind direction measurement bins increases from top to bottom. Courtesy of Barthelmie et al. [4].

Barthelmie et al. [4] carried out a comparison of wind farm models with experimental data from the Horns Rev wind farm, to examine their ability to predict power losses due to wake effects. As shown in Figure 4.14, they found that models which used $k - \varepsilon$ turbulence closure schemes, such as WAKEFARM (based upon UPMWAKE), CENER, and NTUA, tended to overpredict wake losses, particularly for narrow wind direction measurement bins.

4.5.2   LES Models

4.5.2.1   *Single Turbines.*   As LES requires significantly greater computational resources than RANS methods, its introduction into wind turbine and wind farm modeling has only recently been practicable due to recent advances

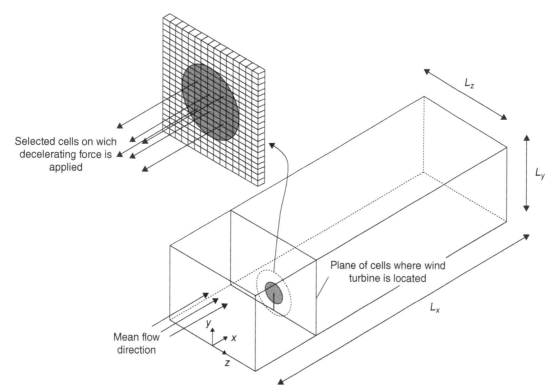

Selected cells on wich
decelerating force is
applied

Plane of cells where wind
turbine is located

Mean flow
direction

FIGURE 4.15   Computational domain used in simulations by Jimenez [78]. The dimensions of the domain are 34.9D × 5.6D × 10.7D. Log law of wall was applied to the rough bottom surface.

in affordable computer power. As large-scale turbulence is resolved explicitly, it potentially affords greater accuracy than RANS models, where both small- and large-scale turbulence are not.

An early important study of the application of LES to modeling wind turbines by Jiminez et al. [78] placed a simple thrust actuator disc within cuboid domain, as shown in Figure 4.15. A dynamic Smagorinsky model (Germano et al. [79]) was used to compute C_s, so that it became a function of space and time.

The axial force exerted by the disc on the flow was specified as

$$f_x = -C_T \frac{1}{2} A u_0 \qquad (4.43)$$

where C_T is the specified thrust coefficient, A the cross-sectional area of the rotor disc, and u_0 the freestream flow speed upwind. Simulation results were compared with measurement data from Sexbierum wind farm, a sample of which is shown in Figure 4.16. This work was later extended for spectral analysis and further comparison with measurements of wind farms [80].

Despite the simplicity of the turbine model, LES was shown to have good qualitative agreement with experimental measurements of turbulence, demonstrating that it could be

a viable technique for wind turbine wake analysis. Similar wind tunnel experiments and LES wind tunnel simulations to test the effect of including wake rotation in the actuator disc model also showed very good quantitative agreement with the experimental data, as long as the wake rotation was included in the model [81].

A more sophisticated model was used by Troldborg et al. [66], who took the actuator line technique developed by Sørensen and Shen [59] and coupled it with the Ellipsys3D flow solver [82, 83], with LES modeling, using mixed-scale model. Simulations ran a cuboid wind tunnel type structure; the fully developed downstream wakes are shown in Figure 4.17.

On a larger scale, Creech et al. [7] extended an actuator volume turbine model of wind and tidal turbine simulations using LES [84] to develop a hybrid blade element/CFD model to simulate a single turbine in a large domain with irregular terrain at Myres Hill, using a finite element CFD with LES [85] for resolving the flow. Rather than having any notion of individual blades as in the actuator line method above, the forces exerted by the blades on the flow varied radially, but were azimuthally smoothed in concentric rings from the hub to the tip. Turbulence was generated in two defined sub-volumes: the main blade section and the tip section as delineated in Figure 4.18.

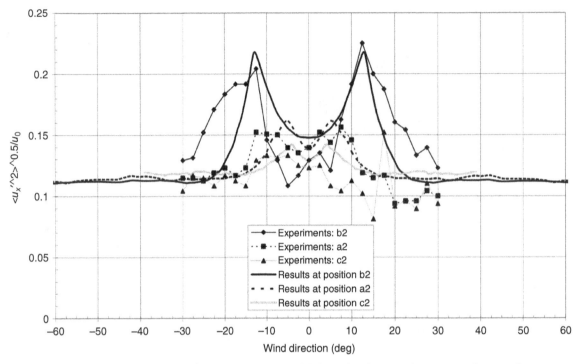

FIGURE 4.16 Turbulence intensity as it varies with flow direction at a downstream distance of 2.5D. Courtesy of Jimenez [78].

LIDAR measurements were taken from the physical site for comparison; good qualitative agreement was found in comparative wake decay rates, as seen in Figure 4.19.

4.5.2.2 Wind Farms. Due to its virtue of handling large-scale turbulence explicitly, LES has recently been used to study the fluid dynamics of wind farms. In their paper, Calaf et al. [54] compared LES models to study the effects of large wind farms on the atmospheric boundary layer (see Figure 4.20). Here, they use thrust (or "drag") discs to model the wind turbines in an array, to develop the idea of a wind turbine array boundary layer (WTABL), conceptually dividing the flow into two layers: one below and one above the turbines' hub height. This allowed them to examine the downward transport of kinetic energy toward the turbines, to parameterize the effective roughness generated by the action of the turbines on the upper layer, in a similar manner to that of Lettau [86].

Lu and Porté-Agel [87] developed a model based upon Sørensen and Shen's actuator line technique [59], with rotor RPM parameterized by freestream wind speed. Rather than model the entire wind farm explicitly, Lu et al. applied periodic boundary conditions horizontally to the domain in Figure 4.21, so that in effect a wind farm of infinite extent could be simulated. The domain was extended downstream by distances of 5D and 8D (see Figure 4.22), at the expense of not modeling the development of the flow from the unperturbed freestream to the developed flow within the farm.

In their findings, they report that while closer spacing of turbines results in greater power extraction, it also increases the atmospheric boundary layer height and overall turbulence intensity. In 2011, Porté-Agel et al. [88] compared three turbine models in the simulation with wake measurements from an actual wind farm: an actuator disc model with no rotation (ADM-NR), actuator disc with rotation (ADM-R), and an actuator line model (ALM). The latter two produced good agreements with SODAR data, as well as having higher turbulence intensity measurements when compared to ADM-NR. ADM-NR consistently under-predicted wake deficits and consequently over-predicted power outputs of downwind turbines. The ADM-R and ALM techniques produced power reductions within 2% of actual SCADA data, suggesting that wake rotation and associated turbulence has a key role in wake diffusion.

4.6 DISCUSSION AND CONCLUSIONS

4.6.1 Comparison of CFD methods

This short overview of the various methods used to model wind turbine wakes shows there are many approaches being used where all have their respective strengths and weaknesses.

The models based on the vorticity are attractive because they reflect the fact that the action of the turbine blades is through the bound vortex induced by the aerofoil shape

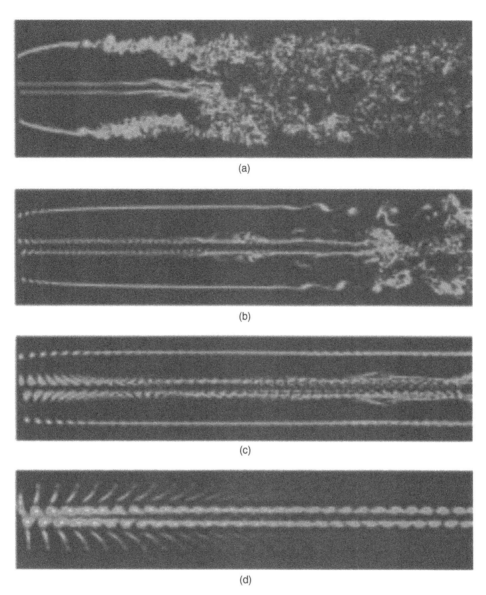

FIGURE 4.17   Vorticity plot of downstream wake of Troldberg [65] for upstream flow speeds of
(a) $u_0 = 6$ m/s, (b) $u_0 = 10$ m/s, (c) $u_0 = 14$ m/s, and (d) $u_0 = 22$ m/s.

of the blade and because they easily describe the transport of vortex features shed by the turbine blades. Their weaknesses, however, are that they cannot easily incorporate solid boundaries, especially in complex geometries, and that they become computationally very expensive and unstable as the complexity of the system increases to allow for free wake development. This is largely due to the need for either solving three Poisson equations—rather than one as for the momentum–pressure formulation—or for integrating Biot–Savart's law over a large domain for each vortex. Considering that this is likely to be a key element for full wind farm modeling, this will be a considerable challenge to a widespread adoption of this approach.

Models based on the momentum and pressure formulation of the Navier–Stokes equations are, in a way, more straightforward as the momentum equation is based on the local momentum balance and only the pressure has to be solved through a single Poisson equation, for which a large number of well-tested solvers exist. However, since the forces on the turbine blades and their reactions in the fluid are due to surface forces, their direct representation for a wind farm would require a model resolution well beyond any imaginable computing resource and a time-step so small to make any such simulation take an extremely long time. This has led to the various approaches of replacing the actual surface forces by other representations, such as the induced velocities from the vorticity of the aerofoils, or by approximating the surface forces by volume forces estimated from actuator disc theory or boundary element momentum theory approaches.

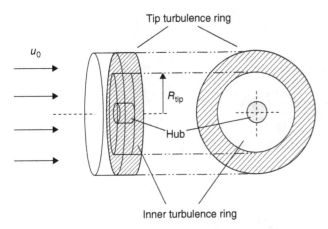

FIGURE 4.18   The tip turbulence rings in the actuator volume model by Creech et al. [7].

Once the forces are parameterized, their effect on the flow through the momentum equation is calculated on either a structured or unstructured mesh or grid. While a structured mesh is computationally more effective, it is less able to optimize the resolution to spatially very diverse geometries and flow fields. Unstructured meshes, on the other hand, can concentrate model resolution in spaces where it is needed, either through careful design of a fixed unstructured mesh or through an adaptive mesh which adjusts to the local flow field. While this is, in principle, a very powerful technique it requires careful control over the mesh adaptation steps.

The two main contenders in choice of CFD modeling are RANS and large eddy simulation (LES). RANS has long been established as a reliable and stable tool for many engineering situations and, for that reason, is available through a large number of commercial packages. Since RANS solves the equations of motion for the mean flow and a bulk turbulence intensity, it cannot explicitly model the actual turbulent velocities and therefore tends to result in smoothed flow solutions, irrespective of whether traditional RANS or unsteady RANS (URANS) is used. As a consequence, the results should only ever be interpreted as averaged approximations of the true case. That said, for the same reason it is computationally relatively cheap and can easily be used on standard desktop computers. LES solves the actual velocities from the Navier–Stokes equations at the resolved scales and only parameterizes the effect of small-scale effects through a suitably calculated eddy viscosity. This suggests that LES should give a closer representation of the flow but only if all relevant scales are sufficiently resolved by the mesh or grid, and it is not always clear what is "sufficient." It can also be easily seen that the computational costs of LES are much higher than those of RANS, and the concept of adaptive meshing becomes very attractive. However, care must be taken that adaptivity does not allow apparently peripheral flow to be excessively smoothed. For example, if the resolution is very high near the turbines but very coarse in a relatively featureless upstream part of the domain, the coarse resolution far upstream will eliminate most freestream turbulence which the model cannot reinject into that fluid once it approaches the turbines.

One option to reduce the computational cost is to replace the region close to solid boundaries, such as complex terrain, by a function which simulates the effect from surface drag onto the main fluid through a wall function; see, for example, Bou-Zeid et al. [89] or Anderson et al. [90]. However, to provide an adequate wall function for a complex surface, such as the land surrounding a wind farm, is problematic and the benefits of using LES are potentially negated by inadequate representation of the effects of the surface on the wind shear and turbulence characteristics.

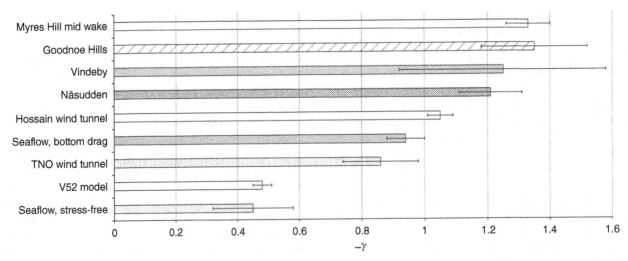

FIGURE 4.19   Decay rate of turbine wakes from both simulation and previous wind tunnel and experimental data. (Creech et al. [7]).

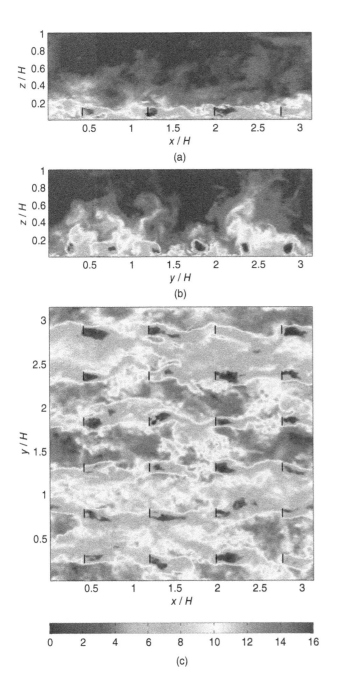

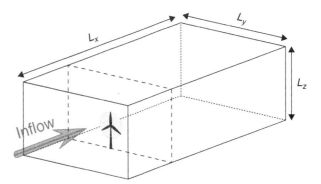

FIGURE 4.21 Schematic of computational domain in Lu and Porté-Agel [87].

modeling is the free flow in the wakes, the fluid away from surfaces is modeled through LES, but in near-wall regions the model switches to RANS. Compared to a typical domain depth of several hundreds of meters or even a few kilometers, the level of the switch from LES to RANS at around 6 m is very small, and the RANS surface model can be regarded as an advanced wall model. Computationally, this approach is marginally more expensive than using a wall function but provides a more rational way of incorporating the forcing from the surface. As this approach is not yet fully explored, it is too early to judge its performance but it has demonstrated that it is worth developing further.

4.6.2   Conclusions

After this survey, it would be tempting to ask "so which one is the best?" The answer would have to depend on the purpose of the wind farm modeling. Would it be to obtain an estimate of an annual electricity production potential for a wind farm with a given wind rose? Would it be to optimize the wind farm layout? Would it be to investigate the blade loading for a particular rotor within a wind farm? Would it be for a fundamental fluid dynamics phenomenon or fluid–structure interaction? Each of these would probably be best investigated with a different approach.

To illustrate the case, let us return to the observed relative performance of downstream turbines in Lillgrund wind farm shown in Figure 4.2 as a mean relative power output against wind direction. That figure suggests that it should be sufficient to use a relatively cheap model and run it for a fairly large number of wind directions, possibly even seeking steady-state solutions, in which case RANS and simple turbine models might be the best. However, if we consider the actual measurements used to generate the mean values in Figure 4.2, a different picture emerges. Figure 4.23 shows the relative power of the second turbine only over a 20° sector where each point represents a valid SCADA measurement from a 3-year measurement period at 10 minutes interval.

FIGURE 4.20 Plots of velocity from LES of a fully developed WTABL case, with black lines representing the modeled turbines: (a) on a vertical plane, from the side; (b) a vertical cross-stream plot 1D downstream of a turbine row; (c) a horizontal plot at hub height. From Calaf et al. [54].

An attempt to combine the strengths of LES and the economy of RANS led to a number of hybrid RANS/LES approaches such as detached eddy simulation (DES) [91,92] or a refinement of it, delayed DES (DDES) [93]. A similar hybrid approach was applied to a wind farm by Bechmann and Sørensen [94]. As the area of interest for wind farm

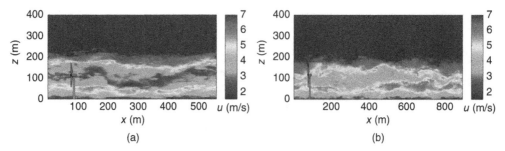

FIGURE 4.22 Vertical slices through x-component of velocity field for (a) 5D and (b) 8D cases. Courtesy of Lu and Porté-Agel [87].

Superimposed on this are the mean and one standard deviation either side of the mean. The dots clearly show that the wind farm performance shows strong fluctuations at each notional wind direction, where it is not known whether the variation arises through inherent turbulence of the wake or through wake meandering caused by upstream turbulence and wind speed changes. In terms of most accurate modeling of the wind farm, a model should reproduce this variability if given appropriate turbulence inlet conditions; this would only be possible with a high-resolution LES model. Similarly, evaluating the forces on wind turbines with a view to investigate reliability or performance, it is important to resolve turbulence appropriately [95], and LES would be required for this. In terms of annual or weekly electricity output, the ability to reproduce the red lines would be sufficient. In that case, RANS might be most appropriate as it models the mean flow and is computationally cheap enough to run many simulations at different wind directions.

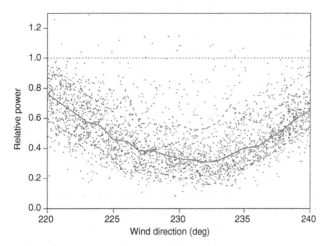

FIGURE 4.23 Scatter plot of the relative power output of the second-row turbine over front turbine when freestream is between the cut-in and rated wind speed, against wind direction. Dots, measurements; solid line, mean; dashed lines, standard deviation. Data source: Vattenfall for Lillgrund wind farm, turbines C07 and C08.

REFERENCES

[1] Dahlberg, Jan-Åke. (2009). Assessment of the Lillgrund Windfarm: power performance, wake effects. Technical report. Vattenfall AB, September.

[2] Steinbuch, M., de Boer, W. W., Bosgra, O. H., Peters, S. A. W. M., Ploeg, J. (1988). Optimal control of wind power plants. *Journal of Wind Engineering and Industrial Aerodynamics*, 27:237–246.

[3] Schepers, J. G.van der Pijl, S. P. (2007). Improved modelling of wake aerodynamics and assessment of new farm control strategies. *Journal of Physics: Conference Series*, 75.

[4] Barthelmie, R. J., Hansen, K., Frandsen, S. T., Rathmann, O., Schepers, J. G., Schlez, W., Phillips, J., Rados, K., Zervos, A., Politis, E. S., Chaviaropoulos, P. K. (2009). Modelling and measuring flow and wind turbine wakes in large wind farms offshore. *Wind Energy*, 12:431–444.

[5] Barthelmie, R. J., Pryor, S. C., Frandsen, S. T., Hansen, K. S., Schepers, J. G., Rados, K., Schlez, W., Neubert, A., Jensen, L. E., Neckelmann, S. (2010). Quantifying the impact of wind turbine wakes on power output at offshore wind farms. *Journal of Atmospheric and Oceanic Technology*, 27(8):1302–1317.

[6] Churchfield, M. J., Lee, S., Moriarty, P. J., Martínez, L. A., Leonardi, S., Vijayakumar, G., Brasseur, J. G. (2012). A large-eddy simulation of wind-plant aerodynamics. *50th AIAA Aerospace Sciences Meeting including the New Horizons Forum and Aerospace Exposition*.

[7] Creech, A. C. W., Früh, W.-G., Clive, P. (2011). Actuator volumes and hr-adaptive methods for 3D simulation of wind turbine wakes and performance. *Wind Energy*, 15(6):847–863.

[8] Sanderse, B. (2009). Aerodynamics of wind turbine wakes. Technical report. Energy Research Centre of the Netherlands.

[9] Adaramola, M. S., Krogstad, P. Å. (2011). Experimental investigation of wake effects on wind turbine performance. *Renewable Energy*, 36(8):2078–2086.

[10] Dahlberg, J. Å., Medici, D. (2003). Potential improvement of wind turbine array efficiency by active wake control (AWC). *Proc. European Wind Energy Conference and Exhibition*, Madrid, Spain, 16–19 June.

[11] Crespo, A., Hernandez, J., Frandsen, S. (1999). Survey of modelling methods for wind turbine wakes and wind farms. *Wind Energy*, 2:1–24.

[12] Larsen, G. C., Crespo, A. (2011). Wind turbine wakes for wind energy. *Wind Energy*, 14(7):797–798.

[13] WAKEBENCH: Benchmarking of Wind Farm Flow Models (2012). Available at: http://www.ieawind.org/summary_page_31.html (accessed May 16, 2012).

[14] Jensen, N. O. (1983). A note on wind generator interaction. Technical report. Risø National Laboratory, Roskilde, Denmark.

[15] Katic, I., Højstrup, J., Jensen, N. O. (1986). A simple model for cluster efficiency. *Proceedings from the European Wind Energy Conference*, Rome, Italy.

[16] Ainslie, J. F. (1988). Calculating the flowfield in the wake of wind turbines. *Journal of Wind Engineering and Industrial Aerodynamics*, 27:213–224.

[17] Barthelmie, R. J., Folkerts, L., Larsen, G. C., Rados, K., Pryor, S. C., Frandsen, S. T., Lange, B., Schepers, G. (2006). Comparison of wake model simulations with offshore wind turbine wake profiles measured by Sodar. *Journal of Atmospheric and Oceanic Technology* 23:888–901.

[18] Mosetti, G., Poloni, C., Diviacco, B. (1994). Optimization of wind turbine positioning in large wind farms by means of a genetic algorithm. *Journal of Wind Engineering and Industrial Aerodynamics*, 97:105–116.

[19] Frandsen, S., Barthelmie, R., Pryor, S., Rathmann, O., Larsen, S., Højstrup, J. (2006). Analytical modelling of wind speed deficit in large offshore wind farms. *Wind Energy*, 9:39–53.

[20] Crespo, A., Manuel, F., Moreno, D., Fraga, E., Hernández, J. (1985). Numerical analysis of wind turbine wakes. *Proc. Delphi Workshop on Wind Energy Applications*, Delphi, pp. 15–25.

[21] Crespo, A., Hernández, J. (1989). Numerical modelling of the flow field in a wind turbine wake. *Proc. 3rd Joint ASCE/ASME Mechanics Conf., Forum on Turbulent Flows*, La Jolla, CA, pp. 121–127.

[22] Beaucage, P., Brower, M., Robinson, N., Alonge, C. (2012). Overview of six commercial and research wake models for large offshore wind farms. *Proceedings EWEA 2012*, Copenhagen, April.

[23] Larsen, G. C., Aa Madsen, H., Thomsen, K., Larsen, T. J. (2008). Wake meandering: a pragmatic approach. *Wind Energy*, 11(4):377–395.

[24] Fletcher, C. A. J. (1991). *Computational Techniques for Fluid Dynamics*, Vol. 2. Springer Series in Computational Physics.

[25] Patankar, S. V.Spalding, D. B. (1972). A calculation procedure for heat, mass and momentum transfer in three-dimensional parabolic flows. *International Journal of Heat Mass Transfer*, 15:1787–1806.

[26] Van Doormal, J. P., Raithby, G. D. (1984). Enhancement of SIMPLE method for predicting incompressible fluid flows. *Journal of Numerical Heat Transfer*, 7:147–163.

[27] Patankar, S. V. (1980). *Numerical Heat Transfer and Fluid Flow*. McGraw-Hill.

[28] Ferziger, J. H., Perić, M. (2002). *Computational Methods for Fluid Dynamics*, 3rd edition. Springer-Verlag.

[29] ANSYS, Inc. (2010). ANSYS CFX Introduction. Technical manual.

[30] Legrand, S., Legat, V., Deleersnijder, E. (2000). Delaunay mesh generation for an unstructured-grid ocean general circulation model. *Ocean Modelling*, 2:17–28.

[31] Nechaev, D., Schrter, J., Yaremchuk, M. (2003). A diagnostic stabilized finite-element ocean circulation model. *Ocean Modelling*, 5(1):37–63.

[32] Pain, C. C., Piggot, M. D., Goddard, A. J. H., Fang, F., Gorman, G. J., Marshall, D. P., Eaton, M. D., Power, P. W., de Oliveira, C. R. E. (2005). Three-dimensional unstructured mesh ocean modelling. *Ocean Modelling*, 10:533.

[33] Canu, D. M., Solidoro, C., Umgiesser, G. (2003). Modelling the responses of the Lagoon of Venice ecosystem to variations in physical forcings. *Ecological Modelling*, 170:265–289.

[34] Chung, T. J. (1978). *Finite Element Analysis in Fluid Dynamics*. McGraw-Hill.

[35] Wilson, C. R. (2009). Modelling multiple-material flows on adaptive unstructured meshes. PhD Thesis, Imperial College London.

[36] Hansen, M. O. L., Sørensen, J. N., Shen, W. Z. (2003). Vorticity–velocity formulation of the 3D Navier–Stokes equations in cylindrical co-ordinates. *International Journal for Numerical Methods in Fluids*, 41:29–45.

[37] Vermeer, L. J., Sorensen, J. N., Crespo, A. (2003). Wind turbine wake aerodynamics. Progress in Aerospace Sciences, 39:467–510.

[38] Hansen, M. O. L. (1994). Vorticity-velocity formulation of the Navier-Stokes equations for aerodynamic flows. PhD Thesis, Department of Fluid Mechanics, Technical University of Denmark.

[39] Zervos, A., Huberson, S., Hemon, A. (1988). Three-dimensional free wake calculation of wind turbine wakes. *Journal of Wind Engineering and Industrial Aerodynamics*, 27:65–76.

[40] Miller, R. H. (1983). The aerodynamic and dynamic analysis of horizontal axis wind turbines. *Journal of Wind Engineering and Industrial Aerodynamics*, 15:329–40.

[41] Øye, S. (1990). A simple vortex model. *Proceedings of the Third IEA Symposium on the Aerodynamics of Wind Turbines*, ETSU, Harwell, pp. 4.1–5.15.

[42] Larsen, J. W., Nielsen, S. R. K., Krenk, S. (2007). Dynamic stall model for wind turbine aerofoils. *Journal of Fluids and Structures*, 23:959–982.

[43] Schreck, S. J., Sørensen, N. N., Robinson, M. C. (2007). Aerodynamic structures and processes in rotationally augmented flow fields. *Wind Energy*, 10:159–178.

[44] Cottet, G. H., Koumoutsako, P. D. (2000). *Vortex Methods: Theory and Practice*. Cambridge University Press, New York.

[45] Coton, F. N., Wang, T. (1999). The prediction of horizontal axis wind turbine performance in yawed flow using an unsteady prescribed wake model. *Proceedings of the Institution of Mechanical Engineers, Part A. Journal of Power and Energy*, 213:33–43.

[46] Coton, F. N., Wang, T., Galbraith, R. A. M. (2002). An examination of key aerodynamic modelling issues raised by the NREL blind comparison. *Wind Energy*, 5:199–212.

[47] Afjeh, A. A., Keith, T. G. (1986). A simplified free wake method for horizontal-axis wind turbine performance prediction. *Transactions of the ASME; Journal of Fluids Engineering*, 108:303–9.

[48] Li, Y., Çalişal, S. M. (2010). A discrete vortex method for simulating a stand-alone tidal-current turbine: modeling and validation. *Journal of Offshore Mechanics and Arctic Engineering*, 132(3):031102

[49] Schmitt, F. G. (2007). About Boussinesq's turbulent viscosity hypothesis: historical remarks and a direct evaluation of its validity. *Comptes Rendus Mécanique*, 335(9–10):617–627.

[50] Launder, B. E., Spalding, D. B. (1974). The numerical computation of turbulent flows. *Computer Methods in Applied Mechanics and Engineering*, 3:269–289.

[51] Wilcox, D. (2008). Formulation of the k-w turbulence model revisited. *AIAA Journal*, 46(11):2823–2838.

[52] Sanderse, B., van der Pijl, S. P., Koren, B. (2011). Review of computational fluid dynamics for wind turbine wake aerodynamics. *Wind Energy*, 14(7):799–819.

[53] Prospathopoulos, J., Politis, E., Rados, K., Chaviaropoulos, P. (2009). Enhanced CFD modelling of wind turbine wakes. *Extended Abstracts for Euromech Colloquium 508 on Wind Turbine Wakes*. European Mechanics Society, Madrid.

[54] Calaf, M., Meneveau, C., Meyers, J. (2010). Large eddy simulation study of fully developed wind-turbine array boundary layers. *Physics of Fluids*, 22(1):1–16.

[55] Meyers, J., Meneveau, C. (2010). Large eddy simulations of large wind-turbine arrays in the atmospheric boundary layer. *48th AIAA Aerospace Sciences Meeting*, Orlando, Florida, January.

[56] Burton, T., Sharpe, D., Jenkins, N., Bossanyi, E. (2006). *Wind Energy Handbook*. Wiley.

[57] Bertagnolio, F., Sørensen, N., Johansen, J., Fugslang, P. (2001). Wind turbine airfoil catalogue. Technical report. Risø National Laboratory, Risø-R-1280(EN).

[58] Masson, C., Smaïli, A., Leclerc, C. (2001). Aerodynamic analysis of HAWTs operating in unsteady conditions. *Wind Energy*, 4:1–22.

[59] Sørensen, J. N., Shen, W. Z. (2002). Numerical modelling of wind turbine wakes. *Journal of Fluids Engineering*, 124:393–399.

[60] Shen, W. Z., Hansen, M. O. L. (2009). Determination of the angle of attack on rotor blades. *Wind Energy*, 12:91–98.

[61] Thiringer, T.Petersson, A. (2005). Control of a variable-speed pitch-regulated wind turbine. Technical report. Chalmers University of Technology.

[62] Jonkman, J., Butterfield, S., Musial, W., Scott, G. (2009). Definition of a 5-MW reference wind turbine for offshore system development. Technical report. National Renewable Energy Laboratory, Rept. NREL/TP-500-38060, Golden, CO, January.

[63] Sørensen, J. N., Shen, W. Z., Munduate, X. (1998). Analysis of wake states by a full-field actuator disc model. *Wind Energy*, 1:73–88.

[64] Troldborg, N., Larsen, G. C., Madsen, H. A., Hansen, K. S., Sorensen, J. N., Mikkelsen, R. (2011). Numerical simulations of wake interaction between two wind turbines at various inflow conditions. *Wind Energy*, 14(7):859–876.

[65] Troldborg, N., Sorensen, J. N., Mikkelsen, R. (2007). Actuator line simulation of wake of wind turbine operating in turbulent inflow. *Journal of Physics Conference Series*, 75:12063–12063. Art. no. 012063

[66] Troldborg, N., Sorensen, J. N., Mikkelsen, R. (2010). Numerical simulations of wake characteristics of a wind turbine in uniform inflow. *Wind Energy*, 13(1):86–99.

[67] Shen, W. Z., Sørensen, J. N.Zhang, J. H. (2007). Actuator surface model for wind turbine flow computations. *Proceedings of European Wind Energy Conference*, Milan.

[68] Shen, W. Z., Zhang, J. H., Sørensen, J. N. (2009). The actuator surface model: a new Navier-Stokes based model for rotor computations. *Journal of Solar Energy Engineering*, 131(1).

[69] Wußow, S., Sitzki, L., Hahm, T. (2007). 3D-simulation of the turbulent wake behind a wind turbine. *Journal of Physics Conference Series*, 75:12036–12036.

[70] Sumner, J.,Sibuet Watters, C., Masson, C. (2010). CFD in wind energy: the virtual, multiscale wind tunnel. *Energies*, 3(5):989–1013.

[71] El Kasmi, A., Masson, C. (2008). An extended model for turbulent flow through horizontal axis wind turbines. *Journal of Wind Engineering*, 96:103–122.

[72] Cabezón, D., Sanz, J., Martí, I., Crespo, A. (2009). CFD modelling of the interaction between the surface boundary layer and rotor wake: comparison of results obtained with different turbulence models and mesh strategies. *Proceedings of EWEC 2009*, Marseilles,

[73] Shih, T.-H., Zhu, J., Lumley, J. L. (1995). A new Reynolds stress algebraic equation model. *Computer Methods in Applied Mechanics and Engineering*, 125:287–302.

[74] Cabezon, D., Migoya, E., Crespo, A. (2011). Comparison of turbulence models for the computational fluid dynamics simulation of wind turbine wakes in the atmospheric boundary layer. *Wind Energy*, 14(7):909–921.

[75] Migoya, E., Crespo, A., Garcia, J., Moreno, F., Manuel, F., Jimenez, A., Costa, A. (2007). Comparative study of the behavior of wind-turbines in a wind farm. *Energy*, 32(10):1871–1885.

[76] Réthore, P. E. (2009). Wind turbine wake in atmospheric turbulence. PhD Thesis, Aalborg University.

[77] Ammara, I., Leclerc, C., Masson, C. (2002). A viscous three-dimensional differential/actuator-disk method for the aerodynamic analysis of wind farms. *Journal of Solar Energy Engineering*, 124(4):345–346.

[78] Jimenez, A., Crespo, A., Migoya, E., Garcia, J. (2007). Advances in large-eddy simulation of a wind turbine wake. *Journal of Physics Conference Series*, 75:12041–12041.

[79] Germano, M., Piomelli, U., Moin, P., Cabot, W.H. (1990). A dynamic subgrid-scale eddy viscosity model. *Proceedings of Summer Workshop*, Center of Turbulence Research, Stanford University.

[80] Jimenez, A., Crespo, A., Migoya, E., Garcia, J. (2008). Large-eddy simulation of spectral coherence in a wind turbine wake. *Environmental Research Letters*, 3(1).

[81] Wu, Y.-T.,Porte-Agel, F. (2011). Large-eddy simulation of wind-turbine wakes: evaluation of turbine parametrisations. *Boundary-Layer Meteorology*, 138(3):345–366.

[82] Michelsen, J. A. (1994). Basis3D – a platform for development of multiblock PDE solvers. Technical report. Department of Fluid Mechanics, Technical University of Denmark, DTU.

[83] Sørensen, N. N. (1995). General purpose flow solver applied to flow over hills. PhD Thesis, Risø-R-827(EN), Risø National Laboratory, Roskilde, Denmark.

[84] Creech, A. C. W. (2009). A three-dimensional numerical model of a horizontal axis, energy extracting turbine. PhD Thesis, Heriot-Watt University (Available at: www.ros.hw.ac.uk).

[85] Piggott, M. D., Pain, C. C, Gorman, G. J., Power, P. W., Goddard, A. J. H. (2004). h, r, and hr adaptivity with applications in numerical ocean modelling. *Ocean Modelling*, 10:95–113.

[86] Lettau, H. (1969). Note on aerodynamic roughness-parameter estimation on the basis of roughness-element description. *Journal of Applied Meteorology*, 8:828.

[87] Lu, H., Porte-Agel, F. (2011). Large-eddy simulation of a very large wind farm in a stable atmospheric boundary layer. *Physics of Fluids*, 23(6).

[88] Porte-Agel, F., Wu, Y.-T., Lu, H., Conzemius, R. J. (2011). Large-eddy simulation of atmospheric boundary layer flow through wind turbines and wind farms. *Journal of Wind Engineering and Industrial Aerodynamics*, 99(4):154–168.

[89] Bou-Zeid, E., Meneveau, C., Parlange, M. B. (2005). A scale-dependent Lagrangian dynamic model for large eddy simulation of complex turbulent flows. *Physics of Fluids*, 17, 2005.

[90] Anderson, W. C., Basu, S., Letchford, W. L. (2007). Comparison of dynamic subgrid-scale models for simulations of neutrally buoyant shear-driven atmospheric boundary layer flows. *Environmental Fluid Mechanics*, 7:195–215.

[91] Spalart, P. R. (2009). Detached-eddy simulation. *Annual Review of Fluid Mechanics*, 41:181–202.

[92] Strelets, M. (2001). Detached eddy simulation of massively separated flows. *AIAA Paper*, 2001-879, 2001.

[93] Sørensen, N. N. (2009). CFD modelling of laminar-turbulent transition for airfoils and rotors using the γ-Re_θ model. *Wind Energy*, 12:715–733.

[94] Bechmann, A., Sørensen, N. N. (2011). Hybrid RANS/LES applied to complex terrain. *Wind Energy*, 14:225–237.

[95] Jimenez, A., Migoya, E., Esteban, M., Gimenez, D., Garcia, J., Crespo, A. (2011). Influence of topography and wakes on wind turbulence: measurements and interpretation of results. *Wind Energy*, 14(7):895–908.

5

FATIGUE FAILURE IN WIND TURBINE BLADES

Juan C. Marin, Alberto Barroso, Federico Paris, and Jose Canas

5.1 INTRODUCTION

With the increase in energy demand throughout the world, the threat of global warming, and the increasing shortage of our fossil fuel reserves, there is an urgent need to develop an alternative energy source that is capable of competing with fossil fuel prices and has a positive environmental impact. Wind power is among the technologically most advanced renewable energy sources and over the past 10 years wind turbines have become a common sight in our landscape and a gradually more essential supply of clean energy. The blades are one of the major constituents of wind turbines and the efficiency of the blade dictates the overall economical performance of the turbine. Therefore, any improvement in the aerodynamic and/or structural design of wind turbine blades is a step forward in making wind turbines more economically viable. Composite materials play an integral role in the design of wind turbine blades but, despite the fact that they are quite flexible to use, there is a need for a deep knowledge of the mechanical behavior of these materials.

Wind turbine as rotary machine is subjected to fatigue phenomena. Therefore, fatigue design of wind turbine components is an interesting topic of research, particularly for the blades that are manufactured in composite materials. The knowledge of fatigue behavior in composite materials is certainly more limited than that in metallic materials. This is due to the more recent development of these materials and to the internal structural complexity. The fatigue analysis approach adopted for composite materials is an extension of the established approach for metallic materials, taking into account the peculiarity of composite laminates [1–11]. A fatigue analysis requires, besides a detailed description of the loads spectrum, a deep knowledge of the fatigue behavior of

the material [5]. In this sense the information which is being generated (see, for instance, DOE/MSU Fatigue Database [12], "Fatigue of materials and components for wind turbine rotor blades" [13], or Refs. [14, 15]), to simulate the behavior under fatigue of the materials used in the building of wind turbine blades, is necessary to advance in the analysis and design of these elements.

The basic method for lifetime prediction includes a counting cycles algorithm (for instance, the rain-flow method), a damage accumulation rule (commonly the Palmgren–Miner rule), and a constant life diagram built on the basis of S–N curves (maximum stress vs. number of cycles) for different R values (ratio between minimum and maximum stress). Improvements of the life prediction can be achieved by modifications of the basic method [3–5]. More refined treatments have been proposed in the last years trying to simulate the fatigue behavior of composite laminates [16–18]. However, in the actual design of blades, the standard regulations [19–21] follow the basic procedure scheme by simplicity. The application of this procedure scheme requires the stress analysis of the blade; this task is mainly accomplished by finite element models (FEM) [6, 9, 22–25], but also alternatively by strength of material-simplified models [26, 27]. An important point in the fatigue analysis procedure is the statistical treatment of the properties that define the fatigue behavior of material, because an appropriate treatment will result in a reliable design [28]. Finally a qualification procedure of the blade design requires full-scale fatigue tests [24, 25, 29].

To illustrate the application of the procedures for fatigue analysis, a study of the fatigue failure that happened in the blades of a 300 kW wind turbine has been presented in this work. These damages, consisting of cracks located in the

Alternative Energy and Shale Gas Encyclopedia, First Edition. Edited by Jay H. Lehr and Jack Keeley.
© 2016 John Wiley & Sons, Inc. Published 2016 by John Wiley & Sons, Inc.

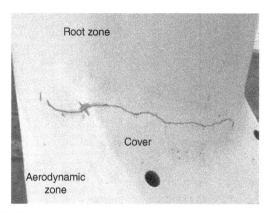

FIGURE 5.1   Location of the cracks on the surface of a blade.

joining zone of the blade with the root, appeared over short periods of time (around 5 years with an average measured number of 14.26×10^6 cycles per year) and systematically in blades in which fatigue loads were more severe. Similar damages appeared also, but over longer periods of time, in blades under more benign fatigue conditions. In any case, these periods were inferior to the design life of the blades (20 years). In this work, the reasons lying behind the damages detected are studied, to be used as a basis for repairing similarly affected blades.

First, a complete inspection of the damaged zone of the blade was carried out, cutting this damaged zone to evaluate the nature of the damage and to observe the inner structure of the zone. The correct correspondence of the structure of the laminates with that established at design as well as the existence of possible defects of manufacture is verified. The results of this visual inspection allow a qualitative estimation of the causes which could originate the appearance of the cracks to be performed. The quantitative justification of the viability of the appearance of the failure observed, in the period of time detected, has been carried out using the simplified model of the Germanischer Lloyd (GL) standard [20].

In order to support the proposal for repair of the damaged blades, several modifications to the original conception of the blade have been studied by means of numerical analysis

(FEM), the influence that these modifications have on the local stress state responsible for the failure being evaluated. Finally, the way to materialize these modifications in the practical repair of the blades is studied.

5.2 DAMAGE INSPECTION IN REAL TURBINE BLADES

A picture of the typical damage found in a blade, in the form of cracks on the external surface, is presented in Figure 5.1. The crack is located in the transition zone between the root of the blade and the zone of aerofoil profile. The cross-sectional section of the blade in the aerodynamic zone is a thin-walled section that reproduces the aerofoil profile and presents a vertical partition (denominated spline) that divides the section into two cells: leading edge and trailing edge (Figure 5.2). The cross section in the root zone is practically cylindrical.

The transition between the root zone and the aerodynamic zone is made in such a way that the cylindrical root geometry is progressively transformed into the cell of the leading edge that constitutes the resistant part of the section. The cell of the trailing edge is closed in the zone of transition by means of an element triangular in shape, constituted by a laminate of low resistance, an element which we will refer to in what follows as cover. As can be observed in Figure 5.1, the crack extends throughout both the union cover–root and the transition root–aerodynamic zone.

In order to have a visual inspection of the failure area, it is divided into slices of 2–3 cm width, as shown in Figures 5.3a and 5.3b, so that the progression of the internal damage can be observed. A detail of the damaged area at slice number 7 is shown in Figure 5.3c, whereas those areas corresponding to slices number 6 (left) and number 5 (right) are shown in Figure 5.3d.

Although the evolution of damage is difficult to establish, the crack observed in the inspection (see Figure 5.1) is formed by several well-differentiated parts: (a) the part that runs through the cover (slices 8 and 9), which affects the whole thickness of a laminate without resistant function;

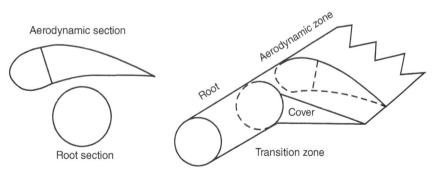

FIGURE 5.2   Scheme of the blade geometry.

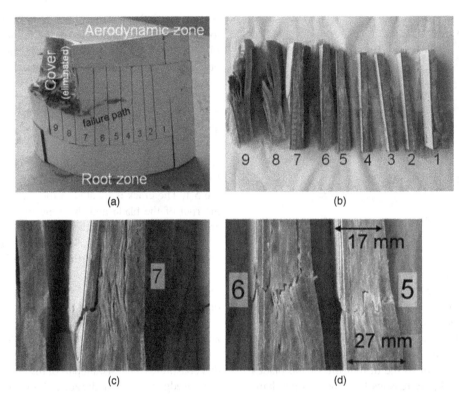

(a)
(b)
(c)
(d)

FIGURE 5.3   Details of the internal damage suffered by the blade.

(b) the continuation of this part on slice 7 which only affects the external layers of the laminate, with breaking of unidirectional fibers and multiple delaminations (as can be observed in Figure 5.3c); (c) the part corresponding to slices 5 and 6 in which the whole thickness is damaged (as observed in Figure 5.3d); and finally (d) slice 4 where the actual crack front is located, as no appreciable damage is observed at the side facing slice 3. Notice that the observed fabrications defects appear at slices 8 and 9 and slices 5 and 6 where the damage starts to develop, being free of initial manufacturing defect.

The following information can be deduced from the inspection of the samples:

- An a posteriori lamination over the gel coat on the external surface and centered in the area of the crack (see Figure 5.3d) can be observed.
- An abrupt change in the thickness of the laminate exists in the area of the crack (see Figures 5.3b and 5.3d for

more detail), the difference of thickness being in some cases (slice 5) up to 10 mm, from 17 to 27 mm, and over a distance of only 150 mm.

- The presence of areas with lack of resin (see Figures 5.3b and 5.3c).
- A debonding is observed between the spline and the aerodynamic walls (when the whole set is looked at from the upper part).

Calcinations of laminate samples from slice 5 were carried out in a muffle oven at 600°C for 90 minutes to obtain the real stacking sequence on both sides of the crack. After the calcination of the resin, the layers defining the laminate are shown in Figure 5.4. The stacking sequence observed after the calcinations is shown in Table 5.1, where MAT indicates layers of short fiber randomly distributed, NUFF represents unidirectional layers oriented at 0° with respect to the longitudinal direction, TRIAX represents triaxial

FIGURE 5.4   Stacking sequence of the laminate.

TABLE 5.1 Laminate obtained in the calcinated samples.

Laminate of root part	GEL	1 MAT	1 TRIAX	13 NUFF	1 BIAX	4 NUFF	1 MAT	1 TRIAX	3 BIAX	1 TRIAX

layers 90/45/–45, BIAX represents biaxial layers 0/90 and/or 45/–45, and GEL denotes the external layer of gel coat.

The most noticeable result is the difference in the number of unidirectional layers (in a proportion of 17 to 10) from the root zone to the aerodynamic area. Also noticeable is the presence of several biaxial 0/90 layers and one triaxial layer inside the root part that do not appear in the aerodynamic part.

Comparing the laminate found in the calcinated samples with the laminate that is described in the lamination manual, an excellent agreement is observed in the laminate of the aerodynamic part, whereas the laminate of the root part presents in the sample several additional layers of NUFF in comparison with that prescribed in the lamination manual. This increment of layers observed may be motivated by the existence of overlap in circumferential direction in the disposition of the layers in the laminate of the root (Figure 5.5b).

5.2.1 Detection of the Causes of Damage

After the visual inspection carried out, the causes most likely to have played a significant role in the origin of the detected damage are the abrupt change of thickness observed, the local geometry of stress concentrator in the transition area, and the observed defects of manufacture of the blade. These factors are discussed separately in the following sections.

5.2.1.1 Abrupt Change of Thickness. The most damaged zone from a structural point of view is the one between slices 5 and 6, in which the damage affects the whole thickness

of the laminate, whereas the crack does not cross the whole thickness in slice 7.

It is in these slices (5 and 6) that, after the inspection carried out in Section 5.2, a more marked transition in thickness can be observed (Figure 5.3). This abrupt change of thickness produces an eccentricity in the transmission of the load, generating bending moments that make the laminate not work uniformly over the whole thickness. Additionally, the unidirectional layers that are present in both laminates suffer misalignment in the longitudinal orientation of the fiber. This misalignment, under compression, can accentuate the failure mechanisms associated to local micro-buckling of the fiber.

Internal discontinuities also appear in the laminate in this area of thickness change. They are provoked by the end of a layer (Figure 5.5a), a circumstance habitually referenced in the literature as "ply drop" [2]. An accumulation of these internal defects could favor the appearance and propagation of greater defects.

5.2.1.2 Local Geometry of Stress Concentrator. The abrupt transition between the root area and the aerodynamic area generates a local geometry of reentrant corner, undesirable from a structural point of view since it acts as local stress concentrator. The root area being the most loaded in the whole blade, the presence of concentrators that amplify the efforts on the laminate results in the existence of critical points for the beginning of failure.

5.2.1.3 Lack of Resin and Debonding Manufacturing Defects. The joints are the weakest points in all structures. If additionally these joints are associated to the presence

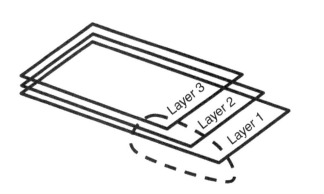

(a) Longitudinal reinforcements

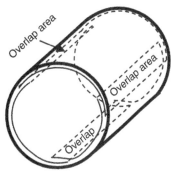

(b) Overlaps pf the circumferential reinforcements

FIGURE 5.5 Changes of thickness motivated by the reinforcements.

of defects, this weakness is consequently accentuated. The corner between the cover, the root, and the aerodynamic area presents manufacture defects associated to the lack of resin (see slice 7) and debonding defects associated to the curing process (see slices 8 and 9).

Although the resin does not have the mechanical properties of the fibers, a laminate with lack of resin does not work as a unique material, its carrying load capacity, mainly under compression, diminishing noticeably.

5.2.1.4 *Evaluation of the Problem.*

Although the reasons set out above do not by themselves fully explain the appearance of the premature failure observed, the combination of the three, in the area of the blade with most stress requirements, can represent sufficient reasons for the premature failure of the blade.

The damage is clearly motivated by a mechanism of fatigue. A prediction of fatigue life is always based on the consideration of a material without defects, the lack of resin, and the observed defects accelerating this failure mechanism. Additionally, the combined effect of the geometry of the concentrator and the abrupt change of thickness amplifies the efforts that act locally in an area which, in addition to suffering the greatest efforts along the whole blade, presents apparent damage.

The quantitative influence on the fatigue life of the blade of each one of these previously described aspects will be evaluated in the following section. The particular objective is to find, with the knowledge we have about the location of the damage, the approximate period of time before its appearance and, as regards the failure mechanism, under what hypothesis it is reasonable to justify quantitatively a fatigue failure such as the one observed.

5.3 EVALUATION OF FATIGUE LIFE

Although the appearance of the damage leads to consider fatigue as the cause of the failure, a classical static analysis was carried out using nominal loads in accordance with GL [20]. The results, not included in this work for the sake of brevity, showed that the stresses under these loads were far from their allowable values and consequently the static loads had no responsibility on the observed failure.

In this section, the estimation of the fatigue life, in order to elucidate if the detected failure can be justified based on the actual configuration of the blade, will be carried out. The simplified procedure proposed in the GL standard [20] (which is one of the most extensive and most specific procedures for the design and certification of these elements) will be used. Although more refined treatments for composite fatigue design (as has been mentioned in Section 5.1) are available, the GL procedure satisfies the requirements of this study.

5.3.1 Calculation of Fatigue Life by Means of Simplified Spectrum According to GL

The GL strength verification adopts the following general expression:

$$S \leq \frac{R_k}{\gamma_{mx}} = R_d \qquad (5.1)$$

where S, which represents the stress associated to a load case, cannot overcome the design strength R_d, which is obtained from decreasing the strength of the material R_k by means of a safety factor that adopts, depending on the case, the values

$$\text{for the static analysis:} \quad \gamma_{ma} = 2.67 \qquad (5.2)$$
$$\text{for the fatigue analysis:} \quad \gamma_{mb} = 1.485 \qquad (5.3)$$

The load spectrum to be used could be obtained from real measurements or using simulation techniques. The simplified spectrum shown in Figure 5.6, which constitutes a conservative option versus the previous ones, will be used. This spectrum is built up using the stresses corresponding to the case N1.0 of GL, called "Basic state power production" which includes aerodynamic, inertia (gravity, centrifugal, and gyroscopic), and functional forces. The range of stresses $\Delta\sigma = |\sigma_{max} - \sigma_{min}|$ is represented versus the number of cycles in logarithmic scale, $\log N$, N_{max} representing an estimation of the blade fatigue life.

Two well-differentiated areas can be clearly detected in Figure 5.6:

Area a where a maximum range $\Delta\sigma_{max} = 1.5\bar{\sigma}$ is assumed, $\bar{\sigma}$ being the average value of stresses originated by the forces associated to load case N1.0. The duration in cycles of this part of the spectrum, for a total life of $N_{max} = 10^x$, is 10^{x-3}.

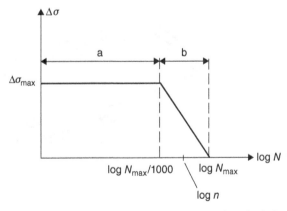

FIGURE 5.6 Simplified spectrum of loads for the calculation of fatigue life.

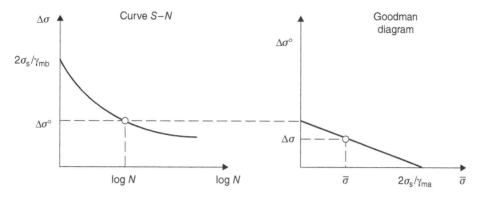

FIGURE 5.7   *S–N* curve and Goodman diagram.

Area b which includes the rest of the cycles up to 10^x, with stress ranges according to the equation

$$\Delta\sigma = 0.5\,\bar{\sigma}\,\log\left(\frac{N_{max}}{n}\right) \text{ for } \log\left(\frac{N_{max}}{1000}\right) \le \log n \le \log N_{max}$$
$$(5.4)$$

The fatigue life calculation, which of course is made using the most unfavorable value of *a*, is based on the *S–N* curve of the laminate and on the Goodman diagram built from it (Figure 5.7). As *S–N* curves are not available for all the different laminates, and to obtain them experimentally is nonviable from a practical point of view, the GL standard proposes the use of expressions (5.5) and (5.6):

$$\Delta\sigma = \frac{2\,\sigma_s}{\gamma_{mb}N^{1/9}} \quad (5.5)$$

$$\Delta\sigma^\circ = \frac{\Delta\sigma}{1 - \dfrac{\bar{\sigma}\,\gamma_{ma}}{\sigma_s}} \quad (5.6)$$

where σ_s is the laminate strength (corresponding to the stress component under consideration), *N* is the number of cycles, $\Delta\sigma = |\sigma_{max} - \sigma_{min}|$ is the stress range, and γ_{ma} and γ_{mb} are, respectively, the static and fatigue safety factors.

The fatigue life can be calculated using these diagrams, evaluating the accumulated damage *D*, by the Miner's rule, as

$$D = \sum_i \frac{n_i}{N_i} \quad (5.7)$$

where n_i is the load cycles number for a stress range *i* and N_i is the allowable load cycles number associated to *i*.

Substituting the expression of the Goodman diagram in the *S–N* curve, the expression that provides N_i for a determined stress range is obtained:

$$N = \left[\frac{2}{\gamma_{mb}}\left(\frac{\sigma_s}{\Delta\sigma} - \frac{\bar{\sigma}\,\gamma_{ma}}{\Delta\sigma}\right)\right]^9 \quad (5.8)$$

The first area of the spectrum (*a*) with a constant stress range contributes to the accumulated damage with n_1/N_1:

$$N_1 = \left[\frac{2}{\gamma_{mb}}\left(\frac{\sigma_s}{1.5\,\bar{\sigma}} - \frac{\gamma_{ma}}{1.5}\right)\right]^9 \quad n_1 = \frac{N_{max}}{1000}$$

$$\frac{n_1}{N_1} = 2.6368\ 10^{-3}\ \frac{N_{max}}{\left(\dfrac{\sigma_s}{\bar{\sigma}} - \gamma_{ma}\right)^9} \quad (5.9)$$

The rest of the spectrum contributes with $\int_{\frac{N_{max}}{1000}}^{N_{max}} \frac{dn}{N}$, where introducing the expression of
N gives the following expression:

$$\int\limits_{\frac{N_{max}}{1000}}^{N_{max}} \frac{dn}{N} = 4.2824 10^{-3}\ \frac{N_{max}}{\left(\dfrac{\sigma_s}{\bar{\sigma}} - \gamma_{ma}\right)^9} \quad (5.10)$$

The complete expression of the accumulated damage is

$$D = 6.9199 10^{-3}\ \frac{N_{max}}{\left(\dfrac{\sigma_s}{\bar{\sigma}} - 2.67\right)^9} \quad (5.11)$$

an expression which allows the fatigue life corresponding to an accumulated damage equal to $D = 1$ to be calculated. The measured relationship between the number of cycles and the years of operation for the blades under study is 14.26×10^6 cycles/year. Using this relationship, the equivalence between cycles and years of life for certain periods of time is shown in Table 5.2.

As is deduced from expression (5.11), in accordance with the simplified treatment adopted, the prediction of fatigue

TABLE 5.2   Equivalence between number of cycles and years.

No. of cycles	0.43×10^8	0.57×10^8	0.71×10^8
Years	3	4	5

life at a certain point of the blade depends exclusively on the relationship between the strength of the laminate and the value of the average stress existing at this point. Hence, in the following sections, the value of these magnitudes will be analyzed in the area affected by the failure.

5.3.2   Evaluation of the Laminate Strength

It should be pointed out that when speaking of R_k as the characteristic strength of the material this refers to the complete laminate. The values of R_k ought to be obtained by testing the different laminates of the blade. This is not practical, due first of all to the number of different stacking sequences of the blade and second to the fact that the failure load of most laminates exceeds the load capacity of most conventional testing machines. As an alternative, an empiric rule has been used to evaluate the strength of the laminate working from the strengths of the component layers. This approach has been contrasted with experimental results and with models of progressive failure [30] by means of degradation properties [31, 32], a good agreement being obtained. The expression used is

$$\sigma_{\text{allowable}\_i} = \frac{\sum\limits_{n} \sigma_{\text{allowable}\_n}\, t_{ni}}{t_i} \qquad (5.12)$$

where $\sigma_{\text{allowable}\_i}$ is the strength of the ith laminate, $\sigma_{\text{allowable}-n}$ is the strength of a layer or a group of layers (the summation extends in to all layers, or groups, with different strengths: NUFF, DDB, biaxial 0/90, MAT, and balsa wood), t_{ni} is the thickness of material n present in the laminate, and finally t_i is the total thickness of the laminate.

Using expression (5.12), with the stacking sequence deduced after the calcination tests (Section 5.2), and using the allowable values (evaluated experimentally) of each one of the component layers, as well as its thickness, the allowable stress of the laminate corresponding to the section of smaller thickness of slice 5 (damaged zone) can be evaluated. The allowable (tensile strength) of each layer appears in Table 5.3. Based on these allowables, a value of 417.1 MPa is obtained for the allowable of the laminate.

It is necessary to consider the appropriate strength value of the laminate to be used in the analysis. To this end, it has to be pointed out that although the predominant form of work

in the area in question is under tension (98%, experimentally measured), a certain percentage also takes place under compression (2%). This fact, knowing the different fatigue behavior under tension and compression, will be taken into consideration in the following.

A series of compression tests on unidirectional (NUFF) specimens, which are the component that predominantly characterizes the resistant behavior of the laminates of the blade, were carried out. The specimens were made of a laminate of glass fiber vinylester resin with three layers of NUFF 1450, the total thickness being 3.99 mm. The tests were carried out in an INSTRON 8801 (100 KN). The tested specimens have experienced a failure type being associated to a phenomenon of micro-buckling of the fibers, which is favored by the longitudinal waving of the tows of the fibers present in these materials.

The measured average value of the compression strength is 250.9 MPa, presenting a typical deviation of 23.7 and a coefficient of variation of 9.45%. Starting from these values, and applying the statistical treatment that recommends the GL standard, the characteristic value of the compression strength of the NUFF layers would be 189.4 MPa. Therefore, the relationship between the characteristic values of the tensile and compressive strength of the unidirectional (NUFF) layers is 2.65.

Given the predominant character of the unidirectional layers in the resistance in longitudinal direction of the laminate, it would be acceptable to suppose that the relationship between the strength in tension and compression of the laminate follows the same proportion as that corresponding to the layers of NUFF. Based on this hypothesis we can estimate the value of the characteristic compression strength of the laminate under question, which would be 157.4 MPa.

Bearing in mind the fact that part of the life of the element works in tension and another part in compression, the correct treatment would involve using the strengths to tension and compression, respectively, for the corresponding parts of the life of the element. In this case, due to the fact that a simplified spectrum, where it is not possible to distinguish between work in tension and compression, will be used, it has been considered appropriate to modify the value of the allowable stress to be used for the fatigue calculation. To this end, the allowable stress of the laminate will be evaluated as the average value between the tension (417.1 MPa) value and

TABLE 5.3   Allowable values (tensile strength), thicknesses, and properties of the different types of layers.

Material	$\sigma_{\text{allowable}}$ (MPa)	Thickness (mm)	E_{11} (GPa)	G_{22} (GPa)	G_{12} (GPa)	v_{12}
NUFF (unidirectional)	501.3	$1.333 \times 10 = 13.33$	34.5	9.34	2.7	0.35
DDB (triaxial 90/45/−45)	74.1	$1.034 + 0.554 \times 2 = 2.142$	27.1	7.3	2.1	0.35
DB (biaxial +45/−45)	112.5	$0.314 \times 2 = 0.628$	27.1	7.3	2.1	0.35
MAT (random)	124.9	$0.460 + 0.214 = 0.674$	9.78	9.78	3.7	0.32

that of compression (157.4 MPa) weighted by the respective work percentages, that is $\sigma_{\text{allowable}} = 411.9$ MPa. It should be pointed out that this small variation in the allowable stress, which could be considered worthless to static effects, has a considerable influence on the fatigue life prediction.

Only reference to normal stresses has been made in the former discussion due to the fact that shear stresses are almost negligible in the zone under consideration, the laminate having in any case [+45/−45] layers (BIAX) to support these stresses.

5.3.3 Evaluation of the Average Stress of the Laminate

The value of the nominal average stress at the neighborhood of the affected zone has been evaluated by means of a model based on the strength of materials theory [26, 27] that considers the characteristics of the laminates (by means of the classic theory of laminates) and the geometric configuration of the section. The load hypothesis considered is case N1.0 of GL standard. The value of the obtained nominal average stress is $\bar{\sigma} = 10.77$ MPa. This nominal value can be affected by the change in thickness of the laminate and by the existing concentrator in the transition zone.

The difference in thickness between the adjacent laminates produces an eccentricity of the transferred load which causes the appearance of a bending moment that amplifies the longitudinal normal stress. This effect will be evaluated using the scheme shown in Figure 5.8.

The moment $N_x\,d$ created by the eccentricity of the load must be absorbed by the laminates. The most unfavorable case is that in which the laminate of smaller thickness absorbs it completely, the maximum normal stress being

$$\sigma_{\max} = \frac{N_x}{e} + \frac{N_x d}{\frac{1}{12}e^3}\frac{e}{2} = 2.57\frac{N_x}{e} \qquad (5.13)$$

If, on the other hand, the moment is distributed in the laminates proportional to their bending stiffness, the minimum value of normal stress for each one of the ends is obtained, its expression being

$$\sigma_{\max} = \frac{N_x}{e} + \frac{1}{6}\frac{N_x d}{\frac{1}{12}e^3}\frac{e}{2} = 1.26\frac{N_x}{e} \qquad (5.14)$$

Therefore, the effect of the change in thickness on the laminates involves an amplification of the normal stress,

whose effect can be quantified by a factor, its value being reasonably between 1.26 and 2.57. With the nominal stress $\bar{\sigma} = 10.77$ MPa and applying the obtained factors, the value of the normal stress would be in the range between 13.6 and 27.7 MPa.

Due to the geometric configuration of the design of the blade, the point under study is within the zone of influence of the concentrator described in Section 5.2.1.2, so that the total normal stress would be affected by a concentration factor associated to this geometry. This factor depends on the real configuration of the geometric detail existing at each blade and therefore on the manufacture process. The total normal longitudinal stress is then calculated by multiplying the nominal average stress by the factor of amplification due to bending f_{bend} and by the factor of amplification due to the concentrator f_{conc} as

$$\bar{\sigma}_{\text{real}} = \bar{\sigma} f_{\text{bend}} f_{\text{conc}} \qquad (5.15)$$

Finally, it is necessary to take into consideration that when the laminate is going to suffer the failure of the resistant layer, the external layer presents a crack, motivated by the effect of the concentrator, described in Section 5.2.1.2, on the weak laminate of the cover, as is indicated in Figure 5.9a.

The effect of this superficial crack on the stress state is double. First, the value of the nominal average stress is increased by the loss of thickness implied by the existence of cracked layers, varying from $\bar{\sigma} = 10.77$ to 11.82 MPa. Second, it would imply, strictly speaking, an intensification of the stresses at the tip of the crack. This effect will, for simplicity, be taken as a concentration of the stresses, being included within the factor f_{conc} of expression (5.15).

5.3.4 Evaluation of Fatigue Life

Once the previous considerations have been made, with regard to the laminate strength and the normal longitudinal average stress appearing at the zone of study, an estimation of the fatigue life can be carried out.

Taking expression (5.11), which evaluates the fatigue life, and introducing $\sigma_s = 411.9$ MPa (see Section 5.3.2) and $N_{\max} = 0.71 \times 10^8$, corresponding to a life of 5 years (see Table 5.2), which is approximately the observed life of the blade under study, the values of the average stress that would cause the fatigue failure, that is $D = 1$, can be estimated. The normal longitudinal average stress obtained by means of the described procedure is $\bar{\sigma}_{\text{real}} = 59.2$ MPa.

FIGURE 5.8   Evaluation of the influence of the change in thickness.

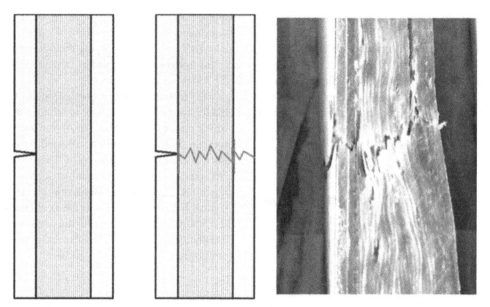

FIGURE 5.9 Scheme of the progression of the crack in the laminate: (a) crack in the superficial layer, (b) crack in the resistant layer, and (c) view of the real crack.

Using expression (5.15), with an average stress of $\bar{\sigma} = 11.82$ MPa, and knowing that the reasonable range of variation of the factor f_{bend} is from 1.26 to 2.57, the factor f_{conc} (due to the geometric concentrator and to the presence of the superficial crack) that causes a real average stress of 59.2 MPa would then be between 1.95 and 3.97.

The obtained values are of a similar order of magnitude to those associated to a geometric concentrator like the one appearing at the root of the blade. In order to evaluate quantitatively the range of values of this concentrator, solutions of similar problems (see Figure 5.10a) can be found in the literature ([33, pp. 286–287]. In particular, two problems with

similar geometry to the real concentrator have been taken, corresponding to tension (Figure 5.10c) and bending (Figure 5.10d) loads.

The parameters on which the stress concentration factor depends can be estimated, for the dimensions of the blade (Figure 5.10b), within the following ranges:

$$15 < h/r < 20; \ 0.3 < 2h/D < 0.4 \qquad (5.16)$$

where r is the radius of curvature, D and d are the characteristic dimensions at both sides of the concentrator, and h is the difference between D and d.

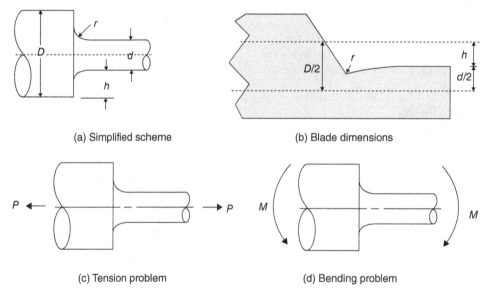

(a) Simplified scheme

(b) Blade dimensions

(c) Tension problem

(d) Bending problem

FIGURE 5.10 Geometric scheme of the concentrator.

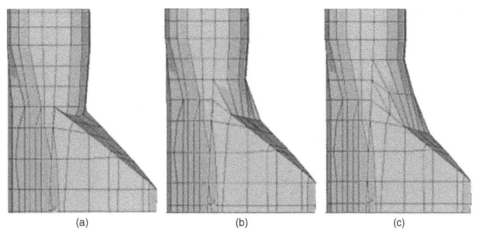

FIGURE 5.11   (a) Original geometry, (b) modified geometry 1, and (c) modified geometry 2.

Using the previous values for the calculation of the stress concentration factor (f_{conc}), the following ranges for the cases of tension and bending are obtained:

$$\begin{aligned} \text{Tension: } & 3.01 < f_{conc} < 3.59 \\ \text{Bending: } & 2.94 < f_{conc} < 3.60 \end{aligned} \quad (5.17)$$

The range of values calculated for f_{conc} is, as can be observed, within the interval of values previously obtained (1.95–3.97), values that can reasonably cause the fatigue failure in the period of time considered.

Additionally, using the finite elements model (which will be detailed later on) of the original configuration that will be used in the analysis of the modifications to be performed on the blades, an estimation of the longitudinal average stress in the zone under study can be obtained, since this point is relatively far from the corner (around 10 cm). The value of the longitudinal average stress at the point considered is 26.43 MPa.

If this value is compared with the nominal average stress previously evaluated $\bar{\sigma} = 10.77$ MPa (considering that in the numerical model the existence of cracked layers is not contemplated), a stress amplification of 2.45 is obtained, which is within the interval of values previously found (1.95–3.97). This fact corroborates that the actual stress level could reasonably cause the fatigue failure in the period of time considered.

5.4   FINITE ELEMENT MODEL OF THE BLADE

Once the causes of the damage have been detected, the design of a suitable repair to stitch the existing cracks in the damaged blades and to relax the stress state in order to prevent similar failures constitutes the next step in the study.

In order to diminish the stress state in the area where the failure appeared, the influence that a variation in the local geometry of the concentrator has on the stress state, and therefore on an improvement of the fatigue life, will be studied. This analysis will be carried out by means of finite elements, using ANSYS [34], and attempts to modify the geometry of the reentrant corner that appears at the beginning of the aerodynamic zone, smoothing the angle of this corner. This modification follows the present tendency in the design of wind turbine blades, in which this transition is made much more smoothly than is the case of the blade under consideration.

Two configurations have been proposed to smooth the geometry of the above-described concentrator. In Figure 5.11, by means of FEM, a detail of the zone under study is shown in its original conception, (a), and in two modified versions, (b) and (c), susceptible of being easily implemented in the blade.

The numerical model of the original geometry is described in detail in Section 5.4.1, whereas those associated to the two proposed modifications are described in Section 5.4.2. A summary of the obtained results will be presented in Section 5.4.3.

5.4.1   Original Geometry

The model performed by ANSYS uses shell elements (shell99), which allow orthotropic materials to be used and the stacking sequence of the laminates (up to 100 layers) to be defined. Shell99 is a quadrilateral element of eight nodes, four in the vertices and four at the midpoints of each edge. Each node has 6 degrees of freedom (three displacements and three rotations). The model consists of 5380 nodes and 2027 elements, more than 500 types of different stacking sequences having been defined. The blade has been divided into 64 sections along the longitudinal direction. The nodes at the first section have the displacements restricted, so that the blade is clamped at the root end. This model has been

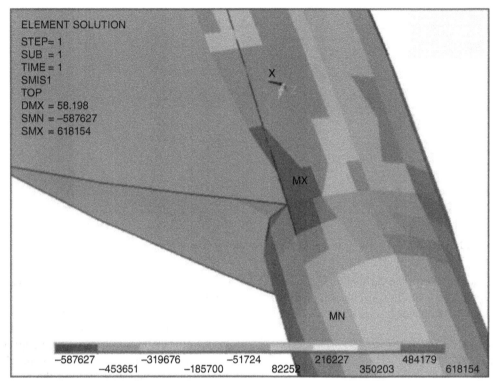

FIGURE 5.12 N_x efforts in the original configuration of the blade.

previously validated by means of comparison with analytical and experimental results [30, 35].

The load case considered for this analysis is N1.0 defined in GL, corresponding to a regular power production, the verifications of fatigue life having been made with this case. The study carried out is a static analysis of first order (without considering large displacements).

The longitudinal normal efforts per unit length (the resultant of the longitudinal normal stresses through the thickness of the laminate) will be taken as a reference to compare the different configurations considered, which from now on will be denoted by N_x.

The values of N_x obtained for the load case N1.0 and for the original geometry of the blade are represented in Figure 5.12, for the face in tension of the blade, where the damages were observed. The zone represented is the one where N_x reaches the highest values, due to the presence of the concentrator. It has to be clarified that, quantitatively speaking, the values of N_x calculated have meaning only for comparison purposes between similar meshes, since the theoretical nominal value of N_x at the vertex of the re-entrant corner is infinite, the values obtained being strongly dependent on the discretization carried out at the neighborhood of the corner.

It can also be observed that the trailing edge zone, within the aerodynamic part of the blade, does not transmit significant loads in the longitudinal direction, in comparison with the zone of the leading edge. For this reason there are almost

no unidirectional laminas oriented longitudinally in the trailing edge. This is an important fact to be taken into account when modifying the local geometry of the corner.

5.4.2 Analysis of Modifications

In order to release the stress state in the zone of influence of the concentrator, a lateral triangle of resistant material that smoothes the reentrant corner (see Figure 5.11b) in the zone where the crack appeared will be introduced. This modification obviously goes along with the modification of the elements that connect the root with the cover. In detail, the modification of the original model consisted of

1. The creation of the elements that define the lateral triangles
2. The redefinition of the stacking sequence of the elements of the cover that is in contact with the root
3. The modification of the laminates in the elements adjacent to the triangular lateral element in the longitudinal direction

In order to have a triangle with suitable mechanical behavior, a similar laminate to that existing in the nearest elements at the root has been considered. The structure of the laminate of the triangle is described in Table 5.4.

TABLE 5.4   Triangle laminate.

Lamina	Weight by unit area (g/m^2)
1 MAT	200
3 NUFF (0°)	1450
1 BIAX (±45°)	300
3 NUFF (0°)	1450
1 BIAX (±45°)	300
6 NUFF (0°)	1450
1 MAT	400

TABLE 5.5   Laminate of the cover.

Lamina	Weight by unit area (g/m^2)
1 MAT	200
3 NUFF (0°)	1450
1 BIAX (±45°)	300
2 NUFF (0°)	1450
1 MAT	200

In the damaged area where in the repaired blade two identical laminates will appear superimposed (one corresponding to the original laminate and the other corresponding to the added triangle), only the properties corresponding to the laminate of the added triangle have been considered in the model. This is a conservative hypothesis which does not consider the residual properties of the original laminate which will be affected by some fiber breaking and general damage.

The complete laminate of the triangle has been added in the model to the laminate of the adjacent elements of the non-damaged area (which originally did not contain NUFF) to give continuity to the resistance of the triangle, thus creating an alternative path for the transmission of longitudinal stresses, which find a path of connection with the contiguous laminates that have a significant resistance capacity.

The geometry of the cover is modified to adapt it to the geometry of the lateral triangles. In addition, the laminate of the cover is modified to make it more resistant and to prevent the appearance of the crack observed in the inspected blades. The laminate defined on the elements of the cover is defined in Table 5.5.

The values of N_x obtained for load case N1.0 and for the modified geometry (configuration 1) are shown in Figure 5.13. The modification produces two effects on the stress state: first a reduction in the level of stress at the neighborhood of the corner of the original configuration, and second the redistribution of the stresses introduced throughout the triangle and the adjacent elements in longitudinal direction. These effects will be evaluated quantitatively in Section 5.4.3.

A second modification is considered (configuration 2), the only difference between the two modifications being the dimension of the triangle introduced, which is greater in the modified configuration 2. The stacking sequence of the

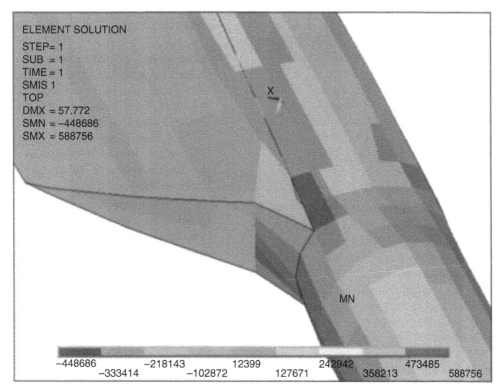

FIGURE 5.13 N_x efforts corresponding to configuration 1.

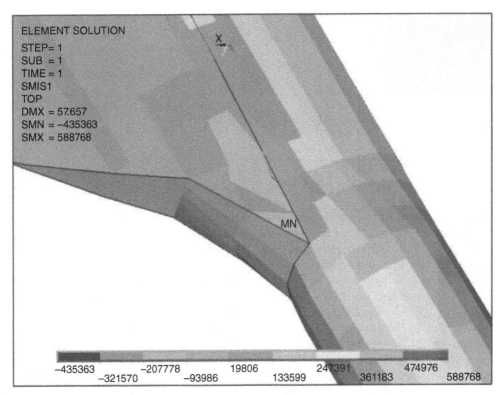

FIGURE 5.14 N_x efforts corresponding to configuration 2.

laminate of the elements that define the triangle responds to that indicated in Table 5.4, whereas the laminate defined on the elements of the cover is that shown in Table 5.5. The modification of the laminates of the adjacent elements now affects a greater number of elements than in the type 1 modification.

The values of N_x obtained for the load case N1.0 and for the modified geometry (configuration 2) are represented in Figure 5.14. Observation of Figure 5.14 leads to similar conclusions as in the case of Figure 5.13, the redistribution of stresses in the elements adjacent to the triangle introduced being even greater than in configuration 1.

5.4.3 Evaluation of Results

To evaluate the influence of the modifications performed in the stress state, the comparison of the results has to be made far enough from the vertex of the corner to leave the results unaffected by the nominal corner singularity. Additionally, the value of N_x must be compared in the area where the damage appeared, this area being represented in Figure 5.15, over the original configuration geometry.

In Figure 5.16 the evolution of the longitudinal average stress σ_x ($\sigma_x = N_x / t$, N_x taken from the centroid of the element and t being the thickness of the laminate) as a function of the distance to the corner is shown for the different configurations. The evolutions corresponding to the elements

placed at both sides (root and blade) of the line of study have been represented, though the observed failure is located on the side of the aerodynamic zone (blade).

From Figure 5.16, the following observations are deduced:

1. Although the qualitative evolution of the stresses in the elements at both sides of the line under study (root and blade) is similar, the difference in thickness between the laminates of these elements (corroborated during the inspection of the blades, see Section 5.2) generates a higher stress level in the elements of the side of the aerodynamic zone of the blade than in the elements of the side next to the root. The thicknesses along the lines under study (root and blade) are listed in Table 5.6.

2. The evolution of the stress in the original configuration quantitatively demonstrates the presence of a stress concentrator. The introduction of the triangles in the modified configurations attenuates this concentration, reducing as well the value of the stress at the damaged zone.

3. Far away from the concentrator, the distribution of stress is not affected significantly in the modified configurations.

4. The stress level found in the zone far from the concentrator (see Figure 5.16) adjusts reasonably well to the value of the previously obtained nominal stress in this zone using the strength of materials approach.

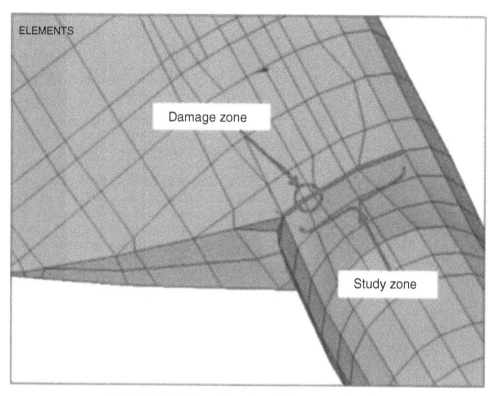

FIGURE 5.15   Line under study and damage zone locations.

5.5   DISCUSSIONS

5.5.1   Fatigue Life Estimation for the Modified Configurations

With a view to evaluating the influence of the obtained stress variations on the fatigue life of the modified configurations of the blade with respect to the original one, the calculation procedure by means of the simplified load spectrum set out in Section 5.3.1 will be used. To do this, the nominal stress that originates the failure by fatigue in a period of 5 years (corresponding to the original configuration) will be considered. This value will be affected by the stress decrements associated to each modified configuration.

The results of the analysis appear in Table 5.7, where the stress decrements between the modified configurations and the original one, as well as the increases in life with respect to the life of 5 years taken as a reference for the original configuration, are shown. The results show the improvement

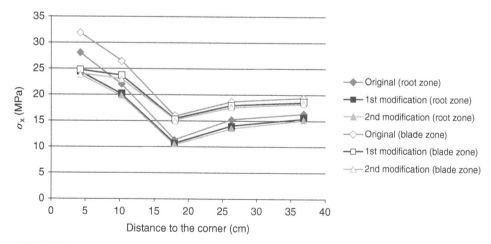

FIGURE 5.16   Evolution of the stress σ_x in the original configuration, configuration 1, and configuration 2.

TABLE 5.6   Thicknesses of the laminates.

Distance to the corner (cm)	Root thickness (cm)	Blade thickness (cm)
4.31	1.99	1.94
10.26	2.18	1.94
17.97	2.37	1.89
26.34	2.13	1.89
36.91	1.99	1.76

TABLE 5.7   Stress variations and fatigue lives for the two modifications of the blade.

Configuration	Δ Stress MPa (%)	Δ Life (years)
Modification 1	−2.73 (10.33%)	4.8
Modification 2	−3.37 (12.75%)	6.6

in fatigue life obtained with the modifications carried out: the greater the size of the triangle, the greater the increase in life, as expected.

5.5.2   Parametric Study

To investigate the influence of the configuration of the laminates used on the increase in fatigue life, variations on the two modified configurations that consist of increasing or decreasing, respectively, the number of unidirectional layers present in the laminates involved in the modification have been studied. The reference laminate configurations are those previously described in Tables 5.4 and 5.5. In particular, variations of ±2 and ±4 unidirectional layers (NUFF 1450 g/m^2) have been considered, the results being shown in Table 5.8.

From the results of Table 5.8 it can be observed that

1. The increase in the number of NUFF layers causes an increase in life, the variation being slightly more pronounced when NUFF layers are eliminated from the reference laminate in comparison with the case where layers are added.
2. As was previously seen, the greatest increase in life was obtained with the greatest size of the triangle. In addition, the greater influence of this effect on fatigue

life in comparison with the variation in the laminate studied can now be observed.

5.5.3   Materialization of the Modifications in the Concentrator Geometry

Due to the fact that the suggested modifications entail the addition of a triangular surface in the concentrator zone, it is necessary to guarantee the structural continuity of this piece with the rest of the blade. This continuity can be guaranteed by using greater dimensions to those of the triangle itself, in order to allow, by means of an overlapping, the bonding over the surface of the blade (see Figure 5.17).

Since the laminate of the triangle contains a considerable number of layers, the appropriate way to build up the laminate requires each lamina to have direct bonding with the surface of the blade, which means that each lamina must overlap the previous one by a certain length along the whole periphery (see Figure 5.17). In this way a progressive transmission of loads takes place from the blade to the triangular surface so that the surface can develop its resistant function.

In relation to the geometry of the reinforcement layers, a geometry in trapezoidal form, as an extension of the triangle, has been chosen.

The overlap lengths chosen for the longitudinal direction (minimum 5 cm) guarantee the correct load transmission to reinforcement layers, so that each lamina can completely develop its resistance capacity in tension by means of the fibers, without being limited by the shear strength of the bonding area with the blade. In transverse direction, where the efforts are much smaller, the overlap lengths (minimum 2 cm) are chosen to generate a smooth transition of the laminates.

5.6   CONCLUSIONS

The inspection of the blade revealed the following damages:

- Damage appearing in the form of a crack that affected the (nonresistant) superficial laminate, extending over the influence zone of the concentrator and with its likely origin at the corner between the cover and the root

TABLE 5.8   Stress and fatigue life variation for the laminate modifications.

Configuration		Reference laminate − 4	Reference laminate − 2	Reference laminate	Reference Laminate + 2	Reference Laminate + 4
Modification 1	Δσ (MPa)	−2.46	−2.61	−2.73	−2.82	−2.89
	Δ Life (years)	4.2	4.6	4.8	5.1	5.3
Modification 2	Δσ (MPa)	−3.03	−3.22	−3.37	−3.48	−3.58
	Δ Life (years)	5.6	6.1	6.6	6.9	7.2

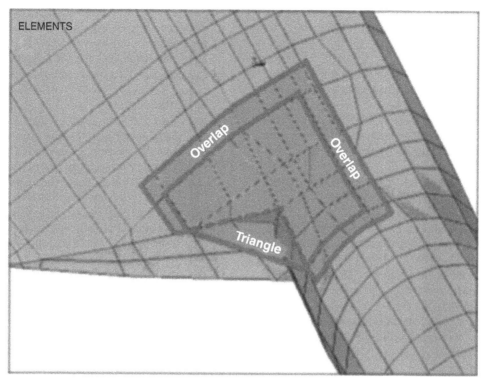

FIGURE 5.17   Scheme of the configuration of reinforcement and the overlaps.

- Damage in the form of a crack that affected the resistant laminate, extending over the zone where there was an abrupt change in the thickness of the laminate
- Lack of bonding, lack of resin, and diverse manufacture defects

The nature of the observed failure seems to be due to a fatigue mechanism. The configuration of the observed cracks seems to indicate that first a superficial crack (Figure 5.9a) appeared, probably around the weakest point (corner between the cover and the root) due to the stress concentration, and that this crack progressed throughout the superficial layer and induced delamination between this superficial layer and the resistant laminate. Later on, in the zone where there was an abrupt change of thickness, the presence of the superficial crack together with the effect of the concentrator and the effect of the change in thickness gave rise to the stress state necessary to generate a crack in the resistant laminate (Figure 5.9b), which completely broke the laminate, as is observed in Figure 5.9c.

By means of the simplified procedure of the standard GL, a calculation of fatigue life for the laminate corresponding to the zone of failure has been carried out. It has been deduced that it is plausible that this laminate reached a fatigue life in the order of 5 years, considering that it was subjected to the combined effect of the presence of the superficial crack, the effect of the geometric concentrator, and the effect of the

change in thickness, all these giving rise to an amplification of the nominal stress at the zone considered. Additionally, the presence of diverse defects of manufacture could, in any case, contribute to a decrement in the fatigue life of the element.

To design a suitable repair, the influence that the geometry of the concentrator that appears in the transition from the root to the aerodynamic zone has on the stress state, and consequently on the fatigue life of the blade, has been studied. It has been found that the inclusion of a triangular surface "filling up" the corner that generates the stress concentration significantly releases the stress state in the zone where the failure has been observed and provides an improvement in the fatigue life of the blade.

The influence of the size of the triangle has been analyzed, as well as the influence of the amount of unidirectional layers in the laminate of these triangles. It has been observed that the size of the triangle is the dominant parameter in the behavior of this type of reinforcement, the greatest increases in fatigue life corresponding to the triangle of greatest size. The variations in life associated to the number of unidirectional layers are of an inferior level.

REFERENCES

[1] Blom, A. F., Svenkvist, P., Thor, S. E. (1990). Fatigue design of large wind energy conversion system and operational experience from the Swedish prototypes. *Journal of Wind Engineering and Industrial Aerodynamics*, 34:45–76.

[2] Sutherland, H. J. (1999). *On the Fatigue Analysis of Wind Turbines.* SAND99-0089, Sandia National Laboratories, Albuquerque, NM.

[3] Nijssen, R. P. L., Van Delft, D. R. V. (2003). Alternative fatigue formulations for variable amplitude loading of fibre composites for wind turbine rotor blades. *European Structural Integrity Society*, 32(c):563–574.

[4] Philippidis, T. P., Vassilopoulos, A. P. (2004). Life prediction methodology for GFRP laminates under spectrum loading. *Composites Part A: Applied Science and Manufacturing*, 35(6):657–666.

[5] Sutherland, H. J., Mandell, J. F. (2004). Effect of mean stress on the damage of wind turbine blades. *ASME Wind Energy Symposium, AIAA/ASME*, pp. 32–44.

[6] Kong, C., Bang, J., Sugiyama, Y. (2005). Structural investigation of composite wind turbine blade considering various load cases and fatigue life. *Energy*, 30:2101–2114.

[7] Brondsted, P., Lilholt, H., Lystrup, A. (2005). Composite materials for wind power turbine blades. *Annual Review of Materials Research*, 35:505–538.

[8] Kensche, C. W. Fatigue of composites for wind turbines. *International Journal of Fatigue*, 28:1363–1374.

[9] Kong, C., Kim, T., Han, D., Sugiyama, Y. (2006). Investigation of fatigue life for a medium scale composite wind turbine blade. *International Journal of Fatigue*, 28:1382–1388.

[10] Hayman, B., Wedel-Heinen, J., Brondsted, P. (2008). Materials challenges in present and future wind energy. *MRS Bulletin*, 33(4):343–353.

[11] Marin, J. C., Barroso, A., Paris, F., Canas, J. (2009). Study of fatigue damage in wind turbine blades. *Engineering Failure Analysis*, 16(2):656–668.

[12] Mandell, J. F., Samborsky, D. D. (1997). *DOE/MSU Composite Material Fatigue Database: Test Methods, Materials and Analysis.* SAND97-3002, Sandia National Laboratories, Albuquerque, NM.

[13] Andersen, S. I., Bach, P. W., Bonnee, W. J. A., Kensche, C. W., Lilholt, H., Lystrup, A., Sys, W. (1996). Fatigue of materials and components for wind turbine rotor blades. *European Commission, Directorate-General XII, Science, Research and Development*, EUR 16684 EN, Brussels.

[14] Philippidis, T. P., Vassilopoulos, A. P. (1999). Fatigue of composite laminates under off-axis loading. *International Journal of Fatigue*, 21(3):253–262.

[15] Mandell, J. F., Samborsky, D. D., Wang, L., Wahl, N. K. (2003). New fatigue data for wind turbine blade materials. *Journal of Solar Energy Engineering*, 125(4):506–514.

[16] Van Paepegem, W., Degrieck, J. (2004). Simulating in-plane fatigue damage in woven glass fibre reinforced composites subject to fully reversed cyclic loading. *Fatigue and Fracture of Engineering Materials and Structures*, 27(12):1197–1208.

[17] Shokrieh, M. M., Rafiee, R. (2006). Simulation of fatigue failure in a full composite wind turbine blade. *Composite Structures*, 74(3):332–342.

[18] Passipoularidis, V. A., Philippidis, T. P., Brondsted, P. (2011). Fatigue life prediction in composites using progressive damage modeling under block and spectrum loading. *International Journal of Fatigue*, 33(2):132–144.

[19] IEC 61400-1 (2005). *Wind Turbines – Part 1, Design Requirements*, 3rd edition. IEC, Geneva, Switzerland.

[20] Germanischer Lloyd (1999). *Rules and Regulations IV – Non Marine Technology, Part I – Wind Energy.* Germanischer Lloyd, Hamburg.

[21] Det Norske Veritas, Riso National Laboratory, (2002). *Guidelines for Design of Wind Turbines.* Det Norske Veritas and Riso National Laboratory, Copenhagen, Denmark.

[22] Bechly, M. E., Clausen, P. D. (1997). Structural design of a composite wind turbine blade using finite element analysis. *Computers & Structures*, 63(3):639–646.

[23] De Goeij, W. C., Van Tooren, M. J. L., Beukers, A. (1999). Implementation of bending-torsion coupling in the design of a wind turbine rotor blade. *Applied Energy*, 63:191–207.

[24] Hahn, F., Kensche, C. W., Paynter, R. J. H., Dutton, A. G., Kildegaard, C., Kosgaard, J. (2002). Design, fatigue test and NDE of a sectional wind turbine rotor blade. *Journal of Thermoplastic Composite Materials*, 15(3):267–277.

[25] Jensen, F. M., Falzon, B. G., Ankersen, J., Stang, H. (2006). Structural testing and numerical simulation of a 34 m composite wind turbine blade. *Composite Structures*, 76:52–61.

[26] Paluch, B. (1993). A software for design and calculation of wind turbine composite rotor blades. *European Community Wind Energy Conf.*, 8–12 March, pp. 559–562.

[27] Fernandes da Silva, G., Marin, J. C., Barroso, A. (2011). Evaluation of shear flow in composite wind turbine blades. *Composite Structures*, 93(7):1832–1841.

[28] Toft, H. S., Sorensen, J. D. (2011). Reliability-based design of wind turbine blades. *Structural Safety*, 33(6):333–342.

[29] Habali, S. M., Saleh, I. A. (2000). Local design, testing and manufacturing of small mixed airfoil wind turbine blades of glass fiber reinforced plastics. Part II: manufacturing of the blade and rotor. *Energy Conversion & Management*, 41:281–298.

[30] Canas, J., Marin, J. C., Barroso, A., Paris, F. (1999). On the use of strength of materials models and finite element models in wind turbine blades design (Sobre el uso de modelos de resistencia de materiales y modelos de elementos finitos en el diseno de palas de aerogenerador). *Proceeding of the MATCOMP-99*, Benalmadena (Spain), pp. 271–278 (in Spanish).

[31] Tsai, S. W. (1988). *Composites Design.* Think Composites, Dayton, OH.

[32] Tan, S. C. (1991). A progressive failure model for composite laminates containing openings. *Journal of Composite Materials*, 25:556–577.

[33] Pilkey, W. D. (1994). *Formulas for Stress, Strain, and Structural Matrices.* John Wiley & Sons Inc., pp. 286–287.

[34] ANSYS (2003). *Swanson Analysis System*, Inc.

[35] Marin, J. C., Barroso, A., Paris, F., Canas, J. (2008). Study of damage and repair of blades of a 300 kW wind turbine. *Energy*, 33(7):1068–1083.

6

FLOATING WIND TURBINES: THE NEW WAVE IN OFFSHORE WIND POWER

Antoine Peiffer and Dominique Roddier

6.1 INTRODUCTION

The onshore wind power industry is already well established and benefits from research and development efforts commenced in the 1960s by different players, including public institutions and private companies, often subsidized by governments. Nowadays, this industry, led by wind turbine manufacturers and project developers, is considered mature. While industry stakeholders continue to install wind farms on land, they are increasingly trying to shift to offshore sites. The reason behind this shift is that compelling onshore wind sites are becoming scarcer and scarcer, and the offshore wind resource is known to be more consistent and could considerably increase the global yield of wind farms. Thus, the offshore wind power industry is clearly on the surge and could be one of the leading sources of renewable energy to meet our future global needs (Czyzewski, 2012). A swift and sustained development of the wind energy sector would support current targets of greenhouse gases emission reductions that are in place in different countries worldwide.

The offshore wind energy industry at its current development stage is relatively limited by water depth and soil constraints. As of 2015, the offshore wind power industry possesses enough knowledge and expertise to deploy wind farms only at shallow water and transitional water sites, where the water depth culminates at 50–60 m. At these water depths, wind turbines are placed on fixed foundations that extend to the sea bottom. The generic foundation types are usually monopiles, tripods, gravity-based, and jackets. The type of foundation is chosen depending on the turbine size, the exact water depth, and the soil properties at the site. Siting wind turbines on fixed foundation, however, is not feasible on loose or soft seabed, which has generated a swelling interest in floating foundations (Briggs, 2010).

This chapter concentrates on the new frontier of the offshore wind power industry, the deep-water areas, where the water depth exceeds 50–60 m. In the future, turbines could be sited very far from shore, independently from water depth and soil conditions on the seabed (Musial and Butterfield, 2004). The only economically feasible means to move further offshore is to install wind turbines on floating foundations generally moored at the site. The aerodynamic loads applied on the wind turbine convert to an overturning moment on the floating structure. Economics show that this moment becomes too high for a fixed foundation to be cost-competitive in water depths exceeding 50–60 m. Wind energy resources from coastal areas where the water depth increases rapidly with the distance from shore or from sites further away from shore could be harvested by such floating wind farms. This transition seems very natural and would also allow the siting of wind turbines where the sea bed is loose. Professor William Heronemus of the University of Massachusetts pioneered the offshore wind power industry in the 1970s by proposing a floating device (see Figure 6.1) with multiple rotors producing either electricity or hydrogen (Heronemus, 1972). The author intends to show the challenges involved with the development of these new-generation floating wind turbines.

In the first section, the transition of the offshore wind power from shallow water to deep-water sites will be assessed as a potential significant part of our future energy mix. Peripheral constraints that affect the siting of floating wind turbines will be examined, including social, environmental, and practical considerations. Then, this chapter will present an overview of the current state of the art in the offshore

Alternative Energy and Shale Gas Encyclopedia, First Edition. Edited by Jay H. Lehr and Jack Keeley.

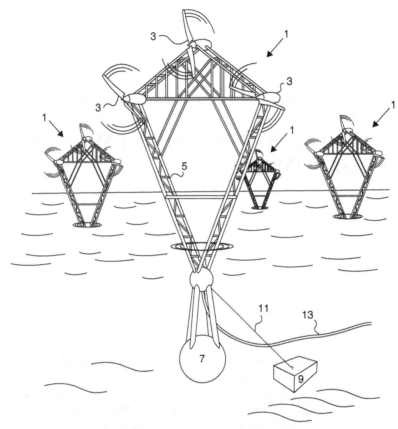

FIGURE 6.1 Pioneering floating wind turbine concept. *Source*: Heronemus et al., patent no.: US 7,075,189 B2, July 11, 2006). The numbers of the figures correspond to the description of the system as written in the original patent.

wind energy and will define the numerous technical and engineering challenges associated with these innovative floating wind turbine designs. Finally, the various generic types of technologies currently under development will be described and the cutting edge of nascent floating wind energy technologies will be discussed.

6.1.1 Peripheral Considerations Affecting Industry Transition

Offshore wind power is one of the fastest growing renewable sources of energy. The fixed offshore energy industry was able to develop considerably, because it managed to meet diverse sociopolitical requirements and constraints at its origin. Many external considerations related to the siting of offshore wind turbines, which could affect the prospects of further development of the industry, are analyzed herein to gauge a potential transition to deeper waters and spot any possible showstoppers noninclusive of technical and financial aspects.

These considerations can be divided into the following three groups:

- Social considerations that include a number of factors related to the society and its activities. These factors are

heavily discussed on the political arena, and hence, are usually well known among citizens.

- Environmental considerations that encompass factors influencing the flora and fauna of an area where a wind turbine is installed. In fact, the current environmental impacts associated with fixed foundations could propel the industry faster toward floating constructions.

- Practical considerations that reflect the current stage of technological development and problems faced by the industry.

6.1.1.1 Social Considerations and Social Acceptance.
The sociopolitical and community response to floating wind turbines might be an issue. Social acceptance linked to land use and the "Not In My Backyard" (NIMBY) effect are both very common for such systems. The visual and noise impacts are a recurrent problem in the onshore and offshore wind energy industry, as local communities (fishermen, beachgoers, etc.) often oppose to the installation of wind turbines at certain sites, for they produce too much noise, alter the local landscape, or modify natural environments. Wind farms also require vast amount of lands to produce a significant amount of electricity, and mar costal landscapes.

This is a good omen for floating wind turbines. The vast ocean offers considerable space for installing offshore wind farms at a very large scale. These wind turbines, however, have to be installed at least 20 miles offshore to appear invisible from shore, based on their estimated height and the curvature of the Earth. The siting flexibility is also an advantage for floating turbines, since water depth is no longer relevant.

Underwater Archeological Sites.   Evidences of shipwrecks and other underwater archeological sites are often found in coastal waters. Again, this is a boon for floating constructions that have a low footprint on the seabed (see next section), and thus should not damage archeological records, if appropriately sited.

Fishing Areas, Shipping, and Aviation Issues.   Both fixed and floating foundations should be located out of shipping lanes and aircraft paths, as well as out of areas used for fishing activities. Coordination with local authorities is necessary for the development of offshore wind farms.

6.1.1.2   *Environmental Considerations.*

Seabed Impact.   For most of the future floating technologies, compared to a fixed foundation, no pilings need to be installed in the seabed, and minimal acoustic disturbances occur during installation. The anchors are non-invasive and can be easily removed. Another positive aspect to keep in mind is that the level of biodiversity is lower in deeper waters, due to lack of sunlight and nutrients in the water. The potential presence of marine sanctuaries should be taken into account during site selection and permitting.

Wildlife Impact.   Wind turbines can be detrimental to the environment, notably to wildlife (avian and marine species). The typical issues include the potential modification of migration patterns, the alteration of natural environments, and the increase in fish populations around foundations. The latter results in the attraction of more birds at the proximity of the site, which therefore raises bird strikes with the tower or rotor. Obviously, potential disruptions have to be assessed and avoided.

At the same time, the foundations may become habitats for certain shellfish, and thus protect certain species. As far as birds are concerned, offshore migration routes and nesting activities should be carefully considered.

6.1.1.3   *Practical Considerations in Offshore Wind Resource* (**Rogers et al., 2000**).   In terms of performance, the offshore wind resource is well known to be more consistent, sustained, and powerful than onshore. Further offshore, fewer ebbs and flows are recorded in the wind speed and direction. Not only higher average wind speeds are measured or predicted, but also lower levels of turbulence and wind

gradients are usually recorded, as the vast ocean presents no obstacles. The electricity production comes out enhanced as capacity factors tend to increase, exceeding sometimes 40% and beating the onshore wind energy industry by about 10 percentage points. Another interesting point is that the wind offshore also tends to blow at the right time of the day, when the electricity demand is usually maximal (Schwartz et al., 2010).

Grid and Transmission Lines.   The wind energy and renewable energy industry across the board suffers from a lack of transmission lines and the need for a more modern and flexible grid to absorb higher rates of intermittent renewable energy. This is highly dependent on the region, but is still considered a global trend. The good news is that about three quarters of worldwide electricity is consumed on the coastal boundaries. Thus, installing wind turbines offshore and further offshore would probably minimize issues linked to grid integration, since most of the transmission lines are already built along the coast. Few new transmission lines would need to be connected and existing transmission lines would merely need to be retrofitted or modernized. Subsequently, the proximity of offshore wind turbine sites and current load centers should allow the grid infrastructure to scale faster to accommodate arrays of new offshore floating wind turbines.

This section presents an overview of many external considerations, which can hinder industry development. A shift of the industry toward floating structures in deeper waters does not seem to exacerbate any of the existing issues associated with the shallow water wind power industry. Some of the issues are even mitigated, which might indicate that a transition to floating turbines is a compelling proposition. If the technological challenges are solved and sufficient cost-competitiveness results, floating wind power could become a major sector of the industry. The next section focuses on the technical challenges to be solved.

6.1.2   Technical Challenges

6.1.2.1   *Current Technological Status: Fixed Foundations for Water Depths Below 50–60 m.*   Currently, offshore wind farms are located in shallow water and transitional water depths from 0 to 60 m. Conventional land-based wind turbines have been marinized by wind turbine manufacturers. The turbine components exposed to the harsh offshore environmental conditions have been beefed up, for example, to resist corrosion. These upgraded wind turbines are placed on concrete bases or steel monopiles that are embedded into the sea bottom and require a buried undersea electrical cable that sends electricity to shore.

Different types of fixed foundation are currently being deployed in the industry, depending on the water depth at the site, the type of soil, and the size of the turbine (Tchou

FIGURE 6.2   Main types of fixed foundation for offshore wind turbines. *Source*: Process Engineering.

and Russell-Smith, 2009). They are described in the order of appearance in Figure 6.2 (from left to right):

1. Gravity base structures made out of concrete or steel are generally limited to very shallow water depths. These structures rely on the heavy weight of the base to stay in place on the seabed. Erosion is a well-known concern for these types of foundations and can increase costs significantly.

2. A monopile base is a single steel column that is an extension of the tower with a transition piece. It is driven 10–20 m into the seabed. This is currently the most popular type of fixed foundation. Monopiles do not require any seabed preparation, but their installations require heavy-duty piling equipment.

3. A tripod is a single steel pile connected to three smaller piles. These piles are either driven into the seabed in a pyramidal fashion or pressed to the ground and suctioned if suction caissons are utilized. This concept is usually more suitable for higher water depths.

4. Steel jackets are usually four-legged structures with a stable latticed construction. They are very suitable foundations for higher water depths up to 50–60 m. One of the drawbacks of this type of structure is its fabrication complexity. Steel jackets are usually piled, but suction caissons (Sukumaran, 1999) can also be used, notably for softer soils (Houlsby and Byrne, 2000).

Piled or gravity foundations are widespread in the industry, but novel designs based on suction bucket technologies are being developed and could be attractive for specific sites.

6.1.2.2   *The Necessity to Transition to Floating Structures Beyond 50–60 m of Water Depth.*
The economic breakpoint at which a fixed foundation becomes more costly than its floating counterpart is driven by the water depth at the site. Wind turbines on fixed foundations become too costly, once the water depth exceeds 50–60 m. Large aerodynamic loads are applied on the wind turbine at the hub height. The dominant aerodynamic load represents the thrust of the wind turbine applied in the direction of the wind, perpendicular to the rotor plane. Viewed from the supporting platform far below the wind turbine hub, the thrust force represents an overturning moment to be withstood. This overturning moment increases with the vertical extension of the foundation, thus increases with the water depth when the structure is fixed to the seabed. The main driver in this case is the amount of steel needed to withstand this overturning moment. At around 50–60 m water depth, the amount of steel required for a fixed foundation to withstand the most critical overturning moment equals or starts exceeding that required for a floating platform. The viability of fixed foundations is thus questioned above this threshold. The border between fixed and floating wind also depends on other aspects, such as installation and fabrication costs, but the amount of steel drives most of the economics.

Sometimes a distinction is made between monopiles/gravity bases versus jackets and tripods, since jackets and tripods are generally suitable for higher water depths. Again, the same economic drivers tend to favor monopiles and gravity bases for water depths inferior to 25 m, while industry leaders opt for jackets in water depths ranging from 25 to 60 m (Figure 6.3).

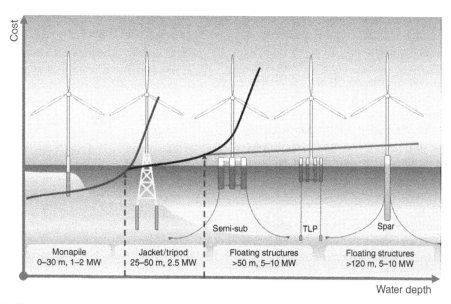

FIGURE 6.3 Technological Progression of offshore wind power with water depth: from shallow water to deep waters (Courtesy of Principle Power). Monopiles on the left-hand side, cost-effective up to 25 m water depth, are then replaced by jackets or tripods. Floating platforms the three platforms on the right-hand side become less costly than jackets or tripods above 50–60 m water depth.

6.1.2.3 Technical Challenges for Floating Wind Turbines.

(Musial et al., 2006) (Butterfield et al., 2007) Floating turbines or turbines placed on floating structures will replace conventional fixed foundations in water depths exceeding 50–60 m. Another level of technical complexity is added to floating foundations accommodating wind turbines. These new types of system must overcome numerous challenges in order to be both technically and financially feasible. This section presents the technical challenges inherent to installing a wind turbine on a floater. Two simple economic aspects are of paramount importance for the viability of floating wind turbine technologies:

(a) The revenues should be maximized, that is, turbine performance should be maximized, or at least comparable to fixed foundations. The maximum amount of electricity should be generated for the longest time possible at the site.

(b) The capital expenditures should be minimized throughout the project lifecycle, that is, costs should be minimized at all stages of the project. For example, any effort to reduce the steel (or other material) weight of the hull or turbine leads directly to a reduction in the cost of energy.

The challenges associated with the design and operations of floating wind turbines are substantial. The floating platform, subject to wave and current loadings, bears a large weight (the wind turbine) and experiences large aerodynamic loads high above the water line. The large masses of the nacelle and rotor considerably raise the center of gravity of the platform compared to conventional floating structures, and the overturning moment created by the aerodynamic thrust at hub height is structurally penalizing (Roddier and Weinstein, 2010).

Motions—Dynamics.   The first challenge associated with floating wind turbines is the minimization of their motions to maintain optimal turbine performance and minimize steel weight. A floating platform supporting a wind turbine will experience motions. In order to keep turbine performance at its best and to mitigate costs, the 6 degree-of-freedom motions have to be minimized as much as possible (Wayman et al., 2006). The lateral motions of the platform usually called surge and sway are the most acceptable degrees of freedom. These planar motions mimic wind turbulence and modify only the apparent wind speed on top of the tower. The most unfavorable motions are the angular motions, rather the pitch and roll motions, and the vertical heave motion. These degrees of freedom result in high undesirable tower-top motions. If not restrained to acceptable limits, these motions could considerably reduce the turbine performance and shrink the structural life of the system. For example, considerable pitch or roll motions could modify the angle between the wind direction and the rotor plan, resulting in energy losses. The size and geometry of the floater matters the most for the platform dynamics. The stability of the floater is usually enhanced by its large size, the spacing between its water-piercing elements, and its large displacement. The developer is faced with a tradeoff between the stability and the size of the floater. The floater must be the most stable and the lightest at the same time (Roddier et al., 2009).

Adaptation of Wind Turbines for Maximum Performance. The conventional horizontal-axis wind turbine in the current offshore configurations might not be able to operate properly on a floater experiencing wave-induced motions. Turbine performance might be reduced significantly. Floater developers and wind turbine manufacturers will have to work hand-in-hand to confirm that the floater motion envelopes abide by the turbine design criteria. The wind turbine hardware and software will need to be verified, because of the large accelerations that occur at the nacelle. Changes might be required on both sides, but ideally the floater technology will only require changes on the turbine software side, meaning that the turbine controller will be tuned. Existing advanced control strategies are already playing an important role in onshore and fixed offshore wind turbines to achieve optimal power production while minimizing the loads on the turbine components. The control objectives thus include the dampening of undesirable structural dynamic responses or the filtering of resonances due to natural wind turbulence or changes in wind speed and direction, and the maximization of power generation. Current control strategies often involving the pitch angle of the turbine blades might prove inadequate and should be redesigned because the turbine is on a floating support (Van der Veen et al., 2012). These active control system schemes could be tweaked and adapted to floating wind turbines to limit wind-induced platform motions and mitigate coupling effects between floater and turbine. The natural periods of response of a floating platform are generally higher than those linked to the flexibility of tower or turbine components, so the risk of dynamic resonance between the floater and the turbine is expected to be small for most platform designs.

If software modifications are not sufficient, hardware modifications might be necessary. Different possible turbine innovations are envisioned in that case:

(a) A reduction of the overall tower weight (the largest weight above water) and tower-top weight would result in a reduction of the supporting platform weight. This weight reduction could stem from a rotor spinning faster than current designs, as long as the noise level remains acceptable. A higher average rotor rotational speed would decrease the torque to be transmitted to the drivetrain, which would reduce its size and weight. This is notably useful in the case of larger wind turbines, since the torque increases more rapidly than the rated power, when the rotor diameter increases. Manufacturers count on the improved reliability of gearless or direct-drive generators to improve efficiency and decrease maintenance costs, but gearless turbines remain heavier than gear-based turbines because direct-drive generators must be bigger to operate at blade speed. The use of permanent magnets in the generator rotor instead of copper coils might be an area to explore for further tower-top weight reduction.

(b) The overall turbine hardware could be modified to become more compliant to large hub motions and accelerations, notably if excessive platform motions alter the lubrication in turbine components or spawn unusual dynamic loading in turbine components.

(c) New lightweight materials could be used to increase flexibility of turbine components, notably for the blades. These flexible components would accommodate large dynamic displacements more robustly. For example, a downwind wind turbine with a more flexible rotor would increase dynamic load compliance and reduce the risk of blade strike with the tower.

All the challenges described in this section might be met for current rotor diameters, but additional complications might arise for the next generation of wind turbines. The main question will be how the current challenges will scale to future wind turbines, which will tend to have larger rotor diameters and more aerodynamically efficient designs.

Design Tool Development. (Passon and Kuhn, 2005) Designers will have to expand already existing codes for bottom-mounted offshore support structures or develop new-generation computer-aided engineering tools to accurately model the system and its response in complex environmental conditions.

The large aerodynamic loads generated by the turbine have a substantial effect on the motion of the substructure, and conversely, motions of the platform generated by waves have a significant impact on the turbine response. Because of the complex responses of both the turbine and the hull, it is extremely difficult to "decouple" the hydrodynamic response (Newman, 1977) from the "aerodynamic" response. A fully coupled simulation tool is the most expedient way to determine accurately the system responses in order to achieve an optimized design (Luo et al., 2012). On top of that, complex mooring system dynamics, structural dynamic effects due to the flexibility of the platform and turbine components, and the implemented turbine control schemes should also be incorporated into the computer model (Jonkman and Sclavounos, 2006).

The challenge lies in both the prediction and validation of extreme loads and cyclic loads (fatigue) under simultaneous incoming wind and waves. Load cases that accurately simulate all possible floating wind turbine design situations with representative metocean conditions are defined during the iterative design process. Time-domain simulations are typically run to account for the nonlinearities present in the aerodynamics and hydrodynamics (Jonkman and Buhl, 2007).

Engineering Standards. Designers currently have to adapt existing standards for offshore fixed wind turbines and oil and gas platforms. The future availability of design codes, engineering guidelines, or industry standards will considerably improve the integrity of floating wind turbines. New engineering guidelines specifically on floating wind turbines are currently being developed by major certification and classification societies, such as the American Bureau of Shipping (ABS) or Stiftelsen Det Norske Veritas (DNV).

Structural Design. Linked to the platform motion minimizations is the structural design of the system. Structural optimizations should be carried out to reduce overall steel weight, while the large overturning moments created by the wind thrust should be transferred throughout the structure. The fatigue life could be a potential issue, as the structure is subject to cyclic loading at different frequencies that could add up quickly until a crack appears, propagates, and jeopardizes the integrity of the system. The interface between the turbine and the floater could be an issue, as different components of the system vibrate at different frequencies. Even if the risks are low and can be mitigated during design, when these components are assembled on the same structure, vibrations could be compounded at certain frequencies, due to unpredicted environmental loading, which could lead to different failures.

Fabrication and Onshore Infrastructure. The objective for fabrication is to optimize the platform for steel weight and provide the most inexpensive system. This is a challenge in that the fabricator constraints must be considered during the design phase to reduce any kind of fabrication complexities that could send costs to the roof. Contrary to other related industries, serial production is envisioned here for a setup of floating wind turbines in farms. All floaters for a given farm at a given site will be identical, allowing for serial production of the units. The components can be prefabricated at multiple locations, taking further advantage of fabrication economies of scale, and the final assembly can take place at the final site. The challenge lies in the optimization of the supply chain for low-cost high-volume production. Another challenge might be the lack of onshore infrastructure for the construction of such a large system. Construction facilities such as shipyards, docks, dry-docks, and staging areas in harbors are required. Different types of ships such as barges with cranes, anchor-handling tug supply (AHTS) vessels, electrical cable-laying vessels, and harbor tug boats might also be needed with enough capacity and appropriate draft. Potential wind farm sites in deep water might be far away from the required infrastructure, which induce higher project costs. The availability of farm interconnection to the local grid could impact costs as well.

Load-Out and Installation. The installation of floating wind turbines is another challenging matter. Generally, these platforms are fabricated in a shipyard. They are then loaded out where the final assembly takes place and towed to site. Any floating wind turbine that is stable at shallow draft with a fully assembled turbine will circumvent the need for complex and expensive offshore assembly operations. Only small harbor tugboats would be required for their towing to site (Figure 6.4).

At the site, the platform is hooked up to its preinstalled electrical cable and mooring lines. These installation operations are considerable and take place only in certain weather windows. To minimize costs, the golden rule is that all offshore operations involving different vessels should be kept to their minimum. The designer should focus on simplicity and use the cheapest nonspecialized vessels available.

Weather window tolerance is another aspect to consider. This is also linked to the accessibility of the platform in rough weather conditions once at the site. Offshore operations should be feasible in the largest sea states possible, in order to minimize weather-dependent windows that cost time and money. A floating wind turbine should ideally be loaded out and installed in a broad range of weather conditions, because poor weather conditions can cause delays. Specialized vessels are usually contracted with a charter fee on a day-to-day basis. These delays result in an increase of installation costs, since crews and vessels are contracted but inactive until weather improves.

Any floating wind turbine stable at shallow draft when fully assembled will save costs, since pre-commissioning of the turbine and platform can take place in the fabrication facility before float-out. This is also an advantage for long-term maintenance and decommissioning, if the platform has to be towed back into port.

Seakeeping—Mooring System. Another challenge will be the seakeeping of the platform at the site. Different types of mooring systems are already in use in the oil and gas industry and could be adapted to floating wind turbines. The mooring lines and anchors selected during the design phase will depend on the water depth and loads on the system during extreme events, as well as cyclic events inducing fatigue. Catenary mooring lines connected to drag-embedment anchors seem to be suitable for a variety of soil conditions, but long length of lines might be required to keep the load horizontal at the anchor point. Vertically loaded anchors, which require shorter taut mooring lines, could also be used depending on the choice of the designer, but seem to be a more expensive option, because of the higher loads to be withstood.

6.1.3 Generic Floating Wind Platform Concepts (Henderson et al., 2004)

The nascent offshore floating wind power industry takes advantages of legacies and expertise bequeathed by two proven industries: the wind power and offshore oil and gas

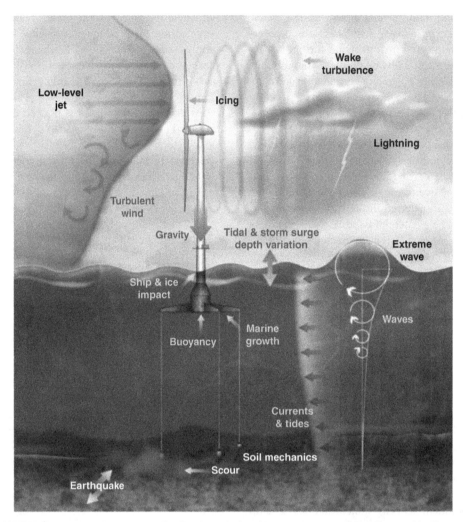

FIGURE 6.4   Loading sources for floating wind turbines. *Source*: National Renewable Energy Laboratory (NREL), Boulder, USA.

industries. Thus, the fast development of the industry lies in the synergies between established technologies from both industries. On the one hand, floating structures deployed in the oil and gas industry date back from the end of the 1970s. The oil and gas industry was in a similar situation at the time, when fields were discovered in very deep waters. The only solution was to develop floating platforms to harness these enormous pockets of energy. The oil and gas industry gained maturity over time for the development of floating platforms and learned the lessons of its numerous failures. Likewise, the offshore floating wind energy industry is being developed to harness stronger wind resources. New technology developments are based on the identification of resources that can be captured.

On the other hand, wind turbines at a commercial scale have been installed for at least 20–30 years and have a proven track record. The challenge for floating platform designers is to develop adequate turbine-agnostic floaters, that is, floaters able to accommodate these already existing

wind turbine technologies with the minimum amount of design modifications.

Oil and gas platforms expertise can be used to design floating wind turbines, but numerous differences exist between the two applications (Musial et al., 2004):

(a) Oil and gas platforms are usually single structures deployed for a specific project with clear applications. The topside equipment of a floating wind turbine will be similar for a given farm. It changes for an oil rig depending on the application.

(b) Oil rigs are large structures that are usually very stable and experience small wave-induced motions for drilling reasons. Offshore wind turbines will be smaller and less stable structures for economic reasons.

(c) The cost of an oil rig is offset very quickly by the large revenues coming from the extraction of abundant and energy-dense oil or gas at the field. The economics

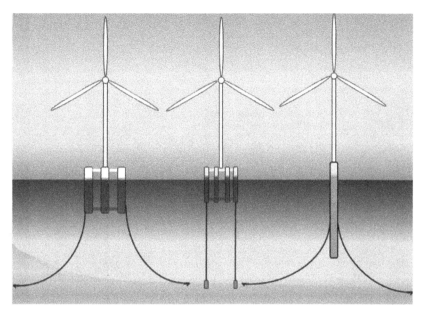

FIGURE 6.5   Generic floating wind turbines concepts. From left to right, water-plane inertia based platform (or semi-submersible platform), mooring-stabilized platform (or tension-leg platform), and gravity-based platform (or spar cylinder) (Courtesy of Principle Power).

is different for floating wind turbines that can only harness the wind.

(d) The workforce is often housed on oil rigs, whereas floating wind turbines will be unmanned platforms. The high safety factors currently employed in the oil and gas industry will have to be adapted and lowered for a floating wind turbine application.

By drawing on the synergies with oil and gas platform technology, while taking into account the design requirements of conventional wind turbine, developers have been developing floating platform supporting wind turbines that fall into three different groups (Henderson and Vugts, 2001) based on the strategy employed to stabilize the platform under static loading conditions (Figures 6.5 and 6.6):

1. The first group can be qualified as *water-plane inertia based*, since the static stability of this group is highly dependent on the inertia of the water-plane area. The floater surface in contact with the water is generally large and spread out. Conventional semi-submersible platforms, composed of several columns interconnected by truss members, are typical examples. Semi-submersible platforms present the highest wave-induced motions and the largest amount of structure present above the water surface. They are suitable for all water depths above 50 m, but have one of the largest footprints among concepts, due to their catenary mooring systems. The main advantage of the semi-submersible appears to be its ease of installation, commissioning, and maintainability,

since the system is fully assembled at quayside already pre-commissioned, can be towed to site due to its stability at low draft, and can be easily decommissioned or sent back to port for overhaul (Roddier et al., 2010).

2. The second group is inherently unstable and requires external restoring forces provided by the mooring line tensions to float upright. This kind of platform can be qualified as *mooring-stabilized* and is called a tension-leg platform (TLP), since it is moored to the seabed by vertical tendons with a high pretension load. A TLP is usually composed of a square pontoon with columns and a topside deck. TLPs are the most stable of the concepts due to the high tension in the mooring lines, and thus usually present very low heave, roll, and pitch motions, which is exactly what is needed for existing wind turbine technology (Matha, 2009). A potential frequency coupling between the mooring lines and the tower should be carefully assessed during design. The installation of the platform and its mooring system, however, is a complex operation, especially when vertically loaded anchors are used. TLPs are also water-depth sensitive, due to the taut mooring system. TLPs are likely to be smaller than any of the other designs and will have the smallest footprint, since the mooring lines are vertical. The tidal variations modify the mean tension in tendon tension, which makes TLPs non-deployable in areas of large tides.

3. The third and last group can be qualified as *gravity-based*. The center of gravity of the platform is tuned to be as low as possible, which means that the hull will be very deep below the waterline. The center of buoyancy

FIGURE 6.6   Generic floating wind turbines concepts. From left to right: semi-submersible (Courtesy of Principle Power), tension-leg (Courtesy of Glosten Associates), and spar (Courtesy of Statoil Hydro) platforms.

of such platforms is usually above the center of gravity, which means that static stability is always assured. That is typically the case of a spar, a slender deep-drafted floating cylinder. Generally, spars have the simplest design of all concepts. Pitch and roll motions can be significant since the water-plane area is low, but the heave motion is nearly not apparent for such systems due to their deep drafts. In terms of possible deployment, spars are limited to certain suitable sites with a minimum water depth that has to be greater than the draft of the spar. Similarly, the final assembly of the wind turbine is not feasible at quayside and usually occurs in sheltered waters (Thiagarajan and Dagher, 2012).

6.2   CONCLUSION

This chapter intended to present the technological state of the art in the offshore wind energy industry and the prospects of development for future floating wind turbine technologies that would be independent of water depth.

Different peripheral hurdles that affect the siting of floating wind turbines in deep water have been described at first, but no social, environmental, or practical showstopper was discovered for the offshore wind energy industry to transition to deeper waters. This augurs well for floating wind turbines, which represent the next logical step for the industry. Many technical challenges, however, are still to be solved, if the wind energy sector is willing to break loose from water depth dependence. Among these technical challenges, the wave and wind-induced motions of the floaters are important elements to consider, if conventional

horizontal-axis wind turbines are to be used without significant modifications. For all that, at the end of the day, the most cost-effective technology along the entire value chain will be the victorious one. It is therefore important to carefully consider cost optimizations at each stage of a floating wind farm project, from design, fabrication, and installation, through commissioning, operation, and maintenance, to decommissioning.

Recently, a significant amount of government funding has been allocated to prototype development. The learning from recent and future deployments of full-scale prototype units in the ocean will benefit the entire industry and help develop the necessary infrastructure, so that the main players can be one step close to commercial reality.

REFERENCES

Briggs, C. (2010). Review of the Suitability of Floating Wind Turbines in the Gulf of Maine, dissertation undertaken in conjunction with SgurrEnergy.

Butterfield, S., Musial, W., Jonkman, J., Sclavounos, P. (2007). Engineering challenges for floating offshore wind turbines. Conference Paper NREL/CP-500-38776.

Czyzewski, A. (2012). Wind energy gets serial. *Process Engineering*. Available at: http://processengineering.theengineer.co.uk/in-depth/the-big-story/wind-energy-getsserial/1012449.article.

Henderson, A. R., Zaaijer, M. B., Huijsmans, R., van Hees, M., Snijders, E., Wijnants, G. H., Wolf, M. J. (2004). Floating windfarms for shallow offshore sites. Proceedings of the 14th ISOPE Conference, Toulon, France.

Henderson, A. R., Vugts, J. H. (2001). *Prospects For Floating Offshore Wind Energy*. Delft University of Technology.

Heronemus, W. E. (1972). Pollution-free energy from offshore winds. Preprints, *8th Annual Conference and Exposition, Marine Technology Society*, September 11–13, 1972, Washington, DC.

Houlsby, G. T., Byrne, B. W. (2000). Suction caisson foundations for offshore wind turbines and anemometer masts. *Wind Engineering*, 24(4):249–255.

Jonkman, J. M., Sclavounos, P. D. (2006). Development and verification of fully coupled aero-elastic and hydrodynamic models for offshore wind turbines. NREL Report No. CP-500-39066, 24 pp.

Jonkman, J. M., Buhl, M. L. (2007). Loads analysis of a floating offshore wind turbine using fully coupled simulation. Conference Paper NREL/CP-500-41714.

Luo, N., Pacheco, L., Vidal, Y., Li, H. (2012). Smart structural control strategies for offshore wind power generation with floating wind turbines. *International Conference on Renewable Energies and Power Quality*, Santiago de Compostela.

Matha, D. (2010). Model development and loads analysis of an offshore wind turbine on a tension leg platform, with a comparison to other floating turbine concepts. Subcontract Report NREL/SR-500-45891.

Musial, W. D., Butterfield, S. (2004). Future for offshore wind energy in the United States. NREL/CP-500–36313. *Energy Ocean Proceedings*, Palm Beach, FL.

Musial, W., Butterfield, S., Ram, B. (2006). Energy from offshore wind. *Offshore Technology Conf.*, Houston.

Musial, W. D., Butterfield, S., Boone, A. (2004). Feasibility of floating platform systems for wind turbines. NREL/CP-500-34874. *23rd ASME Wind Energy Symposium Proceedings*, Reno, Nevada.

Newman, J. N. (1977). *Marine Hydrodynamics*. The MIT Press, Cambridge, MA, ISBN 0-262–14026-8.

Passon, P., Kuhn, M. (2005). *State-of-the-Art and Development Needs of Simulation Codes for Offshore Wind Turbines*. University of Stuttgart.

Roddier, D., Cermelli, C., Weinstein, A. (2009). WindFloat: a floating foundation for offshore wind turbines. Part I: design basis and qualification process. *Proceedings of OMAE 2009, 28th International Conference on Offshore Mechanics and Arctic Engineering*, Honolulu, HI, USA, May 31.

Roddier, D., Cermelli, C., Aubault, A., Weinstein, A. (2010). WindFloat: a floating foundation for offshore wind turbines. *Journal of Renewable and Sustainable Energy*, 2:033104.

Roddier, D., Weinstein, J. (2010). Floating wind turbines. *Mechanical Engineering*, 132.4:28–32.

Rogers, A. L., Manwell, J. F., McGowan, J. G., Ellis, A. F., Abdulwahid, U., LaCroix, A. (2000). A fresh look at offshore wind opportunities in Massachusetts. *Proc. AWEA Annual Conference*.

Schwartz, M., Heimiller, D., Haymes, S., Musial, W. (2010). Assessment of offshore wind energy resources for the United States", Technical Report NREL/TP-500-45889.

Sukumaran, B. (1999). Suction caisson anchors – a better option for deep water applications. Rowan University.

Thiagarajan, K. P., Dagher, H. J. (2012). State-of-the-art review of floating platform concepts for offshore wind energy generation. *Proceedings of OMAE 2012, 31st International Conference on Offshore Mechanics and Arctic Engineering*, Rio de Janeiro, Brazil.

Tchou, J., Russell-Smith, S. (2009). San Francisco offshore wind farm analysis, Preliminary Report, Stanford Solar and Wind Energy Project, Stanford University, 2009.

Van der Veen, G. J., Couchman, I. J., Bowyer, R. O. (2012). Control of floating wind turbines, *Proceedings of the 2012 American Control Conference*, Montreal, Canada.

Wayman, E. N., Sclavounos, P. D., Butterfield, S., Jason, J., Musial, W. (2006). Coupled dynamic modeling of floating wind turbine systems, *Offshore Technology Conf.*

7

WIND POWER—AEOLE TURNS MARINE

ROGER H. CHARLIER AND ALEXANDRE C. THYS

7.1 INTRODUCTION

Of all the ocean energies, marine winds have known the most important development during the last decades. A "renewable" easy to harness, it required only relatively modest capital investments. Sites are abundant, and a judicious choice permits to dampen the objections voiced because of the noise they cause. Danger to migrating birds remains to be further minimized. Tidal, wave "farms" are a well-established concept, but marine wind "farms" have already been implanted in numerous locations particularly in Northern and Western Europe. The environmental-linked objections are being raised, spurring engineers to devise new approaches.

Winds have been tapped for their energy and the traditional windmill, now a quaint ancestor of the folkloric holdover of bygone times, is nevertheless the ancestor of the wind turbine, now a familiar sight of the twenty-first century landscape. In contrast to most other alternative renewables, winds are harnessed both on land and at sea. It is not so long ago that "Aeole took to the sea." Most ocean energies require engineering developments to be harnessed to produce electricity. Marine winds and the tides are the exceptions; both have forerunners: the first have probably been inspired by the millennia-old wind mills, the second find an ancestor in the tide mills, and perhaps the boat [river] mills as well, that came into use centuries ago. Both remained in use well into the twentieth century. Because of the absence of such obstacles as trees, hills, and mountains, marine winds are more powerful than continental winds but also far more destructive in case of storm. And another negative aspect of wind energy is lack of steadiness as winds vary in velocity and direction. In that they differ from tides that are regular, and exceptional tides are few.

Already 5000 years ago Egyptians were using wind power to transport most of their merchandise between Upper and Lower Egypt. In 700 BC the Babylonians used static windmills for irrigation and were among the first to replace the strength of men and animals with mechanical power. About 200 years BC the Persians pioneered the use of windmills to crush olives and to grind grain. And in the Lowlands wind mills were used from the fourteenth century onward to drain the *polders* at the mouth of the Rhine River. Wind mills are a common sight in Greece. Closer to the present, wind pumps were erected worldwide in rural regions, especially in North America and Australia, to pump deep groundwater up to the surface for land irrigation and cattle grazing. In the early twentieth century about 5 million such pumps were estimated at work, solely in the United States. Old Stover wind mills provided electricity in rural America.

7.2 THE WIND TURBINE

The first electricity generating wind turbine was a battery charging machine installed in July 1887 by James Blyth to light his holiday home in Marykirk, Scotland. On his heels inventor Charles F. Brush built the first automatically operated wind turbine for electricity production in Cleveland, Ohio. Blyth's turbine was considered uneconomical but not so for electricity generation in countries with widely scattered populations. In Denmark by 1900, there were about 2500 windmills for mechanical loads such as pumps and mills, producing an estimated combined peak power of about 30 MW. The largest machines were on 24-m towers with four-bladed 23-m diameter rotors. By 1908 there were 72 wind-driven electric generators operating in the United States

Alternative Energy and Shale Gas Encyclopedia, First Edition. Edited by Jay H. Lehr and Jack Keeley.
© 2016 John Wiley & Sons, Inc. Published 2016 by John Wiley & Sons, Inc.

producing from 5 to 25 kW and by 1913 American wind-mill makers were producing 100,000 farm windmills a year, mostly for water pumping. By the 1930s, windmills for electricity were common on farms, mostly in the United States. High-tensile steel was cheap, and windmills were placed atop prefabricated open steel lattice towers.

The URSS, United States, and United Kingdom were first in operating wind turbines. A forerunner of modern horizontal-axis wind generators was in service at Yalta (USSR) in 1931. The 100 kW generator on a 30-m tower was connected to the local 6.3 kV distribution system; its annual capacity factor of 32% did not differ much from that of current wind machines. In the fall of 1941, the Smith-Putnam wind turbine, the first megawatt-class wind turbine was synchronized to a utility grid in Vermont, but it only ran for 1100 hours before suffering a critical failure; it was not repaired because of wartime shortage of materials.

John Brown & Company operated the first utility grid-connected wind turbine in the United Kingdom in 1951 in the Orkney Islands.

The three primary types of turbines are the VAWT Savonius, HAWT towered, and VAWT towered (Figure 7.1). Wind turbines can rotate about either a horizontal or a vertical axis. The device converts kinetic energy from the wind [*wind energy*], into mechanical energy [wind power]. The majority of wind turbines use a horizontal axis. The vertical axis turbines or VAWT move independently from the wind's direction. The life span of a turbine is about 20 years or just a bit less than conventional generating plants and far less than that of tidal power stations. Some 16,500 turbines of 1.5 MW have been installed. The sharing of components has ensured consistent workhorse reliability. Two of the largest wind farms belong to the 2.5 MW class. The *Offshore wind turbine* provides reliable, cost-effective, high-performance solution; the General Electric model replaced the costly gearboxes with reliable slow speed components specifically designed for the offshore environment and can be repaired *in situ* thereby minimizing operating costs. Nevertheless the offshore winds remain a still-to-be-overcome hurdle to development.

Foundations of the modern wind turbine for the generation of electricity were laid more than a century ago by inventors like *Poul la Court* and *Johannes Juul* (Denmark) and *Ulrich Hütter* (Germany) (Figure 7.2).

In the United States, wind energy conversion system (WECS) () made a timid appearance, after the first oil crisis, such systems were spurred on by ever climbing prices of fossil fuels and the need to reduce carbon dioxide emanations. Technical problems were rather rapidly solved and the first energy captors were erected on land and mountain tops, away from human habitat. Complaints from people living close by clusters of wind turbines—a.k.a wind farms—were limited. California was the first region in the world to make huge investments in wind energy. The largest onshore wind farm is located in *Altamont Pass* (CA). More than 7300 turbines

were installed since 1981 on an area of 140 km^2. Those turbines with a power capacity varying between 100 and 750 kW generate about 1.3 million MWh a year, representing about the yearly electricity consumption of nearly 500,000 people. Another huge wind farm is located in the neighborhood of *Tehachapi Pass,* California. Although made up of fewer turbines it has a yearly production of 1.4 million MWh.

After reaching a production of 1000 MW in 1985, it took more than a decade for wind-produced electricity to reach the 2000 MW mark. In 1999 the *US Department of Energy* started the *Wind Powering America Initiative Program* (WPA) to increase the use of clean wind energy in the United States spread over the next two decades. One of its main objectives was to meet 5% of the nation's energy needs with wind power by 2020. Thus a total capacity of 10,000 MW was to be installed by the end of the last millennium and 80,000 MW by 2020 (provided consumption increase remained steady, which it has).

The total installed wind generation in the United States already had by 2005 reached 9149 MW producing approximately 28,000 MWh of electricity, equivalent to the consumption of 2.3 million "average" US households. For 2005, wind power capacity increased by 2431 MW or 36%, and, the following year, installations topped 3000 MW. Of the 30 states with more than 100 MW of generating capacity by 2010, by 2005 only 15 had succeeded, with frontrunners California (2150 MW), Texas (1995 MW), Iowa (836 MW), Minnesota (744 MW), and Oklahoma (475 MW) (Figure 7.3).

7.3 UNITED STATES AND EUROPEAN UNION

The United States pioneered in the field of electricity production using wind turbines, with, until 1991, nearly 80% of the world wind power capacity in California. Since the early 1990s Europe has been catching up and by June 2000 Germany had already reached a production close to 5000 MW, or one-third of the world's total wind power capacity. Member States of the European Union are bound to respect the directives enunciated in the *Green Paper* of November 29, 2000, toward a European strategy for the security of energy supply and the *White Paper* laying down a community strategy and action plan for renewable energy resources. The objective was to increase the share of renewable energy resources in the entire European Union (EU) to a minimum penetration of 12% of gross energy consumption by 2010 and to 21% of electricity generation; in 1995 those levels were, respectively, 5.3% and 13.8%. Reference values were attributed to each Member State for the fixing of national indicative targets for electricity produced from renewable energy sources. For this purpose the different development potentials of each Member State were taken into account and, for instance, smaller countries, like

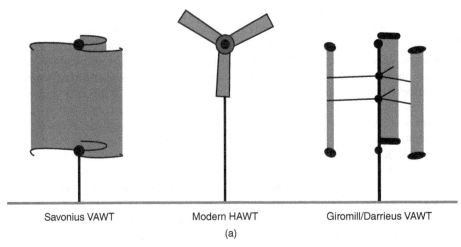

Savonius VAWT   Modern HAWT   Giromill/Darrieus VAWT

(a)

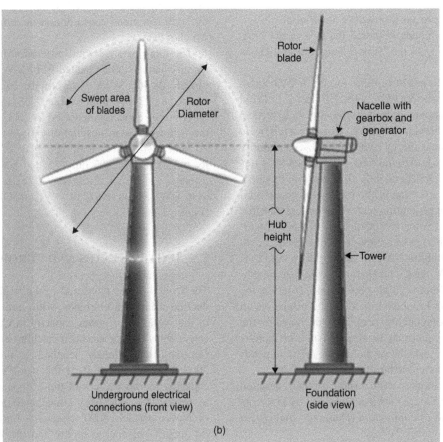

(b)

FIGURE 7.1   (a) Primary types of turbines. *Source*: Wikipedia Encyclopedia reproduced drawing. (b) Drawing of the rotor and blades of a wind turbine, courtesy of ESN 350 × 367. *Source*: European Wind Association (EWA).

Belgium and Luxembourg, had been given a target of only 6% as compared to Austria's 78%.

Those programs aimed to reduce energy dependence of the EU and to decrease CO_2 emission levels, at a time when renewable energy sources accounted only for about 6% of the Union's total internal energy consumption, and implied doubling the ratio. Besides the installation of solar thermal applications, small-scale hydropower stations, photovoltaic systems, and biomass installations, the action plan of the *White Paper* promotes the use of wind energy and the construction of large wind farms with a total capacity of 10,000 MW.

FIGURE 7.2   Prof. Poul la Court in his machine room.

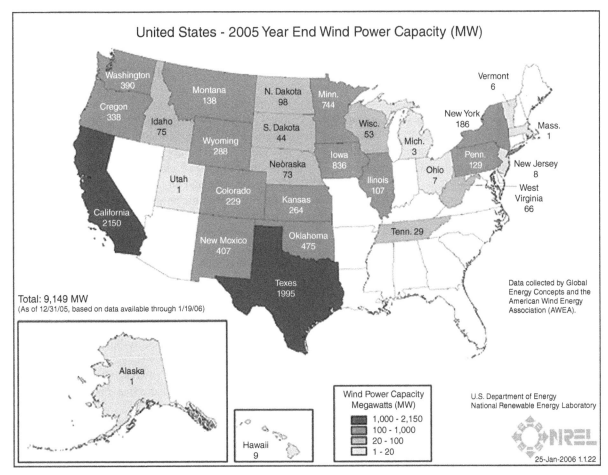

FIGURE 7.3   US wind power capacity in 2005. Data collected by Global Energy Concepts and American Wind Energy Association.

Europe soon led the market with more than 40,500 MW of installed capacity, providing nearly 3% of the EU's electricity consumption in an average "wind year," accounting thereby for about 69% of wind power generating capacity in the world (59,084 MW).

Germany soon ranked first with the highest number of wind turbines (17,592), producing nearly 18,500 MW of power, but Spain with 10,028 MW generates the most electricity from wind (20,236 GWh) and Denmark with 8.5% of its electricity produced by wind has by far the highest wind energy share in total electricity production; it intends to achieve about 50% by 2030 and already has the most important offshore wind capacity (423 MW).

Globally, 11,531 additional megawatts were installed in 2005 to 8207 MW in 2004, a 40.5% increment. The Republic of South Korea achieved with 233% the highest capacity growth, followed by Portugal (100%), Australia (86%), Norway (69%), and Ireland (58%). South Korea produces over 100 MW, Taiwan 87 MW, and the Philippines 25 MW. The Canadian capacity grew with 53% to reach a potential of 683 MW. In the 20 member countries of the *IEA*[1] *Wind Implementing Agreement*, representing 87% of worldwide wind generating capacity, the share of wind generation as percentage of the national electric demand increased to 1.2% in 2005 from a trifle 0.2% 10 years earlier.

With a capacity exceeding 4430 MW, India holds the fourth place among the top-10 wind generating countries on the heels of Germany, Spain, and the United States. Next are Denmark, Italy, the UK, and China (1260 MW) whose market was boosted in anticipation of the country's new *Renewable Energy Law*. Japan ranks tenth with a potential of 1078 MW; among the major producers ranks the Wakamatsu wind farm (*Kitakyushu*). Having a total installed potential of about 7000 MW in 2005, Asia is thus also meeting with a strong growth of 46%.

Expansion of wind energy harnessing has been modest in Latin America and Caribbean countries but a total wind energy capacity of 213 MW has been surpassed with Costa Rica, Caribbean, Brazil, Argentina, and Colombia having a potential ranging between 20 and 71 MW.

Australia is the biggest producer with 708 MW, "down under"; New Zealand has a potential of 169 KW, while some Pacific islands can produce up to 12 MW.

The African market shows promise. Egypt has a capacity of 145 MW and had planned, in mid-decade, to reach 850 MW of wind power by 2010. Morocco and Tunisia (20 MW) were aiming, respectively, for 64 MW and 20 MW. The first commercial wind farm in South Africa was sited near Darling on the west coast, about 75 km from Cape Town.

More expensive to build, less environmentally challenging, more vulnerable to the weather conditions, but also more

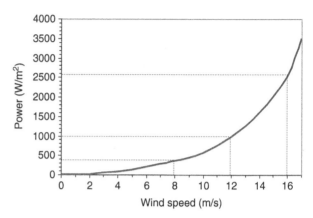

FIGURE 7.4 Ratio between wind speed and produced power per square meter.

productive, offshore wind farms have been the vogue for several years. Marine winds are on average much stronger than continental ones and an offshore location for wind turbines can result in more energy generated than onshore. This difference in intensity is even amplified since the quantity of energy produced by a wind turbine with regard to wind speed does not increase linearly but according to an exponential curve (Figure 7.4).

7.4 NOT IN MY BACKYARD

In some instances wind turbines were placed near population centra and people living close by started complaining about noise as a nuisance; obviously the towering structures were not free from environmental impact. If wind-produced electricity had fired the enthusiasm of the consumer, it did not please everyone and an attitude of "yes, I'm for it, but not in my backyard" prevailed in many locations. Best known is perhaps the Cape Cod (Massachusetts) project that was delayed for years by being held up in court litigation, similar to the offshore wind farm of Knokke-Heist, Belgium. Nevertheless both projects have been implemented (2012). Yet, environmental objections did not stem the determination of some countries to replace by wind, centrals burning coal, oil, or nuclear products. Locations on the coast became favored and better offshore sites were selected. Where installations involving a few wind turbines were first implanted, builders passed to sites where large numbers of turbines were installed. The *Utgrunden* (Sweden) "farm" in the Baltic was inaugurated in 2000 and mark as one of the first offshore "wind farms."

Proliferation of marine wind turbines affected especially the Northern and Western Europe, for example, Sweden, Denmark, Germany, Scotland, and the Netherlands (Figure 7.5). The wind turbines business exploded and pages of such periodicals as *Renewable Energy World* and *Renewable*

---

[1] International Energy Agency.

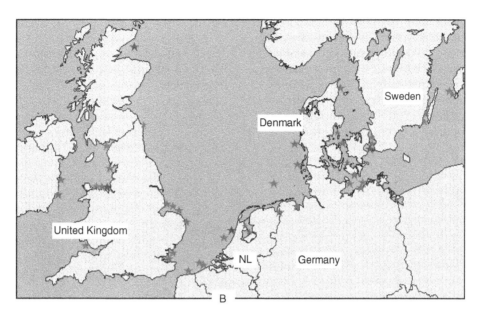

FIGURE 7.5   Map of existing and planned wind farms in Northwest Europe. Red * indicate already built wind turbines; purple * represent built small wind turbines; blue * those under construction in 2005 and grey * planned ones. *Source*: Offshore Wind Energy Europe.

and Sustainable Energy Review became replete with articles and papers on tapping marine energy to produce electrical power. Pylons have become taller and turbines larger.

7.5   MARKET "EXPLOSION" AND ENVIRONMENTAL OBJECTIONS

In 2005 the *IEA Wind countries* reported that more than 686 MW of the wind power capacity was located offshore. In 2004 it had been 578 MW, thus an increment of 19% in a single year had materialized. Denmark had the highest offshore installed wind capacity (423 MW), followed by the United Kingdom (213.8 MW), Ireland (25 MW), and Sweden (23 MW). Japan is trailing with an offshore capacity of 1.2 MW. Germany's expansion slowed down due to scarcity of adequate wind locations onshore and the few offshore wind farm projects. By the end of 2005, German federal authorities approved the construction of 11 pilot wind farms along the German coast consisting of 777 turbines and with a capacity of 3331 MW.

Objections to noise and ruined aesthetics continued unabated and, for instance, in Belgium plans for an offshore row of air turbines at 12 km (7.5 miles) from the coast, offshore Knokke, was held up for years because the view of the horizon would be spoiled by what in fact would look to the human eye at such distance as merely matchsticks. The project would have increased the wind power energy in Belgium from 12 to 613 MW. Knokke's economy being tourism centered took, for a while, precedence over power. Everyone applauded the harnessing of marine winds but nobody

wanted the pylons in his "backyard." Not even on an artificial island. Like for Cape Cod, good sense had the last word. A claim was made that by 2013 the largest wind farm in the world will have been established off the Belgian shore.

Unlike in Europe, no offshore wind farm had been erected by 2006 in the United States. But the success of the European approach to wind power proves to be a strong incentive. To alleviate the energy crisis by means of more powerful marine wind turbines so impressed the Americans that placing turbines on Georges Bank off the coast of Maine was considered. Another project, *Cape Wind*, sited on Horseshoe Shoal off the coast of Cape Cod (Massachusetts) is America's first big offshore wind farm. (130 offshore wind turbines, each with a capacity of 3.6 MW). The closest turbine of this controversial project will be over 5 miles (8.3 km) away from the beaches. Meanwhile, the country's first coastal wind farm became operational in December 2005 in New Jersey. The *Jersey-Atlantic Wind Farm* in Atlantic City consists of five 1.5-MW turbines. Other projects in Northern America are *Bluewater Wind* and *LIPA Offshore Wind Park* (both near Long Island), *Nai Kun* (Hecate Strait, British Columbia, Canada), and locations off the coast of Texas. Further plans and listings are given in Charlier (2011).

Virtually all megawatt-scale wind turbines use pitch control; the purposes of pitch control are to limit the maximum thrust and torque and to give controllability to stop the rotor safely in a fast flow without having to rely on a brake. Current trends are toward taller towers and higher turbine capacity and indeed towers reach 90 m and turbines 2.5 MW; heights of 100 m tubular steel towers and 8 MW turbines are being considered, motivated by efforts to lower infrastructure costs:

fewer towers with larger capacity are sought after in projects. Offshore minimum turbines are 5 MW sized. The American Wind Energy Association (AWEA) equates the 8500 MW out nine medium-sized nuclear power of wind power added in 2009 to 8.5 GW, the equivalent of about nine medium-sized nuclear power plants. The success of the wind industry does not obliterate the fact that it is still in the throes of solving numerous structural problems. (Charlier, 2011).

7.6 ENVIRONMENTAL IMPACT

A new environmental issue has arisen and has already put a halt in several countries to the implant of more air turbines. Indeed concern has been voiced that the machines may cause hecatombs of birds particularly during migration seasons. In his article "How many dead birds equal a dead fish equals an oil spill?" Robert L. Bradley, Jr. claimed that each megawatt of installed US capacity is the equivalent of 2.5 million bird deaths across the entire capacity of the United States. This extrapolation was, maybe, applicable for the first generation of fast rotating turbines like those of *Altamont Pass, CA*. Modern turbines have a much slower motion and do not represent a similar danger for migrating birds.

An answer to noise, migrating birds routes, aesthetics has perhaps been found, at least *in partim* siting of the marine wind turbines on floating—and movable—platforms. The design is ready and the construction on the books.

According to *Power Magazine* European offshore wind turbine capacity grew by 51% in 2010. There are currently 830 offshore wind turbines now installed and connected to the grid, totaling 2,063 MW in 39 wind farms in nine European countries. Some 17 offshore wind farms are under construction in Europe, totaling more than 3,500 MW, with just under half being constructed in UK waters. In addition, a further 52 offshore wind farms have won full consent in European waters, totaling more than 16,000 MW, with just over half of this capacity planned in Germany. And the growth rates are impressive: In 2009, the market grew by 54% compared to 2008. In 2010 the market grew by an even more considerable 75% compared to the previous year.

EWEA has a target of reaching 230 GW of wind power by 2020 which will include 40 GW of offshore power. This is a challenging but very feasible goal. By 2030, just ten years later, we envisage some 150 GW of offshore wind power. If all offshore wind projects in their various stages of planning are added up, there are already some 100 GW of offshore wind projects in addition to existing farms. If these become fully functioning wind farms, they would produce 10% of the EU's electricity while also avoiding 200 million tonnes of CO_2 emissions each year.

Europe must now use the coming decade to prepare for the large-scale exploitation of its biggest indigenous energy resource – offshore wind – overcoming the seemingly significant obstacles – including underwater electricity grids and cables, building the harbors and barges capable of facilitating the installation of offshore wind farms – in the path of its development.

7.7 CONCLUSIONS

The major scientific and technological problems related to the further development of wind energy can be summarized as follows: the cost of wind-generated power is strongly related to turbine size. Increasing energy capture can be obtained by improving the design, the manufacturing material (lighter and stronger), and performance of wind turbines. Longer durability and operational life will contribute to system efficiency.

It will be difficult to exploit wind energy as stand-alone energy generation system. The lack of grid connection is a very important restrictive factor for a larger penetration of wind farms in isolated areas and offshore. In 2004, the European Union financed two electrolysis pilot units supplied by wind power generators, one in Greece, the other on the Canary Islands.

Marine and land winds have a role beyond contributing to energy supply: reduction of CO_2 emission in the atmosphere and its impact on global warming. Furthermore wind energy will improve the security of energy supply, reduce hydrocarbon price volatility, and even stimulate economic development (job creation, industrial development, research, etc.). Stimulating offshore wind energy is an agent in solving the problem of hindrances that hamper the full exploitation of onshore locations.

Recent statistical reports, news releases, and press announcements indicate a continued expansion of the wind energy industry and particularly great progress in the offshore area; currently 830 offshore wind turbines are now installed and connected to the grid, totaling 2063 MW in 39 wind farms in nine European countries. If in 2009, the market grew by 54% compared to 2008, in 2010 growth reached 75% compared to the previous year. EWEA has a 230 GW target for 2020 to include 40 GW of offshore power. By 2030, 150 GW of offshore wind power should be captured. Totaling this up, it means some 100 GW of offshore wind projects in addition to existing farms, 10% of the EU's electricity, and 200 million tons of CO_2 less in annual emissions. Europe must in the upcoming decade overcome significant obstacles, among which are underwater electricity grids and cables, building harbors and barges capable of installing offshore wind farms.

The "Request for Interest" issued recently by the Obama administration launched a process that hopefully may lead to up to 4000 MW of wind energy installed far off US shores.

This would probably be sufficient electricity to power approximately 1.7 million households and enough to take this US industry from its current "adolescence" to maturity.

7.8 MARINE ENERGY2

Pike Research published in the second quarter of 2009 that "The high capital costs associated with renewable energy technologies is largely avoided with marine kinetics. The capital costs of marine renewable energy systems will be 50 to 100 times smaller than investments required to create the same amount of electricity from either wind or solar." Gulf Stream Turbines LLC in Fort Lauderdale has a patented submersible turbine, similar to a wind turbine in price but is 4–10 times as effective as wind or solar in its capacity factor and has no site preparation, has no CO_2 emissions, is invisible, causes no harm to marine life, operates at 85% capacity factors equal to fossil fuel plants but no fuel costs and each turbine has a carbon offset of 10,000 tons of CO_2 each year. If FPL Group realizes that 10,000 turbines can supply the state with 8.94 GW of electricity and at the Euro Carbon Offset futures price of January 2011 is US $30 a ton, that is, US $3 billion in CO_2 offsets alone not counting the revenues of electricity or hydrogen that can be realized by 2012—2013. Rowen Negrin, President of Gulf Stream Turbines LLC, would like to begin a dialogue with FPL Group to explore bringing this technology to market. He can be contacted at rowen@gulfstreamturbines.com or 954 907 2254 or visit www.gulfstreamturbines.com.

REFERENCES

[1] Bradley, R.Jr. (1997) *Renewable Energy – Why Renewable Energy Is Not Cheap and Not Green*. National Center for Policy Analysis.
[2] European Commission. (2005) European Wind Energy at the Dawn of the 21st Century. EUR 21351.
[3] European Commission. (2001) Activities of the European Union, summaries of legislation. *Renewable Energy: White Paper Laying Down a Community Strategy and Action Plan*.
[4] International Energy Agency. (2006) IEA Wind Energy Annual Report 2005. RCN: 26041.
[5] Thys, A. (2003b) *De uitbating van alternatieve energie- en mineraalbronnen uit de oceanen*. Vrije Universiteit Brussel, Brussels.
[6] Thys, A. (2003a) Capturing marine wind energy: juridical and political complexities in Belgium within the framework of the European Union. *Proceedings of PIM XXX Convocation*, Kiev, Ukraine.
[7] US Department of Energy. (2006) Energy efficiency and renewable energy. *Wind & Hydropower Technologies Program*.
[8] Charlier, R. H. (2011). Calling upon Neptune: ocean energies as renewables. In V.Bacescu and R. B.Cathcart (eds), *Macroengineering Seawater in Unique Environments. Arid Lowlands and Water Bodies*. Springer, Heidelberg-Berlin, p. 597.

8

IMPACTS OF WIND FARMS ON WEATHER AND CLIMATE AT LOCAL AND GLOBAL SCALES

JUSTIN J. TRAITEUR AND SOMNATH BAIDYA ROY

With the projected increase in the size and number of large utility-scale wind farms, their interaction with the atmosphere is of great importance for the responsible implementation of these power generation systems. This interaction can have significant effects on weather and climate at local, regional, and global scales and therefore must be studied in order to make wind power a viable solution to the climate change problem. Wind farms interact with the atmosphere because they act as both a sink of the wind's kinetic energy and a source of turbulence induced by the rotor blades. Numerical modeling studies have hypothesized that vertical mixing induced by the rotor blades can cause changes in the vertical structure of the atmosphere downwind (Baidya Roy et al., 2004). This hypothesis has subsequently been validated using field data from a wind farm. These data show that the presence of the wind farm can generate a warming or cooling signal of several degree Celsius for areas in and downwind of the wind farm. The sign of the temperature change was strongly dependent on the lower atmospheric thermodynamic stability, that is, whether temperature increases or decreases with height (Baidya Roy and Traiteur, 2010).

A large number of numerical modeling studies have pointed to the possibility of climatic changes due to large wind farms at the local, regional, and global scale. (Keith et al., 2004; Barrie and Kirk-Davidoff, 2010; Kirk-Davidoff and Keith, 2008; Wang and Prinn, 2010) While these climatic changes are generally smaller in comparison to those induced by an increase in green house gases, they should be given no less concern. Knowledge of these possible changes is critical to being responsible implementers of this exciting and promising technology. This overview aims at summarizing recent studies on the possible atmospheric impacts large utility-scale wind farms can have on climate and weather.

8.1 OBSERVED IMPACTS

While data on wind speed and turbulence in and around wind farm sites are generally available in the public domain, information on other meteorological variables is not. Baidya Roy and Traiteur (2010) is the only study that uses observations to explore the possible impacts of wind farms. The data for this study were obtained from a meteorological field campaign conducted at a wind farm in San Gorgonio, California, in the summer months of 1989. In this field campaign, temperature data were collected at both an upwind and downwind meteorological tower site. The wind farm consisted of 23-m-tall turbines with 8.5-m-long rotor blades arranged in 41 rows that were spaced 120 m apart. This wind farm is of a smaller scale, with respect to area and turbine height, than most currently implemented wind farms but the results and conclusions can be applied to those as well.

Data from the field campaign show that during the night and early morning hours near-surface air temperatures downwind of the wind farm are warmer than upwind regions. During most of the day this signal is reversed (Figure 8.1). These results are statistically significant according to the results of a Mann–Whitney Rank Sum Test (Baidya Roy and Traiteur, 2010). Data from the nearby Edwards Air Force Base showed that this area is statically stable at night and unstable during the day. Therefore, the wind farm generally exhibits a warming effect on near-surface air temperatures

Alternative Energy and Shale Gas Encyclopedia, First Edition. Edited by Jay H. Lehr and Jack Keeley.
© 2016 John Wiley & Sons, Inc. Published 2016 by John Wiley & Sons, Inc.

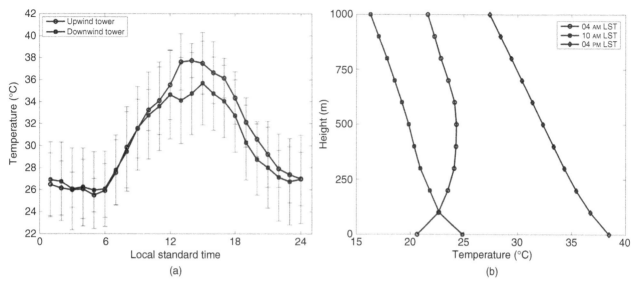

FIGURE 8.1 Observed temperatures in and near the San Gorgonio wind farm: (a) near-surface air temperature patterns at the San Gorgonio wind farm during the field campaign, (b) vertical profiles of air temperature at Edwards Air Force Base during June–August 1989.

during the night when the environment is stable and a cooling effect during the day when the environment is unstable.

8.2 HOW WIND TURBINES INTERACT WITH THE ATMOSPHERE

A wind turbine works by extracting kinetic energy from the wind and converting that to electrical energy. This extraction of energy occurs mainly in the lower 1 km of the atmosphere where the turbines are located (Sta Maria and Jacobson, 2009). The amount of energy extracted from the flow depends on the wind speed and the size and performance of the turbine. A byproduct of this energy extraction is the introduction of mechanically generated turbulence from the rotor blades. Apart from affecting profiles of temperature and moisture, this turbulence can have negative effects on power production and cause structural fatigue for subsequent turbines located downwind.

When a parcel of air approaches a wind turbine, the air converges and causes the parcel's velocity to decrease and pressure to increase. At the rotor blade, a sharp gradient in air pressure occurs along the blades and causes the rotors to turn (Gómez-Elvira et al., 2005). Initially, small-scale turbulence is generated from small eddies formed from the rotors' circulation and strong pressure gradient along the blades. These small eddies then propagate downstream with the wind. The region behind the rotor experiences a reduction in wind speed with the largest reduction at hub-height (Magnusson and Smedman, 1999). A decrease of over 50% in wind speed commonly occurs directly behind the turbine (Porté-Agel et al., 2011). This region of reduced wind speed

is commonly known as a wake and its fundamental characteristics are determined by the environmental atmospheric stability, change of the wind with height (shear), and hub-height wind speed (Barthelmie and Jensen, 2010).

Downstream of the rotor, the wake expands and pressures in this region increase. This is due to turbulent mixing between the slower wind speeds within the wake and faster wind speeds outside of the wake. This is the main cause of pressure increases in the wake and results in the circular ring to expand outwards. Additional turbulent mixing results in the wake velocity to increase until it eventually comes into equilibrium with the ambient flow (Gómez-Elvira et al., 2005).

The end of the near-wake region is usually found between two to five rotor diameters downstream. After the near-wake region a transition region is prevalent before leading into the far-wake region (Gómez-Elvira et al., 2005). Numerical modeling studies have suggested that the far-wake region can extend upwards of 20 rotor diameters downwind. If subsequent turbines are located in the far-wake region of an upwind turbine, a reduction in wind speed and increase in turbulence can occur (Porté-Agel et al., 2011). Since power production is proportional to the cube of wind speed a small decrease in wind can yield a large decrease in power production for wake-encompassed turbines. In addition, an increase in turbulence incident on a wind turbine can cause damage to the rotors for highly turbulent environments (Hand et al., 2003). The outcome of such a situation is decreased power production and increased maintenance costs.

Outside of the observational study done in San Gorgonio, California, most studies on the interaction of wind farms with the environment stem from numerical modeling studies. As stated earlier, wind farms act as an elevated sink of energy in the flow and source of turbulence. As air approaches a large utility-scale wind farm it slows due to the added friction of the turbines and extraction of energy from the flow. In most cases this decrease in wind velocity at hub-height is compensated for by small-scale transfer of momentum from layers above and below that level. This increased turbulence is usually produced by mechanically induced turbulence from the wind turbines and the shear between the wake and ambient flow (Peterson et al., 1998). This mechanically induced turbulence then propagates downwind with the wind and can influence vertical profiles of temperature and moisture and surface fluxes of sensible and latent heat. This may also change the location of the formation of clouds and affect surface temperatures downwind (Wang and Prinn, 2010). This topic will be discussed in more detail later in Section 8.4.

8.3 HOW WIND FARMS ARE REPRESENTED IN WEATHER AND CLIMATE MODELS

Numerical weather prediction (NWP) and climate models solve a set of three-dimensional, compressible, non-hydrostatic, dynamic and thermodynamic equations and a set of microphysics equations to predict how the atmosphere will evolve through time. The models are initialized by observations taken in situ and remotely. NWP models generally have horizontal resolutions ranging from 1 to 40 km while climate models typically are run at a resolution of 100–200 km. Atmospheric processes that are smaller than the model resolution are represented as subgrid-scale parameterizations. Examples of commonly parameterized atmospheric processes are small-scale turbulence, cumulus convection, solar (shortwave) radiation, and infrared (longwave) radiation just to name a few (Dudhia, 1989; Kain and Fritsch, 1993; Mlawer et al., 1997). A process or a phenomenon can be explicitly resolved by an NWP model if it is at least as large as two model grid points. Since current turbine blades are usually between 100 and 125 m in diameter, numerical models would have to have a horizontal resolution between 50 and 60 m to be able to resolve these features. Currently available computing power is not strong enough to run NWP or climate models at such high resolution. Hence, wind turbines can be represented in climate and NWP models only as subgrid parameterizations. The simplest and most common approach to parameterizing the aerodynamic effects of wind farms is to increase surface roughness (Frandsen, 1992; Crespo et al., 1999; Keith et al., 2004). This makes perfect sense because wind farms are sources of added friction and hence act like a surface drag on the flow. In some cases, researchers modify the surface drag coefficients by modifying the roughness length or by implementing an explicit drag scheme in the model. This drag is implemented in the lowest model layers where wind farms typically reside (Keith et al.,

2004). Another method is to use model parameterizations that are similar to wind turbines. For example, by changing the height coefficients of grass, shrubs, or trees up to a typical level of rotor height, numerical modelers can simulate the drag effects of wind farms on the lower model-level wind speeds (Wang and Prinn, 2010).

Baidya Roy et al. (2004) and Baidya Roy and Traiteur (2010) used an alternative approach to model the atmospheric processes induced by the presence of a wind farm. They assumed that since a rotor extracts energy from the flow and generates turbulence in its wake, the rotor can be assumed to be an elevated massless sink of resolved kinetic energy and a source of turbulent kinetic energy (TKE). This is more realistic than the roughness length-based approach because the drag is not at the surface but at the hub-height of the turbine. A major drawback of such a parameterization is that it assumes a constant coefficient of performance. The coefficient of performance is essentially the amount of energy extracted from the flow compared to the amount of energy within the flow. Since the coefficient of performance (C_p) of a wind turbine is a nonlinear function of wind speed, this parameterization scheme is unlikely to produce an extremely accurate estimate of the impacts of wind farms. However, Baidya Roy (2011) has resolved this problem by making the C_p a function of wind speed based on data from the Gamesa G80-2.0 MW turbine.

8.4 IMPACTS OF WIND FARMS ON LOCAL METEOROLOGY

Recently, interest has been growing on the possible impacts wind farms could have on changes in local weather and climate. Observations and computer simulations show that wind farms can significantly change the vertical profiles and surface–atmosphere fluxes of momentum, moisture, and energy. These effects can have substantial impacts on local agriculture, economy, and society in general. Therefore, it is important to provide both a qualitative and a quantitative assessment of these impacts for the responsible implementation of this technology.

8.4.1 Study 1: Baidya Roy et al. (2004)

Baidya Roy et al. (2004) used the Regional Atmospheric Modeling System (RAMS) (Pielke et al., 1992; Cotton et al., 2003) to simulate the local meteorological impacts of a hypothetical wind farm in Oklahoma during the summer of 1995. The hypothetical wind farm consisted of a 100 × 100 array of wind turbines equally spaced 1 km apart. Each turbine was 100 m tall with 100 m diameter rotor blades. The RAMS model was run for 15 days starting on July 1, 1995, at 1200 UTC (Universal Time). The modeled domain consisted of three nested grids with the finest having a horizontal

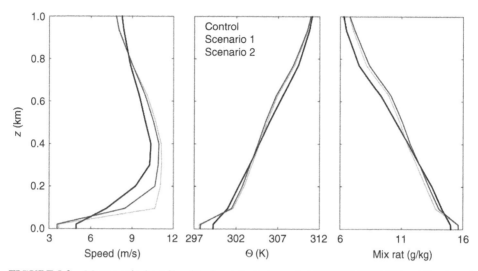

FIGURE 8.2   Mean vertical profile of horizontal wind speed at 0900 UT (0300 LT) and θ and total water-mixing ratio at 1200 UT (0600 LT) over the wind farm.

resolution of 2 km. The weather during this time period varied from strong precipitation events early in the study period to drier conditions for the remainder of the period. The research consisted of a control simulation and two experiment scenarios. Scenario 1 parameterized the turbines as just a sink of kinetic energy from the flow while scenario 2 parameterized the turbine as both a sink of kinetic energy and a source of TKE.

The control run has the highest hub-height wind speeds due to the absence of added friction of the wind farms. Scenario 2 wind speeds are the lowest due to the added friction of the wind farm and the additional transformation of kinetic energy to TKE in the flow. Differences in wind speeds between the scenarios are larger in magnitude in the early morning hours (0300 LT) due to the presence of the nocturnal low-level jet (Bonner, 1968) (Figure 8.2). The nocturnal low-level jet is a fast-moving ribbon of air usually found above a nocturnal capping inversion over the Great Plains regions (Stull, 1988). It is quite common for this jet to flow at the same level as hub-height with weaker flow above and weaker surface winds below. Figure 8.2 shows the vertical profile of wind speed spatially averaged over the wind farm and temporally averaged over the 0300 local time outputs for each day. The presence of the low-level jet is obvious with much stronger winds approximately 100 m above the surface. In scenario 1, a weakening of the flow is evident up to 400 m with the largest weakening of the flow at hub-height. In scenario 2 the low-level jet is further weakened due to the added transformation of kinetic energy from the flow to TKE. Interestingly, scenario 2 shows faster surface winds and winds around 1 km in height. This phenomenon is caused by the added turbulence that mixes faster moving air within the nocturnal low-level jet to levels both above and below hub-height.

Figure 8.2 also shows the vertical profile of potential temperature over the wind farm averaged at 0300 LT. Potential temperature (θ) is the temperature a parcel of air would be if the parcel was moved, and therefore compressed, adiabatically down to 1000 hPa. The reason potential temperature is used instead of temperature is that the impacts of pressure on temperature are eliminated. In Figure 8.2 there is little difference between the control run and scenario 1 at this time. Interestingly, scenario 2 displays a warming signal in the early morning hours compared to scenario 1. During the early morning hours the boundary layer generally exhibits a strong stable stratification (temperature increases with height). Eddies generated by the added turbulence parameterization in scenario 2 causes an increase in vertical mixing, which brings warmer air down to the surface and cooler air up. This leads to a net warming of the surface and cooling above hub-height. This signal is not seen during the daytime because strong surface heating causes the boundary layer to be generally well mixed. In some instances where the atmosphere is unstable, θ decreases with height, a net cooling occurs due to the mixing of cooler air downward to the surface and warmer air upwards from the surface. Scenario 2 also impacts the total water-mixing ratio as seen in Figure 8.2. With the implementation of scenario 2, a drying occurs near the surface and is strongest during the early morning hours of the dry period. Again turbulent eddies mix drier air from hub-height downwards to the surface and moister air upwards.

With the change in surface temperature and moisture induced by the presence of the wind farm, subsequent changes in surfaces fluxes of heat and moisture will also occur. A reduction in the sensible heat flux occurs in scenario 2 on the order of several tens of watts per square meter. During the early morning hours the ground is usually cooler than the air above. This setup equates to a negative thermal ground–atmosphere gradient. The increase of near-surface θ induced by the rotor-generated mixing causes this gradient to become even more negative. This leads to more heat being transferred into the ground. In addition, the drying of the near-surface air in scenario 2 causes the positive land–atmosphere moisture gradient to increase. This increase in the land–atmosphere moisture gradient causes an increase in surface evaporation and net drying of the surface (Baidya Roy et al., 2004).

8.4.2 Study 2: Baidya Roy and Traiteur (2010)

In this study, Baidya Roy and Traiteur (2010) again used the RAMS to simulate the impacts of wind farms on surface temperatures. This is the first and only study to use observations to evaluate the performance of numerical models in simulating the effects of wind farms on temperature. The simulated wind farm was a 7 × 3 array of 100-m-tall wind turbines with 100 m diameter rotor blades. Each turbine was spaced 10 rotor diameters apart in both the north–south and the east–west directions. These turbines were again represented as both an elevated sink of kinetic energy and a source of TKE. The model was initialized with atmospheric sounding data for February 1, May 1, August 1, and November 1, 2009, from 21 rawinsonde stations in the western United States. A pair of 1-hour-long simulations was conducted with each sounding. The pair of simulations included a control case with no wind turbine and a wind farm case with the aforementioned wind turbine parameterization. Most importantly, the results of the simulations were consistent with observations taken from the meteorological field campaign conducted in San Gorgonio wind farm from June 18 to August 9, 1989.

Figure 8.3 shows a scatter plot of the relationship between the 0 and 300 m θ lapse rate and the change in the near-surface air temperature. The results were similar to the results of Baidya Roy et al. (2004) where the turbines caused a warming under positive lapse rates, that is, θ increases with height, and cooling under negative lapse rates. These results match closely to the observations taken in and near the San Gorgonio wind farm. During the early morning hours the presence of the wind farm causes a net warming downwind while during the day the wind farm causes a net cooling downwind (Figure 8.1). In addition to the relationship of the θ lapse rate to changes in near-surface air temperature, hub-height wind speed also plays a critical role. The change in surface air temperature downwind peaks at wind speeds of approximately 12 m/s at hub-height and becomes approximately zero for wind speed less than 2 m/s or greater than 20 m/s. At speeds less than 2 m/s, the turbine is usually below the cut-in speed, so they do not turn. At or above 20 m/s the rotors stop working in order to protect the structural integrity of the turbine. In addition, wind speeds above 20 m/s usually possess a high amount of ambient environmental turbulence resulting

in large amounts of vertical mixing. Figure 8.3c shows the relation of ambient environmental TKE with the change in surface air temperature. The largest change in temperature near the surface usually occurs when ambient TKE is small. In other words, if the environment is highly turbulent the atmospheric boundary layer is generally well mixed and any additional mixing by mechanically generated turbulence by the rotor blades will have a minimal impact on surface air temperatures.

Atmospheric turbulence is a strong indicator of the surface kinetic energy dissipation rate. The surface kinetic energy dissipation rate is just the rate at which small-scale turbulence breaks down to molecular thermal energy or heat. Figure 8.3d shows the relationship between surface kinetic energy dissipation rate and change in surface air temperature. When the environmental surface kinetic energy dissipation rate is weak the impact of the rotors on surface air temperature is large and vice versa. At dissipation rates larger than approximately 2.7 W/m² the impact of rotors starts to decrease and becomes almost zero at rates larger than 6.7 W/m². Therefore, placing turbines in areas where kinetic energy dissipation rates are larger than 2.7 W/m² can minimize the environmental impact of the wind farm.

8.4.3 Study 3: Baidya Roy (2011)

Baidya Roy (2011) investigated the impacts of wind farms on surface–atmosphere moisture and energy fluxes. The RAMS was used again to simulate a 7 × 3 array of 100-m-tall turbines with 80 m diameter rotor blades spaced 1 km apart. The rotors were again parameterized as an elevated sink of kinetic energy and a source of TKE. A total of 306 simulations were conducted and again each was initialized with atmospheric sounding data from 21 rawinsonde stations in the western United States. The model configuration was exactly similar to Baidya Roy and Traiteur (2010) except for one fundamental difference. In this study, the C_p of wind turbines was expressed as a function of wind speed using data from a Gamesa G80-2.0 MW turbine. Thus the wind turbine parameterization employed in this study can be considered to be more realistic than the previous two papers.

Atmospheric stability was estimated from the equivalent potential temperature (θ_e) lapse between 0 and 350 m (similar to θ with an added moisture term). The relatively shallow layer was considered because it is the most likely layer to be impacted by the presence of a wind farm. Similar to Baidya Roy and Traiteur (2010), the wind farm exhibits a cooling under negative θ_e lapse rates and warming under positive θ_e lapse rates. Lapse rates of total water-mixing ratio (R) were also calculated during this study. A positive R lapse rate means humidity increases with height. In this situation turbulent mixing mixes relatively moist air down and relatively dry air up. This leads to a moistening of the near surface. If the R lapse rate is negative, mixing by the rotors will mix

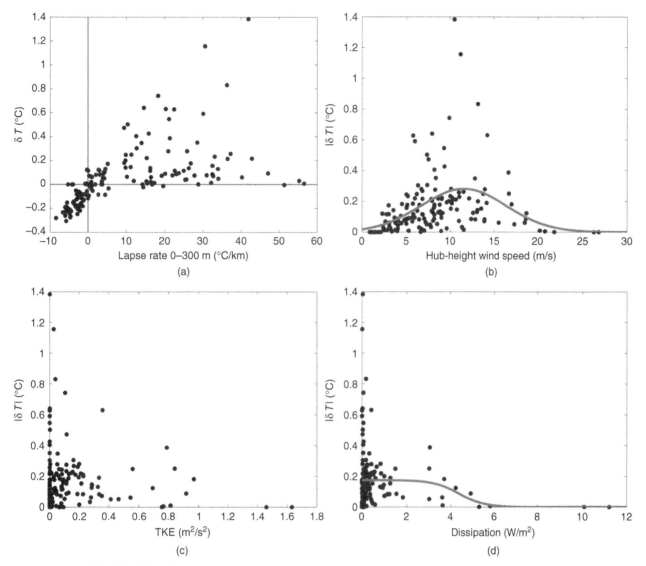

FIGURE 8.3   Simulated change in near-surface air temperatures within the wind farm plotted as a function of (a) 0–300 potential temperature lapse rate at the beginning of the simulations, (b) mean background hub-height (100 m) wind speed, (c) mean background lower ABL (0–300 m) TKE, and (d) mean background surface KE dissipation rate. The variables plotted on the abscissa are from the control simulations, whereas the variables on the ordinate are the difference between the control and the wind farm simulations. The temperature change within the wind farm, wind speed, TKE, and surface KE dissipation rates are averaged over the entire 1-hour-long simulation period for the 21 grid cells containing the turbines.

dry air down and moist air up. This will lead to a drying near the surface. Changes in surface temperature or moisture also have large impacts on surface sensible and latent heat fluxes. The temperature and moisture gradient between the ground and overlying air affects the magnitude of either the sensible or latent heat flux. For example, in early morning hours the ground is relatively cool and the overlying atmosphere stable. Mixing from the turbine causes warmer and drier air to be mixed down to the surface. The temperature gradient between the ground and overlying atmosphere will

now become more negative, which will decrease the magnitude of the sensible heat flux. This results in the warming of the ground and cooling of the overlying air. The latent heat flux will increase due to relatively moist surface air being displaced by dry air from above. In this situation evaporation from the surface will cause the ground to cool.

A sensitivity study was also conducted regarding the relationship between the length of the wind farm and the downwind location where the impacts are largest. Five simulations were conducted with wind farm lengths of 5, 10, 15,

20, and 25 km compared to the previous 7 km length. The 12 UTC August 1, 2009, sounding from Glasgow, Montana, and the 12 UTC November 1, 2008, sounding from Quillayute, Washington, initialized the simulations. Both soundings had similar θ_e lapse rates but the Glasgow sounding exhibited mean hub-height wind speeds twice as strong as the Quillayute sounding. For the Glasgow case, the downwind distance where the largest impacts are reached occurs around 15 km from the upstream edge of the wind farm. As the length of the wind farm increases past 15 km the magnitude of the impacts remained relatively constant. For the Quillayute case, the peak in impact magnitude occurs for a wind farm length of 10 km and remains relatively constant as the length of the wind farm increases. This result points to the dependence of the turbine hub-height wind speed on surface impacts downwind. Larger hub-height wind speeds mean that the peak impacts of the wind farm will be felt farther downwind compared to weaker hub-height wind speeds.

8.5   IMPACTS OF WIND FARMS ON REGIONAL AND GLOBAL CLIMATE

Wind farms can also have impacts on weather and climate far removed from their geographical location. These impacts are caused by the interaction of nonlinear atmospheric processes that can substantially change regional and/or global climate.

8.5.1   Study 4: Keith et al. (2004)

This study explored the possible impacts of the implementation of three approximately continent-sized wind farms on regional and global climate. Two general circulation models (GCMs), the National Center for Atmospheric Research (NCAR) GCM and Geophysical Fluid Dynamics Laboratory (GFDL) GCM, were utilized in order to validate their findings. To simulate the added drag of the wind farm, two parameterization schemes were employed. The first used perturbations to the drag coefficients over the installation regions. These perturbed drag coefficients were calculated by changing the surface roughness lengths over the wind farm areas. The other parameterization used an explicit drag scheme added to the model physics package in the two lowest model layers. This was used to represent a spatial density of 2.8 turbines per square kilometer with each having 100 m hub-heights and 100 m rotor diameters. The NCAR and GFDL GCMs used model horizontal resolutions of 2.8° × 2.8° and 2.0° ×y 2.5°, respectively, and diagnosed sea-surface temperature using climatology.

Results pointed to a near-negligible impact on globally average mean surface air temperatures but some wind farm installation regions experienced temperature changes of ±2°C in certain seasons (Figure 8.4). In winter, a cooling/warming signal is shown for the Europe/central North America regions, respectively. In winter, a stronger poleward temperature gradient and stronger mid-latitude storms cause stronger temperature advection in the Northern Hemisphere (Min et al., 1982; McCabe et al., 2001). The magnitude of temperature advection depends on the strength of the temperature gradient multiplied by the wind speed perpendicular to this gradient. Since wind farms typically slow surface layer winds, by their added friction, temperature advection will also decrease. Therefore, the temperature signal over the installation regions is likely due to the change of the poleward transport of heat in that location. It is important to note that unlike the previous three studies, the wind farm parameterization did not include the effect of added turbulence induced by the rotor blades.

Atmospheric general circulations patterns were also changed by the implementation of these large wind farms. With a 10% increase of surface drag over land areas, mid-latitude average wind speeds decreased, which caused the jet stream to shift poleward. This northward displacement of the location of the jet stream causes storms to track closer to the polar regions. Subsequently, this causes the high latitudes to cool and low latitudes to warm. Since the largest observed temperature increase caused by increases in CO_2 has been in polar regions, cooling of high latitudes will tend to reduce this response.

8.5.2   Study 5: Kirk-Davidoff and Keith (2008)

Kirk-Davidoff and Keith (2008) looked at the dynamical atmospheric processes that are responsible for changes in surface temperature, wind, and cloud fraction. The authors first employed a simplified aquaplanet GCM with isolated high-surface roughness regions to represent the added drag of a large utility-scale wind farm. The aquaplanet GCM was used to simplify the wind farm's impacts on the atmosphere without the influences of complex terrain seen in other GCMs. This simplified model gave valuable insight into the interpretation of the results obtained from a more realistic representation of Earth's surface using the NCAR CAM3.1 GCM. The wind farms were parameterized by perturbations in surface roughness in central North America, Europe, and East Asia, similar to Keith et al. (2004).

With the implementation of the three wind farm installation regions, results on surface air temperatures were quite similar to the results of Keith et al. (2004). In North America, regions in and near to the installation experienced strong warming upwards of 0.5°C (Figure 8.5). In Europe a slight cooling is seen in the installation region with strong cooling northeast and moderate warming to the southeast. Similarly, East Asia experiences slight cooling in the installation region with stronger cooling to the northeast, while warming to the southeast was limited due to sea-surface temperatures being held fixed in the GCM. Also, polar regions exhibited a slight cooling response (0.15°C) due to the

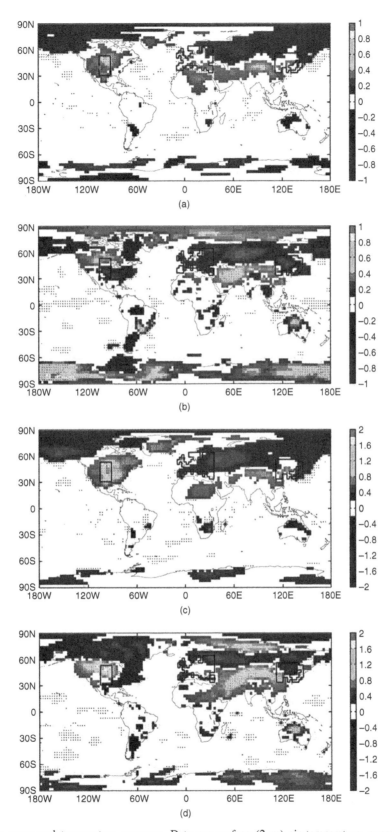

FIGURE 8.4 Wind farm array and temperature response. Data are surface (2 m) air temperature, experiment minus control. Drag perturbation, δCD, was 0.005 over the wind farm array outlined in black. Points that are significant at $P > 0.9$ by using a binary t test on annual/seasonal means are indicated ($\times$). NCAR data are 37 years of perturbed run composed of two runs with differing initial conditions and 108 years of control composed of five independent runs. GFDL perturbed and control runs are both 20 years long. NCAR (a) and GFDL (b) annual means are given, as well as NCAR (c) and GFDL (d) winter (December–February) means.

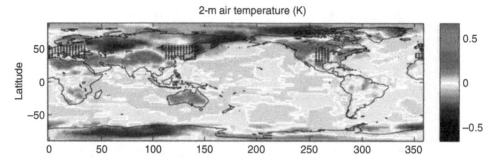

2-m air temperature (K)

FIGURE 8.5   Full-model 2-m air temperature anomalies. Grid points with the * indicate locations of increased surface roughness intended to simulate wind turbine installations.

decreased meridional flux of energy poleward induced by the surface roughness increase. This meridional energy flux is controlled by the placement, track, and strength of storms in the mid- and high latitudes. This change in general circulation is directly correlated to wind velocity changes through a large depth of the atmosphere (Figure 8.6a–8.6d). Near the surface, wind speeds in the installation regions are decreased due to the added drag imposed by the surface roughness increase. This slowing of the wind in the installation region causes an area of convergence upwind and divergence downwind. Not only are wind anomalies in the north–south and east–west directions limited to near the surface but extend several kilometers in height. This points to changes in atmospheric general circulation caused by the surface roughness perturbations.

Examining changes in surface heat fluxes gives more insight into the different signs of surface temperature changes brought on by the areas of surface roughness increase. North America and East Asia show similarities in the reduction of the downward flux of solar radiation, upward flux of surface infrared radiation and sensible heat, and an increase in latent heat flux and the meridional wind component despite having oppositely signed temperature anomalies (Figure 8.7a–8.7d). Therefore, for changes in surface temperature to be oppositely signed in these regions the magnitudes of the flux changes must be different.

The reasons why some areas portray a warming signal while others do not are due to a complex interaction between different atmospheric processes. Starting at the surface, by increasing surface roughness, surface wind speeds inside the installation regions decrease but energy fluxes from the surface increase due to the added turbulence in the flow. By changing surface wind speed, possible changes to the winds aloft can occur. Changing either the speed or direction in the winds at the surface and aloft can affect the sign and magnitude of temperature advection in and downstream of the installation regions. Again, this results from the change in poleward energy transport. Subsequently, if warm air advection is occurring then temperatures aloft will

warm. This will tend to decrease the magnitude of upwelling infrared radiation from the surface that escapes to space. In addition, warm air advection will also decrease sensible heat fluxes from the surface due to decreasing difference between the temperature of the ground and overlying air. The latent heat flux will increase due to increased evaporation from the displacement of cooler air by warmer air. Warm air advection is commonly responsible for cloud formation. Changing the amount of clouds, especially low clouds, can decrease upwelling infrared radiation and downwelling solar radiation. Decreasing upwelling infrared radiation results in a net warming of the surface while decreasing downwelling solar radiation results in a net cooling of the surface. Also, by slowing winds inside the installation regions, air is allowed to converge in the upwind areas and diverge in the downwind areas. Since convergence generally results in upward vertical motion, clouds can form as a result. The sign and magnitude of the temperature response at the location of interest is dependent on the interaction of these complex atmospheric processes.

8.5.3   Study 6: Wang and Prinn (2010)

In this study, Wang and Prinn (2010) used the Community Climate Model Version 3 GCM of the US NCAR (Collins et al., 2006). This model is a fully coupled atmosphere–ocean–land model with a mixed layer ocean option and is used to represent the impact of using wind turbines to meet 10% or more of global energy demand in 2100. The model used a horizontal grid spacing of 2.8° × 2.8° and contained 18 discrete vertical levels. A 7-member model ensemble was developed with each individual member run for 60 years. Four members used different parameterization schemes to represent the wind turbine effects over land, two to represent installation of wind farms in ocean depths shallower than 200 m between 60°S and 74°N and a control. Similar to previous studies, wind turbine parameterizations over land included changing the model surface roughness and/or displacement height coefficients. Over the coastal installation areas, adding

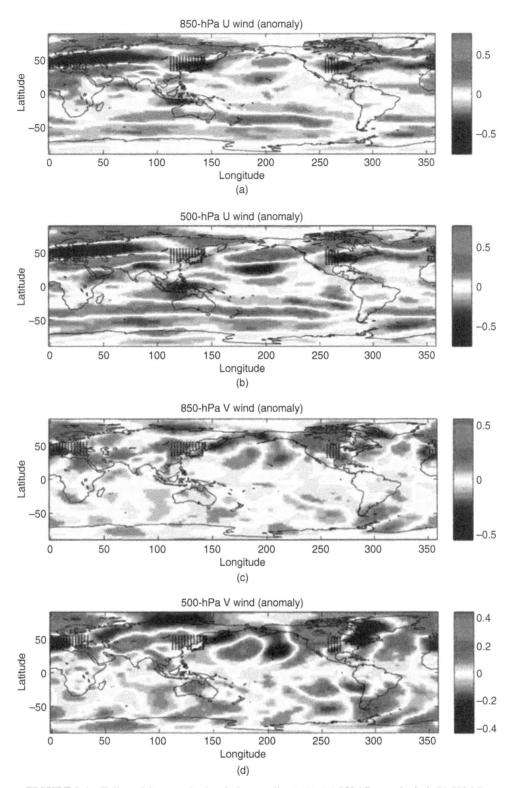

FIGURE 8.6   Full-model tropospheric wind anomalies (m/s): (a) 850-hPa zonal wind, (b) 500-hPa zonal wind, (c) 850-hPa meridional wind, (d) 500-hPa meridional wind.

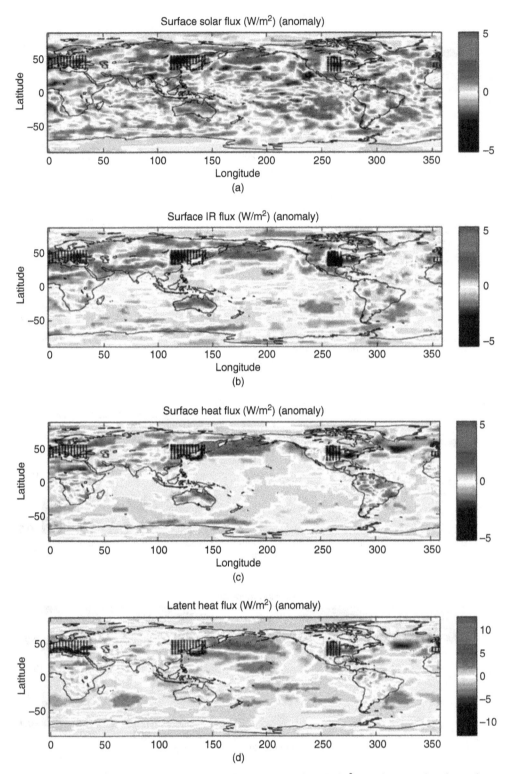

FIGURE 8.7   Full-model results for the surface energy budget (W/m^2): (a) downward surface solar radiative flux, (b) upward surface IR radiative flux, (c) upward sensible heat flux, (d) upward latent heat flux.

an additional surface drag component to the ocean surface parameterized the added friction imposed by the presence of the wind farm. The land-based wind farm was primarily located in regions covered by grass and shrubs and covered an area approximately 58 million km². In addition, coastal wind farm installations totaled over 10 million km².

As seen in previous studies, changes of the surface properties brought on by the presence of the wind farm increase surface friction and slow local surface winds. A moderate change in surface roughness over the installations produces an average surface temperature increase of 1°C over the land-based installation regions (Run L). This member also forecasts a global average increase in surface temperature of approximately 0.15°C. Figure 8.8b shows the temperature change in the lowest model layer (~30 m) globally due to land-based installations. Over land areas a warming signal is observed near the surface while this is opposite over the ocean areas, especially the polar areas. Similar to findings of Baidya Roy et al. (2004), the largest reduction in wind speeds occur at hub-height and tends to reduce the near-surface vertical turbulent transport of energy and therefore warms the surface. This warming also tends to spread beyond the location of the wind farm installations. These effects are likely due to dynamic atmospheric systems such as Rossby wave-train formations (Kirk-Davidoff and Keith, 2008). The existence of these Rossby waves affects global distributions of cloud cover and precipitation. Convective precipitation, in general, is reduced in the Northern Hemisphere and enhanced in the Southern Hemisphere. This is due to the shifting of the Hadley circulation due to changes in surface roughness induced by the presence of the wind turbines (Kirk-Davidoff and Keith, 2008). Changes in the large-scale precipitation are also evident. Mid-latitude cyclones are predominately responsible for this precipitation type and any changes are a result of the change in intensity or mean storm track.

Coastal-based wind farms also affect temperatures in and downwind of the installation regions (Figure 8.8c). Generally, the ocean is much smoother than most land areas due to the absence of vegetation and terrain. This equates to a lower surface roughness length value and subsequent level of ambient turbulence. A substantial increase in this surface roughness length over coastal installation areas will generate much stronger turbulence which enhances ocean–atmosphere heat fluxes. The increase in the ocean–atmosphere heat flux tends to cool the ocean surface over the installation regions. Similar to the land-based runs, temperature changes also occur in areas far removed from the installation regions.

8.5.4   Study 7: Barrie and Kirk-Davidoff (2010)

Barrie and Kirk-Davidoff (2010) used the NCAR Community Atmosphere Model 3.0 (CAM 3.0) (Collins et al., 2006) to simulate the impact of a continent-sized wind farm on global weather patterns. The wind farm stretched from the central United States to south central Canada and occupied 23% of the total North American landmass. Surface roughness was diagnosed by converting an unused Plant Functional Type (PFT), which is used to parameterize the effects of vegetation on land surface characteristics in CAM 3.0, into a wind farm subtype. This hybrid wind farm PFT has a canopy height of 156 m, roughness length of 3.35 m, and a displacement height of 0 m. The wind farm experiment used a roughness length that was reduced by 83% to simulate the minimal drag of a wind turbine profile, where the face of the turbine is turned away from the wind. This is consistent with Vermeer (2003) that a single roughness length value can be used to parameterize an array of wind turbines. Seventy-two 1-month-long case studies were created in branch mode over a 6-year study period holding sea-surface temperatures fixed. This was done in order to simulate the impacts from a sudden decrease in surface roughness caused by the turning of the wind turbine away from the wind.

Figure 8.9 highlights the mean differences between the experiment runs and control runs for the westerly wind field in the lowest model level (approximately 80 m in elevation). The area demarcated by the rectangular box is where the hypothetical wind farm resides. In the wind farm domain, a slowing of the westerly winds is present while winds in northern Canada and western Europe exhibit acceleration in the westerly winds. The largest change in wind and temperature during the first few days following a decrease in magnitude of the surface roughness, when the rotors are turned parallel to the wind direction, was observed primarily within the wind farm. The magnitude and scale of these changes were related to the weather over the region at that time. Over time the anomalies propagated downstream where they eventually reached the North Atlantic. Due to the nonlinearity of the interaction of the anomalies with the North Atlantic storm track, the anomalies grew in magnitude until they exceeded the magnitude of the direct impact of the wind farm. At $t + 5$ days the effect of the wind farm is most significant in the North Atlantic. After 1 week the impacts reach the North Pacific.

The impact of the continent-sized wind farm also produces large instabilities downstream a few days after the roughness length perturbation. Four and half days after the roughness length perturbation, an atmospheric wave-train structure is evident downstream of the hypothetical wind farm. This wave structure is directly related to changes in the baroclinic and barotropic structure of the atmosphere induced by the presence of the wind farm. In one case an anomalously strong upper level high-pressure system is present in the central North Atlantic 4.5 days after the roughness length perturbation. This is clear signal that large-scale wind farms can affect global weather patterns far removed from their geographical location.

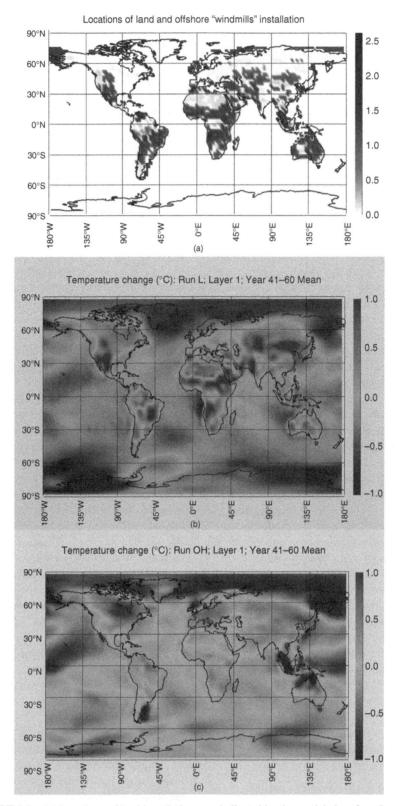

FIGURE 8.8   (a) Locations of land installations are indicated by the modeled surface drag coefficient (nondimensional) averaged over the final 20 years of the 60-year Run L. The drag coefficients
have been scaled by a factor or 1000. Also shown are the locations of the offshore installation regions
where the ocean depth is shallower than 200 m. (b) Temperature changes (Run L minus reference) in
the lowest model layer resulting from large-scale deployment of wind turbines over land sufficient to
generate 158 EJ/year of electrical power. (c) Same as (b) but for run OH (coastal installations).

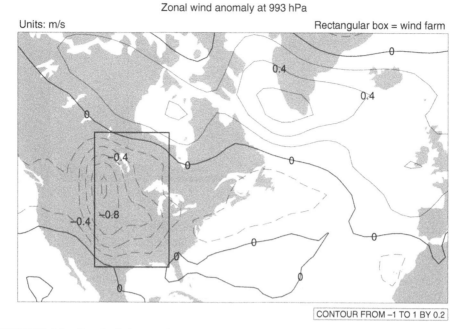

FIGURE 8.9 Zonal wind anomaly at 993 hPa. The mean difference in the eastward wind in the lowest model level between the control and perturbed model runs highlights regions of atmospheric modification. The wind farm is located within the rectangular box over the central United States and central Canada. Areas of the wind farm located over water are masked out during the model runs.

8.5.5 New Study Presented at the 2011 American Meteorological Society Annual Conference

A new study was presented at the 2011 AMS Annual Conference that represents the state of the art in wind farm impact assessment. It is important to remember that this study has not been peer-reviewed and their results/conclusions are subject to change.

A study led by David Sherman from the University of Oklahoma researched the effect that a large utility-scale wind farm would have on changes in weather systems and their associated precipitation patterns. The study focused on the eastern two-thirds of the United States during the warm season (May 1 to August 31). The hypothetical wind farm stretched from the Texas panhandle to northern Nebraska and encompasses 228,375 Bonus 2.0 MW wind turbines. This installation's capacity (450 GW) would be able to meet the US Department of Energy's (DOE's) goal to derive 20% of US electricity demand through wind power by 2030. The simulation was run for 62 years in order to establish the warm season precipitation climatology for the studied domain. Results from the study show that for an individual year, large, but random, changes in precipitation events occurred due to the presence of the wind farm. These changes are caused by either a shifting or strengthening of individual convective systems. When these precipitation anomalies were averaged out over 62 years there appeared to be little change in the precipitation climatology over the domain. This suggests that large utility-scale wind farms can randomly affect precipitation patterns downstream for a given year but do not affect the large-scale precipitation climatology (Sherman, 2011).

8.6 MINIMIZING IMPACTS

As shown, wind farms can have large impacts on surface temperature, moisture, energy fluxes, and atmospheric general circulation patterns which can affect weather and climate on local, regional, and global scales. Therefore, any attempt to minimize these impacts should be investigated. Baidya Roy and Traiteur (2010) devised an approach to minimize the effect of wind farms on changes in surface air temperatures. Two solutions were developed, designing rotors that produce less turbulence and placing wind farms in moderately turbulent environments.

The impact of wind farms on surface temperatures depends on the turbulence generated by the turbines. Turbines that generate less turbulence tend to generate lower impacts (Figure 8.3c). Low-turbulence turbines also have another advantage. Turbines consume kinetic energy to generate turbulence. Low-turbulence rotors consume less kinetic energy, leaving more energy for downwind rotors. Thus, this technology can be a win–win solution that maximizes power generation while minimizing impacts.

Another option to minimizing the impact of wind farms would be to place wind farms in region where background

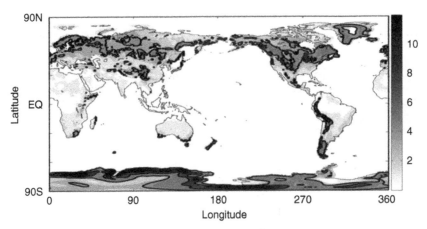

FIGURE 8.10   Mean surface KE dissipation rate (W/m²) for the 1979–2004 period. Regions demarcated by the black line (3 W/m² contour) are ideal for low-impact wind farms.

turbulence is high. The surface kinetic energy dissipation rate is directly related to the environmental turbulence. This is because turbulence breaks down large-scale flow into successively smaller and smaller eddies. Figure 8.10 shows the surface KE dissipation rates averaged over the 1979–2004 period calculated from the JRA25 data set (Onogi et al., 2007). Impacts of wind farms decrease as surface kinetic energy dissipation rates become larger than 2.7 W/m² (Figure 8.3d). These regions of high-surface kinetic energy dissipation rates are areas where the installation of wind farms will have a low impact on surface air temperatures. Regions with high topography such as the Himalayas, Rockies, and Andes have high-surface kinetic energy dissipation rates but are inaccessible to wind energy companies. Fortunately, large areas of the Great Plains, northern Europe, and north/central Asia have relatively high-surface kinetic energy dissipation rates and are ideal locations for minimizing the impacts of wind farms.

8.7   CONCLUSIONS AND DISCUSSIONS

This review presents several pioneering papers describing the local and global impacts of implementing large utility-scale wind farms. The projected accelerated growth of the wind power industry will need studies like these to lead the way in the responsible execution of providing clean, renewable, and inexpensive energy for the global economy.

Wind farms change the vertical structure of the atmosphere, impacting the sign and magnitude of surface energy fluxes. These fluxes have large impacts on surface temperature and moisture. In many cases, these changes tend to produce surface warming and drying in and near the installation region. This can have advantages and disadvantages for the agriculture industry since low-level warming can limit the formation of damaging frosts while drying can increase the chances of crop failure.

Globally, wind farms have the possibility of changing weather patterns and climate far removed from their geographical location. The impact of a large utility-scale wind farm on global average temperatures appears to be negligible, but in certain regions seasonal average temperatures can change several degree Celsius. Wind farms add friction to the flow, which slows mid-latitude wind speeds. Slowing of winds in the mid-latitudes causes the location of the jet stream and subsequent storm track to shift poleward. This can impact the magnitude of large-scale temperature advection affecting the formation of clouds and precipitation.

Fortunately, knowledge on how wind farms interact with the atmosphere has resulted in the development of strategies to minimize their impact. Designing wind turbines that produce less mechanically generated turbulence will reduce changes in surface temperature and moisture. In addition, placing wind farms in areas with relatively high amounts of environmental turbulence reduces their impacts even more. Knowledge of the full scope of these impacts is essential to being responsible advocates of this promising technology.

REFERENCES

Baidya Roy, S. (2011). Simulating impacts of wind farms on local hydrometeorology. *J. Wind Eng. Ind. Aerodyn.,* doi: 10.1016/j.jweia.2010.12.013.

Baidya Roy, S., Traiteur, J. (2010). Impacts of wind farms of surface air temperatures. *P. Natl. Acad. Sci.,* 107:17899–17904.

Baidya Roy, S., Pacala, S. W., Waldo, R. L. (2004). Can large wind farms affect local meteorology. *J. Geophys. Res. Atmos.,* 109:D19101.

Barrie, D. B., Kirk-Davidoff, D. B. (2010). Weather response to a large wind turbine array. *Atmos. Chem. Phys.,* 10:769–775.

Berthelmie, R. J., Jensen, L. E. (2010). Evaluation of power losses due to wind turbine wakes at the Nysted offshore wind farm. *Wind Energy,* 13:573–586.

Bonner, W. D. (1968). Climatology of the low level jet. *Mon. Weather Rev.*, 96:833–850.

Collins, W. D., Bitz, C. M., Blackmon, M. L., et al. (2006). The community climate system model version 3 (CCSM3). *J. Clim.*, 19:2122–2143.

Cotton, W., et al. (2003). RAMS 2001: current status and future directions. *Meteorol. Atmos. Phys.*, 82:5–29.

Crespo, H. G. (1999). Survey of modeling methods for wind turbine wakes and wind farms. *Wind Energy*, 2:1–24.

Dudhia, J. (1989). Numerical study of convection observed during the winter monsoon experiment using a mesoscale two-dimensional model. *J. Atmos. Sci.*, 46:3077–3107.

Frendsen, S. (1992). On the wind speed reduction in the center of cluster of wind turbines. *J. Wind Eng. Ind. Aerodyn.*, 39:251–256.

Gómez-Elvira, R., Crespo, A., Migoya, E., Manuel, F., Hernández, J. (2005). Anisotropy of turbulence in wind turbine wakes. *J. Wind Eng.*, 93:797–814.

Hand, M. M., Robinson, M. C., Kelley, N. D., Balas, M. J. (2003). Identification of wind turbine response to turbulent inflow structures. *Proceedings of the 4th ASME JSME Joint Fluids Engineering Conference*, Renewable Energy Symposium, July 6–10, Honolulu, Hawaii, ASME, New York, NY. NREL/CP-500-33465.

Kain, J., Fritsch, M. (1993). *Convective parameterization for mesoscale models: the Kain-Fritsch scheme. The Representation of Cumulus Convection in Numerical Models*. Meteor. Monogr. No. 24. American Meteorological Society, Boston, pp. 165–170.

Keith, D. W., DeCarolis, J. F., Denkenberger, D. C., Lenschow, D. H., Malyshev, S. L., Pacala, S. N., Rasch, P. J. (2004). The influence of large-scale wind power on global climate. *P. Natl. Acad. Sci.*, 101:16115–16120.

Kirk-Davidoff, D. B., Keith, D. (2008). On the climate impact of surface roughness anomalies. *J. Atmos. Sci.*, 85:2215–2234.

Magnusson, M., Smedman, A.-S. (1999). Air flow behind wind turbines, *J. Wind Eng. Ind. Aerodyn.*, 80:169–189.

McCabe, G. J., Clark, M. P., Serreze, M. C. (2001). Trends in Northern Hemisphere surface cyclone frequency and intensity. *J. Clim.*, 14:2763–2768.

Min, K. D., Horn, H. L. (1982). Available potential energy in the northern hemisphere during the FGGE year. *Tellus*, 34:526–539.

Mlawer, E. J., Taubman, S. J., Brown, P. D., Iacono, M. J., Clough, S. A. (1997). Radiative transfer for inhomogeneous atmospheres: RRTM, a validated k-correlated model for the long-wave. *J. Geophys. Res.*, 102, 16663–16682.

Onogi, K., et al. (2007). The JRA25 reanalyses. *J. Meteorol. Soc. Jpn*, 85, 369–432.

Peterson, E. L., Mortensen, N. G., Landberg, L., Hojstrup J., Frank, H. P. (1998). Wind power meteorology. Part 1: climate and turbulence. *Wind Energy*, 1:25–45.

Pielke, R. A., Cotton, W. R., Walko, R. L., Tremback, C. J., Nicholls, M. E., Moran, M. D., Wesley, D. A., Lee, T. J., Copeland, J. H. (1992). A comprehensive meteorological modeling system – RAMS. *Meteorol. Atmos. Phys.*, 49:69–91.

Porté-Agel, F., Wu, Y. T., Hao, L., Conzemius, R. J. (2011). Large-eddy simulation of atmospheric boundary layer flow through turbines and wind farms. *J Wind Eng. Ind. Aerodyn.*, 99:154–168.

Sherman, D. B. (2011). A WRF simulation of the effect of a large wind farm on 40 years of precipitation in the eastern United States. Oral presentation at the 91st American Meteorological Society Annual Meeting, January 2011, Seattle, WA.

Sta Maria, M. R. V., Jacobson, M. Z. (2009). Investigating the effect of large wind farms on energy in the atmosphere. *Energies*, 2:816–838.

Stull, R. B. (1988). *An Introduction to Boundary Layer Meteorology*. Kluwer Academic, Norwell, MA, 666 pp.

Vermeer, L. J., Sorensen, J. N., Crespo, A. (2003). Wind turbine wake aerodynamics. *Prog. Aerosp. Sci.*, 39:467–510.

Wang, C., Prinn, R. G. (2010). Potential climatic impacts and reliability of very large-scale wind farms. *Atmos. Chem. Phys.*, 10:2053–2061.

9

POWER CURVES AND TURBULENT FLOW CHARACTERISTICS OF VERTICAL AXIS WIND TURBINES

KEVIN POPE AND GREG F. NATERER

9.1 RESIDENTIAL AND SMALL BUSINESS WIND POWER

Combustion of vast quantities of fossil fuels worldwide for power production is responsible for numerous environmental problems. The emissions from hydrocarbon combustion contain pollution agents including NOx, SOx, CO and CO2. These chemicals are connected to a variety of environmental degradation problems, including acid rain, smog, and climate change. There is an urgent need to alter the current consumption and production patterns that rely on fossil fuels. Wind power can provide a sustainable contribution to society's energy needs, while reducing our needs for fossil fuels.

Significant evidence exists that the environmental impact of climate change is rapidly gaining momentum and its alleviation has paramount importance. Since the combustion of fossil fuels, and the subsequent release of carbon dioxide (CO_2) emissions, is a primary cause of climate change, reducing these polluting emissions is crucial to sustainability. Driven by its emission-free operation, the capacity of wind power has expanded rapidly to make a significant contribution to global electricity generation.

Despite rapid developments in wind turbine technology, both in terms of installed capacity and power output, less than 5% of global energy generation can be attributed to wind power [1]. Several different aspects of wind turbine development can be promoted to advance growth by reducing costs and encouraging public acceptance. Currently, the vast majority of installed wind power capacity is from commercial wind farms consisting of several large wind turbines.

However, a significant growth potential exists from small installations distributed among various residential and small business locations.

Increasing wind power utilization will mitigate CO_2 emissions by reducing the reliance on fossil fuel energy production and the subsequent releases that come from the combustion of these fuels. Due to the large environmental costs associated with these pollutants, many concerns have risen about the use of coal. Currently, wind power in city centers is limited to a few isolated installations, and a significant potential exists for CO_2 mitigation from wind power in cities. Limiting the use of fossil fuels will reduce the CO_2 emitted during power generation.

Many deterrents from wind power generation can be avoided with small wind turbines installed at or near the location of power usage (on-site). With many small installations distributed throughout residential and small business locations, the variable output of electrical generation is also distributed throughout many locations, lessening the disconnect of supply and demand, while limiting transmission, installation, and land costs [2–6]. Remote wind turbine operation requires the construction, installation, and maintenance of appreciable amounts of auxiliary systems, reducing the economic and environmental benefits of the installation.

Residential and small business wind (RSBW) power installations offer opportunities for clean power generation in conditions where typical, large turbine systems are ineffective. Typical wind turbine designs do not operate well in regions of high turbulence, such as residential and

Alternative Energy and Shale Gas Encyclopedia, First Edition. Edited by Jay H. Lehr and Jack Keeley.
© 2016 John Wiley & Sons, Inc. Published 2016 by John Wiley & Sons, Inc.

business areas [7]. The yaw system and airfoil designs cannot compensate for the highly variable wind direction and intensity, caused by the interaction between the urban structures and incoming wind. Fluctuations in wind direction can cause excessive power fluctuations, thereby decreasing the stability, fatigue strength, and electrical stability of a wind turbine, which increases fatigue fluctuations on the shaft, reduces longevity, and increases maintenance costs [8].

The turbulence associated with urban structures diverting wind streams was experimentally investigated by placing various structures in a wind tunnel [9]. The tests demonstrated two-scale turbulence behavior at the top of the canopy region, suggesting that separated shear layers, intermingled with large-scale eddies, generate small-scale and unpredictable turbulence. Small-scale turbulence offers severe operational challenges for a wind turbine because it causes highly variable wind direction and intensity. Thus, urban installations, especially if mounted on an urban rooftop, offer difficult operating demands for a small wind turbine. In other wind tunnel experiments [10], the flows in the canyon-like structures developed from several buildings built on both sides of an urban street (urban canyons), which demonstrated that with other obstructions included on the streets, the turbulence increases by 50% to 200%, compared to an urban canyon with no street obstructions. These wind conditions provide harsh demands for wind power systems and can cause many turbine designs to be ineffective for efficient power production.

Unstable and turbulent winds can further deteriorate in residential and commercial centers as urban structures radiate heat and divert wind. Thermal gradients are developed, causing highly unstable and unpredictable flow fields [11]. Furthermore, during high wind speeds and cold weather conditions, the fluctuations in power output are intensified [12].

Obstacles of operating a wind turbine in turbulent conditions are closely related to the start-up requirements of a wind turbine. With a small installation, low cost and maintenance restrictions usually require a turbine with self start-up operation, as opposed to rotating the rotor to the required rotational velocity with a motor. Typically, urban areas have reduced average wind speeds compared to remote locations, which can cause a typical wind turbine to remain stationary for 21% of the year [13]. At start-up, the low relative velocity between the rotating blades and oncoming wind require an increased pitch of the blades to provide an optimal lift from the airfoils. However, without variable pitch blades, the angle of attack of the rotor blades at start-up does not coincide with the low rotational speeds, significantly reducing the rotational forces, which can cause an appreciable delay in the rotor's acceleration. Wright and Wood determined that turbines begin rotating at 4.6 m/s on average, but this varied between 2.5 and 7.0 m/s [14].

9.2 SMALL WIND TURBINE DESIGNS

RSBW power provides promising potential to supplement current wind power growth; however, typical turbine designs often cannot sufficiently provide efficient power production to justify the cost of installation and maintenance. Many municipal and national legislatures are providing attractive incentives for RSBW development [15], though, it is essential that the incentives provide sensible alternatives for effective and efficient power production. With the plethora of turbine designs currently available, wind turbine installations should be carefully selected to provide the best compromise between complexity (cost) and power production (capacity factor).

A wide variety of wind turbines are available for RSBW installations providing various benefits and drawbacks. Typically, wind turbines are categorized by their axis of rotation, that is, horizontal or vertical axis wind turbines (VAWTs and HAWTs). This design feature is important for mechanical properties, operational requirements, and design issues. There are differences in structural dynamics, control systems, maintenance, manufacturing, and electrical equipment.

HAWTs provide the highest efficiencies and represent the majority of current installed wind power capacity. The relative motion of turbine rotor rotation and wind speed generate a rotational force as the wind flows over the rotating airfoils. The major divisions of this category can be represented by variable or fixed pitch blades and active or passive yaw mechanisms. Without variable pitch blades, the approach angle (i.e., the angle of the leading edge of the rotor airfoil exposed to the approaching wind) is optimized to provide maximum lift for a particular range of wind speeds, limiting the turbine's effectiveness at other wind speeds, particularly during start-up. Active pitch blades can be controlled so the pitch is optimized throughout various wind speeds, but this increases the cost and complexity of the system. Without an active yaw mechanism, the turbine blades rotate behind the support tower, which reduces the turbine's efficiency. Whereas, an active yaw mechanism directs the turbine into the wind, allowing the blades to rotate in front of the support tower, improving performance (and increasing cost and complexity). The drag on a HAWT's blades can reduce the torque and increase fatigue stresses on the rotating shaft, although it acts perpendicular to the rotational direction [8]. This popular design is generally ineffective in turbulent winds, as they require stable wind direction and velocity for efficient operation. For this reason, typical wind farms are located in remote, high wind locations.

VAWTs are commonly categorized by their primary rotation force, that is, lift or drag designs. Although many VAWTs utilize both lift and drag forces, in different amounts at various locations of rotation, lift designs can generally be identified by the use of airfoils in the rotor geometry. Typically, a lift force wind turbine can achieve higher efficiencies compared to a drag force design. A lift VAWT has vertical airfoils, either curved (Darrieus) or straight (H-rotor) blades, and are designed to provide consistent lift force in the direction of rotation, throughout the rotation, by considering the relative speed achieved between the wind and rotor rotation.

Designs that utilize a drag force typically capture rotational forces that are highly variable during each rotation, as the angle between the rotor blades and incoming wind changes. Although drag VAWTs typically have reduced maximum efficiencies, compared to lift turbines, the simple construction, low cost, self-starting, and quiet operation are among the many benefits of the design. Typically, a drag turbine is modular and can be designed in stages, that is, two turbines are stacked, one on top of the other, and generally set at varying angles to stagger the peaks in force produced in each stage, thereby providing an overall power output with reduced variability. The magnitude and direction of lift and drag forces can be highly variable, and the magnitude of the lift force is significantly reduced because effective airfoil geometries are not used. The tangential velocity of the rotor blades cannot exceed the speed of the wind (i.e., tip speed ratio (TSR) $\not> 1$), because the blade would be pushing the wind, as a result the power output is limited.

A Savonius wind turbine is a simple and effective drag VAWT. This design consists of opposing and offset concaved blades. A gap between the blades reduces the back pressure on the returning blade and improves efficiency. This design has exhibited competitive efficiencies for a drag wind turbine. The results of a series of 16 wind tunnel experiments [16] investigating Savonius turbines with two or three blades, two or three stages (levels), and semicircular or twisted blades suggest the best design includes two stages and two twisted rotor blades. A maximum power coefficient of 0.32 was achieved. The result was validated by a series of wind tunnel tests [17], whereby Savonius turbines were compared by studying the effect of rotor overlap. A maximum power coefficient of 0.37 was achieved. This compares well with published data for power coefficient of a small HAWT. For example, a micro HAWT with a rotor diameter of 0.5 m was tested for an intended purpose of urban use [18] and achieved a power coefficient of 0.36.

Current researchers are investigating the performance and design of various hybrid VAWTs [19,20], whereby the properties of a drag and lift turbine are combined to exploit the benefits of each design. Recent tests of a hybrid VAWT included Savonius (drag) rotors attached to the same shaft as H-rotor (lift) rotors and achieved a maximum power coefficient of 0.51 [19]. When the wind intensity fluctuates rapidly, a hybrid VAWT can perform better by achieving a higher capacity factor (the fraction that the turbine operates at full capacity), by utilizing the start-up properties of the drag turbine to reduce start-up speed, while utilizing the lift rotors to achieve higher TSR (and higher efficiencies in strong winds) [21].

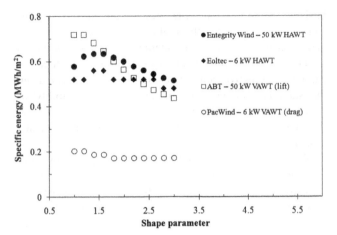

FIGURE 9.1 Specific energy with changes in wind distribution for various wind turbines.

Figure 9.1 compares the specific energy distributions of four different wind turbines, for 1 year. This figure uses wind data from RETSCREEN software [22], based on monthly average wind speeds for Toronto, Canada. A Weibull distribution, equation (9.1) [23], is assumed for the wind velocity distribution:

$$p(U) = \frac{k_w}{c}\left(\frac{U}{c}\right)^{k_w-1}\exp\left[-\left(\frac{U}{c}\right)^{k_w}\right] \quad (9.1)$$

In this equation, k_w represents the shape parameter. A low value of k_w represents wind conditions that can be characterized by an exponential decay function, where a high value of k_w represents conditions that approach a normal distribution. The variable c is the scaling factor, determined analytically by [20]

$$c = \frac{\bar{U}}{\Gamma(1+1/k)} \quad (9.2)$$

The commercial turbines in the plot are the Entegrity Wind Systems (50 kW HAWT) [24], Eoltec (6 kW HAWT) [25], Air Breeze Technologies (ABT) (50 kW H-type VAWT) [26], and the PacWind (6 kW Savonius VAWT) [27]. The characteristic curve that is unique to each turbine, including cut-in and cut-out limits, along with the calculated wind probability functions, is used to calculate the specific power coefficients [28]. The ABT VAWT displays a common feature among Darrieus turbines, whereby they have a lowered cut-in speed compared to HAWTs. These VAWTs have particular advantages when the wind speed distributions have a low shape factor. Such sites coincide with wind speed distributions that approach an exponential decay function. At such conditions, a VAWT can achieve a significantly higher capacity factor than a HAWT. This opposes the high values of k_w that approach a normal distribution of wind velocities about the mean wind speed, where the HAWT design is favored. There is a relatively low efficiency shown by the drag-type turbines in Figure 9.1. The advantages of drag VAWTs need to be better utilized, while the disadvantages need to be further reduced for them to make a significant contribution to the wind energy market.

VAWTs offer several advantages over HAWTs for RSBW, including eliminating the need for a yaw mechanism, pitch regulation, and complicated gearbox, limiting the number of mechanical parts, as well as the complexity, thereby reducing maintenance and installation expenses [6]. VAWTs offer better start-up capabilities and greater flexibility in design and suitability for installation and operating conditions.

9.3 DEVELOPING A VARIABLE-GEOMETRY POWER CORRELATION FOR A VERTICAL AXIS WIND TURBINE

In this section, a dimensional analysis of VAWT performance is presented, in terms of operational characteristics and geometrical properties. While developing a small wind turbine, many independent design variables will affect the turbine performance. Accurately predicting the effect of various changes to a design can be difficult and cumbersome. Experimental prototypes and CFD simulations can be effective techniques to measure the effect on performance of various changes to a turbine's design. However, attempting this technique to bring a turbine from initial conceptualization to commercial production can need an unmanageable number of different designs. Building a specific experimental prototype or CFD simulation for every possible design alternative can be costly and resource intensive. Developing a design correlation can significantly reduce the required number of prototype turbines or CFD simulations needed to effectively develop an efficient small VAWT design by directing designers to the most promising design parameters for a particular set of operating demands.

The power coefficient of a wind turbine, in terms of the TSR, can be represented by the following general form:

$$C_p = a(\text{TSR})^2 + b(\text{TSR}) + c \qquad (9.3)$$

where a maximum C_p is achieved at a unique TSR for each turbine design at a particular set of operating conditions. This maximum is consistent, regardless of wind speed. Typically, complicated flow fields and nonlinear behavior are difficult to predict and can cause undesirable operational effects. In practice, the power coefficient depends on additional design and flow variables which are highly interconnected. Experimental data combined with dimensionless variables can be used to develop convenient correlations that summarize the main trends resulting from complicated flow fields. For example,

experimental data of droplet flow impingement on a wind turbine airfoil was shown to collapse consistently onto a single normalized correlation of $Nu/Pr^{1/3}$ at varying Prandtl numbers [29]. Furthermore, the functional form of the Hilpert correlation can effectively accommodate measured data for a wind turbine airfoil over a range of Reynolds numbers [30]. This chapter describes a method whereby additional geometrical variables are combined with C_p and TSR to produce a power coefficient correlation that reduces the complexity of analyses and helps facilitate practical design and development of small VAWTs [31].

Considering a general VAWT design (with stators), nine key variables can be identified to represent performance for different geometries and operating conditions. The variables include the power output (Φ), rotor velocity (Ω), rotor radius (R), freestream velocity (V), air density (ρ), turbine height (H), turbine width (W), stator spacing (σ), and stator angle (θ). The power output depends on these variables as follows: $\Phi = f(\Omega, V, \rho, R, H, W, \sigma, \theta)$. Dimensions of these variables can be represented by the base dimensions of mass (M), length (L), and time (T), as follows: $\Phi \equiv M^0 \cdot L^2 \cdot T^{-3}$, $\Omega \equiv M^0 \cdot L^0 \cdot T^{-1}$, $\rho \equiv M^1 \cdot L^{-3} \cdot T^0$, $V \equiv M^0 \cdot L^1 \cdot T^{-1}$, $R \equiv M^0 \cdot L^1 \cdot T^0$, $H \equiv M^0 \cdot L^1 \cdot T^0$, $W \equiv M^0 \cdot L^1 \cdot T^0$, $\sigma \equiv M^0 \cdot L^1 \cdot T^0$, and $\theta \equiv M^0 \cdot L^0 \cdot T^0$. From these nine variables and three-dimensional units, the Buckingham–Pi theorem states that six dimensionless Pi terms are sufficient to represent and predict the turbine's performance. The reference variables, ρ, V, and R, lead to the following three Pi variables, as determined by the Buckingham–Pi theorem [32]:

$$\Pi_1 = \frac{\Phi}{\rho \cdot V^3 \cdot R^2} \qquad (9.4)$$

$$\Pi_2 = \frac{R}{H} \qquad (9.5)$$

$$\Pi_3 = \frac{R}{W} \qquad (9.6)$$

Combining these three terms results in the following power coefficient (C_p): $C_p = 2 \cdot \Pi_1 \cdot \Pi_2 \cdot \Pi_3 = \dfrac{\Phi}{1/2 \cdot \rho \cdot H \cdot W \cdot V^3}$. The fourth and fifth Pi terms are additional terms for representing changes in blade geometry:

$$\Pi_4 = \frac{\sigma}{R} \qquad (9.7)$$

$$\Pi_5 = \theta \qquad (9.8)$$

The last Pi term is represented in terms of R, Ω, and V. It is equivalent to the variable TSR, a common dimensionless parameter for wind power analysis:

$$\Pi_6 = \text{TSR} = \frac{R \cdot \Omega}{V} \qquad (9.9)$$

These terms are then combined such that $2 \cdot \Pi_1 \cdot \Pi_2 \cdot \Pi_3 = \phi\left(\Pi_4, \Pi_5, \Pi_6\right)$, which can be reduced to

$$C_\mathrm{p} = \varphi\left(\frac{\sigma}{R}, \theta, \mathrm{TSR}\right) \qquad (9.10)$$

where φ can be determined through prototype experiments or numerical predictions. Further design parameters can be added to equation (9.10), to represent important design features of different turbine conceptualizations. This expression can lead to correlations that provide useful insight into a turbine's geometric characteristics and offer helpful information for design considerations and power improvements.

9.4   FORMULATION OF FLUID FLOW FOR NUMERICAL PREDICTIONS OF SMALL WIND TURBINES

Numerical predictions provide a proficient technique to limit the number of prototype designs required to effectively develop an efficient small wind turbine. Solutions can include sliding mesh or rotating reference frame formulations to represent the motion of the turbine rotor blades. The governing equations, boundary conditions, and solution formulation for numerical predictions of a small VAWT are discussed in this section.

9.4.1   Governing Equations

The governing equations are the incompressible form of the Navier–Stokes equations. The conventional and most widely employed solution approach for turbulence modeling involves time averaging. As a consequence of this averaging, a number of turbulent stresses are introduced into the solution, equations must be defined to represent these terms. As a result of the approximations needed to solve the equations, many models have been developed. Each offers unique advantages for the particular parameters for which it was developed [33]. For the numerical predictions, a rotating frame adaptation of the governing Navier–Stokes equations is solved. The governing equations for fluid flow include the conservation of mass (9.11) and momentum equations (9.12). The conservation of momentum contains two acceleration terms that represent the rotation, including the Coriolis acceleration, defined as $2\vec{\omega} \times \vec{v}_\mathrm{r}$, and the centripetal acceleration described by $\vec{\omega} \times \vec{\omega} \times \vec{r}$:

$$\frac{\partial \rho}{\partial t} + \nabla \cdot \rho\vec{v}_\mathrm{r} = 0 \qquad (9.11)$$

$$\frac{\partial}{\partial t}\left(\rho\vec{v}_\mathrm{r}\right) + \nabla \cdot \left(\rho\vec{v}_\mathrm{r}\vec{v}_\mathrm{r}\right) + \rho\left(2\vec{\omega} \times \vec{v}_\mathrm{r} + \vec{\omega} \times \vec{\omega} \times \vec{r}\right)$$
$$= -\nabla p + \nabla\overline{\overline{\tau}}_\mathrm{r} + \vec{F} \qquad (9.12)$$

In these equations, $\vec{r}$ is the radial position from the origin of the rotating domain, $\vec{\omega}$ is the angular velocity of the rotor domain, $\vec{v}_\mathrm{r}$ is the relative velocity, p is the static pressure, $\overline{\overline{\tau}}$ is the stress tensor, and $\vec{F}$ is the external body forces [34]. A mixed formulation of cylindrical and Cartesian coordinates are adopted, in order to simulate the two separate regions of the domain, that is, rotating turbine section and external incoming flow, respectively.

For the standard k–ε model, equations (9.13) and (9.14) are used to simulate turbulence in the flow field. This is a widely used model that provides reasonable accuracy and a robust ability to represent a wide range of flow regimes. It is a two-equation model that includes turbulent kinetic energy, k, and dissipation rate, ε, as follows:

$$\frac{\partial}{\partial t}\left(\rho k\right) + \frac{\partial}{\partial x_i}\left(\rho k u_i\right) = \frac{\partial}{\partial x_j}\left[\left(\mu + \frac{\mu_\mathrm{t}}{\sigma_\mathrm{k}}\right)\frac{\partial k}{\partial x_j}\right]$$
$$+ G_\mathrm{k} + G_\mathrm{b} - \rho\varepsilon - Y_\mathrm{M} + S_\mathrm{k} \qquad (9.13)$$

$$\frac{\partial}{\partial t}\left(\rho\varepsilon\right) + \frac{\partial}{\partial x_i}\left(\rho\varepsilon u_i\right) = \frac{\partial}{\partial x_j}\left[\left(\mu + \frac{\mu_\mathrm{t}}{\sigma_\varepsilon}\right)\frac{\partial\varepsilon}{\partial x_j}\right]$$
$$+ C_{1\varepsilon}\frac{\varepsilon}{k}\left(G_\mathrm{k} + C_{3\varepsilon}G_\mathrm{b}\right) - C_{2\varepsilon}\rho\frac{\varepsilon^2}{k} + S_\varepsilon \qquad (9.14)$$

In this model, G_k represents the generation of turbulence kinetic energy due to mean velocity gradients, G_b describes the generation of turbulence kinetic energy due to buoyancy, and Y_M represents the contribution of the fluctuating dilatation to the overall dissipation rate. The variables σ_k and σ_ε are the turbulent Prandtl numbers for k and ε, with values of 1.0 and 1.3, respectively. The constants $C_{1\varepsilon}$ and $C_{2\varepsilon}$ are $C_{1\varepsilon} = 1.44$ and $C_{2\varepsilon} = 1.92$. The turbulent (or eddy) viscosity, μ_t, is computed by combining k and ε such that

$$\mu_\mathrm{t} = \rho C_\mu \frac{k^2}{\varepsilon} \qquad (9.15)$$

where C_μ is a constant 0.09 [34]. These equations can predict air flow through the turbine and subsequently improve its performance.

The turbulence model requires the turbulence intensity and turbulence viscosity ratio to be used for turbulence estimation. The turbulence intensity (I) is defined as the ratio of the root-mean-square of the velocity fluctuation (u') to the mean freestream velocity (u_avg), described by equation (9.16). The turbulence viscosity ratio is defined as the ratio of turbulent viscosity (μ_t) to molecular viscosity (μ) in equation (9.17):

$$I \equiv \frac{u'}{u_\mathrm{avg}} \qquad (9.16)$$

$$\text{Turbulent viscosity ratio} \equiv \frac{\mu_\mathrm{t}}{\mu} \qquad (9.17)$$

These variables specify the following inlet and outlet boundary conditions.

9.4.2 Boundary Conditions

The mass flow rate into the solution domain is calculated by

$$\dot{m} = \int \rho \vec{v} \cdot d\vec{A} \qquad (9.18)$$

where $\dot{m}$ is the mass flow rate. The air density is a constant, 1.225 kg/m³, and the magnitude of the wind velocity, $\vec{v}$, is 12.5 m/s [34]. At the velocity inlet, the turbulence intensity is assumed to be 10% [35]. The turbulent viscosity ratio is assumed as 1, to represent external flows. The face pressure (P_f) is resolved from the boundary condition of the pressure outlet used to represent the exit boundary of the solution domain. The face pressure at each cell is calculated from equation (9.19), while other conditions are extrapolated from the interior of the solution domain [34]:

$$P_f = 0.5 \left(P_c + P_e \right) \qquad (9.19)$$

In this equation, P_c is the interior cell pressure neighboring the exit face and P_e is the pressure at the exit, estimated at standard atmospheric pressure of 101,325 Pa. The turbulence intensity is estimated as 12% to account for increased turbulence caused by flow interaction with the turbine. As with the velocity inlet, the backflow turbulence viscosity ratio is assumed as 1.

The standard wall functions U^* are used to represent the blades and outer boundary walls in the solution. The law of the wall [34] for the mean velocity yields

$$U^* = \frac{1}{K} \ln(E y^*) \qquad (9.20)$$

such that

$$U^* \equiv \frac{U_P C_\mu^{1/4} k_P^{1/2}}{\tau_\omega / \rho} \qquad (9.21)$$

$$y^* \equiv \frac{\rho C_\mu^{1/4} k_P^{1/2} y_P}{\mu} \qquad (9.22)$$

where E is an empirical constant equal to 9.793, K is the von Kármán constant ($K = 0.4187$), U_p is the mean velocity of the fluid at point P, k_p is the turbulent kinetic energy at point P, y_p is the distance from point P to the wall, and μ is the dynamic viscosity of air (1.7894×10^{-05} kg/m-s).

9.5 POWER CORRELATION FOR A UNIQUE VERTICAL AXIS WIND TURBINE

In this section, numerical predictions are used to develop a design correlation for a unique VAWT configuration. As illustrated in Figure 9.2, consider a specific design represented by the Zephyr vertical axis wind turbine (ZVWT). This design utilizes stationary stator blades to direct wind to the rotor blades. The blade spacing (σ) and blade angle (θ) are used to represent geometrical changes to the stator blades.

The numerical formulation represents the turbulence and rotor motion with the standard k–ε turbulence model and a moving reference frame, respectively. The details of the numerical formulation can be found in Ref. [36]. With the MRF formulation, only one position of the rotor, relative to the stator, is simulated. Therefore, the flow field along the interface should not be strongly dependent on the rotor position [32]. The design of this turbine suggests that there is a relatively high degree of rotor–stator interaction. However, this problem is overcome by rotating the stator, relative to the wind direction, three times for each of the geometries. A range of predicted C_p values (calculated by the average of three MRF predictions) at three different wind speeds agrees well with an experimental correlation [37], for the same turbine design. The experimental correlation was obtained through a series of wind tunnel tests on a full-scale VAWT prototype. This close agreement with past experimental results provides useful validation of the numerical predictions.

Figure 9.3 illustrates the predicted ZVWT performance curves for various configurations. In order to develop the power correlation, 16 geometrical configurations are generated. Each of the turbine geometries is represented at

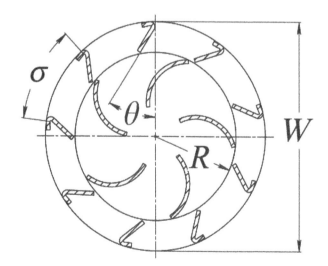

FIGURE 9.2 Geometrical variables of a Zephyr vertical axis wind turbine.

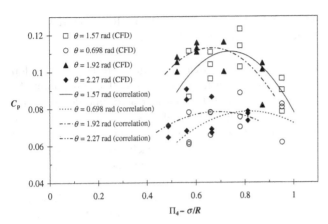

FIGURE 9.3   Results of C_p for different values of Π_4 (TSR = 0.34 rad/s).

three different orientations relative to the wind direction. The orientations are made to rotate the stator by $\pi/3$ radians. A detailed list of geometrical variables is presented in Table 9.1. The average C_p values at varying magnitudes of Π_4 are used for the dimensional analysis. The average C_p values are then used to generate four distinct power curves. These power curves represent discrete values of θ, specifically 0.698, 1.57, 1.92, and 2.27 radians. Each of these curves is discussed in the four cases below.

9.5.1   Case 1: $\theta = 1.57$ Radians

This power curve represents the most recent ZVWT design. This design has been adapted for improved performance with both numerical and experimental studies [36, 37]. The value of θ represents a design where the stator makes a $\pi/4$ planar angle with the turbine's centerline. This angle appears to provide a useful balance between flow stream diversions, while maintaining a low degree of back pressure in front of the turbine. This case will represent the base curve for the subsequent analysis. Geometries in this case have a rotor radius (R) of 0.28 m. A constraint of $\Omega = 15$ rad/s is maintained for the rotational velocity of the rotor subdomain. The turbine width (W) is maintained at 0.762 m and the freestream velocity at 12.5 m/s, throughout the analysis.

Previous experiments have investigated the performance of a ZVWT with nine stator blades [37]. However, recent numerical predictions have indicated that increasing the number of stator blades will improve turbine performance [35]. Therefore, in the four cases, 9, 11, 13, and 15 stator blades are represented. The result is four finite values of σ, specifically 0.27, 0.22, 0.18, and 0.16 m, respectively. The power curve for this case is obtained by curve fitting the CFD data, leading to the following result:

$$C_p = -0.4744\Pi_4{}^2 + 0.6992\Pi_4 - 0.1466 \quad (9.23)$$

This applies to geometries whereby the dimensionless variable Π_4 lies between 0.57 and 0.95. The variability in the data set is calculated by R^2. This coefficient of determination (R^2) for the curve is 0.89, which indicates low variability.

9.5.2   Case 2: $\theta = 0.698$ Radians

This power curve represents the lowest value of θ in the analysis. The low stator angle in this configuration does not effectively divert the flow stream and it produces a relatively low power curve. The geometries in this case are maintained at an identical rotor radius, as case 1, retaining equivalent Π_4 and Ω values for the numerical results, as with case 1. The power curve for this case is represented by

$$C_p = -0.1901\Pi_4{}^2 + 0.3105\Pi_4 - 0.0483 \quad (9.24)$$

for geometries where the dimensionless variable Π_4 lies between 0.57 and 0.95. The resultant R^2 value for this curve is 0.99, which demonstrates extremely low variability. At this low value of θ, the solution accuracy decreases with lower values of Π_4. This is likely caused by the combination of low stator angle and large spacing, which significantly diminishes the stator cage effects. The correlation's accuracy diminishes for this combination of very low θ and high Π_4.

9.5.3   Case 3: $\theta = 1.92$ Radians

The power curve associated with this configuration produces the highest C_p of all the predicted power curves. To maintain a constant turbine radius, the increased stator angle associated with this case imposes a geometrical constraint on the minimum length of R. The radius length is set to 0.3028 m. This change from case 1 shifts the power curve to the left. To maintain a constant TSR between different cases, Ω for the simulations in this curve is set to 13.75 rad/s. Based on CFD data, the power curve for this case is correlated by

$$C_p = -0.3949\Pi_4{}^2 + 0.5242\Pi_4 - 0.061 \quad (9.25)$$

for geometries where the dimensionless variable Π_4 lies between 0.52 and 0.87. The resultant R^2 value for this curve is 0.97, which indicates very low variability.

9.5.4   Case 4: $\theta = 2.27$ Radians

This power curve represents the highest value of θ in the analysis. The high stator angle in this configuration produces a large amount of back pressure and it captures relatively little flow through the turbine. These factors explain the low power curve associated with this case. A radius length of 0.3277 m is simulated, further shifting the power curve to the left. For this curve, Ω is maintained at a constant of

TABLE 9.1   Significant variables.

Geometry	Orientation (°)	Rotor radius [R] (m)	Number of stators	Angle [θ] (°)	Rotor velocity [ω] (rad/s)	Moment (N-m)
Case 1						
1		0.279	9	45	15	4.465
	13.3	0.279	9	45	15	4.178
	26.6	0.279	9	45	15	3.629
1 – average		0.279	9	45	15	4.091
2		0.279	11	45	15	5.281
	10.9	0.279	11	45	15	5.708
	21.8	0.279	11	45	15	4.756
2 – average		0.279	11	45	15	5.248
3		0.279	13	45	15	4.46
	9.23	0.279	13	45	15	4.818
	18.5	0.279	13	45	15	5.128
3 – average		0.279	13	45	15	4.802
4		0.279	15	45	15	5.154
	8	0.279	15	45	15	4.014
	16	0.279	15	45	15	4.717
4 – average		0.279	15	45	15	4.628
Case 2						
5		0.279	9	20	15	2.86
	13.3	0.279	9	20	15	3.743
	26.7	0.279	9	20	15	3.823
5 – average		0.279	9	20	15	3.475
6		0.279	11	20	15	3.499
	10.9	0.279	11	20	15	3.25
	21.8	0.279	11	20	15	4.105
6 – average		0.279	11	20	15	3.618
7		0.279	13	20	15	3.043
	9.23	0.279	13	20	15	3.609
	18.5	0.279	13	20	15	3.619
7 – average		0.279	13	20	15	3.424
8		0.279	15	20	15	2.821
	8	0.279	15	20	15	2.866
	16	0.279	15	20	15	3.619
8 – average		0.279	15	20	15	3.102
Case 3						
9		0.305	9	55	13.75	5.288
	13.3	0.305	9	55	13.75	5.118
	26.7	0.305	9	55	13.75	4.135
9 – average		0.305	9	55	13.75	4.847
10		0.305	11	55	13.75	5.062
	10.9	0.305	11	55	13.75	5.866
	21.8	0.305	11	55	13.75	5.860
10 – average		0.305	11	55	13.75	5.596
11		0.305	13	55	13.75	5.595
	9.23	0.305	13	55	13.75	5.706
	18.5	0.305	13	55	13.75	5.848
11 – average		0.305	13	55	13.75	5.716
12		0.305	15	55	13.75	5.458
	8	0.305	15	55	13.75	5.056
	16	0.305	15	55	13.75	5.318
12 – average		0.305	15	55	13.75	5.277
Case 4						
13		0.328	9	65	12.79	4.200
	13.3	0.328	9	65	12.79	3.994
	26.7	0.328	9	65	12.79	4.270

(continued)

TABLE 9.1   (*Continued*)

Geometry	Orientation (°)	Rotor radius [R] (m)	Number of stators	Angle [θ] (°)	Rotor velocity [ω] (rad/s)	Moment (N-m)
13 – average		0.328	9	65	12.79	4.155
14		0.328	11	65	12.79	3.636
	10.9	0.328	11	65	12.79	4.6961
	21.8	0.328	11	65	12.79	3.743
14 – average		0.328	11	65	12.79	4.025
15		0.328	13	65	12.79	3.696
	9.23	0.328	13	65	12.79	4.923
	18.5	0.328	13	65	12.79	4.6169
15 – average		0.328	13	65	12.79	4.412
16		0.328	15	65	12.79	3.506
	8	0.328	15	65	12.79	3.826
	16	0.328	15	65	12.79	3.857
16 – average		0.328	15	65	12.79	3.730

12.79 rad/s. From curve fitting of CFD data, the power curve can be represented by

$$C_p = -0.1762\Pi_4{}^2 + 0.2424\Pi_4 - 0.005 \quad (9.26)$$

for geometries where the dimensionless variable Π_4 lies between 0.49 and 0.81. The resultant R^2 value for this curve is 0.31, which indicates a relatively high degree of variability. It is evident that the correlation has less accuracy as the value of θ approaches 2.27 radians.

In efforts to collapse all of the previous results into a single correlation, the curves associated with each of the four cases are normalized in the dimensionless plane $[\tilde{C}_p, \tilde{\Pi}_4]$ according to the following function:

$$\tilde{C}_P(f) = -0.4744 \cdot \tilde{\Pi}_4(g)^2 + 0.6992 \cdot \tilde{\Pi}_4(g) - 0.1466 \quad (9.27)$$

where $\tilde{C}_p(f) = C_p + f(\theta)$ and $\tilde{\Pi}_4(g) = \Pi_4 + g(\theta)$. The four curves collapse onto a single plain, as shown in Figure 9.4. As previously discussed, the curve representing $\theta = 2.27$ radians provides a moderate degree of accuracy, and less accuracy with a high Π_4 value ($\Pi_4 = 0.81$). Also, a combination of low angle ($\theta = 0.698$) and high Π_4 ($\Pi_4 = 0.95$) reduces the accuracy of the correlation. These outlier points are removed, because they lie outside of the range of applicability of the correlation (see Figure 9.5).

The normalization coefficient $f(\theta)$ can be represented as follows:

$$f(\theta) = 0.0575\theta^2 - 0.1746\theta + 0.127 \quad (9.28)$$

where minimal variability is exhibited compared with the numerical prediction ($R^2 = 0.91$). The normalization coefficient $g(\theta)$ can be represented by

$$g(\theta) = 0.0825\theta^2 - 0.1782\theta + 0.0833 \quad (9.29)$$

which shows minimal variability with the numerical prediction ($R^2 = 0.98$). The overlapping data points at $\theta = 1.57$ radians coincide with the base case configuration (case 1). All power curves are collapsed onto a single normalized power curve, resulting in a null value for $f(1.57)$ and $g(1.57)$. The general correlation becomes

$$C_P(\Pi_4, \theta) = -0.4744 \cdot [\Pi_4 + g(\theta)]^2$$
$$+0.6992 \cdot [\Pi_4 + g(\theta)] - 0.1466 - f(\theta) \quad (9.30)$$

where $f(\theta)$ and $g(\theta)$ are described by equations (9.28) and (9.29), respectively.

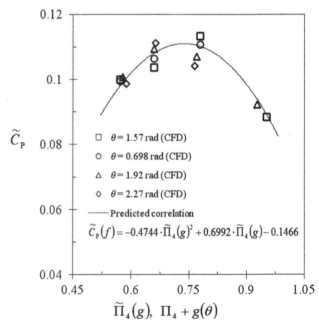

FIGURE 9.4   Comparison of numerical and predicted values, normalized to the plane $|\tilde{C}_p, \tilde{\Pi}_4|$.

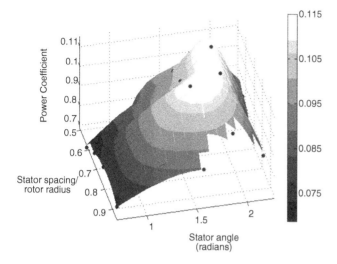

FIGURE 9.5 Surface contours relating the changes of C_p, Π_4, and θ.

Three additional numerical predictions are compared with the correlation described by Eq. (30), confirming the models ability to predict results that are not in the CFD simulated conditions. The first of these configurations provides a verification of the method's ability to provide accurate predictions for the turbine at $\theta = 2.09$ radians. For this prediction, $R = 0.32$ m, $\sigma = 0.20$ m, and $\Omega = 13.2$ rad/s, resulting in $\Pi_4 = 0.63$. The dimensionless predictions provide a useful comparison with the numerical predictions for this configuration. The error between the numerical prediction and the newly developed correlation is 4.4%. The second point maintains $\theta = 2.09$ radians, $R = 0.32$ m, and $\Omega = 13.2$ rad/s, but σ is changed to 0.27 m. This selection of variables represents a Π_4 term of 0.84, which is significantly different than the previous Π_4 value of 0.63. The error is 5.8% compared to the numerical predictions. Finally, the point at $\theta = 1.15$ radians illustrates the model's ability to predict the performance for turbine configurations in the lower range of θ. For this prediction, $R = 0.28$ m, $\sigma = 0.17$ m, and $\Omega = 15$ rad/s. Again the correlation described by equation (9.30) provides a prediction with close agreement with the turbine's performance predicted by CFD simulations. The associated error is only 2.9%.

By developing a general correlation that predicts the ZVWT performance, further design improvements can be undertaken without the need for time-consuming CFD predictions for each case. The correlation provides a useful design tool for adapting the turbine conditions and operating requirements, specific to a particular turbine location. It becomes easier to quickly predict how changes to the turbine's essential design features will impact its performance with a reasonable degree of accuracy. The correlation can also be useful for developing an optimum turbine design, while limiting the need for extensive CFD simulations and prototype experimentation for each turbine configuration.

9.6   CONCLUSIONS

RSBW offer a promising potential to supplement the growth in installed capacity of wind power. However, RSBW installations require small wind turbines to operate in highly variable and often harsh conditions for effective power generation. To increase the effectiveness of RSBW installations, turbines should be designed to operate in different climatic conditions at various locations. Developing a design correlation can be an effective method to supplement the design of a small wind turbine to achieve high performance at different sites and operating conditions. The correlation technique presented in this chapter predicts the power coefficient in terms of not only dimensionless variables including C_p and TSR, but also turbine-specific geometrical variables. It is a robust correlation that can be extended to a variety of other geometries, making it suitable for a wide range of VAWT designs. Its capability was tested on a prototype Zephyr VAWT. The correlation can predict the turbine power with varying stator angles, stator spacing, rotor length, rotor velocity, wind speed, air density, and aspect ratio. The techniques presented in this chapter offer a useful design tool for understanding and improving small wind turbine performance.

ACKNOWLEDGMENTS

The authors express their grateful appreciation to the Natural Sciences and Engineering Research Council of Canada (NSERC) for their financial support.

Nomenclature

c	Scaling factor (dimensionless)
C_p	Power coefficient (dimensionless)
E	Empirical constant (dimensionless)
H	Turbine height (m)
I	Turbulence intensity (dimensionless)
k_w	Weibull shape parameter (dimensionless)
k_p	Turbulence kinetic energy at point P (J)
K	Kármán constant (dimensionless)
L	Length (m)
M	Mass (kg)
$\dot{m}$	Mass flow rate (kg/s)
P_f	Face pressure (Pa)
P_c	Cell pressure (Pa)
P_e	Exit pressure (Pa)
$\vec{r}_0$	Distance to origin of rotating system (m)
R	Rotor radius (m)
R^2	Coefficient of determination (dimensionless)
T	Time (s)
TSR	Tip speed ratio (dimensionless)
$\overline{\overline{\tau}}_r$	Viscous stress (N/m^2)
$\vec{u}_r$	Whirl velocity (m/s)

U_P	Mean fluid velocity at point P (m/s)
V	Mean wind velocity (m/s)
$\vec{v}_\mathrm{r}$	Relative velocity (m/s)
$\vec{v}$	Absolute velocity (m/s)
W	Turbine width (m)
y_P	Distance from point P to the wall (m)

Greek

θ	Blade angle (rad)
Φ	Power (W)
μ	Dynamic viscosity of air (Pa-s)
ρ	Air density (kg/m^3)
σ	Blade spacing (m)
$\vec{\omega}$	Angular velocity relative to stationary frame (rad/s)
Ω	Rotor velocity (rad/s)
τ_w	Wall shear stress (Pa)

REFERENCES

[1] WWEA. (2014). 2014 Half year report. World Wind Energy Association, Bonn, Germany.

[2] McDowall, J. (2006). Integrating energy storage with wind power in weak electricity grids. *Journal of Power Sources*, 162(2):959–964.

[3] Barton,J. P., Infield, D. G. (2006). A probabilistic method for calculating the usefulness of a store with finite energy capacity for smoothing electricity generation from wind and solar power. *Journal of Power Sources*, 162(2):943–948.

[4] Wang, L., Lee, D. J., Lee, W. J., Chen, Z. (2008). Analysis of a novel autonomous marine hybrid power generation/energy storage system with a high-voltage direct current link. *Journal of Power Sources*, 185(2):1284–1292.

[5] Kelouwani, S., Agbossou, K., Chahine, R. (2005). Model for energy conversion in renewable energy system with hydrogen storage. *Journal of Power Sources*, 140(2):392–399.

[6] Eriksson, S., Bernhoff, H., Leijon, M. (2008). Evaluation of different turbine concepts for wind power. *Renewable and Sustainable Energy Reviews*, 12:1419–1434.

[7] Riegler, H. (2003). HAWT versus VAWT: small VAWTs find a clear niche. *Refocus*, 4:44–46.

[8] Akhmatov, V. (2007). Influence of wind direction on intense power fluctuations in large offshore windfarms in the North Sea. *Wind Engineering*, 31(1):59–64.

[9] Castro, I. P., Cheng, H., Reynolds, R. (2006). Turbulence over urban-type roughness: deductions from wind-tunnel measurements. *Boundary-Layer Meteorology*, 118:109–131.

[10] Gayev, Y. A., Savory, E. (1999). Influence of street obstructions on flow processes within urban canyons. *Journal of Wind Engineering and Industrial Aerodynamics*, 82:89–103.

[11] Rohatgi, J., Barbezier, G. (1999). Wind turbulence and atmospheric stability – their effect on wind turbine output. *Renewable Energy*, 16:908–911.

[12] Sahin, A. D., Dincer, I., Rosen, M. A. (2006). Thermodynamic analysis of wind energy. *International Journal of Energy Research*, 30:553–566.

[13] Center for Sustainable Energy. (2003). *Ealing Urban Wind Study*. Ealing Borough Council Urban Wind Study, The CREATE Centre, Bristol.

[14] Wright, A. K., Wood, D. H. (2004). The starting and low wind speed behaviour of a small horizontal axis wind turbine. *Journal of Wind Engineering and Industrial Aerodynamics*, 92:1265–1279.

[15] Heagle, A. L. B., Naterer, G. F., Pope, K. (2011). Small wind turbine energy policies for residential and small business usage in Ontario, Canada. *Energy Policy*, 39:1988–1999.

[16] Saha, U. (2008). Optimum design configuration of Savonius rotor through wind tunnel experiments. *Journal of Wind Engineering and Industrial Aerodynamics*, 96(8):1359–1375.

[17] Biswas, A., Gupta, R., Sharma, K. K. (2007). Experimental investigation of overlap and blockage effects on three-bucket Savonius rotors. *Wind Engineering*, 31(5):313–368.

[18] Hiraharaa, H., Hossainb, M. Z., Kawahashia, M., Nonomurac, Y. (2005). Testing basic performance of a very small wind turbine designed for multi-purposes. *Renewable Energy*, 30:1279–1297.

[19] Gupta, R., Biswas, A., Sharma, K. K. (2008). Comparative study of a three-bucket Savonius rotor with a combined three-bucket Savonius-three-bladed Darrieus rotor. *Renewable Energy*, 33:1974–1981.

[20] Justus, C. G. (1978). *Winds and Wind System Performance*. Franklin Institute Press, Philadelphia, PA.

[21] Alé, J. A., Petry, M. R., Garcia, S. B., Simioni, G. C., Konzen, G. (2007). Performance evaluation of the next generation of small vertical axis wind turbine. *European Wind Energy Conference & Exhibition*, May 7–10, MIC - Milano Convention Centre, Milan, Italy, 2007.

[22] RETScreen International, Empowering Cleaner Energy Decisions. (2008). *RETScreen Clean Energy Project Analysis Software*. Natural Resources Canada, Ottawa, Canada.

[23] Weibull, W. (1951). A statistical distribution function of wide applicability. *Journal of Applied Mechanics - Transactions of the ASME*, 18(3):293–297.

[24] Entegrity Wind Systems Inc. (2008). *EW50 Specifications EW 50 Turbine*. Entegrity Wind Systems Inc, Boulder, CO.

[25] Eoltec (2003). *Eoltec Scirocco 5.5 - 6000 Wind Turbine Performances*. Scirocco High Efficiency Ø5.6m/6kW Professional Wind Turbine. Eoltec, Nice Cedex, France.

[26] Air Breeze Technologies (2008). *Air Breeze VAWT 50kW Air Breeze VAWT technical Data Specifications & Power Curves*. Create Energy with the Innovation of Vertical Axis Wind Turbines. Air Breeze Technologies.

[27] PacWind Inc. (2008). *SeaHawk Owners and Installation Manual*. Alpha Series, 500W SeaHawk Vertical Axis Wind Turbine MorningStar TriStar-60. PacWind Inc, Torrance, CA, USA.

[28] Pope, K., Naterer, G. F. (2009). Potential of Carbon Mitigation by Vertical Axis Wind Turbines in Urban Regions. *Proceedings of 2nd Climate Change Technology Conference*, May 12–15, Hamilton, Canada, 2009.

[29] Wang, X., Bibeau, E., Naterer, G. F. (2007). Experimental correlation of forced convection heat transfer from a NACA airfoil. *Experimental Thermal and Fluid Science*, 31:1073–1082.

[30] Wang, X., Naterer, G. F., Bibeau, E. (2007). Convective droplet impact and heat transfer from a NACA airfoil. *Journal of Thermophysics and Heat Transfer*, 21(3):536–542.

[31] Pope, K., Naterer, G. F., Dincer, I., Tsang, E. (2011). Power correlation for vertical axis wind turbines with varying geometries. *International Journal of Energy Research*, 35(5):423–435.

[32] Munson, B. R., Young, D. F., Okiishi, T. E. (2006). *Fundamentals of Fluid Mechanics*, 5th edition. John Wiley and Sons, Jefferson City, MO, USA.

[33] Gosman, A. D. (1999). Developments in CFD for industrial and environmental applications in wind engineering. *Journal of Wind Engineering and Industrial Aerodynamics*, 81:21–39.

[34] ANSYS Inc. (2006). *Fluent 6.3 Users Guide*. ANSYS Inc, Canonsburg, PA, USA.

[35] Veers, P. S., Winterstein, S. R. (1998). Application of measured loads to wind turbine fatigue and reliability analysis. *Journal of Solar Energy Engineering*, 120(4):233–239.

[36] Pope, K., Naterer, G. F., Tsang, E. (2008). Effects of rotor-stator geometry on vertical axis wind turbine performance. *CSME 2008 Forum*, June 5–8, Ottawa, Canada, 2008.

[37] Hunt, G., Savory, E. (2006). *Report on Wind Tunnel Tests of the Performance of Vertical Axis Turbines*. Department of Mechanical and Materials Engineering University of Western Ontario, London, Canada.

10

WINDMILL BRAKE STATE MODELS USED IN PREDICTING WIND TURBINE PERFORMANCE

Panu Pratumnopharat and Pak Sing Leung

10.1 BACKGROUND

In predicting performance of wind turbines, the blade element momentum (BEM) theory is still commonly used by wind turbine designers and researchers. Despite the fact that more sophisticated approaches are available, the BEM theory has significant advantages in computational speed [1] and ease of implementation. Historically, Glauert [2] originated the basic concepts of aerodynamic analysis of airscrew propellers and windmills. In 1974, Wilson et al. [3–4] extended Glauert's work for application to wind turbines and presented a step-by-step procedure for calculating performance characteristics of wind turbines. The BEM theory is based on one-dimensional momentum theory and two-dimensional blade element theory, and then iterative solutions are obtained for the axial and tangential induction factors.

In recent years, researchers have optimized and modified the BEM calculation to provide more accurate results. There are some differences among several BEM codes because several strategies are used to solve the nonlinear equations. Many corrections, such as tip and hub loss model and windmill brake state model, have been proposed to increase the precision of prediction. Windmill brake state model is the one to be considered because the BEM calculation often calculates the value of thrust coefficient in this state of operation. Referring to Stoddard's work in 1977 [5], the behavior of wind turbine rotors was correlated with known experimental data for helicopters reported by Glauert in 1926 [6]. Stoddard's work showed that thrust coefficient predicted by momentum theory deviated dramatically from the Glauert's experimental data when the value of axial induction factor is greater than 0.5. To solve this problem and to increase the accuracy of the prediction, several researchers applied tip loss effect to the thrust coefficient equation and then proposed windmill break state models. The problem of interest is which windmill break state model is suitable for the wind turbine model being simulated.

10.2 BLADE ELEMENT MOMENTUM THEORY

The basic concept of aerodynamic analysis is based on Glauert's airscrew theory [2]. The development of the airscrew theory followed two independent lines of thought, which may conveniently be called the *momentum* theory and the *blade element* theory. Wilson et al. [3–4] extended Glauert's airscrew theory for application to wind turbines and presented a step-by-step method for calculating performance characteristics of wind turbines. Momentum theory refers to an analysis of thrust and torque by applying the conservation of the linear and angular momentum to an annulus control volume [7]. Blade element theory refers to an analysis of aerodynamic forces at a section of the blade as a function of blade geometry. The results of these theories can be combined into what is known as *blade element momentum* (BEM) theory. Consequently, the BEM theory can be used to relate blade shape to the ability of the rotor in extracting power from the wind.

In momentum analysis, the wind turbine blade rotor is assumed to have a large number of blades so that it becomes effectively an actuator disc; that is, the number of blades, the twist and chord distribution, and the characteristic of the airfoils are not taken into account. It is further assumed that the thrust is uniformly distributed over the disc. The

Alternative Energy and Shale Gas Encyclopedia, First Edition. Edited by Jay H. Lehr and Jack Keeley.
© 2016 John Wiley & Sons, Inc. Published 2016 by John Wiley & Sons, Inc.

results of applying the conservation of linear momentum and angular momentum are the differential thrust and torque expressions written of the axial and tangential induction factors. In fact, the pressure on the suction side of the blade is lower than that on the pressure side, so the wind tends to flow around the tip from lower to upper surface. This effect reduces lift and power generated near the tip of the blade. A number of methods for correcting tip loss effect are proposed. The most straightforward one is Prandtl's tip and hub loss correction models [3]. In 2002, Prandtl's tip loss correction model was improved by Xu and Sankar [8]. The up-to-date tip loss correction model was proposed by Shen et al. in 2005 [9]. In addition, tip and hub loss factors can be combined to the total loss factor. In order to make BEM code more realistic in prediction, the total loss factor is introduced into the differential thrust and torque expressions derived from momentum theory.

In blade element analysis, the blade is assumed to be divided into a number of elements along the blade span. While the blade is rotating, each element sweeps and forms an annular shape. Basically, the blade element theory is based on an assumption that there is no aerodynamic interaction between the adjacent elements, so the thrust and torque on each blade element can be determined by the lift and drag coefficients.

When the thrust equations from momentum theory and blade element theory are equated and some algebraic manipulation performed, an expression of the axial induction factor is obtained. Similarly, an expression of the tangential induction factor is obtained by equating the torque equations from momentum theory and blade element theory.

10.3 WINDMILL BRAKE STATE MODELS

In 1926, Glauert [6] reported that the experimental results showing thrust coefficient equation are not valid if the axial induction factor exceeds 0.4 approximately. *Glauert's empirical formula* [10] was reported in a quadratic form shown in Figure 10.1 and gives the same value as momentum theory when a is 0.4, as well as the same slope of the curve, and yields $C_T = 2$ at $a = 1$. However, the total loss factor is not taken into account. In the BEM calculation, a numerical problem (gap between the momentum theory curve and Glauert's empirical curve) occurs when the total loss correction factor is applied to the thrust equation obtained from the classical momentum theory. This gap causes a discontinuity when a computer is used to iterate for the new value of axial induction factor.

In 1981, Hibbs and Radkey [10] modified Glauert's empirical formula. This modification just lowers the entire parabolic curve, that is, the gap problem is still not completely eliminated as shown in Figure 10.1.

At present, six different windmill brake state models have been proposed to solve the discontinuity problem. They are Glauert's characteristic equation, classical momentum brake state model, advanced brake state model, Wilson and Walker model, modified advanced brake state model, and Shen's correction.

10.3.1 Glauert's Characteristic Equation

Glauert's characteristic equation [11–13] is a cubic equation.

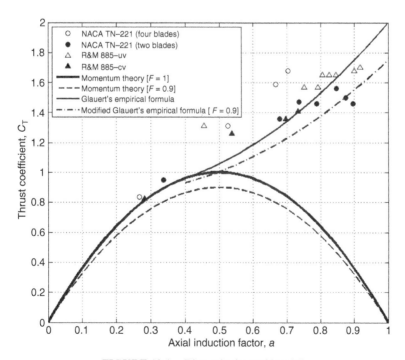

FIGURE 10.1 Discontinuity problem [4].

10.3.2 Classical Momentum Brake State Model

In 1974, Wilson and Lissaman [3] developed a wind turbine analysis computer code named PROP and used *Wilson
model* as a windmill brake state model. Later, in 1976, Wilson, Lissaman, and Walker [14] reported the state of the art
of performance prediction methods for both horizontal and
vertical wind turbines and suggested using Wilson model to
predict the wind turbine performance. From 1981 to 1983,
Hibbs and Radkey [10] updated the version of the PROP
code and renamed the Wilson model as *classical momentum
brake state model*. At present, this model is being used by
WT_Perf code [14] under the name *classical brake model*.

10.3.3 Advanced Brake State Model

Referring to the Glauert's empirical formula [6], this equation
was called advanced brake state model by Hibbs and Radkey
[10]. Unfortunately, it cannot completely eliminate the gab
problem. In 1997, Buhl [14–15] simplified the calculation of
advanced brake state model implemented in WT_Perf code
by replacing the original advanced brake state model with a
simple parabola.

10.3.4 Wilson and Walker Model

Wilson [15] reported the Wilson and Walker model which
was published in 1984. This model is a straight line derived
by the first order of Taylor series. Wilson quoted that this
model well agrees with vortex theory calculations and produces good results between the calculated and measured performance and loads. Also, this model has been found in
Hansen's book [11].

10.3.5 Modified Advanced Brake State Model

In 1998, Buhl [14] modified the use of Glauert's empirical
formula implemented in WT_Perf code so that the coefficients of the quadratic are a function of the losses. Buhl did
not name this relation, so it might be called the *modified
advanced brake state model*.

10.3.6 Shen's Correction

In 2005, Shen et al. [9] proposed a new tip loss correction
model to predict the physical behavior in the proximity of
the tip. In this work, the local thrust coefficient is replaced
by a linear relation when the value of axial induction factor
becomes greater than a critical value.

10.4 COMPARISON OF WINDMILL BRAKE STATE MODELS

In this section, the power curve of AWT-27/P4 wind turbine
model [17] is used as a baseline. Each windmill brake state
model is implemented in aerodynamic code. The predicted
and measured power curves are compared and shown in Table
10.1 and Figure 10.2 [18].

Based on the BEM theory, the rotor power curves predicted by using six different windmill brake state models are
compared to the measured rotor power curve of AWT-27/P4
wind turbine.

Considering Figure 10.2, the predicted power curves in
the range from 5 to 8 m/s can be classified into three groups
according to the predicted power level. The first group,

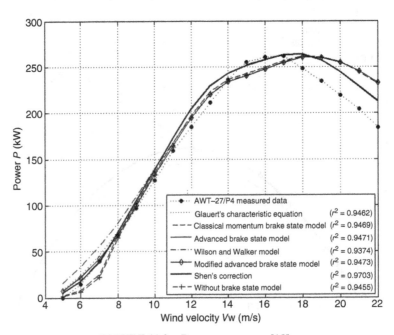

FIGURE 10.2 Rotor power curve [18].

TABLE 10.1   Annual energy production (AEP) at 8.5 m/s annual average wind speed at site [18].

Windmill brake state models	AEP (kWh)
AWT-27/P4 measured data	792,780
Glauert's characteristic equation	821,722
Classical momentum brake state model	793,399
Advanced brake state model	812,563
Wilson and Walker model	859,510
Modified advanced brake state model	811,306
Shen's correction	842,475
Without brake state model	776,908

which consists of the classical momentum brake state model (dashed line) and without brake state model (dash-plus line), predicts lower power level than the measured one. The second group predicts higher power level than the measured one. The windmill brake state model in this second group is Wilson and Walker model (dash-dot line). The last group, which consists of Glauert's characteristic equation (dotted line), advanced brake state model (solid line), modified advanced brake state model (solid line with diamond), and Shen's correction (thick solid line), predicts the power level close to the measured one. In addition, it is noted from Figure 10.2 that the power curve predicted by using the advanced brake state model nearly coincides with that predicted by using the Glauert's characteristic equation. These two curves overlap each other. Finally, the variation between all of the predicted results and the measured AEP values listed in Table 10.1 is less than 9% (for an 8.5 m/s annual average wind speed at site).

REFERENCES

[1] Clifton-Smith, M. J. (2009). Wind turbine blade optimization with tip loss corrections. *Wind Engineering*, 33:477–496.

[2] Glauert, H. (1935). Airplane propellers. In: *Aerodynamic Theory*. Durand, W. F. (editor). Vol. IV Division L, Berlin-Julius Springer, pp. 169–360.

[3] Wilson, R. E., Lissaman, P. B. S. (1974). *Applied Aerodynamics of Wind Power Machines*. The National Science Foundation, 1974, pp. 1–109.

[4] Wilson, R. E., Lissaman, P. B. S., Walker, S. N. (1976). Aerodynamic performance of wind turbines, pp. 1–164 [Online]. Available at: http://wind.nrel.gov/designcodes/papers/Wilson LissamanWalker_AerodynamicPerformanceOfWindTurbines (1976).pdf (Accessed: 19th December 2009).

[5] Stoddard, F. S. (1977). Momentum theory and flow states for windmill. *Wind Technology Journal*, 1:3–9.

[6] Glauert, H. (1926). The analysis of experimental results in the windmill brake and vortex ring states of an airscrew. Reports and Memoranda, No. 1026, His Majesty's Stationery Office, pp. 1–8.

[7] Manwell, J. F., McGowan, J. G., Rogers, A. L. (2002). *Wind Energy Explained: Theory, Design and Application*. John Wiley & Sons Inc.

[8] Xu, G., Sankar, L. N. (2002). Development of engineering aerodynamics models using a viscous flow methodology on the NREL phase VI rotor. *Wind Energy*, 5:171–183 (doi: 10.1002/we.73).

[9] Shen, W. Z., Mikkelsen, R., Sorensen, J. N., Bak, C. (2005). Tip loss corrections for wind turbine computations. *Wind Energy*, 8:457–475 (doi: 10.1002/we.153).

[10] Hibbs, B., Radkey, R. L. (1983). Calculating rotor performance with the revised 'PROP' computer code. AeroVironment, Inc., Report No. PFN-13470W, pp. 1–B13.

[11] Hansen, M. O. L. (2008). *Aerodynamics of Wind Turbines*, 2nd edition. Earthscan, London.

[12] Bak, C., Fuglsang, P., Sørensen, N. N., Madsen, H. A., Shen, W. Z., Sørensen, J. N. (1999). Airfoil characteristics for wind turbines, Risø National Laboratory, Report No. Risø-R-1065(EN), March, pp. 1–51.

[13] Mejia, J. M., Chejne, F., Smith, R., Rodriguez, L. F., Fernandez, O., Dyner, I. (2006). Simulation of wind energy output at Guajira, Colombia. *Renewable Energy*, 31:383–399 (doi:10.1016/j.renene.2005.03.014).

[14] Buhl, M. L. Jr. (2004). NWTC design code (WT-Perf). National Wind Energy Center. Last modified 21st June 2004 [Online]. Available from: http://wind.nrel.gov/designcodes/ designcodes/simulators/wt_perf/ (Accessed 20th October 2009).

[15] Buhl, M. L., Jr. (2005). A new empirical relationship between thrust coefficient and induction factor for the turbulent windmill state. National Renewable Energy Laboratory, Technical report NREL/TP-500-36834, August, pp. 1–7.

[16] Wilson, R. E. (1998). Aerodynamic behavior of wind turbines. In: *Wind Turbine Technology: Fundamental Concepts of Wind Turbine Engineering*. Spera, D. A. (editor). ASME Press, pp. 215–282.

[17] Poore, R. (2000). NWTC AWT-26 research and retrofit project summary of AWT-26/27 turbine research and development. National Renewable Energy Laboratory, Subcontract report NREL/SR-500-26926, January, pp. 1–51.

[18] Pratumnopharat, P., Leung, P. S. (2011). Validation of windmill brake state models used by blade element momentum calculation. *Renewable Energy*, 36:3222–3227 (doi: 10.1016/j.rene.2011.03.027).

11

LIGHTNING PROTECTION OF WIND TURBINES AND ASSOCIATED PHENOMENA

PETAR SARAJCEV

11.1 INTRODUCTION

Wind farms are probably the single most significant contributors to the production of "green" renewable electrical energy, with additional capacity increases planned for the near future. They are often constructed with dozens to several tens of individual wind turbines (WTs) of ever-increasing power, both onshore and offshore. Modern WT generators are already in the several megawatts range, with a possibility for further expansion. WTs are often spread across several square kilometers, covering large surface areas. They are also often mutually interconnected by buried medium-voltage (MV) cable networks (and sometimes overhead transmission lines, or combination thereof). Hence, an MV electrical grid is created for collecting the produced power from individual WTs and transmitting it to the high-voltage (HV) electrical power system. This gives rise to the complex, mutually interconnected, electrical system.

Wind farms are often located in regions which might span across mountain ridges or in other areas of elevated ground (plates), where the wind resource is significant. However, these locations in turn often coincide with high keraunic levels, which has been confirmed by NASA report, among others. Additionally, these areas could, at the same time, have relatively high soil resistivity (e.g., often in excess of thousand Ω m in some parts of southern Europe). Due to the fact that WTs present very tall and isolated objects, exposed to direct lightning strikes, they present extremely vulnerable structures, and in fact tend to get struck by lightning very often. This could have adverse effects on the performance of individual WTs (i.e., their production capability) and on the wind farm as a whole, particularly in terms of the lost revenue. The situation tends to be further exacerbated by the combination of high WT lightning incidence, when combined with high soil resistivity.

The lightning and overvoltage protection analysis of wind farms occupy prominent positions within the design considerations associated with these projects. This includes scrutiny of the various aspects of this problem, from the design of the WT grounding system (in terms of both the dangerous voltages associated with the single-pole short-circuit currents and its lightning-associated transient behavior) to the design of the lightning protection system (LPS) and selection of metal-oxide surge arresters. The topics introduced span quite an extensive field of study and it would be almost impossible, if not quite difficult, to elaborate here, to any significant extent, on these various aspects. To complicate matters even further, some of these aspects are mutually correlated and interact in complex ways; for example, WT grounding system transient behavior has important influence on the subsequently created lightning-related overvoltage (and to some extent on its propagation in the wind farm), that is, the so-called "backsurge" phenomenon. Having said that, we will here concentrate our efforts on the lightning protection of WTs, with brief excursion into the field of overvoltage protection of WTs and wind farms in general.

This chapter is arranged as follows. In Section 11.2, a brief introduction into the statistical depiction of the lightning stroke (i.e., lightning current waveshape) is provided, along with existing statistical correlations between various parameters in question. Here, some basic considerations, relating to statistical distribution of lightning current incident to WTs, are presented and elaborated upon. Next, in Section 11.3, a lightning protection philosophy, as provided by the

Alternative Energy and Shale Gas Encyclopedia, First Edition. Edited by Jay H. Lehr and Jack Keeley.

relevant IEC 61400-24 standard (which is in turn based on IEC 62305) is briefly introduced. Some aspects of this standard are elaborated upon, particularly those pertaining to the assessment of the WT exposure to direct lightning strikes, due to their relevancy in the subsequent overvoltage protection analysis, with emphasis on the estimation of the WT effective height and collection area. Also, procedure for the assessment of the WT service line lightning exposure, in high resistivity soils, is briefly elaborated upon. In Section 11.4, an introduction into the problems of the overvoltage protection analysis of the wind farm electrical systems is provided. The analysis itself rests heavily on the sophisticated numerical simulations of transient overvoltages (their propagation, attenuation, reflection, and refraction) in these complex systems. They are often carried out by some of the EMTP-like software packages. Section 11.5 provides some concluding remarks.

11.2 STATISTICAL PARAMETERS OF LIGHTNING CURRENTS

Lightning is a rather complex and stochastic natural phenomenon which, after more than a century of intensive scholarly research, still attracts significant interest from numerous researchers around the world. A reader interested in various aspects of the lightning is referred to the following two very different and truly seminal books on lightning physics and its various effects [1–3].

Generally speaking, lightning is classified as upward and downward, which bears particular significance in relation to tall and exposed structures. These two forms of lightning can be further subdivided into positive and negative polarity, with the so-called polarity being that of the charge transferred from the cloud to the ground. There is also the bipolar lightning, although it will not be treated here. Significant differences exist between downward and upward lightning flashes. Furthermore, it needs to be emphasized that, generally speaking, approximately 90% of all downward lightning strokes are negative in polarity, while the remaining 10% of the lightning strokes are positive. However, it should also be mentioned that this proportion is obtained from measurements on sites generally not exceeding some 1000 meters above the sea level [4]. Golde mentions in [4] several reports from which it could be argued that on higher elevations this proportion could be significantly altered, with propensity toward the positive strikes. This also bears significance in relation to wind farm sites, which could occupy locations on elevated plates, as previously mentioned.

A detailed explanation and a complete classification of all lightning types was first provided by Berger in [5]. Additionally, an excellent and general overview of the most recent progress in lightning research, accompanied by a vast survey of the associated literature, is given by Rakov and Rachidi [6].

Lightning in general can be seen as having a current wave with a distinct waveshape, described by the four basic parameters: (1) amplitude, (2) duration of the current front, (3) duration of the current wave, and (4) polarity. Engineering applications also make use of lightning current charge transferred during a lightning flash and a so-called action integral, all of which are statistical parameters; for example, see [3]. Furthermore, parameters of the subsequent lightning stroke are of importance in engineering applications as well; the lightning flash is composed of several strokes, where the number of strokes in a flash is often termed flash multiplicity. This holds for the downward (negative) lightning. Upward lightning has different properties and is often characterized by the so-called continuing current which precedes a main current peak. The occurrence of lightning is influenced by several different factors, such as the charge structure of the thunderclouds, the height of the cloud base above the ground surface, the altitude of the site (above the sea level), climatological and topographical effects, and so on.

Due to the stochastic nature of lightning, it needs to be described in terms of statistical distributions of its aforementioned parameters. It has been established that lightning current parameters (e.g., amplitude, front duration/front steepness, and wave duration) obey the log-normal distribution, although it has been questioned by Golde and others whether this is true for lower and upper parts of this distribution, that is, very low and very high values [1]. Additionally, Mousa pointed out several important difficulties and potential pitfalls associated with the scrutiny of the lightning measurements databases, as described in [7].

To complicate matters even further, there is a correlation between some of the statistical variables defining the lightning current waveshape, for example, there is a statistically significant correlation between lightning current amplitude and a front steepness (and/or front duration). This necessitates usage of not only the joint but the conditional probabilities as well. Inherently, correlation coefficient between these statistical variables features prominently in defining the lightning current waveshape. Additionally, there is significant dispersion among statistical parameters defining the log-normal distribution of lightning currents, owing to various influences (both natural and artificial).

Incidence of lightning strikes to modern WTs is still under scrutiny of researchers around the world, although some recommendations exist, as will be mentioned later in accordance with the currently relevant IEC recommendation. However, this rather complex task is further exacerbated by several peculiarities (still under scrutiny) pertaining to the WT lightning incidence (and not found in other similar structures), as will be pointed out later. The intensified research of the WT lightning incidence has been fuelled by the extensive growth of the wind farms and their increasing significance as power producers throughout the world. Importance of this research will probably only increase in the future, as the wind farms are being compelled to occupy less favorable locations (due to the fact that the best ones get occupied first).

11.2.1 Log-Normal Statistical Distribution

It has been a well-established fact that probability distributions of the lightning current parameters follow a log-normal distribution. Statistical variation of the logarithm of a random variable from the log-normal distribution follows the normal (i.e., Gaussian) distribution. In that case, the probability density function (PDF) of the random variable from the log-normal distribution, for example, lightning current amplitude termed I, is given by the following expression [8]:

$$p(I) = \frac{\exp\left[-\frac{(\ln I - \ln I_\mu)^2}{2\sigma_{\ln I}^2}\right]}{\sqrt{2\pi} I \sigma_{\ln I}} \qquad (11.1)$$

where I_μ represents the median value and $\sigma_{\ln I}$ represents the associated standard deviation of the log-normal distribution at hand.

Additionally, statistical depiction of the lightning current parameters is aided by the cumulative distribution function (CDF), which provides a probability that a certain value of random variable (e.g., amplitude) will be exceeded. Hence, the CDF of the lightning current amplitude, which in fact provides a probability that the amplitude I will exceed some arbitrary value I_0 (of the associated log-normal distribution), is obtained from (11.1), by change of variables

$$u_0 = \frac{\ln I_0 - \ln I_\mu}{\sqrt{2}\sigma_{\ln I}} \qquad (11.2)$$

and integrating (11.1) between this new variable I_0 and ∞, which in turn yields [8]

$$P(I \geq I_0) = \frac{1}{\sqrt{\pi}} \int_{u_0}^{\infty} e^{-u^2} du = 0.5 \, \text{erfc}(u_0) \qquad (11.3)$$

where erfc is the complementary error function defined by the following expression:

$$\text{erfc}(u_0) = 1 - \frac{2}{\sqrt{\pi}} \sum_{n=0}^{\infty} \frac{(-1)^n u_0^{2n+1}}{n!(2n+1)} \qquad (11.4)$$

The expression (11.3) holds for the negative and positive polarity (first and subsequent) upward and downward lightning strikes. The median value (I_μ) and standard deviation ($\sigma_{\ln I}$) of the lightning current amplitudes from (11.1) and (11.3) are defined in, for example, [3, 8, 9], and by many others, for the first negative, subsequent negative, and positive polarity lightning strikes.

It needs to be further emphasized that these parameters of the log-normal distribution (median value and standard deviation) are not, unfortunately, uniquely defined for the entire world. Significant differences exist between different geographical locations around the globe, with additional differences arising between different regions, often influenced by meteorological and other parameters. Also, there are appreciable seasonal differences in some regions, such as, for example, the infamous winter lightning in Japan [10]. Naturally occurring differences in lightning activity often arise from different climatological influences (e.g., tropical thunderstorms, monsoons), producing different lightning patterns during different seasons of the year. Significant differences in lightning activity have been observed even on a diurnal basis, especially in the tropical regions, for example, [11].

Furthermore, differences also arise between statistical data provided by different researchers due to inherent differences in measuring equipment and other factors, such as the sample size, differences in the positions of sensors on measuring towers (top or bottom of the tower), and different trigger thresholds, which subsequently influence the parameters derived for the log-normal distribution [6]. Additionally, parameters of lightning currents associated with rocket-triggered lightning differ significantly from those of natural downward lightning, which further complicates the process of gathering the lightning data [3]. Namely, rocket-triggered lightning currents could be (approximately) related to the naturally occurring lightning currents associated with the upward-initiated lightning strikes.

Therefore, it could be concluded from the above-stated facts that a definition of the median value and standard deviation for the log-normal distribution of the lightning current parameters is plagued with various difficulties, and is still a matter of ongoing research around the world. This can be stated equally well for both downward and upward lightning flashes. However, a rather extensive coverage of the lightning activity by the usage of the lightning detection networks (e.g., NLDN in North America and LINET in Europe) will surely remedy at least some of these lightning-associated ambiguities in the near future. This will probably have particular influence in regard to the lightning currents incident to wind parks (including offshore ones to some extent).

11.2.2 Downward Lightning

Downward (negative) lightning has been traditionally a major concern of numerous research papers, due to the fact that the lightning incidence of HV (overhead) transmission lines is influenced (almost) exclusively by this type of lightning strike. There is abundant literature on this subject.

We are here primarily concerned with the lightning current amplitudes. These are in fact currents associated with the so-called return stroke phase of the (negative) downward

TABLE 11.1   Parameters of the log-normal distribution of the first negative downward lightning current amplitude [12].

I_μ (kA)	$\sigma_{\ln I}$	Comment
30.1	0.76	[9]
31.1	0.48	[8, 13]
39.2	0.76	Japan [14]
30.0	0.26	[15]
61.1	1.34	CIGRE ($I \le 20$ kA) [9]
33.3	0.61	CIGRE ($I > 20$ kA) [9]

lightning flash [3]. Hence, in order to appreciate the differences between log-normal distributions of lightning current parameters, for example, amplitudes, a brief summary aggregated from several different sources is presented in Table 11.1 [12].

It can be readily appreciated from the data presented in this table that there is obviously some dispersion between these log-normal distributions, owing to the adopted median values and standard deviations.

The above-introduced CDF of the log-normal distribution for the first negative downward current amplitudes is often approximated with a well-known so-called Anderson expression, for example, [16]:

$$P(I) = \frac{1}{1 + \left(\frac{I}{31}\right)^{2.6}} \qquad (11.5)$$

which provides a probability that the lightning current amplitude I (kA) will be exceeded. This expression has been adopted by the IEEE for the representation of downward negative lightning current distribution. On the other hand, as mentioned in Table 11.1, CIGRE recommends different distribution. The CIGRE distribution, as can be nicely observed from Table 11.1, is divided into two sections (at the lightning current amplitude of 20 kA). This has been traditionally done in order to better assess the shielding effects of HV (overhead) transmission lines (i.e., in the analysis of the so-called shielding failure flashover rate (SFFOR) for transmission lines).

In regard to the downward lightning strike, a so-called subsequent lightning stroke current has (particular) importance in some lightning-related overvoltage studies, due to the fact that it is associated with higher front steepness (although it has lower amplitudes), which in turn (sometimes) produces higher overvoltages on the incident electrical equipment. Hence, according to [8], the following relation could be utilized to approximate the CDF of the (negative) subsequent downward lightning stroke current amplitudes:

$$P(I) = \frac{1}{1 + \left(\frac{I}{12}\right)^{2.7}} \qquad (11.6)$$

which in fact provides a probability that the lightning current amplitude I (kA) will be exceeded. This expression is similar in form to the Anderson approximation, given by (11.5), and is easy to implement and use, which cannot be readily said of expression (11.3).

11.2.3   Upward Lightning

The upward-initiated lightning strikes (i.e., upward lightning flashes) are generally associated with tall structures, that is, communication towers, masts, chimneys, as well as with modern WTs. These lightning strikes would not have been there if the tower/structure were not present. Therefore, in such cases, one could argue that the presence of tower/structure (e.g., WT) causes an increase in the number of lightning strikes to the associated area where that structure is situated.

In accordance with this, it has been observed that towers and other structures having heights below 100 m (i.e., HV transmission lines) do not normally experience upward-initiated lightning strikes, while towers/structures in excess of 500 m will virtually always experience upward-initiated lightning strikes, for example, see [3]. Hence, the new-generation WTs, which are well in excess of 150 m in height (hub height plus one blade length), will certainly experience upward lightning strikes to some extent. It has been shown by Eriksson that the relative percentage of upward-initiated lighting strikes, in relation to the total lightning incidence of a free-standing structure, is a function of the structure's so-called "effective height" [17]. This will be somewhat elaborated upon later.

Median lightning current amplitudes, associated with upward (negative) lightning, significantly differ from those of downward (negative) lightning. For example, median current of negative upward-initiated lightning to Gaisberg tower (100 m tall and located on a mountain 1287 m above the sea level) is estimated to be 9.2 kA [18]. Furthermore, it has been shown by many researchers that median current amplitudes associated with upward-initiated lightning have similar values to those of subsequent (negative) return strokes and to those of rocket-triggered lightning [3, 19]. These currents generally have median values in the range between 8 and 14 kA, according to [3]. Again, of major concern is the lightning current amplitude associated with the upward (negative) lightning flash. Hence, it could be noted here that the expression (11.6), introduced above for the subsequent negative downward lightning, could be utilized for the upward lightning as well.

Interestingly, measurements carried out on instrumented towers show that the upward lightning current amplitudes obtained at the top and bottom of the tall tower are different. It has been subsequently confirmed that this is due to the transient processes initiated in the tower as a consequence

TABLE 11.2 Parameters of the log-normal distribution of the (negative) upward lightning current amplitude [12].

I_μ (kA)	$\sigma_{\ln I}$	Comment
9.2	0.25	Gaisberg tower [17]
10	0.66	Berger [3]
13	0.55	Rocket triggered [3, 21]
12	0.57	Rocket triggered [3, 22]
9.8	5.6	Rocket triggered [3]

TABLE 11.3 Parameters of the log-normal distribution of the negative downward lightning current front duration [12].

$t_{f\mu}$ (µs)	$\sigma_{\ln t}$	Comment
3.83	0.55	[8, 13]
5.5	0.70	[8]
4.5	0.37	Japan [14]

of the lightning strike. In fact, the tower behaves as a transmission line, with reflections of the traveling (lightning) current/voltage waves occurring at points of surge impedance discontinuity, for example, see [20]. In this regard, lightning current measurements should be carried out on top of the towers or subsequently compensated appropriately. This could bear some significance in relation to the lightning current measurements carried out on WTs.

In order to appreciate the differences between the log-normal distributions of upward-initiated (negative) lightning currents, compared to those of rocket-triggered lightning, data aggregated from several different sources are presented in Table 11.2 [12].

It can be appreciated from the data presented in Table 11.2 that there is again some dispersion between the log-normal distributions of upward lightning, owing to the adopted median values and standard deviations. Furthermore, they somewhat differ from the log-normal distributions obtained from rocket-triggered lightning data as well.

11.2.4 Statistical Correlation between Lightning Current Parameters

It has already been mentioned that a statistically significant correlation has been established between some of the parameters defining the lightning current waveshape of the downward lightning (e.g., between the amplitude and front duration, as well as between the amplitude and front steepness). In order to express this correlation of lightning current parameters, one needs to employ a joint (and conditional) PDF. This joint PDF, in case of the lightning current amplitude (I) and front duration (t_f), can be described by the following relation [8]:

$$p(I, t_f) = \frac{\exp\left[-\frac{f_1 - f_2 + f_3}{2(1-\rho_c^2)}\right]}{2\pi I t_f \sigma_{\ln I} \sigma_{\ln t_f} \sqrt{1-\rho_c^2}} \quad (11.7)$$

with the introduction of

$$f_1 = \left(\frac{\ln I - \ln I_\mu}{\sigma_{\ln I}}\right)^2 \quad (11.8)$$

$$f_2 = 2\rho_c \left(\frac{\ln I - \ln I_\mu}{\sigma_{\ln I}}\right)\left(\frac{\ln t_f - \ln t_{f\mu}}{\sigma_{\ln t}}\right) \quad (11.9)$$

$$f_3 = \left(\frac{\ln t_f - \ln t_{f\mu}}{\sigma_{\ln t}}\right)^2 \quad (11.10)$$

Here, the newly introduced variables $t_{f\mu}$ and $\sigma_{\ln t}$ represent the median value and standard deviation for the log-normal distribution of the lightning current front duration, respectively, again for both the first negative and positive polarity lightning. Coefficient ρ_c in the above equations represents a correlation coefficient between these statistical variables, signifying here lightning current amplitude and front duration.

It needs to be accentuated that there is some dispersion between the median value and standard deviation of the log-normal distributions of current front durations, as has been the case with current amplitudes. The reasons for this dispersion are in fact the same as those influencing the lightning current amplitude, as mentioned above, which also influence other lightning parameters (e.g., transferred charge and action integral). Hence, Table 11.3 summarizes several different median values and standard deviations of the log-normal distribution of the lightning current front duration, as obtained by different researchers (mainly from HV transmission line measurements) [12].

It can be appreciated from the data presented in this table that the median value of the negative downward lightning current front duration—which is in fact significantly responsible for the lightning-associated overvoltage stress on the incident electrical equipment—is in the order of several microseconds.

Additionally, correlation coefficient between these two statistical variables (i.e., amplitude and front duration) varies somewhat [3, 8]; also, it is different from the correlation coefficient found between lightning current amplitude and front steepness. Hence, it ought to be specifically provided for each combination of the parameters of the associated log-normal distributions. The problem is thus further exacerbated by this fact, owing to the (appreciable) influence which the correlation coefficient bears in the subsequent lightning-related analysis.

The conditional CDF of the two mutually dependent lightning stroke current parameters, for example, amplitude and front duration, both drawn from the log-normal distribution, is given by the following expression [8]:

$$P(I \geq I_k | t = t_f) = 0.5 \, \mathrm{erfc}(u_0) \quad (11.11)$$

with variable u_0 now given by the expression

$$u_0 = \frac{\ln I_k - b}{\sqrt{2}\sigma} \quad (11.12)$$

where

$$b = \ln I_\mu + \rho_c \frac{\sigma_{\ln I}}{\sigma_{\ln t}} \left(\ln t_f - \ln t_{f\mu} \right) \quad (11.13)$$

$$\sigma = \sigma_{\ln I} \sqrt{1 - \rho_c^2} \quad (11.14)$$

In order to emphasize the influence of the current front duration on the distribution of the lightning current amplitudes, the (conditional) CDF obtained from equations (11.11) to (11.14), with several different lightning current front durations, is graphically displayed in Figure 11.1. Here, six consecutive current front durations were selected, ranging from 1 to 6 μs. The following values of the underlying statistical distributions were utilized: $I_\mu = 30.1$ kA, $\sigma_{\ln I} = 0.76$, $t_{f\mu} = 3.83$ μs, $\sigma_{\ln t} = 0.55$, and $\rho_c = 0.47$. The so-called Anderson expression (11.5) is also depicted in this figure, for the purpose of comparison. It is quite noticeable from this figure that the lightning current amplitudes having shorter front duration are less probable than those having longer front duration. This is important since the lightning currents with

shorter front durations (i.e., those having steeper front) produce larger overvoltages on the associated electrical equipment than those associated with longer front duration.

It also needs to be stated that, apparently, there is no statistical correlation between lightning current amplitude and wave duration, which means that they are treated as two independent statistical variables [8]. Hence, this case of the joint PDF, for example, again between lightning current amplitude and front duration, could be defined as follows:

$$p(I, t_h) = p(I) p(t_h) \quad (11.15)$$

Furthermore, this means that the following equation for expressing the joint cumulative probability holds, that is, the joint CDF [8]:

$$P(I \geq I_k, \, t \geq t_h) = 0.25 \, \mathrm{erfc}(u_{0I}) \mathrm{erfc}(u_{0t}) \quad (11.16)$$

with

$$u_{0I} = \frac{\ln(I_k) - \ln(I_\mu)}{\sqrt{2}\sigma_{\ln I}} \quad (11.17)$$

$$u_{0t} = \frac{\ln(t_h) - \ln(t_{h\mu})}{\sqrt{2}\sigma_{\ln t_h}} \quad (11.18)$$

Here, the newly introduced values $t_{h\mu}$ and $\sigma_{\ln t_h}$ represent the median value and standard deviation for the log-normal distribution of the lightning current wave duration. According to [7, 12], the following parameters are often utilized: $t_{h\mu} = 77.5$ μs and $\sigma_{\ln t_h} = 0.58$.

11.2.5 Distribution of Lightning Current Amplitudes Incident to Wind Turbines

In regard to the lightning current log-normal distributions pertaining to WTs, it needs to be accentuated that, apparently, there is no such distribution publicly available (at least as far as the author is informed), obtained from the long-term and large-scale wind farm measurements. These kinds of measurements, carried out on a number of wind farms (onshore and offshore), are currently underway in different parts of the world, for example, [22–24]. It is believed that they could produce a lightning current distribution better suited, that is, specific, to the WT lightning incidence.

Notwithstanding that, some preliminary measurement results obtained on wind farms do exist, such as those reported in [23, 24]. In accordance with [23], measurements carried out on wind farms revealed that the median value of the lightning current amplitudes, incident to WTs, lies between 6 and 10 kA, which is significantly below the 31 kA value associated with downward negative lightning strikes. Furthermore, lightning current amplitudes measured on WT

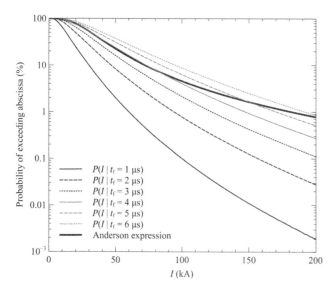

FIGURE 11.1 Joint conditional CDF of the log-normal distribution for the first negative downward lightning current amplitudes having six different front durations.

blades are associated with lower values than those measured on the wind vane.

Hence, in order to put things into perspective, Figure 11.2 depicts the actual measurement data, gathered from real wind farms, and compared with data predicted by IEC 61400-24. This figure, obtained from the data presented in [23], represents a cumulative distribution of lightning strikes exceeding different current amplitudes, being incident to WTs. It is clear from this figure that there is a discrepancy between the IEC-proposed values and the actual measurement data, particularly in the high-amplitude region.

Due to the (significant) number of upward-initiated lightning strikes seen in WTs, there is a clear tendency that the statistical distribution of lightning current amplitudes—pertinent to the WT lightning incidence—will be associated with somewhat lower median values, as has already been observed from several wind farm measurements, for example, [23, 24].

It has been stated that the median current amplitudes associated with upward-initiated lightning have values in the range between 8 and 14 kA, while that of the downward negative lightning are associated with amplitudes generally having a median value of 31 kA. This latter value is often used in lightning-related studies of HV transmission lines, with associated log-normal distribution or Anderson's expression. However, the appropriateness of this distribution for the assessment of the WT lightning incidence is somewhat questionable, as already noticed by several researchers in this field.

Hence, it would seem necessary to derive the unique set of parameters for the log-normal distribution tailored specifically for the analysis of WT lightning incidence, which would account for both upward and downward lightning strikes (and their relative proportion; see also Figure 11.2). Furthermore,

other lightning current parameters (front duration and/or steepness, wave duration, charge transferred, action integral, etc.), associated with WT lightning incidence, should be derived from measurements carried out on actual wind farms. They would certainly differ from those provided for the HV transmission lines. However, the current IEC 61400-24 apparently neglects to carry out full consideration of these implications (beyond the general observations), probably to some extent due to the fact that it draws extensively from the more general IEC 62305.

11.3 LIGHTNING PROTECTION OF WIND TURBINES

Lightning protection philosophy of WTs, according to IEC 61400-24 [25], is based on the assessment of various risks involved in the potential damages related to lightning strikes associated with the WT (direct and nearby lightning strikes to WT installation). This philosophy is inherited from IEC 62305 [26] and adapted to WT installations accordingly. The risk of damage itself can be considered as a sum of several risk components, where each component may be expressed by the following general equation [25]:

$$R_x = N_x P_x L_x \qquad (11.19)$$

with N_x the number of dangerous events per annum (year^{-1}), P_x the probability of damage to the WT which is a function of various protection measures, and L_x the consequent loss. This basic equation is used for assessing the risk of damage based on the probability of damage of various types and the consequent losses, in accordance with the profuse explanations provided in IEC 61400-24. In the case when the risk is found to be too high, protection measures have to be applied, as necessary, in order to reduce the risk to less than the tolerable risk (which in turn may be stipulated by the authorities); see [25].

Hence, in accordance with (11.19), one first needs to assess the WT lightning exposure, also called the WT lightning incidence, in regard to the direct and indirect lightning strikes (accounting for the service lines as well, i.e., electric power lines and possibly telecommunication lines). Then, following the methodology described in IEC 61400-24, one can assess the necessary probabilities of various types of damages needed to be accounted for, as well as consequential losses. The procedure for this assessment is rather straightforward and is presented in IEC 61400-24 in every detail; see [25] for more information. Finally, due to the fact that IEC 61400-24 rests heavily on IEC 62305, the interested reader is advised to consult IEC 62305-2 [26], for more information regarding the risk assessment procedure.

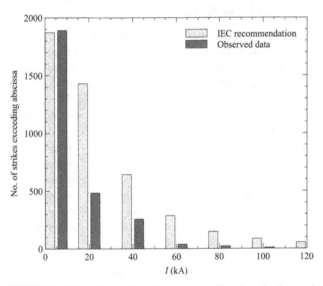

FIGURE 11.2 Comparison between predicted and observed lightning current amplitudes incident to wind turbines.

We will henceforth concentrate on the assessment of the WT lightning exposure, particularly in relation to the direct lightning strikes which are of utmost concern (in terms of both lightning and overvoltage protection of WTs and wind farms in general).

11.3.1 Wind Turbine Lightning Exposure Assessment

11.3.1.1 General Considerations. The problem of determining the WT lightning exposure, that is, the lightning incidence of WTs, is itself exacerbated by several peculiarities, some of them associated with tall structures in general, and some specific to WTs in particular. One such feature is inherent to all tall and isolated structures (such as communication towers, masts, and chimneys, as well as WTs), which is associated with increased lightning "attractiveness" of the structure. Namely, when the stepped leader propagates toward the ground, the field at the grounded objects gradually increases until it reaches a critical value, upon which an upward connecting leader is initiated from the structure, toward the descending stepped leader. The inception of the upward leader from the tip of the WT blades is largely dependent on the level of field enhancement at the blade tips, predominantly governed by the geometry, the proximity of other structures, the background electric field due to the thunderstorm cloud, and the descending stepped leader charge, for example, see [27–29] for more information. In this regard, a phenomenon of the "effective height" for the structures (i.e., WTs) has been introduced, as will be elaborated upon later.

In principle, a tall tower (or a WT) on flat ground tends to collect to itself the lightning flashes that would otherwise have struck the nearby ground. Furthermore, it also produces, if certain conditions are satisfied (as will be elaborated upon later), upward lightning flashes. When dealing with a WT lightning stroke incidence one should account for both the downward- and upward-initiated lightning strikes, although this might be difficult, especially in view of the scarcity of lightning measurement data obtained on modern WTs. Furthermore, local terrain topology can to an extent influence the upward-initiated lightning, mainly due to conditions related to the ambient electrical field, influenced by the thundercloud and WT structure [28, 29].

Additionally, the fact that the blades in WTs rotate introduces new effects and complexity, specific to the WTs, which is not found in other tall structures such as communication towers. Namely, it is known from the rocket-triggered lightning experiments that a fast upward-moving object connected to the ground can initiate lightning. It has been speculated for some time that upward moving WT blades could initiate upward lightning. Hence, a number of similarities found between rocket-triggered lightning and tall WTs with their rotating blades suggest that the WTs may trigger their own (upward) lightning [30]. However, this has not yet been confirmed. It remains to be seen whether this correlation between rocket launch and rotating blades could be fully established and experimentally corroborated. If corroborated, it would significantly alter the way in which the WT lightning exposure has been looked upon.

11.3.1.2 Lightning Exposure According to IEC 61400-24. The frequency of lightning affecting a WT, according to IEC 61400-24, is based on the methodology presented in IEC 62305-2. It starts by assessing the so-called local annual average ground flash density (in km^{-1} year^{-1}), from the following empirical expression [25]:

$$N_g = 0.1T_d \tag{11.20}$$

where N_g is the annual average ground flash density and T_d is the number of thunderstorm days per year (in year^{-1}) obtained from the isokeraunic maps (typically available from the national weather bureaus). It should be mentioned that the (annual average) ground flash density can be obtained from the following expression as well [16]:

$$N_g = 0.04T_d^{1.25} \tag{11.21}$$

although IEC 61400-24 recommends expression (11.20). Both of these expressions have been obtained by some sort of "fitting" of the measurements data gathered from different regions in the world. For example, relation (11.21) is based on the regression equation relating the logarithm of 5-year average value of N_g (measured with 10 kHz lightning flash counters at 62 locations in South Africa) and the logarithm of the value T_d (as reported by the corresponding weather stations); for example, see [3]. There are also other similar expressions, relating thunderstorm hours (T_h) with the annual average ground flash density. It is considered that this parameter (T_h) is potentially more closely related to the lightning incidence than T_d and might have precedence [3]. This stems from the fact that the parameter T_d (unlike parameter T_h) does not distinguish between a small thunderstorm producing a few lightning flashes in tens of minutes and a large storm lasting for several hours and producing hundreds of flashes.

Furthermore, nowadays with lightning detection networks covering entire continents (possibly further combined with satellite-based lightning detectors), it becomes possible to directly "measure" the annual average ground flash density; for example, see [3] for more information. This might be of importance for offshore wind farms, where national weather bureaus' information on lightning activity could be scarce or even deficient. Particularly the high-resolution maps of annual ground flash density, derived by this approach, could be of significant value, for example, in the design phase of the wind farm overvoltage protection (and surge arrester selection). From all of the above stated, there is apparently considerable "room for improvement" in the estimation of this basic parameter (N_g), which stands as a foundation of the subsequent lightning incidence estimation procedure. Taking into account the severe exposure of WTs, coupled with the importance of the associated backsurge phenomenon, the proper assessment of the parameter N_g becomes more important here than with say HV transmission lines (which anyway may span hundreds of kilometers while traversing through very different terrain).

The average annual number of lightning-related events that may endanger in some way a wind turbine, in relation to the risk assessment of equation (11.19), may be separated into the following five components [25]: N_D due to direct lightning strikes to the wind turbine, N_M due to lightning strikes near the WT (within 250 m), N_L due to lightning strikes to the service lines connecting the WT (i.e., electric power or communication cables), N_l due to lightning strikes near the service lines connecting the WT, N_{Db} due to lightning strikes to a WT or another structure at the other end of the service lines connecting the WT in question.

The lightning incidence of WT (i.e., frequency of lightning directly affecting a WT), according to IEC 61400-24, can be assessed as

$$N_D = N_g A_d C_d \times 10^{-6} \qquad (11.22)$$

where A_d (m²) is the collection area of lightning flashes to the WT and C_d is the so-called environmental (i.e., orographic) factor. Appropriate values for this factor are $C_d = 1$ for WTs positioned on flat land and $C_d = 2$ for WTs on a hill or a mountain ridge. Additionally, IEC 61400-24 recommends for the WTs placed at locations known to be very exposed to lightning in general, or to winter lightning in particular, to be assigned higher environmental factor C_d (supposedly to consider the upward lightning being triggered under such conditions). Furthermore, according to IEC 61400-24, WTs placed offshore may have to be assigned an environmental factor C_d of 3–5 to get a realistic estimate of the frequency of lightning attachment [25].

The collection area, sometimes also termed "electrical shadow" area, of a structure in general is defined as an area of ground surface which has the same annual frequency of lightning ground flashes as the structure itself. It is based on the so-called attractive radius concept for downward lightning, which will be elaborated upon somewhat later on. According to IEC 61400-24, for isolated structures, such as the WT, the equivalent collection area is the area enclosed with a border line obtained from the intersection between the ground surface and a straight line with a 1:3 slope which passes from the upper parts of the structure (touching it there) and rotating around it, see [25, 26]. This suggests that the collection area of direct lightning flashes to the WT is determined by the expression [25]

$$A_d = 9\pi H^2 \qquad (11.23)$$

According to IEC 61400-24, it is recommended that all WTs are modeled as a tall mast with a height equal to the hub height plus one rotor radius. This is recommended for WTs with any type of blades including blades made solely from nonconductive material such as glass fiber-reinforced plastic. Figure 11.3 shows the collection area produced by a WT placed on a flat ground, accounting for the connection line to the incident structure at the other end (which could be another WT or a transformer station) [25].

From this figure it is apparent that the collection area A_d assigned to the WT equals the surface area of a gray circle, whose radius amounts to three times the height of the WT (as defined above) on a flat terrain. If the WT is positioned on a hill or a ridge, effective height is introduced instead of the actual height, when determining the collection area [25]. Hence, for WTs positioned in exposed locations IEC 61400-24 directs user to consult the IEC 62305-2, which in turn provides further guidance on the treatment of structures in complex terrain or in proximity to other structures. However, we will return with some more details to the treatment of effective height of exposed structures later on.

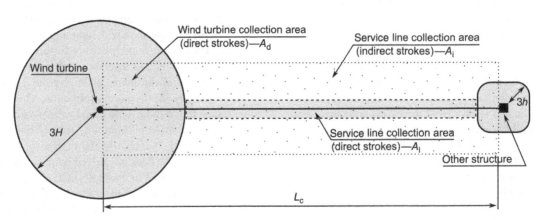

FIGURE 11.3   Collection area produced by a wind turbine placed on a flat ground in accordance with IEC 61400-24:2010.

Additionally, WTs may be endangered by the nearby lightning (in terms of the radiated electromagnetic field effects). The annual number of these dangerous events could be estimated, according to [25], by the following expression:

$$N_{\mathrm{M}} = N_{\mathrm{g}}(A_{\mathrm{m}} - A_{\mathrm{d}}C_{\mathrm{d}}) \times 10^{-6} \qquad (11.24)$$

where the newly introduced variable A_{m} (in m^2) determines a collection area for lightning strikes near the WT, which is the area within a distance of 250 m according to [25]. This area is not depicted in Figure 11.3 for convenience.

Furthermore, IEC 61400-24 views necessary considerations regarding the lightning effects to connecting electrical power (and communication) lines, interconnecting WTs in the wind park and to the HV transformer station (switchyard). Hence, according to [25], the number of lightning strikes to a connecting transmission line can be accessed in the same way as proposed in IEC 62305-2 [25, 26]:

$$N_{\mathrm{L}} = N_{\mathrm{g}}C_{\mathrm{d}}C_{\mathrm{t}}A_{\mathrm{l}} \times 10^{-6} \qquad (11.25)$$

where A_{l} is the collection area of lightning strikes to transmission line (m^2), C_{d} the location factor (assumed equal to 1 in flat areas, 2 in hilly terrain, and 0 for submarine cables), C_{t} the transformer factor ($C_{\mathrm{t}} = 1$ if there is no transformer and $C_{\mathrm{t}} = 0.2$ if there is one between the point of lightning attachment and WT). The collection area of the transmission line is also graphically depicted in Figure 11.3 (designated A_{l}). The collection area of lightning strikes to transmission line (either overhead or buried), in accordance with IEC 61400-24, is provided by the following expression [25]:

$$A_{\mathrm{l}} = \begin{cases} 6h_{\mathrm{c}} \left[L_{\mathrm{c}} - 3(H + h) \right] & \text{for overhead line} \\ \sqrt{\rho} \left[L_{\mathrm{c}} - 3(H + h) \right] & \text{for buried cable line} \end{cases} \qquad (11.26)$$

with ρ being the relative soil resistivity (Ω m), where the maximum value of 500 Ω m is recommended to be assumed, and h_{c} the height of the overhead transmission line (m). However, it is not explained why this particular value of the maximum soil resistivity has been recommended.

The number of lightning strikes near the transmission line (i.e., close enough to affect the line) can be assessed as follows [25, 26]:

$$N_{\mathrm{l}} = N_{\mathrm{g}}C_{\mathrm{e}}C_{\mathrm{t}}A_{\mathrm{i}} \times 10^{-6} \qquad (11.27)$$

where A_{i} is the collection area of lightning strikes near the transmission line (m^2) and C_{e} the environmental factor assumed equal to one for rural areas. The collection area denoted by A_{i} is depicted in Figure 11.3 and is assumed equal to $1000 \cdot L_{\mathrm{c}}$ for overhead lines and $25\sqrt{\rho}L_{\mathrm{c}}$ for buried cables, where L_{c} is the line length in meters (maximum length of 1000 m could be assumed) [25]. Again, the maximum value

of 500 Ω m for the soil resistivity could be assumed in accordance with [25]. Some additional comments on the collection area for the buried cables will be provided later on.

It can be seen from the relations (11.22) to (11.27) that the annual number of dangerous events, estimated by IEC 61400-24, is closely related to the appropriate collection areas, some of which are related to the attractive radius concept, while others have been generally determined (mostly empirically) from the analysis and data gathered on different structures. Several empirically determined factors also influence the estimation procedure, for example, environmental factor and transformer factor. This collection area approach has in fact been inherited and adapted from IEC 62305, which has traditionally been predominantly engaged with HV overhead transmission line (and switchyard/transformer station) lightning protection problems. Some additional comments on the collection area will be provided later on.

11.3.2 Lightning Currents Associated with Dangerous Events

The above-estimated annual number of dangerous events, provided by relations (11.22), (11.23), (11.24), (11.25), (11.26), and (11.27), does not contain information regarding the associated lightning current parameters which are inherent in these events, that is, it does not correlate dangerous events with associated probability that some lightning current amplitude (or some other parameter, e.g., transferred charge) will be attained and/or surpassed. This correlation—particularly regarding the lightning current amplitudes—becomes important in the overvoltage protection analysis, with particular emphasis given to the so-called backsurge phenomenon, and further bears significance in the selection procedure for the metal-oxide surge arresters, for example, see [10, 31].

In order to account for this mentioned correlation, one would generally proceed with the following relationship:

$$n_{\mathrm{D}} = N_{\mathrm{D}}P(I_0)\tau \qquad (11.28)$$

where n_{D} is the number of direct lightning strikes to WT, within the specified time window in years, having lightning currents equal to or exceeding some a priori selected value I_0 (kA); $P(I_0)$ the CDF of the log-normal distribution of lightning current amplitudes to WTs, determining the probability with which a current amplitude I_0 (kA) will be exceeded; N_{D} the previously determined number of dangerous events due to direct lightning strikes to the WT; and τ the time window of interest (years).

Moreover, joint and/or conditional CDF could be used instead of $P(I_0)$ in (11.28), which would further allow definition of lightning current parameters incident to WTs (e.g., amplitude and front duration). This is important due to the

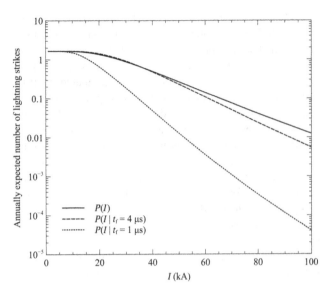

FIGURE 11.4 Annually expected number of direct lightning strikes to wind turbine as a function of the lightning current amplitude which would be exceeded in the event.

fact that lightning current front duration (and/or front steepness) has important influence on the subsequently obtained overvoltage stress on the associated electrical equipment (in relation to the backsurge phenomenon), for example, [10]. However, difficulty with the application of the expression (11.28)—regardless of the CDF used—stems from the fact that the CDF itself, of the lightning current distribution to WTs, is not known (at least one that would account for both the downward and upward lightning strikes and their mutual proportion). Furthermore, the definition of this distribution would be faced with similar problems introduced for the other distributions already presented in Section 11.2.

Notwithstanding that, it could be argued that the large lightning current amplitudes associated with short front durations—which produce severe overvoltages on associated electrical equipment—generally have lower probability of occurrence than the same current amplitudes with longer front durations. Hence, let us take a 120 m high WT on flat terrain ($C_d = 1$) and estimate the annual number of lightning strikes having (or exceeding) some current amplitude, in accordance with (11.28), (11.22), (11.23), (11.20), and data provided in Section 11.2. A factor of 0.9 is included in equation (11.28) in order to account for the fact that approximately 90% of lightning strikes are assumed to be of negative polarity. Figure 11.4 provides (on a semi-log scale) a graphical representation of the obtained results, using different log-normal distributions, as discussed in Section 11.2. The following data were implemented: $P(I)$ with $I_\mu = 31.1$ kA and $\sigma_{\ln I} = 0.48$; $P(I|t_f = 1$ μs$)$ and $P(I|t_f = 4$ μs$)$ with $I_\mu = 31.1$ kA and $\sigma_{\ln I} = 0.48$, $t_{f\mu} = 3.83$ μs, $\sigma_{\ln f} = 0.55$, and $\rho = 0.47$. The annual ground flash density has been computed with expression (11.20), while assuming $T_d = 45$.

An effect of introducing the correlation between the lightning current amplitude and front duration could be clearly seen from this figure, manifested in significant reduction to the annually expected number of dangerous events associated with arbitrarily selected value of lightning current amplitude.

On the other hand, since the CDFs for the downward and upward distributions of lightning current amplitudes are independently known (see Section 11.2), it would seem possible—as a first approximation—to take them into account separately. However, one needs the proportion of downward to upward lightning strikes into the WT in order to proceed. Fortuitously, it happens that Eriksson produced the following empirical relation which relates a tall tower's height to the percentage of upward-initiated lightning strikes [17]:

$$p_u = 52.8\ln(H) - 230 \text{in}(\%) \qquad (11.29)$$

with H (m) being here the (effective) height of the structure (i.e., tower). This expression is valid for tower heights ranging from 78 m up to some 540 m. It shows that the proportion of upward-initiated lightning strikes becomes a dominant feature with tall towers (completely dominating for towers having height in excess of some 500 m) [17]. Hence, combining (11.28) and (11.29), while accounting for the proportion of upward to downward lightning strikes at the same time, yields

$$n_D(I_0) = N_g A_d C_d$$
$$\times \left[0.9\frac{100 - p_u}{100}P_{\text{down}}(I_0) + \frac{p_u}{100 - p_u}P_{\text{up}}(I_0) \right] \tau$$
$$(11.30)$$

where $P_{\text{down}}(I_0)$ is the CDF of the log-normal distribution of downward negative lightning current amplitudes (from Section 11.2.2), while $P_{\text{up}}(I_0)$ is the CDF of the log-normal distribution of upward-initiated lightning current amplitudes (from Section 11.2.3); see also [12, 31]. In the above expression, A_d is the WT collection area provided by (11.23). We are here ignoring the positive lightning strikes (assuming 10% of them), although they could be accounted for, which would be particularly important in those situations where their relative proportion is higher than generally assumed. It is important to mention here that the expression (11.30) is simple to use, although it neglects to consider the complex physical processes associated with upward lightning (see also the next section) [28, 29].

Figure 11.5 depicts the annually expected number of direct lightning strikes to WTs, having two different heights, as a function of the lightning current amplitude which would be exceeded in the event. The annual ground flash density in (11.30) has been computed with expression (11.20), while

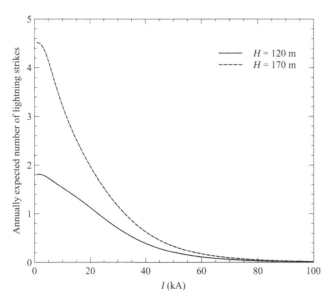

FIGURE 11.5 Annually expected number of direct lightning strikes to wind turbine as a function of the lightning current amplitude (downward and upward lightning combined).

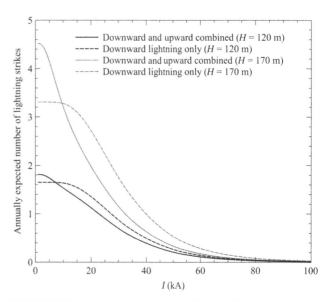

FIGURE 11.6 Comparison between the annually expected number of WT lightning when using only downward and when using both downward and upward lightning.

assuming $T_d = 45$. It is also assumed that $C_d = 1$ (WTs positioned on flat terrain). The following parameters of the log-normal distribution for the downward negative lightning current amplitudes are employed: $I_\mu = 31.1$, $\sigma_{\ln I} = 0.48$. Also, the following parameters of the log-normal distribution for the upward lightning current amplitudes are employed: $I_\mu = 10$, $\sigma_{\ln I} = 0.66$.

It is again clearly evident that the dangerous events having high current amplitudes are less probable than those having lower lightning current amplitudes. This is important from the perspective of the overvoltage protection analysis, for example, [10, 31]. Moreover, according to this figure, a dangerous event associated with lightning current amplitude in excess of, for example, 20 kA would happen approximately once every year for the shorter (i.e., 120 m tall) WT. The situation would be further exacerbated in those cases where WT is being positioned in lightning-exposed location (e.g., on mountaintops).

Furthermore, a comparison between the annually expected numbers of lightning strikes—having (or exceeding) lightning current amplitudes of arbitrary front duration—for 120 and 170 m tall WTs is presented in Figure 11.6. Influence of the upward lightning is here seen as a decrease in the annually expected number of lightning strikes having larger current amplitudes, which is more pronounced for the taller WT, as should be (generally) expected.

Finally, consider the 120 m tall WT again but assume that it has an effective height of 200 m (see also the next section); now Figure 11.7 depicts the annually expected number of upward and downward lightning strikes into this WT (individually), as a function of the lightning current amplitude

(which will be attained or exceeded). In accordance with (11.29), for the WT with effective height of 200 m, approximately 50% of all lightning strikes will belong to upward lightning.

What is important to emphasize here is the fact that, according to (11.29), tall structures experience predominantly upward lightning strikes, which are in turn associated with lower (lightning current) amplitudes (e.g., see data for the Gaisberg tower in Section 11.2.3). This can be readily

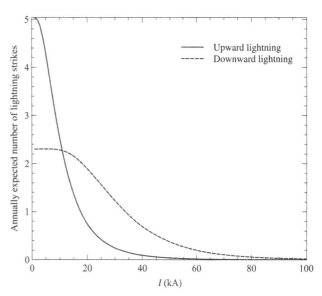

FIGURE 11.7 Annually expected number of upward and downward lightning strikes to the exposed wind turbine.

observed from the parameters of the associated log-normal distributions (again observe Section 11.2.3). The same trend is present with free-standing structures on mountaintops and other elevated areas exposed to lightning (e.g., the famous Berger's 70 m tall twin towers on Mount San Salvatore). However, this feature could not be fully reproduced by the simple analysis presented heretofore. More elaborate analysis is needed, probably in line with that presented by Rizk in, for example, [28,29].

It ought to be mentioned, at last, that the time window featuring in relation (11.28) could be loosely related to the life span of the wind farm, thus providing a measure for the techno-economic optimization of the overvoltage protection selection procedure (i.e., selection of metal-oxide surge arresters); for example, see [31] for more information. Namely, one can derive the lightning current amplitude which should (statistically speaking) strike the WTs in the wind farm at hand, approximately only once during the a priori selected time window, and subsequently use this current (possibly with its associated parameters, i.e., front duration, wave duration) to numerically optimize the wind farm overvoltage protection selection procedure.

11.3.3   On the Effective Height of Wind Turbines

It is quite evident that the expected annual number of direct lightning strikes to WT, provided by the expression (11.22), depends heavily on the collection area A_d, which in turn—at least for WTs positioned in complex terrain—depends on the so-called effective height of the WT. The treatment of the WT effective height, as adopted by IEC 61400-24 (which has been inherited from IEC 62305-2), is somewhat rudimentary in the sense that it lacks regard for the underlying physical picture of the phenomenon.

The effective height is generally seen as a height ascribed to the fictitious structure on the flat ground that would experience the same lightning incidence as an actual structure of some (lower) height exposed on top of the hill. A group of prominent authors presented a study on the effective height of tall towers (or masts) positioned in exposed locations (such as the hill top), in [32]. This study draws heavily from the in-depth analysis of tall structures in lightning-exposed locations provided by Rizk in [28, 29], and could arguably be applied to WTs in the same environmental conditions. Namely, Rizk elaborated on the detailed physical picture of the phenomenon associated with the conditions necessary for the lightning attachment to tall structures in exposed locations (accounting for both downward- and upward-initiated lightning strikes); see [28,29]. From his treatment, authors of [32] further obtained means for defining the effective height of tall structures in exposed locations, which will be briefly presented hereafter.

The effective height (H_{ef}) of the WT positioned on top of the hill, which is assumed to be hemispherical in shape, can be estimated from the following relation [32]:

$$H_{ef} = (k_1 + k_2 E_g)^{-k_3} \qquad (11.31)$$

where $k_1 = 5.87 \times 10^{-3}$, $k_2 = 2.04 \times 10^{-6}$, and $k_3 = 1.3941$ are constants, while E_g (kV/m) is the ambient uniform electric field required for the initiation of the upward lighting flash [28]. This field can be obtained from the following equation [28, 32]:

$$E_g = \frac{U_i}{(H + a)\left[1 - \frac{a^3}{(H+a)^3}\right]} \qquad (11.32)$$

with H being the actual height of the WT (hub height plus rotor radius) in meters, while "a" represents the equivalent radius of the hemispherical mountain base, also in meters. Finally, U_i in the above expression represents the electric potential at the tip of the WT blade, for the WT positioned on top of the mountain and immersed in the ambient uniform electric field of E_g (kV/m). It can be determined from the criteria for the occurrence of upward lightning flash as follows [32]:

$$U_i \geq \frac{1556}{1 + \frac{7.78}{R}} + x_0 E_\infty \ln\left(\frac{E_i}{E_\infty}\right) \qquad (11.33)$$

where x_0 is the parameter proportional to the upward leader speed, assumed by Rizk to be 5 m; E_i the minimum positive streamer gradient which is assumed to be about 400 kV/m; E_∞ the final quasi-stationary leader gradient assumed to be about 3 kV/m; and R the geometric parameter of the (idealized) hemisphere-shaped mountain, analytically determined by Rizk in [28] and provided by the following analytical expression:

$$R = \frac{2a(H + a)}{1 + \frac{2a(H + a)}{(H+a)^2 - a^2} - \frac{2a(H + a)}{(H+a)^2 + a^2}} \qquad (11.34)$$

Additionally, there is a minimum value of the uniform background electric field, below which no upward lightning flash will occur, regardless of the structure (i.e., WT) height. This implies according to [28] and [32] that the background uniform electric field (in which the structure is immersed) has to exceed some 3 kV/m in order for a complete upward flash to become possible.

In accordance with the above-presented theory, Figure 11.8 graphically depicts WT effective height (when positioned on top of the hemispherical mountain) as a function of different mountain base radiuses/heights. Two WTs with

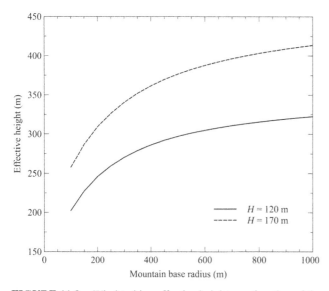

FIGURE 11.8 Wind turbine effective height as a function of the hemispherical mountain base radius.

actual heights of 120 and 170 m have been selected for this example.

It can be observed from this figure that the relation between the WT actual height and its effective height is non-linear and is affected by the hemispherical mountain base radius (i.e., mountain height), along with other influential parameters. It has been acknowledged in [32] that the influence of the final quasi-stationary (lightning) leader gradient has negligible effect on the effective height, as well as the minimum positive streamer gradient. On the other hand, upward positive (lightning) leader velocity has noticeable effect on the effective height of the structures positioned on top of the (hemispherical) mountains [32]. The same can be said of the influence pertaining to the relationship between the semi-minor axis and semi-major axis of the prolate spheroid ellipsoid mountain shape (not treated here); see [28, 32].

It is evident from the presented analysis that the effective height (regardless of its exact definition) is not equal to the simple addition of structure's height to the height of the hill (above the surrounding terrain), but that the relationship is far more complex. This fact ought to be borne in mind while assessing the collection area of WTs in complex terrain.

Finally, notwithstanding the sophisticated analysis of the conditions for the initiation of the upward lightning strikes to exposed structures that has been introduced, some variation in the computed effective height is still to be expected, due to the uncertainty present within influential parameters (e.g., such as the upward leader velocity). Certainly, more research—perhaps on the lines proposed by Rizk in [28, 29]—ought to be taken in order to better assess the effective height of WTs in exposed locations, as well as to assess the lightning incidence of WTs in general.

11.3.4 On the Collection Area of Wind Turbines

Apart from the above-discussed effect of the WT effective height, there is another aspect, regarding the WT lightning incidence, which ought to be elaborated upon. That is the definition of the WT collection area itself, and its estimation procedure. Namely, it has been recognized that the collection area for a free-standing structure is the area of ground surface which has the same annual frequency of lightning ground flashes as the structure itself. It has been traditionally derived by means of the attractive radius concept, for example, [17, 28, 29, 33–35].

It ought to be mentioned here that it is important to make a distinction between striking distance and attractive radius. The striking distance is defined as the distance between the object to be struck and the tip of the downward-moving leader at the instant that the connecting (upward) leader is initiated from the object, for example, [33]. It depends solely on the charge of the downward leader and hence return stroke (i.e., lightning stroke) current. On the other hand, a calculation of the attractive radius (which in turn yields collection area) takes into account the geometry of the object to be struck. This is done by including physical considerations of the ionization process of streamer and leader development [33]. As a very crude approximation, the striking distance can be used in place of the attractive radius, although it should be noted that, in general, striking distance is larger than attractive radius [33, 34]. Furthermore, it has been determined in [35] that the striking distance to structures in high-altitude sites (which could have direct influence on some wind farms) increases significantly, causing the corresponding increase in the risk of lightning strikes.

Collection area of the WT, in accordance with IEC 61400-24 [25] is provided by the expression (11.23) and it depends solely on the WT height (hub height plus blade length). Some authors have argued that the attractive radius, featuring prominently in the collection area definition, is a function of the lightning current amplitude as well as the structure's height, for example, [17, 28, 29, 35]. Considering this fact, the following expression could be utilized for the computation of free-standing structure's (i.e., WT's) collection area [33, 34, 36]:

$$A = \pi \int_0^\infty R^2(H, I) p(I) \mathrm{d}I \qquad (11.35)$$

where $R(H,I)$ is the attractive radius of the structure as a function of the structure's height and lightning current amplitude and $p(I)$ the PDF for the current amplitude distribution provided by (11.1). By considering several different relations for attractive radiuses [33, 34], namely,

– Eriksson:

$$R_1 = 0.84 H^{1.2} I^{0.74} \qquad (11.36)$$

– Petrov et al.:

$$R_2 = 0.56\left[(H+15)I\right]^{2/3} \tag{11.37}$$

– Rizk:

$$R_3 = 25.9H^{0.48} \tag{11.38}$$

and by using the median value of 31 kA and standard deviation of 0.7368 for the lightning current amplitude distribution of (11.1), relation (11.35) provides the following collection areas (in m^2) [33, 34]:

– Eriksson:

$$A_1 = 640H^{1.2} \tag{11.39}$$

– Petrov et al.:

$$A_2 = 153(H+15)^{4/3} \tag{11.40}$$

– Rizk:

$$A_3 = 2106H^{0.96} \tag{11.41}$$

In order to put things into perspective, Figure 11.9 provides a graphical relationship between the WT height (ranging from 60 to 120 m) and its collection area (in km^2) computed by IEC 61400-24 recommended expression (11.23) and those of Eriksson (11.39), Petrov et al. (11.40), and Rizk (11.41). It is important to mention that the upward lightning strikes have not been accounted for, which is consistent with IEC 61400-24 and [33] in defining the collection area. Additional information is provided in [34, 36].

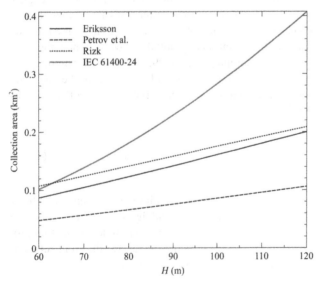

FIGURE 11.9 Wind turbine collection area as a function of its height with different attractive radiuses for lightning incidence.

It can be observed from this figure that the collection area determined by IEC 61400-24 recommended expression is larger than those provided by other authors. Hence, it would further be interesting to compare the WT lightning incidence obtained from IEC 61400-24 standard [25] with the actual data gathered on wind farms (both onshore and offshore). Some discrepancy is surely to be expected, as pointed out in, for example, [23, 36].

11.3.5 On the Lightning Incidence of Buried Cables

Similar to the previously carried discussion (see Section 11.3.2), one would need to account for the lightning current parameters (at least the amplitude), which is incident to the buried cables, now in relation to the collection area A_i in the expression (11.27). This could follow the general form provided by the expression (11.28), where now $P(I_0)$ would be the CDF of the downward (negative) lightning strikes, which are the only strikes associated with the lightning incidence to flat ground. However, log-normal distribution of lightning current amplitudes incident to flat ground differs somewhat from that generally employed (see Section 11.2), which has been traditionally seen as pertaining to the HV transmission lines (employed by IEC 61400-24 as well), for example, see [15] for more information.

It has been established that in the highly resistive soils, lightning currents could propagate for considerable distances in the ground (where the propagation distance increases with soil resistivity and lightning current amplitude), thus posing a threat of terminating on the buried cables. This could in turn produce damage on their outer insulation (i.e., cable jacket), for example, see [37, 38]. Hence, IEC 61400-24 provides means for determining the annual number of dangerous events that can endanger the buried cables, as presented by the relation (11.27). However, it would be additionally beneficial—as just stated—to further account for the lightning current parameters associated with such dangerous events. One possible approach to dealing with that issue is presented in [38].

By following Sunde's approach from [39], one could use the following expression in order to account for the number of lightning strikes to buried cable, per kilometer of cable length per year [36]:

$$n_l = N_g \times 0.9 \times 10^{-3}$$
$$\times \int_0^\infty \mathrm{erfc}\left\{\frac{1}{\sqrt{2}\sigma_{\ln I}}\left[\ln\left(\frac{r^2}{\rho k^2}\right) - \ln(I_\mu)\right]\right\}dr \tag{11.42}$$

where I_μ and $\sigma_{\ln I}$ represent the median value and associated standard deviation of the log-normal distribution of

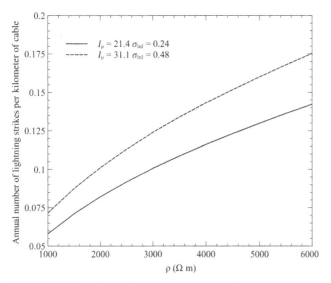

FIGURE 11.10   Annually expected number of lighting strikes per kilometer of cable as a function of the soil resistivity.

downward (negative) lightning strikes, respectively, while factor 0.9 is introduced to account for the fact that, generally speaking, 90% of downward lightning strikes are negative in polarity. Coefficient k in the above expression is determined by Sunde in [39], which for highly resistive soils that are of particular concern (i.e., $\rho \geq 1000\ \Omega$ m) has a value of 0.047. This expression accounts for the lightning current amplitudes, while at the same time providing the estimated lightning incidence of buried cables, which is not the case for the expression (11.27).

In order to accentuate the importance of the parameters associated with the log-normal distribution of downward negative lightning strikes, during the assessment of the buried cable lightning incidence, Figure 11.10 depicts the estimated annual number of lighting strikes per kilometer of cable as a function of the soil resistivity. The computation has been carried out for two different sets of parameters of the log-normal distributions of downward negative lightning currents, one in accordance with the data provided in Section 11.2 and the other in accordance with the analysis presented in [15]. The set of parameters from [15] is associated with lightning strikes to "flat ground" and is derived from lightning measurements on instrumented towers, which makes it suitable for this application. It should be mentioned that the annual ground flash density in (11.42) has been computed with expression (11.21), while assuming $T_d = 45$.

It can be observed from this figure that the selection of parameters for the log-normal distribution of downward (negative) lightning currents holds importance in the assessment of the buried cables lightning incidence. It is desirable, that is recommended, to implement the results from Figure 11.10 produced by the distribution of [15].

11.3.6   On the Wind Turbine Lightning Attachment Points

IEC 61400-24 defines four lightning protection levels (LPLs), I to IV, with four types of relevant protection measures for the design of LPS [25]. For each LPL a set of maximum and minimum lightning current parameters is fixed. The maximum values relevant to LPL-I will not be exceeded with a probability of 99%; they are reduced to 75% for LPL-II and further to 50% for LPL-III/IV. The minimum values of lightning current parameters for different mentioned LPLs are used to derive the so-called rolling sphere radius, in order to subsequently define the lightning protection zones (LPZs) which cannot be reached by direct lightning strikes. The method by which this is accomplished is called the rolling sphere method (RSM) [25]. These currents are 3 kA for LPL-I, 5 kA for LPL-II, 10 kA for LPL-III, and 16 kA for LPL-IV, which translates (according to the electrogeometric model of lightning attachment) to the following radiuses of the associated rolling spheres: 20 m for LPL-I, 30 m for LPL-II, 45 m for LPL-III, and 60 m for LPL-IV [25]. This sphere is in fact rolled along the WT structure, thus identifying potential points of lightning attachment (and associated areas, or protection zones); see [40] for an example of its application.

The IEC 61400-24 standard further recommends that all WT subcomponents should be protected according to LPL-I unless it is shown and demonstrated by the risk analysis that a lower level is adequate [25]. Furthermore, the application of the RSM to the WT structure identifies, as mentioned above, certain parts which are exposed to direct lightning strikes, termed LPZ 0_A according to the mentioned LPZ concept, as well as other areas protected by the structure itself, termed LPZ 0_B; see [25] for more information. Additionally, inner wind turbine zones could be subdivided into more LPZs (e.g., LPZ 1 and LPZ 2) with higher protection level and lower lightning parameter levels; see Annex E of [25] for more information.

Basically, in order to determine the WT LPZs one needs to assess the possible lightning attachment points, using either RSM or some other method. Namely, different parts of the WT structure could be associated with different possibilities of lightning attachment, in relation to both upward and downward lightning incidence. There are certain parts of the WT structure which are more susceptible to lightning attachment (as past experience has shown), such as the blades (particularly their tips). Hence, nowadays all new blade designs incorporate special lightning attachment elements, that is, receptors and down-conductors, as part of the lightning protection system, in accordance with IEC 61400-24 [25]. Another recognized lightning-endangered WT element is the wind wane (installed on the hub), which is also usually shielded from the direct lightning strike.

However, mathematically/geometrically or even purely numerically determining the lightning vulnerability points (i.e., lightning attachment points) on wind turbines is not an easy task. As already mentioned, conforming to IEC 62305, IEC 61400-24 implements the RSM. This method has been traditionally applied to the HV switchyards, transformer stations, and (residential and other) buildings of various shapes and sizes. Some authors maintain that this is a rather crude method, which hides our insufficient understanding of the lightning attachment process, but also appears to be the best tool so far, for the design and the positioning of the air terminals (at least in terms of the simplicity of its application). On the other hand, some authors have provided more sophisticated (purely numerical) methods for assessing the WT lightning attachment points, for example, [41–43]. They provide means for determining the exact positioning (and design) of receptors on blades and other parts of the WT structure and seem to be superior, in that respect, to the far-simpler RSM. However, wider usage of these methods is somewhat limited due to their numerical complexity and difficulty of application [41–43].

Certainly, more research is warranted in this area (in order to assure a high degree of lightning protection of WTs), particularly in regard to the numerical methods suitable for assessing the lightning attachment points (i.e., vulnerable points) on the WT structure. This is especially important, as already mentioned, for the exact positioning (and design) of the lightning receptors on the blades, which are found to be particularly vulnerable to direct lightning strikes. The approach already set forth in this respect in, for example, [41] and [42] features a finite element analysis, by means of extending a sophisticated commercial CAD software package.

11.4 SOME CONCERNS REGARDING THE OVERVOLTAGE PROTECTION OF WIND FARMS

Overvoltage protection analysis of the wind farm electrical systems has often been tackled by means of the well-known ElectroMagnetic Transients Program (EMTP) software package, for example, [37, 44, 45]. This software package has been, in fact, widely used since the 1980s, in conjunction with the increased usage of digital computers, in lightning and switching surge analysis of HV electrical power systems. The main concern in its application in the overvoltage protection analysis of wind farm electrical systems is the development and proper usage of system component models, along with their mutual interconnection and interaction. Hence, from the modeling viewpoint, it could be stated that the main problem in (lightning) overvoltage analysis of wind farms could be seen in the proper application of the appropriate (EMTP-type) models for each of the individual components of the wind farm electrical system. An excellent general treatise of various power system component models,

accompanied with detailed explanations on the range of their applicability, has been given in [46].

There are several key differences, which ought to be accounted for when studying transients, between wind farms and conventional power plants in terms of their layout and the number and type of the electrical power components used. They range from the unconventional design of the WT generators, additionally having dry-type "step-up" transformers, to the extensive use of different types of cables in wind farms (single-core, three-core, four-core cables). Additionally, the extensive implementation of the vacuum circuit breakers and their rather frequent usage (compared with typical distribution system) certainly contribute to the transient phenomena not commonly experienced in conventional power plants, for example, [47].

We will hereafter mainly concentrate on the wind farm lightning surge overvoltages, often associated with a back-surge phenomenon. Hence, in accordance with [37, 45–47], the EMTP model of a wind farm electrical system—for lightning surge studies—could be decomposed into the following main parts: (1) model of the lightning surge current; (2) model of the lightning surge channel; (3) model of the WT which includes models of WT blades and tower, and possibly other (internal) electrical connections, such as the purposefully employed connections within the hub for circumventing the crucial components (i.e., turning gears) from lightning currents; (4) model of the WT grounding system (possibly accounting for the soil ionization and other specifics, e.g., steel-reinforced concrete foundations); (5) model of the MV cable lines; (6) metal-oxide surge arrester model; (7) model of the surge arrester connecting leads; (8) transformer model (which, if needed, includes models for studying lightning surge transference between its windings), and (9) models of other electrical equipment if needed, for example, WT generator model (i.e., model for its stator winding in lightning-related studies). Here models of the vacuum circuit breakers (which are the prominent feature in wind farms) could be mentioned as well, which are needed for studying the switching transient overvoltages in wind farm electrical systems [47]. More details on this subject can be found in, for example, [37, 45–48] and in references cited therein.

It can be seen that there is quite a number of components, which need to be modeled for high frequencies associated with lightning transient analysis. Furthermore, some of the above-mentioned components have several individual and influential factors which determine their transient behavior, for example, frequency dependence, nonlinear behavior, and influence of stray parameters. This is further exacerbated by several ambiguities associated with, for example, transformer and generator models for lightning transients analysis, produced mainly by the unavailability of the geometry and/or measurement data on these components, which stems from the manufacturers' proprietary concerns. Notwithstanding that, some of the components (e.g., transformer, WT grounding system) are associated with several different models, having varying degrees of sophistication, and further associated with specific constraints which establish their valid range of applications.

Furthermore, wind farm topology and other factors influence its overvoltage protection selection, for example, see [37, 48–52]. On top of that, there are several different arrangements of WT systems: generators producing power at the MV level, for example, 12 kV, thus removing the need for the step-up transformer; generators producing power at low voltage level, for example, 600 V, thus having a step-up transformer, which could be located in the WT tower base or in the adjacent housing. Likewise, recent positioning of the step-up transformers within the hub (with some WTs in the several megawatts range) high above the grounding system, and not in the bottom of the tower, could have influence on its overvoltage protection. Additionally, different types of WT generators have been employed, for example, doubly fed induction generators and full-converter generators. Consequently, at least some of these aspects should be given proper account when analyzing the overvoltage protection measures for a wind farm project, particularly from the viewpoint of the lightning-associated surges.

A prominent feature of wind farm overvoltage protection is the aforementioned backflow surge, initiated by the direct lightning strike to the WT, which needs to be accounted for and appropriately addressed [10, 37, 45–52]. It stems from the fact that in case of direct lightning strike to the WT (e.g., WT blade), (overvoltage) traveling wave is created on the WT tower, which propagates into the grounding system, producing transient ground potential rise. Due to the fact that the surge arresters (installed at this stricken WT) are connected to this grounding system, they incidentally bring this transient overvoltage from the WT grounding system onto the phase conductors. This overvoltage subsequently propagates through the wind farm electrical network. It has been found that those WTs, which are positioned at the electrical ends of the wind farm system, have the most adverse conditions regarding the lightning-related overvoltages and need special treatment, in the design of their grounding system, as well as in designing their overvoltage protection, for example, [37, 45, 48].

Metal-oxide surge arresters provide the crucial means of surge protection for wind farm equipment, both from switching and from lightning overvoltages [37, 53, 54]. The main concern in wind farm overvoltage protection comes from the lightning overvoltages, which are the consequences of direct lightning strikes to the associated WTs. Transient behavior of the internal electrical system of the WT could have influence on its overvoltage protection selection. Hence, detailed models of the wind farms (often including models of the WT internal elements) are sometimes needed to facilitate proper numerical verification of the metal-oxide surge

arrester capabilities in protecting the wind farm electrical equipment.

Furthermore, the energy capability of the metal-oxide surge arresters depends on both the switching and lightning overvoltages. It could be argued that the energy capability required of surge arresters installed in wind farms might be defined by the lightning surges (i.e., backsurge), unlike with the surge arresters installed in typical MV distribution networks (where switching surges could dominate), although this might not be the case in general (and depends on wind farm topology and other factors). Hence, a selection of energy capability, in regard to the lightning overvoltages, for the wind farm metal-oxide surge arresters is of particular importance. A statistical approach could be employed in determining the metal-oxide surge arrester lightning surge energy capabilities. Further details on this topic could be found in, for example, [53]. Additionally, study of the wind farm switching surges is of importance, particularly where vacuum circuit breakers are used in combination with extensive wind farm MV cable networks, for example, see [47].

11.5 CONCLUDING REMARKS

Several issues pertinent to the lightning and overvoltage protection of WTs and wind farms in general—as already thus far emphasized on various occasions—are still under scrutiny by many researchers in this field.

One of those issues is related to the assessment of WT lightning incidence, be it through the collection area method (i.e., electric shadow approach) or something else. Namely, some ambiguity is still present in connection with the appropriateness of the method recommended by IEC 61400-24, particularly regarding the treatment of the upward lightning, which gains importance with increasing height of WTs. In relation to this, there is a need to better account for the effective height of WTs or relate upward to downward lightning incidence in some other way, accounting for the underlying physical phenomenon at work, regarding the conditions necessary for the initiation of the upward lightning. The work produced by Rizk in [28, 29] could possibly point the way forward in this matter.

In close connection with the WT lightning incidence is the definition of the lightning current parameters incident to wind turbines, that is, lightning current amplitudes (and other parameters) that will be attained and/or exceeded for treated dangerous events. These parameters are, on one hand, influenced by the geographical location and orographic factors (different for onshore and offshore wind farms) and, on the other hand, by the wind turbine structure itself (and possibly by the rotation of its blades). Hence, there is a clear need, as emphasized in, for example, [23, 24], to define a consistent set of lightning current parameters (e.g., amplitude, front steepness, wave duration), which would be associated with lightning currents incident to WTs. They should take into account both upward and downward lightning strikes and their relative proportion (as far as possible). This could be achieved by the lightning current measurements, as in [24], carried out on a number of wind farms in different parts of the world and possibly further combined with lightning data produced by the lightning detection networks. These parameters would be applied for the overvoltage protection analysis of wind farms (e.g., backsurge phenomenon) and selection of metal-oxide surge arrester parameters.

Other issues pertinent to the analysis and design of the overvoltage protection for wind farms are associated with modeling different components of this electrical system (formed by connecting WTs) for high frequencies and transients. Some difficulties here stem from the fact that appropriate models for some of the components are rather difficult to produce, due to either the lack of input data or the deficiencies in the knowledge of the component's behavior during transients.

The former type of difficulty would be associated with, for example, transformer model for lightning surge transference analysis, where the model needs to be either assembled from the measurements carried out on the actual transformer or produced from the analysis of the exact and detailed transformer internal geometry. This is rather difficult to achieve (sometimes even impossible), primarily due to manufacturers' proprietary concerns. This problem could be alleviated if the manufacturers would provide "black box" models of their transformers, suitable for lightning and/or switching transient analysis; this would not necessitate revealing their cherished proprietary information.

The latter type of difficulty mentioned would be associated with, for example, constructing the detailed model of WT grounding system in lightning transient analysis, particularly if one wishes to include soil ionization and further take into account the influence of the steel-reinforced concrete foundation. More work is still needed on this topic, where analysis carried out in, for example, [55] could point the way forward.

REFERENCES

[1] Golde, R. H. (editor). (1977). *Lightning: Volume 1 – Physics of Lightning*. Academic Press, London.

[2] Golde, R. H. (editor). (1977). *Lightning: Volume 2 – Lightning Protection*. Academic Press, London.

[3] Rakov, V. A., Uman, M. A. (2003). *Lightning: Physics and Effects*. Cambridge University Press, Cambridge.

[4] Golde, R. H. (1977). Lightning currents and related parameters. In: *Lightning: Volume 1 – Physics of Lightning*, Golde, R. H. (editor). Academic Press, London, pp. 309–350.

[5] Berger, K. (1977). The earth flash. In: *Lightning: Volume 1 – Physics of Lightning*, Golde, R. H. (editor). Academic Press, London, pp. 134–185.

[6] Rakov, V. A., Rachidi, F. (2009). Overview of recent progress in lightning research and lightning protection. *IEEE Transactions on Electromagnetic Compatibility*, 51(3):428–442.

[7] Mousa, A. M. (1994). The frequency distribution of the amplitudes of lightning currents. *IEEE Task Force on Parameters of Lightning Strokes*, July, San Francisco.

[8] Lightning and Insulator Subcommittee of the T&D Committee (2005). Parameters of lightning strokes: a review. *IEEE Transactions on Power Delivery*, 20(1):346–358.

[9] IEC 62305-1. (2006). *Protection against Lightning – Part 1: General Principles*. International Electrotechnical Commission, Geneva.

[10] Yasuda, Y., Uno, N., Kobayashi, H., Funabashi, T. (2008). Surge analysis on wind farm when winter lightning strikes. *IEEE Transactions on Energy Conversion*, 23(1):257–262.

[11] Pierce, E. T. (1977). Lightning warning and avoidance. In: *Lightning: Volume 2 – Physics of Lightning*, Golde, R. H. (editor). Academic Press, London, pp. 497–507.

[12] Sarajčev, P. A. (2010). Review of statistics on lightning currents in relation to the lightning incidence of new-generation wind turbines. *International Review of Electrical Engineering*, 5(3):1285–1296.

[13] Anderson, R. B., Eriksson, A. J. (1980). Lightning parameters for engineering applications. *Electra*, 69:65–102.

[14] Narita, T., Yamada, T., Mochizuki, A., Zaima, E., Ishii, M. (2000). Observation of current waveshapes of lightning strokes on transmission towers. *IEEE Transactions on Power Delivery*, 15(1):429–435.

[15] Borghetti, A., Nucci, C. A., Paolone, M. (2004). Estimation of the statistical distributions of lightning current parameters at ground level from the data recorded by instrumented towers. *IEEE Transactions on Power Delivery*, 19(3):1400–1409.

[16] IEEE WG. (1985). A simplified method for estimating lightning performance of transmission lines. *IEEE Transactions on Power Apparatus and Systems*, 104(4):919–932.

[17] Eriksson, A. J. (1987). The incidence of lightning strikes to power lines. *IEEE Transactions on Power Delivery*, 2(3):859–870.

[18] Diendorfer, G., Pichler, H., Mair, M. (2009). Some parameters of negative upward-initiated lightning to the Gaisberg tower (2000-2007). *IEEE Transactions on Electromagnetic Compatibility*, 51(3):443–452.

[19] Rakov, V. A. (2001). Transient response of a tall object to lightning. *IEEE Transactions on Electromagnetic Compatibility*, 43(4):654–661.

[20] Guerrieri, S., Nucci, C. A., Rachidi, F., Rubinstein, M. (1998). On the influence of elevated strike objects on directly measured and indirectly estimated lightning currents. *IEEE Transactions on Power Delivery*, 13(4):1543–1555.

[21] Berger, K., Anderson, R., Kroninger, H. (1975). Parameters of lightning flashes. *Electra*, 41:23–33.

[22] Rodrigues, R. B. et al. (2008). An investigation over the lightning location system in for wind turbine protection development. *Power and Energy Society General Meeting— Conversion and Delivery of Electrical Energy in the 21st Century*. DOI: 10.1109/PES.2008.4596513

[23] Peesapati, V., Cotton, I. (2009). Lightning protection of wind turbines—a comparison of lightning data & IEC 61400-24. *International Conference on Sustainable Power Generation and Supply (SUPERGEN '09)*, April 6–7, Nanjing, China.

[24] Asakawa, A., Shindo, T., Yokoyama, S., Hyodo, H. (2010). Direct lightning hits on wind turbines in winter season: lightning observation results for wind turbines at Nikaho wind park in winter. *IEEJ Transactions on Electrical and Electronic Engineering*, 5:14–20.

[25] IEC 61400-24. (2010). *Wind Turbine Generator Systems – Part 24: Lightning Protection*, Edition 1.0. International Electrotechnical Commission, Geneva.

[26] IEC 62305-2. (2010). *Protection against Lightning – Part 2: Risk Management*, Edition 2.0. International Electrotechnical Commission, Geneva.

[27] Theethayi, N., Thottappillil, R. (2007). Some issues concerning lighting strikes to communication towers. *Journal of Electrostatics*, 65:689–703.

[28] Rizk, F. (1994). Modeling of lightning incidence to tall structures – part I: theory. *IEEE Transactions on Power Delivery*, 9(1):162–171.

[29] Rizk, F. (1994). Modeling of lightning incidence to tall structures – part II: application. *IEEE Transactions on Power Delivery*, 9(1):172–193.

[30] Rachidi, F., et al. (2008) A review of current issues in lightning protection of new-generation wind turbine blades. *IEEE Transactions on Industrial Electronics*, 55(6):2489–2496.

[31] Sarajčev, P., Goić, R. (2012). Assessment of lightning current parameters suitable for wind turbine overvoltage protection analysis. *Wind Energy*, 15(4):627–644.

[32] Zhou, H., Theethayi, N., Diendorfer, G., Thottappillil, R., Rakov, V. A. (2010). On estimation of the effective height of towers on mountaintops in lightning incidence studies. *Journal of Electrostatics*, 68:415–418.

[33] D'Alessandro, F., Petrov, N. I. (2006). Field study on the interception efficiency of lightning protection systems and comparison with models. *Proceedings of the Royal Society A*, 462:1365–1386.

[34] Petrov, N. I., D'Alessandro, F. (2002). Assessment of protection system positioning and models using observations of lightning strikes to structures. *Proceedings of the Royal Society A*, 458:723–742.

[35] Petrov, N. I., Waters, R. T. (1995). Determination of the striking distance of lightning to earthed structures. *Proceedings of the Royal Society A*, 450:589–601.

[36] Chang, G. W., Huang, H. M., Lin, Y. T. (2012). On lightning study for wind generation systems. *CIGRE Session 44*, August 26–31, Paris, France.

[37] Sarajčev, P., Goić, R. (2011). A review of current issues in state-of-art of wind farm overvoltage protection. *Energies*, Special Issue: Wind Energy 2011; 4(4):644–668.

[38] Sarajčev, P., Goić, R. (2010). Lightning incidence of wind farm underground power cable networks. *International Review of Electrical Engineering*, 5(3):1244–1254.

[39] Sunde, E. D. (1949). *Earth Conduction Effects in Transmission Systems*. D. Van Nostrand Company, Inc., New York.

[40] Rodrigues, R. B., Mendes, V. M. F., Catalao, J. P. S. (2009). Estimation of lightning vulnerability points on wind power plants using the rolling sphere method. *Journal of Electrostatics*, 67:774–780.

[41] Bertelsen, K., Erichsen, H. V., Skov Jensen, M. V. R., Madsen, S. F. (2007). Application of numerical models to determine lightning attachment points on wind turbines. *Proceedings of the International Conference on Lightning and Static Electricity*, August 28–31, Paris, France.

[42] Madsen, S. F., Erichsen, H. V. (2008). Improvements of numerical models to determine lightning attachment points on wind turbines. *Proceedings of the 29th International Conference on Lightning Protection*, June 23–26, Uppsala, Sweden.

[43] Peesapati, V., Cotton, I. (2009). Lightning protection of wind turbines – a comparison of real lightning strike data and finite element lightning attachment analysis. *International Conference on Sustainable Power Generation and Supply (SUPERGEN '09)*, April 6–7, Nanjing, China.

[44] Dommel, H. W. (1992). *Electromagnetic Transients Program Theory Book*. MicroTran Power System Analysis Corporation, Vancouver.

[45] Sarajčev, P., Goić, R. (2010). An EMTP model for lightning surge analysis of wind farms. *International Review of Modeling and Simulation*, 3(1):70–81.

[46] Martinez-Velasco, J. A. (editor). (2010). *Power System Transients: Parameter Determination*. CRC Press, Boca Raton.

[47] Badrzaleh, B., Hogdahl, M., Isabegovic, E. (2011). Transients in wind power plants – part I: modeling methodology and validation. *IEEE Industry Applications Society Annual Meeting*, October 9–13, Orlando, FL.

[48] Yasuda, Y., Funabashi, T. (2007). Analysis on back-flow surge in wind farms. *Proceedings of the International Conference on Power Systems Transients (IPST '07)*, June 4–7, Lyon, France.

[49] Vahidi, B., Yazdanpanahi, H. (2009). The effect of wind farm to AC grid connection type on overvoltages due to lightning. *International Review of Modeling and Simulation*, 2(5):520–524.

[50] Vahidi, B., Mousavi, O. A., Hosseinian, S. H. (2007). Lightning overvoltage analysis in wind farm. *Proceedings of the International Conference TENCON 2007*, October 30–December 2, Taipei, Taiwan.

[51] Asuda, O., Funabashi, T. (2004). Transient analysis on wind farm suffered from lightning. *Proceedings of the 39th International Universities Power Engineering Conference*, September 6–8, Bristol, UK.

[52] Rodrigues, R. B., Mendes, V. M. F., Catalao, J. P. S. (2010). EMTP-RV analysis of lightning surges on wind turbines. *Proceedings of the International Conference on Renewable Energies and Quality (ICREPQ'10)*, March 23–25, Granada, Spain.

[53] Sarajčev, P. (2010). Wind farm surge arresters energy capability and risk of failure analysis. *International Review of Modeling and Simulation*, 3(5):926–937.

[54] Wailing, R. A. (2008). Overvoltage protection and arrester selection for large wind plants. *Proceedings of IEEE/PES Transmission & Distribution Conference & Exposition*, April 21–24, Chicago, IL.

[55] Yasuda, Y., Toshiaki, F. (2011). Electromagnetic calculation of a wind turbine earthing system. In: *Wind Turbines*, Al-Bahadly, I. (editor). InTech, Shangai, pp. 507–528.

12

WIND TURBINE WAKE MODELING—POSSIBILITIES WITH ACTUATOR LINE/DISC APPROACHES

STEFAN IVANELL AND ROBERT MIKKELSEN

12.1 INTRODUCTION

Today, wind turbines are often placed close together in small clusters or in large wind farms. This means that they operate in the wakes of each other and that they, depending on the wind direction, can be subject to inflow conditions dominated by vortical structures created by upstream turbines. This reduces the power performance for the individual wind turbine and decreases the lifetime of the rotors [1]. Thus, there is obvious need for understanding and modeling the wake behavior of wind turbines. Wakes behind wind turbines can be distinctly divided into the near- and the far-wake regions. The study of near-wake aerodynamics concerns the description of the vortices in the wake and their relation to the blade loading and inflow conditions. The near wake is followed by the far wake where the focus is put on the influence of wind turbines in farm situations, and the modeling of the actual rotor is less important. In the far wake, turbulence models, wake interaction, and topographic effects are of primary interest.

This study illustrates the differences between results from actuator disc (ACD) and actuator line (ACL) methods in the near and far wake downstream of a wind turbine rotor. Here the near wake is defined as the area just behind the rotor, where the properties of the rotor can be discriminated by the vortices in the wake. Thus the number of blades, the blade aerodynamics, and the tip/root vortices are intrinsic elements of the analysis.

Most experiments on near-wake structures of wind turbine have been performed at rather low Reynolds number (in the following, the Reynolds number is based on blade chord and rotational speed). For an extensive review of the work performed in this area, we refer to Vermeer et al. [2]. For far-wake studies we refer to the survey of Crespo et al. [3]. Early experiments showing the pairing of the vortex spirals downstream in the wake were carried out by Alfredsson and Dahlberg [4,5]. Later experiments based on various techniques, such as flow visualizations, hot-wire anemometry, particle image velocimetry (PIV), and laser Doppler anemometry (LDA), are due to Anderson et al. [6], Savino and Nyland [7], Anderson et al. [8], Eggleston and Starcher [9], Vermeer [10–13], Ebert and Wood [14–16], Whale et al. [17], Hand et al. [18], Shimizu and Kamada [19], Medici [20], and Massouh and Dobrev [21]. As an example, extensive measurements were carried out in the wake flow field behind a small rotor (0.18 m) by Medici [20, 22]. Medici used both two-component hot-wire anemometry and PIV to map the flow field downstream as well as upstream the turbine. From the study it was found that the wake is meandering with a frequency corresponding to the Strouhal number. Recently, the activities in this area have increased to a great extent.

Experiments with Reynolds number higher than 300,000 have been carried out by de Vries [23], Anderson et al. [6], Shimizu and Kamada [19], Schreck [24], and Hand et al. [18]. Recently, the European experiment MEXICO (model rotor experiments under controlled conditions) was carried out in a DNW (German-Dutch Wind tunnels) wind tunnel in the Netherlands. The test section of the MEXICO experiment was 9.5×9.5 m, the rotor to tunnel area ratio was 1:3.8, and the Reynolds number was $\approx 600,000$ at 75% radius. Both pressure sensors and PIV were used. The PIV technique was used to resolve chosen areas of the flow field, including the

Alternative Energy and Shale Gas Encyclopedia, First Edition. Edited by Jay H. Lehr and Jack Keeley.
© 2016 John Wiley & Sons, Inc. Published 2016 by John Wiley & Sons, Inc.

wake. With this experiment it is, for the first time, possible to trace the circulation and position of the tip vortices downstream as a function of azimuthal angle. The wake behind the MEXICO rotor has been studied by Lutz et al. [25] and Nilsson et al. [26], among others within the MEXNEXT project.

There is only one known experiment with a full-scale rotor (radius $\approx$ 5 m): the NREL Unsteady Aerodynamic Experiment in the NASA-Ames wind tunnel [18, 24]. The test section of the NREL experiment was 24.4 × 36.6 m, the rotor to tunnel area ratio was 1:10.8, and the Reynolds number was approximately equal to 1,000,000. The emphasis of the project was, however, not on wake measurements.

To understand the basic aerodynamics of wakes and to analyze experimental data it is necessary to perform numerical simulations, either by vortex wake models or by full field Navier–Stokes CFD (computational fluid dynamics) methods. Recent wake studies based on various CFD techniques have been presented by Sørensen et al. [27], Wufiow et al. [28], Walther et al. [29], Watters and Masson [30], Jimenez et al. [31], Zahle and Sørensen [32], Mikkelsen et al. [33], Troldborg et al. [34], and Nilsson et al. [35], among others. The main problem of utilizing CFD methods for simulating wake flows, however, is the need for resolving many different length scales, ranging from the thickness of the blade boundary layer to the diameter of the rotor. When considering wind turbine clusters or complex terrain conditions it is also important to handle length scales covering the distance between the turbines and the size of the atmospheric boundary layer. In practice it is necessary to focus on specific parts of the problem and/or to make some simplifying assumptions regarding the detailed behavior of the problem. Typically, models coping with clusters of wind turbines are divided into models using parabolized Navier–Stokes equations or models in which the full Navier–Stokes equations are combined with ACDs to represent the loading. The main limitation in ACD methods is that these methods distribute the forces evenly in the tangential direction of the ACD. The influence of the blades is therefore taken as an integrated quantity in the azimuthal direction. To circumvent this problem an extended three-dimensional (3D) method, the actuator line method, was introduced by Sørensen and Shen [36], in which body forces are distributed along lines representing each of the blades. In this chapter we will illustrate results from both ACD and ACL simulations and how these techniques can be used to study the wake structure.

12.2 NUMERICAL MODEL

This section deals with the numerical method and setup, including grid designs and a description of the rotor configuration used for the simulations.

12.2.1 Solver and Numerical Method

The EllipSys3D code is a general-purpose 3D solver developed by Sørensen [37] and Michelsen [38, 39]. The flow solver is based on a finite-volume discretization of the Navier–Stokes equations in general curvilinear coordinates using multi-block topology. The code is formulated in primitive variables, that is, pressure and velocity variables, in a collocated storage arrangement. Rhie/Chow interpolation is used to avoid odd/even pressure decoupling and the main solver is based on multigrid techniques.

The computations are carried out as large eddy simulations (LES) employing the mixed sub-grid-scale model developed by Ta Phuoc [40]. This model exploits the advantage of a closure combining vorticity and turbulent kinetic energy. In this model the vorticity is derived directly from the filtered variables whereas the turbulent kinetic energy is determined by the use of a test filter that is twice as coarse as the computational grid. For more details about the mixed scale model we refer to the text book by Sagaut [41].

Using the coordinate directions (x_1, x_2, x_3), the Navier–Stokes equations are formulated as

$$\frac{\partial u_i}{\partial t} + \frac{\partial u_i u_j}{\partial x_j} = -\frac{1}{\rho}\frac{\partial p}{\partial x_i} + f_{\text{body},i} + f_{\text{c},i} + \frac{\partial}{\partial x_j}$$
$$\left[(v + v_{\text{t}}) \left(\frac{\partial u_i}{\partial u_j} + \frac{\partial u_j}{\partial x_i} \right) \right], \frac{\partial u_i}{\partial x_i} = 0 \quad (12.1)$$

where u_i is the velocity vector, p is the pressure, t is the time, and p is the density of air. f_{body} represents the forces acting on the blades, f_{c} the Coriolis force, v the kinematic viscosity, and v_{t} the eddy viscosity coefficient using the LES method. The numerical method uses a blend of third-order QUICK (10%) and fourth-order CDS (90%) difference scheme for the convective terms and second-order central difference scheme for the remaining diffusive terms.

When using body forces to represent the blades, the viscosity parameter does not need to be exactly equal to the viscosity of air. Using the actual viscosity would result in a Reynolds number, based on inflow velocity and rotor radius, of several millions, and not possible to compute on present-day computer configurations. It should be noted that this type of computation is always a trade-off between accuracy and available computer resources. The important thing in this study, however, is that the aerodynamic loading of the rotor is well captured in order to generate a realistic wake. Mikkelsen [42] and Ivanell [43] have shown that results are independent of the Reynolds number, provided a sufficiently high number, which is in the order of 10^4, has been reached.

12.2.2 Aerofoil Data

The ACL and ACD methods use tabulated aerofoil data taken from 2D measurements and modified in order to agree with experimental data. Therefore, the method is based on tabulated aerofoil data from which C_L and C_D are functions of a. As a consequence, the accuracy of the computed loading of the rotor depends on the quality of the experimental data. Data from the Tjaereborg turbine were used for all simulations presented here. The blade profiles consisted of NACA 4412-43 sections with a blade length of 29 m giving a rotor diameter of 61 m. The chord length was 0.9 m at the tip, increasing linearly to 3.3 m at hub radius 6 m. The blades are twisted 1° per 3 m. The rotor solidity was 5.9%. Simulations were performed with tip speed ratios of 5.05, 5.89, 7.07, 8.84, and 11.79 with corresponding pitch angles of 2.00°, 0.61°, 0.54°, 0.51°, and 0.55°. A tip speed ratio of 7.07 corresponds to optimum performance at a free stream wind speed of 10 m/s. In this work, however, all variables are made dimensionless, such that the actual rotor configuration merely serves to produce a realistic load or circulation distribution.

12.2.3 Actuator Line Method

In the simulations we employ the ACL method, developed previously by Sørensen and Shen [36]. In this technique the loading and the actual geometry of the rotor blades are replaced by body forces that are distributed along three straight lines, each representing the blades of the rotor (see Figure 12.1). A full CFD simulation would require a great number of nodes at the blades to resolve the boundary layer. With the ACL method, the number of node points at the blades is greatly reduced and instead the focus is put on the resolution of the vortical structures in the wake. This

also makes the grid design easier and more efficient. The method was originally developed using a vorticity-based formulation of the Navier–Stokes equations [36] and was later implemented by Mikkelsen [42] into the pressure–velocity code EllipSys3D. Because of numerical discontinuities near the tips the body forces are distributed among neighboring node points in a Gaussian manner. This is done by taking the convolution of the computed load $\mathbf{f}_{r\theta z}$ and the regularization kernel η_ε:

$$f_b = f_{r\theta z}^b * \eta_\varepsilon \tag{12.2}$$

where the regularization kernel is defined as (see Sørensen and Shen [36])

$$\eta_\varepsilon(r) = \frac{1}{\varepsilon^3 \pi^{3/2}} e^{-(p/\varepsilon)^2} \tag{12.3}$$

where p is the distance between the cell-centered grid points and points located on the ACL. Mikkelsen discovered that using a 3D Gaussian smoothing results in inconsistencies near the tip region [42]. Therefore, a 2D Gaussian distribution is used on a 2D plane orthogonal to the ACL; see Figure 12.1. This smearing of the forces is done globally; that is, every node point at a plane orthogonal to the ACL is affected. The effect of the Gaussian function, however, is only perceptible in the vicinity of the ACLs.

The Gaussian distribution is controlled by the parameter ε. The choice of e will affect the numerical discontinuity at the tip, as a result of the 2D distribution. The value of e is typically in the order of 1–3 cell lengths and is, in this study, set to 1. The choice and influence of ε has been studied in detail by Ivanell [43].

When a simulation starts, an initial velocity, corresponding to the free stream velocity, is introduced into the entire flow field. Combining local velocities with the azimuthal velocity from the rotation of the ACLs, local angles of attack at the blade positions are extracted. From tabulated aerofoil data the local forces acting on the blades are determined, and distributions of body forces are subsequently computed and set into the equations. The equations are then solved by iteration in time until they reach either a steady state or a limiting time-dependent state.

12.2.4 Actuator Disc Method

In these simulations we employ an extension of the Froude ACD method. In this technique the loading and the actual geometry of the rotor blades are replaced by body forces distributed on a 3D ACD. These forces are based on the local flow condition across the 3D disc using tabulated aerofoil data and distributed in the computational domain in the direction normal to the disc.

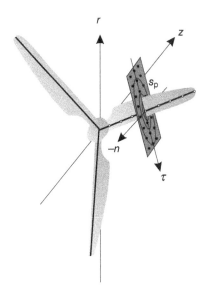

FIGURE 12.1 The actuator line concept. Each blade is represented by a line with J_{ACL} number of points. Forces at each actuator point are Gaussian distributed at all points, s_p, positioned at a plane orthogonal to the actuator line. The plane is infinite.

This numerical approach to the ACD concept has been developed by Sørensen and colleagues [44–46]. The main idea is to determine the flow past a rotor without resolving the boundary layer on the blades. With the ACD method, the number of node points at the blades is greatly reduced, and the focus is put on the resolution of the wake behind the turbines. This also makes the grid design easier and more efficient. The method was originally developed using a vorticity-based formulation of the Navier–Stokes equations [36] and was later implemented by Mikkelsen [42] into the in-house-developed EllipSys3D code.

To avoid singular behavior the body forces are smeared among neighboring node points using a Gaussian distribution. The influence of that smearing parameter has been studied in a separate technical report and will therefore not be presented in detail here [43].

12.2.5   Atmospheric Boundary Layer

The averaged wind shear conditions that the turbines operate in are imposed at the inflow boundary. However, the numerical ability of flow solvers generally does not preserve this desired wind shear profile through the entire domain. In order to impose any wind shear profile, which may include directional changes with height, the flow field is initially generated by applying momentum sources everywhere in the domain; see Mikkelsen et al. [47]. The magnitude of the computed momentum sources is generally very small. A power law wind shear profile with an exponent of 0.15 is chosen.

The influence of atmospheric turbulence is simulated by using a technique where turbulent fluctuations are imported in a 2D plane upstream the rotor from a pre-generated field which the flow solver then convects downstream toward the ACDs. All pre-generated turbulent fields are generated by the Mann model [48–50]. The simulations assume sea conditions defined by Charnock's relation [51]. The turbulence intensity (u'/U) is equal to approximately 6%. The use of imposed

atmospheric turbulence by introducing time-varying body forces in combination with the EllipSys3D code was recently implemented by Troldborg et al. [34]. For more details about the method we refer to Troldborg et al. [34].

12.2.6   Grid, Inflow, and Boundary Conditions

12.2.6.1   Actuator Line. The employed computing code is specially designed to run fast on parallel computers. However, a drawback is that it only handles blocks with the same number of nodes on each block edge or side. The computations were performed on two different mesh designs. Both structures use periodic boundary conditions in the azimuthal direction. Hence, the mesh size is reduced to one-third of a full rotor domain, corresponding to a 120° slice. All simulations are based on steady-state calculations and assume zero yaw. The ACL is fixed in the mesh. Rotation is therefore introduced by boundary conditions. The first grid design, which is mainly used for validation of the method, is based on five blocks distributed after each other axially. The second design, used for simulations with higher resolution, is based on 40 blocks that are distributed in both axial and radial directions. Figure 12.2 shows how the two meshes are designed. Both meshes were created to be able to capture large gradients in the wake.

In earlier studies the accuracy of the ACL method has been checked against measurements of a 500 kW Nordtank wind turbine using a vorticity formulation of the Navier–Stokes equations and an effective Reynolds number of 5000 [36]. In this work we validate the implementation of the ACL technique in the EllipSys3D code by computing the power yield of the Tjaereborg turbine at a wind speed of 10 m/s, corresponding to a tip speed ratio of about 7.

The simulations were performed on three different grids, consisting of 64, 80, and 96 node points at each side of the blocks in the 5-block topology, and at six different Reynolds numbers, ranging from 10 to 50,000.

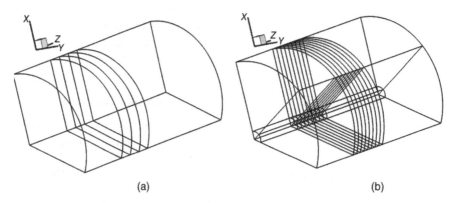

(a)   (b)

FIGURE 12.2   Schematic mesh topology: (a) 5 block mesh; (b) 40 block mesh. The radius of the block closest to the center axis corresponds to 1.25 rotor radii.

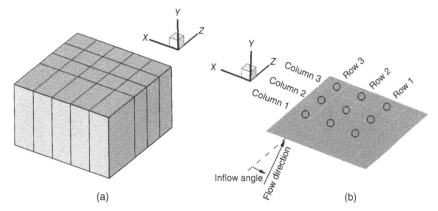

FIGURE 12.3   (a) The block design used for simulations of nine turbines with wall condition at $y = 0$; (b) The definition of the inflow angle and order of row and column numbers.

12.2.6.2 Actuator Disc. A 20 block mesh was designed to be used for simulations with 9 turbines. Figure 12.3a shows the 20 block mesh. The 12 block mesh was only used for a grid dependence investigation. The turbine spacing was 4 radii in both directions in the 20 block mesh. The 20 block mesh structure contained 64 nodes on each block side resulting in $64^3 \times 20 \approx 5.2 \times 10^6$ nodes. This resolution results in the radius being resolved by 10 nodes. In the grid-dependency study, up to 80 mesh points on each block side was tested.

In the mesh design the node point distribution is smeared toward the inlet in flow direction, but equidistant from about two turbine radius in front of the turbine and backwards to the outlet. The mesh also includes a wall boundary condition representing the ground surface. The mesh points are therefore distributed toward that surface.

The side boundaries are set to periodic conditions. The outlet is set to convective boundary conditions allowing vortices to cross. The inlet contains a prescribed boundary layer with a free stream velocity normalized to represent 10 m/s at a hub height of 1.9 rotor radii and a logarithmic wind profile.

A local grid is introduced at which body forces representing the loading are introduced. The body forces are determined by computing local angles of attack and using tabulated aerofoil coefficients. The grid is designed as a polar grid with 21 node points in the radial direction and 64 node points in the azimuthal direction.

We use the notation of a *column* for turbines aligned perpendicular to the wind direction and *row* for turbines aligned parallel to the wind direction; see Figure 12.3b.

12.3 RESULT

12.3.1 Actuator Line

In this work the analysis is mainly focused on understanding and analyzing the structure and position of the vortex system in the wake. From the computations it was found that the

vortex core, as expected, originates close to the tip of the blade at a slightly smaller radial position than the blade tip.

To identify the trajectories of the tip vortices we compute and plot iso-vorticity contours. This is shown in Figure 12.4 where the spiral structure of the tip vortices is clearly seen. In the figure the vertical plane shows the pressure distribution and the horizontal plane shows the streamwise velocity distribution. Figures 12.5 and 12.6 show the pressure and streamwise velocity in the wake more in detail. Figure 12.7 illustrates the axial and azimuthal development in the wake, where RP corresponds to rotor position, $\langle U_z \rangle$ and $\langle U_\theta \rangle$ the azimuthal average of the flow in axial and azimuthal directions, and a and a' the axial and azimuthal induction factors. The wake development is further illustrated in Figures 12.8 and 12.9.

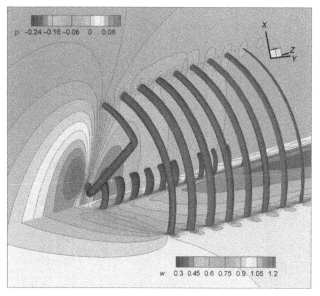

FIGURE 12.4 $x = 0$ plane, pressure distribution; $y = 0$ plane, streamwise velocity; iso-surface, constant vorticity.

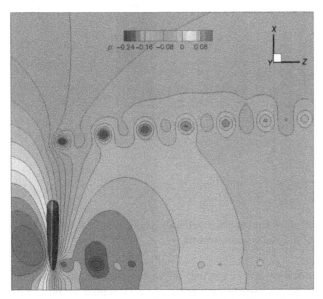

FIGURE 12.5 Pressure distribution for the 96 point mesh simulation 30° after blade passage.

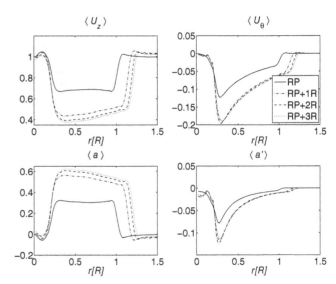

FIGURE 12.7 Velocities and interference factors as functions of radius at different axial positions.

12.3.2 Actuator Disc

Figure 12.10 shows a 3D plot of the flow field from simulations with the 20 block mesh and inflow angles of 0°, 15°, and 30°, both with and without turbulent inflow conditions. The wake structure is identified by an iso-surface of the vorticity on which the pressure is plotted as color contours. Blue color corresponds to low-pressure regions and red to high-pressure regions. The turbine position can therefore be identified by the low-pressure regions, which appear behind the turbines in the figure. The result clearly shows

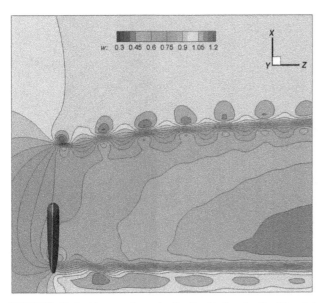

FIGURE 12.6 Streamwise velocity distribution for the 96 point mesh simulation 30° after blade passage.

that the flow strongly depends on the turbulent inflow condition and thereby also the production which has been shown by Ivanell [43]. It also strongly depends on the turbulent inflow condition. Only one turbulence intensity, corresponding to sea conditions, has been simulated here. Simulations with land conditions are expected to result in even greater changes.

Figure 12.11 shows the same cases as Figure 12.10 but now illustrating a 2D plane at hub height. The color contours show the velocity in the flow direction. The values have been normalized and 1 therefore corresponds to a free stream velocity of 10 m/s. The result indicates that the turbulent inflow condition clearly affects the breakup of the flow through the wind farm. It can be noted that the breakup of individual wakes strongly depends on the turbulent inflow. It is also possible to recognize that the flow in between the turbines shows a stronger tendency of acceleration in the cases of turbulent inflow.

No comparison of the data fields from ACL and ACD has been shown here, mainly because these two methods have been simulated on two different grid structures and the comparison would therefore not be justified. The aim here has been to illustrate the principal difference of the results from ACD and ACL methods. However, detailed comparisons have been carried out by Mikkelsen [42].

12.4 DISCUSSION AND CONCLUSIONS

12.4.1 Actuator Line

The behavior of near wakes behind wind turbines was studied by combining numerical solutions of the steady

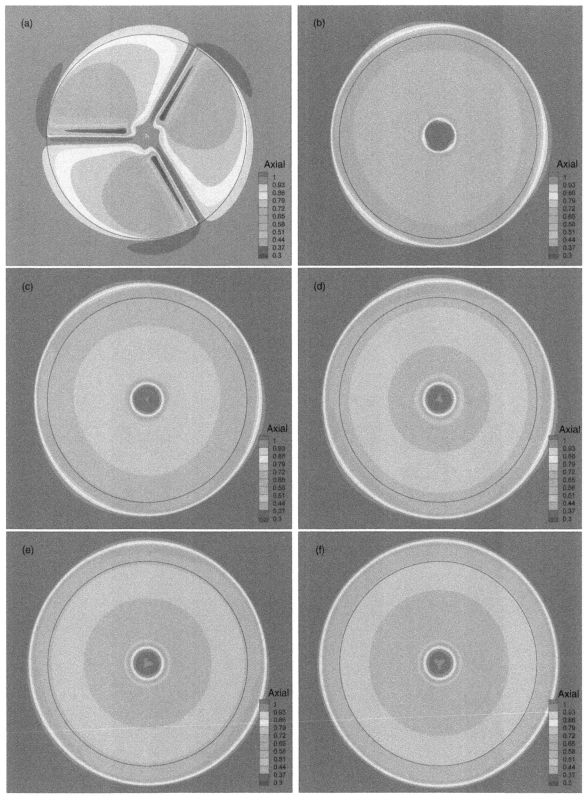

FIGURE 12.8   Axial velocities at different axial positions at and behind the rotor. The levels are normalized with free stream velocity. (a) Rotor position, (b) 0.5R behind turbine position, (c) 1R behind turbine position, (d) 1.5R behind turbine position, (e) 2R behind turbine position, (f) 2.5R behind turbine position.

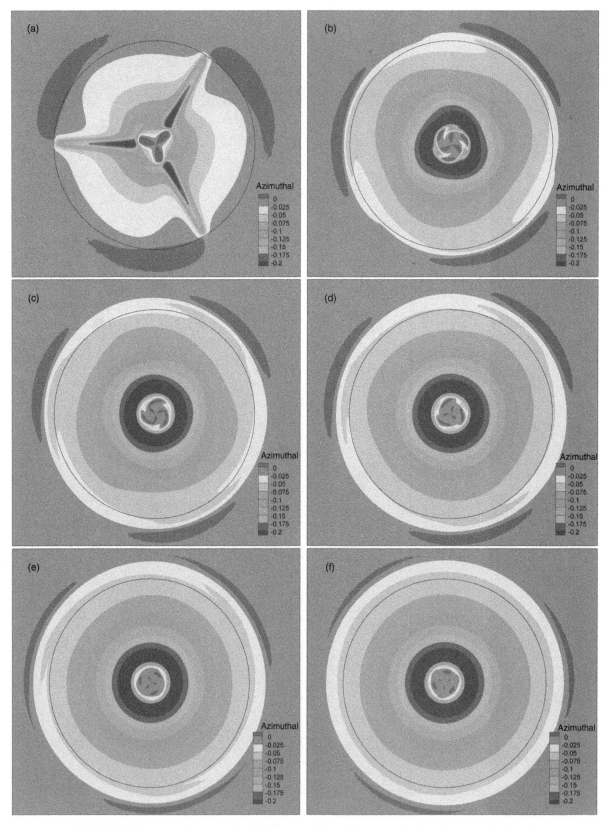

FIGURE 12.9 Azimuthal velocities at different axial positions at and behind the rotor. The levels are normalized with free stream velocity. (a) Rotor position, (b) 0.5*R* behind turbine position, (c) 1*R* behind turbine position, (d) 1.5*R* behind turbine position, (e) 2*R* behind turbine position, (f) 2.5*R* behind turbine position.

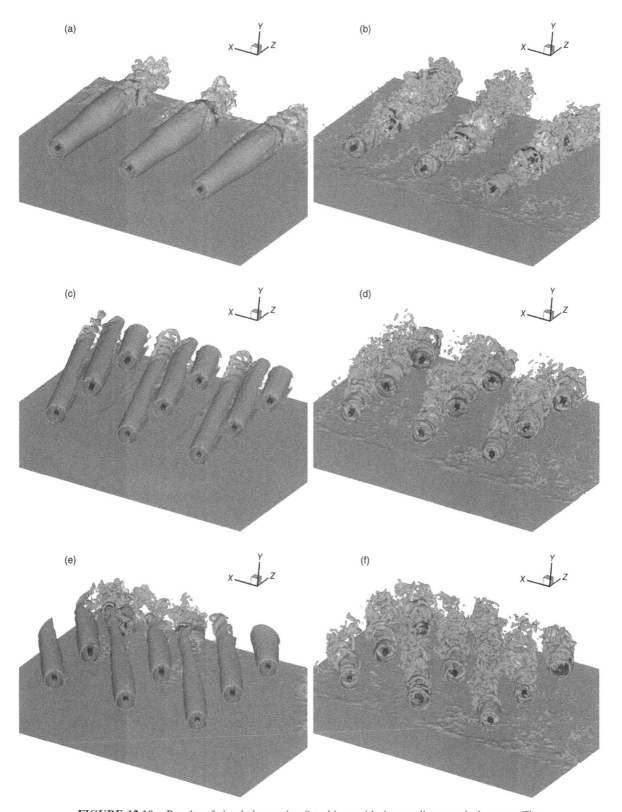

FIGURE 12.10   Results of simulations using 9 turbines with 4 rotor diameters in between. The wake structure is identified by an iso-surface of the vorticity. What appears to be a ground surface is the same iso-surface level illustrating the wake structures. The pressure is identified by color contours at the iso-surface of the vorticity. (a, b) The result at an inflow angle of 0°, (c, d) at an inflow angle of 15°, and (e, f) at an inflow angle of 30°. The left column shows the result without turbulent inflow and the right column the result with turbulent inflow conditions.

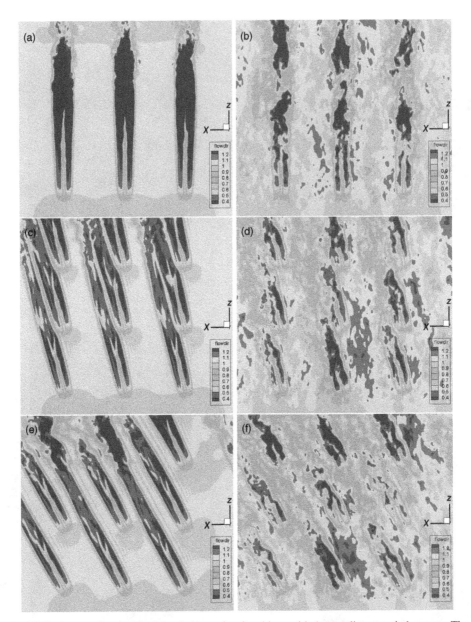

FIGURE 12.11 The results of simulations using 9 turbines with 4 rotor diameters in between. The figure shows a 2D plane at hub height. The downstream velocity is identified by color contours at the plane. (a, b) The result at an inflow angle of 0°, (c, d) at an inflow angle of 15°, and (e, f) at an inflow angle of 30°. The left column shows the result without turbulent inflow and the right column the result with turbulent inflow conditions.

Navier–Stokes equations with the ACL methodology. In the ACL method the blades are represented by lines along which body forces representing the loading are introduced. The body forces are determined by computing local angles of attack using tabulated aerofoil coefficients.

The results from the simulations indicate that the averaged velocity fields in planes perpendicular to the main flow direction essentially consist of a constant axial flow component and an azimuthal velocity following approximately a $1/r$ behavior. This is in accordance with the resulting

vortex system that is formed by stable and distinct tip and root vortices.

The study gives a conceptional picture of differences and possibilities when using body force approaches. The ACL method can clearly be used for studies of more fundamental understanding of the wake. Studies using the ACL method have been performed by Ivanell et al. [52, 53].

The ACL method is, however, computationally demanding for large wind farms. For such applications, the ACD method is more suitable.

12.4.2 Actuator Disc

The ACD simulations clearly indicate, as expected, that turbulent inflow has a strong impact on the result and leads to a more complex flow structure. The study also indicates that the ACD method is suitable for studies of large wind farms. The method has been verified to correspond to measured production data in Horns Rev [43].

ACKNOWLEDGMENTS

The work has been carried out with the support of the Nordic Consortium on Optimization and Control of Wind Farms financed by the Swedish research program Vindforsk. Computer time was granted by the Swedish National Infrastructure for Computing (SNIC).

REFERENCES

[1] Cleijne, J. W. (2003). Results of Sexbierum wind farm; single wake measurements. Technical Report TNO-Report 93-082.

[2] Vermeer, L. J., Sørensen, J. N., Crespo, A. (2003). Wind turbine wake aerodynamics. *Progress in Aerospace Science*, 39:467–510, Division L.

[3] Crespo, A., Hernandez, J., Frandsen, S. (1998). Survey of modelling methods for wind turbine wakes and wind farms. *Journal of Wind Energy*, 2:1–24.

[4] Alfredsson, P. H., Dahlberg, J.-A. (1979). A preliminary wind tunnel study of windmill wake dispersion in various flow conditions. Technical Report Technical Note AU-1499 Part 7, Swedish Defence Research Agency, Stockholm, Sweden.

[5] Alfredsson, P. H., Dahlberg, J.- A. (1981). Measurements of wake intersection effects on power output from small wind turbine models. Technical Report Technical Note HU-2189, part 5, Swedish Defence Research Agency, Stockholm, Sweden.

[6] Anderson, M. B., Milborrow, D. J., Ross, N. J. (1982). Performance and wake measurements on a 3 m diameter horizontal axis wind turbine. Comparison of theory, wind tunnel and field test data. Technical Report Technical Note HU-2189, part 5, University of Cambridge, Department of Physics, Cavendish Lab., Cambridge, UK.

[7] Savino, J. M., Nyland, T. W. (1985). Wind turbine flow visualization studies. In: *Proceedings of the Windpower '85 Conference*, Washington DC. American Wind Energy Association.

[8] Anderson, C. G., Niven, A. J., Jamieson, P., Knight, R. R., Milborrow, D. J. (1987). Flow visualization on rotating blades. In: *9th British Wind Energy Association Conference*.

[9] Eggleston, D. M., Starcher, K. (1990). Comparative studies of the aerodynamics of several wind turbines using flow visualization. *Journal of Solar Energy Engineering*, 112(39):301–309. Division L.

[10] Vermeer, N. J. (1988). Velocity measurements in the near wake of a model rotor (in Dutch). In: *Fourth Dutch National Wind Energy Conference*, Noordwijkerhout, The Netherlands.

[11] Vermeer, N. J. (1989). Velocity measurements in the near wake of a model rotor. In: *European Wind Energy Conference 1989*, Glasgow, UK.

[12] Vermeer, N. J., van Bussel, G. J. W. (1989). Velocity measurements in the near wake of a model rotor and comparison with theoretical results. In: *Fifteenth European Rotorcraft Forum*, Amsterdam, The Netherlands.

[13] Vermeer, L. J. (2001). A review of wind turbine wake research at Tudelft. In: *A Collection of the 2001 ASME Wind Energy Symposium Technical Papers*, New York.

[14] Ebert, P. R., Wood, D. H. (1997). The near wake of a model horizontal-axis windturbine-i. Experimental arrangement and initial results. *Renewable Energy*, 12(3):225–243.

[15] Ebert, P. R., Wood, D. H. (1999). The near wake of a model horizontal-axis windturbine-ii. General features of the three-dimensional flowfield. *Renewable Energy*, 18:513–534.

[16] Ebert, P. R., Wood, D. H. (2001). The near wake of a model horizontal-axis windturbine-iii. Properties of the tip and root vortices. *Renewable Energy*, 22:461–472.

[17] Whale, J., Anderson, C. G., Bareiss, R., Wagner, S. (1996). A study of the near wake structure of a wind turbine comparing measurements from laboratory and full-scale experiments. *Solar Energy*, 56(6):621–633.

[18] Hand, M., Simms, D., Fingersh, L., Jager, D., Cotrell, J., Schreck, S., Larwood, S. (2001). Unsteady aerodynamic experiment phase vi: wind tunnel test configuration and available data campaigns. Technical Report NREL/TP- 500-29955, NREL.

[19] Shimizu, Y., Kamada, Y. (2001). Studies on a horizontal axis wind turbine with passive pitch-flap mechanism (performance and flow analysis around wind turbine). *Journal of Fluid Engineering*, 123:516–522.

[20] Medici, D. (2005). *Experimental Studies of Wind Turbine Wakes - Power Optimisation and Meandering*. PhD thesis, Department of Mechanics, The Royal Institute of Technology, KTH. ISRN KTH/MEK/TR-05/19-SE.

[21] Massouh, F., Dobrev, I. (2007). Exploration of the vortex wake behind of wind turbine rotor. *Journal of Physics: Conference Series, The Science of Making Torque from Wind*, 75: 012036.

[22] Medici, D., Ivanell, S., Dahlberg, J.-A., Alfredsson, P. H. (2011). The upstream flow of a wind turbine: blockage effect. *Journal of Wind Energy*, 14:691697.

[23] de Vries, O. (1979). Wind-tunnel tests on a model of two-bladed horizontal axis wind turbine and evaluation of an aerodynamic performance calculation method. Technical Report NLR TR 79071 L, NLR, Amsterdam.

[24] Schreck, S. (2002). The NREL full-scale wind tunnel experiment. Introduction to the special issue. *Wind Energy*, 5:77–84.

[25] Lutz, T., Meister, K., Kramer, E. (2011). Near wake studies of the Mexico rotor. In: *EWEC*, Brussels, Belgium.

[26] Nilsson, K., Shen, W. Z., Sørensen, J. N., Breton, S.-P., Ivanell, S. (2014). Validation of the actuator line method using near wake measurements of the Mexico rotor. *Journal of Wind Energy*, DOI:10.1002/we.1714.

[27] Sørensen, J. N., Mikkelsen, R., Troldborg, N. (2007). Simulation and modelling of turbulence in wind farms. In: *EWEC 2007*, Milan, Italy. European Wind Energy Association.

[28] Wußow, S., Sitzki, L., Hahm, T. (2007). 3d simulation of the turbulent wake behind a wind turbine. *Journal of Physics: Conference Series, The Science of Making Torque from Wind*, 75:012033.

[29] Walther, J. H., Guenot, M., Machefaux, E., Rasmussen, J. T., Chatelain, P., Okulov, V. L., Sørensen, J. N. (2007). A numerical study of the stability of vortices using vortex methods. *Journal of Physics: Conference Series, The Science of Making Torque from Wind*, 75:012034.

[30] Watters, C. S., Masson, C. (2007). Recent advantages in modeling of wind turbine wake vortical structure using a differential actuator disc theory. *Journal of Physics: Conference Series, The Science of Making Torque from Wind*, 75:012037.

[31] Jimenez, A., Crespo, A., Migoya, E., Garcia, J. (2007). Advances in large-eddy simulations of a wind turbine wake. *Journal of Physics: Conference Series, The Science of Making Torque from Wind*, 75:012041.

[32] Zahle, F., Sørensen, N. N. (2007). On the influence of far-wake resolution on wind turbine flow simulations. *Journal of Physics: Conference Series, The Science of Making Torque from Wind*, 75:012042.

[33] Mikkelsen, R., Sørensen, J. N., Oye, S., Troldborg, N. (2007). Analysis of power enhancement for a row of wind turbines using the actuator line technique. *Journal of Physics: Conference Series, The Science of Making Torque from Wind*, 75:012044.

[34] Troldborg, N., Sørensen, J. N., Mikkelsen, R. (2007). Actuator line simulation of wake of wind turbine operating in turbulent inflow. *Journal of Physics: Conference Series, The Science of Making Torque from Wind*, 75:012063.

[35] Nilsson, K., Ivanell, S., Hansen, K. S., Mikkelsen, R., Sørensen, J. N., Breton, S.-P., Henningson, D. (2014). Large-eddy simulations of the Lillgrund wind farm. *Journal of Wind Energy*, DOI:10.1002/we.1707.

[36] Sørensen, J. N., Shen, W. Z. (2002). Numerical modeling of wind turbine wakes. *Journal of Fluid Engineering*, 124.

[37] Sørensen, N. N. (1995). *General Purpose Flow Solver Applied to Flow over Hills*. PhD thesis, Risø National Laboratory, Roskilde, Denmark.

[38] Michelsen, J. A. (1992). Basis3d - a platform for development of multiblock pde solvers. Technical Report AFM 92-06, Department of Fluid Mechanics, Technical University of Denmark, DTU.

[39] Michelsen, J. A. (1994). Block structured multigrid solution of 2d and 3d elliptic pde's. Technical Report AFM 94-06, Department of Fluid Mechanics, Technical University of Denmark, DTU.

[40] Ta Phuoc, L. (1994). Modeles de sous maille appliques aux ecoulements instationnaires decolles. In: *Proceedings of the DRET Conference: 'Aerodynamique Instationnaire Turbulents - Aspects Numeriques et Experimentaux*.

[41] Sagaut, P. (2006). *Large Eddy Simulation for Incompressible Flow*. Springer, 3rd edition.

[42] Mikkelsen, R. (2003). *Actuator Disc Methods Applied to Wind Turbines*. PhD thesis, Department of Fluid Mechanics, Technical University of Denmark, DTU.

[43] Ivanell, S. (2009). Numerical computations of wind turbine wakes. Technical Report PhD thesis, ISSN 0347-467X, ISBN 978-91-7415-216-6, Stockholm, Sweden.

[44] Sørensen, J. N., Myken, A. (1992). Unsteady actuator disc model for horizontal axis wind turbine. *Journal of Wind Engineering and Industry Aerodynamics*, 39.

[45] Sørensen, J. N., Kock, C. W. (1995). A model for unsteady rotor aerodynamics. *Journal of Wind Energy and Industry Aerodynamic*, 58:259–275.

[46] Sørensen, J. N., Shen, W. Z., Munduate, X. (1998). Analysis of wake states by a full-field actuator disc model. *Wind Energy*, 1:73–88.

[47] Mikkelsen, R., Sørensen, J. N., Troldborg, N. (2007). Prescribed wind shear modelling with the actuator line technique. In: *Proceedings of EWEC 2007, European Wind Energy Conference*, Milano, Italy.

[48] Mann, J. (1998). Wind field simulation. *Probabilistic Engineering Mechanics*, 13(4):269–282.

[49] Mann, J. (1994). The spatial structure of neutral atmospheric surface-layer turbulence. *Journal of Fluid Mechanics*, 273:141–168.

[50] Mann, J., Ott, S., Hoffmann Jørgensen, B., Frank, H. P. (2002). Wasp engineering 2000. Technical Report Risø-R-1356(EN).

[51] Charnock, H. (1955). Wind stress on water surface. *Quarterly Journal of the Royal Meteorological Society*, 81:639–640.

[52] Ivanell, S., Sørensen, J. N., Mikkelsen, R., Henningson, D. (2009). Analysis of numerically generated wake structures. *Wind Energy*, 12(1):63–80.

[53] Ivanell, S., Mikkelsen, R., Sørensen, J. N., Henningson, D. (2010). Stability analysis of the tip vortices of a wind turbine. *Wind Energy*, 12(8):705–715.

13

RANDOM CASCADE MODEL FOR SURFACE WIND SPEED

R. Baile and J. F. Muzy

13.1 INTRODUCTION

The fast growth of wind energy technology shows that more and more countries attach importance to this renewable resource. However, the energy production is strongly dependent on the wind, which is volatile, and is consequently characterized by a large amount of uncertainty. Therefore, a reliable description of wind speed statistical features is essential. First of all, in order to evaluate wind power resource and system production or to optimize plant scheduling, precise estimation of wind speed probability distribution laws is of great interest [1, 2]. Moreover, accurate wind speed predictions and thereby forecast of power output of wind farms in the next hours up to the next days are necessary for the energy suppliers to optimize wind integration into a power system through, for example, economic dispatch, reserve allocation, or power exchanges with neighbor system [3].

Recent empirical findings [4, 5] suggest the existence of some cascading process in the atmospheric mesoscale range, reflecting an energy transfer process similar to the one well known in fully developed turbulence regime [6]. These observations led us to build a stochastic model for wind speed fluctuations, relying on a random cascade. More precisely, we consider that wind speed can be described by an autoregressive seasonal time series model where the noise term is "multifractal," that is, associated with a random cascade. In this contribution, we provide a brief survey of this model called "M-Rice" and first introduced in [7]. The "M-Rice" model can be of great interest in wind power industry since it can provide solutions to various problems related to wind resource as the issues mentioned above. On the one hand, this new model is proposed for estimating wind speed frequency distributions [8]. This leads to wind speed distributed according to M-Rice probability density function (pdf), that is, a Rice distribution multiplicatively convolved with a normal law. On the other hand, the "M-Rice" model also performs well in short-term forecasting (1–12 hours ahead) of surface layer wind speed. Indeed, a better modeling of the noise term based on cascade process enhances the forecast [7].

13.2 THE "M-RICE" TIME SERIES MODEL OF WIND SPEED

The "M-Rice" model is a time series model of wind speed components $V_x(t)$ and $V_y(t)$. It is formulated as a seasonal autoregressive process where errors are given by a (seasonal) continuous cascade. Before introducing this model, let us briefly enumerate all the empirical observations over which the model is based.

Subsequently, $V(t)$ will denote the modulus of the wind speed horizontal vector while $V_x(t)$ and $V_y(t)$ will stand for its two components along arbitrary orthogonal axes x and y. We have by definition

$$V(t) = \sqrt{V_x(t)^2 + V_y(t)^2}$$

13.2.1 Wind Speed Empirical Observations

13.2.1.1 Wind Speed Basic Statistical Properties. In order to construct a parsimonious stochastic model of wind variations, we have chosen to study V_x and V_y separately, since a modeling of the wind dynamics in polar coordinates (modulus and direction) would be cumbersome and more difficult to handle using Gaussian processes (see next section).

Alternative Energy and Shale Gas Encyclopedia, First Edition. Edited by Jay H. Lehr and Jack Keeley.
© 2016 John Wiley & Sons, Inc. Published 2016 by John Wiley & Sons, Inc.

Moreover, since there is no well-defined wind direction with a small "turbulent rate," components along longitudinal and transverse directions are meaningless and we have preferred to focus on components along arbitrary fixed directions.

The power spectrum is one of the most common tools for analyzing random processes. Since the pioneering work of Ven der Hoven [9, 10], the shape of a typical atmospheric wind speed spectrum (spectrum of V, V_x, or V_y) in the atmospheric boundary layer is still a matter of debate. It is relatively well admitted that it possesses two regimes separated by a low-energy valley called the "spectral gap" located at frequencies around a few minutes. This gap separates the microscale regime, where turbulent motions take place, from the mesoscale range. At the mesoscale range, the first striking feature of a wind speed power spectrum is the main peaks associated with diurnal oscillations. Up to the presence of these peaks, the spectrum can be represented by a decreasing function that connects the flat low-frequency behavior to the high-frequency spectral gap. The exact shape of the spectrum in the (intermediate) mesoscale range is unknown but it can be modeled by a power law $P(f) \sim f^{-\beta}$ with an exponent β around 1.5–2. One does not expect, in the mesoscale range, the same level of universality of the speed statistics at different sites as in the microscale (turbulence) range [6]. Notably the value of the exponent can depend on local orographic and atmospheric conditions, height above ground level, and so on [11]. The main power spectrum features can be alternatively observed through the behavior of the correlation function of the wind speed components. Let us note that the correlation decreases exponentially fast but with a large correlation time: the wind speed can remain correlated with a correlation greater than 10% up to lags of 5 days. Wind component autocorrelations and cross-correlations can be easily described within the framework of linear time series models like ARMA modeling [12].

"M-Rice" model relies on (seasonal) ARMA processes and accounts for the non-Gaussian observed statistics through the nature of the noise term that will be given, as explained in the next section, by a multifractal process.

13.2.1.2 Nonlinear Statistical Properties: Non-Gaussian Fluctuations and Magnitude Long-Range Correlations.

When referring to the non-Gaussian nature of wind speed statistics, one has to be precise since wind speed amplitudes are obviously not normally distributed. Indeed, even in the case when V_x and V_y are Gaussian, the wind speed modulus pdf is a Rayleigh distribution (or a Rice distribution if the components have a nonzero mean), a particular case of the Weibull family. This is probably the main reason why Weibull is the most commonly considered distribution in order to reproduce the pdf of wind amplitudes [13].

As discussed in the previous section, a modeling of both components V_x and V_y allows one to reproduce the

observed (partial) correlation functions within the framework of ARMA models. However, even in this context, non-Gaussian statistics are observed: if one studies the distribution of the prediction errors of these models (or simply the distribution of wind speed component variations), it appears that their pdfs are characterized by stretched exponential tails very similar to the distribution of velocity increments at small scales in fully developed turbulent flows [6, 7]. As explained in [14], such a shape is a characteristic of the law associated with the small-scale fluctuations of a log-normal random cascade model (see Section 13.2.2.2). Another striking feature of wind series is that the amplitude of the error noise is long-range correlated. This is another property that has been observed in turbulence [15, 16]. More precisely, if one defines the local "magnitude" as $v(t) = \frac{1}{2}\ln\left(\rho_x(t)^2\right)$ or $v(t) = \frac{1}{2}\ln\left(\rho_y(t)^2\right)$ or

$$ v(t) = \frac{1}{2}\ln\left(\rho_x(t)^2 + \rho_y(t)^2\right) \tag{13.1} $$

where ρ_x and ρ_y are the noise terms associated with a linear prediction of V_x and V_y[1] then, as illustrated in Figure 13.1, the empirical covariance of v can be fitted as

$$ \mathrm{Cov}\left[v(t), v(t+\tau)\right] \simeq \beta^2\ln^2\left(\frac{\tau}{T}\right) \tag{13.2} $$

Indeed, if one represents the square root covariance as a function of the logarithm of the lag τ one observes a straight line. In Figure 13.1, we have plotted the magnitude covariances estimated for different sites in Corsica and Netherlands (see caption). It can be observed that the parameters β^2 (slope of the curves) and T (time lag where correlations vanish) are very close for all site data. As explained in [4, 7] and briefly reviewed in Section 13.2.2.2, these properties are intimately related to random cascade processes.

13.2.2 Building the Model

13.2.2.1 Construction of the Seasonal Autoregressive Part.
As previously mentioned, $V_x(t)$ and $V_y(t)$ both contain additive seasonal components; that is, they can be written as

$$ V_{x,y}(t) = S_{x,y}(t) + V_{x,y}^S(t) \tag{13.3} $$

where $S_{x,y}(t)$ represents the deterministic diurnal oscillations (that vary with the season) and $V_{x,y}^S(t)$ the "deseasonalized" wind speed components. Since the seasonality is caused by the variation of the sun position during the day, $S_x(t)$ and $S_y(t)$ are almost daily periodic functions, with a period shape that changes according to the considered season in the year.

---

[1]We would obtain the same results with $v_s(t) = \frac{1}{2}\ln(\delta_s V_x^2 + \delta_s V_y^2)$ where s is small scale and $\delta_s F(t) = F(t+s) - F(t)$.

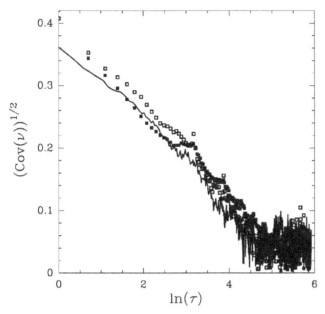

FIGURE 13.1 Square root of the wind magnitude covariance as a function of the log of the lag (time units are hours). The solid line corresponds to a 1 minute sampled data series (20 minutes average) from Corsica. The symbol (□) curve represents the average over seven 1 hour sampled data series from Corsica, and the symbol (■) curve, the mean of three 1 hour sampled data series from Netherlands. Let us note that the same decreasing linear function (that becomes zero above some lag T) is obtained for all sites individually. All the slopes are close to each other and the "correlation" scale (integral scale) is roughly the same for all sites (see [4]).

In order to determine this shape, we therefore have to perform a "local" estimation. For that purpose, we use a standard methodology described in [17] (see also [18] for similar methodology applied to wind power in the tool called WPPT): each seasonal component $S_x(t)$ and $S_y(t)$ (denoted as $S(t)$) is described by m Fourier modes of period 1 *day* ($D = 24$ samples for hourly data):

$$S(t) = \alpha_0 + \sum_{k=1}^{m} \left[\eta_{1,k} \sin\left(\frac{2k\pi t}{D}\right) + \eta_{2,k} \cos\left(\frac{2k\pi t}{D}\right) \right] \quad (13.4)$$

Because of the yearly variation in the seasonality, the coefficients $\{\eta_{i,k}\}_{i=1,2;k=1...m}$ depend a priori on the day d and the local estimation simply consists of using the least squared method associated with a local exponential moving average:

$$\{\eta_{i,k}\}_{i=1,2;k=1...m}(d) =$$

$$\arg\min\left\{ \sum_{yy=1}^{Y} \sum_{j} \psi^{|d-j|} \sum_{t=0}^{D-1} \left[V_{x,y}(yy,j,t) - S(t) \right]^2 \right\}$$

$$(13.5)$$

where Y is the number of available years in the data series, $V_{x,y}(yy,j,t)$ represent the wind speed components at year yy,

day j, and "hour" t. ψ is an exponential discount factor chosen so that $\frac{-1}{\ln(\psi)} \simeq 10$ days ($\psi = 0.9$).

In order to account for the linear correlations and cross-correlations of the stationary parts $V_x^S(t)$ and $V_y^S(t)$, we have considered the class of bivariate ARMA processes. The study of partial autocorrelation (PACF) and cross-correlation functions or procedure like the Akaike Information Criterion (AIC) suggests that an AR of order 2 is appropriate to fit the observations.

Finally, we are led to the following simple model for the deseasonalized wind components:

$$\begin{cases} V_x^S(t+1) = \sum_{k=0}^{1} \left(\gamma_{xx}(k) V_x^S(t-k) + \gamma_{xy}(k) V_y^S(t-k) \right) \\ \qquad\qquad + \rho_x(t+1) \\ V_y^S(t+1) = \sum_{k=0}^{1} \left(\gamma_{yy}(k) V_y^S(t-k) + \gamma_{yx}(k) V_x^S(t-k) \right) \\ \qquad\qquad + \rho_y(t+1) \end{cases}$$

$$(13.6)$$

where $\rho_{x,y}(t)$ represents the noise terms which will be modeled as a log-normal continuous cascade (see next section) and $\gamma_{xx}(k)$, $\gamma_{yy}(k)$, $\gamma_{xy}(k)$ and $\gamma_{yx}(k)$ ($k = 1,2$) are the AR coefficients.

13.2.2.2 Random Cascade Model for Wind Speeds.

Let us briefly overview random cascade to show that empirical observations reported in Figure 13.1 are intimately related to the existence of such a process. Discrete random multiplicative cascades [19–21] were originally introduced as models for the energy cascade in fully developed turbulence. In the simplest case, these objects are positive fields (measures) whose construction involves a recursive procedure along a dyadic tree: the cascading process starts at a large "integral" time scale T where the measure is uniformly spread (meaning that the density is constant). One then splits this interval into two equal parts over which the densities are obtained by multiplying the "father" density by two (positive) i.i.d. random factors $W_1 = e^{\kappa_1}$ and $W_2 = e^{\kappa_2}$. Each of these two subintervals is again cut into two equal parts and the process is repeated infinitely. At construction step n, the dyadic intervals have a size $T2^{-n}$ and their measure denoted σ_n^2 is simply $\sigma_n^2 = \sigma_0^2 \prod W_i = 2^{-n} e^{\sum \kappa_i}$, where all the $W_k = e^{\kappa_k}$ are i.i.d. such that $E(W) = 1$. If the random variables κ are Gaussian, then the corresponding model is log-normal and its scaling properties are easy to control (see, e.g., [22] and references therein for more details). Let us notice that nonpositive fields like, for example, the velocity field in developed turbulence, can be simply derived from the construction of the measure by considering that $\sigma^2(t)$ is the (stochastic) variance of a Brownian motion (or another Gaussian process), that is, $\delta_\tau X(t) = \sqrt{\sigma^2(\tau)}\varepsilon$, where ε is a Gaussian random noise. Such "grid-bounded" cascades, though simple, do not, however, provide a satisfying model for a stationary physical process such as wind temporal variations. Indeed, they are built on a fixed time interval $[0, T]$, are not causal, and are not stationary. Moreover, they involve an arbitrary fixed scale ratio (2 in the dyadic case). Very recently, several constructions have been proposed to generalize discrete cascades to stationary, causal, and continuous processes [22,23]. We will not enter into details but if one calls the magnitude process $\omega(t) = \sum_i \kappa_i$, then in the log-normal case, ω is a Gaussian process characterized by its covariance function. If one notices that the tree-like structure underlying the discrete construction implies a logarithmic correlation function, then one can naturally define the log-normal continuous cascade as follows [23,24]:

$$\sigma_s^2(t) = e^{2\omega_s(t)} \quad (13.7)$$

where $\omega_s(t)$ is a stationary Gaussian process of covariance defined by

$$\text{Cov}[\omega_s(t), \omega_s(t+\tau)] = \lambda^2 \ln\left(\frac{T}{s+\tau}\right) \quad (13.8)$$

Here T and λ^2 are two parameters that correspond, respectively, to the integral scale (correlation length analog to the time scale where cascading process starts) and the intermittency coefficient (which quantifies the degree of occurrence of strong events in the process). The parameter s is a time sampling parameter that can be chosen arbitrarily small (since the weak limit $s \to 0$ of the process exists [22, 25]). It can be proven that such a process is the continuous equivalent of discrete random cascades. Therefore, according to this picture, a continuous cascade is nothing but a stochastic process $X(t)$ whose variations can be written as

$$\delta X(t) = e^{\omega_s(t)} \varepsilon(t) \quad (13.9)$$

where ε is a Gaussian noise and the magnitude ω has a covariance correlated as a logarithmic function (equation (13.8)). It can be shown that, unlike Brownian motion, the scaling properties of such processes are not characterized by a single scaling exponent but by a multifractal spectrum. In that respect, random cascades are also called multifractal processes (see [4, 6] for more details).

In order to link these considerations with previous observed features for wind data, let us remark that wind fluctuations at a fixed spatial location result from two types of stochastic variations: first, the spatial fluctuations at a fixed time (Eulerian) and then the temporal fluctuations for a fixed fluid element (Lagrangian). Since there is no strong mean velocity and Taylor frozen hypothesis cannot be invoked, both Lagrangian and Eulerian variations have to be taken into account. In [26], Castaing shows that if one supposes a continuous cascade paradigm (equation (13.8)) for both Eulerian and Lagrangian fields, then the magnitude correlation function at a fixed location should behave like a squared logarithmic function:

$$\text{Cov}[\omega_s(t), \omega_s(t+\tau)] = \beta^2 \ln^2\left(\frac{T}{s+\tau}\right) \quad (13.10)$$

where the coefficient β^2 depends on both Lagrangian and Eulerian intermittency coefficients. Let us note this is precisely the behavior that has been observed in real data (see Figure 13.1 and [4, 7] for more details). Therefore, the residual variance of the errors $\rho_x(t)$ and $\rho_y(t)$ associated with the linear models of $V_x(t)$ and $V_y(t)$ (see equation (13.6)) can be both defined as in equation (13.9):

$$\begin{cases} \rho_x(t) = e^{\omega^s(t)+M_s(t)}\varepsilon_x(t) = e^{\omega(t)}\varepsilon_x(t) \\ \rho_y(t) = e^{\omega^s(t)+M_s(t)}\varepsilon_y(t) = e^{\omega(t)}\varepsilon_y(t) \end{cases} \quad (13.11)$$

where $\varepsilon_x(t)$ and $\varepsilon_y(t)$ are independent realizations of white Gaussian noise, $M_s(t)$ is a deterministic function that represents a multiplicative seasonality of the noise amplitude, and $\omega^s(t)$ is a zero mean stationary Gaussian sequence independent of $\varepsilon(t)$, with a squared log covariance as described in Section 13.2.2.2 (equation (13.10)). In practice, one computes $2v(t) = \ln(\rho_x^2(t) + \rho_y^2(t)) = 2\omega^s(t) + 2M_s(t) + \ln(\varepsilon_x^2(t) + \varepsilon_y^2(t))$ and since the mean and variance of $\ln Z(t) = \ln(\varepsilon_x^2(t) + \varepsilon_y^2(t))$ are known, one can obtain $M_s(t)$ along the same line as we have estimated $S_{x,y}(t)$ (equations (13.4) and (13.5)). A generalized method of moments [22] applied to the sample covariance of $v(t)$ allows one to evaluate the parameters β^2 and T of equation (13.10), defining the Gaussian process $\omega^s(t)$.

13.3 APPLICATION TO WIND SPEED PROBABILITY DISTRIBUTION ESTIMATION

As recalled in the introduction, wind speed and consequently wind energy production are characterized by a large amount of randomness. A precise estimation of wind speed probability distribution law is particularly necessary to evaluate, among others, the resource at one site. These distributions are commonly described by the two-parameter Weibull distribution [2, 27–33]. Even if, in some cases, other laws can provide a better fit of the data (like, e.g., Gamma distribution [34], three-parameter Weibull model [1], Rayleigh distribution [29, 35], bivariate normal distribution [36, 37], gamma distribution [34], elliptical bivariate distribution [38]), Weibull distribution is considered by the community as the reference. It is the most widely used and accepted velocity law in the wind energy literature. Indeed, for the scientific community, its popularity is not only due to goodness of fit. First of all, one can underline its simplicity with the fact that there are only two parameters (against, e.g., five for a bivariate normal distribution). Moreover, if the Weibull distribution is fitted for wind speed at one height, distributions at other heights can be deduced from the power law wind profile (see [39]). At last, if wind speed follows the Weibull distribution with parameters λ and k, then the distribution of wind speed cubed, proportional to wind power, follows the Weibull distribution with parameters $k/3$ and λ [2].

A nonnegative random variable V is Weibull distributed if its pdf is defined by

$$f_{\mathrm{w}}(\lambda,k)[V] = \frac{k}{\lambda}\left(\frac{V}{\lambda}\right)^{k-1}\mathrm{e}^{-\left(\frac{V}{\lambda}\right)^k}, \ (k,\lambda > 0) \quad (13.12)$$

where k is a dimensionless shape factor and λ is the scale parameter of the function, with the same dimension as V.

In [8], we aimed at studying the velocity probability distribution function associated with the "M-Rice" model described above. Since the velocity modulus has a Rice distribution conditionally to the stochastic variance, the resulting unconditional law is simply a Rice law multiplicatively convolved with a log-normal law. Let us now explain how this wind speed time series model allows one to build the M-Rice distribution.

13.3.1 M-Rice Distribution

The M-Rice distribution of wind speed amplitude V is based on its decomposition into V_x and V_y and an assumption on the distributions of V_x and V_y. Let us note that the description of wind speed distribution in terms of the $V_{x,y}$ component distributions has already been considered in the literature [29, 35–38, 40]. For instance, in the simple case where V_x and V_y are centered Gaussian processes, the resulting distribution of $V = \sqrt{V_x^2 + V_y^2}$ is a Rayleigh distribution, a particular case of the Weibull one. The Rice distribution [41] is a generalization of the Rayleigh distribution and accounts for non-centered Gaussian components. $V \sim \mathrm{Rice}(v,\sigma)$ has a Rice distribution of parameters $v \geq 0$ and $\sigma \geq 0$ if $V = \sqrt{V_x^2 + V_y^2}$, where $V_x \sim N\left(v\cos\theta,\sigma^2\right)$ and $V_y \sim N\left(v\sin\theta,\sigma^2\right)$ are two independent normal distributions of respective mean $v\cos\theta$ and $v\sin\theta$ and of standard deviation σ (θ being a real number). In other words, V is Rice distributed if V_x and V_y can be written as

$$\begin{cases} V_x = v\cos\theta + \sigma\varepsilon_1 \\ V_y = v\sin\theta + \sigma\varepsilon_2 \end{cases} \quad (13.13)$$

where ε_1 and ε_2 are standardized normal random variables. If one denotes by $R(v,\sigma)[V]$ the Rice probability density of a nonnegative random variable V, its expression reads

$$R(v,\sigma)[V] = \frac{V}{\sigma^2}\exp\left(-\frac{V^2+v^2}{2\sigma^2}\right)I_0\left(\frac{Vv}{\sigma^2}\right) \quad (13.14)$$

where $I_0(x)$ is the order 0 Bessel function [42].

According to equations (13.3), (13.6), and (13.11), $V_x(t)$ (or $V_y(t)$) can be written as

$$V_x(t) - S_x(t) = \sum_{i=0}^{+\infty}\alpha_x^i\rho_x(t-i) = \sum_{i=0}^{+\infty}\alpha_x^i\mathrm{e}^{\omega(t-i)}\varepsilon_x(t-i) \quad (13.15)$$

If $\omega(t-i)$ are fixed, it results that

$$V_x(t) - S_x(t) \underset{\mathrm{law}}{=} \sqrt{\sum_i \alpha_x^{2i}\mathrm{e}^{2\omega(t-i)}}\varepsilon$$

where ε is a standardized Gaussian random variable. Along the same approach as the one proposed in [22], it can be shown that if intermittency coefficient β^2, defined in equation (13.10), is sufficiently small, then

$$\sum_{i=0}^{+\infty}\alpha_x^{2i}\mathrm{e}^{2\omega(t-i)} \underset{\mathrm{law}}{\simeq} \mathrm{e}^{2\Omega} \quad (13.16)$$

where

$$\Omega = \sum_k \alpha_x^{2k}\omega(k) \quad (13.17)$$

is also a Gaussian random variable, independent of ε, whose mean and variance can be deduced from those of ω. It results finally from equations (13.6) and (13.16) that

$$V_{x,y} \underset{\mathrm{law}}{=} S_{x,y} + \mathrm{e}^\Omega\varepsilon_{x,y} \quad (13.18)$$

By analogy between equations (13.13) and (13.18), it turns out that, at fixed Ω, the velocity V has a Rice distribution of parameters

$$\begin{cases} v = \sqrt{\bar{S}_x^2 + \bar{S}_y^2} \\ \sigma = \mathrm{e}^\Omega \end{cases} \quad (13.19)$$

where $\bar{S}_{x,y}$ is the mean of $V_{x,y}$ seasonal component. In fact, we have seen that in our model, Ω (i.e., $\ln\sigma$) is not fixed but "fluctuates" as a Gaussian random process. Let $P(\Omega)$ be the unconditional Gaussian law of Ω, of mean $\hat{\Omega}$ and variance s_Ω:

$$P(\Omega) = \frac{\mathrm{e}^{-\frac{(\Omega-\hat{\Omega})^2}{2s_\Omega}}}{\sqrt{2\pi s_\Omega}} \quad (13.20)$$

From equation (13.18), the velocity distribution, called M-Rice, can be obtained as a "geometric convolution" between $P(\Omega)$ and the Rice distribution [14]:

$$f_{\mathrm{mr}}\left(v,\hat{\Omega},s_\Omega\right)[V] = \int P(\Omega)R(v,\mathrm{e}^\Omega)[V]\,\mathrm{d}\Omega \quad (13.21)$$

This quantity can be easily evaluated numerically by a Gaussian quadrature approximation of the integral [43]. M-Rice distribution is then a Rice distribution whose standard deviation σ is a log-normal random variable.

Thus, the M-Rice distribution is characterized by three parameters. The value of v can be directly estimated as follows:

$$v = \sqrt{\bar{V}_x^2 + \bar{V}_y^2} = \sqrt{\bar{S}_x^2 + \bar{S}_y^2} \qquad (13.22)$$

where $\bar{V}_x$ and $\bar{V}_y$ are, respectively, the mean values of V_x and V_y. Let us note that a nonzero value of v means that there is a preferred velocity direction in the data, that is, there exists a non-vanishing mean velocity vector. The two remaining parameters are the mean and the variance of Ω, denoted as $\hat{\Omega}$ and s_Ω. The (exponential of the) mean value of Ω is simply an overall scale factor of the distribution while the variance s_Ω controls its shape: the greater s_Ω, the wider the M-Rice right tail. In theory, these parameters could be computed using mean and variance of ω according to equation (13.17). However, a simpler and more direct method is to use a method of moments. $\hat{\Omega}$ and s_Ω can be determined from the expressions of Rice order 2 and 4 moments, $\mu_2 = 2\sigma^2 + v^2$ and $\mu_4 = 8\sigma^4 + 8\sigma^2 v^2 + v^4$ where, from equation (13.19), $\sigma^2 = e^{2\hat{\Omega}+2s_\Omega}$ and $\sigma^4 = e^{4\hat{\Omega}+8s_\Omega}$. Therefore, the two unknown parameters can then be evaluated as

$$\hat{\Omega} = \ln\sigma^2 - \frac{1}{4}\ln\sigma^4, s_\Omega = \frac{1}{4}\left(\ln\sigma^4 - 2\ln\sigma^2\right) \quad (13.23)$$

where $\sigma^2 = \left(\mu_2 - v^2\right)/2$ and $\sigma^4 = \left(\mu_4 - 8\sigma^2 v^2 - v^4\right)/8$ only depend on known parameters.

13.3.2   Correlations between Weibull and M-Rice Parameters

As it has been shown in [8], there are correlations between Weibull and M-Rice parameters. In order to determine the precise relationship between M-Rice and Weibull parameters, let us notice that the nth moment associated with Weibull distribution can be written as $\mu_n = \lambda^n \Gamma(1 + n/k)$ with $\Gamma(z) = \int_0^\infty x^{z-1}e^{-x}dx$. Since M-Rice parameters are calculated from 2 and 4 order moments, we can write, from equation 13.23

$$\begin{cases} \hat{\Omega} = \ln\left(\frac{\lambda^2\Gamma\left(1+\frac{2}{k}\right)-v^2}{2}\right) \\ \quad -\frac{1}{4}\ln\left(\frac{\lambda^4\Gamma\left(1+\frac{4}{k}\right)-4\lambda^2 v^2\Gamma\left(1+\frac{2}{k}\right)+3v^4}{8}\right) \\ s_\Omega = \frac{1}{4}\ln\left(\frac{\lambda^4\Gamma\left(1+\frac{4}{k}\right)-4\lambda^2 v^2\Gamma\left(1+\frac{2}{k}\right)+3v^4}{8}\right) \\ \quad -\frac{1}{2}\ln\left(\frac{\lambda^2\Gamma\left(1+\frac{2}{k}\right)-v^2}{2}\right) \end{cases} \quad (13.24)$$

These relations provide the link between the two distributions and allow one to interpret the usual Weibull parameters within the framework of our model. Indeed, one can directly check in equation (13.24) that, within our model, when $v = 0$ and $s_\Omega = 0$, one recovers a Rayleigh distribution ($k = 2$) with a scale parameter $\lambda = \sqrt{2}e^{\hat{\Omega}}$. By performing a series expansion around $v = 0$ and $s_\Omega = 0$, we find (after some cumbersome algebra)

$$\begin{cases} \lambda = \sqrt{2}e^{\hat{\Omega}} + 8e^{2\hat{\Omega}}s_\Omega(\gamma - 1) + \frac{v^2}{2}(\gamma - 1) \\ \quad + O\left(v^4\right) + O\left(s_\Omega^2\right) \\ k = 2 + e^{-2\hat{\Omega}}v^2 - 8s_\Omega + O\left(v^4\right) + O\left(s_\Omega^2\right) \end{cases} \quad (13.25)$$

One sees that values of $k > 2$ are due to the existence of a strong preferred velocity direction (v large) while k values smaller than 2 indicate a strong intermittency (s_Ω large).

13.3.3   Wind Speed Distribution Law Estimation

Figure 13.2 illustrates the pdf of the wind speed estimated at Calvi (Corsica) with the two models. One can see that the pdf shapes from Weibull and M-Rice models are quite close and describe relatively well the data. In order to quantitatively compare the two models, we studied wind speed moment estimation and directly confronted the pdfs by resorting to several goodness-of-fit tests [8].

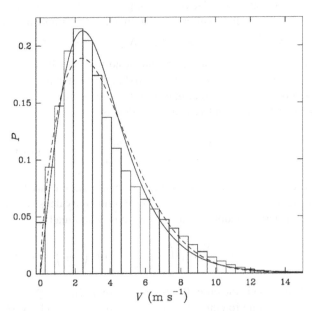

FIGURE 13.2   Probability density function (pdf) of wind speed amplitude in Calvi (Corsica). The histogram represents the pdf based on data set, the dashed line (respectively, the solid line) corresponds to pdf deduced from Weibull model (respectively, M-Rice model).

Thus, it turns out that this model can be used as an alternative to standard Weibull law since fitting wind speed modulus distribution with the M-Rice model has been shown to be as efficient as the Weibull one. Moreover, M-Rice seems to provide a better fit of tail events. Nevertheless, our goal was not to show that M-Rice model could fit wind data better than Weibull since the popularity of the Weibull distribution is not only due to goodness of fit. Even if our distribution could be parameterized in order to have the same advantages as the Weibull one, we aim to highlight the possibility to explain Weibull (and then wind) distribution shape using our M-Rice distribution, whose parameters (related to Weibull ones) are directly deduced from a model of the intermittent dynamics of wind fluctuations, as shown in Section 13.3.2. In that respect, M-Rice description can be considered as complementary to the Weibull one. Our hypothesis is then that wind speed can be described by a Rice distribution associated with a dilation factor, Rice being a Rayleigh distribution (particular case of Weibull with $k = 2$) with an additional constant.

13.4   APPLICATION TO SHORT-TERM FORECASTING

As previously mentioned, because of wind speed volatility, accurate wind speed predictions are necessary for the energy suppliers to optimize wind integration into a power system. The prediction horizon of interest depends on the size of the system [3]: for interconnected systems, it is typically 48–72 hours. For isolated systems (e.g., islands), the horizon of interest can vary from 1 to a few hours. Indeed, the impact of wind variability is higher in non-interconnected systems, and real-time scheduling of electricity generation is necessary: the forecast allows one to plan a generation schedule in order to be able to respond to a potential rapid change in available wind power output. Another application for wind power forecasting with a 1–12 hours horizon is for generators to optimize their bidding strategies in some electricity markets where bids can be made with a short lead time [44, 45].

For that purpose, many efforts have been made by the scientific community over several years in order to design models that allow one to produce sufficiently good forecasts. As reviewed, for example, in [46], there are mainly two families of approaches. The "physical" models rely upon physical considerations leading to atmospheric models that provide a "numerical weather prediction" (NWP) system. For very short prediction horizons, one often prefers "statistical" approaches that mainly consist of designing stochastic models or using methods of time series analysis, calibrated on historical data or other explanatory variables (like the output of a physical model).

The method we propose, based on M-Rice model, is part of statistical approaches since it uses only recent wind speed observations as inputs. We aim at performing predictions of the wind speed over horizons extending from 1 to 12 hours (i.e., as advocated previously, potentially useful in isolated power systems or some electricity markets). Indeed, as mentioned above, statistical methods are pertinent at very short term horizons (up to 6 hours) while at longer horizons, approaches based on NWP turn out to be more efficient

13.4.1   H-Step Forward Prediction

Our goal is to predict wind speed amplitude $V(t) = \sqrt{V_x(t)^2 + V_y(t)^2}$ at different horizons of time (from 1 to 12 hours). Since the (conditional) law of the wind speed modulus is not Gaussian, the "best" prediction depends, in general, on the type of error one wants to minimize. In theory, since the multifractal AR model we have introduced provides the full conditional law of each wind speed component, one should be able to solve various forecasting problems corresponding to different kinds of error to minimize. For the sake of simplicity, we will only estimate the conditional mean of V, denoted as $E(V|t)$ in the following, that is, the predictor which minimizes the mean square error.

Let $\hat{V}^S_{x,y}(t, h)$ (resp. $\hat{V}_{x,y}(t, h)$) be the best linear predictors of $V^S_{x,y}(t + h)$ (resp. $V_{x,y}(t + h)$), at time t and horizon h, that is, from the definition of the model:

$$\hat{V}^S_{x,y}(t, h) = E[V^S_{x,y}(t + h)\,|t]$$
$$\hat{V}_{x,y}(t, h) = \hat{V}^S_{x,y}(t, h) + S_{x,y}(t + h)$$

These predictors are easy to compute: since the linear part of our model reduces to a vector AR(2) model (equation (13.6)), h iterations of the model provide the linear coefficients. Indeed, equation (13.6) can be rewritten in a vector form:

$$\mathbf{V}^S(t + 1) = \mathcal{A}\mathbf{V}^S(t) = e(t + 1) \qquad (13.26)$$

where the vectors $\mathbf{V}^S(t)$ and $\mathbf{e}(t)$ are defined by

$$\mathbf{V}^S(t) = \begin{pmatrix} V^S_x(t) \\ V^S_y(t) \\ V^S_x(t-1) \\ V^S_y(t-1) \end{pmatrix}, \quad e(t) = \begin{pmatrix} \rho_x(t) \\ \rho_y(t) \\ 0 \\ 0 \end{pmatrix} \qquad (13.27)$$

and the matrix $\mathcal{A}$ reads

$$\mathcal{A} = \begin{pmatrix} \gamma_{xx}(0) & \gamma_{xy}(0) & \gamma_{xx}(1) & \gamma_{xy}(1) \\ \gamma_{yx}(0) & \gamma_{yy}(0) & \gamma_{yx}(1) & \gamma_{yy}(1) \\ 1 & 0 & 0 & 0 \\ 0 & 1 & 0 & 0 \end{pmatrix} \qquad (13.28)$$

When one considers a horizon h, the iteration of equation (13.26) gives

$$\mathbf{V^S}(t+h) = \mathcal{A}^h \mathbf{V^S}(t) + \sum_{k=0}^{h-1} \mathcal{A}^k \mathbf{e}(t+h-k) \\ = \mathcal{A}^h \mathbf{V^S}(t) + \mathbf{e^{(h)}}(t+h) \qquad (13.29)$$

According to this representation, $\hat{V}_{x,y}^S(t,h)$ corresponds to the first two components of $\mathcal{A}^h \mathbf{V^S}(t)$. From equations (13.11) and (13.27), the components of the noise vector, in the RHS of the previous equation, can be written as

$$e_{x,y}^{(h)}(t+h) = \sum_k a_k e_{x,y}^{\omega(t+h-k)}(t+h-k) \qquad (13.30)$$

where the constants a_k can be deduced from the $\mathcal{A}$ coefficients. Moreover, by considering, as shown in [47], $\varepsilon(t)e^{\omega(t)}$ is quasi-stable with respect to linear combinations, we have

$$e_{x,y}^{(h)}(t+h) \underset{\text{law}}{=} e^{\omega^{(h)}}(t+h)_{x,y}^{(h)}(t+h) \qquad (13.31)$$

where $\varepsilon^{(h)}$ is a standardized Gaussian noise and $\omega^{(h)}$ is also Gaussian, at fixed h, with the same covariance as $\omega(t)$ for lags greater than h (equation (13.10)). equations (13.29) and (13.31) show that the model conserves the same shape for all prediction horizons:

$$V_{x,y}^S(t+h) = \hat{V}_{x,y}^S(t,h) + e^{\omega^{(h)}(t+h)}\varepsilon_{x,y^{(h)}}(t+h) \qquad (13.32)$$

This property is of great practical interest because, whatever the horizon h, at fixed value of $\omega^{(h)}(t+h)$, the law of the wind speed modulus $V(t+h)$ is a Rice distribution [41] of parameters $r = \sqrt{\hat{V}_x^2(t,h) + \hat{V}_y^2(t,h)}$ and $\sigma^2 = e^{2\omega^{(h)}(t+h)}$. More specifically, let $M_R(r,\sigma^2)$ be the mean value of a Rice distribution, that is,

$$M_R(r,\sigma^2) = \sigma\sqrt{\frac{\pi}{2}}L_{1/2}\left(-\frac{r^2}{2\sigma^2}\right) \qquad (13.33)$$

(where $L_{1/2}(x)$ is the order 1/2 Laguerre polynomial), and $P_h(\omega|t)$ the conditional Gaussian law of $\omega^{(h)}(t+h)$. The conditional wind speed value at horizon h is then

$$E(V(t+h)|t) = \int P_h(\omega|t)M_R\left(r, e^{2\omega}|t\right)d\omega \qquad (13.34)$$

This quantity can be evaluated numerically by a Gaussian quadrature approximation of the Gaussian integral [43]. The conditional law of $\omega^{(h)}(t+h)$ is a normal law whose mean, $\hat{\omega}^{(h)}(t+h)$, and variance, $s_\omega^{(h)}(t+h)$, can be computed using the known mean and covariance of $\omega^{(h)}$. $\hat{\omega}^{(h)}(t+h)$ is nothing

but the best linear predictor of $\omega(t+h)$ at time t and horizon h, that is,

$$\hat{\omega}^{(h)}(t+h) = M_S^{(h)}(t+h) + \sum_{k=0}^{T-1}\alpha_k\omega^{s(h)}(t-k) \qquad (13.35)$$

where the filter size T and the coefficients α_k are obtained from the shape of the covariance function of $\omega^{s(h)}$ (equation (13.10)). If one denotes $C_{ij}^{(h)} = \text{Cov}\left[\omega^{s(h)}(t), \omega^{s(h)}(t+|j-i|)\right]$ and $\zeta_k^{(h)} = \text{Cov}\left[\omega^{s(h)}(t), \omega^{s(h)}(t+k+h)\right]$, then

$$\alpha_k = \sum_j \left[C_{kj}^{(h)}\right]^{-1}\zeta_j^{(h)} \qquad (13.36)$$

Let us end this section by noticing that the alternative predictor

$$\hat{V}(t,h) = \sqrt{E\left(V^2(t+h)|t\right)} \qquad (13.37)$$

which, after a little algebra, reduces to

$$\hat{V}(t,h) = \sqrt{\hat{V}_x(t,h)^2 + \hat{V}_y(t,h)^2 + 2e^{2\hat{\omega}^{(h)}(t+h)+2s_\omega^{(h)}(t+h)}} \qquad (13.38)$$

provides performances relatively close to the former "Rice" predictor.

13.4.2 Forecasting Performances

In [7], we have addressed the problem of short-term wind speed forecasting using the "M-Rice model." It has been applied to forecast hourly wind speed data up to 12 hours ahead. The obtained results showed that the proposed method is more accurate than the standard reference models, in regard to minimizing the root mean square error or mean absolute error. Let us note that when the horizon increases, the performances of each model decrease, but the relative accuracy of our model becomes more and more significant (average improvement of, respectively, 10% and 26% as compared to persistence[2] at 1 and 6 hours forecasting horizons). Moreover, by comparing the performances of our model with and without the multifractal noise, we have established that the former leads to a systematic improvement of the forecast. Furthermore, a systematic improvement in the prediction accuracy has been observed when one uses better

---

[2]Persistence is the most commonly used reference predictor. According to Giebel [49], for short prediction horizons (from few minutes to hours), this model is the benchmark all other prediction models have to beat. It consists in a simple martingale hypothesis according to which future wind speed at horizon h will be the same as the present observed value: $\hat{V}(t+h|t) = V(t)$.

resolved input data: the finer the sampling rate, the better the forecast. This result highlights the importance of having high-frequency data to enhance the forecast quality. Let us notice that our approach can be improved by considering, for instance, its natural multivariate generalization. This may allow us to describe the joint wind variations at different locations. Let us also mention that our time series cascade model is able to provide conditional wind speed probability distributions and therefore addresses questions related to risk management (e.g., to predict the probability that a given strong or weak wind regime will occur during the next few hours [48]). It enables to expand the knowledge on wind speed at horizon we are interested in.

REFERENCES

[1] Stewart, D. A.,Essenwanger, O. M. (1978). Frequency distribution of wind speed near the surface. *Journal of Applied Meteorology*, 17:1633–1642.

[2] Hennessey, J. P.,Jr. (1977). Some aspects of wind power statistics. *Journal of Applied Meteorology*, 16(2):119–128.

[3] Kariniotakis, G., Pinson, P., Siebert, N., Giebel, G., Barthelmie, R. (2004). The state of the art in short term prediction of wind power - from an offshore perspective. Brest, France. Proceedings, Ocean Energy Conference ADEME-IFREMER.

[4] Muzy, J. F., Baile, R., Poggi, P. (2010). Intermittency of surface layer wind velocity series in the mesoscale range. *Physical Review E*, 81:056308.

[5] Baile, R., Muzy, J. F. (2010). Spatial intermittency of surface layer wind fluctuations at mesoscale range. *Physical Review Letters*, 105:254501.

[6] Frisch, U. (1995). *Turbulence*. Cambridge University Press, Cambridge.

[7] Baile, R., Muzy, J. F., Poggi, P. (2011). Short term forecasting of surface layer wind speed using a continuous random cascade model. *Wind Energy*, 14:719–734.

[8] Baile, R., Muzy, J. F., Poggi, P. (2011). An m-Rice wind speed frequency distribution. *Wind Energy*, 14:735–748.

[9] Van der Hoven, I. (1957). Power spectrum of horizontal wind speed in the frequency range from 0.0007 to 900 cycles per hour. *Journal of Meteorology*, 14:160–164.

[10] Oort, A. H., Taylor, A. (1969). On the kinetic energy spectrum near the ground. *Monthly Weather Review*, 97:623–636.

[11] Lauren, M. K., Menabde, M., Seed, A. W., Austin, G. L. (1999). Characterisation and simulation of the multiscaling properties of the energy containing scales of horizontal surface layer winds. *Boundary-Layer Meteorology*, 90:21–46.

[12] Box, G. E. P., Jenkins, G. M. (1976). Time series analysis: forecasting and control. Wiley series in probability and statistics.

[13] Burton, T., Sharpe, D., Jenkins, N., Bossanyi, E. (2001). *Wind Energy Handbook*. Wiley, Chichester, England.

[14] Castaing, B., Gagne, Y., Hopfinger, E. (1990). Velocity probability density functions of high Reynolds number turbulence. *Physica D*, 46:177–200.

[15] Delour, J., Muzy, J. F., Arneodo, A. (2001). Intermittency of 1d velocity spatial profiles in turbulence: a magnitude cumulant analysis. *The European Physical Journal B*, 23:243–248.

[16] Delour, J. (2001). Processus aleatoires auto-simimaires: applications en turbulence et en finance. PhD thesis, Universite de Bordeaux I, Pessac, France.

[17] Abraham, B., Ledolter, J. (1983). *Statistical Methods for Forecasting*. John Wiley & Sons, New York.

[18] Madsen, H., Nielsen, H. A., Nielsen, T. S. (2005). A tool for predicting the wind power production of off-shore wind plants. In: Proceedings of the Offshore Wind Conference and Exhibition, Copenhagen, Denmark.

[19] Mandelbrot, B. B. (1974). Intermittent turbulence in self-similar cascades: divergence of high moments and dimension of the carrier. *Journal of Fluid Mechanics*, 62:331–358.

[20] Kahane, J. P., Peyriere, J. (1976). Sur certaines martingales de Benoit Mandelbrot. *Advances in Mathematics*, 22:131–145.

[21] Molchan, G. M. (1996). Scaling exponents and multifractal dimensions for independent random cascades. *Communication in Mathematical Physics*, 179:681–702.

[22] Bacry, E., Kozhemyak, A., Muzy, J. F. (2008). Continuous cascade models for asset returns. *Journal of Economic Dynamics and Control*, 32:156–199.

[23] Muzy, J. F., Delour, J., Bacry, E. (2000). Modelling fluctuations of financial time series: from cascade process to stochastic volatility model. *The European Physical Journal B*, 17:537–548.

[24] Arneodo, A., Muzy, J. F., Sornette, D. (1998). "Direct" causal cascade in the stock market. *European Physical Journal B*, 2:277–282.

[25] Bacry, E., Muzy, J. F. (2003). Log-infinitely divisible multifractal process. *Communications in Mathematical Physics*, 236:449–475.

[26] Castaing, B. (2002). Lagrangian and Eulerian velocity intermittency. *The European Physical Journal B*, 29:357–358.

[27] Justus, C. G., Hargraves, W. R., Yalcin, A. (1976). National assessment of potential output from wind-powered generators. *Journal of Applied Meteorology*, 15(7):673–678.

[28] Justus, C. G., Hargraves, W. R., Mikhail, A., Graber, D. (1978). Methods for estimating wind speed frequency distributions. *Journal of Applied Meteorology*, 17(3):350–353.

[29] Hennessey, J. P.,Jr. (1978). A comparison of the Weibull and Rayleigh distributions for estimating wind power potential. *Wind Engineering*, 2(3):156–164.

[30] Takle, E. S., Brown, J. M. (1978). Note on the use of Weibull statistics to characterize wind speed data. *Journal of Applied Meteorology*, 17(4):556–559.

[31] Takle, E. S., Davis, W. M. (1978). Characteristics of wind and wind energy in Iowa. *Iowa State Journal Research*, 52:313–339.

[32] Petersen, E. L., Troen, I., Frandsen, S., Hedegard, K. (1981). *Wind atlas for Denmark. A rational method for wind energy siting. Risø-R-428.* Risø National Laboratory, Roskilde, 1981.

[33] Conradsen, K., Nielsen, L. B. (1984). Review of Weibull statistics for estimation of wind speed distributions. *Journal of Applied Meteorology*, 23:1173–1183.

[34] Sherlock, R. H. (1951). Analyzing winds for frequency and duration. *Meteorological Monographs*, 4:72–79.

[35] Baynes, C. J. (1974). The statistics of strong winds for engineering applications. PhD thesis, Faculty of Engineering Science, University of Western Ontario.

[36] Crutcher, H. L., Baer, L. (1962). Computations from elliptical wind distribution statistics. *Journal of Applied Meteorology*, 14:1512–1520.

[37] Brooks, C. E. P., Carruthers, N. (1953). *Handbook of Statistical Methods in Meteorology.* London (H. M. S. O.)

[38] Essenwanger, O. M. (1976). *Applied Statistics in Atmospheric Science, Part A.* Frequencies and curve fitting. Amsterdam, Netherlands. Elsevier scientific publishing company.

[39] Justus, C. G., Mikhail, A. (1976). Height variation of wind speed and wind distribution statistics. *Geophysical Research Letters*, 3:261–264.

[40] Chou, K. C., Corotis, R. B. (1983). Generalized wind speed probability distribution. *Journal of Engineering Mechanics*, 109:14–29.

[41] Rice, S. O. (1945). Mathematical analysis of random noise. *Bell System Technical Journal*, 24:46–156.

[42] Abramowitz, M., Stegun, I. A. (1968). *Handbook of Mathematical Functions, Applied Mathematics Series.* Dover Publications, New York.

[43] Press, W. H., Teukkolsky, S. A., Vetterling, W. T., Flannery, B. P. (1988). *Numerical Recipes in C.* Cambridge University Press, Cambridge.

[44] Bathurst, G. N., Weatherill, J., Strbac, G. (2002). Trading wind generation in short term energy markets. *IEEE Transactions on Power Systems*, 17:782–789.

[45] AEMO, cited 2010 [http://www.aemo.com.au/corporate/0000-0262.pdf].

[46] Giebel, G., Kariniotakis, G., Brownsword, R. (2003). State-of-the-art on methods and software tools for short-term prediction of wind energy production. In: Proceedings of the EWEC, Madrid, Spain.

[47] Bacry, E., Kozhemyak, A., Muzy, J. F. (2013). Log-normal continuous cascades: aggregation properties and estimation. Application to financial time-series. *Quantitative finance*, 13:795–818.

[48] Baile, R., Muzy, J. F., Poggi, P. (2010). Intermittency model for surface layer wind speed fluctuations. Applications to short term forecasting and calibration of the wind resource. In: EWEC, Warsaw, Poland.

[49] Giebel, G., Kariniotakis, G. N., Brownsword, R. (2003). The state-of-the-art in short-term prediction of wind power from a Danish perspective. In: Workshop on Large-Scale Integration of Wind Power and Transmission Networks for Offshore Wind Farms, Billund, Denmark.

14

WIND POWER BUDGET

Hugo Abi Karam

INTRODUCTION

In this chapter, we explore some equations of the wind power budget in the surface boundary layer (SBL). Key concepts are selected from references of the dynamics and thermodynamics of the atmosphere to give the reader just an overview about the subject. Concepts of shallow and deep convection are presented related to enough accurate approximations of the momentum equation, that is, perturbation approximations to momentum and energetic budget from Dutton and Fichtl. Among the select concepts, some derivation about the Monin–Obukhov similarity theory applicable to obtain the vertical structure of the horizontal mean wind in the SBL was included. Some comments are given about the Betz law and atmospheric boundary-layer adjustment over hills in function of the Froude number.

14.1 WIND BUDGET EQUATION IN THE SURFACE BOUNDARY LAYER (SBL)

The interaction between the atmospheric boundary layer (ABL) and the Earth's surface is very important in providing the energy to atmospheric circulations. If the nature is considered as a steam engine then the mechanical work may be written by

$$W = Q \left(\frac{T_{hot} - T_{cold}}{T_{hot}} \right) \qquad (14.1)$$

where Q is the rate of heat input and T_{hot} and T_{cold} the beginning and end step temperatures in the atmospheric system, for instance, the circulation flow in hurricanes (Emanuel, 2005).

$\eta = \left(T_{hot} - T_{cold}/T_{hot} \right)$ is the thermodynamic efficiency. The rate of dissipation of wind kinetic energy per unit area of the surface due to friction is given by

$$D = C_D \rho V^3 \qquad (14.2)$$

where V is the wind speed, ρ is the density of the air, and C_D is the drag coefficient. In accord with Emanuel (2005), the total rate of heat input per unit area is given by

$$Q = C_K \rho V E + C_D \rho V^3 \qquad (14.3)$$

where C_K is the enthalpy exchange coefficient and E is the evaporative potential of the sea surface. We can rewrite the work as follows:

$$W = \left(C_K \rho V E + C_D \rho V^3 \right) \left(\frac{T_{hot} - T_{cold}}{T_{hot}} \right) \qquad (14.4)$$

If the work is exactly equal to the dissipation, we obtain a formula for the maximum wind speed (in deep convective cell of a hurricane) in function of the thermodynamic efficiency,

$$V_{max} \approx \sqrt{\eta E} \qquad (14.5)$$

Thus, intense deep convection in the atmosphere comes close to the Carnot's heat engine, driven by the surface heat flux. A better intensity metric is the power dissipation index (PDI), a measure of the total frictional dissipation of kinetic energy in the hurricane boundary layer over lifetime of the storm (τ):

$$PDI = \int_0^\tau V_{max}^3 \, dt \qquad (14.6)$$

Alternative Energy and Shale Gas Encyclopedia, First Edition. Edited by Jay H. Lehr and Jack Keeley.
© 2016 John Wiley & Sons, Inc. Published 2016 by John Wiley & Sons, Inc.

The boundary layer is the lower layer of the atmosphere where the fluxes are turbulent in nature.

Dutton and Fichtl (1969) showed that the motions in the atmosphere can be interpreted in terms of shallow and deep convection, depending on the vertical scale of the motion. In these cases, the approximate equations of motion (momentum budget) become

$$\partial \vec{v} \partial t + \vec{v} \dot{\nabla} \vec{v} \approx - \alpha_0 \nabla p' + g \frac{\alpha'}{\alpha_0} \vec{k} \qquad (14.7)$$

On the left side of the equation is the total derivate of the velocity vector and on the right side of the equation the pressure perturbation force and the buoyancy term written to the perturbation of the specific volume of the air, or effective gravity, g being the acceleration of gravity and the index $(')$ referring to perturbation in relation to the horizontally concept basic (or reference) state of the atmosphere in this case, for instance, α_0 represents the basic state of the specific volume of air in the atmosphere. The form of energy of the momentum equation is very useful in atmospheric studies. In particular, Dutton and Fichtl (1969) showed that the rate of transformation (C) from *available potential energy* (A) to *kinetic energy* (K) is given by

$$C(A,K) = \int_V \rho_0 w \left(g \frac{\alpha'}{\alpha_0} + \frac{p'}{\rho_0 H_\alpha} \right) dV \qquad (14.8)$$

for deep convection, w being the vertical component of velocity and H_α the height scale of circulation $(H_\alpha \approx 8\,\text{km})$. Shallow convection circulation is characterized by $H_\alpha \ll 8\,\text{km}$ in the Earth's atmosphere and can be associated with circulations in the ABL, like sea-land breezes (Dutton and Fichtl, 1969).

14.2 MONIN–OBUKHOV SIMILARITY THEORY AND MEAN WIND PROFILES IN THE SBL

14.2.1 Vaschy–Buckingham Formalism (the PI Theorem)

Since the pioneering work of Rayleigh (1915) and Vaschy (1890, 1896) about the method of dimensions, the similarity theory became known such that physical laws can be written in the form of a multiplication of variables, under different powers, particularly in the dimensionless form. Thus,

$$\prod_{i=1}^p X_i^{a_i} = 1 \qquad (14.9)$$

where $F = F(X_i)$ is a power function that depends on a set of p variables, x_i. From the partial derivatives of the multiplicand

for each variable we can obtain a physical interpretation of the exponents of physical laws, that is,

$$b_{ji} = \frac{X_i}{X_j} \frac{\partial X_j}{\partial X_i} = \frac{\partial \ln |X_j|}{\partial \ln |X_i|} \quad \text{or} \quad \partial \ln |X_j| = b_{ji} \partial \ln |X_i| \qquad (14.10)$$

if $i \neq j$ and $b_{ii} = 1$ if $i = j$. Therefore, the exponent of the positive dimensionless variables can be considered components of the gradient of the logarithm of the variables, or the given law of scales (logarithmic) can be considered dimensions (fractals). In practice, to obtain the dimensions, logarithmic conversion of the multiplicand is applied to obtain the expression as a corresponding sum. Consequently,

$$a_1 \ln X_1 + a_2 \ln X_2 + a_3 \ln X_3 + \cdots + a_{p-1} \ln X_{p-1} + a_p \ln X_p = 0 \qquad (14.11)$$

For the pth variable it is written as

$$\ln X_p = \frac{1}{a_p} \sum_{i=1}^{p-1} a_i \ln X_i \qquad (14.12)$$

Using the notation $x_i = \ln X_i$ and $A_i = a_i / a_p$, the equation can be rewritten for $i = 1, 2, 3, \ldots, p-1$ as

$$x_p = \sum_{i=1}^{p-1} A_i x_i \qquad (14.13)$$

Let us consider $z_i = \dfrac{x_i - \bar{x}}{\sigma_x}$ to obtain the scalar product

$$z_p = \sum_{i=1}^{p-1} b_i z_i \qquad (14.14)$$

Here b_i are the correlations between variables z_i and variable z_p for $i = 1, 2, 3, \ldots, p-1$. The association with a multiple linear regression problem becomes apparent. The weights b_i are the regression correlations of the variables predicted in relation to the predictors (Cooley and Lohnes, 1971). To obtain the weights of the regression, it is common to employ the method of least squares. A cost function is defined to minimize the mean squared errors of preditantes, e_i^2. Thus

$$J = \frac{1}{p} \sum_{i=1}^p e_i^2 \qquad (14.15)$$

where $e_i = z_i - \bar{z}$. Deriving for each b_i and equating to zero, we obtain a system of p equations as a function of $p - 1$ unknowns. That is,

$$1b_1 + r_{12}b_2 + r_{13}b_3 + \cdots + r_{1p-1}b_{p-1} + r_{1p} = 0$$
$$r_{21}b_1 + 1b_2 + r_{23}b_3 + \cdots + r_{2p-1}b_{p-1} + r_{2p} = 0$$
$$r_{31}b_1 + r_{31}b_2 + 1b_3 + \cdots + r_{3p-1}b_{p-1} + r_{3p} = 0$$
$$\cdots = \cdots$$
$$r_{p-1,1}b_1 + r_{p-1,2}b_2 + r_{p-1,3}b_3 + \cdots + 1b_{p-1} + r_{p-1,p} = 0$$

$$(14.16)$$

Here r_{ij} are the correlation coefficients between the variables z_i and z_j. The complete set of correlations between p variables forms a square symmetric matrix of size $p \times p$. In matrix notation the system can be rewritten as

$$R_{p \times p} \cdot B_{p \times 1} = 0 \qquad (14.17)$$

The correlation matrix can be written in the form of an augmented matrix, comprising four subarrays,

$$R = \begin{bmatrix} R_{11} & R_{12} \\ R_{21} & I \end{bmatrix} = \begin{bmatrix} 1 & r_{12} & r_{13} & \cdots & r_{1p-1} & r_{1p} \\ r_{21} & 1 & r_{23} & \cdots & r_{2p-1} & r_{2p} \\ r_{31} & r_{32} & 1 & \cdots & r_{3p-1} & r_{3p} \\ \cdots & \cdots & \cdots & \cdots & \cdots & \cdots \\ r_{p-11} & r_{p-12} & r_{p-13} & \cdots & 1 & r_{p-1p} \\ \cdots & \cdots & \cdots & \cdots & \cdots & \cdots \\ r_{p1} & r_{p2} & r_{p3} & \cdots & r_{pp-1} & 1 \end{bmatrix}$$

$$(14.18)$$

where R_{11}, R_{12}, R_{21}, and I are the submatrices of R. $I = I_{q \times q}$ is the identity matrix of size $q \times q$; there are $q \leq p$ rows that are being linearly dependent on the first $(p - q)$ lines of R. Note that q is equal to the rank of the dimensional array, equal to the number of variables p minus the number of basic dimensions $(p - q)$. Here R_{11} is the matrix of variance–covariance or dispersion comprising correlation coefficients and $R_{12} = R_{21}^T$ is associated with a matrix of correlations of predictors or unknowns. The order of R_{11} is $p - 1$. The solution matrix of weights is given by

$$B_{q \times p} = -R_{11} \cdot R_{12} \qquad (14.19)$$

corresponding to the kernel of the equation of a singular decomposition, $R \cdot B = 0$. The kernel corresponds to exponents or dimensions of the physical law that can be easily obtained from the matrix R, which can be rearranged into a canonical echelon form. In this case R_{11} becomes the identity matrix, R_{21} are linear combinations of R_{11} rows, and $B_{q \times p} = -R_{11} \cdot R_{12}$ are the exponents of the set of independent variables. In the practice, it is sufficient obtaining a

diagonal staggered matrix by the pivot method, in order to obtain the identity matrix $I_{q \times q}$, where q is the number of basic dimensions. See also applications in different contexts by Cooley and Lohnes (1971) and Carneiro (1996).

14.2.2 Application of the Matrix Formalism of Vaschy–Buckingham

To illustrate the application of the theory of similarity to a problem based on the relationship between the three variables, $\text{var} = \{\partial d / \partial x', x', \bar{d}\}$, and only two basic units, $\dim = \{L, T\}$, we write the dimensional matrix as

$$\begin{bmatrix} \text{var/dim} & \dfrac{\partial d}{\partial x'} & x' & \bar{d} \\ L & 0 & 1 & 1 \\ T & 1 & -1 & 0 \end{bmatrix} \qquad (14.20)$$

The array has two rows, associated with two basic dimensions, and three columns corresponding to the three variables. Therefore, the number of linearly independent rows (LI) is equal to two. The matrix of the augmented dimensional array can be written as

$$R = \begin{bmatrix} R_{11} & R_{21} \\ R_{21} & R_{22} \end{bmatrix} = \begin{bmatrix} 0 & 1 & 1 \\ 1 & -1 & 0 \end{bmatrix} \qquad (14.21)$$

Its reduced echelon form is also called the canonical form being associated with kernel E such as $R \cdot E = \mathbf{0}$, whereas the irreducible submatrix R_{11} must have a nonzero inverse so the nontrivial solution can exist. The submatrices R_{21} and $R_{22} = I_{1 \times 1}$ are simple linear combinations and permutations of the rows of submatrices R_{11} and R_{21} (Louis Brans: The Pi-Theorem of Dimensional Analysis, op. cit. Carneiro in 1996). Reducing R to the canonical form we obtain

$$R\,(\text{canonical}) = \begin{bmatrix} 1 & 0 & +1 \\ 0 & 1 & +1 \\ 0 & 0 & 1 \end{bmatrix} \qquad (14.22)$$

Note the identity submatrix on the upper left corner and the last column correspond to the elements of E, or to the exponent of the equation. Therefore, in association with the last column of R we obtain

$$\left(-B_{p,p-q}\right)^T = \left(D_{11}^{-1} \cdot D_{12}\right)^T = \begin{bmatrix} +1 \\ +1 \end{bmatrix} \qquad (14.23)$$

Therefore, with the exponents $(-1, -1, 1)$ we can write the dimensionless group as

$$\left(\frac{\partial d}{\partial x'}\right)^{-1} (x')^{-1} (\bar{d})^{+1} = 1 \geq \Pi = \frac{1}{\bar{d}} \left(\frac{\partial d}{\partial x'}\right) x'$$
$$= \frac{\partial D}{\partial \ln x'} = 1 = cte\,(m) \qquad (14.24)$$

Finally, the quadratic multiple correlation coefficient can be obtained by calculating $R^2 = B^T \cdot R_{12}$.

14.2.3 Momentum Similarity in the Prandtl Layer

Let us determine the gradient wind in the SBL under the neutral condition $\partial\theta/\partial z = 0$. We also assume stationary and horizontally homogeneous conditions. Prandtl found that the vertical gradient of the moment in neutral SBL depends only on the shear stress $\tau_0 = \rho u *^2$ in units of $(\mathrm{kg\,m^{-1}\,s^{-2}})$ and height (m). Therefore, the physical variables of the problem are

$$F\left[\frac{\partial u}{\partial x}, z, u^*\right] = 0 \qquad (14.25)$$

Consequently,

$$\frac{\partial u}{\partial x} - G\left(z, u^*\right) = 0 \qquad (14.26)$$

There are just three variables $\left\{\frac{\partial U}{\partial z}; z; u^*\right\}$, $n_v = 3$, and two basic dimensions $n_d = 2$, L being the length and T the time.

Thus, by Buckingham's Pi theorem, the number of dimensionless groups is equal to one, since $n_g = n_v - n_d = 3 - 2 = 1$. This implies that the dimensionless group to be formed, Π_1, must be a constant. In this case, the dimensionless group is also a constant function, not dependent on other variables, $\Pi_1 = $ constant. We can write

$$\Pi_1 = \frac{\frac{\partial U^\alpha}{\partial z}}{z^\beta (u^*)^\gamma} = A = cte. \qquad (14.27)$$

Note that the constant can take any real value a priori, but its dimension is always one. For this group to be dimensionless it is necessary that

$$[\Pi_1] = \frac{\left[\frac{\partial U}{\partial z}\right]^\alpha}{[z]^\beta [(u^*)]^\gamma} = [A] = 1 \qquad (14.28)$$

namely,

$$[\Pi_1] = \frac{\left[\mathrm{s^{-1}}\right]^\alpha}{[\mathrm{m}]^\beta \left[(\mathrm{ms^{-1}})\right]^\gamma} = 1\,\mathrm{or}\,[\Pi_1]$$

$$= \mathrm{m}^{-\beta-\gamma}\mathrm{s}^{-\alpha+\gamma} = 1 = \mathrm{m}^0\mathrm{s}^0 \qquad (14.29)$$

Thus it has a linear system of two equations and two unknowns:

$$\begin{aligned} -\beta - \gamma &= 0 \\ -\alpha + \gamma &= 0 \end{aligned} \qquad (14.30)$$

It follows that $\beta = -\alpha, \gamma = \alpha$. After replacing

$$[\Pi_1] = \frac{\left[\mathrm{s^{-1}}\right]^\alpha}{[\mathrm{m}]^{-\alpha} \left[(\mathrm{ms^{-1}})\right]^\alpha} = 1 \qquad (14.31)$$

Therefore

$$[\Pi_1] = \frac{\left[\frac{\partial U}{\partial z}\right]^\alpha [z]^\alpha}{[(u^*)]^\alpha} = 1\,\mathrm{or}\,\frac{\left[\frac{\partial U}{\partial z}\right][z]}{[(u^*)]} = 1\frac{1}{\alpha} = 1 \qquad (14.32)$$
$$\alpha = 1$$

Finally, we can write the dimensional form:

$$\Pi_1 = \frac{z}{u^*}\frac{\partial U}{\partial z} = A = cte. \qquad (14.33)$$

The estimated value of A is 5/2; therefore, its inverse, known as Von Karman constant (κ), is approximately equal to 0.4.

14.2.4 Theory of Similarity of Monin–Obukhov for the Surface Boundary Layer (Under Nonneutral Conditions)

In 1954, Monin–Obukhov proposed a theory of similarity to the stratified layer or to a surface layer having a vertical temperature gradient. In this case, the vertical gradients of momentum and temperature

$$\frac{\partial u}{\partial z}, \frac{\partial \theta}{\partial z} \qquad (14.34)$$

can be determined on the basis of five physical quantities, which are

$$z$$
$$\rho_0$$
$$\tau_0 = \sqrt{\left(\overline{u'w'}\right)_0^2 + \left(\overline{v'w'}\right)_0^2} \qquad (14.35)$$
$$Q_0 = \left(\overline{\theta'w'}\right)_0$$
$$\beta = g/T_0$$

We note that the characteristics of turbulence to a height z should depend only on five quantities. In this case, since there are four independent dimensions—length, time, mass, and temperature, we can formulate a single dimensionless combination of these five independent quantities. Following Sorbjan (1989) (see Monin and Yaglom, 1971 for a detailed review) we found that this combination is given by

$$\zeta = z/L \qquad (14.36)$$

where

$$L = u^{*3} \left(\kappa \frac{g}{T_0} \frac{Q_0}{c_p \rho_0} \right)^{-1} \qquad (14.37)$$

Since the Von Karman constant was first introduced in the equation of L by Obukhov in 1946 it has been maintained on subsequent works by tradition (Monin and Yaglom, 1971). With the six significant variables we can define scales of velocity, temperature, and length, respectively,

$$u^* = \rho_0 \tau_0^2$$

$$T^* = \frac{-Q_0}{\kappa u^*} \qquad (14.38)$$

$$L = \frac{u^{*2}}{\beta \kappa T^*}$$

These scales-stratified characteristics of SBL, with basic units, take the place of the variables considered first. Note that we can use them since they are dependent on earlier amounts. Really, the reduction to the smallest number of dimensions is a goal and also an advantage of the dimensional analysis. So the independent variables are (z, u^*, T^*, L). The vertical gradients are the following functions:

$$\frac{\partial u}{\partial z} = f_1(z, u^*, L)$$

$$\frac{\partial \theta}{\partial z} = f_2(z, T^*, L) \qquad (14.39)$$

In this case, the number of independent variables is three, (z, u^*, L) for $\partial u/\partial z$ and (z, T^*, L) for $\partial \theta/\partial z$. The number of basic dimensions is also equal to two $\{L = \text{length}, T = \text{time}\}$. On the other hand, the number of basic dimensions considering $\{L = \text{length}, K = \text{temperature}\}$ is also two; therefore, the number of independent dimensionless group is one, in this case also. This implies that we have two dimensionless groups, one for each vertical gradient, called Π_0 and Π_1. First we write

$$\Pi_0 = \frac{\left(\frac{\partial u}{\partial z} \right)}{(u^*)^{\alpha_1} (L)^{\alpha_2}} = \phi \left(\Pi_1 \right) \wedge \Pi_1 = \frac{z}{L^{\alpha_3}} \qquad (14.40)$$

For the group to be dimensionless it is sufficient that

$$(\alpha_1, \alpha_2, \alpha_3) = (1, -1, 1) \qquad (14.41)$$

Thus,

$$\Pi_0 = \frac{L}{u^*} \left(\frac{\partial u}{\partial z} \right) \wedge \Pi_1 = \frac{z}{L} \qquad (14.42)$$

The empirical relationship may be written generically as

$$\Pi_0 = \frac{L}{u^*} \left(\frac{\partial u}{\partial z} \right) = \varphi \left(\Pi_1 \right) = g \left(\frac{z}{L} \right) \qquad (14.43)$$

Similarly, we obtain

$$\frac{L}{u^*} \left(\frac{\partial u}{\partial z} \right) = g \left(\frac{z}{L} \right)$$

$$\frac{L}{T^*} \left(\frac{\partial \theta}{\partial z} \right) = g_1 \left(\frac{z}{L} \right) \qquad (14.44)$$

By tradition, we have included a constant of Von Karman in the equations. Additionally, consider $\zeta = (z/L)$, the stability parameter in the equations. Thus,

$$\frac{\kappa z}{u^*} \left(\frac{\partial u}{\partial z} \right) = \zeta g(\zeta) = \phi(\zeta)$$

$$\frac{\kappa z}{T^*} \left(\frac{\partial \theta}{\partial z} \right) = \zeta g_1(\zeta) = \phi_1(\zeta) \qquad (14.45)$$

However, it is noted that

$$\phi(\zeta) = \zeta g(\zeta) = \zeta \Psi'(\zeta) = \zeta \frac{\partial \Psi}{\partial \zeta} \qquad (14.46)$$

Accordingly,

$$g(\zeta) = \frac{\partial \Psi(\zeta)}{\partial \zeta} \qquad (14.47)$$

The profiles of temperature and wind speed at the SBL can be obtained by integrating the gradient equations derived above (Monin and Yaglom, 1971) to obtain

$$u(z_2) - u(z_1) = \frac{u^*}{\kappa} \left[\Psi \left(\frac{z_2}{L} \right) - \Psi \left(\frac{z_1}{L} \right) \right]$$

$$\theta(z_2) - \theta(z_1) = \frac{T^*}{\kappa} \left[\Psi_1 \left(\frac{z_2}{L} \right) - \Psi_1 \left(\frac{z_1}{L} \right) \right] \qquad (14.48)$$

where

$$\left[\Psi \left(\frac{z_2}{L} \right) - \Psi \left(\frac{z_1}{L} \right) \right] = \int_{z_1/L}^{z_2/L} \left(\frac{\phi(\zeta) - 1}{\zeta} \right)$$

$$\left[\Psi_1 \left(\frac{z_2}{L} \right) - \Psi_1 \left(\frac{z_1}{L} \right) \right] = \int_{z_1/L}^{z_2/L} \left(\frac{\phi_1(\zeta) - 1}{\zeta} \right) \qquad (14.49)$$

The empirical universal functions of the Monin–Obukhov similarity theory are reviewed in the literature of micrometeorology, for instance, Businger et al. (1971), Benoit (1977), Panofsky and Dutton (1984), Stull (1988), and Kaimal and Finnigan (1994).

14.3 MAXIMUM EFFICIENCY OF A WIND TURBINE (BETZ'S LAW)

The maximum efficiency of a wind turbine (Betz's law) is limited to approximately 16/27 (0.593). This is a consequence of the restriction in the reduction of the wind momentum after crossing the turbine. Real turbines operate at efficiencies still lower than the Betz limit. Note that the density of water is 1000 times greater than the density of air. Thus, the increase of real efficiency toward the Betz limit should depend more on the fluid density application, for instance, some turbines could be immersed in sea water currents, than in the geometric reshape of the turbines.

14.4 FROUDE NUMBER AND BOUNDARY LAYER AIRFLOW ADJUSTMENT OVER COMPLEX TERRAIN

In the case of airflow adjustment over a hilly terrain, the Froude number (Fr) is just the rate between the natural wavelength of the vertical oscillation of air and the width of the hill. Under stable stratified conditions, the flow can develop blocking, lee waves, rotors, cavities, boundary layer separations, and downwind turbulent wake. Under convective conditions in the ABL, a mixture layer (ML) develops along the sunny hours, in response to the positive input of surface heat flux. In this case (convective), the shallow water equation model may give some clues about the airflow adjustment of the ML, and the Bernoulli effect can accelerate the flow over the hill. In this case, Fr may be interpreted as the ratio that compares the mean flow velocity in the ML and the phase velocity of the external gravity wave, propagating in the ML top interface (Markowski and Richardson, 2010; Stull, 1988). Under convective conditions, the adjustment of the flow can induce horizontal acceleration over the hill (Bernoulli effect), hydraulic jump, and Karman's vortices.

REFERENCES

Benoit, R. (1977). On the integral of the surface layer profile-gradient functions. *Journal of Applied Meteorology*, 16:859–860.

Businger, J. A., Wyngaard, J. C., Izumi, Y., Bradley, E. F. (1971). Flux-profile relationships in atmospheric surface layer; *Journal of the Atmospheric Science*, 28:181–189.

Carneiro, F. L. (1996). *Análise dimensional e teoria da semelhança e dos modelos físicos*. 2a Edição. Ed. UFRJ, ISBN 85-7108-077-1.

Cooley, W. W., Lohnes, P. R. (1971). *Multivariate Data Analysis*. John Wiley & Sons, New York, 364 pp., ISBN 0471170607, 9780471170600.

Dutton, J. A., Fichtl, G. H. (1969). Approximate equations of motion for gases and liquids. *Journal of the Atmospheric Sciences*, 26:241–254.

Emanuel, K. (2005). *Divine Wind – The History and Science of Hurricanes*. Oxford University Press, 285 pp., ISBN 0-19-514941-6.

Kaimal, J. C., Finnigan, J. J. (1994). *Atmospheric Boundary Layer Flows*, Oxford University Press, New York, Oxford, 289 pp.

Markowski, P., Richardson, Y. (2010). *Mesoscale Meteorology in Midlatitudes (Advancing Weather and Climate Science)*. 1st edition (March 2, 2010). Wiley, 430 p., ISBN-10: 0470742135, ISBN-13: 978-0470742136.

Monin, A. S., Yaglom, A. M. (1971). *Statistical Fluid Mechanics: Mechanics of Turbulence*. Vol. I. MIT Press, Cambridge, MA, 769 pp.

Panofsky, H. A., Dutton, J. A. (1984) *Atmospheric Turbulence – Models and Methods for Engineering Applications*. John Wiley & Sons, New York, 397 pp.

Rayleigh, Lord (1915). The principle of Similitude. *Nature*, 95:66.

Sorbjan, Z. (1989). *Structure of the Atmospheric Boundary Layer*. Prentice Hall, Upper Saddle River, NJ, 317 pp.

Stull, R. B. (1988). *An Introduction to Boundary Layer Meteorology*. Kluwer Academic Publishers, Dordrecht (Netherlands), 647 pp.

Vaschy, A. (1890). Traité d'életricité et de magnétisme. Volume 1 – Expos des Phènomemes elètriques et Magnètiques. Maison a Liège (1890). Volume 2 – théorie et applications, instruments et méthodes de mesure électrique. Baudry. Digitalized by the Universidade de Wisconsin – Madison in 2007.

Vaschy, A. (1896). Théorie de l'électricité: Exposé des phénomènes électriques et magnétiques fondé uniquement sur l'expérience et le raisonnement. Librairie polytechnique, Baudry et cie, 1896, 340 p. Digitalized by the Universidade de Wisconsin – Madison, Dec. 2, 2009.

15

IDENTIFICATION OF WIND TURBINES IN CLOSED-LOOP OPERATION IN THE PRESENCE OF THREE-DIMENSIONAL TURBULENCE WIND SPEED: TORQUE DEMAND TO MEASURED GENERATOR SPEED LOOP

MIKEL IRIBAS-LATOUR AND ION-DORÉ LANDAU

15.1 INTRODUCTION AND MOTIVATION OF THE WORK

15.1.1 WT Control Design and Implementation: State of the Art

It is well known in the literature [1–10], [11–15] that control loops and its parameters are critical from the point of view of generated loads in a wind turbine (WT). The increase in size and power of newer WT designs makes the control more important in terms of loads and stabilization. The basic tool for the design of a controller is the linear model of the plant to be controlled in various points of operation [16]. For design purposes, tuning of control loop parameters is a relatively well-solved problem in the wind sector [4–10], since linear models based on linearization of nonlinear aeroelastic codes are generally used [17–19]. On the other hand, tuning the controllers for WT on site is in many occasions time consuming, inaccurate, and does not cope with the time-varying characteristics of the hold system. Moreover, it seems to be very common in the WT sector that control parameters tuned during WT design and used for the certification process do not exactly correspond to those finally implemented in the real WT [10]. Then, the adequacy of the linearized models [17–19] can be criticized, since in many occasions, the control parameters are not subsequently used in the real WT [10]. In that case, how are these parameters tuned in real WT? Most of the time, the

final on-site tuning is based on the experience of the control engineer who proceeds according to rules of thumb or on a trial and error basis. There is absolutely no evidence that this tuning is close to the optimal controller that can be obtained with model-based control. Therefore, there is a need for obtaining relevant and accurate models on-site that can effectively be used for controller tuning.

Consequently, although big efforts in modeling WT have been done [1, 17–25], the search of a procedure to obtain reliable linear models for control design seems a pertinent research task. Open-loop system identification techniques can be used for obtaining on-site realistic linear models for control purposes [26]. Different approaches have been tried in the frame of WT identification, but they usually look for different purposes. In Ref. [27], an experimental approach for finding aeroelastic damping is introduced. In Refs. [28] and [29], the identification is divided between a linear model for drive train identification and nonlinear model for the aerodynamical effects, with extra measurements. In Ref. [30], a full transfer function from torque demand to generator speed is obtained based on open-loop identification algorithms.

Since it is difficult and dangerous to operate WT in open loop, identification in closed-loop operation should be considered. A first approach for solving this problem based on closed-loop identification algorithms was presented in Ref. [31]. However, this procedure had a practical limitation since constant wind speed was supposed. In this chapter, this limitation is overcome for the identification of the transfer

Alternative Energy and Shale Gas Encyclopedia, First Edition. Edited by Jay H. Lehr and Jack Keeley.
© 2016 John Wiley & Sons, Inc. Published 2016 by John Wiley & Sons, Inc.

function for generator torque demand to measured generator speed. Accurate identification is made possible in the presence of three-dimensional turbulence wind speed.

15.1.2   The Protocol for Obtaining Reliable Linear Models from Experimental Data

System identification in open-loop operation is a very powerful technology for obtaining linear models from experimental data for simple to complicated plants (like DC motor, thermal motors, active suspensions, flexible mechanical structures, etc.). The open-loop identification techniques are well known and have been extensively used [26]. These models have been successfully used for control design purposes for many years in different applications. The basics of open-loop identification consist of having access to the source of energy that controls the behavior of the system operating in open loop. Applying enough exciting signal (in terms of frequency content), the input and the output data of the system are collected. After some data treatment (filtering, removing DC components, etc.) parametric or nonparametric identification algorithms are used to obtain linear models that should be validated (statistical validation) prior to their use for control design [26, 32].

However, this technique has two main drawbacks. First, it can be complicated or dangerous to identify in open-loop operation when the system to be identified is unstable, when it has an integrator, or when an important drift of the operating point occurs; and second, when the source of energy cannot be manipulated by the user [33]. In the case of WTs, these drawbacks appear. It can be dangerous to operate with the WT in open loop for safety and integrity reasons. Big excursions to different operational points are very common because of wind turbulences. On the other hand, the main problem for the implementation of an open-loop experiment is that the wind energy, the energy source of the system, cannot be manipulated by the control engineer. For these reasons, open-loop system identification techniques cannot be realistically used in the field of WT.

In order to overcome the aforementioned problems, techniques for identification in closed-loop operation [33–35] should be used for obtaining realistic linear models based on experiments performed on a real WT or even on nonlinear aeroelastic simulators [31, 36]. These models can help to improve on-site controller tuning and is a step further toward implementation of new control strategies. This technique can also be used for control loop maintenance. There is also another reason for considering the use of these techniques. It was shown and proven experimentally [37, 38] that if the objective is to identify a model for control design, the models identified in closed-loop operation with appropriate algorithms are better in terms of control results than the models identified in open loop. While identification in closed loop has been addressed since the 1970s [35], only in the second half of the 1990s have efficient methods been

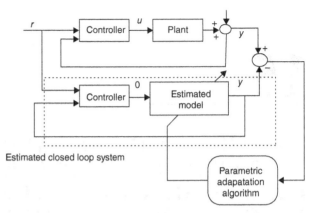

FIGURE 15.1 Principle of identification in closed-loop operation.

developed for dealing with experimental data acquired in closed loop [33–35].

A first application of identification in closed-loop operation of a WT was presented in Ref. [31]. Results based on data obtained from nonlinear aeroelastic simulators were presented and discussed in comparison with models obtained by linearization techniques. However, this procedure and model was obtained in the absence of wind turbulences. This chapter deals with a modification of the procedure in order to be able to identify reliable linear models for control purposes in the presence of three-dimensional wind turbulences. More precisely, the identification of the torque loop for a variable speed-pitch controlled WT is presented. This new feature gives the capability to use this procedure for the control tuning based on experimental data from a real WT.

The principle of identification in closed loop is illustrated in Figure 15.1, for the case when the excitation signal is superposed on the reference, but similar figure can be drawn when the excitation signal is superposed on the output of the controller [34]. The objective of identification in closed-loop operation is to find the estimated plant model, G, which in feedback with the same controller used on the real plant, gives the best model for the real closed-loop system. To achieve this, basically, the error between the true closed-loop system and the estimated closed-loop system is fed into an adaptation algorithm that updates the parameters of the estimated plant model. The big difficulties come from the correlations of the output and input noise because of feedback. Specific algorithms have been developed, which take advantage of this configuration but are not sensitive to input/output noise correlation [34].

15.2   SELECTION OF THE EXPERIMENT FOR IDENTIFYING A DATA-BASED MODEL

The objective of this chapter is to show a procedure for obtaining a reliable transfer function of the torque demand to measured generator speed for control design purposes in the presence of wind turbulences at low wind speeds. Although

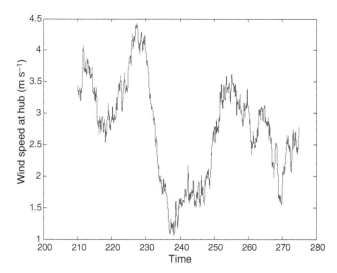

FIGURE 15.2 Three-dimensional turbulence wind. Nominal wind measured at hub position.

the model has been obtained for this particular case, it has been proved that the designed control loop based on this identified linear model is also valid for mean wind speeds cases from 3 to 25 m s^{-1}.

In order to identify linear models for control purposes, a set of data acquired on a plant is needed. For this work, a three blades-variable speed-pitch controlled WT has been chosen. The data for the experiment have been obtained from its nonlinear model in Bladed® [17]. For the identification of the transfer function from the generator torque demand to measured generator speed, a standard fatigue case is used.

Figure 15.2 shows the nominal wind speed measured at the hub position. An important variance can be seen. The turbulence intensity for lateral, vertical, and longitudinal wind is between 20% and 35%. For the case of this set of data, the longitudinal mean speed of the wind is 2.7 m s^{-1}. At such low wind speed, the WT is controlled through the torque demand of the generator. Sometimes, small changes in pitch angle are implemented in order to improve power production at low wind speeds. However, in order to avoid problems concerning multi-inputs, multi-outputs system control-identification problems, the pitch is held constant during the identification procedure. As will be shown in Section 15.4, this does not affect the quality of the model identified for control design purpose.

15.3 THE PROCEDURE FOR WT IDENTIFICATION OPERATING IN CLOSED LOOP: THE TORQUE LOOP

15.3.1 Control Loop Configuration

Identification in closed-loop operation is done in discrete time. Therefore, a correct sampling frequency must be

chosen for both control and identification purposes. It is a commonly accepted idea that "with a higher sampling frequency, a better controller is obtained." However, this is not always true. The sampling frequency should be chosen not on the available computer speed, but in relation with the dynamics of the system to be controlled and identified, as well as the disturbance signals, measurement noise, uncontrolled body-bending mode signals detected by system sensors, and the inherent delay. In essence, the primary influence on sampling time is the bandwidth characteristics of these signals [16]. In this case, several sampling frequencies have been used, but 80 Hz is the final frequency chosen. Although this frequency seems too high based on identified dynamics, this frequency has been chosen for convenience in order to present a comparison with existing controllers operating with a sampling frequency of 80 Hz (see Section 15.5).

It is well known [1] that in a variable-speed WT operating at constant generator torque there is very little damping for the drive train mode, since torque does not vary with generator speed. The very low damping can lead to large torque oscillations at the gearbox. Although it may be possible to provide some damping mechanically, there is a cost associated to this. Another solution, which is very common in variable-speed WT, is to modify the generator torque control to provide some damping. More precisely, it is used to damp the drive train and, in some cases, the side-to-side tower mode and the in-plane rotor mode. A classical and well-known control scheme for controlling a variable wind speed-pitch controlled WT is shown in Figure 15.3 [1–3].

In order to be able to perform an experiment for closed-loop identification, two options are available for introducing the excitation signal. This can be superposed either on the reference or on the output of the controller. In this case, the second option has been chosen, so that the final scheme for the closed-loop identification experiment is represented in Figure 15.4. This new scheme plays the role of the control torque in feedback with the measured generator speed shown in the scheme of Figure 15.3. Of course, only power regulation will be possible during the experiment. However, if the experiment is under control, this should not be a big

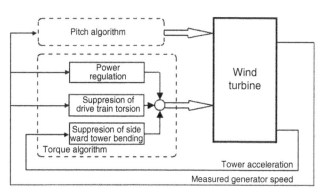

FIGURE 15.3 Control loops for a wind turbine.

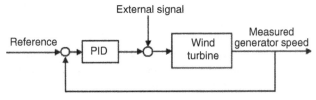

FIGURE 15.4 Closed-loop scheme used.

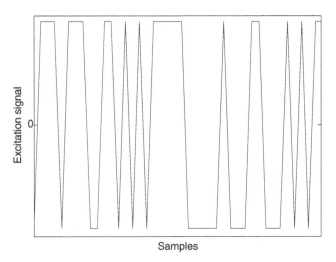

FIGURE 15.5 Small set of the input signal.

problem, since the length of the experiment chosen is 60 seconds (and this duration can be even reduced). This means that the drive train torsion mode will not be damped during the experiment, which will allow a better model identification in this frequency region [33].

The main objective of this study is to prove that the procedure can be used for in-site control design and tuning, so a step-by-step procedure is needed. The methodology can also be extended (with some modifications) to the case when the drive train torsion mode is already damped. It is important to note that identification in closed loop can be applied iteratively for each control loop design. By repeating the identification in closed loop and redesigning at each step, the obtained controller can improve the performances of the control system [37, 38]. This approach can be considered for this application but it is beyond the specific objective of this chapter.

15.3.2 Excitation Signal Design

The design of the excitation signal for system identification is crucial. On one hand, the dynamics to be identified should be excited, and on the other hand, care should be taken because there can be important undamped modes that need to be identified, but should not be amplified in order to avoid damages in the WT. For a correct design of this input signal, the detailed model of the design process of the WT can be of certain interest. Stability and integrity reasons are critical for the choice of closed-loop identification technique, since this technique can be used while stability is guaranteed because the control loop is always active. The stability and integrity of the WT should not be threatened; therefore, special care must be taken at undamped modes of the tower, drive train, flapwise, edgewise, bending of the blades, and so on. So for this case, the a priori information about the drive train torsion frequency or the side-to-side and in-plane frequencies should be taken into account. The frequency shape and amplitude of the input signal should be designed according to the natural frequencies associated to the dynamics of the WT.

An important set of different combinations of input signals has been tried: pseudorandom binary signals (PRBS), step signals, chirp signals, sinusoids, square signals, and so on. Finally, a PRBS has been selected [33, 36]. The PRBS is generated by means of shift registers with feedback (implemented in software). The number of cells in the shift register

used was $n = 11$, which generates a PRBS with a length of 2047 samples ($2^N - 1$). With this PRBS configuration, the lowest frequency that can be identified using one PRBS sequence is 3.63 Hz [33]. Then, in order to try to get a better identification at lower frequencies, the sequence of 2048 samples has been repeated once, in order to enlarge the frequency content at low frequencies, moving to almost 60 seconds of experiment length. So a better identification at the interesting frequency range should be obtained. It is important to emphasize that the amplitude of this input signal does not need to be significant. For this case, the input signal is around 10% of the mean value of the generator demand during the experiment. This 10% corresponds with the 0.25% of the nominal torque of the generator. This means that the amount of energy that is needed for doing identification in closed-loop operation is very small and that the stress on the system is negligible. In Figure 15.5, the input excitation signal has been plotted for a period of 1 second.

15.3.3 Output Data of the Experiment

Once the benchmark has been completely defined, the experimental conditions fixed, and the input excitation signal designed, the experiment is performed in Bladed® for a three-dimensional turbulence wind speed fatigue case. At this point, the only decision that should be taken is when the experiment should be done. Since this procedure is designed for identification of torque loop, and no interaction with pitch control is desired, the experiment should be done in the presence of low wind speed, which in addition is interesting from the point of view of WT integrity. For a real WT case, information from the anemometer and power production would be needed. For a simulation environment, this is not critical, since the wind speed as well as the behavior of the system is known a priori.

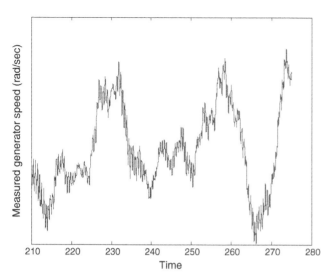

FIGURE 15.6 Output signal. Measured generator speed.

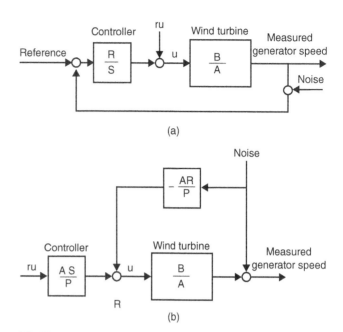

FIGURE 15.8 (a) Identification scheme in closed loop and (b) equivalent scheme.

15.3.4 Identification Algorithms

The objective of system identification in closed-loop operation is to search for a plant model, which in feedback with the controller operating on the true plant will lead to a closed-loop transfer function (sensitivity function), which is as close as possible to that of the real closed-loop system. Of course, this is possible assuming that the real system is linear around an operating point for small signal variations. One has to note that the system is able to stay around this operating point because of the effect of the acting controller.

Let us first try to explain why using identification in closed-loop operation is potentially possible to obtain a better model than using open-loop identification and what the difficulties encountered are. Consider Figure 15.8a, where the controller is implemented in discrete time (polynomial controller $K = R/S$) and the plant model is characterized by a discrete time transfer function $G = B/A$. An equivalent scheme is presented in Figure 15.8b (assuming that the reference is constant). The effective plant input u corresponds to the external excitation r_u filtered through AS/P, which is nothing else than the output sensitivity function S_{yp} (the transfer function from r_u to u is identical with the transfer function from the output disturbance to the output). The output sensitivity function is given in equation (15.1). Its magnitude has a maximum in the frequency regions close to the critical point in the Nyquist plane [16]. Therefore, the frequency spectrum of the effective input applied to the plant will be enhanced in these frequency zones. As a consequence, the quality of the identified model in these critical frequency regions for stability and performance will be improved. It has been shown [3]

The time domain output of the experiment is plotted in Figure 15.6. It can easily be observed that there is an important ripple of high frequency that corresponds to the different undamped modes of the drive train and tower side-to-side modes. It is also possible to realize that an important low-frequency content appears in the output because of the different excitation signals of 3P and 6P, as well as because of the effect of the three-dimensional turbulence of the wind. It is easy to realize that there is a big correlation between the measured generator speed and the nominal wind speed measured at hub position (see Figure 15.2). All off this frequency content can be seen in the fast Fourier transformation (FFT) plot of the output in Figure 15.7.

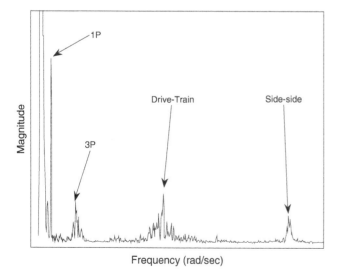

FIGURE 15.7 FFT of the measured generator speed signal.

that the plant identification operating in closed loop, provided that appropriate identification algorithms are used, leads in general to better models for control design than open-loop identification [37, 38]:

$$S_{\text{yp}} = \frac{AS}{AS + BR} \tag{15.1}$$

Unfortunately, in the meantime, the feedback introduces a correlation between the measurement noise and the plant input. This leads to an important bias on the estimated parameters if one would like to identify the plant model with open-loop identification techniques based on uncorrelation. One may expect that the open-loop techniques based on the whitening of the prediction error will still provide good results in closed-loop operation. However, as a consequence of feedback, interdependence between the noise model and the plant model occurs and the parameter estimation will also be biased [33, 34].

Therefore, for a correct identification in closed-loop operation, one needs identification methods that take advantage of the "improved" characteristics of the effective excitation signal applied to the plant input but are not affected by the noise correlation introduced by the feedback. This can never be obtained with classical open-loop identification algorithms [33, 34]. An efficient solution for this problem is provided by the family of "closed-loop output error" (CLOE) identification algorithms [34].

Of great importance is the selection of the model structure to be identified. At this step, the noise model is the main decision to be made. In Ref. [31], an auto regressive moving average model (ARMAX) structure was selected (equation (15.2)). In order to be able to identify an ARMAX model in closed-loop operation, the extended CLOE identification algorithm method was used [34]:

$$A\left(q^{-1}\right) y(t) = q^{-d} B\left(q^{-1}\right) u(t) + C\left(q^{-1}\right) e(t) \tag{15.2}$$

The structure of the noise model depends on the disturbance that appears on the system. In the framework of WT, there are disturbances of 1P, 3P, 6P, ..., nP [1]. Also there are several frequencies involved and they are usually very narrow band disturbances. This set of disturbance is much stronger in the measured generator speed for the case of three-dimensional turbulence wind speed (see Figure 15.7).

Once model identification is done, validation tests should be performed in order to qualify the model control design. The validation tests are always needed for an identified model, obtained in open loop or in closed loop. The validation tests depend on the identification technique used and on the algorithm selected [26, 32, 33, 35]. The validation tests in closed loop help to decide which model is correct and which one is not correct. In addition, the validation tests help to decide among several validated models which one is the best. At the end of the identification process, there will be one model that will be better in terms of the validation test applied to the set of identified models. For the case of identification in closed-loop operation, the objective of the model validation in closed loop is to find what plant model combined with the current controller provides the best prediction of the behavior of the true closed-loop system. The validation tests applied are [33]

- *Uncorrelation test.* A statistical uncorrelation test between the residues (difference of the real output of the plant and the estimated output of the identified plant with CLOE identification algorithms, in closed loop with the controller) and the predicted output of the model is done (see Figure 15.1). This test checks unbiased estimation of the model parameters, as well as the uncorrelation between the CLOE and the external excitation signal. The procedure involves computation of normalized cross-correlations that should be below a certain level (see Ref. [11] for the definition of the threshold level).

- *Closeness of closed-loop poles.* The poles of the identified closed-loop system[1] and the poles of the computed closed-loop transfer function[2] are compared.

- *Vinnicombe gap.* This is a number that quantifies the difference between the identified sensitivity transfer function and the computed sensitivity transfer function [39].

- *Frequency validation.* It consists of the comparison of the magnitude Bode plots of the computed closed-loop transfer function and the identified closed-loop transfer function.

- *Time domain validation in closed loop.* Here the real output and the estimated output of the identified plant in closed loop with the controller are plotted (however, this test is not very significant when the level of the noise is high).

All these tests will usually give an indication of the quality of the identified model.

Once the algorithm to be used is selected, an important issue is the search for the appropriate orders, numerator, denominator of plant and noise models, as well as the delay of the plant. This iterative search for the best model structure is done using the validation tests on the corresponding identified models.

---

[1] Sensitivity (closed-loop) transfer function identified with an open-loop algorithm, taking into account the plant and controller orders.
[2] Sensitivity (closed-loop) transfer function computed with the model identified in closed-loop operation and the actual controller running in the experiment.

15.4 MODEL IDENTIFICATION RESULTS

From Figure 15.9a, the frequency domain validation (linear frequency scale and log-normalized magnitude scale) can be seen. Both sensitivity functions, the identified and the computed, are quite close until 0.15 normalized frequency (i.e., 12 Hz), where a ripple begins. Also, a small difference seems to appear at low frequencies, where there is a bit of difference between the identified sensitivity function and the computed one. This may come from the low energy of the input excitation signal at very low frequencies explained in Section 15.3.2. The differences at high frequencies are less important since they are out of the control bandwidth. In Figure 15.9b, the closeness poles test is depicted. Here it can be seen that poles of the identified sensitivity transfer function and the computed sensitivity transfer function are very close, especially within the bandwidth of the control loop. This means that the estimated plant model is correct since the two systems use the same controller. On the other hand, the Vinnicombe distance between the identified and the computed sensitivity transfer functions is 0.0474, which is a very low value. The statistical test has been also done. To do this, normalized cross-correlations have been computed as indicated earlier. This test has been passed by the model, since all values of the normalized cross-correlations are below the threshold level of 0.15 [33]. Then, it can be concluded that the estimated parameters of the model identified with CLOE algorithms are correct and that the model can be used safely for controller design. The time domain validation test is omitted since, although it can be plotted, the plot does not take into account the noise model, which in addition has different

converge properties, and its view does not give additional information to the validation process.

At this point, one has three models for the plant: (i) the model identified in closed operation in the presence of three-dimensional turbulence wind speed; (ii) the model identified at constant wind speed [31]; and (iii) the linearized model obtained from Bladed® at 3 m s^{-1}. These control models have been obtained by different methods but from the same model and simulation tool. Therefore, one can consider a comparative validation.

The linearized model is not a good model for control design, since even if the controller used during closed-loop identification stabilizes the true plant, the nonlinear Bladed® model does not stabilize the linearized model of the plant (there are certainly numerically problems inherent to the method used to generate this model).

The results for the validation of the model identified in Ref. [31] are shown in Figure 15.10. The results are close to those obtained for the model identified in the presence of wind turbulence. A slightly higher difference can be noticed at low frequencies between the computed closed-loop transfer function and the identified closed-loop transfer function, but the computed and identified closed-loop poles look to be as close as for the case of the model identified in the presence of wind turbulence. The Vinnicombe distance is 0.0412, which is also very close to the value obtained for the model identified in the presence of wind turbulence (differences below 0.01 are absolutely not significant).

Therefore, despite that the model identified in this paper has been obtained in the presence of wind turbulence, it is as good as the model identified at constant wind speed.

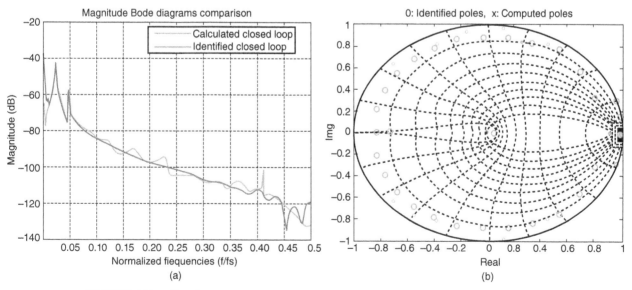

FIGURE 15.9 Closed-loop validation tests for identified model for the case of three-dimensional turbulence (Vinnicombe gap = 0.0474). (a) Frequency domain validation and (b) closed-loop poles closeness.

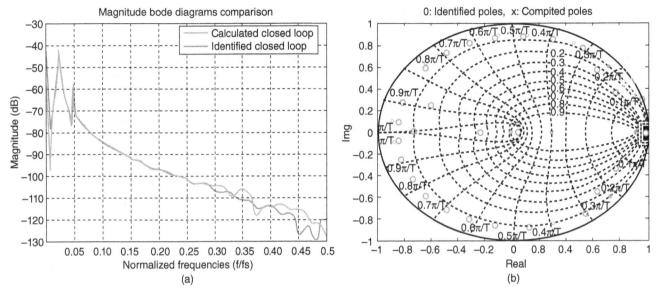

FIGURE 15.10   Closed-loop validation tests for the model identified at constant wind speed (Vinnicombe gap = 0.0412). (a) Frequency domain validation and (b) closed-loop poles closeness.

The magnitude versus frequency characteristics of the three models have been plotted in Figure 15.11. The model identified in closed operation in the presence of three-dimensional turbulence wind speed is given as solid line, the model identified at constant wind speed [31] as dash line, and the linearized model obtained from Bladed® at 3 m s^{-1} as dash dot line. The differences between the models can be summarized as follows:

- A positive slope is observed at lowest frequencies in the linearized model, while a constant gain is obtained for the identified models.

- The resonances (complex poles) and antiresonance (complex zeros) are almost at the same frequencies for the model from Ref. [31] and the linearized model, but the dampings are different.

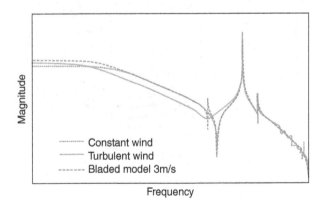

FIGURE 15.11   Linear models. Solid, model identified in presence of disturbance; dash, model identified at constant wind speed; dash dot, linear model obtained from Bladed® at 3 m s^{-1}.

- The model identified in the presence of wind turbulence shows the same resonances and antiresonance as the model from Ref. [31] except in the medium frequencies range where the effect of a close resonance and antiresonance is smoothed (which is good for control design purposes).

The final conclusion that can be drawn from the comparative model validation in closed loop is that the model identified in closed loop in the presence of wind turbulence is as good as the one identified at constant wind speed (which is not a very realistic situation), and both are better than the model obtained by linearization. It is important to mention, however, that this conclusion is valid for models to be used for control design that capture only the features necessary for this task (they do not replace a physical model).

15.5   CLOSED-LOOP PERFORMANCE

The purpose of the identified models is to tune controllers. In order to conclude which is the best model for controller design and tuning, this section presents controller performance in the torque control loop using controllers designed on the basis of different models. In WT design, the most important objectives are always increase of power production and reduction of loads. If these objectives are met, the always present objective of reducing the cost of wind energy production will be satisfied. For the case of variable-speed WTs with double feed generator, the gearbox loads are one of the most important variables to be analyzed in the torque loop, since it is well known that the gearbox failure is one

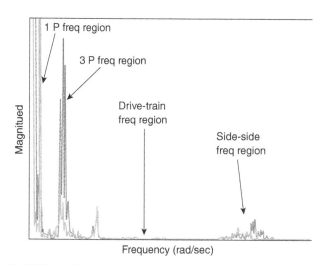

FIGURE 15.12 PSD of the gear box torque for (a) on dots, design based on linearized model from aeroelastic model and (b) on solid, designed on identified models obtained in closed-loop operation.

of the major WT maintenance problems. So a reduction in gearbox loads will allow much more wind energy production because of WT availability, together with less maintenance costs. The spectral analysis of gearbox torque is a good measurement of loads in torque loop.

Using the model identified with CLOE algorithms, one controller has been designed. On the other hand, another controller is designed based initially on linear models from Bladed®, and later, this controller is "fine tuned" based on nonlinear simulation in Bladed® in order to get a better result in terms of gearbox loads. Then, both controllers have been tested on the Bladed® model. Figure 15.12 shows the power spectral density of the gearbox torque for both cases. The performance obtained with the controller designed using the model identified in closed loop is plotted in solid. The performance obtained with the controller designed using the linear model from Bladed® is plotted in dots.

For the normal production case plotted (wind at 4 m s^{-1}), the drive train mode is totally damped by both controllers. However, in the frequency region close to the side-to-side mode, a reduction of the peak can be observed when using the controller designed on the basis of the closed-loop identified model. The 6P disturbance is almost the same for both controllers, but an important reduction in 3P is obtained, as well as for 1P disturbance.

So for the control design point of view, it can be concluded that the results obtained with a controller designed with the linear model identified in closed loop are better than the results obtained with the controller designed using the linearized model from Bladed® and "fine tuning" based on nonlinear simulations. This gives the main idea of this work, which is that identification in closed loop can provide a reliable model for control design purposes (as well as for on-site controller tuning).

15.6 DISCUSSION OF THE RESULTS

The possibilities offered by the identification of WT models operating in closed loop using experimental data for obtaining improved models for controller design leading to better performances have been explored. A procedure to be used in the context of WT has been developed and tested. This procedure can be applied to a simulation tool or to a real WT. This fact is especially important since this provides control engineers with a tool that makes it possible to obtain reliable linear models of each WT. These models will give the challenge to design and tune better and more reliable controllers in a shorter period of time.

It is important to point out that the challenge of providing reliable linear models is also important from the reliability of actual stability margins and robust stability. The linearized models [17–19] obtained from aeroelastic tools have been very useful from the initial design point of view. However, the robustness analysis based on these models is not trustworthy. Robustness margins have to be increased, giving more conservative controllers. On the other hand, with reliable linear models, a better controller can be designed in terms of performance, while in the mean time realistic robustness margins will be assured.

Another important challenge appears with this technique. This is related to the possibility of moving to more advanced controllers. Identification in closed loop followed by the redesign of the controller can be considered as a first step toward implementation of an adaptive control scheme [38]. It is a fact that it is not possible to characterize a full WT with one or two linear models, since big changes occur in a WT in terms of operation point. Therefore, the controller should be adapted in order to cope with large variations of the plant model parameters. This appears clearly in the pitch loop, where adaptive solutions are proposed based on gain scheduling [1, 40]. The gain scheduling is probably the simplest adaptive control realization.

It is an "open-loop"-type adaptive system that relies on a perfect correlation between the measure of the operating point and the associated model. What is needed is a "performance"-oriented adaptive control that will modify the controller in order to meet the performances. This, in one way or another, will require identification in closed loop [38]. Therefore, identification of WT models in closed-loop operation can become the first step toward implementation of more advanced adaptive control techniques (performance oriented).

15.7 CONCLUSION AND FUTURE WORK

The procedure presented in Ref. [31] for identification of WT operating in closed loop has been extended for a more general and realistic environment, namely the three-dimensional turbulence winds. It has been demonstrated that this procedure can be used in full aeroelastic and nonlinear simulators, as well as in real WTs, since the stability of the WT is guaranteed by feedback controllers.

The procedure is presented for the case of identification of the transfer between the generator torque and the measured generator speed. The reasons that make useless the open-loop identification algorithms and justify the proposed approach have been explained. The efficiency of the procedure, algorithms, and identified models has been demonstrated through the validation test and a controller design example. The suitability of the identified model has also been proven based on the FFT in the gearbox torque.

Application of this approach on an existing WT is considered for the near future.

Future work will have to focus on the way to identify other WT loops. In terms of new control developments, an online closed identification would be implemented in order to move toward adaptive controllers.

ACKNOWLEDGMENT

This work has been made possible by the financial support of CENER.

NOMENCLATURE

WT	wind turbine
CLOE	closed-loop output error identification algorithm
FCLOE	filtered closed-loop output error identification algorithm
AFCLOE	adaptive gain filtered closed-loop output error identification algorithm
XCLOE	extended closed-loop output error identification algorithm
GLS	generalized least squares identification algorithm
PID	proportional integral derivative controller
ARMAX	auto regressive moving average model
ARARX	auto regressive auto regressive model
PRBS	pseudorandom binary signal
1P	disturbance which corresponds to the frequency of the time that one blade spends in one revolution
3P	disturbance at three times the frequency 1P
6P	disturbance at six times the frequency 1P
nP	disturbances at n times the frequency 1P
MIMO	multi-inputs, multi-outputs system
FFT	fast Fourier transformation
$A(q^{-1}), B(q^{-1}),$	polynomials in the delay operator q^{-1}
$C(q^{-1}), D(q^{-1}),$	
$R(q^{-1}), S(q^{-1})$	
$A(z^{-1}), B(z^{-1}),$	polynomials in the complex variable z^{-1}
$C(z^{-1}), D(z^{-1}),$	
$R(z^{-1}), S(z^{-1})$	

REFERENCES

[1] Burton, T., Sharpe, D., Jenkins, N., Bossanyi, E. (2001). *Wind Energy Handbook*. Wiley, Chichester, UK.

[2] Bossanyi, E. (2000). The design of closed loop controllers for wind turbines. *Wind Energy*, 3(3):149–163.

[3] Van der Hoof, E. L., Schaak, P., Van Engelen, T. G. Wind turbine control algorithms. Dowec-F1w1-eh-03-094/0.

[4] Bossanyi, E. (2002). Individual blade pitch control for load reduction. Technical Report. *Wind Energy*, 6:119–128.

[5] Bossanyi, E. (2003). Wind turbine control for load reduction. Technical Report. *Wind Energy*, 6:229–244.

[6] Bossanyi, E. (2005). Further load reductions with individual pitch control. Technical Report. *Wind Energy*, 8:481–485.

[7] Bossanyi, E. (2004). Developments in individual blade pitch control. *Proceedings of the Special Topic Conference: The Science of Making Torque from Wind*, April 19–21, Delft, The Netherlands.

[8] Leithead, W. E., Dominguez, S. (2006). Coordinated control design for wind turbine control systems. Technical Report, University of Strathclyde, Glasgow.

[9] Leithead, W. E., Dominguez, S. (2005). Controller design for the cancellation of the tower fore-aft mode in a wind turbine. *Proceedings of the 44th IEEE Conference on Decision and Control, and the European Control Conference*.

[10] Johnson, K. E. (2004). Adaptive torque control of variable speed wind turbine. NREL/TP-500-36265.

[11] Iribas, M. (2007). Identification in closed loop in the presence of narrow band disturbances. *CENER Technical Note*.

[12] Mann, J. (1998). Wind field simulation. *Probabilistic Engineering Mechanics*, 13:269–282.

[13] Chapman, J. (2000). *Mdquinas Electricas*. McGraw-Hill, Oxford.

[14] Bongers, P. M. M. (1994). Modeling and identification of flexible wind turbines and factorizational approach to robust control. PhD Thesis, Delft University of Technology.

[15] Pitt, D. M., Peters, D. A. (1981). Theoretical prediction of dynamic inflow derivatives. *Vertica*, 5:21–34.

[16] D'Azzo, J., Houpis, C. H. (1975). *Linear Control System Analysis and Design, Conventional and Modern*. McGraw Hill, New York.

[17] Garrad Hassan and Partners Ltd. (2009). Bladed. [Online]. Available at: http://www.garradhassan.com (Accessed January 23, 2009).

[18] FAST. NREL. (2009). [Online]. Available at: http://wind.nrel.gov/designcodes./ (Accessed January 23, 2009).

[19] Miranda, E., Iribas, M. (2009). HMC, CENER. Available at: http://www.cener.com (Accessed January 23, 2009).

[20] Moolenar, P. D. (2004). *Cost-effective Design & Operation of Variable Speed Wind Turbines: Closing the Gap between Control Engineering & the Wind Engineering Community.* Delft University Press, Delft, The Netherlands.

[21] Rodriguez Amenedo, J. L., Burgos Diaz, J. C., Arnalte Gomez, S. (2003). *Sistemas Eolicos de Production de Energia Electrica.* Rueda, Madrid.

[22] Leithead, W. E., Rogers, M. C. M. (1996). Drive-train characteristics of constant speed HAWT's: Part I. Representation by simple dynamic models. *Wind Engineering*, 20:149–174.

[23] Leithead, W. E., Rogers, M. C. M. (1996). Drive-train characteristics of constant speed HAWT's: Part II. Simple characterisation of dynamics. *Wind Engineering*, 20:175–201.

[24] Sanz Feito, J. (2002). *Mdquinas Electricas.* Pearson Education, Madrid.

[25] Leishman, J. G., Beddoes, T. S. (1989). A semi-empirical model for dynamic stall. *Journal of the American Helicopter Society,* 34:3–17.

[26] Ljung, L. (1999). *System Identification*, 2nd edition. Prentice Hall, New Jersey.

[27] Hansen, M. H., Thomsen, K., Fuglsang, P. (2006). Two methods for estimating aeroelastic damping of operational wind turbine modes from experiments. *Wind Energy*, 9:179–191.

[28] Leithead, W. E., Hardan, F., Leith, D. J. (2003). Identification of aerodynamics and drive train dynamics for a variable speed wind turbine. *Proc. European Wind Energy Conference*, Madrid.

[29] Leithead, W. E., Zhang, Y., Neo, K. S. (2005). Wind turbine rotor acceleration: identification using Gaussian regression. *Proceedings of the 2nd International Conference on Informatics in Control, Automation and Robotics, ICINCO 2005,*

September 14–17, 2005, Barcelona, Spain. Available at: http://www.hamilton.ie/systemsmodelling/papers/leithead_wind_rotoraccel.pdf.

[30] Novak, P., Ekelund, T., Jovik, I., Schmidtbauer, B. (1995). Modeling and control of variable speed wind turbine, drive-system dynamics. *IEEE Control Systems*, 15(4):28–38.

[31] Iribas, M. (2007). Identification of wind turbines in closed loop operation. Torque demand to measured generator speed loop. *Proceedings of the EAWE Conference,* Pamplona.

[32] The Mathworks. (2009). Identification Toolbox. Matlab. [Online]. Available at: http://www.mathworks.com. (Accessed January 23, 2009).

[33] Landau, I.-D., Zito, G. (2006). *Digital Control System.* Springer, London.

[34] Landau, I.-D., Karimi, A. (1997). Recursive algorithms for identification in closed loop a unified approach and evaluation. *Automatica*, 33:1499–1523.

[35] Van den Hof, P. (1998). Closed loop issues in system identification. *Annual Reviews in Control*, 22:173–186.

[36] Iribas, M. (2006). Closed loop identification for wind turbines. CENER Ref IN-08.00289.

[37] Langer, J., Landau, I.-D. (1996). Improvement of robust digital control by identification in the closed loop. Application to a 360° flexible arm. *Control Engineering Practice*, 4:1637–1646.

[38] Landau, I.-D. (1999). From robust control to adaptive control. *Control Engineering Practice*, 7:1113–1124.

[39] Vinnicombe, G. (1993). Frequency domain uncertainty and the graph topology. *IEEE Transactions on Automatic Control*, 38:1371–1383.

[40] Leith, D. J., Leithead, W. E. (1996). Appropriate realization of gain-scheduled controllers with application to wind turbine regulation. *International Journal of Control*, 65:223–248.

16

IDENTIFICATION IN CLOSED-LOOP OPERATION OF MODELS FOR COLLECTIVE PITCH ROBUST CONTROLLER DESIGN

Mikel Iribas-Latour and Ion-Doré Landau

16.1 INTRODUCTION

Better models lead to better controllers. This is a very well-known axiom in the control engineering world. It is also well known that for control purposes, the best model for control design is not necessarily the most complicated and detailed one. Wind turbine (WT) control is not an exception. Actually, for control design purposes, the most used representation of a system is through a set of linear time-invariant (LTI) models. Of course, more sophisticated nonlinear models are needed for control testing but not for controller design.

It is well known [1–11] that control loops and its parameters are critical from generated loads point of view. Offshore challenges such as uncertainty, size, and power increase make the control even more important in terms of loads and stabilization. The linear models in several points of operation are the basic tool for controller design. Tuning of control loop parameters is a relatively well-solved problem in the wind sector [1–13] since linear models based on linearization of nonlinear aeroelastic codes [14–16] are generally used. However, it seems to be very common that control parameters tuned during WT design, and used for the certification process, do not exactly correspond to those finally implemented in the real WT. [10] Then the pertinence of these linearized models can be criticized since in many occasions the control parameters are not subsequently used in the real WT. Then one question appears. How are these controllers tuned in real WT? Most of the times, the final on-site tuning is based on the experience of the control engineer who proceeds according to rules of thumb or on a trial and error basis. This approach,

in many situations, makes the controller tuning very time consuming, inaccurate, and unable to cope with the nonlinear characteristics of the whole system. Even more, there is absolutely no evidence about the robustness of these tuned controllers. Therefore, there is a need for obtaining relevant and accurate models on site that can effectively be used for controller tuning.

Consequently, although big efforts in modeling WT [17–28] as well as for getting linearized models [14–16] were performed, the search of a procedure to obtain reliable control-oriented linear models (COLM) for control design seems a pertinent research task. Open-loop system identification techniques [29] can be used for obtaining on-site realistic COLM. Open-loop identification is probably the most extended technique for obtaining empirical models from experimental data. This technique is well known and has proven its goodness in a number of fields and applications. However, this technique should be avoided for the extraction of the empirical model of WTs in practice for different reasons that are commented in Section 16.3.

Different approaches have been tried in the frame of WT identification, but they usually look for different purposes. In Hansen et al., [30] an experimental approach for finding aeroelastic damping is introduced. In Leithead et al., [31,32] the identification is divided between a linear model, for drive train identification, and nonlinear model for the aerodynamical effects, with extra measurements. In Novak et al., [33] a full transfer function from torque demand to generator speed is obtained on the basis of open-loop identification algorithms.

Alternative Energy and Shale Gas Encyclopedia, First Edition. Edited by Jay H. Lehr and Jack Keeley.
© 2016 John Wiley & Sons, Inc. Published 2016 by John Wiley & Sons, Inc.

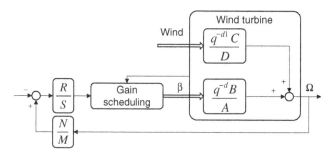

FIGURE 16.1 Classical collective pitch control loop representation.

The use of a controller during the execution of identification experiments warranties not only the stability and integrity of the WT but also the operation around a fixed operational point. Therefore, it is important to study the possibility of WT identification in closed-loop operation. [34–38] Using these new techniques of identification in closed loop is probably the best option for solving the problems coming from the nature of the WTs and open-loop identification. [39] Closed-loop output error (CLOE) algorithms [37, 38] have shown their capabilities for extracting reliable linear models for control design purposes in different applications while the system is operated in closed loop. [40] These algorithms are, however, designed to operate only in the presence of LTI controllers.

In variable speed–variable regulated WT, collective pitch control loop regulates the generator speed when rated power is reached (Figure 16.1), while is inactive below. This loop usually has a nonlinear structure. One of the most common approaches is to use an LTI controller with a gain, whose value depends on the state of the system, known as "gain scheduling". So, appropriate algorithms have to be developed to do identification in closed-loop operation in the presence of this kind of nonlinear controllers (NLCs). [41].

The first objective of this chapter is to develop an algorithm for plant model identification in closed-loop operation when the used controller is nonlinear.

The second objective of this chapter is to show how this new algorithm can be used to obtain a set of COLM for representing the dynamics between the collective pitch demand and the measured generator speed, around various operating points. Of course, linear behavior of the system is assumed during the experiments developed in Bladed. [14] The experimental protocol designed uses a linear controller with gain scheduling, however. The authors consider that this is the most realistic and safe approach for WT commissioning, since these controllers are available from the WT design process.

The third objective of this chapter is to analyze the properties of the various identified models. It will be shown that their properties vary drastically with the wind speed. These

models present zeros that move from outside the unit to inside the unit circle, while wind speed increases.

Since a set of identified COLM around various points of operation is available, at least two approaches are possible to design the pitch controller:

- Design a NLC with gain scheduling, that is, use the identified models and keep the control structure concept.
- Design a single fix robust linear controller for the collective pitch control loop, which achieves the desired performances, for all the identified COLM.

Then the fourth objective of this chapter is to show that indeed it is possible to replace the linear controller with gain scheduling, by a single fix linear controller for the same level of performance, provided that a set of reliable identified COLM is available and a robust controller design technique is used. A comparative simulation of the WT, with wind speeds from 8 to 22 m s^{-1}, demonstrates the equivalent performance in terms of rotor speed and collective pitch demand for both controllers. In addition, for the fix controller, the robust stability is guaranteed.

This chapter is organized as follows. In Section 16.2, a brief review of physical model and the control of generator speed in rated power by the collective pitch is presented. In Section 16.3, the linearized models coming from aeroelastic codes are evaluated, the state of the art on identification is commented, and the new algorithm of CLOE family for the identification in closed-loop operation in the presence of linear controller with gain scheduling is presented. In Section 16.4, the simulations carried on for obtaining data for closed-loop identification are presented. Section 16.5 presents the obtained models and the validation through a control performance on a realistic simulation. Concluding remarks and future direction of research are in Section 16.6.

16.2 COLLECTIVE PITCH CONTROL LOOP

16.2.1 Collective Pitch Demand to Generator Speed, Physical System Overview

The dynamic from collective pitch demand to generator speed is complex. It has been extensively studied. [2, 17–21] Many important details should be taken into account, such as aerodynamics, blade's and tower's structural dynamics, and drive train. Many efforts in describing the physics were performed in these fields. Here, only some remarks concerning the main control problems will be briefly commented.

There exists a very well-known nonlinear relationship between the extracted power, P, from the wind and the wind speed, V. Such relation is given in equation (16.1), where ρ is the air density. The extracted power depends linearly on the area of the rotor characterized by the rotor's radius, R. The term C_p in equation (16.1) deals with the characterization of the aerodynamical performance of the rotor blades. Taking into account that common rated wind speeds are between 10 and 12 m s^{-1} and that common cut-out wind speeds are between 20 and 30 m s^{-1}, there is an important nonlinear effect that should be taken into account by the controller in this range of wind speeds.

$$P = \frac{1}{2}\rho\pi R^2 V^3 C_p(\lambda, \beta) \qquad (16.1)$$

Active blade pitch control can be carried out in two opposite directions: to stall or to feathering. The most popular concept is pitching to feathering. In this case, for the operating pitch angle range, there is a linear relation between the angle of attack and the lift coefficient, C_l (see Figure 16.2). If the pitch angle increases, the angle of attack decreases. If the wind speed increases, the angle of attack increases. Normally, the control loop is designed to operate at this linear range, avoiding problems coming from stall operating at higher angles of attack, where higher loads and uncertainties appear and the linear relationship disappears. [17]

Another crucial effect, in the collective pitch demand to generator speed dynamics, is the coupling of the regulation speed loop with blade's and tower's structural dynamics. Changing pitch angle causes variations in both aerodynamical torque and thrust. This means a direct coupling between the rotor speed excursion and the structural dynamics. Torque variations may come not only from wind variations but also from blades and tower motions induced by the aerodynamical thrust changes. This well-known phenomenon may introduce non-minimum phase behavior in the control loop. This effect depends on the mass distribution and the operational conditions. [17, 20]

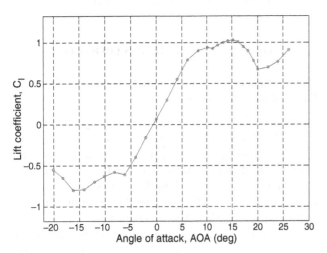

FIGURE 16.2 C_l curve for S809, Reynolds 1.000.000. *Source*: NREL.

Other features of the collective pitch demand loop, such as wind shear, gravity loads, wind turbulence, tower shadow, or imbalanced rotor, should also be taken into account. These disturbances are related with the time per revolution of the rotor, leading to the well-known nP disturbances. The effects of these disturbances on the controlled output are very powerful. Actually, it has been shown [42] that this effect on the output is sometimes more powerful than the effect of the controlled input demand. The effect of these disturbances, when using a simple proportional integral (PI) controller, can cause unwanted activity of the pitch actuator at these frequencies. This can cause an amplification of these disturbances, which can destabilize the turbine if they are not taken into account in the controller loop design.

Note also that the undamped torsion mode of the drive train is always present in the collective pitch dynamics in open loop. However, this is usually damped in the torque loop. [43] So if this torsion mode is correctly damped, it is invisible in the collective pitch control loop.

16.2.2 Classical Collective Pitch Control Algorithm

For variable speed WT configuration, the collective pitch control loop is active at medium and high wind speeds. The main objective of the collective pitch loop is to modify the pitch angle of the blades to regulate the rotor speed, while trying to reduce mechanical loads.

The most extended solution to regulate the generator speed at rated power with collective pitch demand is shown in Figure 16.1. A polynomial controller, R/S, combined with a gain scheduling is commonly used. According to the dominant dynamics, low-order controllers such as PI or proportional integral derivative (PID) are very popular. This could be correct because the dynamic of the rotor speed is dominated by its huge inertia at lower frequencies. In addition, PI and PID are very well-known controllers that can be tuned on site, although probably their tuning is far from being optimal.

However, a simple PID cannot deal with the nonlinear aerodynamical effects coming from the aerodynamics (equation (16.1)). There is certainly an important change in the gain of the collective pitch to generator speed model coming from the wind speed variations. In addition, the PID cannot take care of the nonlinearities coming from the coupling between aerodynamics and structural dynamics. These phenomena are important since they oblige to reduce controller's bandwidth. Then simple controllers like PI or PID are modified to accommodate these nonlinearities. A common solution is to implement a gain scheduling on the basis of indirect measurements of wind speed. [1, 2, 7] The gain scheduling introduces a nonlinear behavior to the controller.

It is also very common to use filters in the feedback loop of the collective pitch control loop. These filters, summarized in Figure 16.1 as N/M, may have different purposes. Avoiding the excitation by some disturbances and reducing collective pitch demand activity at high frequencies are typical objectives of these filters.

In Figure 16.1, the plant model is defined as the transfer operator:

$$\frac{\text{Measured generator speed}}{\text{Collective pitch demand}}$$

$$= \frac{B(q^{-1})}{A(q^{-1})} = \frac{q^{-d}\left[b_0 + b_1 q^{-1} + \cdots + b_{nb} q^{-nb}\right]}{1 + a_1 q^{-1} + \cdots + a_{na} q^{-na}} \quad (16.2)$$

where d represents the integer delay between the pitch actuation and the generator speed (in terms of sampling periods), and $B(q^{-1})$ and $A(q^{-1})$ represent the numerator and the denominator of the collective pitch to generator speed model.[1]

In Figure 16.1, the full collective pitch control loop is also represented where *gain scheduling* represents the nonlinear part of the controller; $R(q^{-1})$ and $S(q^{-1})$ represent the numerator and the denominator of the linear part of the controller; $N(q^{-1})$ and $M(q^{-1})$ represent the numerator and the denominator of the filters in the feedback loop; d_1, $C(q^{-1})$, and $D(q^{-1})$ represent the disturbance model of the pitch loop; Ω is the measured generator speed; β is the collective pitch demand; and the feedback loop uses positive feedback, since the system has negative gain.

The collective pitch control loop presented is a commonly used solution for regulating generator speed at rated power. Although this control scheme can be more complicated, this one is representative enough of the actual state of the art.[2]

The model of the system changes with wind conditions. Then models for various operation points are needed to have a good representation of the WT dynamics at different wind speed conditions. Therefore, an identification of these models is necessary. This has to be performed in closed-loop operation, in the presence of the predesigned controller with gain scheduling.

16.3 CLOSED-LOOP OUTPUT ERROR IDENTIFICATION ALGORITHM IN THE PRESENCE OF NONLINEAR CONTROLLERS

16.3.1 Linearized Models Coming from Wind Turbine Aeroelastic Code

The linearized models obtained from aeroelastic codes were a significant advance for understanding the main control loop dynamics and couplings. Of course, they help in the

design process of model-based controllers. These models have shown a number of drawbacks; however,

- The operational point for these models, constant wind speed for every point of the rotor area, may be very far from the real operational conditions. This is especially true for WT operating in complex terrains. Effects like wind shear, turbulence, or tower shadow also move the real operating condition far from constant wind speed for the whole rotor. [14–16]

- The routines that deliver the models do not always converge to a periodic state solution, which is commonly used as operating point for obtaining the linearized models.

- The high order of the obtained models (between 30 and 40) is unnecessarily high for describing the main plant dynamics for control design purposes.

- It is common that nonobservable modes appear in the linearized models, which, in addition, make the models numerically ill conditioned.

- Numerical errors caused by linearization techniques introduce unrealistic dynamics, especially at low frequencies. This is especially noticeable in phase plots and makes the stability analysis of the designed controllers unreliable.

- The modeling errors on aeroelastic codes or wrong parameterization of the models can enlarge the gap between theoretical models and the real dynamic behavior of the WT.

A consequence of the aforementioned problems is that sometimes the controllers designed on the basis of these linearized models do not provide the expected performance on real WT. The consequences in practice are as follows:

- A new controller should be designed to stabilize the real WT.

  If the controller designed with the linearized models does not work properly, no matter what the reason is, this controller needs to be redesigned. Sometimes, small changes in the parameters need to be carried out, but in a number of situations, important deviations from the original controller's parameters may occur. [10]

- Model-based controller design is no more possible for the real operating controller.

  As long as a reliable model is not available, the redesign of the controller is commonly based on the expertise of the control engineer, who usually applies rules of thumb. This can be very time consuming and introduce many doubts about the optimality or robustness of the controller.

  In addition, it is not possible to compute the basic robustness margins, which allow to assess the

---

[1] q^{-1} is the unit delay operator: $q^{-1} y(t) = y(t-1)$.
[2] More complicated nonlinear controllers or multivariable control schemes for reducing or alleviating some of the difficulties commented in Section 16.2.1 have been proposed. These more advanced control schemes are out of the scope of this chapter.

robustness of the controller, without a reliable model. Then it is not possible to assess the robustness of the used controller. This is probably the major problem, since this may drive the system to instability under certain circumstances, such as disturbances, nonlinear effects, or uncertainties.

- The controller used for certification of WT loads and consequently used in the mechanical design of each component is no more used in real WT.

  Certification loads are computed for 20 years of WT operation for both extreme and fatigue loads. Fatigue loads, where control plays an important role, are strongly affected by a wrong controller parameterization. If there is an important deviation from the performance of the certified controller, this may translate into a deviation between estimated—certified—loads and real operating loads. This is a drawback from the long-term WT integrity point of view. Even more, this means that there is no certainty about the loads that will be generated in the WT.

From these drawbacks, it can be concluded that the linearized models coming from WT simulators are a good qualitative representation of the WT dynamics, although they are not an accurate enough representation of the real operating WT dynamics. Then they cannot be used for in-site control tuning if the designed control fails on the real WT. One also concludes that a procedure is needed to obtain COLM of each WT to obtain a robust controller that will approach an optimal behavior.

16.3.2   State of the Art for Obtaining Linear Models from Experimental Data for Wind Turbine

In many fields, the most extended technique for obtaining linear models for control purposes is the open-loop system identification. [29] However, these techniques are too risky and should be avoided in the frame of WT since

- The source of energy that governs the WT, the wind, is not under control. Then it is hard to set an open-loop identification experiment.
- Since the wind speed is not under control, it is not possible to fix an operating point representative of the real operating condition of the WT.
- Operating a WT in open loop can be very risky, can create excessive loads and the possibility of WT instability, or, in the worst case, can collapse the WT.

To avoid some of these inconveniences and restrictions, the identification of WT in closed-loop operation has been considered. [35–38]

Identification in closed-loop operation was initially discarded in the 1980s. The available algorithms for open-loop identification were unable to produce relevant models when used in closed-loop operation. These algorithms suppose that there is no correlation between the input and output data. However, if the experimental data are obtained in closed-loop operation, there is a strong correlation between the measured noise and the real input to the plant. A number of "indirect" techniques have been developed to try to overcome this correlation between input and output data. One of these use the "instrumental variable"[3] concept and requires several steps. This technique was used by Novak et al. [33] during the 1990s in the frame of WT identification.

Newer developments for identification in closed-loop operation opened the possibility to obtain relevant COLM in one step from experimental data acquired in closed-loop operation. WT identification, when operating in closed loop, has been studied in the past years, and different approaches have been proposed.

In van Windergen et al., [44] a linear parameter with varying identification on repetitive sequences and subspace identification in closed loop has been applied to a WT. However, quite long experiments (50 minutes) are required, which may be a serious problem in practice. A different approach for WT identification in closed-loop operation was presented in Iribas, [45] where CLOE algorithms and heavy filtering were used for constant wind speeds. In Iribas and Landau, [43] the limitation of constant wind speed was removed. CLOE algorithms succeed in identifying the transfer function between torque demand to measured generator speed, in the presence of three-dimensional turbulent wind speed. This solution was able to identify the WT model in the presence of linear controllers, filters, and also linear drive train dampers.

This chapter is an extension of the methodology used in Iribas [45] and Iribas and Landau [43] to the collective pitch control loop. The existing CLOE algorithms are not applicable since in the collective pitch control loop, one often uses a NLC. In this chapter, a new CLOE algorithm is proposed. This allows the identification of the transfer operator for collective pitch demand to measured generator speed, when a NLC is being used in the loop. Of course, the identification is made possible in the presence of three-dimensional turbulence wind speed and without the measurement of the wind.

16.3.3   The New Identification Algorithm

The objective of CLOE algorithms is to identify a plant model that, in feedback with the actual controller, gives a closed-loop transfer function as close as possible to the real operating one. The algorithm developed for the identification of the

---

[3]An "instrumental variable" is an auxiliary variable that is correlated with the true variable and uncorrelated with the noise.

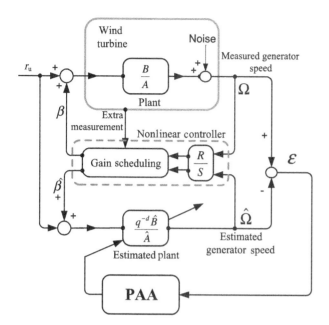

FIGURE 16.3 The new closed-loop identification algorithm scheme used for systems controlled with nonlinear time-varying controller. PAA, parameter adaptation algorithm.

collective pitch control loop, shown in Figure 16.3, is an extension of CLOE family algorithms for this case of NLC.

This approach is applied to the collective pitch control loop of a WT. The objective is to have a transfer function from the collective pitch demand to generator speed, based on experimental data, while operating in closed loop. Suppose the real generated speed, Ω, and the estimated generator speed, $\hat{\Omega}$, are given by equations (16.3) and (16.4), where β is the collective pitch angle demand, e is the white noise, and ∂ the predicted error:

$$\Omega(t) = \frac{z^{-d}B(q^{-1})}{A(q^{-1})}\beta(t) + \frac{z^{-d1}C(q^{-1})}{D(q^{-1})}e(t+1) \quad (16.3)$$

$$\hat{\Omega}(t) = \frac{z^{-d}\hat{B}(q^{-1})}{\hat{A}(q^{-1})}\hat{\beta}(t) + \frac{z^{-d1}\hat{C}(q^{-1})}{\hat{D}(q^{-1})}\varepsilon(t) \quad (16.4)$$

where

$$A(q^{-1}) = 1 + a_1 q^{-1} + \cdots + a_{nA}q^{-na} \quad (16.5)$$

$$B(q^{-1}) = b_1 q^{-1} + \cdots + b_{nB}q^{-nb} \quad (16.6)$$

$$\hat{A}(q^{-1}) = 1 + \hat{a}_1 q^{-1} + \cdots + \hat{a}_{nA}q^{-na} \quad (16.7)$$

$$\hat{B}(q^{-1}) = \hat{b}_1 q^{-1} + \cdots + \hat{b}_{nB}q^{-nb} \quad (16.8)$$

characterize the real and estimated models of the transfer function between collective pitch demand and generator speed. The real measured output of the system, equation (16.3), and the estimated plant output, equation (16.4), are used to compute the closed-loop error, $\partial_{cl} = \Omega - \hat{\Omega}$ (see Figure 16.3).

The input excitation signal can be applied in the reference signal or at the output of the controller. Since this is a regulation problem, the input excitation signal, r_u, is applied at the output of the controller (see Figure 16.3). The estimated collective pitch demand and the real collective pitch demand are both computed with the real operating NLC. The first gets into the estimated model, and the second gets into the real operating system. The closed-loop error, ∂_{cl} is used by the parameter adaptation algorithm (PAA), which recursively estimates the parameters of the estimated plant model.

The estimated parameters can be organized in a vector of parameters θ to be estimated. The measurements can also be arranged in a measurements vector, Φ, and used in the PAA. The PAA algorithm is defined by equations (16.9), (16.10), (16.11), and (16.12), a recursive algorithm inspired by the recursive least squares. Here, ε_{cl}^0 is the a priori prediction error; this means it is computed before measurement for time t + 1 is available. The prediction error ϵ_{cl} is computed once this measurement is available. F is a gain, and gives an idea of the magnitude of the change in the estimation of the parameters, and can be manipulated to weigh the measurements in time by different parameterization of λ_1 and λ_2. For details on PAA, see Verhaegen and Verdult [36] and Landau and Karimi [37]:

$$\hat{\theta}(t+1) = \hat{\theta}(t) + F(t)\phi(t)\varepsilon(t+1) \quad (16.9)$$

$$\varepsilon(t+1) = \frac{\varepsilon^0(t+1)}{1 + \phi^T(t)F(t)\phi(t)} \quad (16.10)$$

$$\varepsilon^0(t+1) = w_1(t+1) - \hat{\theta}(t)\phi(t) \quad (16.11)$$

$$F(t+1) = \frac{1}{\lambda_1(t)}\left[F(t) - \frac{F(t)\phi(t)\phi^T(t)F(t)}{\frac{\lambda_1(t)}{\lambda_2(t)} + \phi^T(t)F(t)\phi(t)} \right] \quad (16.12)$$

The novelty of the scheme is the use of the NLC used in the real WT, since previous approaches of CLOE algorithms supposed LTI controllers. Then the input to the real plant will be

$$\beta(t+1) = f(\text{Gain scheduling}, R/S, N/M, \Omega(t), ru) \quad (16.13)$$

and the estimated input to the estimated plant model will be

$$\hat{\beta}(t+1) = f(\text{Gain scheduling}, R/S, N/M, \hat{\Omega}(t), ru) \quad (16.14)$$

16.4 CLOSED-LOOP IDENTIFICATION SIMULATIONS FOR COLLECTIVE PITCH CONTROL LOOP

The WT model used is a 2 MW variable speed, variable pitch to feather regulated, with gearbox, 80 m tower height, and

80 m rotor diameter. The WT model uses two modes for modeling the tower fore aft and side-to-side dynamics. In addition, the WT model uses three modes for modeling the in-plane and out-of-plane dynamics. The WT is modeled in Bladed®, where the simulated experiments are carried out.

The simulated experiments are developed at rated power and rated speed. In this WT conditions, the collective pitch control regulates the WT. Various turbulent wind speeds are used to obtain models for different operating conditions. Operating in closed-loop warranties the integrity of the WT, as well as the operation around a specified operating point. The used controller should always be the best available at each time, although this would be far from being an optimum. The procedure can be applied iteratively. [38]

This approach supposes a linear behavior of the WT only during the experiment. Of course, this cannot be warranted before the closed-loop identification experiment is carried out. It will mainly depend on the behavior of the full WT, its controller, and, of course, the behavior of the wind. So, experimental data in a real system should be evaluated to warranty that the system has operated close to a fixed operating point.

16.4.1 Simulations Design

16.4.1.1 The Operational Conditions.
It is well known that one of the most complicated problems for system identification in open loop is to maintain the operational conditions during the experiments. For the case of WT identification, where the source of energy is not under control, this problem is even harder. However, operating in closed loop will ensure that the input and the output of the plant will remain relatively close to the defined operational conditions, even in the case that the controller is not optimum.

Although the wind is not predictable and can suffer big variations during the experiment, the stability of the WT is always warranted since the system is always working in closed loop with a controller that stabilizes the WT. In addition, as long as the wind can be considered as a disturbance at the output of the plant, the wind excursions will not be a problem for plant identification provided that the controller is able to keep the rotor speed at each rated value. However, it is true that if the wind changes a lot during the experiment, the identified model should be used with care or the experiment should be repeated. For this reason, techniques that require short experiments, like the one presented here, are more appropriate.

16.4.1.2 Design of the Input Excitation Signal.
Several simulations for various wind speeds should be considered. Several input excitation signals may be designed for different wind conditions. In a simulation environment, where the wind speed is known a priori, it is easy to select when the input signal should be applied. However, for a real operating WT, where only previous WT states are known, it will

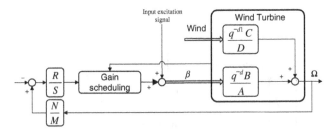

FIGURE 16.4 Pitch control loop for closed-loop identification simulations.

probably be necessary to repeat the experiment a number of times.

The input excitation signal can be applied at the demanded generator speed or at the output of the controller. Depending on the selected solution, the input design will be different. For the case of the collective pitch control loop, the input excitation signal has been applied at the output of the controller, as can be seen in Figure 16.4.

The stability and integrity of the WT should not be threatened by the excitation signal, which should be correctly designed, in terms of amplitude and frequency content.

The design of the input excitation signal is critical for a correct identification of the plant. On one hand, the dynamics to be identified should be excited. On the other hand, to avoid damages in the WT, one should not amplify important undamped modes that do not need to be identified. For a correct design of this input signal, the complete model of the WT is of interest, since it allows testing the feasibility of the experiment. Preliminary trials on simulation tools before developing experiments in a real WT should be considered.

On the basis of the simulations carried out in Bladed®, a good input excitation signal for the considered WT can be seen in Figure 16.5. The duration of the experiment is 54 seconds with a total number of 1080 samples for each experiment (sampling frequency: 20 Hz). However, it is probable that a longer experiment would be needed in a real WT because of the presence of a higher level of measurement noise, that is, a lower signal to noise ratio. The input excitation signal designed is presented in time domain in Figure 16.5a, and its frequency content is presented in Figure 16.5b.

It is important to see in Figure 16.5a that a low energy excitation is used for the identification in closed-loop operation. Amplitude of 1°, 0.01745 rad, is being used for these simulation experiments of identification in closed-loop operation. This amplitude is demanded to the pitch actuators that will rotate the blade at the desired value.

16.4.2 Simulated Experimental Data

Several simulations have been carried out for the identification of WT models from collective pitch demand to generator speed. The selected operational points were characterized by

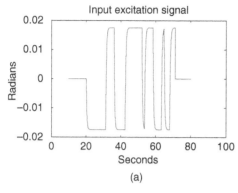 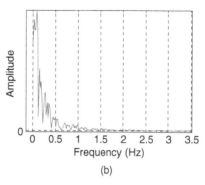

(a) (b)

FIGURE 16.5 Input excitation signal example. (a) Time domain and (b) fast Fourier transform of input excitation signal.

the mean wind speed at hub position during the execution of the simulations (see Figure 16.6 and Table 16.1 for details). The data of the mean wind speed are only used for naming the models that were identified. As can be seen in Table 16.1, the used winds are highly turbulent to be as close as possible to real operating wind conditions.

16.4.3 Simulated Experimental Data Analysis

Before proceeding to the identification, a preliminary analysis of the acquired data is necessary. It is important to focus on the evolution of the operational conditions of the plant during the experiment. The identification procedure may fail if there are important variations in the operational conditions during the execution of the experiment. Although the active control keeps the rotor speed close to its rated value during the experiment, the nonlinearity of the system can make the identified LTI model unreliable for the pitch controller design.

TABLE 16.1 Wind characteristics at hub during simulations.

Case (m s^{-1})	Standard deviation (m s^{-1})	Turbulence intensity (%)
14	1.66	11.58
16	2.57	15.29
17	1.68	9.76
20	2.58	13.07
22	2.69	11.29
24	3.15	12.13

For data coming from a simulation aeroelastic code, it is easy to see the incident wind. This could not be an easy option for data coming from real WT. However, there are several indirect measurements of the operational point that can be used. One may be the evolution of the controller gain during the experiment. This has been plotted in Figure 16.7 for the duration of developed closed-loop identification simulations. The obtained gain variations, for each simulation, have not been a problem for model identification. As was expected a

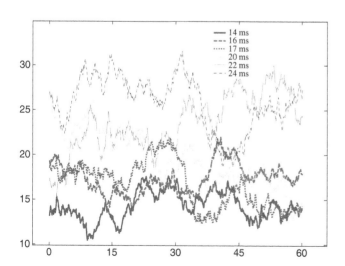

FIGURE 16.6 Wind conditions at hub during closed-loop identification simulations.

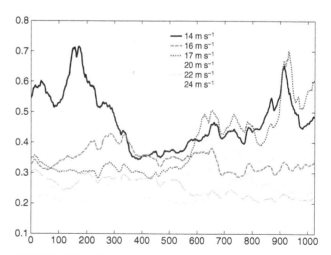

FIGURE 16.7 Gain variations during the set of closed-loop identification simulations.

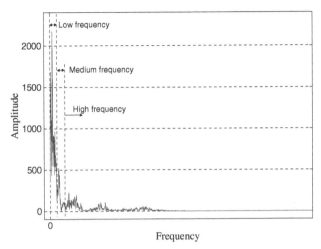

FIGURE 16.8 Fast Fourier transform for the measured generator speed for the identification case at 22 m s^{-1}.

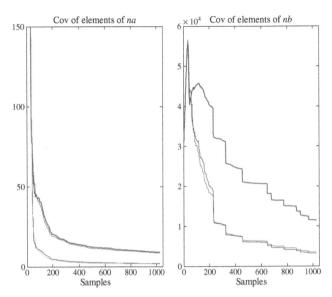

FIGURE 16.9 Covariance evolution of the identified parameters for the identification case at 14 m s^{-1}.

priori, the mean value of the gain and its deviations are higher for the cases of lower wind speed. Of course, it should be pointed that time evolution of this gain depends a lot on wind excursions during the experiment.

It is also important to check the effect that the input excitation signal has on the full WT. A little amount of energy is introduced in the system through this input signal. However, if the signal is not correctly designed, it can produce undesired amplification of different vibration modes. A frequency domain analysis of different variables can give a precise idea of any possible dynamic amplification, in case something unwanted would happen. Such an example is the fast Fourier transform of the measured generator speed during data acquisition for identification at 22 m s^{-1} shown in Figure 16.8. If a frequency correlation between Figure 16.8 and the WT component modes is performed, it can be concluded that no dynamic amplification appears during the experiment for main component's modes. Then one can conclude that the frequency content of measured data comes from the plant dynamics and disturbances, without the effect of dynamic amplification of any mode. Similar results can be obtained for all the identified cases.

16.5   IDENTIFIED MODELS

16.5.1   Algorithm Performance

The performances of the identification algorithms are usually evaluated by the so-called "model validation techniques," which give clear indication if the identified models can be used for controller design or not. If the model is validated, then it is used for controller design. Unfortunately, all the available techniques for validation of models identified in closed loop with CLOE algorithms make the assumption that

the controller is LTI, which is not the case for the pitch loop. Therefore, the evaluation of the quality of the identified models can be carried out only by evaluating the performances of the control system using a controller designed on the basis of the identified models.

However, some indications upon the quality of the identified model can be obtained before using the model for controller design. One of these indicators is the covariance matrix of the estimated parameters. It is possible to analyze the evolution of the covariance of each parameter. This will tell if the experiment was long enough and if the frequency content of the excitation signal is rich enough in terms of frequency content. In Figure 16.9, the evolution of the covariance of each element of the identified model is plotted as a function of the number of samples during the identification of the model for the 14 m s^{-1} case, where na is the number of elements of identified denominator and nb is the number of parameters of identified numerator (see equations (16.5), (16.6), (16.7), and (16.8)).

In Figure 16.9, it can be observed how the elements of the covariance matrix of the identified parameters decrease with the number of samples. This figure also indicates that there is not instability in the algorithm during parameter estimation. It can also be seen that the speed of convergence is faster for the elements of the denominator, and its value is much smaller than the ones for the numerator parameters. The convergence of the numerator's elements is slower, although it seems that both numerator and denominator parameters converge toward a fixed value. In case these values would not show a clear convergence, the duration of the experiment should be augmented, provided the correct model structure is used.

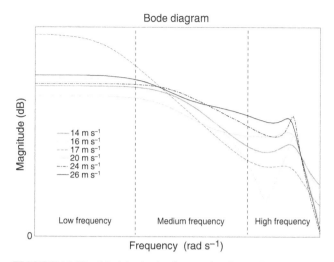

FIGURE 16.10 Models obtained supposing linear time-invariant controller in closed-loop operation.

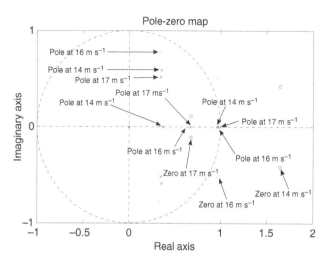

FIGURE 16.11 Pole–zero map for models that cross from non-minimum phase to minimum phase plant depending on wind speed velocity.

16.5.2 Analysis of the Identified Models

All the identified models presented here have the same number of poles and zeros. Actually, four poles and two zeros and a delay of two samples have been a good choice for all the analyzed cases. It is important to emphasize that good results have been obtained with the same plant structure for all the identification cases, that is, the change in the operational conditions influence the value of the parameters but not the complexity of the model.

The frequency characteristics are summarized in Figure 16.10. The variations of the dynamic characteristics coming from the different operational conditions can be seen, and they are summarized as follows:

- The models have different static gains. However, taken into account the length of the experiment, the DC gain of the identified models should be taken with care. It is not possible to identify the dynamics at frequencies below twice the inverse of the length of the experiment, in this case 0.035 Hz. It is important to mention that the existence of a pure integrator in the controller would minimize the importance of the knowledge of the exact DC gain value, provided that the identified DC gain is close to the real one.

- At medium frequencies, in general, similar slopes are identified. The behavior of this frequency region is clearly defined by the dominant dynamic of the rotor inertia.

- Probably, the most relevant differences are at high frequencies, where important differences in the resonance and antiresonance are clearly shown in Figure 16.10. These changes in the frequency characteristics are due to the coupling between the structural dynamics and the collective pitch control loop.

This nonlinear behavior of the system can also be illustrated by the pole–zero map of three different linear models obtained at 14, 16, and 17 m s^{-1}, shown in Figure 16.11. For this particular WT, it can be seen how at certain wind speeds between 16 and 17 m s^{-1} the systems pass from being non-minimum phase to minimum phase.[4] Here, one can observe the effect of the wind speed upon its dynamic properties of the system. Not only the parameters of the model but also the system dynamic properties changed.

It should also be noted that the drive train mode is not identified, as was expected from frequency domain analysis of the measured generator speed shown in Figure 16.8. This is because the frequency analysis of the output does not show any amplification of the drive train mode and because the damper in the torque loop is active during the execution of the simulations.

16.5.3 Validation Through Collective Pitch Controller Design and Performance Evaluation

The existing validation tests for models identified with CLOE cannot be applied since these validation methods suppose linear fix controllers. The model validation can only be performed through the following procedure:

- Design a controller based on the identified models.
- Test the controller.

---

[4] In discrete time models, zeros outside the unit circle with positive real part characterize non-minimum phase behavior, whereas zeros inside the unit circle characterize minimum phase behavior. Zeros outside the unit circle with negative real part may also come from fractional delay in the system, but this is not the case here.

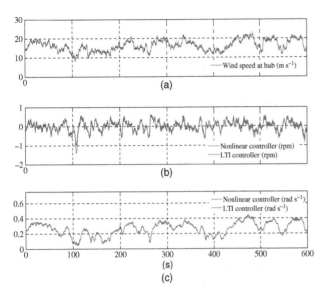

FIGURE 16.12 Comparison between nonlinear controller and a robust linear controller for 600 seconds simulation for high turbulence wind speed: (a) wind speed; (b) rotor speed (normalized); (c) pitch angle.

- Analyze and compare loop performance with different controllers: classical LTI controller with gain scheduling based on linearized models and the robust linear controller based on identified models.

Therefore, a controller was designed on the basis of linearized models coming from Bladed. A set of linearized models from 12 to 25 m s^{-1} wind speed has been used. The design of the PID and of the gain scheduling was performed to obtain similar control performance for each linearized model.

For the case of the single controller designed on the basis of the identified models, a different approach was considered. A robust controller design was developed on the basis of pole placement technique. This controller design gave good performance for all identified models.[5]

The designed LTI robust controller without any nonlinearity is tested against the aeroelastic code model, and its robustness is proved through a full time domain simulation. The comparison of performances between the two controllers is shown in Figure 16.12.

Figure 16.12a shows the wind speed for a simulation of 600 seconds. As it can be observed, the wind speed is moving from 8 m s^{-1} to more than 22 m s^{-1}. In addition, in Figure 16.12b, it can be observed that there is no major difference in the rotor speed error when using the NLC and when using the robust LTI controller. Finally, in Figure 16.12c, the pitch angle demand shows similar activity.

[5]These techniques cannot be applied to the Bladed linearized models since these models feature nonobservable (or almost nonobservable) modes and they are numerically ill conditioned.

16.6   CONCLUSIONS AND FUTURE WORK

A new algorithm for identifying linear models, while a NLC is running in the feedback loop, has been presented. This technique has been applied to obtain a set of models from the collective pitch demand to generator speed, at six different mean wind speeds between 14 and 26 m s^{-1}.

The obtained models indicate a correlation between physical knowledge and identified models, in the sense that they show DC gain variations, low-frequency dominant dynamics, and transition from non-minimum to minimum phase behavior with operational conditions.

The obtained models are used for a single robust LTI controller design. Its performance is compared with that of a NLC using a PID controller with gain scheduling. Similar control performances are obtained for both approaches, but the linear robust controller is definitely simpler to implement.

The presented procedure and algorithm gives a new tool that makes possible to do model-based WT control tuning in site, when controllers designed on theoretical model do not work properly. This procedure avoids dangerous tuning methods based on rules of thumb and makes possible to check the robustness of the used controllers prior to their use since reliable COLM are now available.

Specific real tests for applying this algorithm will soon be developed at NREL's CART2 and CART3 in Boulder, Colorado.

The move toward iterative identification in closed loop and controller redesign, once the LTI controller is implemented, is an interesting option for future studies.

Future developments will deal with the identification of multi-input, multi-output systems in the frame of individual pitch control. Important uncertainties about flapwise couplings exist, and closed-loop identification of the involved loops seems a good tool to analyze this and other effects.

APPENDIX A. MODEL COMPARISON: IDENTIFIED MODELS VERSUS LINEARIZED MODELS

There exists the possibility to compare the obtained identified model with the linearized model obtained from Bladed®. This comparison should be performed with care, since the origin of the models is totally different. Of course, similar characteristics should be shown, but a perfect match between these models should not be expected.

Then it should be kept in mind that there are important differences in the procedure to obtain both models:

- The procedure to obtain linearized models uses constant wind speeds, whereas identified models use data coming from aeroelastic simulations with three-dimensional turbulent wind speeds.

- The procedure to obtain linearized models removes nonlinearities in the linearization procedure, especially those coming from the actuators, whereas the identified model uses the most detailed description of the WT.

- Azimuth dependence and certain other features such as tower shadow, upwind turbine wake, yaw motion, gravity loads, safety factors, or imbalances are not used for obtaining linearized model, whereas identified models use the most realistic representation of the WT dynamics and aerodynamic characteristics.

- The linearized models are obtained in open-loop simulation, whereas identified models are obtained in closed-loop operation. This is important since, for example, the drive train mode is always present in the linearized model, but in the identified models, the controller is active and damps this mode that it no more appears in the identified models. There are other important differences, like the way the operating point is obtained. In closed-loop operation, this is carried out by the active feedback control, which is not the case for the linearization process of Bladed.

For the case of 16 m s^{-1} wind, the magnitude Bode plots of the identified model (solid line) and the linearized model (dotted line) are plotted in Figure A.1. To be able to make an easier comparison, the linearized models have been modified to include the drive train damper, active in the identification procedure. It is easy to see that even the drive train damper has been included in the linearized model, both models looks different in some frequency regions. At this point, it is important to remember that the identified model is of order 4, whereas the linearized is of order 40, which means much more slope changes are necessary in this model.

If models are analyzed in ranges of frequency, it is clear that at low frequency there are differences at zero frequency, where linearized models suggest a pure derivator, whereas a flat gain is obtained in the identified model. In addition, once the slope of the linearized models moves to zero, there is an important gain difference. However, at medium frequencies, the models converge to similar values, with the same gain and slope around 1 rad s^{-1}. Clearly, both models represent similarly the big influence of the rotor inertia in the WT dynamics. Although frequency moves to higher values, deviations are again evident between both models. But it can be observed that the linearized models show similar slope changes at the resonance of the identified model, which coincides with a peak in the linearized model.

Once again, it is not the comparison of the models that is important but the performances that can be obtained with the controller designed on the basis of these models.

NOMENCLATURE

R	rotor diameter
V	wind speed
λ	tip speed ratio
ρ	air density
C_p	power coefficient
C_1	aerodynamical lift coefficient
Ω	generator speed
$\hat{\Omega}$	estimated generator speed
β	collective pitch angle demand
$\hat{\beta}$	estimated collective pitch angle demand
$\in^0$	predictor error a priori
$\in$	predictor error
E	white noise
Ru	input excitation signal
θ	vector of parameters to be estimated
Φ	measurements vector
R/S	linear controller of numerator R and denominator S
B/A	linear transfer function, with numerator B and denominator A
N/M	linear filter with numerator N and denominator M
1P	disturbance that corresponds to the frequency of one rotor revolution time
nP	disturbances at n times the frequency 1P
$A(q^{-1}), B(q^{-1}),$ $C(q^{-1}), D(q^{-1}),$ $R(q^{-1}), S(q^{-1})$	polynomials in the delay operator q^{-1} $(q-1y(t) = y(t-1))$
$A(z^{-1}), B(z^{-1}),$ $C(z^{-1}), D(z^{-1}),$ $R(z^{-1}), S(z^{-1})$	polynomials in the complex variable z^{-1}
T	normalized sampling time $(= t/T_s$ where T_s is the sampling period)

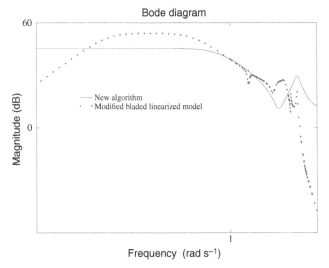

FIGURE A.1 Linear models for 16 m s^{-1} wind speed identified with the new algorithm and the linearized model with damper.

ABBREVIATIONS

WT   wind turbine
COLM   control-oriented linear models
LTI   linear time invariant
CLOE   closed-loop output error
PI   proportional integral controller
PID   proportional integral derivative controller
NLC   nonlinear controller
PAA   parameter adaptation algorithm

REFERENCES

[1] Bossanyi, E. (2000). The design of closed loop controllers for wind turbines. *Wind Energy*, 3(3):149–163.

[2] Bianchi, F. D., Battista, H. de., Mantz, R. J. (2007). *Wind Turbine Control Systems. Principles, Modelling and Gain Scheduling Design*. Springer, London.

[3] Bossanyi, E. (2002). Individual blade pitch control for load reduction. Technical Report. *Wind Energy*, 6:119–128.

[4] Bossanyi, E. (2003). Wind turbine control for load reduction. Technical Report. *Wind Energy*, 6:229–244.

[5] Bossanyi, E. (2004). Developments in individual blade pitch control. *Special Topic Conference. The Science of Making Torque from Wind*, April 19–21, Delft, The Netherlands.

[6] Bossanyi, E. (2005). Further load reductions with individual pitch control. Technical Report. *Wind Energy*, 8:481–485.

[7] Leith, D. J., Leithead, W. E. (1996). Appropriate realization of gain-scheduled controllers with application to wind turbine regulation. *International Journal of Control*, 65(2):223–248.

[8] Leithead, W. E., Dominguez, S. (2006). Coordinated control design for wind turbine control systems. Technical report, University of Strathclyde, Glasgow.

[9] Leithead, W. E., Dominguez, S. (2005). Controller design for the cancellation of the tower fore-aft mode in a wind turbine. *Proceedings of the 44th IEEE Conference on Decision and Control, and the European Control Conference*.

[10] Johnson, K. E. (2004). Adaptive torque control of variable speed wind turbine. NREL/TP-500-36265.

[11] Van der Hoof, E. L., Schaak, P., Van Engelen, T. G. Wind turbine control algorithms. Dowec-F1w1-eh-03-094/0.

[12] D'Azzo, J., Houpis, C. H. (1975). *Linear Control System Analysis and Design, Conventional and Modern*. McGraw-Hill, New York.

[13] Landau, I. D. (1999). From robust control to adaptive control. *Control Engineering Practice*, 7(10):1113–1124.

[14] Garrad Hassan and Partners. Bladed. Available at: www.garradhassan.com.

[15] FAST. NREL. Available at: http://wind.nrel.gov/design codes/.

[16] Miranda, E., Iribas, M. HMC, CENER. Available at: www.cener.com.

[17] Burton, T., Sharpe, D., Jenkins, N., Bossanyi, E. (2001). *Wind Energy Handbook*. Wiley, Chichester.

[18] Leithead, W. E., Rogers, Y. M. C. M. (1996). Drive-train characteristics of constant speed HAWT's: part I. Representation by simple dynamic models. *Wind Engineering*, 20(3):149–174.

[19] Leithead, W. E., Rogers, Y. M. C. M. (1996). Drive-train characteristics of constant speed HAWT's: part II. Simple characterisation of dynamics. *Wind Engineering*, 20(3):175–201.

[20] Bongers, P. M. M. (1994). Modeling and identification of flexible wind turbines and factorizational approach to robust control. PhD Thesis, Delft University of Technology.

[21] Suryanarayanan, S., Dixit, A. (2005). On the dynamics of the pitch control loop in horizontal-axis large wind turbines. *American Control Conference*, pp. 686–690.

[22] Mann, J. (1998). Wind field simulation. *Probabilistic Engineering Mechanics, Elsevier* 13(4):269–282.

[23] Pitt, D. M., Peters, D. A. (1981). Theoretical prediction of dynamic inflow derivatives. *Vertica*, 5:21–34.

[24] Leishman, J. G., Beddoes, T. S. (1989). A semi-empirical model for dynamic stall. *Journal of the American Helicopter Society*, 34:3–17.

[25] Moolenar, P. D. (2004). *Cost-effective Design & Operation of Variable Speed Wind Turbines: Closing the Gap between Control Engineering & the Wind Engineering Community*. Delft University Press, Delft.

[26] Rodriguez Amenedo, J. L., Burgos Diaz, J. C., Arnalte Gomez, S. (2003). *Sistemas eolicos deproduccion de energia electrica Ed*. Rueda Madrid, Madrid.

[27] Sanz Feito, J. (2002). *Mdquinas Electricas*. Prentice Hall, Madrid.

[28] Chapman, J. (2000). *Mdquinas Electricas*. McGraw-Hill, Madrid.

[29] Ljung, L. (1999). *System Identification*, 2nd edition Prentice Hall, Upper Saddle River, NJ.

[30] Hansen, M. H., Thomsen, K., Fuglsang, P. (2006). Two methods for estimating aeroelastic damping of operational wind turbine modes from experiments. *Wind Energy*, 9:179–191.

[31] Leithead, W. E., Hardan, F., Leith, D. J. (2003). Poster session presented at European Wind Energy Conference and Exhibition, Madrid, Identification of aerodynamics and drive train dynamics for a variable speed wind turbine.

[32] Leithead, W. E., Zhang, Y., Neo, K. S. (2005). Wind turbine rotor acceleration: identification using Gaussian regression, ICINCO, Barcelona.

[33] Novak, P., Ekelund, T., Jovik, I., Schmidtbauer, B. (1995). Modeling and control of variable speed wind turbine, drive-system dynamics. *IEEE Control Systems*, 14(4):28–38. ISSN: 1066-033X.

[34] Hof, P.Van den, Schrana, R. (1995). Identification and control—closed loop issues. *Automatica*, 31:1751–1770.

[35] Hof, P.Van den. (1998). Closed loop issues in system identification. *Annual Reviews in Control*, 22:173–196.

[36] Verhaegen, M., Verdult, V. (2007). *Filtering and System Identification: A Least Squares Approach*. Cambridge University Press, Cambridge.

[37] Landau, I.-.D., Karimi, A. (1997). Recursive algorithms for identification in closed loop—a unified approach and evaluation. *Automatica*, 33(8):1499–1523.

[38] Landau, I.-D., Zito, G. (2006). *Digital Control System*. Springer, London.

[39] Iribas, M. (2007). Identification in closed loop in the presence of narrow band disturbances. *CENER Technical Note*.

[40] Langer, J., Landau, I. D. (1996). Improvement of robust digital control by identification in the closed loop. Application to a 360° flexible arm. *Control Engineering Practice*, 4(12):1637–1646.

[41] Iribas, M., Landau, I. D. (2009). Closed loop identification of wind turbines models for pitch control. *Mediterranean Control Conference*.

[42] Iribas, M. (2006). Closed loop identification for wind turbines. CENER RefIN-08.00289.

[43] Iribas, M., Landau, I. D. (2009). Identification of wind turbines in closed loop operation in the presence of three dimensional turbulence wind speed. *Wind Energy Journal*, 12:660–675.

[44] van Windergen, J. W., Houtzager, I., Felici, F., Verhaegen, M. (2009). Closed loop identification of time-varying dynamics of variable wind turbines. *International Journal of Robust and Nonlinear Control*, 19:4–21.

[45] Iribas, M. (2007). Identification of wind turbines in closed loop operation. Torque demand to measured generator speed loop. *EAWE Conference*, Pamplona.

17

WIND BASICS—ENERGY FROM MOVING AIR

17.1 WIND BASICS

17.1.1 Energy from Moving Air

Wind is simply air in motion. It is caused by the uneven heating of the Earth's surface by the sun. Because the Earth's surface is made of very different types of land and water, it absorbs the sun's heat at different rates. One example of this uneven heating can be found in the daily wind cycle.

17.1.2 The Daily Wind Cycle

During the day, the air above the land heats up more quickly than the air over water. The warm air over the land expands and rises, and the heavier, cooler air rushes in to take its place, creating wind. At night, the winds are reversed because the air cools more rapidly over land than over water.

In the same way, the atmospheric winds that circle the earth are created because the land near the Earth's equator is heated more by the sun than the land near the North and South poles.

17.1.3 Wind Energy for Electricity Generation

Today, wind energy is mainly used to generate electricity, although water pumping windmills were once used throughout the United States.

17.2 ELECTRICITY GENERATION FROM WIND

17.2.1 How Wind Turbines Work

Like old-fashioned windmills, today's wind machines (also called wind turbines) use blades to collect the wind's kinetic energy. The wind flows over the blades creating lift, like the effect on airplane wings, which causes them to turn. The blades are connected to a drive shaft that turns an electric generator to produce electricity.

With the new wind machines, there is still the problem of what to do when the wind is not blowing. At those times, other types of power plants must be used to make electricity.

17.2.2 Wind Production

In 2012, wind turbines in the United States generated about 3% of total US electricity generation. Although this is a small share of the country's total electricity production, it was equal to the annual electricity use of about 12 million households.

The amount of electricity generated from wind has grown significantly in recent years. Generation from wind in the United States increased from about 6 billion kWh in 2000 to about 140 billion kWh in 2012.

New technologies have decreased the cost of producing electricity from wind, and growth in wind power has been encouraged by tax breaks for renewable energy and green pricing programs. Many utilities around the country offer green pricing options that allow customers the choice to pay more for electricity that comes from renewable sources to support new technologies.

17.3 WHERE WIND IS HARNESSED

17.3.1 Wind Power Plants Require Careful Planning

Operating a wind power plant is not as simple as just building a windmill in a windy place. Wind plant owners must

Alternative Energy and Shale Gas Encyclopedia, First Edition. Edited by Jay H. Lehr and Jack Keeley.
© 2016 John Wiley & Sons, Inc. Published 2016 by John Wiley & Sons, Inc.

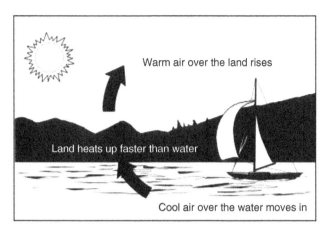

FIGURE 17.1 How uneven heating of water and land causes wind. *Source*: National Energy Education Development Project (public domain).

carefully plan where to locate their machines. It is important to consider how fast and how much the wind blows at the site.

As a rule, wind speed increases with altitude and over open areas that have no windbreaks. Good sites for wind plants are the tops of smooth, rounded hills, open plains or shorelines, and mountain gaps that produce wind funneling.

17.3.2   Wind Speed Is Not the Same Across the Country

Wind speed varies throughout the United States. It also varies from season to season. In Tehachapi, California, the wind blows more from April through October than it does in the winter. This is because of the extreme heating of the Mojave

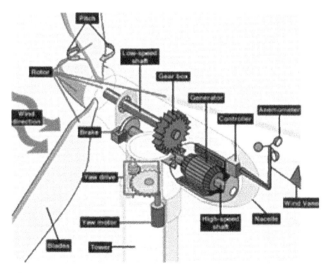

FIGURE 17.2 Diagram of windmill workings. *Source*: National Renewable Energy Laboratory, US Department of Energy (Public Domain).

Desert during the summer months. The hot air over the desert rises, and the cooler, denser air above the Pacific Ocean rushes through the Tehachapi mountain pass to take its place. In a state like Montana, on the other hand, the wind blows more during the winter.

Fortunately, these seasonal variations are a good match for the electricity demands of the regions. In California, people use more electricity during the summer for air conditioners. In Montana, people use more electricity during the winter.

17.3.3   Major Wind Power Locations

Large wind turbines (sometimes called wind machines) generated electricity in 36 different states in 2011. The top five states with the largest generation of electricity from wind were Texas, Iowa, California, Minnesota, and Illinois.

17.3.4   International Wind Power

In 2009, most of the wind power plants in the world were located in Europe and in the United States where government programs have helped support wind power development. The United States ranked first in the world in wind power generation, followed by Germany, Spain, China, and India. Denmark ranked ninth in the world in wind power generation, but generated about 19% of its electricity from wind, the largest share of any country.[1]

17.3.5   Offshore Wind Power

Conditions are well suited along much of the coasts of the United States to use wind energy. However, there are people who oppose putting turbines just offshore, near the coastlines, because they think the wind turbines will spoil the view of the ocean. There is a plan to build an offshore wind plant off the coast of Cape Cod, Massachusetts.

Wind is a renewable energy source that does not pollute, so some people see it as a good alternative to fossil fuels.

17.4   TYPES OF WIND TURBINES

There are two types of wind machines (turbines) used today, based on the direction of the rotating shaft (axis): horizontal-axis wind machines and vertical-axis wind machines. The size of wind machines varies widely. Small turbines used to power a single home or business may have a capacity of less than 100 kW. Some large commercial-sized turbines may have a capacity of 5 million watts, or 5 MW. Larger turbines are often grouped together into wind farms that provide power to the electrical grid.

---

[1] 2008 is most recent year data on capacity available as of June 29, 2011.

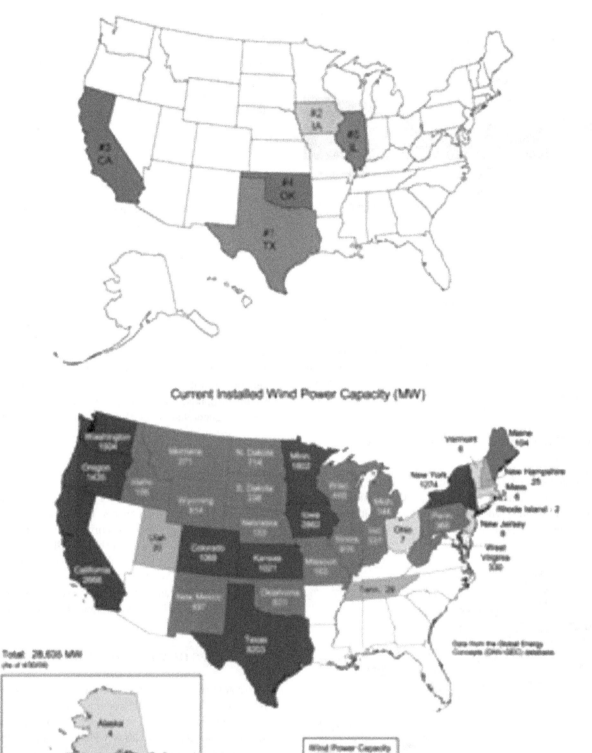

FIGURE 17.3 Map of US wind capacity. *Note*: See progress of installed wind capacity between 1999 and 2012. *Source*: (a) US Energy Information Administration, *Electric Power Monthly*, Table 1.17.B (February 2013). (b) National Renewable Energy Laboratory, US Department of Energy (Public Domain).

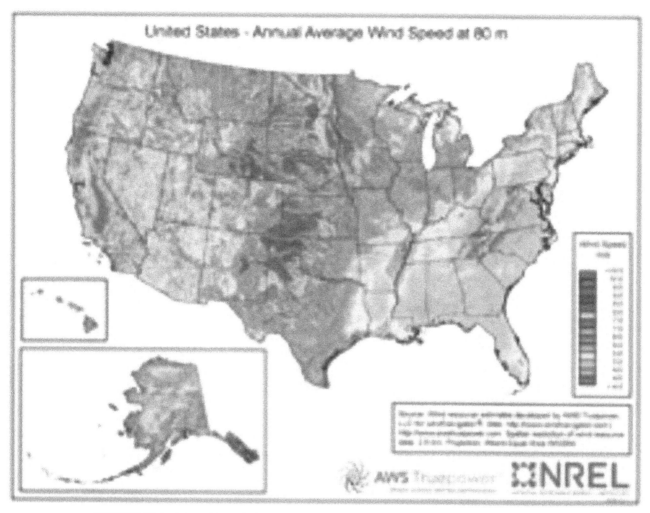

FIGURE 17.4   Map of US wind resources. *Source*: National Renewable Energy Laboratory, US Department of Energy (Public Domain).

17.4.1   Horizontal-Axis Turbines Look Like Windmills

Most wind machines being used today are the horizontal-axis type. Horizontal-axis wind machines have blades like airplane propellers. A typical horizontal wind machine stands as tall as a 20-story building and has three blades that span 200 ft across. The largest wind machines in the world have blades longer than a football field. Wind machines stand tall and wide to capture more wind.

17.4.2   Vertical-Axis Turbines Look Like Egg Beaters

Vertical-axis wind machines have blades that go from top to bottom. The most common type—the Darrieus wind turbine, named after the French engineer Georges Darrieus who patented the design in 1931—looks like a giant, two-bladed egg beater. This type of vertical wind machine typically stands 100 ft tall and 50 ft wide. Vertical-axis wind machines make up only a very small share of the wind machines used today.

17.4.3   Wind Power Plants Produce Electricity

Wind power plants, or wind farms, as they are sometimes called, are clusters of wind machines used to produce electricity. A wind farm usually has dozens of wind machines scattered over a large area. The world's largest wind farm, the Horse Hollow Wind Energy Center in Texas, has 421 wind turbines that generate enough electricity to power 220,000 homes per year.

Many wind plants are not owned by public utility companies. Instead, they are owned and operated by business people who sell the electricity produced on the wind farm to electric utilities. These private companies are known as independent power producers.

FIGURE 17.5 Wind turbines in the ocean. *Source*: US Energy Information Administration, *Electric Power Monthly*, Table 1.17.B (February 2013).

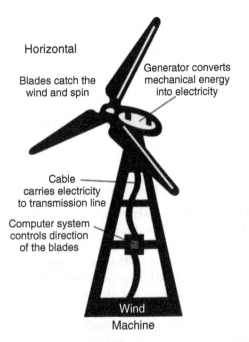

FIGURE 17.6 Horizontal-axis wind machine. *Source*: National Energy Education Development Project (public domain).

FIGURE 17.7 Darrieus vertical-axis wind turbine in Martigny, Switzerland. *Source*: Lysippos, Wikimedia Commons author (GNU free documentation license) (public domain).

FIGURE 17.8 Traditional Dutch-type windmill.

FIGURE 17.9 Wind turbines at the Cerro Gordo Project, west of Mason City, Iowa. *Source*: National Renewable Energy Laboratory (public domain).

17.5 HISTORY OF WIND POWER

17.5.1 The Oldest Windmills Were in Ancient Persia

Since early recorded history, people have been harnessing the energy of the wind. Wind energy propelled boats along the Nile River as early as 5000 BC. By 200 BC, simple windmills in China were pumping water, while vertical-axis windmills with woven reed sails were grinding grain in Persia and the Middle East.

New ways of using the energy of the wind eventually spread around the world. By the 11th century, people in the Middle East were using windmills extensively for food production; returning merchants and crusaders carried this idea back to Europe. The Dutch refined the windmill and adapted it for draining lakes and marshes in the Rhine River Delta. When settlers took this technology to the New World in the late 19th century, they began using windmills to pump water for farms and ranches and, later, to generate electricity for homes and industry.

American colonists used windmills to grind wheat and corn, to pump water, and to cut wood at sawmills. As late as the 1920s, Americans used small windmills to generate electricity in rural areas without electric service. When power lines began to transport electricity to rural areas in the 1930s, local windmills were used less and less, though they can still be seen on some Western ranches.

17.5.2 Windmills Make a Comeback in the Wake of Oil Shortages

The oil shortages of the 1970s changed the energy picture for the country and the world. It created an interest in alternative energy sources, paving the way for the reentry of the windmill to generate electricity. In the early 1980s, wind energy really took off in California, partly because of state policies that encouraged renewable energy sources.

In the 1970s, oil shortages pushed the development of alternative energy sources. In the 1990s, the push came from a renewed concern for the environment in response to scientific studies indicating potential changes to the global climate if the use of fossil fuels continues to increase. Wind energy is an economical power resource in many areas of the country.

Growing concern about emissions from fossil fuel generation, increased government support, and higher costs for fossil fuels (especially natural gas and coal) have helped wind power capacity in the United States to grow substantially over the past 10 years.

17.6 WIND ENERGY AND THE ENVIRONMENT

17.6.1 Wind: A Clean Source of Energy

Wind is a clean source of energy, and overall, the use of wind for energy has fewer environmental impacts than using many other energy sources. Wind turbines (often called windmills) do not release emissions that pollute the air or water (with rare exceptions) and they do not require water for cooling. They may also reduce the amount of electricity generated from fossil fuels and therefore reduce the amount of air pollution, carbon dioxide emissions, and water use of fossil fuel power plants.

A wind turbine has a small physical footprint relative to the amount of electricity it can produce. Many wind projects, sometimes called wind farms, are located on farm, grazing, and forest land. The extra income from the turbines may allow farmers and ranchers to stay in business and keep their property from being developed for other uses. For example, wind power projects have been proposed as alternatives to mountain top removal coal mining projects in the Appalachian Mountains of the United States. Offshore wind turbines on lakes or the ocean may have smaller environmental impacts than turbines on land.

17.7 DRAWBACKS OF WIND TURBINES

Wind turbines do have negative impacts on the environment, but the negative impacts have to be balanced with our need for electricity and the overall lower environmental impact of using wind for energy relative to other sources of energy to make electricity.

Modern wind turbines are very large machines, and some people do not like their visual impact on the landscape. A few wind turbines have caught on fire, and some have leaked lubricating fluids, though this is relatively rare. Some people do not like the sound that wind turbine blades make. Some types of wind turbines and wind projects cause bird and bat deaths. These deaths may contribute to declines in species that are also being affected by other human-related impacts. Many birds are killed from collisions with vehicles and buildings, by house cats and hunters, and by pesticides. Their natural habitats may be altered or destroyed by human development and by the changes in the climate that most scientists believe are caused by greenhouse gases emissions from human activities (which wind energy use can help reduce). The wind energy industry and the US government are researching ways to reduce the impact of wind turbines on birds and bats.

Most wind power projects on land also require service roads that add to their physical impact on the environment. Making the metals and other materials in wind turbines and the concrete for their foundations requires the use of energy, which may be from fossil fuels. Some studies have shown that wind turbines produce much more clean electricity over their operating life than the equivalent amount of energy used to make and install them.

18

WIND—CHRONOLOGICAL DEVELOPMENT

Public Domain

500–900 AD	The first windmills were developed in Persia for pumping water and grinding grain.
About 1300	The first horizontal-axis windmills (like a pinwheel) appeared in Western Europe.
1850s	Daniel Halladay and John Burnham worked to build and sell the Halladay Windmill, designed for the American West. It had an open tower design and thin wooden blades. They also started the U.S. Wind Engine Company.
Late 1880s	Thomas O. Perry conducted over 5000 wind experiments trying to build a better windmill. He invented the mathematical windmill, which used gears to reduce the rotational speed of the blades. This design had greater lifting power and smoother pumping action, and the windmill could operate in lighter winds. Perry also started the Aermotor Company with LaVerne Noyes.
	The development of steel blades made windmills more efficient. Six million windmills sprang up across America as settlers moved west. Homesteaders purchased windmills from catalogs or traveling salesmen or, otherwise, built their own. Mills were used to pump water, shell corn, saw wood, and mill grain.
1888	Charles F. Brush used the first large windmill to generate electricity in Cleveland, Ohio. Windmills that produce electricity started to be called wind turbines. In later years, General Electric acquired Brush's company, Brush Electric Co.
1893	In Chicago, Illinois, the World's Columbian Exposition (also known as the Chicago World's Fair) highlighted 15 windmill companies that showcased their goods.
Early 1900s	Windmills in California pumped saltwater to evaporate ponds. This provided gold miners with salt.
1941	For several months during World War II, the Smith-Putnam wind turbine supplied power to the local community at "Grandpa's Knob," a hilltop near Rutland, Vermont. Its blades were 53 m (175 ft) in diameter.
1943	The Smith-Putnam wind turbine broke down, and the machine was shut down.
1945	The Smith-Putnam machine was restarted, but small cracks in the blade caused one blade to break; the turbine was shut down forever.
1950s	Most windmill companies in the United States went out of business.
1973	The Organization of Petroleum Exporting Countries (OPEC) oil embargo caused the prices of oil to rise sharply. High oil prices increased interest in other energy sources, such as wind energy.
1974–1982	With funding from the National Science Foundation and the US Department of Energy (DOE), the National Aeronautics and Space Administration (NASA) led an effort to increase wind power technology at the Lewis Research Center in Cleveland, Ohio. NASA developed 13 experimental wind turbines with four major designs: 1. the MOD-0A (200 kW); 2. the MOD-1 (2 MW, the first turbine in 1979 over 1 MW); 3. the MOD-2 (2.5 MW); 4. the MOD-5B (3.2 MW).

(continued)

Alternative Energy and Shale Gas Encyclopedia, First Edition. Edited by Jay H. Lehr and Jack Keeley.
© 2016 John Wiley & Sons, Inc. Published 2016 by John Wiley & Sons, Inc.

1978	Congress passed the Public Utility Regulatory Policies Act (PURPA) of 1978 to encourage the use of renewable energy and cogeneration facilities (plants that have another purpose besides producing electricity). PURPA requires utility companies to buy extra electricity from renewable and cogeneration facilities that meet certain qualifications, called qualifying facilities (QFs). The amount that a utility pays a QF must be equal to the cost that it would have taken the utility to produce the same amount of electricity, called the avoided cost.
1979	The first wind turbine rated over 1 MW (MOD-1) began operating; MOD-1 had a 2-MW capacity rating. The cost of electricity from wind generation was about 40 cents per kWh.
1980	The Crude Oil Windfall Profits Tax Act of 1980 further increased tax credits for businesses that used renewable energy. The Federal tax credit for wind energy reached 25%, rewarding those businesses choosing to use renewable energy.
1983	Because of a need for more electricity, California began using a contract system that allowed certain renewable and cogeneration facilities (or in other words, QFs) to lock into rates that would make electricity generated from renewable technologies, like wind farms and geothermal plants, more cost competitive. Prices were based on the costs saved by not building planned coal plants.
1985	Many wind turbines were installed in California in the early 1980s to help meet growing electricity needs and to take advantage of government incentives. By 1985, California wind capacity exceeded 1000 MW, enough power to supply 250,000 homes. These wind turbines were very inefficient.
1987	The MOD-5B was the largest wind turbine operating in the world—with a rotor diameter of nearly 100 m (330 ft) and a rated power of 3.2 MW.
1988	Many of the hastily installed turbines of the early 1980s were removed and later replaced with more reliable models.
1989	Throughout the 1980s, DOE funding for wind power research and development declined, reaching its low point in 1989.
1990	More than 2200 MW of wind energy capacity was installed in California—more than half of the world's capacity at the time.
1992	The Energy Policy Act of 1992 called for increased energy efficiency and renewable energy use and authorized a production tax credit of 1.5 cents per kWh for wind-generated electricity. It also reformed the Public Utility Holding Company Act to help make smaller utility companies more able to compete with larger ones.
1993	US windpower developed one of the first commercially available variable-speed wind turbines, the 33M-VS. The development was completed over 5 years, with the final prototype tests completed in 1992. The $20 million project was funded mostly by US windpower, but also involved Electric Power Research Institute (EPRI), Pacific Gas & Electric, and Niagara Mohawk Power Company.
1995	In a ruling against the California Public Utility Commission, the Federal Energy Regulatory Commission (FERC) refused to allow utilities to pay qualifying renewable facilities (QFs) rates that were higher than the utilities' avoided cost, the amount that it would cost the utility to produce the same amount of electricity. The DOE's Wind Energy Program lowered technology costs. DOE's advanced turbine program led to new turbines with energy costs of 5 cents per kWh of electricity generated.
Mid-1990s	Ten-year Standard Offer contracts written during the mid-1980s (at rates of 6 cents per kWh and higher) began to expire. The new contract rates reflected a much lower avoided cost of about 3 cents per kWh and created financial hardships for most qualifying renewable and cogeneration facilities (QFs). Kenetech, the producer of most of the US-made wind generators, faced financial difficulties; it sold off most of its assets and stopped making wind generators.
1999	Wind-generated electricity reached the 2000 MW mark.
1999–2000	Installed capacity of wind-powered, electricity-generating equipment exceeded 2500 MW. Contracts for new wind farms continued to be signed. The cost of electricity from wind generation was from 4 to 6 cents per kWh.
2003	Installed capacity of wind-powered, electricity-generating equipment was 4685 MW as of January 21.
2004	The cost of electricity from wind generation was 3 to 4.5 cents per kWh.
2005	The Energy Policy Act of 2005 strengthened incentives for wind and other renewable energy sources.
2006	DOE's budget for wind subsidies was about $500 million—about 10 times as much as the 1978 level.
2007	Wind power provided 5% of the renewable energy used in the United States. US wind power produced enough electricity, on average, to power the equivalent of over 2.5 million homes. Installed capacity of wind-powered, electricity-generating equipment was 13,885 MW as of September 30—more than four times the capacity in 2000.

Last Revised: September 2008.
Sources: Energy Information Administration, Energy in Brief: How much renewable energy do we use? August 2008
Office of Energy Efficiency and Renewable Energy, US Department of Energy, Wind & Hydropower Technologies Program, March 2008; American Wind Association, *Wind Energy Projects Throughout the United States*, March 2008.

PART II

SOLAR

19

SOLAR AIR CONDITIONING

Winston Garcia-Gabin and Darine Zambrano

19.1 INTRODUCTION

The need to increase the use of renewable energy sources is an issue that has received increased attention during the last decades. Air conditioners, for commercial and residential services, are widely used around the world, excluding only some regions because of their latitudes or countries due to the scarcity of economic resources.

According to the figures reported by [1], the energy demand for air conditioning is projected to increase rapidly over the 2000–2100 period, mostly driven by income growth. The associated CO_2 emission for both heating and cooling is expected to increase from 0.8 GtC in 2000 to 2.2 GtC in 2100, that is, about 12% of total CO_2 emissions from worldwide energy consumption. Solar air conditioning systems are a technology that allows the emissions to be reduced.

The sun is a very important source of energy; all of us know about the importance of the sun in our lives. There are a lot of applications for solar energy (see, e.g., [2]); however, its use for solar air conditioning is not very popular. Solar energy is a potential choice as energy source to deal with adverse effects to the environment and also because of its availability in countries and in seasons where air conditioning is much needed.

The production of cold by solar energy is one of the most attractive applications; for our physical comfort and for many other purposes cold is needed when the sun is shining. So, in some way, the demand matches the supply and thus storage problems are reduced. For this reason, and also considering that air conditioning does not need very low temperatures, interest in this field is increasing throughout the world. Although there are a large number of theoretical studies and

projects, very few operating solar air condition systems are available so far.

A particularly interesting application is the air conditioning in buildings, because in many cases there is an overlap between the peak cooling demand and the available solar energy, as well as air conditioning being a significant part of the electrical demand in buildings (around 35%). The use of solar energy for heating and cooling of buildings is especially attractive in cases that require the use of both these kinds of conditioning in winter and summer, respectively, and hence solar collectors are used throughout the year. The incorporation of heat and cold storage systems will be needed when there is no overlap between the air conditioning requirements and the high solar irradiation.

19.1.1 Air Conditioner

The necessity for humans to modify air conditions began in Roman times, when houses were designed to allow water circulation inside the walls of some houses in order to chill the rooms. In general, an air conditioner refers to an appliance or system that is able to modify the temperature and humidity of the air within a specific area by using cooling, heating, ventilation, and so on. In the following, we refer to air conditioner as the system that allows temperature regulation, mainly by cooling the air. In modern times, air conditioners are a popular technology, which is being used by more and more people. There are several kinds of air conditioners, which differ mainly in the working principle and the energy source. These can be operated by using solar energy, by generating electricity with photovoltaic panels, or by warming a liquid with solar collectors. An interesting introduction about

Alternative Energy and Shale Gas Encyclopedia, First Edition. Edited by Jay H. Lehr and Jack Keeley.
© 2016 John Wiley & Sons, Inc. Published 2016 by John Wiley & Sons, Inc.

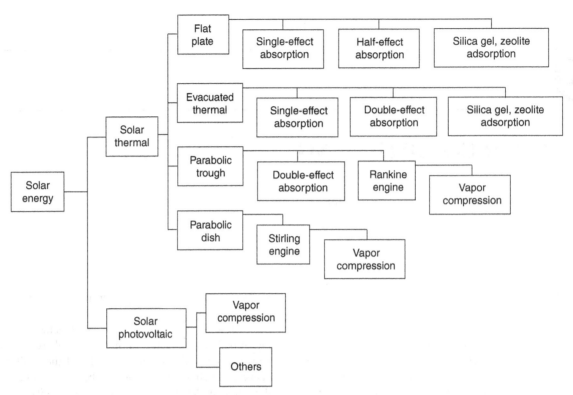

FIGURE 19.1 Solar cooling technology diagram.

the different options for producing cold by means of solar energy is presented in [3]. Different methods for producing cold using solar radiation as the primary source of energy are presented in [4, 5]. Figure 19.1 schematizes the most widespread options for solar cooling production. This chapter focuses on sorption (absorption and adsorption) air conditioning systems, and mainly on solar absorption systems.

19.2 REFRIGERATION PROCESS

The air refrigeration process consists in removal of heat from air. The general idea can be understood with the vapor compression cycle. It uses a condensable fluid called refrigerant, which is cyclically converted from a gas to a liquid and then to a gas. Heat is removed from the air inside of a room to transfer it to the outside air. The main components are an evaporator, a compressor, and a condenser. The refrigerant's molecules are packed in at constant entropy increasing the temperature, pressure, and density of the refrigerant. Thus, the working fluid leaves the compressor as a superheated, high-pressure gas and flows into the condenser. The condenser has metal fins all around it that allows superheated vapor to transfer heat to the surrounding outdoor air. As a result, the condenser first cools and removes the superheat and then condenses the vapor, causing the working fluid to transform into a liquid by removing additional heat at constant pressure and

temperature. When the refrigerant fluid leaves the condenser, its temperature has almost dropped back down to the outside temperature but it is now a liquid rather than a gas. The liquid refrigerant goes through the evaporator where its pressure abruptly decreases. This pressure drop causes liquid flash evaporation. To produce this evaporation, it extracts heat from the air around the evaporator because that heat is needed to separate the molecules of the refrigerant. The metal fins of the evaporator allow exchanging thermal energy with the surrounding air. A fan blows across the evaporator tubes chilling the warm air when it flows to the room being refrigerated. The refrigerant returns to the compressor inlet as a cool, low-pressure gas to complete the thermodynamic cycle and begins its trip all over again. In summary, the refrigerant releases heat into the outside air and absorbs heat from the inside air.

A key factor to describe the operation of a thermal system is the coefficient of performance (COP), which is defined as the cold produced per unit of heat. Vapor absorption cycle has a lower COP than vapor compression cycle. However, when heat is obtained from solar thermal energy, it is a profitable way for air conditioning. The absorption cycle is similar to the compression cycle, except that it does not have a compressor raising the pressure of the refrigerant vapor. In the vapor absorption cycle, an absorber that dissolves the refrigerant in a liquid substitutes the compressor. The typical combinations (refrigerant–absorbent) are ammonia–water and water–lithium bromide. First, the refrigerant is driven off the absorbent solution using solar thermal heat. Then, the refrigerant under low pressure is evaporated from the coils that are being chilled. The refrigerant is absorbed by an absorbent–refrigerant solution. Solar single-effect absorption systems have been commonly used despite their low COP and high initial cost [6]. They can operate at low temperatures that can be provided by flat plate solar collectors, which are cheaper and easier to operate than concentrating solar systems, and can provide cool water at the range needed for buildings and residential conditioning. On the other hand, photovoltaic driven systems have been too expensive with respect to the solar thermal ones. This trend is changing now, mainly due to the lower cost of the photovoltaic systems. Economic analyses for different solar cooling systems are presented in [7–9].

19.3 WORKING PRINCIPLE OF SOLAR SORPTION AIR CONDITIONERS

19.3.1 Desiccant Systems

A desiccant system is a form of dehumidification air conditioning. This system is considered as open sorption cycle, where the refrigerant is discarded from the system after providing the cooling effect. It is based on desiccant materials that attract moisture due to differences in vapor pressure; also, they can be dried or regenerated by adding heat supplied by solar thermal energy. Then, a cooling system is used to lower the temperature to the comfort value required at the occupied spaces. Thus, the desiccant system is combined with a conventional air conditioning system in which the desiccant removes humidity and the air conditioner reduces air temperature. In this configuration, the desiccant system reduces the work of the cooling system, because it is used only for lowering the temperature allowing an efficient cooling cycle. The desiccant materials can work in solid state (silica gel or zeolite) or liquid state (a mix of calcium chloride with water) to absorb water directly from the air. The advantage of liquid desiccant is that it can be used directly in cooling and dehumidification processes, resulting in a COP of 1.2, which is better than the COP obtained with solid materials.

19.3.2 Adsorption Systems

The adsorption systems produce chilled water for the air conditioning system based on adsorption/desorption process. They mainly include four processes (evaporation, adsorption/desorption, and condensation) working in two phases. During the first phase of the operation the refrigerant is evaporated at low pressure and low temperature in the evaporator, and heat is adsorbed by the refrigerant. Thus, the desired refrigeration effect is achieved. At this phase, the sensible heat and the adsorption heat are consumed by a cooling medium, which is usually water or air. In the second phase of the operation, also called regeneration, the refrigerant is condensed at the condenser and heat is released into the environment. The heat necessary for the regeneration process is supplied by thermal solar energy. In the adsorption systems, zeolite–water and silica gel–water are used for cold storage. This technology has the disadvantages of low specific cooling power and the lowest COP of all known solar air cooling systems, between 0.02 and 0.6.

19.3.3 Absorption Systems

The absorption refrigeration machines work much like a compression refrigerators, except they do not have a mechanical compressor. Instead, the evaporated refrigerant goes from the evaporator to the absorber, where it is diluted with an absorbent solution. The diluted solution is driven by a pump to the generator (note that pumps are more efficient than compressors), wherein the absorbent solution is regenerated by supplying external heat, so that the refrigerant is vaporized again, at a higher temperature and higher pressure. The working fluid in the absorption refrigeration system is a binary solution consisting of refrigerant and absorbent. In the absorption process, the binary solution absorbs the refrigerant vapor from a refrigerant liquid, and as a consequence, its pressure and temperature are reduced. The binary solution cannot absorb more vapor due to its saturation; for this reason, the refrigerant is separated from the diluted solution. This separation is done by heat input with a high pressure in the solution. In [10], there is a review of the refrigeration technology by absorption. A variety of working fluids are used; the two most common absorbent–refrigerant pairs are LiBr–water and water–ammonia. Single-effect absorption systems are limited in COP to about 0.7 for LiBr–water and 0.6 for ammonia–water. Double-effect absorption machines require higher input temperature than the single-effect ones. As a consequence, they need huge solar collector area to supply the heat necessary for their operation. The solar collectors' surface can be reduced by using systems with an improved COP, which can be achieved using heat sources with high temperatures. Evacuated tubes are used to obtain high temperatures. In this case, the high cost of the absorption machine and the solar collectors should be considered.

19.4 A DEEPER INSIGHT INTO SOLAR ABSORPTION AIR CONDITIONING

19.4.1 An Overview

Since 1970s, solar refrigeration and cooling technologies have received great interest. This interest was promoted by fuel prices as today, and nowadays there is also the need to use energies different from fossil energy and to reduce harmful emissions. Since then, many researchers have invested large effort in this field, and many significant works have been published. A comprehensive and pioneering survey of comparison by simulating different solar-powered absorption air conditioning systems is given in [11]. A survey about the development of the technologies for solar cooling and air conditioning is provided in [12]. An overview of this technology in Europe is presented in [13]. The authors in [4] make a review on different projects and solar refrigeration technologies applied in Latin America. A review of the state of the art about solar-assisted air conditioning of building is given in [14]. The main requirements on control schemes for thermal solar systems in Europe can be found in [13].

Solar technology has attracted interest from various research groups (e.g., [15–17]). The refrigeration technology based on cycles of absorption and adsorption, with single and multiple effects, is being continuously researched to improve the performance, lower the costs, and reduce the value of the hot water temperature required for operation. The absorption systems offer an exciting alternative to conventional compression chillers, because in these systems the main source of energy is heat rather than mechanical power. An evaluation of absorption refrigeration systems can be found in [18, 19]. Several air conditioning systems that use solar energy absorption machines-based cooling technologies and thermal technologies are presented in detail in [20]. This work reviews the operation of various air conditioning systems that use solar energy absorption machines with lithium bromide and water as the energy exchange fluid, since this combination is considered the most suitable to achieve the needs of air conditioning applications using solar energy. For air conditioning applications in which cold water is needed with more than 5°C, the combination of water (H_2O) as a refrigerant and lithium bromide (LiBr) as absorbent is the most common refrigeration cycle. Using ammonia as absorbent is more suitable for industrial cooling, since water can be cooled to temperatures less than −10°C.

19.4.2 Currently Available Technology

Many studies about real applications of solar air conditioning [13, 21, 22] have shown favorable results of feasibility. These studies will encourage the spread of these technologies, as they are more environmental friendly than the conventional ones and are commercially available.

The single-effect absorption chiller is mainly used for building cooling loads. Single-effect chillers can operate with hot water at lower temperature range than the double-effect absorption chiller. Double-effect chillers have a higher COP. Although double-effect chillers are more efficient than the single-effect machines they are obviously more expensive to purchase.

Traditionally, absorption machines have had a large cooling capacity (thermal power outputs from 100 kW to 5 MW, or even more); for a long time very few systems were available in a range of cooling capacity below 100 kW. However, a trend can be observed in the manufacturing market toward small absorption machines with low cooling capacity (20 kW or less). There are small absorption machines that can run with hot water at temperatures that can be achieved with the use of solar collectors.

When solar energy is used to drive the absorption chillers, the initial cost is increased significantly. Under these conditions solar cooling system costs are higher than conventional system, as reported in [13]. According to [3], the initial costs can be reduced by using low-temperature solar collectors. This is a way to improve the chillers' efficiency when they work with low temperatures.

Commercially available absorption chillers for air conditioning applications usually operate with a solution of lithium bromide in water and use steam or hot water as heat source. In the market, three types of chillers are available, the single-, the double-, and the triple-effect absorption systems, with typical cooling COPs of 0.7, 1.2, and 1.7, respectively. The most widely used technology is the single-effect system. The market is dominated mainly by United States and Japan; Yazaki, Carrier, Trane, and Daikin are some of the main manufacturers.

19.4.3 A Solar Absorption Air Conditioning Pilot Plant

A solar air conditioning plant located at Seville (Spain) is used as an application example of this technology. The plant consists of a solar field that produces hot water which feeds an absorption machine which generates chilled water and injects it into the air conditioning system, achieving a cooling power of 35 kW. The plant operates with three different thermal sources: solar energy, stored hot water, and gas heater, which can be combined or used independently. As can be seen in Figure 19.2, its main components are the solar system, auxiliary energy system, thermal load system, and cooling system based on adsorption cycle. The solar air conditioning plant configuration needs to be handled by a hybrid controller (e.g., [23, 24]), manipulating valves and pumps to allow the selection of the components to optimize the energy flow. For a detailed description see [25].

The primary energy source is the solar irradiation, which is used by the solar collectors to increase the temperature of the circulating water. The solar field is composed of 151 m^2 of flat collectors that work in the range of 60–100°C and supply a nominal power of 50 kW. A specific feature of the flat plate collectors is that they do not have any kind of concentration of energy incident. Flat collectors capture both the direct and diffuse radiation; however, they cannot track the position of the sun throughout the day. The hot

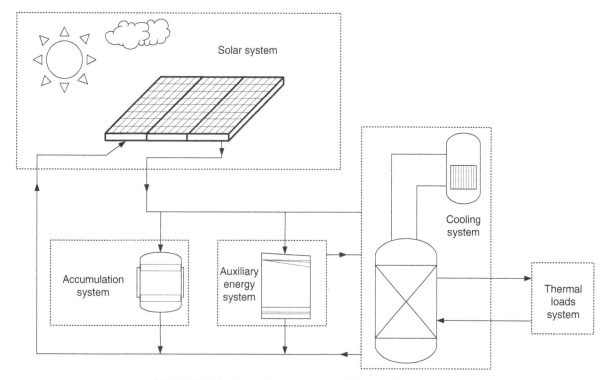

FIGURE 19.2   Solar absorption air conditioning plant diagram.

water flows to the cooling system and/or the accumulation system, which is composed of two isolated tanks. Also, the hot water can be combined with the water heated by the auxiliary energy system (a gas heater). Figure 19.3 shows the key components of the solar plant: the solar collectors, the tanks, and the absorption machine.

The cooling system is an absorption machine that works with water as cooling fluid and a lithium bromide/water solution (LiBr–H_2O) as absorbent fluid. The operating procedure of the machine requires that its inlet temperature is within the range of 75–100°C for chilled water production.

The machine has four different circuits: evaporator, generator, condenser, and absorber, where the energy exchanges that produce chilled water take place. The critical variables that have influence on operation are condenser temperature, which establishes the condenser and generator pressures; evaporator temperature, which establishes the evaporator and absorber pressures; generator temperature, which together with the condenser pressure fixes the solution concentration that leaves the generator; and absorber temperature, which together with the evaporator pressure fixes the solution concentration that enters the generator. The absorption

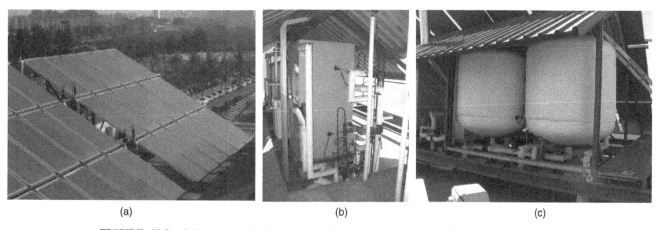

FIGURE 19.3   Solar, accumulation, and cooling systems: (a) solar collectors, (b) absorption machine, (c) tanks.

TABLE 19.1 Technical specifications of the absorption machine Yazaki WFC-10.

Circuit	Thermal power (kW)	Flow (L/h)	Input temperature (°C)	Output temperature (°C)
Evaporator	34.9	6,012	14	9
Condenser	84.8	14,580	29.5	34.5
Generator	49.8	8,568	88	83

equipment used in the air conditioning solar plant is AROCE Yazaki, model WFC-10 [26], which has the technical specifications shown in Table 19.1. The operating cycle of this absorption machine can be explained as follows. The water vapor, which comes from the evaporator, is absorbed by the solution of LiBr–H_2O in the absorber. Then, the saturated solution is driven to the generator. Inside the generator, the solution is heated and a portion of water evaporates from the boiling solution making the solution more concentrated. This concentrated solution is driven to the absorber again. As the temperature of the solution that comes from the generator is higher than the absorber temperature, and the solution going to the generator needs to be warmed up, a heat exchanger is used to preheat the diluted solution with the concentrated solution. In the condenser, the water vapor produced by the generator is condensed. From the condenser, the coolant goes through a pressure limiter to the evaporator, where water evaporation creates the cooling effect. Evaporation is produced because the solution vapor pressure in the absorber is lower than water temperature at the evaporator temperature. There are two pressure levels in the cycle; the first is a high pressure in the generator and the condenser, which is defined by condenser temperature. The second is a low pressure in the evaporator and the absorber, which depends on evaporator temperature. This model does not use internal pumps. The processes of evaporation and condensation are first-order transitional phases. These transitions are characterized by the fact that temperature and pressure remain constant during transition, while the entropy and volume change.

There are many factors affecting the performance of the absorption machine. For example, heat rejection is a critical factor in the COP of the absorption machine. In [27], COPs lower than the nominal values were obtained due to the low flow rate of the water coming from an aqueduct in the cooling of the absorption machine. Another common problem in the absorption machine with LiBr is the crystallization due to excessive temperature in the absorber. The COP is a function of several factors, such as chilled water temperature, the water temperature of the heat rejection system, and the temperature of the generator. Under normal operating conditions, single-effect absorption machines need inlet temperatures in the generator between 80°C and 100°C to achieve a COP around 0.7. In [27], it is shown that during the transient, the absorber

temperature has drastic changes. First, the temperature rises rapidly reaching 45°C, then after a few minutes, decreases rapidly to 30–35°C. This is produced by how the salt is mixed in the absorber.

19.4.3.1 Energetic Balance. The energetic balance of this plant was analyzed based on the performance test published in [25]. The results showed that the absorption machine can be driven mostly by solar energy when solar irradiation is higher than 700 W/m^2.

Besides, solar energy can contribute to reduce the auxiliary energy use when it is higher than 600 W/m^2. The solar energy was used by driving the absorption machine for more than 5 hours, of which for 3 hours the absorption machine was used without combination with auxiliary energy. Notice that for this test, there were no control algorithms optimizing the energy efficiency; so these results could have been better than the reported ones. One important point is that there is a significant coincidence between the working hours, when air conditioning is needed, and the absorption machine being driven by solar energy.

19.5 FINAL REMARKS: ADVANTAGES AND MAIN DRAWBACKS

Air conditioners are part of the everyday appliances in many countries, for residential, commercial, and industrial end uses. The benefits of air conditioners are clearly related with comfort and also with generating suitable air conditions, for example, for medical environments, working conditions of electronic appliances, drugs, and food conservation. There are many technologies well developed to this end. So, the main goal of the research in this field is to find new alternatives to improve the energy efficiency and the environmental effects.

Solar air conditioning appears as a suitable technology to reduce the environmental negative effect of the traditional ones, due to the use of renewable energy. However, this technology has some clear drawbacks, as those imposed by unfavorable climates, such as insufficient solar radiation or high cloudiness. Also, there are economic issues, because it implies a high initial investment. For this reason, nowadays, they are not considered to be competitive with conventional gas or electric air conditioners. Also, from an energetic point of view, they are not competitive with conventional systems (e.g., absorption air conditioners [20]). However, these technologies are continuously being developed, in order to reduce costs and to improve competitiveness. This could be helpful in extending their number of applications.

One advantage is that these systems allow the combination of solar energy with other sources of energy, that is, gas. It can be used in combination with advanced control schemes to deal with the environmental disturbance and improve the energy efficiency by selecting the best combination between energy sources. Another way to improve global efficiency is to use systems with improved COP; however, they require a higher temperature heat source which can be met by using concentrating collectors or evacuated tubes.

REFERENCES

[1] Isaac, M., and van Vuuren, D. P. (2009). Modeling global residential sector energy demand for heating and air conditioning in the context of climate change. *Energy Policy*, 37(2):507–521.

[2] Thirugnanasambandam, M., Iniyan, S., and Goic, R. (2010). A review of solar thermal technologies. *Renewable and Sustainable Energy Reviews*, 14(1):312–322.

[3] Kim, D. S., and Infante Ferreira, C. A. (2008). Solar refrigeration options—a state-of-the-art review. *International Journal of Refrigeration*, 31(1):3–15.

[4] Best, R., and Pilatowsky, I. (1998). Solar assisted cooling with sorption systems: status of the research in Mexico and Latin America. *International Journal of Refrigeration*, 21(2):150–159.

[5] Sayigh, A. A. M. (1992). *Solar Air Conditioning and Refrigeration*. Pergamon Press, New York.

[6] Weber, C., Berger, M., Mehling, F., Heinrich, A., and Nunez, T. (2014). Solar cooling with water–ammonia absorption chillers and concentrating solar collector—operational experience. *International Journal of Refrigeration*, 39:57–76.

[7] Ferreira, C. I., and Kim, D.-S. (2014). Techno-economic review of solar cooling technologies based on location-specific data. *International Journal of Refrigeration*, 39:23–37.

[8] Lazzarin, R. M. (2014). Solar cooling: PV or thermal? A thermodynamic and economical analysis. *International Journal of Refrigeration*, 39:38–47.

[9] Otanicar, T., Taylor, R. A., and Phelan, P. E. (2012). Prospects for solar cooling—an economic and environmental assessment. *Solar Energy*, 86:1287–1299.

[10] Srikhirin, P., Aphornratana, S., and Chungpaibul-patana, S. (2001). A review of absorption refrigeration technologies. *Renewable and Sustainable Energy Reviews*, 5:343–372.

[11] Wilbur, P. J., and Mancini, T. R. (1976). A comparison of solar absorption air conditioning systems. *Solar Energy*, 18:569–576.

[12] Grossman, G., and Johannsen, A. (1981). Solar cooling and air conditioning. *Progress in Energy and Combustion Science*, 7(3):185–228.

[13] Balaras, C. A., Grossman, G., Henning, H.-M., Infante Ferreira, C. A., Podesser, E., Wang, L., and Wiemken, E. (2007). Solar air conditioning in Europe—an overview. *Renewable and Sustainable Energy Reviews*, 11(2):299–314.

[14] Henning, H.-M. (2007). Solar assisted air conditioning of building—an overview. *Applied Thermal Engineering*, 27(10):1734–1749.

[15] Lamp, P., and Ziegler, F. (1998). European research on solar-assisted air conditioning. *International Journal of Refrigeration*, 21(2):89–99.

[16] Florides, G. A., Kalogirou, S. A., Tassou, S. A., and Wrobel, L. C. (2003). Design and construction of a $LiBr$—water absorption machine. *Energy Conversion and Management*, 44:2483–2508.

[17] Fan, Y., Luo, L., and Souyri, B. (2007). Review of solar sorption refrigeration technologies: development and applications. *Renewable and Sustainable Energy Reviews*, 11:1758–1775.

[18] Ayyash, S., Suri, R. K., and Al-Shami, H. (1985). Performance results of solar absorption cooling installation. *International Journal of Refrigeration*, 8(3):177–183.

[19] Casals, X. G. (2006). Solar absorption cooling in Spain: perspectives and outcomes from the simulation of recent installations. *Renewable Energy*, 31(9):1371–1389.

[20] Li, Z. F., and Sumathy, K. (2000). Technology development in the solar absorption air-conditioning systems. *Renewable and Sustainable Energy Reviews*, 4(3):267–293.

[21] da Silva, M. M. (2005). Solar air conditioning potential for offices in Lisbon. Master's thesis, University of Strathclyde.

[22] Henning, H. M. (2004). *Solar Assisted Air-Conditioning in Buildings: A Handbook for Planners*, 1st edition. Springer, New York.

[23] Zambrano, D., and Garcia-Gabin, W. (2008). Hierarchical control of a hybrid solar air conditioning plant. *European Journal of Control*, 14(6):464–483.

[24] Zambrano, D., Garcia-Gabin, W., and Camacho, E. F. (2010). Application of a transition graph-based predictive algorithm to a solar air conditioning plant. *IEEE Transactions on Control Systems Technology*, 18(5):1162–1171.

[25] Zambrano, D., Bordons, C., Garcia-Gabin, W., and Eduardo, F. C. (2007). Model development and validation of a solar cooling plant. *International Journal of Refrigeration*, 31(2):315–327.

[26] Yazaki. *Water fired chiller/chiller-heater*.

[27] Asdrubali, F., and Grignaffini, S. (2005). Experimental evaluation of the performances of a H_2O–$LiBr$ absorption refrigerator under different service conditions. *International Journal of Refrigeration*, 28:489–497.

20

ENERGY PERFORMANCE OF HYBRID COGENERATION VERSUS SIDE-BY-SIDE SOLAR WATER HEATING AND PHOTOVOLTAIC FOR SUBTROPICAL BUILDING APPLICATION

Tin-Tai Chow, Ka-Kui Tse, and Norman Tse

20.1 HYBRID SOLAR VERSUS SIDE-BY-SIDE SYSTEMS

20.1.1 Introduction of Hybrid Solar Collectors

The market of solar products is expected to grow continuously as long as carbon reduction remains a global issue. For the building sector, solar power generation is commonly adopted in low-carbon buildings, and there are more and more interests in zero carbon or even positive carbon building designs. The target of maximizing renewable power outputs favors the use of hybrid solar systems. A hybrid photovoltaic/thermal (PV/T) system combines photovoltaic (PV) and solar thermal (ST) devices together to form an integral cogeneration system. Both electricity and thermal energy are generated through a single hybrid unit [1]. An example of the PV/T collector incorporated with front glazing is shown in Figure 20.1. The presence of front glazing favors the capture of solar heat and strengthens the thermosyphon water circulation. With the bonding of solar cells on the thermal absorber surface, electricity is produced simultaneously. A hybrid unit thus has higher energy output per unit collector surface area [2].

PV/T system with flat plate thermal collector design is well applicable to buildings with low-temperature fluid heating demands. It is good for water preheating when higher water temperature is in need. In the warm climate cities, water-based ST system has wide applications since air heating in buildings is seldom required. The higher energy output density makes it especially suitable for use in multistory buildings, where the available space is inadequate for accommodating the quantity of solar collectors in meeting the overall building demand.

In solar physics, each PV cell type has its threshold photon energy corresponding to the particular energy band gap. Below this threshold, electricity conversion will not take place. Photons of longer wavelength do not generate electron–hole pairs, but instead dissipate as heat. For a PV module exposed to sunlight, the heat accumulation leads to its temperature increase and drop in cell efficiency. The drop is typically 0.4% per degree Celsius temperature rise for crystalline silicon cells. In this regard, the PV/T system is advantageous in that the liquid cooling effect lowers the PV operating temperature. As a result, the electricity yield is improved.

20.1.2 Real Building Studying Approach

Theoretical and experimental studies of PV/T were documented as early as in mid-1970s [3]. Despite the technical validity being concluded 40 years ago, it is only in recent years that it has attracted worldwide attention. Although the research and development work is escalating, those academic publications mentioning real project applications remain the minority [4]. For the commercial application of water-based PV/T collectors in real buildings, the cooling effect is climate dependent and varies with the hot water usage pattern. As a matter of fact, the merit of the cogeneration as compared to

Alternative Energy and Shale Gas Encyclopedia, First Edition. Edited by Jay H. Lehr and Jack Keeley.
© 2016 John Wiley & Sons, Inc. Published 2016 by John Wiley & Sons, Inc.

FIGURE 20.1   A glazed water-based PV/T collector connected for thermosyphon flow.

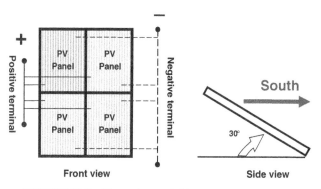

FIGURE 20.2   Front view and side view of the PV array.

the side-by-side system at a specific location is best examined through real case studies. The findings with respect to different climate zones and from a batch of building types having different functional usage will constitute an overall picture about the further R&D strategies in need. In view of these, an office building in Hong Kong was selected for comparative performance evaluation of the side-by-side installation against the hybrid cogeneration.

An experimental side-by-side PV+ST system was installed at the rooftop of an existing 10-storey office building at our university campus. By that time this building was under major retrofitting work. The ST subsystem provides warm/hot water to a number of pantries. Electricity generated from the PV subsystem serves a group of exit sign boxes of the building. The LED lighting of these exit sign boxes indicate the locations of the exit doors for emergency evacuation. Theoretical models of the required services system components were derived making reference to the physical plant running conditions. Further on, based on these mathematical models the computer simulation program of this study was developed. Concerning the real system performance, on-site operation data of the two subsystems, as well as the climatic data were monitored by built-in sensors and instruments. Their time-averaged values were periodically recorded by a data-logging station. The measured data were primarily used for verifying the accuracy and robustness of the self-developed simulation programs. Along this line, the well-validated models of the PV subsystem and the ST subsystem were integrated to form a hypothetical PV/T hybrid system model. With the availability of the three system

models and the typical meteorological year (TMY) data of Hong Kong, year-round performance simulations were done for each of the PV and ST subsystems, and for the hybrid PV/T system as well. In doing so, the physical scale of the services systems was adjusted to include all pantries and LED exit sign boxes as in the real building. These simulation results were then finally compared and evaluated.

20.2   EXPERIMENTAL SYSTEM DESCRIPTION

20.2.1   Photovoltaic Subsystem

Figure 20.2 shows the arrangement of the PV array. This was formed by four numbers of 220 W polycrystalline silicon PV panels in parallel connection. Figure 20.3 shows an indicative circuit diagram of the PV subsystem. Table 20.1 lists the basic parameters of a single PV panel, whereas those of the balance of system (BOS) components, that is, the storage battery, inverter, and charging controller, are given in Table 20.2.

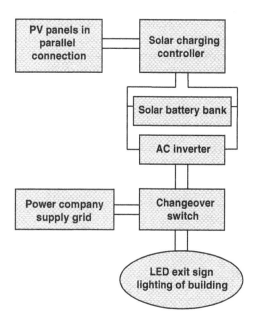

FIGURE 20.3   Brief circuit diagram of the PV subsystem.

TABLE 20.1   Physical parameters of individual PV panel.

Solar cell type	Polycrystalline silicon (p-si)
Nominal maximum power	220 W
Nominal maximum power voltage	29 V
Nominal maximum power current	7.57 A
Nominal open circuit voltage	36.4 V
Nominal short circuit current	8.19 A
Number of cells	60 connected in series
Size of the modules	1.643×0.99 m (L $\times$ W)
Size of each solar cell	0.156 m $\times 0.156$ m
Nominal cell efficiency	13% at STC

The battery bank consists of two lead-acid batteries which are connected in series, giving a storage capacity of 6 kWh. When the electricity generated by the PV array is in excess of the power demand of the inverter and the LED light system, the surplus will be charged into the battery bank. This process is regulated by the charging controller. At a time when the solar power generation is insufficient, make-up current will be discharged out of the batteries to top up the total load demand. The inverter is to convert the 24 V DC input power to 220 V AC output power to match the nominal operating voltage of the group of LED lights. The changeover switch interchanges the power supply of the LED lights between the PV subsystem and the main supply grid, depending on the terminal voltage level (as an indicator of the energy storage level) of the battery bank.

20.2.2   Solar Thermal Subsystem

Figure 20.4 shows the pipe arrangement of the ST subsystem. This arrangement is known as an indirect thermosyphon solar water heating system, in that two water circuits exist: (i) the thermosyphon fluid circulation between the ST collectors and the insulated storage tank, and (ii) the potable water circuit through which the water from the city main is warmed through a heat exchange coil and then discharged out of the storage tank.

TABLE 20.2   Physical parameters of the PV subsystem components.

Battery	Type	Lead-acid
	Nominal voltage	12 V
	Nominal capacity	250 AH
Inverter	Type	24 V DC to 220 V AC
	Conversion efficiency	90%
Solar charging controller	Maximum current load	45 A
	System voltage	12–48 V
	Self-consumption current	< 20 mA

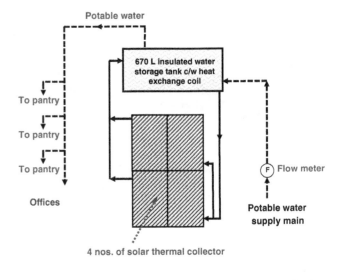

FIGURE 20.4   Indicative piping arrangement of the solar water heating subsystem.

In the first circuit, in principle the fluid (like water) circulation is induced by the natural buoyant force due to the fluid density difference at the two legs. Hence, no circulation pump is required. The water tank serves the dual purposes of water heating and storage. The entire ST collector array is assembled by four numbers of flat plate-and-tube-type solar collectors. Table 20.3 shows the physical parameters of this ST subsystem.

20.3   MATHEMATICAL MODELING

20.3.1   Electrical Subsystem Model

20.3.1.1   PV Array.   A PV panel can be modeled by the single-diode five-parameter approach [5]. Based on this the current–voltage (I–V) is expressed as

$$I = I_{pv} - I_0 \left[\exp\left(\frac{V + R_s I}{V_t a} \right) - 1 \right] - \frac{V + R_s I}{R_p} \qquad (20.1)$$

where I_{pv} is the photovoltaic current, I_0 is the saturation current, V_t is the thermal voltage, R_s is the series resistance, R_p is the shunt (parallel) resistance, and a is the diode ideality constant.

For a PV panel with high shunt (parallel) resistance R_p and negligibly small series resistance R_s, equation (20.1) can be simplified to [6]

$$I = I_{pv} - I_0 \left[\exp\left(\frac{V}{V_t a} \right) - 1 \right] \qquad (20.2)$$

TABLE 20.3 Design parameters of the thermosyphon solar water heating subsystem.

Solar collector		
Front glazing	Material	Low-iron glass
	Effective area	1.91 m^2
	Thickness	0.0032 m
	Air gap underneath	0.025 m
Thermal absorber	Material	Copper
	Dimension	1.84 × 0.95 m (L × W)
	Base plate thickness	0.002 m
	Number of riser tubes	8
Back thermal insulation	Material	Fiber glass
	Thickness	0.05 m
Insulated water storage tank	Type	Horizontal cylindrical
	Water storage capacity	670 L
	Thermal insulation material	Polyurethane
	Insulation thickness	0.05 m
	No. of headers of heat exchange coil set	2
	No. of parallel connected tube rings	28

If a PV array is composed of N numbers of rows and each row carries m numbers of parallel panels, in each of which n numbers of cells are connected in series, the maximum power output derived from the five-parameter model is given by

$$P_{max} = \{V_{max}I_{max}\}mN$$
$$= \left\{ V_{max}I_{pv} - V_{max}I_o \left[\exp\left(\frac{V_{max} + R_s I_{max}}{V_t a} \right) - 1 \right] - \frac{V_{max}^2 + V_{max}R_s I_{max}}{R_p} \right\} mN \quad (20.3)$$

And further based on the simplified equation (20.2), equation (20.3) is reduced to

$$P_{max} = \{V_{max}I_{max}\}mN$$
$$= \left\{ V_{max}I_{pv} - V_{max}I_o \left[\exp\left(\frac{V_{max}}{V_t a} \right) - 1 \right] \right\} mN \quad (20.4)$$

For an unshaded PV panel exposed to instantaneous solar irradiation level Q, the light-generated current (photocurrent) is given by

$$I_L = I_{pv} = \frac{Q}{Q_{ref}} I_{sc} \quad (20.5)$$

where Q_{ref} is the reference solar irradiation level of 1000 W/m^2.

20.3.1.2 BOS Components. Figure 20.5 shows the electrical system model, in which the current flowing route from the power generation part (i.e., the PV array) to the storage part (i.e., the battery bank) and to the final power loading part (i.e., the inverter + LED system) is indicated.

(a) PV array output

The power output of the PV array is given by

$$P_{pv} = V_{pv}I_{pv} \quad (20.6)$$

(b) Solar charging controller

The power delivered from the controller is

$$P_C = V_{pv}(I_{pv} - I_{c,self}) \quad (20.7)$$

where the self-consumption current $I_{c,self}$ of the controller can be obtained from equipment catalog, and is 20 mA in this case study.

(c) LED exit sign lights

The total power demand P_{LED} of 36 numbers of LED lights in the exit sign boxes is 136 W round the clock.

(d) 24 V DC to 220 V AC inverter

Based on the manufacturer data and the recommendation in Ref. [7], the efficiency of the DC to AC inverter in this study is

$$\eta_{INV} = \frac{V_{ac}I_{ac}\cos\phi}{V_{dc}I_{dc}} = 0.9 \quad (20.8)$$

where ϕ is the phase angle. Since the LED system is simple and carries no induction motor,

$$\cos\phi \approx 1.$$

(e) Solar battery bank

When the available power P_C from the controller is in surplus of the power consumption of the inverter plus LED system, the power delivered to the battery is given by

$$P_B = P_C - P_{INV+LED} \quad (20.9)$$

and

$$P_B = V_B I_B \quad (20.10)$$

The state of charge (SOC) of a lead-acid rechargeable battery during the charging process is given by [8]

$$SOC(t+1) = SOC(t)[1 - \sigma(t)] + \frac{I_B(t)dt\,\eta_c(t)}{C_B} \quad (20.11)$$

where C_B is the nominal capacity of the battery and $\sigma(t)$ is the hourly self-discharge rate—a variable which depends on

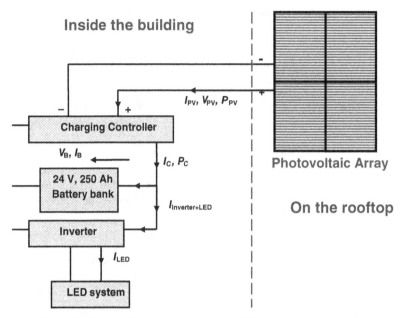

FIGURE 20.5   Indicative diagram for the PV electrical system model.

the amount of charge accumulated in the battery and also on its state of health [9]. A proposed value of 0.02% is used in the study.

The charging efficiency is determined by the battery working conditions as well as the charging current. It can be shown that

$$\eta_c(t) = 1 - \exp\left\{\frac{a_o[\text{SOC}(t) - 1]}{\dfrac{I_B(t)}{I_{10}} + b_o}\right\} \quad (20.12)$$

where a_o, b_o, and I_{10}, are the parameters subjected to the working conditions of the battery. In general, the charging efficiency $\eta_c(t)$ is around 0.65–0.85, and equal to 1.0 during discharge [9]. In this study, an averaged value of 0.75 was used in the numerical simulation.

During the discharging process, the battery SOC is given by

$$\text{SOC}(t + 1) = \text{SOC}(t)[1 - \sigma(t)] + \frac{I_B(t)\text{d}t}{C_B} \quad (20.13)$$

Here the charging efficiency term, $\eta_c(t)$, is removed. The battery current $I_B(t)$ is flowing out of the battery and hence is negative in value.

In addition to the determination of SOC, a definite mathematical relationship of the voltage, current, and SOC should be made available in order to model this electrical

subsystem well. Readers can refer to Refs. [10, 11] for detailed information about this relationship.

The terminal voltage (i.e., the floating charge voltage) of a battery can be represented by the sum of its open circuit voltage E_{oc} and the voltage drop across the battery internal resistance R_B, that is,

$$V_B = E_{oc} + I_B R_B \quad (20.14)$$

The open circuit voltage has a logarithmic relationship with the battery SOC, in that

$$E_{OC} = \text{VF} + \text{VS}\,\log(\text{SOC}) \quad (20.15)$$

where VF is the full charge rest voltage and VS is an empirical constant. VF and VS carry different numerical values at the charging and discharging processes.

The instantaneous power loss owing to the self-loss of battery and during the charging process is given by

$$P_{\text{loss}}(t) = V_B(t)\{\sigma(t)\text{SOC}(t)C_B + I_B(t)[1 - \eta_c(t)]\} \quad (20.16)$$

and the corresponding energy losses over a time period t is then

$$E_{\text{loss}} = \int_0^t V_B(t)\{\sigma(t)\text{SOC}(t)C_B + I_B(t)[1 - \eta_c(t)]\}\text{d}t \quad (20.17)$$

Within a short period of time dt, the above power losses as well as the energy losses can be represented by their averaged values, that is,

$$\overline{P_{\text{loss}}} = \overline{V_{\text{B}}} \left[\sigma \, \overline{\text{SOC}} \, C_{\text{B}} + \overline{I_{\text{B}}}(1 - \eta_{\text{c}}) \right] \quad (20.18)$$

$$\overline{E_{\text{loss}}} = \overline{V_{\text{B}}} \left[\sigma \, \overline{\text{SOC}} \, C_{\text{B}} + \overline{I_{\text{B}}}(1 - \eta_{\text{c}}) \right] \mathrm{d}t \quad (20.19)$$

where $\overline{V_{\text{B}}}$ is the averaged battery voltage, $\overline{I_{\text{B}}}$ is the averaged charging current, and $\overline{\text{SOC}}$ is the averaged SOC of the battery over the time period $\mathrm{d}t$.

During the discharging process, the second terms at the right hand side of equations (20.16), (20.17), (20.18), and (20.19) are reduced to zero.

20.3.2 Solar Thermal Subsystem Model

The flat plate-and-tube-type solar thermal collector can be modeled by the finite difference control volume method [12]. The whole collector is tactfully divided into three subcomponents. Each of the subcomponents is assumed to have uniform physical properties and working conditions. More details about the thermal collector model are given in Section 20.4.3 below.

The modeling technique for the fluid flow in the thermosyphon loop is based on the work of Koffi et al. [13]. By balancing the buoyancy pressure due to the density difference of water inside the collector system against the pressure losses due to fiction within the entire thermosyphon loop, the following quadratic equation can be obtained:

$$\frac{8k\dot{m}_{\text{f}}^2}{\rho_{\text{m}}\pi^2 d_{\text{ex}}^4} + (1 + \phi)\frac{128 v_{\text{m}} L_{\text{cl}} \dot{m}_{\text{f}}}{\pi N d_{\text{cl}}^4} - g\rho_{\text{m}}\beta' \sin(\theta)$$
$$\frac{L_{\text{cl}}}{2}(T_{\text{out}} - T_{\text{in}}) = 0 \quad (20.20)$$

where k is the singular pressure coefficient; N is the total number of riser tubes; d_{cl} and L_{cl} are the tube diameter and length at the thermal collector, respectively; θ is the collector tilt angle; d_{ex} is the tube diameter of the heat exchange coil in tank; ρ, v, and β' are the density, kinematic viscosity, and coefficient of thermal expansion of water, respectively; and T_{in} and T_{out} are the temperatures of fluid that enters and leaves the collector, respectively.

By solving the above quadratic equation, the instantaneous mass flow rate $\dot{m}_{\text{f}}$ of water inside the thermosyphon loop can be obtained.

The efficiency of a flat plate thermal collector can be expressed as [14]

$$\eta = F_{\text{R}}(\tau\alpha)_\theta - F_{\text{R}}U_{\text{L}}(T_{\text{in}} - T_{\text{a}})/Q_\theta \quad (20.21)$$

where F_{R} is the collector heat removal factor, $(\tau\alpha)_\theta$ is the transmittance-absorptance product of the collector surface at a tilted angle of θ, U_{L} is the upward heat loss coefficient, T_{a}

is the ambient air temperature, and Q_θ is the incident solar irradiation on a surface with tilted angle θ.

For the PV/T and ST systems, the transient energy balance equation of the insulated storage water tank at temperature T_{tk} can be expressed as

$$\rho_{\text{tk}}V_{\text{tk}}(T_{\text{tk}})C_{\text{p}}\frac{\mathrm{d}T_{\text{tk}}}{\mathrm{d}t} = \dot{m}_{\text{f}}C_{\text{p}}(T_{\text{f,in}} - T_{\text{f,out}})$$
$$+ \dot{m}_{\text{p}}C_{\text{p}}(T_{\text{p,in}} - T_{\text{p,out}})$$
$$+ U_{\text{tk}}A_{\text{tk}}(T_{\text{a}} - T_{\text{tk}}) \quad (20.22)$$

where $\dot{m}_{\text{f}}$ is the potable water mass flow rate, U_{tk} and A_{tk} are, respectively, the heat loss coefficient and inside surface area of the tank body.

20.4 NUMERICAL MODEL DEVELOPMENT AND VERIFICATION

20.4.1 PV Subsystem

20.4.1.1 PV Array. The following discussions illustrate the importance of verifying the numerical models before their use in real cases. When the series resistive effect is negligible in the PV array model, equation (20.4) can be applied to determine the maximum PV power output. With the use of the parameters in Table 20.1, the ideal single-diode model generates the family of current–voltage (I–V) characteristic curves as shown in Figure 20.6a and the power–voltage (P–V) characteristic curves as in Figure 20.6b. These are for the standard panel working temperature of 25°C. It can be seen that under the reference 1000 W/m² solar irradiation level, the maximum power output of the PV array is 880 W. This is equal to the total nominal power of the four PV panels (at 220 W each). The maximum power voltage and current are 29 V and 30.28 A (i.e., 4 × 7.57A), respectively. The maximum power, voltage, and current determined by the numerical model well match the nominal values quoted in the product catalog.

The actual performance of the PV panels as installed in this campus building was determined through quality tests at the system commissioning stage. The performance data are listed in Table 20.4. Based on these, the actual I–V and P–V curves are generated and shown in Figures 20.7a and 20.7b. Great discrepancy can be observed in these two graphical plots against the corresponding ones in Figures 20.6a and 20.7b. For instance, at 1000 W/m² irradiation level, the maximum power output reduces to 780 W. This occurs when $V = 26$ V and $I = 30$ A. This outcome indicates that the use of the simplified equation (20.4) is inappropriate for this real building project. Instead, equation (20.3) from the five-parameter model should be used.

For better illustration, the experimental measurements and simulation results of power output of the PV array at 36°C

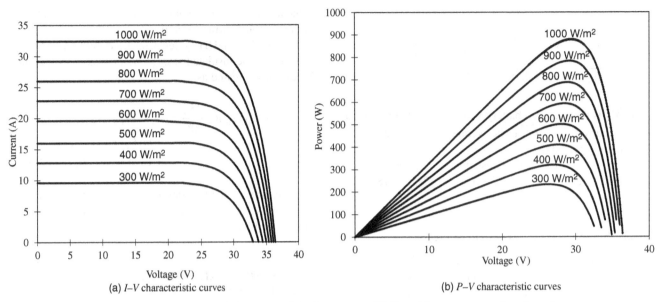

(a) *I–V* characteristic curves

(b) *P–V* characteristic curves

FIGURE 20.6 Characteristic curves of the PV array at 25°C working temperature based on the ideal single-diode model without resistive effect.

TABLE 20.4 **Actual performance parameters of the (PV) panels in a quality test before handover.**

Open circuit voltage	35.4 V
Short circuit current	8.55 A
Shunt (parallel) resistance	166 ohms
Series resistance	0.584 ohms

working temperature are compared in Figure 20.8. The simulation results from both the ideal single-diode model using equation (20.4) and the single-diode five-parameter model using equation (20.3) are displayed. It can be seen that the results obtained from the five-parameter model align well with the experimental results. The computations were based on the model parameters listed in Table 20.5. It is worth to notice that according to Figure 20.8 the power output at 1000 W/m² incident solar irradiation is around 547 W. This is different from the prediction of around 760 W by the five-parameter model shown in Figure 20.9. Such further reduction can be owing to a shift of the operating point in the *I–V* curve. As the nominal voltage of the battery bank is 24 V, during the charging process in daytime the battery voltage normally floats between 25 and 28V. Referring to Ref. [15], the PV module output voltage has to match the

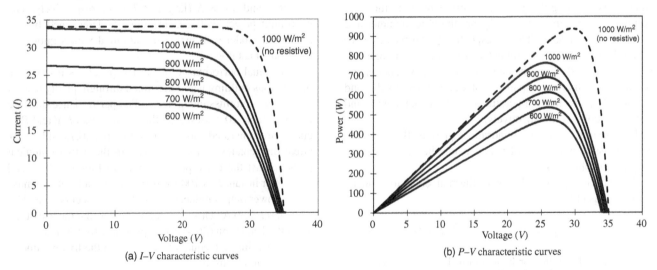

(a) *I–V* characteristic curves

(b) *P–V* characteristic curves

FIGURE 20.7 Characteristic curves of the PV array at 25°C working temperature based on the single-diode model with resistive effect.

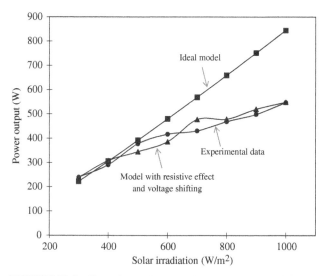

FIGURE 20.8 Experimental versus simulation results of the PV array power output at 36°C working temperature.

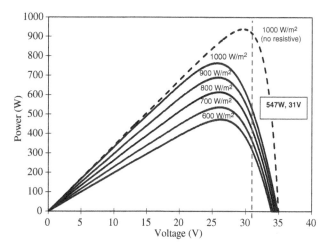

FIGURE 20.9 P–V characteristic curves of the PV array at 36°C working temperature based on the single-diode model with resistive effect.

critical voltage region of the battery. The voltage drops in the blocking diode and in the wiring resistance also have to be taken into account. Therefore, the PV output voltage has to be sufficiently higher than the battery nominal voltage. In the above case, the operating point of the PV array shifts to 31 V. So the power output further drops to 547 W.

Figure 20.10 shows the hourly distribution of electricity output of the PV subsystem on a sunny day. It can be seen that the daily profile obtained from the simulation results agrees well with the experimental data. Hence, good quality of the self-developed simulation program of the PV subsystem has been verified.

20.4.1.2 Battery and LED System. Figure 20.11 presents the daily performance of the BOS in seven consecutive days. The uppermost line indicates the variation of daily electrical energy conversion from the solar irradiance. The two in the

middle are the actual and simulated time duration within a day that the LED lights are powered by the PV subsystem. The line at the bottom end shows the daily power loss of the system. Included in the loss term are the electrical energy loss during the charging process, the self-discharge loss of the battery bank, and the loss due to the operation of the inverter. It can be seen that the predicted time duration from simulation is in good agreement with the experimental record. These are in the range of 7.5 hours to 19.5 hours, depending on the solar availability within the week.

20.4.2 Solar Thermal Subsystem

Figure 20.12 shows the model verification of water temperature in tank, as a result of thermal energy exchange. Here

TABLE 20.5 Simulation parameters for the battery model, inverter, and the LED system.

Battery	Parameters	Charging mode	Discharging mode
Open circuit voltage	VF	13.44 V	12.86 V
	VS	0.699 V	0.712 V
Internal resistance	r_1	0.0219 Ω	0.0039 Ω
	r_2	0.0565Ω	0.0335 Ω
	r_3	85.638 Ω^{-1}	2.769 Ω^{-1}
	r_4	72.670 Ω^{-1}	−102.99 Ω^{-1}
Capacity	207 AH		
Inverter	Efficiency	0.83	
LED system	Power	136 W	

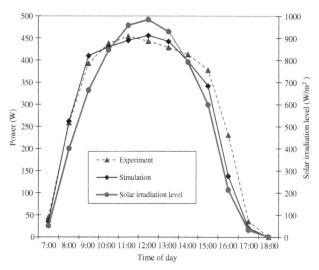

FIGURE 20.10 Comparison of the experimental and simulation results of the PV array hourly power output on a sunny day.

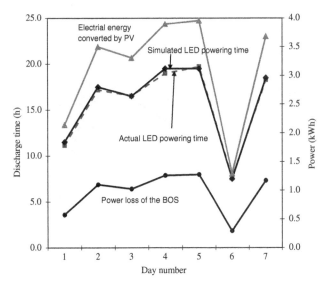

FIGURE 20.11   Experimental and simulation performance data of the components of the PV system for seven consecutive days.

the results of two consecutive days for the water temperature at the upper portion of the stratified tank are shown. Since the ST subsystem model has included the thermal loss of the storage tank during the nighttime, both daytime and nighttime simulation results were obtained. It can be seen that the deviations at any time instant between the two temperature lines are small. The water temperature rise at the upper layer during the daytime can reach 20°C while the thermal loss at night can be a 10°C drop. Good quality in model prediction is again demonstrated.

20.4.3   PV/T System Model Development

After the successful verification of the side-by-side simulation models, they were integrated to develop a corresponding hybrid PV/T system model. In principle, all physical features

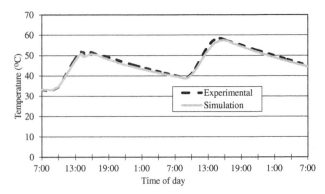

FIGURE 20.12   Thermal model validation—comparison of upper layer water temperature in storage tank.

of the PV/T system are kept the same as in the thermosyphon solar water heating system, except that the ST collectors are replaced by the water-based PV/T collectors. Hence their system thermal models are virtually the same. Figure 20.13 shows the cross-section view of the flat plate-and-tube-type PV/T collector. The PV encapsulation and the riser tubes are at the opposite side of the thermal absorber plate.

In our finite difference model, the PV encapsulation and the thermal absorber plate are treated as one single node. Hence, the PV/T collector is divided into three nodes, namely (i) the front glazing, (ii) the PV encapsulation plus thermal absorber, and (iii) the fluid (water) inside the copper tubing. The transient energy balanced equation of the (PV encapsulation + thermal absorber) as one element is expressed as

$$\rho_{ab}\delta_{ab}C_{ab}\frac{dT_{ab}}{dt} = Q\alpha_{ab} - E(Q,T_{ab}) + (h_{ab-g} + h_{r,ab-g})$$
$$\times (T_g - T_{ab})$$
$$+ h_{ab-f}A_{f,1}(T_f - T_{ab}) + \frac{T_a - T_{ab}}{R_{in}}$$

$$(20.23)$$

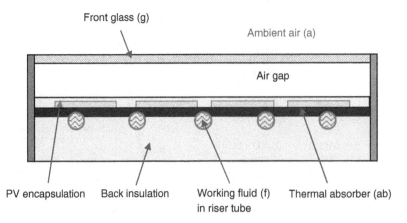

FIGURE 20.13   Cross-section view of hybrid photovoltaic/thermal collector.

where the term $E(Q, T_{ab})$ is the instantaneous electrical power generated per unit collector surface area. This is dependent on the incident solar irradiation level Q and the absorber temperature T_{ab}, and can be omitted in the ST collector model. $A_{f,1}$ is the area ratio between the wet surface of tubular water flow and the flat absorber plate.

The transient energy balanced equations that represent the front glazing node "g" and the working fluid node "f" are given by equations (20.27) and (20.28) as follows [16]:

$$\rho_g \delta_g C_g \frac{dT_g}{dt} = Q\alpha_g + (h_{wind} + h_{r.g-a})(T_a - T_g) + (h_{ab-g} + h_{r,ab-g})(T_{ab} - T_g) \qquad (20.24)$$

$$\rho_f C_f \frac{dT_f}{dt} = (h_{ab-f})A_{f,2}(T_{ab} - T_f) - \rho_f u_f C_f \frac{\partial T_f}{\partial L} \qquad (20.25)$$

where h_{wind} is the wind-induced convective heat transfer coefficient, u_f is the flow velocity of water along tube direction L, and $A_{f,2}$ is the area ratio between the wet surface of tubular water flow and the tube cross-section.

20.5   CASE STUDY FORMULATION

20.5.1   Cases Description

With the developed computer programs of the PV subsystem, ST subsystem, and the hybrid PV/T system, full-scale building energy simulation can be performed. Here the scope of services is extended to include all LED exit sign boxes and office pantries of this building block, that is,

(i) the total number of LED exit sign lights increased from 36 to 83; and

(ii) the total number of pantries being served increased from 5 to 10.

The annual energy performance between PV + ST and PV/T were compared making use of the following four hypothetical cases:

Case 1: Side-by-side system 1 (PV + ST 1): this includes a PV subsystem consisting of eight PV panels (with 13.0 m² overall panel area) and a ST subsystem consisting of eight ST collectors (with 15.28 m² overall collector area).

Case 2: PV/T system 1 (PVT 1): the PV/T system consisted of eight water-based PV/T collectors (with 15.28 m² overall collector area).

Case 3: Side-by-side system 2 (PV + ST 2): this includes a PV subsystem consisting of eight PV panels (with 13.0 m² overall panel area) and a ST subsystem consisting of eight ST collectors (with 13.0 m² overall

collector area); compared to the ST collector in case 1 each ST collector is shortened by 0.3 m in length.

Case 4: PV/T system 2 (PVT 2): the PV/T system consisted of nine water-based PV/T collectors (with 17.19 m² overall collector area).

In the above cases, the nominal cell conversion efficiency of the PV panels is 13% at STC. The ST collectors are of flat plate-and-tube design as in the experimental study. In principle, the solar collector systems of cases 1 and 2 are worked out based on the emphasis on the similarity in material use in the solar collectors. The systems of cases 3 and 4 are designed under the emphasis on similarity in annual electricity savings for the operation of the LED exit sign lights and hot water provision at the pantries. So the comparison of system performance can be done in two pairs: case 1 versus case 2 and case 3 versus case 4.

20.5.2   Electrical and Hot Water Services Systems

The electrical and water heating system provisions in the above four cases were modified to match the expanded scope of services. Accordingly, the number of solar battery has been increased from 2 to 4. Same as before, the resistive effect and the voltage shifting effect due to the mismatch of the operation voltages between the PV array and the solar battery are included. A constant power conversion efficiency of 90% has been used for the inverter operation.

Zero hot water demand is assumed during Saturday, Sunday, and public holidays. All working days are having hot water consumption profile as shown in Figure 20.14. This is constructed based on the hot water demand histogram for office building described in Ref. [17]. The total daily hot water consumption is 520 L. This is around 81% of the storage tank net water holding capacity of 640 L. Electric water heaters of 95% assumed energy conversion efficiency are provided at the pantries for final water heating. This allows a full comparison of the overall electricity saving of the combined system services.

20.5.3   Environmental Parameters

The Typical Meteorological Year (TMY) data file of Hong Kong available at the US Department of Energy website [18] was used in all year-round simulation runs. The hourly solar irradiation falling on the solar collectors are computed using the Perez's tilted surface model [19].

20.6   RESULTS AND DISCUSSION

Table 20.6 summarizes the simulation results of the above four cases.

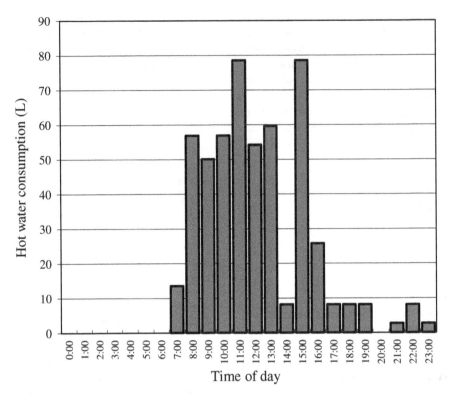

FIGURE 20.14 Hot water consumption histogram for year-round performance simulation.

20.6.1 Case 1 versus Case 2

The commonality between case 1 and case 2 is that the material usages for the constructions of the solar energy systems are relatively the same. For case 1 (PV + ST 1), the annual electrical output from the PV array is 1614.6 kWh. The monthly averaged daily conversion efficiency is ranged from 9.52% to 10.12%. The annual servicing hour of the LED exit sign lightings by the PV subsystem (with battery backup) is 3989 hours. This is equivalent to 1250.8 kWh electricity saving. The system loss is therefore 363.8 kWh. Hence, the overall efficiency of the electrical subsystem is 77.5%. With the ST subsystem, the annual thermal energy saved is 11545.8 MJ. The corresponding electricity saving in water heating is 3376 kWh. Hence in case 1, the overall annual electricity saving for the combined services is 4626.8 kWh.

For case 2 (PV/T 1), the annual electrical power output from eight numbers of PV/T collectors is 1338.8 kWh. The monthly averaged daily conversion efficiency from solar energy to electricity is ranged from 7.76% to 8.73%. These figures are lower than those in case 1 primarily because of the presence of the front glazing and air gap in the PV/T collectors. The actual amount of solar irradiance falling on the surface of the PV encapsulation is reduced as compared with the plain PV subsystem in case 1. The corresponding annual thermal energy gain of the hybrid system is 9621 MJ which is also lower than in case 1. Since a part of the incoming solar energy is converted into electricity by the solar cells, the heat absorbed at the thermal absorber is reduced. In case 2, 2813 kWh of electricity can be saved for water heating purpose. The overall annual electricity saving for the LED lights and water heating combined is 15.9% less than that in case 1.

TABLE 20.6 Summary of annual energy performance and electricity saving for different system designs.

Case	No. of PV panels	No. of ST collectors (area of each collector)	No. of PV/T collectors collectors	Annual PV output (kWh)	Annual LED electricity saving (kWh)	Annual thermal gain (MJ)	Annual electricity saving for water heating (kWh)	Overall electricity saving (kWh)
1	8	8 (1.91 m²)	0	1614.6	1250.8	11546	3376	4626.8
2	0	0	8	1338.8	1078.1	9621	2813	3891.1
3	8	8 (1.62 m²)	0	1614.6	1250.8	9747	2850	4100.8
4	0	0	9	1504.8	1185.9	9763	2855	4040.9

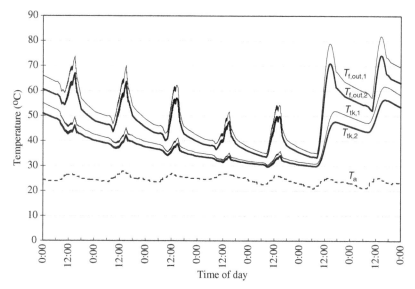

FIGURE 20.15   Case 1 versus case 2: comparison of water tank temperature profiles of the side-by-side and PV/T systems in a typical week.

Selected temperature profiles for cases 1 and 2 during a 7-day period are shown in Figure 20.15. The plots include

(i) Average water temperatures in stratified storage tank ($T_{tk,1}$ and $T_{tk,2}$);

(ii) Temperatures of hot water delivery to occupants ($T_{f,out,1}$ and $T_{f,out,2}$); and

(iii) Outdoor air temperature (T_a).

The last 2 days shown in this figure are Saturday and Sunday. Hence there is no water consumption on these 2 days and the four temperature profiles as shown are completely smooth. The toughs of the water temperature profiles of the first five consecutive days represent the peak hot water consumption rates within that time interval. It can be seen that for case 2, the water temperature levels are several degrees Celsius lower than the ST subsystem in case 1. This is because a portion of the incident solar irradiance is converted into electricity. Since the PV conversion efficiency is around 8–9%, the drop in hot water temperature delivery of the PV/T system in case 2 is not severe compared to the ST subsystem in case 1.

The simulation results show that there are minor reductions of both electrical and thermal energy output of the PV/T system (case 2) as compared to the side-by-side system (case 1). The electrical energy output is reduced by 17.1% while the thermal energy output is reduced by 16.7%. Compared with the conventional system design without solar application, the overall electricity saving is reduced from 4626.8 kWh (for case 1) to 3891.1 kWh (for case 2), and hence an overall reduction by 15.9%. However, in terms of unit collector surface area, the annual electricity saving of the PV/T

system is 254.6 kWh/m² while that of the side-by-side system is only 163.6 kWh/m². The percentage of improvement is 35.7% which is substantial.

20.6.2   Case 3 versus Case 4

In case 3 (PV + ST 2), the size and the numbers of the plain PV panels are the same as those in case 1. The number of ST collectors remains the same, but the length of each ST collector is shortened by 0.3 m. The design of the side-by-side system aims to provide the same amount of annual electricity saving as compared to a similar PV/T system (case 4). The performance of the PV subsystem in this case is the same as in case 1. Due to a reduction of the ST collector area, the annual thermal energy gain reduces to 9747 MJ. The equivalent annual electricity consumption by the electric water heater is 2850 kWh. The overall annual electricity saving is 4100.8 kWh. The annual electricity saving per unit solar collector area reaches 157.7 kWh/m².

Case 4 is a hybrid PV/T system with nine numbers of PV/T collectors. This is to achieve the criteria of equivalent amount of annual electricity saving for case 4 as compared to case 3. Here the size of the PV/T collector is taken the same as the ST and PV/T collectors in cases 1 and 2. The overall annual electricity saving by adopting this PV/T system is 4040.9 kWh. The difference of electricity saving between case 3 and case 4 is therefore less than 1.5%. Since there is one more PV/T collector in case 4 than in case 2, the annual electricity output of the hybrid system in case 4 increases from 1338.8 to 1504.8 kWh. However, there is a vey minor reduction of the PV conversion efficiency. The monthly averaged daily conversion efficiency of the PV encapsulation in this PV/T system is ranged from 7.73% to 8.72%. As there is one extra PV/T collector in case 4, the ratio of water storage tank capacity to collector area slightly decreases. When the daily hot water drawing pattern and the amount of hot water consumption in both cases 2 and 4 remain the same, the working temperature of the PV/T collectors in case 4 is slightly higher throughout the year. This has a negative effect on the energy performance. The efficiency of the electrical subsystem is 78.8%. Since the electricity output from the PV/T system is lower, there is less extra power current charging into the battery bank. The power loss in the battery charging process is reduced. Thus the electrical energy loss in the BOS is only 21.2%. The annual thermal energy gain is 9763 MJ. This gives the electricity saving of 2855 kWh for hot water production. The overall electricity saving on unit collector area bases is 235.1 kWh/m^2, which is close to case 1 situation. Comparing with case 3, the improvement of electricity saving per unit collector area is 32.9%.

20.7   CONCLUSION

Generally speaking, it is the numerical analysis of system performance based on real design cases and typical year-round weather data that is more representative to evaluate and compare the long-term energy performance of different renewable energy systems. In doing so, satisfactory verification of the numerical models is crucial. In this study, numerical model of an electrical PV system was first developed. Good quality of the single-diode five-parameter model in use was successfully verified by the measured data obtained from an experimental setup installed at a real office building. Similarly, an indirect thermosyphon solar water heating system model was developed and also successfully validated by the experimental results. A new hypothetical PV/T system model was then developed by the integration of these two reliable computer models.

The findings of the comparative case studies support that the use of PV/T system is able to improve the overall energy output intensity. Based on the design principle of having similarity in collector material usage, the PV/T system is found to be able to increase the overall electricity saving by around 36% on unit collector area basis, comparing with the side-by-side provision. Nevertheless, the PV/T system does not have an advantage over the side-by-side system in the overall quantities of electricity and thermal energy outputs. On the other hand, based on the design principle of having relatively the same annual electricity saving, the PV/T system is again found superior than the side-by-side option in overall electricity saving per unit collector area. The improvement is around 33%. The results of this case study support that in the subtropical climate zone, the water-based PV/T system embraces very good potential in maximizing the renewable energy output in a reserved area for solar collector

installation. This can be an excellent choice in the era of hunting for optimal sustainable building design solutions. More R&D work is deserved to make this a mature technology.

20.8   NOMENCLATURE

a	diode ideality constant, –
A_f	area ratio, -
C	specific heat capacity, J/(kg K)
C_B	nominal capacity of battery bank, Ah
d	diameter, m
E	instantaneous power per unit area, W/m^2
E_{OC}	open circuit voltage of battery, V
F_R	collector heat removal factor, -
g	gravitational constant, m/s^2
h	heat transfer coefficient, W/(m^{2}K)
I	current, A
k	singular pressure coefficient, –
L	length, m
m	number of module in parallel, –
$\dot{m}$	mass flow rate, kg/s
N	number of PV array / riser tube, –
n	number of cell connected in series, –
P	power, W
Q	solar irradiation, W/m^2
R	electrical resistance, Ω
R	thermal resistance, m^{2}K/W
SOC	state of charge of battery, –
T	temperature, K
t	time, s
U	heat loss coefficient, W/(m^{2}K)
u	flow velocity, m/s
V	voltage, V
V	volume, m^3
VF	full charge rest voltage, V
VS	empirical constant, V
Greek	
α	absorptance coefficient, –
β'	coefficient of thermal expansion, 1/K
δ	thickness, m
η	collector efficiency, –
η_c	battery charging efficiency, –
θ	collector tilted angle, degree
ρ	density, m^2/s
σ	battery self-discharge rate, –
$\tau\alpha$	transmittance-absorptance product, –
υ	kinematic viscosity, m^2/s
φ	phase angle, degree
ϕ	pressure loss difference ratio, –
Subscripts	
a	ambient
ab	thermal absorber + PV encapsulation
ac	alternating current

B	Battery
C	solar charging controller
c	PV cell
c,self	charging controller self-consumption
cl	solar thermal collector
dc	direct current
ex	heat exchange coil
f	fluid
f,in	fluid inlet of the tank
f,out	fluid outlet of the tank
g	front glass
in	thermal insulation; thermal collector inlet
INV	inverter
L	light generated
LED	light emit diode
m	mean
max	maximum power
o	saturation; charging parameter
oc	open circuit
out	thermal collector outlet
p	parallel; potable
pv	photovoltaic
r	radiation
ref	reference
s	series
sc	short circuit
t	thermal
tk	tank
x–y	between object "x" to object "y"

REFERENCES

[1] IEA. (2002). Photovoltaic/thermal solar energy system: status of the technology and roadmap for future development. Task 7 Report, International Energy Agency, PVPS T7-10.

[2] Chow, T. T., He, W., Ji, J., Chan, ALS. (2007). Performance evaluation of photovoltaic-thermosyphon system for subtropical climate application. *Solar Energy,* 81:123–130.

[3] Wolf, M. (1976). Performance analysis of combined heating and photovoltaic power systems for residences. *Energy Conversion,* 16:79–90.

[4] Chow, T. T. (2010). A review on photovoltaic/thermal hybrid solar technology. *Applied Energy,* 87:365–379.

[5] Rauschenbach, H. S. (1980). *Solar Cell Array Handbook.* Van Nostrand Reinhold, New York.

[6] Villalva, M. G., Gazoli, J. R., Filho, E. R. (2009). Comprehensive approach to modeling and simulation of photovoltaic arrays. *IEEE Transactions on Power Electronics,* 24:1198–1208.

[7] Hansen, A. D., Sorensen, P., Hansen, L. H., Bindner, H. (2000). *Model for a Stand-Alone PV System.* Riso National Laboratory, Roskilde.

[8] Rajapakse, A., Chungpibulpatana, S. (1994). Dynamic simulation of a photovoltaic refrigeration system. *RERIC International Energy Journal,* 16:67–101.

[9] Guash, D., Silvestre, S. (2003). Dynamic battery model for photovoltaic applications. *Progress in Photovoltaics: Research and Applications,* 11:193–206.

[10] Smieee, Z. A. Development of an electrical model for a PV/battery system for performance prediction. *Renewable Energy,* 15:531–534.

[11] Sukamongkol, Y., Chungpaibulpayana, S., Ongsakul, W. (2002). A simulation model for predicting the performance of a solar photovoltaic system with alternating current loads. *Renewable Energy,* 27:237–258.

[12] Chow, T. T. (2003). Performance analysis of photovoltaic thermal collector by explicit dynamic model. *Solar Energy,* 75(2):143–152.

[13] Koffi, P. M. E., Andoh, H. Y., Gbaha, P., Toure, S.,Ado, G. (2009). Theoretical and experimental study of solar water heater with internal exchanger using thermosiphon system. *Energy Conversion and Management,* 49:2279–2290.

[14] Duffie, J. A., Beckman, W. A. (1991). *Solar Engineering of Thermal Processes.* 2nd edition. Wiley, New York.

[15] Lasnier, F., Ang, T. G. (1990). *Photovoltaic Engineering Handbook.* Adam Hilger, New York.

[16] Chow, T. T., He, W.,Ji, J. (2006). Hybrid photovoltaic thermosyphon water heat system for residential application. *Solar Energy,* 80:298–306.

[17] CIBSE Guide B. (1986). *Installation and Equipment Data.* The Chartered Institution of Building Services Engineers, London.

[18] Energy Efficiency & Renewable Energy, Department of Energy, United States. Available at: http://www.eere.energy.gov/buildings/energyplus/cfm/weather_data.cfm.

[19] Perez, R., Ineichen, P., Seals, R., Michalsky, J., Stewart, R. (1990). Modeling daylight availability and irradiance components from direct and global irradiance. *Solar Energy,* 44(5):271–289.

21

POLYCRYSTALLINE SILICON FOR THIN FILM SOLAR CELLS

Nicolás Budini, Roberto D. Arce, Román H. Buitrago, and Javier A. Schmidt

21.1 INTRODUCTION

Solar energy, as well as other renewable energy sources, appears as a promising candidate to capture an important proportion of the world energetic matrix. Indeed, solar energy conversion and photovoltaic electricity consumption have been increasing in the last decades at high rates. At the same time, research in this field has also increased, giving rise to new technologies. One of the main driving forces for research in this field is the need to overcome the current compromise existing between conversion efficiency and production cost of solar cells. The latter is hard to reduce while maintaining, or even without decreasing, the former. However, present research intends to force this restriction to the limit of developing low-cost materials and processes leading to higher conversion efficiencies. In Section 21.2, we present some aspects of the current state of photovoltaic technology so as to set the place that polycrystalline silicon solar cells occupy. Silicon-based solar cells, which represent an important proportion of commercially available devices, are shortly addressed in Section 21.3. Several available processes to obtain polycrystalline silicon, which can be classified basically by the temperature range involved in the overall process, are presented in Sections 21.4 and 21.5. This chapter concludes with a summary of the discussed items in Section 21.6.

21.2 TRENDS IN PHOTOVOLTAICS

The need for development of alternative and renewable energy sources has turned photovoltaics (PV) one of the most promising technologies in this field. The main characteristic and great advantage of solar energy is the enormous and widely available power that our star offers us. For practical purposes we can consider the Sun as an "infinite" source of energy, since it will keep radiating for billions of years. Therefore, it is well worth taking advantage of that power to satisfy, at least partially, the requirements of our nowadays lifestyle, aiding and complementing the present and long-term worldwide energetic matrix.

Great efforts have been made in the last years in order to reduce the production costs and, at the same time, to improve the conversion efficiency of PV devices. This challenge has led to a classification of the different stages of technological innovations from the point of view of design of the solar cell and also of the materials involved. Actually, there are three well-established *generations* in the field of PV devices. The *first generation* makes reference mainly to silicon (Si) or gallium arsenide (GaAs) wafer-based, single-junction, solar cells. Typical efficiencies for Si solar cells are in the range between 15% and 18% at module level and go up to 24.7% at laboratory level [1]. GaAs cells have slightly higher efficiencies, reaching 26–29% at laboratory level and averaging 20% at module level [2,3]. However, the production costs for both types are relatively high. The *second generation* is based on the thickness reduction and optimization of single-junction solar cells. It is also known as "thin film solar cell technology," which is thought to be suitable for the mid-term future of PV [4]. At the laboratory level, efficiencies as high as 20% have been reached so far [5–7] while theoretical considerations fix an upper limiting value of 31% under an illumination of one sun, extendable to a mere 41% under extreme sunlight concentration (46,200 suns) [8]. Amorphous silicon (a-Si), polycrystalline silicon (poly-Si), copper indium diselenide (CIS), and cadmium telluride (CdTe) solar cells

Alternative Energy and Shale Gas Encyclopedia, First Edition. Edited by Jay H. Lehr and Jack Keeley.
© 2016 John Wiley & Sons, Inc. Published 2016 by John Wiley & Sons, Inc.

are the principal contenders at this level. A *third generation* of PV devices is still emerging. Its distinctive feature is the utilization of novel polymers or dye-sensitized materials for a better capture of incident sunlight, together with multi-junction thin film structures which would allow overcoming the current efficiency limit. The estimated enhanced range of conversion efficiencies for this kind of devices is 30–60%, keeping low production costs. Current research concerning third-generation solar cells is being made over a wide variety of new approaches, such as tandem or quantum-well cells, which make use of processes like multiple energy thresholds, multiple electron–hole pairs creation, and hot carriers excitation [9]. In the rest of the chapter we will be centered in second-generation solar cells made of polycrystalline silicon.

21.3 SILICON-BASED SOLAR CELLS

Nowadays, first- and second-generation solar cells coexist at the industrial level with an incipient inclusion of third-generation devices. In this scenario, crystalline silicon (c-Si, including single-crystalline and multi-crystalline) widely dominates the market, despite being an expensive material, with 80–90% of the worldwide solar cell production based on wafer technology [10, 11]. This is due to the fact that silicon has plenty of advantages compared to other materials. First of all, being a nontoxic element, its notable abundance is of great importance. Silicon composes a fraction of 30% of the Earth's crust, possessing excellent electronic, chemical, and mechanical properties. Moreover, it has a bandgap of 1.1 eV, which is almost perfectly matched to the solar spectrum impinging on the terrestrial surface (see Figure 21.1). Furthermore, it gives long-term stable solar cells and also offers the vast experience of processing and production acquired from the microelectronics industry. However, the need to reduce the costs of PV energy forces toward a reduction in the production costs. To achieve this goal, other approaches using less material and lowering the processing temperature are needed.

Several possibilities have emerged during the second-generation stage as powerful substitutes, lowering drastically the cost/throughput relation. Most of these alternatives are based on silicon, anyway, but they take advantage of other methods of production and processing which are economically more convenient. Such is the case of amorphous (a-Si), nanocrystalline (nc-Si), microcrystalline (μc-Si), and polycrystalline (poly-Si) silicon. Extensive investigation has been made concerning these materials [12] since they can be deposited over different substrates by several methods and, as a remarkable characteristic from the point of view of cost reduction, in the form of thin films. The prefixes nano-, micro-, poly-, and so on refer to the same kind of material, consisting in crystalline domains (grains) separated by

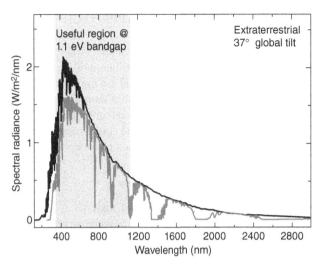

FIGURE 21.1   Solar spectrum reaching Earth. Incoming energy from the Sun (darker line) is affected by absorption as it travels through the atmosphere (lighter line). This spectrum corresponds to radiance at sea level, received at a tilt angle of 37° toward the equator, that is, facing the Sun. The approximate useful portion of the spectrum for silicon solar cells is highlighted with a dark band.

an amorphous and defective interface (grain boundary). The size of these grains determines which one of the prefixes should be used to design the material in question. A commonly used classification is given by

- amorphous: no grains;
- nanocrystalline: grain size of some nanometers in diameter;
- polymorphous: nanograins embedded within an amorphous matrix;
- microcrystalline: grain size between 100 nm and 1 μm;
- polycrystalline: grain size between 1 μm and 1 mm;
- semi-crystalline: grain size above 1 mm;
- single crystalline: no grain boundaries.

An important factor to take into account is that the greater the grain size the better the material's properties.

Generally, the starting material is deposited in an amorphous state and is subsequently crystallized by means of thermal annealing. However, some methods allow growing an already crystallized thin film. The main difference between both approaches is basically the temperature range that is maintained during deposition. Thus we can distinguish two approaches to obtain nano-, micro-, or poly-Si thin films, namely high-temperature and low-temperature approaches. Any of these processes allows obtaining a complete single-junction polycrystalline thin film solar cell by suitably controlling the dopant concentration in each layer.

21.4 HIGH-TEMPERATURE APPROACH

To obtain a crystallized (nano- or microcrystalline) silicon thin film, the deposition temperature should be above 700°C (in some cases as high as 1200°C). These high temperatures give the atoms sufficient energy to allow them to distribute in the form of crystalline domains (grains) onto the supporting substrate. Among these methods, one could mention chemical vapor deposition (CVD) and liquid phase epitaxy (LPE).

21.4.1 Chemical Vapor Deposition

CVD is a process by which a solid material is deposited through a chemical reaction from gaseous reactants. The silicon-containing gas source is usually silane (SiH_4) or a chlorinated silane, like dichlorosilane (SiH_2Cl_2), trichlorosilane (TCS, $SiHCl_3$), or tetrachlorosilane ($SiCl_4$). The energy necessary to initiate the chemical reaction can be provided thermally ("thermal CVD" or simply "CVD"), by a plasma ("PE-CVD"), by a heated filament catalyzer ("hotwire CVD"), or by making use of the cyclotron resonance frequency ("ECR-CVD"), among others. The resulting molecules diffuse and react on the surface of the heated substrate and subsequently incorporate on the growing film. A scheme of a basic CVD reactor is shown in Figure 21.2.

Thermal CVD offers the advantages of providing a high throughput, high growth rate, and high chemical yield of the silicon precursor. Moreover, atmospheric pressure CVD (AP-CVD) has the advantage of a simple setup. From the point of view of cost, the most suitable gas precursor is TCS, which is a by-product of the industrial silicon production. $SiHCl_3$ can be obtained by hydrochlorination of metallurgical grade silicon and can be easily purified by distillation.

Poly-Si deposition by thermal CVD from $SiHCl_3$ is usually performed in a temperature range of 900–1200°C, although lower deposition temperatures have also been reported [13]. Doping can be performed by diluting a suitable gas in the precursor gas. In Ref. [14], AP-CVD was

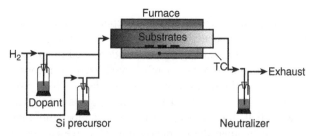

FIGURE 21.2 Simple scheme of a basic thermal CVD reactor. The carrier gas (H_2) drags the precursor vapor (usually SiHCh) together with the dopant, if necessary. The furnace provides thermal energy to the system (~900–1200°C) needed for decomposition of the species, which deposit onto the substrates. The remaining by-product of the reaction is neutralized (e.g., with aqueous NaOH) before being exhausted.

used to thicken a poly-Si seed layer. The deposition temperature was 1130°C with TCS as precursor, and the solar cells made by this method achieved a maximum efficiency of 6.4%. Recently, monocrystalline-silicon seed layers were prepared on transparent glass-ceramic substrates by using a patented process that consists of anodic bonding and an implant-induced separation [15]. These layers were then epitaxially thickened by thermal CVD, obtaining simple solar cell structures with efficiencies of up to 7.5% [15].

21.4.2 Liquid Phase Epitaxy

The LPE method [16] allows growing polycrystalline layers of silicon directly from the melt on suitable solid substrates. This is achieved at temperatures which are well below the melting point of the deposited semiconductor. Generally, the semiconductor to be deposited is dissolved in the melt of another material at conditions close enough to the equilibrium point between dissolution and deposition. In this manner the semiconductor layer grows uniformly and slowly on the substrate forming polycrystals. Doping can be controlled by adding a specific concentration of impurity atoms in the same melt. The quality of the resulting films is high.

21.4.3 Zone Melting Recrystallization

Zone melting recrystallization (ZMR) is quite a different method in the sense that the material has been previously deposited amorphous or nanocrystalline and is then recrystallized to increase the grain size [17]. The method consists in creating a narrow molten zone (recall that silicon melts at ~1400°C) that is scanned across the film at a constant velocity. Different methods can be used to generate the molten zone, like an electron beam, a laser, a stripe heated by an electric current, or the light from incoherent lamps. The material behind the molten zone cools down and solidifies forming crystals (see Figure 21.3). The size of the crystals, which can reach up to some millimeters (μc-Si), depends strongly on the scanning velocity.

21.5 LOW-TEMPERATURE APPROACH

In particular, the possibility of using glass substrates introduces an important cost reduction, being one of the advantages of thin film solar cells. An important requirement that the chosen substrate must fulfill is to tolerate the temperatures used during processing (deposition and/or annealing), which are usually lower than 600°C. Borosilicate, aluminosilicate, and alkali-free glasses are among the most commonly used substrates. Some particular processes (e.g., defect annealing or doping) may need higher temperatures (~900°C), which are almost at the limit or even higher than the softening point

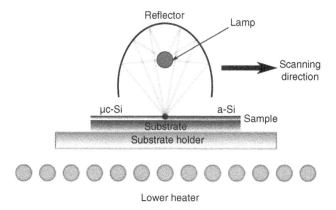

FIGURE 21.3   Zone melting recrystallization consists of locally melting the sample. After cooling down, as the molten zone advances in a specific scanning direction, pc-Si is obtained.

of commercially available and cost-effective glasses. However, there are some possible ways to overcome this inconvenience as is the case of rapid thermal annealing (RTA), which allows the rise and fall of the temperature of the sample at high rates without heating the glass substrate in excess.

21.5.1   Solid Phase Crystallization

This approach consists in depositing an amorphous film, which is subsequently crystallized through an isothermal annealing at temperatures below 700°C. Solid phase crystallization (SPC), the simplest and at the same time one of the first techniques explored to crystallize a-Si [18], is the direct annealing of the material for a sufficiently long time. The temperature used is generally in the range between 550°C and 700°C, and most commonly around 600°C. This temperature is chosen as a compromise between high temperatures that produce faster crystallization but smaller grains, and low temperatures that produce larger grains but needing longer annealing times. The method is simple and straightforward, but the main drawbacks are the need for long annealing periods, generally tens of hours [19], and the relatively small grain size obtained, which is not larger than 1.5 μm [20]. In spite of these limitations, it is the basic process used to produce commercial poly-Si solar cells at the CSG Solar company with a great performance [21, 22].

21.5.2   Metal-Induced Crystallization

By adding some metals to a-Si, the crystallization temperature can be reduced in comparison to SPC, leading to a larger grain size with a good uniformity. This is the innovation introduced by the technique named metal-induced crystallization (MIC). Some metals like aluminum, indium, gold, or antimony form eutectics with silicon, lowering the crystallization temperature. Other metals like nickel, palladium,

or titanium form silicides. This method has been intensively investigated lately and it proved to be successful in terms of enlarging the grain size and lowering the annealing temperature to attain crystallization [23, 24]. Both characteristics go also in the direction of costs reduction as is desired. By using this method, an increase in grain size of over one order of magnitude as compared to the grain size obtained by direct SPC of a-Si can be achieved. The method is based on the capability of some metals to form eutectics or silicides with silicon. Whether the former or the latter occurs depends on the metal used as the inductor. Several metals were studied so far, Al and Ni being the most interesting and, hence, those that have been more thoroughly investigated. Aluminum forms an eutectic with silicon at relatively low temperatures (< 600°C), while nickel forms nickel disilicide ($NiSi_2$) at temperatures around 350°C.

The main drawback of MIC in general is the metal contamination that may remain after crystallization. Due to the fact that the desired outcome of this method is a good quality poly-Si, metal impurities are thus detrimental and should be avoided. Taking account of MIC's advantages, concerning grain sizes and lower annealing temperatures, it is reasonable to develop suitable etching processes for applying them as impurity removals without affecting the overall quality of the poly-Si obtained. This aspect has been addressed and investigated so far [25].

21.5.2.1   Aluminum-Induced   Crystallization. The aluminum-induced crystallization (AIC) method has been extensively studied for the production of thin film silicon solar cells [26, 27]. Although large grains are obtained, the metal contamination problem is hard to overcome. Aluminum is a shallow acceptor in silicon and hence the layers result strongly p-doped, with aluminum contents as high as 1×10^{19} cm^{-3} [28]. Recombination processes at defect centers, introduced by metal impurities, reduce the carriers' lifetime. Moreover, solar cells usually present a low parallel resistance due to the segregation of aluminum atoms in the grain boundaries. Therefore, layers prepared by AIC are commonly used as the back surface field (BSF) of the solar cell, but are not suitable to act as the absorber layer.

The physical process by which AIC acts is based in an exchange between an a-Si layer in contact with an Al layer during annealing (see Figure 21.4), resulting in a poly-Si

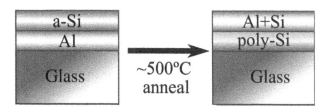

FIGURE 21.4   Layer exchange process in Al-induced crystallization at temperatures below the eutectic point (~400°C).

film with an Al + Si capping layer. This process was named aluminum-induced layer exchange (ALILE) [10]. The driving force during crystallization is the ability of aluminum atoms to diffuse into the a-Si structure producing a change in the Si–Si covalent bonds at the Si/Al interfaces due to the free electrons of the metallic phase [23]. Final grain sizes are generally in the order of 20–30 μm.

21.5.2.2 Nickel-Induced Crystallization.

The nickel-induced crystallization (NIC) technique has been thoroughly studied for the production of thin film transistors (TFTs) for liquid crystal displays [29, 30]. When crystallization advances in the lateral direction, starting from a Ni electrode, the process is named metal-induced lateral crystallization (MILC) [31, 32]. Nickel silicides can be formed at temperatures around 350°C and, in this way, it is possible to obtain poly-Si at temperatures as low as 480°C [33]. Most of the research concerning NIC and MILC for TFT applications is done on thin samples, generally with film thicknesses that lay below 0.1 μm. Since we are concerned with solar cell applications, considerably thicker layers are needed.

The crystallization of a-Si by NIC is mediated by the formation of $NiSi_2(111)$ octahedral precipitates, having a very similar lattice parameter ($\sim$99.6%) to that of c-Si(111) [24]. A simple diagram of the process can be seen in Figure 21.5. The detailed mechanism by which the crystallization is induced by Ni is still under investigation. However, it is thought that at a first stage of annealing, Ni atoms diffuse over the a-Si surface until they form $NiSi_2$ nuclei. For this to accomplish, a minimum surface density of Ni atoms and a critical size of the silicide nuclei are required [34]. In a second stage, these $NiSi_2$ precipitates migrate through the a-Si volume leaving behind c-Si, until grain boundaries collide with those of neighboring grains. Due to the nature of the growing process, the Ni concentration inside the grains is relatively low. It has been demonstrated by time-of-flight secondary ion mass spectroscopy (TOF-SIMS) analysis that after full crystallization the nickel concentration at grain boundaries is much higher than inside the grains. This proves that Ni atoms migrate during the grain growth, accumulating at grain boundaries [35]. NIC has two important features. First, the required quantity of Ni atoms per unit of area is really small for crystallization to proceed and, second, the final grain size (depending upon Ni density) can be larger than that obtained by AIC. In a recent work it has been demonstrated that in a-Si

samples with sputtered Ni surface densities lower than 1×10^{15} at./cm^2, disk-shaped grains with a diameter in excess of 100 μm can be obtained [36,37]. It is important to remark that those values of surface densities are equivalent to depositing less than a monolayer of Ni atoms.

There are several ways to deposit Ni atoms onto the a-Si layer, such as ion sputtering of a Ni target [24, 36, 37], ink-jet printing of an appropriate Ni-containing solution [38,39], and electroless deposition [40,41], among other possibilities.

21.5.3 Laser-Induced Crystallization

Laser-induced crystallization (LIC) makes use of the local heating that can be achieved by a laser beam when impinging on an a-Si thin film [42]. The rapid increase of temperature provides the atoms with enough energy to nucleate and form crystalline grains without heating the underlying substrate beyond some upper temperature limit. Generally, pulsed excimer or green lasers (like frequency-doubled Nd:YAG, $\lambda = 532$ nm) are used for heating. If an excimer laser is used the technique generally takes the name of excimer laser annealing (ELA) [43,44]. The flux of photons should be controlled in order to induce nucleation and crystallization in the entire film thickness without producing a widespread melting. As the laser is focused on the a-Si film, small regions are heated until they melt and they are allowed to cool subsequently. It is also important to control the process in such a way that the laser melts the silicon film through its entire thickness, without damaging the substrate. The advantage of this method is that heating can be performed at high rates and thus crystallization is quickly achieved. However, some drawbacks are a poor uniformity due to the laser focusing and high production costs due to the elevated laser power needed to achieve the required melting temperature.

21.5.4 Ion-Assisted Deposition

In the ion-assisted deposition (IAD) technique the evaporation of some material onto a substrate is performed while it is simultaneously irradiated with a low-energy high-flux ion source [45]. The constant impact of the ions during the deposition of the film modifies the properties of the resulting material. If the deposited film is to be of the same atomic composition of the evaporated substance, then some inert gas like argon should be used for ion production. Otherwise,

FIGURE 21.5 Nickel-induced crystallization of a-Si. Ni atoms are deposited onto the amorphous semiconductor which is subsequently annealed at temperatures below 600°C. After crystallization has succeeded, a poly-Si with segregated Ni is obtained.

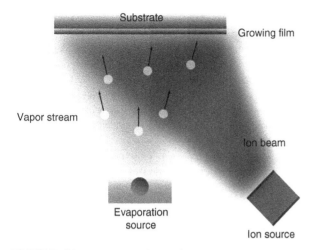

FIGURE 21.6 Scheme of the ion-assisted deposition. The growing film is composed of the evaporated material. The ion beam allows enhancing the deposition, gaining control over film properties.

some reactive gases could be added in order to obtain a different compound. In such case, the method is called reactive ion-assisted deposition. Here, compound formation depends upon a suitable supply of reactant, collisions between reactant species, and reaction between colliding species. The reactivity of the different species can be further enhanced by ionization through an electrical discharge during evaporation and the process is then classified as activated reactive evaporation (ARE) [46]. Ions can be produced by means of a plasma or by means of bombardment with an ion gun. The latter allows a greater control over film properties. A schematic diagram of the deposition process is presented in Figure 21.6.

21.6 CONCLUSION

Summarizing, electricity generation by means of solar energy conversion is becoming an extremely important member inside the worldwide energetic matrix. Research is being intensively directed to costs reduction and several materials have appeared in scene, each of them having different benefits, characteristics, and drawbacks. Copper indium gallium diselenide (CIGS) thin film solar cells reached the top-record module efficiency of 20% for single-junction devices, pertaining to the *second* generation. The main problem of this technology is the utilization of contaminating and not-so-abundant materials. This turns poly-Si in the form of thin films to be the most promising material at mid- to long term, with module efficiencies to date in the order of 10% (CSG Solar). However, *third-generation* solar cells are emerging as good companions for the long-term perspective, elevating the efficiencies to above the c-Si limit of 31–41% and further reducing production costs.

REFERENCES

[1] Green, M. A., Jianhua, Z., Wang, A., Wenham, S. R. (1999). Very high efficiency silicon solar cells-science and technology. *IEEE Transactions on Electron Devices*, 46(10):1940–1947.

[2] Bauhuis, G. J., Mulder, P., Haverkamp, E. J., Huijben, J. C. C. M., Schermer, J. J. (2009). 26.1% thin-film GaAs solar cell using epitaxial lift-off. *Solar Energy Materials and Solar Cells*, 93:1488–1491.

[3] Green, M. A., Emery, K., Hishikawa, Y., Warta, W. (2010). Solar cell efficiency tables (version 36). *Progress in Photovoltaics: Research and Applications*, 18:346–352.

[4] Green, M. A. (2004). Recent developments in photovoltaics. *Solar Energy*, 76:3–8.

[5] Ramanathan, K., Teeter, G., Keane, J. C., Noufi, R. (2005). Properties of high-efficiency CuInGaSe2 thin film solar cells. *Thin Solid Films*, 480–481:499–502.

[6] Jackson, P., Wurz, R., Rau, U., Mattheis, J., Kurth, M., Schlotzer, T., Bilger, G., Werner, J. H. (2007). High quality baseline for high efficiency, Cu(In1 − x, Gax)Se2 solar cells. *Progress in Photovoltaics: Research and Applications*, 15(6):507–519.

[7] Repins, I., Contreras, M., Romero, M., Yan, Y., Metzger, W., Li, J., Johnston, S., Egaas, B., DeHart, C., Scharf, J., McCandless, B. E., Noufi, R. (2008). *Photovoltaic Specialists Conference, 2008. PVSC '08. 33rd IEEE*, May 11–16, 2008, pp. 1–6.

[8] Shockley, W., Queisser, H. J. (1961). Detailed balance limit of efficiency of p-n junction solar cells. *Journal of Applied Physics*, 32:510.

[9] Green, M. A. (2005). *Third Generation Photovoltaics: Advanced Solar Energy Conversion*. Springer Series in Photonics.

[10] Gall, S., Schneider, J., Klein, J., Hiibener, K., Muske, M., Rau, B., Conrad, E., Sieber, I., Petter, K., Lips, K., Stoger-Pollach, M., Schattschneider, P., Fuhs, W. (2006). Large-grained polycrystalline silicon on glass for thin-film solar cells. *Thin Solid Films*, 511–512:7–14.

[11] European Photovoltaic Industry Association (2010). Global market outlook for photovoltaic until 2014. Available at: www.epia.org.

[12] Beaucarne, G. (2007). Silicon thin-film solar cells. *Advances in OptoElectronics*, vol. 2007, Article ID 36970, 12 pages.

[13] Benvenuto, A., Buitrago, R. H., Schmidt, J. A. (2010). *Proc. 25th European Photovoltaic Solar Energy Conference and Exhibition, WIP*, Munich, pp. 3584–3587.

[14] Gordon, I., Carnel, L., Van Gestel, D., Beaucarne, G., Poortmans, J., Pinckney, L., Mayolet, A., (2007). *Proc. 22nd European Photovoltaic Solar Energy Conf., WIP*, Munich, pp. 1993–1996.

[15] Gordon, I., Vallon, S., Mayolet, A., Beaucarne, G., Poortmans, J. (2010). Thin-film monocrystalline-silicon solar cells made by a seed layer approach on glass-ceramic substrates. *Solar Energy Materials & Solar Cells*, 94:381–385.

[16] Kopecek, R., Peter, K., Hötzel, J., Bucher, E. (2000). Structural and electrical properties of silicon epitaxial layers grown by LPE on highly resistive monocrystalline substrates. *Journal of Crystal Growth*, 208(1–4):289–296.

[17] Slaoui, A., Bourdais, S., Beaucarne, G., Poortmans, J., Reber, S. (2002). Polycrystalline silicon solar cells on mullite substrates. *Solar Energy Materials & Solar Cells*, 71(2):245–252.

[18] Iverson, R. B., Reif, R., (1987). Recrystallisation of amorphized poly-Si films on SiO_2: temperature dependence of the crystallisation parameters. *Journal of Applied Physics*, 62:1675–1681.

[19] Matsuyama, T., Terada, N., Baba, T., Sawada, T., Tsuge, S., Wakisaka, K., Tsuda, S. (1996). High-quality polycrystalline silicon thin film prepared by a solid phase crystallization method. *Journal of Non-Crystalline Solids*, 198–200:940–944.

[20] Song, D., Inns, D., Straub, A., Terry, M. L., Campbell, P., Aberle, A. G. (2006). Solid phase crystallized polycrystalline thin-films on glass from evaporated silicon for photovoltaic applications. *Thin Solid Films*, 513:356.

[21] Green, M. A., Basore, P. A., Chang, N., Clugston, D., Egan, R., Evans, R., Hogg, D., Jarnason, S., Keevers, M., Lasswell, P., O'Sullivan, J., Schubert, U., Turner, A., Wenham, S. R., Young, T. (2004). Crystalline silicon on glass (CSG) thin-film solar cell modules. *Solar Energy*, 77:857–863.

[22] Basore, P. A. (2006). *Conf. Record 2006 IEEE 4th World Conf. Photovoltaic Energy Conversion, IEEE*, New York, pp. 2089–2093.

[23] Dimova-Malinovska, D. (2010). Structural and optical properties of poly-Si thin films obtained by aluminium induced crystallization. *Journal of Physics: Conference Series*, 223, 012013.

[24] Schmidt, J. A., Budini, N., Rinaldi, P. A., Arce, R. D., Buitrago, R. H. (2008). Large-grained oriented polycrystalline silicon thin films prepared by nickel-silicide-induced crystallization. *Journal of Crystal Growth*, 311:54–58.

[25] Song, N. K., Kim, M. S., Kim, Y. S., Han, S. H., Joo, S. K. (2007). A study on the properties of polycrystalline silicon crystallized by a Ni seed. *Journal of the Korean Physical Society*, 51(3):1076–1079.

[26] Nast, O., Puzzer, T., Koschier, L. M., Sproul, A. B., Wenham, S. R. (1998). Aluminum-induced crystallization of amorphous silicon on glass substrates above and below the eutectic temperature. *Applied Physics Letters*, 73:3214–3216.

[27] Gall, S., Becker, C., Lee, K. Y., Sontheimer, T., Rech, B. (2010). Growth of polycrystalline silicon on glass for thin-film solar cells. *Journal of Crystal Growth*, 312:1277–1281.

[28] Nast, O., Brehme, S., Neuhaus, D. H., Wenham, S. R. (1999). Polycrystalline silicon thin films on glass by aluminum-induced crystallization. *IEEE Transactions on Electron Devices*, 46:2062–2068.

[29] Zhao, S., Meng, Z., Wu, C., Xiong, S., Wong, M., Kwok, H. S. (2007). Solution-based metal induced crystallized polycrystalline silicon films and thin-film transistors. *Journal of Materials Science: Materials in Electronics*, 18:S117–S121.

[30] Zhang, B., Meng, Z., Zhao, S., Wong, M., Kwok, H. S. (2007). Polysilicon thin film-transistors with uniform and reliable performance using solution-based metal-induced crystallization. *IEEE Transactions on Electron Devices*, 54:1244–1248.

[31] Lee, S. W., Joo, S. K. (1996). Low temperature poly-Si thin-film transistor fabrication by metal-induced lateral crystallization. *IEEE Electron Device Letters*, 17:160–162.

[32] Ma, T., Wong, M. (2002). Dopant and thickness dependence of metal-induced lateral crystallization of amorphous silicon films. *Journal of Applied Physics*, 91:1236–1241.

[33] Miyasaka, M., Makihira, K., Asano, T., Polychroniadis, E., Stoemenos, J. (2002). In situ observation of nickel-induced lateral crystallization of amorphous silicon thin films. *Applied Physics Letters*, 80:944–946.

[34] Hayzelden, C., Batstone, J. L. (1993). Silicide formation and silicide-mediated crystallization of nickel-implanted amorphous silicon thin films. *Journal of Applied Physics*, 73:8279.

[35] Kim, K. H., Oh, J. H., Kim, E. H., Jang, J. (2004). Formation and analysis of disk-shaped grains by Ni-mediated crystallization of amorphous silicon. *Journal of Vacuum Science & Technology A*, 22:2469.

[36] Schmidt, J. A., Budini, N., Rinaldi, P. A., Arce, R. D., Buitrago, R. H. (2009). Nickel-induced crystallization of amorphous silicon. *Journal of Physics: Conference Series*, 167, 012046.

[37] Schmidt, J. A., Budini, N., Arce, R. D., Buitrago, R. H. (2010). Polycrystalline silicon thin films on glass obtained by nickel-induced crystallization of amorphous silicon. *Physica Status Solidi C*, 7(3–4):600–603.

[38] Ishida, Y., Nakagawa, G., Asano, T. (2007). Inkjet printing of nickel nanosized particles for metal-induced crystallization of amorphous silicon. *Japanese Journal of Applied Physics*, 46:6437.

[39] Lee, J. S., Kim, M. S., Kim, D., Kim, Y. M., Moon, J., Joo, S. K. (2009). Fabrication and characterization of low temperature polycrystalline silicon thin film transistors by ink-jet printed nickel-mediated lateral crystallization. *Applied Physics Letters*, 94:122105.

[40] Jeske, M., Schultze, J. W., Thonissen, M., Minder, H. (1995). Electrodeposition of metals into porous silicon. *Thin Solid Films*, 255:63-66.

[41] Liu, Y. M., Sung, Y., Pu, N. W., Chou, Y. H., Yeh, K. C., Ger, M. D. (2008). Electroless deposition of nickel-phosphorous nano-dots for low-temperature crystallization of amorphous silicon. *Thin Solid Films*, 517:727–730.

[42] Choi, T. Y., Hwang, D. J., Grigoropoulos, C. P. (2003). Ultrafast laser-induced crystallization of amorphous silicon films. *Optical Engineering*, 42:3383.

[43] Im, J. S., Kim, H. J., Thompson, M. O. (1993). Phase transformation mechanisms involved in excimer laser crystallization of amorphous silicon films. *Applied Physics Letters*, 63(14):1969–1971.

[44] Chen, Y.-R., Chang, C.-H., Chao, L.-S. (2007). Modeling and experimental analysis in excimer-laser crystallization of a-Si films. *Journal of Crystal Growth*, 303(1):199–202.

[45] Fuhs, W., Gall, S., Rau, B., Schmidt, M., Schneider, J. (2004). A novel route to a polycrystalline silicon thin-film solar cell. *Solar Energy*, 77(6):961–968.

[46] Durandet, A., Boswell, R., McKenzie, D. (1995). New plasma-assisted deposition technique using helicon activated reactive evaporation. *Review of Scientific Instruments*, 66(4):2908–2913.

22

SOLAR BASICS – ENERGY FROM THE SUN

22.1 SOLAR BASICS

22.1.1 Energy from the Sun

The sun has been producing energy for billions of years. Solar energy is the sun's rays (solar radiation) that reach the Earth. This energy can be converted into other forms of energy, such as heat and electricity.

In the 1830s, the British astronomer John Herschel famously used a solar thermal collector box (a device that absorbs sunlight to collect heat) to cook food during an expedition to Africa. Today, people use the sun's energy for lots of things.

22.1.2 Solar Energy Can Be Used for Heat and Electricity

When converted to thermal (or heat) energy, solar energy can be used to:

- Heat water—for use in homes, buildings, or swimming pools.
- Heat spaces—inside homes, greenhouses, and other buildings.
- Heat fluids—to high temperatures to operate a turbine to generate electricity.

Solar energy can be converted to electricity in two ways:

- Photovoltaic (PV devices) or "solar cells" change sunlight directly into electricity. Individual PV cells are grouped into panels and arrays of panels that can be used in a wide range of applications ranging from single small cells that charge calculator and watch batteries, to systems that power single homes, and to large power plants covering many acres.
- Solar thermal/electric power plants generate electricity by concentrating solar energy to heat a fluid and produce steam that is used to power a generator. In 2012, solar thermal power-generating units were the main source of electricity at 12 power plants in the United States:
  - 11 in California
  - 1 in Nevada

The main benefits of solar energy:

- Solar energy systems do not produce air pollutants or carbon-dioxide.
- When located on buildings, they have minimal impact on the environment.

Two limitations of solar energy:

- The amount of sunlight that arrives at the Earth's surface is not constant. It varies depending on location, time of day, time of year, and weather conditions.
- Because the sun does not deliver that much energy to any one place at any one time, a large surface area is required to collect the energy at a useful rate.

22.2 WHERE SOLAR IS FOUND

22.2.1 Solar Energy Is Sunshine

The amount of solar energy that the earth receives each day is many times greater than the total amount of energy

Alternative Energy and Shale Gas Encyclopedia, First Edition. Edited by Jay H. Lehr and Jack Keeley.
© 2016 John Wiley & Sons, Inc. Published 2016 by John Wiley & Sons, Inc.

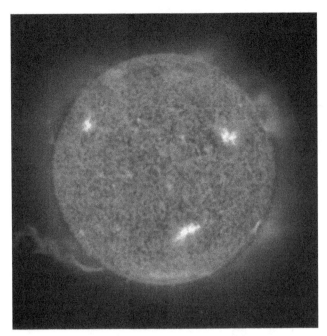

FIGURE 22.1   Radiant energy from the sun has powered life on Earth for many millions of years. From: NASA.

consumed around the world. However, solar energy is a variable and intermittent energy source. The amount and intensity of sunlight varies by location, and weather and climate conditions affect its availability on a daily and seasonal basis. The type and size of a solar energy collection and conversion system determines how much of available solar energy can be converted to useful energy.

22.2.2   Solar Thermal Collectors

Low-temperature solar thermal collectors absorb the sun's heat energy to heat water or air in homes, offices, and other buildings.

22.2.3   Concentrating Collectors

Concentrating solar power technologies use mirrors to reflect and concentrate sunlight onto receivers that collect solar energy and convert it to heat. This thermal energy can then be used to produce high temperature heat or electricity via a steam turbine or heat engine driving a generator.

22.2.4   Photovoltaic Systems

PV cells convert sunlight directly into electricity. PV systems can range from those that provide tiny amounts of power for watches and calculators up to those that provide the amount used by hundreds of homes.

Hundreds of thousands of houses and buildings around the world have PV systems on their roofs. Many multi-megawatt (MW) PV power plants have also been built. Covering 4% of

the world's desert areas with PVs could supply the equivalent of all of the world's electricity. The Gobi Desert alone could supply almost all of the world's total electricity demand.

22.3   SOLAR PHOTOVOLTAIC

22.3.1   PV Cells Convert Sunlight into Electricity

A photovoltaic cell, commonly called a solar cell or PV, is the technology used to convert solar energy directly into electrical power. A PV cell is a nonmechanical device usually made from silicon alloys.

22.3.2   Photons Carry Solar Energy

Sunlight is composed of photons, or particles of solar energy. These photons contain various amounts of energy corresponding to the different wavelengths of the solar spectrum.

When photons strike a PV cell, they may be reflected, pass right through, or be absorbed. Only the absorbed photons provide energy to generate electricity. When enough sunlight (energy) is absorbed by the material (a semiconductor), electrons are dislodged from the material's atoms. Special treatment of the material surface during manufacturing makes the front surface of the cell more receptive to free electrons, so the electrons naturally migrate to the surface.

22.3.3   The Flow of Electricity

When the electrons leave their position, holes are formed. When many electrons, each carrying a negative charge, travel toward the front surface of the cell, the resulting imbalance of charge between the cell's front and back surfaces creates a voltage potential like the negative and positive terminals of a battery. When the two surfaces are connected through an external load, such as an appliance, electricity flows.

22.3.4   How PV Systems Operate

The PV cell is the basic building block of a PV system. Individual cells can vary in size from about 0.5 inches to about 4 inches across. However, one cell produces only 1 or 2 W, which is not enough power for most applications.

To increase power output, cells are electrically connected into a packaged weather-tight module. Modules can be further connected to form an array. The term array refers to the entire generating plant, whether it is made up of one or several thousand modules. The number of modules connected together in an array depends on the amount of power output needed.

22.3.5   Weather Affects PVs

The performance of a PV array is dependent upon sunlight. Climate conditions (such as clouds or fog) have a

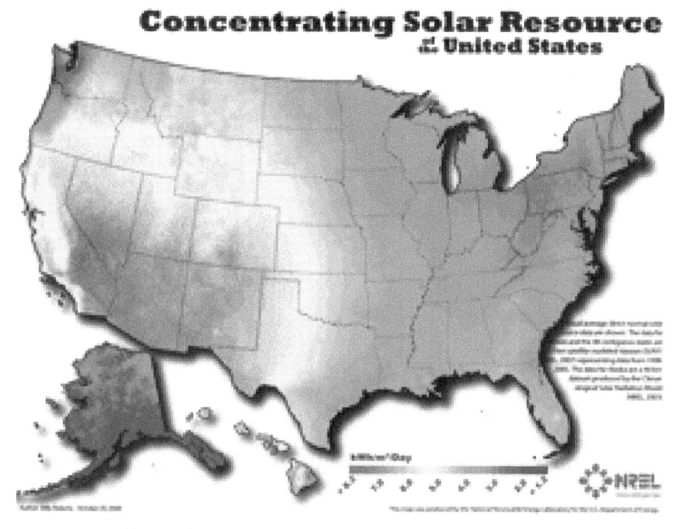

FIGURE 22.2   From: National Renewable Energy Laboratory, U.S. Department of Energy.

significant effect on the amount of solar energy received by a PV array and, in turn, its performance. The efficiency of most commercially available PV modules in converting sunlight to electricity ranges from 5% to 15%. Researchers around the world are trying to achieve efficiencies up to 30%.

22.3.6   Commercial Applications of PV Systems

The success of PV in outer space first generated commercial applications for this technology. The simplest PV systems power many of the small calculators and wrist watches used every day. More complicated systems provide electricity to pump water, power communications equipment, and even provide electricity to our homes.

Some advantages of PV systems:

1. Conversion from sunlight to electricity is direct, so bulky mechanical generator systems are unnecessary.

2. PV arrays can be installed quickly and in any size.

3. The environmental impact is minimal, requiring no water for system cooling and generating no by-products.

PV cells, like batteries, generate direct current (DC), which is generally used for small loads (electronic equipment). When DC from PV cells is used for commercial applications or sold to electric utilities using the electric grid, it must be converted to alternating current (AC) using inverters, solid state devices that convert DC power to AC.

22.3.7   History of the PV Cell

The first practical PV cell was developed in 1954 by Bell Telephone researchers examining the sensitivity of a properly prepared silicon wafer to sunlight. Beginning in the late 1950s, PV cells were used to power US space satellites. PV

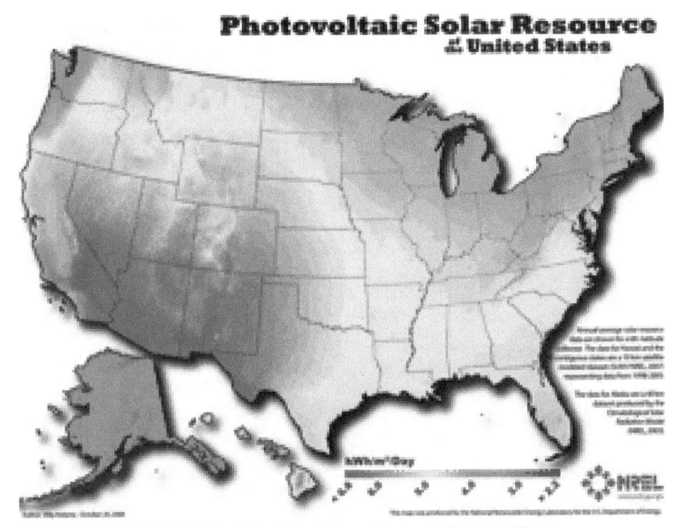

FIGURE 22.3   From: National Renewable Energy Laboratory, U.S. Department of Energy.

cells were next widely used for small consumer electronics like calculators and watches and to provide electricity in remote or "off-grid" locations where there were no electric power lines. Technology advances and government financial incentives have helped to greatly expand PV use since the mid-1990s.

US shipments (includes imports, exports, and domestic shipments) of PV panels (modules) by US industry in 2012 was the equivalent of about 4655 MW, about 245 times greater than the shipments of about 19 MW in 1994[1]. Since about 2004, most of the PV panels installed in the United States have been in "grid-connected" systems on homes, buildings, and central-station power facilities. There are PV products available that can replace conventional roofing materials.

[1]Earliest year for which data are available.

22.4   SOLAR THERMAL POWER PLANTS

22.4.1   Solar Thermal Power Uses Solar Energy Instead of Combustion

Solar thermal power plants use the sun's rays to heat a fluid to very high temperatures. The fluid is then circulated through pipes so it can transfer its heat to water to produce steam. The steam, in turn, is converted into mechanical energy in a turbine and into electricity by a conventional generator coupled to the turbine.

So solar thermal power generation works essentially the same as generation from fossil fuels except that instead of using steam produced from the combustion of fossil fuels, the steam is produced by the heat collected from sunlight. Solar thermal technologies use concentrator systems to achieve the high temperatures needed to heat the fluid.

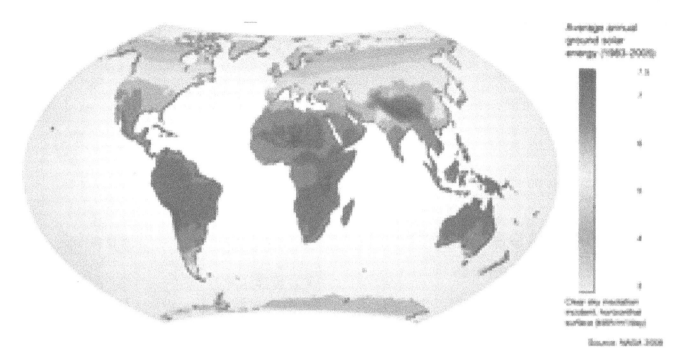

FIGURE 22.4   World map of solar resources. From: United Nations Environment Programme (UNEP), NASA Surface meteorology and Solar Energy (SSE), 2008.

The three main types of solar thermal power systems:

1. Parabolic trough (the most common type of plant)
2. Solar dish
3. Solar power tower

22.4.2   Types of Solar Thermal Power Plants

22.4.2.1   Parabolic Troughs. Parabolic troughs are used in the largest solar power facility in the world located in the Mojave Desert at Kramer Junction, California. This facility has operated since the 1980s and accounts for the majority of solar electricity produced by the electric power sector today. A parabolic trough collector has a long parabolic-shaped reflector that focuses the sun's rays on a receiver pipe located at the focus of the parabola. The collector tilts with the sun as the sun moves from east to west during the day to ensure that the sun is continuously focused on the receiver.

Because of its parabolic shape, a trough can focus the sun at 30–100 times its normal intensity (concentration ratio) on the receiver pipe located along the focal line of the trough, achieving operating temperatures over 750°F.

The "solar field" has many parallel rows of solar parabolic trough collectors aligned on a north–south horizontal axis. A working (heat transfer) fluid is heated as it circulates through the receiver pipes and returns to a series of "heat exchangers" at a central location. Here, the fluid circulates through pipes so it can transfer its heat to water to generate high-pressure, superheated steam. The steam is then fed to a conventional steam turbine and generator to produce electricity. When the hot fluid passes through the heat exchangers, it cools down and is then recirculated through the solar field to heat up again.

The plant is usually designed to operate at full power using solar energy alone, given sufficient solar energy. However, all parabolic trough power plants can use fossil fuel combustion to supplement the solar output during periods of low solar energy, such as on cloudy days.

22.4.2.2   Solar Dish. A solar dish/engine system uses concentrating solar collectors that track the sun, so they always point straight at the sun and concentrate the solar energy at the focal point of the dish. A solar dish's concentration ratio is much higher than a solar trough's, typically over 2000, with a working fluid temperature over 1380°F. The power-generating equipment used with a solar dish can be mounted at the focal point of the dish, making it well suited for remote operations or, as with the solar trough, the energy may be collected from a number of installations and converted to electricity at a central point.

The engine in a solar dish/engine system converts heat to mechanical power by compressing the working fluid when it is cold, heating the compressed working fluid, and then expanding the fluid through a turbine or with a piston to produce work. The engine is coupled to an electric generator to convert the mechanical power to electric power.

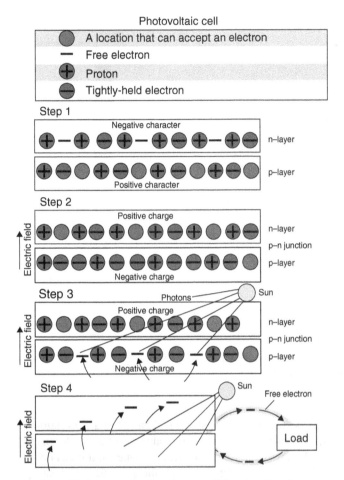

FIGURE 22.5   From: National Energy Education Development Project (public domain).

22.4.2.3 *Solar Power Tower.* A solar power tower, or central receiver, generates electricity from sunlight by focusing concentrated solar energy on a tower-mounted heat exchanger (receiver). This system uses hundreds to thousands of flat, sun-tracking mirrors called heliostats to reflect

FIGURE 22.6   Parabolic trough power plant.

FIGURE 22.7   Solar dish.

and concentrate the sun's energy onto a central receiver tower. The energy can be concentrated as much as 1500 times that of the energy coming in from the sun.

Energy losses from thermal energy transport are minimized because solar energy is being directly transferred by reflection from the heliostats to a single receiver, rather than being moved through a transfer medium to one central location, as with parabolic troughs.

Power towers must be large to be economical. This is a promising technology for large-scale grid-connected power plants. Power towers are in the early stages of development compared with parabolic trough technology.

FIGURE 22.8   Solar power tower. From: National Renewable Energy Laboratory (NREL).

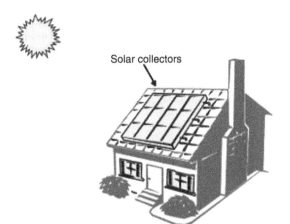

FIGURE 22.9   From: National Energy Education Development Project (public domain).

FIGURE 22.10   From: U.S. Marine Corps photo by Pfc. Jeremiah Handeland/Released (public domain).

The U.S. Department of Energy, along with a number of electric utilities, built and operated a demonstration solar power tower near Barstow, California, during the 1980s and 1990s. Projects from private companies:.

- A 5-MW, two-tower project, built in the Mojave Desert in southern California in 2009
- A 390-MW, three-tower project being built in the Mojave Desert
- A 110-MW project located in Nevada

Learn more about the history of solar power in the Solar Timeline.

22.5   SOLAR THERMAL COLLECTORS

22.5.1   Heating with the Sun's Energy

Solar thermal (heat) energy is often used for heating water in homes and swimming pools and for heating the insides of buildings ("space heating"). Solar space heating systems can be classified as passive oractive.

Passive space heating is what happens to your car on a hot summer day. The sun's rays heat up the inside of your car. In buildings, the air is circulated past a solar heat surface and through the building by convection (meaning that less dense warm air tends to rise, while denser cool air moves downward). No mechanical equipment is needed for passive solar heating.

Active heating systems require a collector to absorb and collect solar radiation. Fans or pumps are used to circulate the heated air or heat absorbing fluid. Active systems often include some type of energy storage system.

22.5.2   Solar Collectors Are Either Nonconcentrating or Concentrating

22.5.2.1   Nonconcentrating collectors. The collector area (the area that intercepts the solar radiation) is the same as the absorber area (the area absorbing the radiation). Flat-plate collectors are the most common type of nonconcentrating collector and are used when temperatures below about 200°F are sufficient. They are often used for heating buildings.

There are many flat-plate collector designs but generally all consist of

- a flat-plate absorber that intercepts and absorbs the solar energy;
- a transparent cover(s) that allows solar energy to pass through but reduces heat loss from the absorber; and
- a heat-transport fluid (air or water) flowing through tubes to remove heat from the absorber, and a heat insulating backing.

22.5.2.2   Concentrating collectors. The area intercepting the solar radiation is greater, sometimes hundreds of times greater, than the absorber area.

22.6   SOLAR ENERGY AND THE ENVIRONMENT

An array of solar PV panels supplies electricity for use at Marine Corps Air Ground Combat Center in Twentynine Palms, California.

Using solar energy produces no air or water pollution and no greenhouse gases, but may have some indirect negative impacts on the environment. For example, there are some toxic materials and chemicals that are used in the manufacturing process of PV cells, which convert sunlight into electricity. Some solar thermal systems use potentially hazardous fluids to transfer heat. US environmental laws regulate the use and disposal of these types of materials.

As with any type of power plant, large solar power plants can affect the environment where they are located. Clearing land for construction of the power plant may have long-term impacts on plant and animal life. They may require water for cleaning solar collectors or concentrators and for cooling turbine-generators. Using ground water from wells may affect the ecosystem in some arid locations. Birds and insects can be killed if they fly into a concentrated beam of sunlight created by a "solar power tower."

23

NASA ARMSTRONG FACT SHEET: SOLAR-POWER RESEARCH

PUBLIC DOMAIN

Since 1980 AeroVironment, Inc. (founded in 1971 by the ultra-light airplane innovator Dr. Paul MacCready) has been experimenting with solar-powered aircraft, often in conjunction with NASA's Dryden Flight Research Center, Edwards, CA. Thus far, AeroVironment, now headquartered in Monrovia, CA, has achieved several altitude records with its Solar Challenger, Pathfinder, and Pathfinder-Plus aircraft. It expects to exceed them with the newer and larger solar-powered Centurion and its successors in NASA's Environmental Research Aircraft and Sensor Technology (ERAST) project, the Centelios and Helios vehicles.

Solar Challenger set an initial altitude record of 14,300 feet. More spectacularly, on July 7, 1981, the solar-powered aircraft flew 163 miles from Corneille-en-Verin Airport north of Paris across the English Channel to Manston Royal Air Force Base south of London, staying aloft 5 hours and 23 minutes.

At the time, AeroVironment was headquartered in Pasadena, CA. Dr. MacCready, a former gliding champion, whose Gossamer Albatross crossed the English Channel using human power in 1979, saw solar power as a way to "help business and government recognize and meet their environmental and energy objectives." MacCready remains the Chairman of the Board at AeroVironment.

1981 from the US government for a classified program to look into the feasibility of long-duration, solar-electric flight above 65,000 feet. The firm designed an airplane designated HALSOL (High-Altitude Solar Energy), built and test flew three subscale models and a final prototype. The HALSOL proved the aerodynamics and structures for the approach, but subsystem technologies, principally for energy storage, were inadequate for the intended mission. The HALSOL was mothballed for 10 years but later evolved into Pathfinder, which first flew at Dryden in 1993 under the auspices of the Ballistic Missile Defense Office for potential use in an antiballistic-missile defense role.

When funding for this program ended, Pathfinder became part of NASA's ERAST project to develop remotely piloted, long-duration aircraft for environmental sampling and sensing at altitudes above 60,000 feet. On September 11, 1995, Pathfinder exceeded Solar Challenger's altitude record for solar-powered aircraft by a long margin when it reached 50,500 feet at Dryden.

After moving its flight operations to the US Navy's Pacific Missile Range Facility (PMRF) on Kauai, HI, in April 1997, an upgraded Pathfinder took advantage of the improved angle of sunlight to set a new record of 71,530 feet for solar-powered aircraft on July 7, 1997.

Further modified with longer wings, improved motors, and more efficient solar array, Pathfinder-Plus (as it was now called) flew to still another record of 80,201 feet at the PMRF on August 6, 1998. In the process, it stayed above 70,000 feet for almost three and a half hours while carrying 68 pounds of test instrumentation and other payload.

23.1 BACKGROUND

The first flight of a solar-powered aircraft took place on November 4, 1974, when the remotely controlled Sunrise II, designed by Robert J. Boucher of AstroFlight, Inc., flew following a launch from a catapult.

Following this event, AeroVironment took on a more ambitious project to design a human-piloted, solar-powered

Alternative Energy and Shale Gas Encyclopedia, First Edition. Edited by Jay H. Lehr and Jack Keeley.

FIGURE 23.1 Pathfinder-Plus with the Hawaiian island of N'ihau in the background.

FIGURE 23.3 Gossamer Albatross II during a test flight at NASA's Dryden Flight Research Center, Edwards, CA. The original Gossamer Albatross was best known for completing the first completely human-powered flight across the English Channel on June 12, 1979. The Albatross II was the backup craft for the Channel flight.

aircraft. The firm initially took the human-powered Gossamer Albatross II and scaled it down to three-quarters of its previous size for solar-powered flight with a human pilot controlling it. This was more easily done because in early 1980 the Gossamer Albatross had participated in a flight research program at NASA Dryden in a program conducted jointly by the Langley and Dryden research centers. Some of the flights were conducted using a small electric motor for power.

The scaled-down aircraft was designated the Gossamer Penguin. It had a 71-foot wingspan compared with the 96-foot span of the Gossamer Albatross. Weighing only 68 lb without a pilot, it had a low power requirement and thus was an excellent test bed for solar power.

AstroFlight, Inc., of Venice, CA, provided the power plant for the Gossamer, Penguin, an Astro-40 electric motor. Robert Boucher, designer of the Sunrise II, served as a key consultant for both this aircraft and the Solar Challenger. The

power source for the initial flights of the Gossamer Penguin consisted of 28 nickel–cadmium batteries, replaced for the solar-powered flights by a panel of 3920 solar cells capable of producing 541 W of power.

The battery-powered flights took place at Shafter Airport near Bakersfield, CA. Dr. MacCready's son Marshall, who was 13 years old and weighed roughly 80 lb, served as the initial pilot for these flights to determine the power required to fly the airplane, optimize the airframe/propulsion system, and train the pilot. He made the first flights on April 7, 1980, and made a brief solar-powered flight on May 18.

The official project pilot was Janice Brown, a Bakersfield school teacher, who weighed in at slightly under 100 lb and was a charter pilot with commercial, instrument, and glider ratings. She checked out in the plane at Shafter and made about 40 flights under battery and solar power there. Wind direction, turbulence, convection, temperature, and radiation at Shafter in mid-summer proved to be less than ideal for Gossamer Penguin because takeoffs required no crosswind and increases in temperature reduced the power output from the solar cells.

Consequently, the project moved to Dryden in late July, although conditions there also were not ideal. Nevertheless, Janice finished the testing, and on August 7, 1980, she flew a public demonstration of the aircraft at Dryden in which it went roughly 1.95 miles in 14 minutes and 21 seconds.

This was significant as the first sustained flight of an aircraft relying solely on direct solar power rather than batteries. It provided the designers with practical experience for developing a more advanced, solar-powered aircraft, since the Gossamer Penguin was fragile and had limited controllability. This necessitated its flying early in the day when there

FIGURE 23.2 Gossamer Penguin in flight above Rogers Dry Lakebed at Edwards, CA, showing the solar panel perpendicular to the wing and facing the sun.

FIGURE 23.4 AstroFlight, Inc., again provided the motor, and the DuPont Company, which produced many of the advanced materials for the Gossamer Albatross, Gossamer Penguin, and Solar Challenger, sponsored the project. Janice Brown remained one of the pilots, but she was joined by the slightly heavier Stephen R. Ptacek (almost 150 pounds), who brought to the project over 4600 hours of flight in a variety of aircraft.

FIGURE 23.5 Pathfinder in flight over Hawaii.

were minimal wind and turbulence levels, but the angle of the sun was also low, requiring a panel for the solar cells that could be tilted toward the sun.

23.2 SOLAR CHALLENGER

Using the specific conclusions derived from their experience with Gossamer Penguin, the AeroVironment engineers designed Solar Challenger, a piloted, solar-powered aircraft strong enough to handle both long and high flights when encountering normal turbulence. As compared with the Penguin's 71-foot wingspan, Solar Challenger had only a 46.5-foot wingspan, but it had a huge horizontal stabilizer and a large enough wing area to accommodate 16,128 solar cells.

Using in-house computer programs, AeroVironment engineers Peter Lissaman and Bart Hibbs designed the unusual wings and stabilizers, which they made flat on top to hold the solar cells. Hibbs developed the aerodynamic design for the propeller with another in-house computer program. The result was a "smooth and docile" aircraft that dropped in a steady, wing-level attitude when stalled and rapidly regained unstalled flight.

The pilots flew the aircraft, first with batteries and then under solar power, at the Santa Susana, Shafter, and El Mirage airports in California before moving to Marana Airpark northwest of Tucson, AZ, in the late 1980 and early 1981. With some modifications, the Solar Challenger showed itself to be an effective aircraft. This was proved to the world during the cross-Channel flight on July 7, 1981, with Ptacek at the controls.

Growing out of the post-1983 development of HALSOL, Pathfinder was modified with additional solar arrays and other upgrades. It was then brought back to Dryden for another series of developmental flights in 1995. On September 11, 1995, Pathfinder reached an altitude of 50,500 feet,

setting a new altitude record for solar-powered aircraft. The National Aeronautic Association presented the NASA-industry team with an award for one of the "10 Most Memorable Record Flights" of 1995.

After additional upgrades and one checkout flight at Dryden in late 1996, Pathfinder was transferred to the US Navy's Pacific Missile Range Facility (PMRF) at Barking Sands, Kauai, HI, in April, 1997. Kauai was chosen as an optimum location for testing the solar-powered Pathfinder due to the high levels of sunlight, available airspace and radio frequencies, and the diversity of terrestrial and coastal ecosystems for validating scientific imaging applications. While in Hawaii, Pathfinder flew seven high-altitude flights from PMRF, one of which reached a world altitude record for propeller-driven as well as solar-powered aircraft of 71,530 feet.

During 1998, the Pathfinder was modified into the longer-winged Pathfinder-Plus configuration. On August 6, 1998, the modified aircraft was flown to a record altitude for propeller-driven aircraft of 80,201 feet on the third of a series of developmental test flights from PMRF on Kauai. The goal

FIGURE 23.6 Pathfinder-Plus flight in Hawaii, June 2002.

FIGURE 23.7 Centurion. The slow-flying Centurion solar-electric flying wing, one of several remotely piloted aircraft developed under NASA's ERAST project, glides in for a landing on Rogers Dry Lake following a test flight at NASA's Dryden Flight Research Center.

of the flights was to validate new solar, aerodynamic, propulsion and systems technology developed for the Pathfinder's successor, the Centurion, which is designed to reach and sustain altitudes in the 100,000-foot range.

Essentially a transitional vehicle between the Pathfinder and the follow-on Centurion, the Pathfinder-Plus is a hybrid of the technology that was employed on Pathfinder and developed for Centurion.

The most noticeable change is the installation of a new 44-foot-long center wing section that incorporates a high-altitude airfoil designed for Centurion. The new section is twice as long as the original Pathfinder center section and increases the overall wingspan of the craft from 98.4 to 121 feet. The new center section is topped by more-efficient silicon solar cells developed by SunPower Corp., Sunnyvale, CA; they can convert 19% of the solar energy they receive to useful electrical energy to power the craft's motors, avionics, and communication systems. That compares with about 14% efficiency for the older solar arrays that cover most of the surface of the middle and outer wing panels from the original Pathfinder. Maximum potential power was boosted from about 7500 W on Pathfinder to about 12,500 W on Pathfinder-Plus.

In addition, the Pathfinder-Plus was powered by eight electric motors, two more than had powered the previous version of Pathfinder. Designed for Centurion, the motors are slightly more efficient than the original Pathfinder motors.

Centurion, like its immediate predecessors Pathfinder and Pathfinder-Plus, is a lightweight, solar-powered, remotely piloted flying wing aircraft that is demonstrating the technology of applying solar power for long-duration, high-altitude flight. It is considered to be a prototype technology demonstrator for a future fleet of solar-powered aircraft that could

stay airborne for weeks or months on scientific sampling and imaging missions or while serving as telecommunications relay platforms.

Although it shares much of the design concepts of the Pathfinder, the Centurion has a wingspan of 206 feet, more than twice the 98-foot span of the original Pathfinder and 70% longer than the Pathfinder-Plus' 121-foot span. At the same time, it maintains the 8-foot chord (front to rear distance) of the Pathfinder wing, giving the Centurion wing an aspect ratio (length-to-chord) of 26:1.

Other visible changes from its predecessor include a modified wing airfoil designed for flight at extreme altitude and four underwing pods to support its landing gear and electronic systems, compared with two such pods on the Pathfinder. The flexible wing is primarily fabricated from carbon fiber and graphite epoxy composites and kevlar. It is built in five sections, a 44-foot-long center section and middle and outer sections just over 40 feet long. All five sections have an identical thickness, that is, 12% of the chord, or about 11.5 inches, with no taper or sweep.

Solar arrays that will cover most of the upper wing surface will provide up to 31 kW of power at high noon on a summer day to power the aircraft's 14 electric motors, avionics, communications, and other electronic systems. Centurion also has a backup lithium battery system that can provide power for between 2 and 5 hours to allow limited-duration flight after dark. Initial low-altitude test flights at Dryden in 1998 are being conducted on battery power alone, prior to installation of the solar cell arrays.

Centurion flies at an airspeed of only 17–21 mph, or about 15–18 knots. Although pitch control is maintained by the use of a full-span 60-segment elevator on the trailing edge of the wing, turns and yaw control are accomplished by applying differential power-slowing down or speeding up the motors on the outboard sections of the wing.

AeroVironment envisions Helios as the ultimate solar aircraft that can offer virtually eternal flights in the stratosphere. It will build upon the technologies developed by Pathfinder and Centurion but will add an energy storage system for nighttime flying. From 25% to 50% larger than Centurion, the Helios will store up to two-thirds of the energy received by its solar array during the day and will use this stored energy to maintain its altitude overnight. Because it will renew its energy every day from the sun, the Helios will have flight endurance limited only by the reliability of its systems, meaning a practical limit of perhaps 6 months on station.

Because of this long duration of flight, the Helios will be extremely economic in operation. However, it will have to be perhaps the most reliable aircraft ever built, with each flight lasting longer than the time between overhaul for a typical, general aviation aircraft. As a consequence, much of the Helios design will involve a minimum of moving parts, high redundancy, low temperatures, and solid-state control

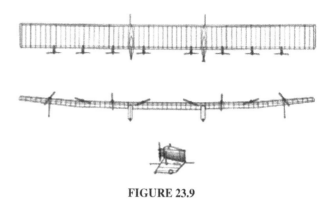

FIGURE 23.9

FIGURE 23.8 Helios. The solar-electric Helios Prototype flying wing near the Hawaiian islands of Niihau and Lehua during its first test flight on solar power from the US Navy's Pacific Missile Range Facility on Kauai, HI, July 14, 2001.

systems. The Helios will also be able to diagnose degradation of its control systems and reconfigure the autopilot while in flight. As a result, AeroVironment expects the Helios to perform as a non-polluting, re-configurable "atmospheric satellite" in the new millennium.

23.3 THE ERAST PROJECT

Centurion is and Helios will be one of a number of remotely piloted aircraft being evaluated under NASA's Environmental Research Aircraft and Sensor Technology (ERAST) project. The ERAST project is one of NASA's initiatives designed to develop the new technologies needed to continue America's leadership in the highly competitive aerospace industry.

The primary focus of ERAST is on the development of slow-flying, remotely operated aircraft that can perform long-duration science missions at very high altitudes above 60,000 feet. These missions could include *in situ* atmospheric sampling, tracking of severe storms, remote sensing for earth sciences studies, hyperspectral imaging for agriculture monitoring, and serving as telecommunications relay platforms. The most extreme mission envisioned for solar-powered aircraft such as the Centurion, Centelios, and Helios would reach altitudes of 100,000 feet.

A parallel effort is developing lightweight, microminiaturized sensors that can be carried by these aircraft. Additional technologies considered by the joint NASA-industry ERAST Alliance include lightweight materials, avionics, aerodynamics, and other forms of propulsion suitable for extreme altitudes and duration.

The ERAST project was sponsored by the Office of Aeronautics and Space Transportation Technology at NASA Headquarters and was managed by the NASA Dryden Flight Research Center. Sensor technology development

was headed by NASA Ames Research Center, Moffett Field, CA.

23.4 AIRCRAFT SPECIFICATIONS

- **Wingspan:** Solar Challenger, 46.5 feet (14.8 m); Pathfinder, 98.4 feet (29.5 m); Pathfinder-Plus, 121 feet (36.3 m); Centurion, 206 feet (61.8 m).
- **Length:** Solar Challenger, 30.3 feet (9.22 m); Pathfinder, Pathfinder-Plus, and Centurion, 12 feet (3.6 m).
- **Wing chord:** Solar Challenger, 5.8 feet (1.78 m); Pathfinder, Pathfinder-Plus, and Centurion, 8 feet (2.4 m).
- **Gross weight:** Solar Challenger, about 336 lb (152.8 kg); Pathfinder, about 560 lb (252 kg); Pathfinder-Plus, about 700 lb (315 kg) Centurion, varies depending on power availability and mission profile; approximately 1900 lb for a mission to 80,000 feet altitude.
- **Payload:** Solar Challenger, weight of pilot, up to 150 lb; Pathfinder, up to 100 lb (45 kg); Pathfinder-Plus, up to 150 lb (67.5 kg) Centurion, varies depending on altitude; about 100 lb to 100,000 feet, 600 lb to 80,000 feet.
- **Airspeed:** Solar Challenger, approximately 25–34 mph cruise; Pathfinder, Pathfinder-Plus, approximately 17–20 mph cruise; Centurion, approximately 17–21 mph cruise.
- **Power:** Arrays of solar cells, max. output: Solar Challenger, 2700 W; Pathfinder, about 7500 W; Pathfinder-Plus, about 12,500 W; Centurion, 31,000 W.
- **Motors:** Solar Challenger, one electric motor, 2.7 kW Pathfinder, six electric motors, 1.25 kW each Pathfinder-Plus, eight electric motors, 1.5 kW maximum each Centurion, 14 electric motors, 2.2 kW each.
- **Manufacturer:** AeroVironment, Inc.
- **Primary materials:** Composites, plastic, foam.

BIBLIOGRAPHY

AeroVironment, Available at: http://www.aerovironment.com/overview/history/history.html and linked sites

Benedek, L. (1985). "Solar energy and the aeronautics industry," translation of "Energrave;a Solar y la Industria Aeronautica," thesis, Olivos, Argentina, October 9 (Washington, DC: NASA TM-77957): 24–33.

Cowley, M. (1981). "Wings in the sun: the evolution of Solar Challenger," *Flight International* (June 13): 1865–1868.

Documents relating to Solar Challenger, Dryden historical reference collection, including 3-view photograph and various background papers and fact sheets.

Dornheim, M. A. (1998). "Pathfinder Plus Exceeds 80,000 Ft.; Centurion Assembled," *Aviation Week & Space Technology* (August 17): 33.

Brown, A. (1998). Fact Sheets on Pathfinder and Centurion, NASA Dryden Flight Research Center, September, and coordinated with ERAST and AeroVironment personnel.

Flittie, K. and Curtin, B. "Pathfinder solar-powered aircraft flight performance," paper presented at AIAA Atmospheric Flight Mechanics Conference & Exhibit, Boston, MA, August 10–12 (AIAA 98-4446).

(1980). "Gossamer Penguin makes first sustained flight," *X-Press*, 23(6): 2.

MacCready, P. B. et al. (1981). "Sun powered aircraft design," paper presented at the AIAA 1981 Annual Meeting and Technical Display, Long Beach, CA (AIAA-81-0916).

Penner, H. (1981). "Sun Worshiper [sic]. MacCready's [sic] Solar Challenger flies over the English Channel", translation of "Sonnenanbeter. McCreadys Solar Challenger flog ber den Armelkanal," *Flug Revue* (No. 9, September): 67–69. (Washington, DC: NASA TM-77327, 1983).

24

SOLAR THERMAL – CHRONOLOGICAL DEVELOPMENT

PUBLIC DOMAIN

1860	Auguste Mouchout (France), a mathematics instructor, was able to convert solar radiation directly into mechanical power.
1878	William Adams (England) constructed a reflector of flat-silvered mirrors arranged in a semicircle. To track the sun's movement, the entire rack was rolled around a semicircular track, projecting the concentrated radiation onto a stationary boiler.
1883	Charles Fritts (United States) built the first genuine solar cell with an efficiency rate between 1% and 2%.
1883–1884	John Ericsson (United States) invented and erected a solar engine that used parabolic trough construction.
1921	Albert Einstein won the 1921 Nobel Prize in Physics for his theories that explained the photoelectric effect.
1947	Energy was scarce during World War II, so passive solar buildings became popular in the United States. Libbey-Owens-Ford Glass Company published a book titled, *Your Solar House,* which profiled 49 of the nation's greatest solar architects.
Mid–1950s	Frank Bridgers (United States) designed the world's first commercial office building that features solar water heating and passive design. The Bridgers-Paxton Building is listed in the National Historic Register as the world's first solar-heated office building.
1969	A "solar furnace" was constructed in Odeillo, France; it featured an eight-story parabolic mirror.
1973	The University of Delaware built "Solar One," a PV/thermal hybrid system. Roof-integrated arrays fed surplus power through a special meter to the utility during the day; power was purchased from the utility at night. In addition to providing electricity, the arrays were like flat-plate thermal collectors; fans blew warm air from over the array to heat storage bins.
1974	The Solar Energy Industries Association (SEIA) was formed. The organization represents the interests of the solar industry and acts as a lobbying group in Washington, DC.
1977	The Solar Energy Research Institute (SERI) was formed (now the National Renewable Energy Laboratory (NREL)), a national laboratory that provides research and development support for solar and photovoltaic technologies.
1978	The Public Utility Regulatory Policies Act (PURPA) of 1978 mandated the purchase of electricity from qualifying facilities that meet certain standards on energy source and efficiency. A 15% energy tax credit was added to an existing 10% investment tax credit, providing incentive for capital investment in solar thermal generation facilities for independent power producers.
1981	California enacted a 25% tax credit for the capital costs of renewable energy systems.
1982	Solar One, a 10-MW central receiver demonstration project, was first operated and established the feasibility of power tower systems. In 1988, the final year of operation, the system achieved an availability of 96%.
1983	California's Standard Offer Contract system provided renewable electric energy systems with a relatively firm, stable market for their output. This system allowed the financing of capital-intensive technologies such as solar thermal-electric. The SEGS I plant (13.8 MW) was installed, the first in a series of Solar Electric Generating Stations (SEGS). SEGS I used solar trough technology to produce steam in a conventional steam turbine generator. Natural gas was used as a supplementary fuel for up to 25% of the heat input.

(continued)

Alternative Energy and Shale Gas Encyclopedia, First Edition. Edited by Jay H. Lehr and Jack Keeley.
© 2016 John Wiley & Sons, Inc. Published 2016 by John Wiley & Sons, Inc.

1984	Advanco and McDonnell Douglas systems demonstrated the potential for the high-efficiency 25-kW solar dish. Dish/engine systems convert the thermal energy in solar radiation to mechanical energy and then to electrical energy — in much the same way that conventional power plants convert thermal energy from combustion of a fossil fuel to electricity. The Sacramento Municipal Utility District commissioned its first 1-MW photovoltaic electricity-generating facility.
1989	Federal regulations that govern the size of solar power plants were modified to increase maximum plant size to 80 MW from 30 MW.
1991	Luz International went bankrupt while building its tenth SEGS plant. SEGS I through IX remained in operation.
1992	A 7.5-kW dish prototype system became operational, using an advanced stretched-membrane concentrator, through a joint venture of Sandia National Laboratories and Cummins Power Generation. The Energy Policy Act of 1992 restored the 10% investment tax credit for independent power producers, using solar technologies.
1994	The first solar dish generator, using a free-piston Stirling engine, was tied to a utility grid. The Corporation for Solar Technology and Renewable Resources, a public corporation, was established to facilitate solar developments at the Nevada Test Site. 3M Company introduced a new silvered plastic film for solar applications.
1995	Federal Energy Regulatory Commission (FERC) prohibits qualifying facility contracts above avoided costs.
2000	A 12-kW solar electric system, in Colorado, was the largest residential installation in the United States to be registered with the U.S. Department of Energy's Million Solar Roofs Initiative. The system provided most of the electricity for the family of eight's 6000-square-foot home.
2001	Home Depot began selling residential solar power systems in three stores in San Diego, California. NASA's solar-powered aircraft, Helios, set a new world altitude record for non-rocket-powered craft: 96,863 feet (more than 18 miles).
2002	Students from the University of Colorado built an energy-efficient solar home for the Solar Decathlon, a competition sponsored by the Department of Energy. Student teams integrated aesthetics and modern conveniences with maximum energy production and optimal efficiency. The houses were transported to the National Mall in Washington, DC, where the student team took first prize overall.
2007	The Technische Universität Darmstadt won the 2007 Solar Decathlon. The team won the architecture, lighting, and engineering contests.

Last Revised: July 2008.
Sources: U.S. Department of Energy, Office of Energy Efficiency and Renewable Energy, "Solar History Timeline:1900s," October 18, 2007.
U.S. Department of Energy, Office of Energy Efficiency and Renewable Energy, "Solar Dish Engine," October 21, 2007.
Smith, C., "History of Solar Power, Revisiting Solar Power's Past," *Technology Review,* July 1995: Solar Power.
National Renewable Energy Laboratory, Feature: "NREL Teams Up with Boeing Spectrolab to Win R&D 100 Award," July 2007.
ASME, "Prime Movers and Power Plants," August 2007.

25

PHOTOVOLTAIC – CHRONOLOGICAL DEVELOPMENT

PUBLIC DOMAIN

A photovoltaic cell, also called a PV or solar cell, is a device that converts light (radiant) energy directly into electrical energy. PV cells are usually made from silicon. The first PV cells were very inefficient, converting less than 1% of radiant energy into electricity. Today, some solar cells have a 40% conversion rate.

1904	Wilhelm Hallwachs (German) discovered that a combination of copper and cuprous oxide was sensitive to light.
1905	Albert Einstein published a paper on the photoelectric effect. He won 1921 Nobel Prize in Physics for these theories.
1950s	Inventors at Bell Labs (Daryl Chapin, Calvin Fuller, and Gerald Pearson) developed a more efficient PV cell (6%) made from silicon. This was the first solar cell capable of generating enough power from the sun to run everyday electrical equipment.
1955	Western Electric began to sell commercial licenses for silicon photovoltaic technologies. Early successful products included PV-powered dollar bill changers and devices that decoded computer punch cards and tape.
1958	Federal support for photovoltaic technology was initially tied to the space program to provide power for the Vanguard satellite.
1973	Spurred by the oil embargo, interest in space applications of photovoltaics grew.
1970s	By the late 1970s, a program for the development of distributed photovoltaics was established by the U.S. Government at the Massachusetts Institute of Technology, focusing on design and demonstration issues for the buildings sector.
1978	The Energy Tax Act of 1978 established a 10% investment tax credit for photovoltaic applications. The Solar Photovoltaic Energy, Research, Development and Demonstration Act of 1978 committed $1.2 billion, over 10 years, to improve photovoltaic production levels, reduce costs, and stimulate private sector purchases. Photovoltaic energy commercialization program accelerated the procurement and installation of photovoltaic systems in Federal facilities.
1980	The Carlisle house (Massachusetts) was completed with participation from MIT, DOE, and Solar Design Associates. It featured the first building-integrated photovoltaic system, passive solar heating and cooling, superinsulation, internal thermal mass, earth-sheltering, daylighting, a roof-integrated solar thermal system, and a 7.5-peak-watt photovoltaic array of polycrystalline modules from Solarex. The Crude Oil Windfall Profit Tax Act of 1980 was enacted, raising the residential tax credit to 40% of the first $10,000 for photovoltaic applications, and the business tax credit to 15%. The Act also extended the credit to the end of 1981. Boeing and Kodak fabricated the first thin-film photovoltaic cells with efficiencies greater than 10%.
1985	The 6-MW Carissa Plains plant was added to Southern California Edison's system. The project was later dismantled.
1989	The Renewable Energy and Energy Efficiency Technology Competitiveness Act of 1989 sought to improve the operational reliability of photovoltaic modules, increase module efficiencies, decrease direct manufacturing costs, and improve electric power production costs. PV for Utility Scale Applications (PVUSA), a national public–private partnership program, was created to assess and demonstrate the viability of utility-scale photovoltaic electric generating systems. PVUSA participants include the DOE and other agencies, the Electric Power Research Institute, the California Energy Commission, and Pacific Gas & Electric (PG&E), and eight other utilities.

(continued)

Alternative Energy and Shale Gas Encyclopedia, First Edition. Edited by Jay H. Lehr and Jack Keeley.
© 2016 John Wiley & Sons, Inc. Published 2016 by John Wiley & Sons, Inc.

1990	Siemens A.G. of Munich, West Germany, acquired California-based ARCO Solar, the world's largest photovoltaic company. The PV Manufacturing Technology (PVMaT) project began. A government-industry research and development partnership between DOE and members of the US photovoltaic industry was designed to improve manufacturing processes, accelerate manufacturing cost reductions for photovoltaic modules, improve commercial product performance, and lay the groundwork for a substantial scale-up of manufacturing capacity.
1992	The University of South Florida fabricated a 15.89% efficient thin-film cell, breaking the 15% barrier for the first time.
1993	Pacific Gas and Electric completed the installation of the first grid-supported photovoltaic system in Kerman, California. The 500-kW system was the first effort aimed at "distributed power," whereby a relatively small amount of power is carefully matched to a specific load and is produced near the point of consumption. New world-record efficiencies in polycrystalline thin film and in single-crystal devices, approaching 16% and 30%, respectively, were achieved in 1993.
1994	The National Renewable Energy Laboratory (NREL) developed a solar cell made of gallium indium phosphide and gallium arsenide; it was the first one of its kind to exceed 30% conversion efficiency.
1995	An Amoco–Enron joint venture announced its intention to use amorphous silicon modules for utility-scale photovoltaic applications.
1998	Subhendu Guha, a scientist noted for his pioneer work in amorphous silicon, led the invention of flexible solar shingles, a roofing material and state-of-the-art technology for converting sunlight into electricity on buildings.
1999	Construction was completed on Four Times Square in New York. The office building had more energy-efficient features than any other commercial skyscraper and included building-integrated photovoltaic panels on the 37th–43rd floors, on the south- and west-facing facades, to produce part of electricity needed for the building. Spectrolab, Inc., and the NREL develop a 32.3% efficient solar cell. The high efficiency resulted from combining three layers of photovoltaic materials into a single cell. Researchers at the NREL developed a record-breaking prototype solar cell that measured 18.8% efficiency, topping the previous record for thin-film cells by more than 1%. Worldwide, installed photovoltaic capacity reached 1000 MW.
2000	First Solar began production at the Perrysburg, Ohio, photovoltaic manufacturing plant. Each year, it could produce enough solar panels to generate 100 MW of power. Astronauts began installing solar panels at the International Space Station, on the largest solar power array deployed in space. Each "wing" of the array consisted of 32,800 solar cells.
2001	BP and BP Solar announced the first BP Connect gasoline retail and convenience store in the United States. The Indianapolis, IN, service station features a solar–electric canopy. The canopy contains translucent photovoltaic modules made of thin-film silicon integrated into glass.
2007	Boeing Spectrolab and the NREL created the High-Efficiency Metamorphic Multijunction Concentrator Solar Cell, or HEMM solar cell, which achieved the highest efficiency level of any photovoltaic device to date. The HEMM solar cell broke the 40% conversion efficiency barrier, making it twice as efficient as a typical silicon cell.

Last Revised: July 2008.
Sources: U.S. Department of Energy, Office of Energy Efficiency and Renewable Energy, "Solar History Timeline: 1900's," January 2008.
U.S. Department of Energy, Office of Energy Efficiency and Renewable Energy, "Solar Cells," October 21, 2007.
Big Frog Mountain, Alternative Energy, "Intro to History of Solar Electric Power," January 2008.

PART III

GEOTHERMAL

26

GEOTHERMAL: HISTORY, CLASSIFICATION, AND UTILIZATION FOR POWER GENERATION

MATHEW C. ANEKE AND MATTHEW C. MENKITI

26.1 INTRODUCTION

Industrialization and world population are the two major factors that drive the global energy demand. Thus, since both are always on the increase, the world energy demand will continuously be on the increase.

Among the different sources of energy, fossil fuel, wind, solar, hydro, geothermal, so on, burning of fossil fuels still remains the most widely used. In 2008, it was estimated that the total global energy consumption was 531×10^{18} J [1]. Out of this, about 86% were produced through the burning of fossil fuels. The projection for 2035 was estimated to be about 812×10^{18} J [2], with fossil fuels still playing a dominant role although there is a remarkable improvement in the use of renewable (Figures 26.1 and 26.2).

The increase in the world population and the enhancement in the social and economic development have also resulted in an increased stress on fossil fuels which are exhaustible. Despite its exhaustible nature, the overdependence on fossil fuels for energy production had resulted in the release of large amount of anthropogenic CO_2, a greenhouse gas, responsible for causing global warming. This detrimental effect on the overdependence on fossil fuels had resulted in a clamor by the world leaders to develop cleaner energy systems in order to ensure the sustainability of our atmospheric, hydrologic, mineral resource, biological, social, and economic systems.

Although no single energy source can meet the global energy demand, however, among the cleaner energy resources, geothermal energy had played an important role in meeting some quota of the world energy demand and is still prominent in meeting the energy demand of the future for sustainable environment.

Geothermal resource is made up of volumes of rock found beneath the earth surface where heat can be harnessed economically [3]. The heat is linked to the internal structure of our planet and the physical processes occurring there [4]. It occurs as a result of the radioactive decay of isotopes of uranium, thorium, potassium, and other radioactive elements contained in the rock [3, 5]. The heat, also known as geothermal energy, is extracted through a carrier phase and continually conducted to the earth surface along the geothermal gradient. The carrier phase or geothermal fluid is usually water and occurs in the form of either liquid or steam.

The geothermal gradient occurs as a result of an increase in the temperature with depth in the earth crust. At depths of over 10,000 m, the average geothermal gradient is about $2.5-3°C/100$ m [4, 6]. As a result of the increase in temperature with depth of the earth crust, heat is transferred by conduction and convection [4] from the deep hotter zones to the shallow colder zones with the tendency to attain an equilibrium condition; however, this equilibrium is never actually attained.

Local and regional geological and tectonic phenomena play a major role in determining the location and quality of a particular resource [7]. For example, regions of higher than normal heat flow are usually associated with tectonic plate boundaries and with areas of geologically recent volcanic events. This is why geothermal resources are usually associated with highly volcanic and plate boundary regions like Iceland, Japan, New Zealand (plate boundaries), Larderello field in Italy (volcanism), therefore, neglecting the exploration of geothermal energy opportunities presented by other regions with neither plate boundary nor volcanic property.

Alternative Energy and Shale Gas Encyclopedia, First Edition. Edited by Jay H. Lehr and Jack Keeley.
© 2016 John Wiley & Sons, Inc. Published 2016 by John Wiley & Sons, Inc.

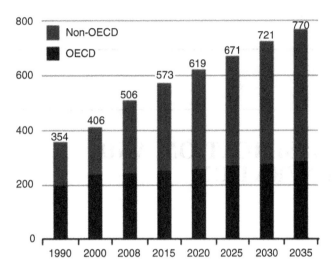

FIGURE 26.1   World energy usage projection (in quadrillion Btu) [2].

The mantle extends from the base of the earth crust for about 2900 km. Studies with the use of seismic waves have proved that the mantle and the crust are solid in nature although there have been cases where localized magma chambers occurs as isolated liquid pockets in both the crust and the upper mantle. The earth's crust and the uppermost part of the mantle together form the lithosphere, the outermost shell of the earth that is relatively rigid and brittle.

The lower boundary of the lithosphere inside the mantle is marked by a particular layer known as the low-velocity zone in which the seismic waves slow down. This zone known as the asthenosphere extends to a depth of about 200 km from the surface and contains rocks which are closer to their melting point than rocks above or below this zone. The lithosphere is believed to be in continual movement, probably due to the underlying mantle convection and plates of brittle lithosphere which moves easily over the asthenosphere.

The core of the earth extends from 2900 to 6370 km with thickness or radius of about 3470 km and density of 10 g/cm^3 at the core–mantle boundary, increasing to 12–13 g/cm^3 at the center of the earth. Its temperature is around 4000°C and pressure at the earth's center is 3.6×10^6 bars.

It has been estimated that about 5×10^{24} kJ of the earth crust's heat energy exists at temperatures above 15°C [3] which is an enormous amount of energy capable of providing the total annual energy need of the world which stands at 519×10^{18} J according to the 2008 estimate [1].

The quantity of geothermal fluid in the rock is a function of its porosity. Rocks found in volcanic regions tend to have higher porosity with values as high as 40% near the surface which decreases to <5% at a depth of several kilometers [3]. Sedimentary rocks like sandstones and mudstones have porosity of about 20%. At any given geothermal resource temperature, the higher the porosity of the rocks, the higher the geothermal fluid contained in them and the higher the extractable heat energy.

So far, only a small fraction of the earth's available geothermal energy has been utilized by mankind, and this occurs at areas in which the geological conditions permit a carrier to transfer the heat from the hot deep zones to or near the earth surface [6]. With technological advancement, it is believed that in the near future innovative techniques will offer new perspectives into extracting geothermal energy from the regions where the geological conditions are not favorable (analogous to what is currently happening in the oil and gas industry).

However, experience had shown that geothermal resources are not limited to those areas mentioned above. There are vast magnitudes of geothermal resources with global distribution most of which are yet untapped, and with the improvement in technological advancement, geothermal energy will contribute a reasonable quota of the world energy demand for centuries to come.

26.2   THE EARTH STRUCTURE

The earth is made up of three concentric zones: crust, mantle, and core (Figure 26.3). The earth crust is the outermost part of the earth with a thickness of about 7 km on the average under the ocean basins, and 20–65 km under the continents [4].

Geothermal resources occur at temperatures which vary from one location to another across the earth surface with values ranging from about 50°C to 350°C [8, 9]. The temperature obtainable in any given location is dependent on the level of radioactive decay occurring in the rocks beneath the surface. Furthermore, the temperature and the nature of the rocks also determine the state and constituent of the geothermal fluid.

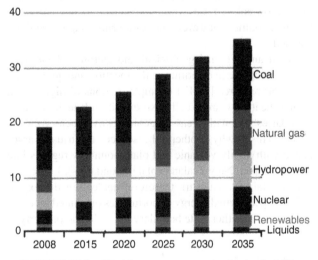

FIGURE 26.2   World energy usage projection by fuel [2].

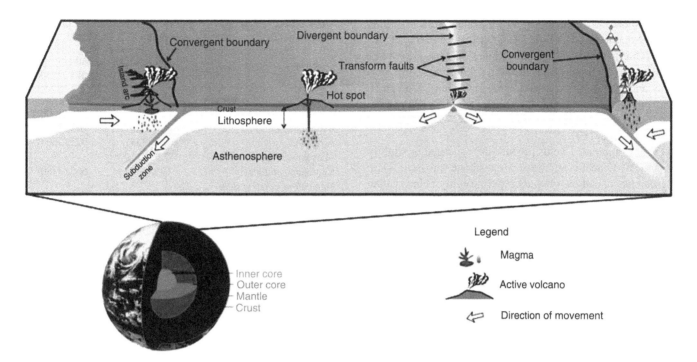

FIGURE 26.3   Layers of earth and typical tectonic settings for geothermal systems.

26.3   NATURE AND CONSTITUENT OF GEOTHERMAL FLUID

Geothermal systems are classified into four different types: hydrothermal, hot dry, geopressured, and magmatic [10]. With the current technological status, the system usually utiliszd is the hydrothermal system. It is believed that in the foreseeable future, advancement in technology will lead to the exploitation of the other three systems.

Hydrothermal systems, also known as the geothermal reservoirs or fields, are usually classified into two categories:

- Water dominated
- Vapour dominated

Water-dominated fields are further subdivided into two categories:

- Hot water fields
- Wet steam fields

As the name implies, the hot water fields produce hot water while the wet steam fields produce a mixture of water and steam with the quality of the steam dependent on the temperature and pressure of the geothermal fluid.

The hot water field produces surface water at temperatures up to 100°C. They usually contain water in the liquid phase and occur as thermal spring on the surface of the earth. In

some cases, their temperature is close to the boiling water temperature. A hot water field is usually considered to have economic interest if the reservoir is found at a depth of less than 2 km with a salt content of the water lower than 60 g/kg and a flow rate of over 150 t/h [10]. Some examples of hot water fields include those of Pannonian basin in Hungary, Paris basin and Aquitanian basin in France, the Po river valley in Italy, Klamath Falls and Chena hot spring in the United States, and Tianjin in China.

On the other hand, wet steam fields contain pressurized water at temperatures higher than 100°C and small quantities of steam in the shallower, lower pressure parts of the reservoir. The continuous phase is the liquid and it controls the pressure inside the reservoir. The steam always occurs in the form of bubbles surrounding the water and has negligible effect on the pressure of the fluid [10]. As a common practice in this type of system, only the steam is used to generate electricity while the liquid water must be removed at the surface in special separators. This type of fields usually manifest in the earth surface in the form of boiling springs and geysers. The produced water often contains large quantities of chemicals which cause severe scaling problems to pipelines and plants. Over 90% of the hydrothermal reservoirs exploited industrially are usually of the wet steam type. Some typical examples of wet steam fields used for power generation include Cerro Prieto, Los Azufres and Los Humeros (Mexico), Momotombo (Nicaragua), Ahuachapan-Chipilapa (El Salvador), Miravalles (Costa Rica), Zunil (Guatamala), Wairakei, Ohaaki, and Kawerau (New Zealand), Salton Sea,

Coso, and Casa Diablo (California), Puna (Hawaii), Soda Lake, Steamboat, and Brady Hot Springs (Nevada), Cove Fort (Utah), Dieng and Salak (Indonesia), Mak-Ban, Tiwi, Tongonan, Palinpinon, and Bac Man (Philippines), Pauzhetskay and Mutnovsky (Russia), Fang (Thailand), Kakkonda, Hatchobaru, and Mori (Japan), Olkaria (Kenya), Krafla (Iceland), Azores (Portugal), Kizildere (Turkey), Latera (Italy), Milos (Greece) [10].

Vapor-dominated fields produce dry saturated or slightly superheated steam at above atmospheric pressure [10]. Researches had shown that their permeability is lower than in wet steam fields. In this field, steam serves as the dominant or the continuous phase and regulates the pressure of the reservoir which is practically constant throughout. The produced steam also contains other gases mainly CO_2 and H_2S. The surface geothermal activity associated with this type of field is similar to that present in wet steam fields. About 50% of the geothermal electric energy generated in the world comes from six vapor-dominated fields: Larderello (Italy), Mt. Amiata (Italy), The Geysers (California), Matsukawa (Japan), Kamojang and Darajat (Indonesia).

Some major examples of water and vapor-dominated reservoirs and their respective geothermal fluid temperatures are shown in Table 26.1.

In summary, of approximately 100 hydrothermal systems investigated so far, less than 10% are vapor dominated, 60% are wet steam fields (water dominated), and 30% produce hot water [11].

Generally, geothermal steam or water is not pure but rather contains some other compounds. The kind of compound depends on the geological formation of the area where the geothermal fluid is formed. The average steam composition of some geothermal field under exploitation is given in Table 26.2. The fluid contains mainly gases such as CO_2, H_2O, HCl, HF, NH3, CH_4, H_2 and occurs in different ranges from field to field. In any given field, the content of these gases tends to decrease with the age of the field as production continues.

26.4 HISTORY OF GEOTHERMAL

Like in many areas of life, the practical application of geothermal energy precedes its scientific research. There are

TABLE 26.1 Reservoir temperature and enthalpies of geothermal fields generating electricity.

Geothermal field	Reservoir temp, C ()max	Enthalpy max kcal/kg (kJ/kg)
Vapor-dominated systems		
The Geysers (USA)	237 (310)	718 (3000)
Larderello (Italy)	200 (420)	742 (3100)
Amiata (Italy)	154 (344)	622 (2600)
Matsukawa (Japan)	220	
Kamojang (Indonesia)	175 (248)	665 (2780)
Water-dominated systems		
Wairakei (New Zealand)	230 (290)	281 (1175)
Broadlands (New Zealand)	280 (326)	401 (1675)
Imperial Valley fields (USA)	160 (370)	239 (1000)
Cerro Prieto (Mexico)	265 (388)	581 (2430)
Los Azufres (Mexico)	175 (300)	646 (2700)
Los Humeros (Mexico)	310 (418)	622 (2600)
Momotombo (Nicaragua)	210 (327)	646 (2700)
Ahuachapan (El Salvador)	210 (240)	660 (2760)
Tiwi (Philippines)	273 (309)	670 (2800)
Mac-Ban (Philippines)	207 (313)	428 (1790)
Hatchobaru (Japan)	218 (308)	538 (2250)
Olkaria (Kenya)	205 (330)	574 (2400)
Krafla (Iceland)	205 (344)	641 (2680)

Source: From Reference 12.

some archaeological evidences which show that the earlier humans probably made use of geothermal water that occurs in natural pools and hot springs for cooking, bathing, keeping warm, and performing some religious activities [5].

Early recorded history shows that Romans, Japanese, Turks, Icelanders, Central Europeans, as well as the New Zealander's used geothermal resources for bathing, cooking, and space heating [5, 13]. Balneology in the early Roman Empire, the middle kingdom of the Chinese, and the Turkish has also been carried out using geothermal hot springs.

The first practical recorded industrial application of geothermal energy was carried out in a chemical extraction process by Francesco Larderel in Larderello region of Italy in 1827 [5,6], where he utilized the heat from boric fluids emerging naturally to extract boric acid from boric hot water rather than burning woods from the rapidly depleting forest (Figure 26.4).

TABLE 26.2 Composition of steam from some geothermal fields.

Constituents (g/kg)	The Geyers (USA)	Larderello (Italy)	Matsukawa (Japan)	Wairakei (New Zealand)	Cerro Prieto (Mexico)
H_2O	995.9	953.2	986.3	997.5	984.3
CO_2	3.3	45.2	12.4	2.3	14.1
H_2S	0.2	0.8	1.2	0.1	1.5
NH_3	0.2	0.2			0.1
$CH_4 + H_2$	0.2	0.3			
Others	0.2	0.3	0.1	0.1	

Source: From Reference 4.

FIGURE 26.4   Covered lagoon used to collect hot boric waters and extract the boric acid [6].

This was followed by the use of geothermal steam to raise liquids in primitive gas lifts and later in reciprocating and centrifugal pumps and winches used in drilling and the local boric acid industry. Between 1910 and 1940, low-pressure steam of Tuscany, Italy, was used to heat the industrial and residential buildings and greenhouses [6]. The world's first geothermal district heating system was started in the 14th century at Chaudes-Aigues in France, while the oldest geothermal district heating project in the United States occurs in Warm Springs Avenue in Boise, Idaho, in 1892 [5]. In 1928, Iceland exploited its first geothermal energy for domestic heating purpose.

Ever since then, the use of geothermal energy for different applications has always been on the increase with the total installed capacity well over 19,500 MW by 2004. With more geothermal energy been discovered and utilized and more countries joining the league of geothermal energy users, this value will always be on the increase.

26.5   CLASSIFICATION OF GEOTHERMAL ENERGY

There is no standard international classification of geothermal resources. Different researchers have adopted different classifications to suit their individual research interest. However, most of the classifications are generally based on temperature.

In some literature [3], they are classified into two groups:

- Low-temperature geothermals <150°C
- High-temperature geothermals >150°C

while in others [6, 8] they are classified as

- Low-temperature geothermals 70–100°C
- Mid-temperature geothermals 100–150°C
- High-temperature geothermals >150°C

There are still other classifications which can be found in literature [14–16].

26.6   UTILIZATION OF GEOTHERMAL ENERGY FOR POWER GENERATION

The use of geothermal energy is generally divided into two categories:

- Direct use
- Electricity use

This division is purely based on the temperature of the geothermal resource.

Direct uses are regarded as those uses where the geothermal fluids are used for any other purpose (domestically, industrially, or commercially) than for power generation. This includes bathing, space heating, district heating, cooling, swimming, balneology, aquaculture, space conditioning, greenhouse heating, snow melting, heat pumps, and so on.

Initially, geothermal resources at temperatures lower than 100°C are usually used for direct application and regarded as not been good enough for power generation. However, with the recent development in geothermal binary power plants

FIGURE 26.5   First geothermal power generation experiment at Larderello, Italy, in 1904 by Prince Ginori Conti [6].

like Organic Rankine Cycle and Kalina Cycle, geothermal energy resources lower than 100°C are currently been utilized for power generation at commercial scale.

The use of geothermal energy for commercial electricity generation has been in existence since 1913 [17] and ever since then there has been a tremendous increase although the increase is small compared with the vast deposit of geothermal resources in the world.

The first electricity generation from geothermal energy started as far back as 1904 at Larderello, Italy [6, 18]. The plant which was developed by Prince Ginori Conti was in experimental scale and makes use of geothermal wet steam to produce pure steam in a heat exchanger which is used to produce power used to light five bulbs using piston engine coupled with a 10-kW dynamo (Figure 26.5).

The success of the experimental work encouraged Ginori to develop the first prototype of a geothermal power plant in 1905. The plant makes use of Cail reciprocating engine coupled to a 20-kW dynamo. In 1908, another reciprocating engine, Neville was coupled to a second 20-KW dynamo. Both were used to provide the electricity need of Larderello industrial plant and some main buildings.

The first commercial geothermal power plant known as Larderello 1 [18] was commissioned in 1913 and generates 250 kW of electricity using a turbine which was driven by pure steam generated using a heat exchanger supplied with a geothermal fluid from two wellheads at 200–250°C (Figure 26.6). The plant was designed and built by Tosi Electromechanical Company and uses a geothermal wellhead pressure of about 3 atm. The power generated from the

FIGURE 26.6   Larderello 1; first commercial geothermal power plant, 250 kW, Larderello, Italy, 1913 [18].

FIGURE 26.7   First direct cycle geothermal power plant at Serrazzano, Italy, 1923 [18].

plant was used to meet the power requirement of the whole Larderello chemical plant and the surrounding villages.

Furthermore, in 1923, two 3.5-MW turbo alternators using indirect cycle were installed. This was followed by the first direct cycle pilot plant installation at Serrazano in 1923. The plant makes use of natural steam from the geothermal resource directly fed to a turbine to generate 23 kW_e (Figure 26.7).

In the late 1920s, more direct cycle plants were installed at Castelnuovo and Larderello with an installation capacity of 600 and 800 kW at the former and 3.5 MW at the latter [18]. The installation capacity increased gradually, and by 1943, the total capacity rose to about 132 MW with 107 MW using the indirect cycle technique. However, the increase was hindered in 1944 as a result of the World War II (WWII) (Figure 26.14) which resulted in the bombing of all the major geothermal power plant installations in the Larderello area except the 23 kW direct cycle plant (Figure 26.8).

Immediately after the WWII, the destroyed geothermal plants were rebuilt and new ones constructed, and by 1950 the installation capacity rose to about 300 MW and ever

since then, the installation capacity of geothermal power plant has continuously been on the increase with the technology penetrating other countries with adequate geothermal resources.

The success of the geothermal power generation in Italy provided opportunity for other countries like Japan, USA, New Zealand, Mexico, Iceland, and so on to tap into the geothermal power sector.

Japan carried out the first geothermal power plant test in 1925 at Beppu, Oita Prefecture, in southern Japan [19]. This was followed by Otake in 1926, but unfortunately, the tests were not successful. In 1966, they developed the first dry steam-based commercial geothermal power plant at Matsukawa with an installation capacity of 23 MW of electricity [18, 20] (Figure 26.9). Nowadays, it is estimated that there are over 535 MW of electricity generation in Japan using geothermal power plant.

In the United States, the first successful geothermal power plant was constructed at Geysers in the early 1930 by John Grant (Figure 26.10). The plant uses 153°C of geothermal steam at a pressure of 4 bar to generate 35 kW of electricity

FIGURE 26.8   Larderello geothermal plants destroyed during WWII [18].

FIGURE 26.9   First geothermal power plant in Japan.

using two reciprocating steam engine-driven turbine generators from General Electric.

This was followed by another plant in 1960, and by 1968, the capacity of the Geysers geothermal field rose to about 82 MW. By 1989, 29 geothermal units had been constructed with an installed capacity of 2098 MW [18]. The total installed capacity in the United States currently stands at over 2534 MW (more than 2000 operating units).

Mexico joined the league of geothermal power producers in 1959 when a dry steam from the Pathé geothermal field was used to generate about 3.5 MW_e. The plant was decommissioned in 1972 but later fields were developed which brought the geothermal installation capacity in Mexico to about 953 MW.

Recent developments and successful operation of early geothermal installations resulted in the rapid increase in geothermal power plant construction all over the world.

In 1975, 30-MW geothermal power plant was installed at Ahuachapan, El Salvador. In 1980, plants were commissioned in Indonesia [21], Turkey, Kenya, the Philippines, and Portugal. Greece, France, and Nicaragua joined in 1985, while Thailand, Argentina, Taiwan, and Australia followed suit in 1990, while the plant in Greece was shut down the same year. Costa Rica started geothermal power production in 1994 [22], while Austria, Guatemala, and Ethiopia came online in 2000 with Argentina shutting down their plant the same year. Germany and Papua New Guinea joined the league in 2004.

Technological advancement in the binary power plant technology (Organic Rankine Cycle and Kalina Cycle) widens the application of geothermal resources for power generation to the lower and mid-temperature regions.

The first recorded low-temperature geothermal power plant known as Paratunka power plant was located at

FIGURE 26.10   First geothermal power plant at The Geysers, USA, early 1930.

FIGURE 26.11   First binary plant using geothermal water at 81°C.

Kamchatka peninsula in eastern Siberia, Russia. The plant makes use of R-12 refrigerant as working fluid and produces power using geothermal water at 81°C (Figure 26.11).

Another binary power plant was commissioned in Austria and uses 110°C of geothermal water to generate about 2.5 MW of power. In Thailand, a 300-kW binary plant using 116°C of geothermal water was installed (Figure 26.12), and the electricity generated was used to replace a diesel generator.

In 2006, a 400-kW binary geothermal power plant was commissioned in Alaska, USA (Figure 26.13). The plant was designed and constructed by the United Technologies and makes use of R134 refrigerant as working fluid with geothermal water at 73.33°C [23].

26.7   GEOTHERMAL TO POWER PRODUCTION FOR SUSTAINABLE ENVIRONMENT: PRESENT AND FUTURE PROSPECTS

The world energy demand had always been on the increase, and one energy source will never be enough to meet this demand. The only way of meeting this demand is through the use of energy mix system which will involve both the use of conventional and renewable energy system. However, the proportion of each portion (conventional and renewable) will vary significantly as time goes by. With the recent interest in safeguarding our environment through the use of renewable energy system (to which geothermal energy belongs), it is a well-established fact that renewable energy systems will play

FIGURE 26.12   Binary power plant, 300 kW, at Fang, Thailand.

FIGURE 26.13   400 kW Chena binary geothermal power plant, Alaska, USA.

a prominent role in the future in meeting the world energy demand.

Ever since the first geothermal installation in 1904, the total installed capacity has always been on the increase except during the WWII when its growth was hindered (Figure 26.14).

In year 2000, the installed capacity of the world geothermal electricity was estimated to be in the neighborhood of 7974 MW$_e$ with generation in that year of 49.3 billion kWh of electricity (Table 26.3). After the compilation of this table by the author, it was also recorded that a further 282 MW$_e$ was also installed that same year bringing the figure to 8256 MW$_e$.

This figure has no doubt been on the increase with many other countries joining the league of geothermal electric producers and others expanding their installations.

In 1973, the total electricity production in the world was estimated to be about 6116 TWh. Out of this, 21.6% were from renewable with the new renewable like biomass, geothermal, wind, solar, and tidal contributing only 0.6%, while 21% came from the hydrothermal energy [24]. In 1998,

TABLE 26.3   Installed geothermal generating capacities in the world in the year 2000 [26].

Country	Installed MWe	GWh generated	Percentage of national capacity	Percentage of national energy
Australia	0.17	0.9	n/a	n/a
China	29.17	100	n/a	n/a
Costa Rica	142	592	7.77	10.21
El Salvador	161	800	15.39	20
Ethiopia	8.52	30.05	1.93	1.85
France	4.2	24.6	n/a	2
Guatemala	33.4	215.9	3.68	3.69
Iceland	170	1138	1304	14.73
Indonesia	589.5	4575	3.04	5.12
Italy	785	4403	1.03	1.68
Japan	546.9	3532	0.23	0.36
Kenya	45	366.47	5.29	8.41
Mexico	755	5681	2.11	3.16
New Zealand	437	2268	5.11	6.08
Nicaragua	70	583	16.99	17.22
Philippines	1909	9181	n/a	21.22
Portugal	16	94	0.21	n/a
Russia	23	85	0.01	0.01
Thailand	0.3	1.8	n/a	n/a
Turkey	20.4	119.73	n/a	n/a
USA	2228	15,470	0.25	0.4
Totals	7974.06	49,264.45		

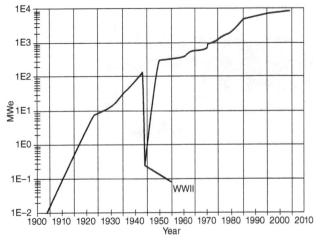

FIGURE 26.14   World Geothermal power generation 1904–2004.

Source: From Reference 26.

the total electricity production in the world was estimated to be 14,411 TWh; out of this, about 20% were from renewable with the new renewable contributing 1.6% of the energy [17, 25]. In 2007, the total electricity generation in the world was 19,771 TWh with the new renewable contributing about 2.6% of the electricity [24].

The trend shows that the world energy demand is continuously on the increase and that the use of renewable energy system in meeting the world energy demand is also on the increase.

Although the use of conventional electricity generation systems which are based on fossil fuels like crude oil, coal, natural gas, so on, still play a dominant role in meeting the world electricity demand, there is a general belief that their share of the world energy will definitely decrease in the future as many world leaders are now promoting the use of clean energy systems in meeting the world energy demand in order to reduce CO_2 emission; a greenhouse gas believed to be responsible for causing global warming.

Although geothermal energy is categorized among the "new renewable" energy systems, it is not a new energy source as had been established in Section 1.3. This long stay of geothermal power technology has made it a mature technology, and this had given the geothermal energy system an advantage over other renewable energy systems.

For example, according to a comprehensive report by The World Energy Assessment [27] which was compiled in year 2000, it was estimated that geothermal energy accounted for 69.6% of the electricity generation from the "new renewables," followed by wind energy with 27.2%. This value confirms the reliability of geothermal plants which can operate at capacity factors in excess of 90%.

Geothermal energy like other renewable energy systems is a clean energy system with little environmental impact. It is independent of weather as opposed to other renewable energy systems like solar, wind, or hydro applications. It has an inherent storage capability and can be used for both base load and peak power plant.

With the recent development in the binary power plant technology, the range of application of geothermal energy systems for electricity generation has also been increased with plants producing electricity from geothermal fluid sources as low as 70°C.

Furthermore, the technological advancement in the development of hot dry rock (HDR) [10] systems, an artificial form of geothermal reservoir will also reposition geothermal power generation as a renewable energy of choice in the future.

HDR systems are man-made reservoirs in rocks that are artificially fractured, and thus any convenient volume of HDR in the earth's crust, at accessible depth, can become an artificial reservoir [28]. A pioneer HDR project has been carried out at Los Alamos, New Mexico, USA, and the result of the project proves its economical viability. The success of the project opened up field experiments in other countries like the United Kingdom, France, Germany, Japan, Sweden, and Australia.

Thus, HDR systems could, however, represent the new frontier especially in areas with no natural hydrothermal system thus widening the application of geothermal energy generation in areas without natural geothermal capability.

Furthermore, studies have shown that the use of HDR with supercritical CO_2 instead of water can also improve the electricity of the HDR system and reduce some operational problems of heat exchanger scaling [29]. It can also create opportunities in carbon capture and storage (CCS) by using the CO_2 captured in the supercritical HDR system for power generation.

26.8 CONCLUSION

The overall installed capacity of geothermal power plants has continuously been on the increase after the WWII (Figure 26.14).

Considering the level of technological advancement in the geothermal power generation technology, it can be regarded as a mature technology, however, more improvement is still been expected especially in the low- to mid-temperature range especially in the development of high-efficiency turbines/expanders which will help to improve the overall efficiency of the cycle and also in the development of both HDR systems and supercritical HDR system which will create geothermal resources in areas where it never existed naturally.

Thus, geothermal energy will play a prominent role in meeting the world energy demand both currently and in the future and is foreseen to be a renewable energy of choice in the future as a result of its stability and sustainable environmental friendliness.

REFERENCES

[1] EIA. (2010). International Energy Statistics. In: *U.S. Energy Information Administration Independent Statistics and Analysis*. Department of Energy, USA.

[2] EIA. (2011). International Energy Outlook. In: *U.S. Energy Information Administration Independent Statistics and Analysis*. Department of Energy, USA.

[3] Thain, I., Reyes, A. G., and Hunt, T. (2006). A practical guide to exploiting low temperature geothermal resources. GNS Science Report 2006/09.

[4] Barbier, E. (1997). Nature and technology of geothermal energy: a review. *Renewable & Sustainable Energy Review*, 1(1/2):1–69.

[5] Lund, J. W., et al. (2008). Characteristics, development and utilization of geothermal resources—a Nordic perspective. *Episodes*, 31(1):140–147.

[6] Dickson, M. H. and Fanelli, M. (2004). *What Is Geothermal Energy?* CNR-Istituto di Geoscience e Georisorse, 1–61.

[7] Mock, J. E., Tester, J. W., and Michael Wright, P. (1997). Geothermal energy from the earth: its potential impact as an environmentally sustainable resource. *Annual Review of Energy and the Environment*, 22:305–356.

[8] Aneke, M., Agnew, B., and Underwood, C. (2011). Performance analysis of the Chena binary geothermal power plant. *Applied Thermal Engineering*, 31(10):1825–1832.

[9] Hettiarachchi, H. D. M., et al. (2006). Optimum design criteria for an organic rankine cycle using low-temperature geothermal heat sources. *Energy*, 32:1698–1706.

[10] Barbier, E. (2002). Geothermal energy technology and current status: an overview. *Renewable & Sustainable Energy Review*, 6:3–65.

[11] Hochstein, M. P. (1990). Classification and assessment of geothermal resources. In: *Small Geothermal Resources—A Guide to Development and Utilisation*, Dickson, M. H. and Fanelli, M. (editors). UNITAR/UNDP Centre on Small Energy Resources, Rome.

[12] Sommaruga, C. and Zan, L. (1995). World geothermal resources- main characteristics and maximum values, Unpublished data.

[13] Hepbasli, A. and Ozgener, L. (2004). Development of geothermal energy utilization in Turkey: a review. *Renewable & Sustainable Energy Review*, 8:433–460.

[14] Sanyal, S. K. (2005). Classification of geothermal systems—a possible scheme. In: *Thirtieth Workshop on Geothermal Reservoir Engineering*. Stanford University, Stanford, California.

[15] Axelsson, G. and Gunnlaugsson, E. (2000). *Background: Geothermal utilization, management and monitoring.* In: Axelsson G and Gunnlaugsson E, (convenors), Long -term monitoring of high and low-enthalpy field under expoitation. World Geothermal Congress 2000, Pre -Congress Course, Kokonoe, Japan, p. 3–10.

[16] Nicholson, K. (1993). *Geothermal Fluids*. Springer, Berlin, XVIII: 264pp.

[17] Fridleifsson, I. B. (2001). Geothermal energy for the benefit of the people. *Renewable & Sustainable Energy Review*, 5:299–312.

[18] Lund, J. W. (2005). 100 Years of geothermal power product. In: *Proceedings, Thirtieth Workshop on Geothermal Reservoir Engineering*. Stanford University, Stanford, California.

[19] Kawazoe, S. (2004). *Geothermal Japan: History and Status of Geothermal Power Development and Production*. International Geothermal Development.

[20] Hanano, M. (2003). Sustainable steam production in the Matsukawa geothermal field, Japan. *Geothermics*, 32:311–324.

[21] Hochstein, M. P. and Sudarman, S. (2008). History of geothermal exploration in indonesia from 1970 to 2000. *Geothermics*, 37:220–266.

[22] Moya, P. and DiPippo, R. (2007). Unit 5 bottoming binary plant at Miravalles geothermal field, Costa Rica: planning, design, performance and impact. *Geothermics*, 36: 63–96.

[23] Holdmann, G. (2007). 400kW geothermal power plant at Chena Hot Springs, Alaska. Final Project Report Prepared for the Alaska Energy Authority, 1–37.

[24] IEA. (2009). *Key World Energy Statistics*. International Energy Agency, 16th September, 2011, available at: www.iea.org

[25] Fridleifsson, I. B. (2003). Status of geothermal energy amongst the world's energy sources. *Geothermics*, 32:379–388.

[26] Huttrer, G. (2001). The status of world geothermal power generation 1995–2000. *Geothermics*, 30:1–27.

[27] WEA. (2000). *World Energy Assessment: Energy and the Challenge of Sustainability*. Prepared by UNDP, UN-DESA and the World Energy Council United Nations Development Programme, New York.

[28] Saito, S., Sakuma, S., and Uchida, T. (1998). Drilling procedures techniques and test results for a 3.7 km deep 500°C exploration well, Kakkonda, Japan. *Geothermics*, 27:573–590.

[29] Brown, D. W. (2000). A hot dry geothermal energy concept utilizing supercritical CO_2 instead of water. In: *Twenty-Fifth Workshop on Geothermal Reservoir Engineering*. Stanford University, Stanford, California.

27

ENHANCED GEOTHERMAL SYSTEMS

Rosemarie Mohais, Chaoshui Xu, Peter A. Dowd, and Martin Hand

27.1 INTRODUCTION

Pliny's *Natural History* and the Aristotelian *Problemata* (Martin, 2011) document the long-standing belief that the Earth's interior is much hotter than the surface. Speculation continued well into the late eighteenth century (Hutton, 1785), but it was only in the mid-nineteenth century that experiments in support of this hypothesis began (Caldecott, 1845; Thomson, 1860). Caldecott's measurements of temperature at depths of 3, 6, and 12 French feet below the Earth's surface prompted the following statement: "The most remarkable circumstance connected with the observations, is the extraordinary excess uniformly observed of the earth temperature above the air temperature" (Caldecott, 1845). It was later determined that the Earth's internal heat came from radioactive sources (Rutherford and Barnes, 1904).

It is easy to relate to Caldecott's excitement when one encounters visible geothermal systems such as hot springs or mud pools. Generally, a geothermal system is characterized by a reservoir capable of storing a large quantity of heat, which is prevented from escaping by some sort of barrier. Different geological conditions give rise to different types of geothermal systems, for example, vapour-dominated, hot water, geopressured, hot dry rock (HDR), and magma (Gupta and Roy, 2007). Where geothermal fluids, vapour and liquid, reach the Earth's surface through cracks or fissures, the energy is readily exploitable as at Larderello in Italy and in many parts of Iceland (Figure 27.1). As can be expected, the associated fluid enthalpy in surficial systems is lower than that of fluids extracted from sub-surface layers, however, the feasibility of heat production from deep-earth environments depends on the presence of fluid at, and in the vicinity of, an accessible drilling depth at the borehole site, and the rocks that contain the fluid being sufficiently permeable to allow flow and, thereby, extraction of heat. As a large portion of the world's hot dry rock geothermal resources is in rocks that are not in contact with an aquifer, the challenge is to engineer systems that will enable the extraction of geothermal energy from hot dry rocks.

27.2 HDR GEOTHERMAL RESOURCES AND ENHANCED GEOTHERMAL SYSTEMS

The US Energy Research and Development Administration describes a HDR geothermal resource as the heat stored in rocks within 10 km of the Earth's surface from which energy cannot be economically produced by natural water or hot steam. The definition limits these resources to those with temperatures less than 650°C so as to exclude magma. The source of heat can be classified as (a) igneous related, (b) upper-mantle related, and (c) local (due to a high concentration of radioactive materials or large-scale faulting) (Gupta and Roy, 2007).

In 1927, John L. Hodgson proposed the possibility of exploiting stores of sub-surface heat by using a downhole heat exchanger. He conjectured that sinking a borehole would cause fragmentation of the rock at the bottom of the hole giving rise to an area of high permeability through which water could be circulated to extract thermal energy (Smith, 1983). Drilling was discouraged by the realisation that, due to the low heat conductivity, the rock temperature near the borehole would reduce significantly with repeated extraction of heat. About 40 years later, the idea was rekindled, somewhat serendipitously (see Brown et al., 2012 for a historical perspective), by a group of scientists in Los Alamos who

Alternative Energy and Shale Gas Encyclopedia, First Edition. Edited by Jay H. Lehr and Jack Keeley.

proposed the extraction of heat from HDR through engineered geothermal systems (EGS). The plan involved drilling two boreholes in impermeable rock and then creating a reservoir to link the injection and production wells, using water as the heat exchange fluid. After an in-depth mathematical study by Harlow and Pracht (1972), a full-scale project was implemented; the principal limitation at that time was drilling costs. Driven perhaps by international competitiveness or possibly a genuine belief in the idea, other countries, such as Japan, Germany, the United Kingdom, France, and Australia also invested in HDR projects, thereby establishing a new industry.

The first step in HDR geothermal energy extraction is fracture stimulation of a zone of high heat-generating granite in order to convert large hot masses of rock into a permeable thermal field (Christopher and Armstead, 1978). Once permeability is established by creating a connected fracture network, cold water or other suitable fluids can be injected in one well and returned via another to the surface as steam. Heat is conducted through the intact rock blocks toward fractures and transported to the production well via convection (within the heated fluid) and advection (the transport of heat by the bulk flow of the fluid). The circulation may be free or partially or totally forced by pumping. The two principal parameters that determine the technical and economic potential of a HDR geothermal resource—the mean rock temperature and the reservoir size—depend on the drilling depth and the effectiveness of the hydro-fracturing process. Although these two variables can be controlled by reservoir engineers, the factors that affect the grade of the geothermal resource, viz., the temperature–depth relationship and the permeability and porosity of the reservoir rocks, are specific to each site. Rummel (2005) estimates that cooling 1 km^3 of hot rock by 100°C will enable operation of a 30-MWe geothermal power plant for 30 years.

The terms Enhanced Geothermal System (EGS) and Engineered Geothermal System (also abbreviated EGS) are used synonymously to describe man-made permeable HDR reservoirs. By strict definition, HDR reservoirs must be fully engineered, dry, and confined systems; on that basis Brown et al. (2012) argue that the Fenton Hill EGS is the only one that can be considered a true HDR system. Under this definition, for cases in which permeability is initially inherent in the rock matrix, fluid is present and there is no clear boundary to the reservoir, the system cannot be described as HDR and the term Hot Wet Rock (HWR) may be more suitable. However, Brown's definition is not always strictly applied in the literature and EGS projects in Japan, Australia, and France, for example, that do not strictly obey Brown's criteria for HDR, are sometimes loosely described as HDR geothermal systems. In this chapter, the term "Enhanced Geothermal System" is used to describe any geothermal system that has been altered in order to improve the quantity and efficiency of heat extraction from sub-surface rocks.

FIGURE 27.1   Geothermal energy is easily accessible in many parts of the world. The Nesjavellir geothermal power plant is one of many in Iceland. Image courtesy of public domain: https://en .wikipedia.org/wiki/File:NesjavellirPowerPlant_edit2.jpg

27.3   EGS TO DATE

The first experiment in HDR technology was at Fenton Hill, in the Jemez Mountains of New Mexico (Figure 27.2), where an initial well 3 km deep intercepted granitic basement rock at 195°C. Between 1974 and 1995, two reservoirs were created, Phase I and Phase II, at depths of 2700 m and 3600 m at temperatures of 180°C and 240°C respectively. These reservoirs were considered to be fully engineered and confined, meaning that they were created in an originally impermeable region and the reservoir of pressurized fluid was totally contained within sealed rock (see Chapter 2, Brown et al., 2012 for further details); as a result, fluid losses were minimal. Hydraulic stimulation was used to open, dilate, and extend a network of pre-existing sealed joints in the hot dry rock. Besides hydraulic stimulation, there also exist several other techniques that can be used to create artificial pathways for fluid flow, for example, thermal and chemical/acid fracturing. Hydraulic fracturing is however a very successful mechanism as illustrated in experiments conducted in Groβ Schönebeck EGS research site, situated in the low permeability region of the North German Basin. In Groβ Schönebeck, cyclic changes of flow rates over 6 days with the incorporation of quartz, led to a fourfold increase in reservoir productivity (Zimmermann et al., 2010). The primary function of the quartz was to prop open existing and new fractures, however, although proppants are commonly used in oil and gas reservoirs, their use in EGS is contestable as they have the potential to dissolve or act as nucleation sites for mineral deposition in fractures, which can cause significant alteration of the flow.

During the fracturing process, seismic and tracer mapping techniques are commonly used to verify the creation of fractured reservoirs (see Section 27.4), which may be of the

FIGURE 27.2 The original Fenton Hill EGS site. From: http://www.lanl.gov/history/photo.php?photo_id=290&story_id=33&page_num=1&row_num=0&photo_num=1

order of cubic kilometer, as in Fenton Hill which was approximately 1 km^3. Locations of the seismic events at Fenton Hill (see Figure 27.3) demonstrated that the seismicity occurred along surfaces that were favourably aligned with the *in situ* stress field to allow for shear slip (Murphy and Fehler, 1986). It is important to understand the *in situ* stress field as it dictates the orientation of tensile fractures, borehole stability, and the state of stress along existing faults (Zimmermann et al., 2010). This allows the reservoir engineer to make informed decisions on well planning such as the distance between wells and the direction of well deviation in relation to the stress field and fracture zones.

The Fenton Hill reservoir achieved a thermal power output of 4 MW for routine production intervals. Closed-loop circulation of fluid in a man-made reservoir from hydraulic stimulation of impermeable granite, meant that nothing was released to the environment other than heat. The designed well-doublet comprising one injection and one production well served as a prototype of a basic sustainable system. Another example is at Groβ Schönebeck, which provides access to geothermal reservoir at about 4100 m. In the Soultz-sous-Forêts EGS in France (see Table 27.1), however, a triplet system is used that comprises one injection well and two production wells to complete the water circulation system. Expansion of production capability may include drilling multiple wells along the line connecting the injection and production wells, but this will only improve a classic HDR system when the stimulated fractures are aligned along this line. In fact, it is possible to enlarge the effective heat transfer area through many vertical wells aligned in various patterns (e.g., hexagonal); the optimum number of these wells depends on the geological characteristics of the reservoir, the target depth, the initial productivity of the reservoir rocks, and the cost of stimulation treatments (Zimmermann et al., 2010).

Ultimately, the thermal output at Fenton Hill was too small for commercial development of the concept. Initially it was believed that drilling costs were the main limitation on the productivity of an EGS, but with experience, practical engineering considerations emerged as critical factors in obtaining a large thermal output. These include the total mass of hot rock at reasonable depth, the grades of the flow path and heat exchanger created in the rock mass, the fluid flow resistance, and the size of the heat exchange surface between the rock and circulating fluid (Baria et al., 1999). There are many other practical difficulties in creating an EGS which are well documented in the literature. For example at The Geysers project (see Table 27.1) in California (Figures 27.4, 27.5 and 27.7), Alta Rock Energy was forced to suspend operations after months of drilling in a highly promising EGS site because penetration into the rock cap overlying the HDR was impossible (Huang and Liu, 2010). Another example is a stuck drill pipe that could not be recovered from a deep well in the Geodynamics lease in the Cooper Basin, Australia (see Table 27.1) (Geodynamics Ltd. Annual Report, 2009). Although these two examples may appear to be simple issues, they were unsolvable industrial-scale problems that most likely resulted in large economic losses for the respective companies. The creation of a larger heat exchange surface

TABLE 27.1 Summary details for geothermal fields identified as EGS sites.

Geothermal field	Location	Time line of activity	Approx. latitude/longitude	Thermal reservoir (lithology)	Tectonics and structural elements	Depth to thermal reservoir	Temperature of thermal reservoir	Potential power output	EGS activity
Cooper Basin (Habanero)	Approximately 8 km SSE of Innamincka, South Australia	2003: Drilling of Habanero 1, 4421 m, 243°C; 2004: Habanero 2, 4459 m, 244°C; 2008: Habanero 3, 4200 m, 242°C 2009: proof of concept, 1 MW pilot plant (Geodynamics 2009) 2012: Drilling of Habanero 4, 3883 m depth on 21 June.	27.85S 140.72E	Carboniferous granitic intrusives	Permo-triassic basin; intracratonic region with evidence for carboniferous compression	3700–4900 m	In excess of 240°C	37 wells to produce an estimated 275 MWe	Hydraulic stimulation of sub-horizontal joints and fractures
Coso Geothermal field	California, USA (Naval Weapons Air Station near China Lake, CA); ~161 km North of Los Angeles, CA	1993: Well 34A-9 drilled to 2985 m, >300°C; 2005: 34-9RD2 re-worked, re-drilled, and stimulated (Foulger et al., 2008)	36.00N 117.75W	Complex, interfingering sequence of mesozoic diorite, granodiorite, and granite	Situated in major volcanic area with 38 rhyolite domes and abundant basalts; transition between regions of strike-slip and extension	Less than 3000 m	In excess of 300°C at depths less than 3000 m	240 MWe	Hydraulic fracturing of existing reservoir
Desert Peak	Nevada, USA; ENE of Reno, NV	2009: Initial research on bottom hole temp. and hydraulic conductivity; 2010: Hydraulic stimulation and petrological studies 2011: Chemical stimulation (Chabora et al., 2012)	39.76N 118.92W	Fault dissected, tertiary volcanics and sedimentary rocks that overlie mesozoic metamorphics	Humboldt structural zone (extensionalnormal faults, strike-slip transfer faults)	762–1280 m	~200°C	9.9 MWe (year 2000)	Hydraulic stimulation

Site	Location	Timeline	Coordinates	Reservoir rock	Structure/stress	Depth	Temperature	Capacity	Notes
Geysers Geothermal Field	California, USA (south of Clear Lake, CA); ~193 km north of San Francisco, CA	*1921–1926*: 8 wells drilled from 47 to 195 m; *1927*: Potential exists for generating 4500 kW but there was no market for steam; *1955*: Geothermal steam development; *1960*: Electricity generation; *1966*: 51 MWe capacity; *1975*: Capacity 502 MWe; (Lipman et al., 1978) *1989*: 2043 MWe; *1980s*: Overdevelopment and prod. decline; *1982*: Injection augmentation *2000*: Production declined to 1000 MW *Expected in 2020*: 625 MW(Sanyal, 2000)	38.8N, 122.8W	Steam reservoir rocks are typically massive greywacke turbidites of the mesozoic Franciscan Fm. Underlain by a 2.4–0.9 Ma silicic batholith (felsite)	Fault-bounded, quasi-extensional region; fractures in greywacke are randomly oriented and sub-horizontal, while in the felsite are oriented NW and are near vertical	60–~3000 m	~40°C at shallow depths to greater than 240°C in the deepest wells	2043 MWe cumulative installed gross capacity in 1989	Recharge reservoir at depths between 2134 and 3048 m
Hijiori, Japan	South edge of the inner Hijiori caldera, Okura Village in Yamagata Prefecture	*1985–1991*: Shallow reservoir created and developed *1991*: 90-day circulation test, 78% of fluid recovered *1992*: Second phase of investigation launched *1992*: Two-layered reservoir created *1995, 1996*: Short-term circulation tests *2000–2001*: Long-term circulation tests *2001–2002*: Dual circulation test *June–August 2002*: Small binary power plant installed	38.60N 140.18E	Granodiorite	Max. compressive stress direction is E–W; tectonic regimes are strikeslip and normal faulting	Upper reservoir at 1800 m; lower at 2200 m	~250–270°C		Hydraulic fracturing and stimulation

(continued)

TABLE 27.1 *(Continued)*

Geothermal field	Location	Time line of activity	Approx. latitude/longitude	Thermal reservoir (lithology)	Tectonics and structural elements	Depth to thermal reservoir	Temperature of thermal reservoir	Potential power output	EGS activity
Landau, Germany	Upper Rhine Graben, 35 kilometer northeast of Soultz	*2005–2006:* Two boreholes drilled in opposite EW directions *2007:* Testing of doublet *2007:* 3.8 MWe ORC plant *2008:* 65 l·s⁻¹ balanced circulation rate	49.2N 8.1E	Permeable carbonates and sandstones (Evans et al., 2012)	Tertiary rift composed of several asymmetric half grabens (Evans et al., 2012)	3.3 km		3 MW (Azim et al., 2010)	Electricity generation (and heat) using an ORC (organic Rankine cycle) plant (Bromley and Mongillo, 2008)
Larderello, Italy	A few kilometer west of Larderello, Tuscany, Italy	*1904:* Discovered; various wells explored 1940, 1950, 1960, 1975 (Celati et al., 1977)	43.25N 10.87E	Upper reservoir has anhydrites and dolomitic limestones; quartzitesand phyllites in lower reservoir	Structural high; series of nappes with predominant ENE vergence.	~4 km	>400°C	547 MWein 1999	Recharge of reservoir by reinjection
Rosemanowes, Quarry, UK	Near Pen-ryn, Cornwall, UK	*1978–1992:* Active phase. Development of a circulation system at 2 km *1985:* Well drilled and stimulated with intermediate viscosity gel; fluid loss ~20%	50.15N 5.1W	Late carboniferous to early Permian Carnmenellis Granite	No major faults outcrop at Earth surface; sub-horizontal joints near Earth surface; two main subvertical joint sets at depth (NE–SW, NW–SE)	Initial borehole depths to 300 m; subsequent depths to 2000 m	80°C at 2000 m (average geothermal gradient of 35°C·km⁻¹)	Not established for power generation	Hydraulic stimulation; explosive stimulation
Soultz–sous–Forêts, France	~ 50 km north of Strasbourg, Alsace	*1984–1987:* Prep. Phase; *1987–1991:* Exploration phase; *1991–1998:* Two well system; *1997–2007:* Three well system; *2007–2008:* Power production unit (Genter et al., 2010)	48.93N 7.88W	Granites	Local horst structure within the extensional tectonics of the Rhine Graben	3500–5000 m	150°C to more than 200°C	6 MWe (year 2005)	Hydraulic fracturing and stimulation

Source: Most details were obtained from Karner (2005).

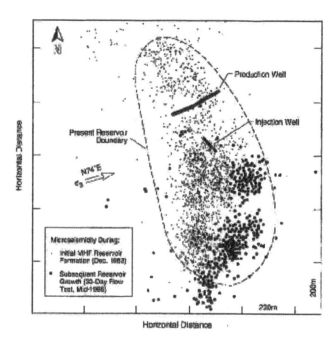

FIGURE 27.3   Plan view of the deeper (Phase II) HDR reservoir at Fenton Hill showing microseismic event locations. From: Fig. 1 of Brown et al. (1999).

in an EGS is not as simple as injecting more water at high injection pressures during hydro-fracturing; this has been demonstrated at Hijiori, Japan, which is located within volcanic calderas and is associated with open, fluid-conductive joints. The pre-existing network of open faults and fractures at Hijiori actually resulted in high water losses in the EGS.

Another issue in EGS is that during the initial exploration and development phases, there may be several negative

FIGURE 27.4   Geothermal power plant at The Geysers near the city of Santa Rosa in northern California. The Geysers area is the largest geothermal development in the world. Photograph by Julie Donnelly-Nolan, USGS.

FIGURE 27.5   The Geysers, California. The picture depicts a drill rig in the NW Geysers drilling the WHS-71 well with flow coming from the adjacent WHS-36 well (located on the same drill pad). A downhole sampler was being run in the WHS-36 well to collect fluid samples at different depths. The photo was taken by Patrick Dobson, Lawrence Berkeley National Laboratory on February 17, 2012.

impacts on the environment, especially from land clearance and the construction of access tracks for geophysical and geochemical measurements. During the production phase of development, there is increased noise and dust levels (Dickson and Fanelli, 2006), but by far, earthquakes or tremors are the most disturbing effect; seismic events occur during high-pressure fluid injection which gives rise to shear failure in favourably oriented natural joints (Kraft et al., 2009).

A pilot EGS project was launched in the Cooper Basin, South Australia, in 2002 by Geodynamics Limited (Weidler, 2005, see Figure 27.6). The site was extensively studied prior to the launch of the project through several exploration wells and seismic traverses from petroleum exploration and production. Geophysical evidence from these studies predicted that the basin is underlain by approximately 1000 km^2 of high heat-producing granites. Habanero No. 1, the first well that was drilled, recorded a temperature of 250°C at 4.4 km depth (Chopra and Wyborn, 2003). Hydraulic fracturing

FIGURE 27.6   (a) The Cooper Basin EGS site is in a remote location in the north-east of South Australia close to Moomba; (b) steam flows out of Habanero in March 2008. Image courtesy of Geodynamics Ltd.

was instigated to enhance the *in situ* permeability between the injection well and a future production well. Analysis of the flow and temperature profiles indicated a possible flow zone at 4254 m depth. Stimulation continued using stepwise increases of injection rates and ceased on December 9, 2003 (Baisch et al., 2006). Micro-earthquakes were common during the stimulation phase, and the frequency of these events was as high as 1500 per day (Baisch et al., 2006). Seismic events were monitored using an eight-element station network comprising three component geophones at boreholes of depths between 70 and 1700 m. Locations of seismic events are commonly used to infer the geometry of the stimulated reservoir; in this case the seismic cloud indicated fractures that are mainly sub-horizontal (see also Section 27.4). Although the hypocentres were distributed in space and time, the migration pattern was unsurprisingly not radially symmetric; the shape of the seismic cloud is believed to be controlled by the fracturing process induced by hydro-mechanical coupling. Analysis of the larger events (> moment magnitude M 1.0) showed significant extension of the seismic cloud, which suggests that some sort of hydraulic barrier was broken by larger events. During the stimulation, events as large as M 3.0 were recorded (Asanuma, 2005) and about 27,000 micro-earthquakes were detected (Baisch et al., 2006).

Despite providing useful information about the extent and orientation of the stimulated reservoir, seismic events during fracture stimulation are often viewed by the general public with much apprehension, since they can be felt at the surface and may cause damage as well as annoyance. The remote, sparsely populated location of the Cooper Basin EGS, means the earthquake disturbances do not have a significant impact, although on the other hand, the remoteness requires significant additional resources to link with the national

electricity grid (Beardsmore, 2005). At the Geysers EGS (Figures 27.4 and 27.5) in California, however, the impact of earthquakes from hydraulic fracturing did not go unnoticed (Figure 27.7), and a planned geothermal power plant project was suspended after undergoing review by the U.S. Department of Energy (Giardini, 2009); plans are in place to re-start the project (Garcia et al., 2012). Another example is the Deep Heat Mining Project initiated in Basel, Switzerland, by Geopower Basel (GPB) in 1996. The city, an industrial centre with a population of over 700,000, is also the site of the largest known earthquake in central-northern Europe (local magnitude M_L 6.5–6.9) which occurred in 1356. No other large event has since been recorded and although during the period 1975–2008, there were only 15 events with $M_L < 2$, authorities made it abundantly clear that tremors arising from hydraulic injection that can be felt at the surface are unacceptable. Using pre-defined thresholds derived from the Soultz-sous-Forêts EGS project (Figure 27.8), which were corrected for local near-surface amplification, a seismic response procedure was agreed between the city of Basel and Geopower Basel. The seismic response procedure was based on three independent parameters: public response, local magnitude M_L, and peak ground velocity; the first of these parameters being the most significant. In October 2006, an injection well was drilled to a depth of 5 km passing through 2.4 km of sedimentary rocks and 2.6 km of granite (Häring et al., 2008). During the first injection, an M_L 2.6 event occurred and injection was stopped. The well was shut in for 5 hours and while preparations were being made to bleed off the well, a magnitude M_L 3.4 event occurred. Fifty-six days and three aftershocks ($> M_L$ 3) later, the project was suspended.

Unlike Basel, the recording of thousands of seismic events during hydraulic stimulation at the Soultz-sous-Forêts site in

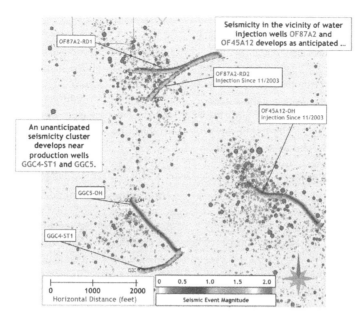

FIGURE 27.7   Map of anticipated and unanticipated seismic events in the Geysers geothermal field over the period January 2002 to April 2009. Reproduced from Hartline et al., (2015) and showing Northern California Earthquake Data Centre tomographic double-difference induced seismicity hypo-centres, injection wells and production wells. The seismic event hypo-centre magnitudes are scaled by size.

the Upper Rhine graben region, France, is regarded primarily as data collection. The main purpose of the data analysis is to determine the spatio-temporal distribution and density of the events in order to estimate the large-scale permeability of the medium. The Soultz experiments began in 1993 and are still ongoing (Genter et al., 2009). In Soultz-2000, hydraulic stimulation produced 7200 seismic events that were located using a borehole and a free surface seismic network. The seismic cloud measured 1500 m in length, 500 m in width,

and 1500 m in height (Delépine et al., 2004). One of the main risks at Soultz is mineral precipitation within fractures. Additional hydraulic and chemical stimulation is used to improve hydraulic connectivity between the wells and the entire network in general (Porter et al., 2009). The stimulation is geared toward the removal of the permeability damage near the well-bore and the minerals (usually calcium carbonate, amorphous silica calcium sulfates) that are deposited through transport and precipitation mechanisms.

FIGURE 27.8   Overview of EGS Soultz-sous-Forêts, France. Soultz is perhaps the most outstanding EGS to date. Photo courtesy of Dr Albert Genter, GEIE Soultz.

27.4 FRACTURE MODELLING FOR EGS

Potential hot dry rock geothermal systems occur in deep underground crystalline rock. The rock matrix (granite) is almost impermeable and the only viable pathway for geothermal flow is through a fracture network. *In-situ* natural fractures in HDR are generally closed or sealed with very low hydraulic conductivity (with micron-scale hydraulic apertures). The creation of a viable geothermal reservoir, appropriate for industrial-scale exploitation, is usually achieved by hydraulically stimulating fractures to create an EGS. The purpose of fracture stimulation is to engineer an effective heat exchange fracture network inside the reservoir that is capable of producing the required flow rate of around $100 \, \mathrm{l} \cdot \mathrm{s}^{-1}$ at an elevated temperature and is sustainable in terms of heat extraction from the ground.

Prior to fracture stimulation, geophysical surveys are conducted to determine whether a high-temperature body exists in the subsurface. In the case of the Cooper Basin in South Australia, a granite body of extent of $\sim 1000 \, \mathrm{km}^2$ was predicted. In November 2003, $20{,}000 \, \mathrm{m}^3$ of water was injected into the 4421-m-deep well, Habanero 1. Hydraulic tests were conducted, and on November 30, 2003, $131 \cdot \mathrm{s}^{-1}$ of water was injected for about 2 days, and injection rates were increased stepwise to $24 \, \mathrm{l} \cdot \mathrm{s}^{-1}$ for 10 days. In order to improve connectivity, the casing was perforated at several intervals between 3994 and 4136 m, and fluid was allowed to enter the main fracture zone (Baisch et al., 2006).

Following fracture stimulation, it is essential to understand the fracture system within the reservoir as it provides the basis for the modeling of flow within, and heat extraction from, the EGS and thus the design and performance assessment of the entire energy production system.

The characterization of rock fracture networks is a very difficult problem not least because accurate field measurement of a single fracture is difficult and measurement of all fractures is impossible. The only way to measure the fracture system accurately is to completely dismantle the fractured rock along fractures surfaces (Dowd et al., 2009), which is obviously impossible for a volume of rock of reasonable size. Therefore, in practice, the whole fracture system is not observable on any meaningful scale, and the only realistic approach is via a stochastic model informed by sparse data and/or by surface analogues. In HDR applications, a realistic solution is even more difficult as the only reference data related to the fracture system are either from borehole logs revealing mainly borehole wall images and/or sparse seismic events kilometers beneath the surface detected during the hydraulic stimulation process.

The stochastic modeling of fracture networks originated in percolation studies (Robinson, 1983; Sahimi, 1993) and its wider application to rock engineering was promoted in the 1980s by the work of several research groups (e.g., Long et al., 1982; Baecher, 1983; Andersson, 1984;

Dershowitz and Einstein, 1988). The principle of stochastic modeling is increasingly accepted as a more realistic alternative to the conventional continuum model (i.e., equivalent porous media (EPM) approach) for the investigation of flow transport in fractured rock masses. In applications to deep underground crystalline rock (e.g., EGS reservoir characterization), a stochastic approach is the only viable means of establishing a model from very sparse, often unreliable data.

27.4.1 Fracture Model Construction

The parameters required to construct a fracture model include location, size (persistence), orientation, and other fracture properties such as aperture and surface roughness. Stochastic modeling of rock fractures is the general approach in which these parameters are treated as spatially correlated random variables with inferred probability distributions. In the simplest case, once the parameter distributions are inferred, the fracture model can be constructed by Monte Carlo simulation.

The first stage of the modeling process is the collection of fracture data for statistical analysis. The most common measurement technique is by scan lines or window surveys of rock outcrops or excavated rock surfaces (Priest and Hudson, 1981; Kulatilake et al., 2003). If drill cores are available, scan line surveys can also be applied to core samples (Zhang and Einstein, 2000; Gupta and Adler, 2006). Additional data can be obtained along the borehole using a down-the-borehole camera or using wireline geophysical logging (Ozkaya and Mattner, 2003). In EGS applications, an important source of data for inferring the characteristics of the fractures in the reservoirs is provided by the locations of detected seismic events (caused by fracture slips or fracture initiation and/or propagation) obtained during the fracture stimulation (Pine and Batchelor, 1984; Brown et al., 1999; Parker and Jupe, 1997) as discussed in the previous section.

Data collected from scan lines and/or window surveys of exposed rock surfaces, core samples, or borehole walls are biased due to sample truncation, censoring, and edge effects (Laslett, 1982; Baecher, 1983) and may require statistical correction before they can be used. These data are essentially 2D and are, therefore, due to the orientation and geometry of the sampling face, also biased samples of the complete, but inaccessible, 3D fracture networks that are being sampled. The biases require correction before the 2D data can be used for 3D modeling (Einstein and Baecher, 1983; Villaescusa and Brown, 1992; La Pointe, 2002; Fouché and Diebolt, 2004). The various published attempts at bias correction have had varying degrees of success and none of them are completely satisfactory. The derivation of 3D fracture model parameters from 2D data is still a very challenging issue, and additional assumptions are in general required to find a solution (Waburton, 1980; Zhang and Einstein, 2000; Xu and Dowd, 2010). In addition, the use of fracture data from analogues, such as

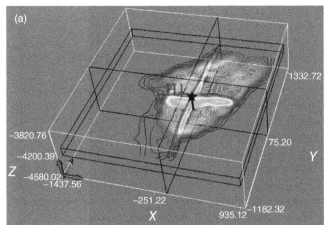 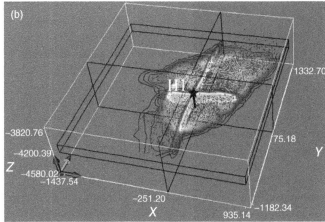

FIGURE 27.9 Inhomogeneous point process to model fracture density of a stimulated fracture reservoir: (a) contours show fracture density model on three cross sections, the crossing point of the two vertical cross sections is the location of Habanero 1 where the fracture density is expected to higher due to the hydraulic stimulation and the sub-horizontal cross section is the lest-square fitted plane cutting through the reservoir; (b) points show a simulated point pattern (representing fracture locations) based on the density model. The density model is the one constructed for the Habanero reservoir.

surface outcrops, is always questionable when used to model deep underground rock fractures, primarily because of the significantly different stress regimes.

Once the distribution parameters are determined, the simulation of fractures is relatively straightforward. Fracture location is the most important distribution parameter of fracture networks and is usually modeled separately from other fracture properties. The most common approach is to use a single point, usually the center of the fracture (e.g., a line in 2D or ellipse in 3D), to represent the location. Geostatistics or a point process can then be used to model the spatial distribution of the fractures. In geostatistical modeling (Young, 1987; Chilès, 1988; Billaux et al., 1989; Viruete et al., 2003), the variable modeled is usually the fracture density or the number of fracture representation points within a unit area (2D) or a unit volume (3D). For a given dataset, the variable is derived from a specified grid system in which the fracture density for each cell is calculated and its spatial correlation modeled. During fracture simulation, fracture density for each grid cell is simulated first on the basis of the geostatistical model, and then the location points are distributed within the cell according to a Poisson rule. In point process modeling, the point intensity model, either parametric or nonparametric, is derived first for the dataset. The intensity model may be a simple Poisson model in which case fractures are uniformly distributed within the volume of interest. It may also be one of a number of nonhomogeneous models including inhomogeneous, cluster, and Cox models. The use of various nonhomogeneous point processes makes the fracture modeling extremely flexible and able to simulate various kinds of fracture patterns. In fact, for HDR applications, the

inhomogeneous model is perhaps the most appropriate, as fracture density close to the stimulation well is expected to be high and the density will reduce gradually with distance from the well; see the example shown in Figure 27.9. A cluster model might also be applicable, as fracture propagation/growth is expected to originate from or around existing fractures and is hence analogous to the formation of a cluster point process model. For detailed descriptions of these models, readers are referred to Xu and Dowd (2010).

Once fracture locations are fixed, the next step is to simulate the geometries of individual fractures. Fracture size and orientation are generally modeled as probability distribution functions (PDF). In the simplest case, the geometrical parameters for a fracture are simulated by Monte Carlo sampling of the corresponding PDF, and these are then used to construct the lines or polygons representing the fractures. Orientations of fractures are most commonly modeled using a Fisher distribution, while sizes of fractures are commonly modeled using lognormal, exponential, or gamma distributions (Priest, 1993). Additional fracture properties (e.g., fracture aperture, surface roughness) can be included in the model if required. These properties are generally described by PDF models and they can also be simulated by Monte Carlo sampling. The term marked point process modeling (MPP) is used to describe the whole simulation process discussed above when the fracture locations are modeled by a point process model and the fracture properties (including geometrical parameters and additional properties) are considered as marks associated with the points. The simulation process is simple if there is no significant multivariate correlation among fracture properties. Correlated MPP (Xu et al., 2007) can be used for cases when the correlation has to be incorporated in the simulation. Simulated fracture models are usually validated by sampling the model (using scan lines or areas) and assessing the extent to which the sampled values conform to the statistical models inferred from the survey data (Kulatilake et al., 2003).

It is a common practice in fracture modeling to divide the fractures into sub-sets according to fracture orientations. This is based on the belief that the fractures formed by the same geological activity will have similar properties, including similar preferential orientations. Classification of sub-sets is aided by fracture orientation rose diagrams in 2D applications and hemispherical projections in 3D applications. Each sub-set is modeled separately, and the final simulation is the simple combination of all independently simulated sub-sets. In the case when fracture sets are spatially correlated, the hierarchical modeling (Lee et al., 1990) or plurigaussian simulation (Dowd et al., 2007; Xu et al., 2006a,b) can be used. In HDR applications, reservoir analogues are, in general, difficult to find and there is, in general, insufficient information for fracture set classification and for reliable parameter estimation. In this case, there is no sub-set division in the prior belief of the fracture model, and some fracture model parameters have to be assumed based on research done in other related areas.

Data comprising the locations of detected seismic events obtained during borehole stimulation of geothermal reservoirs are generally in the form of a point cloud. Although these data contain important information about the *in-situ* fracture network, they are difficult to interpret and have never been used directly for the statistical inference of the distribution parameters of fracture networks in geothermal reservoirs. To our knowledge, the only published, direct uses of these data to date are the estimation of the volume of a geothermal resource that can be reached by wells (Cuenot et al., 2006; Hori et al., 1999; Brown et al., 1999), the matching of simulated fracture locations (Tran and Rahman, 2007), and the estimated permeability for the equivalent porous reservoir model (Xing et al., 2009). There have been a few recent attempts to use the data directly to fit a discrete fracture model to the reservoir (Xu et al., 2011a,b; 2012; Fadakar-A et al., 2011).

27.4.2 Type of Fracture Models

It is impossible to measure accurately a 3D rock fracture network for a volume of rock of reasonable size. The fracture network used for various analyses is commonly derived from a fracture model. Fracture models commonly use planes to represent 3D fractures (or lines for 2D fracture traces).

Fracture models can be classified as either deterministic or stochastic although it is arguable whether a fracture model can be truly deterministic as some fracture properties of a supposedly deterministic model may always remain uncertain. Common deterministic models use parallel fractures with known orientations (usually coinciding with the coordinate system) and uniform fracture density although regular mosaic tessellation models can also be regarded as deterministic (Dershowitz and Einstein, 1988). This type of model is commonly used to represent rock fractures using the EPM approach in dual-porosity or multiple interacting continua models (Pruess et al., 1999) for simulating flow and heat transfer in fractured rock masses. Two examples of such models are illustrated in Figure 27.10. Stochastic models treat all fracture parameters as random variables (Figure 27.10c). Stochastic models can be unconditional or conditional. A conditioned fracture model is generated by incorporating some known data from the fracture network, either surveys or seismic events indicating points on fracture planes.

According to the size characteristics of fractures, fracture models can also be classified as bounded or unbounded. An unbounded fracture model assumes infinite fracture persistence, i.e., infinite lines in 2D or infinite planes in 3D, so that the area or the volume of the rock is divided completely by these fractures into individual blocks (Figure 27.11a). Commonly used models include Poisson line and Poisson plane models (Dershowitz and Einstein, 1988). The rock mass in this case becomes an assemblage of discrete intact blocks containing no internal defects (fractures). Bounded fracture models assume fractures are limited in persistence (Figure 27.11b). Commonly used 3D fracture models include circular disks, elliptical disks (Bacher, 1983), and irregular planar polygons (Xu and Dowd, 2010; Fadakar-A et al., 2011). In these models, rock mass may consist of blocks with various types of fracture intersections. In EGS, a bounded fracture model is more appropriate, as the fracture stimulation seismic point cloud implies that the fractured reservoir is finite in size. There are modifications to Poisson plane models to eliminate the infinite persistence assumption but the artifact of co-planarity in the fracture model is not suitable for an EGS reservoir (Dershowitz and Einstein, 1988). Fracture models can also be classified by the corresponding percolation states (Sahimi, 1993) of their realizations. A percolated fracture model will always produce a percolated fracture network (realization) (Figures 27.12b and 27.12c). For this type of fracture model in EGS, the connectivity index (Xu et al., 2006b) between the injection well and the production well is always 1. On the other hand, the corresponding connectivity index for an unpercolated fracture model will be less than 1. In other words, an unpercolated fracture model does not always produce a realization where the injection and production wells are connected by the fracture network (Figure 27.12a). Obviously in an EGS, the injection and production wells must be well connected for the reservoir to be of any practical use and therefore the corresponding fracture model must be a percolated one. Of course this may simply reflect reality. Stimulating fractures will not necessarily ensure that injection and production wells are connected.

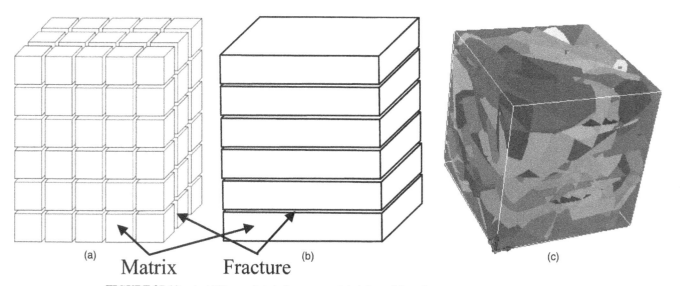

Matrix   Fracture

FIGURE 27.10 (a, b) Deterministic fracture model. Adapted from Pruess et al. (1999) and Carneiro (2009). (c) Stochastic fracture model. Adapted from Xu and Dowd (2010).

27.4.3 Realistic Fracture Modeling for EGS

For EGS, down-hole geophysical logging and seismic events recorded during fracture stimulation are the only measurements of the fracture system. Borehole acoustic or ultrasonic image logging is a direct measurement but the observed fractures are only the traces of intersections between fractures and the borehole walls, and therefore, they reveal very limited information about the entire fracture network. The seismic event cloud reveals the fracture system for the entire reservoir but unfortunately it is extremely difficult to extract the fracture network information from the point cloud.

Fracture stimulation causes planes of existing (naturally occurring) fractures to slip against each other due to the reduction in effective normal stress. Slipping causes misalignment of fracture surface profiles causing lateral dilation, which essentially causes the hydraulic aperture of the fracture to increase (Baisch et al., 2009). In addition, fracture stimulation causes the existing fractures to propagate and creates new fractures. The final product of stimulation is a fractured rock mass that can be exploited to produce heat by using injection and production wells intersecting the reservoir. During fracture stimulation, fracture initiation, propagation, and slipping generate micro-seismic events that can be monitored and their spatial locations determined. The resulting seismic point cloud can be used to determine not just the geographical extent of the HDR reservoir and the amount of

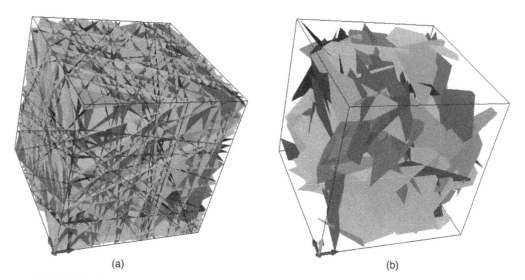

FIGURE 27.11 Unbounded (a) and bounded (b) stochastic fracture models; both have the same number of fractures.

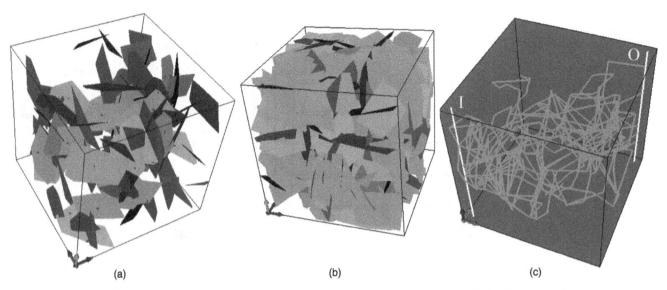

FIGURE 27.12   (a) Unpercolated fracture network where fracture clusters are isolated; percolated fracture network (b) where fracture clusters are interconnected and also connected to both the injection well (I) and production well (O) following the connection paths (c) through fractures shown in (b).

fracturing but also the fracture network within the reservoir (Xu et al., 2011a,b). Establishing the fracture network reservoir model conditioned by this seismic point cloud is critical in creating a more realistic and reliable fracture model for the HDR EGS.

It is reasonable to assume that seismicity caused by either the fracture slipping/propagation or fracture initiation occurs only on the fracture plane. In other words, an accurately conditioned fracture model should pass through these points. The problem now is to construct a fracture model conditioned to the seismic events under the assumption that at least one fracture passes through a seismic event point, i.e., an accurately conditioned fracture model should intersect all seismic event locations. This essentially becomes a stochastic geometry reconstruction problem given a set of point clouds. Reconstruction of a surface from random point clouds is computationally and algorithmically challenging and is an active research area in computer and mathematical sciences (Bercovier et al., 2002). The success of current practice, however, depends critically on close sampling points on the surface, which is usually not an issue as the point clouds are generally obtained from laser scanning or some form of digitizing. For seismic point clouds in geothermal applications, however, samples are very sparse. For a given fracture, only a few points are available, which indicate either the propagation front of the fracture or a point on the fracture surface where shear slip occurs at the time the events are detected. Current methodology is thus not directly applicable to fracture modeling.

Although we use planar polygons to construct the fracture model, in reality, fracture planes are highly tortuous.

It is impossible to represent all fractures by tortuous surfaces using a reasonable number of parameters and some simplification is required. This is the main reason that common approaches use circular discs, elliptical discs, planar polygons, or planes with infinite extents. The validity of this representation is based on the assumption that fracture planes in the fitted model will follow closely the actual tortuous surfaces of the fractures in the reservoir, or in other words, the fitted fracture model is the first-order approximation of reality. Within this framework, even the "best" fitted model will not intersect all seismic points, but the distance of the points to fracture planes can be used as a criterion to assess the goodness-of-fit of the fracture model. It should be noted that the accuracy of the seismicity detection and the inversion process in deriving the locations of seismic points are not a topic of this chapter. Readers interested in this topic should consult Baisch et al. (2006).

Assume there are m conditioning seismic events and n fractures in the fracture model. We introduce a projection distance for each seismic event j ($j = 1, \ldots, m$) as $d_{j;i}$ defined as the distance of point j to fracture plane i which is the shortest distance from point j to all fracture planes. We can parameterize each fracture with parameters (x_i, y_i, $z_i, \alpha_i, \beta_i, \gamma_i$, a_i, b_i), $i = 1, 2, \ldots, n$, where the coordinates of the center of the fracture are given by the point (x_i, y_i, z_i), the orientations are described by three angles: dip direction α_i, dip angle β_I, and rotation angle γ_I, and the sizes of the fractures are described by a major axis a_i and a minor axis b_i of an ellipse containing the polygon (Xu and Dowd, 2010). Fracture F_i can also be expressed in functional form as $\lambda_x^{(i)}x + \lambda_y^{(i)}y + \lambda_z^{(i)}z = \omega^{(i)}$, where ($\lambda_x^{(i)}, \lambda_y^{(i)}, \lambda_z^{(i)}$) is the unit vector normal to the fracture plane and can be calculated from $(\alpha_i, \beta_i, \gamma_i)$. Given a point P_j (x_j, y_j, z_j), not necessarily lying on the plane, the signed orthogonal distance to fracture F_i is defined by

$$d_{j:i} = \begin{cases} \lambda_x^{(i)} x_j + \lambda_y^{(i)} y_j + \lambda_z^{(i)} z_j - \omega^{(i)} & \text{if } P_j^{(F_i)} \in F_i \\ \infty & \text{otherwise} \end{cases} \quad (27.1)$$

where $P_j^{(F_i)}$ is the projection point of P_j on fracture plane F_i. A matching function $\xi(j) \in \{1, 2, \ldots, n\}$ is used to associate each point P_j with one and only one fracture polygon. We impose a simple criterion of minimum distance for this association and by writing $d_{j:i}$ we mean point P_j is associated with fracture polygon F_i and its distance calculated by equation (27.1) is the minimum when compared with distances to any other fracture plane. Point j is then said to be associated with, and only with, fracture plane i. Because of the first-order approximation of the fracture model, the complete set of projection distances $d = \{d_{j:i}\} \forall j$ will exhibit random variation from the fitted fracture planes. A Gaussian model is the most appropriate statistical model to describe random variation, or noise, from a first-order approximation, that is, $d_{j:i} \sim N(0, \sigma^2)$, where σ^2 measures the degree of tortuous variability of fracture surfaces. The likelihood for the set of seismic event points $P = \{P_j\}$ given a set of fitted fractures $F = \{F_i\}$ and a matching function ξ, can then be defined as

$$f\left(\{d_{j:i}\} \mid \theta\right) = \prod_{j=1}^{m} f\left(d_{j:i} \mid \theta\right) = \left(\frac{1}{\sqrt{2\pi}\sigma}\right)^m \prod_{j=1}^{m} e^{-\frac{(d_{j:i})^2}{2\sigma^2}} \quad (27.2)$$

with the set of parameters $\theta = \{(x_i, y_i, z_i, \alpha_i, \beta_i, \gamma_i, a_i, b_i), i = 1, 2, \ldots, n)$. The posterior distribution given P and F is then

$$\pi(\theta \mid P; F) = f\left(\{d_{j:i}\} \mid \theta\right) \cdot \pi(\theta) \int_\infty^\infty f\left(\{d_{j:i}\} \mid \theta\right)$$
$$\cdot \pi(\theta) \cdot d\theta \infty f\left(\{d_{j:i}\} \mid \theta\right) \cdot \pi(\theta) \quad (27.3)$$

and $\pi(\theta)$ is the prior distribution or prior belief for the parameters of the set of fractures F. Note the normalizing constant $\int_\infty^\infty f(\{d_{j:i}\} \mid \theta) \cdot \pi(\theta) \cdot d\theta$ does not usually need to be calculated as the parameters of the posterior distribution $\pi(\theta \mid P; F)$ can be derived from the numerator.

In practice, the number of fractures could be in the order of millions, and the dimension of θ will be in the same order of magnitude, if not higher. The posterior distribution of θ is therefore analytically intractable. One way to derive samples of the parameters θ from the posterior distribution $\pi(\theta \mid P; F)$ is to construct a Markov chain using Markov Chain Monte

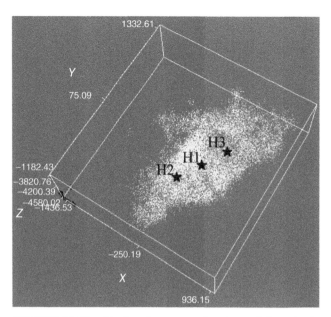

FIGURE 27.13   Absolute hypocentre locations of the seismic events.

Carlo (MCMC) simulation. A Markov chain is generated by Monte Carlo sampling from the posterior distribution which is commonly constructed by the Metropolis–Hastings algorithm using the Monte Carlo acceptance/rejection technique imposed by the Hastings' ratio. For a more detailed description of this algorithm, users are referred to Mardia et al. (2007) and Xu et al. (2011a,b, 2012).

As a demonstration of the application of the fracture fitting technique described above, the seismic event dataset collected during the Habanero 1 fracture stimulation period is used to demonstrate the construction of a conditioned fracture model. Habanero 1 was drilled to a depth of 4421 m, 753 m into the basement rock where the temperature reaches as high as 250°C. It was intended to serve as the injection well for the future pilot electricity generation plant. During the entire fracture stimulation period, a total number of 23,232 seismic events were recorded although 27,000 triggers were reported as discussed above and analyzed, as shown in Figure 27.13 where the location of Habanero 1 (H1) and Habanero 3 (H3) are marked. H3 is intended to serve as the production well for the pilot plant, and the connectivity analyses between H1 and H3 are used to demonstrate the effectiveness of the fracture fitting algorithm described.

Figure 27.14 shows the fracture model fitted to the reservoir after 1M steps of MCMC optimization and a corresponding simulated flow model. Compared with the original model (not shown), the $\sum d_{j:i}^2$ has been reduced from 26,437,002 to 501,862 m^2, fracture orientations have changed from purely random to mostly sub-horizontal, and agree well with earlier analyses (Baisch et al., 2006). The number of connection paths between H1 and H3 has increased from 9135 to

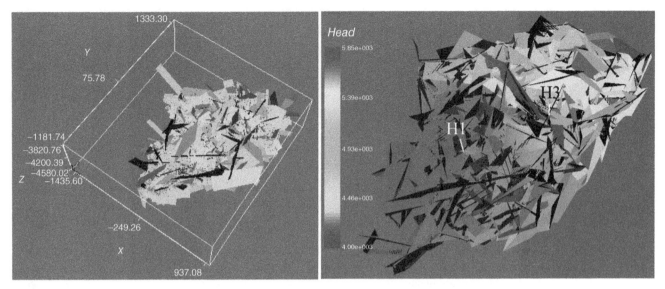

FIGURE 27.14   Habanero fracture model after optimization.

27,293. The number of connection paths between the two wells with distance less than 1 km has increased from 20 to 103, which certainly gives a more realistic number of connections between the two wells due to the fracture stimulation. Using a constant hydraulic aperture of 100 μm for both fracture models and 10^{-6} m$^2 \cdot$ s^{-1} as the kinematic viscosity for the fluid, the flow rate achieved in the original and the optimized model are 1.6 l·s^{-1} and 5.9 l·s^{-1}, which is almost a fourfold increment. These correspond to the reservoir impedance of 9.4 MPa·l^{-1}·s and 2.5 MPa·l^{-1}·s respectively. Clearly the optimized model reflects much better the actual connectivity of the reservoir due to the stimulation. For more detailed analyses, readers are referred to Xu et al. (2012).

27.5   FLUID FLOW MODELING FOR EGS

In addition to the fracture model, the second component of an EGS is its fluid flow and heat extraction characteristics. There are no dedicated flow and heat modeling tools available for EGS applications although flow and heat can be modeled using tools developed for conventional geothermal applications (porous media), hydrogeology modeling, contaminant/radioactive particle transport problems (porous/fractured media), or petroleum engineering. Modeling fluid flow through fractured media is very challenging especially when significant numbers of fractures are involved and the fracture network is complex.

There are essentially four broad approaches to modeling fluid flow through fractured media (Berkowitz, 2002). The EPM approach represents the fractured medium as a porous continuum based on the equivalent hydraulic conductivity of the fracture system. Double and multiple continua models have also been developed as a means of accounting for two or more interacting systems such as a fracture network and a porous rock matrix. This approach is almost exclusively deterministic. The advantage of this approach is efficiency as the overall flow behavior of the entire system can be quickly assessed.

The discrete fracture network (DFN) approach lends itself more naturally to the solution of this problem. Here, each fracture is modeled explicitly, either as a line segment (2D) or as an elliptical disk or polygonal plane (3D). Each fracture is further subdivided into elements so that values of flow variables can be obtained for each element. Numerical methods used include the finite element method (FEM), the boundary element method (BEM), the finite difference method (FDM), and the distinct element method (DEM) (e.g., Jing and Stephansson, 2007) with associated commercial codes including FracMan/Mafic, ConnectFlow, and UDEC/3DEC, none of which are specific to HDR applications.

A third general approach, the stochastic continuum (SC) or fractured continuum (FC) model (Tsang et al. 1996), has been proposed as a means of increasing DFN modeling efficiency for large-scale applications. This is a hybrid of EPM and DFN. The modeled volume is subdivided into smaller blocks, and hydraulic conductivities (permeability tensors) are assigned individually to each block. The most common way of assigning permeability tensors to blocks is by geostatistical simulation using conditioning data from borehole packer tests (e.g., Hamm et al., 2007). A fourth approach is the equivalent pipe networks (EPN) method (Cacas et al., 1990; Dershowitz and Fidelibus, 1999a). In this method, the 3D fracture network model is transformed to a pipe network with equivalent hydraulic conductivity; the fractures are represented as pipes originating and ending at centers of fracture intersection lines.

27.5.1 Flow and Heat Transfer Through Individual Fractures

In the 1970s, the first series of mathematical studies on EGS were developed based on the existing models. EGS research in the USA, UK, France, and the Federal Republic of Germany were thus built on the expertise from the oil industry which sought to idealize an EGS reservoir as a penny-shaped fracture which behaves as an isotropic continuum in a uniform stress field (Abé et al., 1976).

Fluid flow within a single fracture begins by considering the Navier–Stokes and continuity equations. The Navier–Stokes equation can be written as

$$\frac{\partial u}{\partial t} + (u \cdot \nabla) u = F - \frac{1}{\rho} \nabla \rho + \frac{\mu}{\rho} \nabla^2 u \qquad (27.4)$$

Here ρ is the fluid velocity, F is the body force vector per unit mass, p is the pressure, μ is the fluid viscosity, and u is the velocity vector. In the context of Newton's second law of motion, the terms on the left of the Navier–Stokes equation represent the net acceleration of the fluid and the terms on the right represent the net force per unit mass (Zimmerman and Bodvarsson, 1996).

Equation (27.4) can be written together with the equation of continuity to give a closed form system of equations with three velocity components and the pressure term. The equation of continuity for an incompressible fluid is given as

$$\nabla \cdot u = 0 \qquad (27.5)$$

The next step in the analysis involves determining the boundary conditions. The no-slip conditions are commonly used; these conditions state that at the boundary of the fluid and the solid, the velocity vector of the fluid is equal to the velocity of the solid. This implies that at a stationary wall, the normal and tangential components of velocity are zero. Once the boundary conditions are determined, the system of equations can be solved.

In describing fracture permeability, steady state flow conditions are considered and the nonlinear terms in the Navier–Stokes equations are usually omitted. Equation (27.4) then reduces to a much simpler form (equation 27.6):

$$\mu \nabla^2 u - \rho (u \cdot \nabla) u = \nabla p \qquad (27.6)$$

Perhaps the simplest way to model flow through individual fractures is using the parallel plate model (Snow, 1965). In this model, the major assumptions are that the fracture is flat and has a finite size and arbitrary shape, while the reservoir is isotropic, homogeneous rock. By using parallel plate representations for fractures, the solution of the Navier–Stokes equation for Newtonian fluid flow within fractures is linked through the cubic law to Darcy's law, which characterizes fluid flow through porous media. The transmissivity, T, (or effective permeability) of a fracture analyzed in this context can be written as

$$T = \frac{wh^3}{12} \qquad (27.7)$$

This is the expression of the cubic law. Here w is the width of the fracture and h is the fracture aperture. When using the cubic law, it is important to realize the limitations of the model, the most obvious being that a real fracture does not have smooth parallel walls but rather rough walls which make contact with each other at some points (Figures 27.15 and 27.16). Some authors (Zimmerman and Bodvarsson, 1996) have improved on the model to include a varying fracture aperture, where a mean aperture, $<h>$, is defined and T is given by

$$T = \frac{w \langle h \rangle^3}{12} \qquad (27.8)$$

Another improvement to the cubic law is by considering the walls of the fracture as having permeable properties as a result of granular rock properties, as well as the cracks and fissures that may occur during the initial hydro-fracturing process. It is generally accepted that when an impermeable wall confines fluid, a very thin layer of fluid particles adheres to the boundary as fluid flows past the wall. This is the no-slip boundary condition. For permeable walls, "slip boundary conditions" are needed; these are applicable for viscous liquids at low Reynolds number; the Reynolds number is a dimensionless measure of the relative strengths of inertial to viscous forces. Berkowtiz (1989) points out that in the general modeling of fractures, it is assumed that the mass flux normal to the interface is constant, the pressure across the interface is constant, and the tangential velocity across the interface tends to be zero, the no-slip boundary condition. He sums up the effect of slip by saying that the effects of viscous shear in the free fluid propagate across the interface and lead to a transition region within the boundary which, although small, affects the flow profile and volumetric flow rate within the channel. He concluded that the difference in volumetric flow rate within a fracture using no-slip and slip boundary conditions is as much as 19% for the case he examined. Berkowitz modeled the slip boundary conditions using Brinkman's extension to Darcy's law, which includes a macroscopic shear term to account for the transition region.

Although Berman (1953) was the first to investigate the effects of wall porosity in a channel, Beavers and Joseph were the first to account for the material properties of permeable walls in fluid flow problems (Beavers and Joseph, 1967; Beavers et al., 1970). Beavers and Joseph hypothesized that when a viscous fluid in a channel flows parallel to the permeable medium, the effects of viscous shear propagate across

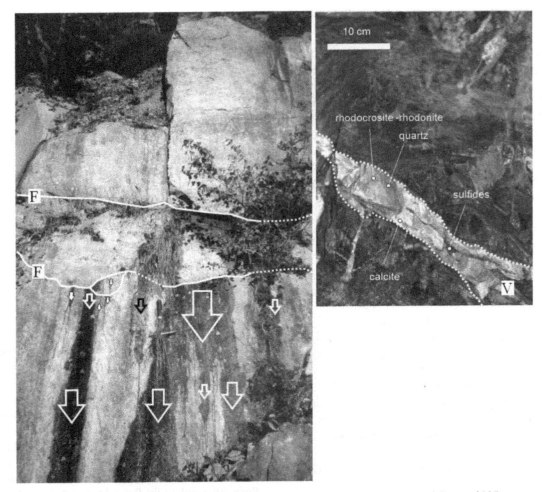

FIGURE 27.15   Fractures in granite showing complexity in fluid flow (Sausse and Genter, 2005).

the interface and result in thin layer of streamwise moving fluid lying just below the permeable layer. This fluid layer is pulled along by the free fluid external to the permeable layer. The tangential component of the velocity of the free fluid,

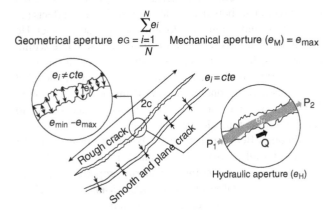

Geometrical aperture $e_G = \dfrac{\sum\limits_{i=1}^{N} e_i}{N}$ Mechanical aperture $(e_M) = e_{max}$

FIGURE 27.16   Comparing a smooth and parallel model of a crack with a rough crack (Sausse and Genter, 2005).

u_f, at the boundary of the permeable material is considerably higher than the mean filter velocity (seepage velocity), u_m, within the permeable body. This is the Beavers–Joseph slip flow hypothesis which is valid for common viscous liquids at low Reynolds number (Neale and Nadar, 1974). Slip generally results from fluid properties, interactions between the fluid and the wall, the shear rate at the wall, and surface roughness (in the case of liquid flows) (Ligrani et al., 2010). Crandall et al. (2010) performed numerical computations of flow within open fractures and confined fractures surrounded by permeable medium using an interface condition and found an increase in flow volume of about 10%.

Apart from Crandall et al. (2010) and Berkowitz (1989), few authors have attempted to apply slip boundary conditions to flow within fractures. An extension of Berkowitz's work was achieved through a more sophisticated model for a fracture with slip at the walls for a 2D case using a similarity solution and the Beavers–Joseph slip boundary conditions (Mohais et al., 2011a,b, 2012). The problem was solved using a similarity parameter in conjunction with a perturbation solution. The coupled fluid flow and heat transfer within the

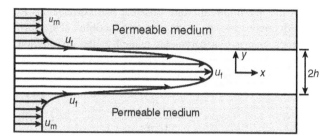

FIGURE 27.17 Velocity profile for coupled parallel flows within a channel and bounding porous medium according to the slip flow hypothesis of Beavers and Joseph (after Beavers and Joseph (1967) and Neale and Nadar (1974)).

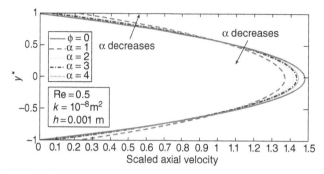

FIGURE 27.18 The property α is unique to each material which comprises the wall of the channel. Experimental results have shown that α ranges from 1 to 4. This figure shows the effect of a changing alpha on the axial velocity in the channel.

free fluid region in the fracture and in the fracture walls were addressed by first solving for the velocity components and then using the solutions in the energy equation. In the fractures, the heat advection of the fluid is balanced by the heat supply at the walls, whereas in the reservoir, heat conduction is governed by the 3D diffusion equation. Deviations from the cubic law occur also at high Reynolds numbers, at the onset of turbulence.

The slip boundary condition can be defined as

$$\frac{\partial u_f}{\partial y} = \frac{\alpha}{\sqrt{k}} \left(u_f - u_m \right) \qquad (27.9)$$

This condition is evaluated at a boundary limit point from the exterior of the fluid (see Figure 27.17). Here k is the wall permeability and α is a dimensionless quantity which characterizes the structure of the permeable material within the boundary region. It can be determined through flow simulations and measurements.

In the model presented in Mohais et al. (2011a,b, 2012), a fractured reservoir was considered at approximately 3–5 km below the surface. Pressurized water of density, ρ, and viscosity, η, was introduced into a single fracture. As the water flows through the fracture, it is heated by the surrounding rock (at approximately 200°C) of which the reservoir is comprised. The heated water can then be extracted via a recovery well located some distance away from the injection well. Neglecting end effects at the entry and exit points and modeling the fracture as a channel with horizontal parallel walls, define the distance between the channel walls to be $2h$. Assume that both walls have a small permeability, k, flow within the permeable region can be ignored above the first order in permeabilities, and the effect of permeability enters through the slip boundary conditions at $y = \pm h$. The nondimensional distance y^* is defined as $y^* = y/h$.

The stream function, ψ, can be written in terms of the entrance velocity u_0, a constant wall velocity v_w, and a similarity function f according to Terrill and Shrestha (1965):

$$\Psi \left(x, y^* \right) = \left(h u_0 - v_w x \right) f \left(y^* \right) \qquad (27.10)$$

The velocity components are

$$u \left(x, y^* \right) = \left(u_0 - \frac{v_w}{h} x \right) f' \left(y^* \right) \qquad (27.11)$$

$$v \left(y^* \right) = v_w f \left(y^* \right) \qquad (27.12)$$

The wall Reynolds number is

$$\mathrm{Re}_w = \frac{v_w h}{v} \qquad (27.13)$$

After some substitutions and manipulations, the Navier–Stokes equation reduces to a third-order equation of the similarity parameter, where C is a constant of integration (see Mohais et al., 2011a,b, 2012 for full details):

$$f''' + \mathrm{Re}_w \left(-f f'' + (f')^2 \right) = C \qquad (27.14)$$

The axial velocity within the channel is affected by α, k, and the height of the channel, h, and the temperature distribution and temperature gradient is affected by k. Axial velocity profiles are presented in Figures 27.18 and 27.19.

Besides the permeable nature of the fracture walls, another deviation in the flow profile arises as a result of asperity contact within the fracture. Asperity contact in the Hijiori EGS system was addressed using the Gangi model (sometimes called the bed of nails model). The basic premise of the Gangi model is that asperities of various heights can be modeled inside two fracture walls separated by a varying width by defining a relationship between fracture aperture and effective stress (using the concept of a bed of nails as an analogy for asperities).

The Hijiori HDR system, located at the edge of a 10,000-year-old 2-km caldera, comprises a shallow reservoir at 250°C, 1800 m, and a deep reservoir at 270°C, 2200 m. Two injection and two production wells were drilled at the site to intercept each reservoir. Field tests were conducted in 1984–2002 to create a heat extraction system. Using core

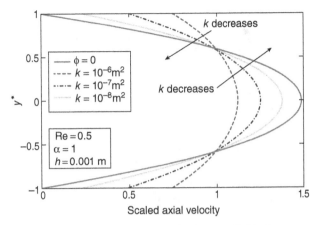

FIGURE 27.19 The effects have also been studied in the context of very high wall permeability. With a decrease in wall permeability, there is a greater maximum axial velocity at the mid-line of the channel.

samples, rock parameters such as thermal conductivity, density, and porosity were measured in the laboratory and the following values were obtained:

- rock permeability $= 1 \times 10^{-16}$ m^2
- rock porosity $= 0.01$
- rock density $= 2700$ kg·m^{-3}
- rock thermal conductivity $= 30$ W·(mK)$^{-1}$
- rock specific heat $= 1.0$ kJ·(kgK)$^{-1}$
- water specific heat $= 4.2$ kJ·(kgK)$^{-1}$

The calculation of temperature at depth was achieved through two basic equations:

(a) at depths < 1500 m,

$$T_1 = 4.564 + 0.263 \,(\text{depth}) \\ - (7.41 \times 10^{-5}) \times (\text{depth})^2 \quad (27.15)$$

(b) at depths ≥ 1500 m,

$$T_2 = 144.739 + 0.0584 \,(\text{depth}) \quad (27.16)$$

The initial pressure at 1300 m depth was 11.410 MPa and the fracture opening pressure at depths greater than 1300 m was calculated (see Tenma et al., 2008 for more details). In the Gangi model, one assumes a 2D X–Z radial model system which includes one fracture. The model provides an equation that relates fracture aperture to effective stress, for a fracture of width W_f and effective stress P,

$$W_f = W_{0f} \left(1 - \frac{p}{p_a}\right)^{\frac{1}{n}} \quad (27.17)$$

W_{0f} is the maximum aperture possible that corresponds to zero effective stress, P_a is the effective modulus of fracture asperities, and n is a parameter related to the distribution of heights of the asperities. Applying Darcy's law using the Gangi model, the permeability of a fracture reduces to

$$k = \frac{w_f^2}{12 \, fl} \quad (27.18)$$

fl is the Lomize's friction factor for flow, which for relative asperity height is given by

$$fl = 1 + 17 \left(\frac{a}{2w_f}\right)^{3/2} \quad (27.19)$$

In the modeling of flow and heat transfer through fractures, it must be emphasized that the models created are based solely on scientific interpretations of fractures. In reality, the fractures exist at such great depths and high pressures that one can only speculate at the true physics that govern the flow processes. Scientific interpretations include the leaky fracture model, where a spatial dependence of leakage rate per unit length decreases linearly from well-bore to fracture tip (Johnson and Gustafson, 1988).

The fracture can also be modeled as a flat open space between two large blocks of homogeneous, isotropic, impermeable rock where the temperature of the rock is governed by conduction, and convective transport enters only through the boundary conditions of the fracture surfaces (Bodvarsson, 1969). In the case of an arbitrarily shaped fracture, as in the Coso geothermal field in central eastern California (Ghassemi et al., 2005), incompressible fluid flow is solved using 2D fracture geometry as follows:

$$\nabla \cdot q(x, y) = -\sum_{i=1}^{n} Q_i \delta \left(x - x_i, y - y_i\right) \quad (27.20)$$

∇ is the divergence operator in 2D, Q_i is the injection rate at well i, which is negative for extraction, (x_i, y_i) is the well location, and δ is the Dirac delta function. Flow in the fracture is driven by a pressure gradient

$$\nabla p(x, y) = -\frac{\pi^2 \mu}{w^3} q(x, y) \quad (27.21)$$

p is the fluid pressure, μ is the fluid viscosity, and w is the fracture width. The boundary condition that is applied at the rim of the planar fracture is

$$\frac{\partial p}{\partial n} = 0 \text{ on } \partial A \quad (27.22)$$

∂A is the rim of the planar fracture and n is the outward normal of ∂A. Numerical techniques such as the FEM are used to solve the flow in the fracture after it is discretized into surface elements.

Ghassemi et al. (2008) extended their earlier study (Ghassemi et al., 2003, 2005) to include the poroelastic and thermoelastic effects of the rock walls, where leak off is possible within the fracture; their numerical studies addressed varying fluid loss and injection rates within the fractures. They also analyzed the pressure profile along the fracture for fluid loss and injection rate ratios after 6 months of injection. Other authors focused on fully coupled systems where the effects of thermal, hydrologic, mechanical, and chemical processes were considered in the models over long-term and short-term periods (Taron and Elsworth, 2009). The rationale behind these models is that chemical reactions and thermal–hydro–mechanical deformation influences fracture permeability and interconnectivity which may lead to increased fluid losses and resistance to fluid flow, which are detrimental effects in EGS.

In the case of the Cooper Basin, thermo-hydraulic analysis was conducted by considering a stacked reservoir model comprised of individual layers of approximately 100 m extent (Vöros and Weidler, 2006). The temperature range of the stacked system at depths of 4.2–5 km was 247–274°C. Two primary modules were designed, one which modeled a central injection well with three production wells at the apices of an equilateral triangle, and the other with a square base, with a central injection well and four production wells at the vertices. These modules were then arranged in a hexagonal (with 24 triangular base modules) and a square pattern (16 square base modules) and numerically analyzed for hydraulic and thermal modeling. The temperature decline was computed for 20 years. Vöros and Weidler (2006) found that the distance between central and outer wells was the dominant parameter of the system. They were able to provide the following estimates: (a) for a total circulation rate of 600 $l \cdot s^{-1}$ per layer, the models with 700 m well spacing a temperature decay will result in the second year with a temperature decline of about 2.5 $K \cdot yr^{-1}$ and (b) for the larger model with a well separation of 1000 m, the total decay will be 12 K after 20 years.

Other numerical solutions for the fluid flow and heat transfer in geothermal reservoirs exist in the literature, see for example: FEM (Bundschuh and Suárez-Arriaga, 2010) and finite difference numerical model for a 3D double porosity system (Sanyal and Butler, 2005). Also, models have been developed where the working fluid in the EGS is carbon dioxide, rather than water (Pruess, 2006). Pruess commented that the main disadvantage of carbon dioxide as a working fluid, viz., its low heat capacity, will be sufficiently compensated for by its other desirable physical properties, including a low viscosity and large expansivity compared to water, and its inability to act as a solvent for many rock minerals at high temperatures.

27.6   CONCLUDING REMARKS

EGS technology is still in its infancy compared to the conventional geothermal industry, but it is rapidly gaining momentum because of the increasing need for clean energy sources. Many experimental projects are currently underway in many parts of the world, and some have demonstrated that although no two EGS behave in exactly the same way, many lessons have been learnt and have been applied successfully in new projects.

The critical test for an EGS is obviously the performance. In a commercial scale reservoir, the three main parameters considered are

- thermal performance
- hydraulic impedance
- water loss

Major EGS projects at Fenton Hill in New Mexico, Falkenberg in Germany, Mayet de Montagne in France, and the Phase I experiments at Rosemanowes in Cornwall, UK, were designed to show that it is possible to circulate water between boreholes linked by a fractured reservoir (Richards et al., 1994). However, none of these projects were capable of commercial operation. For the Rosemanowes EGS project, which began in 1977, the thermal performance was only about 0.4% of the requirements of a commercial system and in 1991, after three phases of experimentation, a review panel concluded that electricity generation from HDR in the UK is not viable in the short–medium term and the EGS project was suspended. On the other hand, projects in Soultz and Landau demonstrate that technically and commercially viable EGS is achievable and realistic; these two systems are fully operating and provide a clean ongoing supply of geothermal energy. In addition, ongoing numerical studies on EGS always show great power generation potential, see for example, the case study on Desert Peak (Butler et al., 2004).

As outlined by Richards et al. (1994), the challenge of EGS is to achieve the following performance parameters for a commercial-scale HDR well-doublet:

1. A 25-year lifetime at a production flow rate of 75 $kg \cdot s^{-1}$ before the outlet temperature declines by 10%. Further, for an initial inlet temperature of 40°C and rock temperature of 200°C, the rate of heat extraction should be ~50 MW.
2. A hydraulic impedance less than 0.1 $MPa \cdot kg^{-1} \cdot s$.
3. Fluid loss of 10% or less.

Once these performance parameters are achieved, EGS will be able to meet the baseload demands of consumers and ultimately replace or supplement other traditional forms of energy.

REFERENCES

Abé, H., Mura, T., and Keer, L. M. (1976). Growth rate of a penny-shaped crack in hydraulic fracturing of rocks, *J. Geophys. Res.*, 81(29), 5335–5340.

Andersson, J., Shapiro, A. M., and Bear, J. (1984). A stochastic model of fractured rock conditioned by measured information, *Water Resour. Res.*, 20, 79–88.

Asanuma, H., Nozaki, H., Niitsuma, H., and Wyborn, D. (2005). Interpretation of microseismic events with larger magnitude collected at the Cooper Basin, Australia, *GRC Trans.*, 29, 87–91.

Azim, R., Amin, S., and Shoeb, A. (2010). Prospect of enhanced geothermal systems in baseload power generation. In: IEEE International Conference on Advanced Management Science, July 9–11, Vol. 3, pp. 176–180.

Baecher, G. B. (1983). Statistical analysis of rock mass fracturing, *Math. Geol.*, 15(2), 329–348.

Baisch, S., Vörös, R., Weidler, R., and Wyborn, D. (2009). Investigation of fault mechanisms during geothermal reservoir stimulation experiments in the Cooper Basin, Australia, *Bull. Seismol. Soc. Am.*, 99(1), 148–158.

Baisch, S., Weidler, R., Vörös, R., Wyborn, D., and de Graaf, L. (2006). Induced seismicity during the stimulation of a geothermal HFR reservoir in the Cooper Basin, Australia, *Bull. Seismol. Soc. Am.*, 96(6), 2242–2256.

Baria, R., Baumgärtner, J., Rummel, F., Pine, R. J., and Sato, Y. (1999). HDR/HWR reservoirs: concepts, understanding and creation, *Geothermics*, 28, 533–552.

Beardsmore, G. R. (2005). Thermal modelling of the hot dry rock geothermal resources beneath GEL99 in the Cooper Basin, South Australia. In: Proceedings of World Geothermal Congress 2005, Antalya, Turkey, April 24–29, 2005.

Beavers, G. S. and Joseph, D. D. (1967). Boundary conditions at a naturally permeable wall, *J. Fluid Mech.*, 30, 197–207.

Beavers, G. S., Sparrow, E. M., and Magnuson, R. A. (1970). Experiments on coupled parallel flows in a channel and bounding porous medium, Trans, *ASME J. Basic Eng.*, 92, 843–848.

Bercovier, M., Luzon, M., and Pavlov, E. (2002). Detecting planar patches in an unorganized set of points in space, *Adv. Comput. Math.*, 17, 153–166.

Berkowitz, B. (1989). Boundary conditions along permeable fracture walls: influence on flow and conductivity, *Water Resour. Res.*, 25(8), 1919–1922.

Berkowitz, B. (2002). Characterizing flow and transport in fractured geological media: a review, *Adv. Water Resour.*, 25, 861–884.

Berman, A. S. (1953). Laminar flow in channels with porous walls, *J. Appl. Phys.*, 24(9), 1232–1235.

Bodvarsson, G. (1969). On temperature of water flowing through fractures, *J. Geophys. Res.*, 74(8), 1987–1992.

Bromley, C. J. and Mongillo, M. A. (2008). Geothermal energy from fractured reservoirs-dealing with induced seismicity, *IEA Open Energy Technol. Bull.*, 48, 7.

Brown, D., DuTeaux, R., Kruger, P., Swenson, D., and Yamaguchi, T. (1999). Fluid circulation and heat extraction from engineered geothermal reservoirs, *Geothermics*, 28, 553–572.

Brown, D. W., Duchane, D. V., Heoken, G., and Hriscu, V. T. (2012). *Mining the Earth's Heat: Hot Dry Rock Geothermal Energy*, Springer.

Bundschuh, J. and Suárez-Arriaga, M. C. (2010). *Introduction to the Numerical Modelling of Groundwater and Geothermal Systems*, CRC Press, London.

Butler, S. J., Snayal, S. K., and Robertson-Tait, A. (2004). A numerical simulation study of the performance of enhanced geothermal systems. In: Proceedings of Thirty-Seventh Workshop on Geothermal Reservoir Engineering Stanford University, Stanford, California, January 26–28, SGP-TR-175.

Cacas, M. C., Ledoux, E., de Marsily, G., Barbreau, A., Calmels, P., Gaillard, B., and Margrita, R. (1990). Modelling fracture flow with a stochastic discrete fracture network: calibration and validation—the transport model, *Water Resour. Res.*, 26(3), 491–500.

Caldecott, J. (1845). Observations on the temperature of the Earth at Trevandrum in Lat. 8° 30′ 32″. *Proc.Roy. Soc. Edinburgh*, 2, 29–31.

Carneiro, J. F. (2009). Numerical simulations on the influence of matrix diffusion to carbon sequestration in double porosity fissured aquifers, *Int. J. Greenhouse Gas Control*, 3, 431–443.

Celati, R., Squarci, P., Stefan, G. C., and Taffi, L. (1977). Study of water levels in Larderello region geothermal wells for reconstruction of reservoir pressure trend, *Geothermics*, 6, 183–198.

Chabora, E., Zemach, E., Spielman, P., Drakos, P., Hickman, S., Lutz, S., Boyle, K., Falconer, A., Robertson-Tait, A., Davatzes, N. C., Rose, P., Majer, E., and Jarpe, S. (2012). Hydraulic stimulation of well 27-15 Desert Peak geothermal filed, Nevada USA. In: Proceedings of Thirty-Seventh Workshop on Geothermal Reservoir Engineering Stanford University, Stanford, California, January 30–Feb 1, SGP-TR-194.

Chilès, J. P. (1988). Fractal and geostatistical methods for modelling of a fracture network, *Math. Geol.*, 20(6), 631–654.

Chopra, P. and D.Wyborn (2003). Australia's first hot dry rock geothermal energy extraction project is up and running in the granite beneath the Cooper Basin, NE South Australia. In: *Magmas to Mineralisation. The Ishihara Symposium*, Blevin, P. Jones, M., and Chappell, B. (editors)., Geoscience Australia, pp. 43–45. Record 2003/14.

Christopher, H. and Armstead, H. (1978). *Geothermal Energy*, E and FN Spon, London.

Crandall, D., Ahmadi, G., and Smith, D. H. (2010). Computational modelling of fluid flow through a fracture in permeable rock, *Transp. Porous Media*, 84, 493–510.

Cuenot, N., Charléty, J., Dorbath, L., and Haessler, H. (2006) Faulting mechanisms and stress regime at the European HDR site of Soultz-sous-Forêts, France, *Geothermics*, 35, 561–575.

Delépine, N., Cuenot, N., Rothert, E., Parotidis, M., Rentsch, S., and Shapiro, S. A. (2004). Characterisation of fluid transport properties of the hot dry rock reservoir Soultz-2000 using induced seismicity, *J. Geophys. Eng.*, 1, 77–83.

Dershowitz, W. S. and Fidelibus, C. (1999). Derivation of equivalent pipe network analogues for three-dimensional discrete fracture networks by boundary element method, *Water Resour. Res.*, 35(9), 2685–2691.

Dershowitz, W. S. and Einstein, H. H. (1988). Characterizing rock joint geometry with joint system models, *Rock Mech. Rock Eng.*, 21, 21–51.

Dickson, M. H. and Fanelli, M. (2006). *Geothermal Energy: Utilization and Technology*, Earthscan.

Dowd, P. A., Xu, C., Mardia, K. V., and Fowell, R. J. (2007). A comparison of methods for the simulation of rock fractures, *Math. Geol.*, 39, 697–714.

Dowd, P. A., Martin, J. A., Xu, C., Fowell, R. J., and Mardia, K. V. (2009). A three-dimensional data set of the fracture network of a granite, *Int. J. Rock Mech. Min. Sci.*, 46(5), 811–818.

Einstein, H. H. and Baecher, G. B. (1983). Probabilistic and statistical methods in engineering geology, *Rock Mech. Rock Eng.*, 16, 39–72.

Evans, K. F., Zappone, A., Kraft, T., Diechmann, N., and Maia F. (2012). A survey of the induced seismic responses to geothermal and CO_2 reservoirs in Europe, *Geothermics*, 41, 30–54.

Fadakar-A, Y., Xu, C., and Dowd, P. A. (2011). A general framework for fracture intersection analysis: algorithms and practical applications. In: AGEC 2011, November 16–18, Melbourne, Australia.

Foulger, G. R., Julian, B. R., and Monastero, F. C. (2008). Seismic monitoring of EGS tests at the Coso Geotehrmal area, California using accurate MEQ locations and full moment tensors. In: Proceedings of Thirty-Third Workshop on Geothermal Reservoir Engineering Stanford University, Stanford, California, January 28–30, SGP-TR-185.

Fouché, O. and Diebolt, J. (2004). Describing the geometry of 3D fracture systems by correcting for linear sampling bias, *Math. Geol.*, 36(1), 33–63.

Garcia, J., Walters, M., Beall, J., Hartline, C., Pingol, A., Pistone, S., and Wright, M. (2012). Overview of the Northwest Geysers EGS demonstration project. In: Proceedings of Thirty-Seventh Workshop on Geothermal Reservoir Engineering Stanford University, Stanford, California, January 30–February 1, SGP-TR-194.

Genter, A., Fritsch, D., Cuenot, N., Baumgärtner, J., and Graff, J-J. (2009). Overview of the current activities of the European EGS Soultz project: from exploration to energy production. In: Proceedings of Thirty-Fourth Workshop on Geothermal Reservoir Engineering, Stanford University, Stanford, California, February 9–11, SGP-TR-187.

Genter, A., Evans, K., Cuenot, N., Fritsch, D., and Saujuan, B. (2010). Contribution of the exploration of deep crystalline fractured reservoir of Soultz to the knowledge of enhanced geothermal systems (EGS), *Comptes. Rendus. Geosci.*, 342(7–8), 502–516.

Ghassemi, A., Tarasovs, S., and Cheng, A. H.-D. (2005). Integral equation solution of extraction-induced thermal stress in enhanced geothermal reservoirs, *Int. J. Numer. Anal. Meth. Geomech.*, 29, 829–844.

Ghassemi, A., Tarasovs, S., and Cheng, A. H.-D. (2003). An integral equation method for modelling three dimensional heat extraction from a fracture in hot dry rock, *Int. J. Numer. Anal. Meth.Geomech.*, 27, 989–1004.

Ghassemi, A., Nygren, A., and Cheng, A. H.-D. (2008). Effects of heat extraction on fracture aperture: a poro-thermoelastic analysis, *Geothermics*, 37, 525–539.

Geodynamics Ltd. (2009). Annual report 2009. http://www.geodynamics.com.au/IRM/Company/ShowPage.aspx?CPID=2273&EID=36841324.

Giardini, D. (2009). Geothermal quake risks must be faced, *Nature*, 462, 848–849.

Gupta, A. K., Adler, P. M. (2006). Stereological analysis of fracture networks along cylindrical galleries, *Math. Geol.*, 38(3), 233–267.

Gupta, H., and Roy, S. (2007). *Geothermal Energy. An Alternative Resource of the 21st Century*. Elsevier.

Hamm, S.-Y., Kim, M.-S., Cheong, J.-Y., Kim, J.-Y., Son, M., and Kim, T. W. (2007). Relationship between hydraulic conductivity and fracture properties estimated from packer tests and borehole data in a fractured granite, *Eng. Geol.*, 92, 73–87.

Häring, M. O., Schnaz, U., Ladner, F., and Dyer, B. C. (2008). Characterisation of the basel 1 enhanced geothermal system, *Geothermics*, 37, 469–495.

Harlow, F. H. and Pracht, W. E. (1972). A theoretical study of geothermal energy extraction, *J. Geophys. Res.*, 77, 7038–7048.

Hartline, C. S., Walters, M. A., and Wright, M. C. (2015). Three-dimensional structural model building, induced seismicity analysis, drilling analysis, and reservoir management at The Geysers Geothermal Field, Northern California. *GRC Transactions*, 39, 603–614.

Hori, Y., Kitano, K., Kaieda, H., and Kiho, K. (1999). Present status of the Ogachi HDR project, Japan, and future plans, *Geothermics*, 28, 637–645.

Huang, S. and Liu, J. (2010). Geothermal energy stuck between a rock and a heat place, *Nature*, 463.

Hutton, J. (1785). Theory of the Earth; or an investigation of the laws observable in the composition, dissolution and restoration of land upon the globe, *Proc. Royal Soc. Edinburgh*, 1, 209–304.

Jing, L. and Stephansson, O. (2007). *Fundamentals of Discrete Element Methods for Rock Engineering—Theory and Applications*, Elsevier.

Johnson, R. E. and Gustafson, C. W. (1988). Leakage losses from a hydraulic fracture and fracture propagation, *Phys. Fluids*, 31(11), 3180–3187.

Karner, S. (2005). Stimulation techniques used in enhanced geothermal systems: perspectives from geomechanics and rock physics. In: Proceedings of Thirtieth Workshop on Geothermal Reservoir Engineering, Stanford University, Stanford, California, January 31–Febuary 2, SGP-TR-176.

Kraft, T., Mai, P. M., Wierner, S., Deichmann, N., Ripperger, J., Kästli, P., Bachman, C., Fäh, D., Wössner, J., and Giardini, D. (2009). Enhanced geothermal systems: mitigating risks in urban areas, *EOS Trans.*, 90(32), 273–280.

Kulatilake, P. H. S. W., Um, J., Wang, M., Escandon, R. F., and Narvaiz, J. (2003). Stochastic fracture geometry modeling in 3D including validations for a part of Arrohead East Tunnel, California, USA, *Eng. Geol.*, 70, 131–155.

La Pointe, P. R. (2002). Derivation of parent fracture population statistics from trace length measurements of fractal fracture populations, *Int. J. Rock Mech. Min. Sci.*, 39, 381–388.

Laslett, G. M. (1982). Censoring and edge effects in areal and line transect sampling of rock joint traces, *Math. Geol.*, 14(2), 125–140.

Ligrani, P., Blanchard, D., and Gale, B. (2010). Slip due to surface roughness for a Newtonian liquid in a viscous microscale pump, *Phys. Fluids*, 22, 0520021-15.

Lipman, S. C., Strobel, C. J., and Gulati, M. S. (1978). Reservoir performance of the Geysers field, *Geothermics*, 7, 209–219.

Lee, J. S., Veneziano, D., and Einstein, H. H. (1990). Hierarchical fracture trace model. In: *Rock Contributions and Challenges, Proceedings of the 31st US Rock Mech. Symposium*, Hustrulid, W. and Johnson G. A. (editors), Balkema, Rotterdam, pp. 261–268.

Long, J. C. S., Remer, J. S., Wilson, C. R., and Witherspoon, P. A. (1982). Porous media equivalents for networks of discontinuous fractures, *Water Resour. Res.*, 18(3), 645–658.

Mardia, K. V., Nyirongo, V. B., Walder, A. N., Xu, C., Dowd, P. A., Fowell, R. J., and Kent, J. T. (2007). Markov chain Monte Carlo implementation of rock fracture modeling, *Math. Geol.*, 39, 355–381.

Martin, C. (2011). *Renaissance Meterology*, John Hopkins University Press, Baltimore, MD.

Mohais, R., Xu, C., and Dowd, P. A. (2011). An analytical model of coupled fluid flow and heat transfer through a fracture with permeable walls in an EGS. In: AGEC, November 2011, Melbourne, Australia.

Mohais, R., Xu, C., and Dowd, P. A. (2011). Fluid flow and heat transfer within a single horizontal channel in an enhanced geothermal system, *ASME J. Heat Transf.*, 133(11), 112603.

Mohais, R., Xu, C., Dowd, P. A., and Hand, M. (2012). Permeability correction factor for fractures with permeable walls, *Geophys. Res. Letts.*, 39, L03403, doi:10.1029/2011GL050519

Murphy, H. and Fehler, M. (1986). Hydraulic fracturing of jointed formations, *Soc. Petrol. Engrs. Int. Mg. Petrol. Eng.*, March 17–20, Beijing, China, SPE paper 14088.

Neale, G. and Nader, W. (1974). Practical significance of Brinkman's extension of Darcy's Law: coupled parallel flows within a channel and a bounding porous medium, *Can. J. Chem. Eng.*, 52, 475–478.

Ozkaya, S. I. and Mattner, J. (2003). Fracture connectivity from fracture intersections in borehole image logs, *Compus. Geosc.*, 29, 143–153.

Parker, R. H. and Jupe, A. (1997). In situ leaching mining and hot dry rock (HDR) geothermal energy technology, *Min. Eng.*, 10(3), 301–308.

Pine, R. J. and Batchelor, A. S. (1984). Downard migration of shearing in jointed rock during hydraulic injections, *Int. J. Rock Mech. Min. Sci.*, 21(5), 249–263.

Portier, S., Vuataz, F-D., Nami, P., Snajuan, B., and Gérard, A. (2009). Chemical stimulation techniques for geothermal wells: experiments in the three-well EGS system at Soultz-sous-Forêts, France, *Geothermics*, 38(4), 349–359.

Priest, S. D. (1993). *Discontinuity Analysis for Rock Engineering*, Chapman & Hall, p. 473.

Priest, S. D. and Hudson, J. A. (1981). Estimation of discontinuity spacing and trace length using scanline surveys, *Int. J. Rock Mech. Min. Sci.*, 18, 181–197.

Pruess, K., Oldenburg, C., and Moridis, G. (1999). TOUGH2 User's Guide, V2.0, Earth Sciences Division, Lawrence Berkeley National Laboratory, University of California, Berkeley, 94720.

Pruess, K. (2006). Enhanced geothermal systems (EGS) using CO_2 as a working fluid—a novel approach for generating renewable energy without simultaneous sequestration of carbon, *Geothermics*, 35, 351–367.

Richards, H. G., Parker, R. H., Green, A. S. P., Jones, R. H., Nicholls, J. D. M., Nicol, D. A. C., Randall, M. M., Richards, S., Stewart, R. C., and Wills-Richards, J. (1994). The performance and characteristics of the experimental hot dry rock reservoir at Rosemanowes, Cornwall (1985–1988), *Geothermics*, 23(2), 73–109.

Robinson, P. C. (1983). Connectivity of fracture systems—a percolation theory approach, *J. Physics A*, 16, 605–614.

Rummel, F. (2005). Heat mining by hot-dry-rock technology. In: Proceedings of International Conference of ECOMINING, Bukharest.

Rutherford, E. and Barnes, H. T. (1904). Heating effect of the radium emanation, *Philos. Mag. Series 6*, 7(38), 202–219.

Sahimi, M. (1993). Flow phenomena in rocks: from continuum models to fractals, percolation, cellular automata and simulated annealing, *Rev. Mod. Phys.*, 65(4), 1393–1534.

Sanyal, S. K. and Butler, S. J. (2005). An analysis of power generation prospects from enhanced geothermal systems. In: Proceedings of World Geothermal Congress, Antalya, Turkey, April 24–29.

Sanyal, S. K. (2000). Forty years of production history at the Geysers geothermal field, California—the lessons learned, *Geotherm. Resour. Counc. Trans.*, 24, 317–323.

Sausse, J. and Genter, A. (2005). Types of permeable fractures in granite, In: Petrophysical properties of crystalline rocks, Harvey, P. K., Brewer, T. S., Pezard, P. A. and Petrov, V. A. (editors), Geological Society, London, *Special Publications*, 240, 1–14.

Smith, M. C. (1983) A history of hot dry rock geothermal energy systems, *J. Volcanol. Geotherm. Res.*, 15, 1–20.

Snow, D. T. (1965). A parallel plate model of fractured permeable media. PhD thesis, University of Berkeley.

Taron, J. and Elsworth, D. (2009). Thermal-hydrologic-mechanical-chemical processes in the evolution of engineered geothermal reservoirs, *Int. J. Rock Mech. Min. Sci.*, 46, 855–864.

Tenma, N., Yamaguchi, Y., and Zyvoloski, G. (2008). The Hijiori hot dry rock test site, Japan. Evaluation and optimization of heat extraction from a two-layered reservoir, *Geothermics*, 37, 19–52.

Terrill, R. M. and Shrestha, G. M. (1965). Laminar flow through parallel and uniformly porous walls of different permeability, *ZAMP*, 16, 470–482.

Thomson, W. (1860). On the reduction of observations of underground temperature, with application to Professor Forbes'

Edinburgh observations, and the continued Carlton Hill series, *Trans. Roy. Soc. Edinburgh*, 4, 342–346.

Tran, N. H. and Rahman, S. S. (2007). Development of hot dry rocks by hydraulic stimulation: natural fracture network simulation, *Theor. Appl. Fract. Mech.*, 47, 77–85.

Tsang, Y. W., Tsang, C. F., Hale, F. V., and Dverstorp, B. (1996). Tracer transport in a stochastic continuum model of fractured media, *Water Resour. Res.*, 32(10), 3077–3092.

Villaescusa, E. and Brown, E. T. (1992). Maximum likelihood estimation of joint size from trace length measurements, *Rock Mech. Rock Eng.*, 25, 67–87.

Viruete, J. E., Carbonell, R., Martí, D., and Pérez-Estaún, A. (2003). 3D stochastic modelling and simulation of fault zones in the Albalá granitic pluton, SW Iberian Variscan Massif, *J. Struct. Geol.*, 25, 1487–1506.

Vöros, R. and Weidler, R. (2006). Numerical thermo-hydraulic stimulation of a large scale power production in the Cooper Basin. Q-Con Report, GDY 018.

Warburton, P. M. (1980). A stereographical interpretation of joint trace data, *Int. J. Rock Mech. Min. Sci.*, 17, 181–190.

Weidler, R. (2005). The Cooper Basin HFR Project 2003/3004: findings, achievements and implications, Technical Report, Geodynamics Ltd.

Xing, H., Zhang, J., Liu, Y., and Mulhaus, H. (2009). Enhanced geothermal reservoir simulation. In: Proceedings of the Australian Geothermal Energy Conference 2009, Brisbane.

Xu, C. and Dowd, P. A. (2010). A new computer code for discrete fracture network modelling, *Comps. Geosc.*, 36, 292–301.

Xu, C., Dowd, P. A., and Mohais, R. (2012). Connectivity analysis of the Habanero enhanced geothermal system. In: Proceedings of the Thirty-Seventh Workshop on Geothermal Reservoir Engineering, Stanford University, Stanford, California.

Xu, C., Dowd, P. A., and Wyborn, D. (2011a). Optimisation of a stochastic rock fracture model using Markov chain Monte Carlo simulation. In: Proceedings of the 35th International Symposium on the Application of Computers and Operations Research in the Minerals Industries, (APCOM2011), Wollongong, Australia.

Xu, C., Dowd, P. A., and Wyborn, D. (2011b). Optimised fracture model for Habanero reservoir. In: Proceedings of the Australian Geothermal Conference 2010, Geoscience Australia, Adelaide.

Xu, C., Dowd, P. A., Mardia, K. V., and Fowell, R. J. (2006a). A flexible true plurigaussian code for spatial facies simulations, *Comps. Geosc.*, 32, 1629–1645.

Xu, C., Dowd, P. A., Mardia, K. V., and Fowell, R. J. (2006b). A connectivity index for discrete fracture networks, *Math. Geol.*, 38, 611–634.

Xu, C., Dowd, P. A., Mardia, K. V., Fowell, R. J., and Taylor, C. C. (2007). Simulating correlated marked point processes, *J. Appl. Stat.*, 34(9), 1125–1134.

Young, D. S. (1987). Indicator Kriging for unit vectors: rock joint orientations, *Math. Geol.*, 19(6), 481–501.

Zhang, L. and Einstein, H. H. (2000). Estimating the intensity of rock discontinuities, *Int. J. Rock Mech. Min. Sci.*, 37, 819–837.

Zimmerman, R. W. and Bodvarsson, G. S. (1996). Hydraulic conductivity of rock fractures, *Trans. Porous. Media*, 23, 1–30.

Zimmermann, G., Moeck, I., and Blöcher, G. (2010). Cyclic water-frac stimulation to develop an enhanced geothermal system (EGS)—conceptual design and experimental results, *Geothermics*, 39, 59–69.

28

THERMODYNAMIC ANALYSIS OF GEOTHERMAL POWER PLANTS

MEHMET KANOGLU AND ALI BOLATTURK

28.1 INTRODUCTION

Geothermal energy is the thermal energy within the earth's interior. It is to some extent a renewable energy source since a geothermal resource usually has a projected life of 30–50 years. The life of a resource may be prolonged by reinjecting the waste fluid, which is the most common method of disposal. Reinjection may also help to maintain reservoir pressure. Over-ground disposal of geothermal fluid has a potential hazard of water pollution of rivers and lakes as well as air pollution (Barbier, 1997).

There are several options for utilizing the thermal energy produced from geothermal energy systems. The most common is base-load electric power generation, followed by direct use in process and space-heating applications. In addition, combined heat and power in cogeneration and hybrid systems, and as a heat source and sink for heat pump applications, are options that offer improved energy savings.

Electricity has been generated from geothermal resources since the early 1960s. Most of the world's geothermal power plants were built in the 1970s and 1980s following the 1973 oil crisis. The urgency to generate electricity from alternative energy sources and the fact that geothermal energy was essentially free lead to nonoptimal plant designs for using geothermal resources. Today, with nearly 10,000 MWe of electricity generated by geothermal worldwide, there are several energy conversion technologies commercially available at various stages of maturity. These include direct steam expansion, single and multistage steam flashing, organic binary Rankine cycles, and two-phase flow expanders. Direct use and heat pump applications are also having an increasing impact, with

a combined, estimated market penetration of about 100,000 MW worldwide (MIT, 2006).

A geothermal well can produce brine (saturated or superheated vapor), wet steam (liquid–vapor mixture), and dry steam (saturated or superheated vapor). Liquid-dominated systems are much more common than vapor-dominated systems and can be produced either as brine or as brine–steam mixture, depending on the pressure maintained on the production system. If the pressure is reduced below the saturation pressure at that temperature, some of the brine will flash, and a two-phase mixture will result. If the pressure is maintained above the saturation pressure, the fluid will remain single phase. High-temperature geothermal resources above 150°C are generally used for power generation. Moderate-temperature (between 90°C and 150°C) and low-temperature (below 90°C) geothermal resources are best suited for direct uses. Some novel designs are proposed to generate electricity from low-temperature resources economically (ASHRAE, 1995; Barbier, 1997). However, geothermal energy is more effective when used directly than when converted to electricity particularly for moderate and low-temperature geothermal resources, since the direct use of geothermal heat for heating and cooling would replace the burning of fossil fuels from which electricity is generated much more efficiently (Kunze, 1979; Kanoglu and Cengel, 1999b; Barbier, 1997).

Most of the world's high-temperature geothermal resources have already been exploited for the generation of electricity. For the geothermal resources above 90°C, only about a quarter are at 150°C or above (ASHRAE, 1995). For most moderate-temperature geothermal resources, generation of electricity is not economical due to the very

Alternative Energy and Shale Gas Encyclopedia, First Edition. Edited by Jay H. Lehr and Jack Keeley.
© 2016 John Wiley & Sons, Inc. Published 2016 by John Wiley & Sons, Inc.

low thermal efficiencies. Some novel designs are proposed to generate electricity from those resources economically (Mohanty and Paloso, 1992; Yuan and Michaelides, 1993). It appears that more research will focus on the methods of economic electric generation from these moderate-temperature geothermal resources.

28.2   GEOTHERMAL CYCLES

The technology for producing power from geothermal resources is well established and there are many geothermal power plants operating worldwide (Barbier, 1997). Depending on the state of the geothermal fluid in the reservoir, different power-producing cycles may be used including direct-steam, flash-steam (single-flash and double-flash), binary and combined flash-binary cycles.

The simplest and cheapest geothermal cycle is dry steam direct-intake noncondensing cycle. Steam from the geothermal well is simply passed through a turbine and exhausted to the atmosphere. Flash steam plants are used to generate power from liquid-dominated resources that are hot enough to flash a significant proportion of the water to steam in surface equipment, either at one or two pressure stages (single-flash or double-flash plants). Steam flows through a steam turbine to produce power while the brine is reinjected back to the ground. Steam exiting the turbine is condensed by cooling water obtained in a cooling tower or a spray bond before being reinjected. Steam ejectors are used in all steam-condensing cycles to keep vacuum conditions (under atmospheric pressure) in the condensers. Ejectors consume some

steam to accomplish their purpose. Binary cycle plants (Figure 28.1) use the geothermal brine from liquid-dominated resources usually below 170°C. These plants operate with a binary working fluid (isobutane, isopentane, R-114, etc.) that has a low boiling temperature in a Rankine cycle. The working fluid is completely vaporized and usually superheated by the geothermal heat in the vaporizer. The vapor expands in the turbine. It is then condensed in an air-cooled condenser or water-cooled condenser before being pumped back to the vaporizer to complete the cycle. Combined flash or binary plants incorporate both in a binary unit and a flashing unit to take advantages associated with both the systems. The liquid portion of the geothermal mixture serves as the input heat for binary cycle, while the steam portion goes through a steam turbine to produce power.

The thermal efficiency of steam plants, defined as the ratio of the net power generated to the energy of the geothermal steam in reservoir or at the plant site, ranges from 10% to 17%. These low percentages are due to geothermal resources being at relatively low temperatures. The thermal efficiency of binary cycle plants, defined as the ratio of the net power generated to the energy of brine in the reservoir or at the plant site, ranges from 2.8% to 5.5%. These percentages are even lower since binary design plants use lower temperature geothermal resources available for power generation. Also, the only cooling medium available for most binary plants is ambient air, which keeps the condenser temperature relatively high throughout the year. Alternatively, a conversion efficiency can be defined for binary plants as the ratio of the net power generated to the heat transferred to the binary fluid in the vaporizer, which is in fact the thermal efficiency

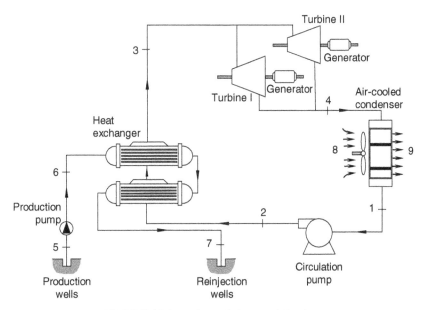

FIGURE 28.1   Schematic layout of the plant.

of the Rankine cycle. It ranges from 6% to 12% (Barbier, 1997). An extensive overview is provided of various energy-based and exergy-based efficiencies used in the analysis of power cycles including geothermal cycles (Kanoglu et al., 2007). Vapor and gas power cycles, cogeneration cycles, and geothermal power cycles are examined, and consideration is given to different cycle designs. The many approaches that can be used to define efficiencies are provided and their implications are discussed.

A binary power plant is most suitable for liquid-dominated low-temperature geothermal resources in the range of 100–170°C. Binary plants have been proven to have greater efficiencies than flashing plants in this temperature range. Performance of a binary geothermal power plant strongly depends on the thermophysical and critical properties of the working fluid selected. Therefore, the selection of the most appropriate working fluid is important to maximize the work output for a specified geothermal resource. A working fluid that was best when a binary plant was first constructed may no longer be the right one if the geothermal resource conditions change. In such cases, it may be necessary to change the working fluid or to modify its composition and/or to change the operating conditions of the plant to re-establish a match between operation and the new resource conditions (Phair, 1994).

28.3 FIRST LAW ANALYSIS

The first law of thermodynamics is the application of the conservation of energy principle to heat and thermodynamic processes. Efficiency traditionally has been primarily defined based on the first law (i.e., energy). Efficiency is one of the most frequently used terms in thermodynamics, and it indicates how well an energy conversion or process is accomplished. Efficiency is also one of the most frequently misused terms in thermodynamics and is often a source of misunderstanding. This is because efficiency is often used without being properly defined first (Cengel and Boles, 2006).

To assist in improving the efficiencies of power plants, their thermodynamic characteristics and performances are usually investigated. Power plants are normally examined using energy analysis but, as pointed out previously, a better understanding is attained when a more complete thermodynamic view is taken, which uses the second law of thermodynamics in conjunction with energy analysis via exergy methods.

Mass and energy balances for any control volume at steady state with negligible kinetic and potential energy changes can be expressed, respectively, by

$$\sum \dot{m}_{in} = \sum \dot{m}_{out} \tag{28.1}$$

$$\dot{Q} + \dot{W} = \sum \dot{m}_{out} h_{out} - \sum \dot{m}_{in} h_{in} \tag{28.2}$$

where the subscripts "in" and "out" represent the inlet and exit states, $\dot{Q}$ and $\dot{W}$ are the net heat and work inputs, $\dot{m}$ is the mass flow rate, and h is the enthalpy.

In general, the thermal efficiency of a geothermal power plant may be expressed as

$$\eta_{th,1} = \frac{\dot{W}_{net,out}}{\dot{m}_{geo}(h_{geo} - h_0)} \tag{28.3}$$

where the expression in the denominator is the energy input to the power plant, which is expressed as the enthalpy of the geothermal water with respect to environment state multiplied by the mass flow rate of geothermal water. Using the states, it becomes

$$\eta_{th,1} = \frac{\dot{W}_{net,out}}{\dot{m}_{geo}(h_5 - h_0)} \tag{28.4}$$

or according to the energy of geothermal water at the heat exchanger inlet:

$$\eta_{th,2} = \frac{\dot{W}_{net,out}}{\dot{m}_{geo}(h_6 - h_0)} \tag{28.5}$$

In equation (28.4), the energy input to the power plant represents the maximum heat the geothermal water can give, and this can only happen when the geothermal water is cooled to the temperature of the environment.

The actual heat input to a geothermal power cycle is less than the term in the denominator of equation (28.4) since part of geothermal water is reinjected back to the ground at a temperature much greater than the temperature of the environment. In this approach, the thermal efficiency is determined from

$$\eta_{th,3} = \frac{\dot{W}_{net,out}}{\dot{Q}_{in}} \tag{28.6}$$

The thermal efficiency may be expressed based on the heat transfer to the binary Rankine cycle (i.e., the heat transfer in the heat exchanger):

$$\eta_{th,binary} = \frac{\dot{W}_{net,out}}{\dot{m}_{geo}(h_6 - h_7)} \tag{28.7}$$

or

$$\eta_{th,binary} = \frac{\dot{W}_{net,out}}{\dot{m}_{binary}(h_3 - h_2)} \tag{28.8}$$

The isentropic efficiency of a turbine is defined as the ratio of the actual work output of the turbine to the work output that would be achieved if the process between the inlet state and exit pressure were isentropic. It is given by

$$\eta_{turb} = \frac{\dot{W}_{turb}}{\dot{W}_{turb,rev}} \qquad (28.9)$$

The isentropic efficiency of a pump is defined as the ratio of the work input required to raise the pressure of a liquid to a specified value in an isentropic manner to the actual work input:

$$\eta_{pump} = \frac{\dot{W}_{pump,rev}}{\dot{W}_{pump}} \qquad (28.10)$$

The throttling process is one for which $h_1 = h_2$. Its objective is to cause a significant pressure drop in the fluid. In flash cycle geothermal power plants, if the geothermal fluid in the reservoir is liquid or a steam–water mixture, it is throttled to decrease its pressure so that a substantial fraction of liquid can be converted to steam. Since throttling is essentially an isentropic process, the first law efficiency can be taken as 100%.

Heat exchangers are devices which facilitate two moving fluid streams to exchange heat without mixing. Condensers, vaporizers, and preheaters are all heat exchange devices used in geothermal power plants. Heat exchanger (or condenser) effectiveness is defined as the ratio of actual heat transfer rate to maximum possible heat transfer rate. It is given by

$$\varepsilon_{heatexc,cond} = \frac{\dot{Q}}{\dot{Q}_{max}} \qquad (28.11)$$

28.4   SECOND LAW ANALYSIS

The maximum possible work as a system undergoes a reversible process from the specified initial state to the state of its environment is called exergy (or availability). Reversible work is defined as the maximum amount of useful work that can be obtained as a system undergoes a process between the specified initial and final states. The difference between the reversible work and the actual work for the same specified initial and final states is called reversibility. Note that the exergy difference between the specified initial and final states also represents reversible work. That is, the exergy change is simply the maximum available work for a process. The second law efficiency is expressed as the ratio of the actual work output and the maximum possible work output (Cengel and Boles, 2006)

Exergy analysis has found increasingly widespread acceptance as a useful tool in the design, assessment, optimization, and improvement of energy systems (Bejan, 1988; Kotas, 1995; Szargut et al., 1988). Determining exergy efficiencies

for an overall system and/or the individual components making up the system constitutes a major part of exergy analysis. A comprehensive analysis of a thermodynamic system includes both energy and exergy analyses in order to obtain a more complete picture of system behavior.

Although exergy analysis can be generally applied to energy and other systems, it appears to be a more powerful tool than energy analysis for power cycles due to the fact that it helps determine the true magnitudes of losses and their causes and locations, and improve the overall system and its components.

Exergy balance for any control volume at steady state can be expressed by

$$\dot{E}_{heat} + \dot{W} = \sum \dot{E}_{out} - \sum \dot{E}_{in} + \dot{I} \qquad (28.12)$$

where $\dot{I}$ is the rate of irreversibility (exergy destruction). Also, $\dot{E}_{heat}$ is the net exergy transfer by heat at the temperature T, which is given by

$$\dot{E}_{heat} = \sum \left(1 - \frac{T_0}{T}\right)\dot{Q} \qquad (28.13)$$

The subscript 0 stands for the restricted dead state.

The specific flow exergy is given by

$$e = h - h_0 - T_0(s - s_0) \qquad (28.14)$$

Multiplying specific exergy by the mass flow rate of the fluid gives the exergy rate

$$\dot{E} = \dot{m}e \qquad (28.15)$$

The exergetic efficiency of a turbine is defined as a measure of how well the stream exergy of the fluid is converted into actual turbine output. Then,

$$\eta_{ex,turb} = \frac{\dot{W}_{turb}}{\dot{W}_{turb,rev}} \qquad (28.16)$$

where $\dot{W}_{turb}$ is the actual turbine power and $\dot{W}_{turb,rev}$ is the reversible turbine power, which is equal to $\dot{W}_{turb} + \dot{I}$. The exergy efficiency of the compressor is defined similarly as

$$\eta_{ex,pump} = \frac{\dot{W}_{pump,rev}}{\dot{W}_{pump}} \qquad (28.17)$$

where $\dot{W}_{pump,rev}$ is the reversible pump power, which is equal to $\dot{W}_{pump} - \dot{I}$. The exergetic efficiencies of a heat exchanger and condenser may be measured by increase in the exergy of the cold stream divided by the decrease in the exergy of the stream (Wark, 1995). Applying this definition to heat exchanger or condenser, we obtain

$$\eta_{ex,heatexc,cond} = \frac{(\dot{E}_{out} - \dot{E}_{in})_{cold}}{(\dot{E}_{in} - \dot{E}_{out})_{hot}} \quad (28.18)$$

where the subscripts cold and hot represent the cold stream and the hot stream, respectively. The difference between the numerator and denominator of equation (28.18) is the exergy destruction in the heat exchanger or condenser. One may take all the exergy given up by the hot fluid in the condenser as part of the exergy destruction for the power plant.

Using the exergy of geothermal water as the input to the plant, the exergy efficiency of a geothermal power plant can be expressed as

$$\eta_{ex,1} = \frac{\dot{W}_{net,out}}{\dot{E}_{in}} = \frac{\dot{W}_{net,out}}{\dot{m}_{geo}[h_5 - h_0 - T_0(s_5 - s_0)]} \quad (28.19)$$

or according to the exergy of geothermal water at the inlet of heat exchanger,

$$\eta_{ex,2} = \frac{\dot{W}_{net,out}}{\dot{E}_6} = \frac{\dot{W}_{net,out}}{\dot{m}_{binary}[h_6 - h_0 - T_0(s_6 - s_0)]} \quad (28.20)$$

For a binary cycle, the exergy efficiency may be defined based on the exergy decrease of geothermal water or exergy increase of the binary working fluid in the heat exchanger. That is,

$$\eta_{ex,binary,1} = \frac{\dot{W}_{net,out}}{\dot{m}_{geo}[h_6 - h_7 - T_0(s_6 - s_7)]} \quad (28.21)$$

$$\eta_{ex,binary,2} = \frac{\dot{W}_{net,out}}{\dot{m}_{binary}[h_3 - h_2 - T_0(s_3 - s_2)]} \quad (28.22)$$

The difference between the denominators of equations (28.21) and (28.22) is the exergy destruction in the heat exchanger.

The total exergy lost in cycle is determined from

$$\dot{I}_{cycle} = \dot{I}_{pump} + \dot{I}_{heatexc.} + \dot{I}_{turb.} + \dot{I}_{cond.} + \dot{I}_{reinj.} \quad (28.23)$$

The total exergy destruction in the plant is the difference between the brine exergy at the heat exchanger inlet and the net power output from cycle:

$$\dot{I}_{plant} = \dot{E}_{in} - \dot{W}_{net,out} \quad (28.24)$$

This includes various exergy losses in the plant components as well as the exergy of brine leaving heat exchanger. One may argue that the exergy of used brine is a recovered exergy, and it should not be considered to be part of the exergy loss. However, the used brine is reinjected back to the ground without any attempt to make use of it.

28.5 AN APPLICATION

Here, we present an application for thermodynamic analysis of an actual binary geothermal power plant. More details can be found in Kanoglu and Bolatturk (2008).

The geothermal power plant analyzed is a binary design plant that generates a yearly average net power output of about 27 MW. The plant consists of two identical units, each having two identical turbines. Only one unit will be considered for the rest of the paper. Schematic of one unit and the properties at various states are given in Figure 28.1 and Table 28.1, respectively. Brine is extracted from five production wells whose average depth is about 160 m. The power plant operates on a liquid-dominated resource at 160°C. The brine passes through the heat exchanger system that consists of a series of counter-flow heat exchangers where heat is transferred to the working (binary) fluid isobutane before the brine is reinjected back to the ground. Isobutane is found superheated at the heat exchanger exit. An equal amount of isobutane flows through each turbine. The mechanical power extracted from the turbines is converted to electrical power in generators. It utilizes dry-air condenser to condense the working fluid, so no fresh water is consumed. Isobutane circulates in a closed cycle, which is based on the Rankine cycle.

The harvested geothermal fluid is saturated liquid at 160°C and 1264 kPa in the reservoir. The heat source for the plant is the flow of geothermal water (brine) entering the plant at 158°C and 609 kPa with a total mass flow rate of 555.9 kg/s. Geothermal fluid remains as a liquid throughout the plant. The brine leaving the heat exchangers is directed to the reinjection wells where it is reinjected back into the ground at 90°C and 423 kPa.

In the plant, mass flow rate 305.6 kg/s of working fluid circulates through the cycle. The working fluid enters the heat exchanger at 13.7°C and leaves after it is evaporated at 128°C and superheated to 146.8°C. The working fluid then passes through the turbines with a total mass flow rate of 305.6 kg/s. It exhausts to an air-cooled condenser at about 79.5°C, where it condenses to a temperature of 11.7°C. Approximately 8580 kg/s air at an ambient temperature of 3°C is required to absorb the heat yielded by the working fluid. This raises the air temperature to 19.4°C. The working fluid is pumped to heat exchanger pressure to complete the Rankine cycle. The isobutane cycle on a T–s diagram is shown in Figure 28.2. It is noted in Figure 28.2 that the saturated

TABLE 28.1   Exergy rates and other properties at various plant locations for one representative unit.

State No	Fluid	Phase	Temperature T (°C)	Pressure P (kPa)	Enthalpy h (kJ/kg)	Entropy s (kJ/kg°C)	Mass flow rate $\dot{m}$(kg/s)	Exergy rate $\dot{E}$(kW)
0	Brine	Dead state	3.0	84	12.6	0.046	–	–
0′	Isobutane	Dead state	3.0	84	207	1.025	–	–
0″	Air	Dead state	3.0	84	276.5	5.672	–	–
1	Isobutane	Comp. liquid	11.7	410	227.5	1.097	305.6	221.3
2	Isobutane	Comp. liquid	13.76	3250	234.3	1.103	305.6	1768
3	Isobutane	Sup. vapor	146.8	3250	760.9	2.544	305.6	41,094
4	Isobutane	Sup. vapor	79.5	410	689.7	2.601	305.6	14,528
5	Brine	Liquid	160.0	1264	675.9	1.942	555.9	77,656
6	Brine	Liquid	158.0	609	666.8	1.923	555.9	75,586
7	Brine	Liquid	90.0	423	377.3	1.193	555.9	26,706
8	Air	Gas	3.0	84	276.5	5.672	8580	0
9	Air	Gas	19.4	84	292.9	5.730	8580	4036

State numbers refer to Figure 28.1.

vapor line of isobutane has positive slope ensuring super-heated vapor state at the turbine outlet. Thus, no moisture is involved in the turbine operation. This is one reason isobutane is a suitable working fluid in binary geothermal power plants.

The heat exchange process between the geothermal brine and working fluid isobutane is shown in Figure 28.3. An energy balance can be made from Figure 28.3 for the heat exchanger as

$$\dot{m}_{geo}(h_6 - h_{pp}) = \dot{m}_{binary}(h_3 - h_{f,binary}) \quad (28.25)$$

and

$$\dot{m}_{geo}(h_{pp} - h_7) = \dot{m}_{binary}(h_{f,binary} - h_2) \quad (28.26)$$

where $\dot{m}_{geo}$ and $\dot{m}_{binary}$ are the mass flow rate of geothermal brine and binary fluid, respectively. $h_{f,binary}$ is the saturated liquid enthalpy of isobutane at the saturated (vaporization) temperature, T_{vap} to be 128°C, and h_{pp} is the enthalpy of the brine at the pinch-point temperature of the brine. Solving these equations for h_{pp}, we determine the corresponding brine pinch-point temperature, T_{pp}, to be 133.9°C. The pinch-point temperature difference ΔT_{pp} is simply the difference between brine pinch-point temperature and the vaporization temperature of isobutane, resulting in 5.9°C.

The exergetic efficiencies and exergy destructions of major plant components and the entire plant are calculated as explained in this section and listed in Table 28.2. All values are for one representative unit. To pinpoint the sites of exergy destruction and quantify those losses, an exergy diagram is given in Figure 28.4. An energy losses diagram is given in Figure 28.5 to provide a comparison with the exergy losses diagram.

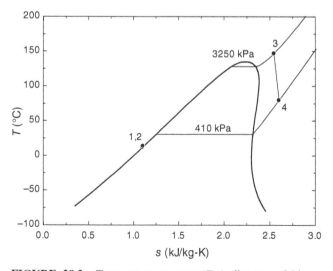

FIGURE 28.2   Temperature–entropy (*T–s*) diagram of binary Rankine cycle.

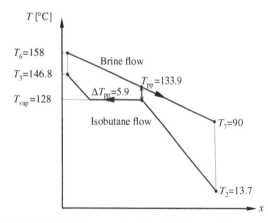

FIGURE 28.3   Diagram showing the heat exchange process between the geothermal brine and the working fluid isobutane in the heat exchanger.

TABLE 28.2   Some exergetic and energetic performance data provided for one representative unit of the plant.

Component	Exergy destruction (kW)	Exergetic efficiency (%)	Heat transfer or power (kW)	Effectiveness or isentropic efficiency (%)
Reinjection well	26,706	–	202,742	–
Heat exchanger	9552	80.5	160,929	47.1
Air-cooled condenser	14,307	28.2	141,271	88.6
Turbine I	2411	81.8	10,872	78.2
Turbine II	2411	81.8	10,872	78.2
Circulation pump	541	74.1	2087	73.4
Parasitic power			3262	
Cycle	51,891	41.7 (equation 28.22)	16,396	10.2 (equation 28.8)
		33.5 (equation 28.21)		10.2 (equation 28.7)
		21.7 (equation 28.20)		4.5 (equation 28.5)
		21.1 (equation 28.19)		4.4 (equation 28.4)

The power output from the turbines is 10,872 kW in Turbine I and 10,872 kW in Turbine II. The pump power requirement for the circulation pumps is calculated to be 2087 kW. The net power output from Rankine cycle then becomes 19,657 kW. It is further estimated based on plant data that about 16.6% of the net power generated in the cycle is consumed by the parasites of unit of plant, which corresponds to 3262 kW (Kanoglu and Cengel, 1999b). Parasitic power includes brine production pumps, condenser fans, and other auxiliaries. Subtracting the parasitic power from the net power generated in the cycle, the net power output becomes 16,396 kW.

In Table 28.1, temperature, pressure, and mass flow rate data for geothermal fluid, working fluid, and air are given according to their state numbers specified in Figure 28.1. Exergy rates are calculated for each state, and listed in Table 28.1. State 0, 0′ and 0″ are the restricted dead states for the geothermal fluid, working fluid and air, respectively. They correspond to an environment temperature of 3°C and an atmospheric pressure of 84 kPa, which were the values measured at the time when the plant data were obtained. For

geothermal fluid, the thermodynamic properties of water are used. By doing so, effects of salts and noncondensable gases that might present in the geothermal brine are neglected. This should not cause any significant error in calculations since their fractions are estimated by the plant management to be small. Thermodynamic properties of working fluid, isobutane, are obtained from a software with built-in thermodynamic property functions (Klein, 2006).

As part of the analysis, we investigate the effects of turbine inlet pressure and temperature and the condenser pressure on exergy and energy efficiencies, the net power output, and the brine reinjection temperature. In order to facilitate this parametric study, we used the given geothermal inlet temperature and flow rate values (158°C, 555.9 kg/s) and the calculated isentropic efficiencies for the turbine (0.782) and pump (0.734) and the pinch-point temperature difference (6°C). The brine temperature at the heat exchanger exit (reinjection temperature) and the mass flow rate of isobutane are the unknown parameters in this analysis. The results of this parametric study are given in Figures 28.6, 28.7, 28.8, 28.9, 28.10, and 28.11.

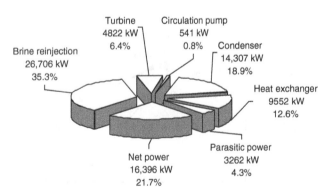

FIGURE 28.4   Exergy losses diagram. Given as the percentages of brine exergy input (75,586 kW), which is taken as the exergy of brine at state 6 in Figure 28.1.

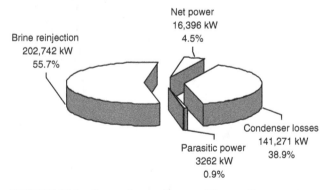

FIGURE 28.5   Energy losses diagram. Given as the percentages of brine energy input (363,671 kW), which is taken as the energy of brine at state 6 in Figure 28.1.

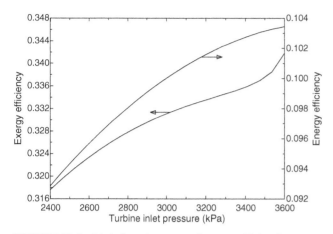

FIGURE 28.6 Variation of exergy and energy efficiencies versus turbine inlet pressure.

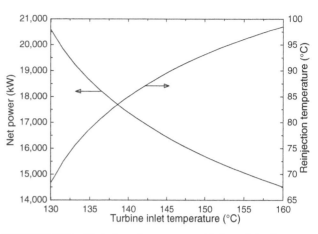

FIGURE 28.9 Variation of net power and brine reinjection temperature versus turbine inlet temperature.

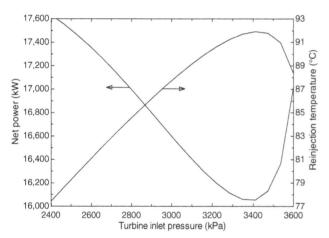

FIGURE 28.7 Variation of net power and brine reinjection temperature versus turbine inlet pressure.

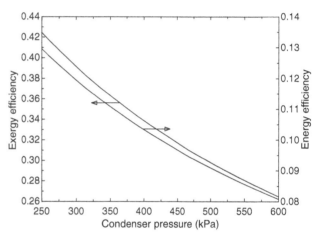

FIGURE 28.10 Variation of exergy and energy efficiencies versus condenser pressure.

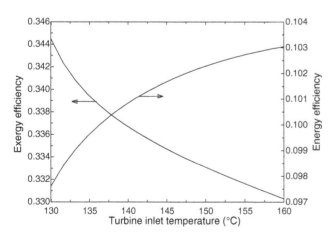

FIGURE 28.8 Variation of exergy and energy efficiencies versus turbine inlet temperature.

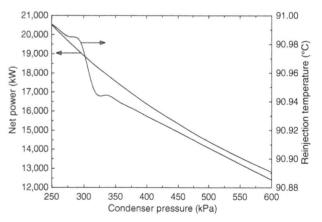

FIGURE 28.11 Variation of net power and brine reinjection temperature versus condenser pressure.

28.6   RESULTS AND DISCUSSION

An investigation of the exergy pie diagram given in Figure 28.4 shows that 74% of the exergy entering to the plant is lost. The remaining 26% is converted to power and 21.7% of this power is used for parasitic load in the plant. The exergetic efficiency of the plant is determined to be 41.7% (equation 28.22) and 33.5% (equation 28.21) based on the exergy input to the isobutene Rankine cycles (i.e., exergy drops of the brine in the heat exchanger), and 21.1% (equation 28.19) and 21.7% (equation 28.20) based on the exergy input to the plant (i.e., exergy of the brine at the reservoir and the heat exchanger inlet, respectively) (Table 28.2). Bodvarsson and Eggers (1972) gives the exergetic efficiencies of a single-flash and a double-flash cycle to be 38.7% and 49.0%, respectively, based on 250°C resource water temperature and 40°C sink temperature. Both values are significantly greater than the value calculated for the binary plant analyzed in this paper. This is expected since additional exergy destruction occurs during the heat exchange between the geothermal and the working fluids in binary plants. DiPippo and Marcille (1984) calculated the exergetic efficiency of an actual binary power plant using a 140°C resource and 10°C sink to be 20% and 33.5% based on the exergy input to the plant and to the Rankine cycle, respectively.

Using low-temperature resources, geothermal power plants generally have low first law efficiencies. Consequently, the first law efficiency of the plant is calculated to be 4.4% (equation 28.4) and 4.5% (equation 28.5) based on the energy input to the plant and 10.2% (equation 28.8) based on the energy input to the isobutene Rankine cycles. This means that more than 90% of the energy of the brine in the reservoir is discarded as waste heat. There is a strong argument here for the use of geothermal resources for direct applications such as district heating instead of power generation when economically feasible. A cogeneration scheme involving power generation and the district heating may also be considered when the used brine is reinjected back to the ground at a relatively high temperature. The energy loss diagram in Figure 28.5 shows that 55.7% of the energy of the brine is reinjected, 38.9% of it is rejected in the condenser, and the remaining is converted to power. Yet it provides no specific information on where the power potentials are lost. This shows the value of an exergy analysis.

The causes of exergy destruction in the plant include heat exchanger losses, turbine-pump losses, the exergy of the brine reinjected, and the exergy of isobutane lost in the condenser. They represent 18.9%, 7.2%, 35.3%, and 12.6% of the brine exergy input, respectively (Figure 28.4). The exergetic efficiencies and effectiveness of heat exchanger are 80.5% and 47.1%, respectively. This exergetic efficiency can be considered to be high and indicates a satisfactory performance of the heat exchange system. In binary geothermal power plants, heat exchangers are important components, and their individual performances considerably affect the overall performance of the plant.

The exergetic efficiency of the turbine is 81.8%, which is reasonable. The exergetic efficiencies of the condensers are in the range of 28.2%, making them the least efficient components in the plant. This is primarily due to the high average temperature difference between the isobutane and the cooling air. The brine is reinjected back to the ground at about 90°C.

For binary geothermal power plants using air as the cooling medium, the condenser temperature varies as the ambient air temperature fluctuates throughout the year and even throughout the day. As a result, the power output decreases by up to 50% from winter to summer (Kanoglu and Cengel, 1999b; Michaelides and Ryder, 1992). Consequently, the exergy destruction rates and percentages at various sites change, this effect being most noticeable in the condenser.

As part of the analysis, we investigate the effect of some operating parameters on exergy and energy efficiencies, the net power output, and the brine reinjection temperature. The energy and exergy efficiencies are those given in equations (28.7) and (28.21), respectively. The effect of turbine inlet pressure on the exergy and energy efficiencies is given in Figure 28.6. Both the exergy and energy efficiencies increase with the turbine inlet pressure. The critical pressure of isobutane is 3640 kPa. As the pressure approaches the critical pressure, the increase in exergy efficiency becomes more dramatic. On the other hand, energy efficiency shows a different trend as the pressure approaches the critical value. Figure 28.7 indicates a pressure of about 3430 kPa at which the net power from the plant is minimized. This is also about the same pressure at which the brine reinjection temperature is a maximum. This means that less heat is picked up from the geothermal brine by the isobutane. For geothermal power plants where the brine leaving the heat exchanger is not used for any purpose and directly reinjected to the ground, maximizing the power output (not the energy or exergy efficiency) is a priority. This is the case for this particular power plant. Note that the brine temperature is high enough for use in district heating systems. This may be explored if there is a residential, commercial, or industrial district in a close distance to the power plant.

The exergy efficiency decreases and the energy efficiency increases with increasing turbine inlet temperature as shown in Figure 28.8. The reason for decreasing trend of exergy efficiency: As the temperature increases, the power potential increases but the power output decreases (Figure 28.9). The reason for decreasing trend in the power output is the decrease in mass flow rate of isobutane. It decreases from 457 kg/s at 130°C to 266 kg/s at 155°C. The reason for increasing trend in energy efficiency: Both the power output and the heat input decrease with increasing turbine inlet temperature while heat input decreases at a greater rate than power output. The reason for decreasing heat input is due to decreasing mass flow of isobutane.

Figures 28.10 and 28.11 show that exergy and energy efficiencies and the net power decrease as the condenser pressure increases. The reinjection temperature remains almost constant with varying condenser pressure. Note that the mass flow rate of isobutane and the rate of heat input to the Rankine cycle remain essentially constant when condenser pressure is changed, since changing condenser pressure only affects part of the heat exchange process described in equation (28.26) whose effect is very small.

28.7 CONCLUSIONS

Thermodynamic analysis of a geothermal power plant can be done using the first and the second laws of thermodynamics. A first law analysis provides energy balance of the system indicating consumed energy and power outputs as well as thermal efficiency. A second law analysis uses the results of energy analysis but improves the analysis by locating the losses. The source temperature of geothermal power plants is typically very low compared to conventional fossil-fuel-fired power plants, and consequently their first law efficiencies are very low. One cannot make a conclusion as to how efficiently a geothermal power plant is operating on the basis of the first law efficiency only. Second law efficiency provides an additional measure to assess performance since it represents how efficiently the work potential of a source is utilized.

The aim of a second law analysis for a power plant is usually to identify and quantify the sites of exergy destruction, so that the directions for any attempt to improve the performance can be identified. This is illustrated in the case study. The investigation of some operating parameters in the cycle on the cycle performance parameters yielded some important insights to the plant operation and heat exchange process, and this information can be used in the design, analysis, and performance improvement of binary geothermal power plants.

28.8 NOMENCLATURE

e	specific exergy, kJ/kg
$\dot{E}$	exergy rate, kW
h	specific enthalpy, kJ/kg
$\dot{I}$	exergy destruction, kW
$\dot{m}$	mass flow rate, kg/s
P	pressure, kPa
$\dot{Q}$	heat flow rate, kW
s	specific entropy, kJ/kgK
T	temperature, °C
T_{vap}	vaporization temperature, °C
T_{pp}	pinch-point temperature, °C
$\dot{W}$	power, kW

ΔT_{pp}	pinch-point temperature difference, °C
ε	effectiveness
η_{ex}	exergetic efficiency
η_{th}	first-law efficiency

Subscripts

0	dead state
binary	isobutane fluid
cold	cold stream
cond	condenser
cycle	cycle
ex	exergy
f	saturated liquid
geo	geothermal fluid
heat exc	heat exchanger
hot	hot stream
in	inlet
max	maximum
out	outlet
plant	plant
pp	pinch-point
pump	pump
rev	reversible
reinj	reinjection
th	thermal
turb	turbine
vap	vaporizer

REFERENCES

ASHRAE. (1995). *ASHRAE Handbook of Applications.* American Society of Heating, Refrigerating, and Air-Conditioning Engineers, Inc., Atlanta, GA.

Barbier, E. (1997). Nature and technology of geothermal energy: a review. *Renewable & Sustainable Energy Reviews*, 1(1/2):1–69.

Bejan, A. (1988). *Advanced Engineering Thermodynamics.* John Wiley & Sons, New York.

Bodvarsson, G and Eggers, D. E. (1972). The exergy of thermal power. *Geothermics*, 1:93–95.

Cengel, Y. A. and Boles, M. A. (2006). *Thermodynamics: An Engineering Approach*, 5th edition. McGraw-Hill, New York.

DiPippo, R. and Marcille, D. F. (1984). Exergy analysis of geothermal power plants. *Geothermal Resources Council Transactions*, 8:47–52.

Kanoglu, M. and Bolatturk, A. (2008). Performance and parametric investigation of a binary geothermal power plant by exergy. *Renewable Energy*, 33(11):2366–2374.

Kanoglu, M. and Cengel, Y. A. (1999a). Improving the performance of an existing binary geothermal power plant: a case study. *Transactions of the ASME, Journal of Energy Resources Technology*, 121(3):196–202.

Kanoglu, M. andCengel, Y. A. (1999b). Economic evaluation of geothermal power generation, heating, and cooling. *Energy*, 24(6):501–509.

Kanoglu, M., Dincer, I., and Rosen, M. A. (2007). Understanding energy and exergy efficiencies for improved energy management in power plants. *Energy Policy*, 35:3967–3978

Klein, S. A. (2006). *Engineering Equation Solver (EES),* Academic Commercial V7.714, F-Chart Software, available at: www.fChart.com (accessed July 29, 2015).

Kotas, T. J. (1995). *The Exergy Method in Thermal Plant Analysis*, 2nd edition. Krieger, Malabar, FL.

Kunze, J. F. (1979). Utilizing geothermal resources below 150°C. *ASME Journal of Energy Resources Technology*, 101: 124–129.

Michaelides, E. E. and Ryder, J. K. (1992). The Influence of seasonal and daily temperature fluctuations on the work produced by geothermal power plants. *International Journal of Energy Systems*, 12(2):68–72.

MIT. (2006). *The future of geothermal energy*: Impact of an enhanced geothermal systems on the united states in the 21st century, Massachusetts Institute of Technology, available at: http://geothermal.inel.gov/publications/future_of_geothermal _energy.pdf (accessed July 29, 2015).

Mohanty, B. and Paloso, G. Jr. (1992). Economic power generation from low-temperature geothermal resources using organic Rankine cycle combined with vapor absorption chiller. *Heat Recovery Systems & CHP*, 12(2):143–158.

Phair, K. A. (1994). Getting the most out of geothermal power. *ASME Mechanical Engineering*, 116(9):76–80.

Szargut, J., Morris, D. R., and Steward, F. R. (1988). *Exergy Analysis of Thermal, Chemical, and Metallurgical Processes*. Hemisphere Publishing Corp., New York.

Wark, K. J. (1995). *Advanced Thermodynamics for Engineers*. McGraw-Hill, New York.

Yuan, Z. and Michaelides, E. E. (1993). Binary-flashing geothermal power plants. *ASME Journal of Energy Resources Technology*, 115:232–236.

29

SUSTAINABILITY ASSESSMENT OF GEOTHERMAL POWER GENERATION

Annette Evans, Vladimir Strezov, and Tim J. Evans

29.1 INTRODUCTION

Electricity produced from geothermal renewable energy sources harnesses the energy contained as heat in the Earth's interior. Electricity was first generated from geothermal heat in Italy in 1904 [1]. Although there are four types of geothermal heat sources available—hydrothermal, hot dry rock, geopressured, and magmatic sources—only hydrothermal heat sources have been commercially proven for electricity generation on a large scale. Hot dry rock power stations, where boreholes are drilled up to 5 km into a geothermal heat source, are currently under development, with Australia showing strong potential for full commercialization of this technology.

Hydrothermal systems of geothermal heat sources are characterized by an underground reservoir of hot water and/or steam and classified by the primary water phase as either water or vapor dominated. Water-dominated fields are further categorized as either hot water or wet steam (water and steam) fields. Vapor-dominated fields differ from wet steam fields in that the former are typically dry saturated or slightly superheated, giving higher heat transfer from depth than wet steam. Vapor-dominated fields have the highest energy yield per unit mass of fluid [2] as the geothermal fluid can be directly used to generate electricity on a low-pressure turbine [3]. Water from high temperature (>230°C) wet reservoirs must first undergo flashing to steam before electricity generation in a low-pressure turbine.

Traditional geothermal power plants require a source at a junction of underground heat, water, and fractured rock. The viability of a reservoir depends primarily on the depth of the reservoir (up to 2 km for a hot water resource or 5 km for a steam resource are generally considered the limits), rock permeability, and the salt content, flow rate, and temperature of the geothermal fluid (water). Enhanced geothermal and hot dry rock system advancements will allow cost-effective exploitation of geothermal resources where the ideal conditions juncture does not exist. They will further allow for drilling and exploitation of geothermal resources at much greater depths [4]. An enhanced geothermal system is an artificially created reservoir to allow for electricity production from geothermal resources that due to lack of fluid and/or permeability would otherwise be deemed uneconomical [5].

Geothermal power is generated by extracting hot fluids from the geothermal well and passing them through a turbine. The cheapest and easiest type of geothermal plant is a direct intake, noncondensing plant where geothermal steam is passed once through a turbine and then emitted to the atmosphere. Condensing plants, where exhaust gases are condensed prior to emission, use significantly less steam per kilowatt hour generated; however, they can only operate where noncondensable gases make up less than 15% of the input geothermal fluid. Flash steam plants operate where input fluid heats are insufficient to generate their own steam. Binary cycle plants operate on hot water fluids (as low as 85°C) by using a secondary, low boiling point fluid such as freon or ammonia through an organic Rankine cycle. These plants operate with very low efficiencies, but the benefit of electricity generation from low-quality geothermal resources makes this method of electricity generation cost-effective [2].

Alternative Energy and Shale Gas Encyclopedia, First Edition. Edited by Jay H. Lehr and Jack Keeley.
© 2016 John Wiley & Sons, Inc. Published 2016 by John Wiley & Sons, Inc.

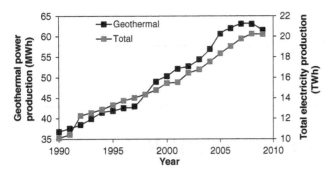

FIGURE 29.1   Historical geothermal and total world power generation [6].

29.2 PRESENT STATUS OF THE GEOTHERMAL POWER

In 2009, 0.3% of the world's total primary energy supply came from geothermal power, with 61.6 TWh of electricity produced [6]. In 2004, there were 24 countries using geothermal power with a combined installed capacity of 8.9 GW [7]. By 2010, this had increased to a globally installed capacity of 10 GW [3]. Geothermal power generation has steadily increased over the last 25 years, with a sharp increase of 50% between 1995 and 2005, as shown in Figure 29.1. The United States is the largest geothermal power generator (26%), followed by Philippines (17%), Indonesia, and Mexico (10% each) [6]. In 2004, geothermal power substantially contributed to the total electricity generated in Tibet (30%), Costa Rica (15%), El Salvador (24%), Iceland (17%), Kenya (19%), and the Philippines (19%) [7]. Around half of the geothermal electricity worldwide is generated from six vapor-dominated fields, Larderello and Mt. Amiata (Italy), The Geysers (USA), Matsukawa (Japan), Kamojang and Darajat (Indonesia) (Barbier, 2002). Geothermal heat is also commonly utilized directly in 72 countries for applications in space heating, agricultural drying, bathing, cooling, and snow melting [8].

Geothermal electricity has an important capability to supply baseload power, available all year round, 24 hours a day, with typical plant lifetimes around 20–30 years [9]. Geothermal plants have high reliability, with capacity factors exceeding 90% [1]. In order to direct future investment, it is necessary to understand the environmental footprint of projected energy growth scenarios, focusing on sustainable energy generation practices.

29.3 SUSTAINABILITY INDICATORS OF GEOTHERMAL POWER

The full environmental footprint accounts for the entire energy chain life cycle, from mining and processing to direct and indirect emissions, waste disposal, and/or recycling. In the assessment of each stage of the chain, key indicators must be identified to allow quantification of environmental impacts. The indicators will be based upon environmental and societal impacts, greenhouse gas emissions, resource depletion, availability of the energy sources, and the value that they add to the economy.

Significant research has already been given to understand the impacts of electricity generation on the environment and economy. Most work seeks to quantify parameters such as emissions [10, 11], energy payback periods, and costs [12]. Several authors have completed full life-cycle assessment (LCA) of individual energy generation technologies [13, 14]. Life-cycle assessment (LCA) is an internationally accepted tool for evaluation of the impact for a product or service. LCA of energy generation technologies allows direct comparison of a range of impacts by breaking them down into relative consequences, that is, effect of wind power generation on migratory birds [15, 16], potential incidence of leukemia clusters surrounding nuclear power plants [17, 18], and so on. There are other methods of assessing sustainability, such as input–output analysis, mass and energy balances, and emergy (embodied energy) accounting; however, LCA is a combination of these tools, providing the most comprehensive method currently available.

Life-cycle assessment as a tool to assess sustainability is not without its limitations, as identified by Bergerson and Lave [19]. It is the responsibility of the analyst to ensure all necessary inputs and outputs are considered and weighted. Gagnon et al. [20] highlighted the fact that LCAs are unable to account for the dual function of hydroelectric dams or the reliability of electricity supply. As with all analysis methods, there is also difficulty attributing full value to more flexible generation options [21].

The most comprehensive examples of previous LCA studies on electricity generation have been produced by Bilek et al. [10], Hondo [11], Gagnon et al. [20], Denholm and Kulcinski [22], Uchiyama [23], and Weisser [24]. These studies used one or more indicators to provide assessment, typically greenhouse gas emissions and possibly energy accounting. Gagnon et al. [20] consider the widest range of indicators of sustainability in their assessment, but avoid consideration of social impacts. The previous studies are limited to only small number of indicators and limited variation of energy generation technologies to gain a full understanding of the sustainability of all of the modern electricity generation.

There is a range of other significantly important indicators that must be considered when evaluating sustainability of energy generation technologies. It is not only the traditional form of the environment that is impacted by electricity generation, the human social and economic environments are also significantly impacted by the choice of production method. The key indicators of sustainability used in this assessment with the main justification for their selection are as follows:

- Price of electricity generation unit must be considered since unfavorable economics are not sustainable.
- Greenhouse gas emissions are increasingly becoming one of the key parameters that define sustainability of energy generation.
- Availability and limitations of each technology must be considered since some technologies or fuels may be heavily resource constrained.
- Efficiency of energy transformation must be known for meaningful comparison. Efficient processes will typically have lower process requirements, capital, and operating costs.
- Less efficient processes may have more significant room for technological advancement and innovation.
- Land use requirements are important as renewable energy technologies are often claimed to compete with agriculturally arable land or to change biodiversity.
- Water consumption is particularly important in arid climates such as Australia. It is not sustainable to have high water consumption and evaporation rates to support the energy generation process when already water shortages are problematic. Previous LCAs often ignore the high water requirements of thermal technologies such as coal when it must be considered.
- Social impacts are important to correctly identify and quantify the human risks and consequences, which will allow better acceptance and understanding of some technologies that are often subject to public objection.

Geothermal power generation has been assessed against each sustainability indicator in detail, across the full life-cycle and subsequently compared to more traditional baseload power generation technologies: coal, natural gas, and nuclear.

29.3.1 Price

In areas of recent volcanic activity from tectonic plate boundaries, geothermal resources are typically located close to the surface and geothermal electricity can be produced at low cost. The price depends significantly on the grade of the resource. High-grade (shallow, hot, and permeable) resources provide electricity at the same cost or cheaper than fossil fuel alternatives. A summary of literature values for the cost of geothermal electricity is shown in Table 29.1. Power can be produced for as low as $0.02/kWh [25, 26]. The life cycle cost of geothermal power generation is heavily affected by the capital-intense nature of the technology, due mostly to the high costs of raw materials such as steel in

TABLE 29.1   Literature prices for geothermal electricity [9, 26–32].

Author	Year	USD/kWh	Comment
Fridleifsson and Freeston	1994	0.04	
Mock et al.	1997	0.05–0.07	High-grade resource
Mock et al.	1997	0.10–0.17	Low-grade resource
Dincer	1999	0.05	
UNDP	2000	0.02–0.10	
Sannner and Busman	2003	0.10	
Hammons	2004	0.05	Large plants
Hammons	2004	0.07	Small plants
IEA	2004	0.05	
IEA	2006	0.06–0.08	
IEA	2007	0.07–0.15	Average current cost
IEA	2007	0.05–0.08	Good new plants
IEA	2007	0.02–0.03	Lowest cost in the United States

the power plant construction and drilling for exploration and production. However, geothermal power generation has low operating costs and a high capacity factor due to operation as a consistent baseload power generator, as reflected in the low minimum costs seen in the table.

29.3.2 Efficiency

Significant amounts of heat are lost during the extraction and processing of the geothermal fluid. For this reason, geothermal power plants do not have high thermal efficiencies, as shown in Table 29.2. The main influence on efficiency is the heat of the geothermal fluid. High-temperature fluids give better heat transfer and offer higher efficiencies. As shown by Williamson [3], the efficiency can range from as low as 5% when temperatures are low (100°C) but increases to over 25% at high temperatures (>300°C).

29.3.3 Greenhouse Gas Emissions

Geothermal vapors have varying amounts of CO_2 present, depending on the site and time in operation. Plants located

TABLE 29.2   Thermal efficiency of geothermal electricity generation [2, 3, 28, 33–35].

Author	Year	Efficiency (%)	Comment
Mock et al.	1997	10–20	< 200°C
Barbier	2002	10–17	
Kanoglu	2002	5.8–8.9	Turkey
Hepbasli	2003	10–17	
DiPippo	2007	13.5	Thermal Efficiency
Williamson	2010	5-> 25	100–300°C

TABLE 29.3 Geothermal power plant greenhouse gas emissions [2, 9, 11, 25, 32, 36–41].

Author	Year	g CO_2/kWh	Comment
Hunt	2000	0–400	
UNDP	2000	13–380	
Barbier	2002	13	Wairakei, NZ
Barbier	2002	33	The Geysers, USA
Barbier	2002	96	Krafla, Iceland
Barbier	2002	175	Cerro Prieto, Mexico
Barbier	2002	272	Tiwi, Philippines
Barbier	2002	380	Larderello, Italy
Bertani and Thain	2002	4–740	Larger plants have lower emissions
Bertani and Thain	2002	122	Weighted average
Brown and Ulgati	2002	655	20 MW
Hammons	2004	82	
Armannsson et al.	2005	152	Krafla, Iceland
Armannsson et al.	2005	181	Svartsengi, Iceland
Armannsson et al.	2005	26	Nesjavellir, Iceland
Hondo	2005	15	Double flash
IEA	2006	27–40	
Pehnt	2006	37.8	
Chatzmouratidis and Pilavachi	2007	18.9	

in volcanic areas tend to have high background emissions, which are not usually measured prior to construction. Geothermal extraction can also lower the naturally occurring emissions elsewhere on the site. It is therefore difficult to allocate emissions to the power plant as the measurements may not be significantly different to overall site emissions during preconstruction [36]. Carbon emissions will also tend to reduce naturally with the age of the site [2]. Major geothermal fields have reportedly between 2.3 and 45.2 g CO_2/kg of fluid [2].

As shown in Table 29.3, literature values for emissions vary greatly from nothing to high at 740 g/kWh. A value of zero should not be considered accurate, as even if all CO_2 were captured during power generation and not emitted, there would be some (although small) emissions associated with the construction and decommissioning of the plant.

Many geothermal power plants use either CO_2 capture for liquefaction and sale [42] or reinject waste fluids into the reservoir to avoid emissions [11].

29.3.4 Water Use

Geothermal power production can consume large quantities of water. Abbasi and Abbasi [43] argued that water consumption is reduced when the source is very hot as more water is converted to steam. While geothermal concentrate can be used as cooling water to reduce overall water consumption, the concentrate is highly polluted and with foul smell.

During a study on power production in Wairakei, NZ, Axtmann [44] found that geothermal plants produce 5.5 times more water vapor than a modern NZ coal-fired power station. Inhaber [45] showed that a geothermal power plant consumes between 12 and 300 m^3 of water per megawatt hour of power production. The maximum value of this figure is over six times higher than for coal, gas, or nuclear power plants.

Significant amounts of water consumption are particularly unacceptable in arid climates, such as Australia. Mock et al. [28] suggested methods to reduce water consumption including total geothermal fluid reinjection, non-evaporative cooling, and general pressure management in closed-loop cycles.

29.3.5 Availability

Geothermal resources at temperatures greater than 150°C are available in over 80 countries around the world, with a potential generating capacity of 11 ± 1.3 PWh/year [46]. Some of the best geothermal resources are in the developing countries, including Ethiopia, Guatemala, Kenya, and Nicaragua [47]. The largest geothermal resources are in the United States, Central America, Indonesia, East Africa, and the Philippines [26]. As hot dry rock technology is further developed and becomes commercial, the potential of geothermal power will significantly increase. It has been estimated that there is in excess of 2.5 PJ of geothermal energy in Australia alone [48].

The availability of geothermal power is enhanced by its independence from weather and climatic conditions, inherent storage capability, and also the ability to provide base or peaking power, unlike most other renewable energy sources [47].

29.3.6 Limitations

The primary limitation to geothermal electricity is the fixed and limited number of appropriate geothermal resources. If the development of hot dry rock technology is successful, many new geothermal sites will be possible.

One of the main problems with geothermal power is the heavily contaminated process water. Geothermal fluid contains significant amounts of hydrogen sulfide, ammonia, mercury, radon, boron [47], cadmium, lead, and arsenic [49]. Since these chemicals are concentrated in the disposal water, they are mostly reinjected in the reservoir. In some older plants disposal waters have caused environmental damage after release into local waterways, for example, the Kizildere plant in Turkey where disposal into the local river, used downstream for irrigation, caused excessive boron accumulation in crops, which is toxic in large doses [50].

Other chemicals of concern present in geothermal fluids are carbonates and silicates that cause scaling, reducing the overall diameter of wells, necessitating regular mechanical cleaning, and reducing production capacity [51].

The reinjection of wastewater in the reservoir also benefits the process by helping to maintain reservoir pressures. Natural reservoir recharge rates are lower than power plant extraction rates, causing the pressure to drop over time. By returning fluids after the heat has been extracted, well pressures may be maintained for prolonged periods of time. It is important that the reinjection sites are carefully selected to avoid short-circuiting, which would allow cooler fluids to be recirculated to the extraction well before sufficient residence time is allowed to bring the cool fluids up to operational temperatures [52].

29.3.7 Land Use

Geothermal power plants generally have a small surface footprint, given that most infrastructure is underground and there is no need for separate mining activities [28]. Bertani [7] gives the average global power density of these plants as 7.4 ± 6 MW/km^2, with smaller values more common, in the range 2–6 MW/km^2. With an average power production of around 6.5 GWh per MW of installed capacity [7], the land use of geothermal power plants ranges between 0.026 and 0.077 m^2/kWh.

Geothermal extraction can cause land subsidence in areas where rock above the geothermal well is not hard. Water-dominated fields cause more subsidence than vapor-dominated fields. Spent fluid reinjection reduces or prevents subsidence; however, it can lead to microseismic activity [2].

29.3.8 Social Impacts

There are several likely adverse environmental impacts associated with geothermal power generation that must be considered. These include surface disturbances, physical effects, such as land subsidence caused by fluid withdrawal, noise, thermal pollution, and release of offensive chemicals [43]. There are large variations from site to site that are also technology dependent. Abbasi and Abbasi [43] also found that highly mineralized water is being carried from cooling towers in California, in turn killing downwind vegetation.

The primary air pollutant of concern from geothermal activity is hydrogen sulfide, H$_2$S. Concentrations of H$_2$S in geothermal exhaust gases range from 0.5 to 6.8 g/kWh, up to approximately half that associated with coal-fired power stations. Geothermal steam gases can also contain ammonia, mercury, boron, and radon as well as hydrocarbons such as methane [2].

The water and the condensate typically contain a variety of toxic chemicals in suspension and solution: arsenic, mercury, lead, zinc, boron, sulfur, carbonates, silica, sulfates, and chlorides. In water-dominated fields, the quantities of wastewater can be very high, between 70 and 400 kg/kWh. Reinjection is the optimal method of disposal, which is beneficial in that it helps to maintain reservoir pressure, extract

additional heat from the rock, and prolong the useful life of the resource, such that the costs of reinjection typically pay for themselves in productivity improvements [2].

Mock et al. [28] acknowledge that many geothermal systems approach emission- and waste-free operation. Binary plants do not emit water or gas into the environment during normal operation [2]. This is aided by the single location required for power production, avoiding the need for mines, pipelines, and waste repositories.

Rybach [53] found that a drawback of reinjection was a potential increase in the frequency (but not severity) of seismic activity. This seems less of a problem than the soil and waterway contamination and increased water use if waste is not reinjected.

29.4 COMPARISON OF SUSTAINABILITY OF GEOTHERMAL POWER WITH CONVENTIONAL POWER GENERATION TECHNOLOGIES

The life-cycle assessment data for geothermal power generation in comparison to nuclear, coal, and gas as a reference point to current market standards are outlined in the following section.

29.4.1 Price

Figure 29.2 shows the average, high, and low prices for geothermal power compared to the electricity produced from coal, gas, and nuclear sources. Geothermal power has the lowest minimum cost ($0.01/kWh), but highest average ($0.068/kWh) and maximum ($0.17/kWh) costs. Nuclear power has the lowest average cost ($0.043/kWh), while coal has the lowest maximum cost ($0.06/kWh). Electricity produced from coal also has the highest minimum cost ($0.04/kWh), giving it the narrowest price range. It can be seen that geothermal power is currently only cost competitive where generation costs are at their lowest level. Any cost higher than average is currently unsustainable.

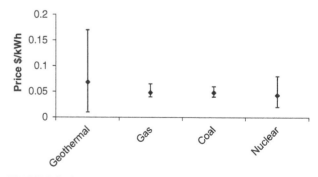

FIGURE 29.2 Comparison of the prices for electricity generation from geothermal, coal, gas, and nuclear energy sources.

TABLE 29.4   Efficiency of each technology.

Technology	Efficiency (%)
Geothermal	10–25
Gas	45–53
Coal	32–45
Nuclear	30–36

TABLE 29.5   Water use for all technologies.

Technology	Water use (kg/kWh)
Geothermal	12–300
Gas	78
Coal	78
Nuclear	107

29.4.2   Efficiency

The efficiency of geothermal power is very low, as shown in Table 29.4. Gas has the highest efficiency at 53%, followed by coal at a maximum efficiency of 45% and nuclear with 36% maximum efficiency. Geothermal efficiencies, even at their maximum potential, are much lower than the other technologies due to the lower temperature and higher gas content of the geothermal fluid, compared to the concentrated saturated steam of other technologies.

29.4.3   Greenhouse Gas Emissions

The minimum, maximum, and average greenhouse gas emissions for geothermal power compared to the other technologies are shown in Figure 29.3. Nuclear power has the lowest minimum (1.8 gCO_2e/kWh), average (16 gCO_2e/kWh), and maximum (65 gCO_2e/kWh) greenhouse gas emissions. Geothermal minimum emissions are only slightly higher than for nuclear with a minimum of 10 gCO_2e/kWh. The minimums shown by gas (440 gCO_2e/kWh) and coal (860 gCO_2e/kWh) are higher than the maximum for nuclear and the average for geothermal. The average (1000 gCO_2e/kWh) and maximum (1340 gCO_2e/kWh) emissions for coal are approximately double that for gas. Geothermal power has the second highest maximum emissions (740 gCO_2e/kWh); however, the low average value (170 gCO_2e/kWh) shows that this is not typically the case.

29.4.4   Water Use

Comparison of the water use for power generation technologies from geothermal, gas, coal, and nuclear energy sources

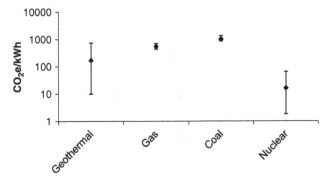

FIGURE 29.3   Comparison of the greenhouse gas emissions for geothermal, coal, gas, and nuclear energy generation technologies.

is shown in Table 29.5. Geothermal power has a broad range of water use, from the lowest to the highest. Gas and coal are more water intensive than some geothermal, while nuclear has a higher water use. For geothermal power generation to operate sustainably in arid climates where water use is critical, plants would have to be designed with the minimum values of water use, making them more water efficient than the other baseload electricity generation options.

29.4.5   Availability

The annual availability of geothermal energy is estimated at 11 PWh/year. For comparison, global energy consumption in 2009 was 20.2 PWh [6]. Recent estimates of fuel reserves under current consumption rates are 107–173 years for black coal, 225 years for brown coal, 37–66 years for natural gas, and 49–70 years for uranium [54, 55, 56]. Coal reserves of significant suitable quantity are found throughout the world, with the largest reserves primarily in the United States (29%), Russia (19%), China (14%), and Australia (9%) with a total global reserve of 826,001 million tonnes [57]. Proven natural gas reserves are the largest in Russia (24%), Iran (16%), and Qatar (13.5%), with a total global resource of 187 trillion cubic meters [57]. Known recoverable uranium resources are located in Australia (31%), Kazakhstan (12%), Canada (9%), and Russia (9%) with 5.4 million tonnes of global available uranium [58].

29.4.6   Limitations

Geothermal power is a renewable energy resource that is able to provide baseload electricity. Once installed and operational, geothermal power plants operate 24 hours a day, providing baseload due to the reliability of the power, with plant life times of around 20 years. The biggest limitation on geothermal installation is identifying and accessing the necessary hydrothermal well at an appropriate depth, with adequate temperature and porosity within the reserve to make power generation economical. It is anticipated that the future success of hot dry rock and enhanced geothermal systems will increase the potential and reduce the limitations of this technology.

Coal is a readily transportable and storable fuel source with abundant reserves and without many practical limitations. This is why it is currently the most popular fuel for power generation worldwide. Natural gas is also easily

TABLE 29.6   Land use for all technologies [20, 59].

Technology	Gagnon et al. (m²/kWh)	Fthenakis and Kim (m/kWh)
Geothermal (Bertani 2005)	0.05	0.05
Gas		0.0003
Coal	0.004	0.0004
Nuclear	0.0005	0.00005

transported and stored; however, current proven reserves are limiting. Nuclear power requires significantly more processing of the ore than coal or gas and is limited in suitability to areas of geographical stability; as highlighted recently in Japan, it is not suited to areas with high risk of strong disturbances such as earthquakes.

29.4.7   Land Use

Table 29.6 shows the comparison of the values for land use of each energy technology. The values listed show a discrepancy of an order of magnitude between the two sources. However, both sources show a similar relative relationship, with nuclear, gas, and coal having the lowest land occupation. The entire geothermal field is measured for the land occupation of geothermal power, despite the power plant occupying a small fraction of this area; therefore, additional users may share the geothermal site.

Complications with measuring land use arise from the inability to assess the functional capacity change of the land before and after power generation.

29.4.8   Social Impacts

There is a direct correlation between electricity access and quality of life, where communities with abundant, constant electricity access enjoy high quality of life. An influential factor in reliable power access is the ability of the generator to obtain a secure source of fuel over a long term at a stable price. Local supplies are more secure than those that have to travel great distance and international relationships can add to long-distance complications. For these reasons, the most secure power generation will be from a fuel that is available within the country of generation in abundant supply.

When comparing between different methods of electricity generation, it should be highlighted that electricity generation causes more air pollution than any other industry; therefore, these emissions are of critical importance. In addition to the greenhouse gas emissions previously discussed, electricity generation often results in significant emissions of NO_X, SO_2, non-methane volatile organic compounds (NMVOC), and particulate matter (PM) as shown in Table 29.7. PM is in the size bracket less than 10 nm to more than 100 mm in diameter. Geothermal power generation typically has the

TABLE 29.7   Non-greenhouse gas emissions for each technology [42].

	NMVOC (mg/kWh)	NOx (mg/kWh)	SO₂ (mg/kWh)	PM (mg/kWh)
Coal/lignite	24	2986	16511	347
Natural gas turbine	118	1477	152	34
Natural gas Combined cycle	118	756	152	6
Nuclear	0	51	27	2
Geothermal	0	280	20	0

lowest non-greenhouse gas emissions, including zero emission of NMVOC and particulates. Only nuclear power generation has comparable emissions to geothermal, including lower NO_X.

All renewable energy sources show societal benefits by providing power with significantly reduced mining requirements than fossil fuels. Geothermal power generation has low environmental impact when wastewater and exhaust gases are reinjected or adequately scrubbed, as is the case in modern plants. The use of geothermal sites for electricity generation increases the frequency but not the severity of geothermal disturbances in the area.

The social impacts of coal-fired power generation include mining-related problems such as land excavation, dust, and noise, while the power stations emit large amounts of criteria pollutants in addition to those discussed above, such as heavy metals, carbon monoxide, and radioactive material.

Natural gas mining occurs in fragile marine environments and gas-fired power plants emit a range of pollutants, as outlined in Table 29.7 as well as carbon monoxide and fugitive emissions. Nuclear power causes impacts from mining, enrichment, and transportation of uranium; however, the biggest unsolved issues surrounding nuclear power generation are the safe, long-term storage of nuclear waste and how to decommission a nuclear power plant that has undergone a meltdown.

29.5   CONCLUSIONS

Geothermal power has significant potential to meet the world's electricity needs, particularly with the advent of advanced geothermal systems. After comparison to electricity generated from coal, natural gas, and nuclear power, it can be seen that geothermal power generation offers better sustainability in the long term. Strengths of geothermal power include the ability to provide baseload electricity, alleviating requirements for fuel transportation and mining during operation. However, costs, CO_2 emissions, water pollution, and water use can be high, so these must be evaluated carefully. The efficiency of geothermal power generation is currently very low, so any improvements in this area will greatly increase the economics of the process. Further investigation of geothermal land use is warranted as there is significant discrepancy between land use data for geothermal and the other technologies considered.

REFERENCES

[1] Fridleifsson, I. B. (2003). Status of geothermal energy amongst the world's energy sources. *Geothermics*, 32:379–388.

[2] Barbier, E. (2002). Geothermal energy technology and current status: an overview. *Renewable & Sustainable Energy Reviews*, 6:3–65.

[3] Williamson, K. H. (2010). Geothermal power: the baseload renewable. In: *Generating Electricity in a Carbon-Constrained World*, Sioshansi, F. P. (Editor).

[4] National Geothermal Collaborative, NGC. (2004). Location of geothermal resources. Available at: http://www.geocollaborative.org/publications/

[5] Energy Efficiency & Renewable Energy, EERE. (2004). Geothermal energy: technologies and costs. Available at: www.eere.energy.gov.

[6] Euromonitor International. (2010). Available at: http://www.euromonitor.com/

[7] Bertani, R. (2005). World geothermal power generation in the period 2001-2005. *Geothermics*, 34:651–690.

[8] International Energy Agency, IEA. (2005). Energy statistics manual. Available at: www.iea.org/

[9] Hammons, T. J. (2004). Geothermal power generation worldwide: global perspective, technology, field experience, and research and development. *Electric Power Components and Systems*, 32:529–553.

[10] Bilek, M., Lenzen, M., Hardy, C., Dey, C. (2006). *Life-Cycle Energy and Greenhouse Gas Emissions of Nuclear Power in Australia*. The University of Sydney.

[11] Hondo, H. (2005). Life cycle GHG emission analysis of power generation systems: Japanese case. *Energy*, 30:2042–2056.

[12] Kammen, D. M., Pacca, S. (2004). Assessing the costs of electricity. *Annual Review of Environment and Resources*, 29:301–344.

[13] Phylipsen, G. J. M., Alsema, E. A. (1995). Environmental life-cycle assessment of multicrystalline silicon solar cell modules. Department of Science, Technology and Society, Utrecht University, The Netherlands.

[14] Vestas Wind Systems, A. S. (2006). *Life Cycle Assessment of Offshore and Onshore Sited Wind Power Plants Based on Vestas V90-3.0MW Turbines*. Vestas, Randers, Denmark.

[15] Schleisner, L. (2000). Life cycle assessment of a wind farm and related externalities. *Renewable Energy*, 20:279–288.

[16] Stewart, G. B., Pullin, A. S., Coles, C. F. (2007). Poor evidence-base for assessment of windfarm impacts on birds. *Environmental Conservation*, 34:1–11.

[17] Clapp, R. W. (2005). Nuclear power and public health. *Environmental Health Perspectives*, 113:A720–A721.

[18] Brown, V. J. (2007). Childhood leukemia in Germany: cluster identified near nuclear power plant. (Science Selections). *Environmental Health Perspectives*, 115:A313.

[19] Bergerson, J., Lave, L. (2002). *A Life Cycle Analysis of Electricity Generation Technologies: Health and Environmental Implications of Alternative Fuels and Technologies*. Carnegie Mellon Electricity Industry Center.

[20] Gagnon, L., Belanger, C., Uchiyama, Y. (2002). Life-cycle assessment of electricity generation options: the status of research in year 2001. *Energy Policy*, 30:1267–1278.

[21] Chapman, C., Ward, S. (1996). Valuing the flexibility of alternative sources of power generation. *Energy Policy*, 24:129–136.

[22] Denholm, P., Kulcinski, G. L. (2004). Life cycle energy requirements and greenhouse gas emissions from large scale energy storage systems. *Energy Conversion and Management*, 45:2153–2172.

[23] Uchiyama, Y. (2007). Life cycle assessment of renewable energy generation technologies. *IEEJ Transactions on Electrical and Electronic Engineering*, 2:44–48.

[24] Weisser, D. (2007). A guide to life-cycle greenhouse gas (GHG) emissions from electric supply technologies. *Energy*, 32:1543–1559.

[25] United Nations Development Program, UNDP. (2000). World energy assessment energy and the challenge of sustainability. Available at: www.undp.org/

[26] International Energy Agency, IEA. (2007). Renewables in global energy supply. Available at: www.iea.org/

[27] Fridleifsson, I. B., Freeston, D. H. (1994). Geothermal energy research and development. *Geothermics*, 23:175–214.

[28] Mock, J. E., Tester, J. W., Wright, P. M. (1997). Geothermal energy from the earth: its potential impact as an environmentally sustainable resource. *Annual Review of Energy and the Environment*, 22:305–356.

[29] Dincer, I. (1999). Environmental impacts of energy. *Energy Policy*, 27:845–854.

[30] Sanner, B., Bussmann, W. (2003). Current status, prospects and economic framework of geothermal power production in Germany. *Geothermics*, 32:429–438.

[31] International Energy Agency, IEA. (2004). World energy outlook 2004. Available at: www.iea.org/

[32] International Energy Agency, IEA. (2006). Geothermal energy annual report 2005. Available at: www.iea.org/

[33] Kanoglu, M. (2002). Exergy analysis of a dual-level binary geothermal power plant. *Geothermics*, 31:709–724.

[34] Hepbasli, A. (2003). Current status of geothermal energy applications in Turkey. *Energy Sources*, 25:667–677.

[35] DiPippo, R. (2007). Ideal thermal efficiency for geothermal binary plants. *Geothermics*, 276–285.

[36] Armannsson, H., Fridriksson, T., Kristjansson, B. R. (2005). CO_2 emissions from geothermal power plants and natural geothermal activity in Iceland. *Geothermics*, 34:286–296.

[37] Hunt, T. M. (2000). Five lectures on environmental effects of geothermal utilization. Geothermal Training Programme 2000, Report 1. United Nations University.

[38] Bertani, R., Thain, I. (2002). Geothermal power generating plant CO_2 emission survey. *IGA News*, 49:1–3.

[39] Brown, M. T., Ulgiati, S. (2002). Emergy evaluations and environmental loading of electricity production systems. *Journal of Cleaner Production*, 10:321–334.

[40] Pehnt, M. (2006). Dynamic life cycle assessment (LCA) of renewable energy technologies. *Renewable Energy*, 31:55–71.

[41] Chatzimouratidis, A. I., Pilavachi, P. A. (2007). Objective and subjective evaluation of power plants and their non-radioactive emissions using the analytic hierarchy process. *Energy Policy*, 35:4027–4038.

[42] Huttrer, G. W. (2001). The status of world geothermal power generation 1995-2000. *Geothermics*, 30:1–27.

[43] Abbasi, S. A., Abbasi, N. (2000). Likely adverse environmental impacts of renewable energy sources. *Applied Energy*, 65:121–144.

[44] Axtmann, R. C. (1975). Environmental impact of a geothermal power plant. *Science*, 187:795–803.

[45] Inhaber, H. (2004). Water use in renewable and conventional electricity production. *Energy Sources*, 26:309–322.

[46] Stefansson, V. (1998). Estimate of the world geothermal potential. *Geothermal Training in Iceland 20th Anniversary Workshop. Reykjavik, United Nations University Geothermal Training Programme.*

[47] Fridleifsson, I. B. (2001). Geothermal energy for the benefit of the people. *Renewable & Sustainable Energy Reviews*, 5:299–312.

[48] Harries, D., McHenry, M., Jennings, P., Thomas, C. (2006). Geothermal energy in Australia. *International Journal of Environmental Studies*, 63:815–821.

[49] Baba, A., Armannsson, H. (2006). Environmental impact of the utilization of geothermal areas. *Energy Sources Part B-Economics Planning and Policy*, 1:267–278.

[50] Simsek, S., Yildirim, N., Gulgor, A. (2005). Developmental and environmental effects of the Kizildere geothermal power project, Turkey. *Geothermics*, 34:234–251.

[51] Gokcen, G., Ozturk, H. K., Hepbasli, A. (2004). Overview of Kizildere geothermal power plant in Turkey. *Energy Conversion and Management*, 45:83–98.

[52] Williamson, K. H., Gunderson, R. P., Hamblin, G. M., Gallup, D. L., Kitz, K. (2001). Geothermal power technology. *Proceedings of the IEEE*, 89:1783–1792.

[53] Rybach, L. (2003). Geothermal energy: sustainability and the environment. *Geothermics*, 32:463–470.

[54] Kruijsdijk, C. P. J. W. V. (2004). Resources, reserves and peaking. In: *Fossil Fuels: Reserves and Alternatives - A Scientific Approach, 2005 Amsterdam*, Geuns, L. C. V., Groen, L. A. (editors). *Earth and Climate Council of the Royal Netherlands Academy of Arts and Sciences and Clingendael International Energy Programme*, 15–18.

[55] World Nuclear Association, WNA. (2006). The new economics of nuclear power. Available at: http://www.world-nuclear.org

[56] Shafiee, S., Topal, E. (2009). When will fossil fuel reserves be diminished? *Energy Policy*, 37:181–189.

[57] BP. (2010). Statistical Review of World Energy 2010 - Excel File [Online]. Available at: http://www.bp.com/statisticalreview

[58] World Nuclear Association, WNA. (2011). Supply of uranium. Available at: http://www.world-nuclear.org

[59] Fthenakis, V., Kim, H. C. (2009). Land use and electricity generation: a life-cycle analysis. *Renewable & Sustainable Energy Reviews*, 13:1465–1474.

30

GEOTHERMAL ENERGY AND ORGANIC RANKINE CYCLE MACHINES

Bertrand F. Tchanche

30.1 INTRODUCTION

Industrialization and economic growth have long been supported by fossil fuels and this has led to the destruction of the environment. The finite nature of the fossil fuels resources brings the fear of the collapse of the civilization and increases international tensions. Adding the exclusion of more than a billion humans living without access to clean and affordable energy, agreement is easily reached to seek for alternative and sustainable energy resources to sustain economic activities, power the increasing billions of the Earth's inhabitants while preserving the environment. Untapped energy resources are now put into contribution. These are solar energy, bioenergy, wind energy, ocean energy, water potential and kinetic energy, and geothermal energy. The next sections focus on exploitation of hydrothermal energy for electricity generation.

30.2 THE GEOTHERMAL RESOURCE

30.2.1 Typology of the Resource

Geothermal energy is defined as thermal energy that takes its origin in the decay of radioactive isotopes and partly from the relic released about 4.5 billion years ago during the Earth's formation. It appears in the form of dry rock, steam, and pressurized water [1]. Classification is performed either by temperature level or by heat transfer mode [2]. High-temperature (> 180°C) systems are associated with recent volcanic activities and mantle hot spot anomalies; intermediate (100–180°C) and low-temperature (< 100°C) systems are found in continental settings and supplied by radioactive isotope decay and water circulation along deep penetrating fault zones. Based on heat transfer mode, convection-dominated (liquid and vapor), conduction-dominated (hot rock and magma), and mixed systems are distinguished. Convective or hydrothermal resources and enhanced geothermal systems (EGS) based on hydraulic stimulations and advanced well configurations are the most exploited geothermal systems. Abandoned oil and gas wells are being investigated in view of their residual heat exploitation [3].

30.2.2 Technologies for Resource Exploitation

Geothermal can be used for electricity generation and applications requiring heat such as district heating, cooling and heating of buildings/processes, and aquaculture [1, 4]. Prince Ginori Conti in 1904/1905 was the first to produce electricity from geothermal fluids in Italy by experimental work. Commercialization followed with a plant of 250 kWe in 1913 at Larderello, Italy [1]. More than 500 geothermal power plants are in operation globally with a total installed capacity close to 11 GW [2]. Geothermal technologies for power are dry steam, single/double-flash, and binary cycles (organic Rankine and Kalina cycles). Dry steam plants are used for vapor- or dry steam-dominated resources, flash systems for moderate temperature and liquid-dominated resources, and binary cycles are well adapted for low-temperature liquid-dominated resources.

Alternative Energy and Shale Gas Encyclopedia, First Edition. Edited by Jay H. Lehr and Jack Keeley.

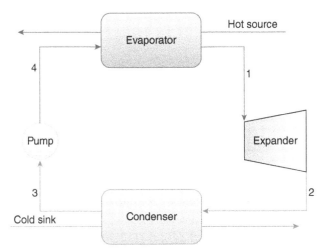

FIGURE 30.1   Schematic of the Rankine cycle.

30.2.3   The Resource Potential

Geothermal wells/heat reservoirs are continually restored by natural heat production. The Earth's heat flow (amount of heat released into space from the interior through a unit area in a unit of time) varies from place to place on the surface and with the time at a particular location. An average value of 65 mW/m^2 was measured on continents and 101 W/m^2 on the ocean floor. The global earth's heat flow rate is estimated at 1400 EJ/year (315 EJ/year emitted by continents)—close to threefold of the world's total energy supply (500 EJ/year) [5]. Global resource assessment carried out at different depths gave 42.67×10^6 EJ/year down to 3 km, 139.5×10^6 down to 5 km, and 403×10^6 EJ/year within 10 km depth [6]. Based on mature geothermal technologies, the global technical potential for electricity generation is within the range of 117.5 EJ (3 km) to 1108.6 EJ (10 km depth) [6].

30.3   ORGANIC RANKINE CYCLE TECHNOLOGIES

30.3.1   The Cycles

The Rankine cycle is made of four connected components: pump, boiler, turbine, and condenser. In large conventional thermal power plants, water is used as the working fluid. Small-scale plants of less than 10 MW are possible with organic fluids like refrigerants, siloxanes, and hydrocarbons, in the so-called "organic Rankine cycles." Organic fluids are the best choice at low (< 150°C) and moderate (150–300°C) temperatures [7, 8]. The theoretical Rankine cycle consists of four processes; see Figure 30.1: (1, 2) isentropic expansion—expansion device, (2, 3) isobaric heat rejection—heat exchanger, (3, 4) isentropic compression—pumping device, and (4, 1) isobaric heat addition—heat exchanger.

A binary plant (Figure 30.2) combines two loops: geothermal fluid loop and an organic fluid closed loop. The heat of the geothermal fluid is transferred to the organic fluid via a set of heat exchangers. The organic working fluid receives heat and evaporates and expands in the power-producing device before the condensation can take place and the fluid returned back to the evaporator by the feed pump. The cooling of the system is assured by air coolers, surface water cooling systems, wet-type cooling towers, or dry-type cooling towers.

For low-temperature fluids below 150°C, it is difficult to implement cost-effective flash steam plants and the binary option is the sole solution. Binary power plants are the most widely used type of geothermal power plant technologies (> 162 units), represent 32.14% of all geothermal units in operation, but have only 4% of generating capacity [4]. Efficiency improvements of 1–5% can be achieved by integration of regenerator or feed liquid heaters or by operation taking place above the critical point in transcritical/supercritical cycles.

Organic dry fluids with their positive slope are suitable for safe and economic operation of the Rankine cycle system. With these fluids, the expansion process happens in the superheated zone. An internal heat exchanger can be integrated into the cycle to increase the cycle thermal efficiency by providing additional heat to the fluid entering the evaporator: stream 2–3 to stream 5–6; see Figure 30.3. This requires high exergy superheated fluid at the turbine outlet. Numerous studies [9, 10] show that dry fluids such as alkanes (pentanes, butanes) and refrigerants R141b, R245fa, R123, and so on

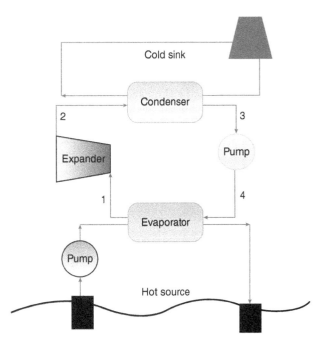

FIGURE 30.2   Sketch of a binary organic geothermal power plant.

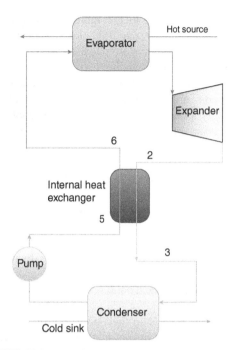

FIGURE 30.3   Rankine cycle with internal heat exchanger.

are good working media for Rankine cycles with integrated regenerator.

In a supercritical Rankine cycle, both heat addition and heat rejection take place at pressures above the critical point. For a transcritical cycle the heat rejection takes place at a subcritical pressure—see Figure 30.4. Transcritical cycles operating with steam yield higher efficiency and require special materials and safety precautions. Transcritical organic Rankine cycles have been investigated by Saleh et al. [10], Karellas et al. [11], Schuster et al. [12], and many others. Efficiencies of transcritical and subcritical organic Rankine cycles are of the same order. Evaporator heat exchangers are more efficient in transcritical organic Rankine cycles due to high-pressure operation leading to high system cost.

Carbon dioxide (CO_2) as working fluid for power generation in nuclear plants, waste heat recovery systems, as well as

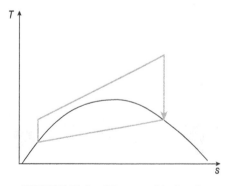

FIGURE 30.4 CO_2 transcritical cycle.

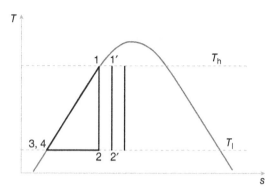

FIGURE 30.5   Schematic diagram of a trilateral flash cycle.

solar thermal power installations has become attractive for many reasons: it is abundant, environmental friendly, non-flammable, nontoxic, cheap, and its thermophysical properties are well known. Chen et al. [13], Zhang et al. [14], Wang et al. [15], and Cayer et al. [16] studied the application of CO_2 in low-temperature transcritical Rankine cycles.

In a common organic Rankine cycle, the expansion process starts in the superheated zone, but it is also possible to have a two-phase expansion provided the expansion device is capable of handling liquid. This is possible at low temperature with screw expanders. The cycle is known as "triangular or tri-lateral cycle" and is depicted in Figure 30.5. T_h and T_l represent the absolute temperature of the heat source and that of the cold sink, respectively. The four-step process is (4–1) isobaric heating (no evaporation!), (1–2) two-phase expansion, (2–3) isobaric condensation, and (3–4) isentropic pumping. The thermodynamics of the concept are presented by Fischer [17]. The possibility of using a trilateral cycle for power generation has long been investigated by many authors. As example, Zamfirescu and Dincer [18] proposed an ammonia–water trilateral Rankine cycle. Their cycle uses no boiler and the saturated liquid at the pump exit is flashed by a positive displacement expander (e.g., reciprocating, centrifugal, rotating vane, and screw- or scroll-type expander). According to their findings, the proposed cycle showed good performance in comparison with Rankine and Kalina cycles. Research on power generation using two-phase expanders and the replacement of throttle valves in large vapor compression systems were carried out at City University, United Kingdom [19, 20], where screw machines were tested in two-phase flows with performance of 70% in expansion mode.

A new concept of flash organic cycle was introduced by Ho et al. [21] with the advantage of reducing exergy losses and increasing heat utilization when recovering heat for power generation. The operating principle is shown in Figure 30.6. In the organic flash cycle a fluid at low pressure (9) is pumped into the heater where it is heated by the hot source stream, and then the vapor exiting the heater (2) is

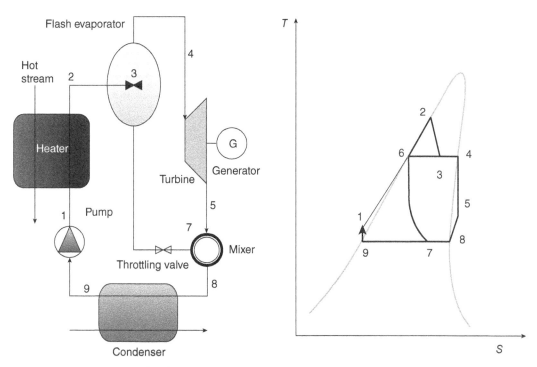

FIGURE 30.6   Sketch of a flash organic cycle [21].

flashed in a separator (3). The high-pressure mixture in the separator is separated into two phases: the liquid (6) is throttled while the saturated vapor (4) is expanded in the turbine. The fluid exiting the throttling valve (7) is mixed with the cold vapor (5) exiting the turbine after expansion. The mixture is further cooled in the condenser and pumped back in the evaporator. The concept is still at an early stage.

30.3.2 Working Fluids

Organic and inorganic fluids are widely used in refrigeration industry and chemical industry and recently they have been applied for power generation and heat recovery in organic Rankine cycles and heat pumps. Few are displayed in Table 30.1. According to the shape of the saturation curve in a T–s diagram it is distinguished [22–24]: wet, isentropic, and dry shapes already well commented in literature. Other classifications are used such as ASHRAE/halocarbons classification and IUPAC/chemical compounds classification. With respect to the latter we have many families among which are

- Alcohols (methanol, ethanol, etc.)
- Inorganic fluids (water (R718), carbon dioxide (R744), and ammonia (R717))
- Hydrocarbons (propane (R290), n-butane (R600), n-pentane (R601), etc.)
  - Chlorofluorocarbons (R12, R13, etc.)
  - Hydrochlorofluorocarbons (R22, etc.)

- Hydrofluorocarbons (R134a, etc.)
- Perfluorocarbons (PP50, etc.)
- Ethers and hydrofluoroethers (HFE7000, RE134, HFE7100, etc.)
- Siloxanes (octamethyltrisiloxane (MDM), hexamethyldisiloxane (MM), etc.)

The choice of working fluids is a critical issue in any organic Rankine cycle-based system. There exist a large number of working fluids but the problem after the operating conditions have been set is to choose the most suitable one for optimal operation. In the absence of a dominant criterion and an appropriate methodology for the selection of the best working fluid, it is better to look at different data concerning the fluids. The data are usually classified into three groups: thermophysical, safety, and environmental data. A good fluid should be the one that has good thermodynamic properties and for which the efficiency of the system is high and the cost low. Selection of working fluids has been an area of intensive research [10, 25–28]. Useful criteria for the selection of suitable working fluids as proposed are summarized below:

- Appropriate low critical temperature and pressure.
- Small specific volume.
- Low viscosity and surface tension.
- High thermal conductivity.

TABLE 30.1 **Organic fluids for organic cycles.**

	Substances	Molecular mass (kg/kmol)	[a]T (°C)[a]	[A]Crlt (°C)	[c]P crlt (bar)	ASHRAE 34 safety group	Atmospheric lifetime (years)	[d]ODP	[c]GWP (100 years)
1	R32	52.02	−51.7	78.11	57.84	A2	4.9	0	675
2	R227ea	170.0	−17.5	102.0	29.5	–	36.5	0	2900
3	R152a	66.05	−24.0	113.3	45.20	A2	1.4	0	124
4	R600a	58.12	−11.7	135	36.47	A3	0.019	0	~ 20
5	R134a	102.03	−26.1	101	40.59	A1	14.0	0	1430
6	R601	72.15	36.1	196.5	33.64	–	0.01	0	~ 20
7	R717	17.03	−33.3	132.3	113.33	B2	0.01	0	< 1
8	R236fa	152.0	−1.4	124.0	32.0	–	209	0	6300
9	R142b	100.5	−9.12	137.11	40.55				
10	R236ea	152.04	6.19	139.29	35.02				
11	R114	170.92	3.6	145.7	32.89	A1	300	0.85	9200
12	RE134	118.0	6.2	147.0	42.28	–	23–24	0	6900
13	R245fa	134.05	15.3	154.1	36.4	B1	8.8	0	820
14	RE347mccl	200.1	34.2	164.6	24.8	–	6.4	0	485
15	RE347	200.1	29.4	160.2	25.5	–	4.9	0	368
16	Novec649	316	49.26	168.66	18.646				
17	HFE7000	200	34	164.46	24.82				
18	SES36	184.53	35.64	177.55	28.49				
19	RE245	150.0	29.2	170	34.20	–	6.5	0	640
20	R245ca	134.05	25.13	174.42	39.25				
21	R123	152.93	27.8	183.7	36.68	B1	1.3	0.020	77
22	R365mfc	148.07	41.4	186.85	32.66			0	825
23	HFE7100	250	61	195.3	22.3		4.1	0	320
24	R290	44.10	−42.1	96.7	42.5	A3	0.041	0	~ 20
25	R141b	116.95	32.0	204.2	42.49	n.a.	9.3	0.120	725

NBP (°C), NBP: Normal boiling point; Tc (°C), Tc: critical temperature; Pc (bar), Pc: critical pressure; ODP (ozone depletion potential); GWP (global warming potential)

- High latent heat and low specific heat.
- Suitable thermal stability.
- Be noncorrosive, nonflammable, nontoxic, and compatible with engine material and lubricating oil.
- In T–s diagram, the saturation curve should have a near-vertical saturated liquid line and a near-vertical vapor line so that most of the heat input should be added during the phase change and only little quantity of moisture should result during the expansion.
- It is convenient to have moderate vapor pressure in the range 0.1–2.5 MPa in the heat exchangers.
- Low ODP, low GWP, and short ALT (atmospheric lifetime).

30.4 NEW ORGANIC RANKINE CYCLE MACHINES FOR GEOTHERMAL APPLICATION

Organic Rankine cycle machines are designed to produce power from 1 kW up to few megawatts. Market survey showed a great diversity of machines in terms of size, expansion device technology, heat source temperature handled, maturity, market availability, type of working media, and

FIGURE 30.7 PureCycle280 (UTC).

cost. ORC machines are available from an increasing number of manufacturers. The machine can be tailored according to the customer's specifications or bought ready from suppliers. Once on the site, the machine is adapted to the heat source.

FIGURE 30.8   ElectraTherm® machine.

In the megawatt power range, conventional turbines are cost effective, and manufacturers are well established as Turboden [29], Adoratec [30], Ormat Inc. [31], and so on while at lower power outputs the lack of cheap turbines renders the technology hardly applicable. Brasz et al. [32], Lemort et al. [33], and Badr et al. [34] suggested to use HVAC components. As an example, a standard 350 ton air-conditioning system was turned into a 200 kW thermal power system and is commercialized under the brand name PureCycle® 280 by United Technologies Corporation (UTC)—see Figure 30.7. Plants based on this technology are in Austin (TX), Danville (IL), and Chena (AK) [32, 35, 36]. Smith and Stosic at City University, United Kingdom [19, 20, 37], converted screw

FIGURE 30.9   Opcon power box.

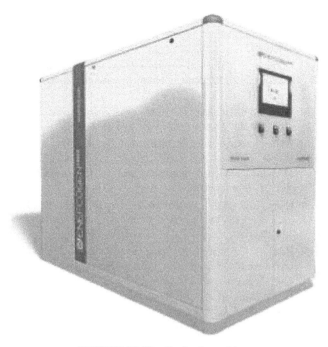

FIGURE 30.10   Eneftech machine.

compressors into expanders; and ElectraTherm [38] and BEP Europe [39] are those manufacturers (Figure 30.8). Lysholm expander is used in Opcon powerbox (Figure 30.9) [40]. Scroll compressors as power-producing device are developed by Eneftech (Figure 30.10) [41].

30.5   CONCLUSION

Various thermodynamic cycles developed to convert low-grade heat into power by exploitation of phase changes of organic fluids have been presented. Organic Rankine cycle was first applied at Paratunka, Russia. In the 1970s during the energy crisis it was pioneered by Ormat Inc. ORC, TFC, and OFC are capable of converting geothermal heat in the range 70–150°C. Plants of more than 400 kW are built based on conventional turbines but for the years to come machines based on innovative expansion devices and with lower power output will surely be made available.

REFERENCES

[1] Lund, J. W. (2007). Characteristics, development and utilization of geothermal resources. *GCH Bulletin*, June.

[2] Goldstein, B. A., Hiriart, G., Tester, J., Bertrani, B., Bromley, R., Gutierrez-Negrin, L., Huenges, C. J., Ragnarson, E. H., Mongillo, A., Muroaka, M. A., Zui, V. I. (2011). Great expectations for geothermal energy to 2100. 36th Workshop on Geothermal Reservoir Engineering, Stanford University, California, January 31–February 2, 2011.

[3] Bu, X., Ma, W., Li, H. (2012). Geothermal energy production utilizing abandoned oil and gas wells. *Renewable Energy*, 41:80–85.

[4] DiPippo, R. (2008). *Geothermal Power Plants: Principles, Applications, Case Studies and Environmental Impact*. Elsevier.

[5] Moriarty, P., Honnery, D. (2009). What energy levels can the Earth sustain? *Energy Policy*, 37:2469–2474.

[6] Edenhofer, O., Sokona, Y., Pichs-Madruga, R. (2011). *Special Report on Renewable Energy Sources and Climate Change Mitigation*. Intergovernmental Panel on Climate Change (IPCC) - Working Group 3.

[7] Hung, T.-C. (2001). Waste heat recovery of organic Rankine cycle using dry fluids. *Energy Conversion and Management*, 42:539–553.

[8] Hung, T. C., Shai, T. Y., Wang, S. K. (1998). A review of organic Rankine cycles (ORCs) for the recovery of low-grade waste heat. *Energy,* 22:661–667.

[9] Dai, Y., Wang, J., Gao, L. (2009). Parametric optimization and comparative study of organic Rankine cycle (ORC) for low grade waste heat recovery. *Energy Conversion and Management*, 50:576–582.

[10] Saleh, B., Koglbauer, G., Wendland, M., Fischer, J. (2007). Working fluids for low-temperature organic Rankine cycles. *Energy*, 32:1210–1221.

[11] Karellas, S., Schuster, A., Leontaritis, A.-D. (2012). Influence of supercritical ORC parameters on plate heat exchanger design. *Applied Thermal Engineering*, 33:70–76.

[12] Schuster, A., Karellas, S., Aumann, R. (2009). Efficiency optimization potential in supercritical organic Rankine cycles. *Energy*, 35:1033–1039.

[13] Chen, H., Yogi Goswami, D., Rahman, M. M., Stefanakos, E. K. (2011). Energetic and exergetic analysis of CO_2- and R32-based transcritical Rankine cycles for low-grade heat conversion. *Applied Energy*, 88:2802–2808.

[14] Zhang, X.-R., Yamaguchi, H., Cao, Y. (2010). Hydrogen production from solar energy powered supercritical cycle using carbon dioxide. *International Journal of Hydrogen Energy*, 35:4925–4932.

[15] Wang, J., Zhao, P., Niu, X., Dai, Y. (2012). Parametric analysis of a new combined cooling, heating and power system with transcritical CO_2 driven by solar energy. *Applied Energy*, 94:58–64.

[16] Cayer, E., Galanis, N., Desilets, M., Nesreddine, H., Roy, P. (2009). Analysis of a carbon dioxide transcritical power cycle using a low temperature source. *Applied Energy*, 86 1055–1063.

[17] Fischer, J. (2011). Comparison of trilateral cycles and organic Rankine cycles. *Energy*, 36:6208–6219.

[18] Zamfirescu, C., Dincer, I. (2008). Thermodynamic analysis of a novel ammonia-water trilateral Rankine cycle. *Thermochimica Acta*, 477:7–15.

[19] Smith, I. K., Stosic, N. (2001). Prospects for energy conversion efficiency improvements by the use of twin screw two-phase expanders. 2nd International Heat Powered Cycles Conference, Paris, France.

[20] Smith, I. K., Stosic, N., Kovacevic, A. (2005). Screw expanders increase output and decrease the cost of geothermal binary power plant systems. Geothermal Resources Council Annual Meeting, Reno, NV, USA, September 25–28, 2005.

[21] Ho, T., Mao, S. S., Greif, R. (2012). Comparison of the organic Flash cycle (OFC) to other advanced vapor cycles for intermediate and high temperature waste heat reclamation and solar thermal energy. *Energy*, 42:213–223.

[22] Badr, O., Probert, S. D., O'Callaghan, P. W. (1985). Selecting a working fluid for a Rankine-cycle engine. *Applied Energy*, 21:1–42.

[23] Tchanche, B. F., Papadakis, G., Lambrinos, G., Frangoudakis, A. (2009). Fluid selection for a low-temperature solar organic Rankine cycle. *Applied Thermal Engineering*, 29:2468–2476.

[24] Mago, P. J., Chamra, L. M., Srinivasan, K., Somayaji, C. (2008). An examination of regenerative organic Rankine cycles using dry fluids. *Applied Thermal Engineering*, 28:998–1007.

[25] Heberle, F., Brüggemann, D. (2010). Exergy based fluid selection for a geothermal Organic Rankine Cycle for combined heat and power generation. *Applied Thermal Engineering*, 30:1326–1332.

[26] Borsukiewicz-Gozdur, A., Nowak, W. (2007). Comparative analysis of natural and synthetic refrigerants in application to low temperature Clausius-Rankine cycle. *Energy*, 32:344–352.

[27] Guo, T., Wang, H. X., Zhang, S. J. (2011). Selection of working fluids for a novel low-temperature geothermally-powered ORC based cogeneration system. *Energy Conversion and Management*, 52:2384–2391.

[28] Heberle, F., Preißinger, M., Brüggemann, D. (2012). Zeotropic mixtures as working fluids in organic Rankine cycles for low-enthalpy geothermal resources. *Renewable Energy*, 37:364–370.

[29] Turboden, http://www.turboden.eu/en/home/index.php, July 2012.

[30] Adoratec, http://www.adoratec.com/productnav.html, September 2012.

[31] Ormat Inc., http://www.ormat.com/, July 2012.

[32] Brasz, J. J., Biederman, B. P., Holdmann, G. (2005). Power production from a moderate temperature - geothermal resource. GRC Annual Meeting, Reno, NV, USA.

[33] Lemort, V., Quoilin, S., Cuevas, C., Lebrun, J. (2009). Testing and modeling a scroll expander integrated into an organic Rankine cycle. *Applied Thermal Engineering*, 29:3094–3102.

[34] Badr, O., Naik, S., O'Callaghan, P. W., Probert, S. D. (1991). Expansion machine for a low power-output steam Rankine-cycle engine. *Applied Energy*, 39:93–116.

[35] Aneke, M., Agnew, B., Underwood, C. (2011). Performance analysis of the Chena binary geothermal power plant. *Applied Thermal Engineering*, 31:1825–1832.

[36] Chena Power LLC (2007). 400 kW geothermal power plant at Chena Hot Springs, Alaska. Final Report for Alaska Energy Authority.

[37] Smith, I.K., Stosic, N., Kovacevic, A. (2001). Power recovery from low cost two-phase expanders. GRC Annual Meeting, San Diego, USA.

[38] ElectraTherm, http://www.electratherm.com, June 2012.

[39] BEP Europe, http://www.bepenergy.com, July 2012.

[40] Opcon, http://www.opcon.se/web/Opcon_Powerbox_2.aspx, June 2012.

[41] Eneftech, http://www.eneftech.com/fr/technologie.php, July 2012.

31

LOW TEMPERATURE GEOTHERMAL ENERGY: GEOSPATIAL AND ECONOMIC INDICATORS

Alberto Gemelli, Adriano Mancini, and Sauro Longhi

31.1 INTRODUCTION TO GEOSPATIAL MODELLING OF THE GEOTHERMAL RESOURCE

The home-heating system exploiting low-temperature geothermal energy (LTGE) has spread at different rates in the world; in some countries its use has grown continuously since the debut more than half a century ago, while in other countries, such as Italy, its use started later and is expected to increase consistently in the coming years for economic contingencies (Renzulli et al., 2009). In regions where the development is rapid and regulated by the market and competitiveness, the investment on LTGE requires a careful prior evaluation of costs and benefits due to the tipically high upfront cost of the plants. In this context, economic and potential models of this resource will be essential tools for investors to support their decisions. Modeling tools shall be integrated into an information system available to all the stakeholders involved in the investments. An informative system is needed that suggests where and how much to invest in LTGE. Compared to other renewable resources, the LTGE has a peculiar feature: it is used in the same production site, whereas for the other renewable resources, including the high-temperature geothermal energy, the energy is converted into electricity to be transported up to medium distances. This implies that the costs and benefits of LTGE must be evaluated in relation to local needs around the production site.

The economy of LTGE is quantified with numerical indicators that can be either calculated for an individual plant or aggregated over an arbitrarily large territory, such as a commune or a province. Aggregated economic indicators are attractive to investors for comparative analysis to decide if investing or not in a site and to administrators for granting funds to encourage the use of renewable energy. In informative systems dedicated to LTGE, any data used to calculate indicators and the indicators themselves are georeferenced. The *Exploratory Spatial Data Analysis* (ESDA) is a data analysis approach that includes statistical procedures and graphical visualization techniques that are applied to the georeferenced data to calculate indicators significant to given decision problems. The spatial indicators are represented on cartograms highlighting spatial relationships between the data. In this chapter the ESDA methods are applied to LTGE data with the support of a software platform specialized in statistical and geospatial computing.

31.2 SPATIAL DISTRIBUTION OF LOW-TEMPERATURE GEOTHERMAL RESOURCE

31.2.1 Geographic Segmentation and Indexing of Resource

The Earth absorbs and retains the energy it receives from the sun, acting as a large exploitable reservoir available a few meters below the surface and about anywhere. The average temperature of soil—even at low depth—remains constant, unlike the atmospheric temperature which varies daily and seasonally. This allows us to extract heat from the ground in winter for domestic heating and dispersing heat in summer for cooling. The exploitation of this natural condition is defined as *LTGE*. The cost of an LTGE system is due mainly to the upfront cost for constructing the piping system of the geothermal heat exchanger that crosses the ground in vertical or horizontal directions (Lund et al., 2004). The front cost is

Alternative Energy and Shale Gas Encyclopedia, First Edition. Edited by Jay H. Lehr and Jack Keeley.
© 2016 John Wiley & Sons, Inc. Published 2016 by John Wiley & Sons, Inc.

proportional to the depth of the probe which in turn depends on the energy requirements it must satisfy. In the following text, with the term *plant* we refer to the heat exchanger subsystem. For Biberacher et al. (2008) the *natural potential* of any renewable resource is never exploited fully, rather it is limited by the status of art of exchanger technology which reduces the potential to the so-called *technical potential;* also local socioeconomic context imposes further limitation reducing the potential to a *realizable potential*, which is actually the one taken into account when energy planning is carried out. A model for LTGE consists of an approximate representation of its potential, calculated at one of these three levels (natural, technical, or realizable level) depending on the application. For example, in Hiremath et al., (2007) the *realizable potential* of renewable resources is calculated by resolving a constrained optimization problem whose objective is to minimize the total annual cost of energy, the dependence on nonlocal resources, and to maximize the overall performance of the production system throughout the region. In the survey of Jebaraj and Iniyan (2006) several techniques for modeling energy demand, investment, and alternative scenarios of energy use are described. We notice the modern approach to energy planning makes use of spatial modeling of resources: this is a consequence of energy production industry moving toward the paradigm of distributed generation in the geographic space. Some study cases can be cited: in Ramachandra and Shruthi (2007), the energy models are spatial and are developed with the support of the *Geographic Information System* (GIS). In Ondreka et al. (2007) the GIS was used to quantify the potential of LTGE in a region of Germany by modeling the thermal properties of rocks within a given depth from the surface. In Ramachandra (2009), a spatial model of energy resources is used as active support to the decision in energy planning.

The basis of spatial data processing systems is the *geospatial information*. For Goodchild et al. (1999) the primitive element of the geographic information is a tuple of values:

$$\langle x, y, z, t, U \rangle$$

where x, y are the coordinates in a given reference system, z is the elevation (or another attribute), t is time, and U is an associated geographic element. For example, a geographic element might be the area of an *administrative commune* delimited by its boundary, while the attribute z can be a numeric value of any physical or socioeconomic variable within the commune. There are several methods to represent computationally the geographic elements. With the vector data structure, the contour of the geographic element is represented by its geometric shape, and a commune is represented by a polygon representing its border. The graphic symbolism of choropleths (Carr et al., 2005) consists in shading the polygons in proportion to the value of attribute z. Once structured, the geospatial information is input to the processing methods

of ESDA. ESDA is a discipline largely applied in analyzing spatial resources and econometrics studies as discussed by Herath and Prato (2006). ESDA has been defined by Anselin (1998) as techniques aimed to (a) describe data; (b) visualize spatial distributions; (c) identify the favorable conditions; (d) discover patterns and spatial associations, clusters, outliers, or unequal conditions of homogeneity; (e) evaluate assumptions and the accuracy of models. The ESDA emphasizes the use of simple and intuitive descriptive methods based on graphic user interfaces. The descriptive statistics is the fundamental component of ESDA (De Smith et al., 2007); more complex methods are also used in ESDA such as the Moran's index which calculates the degree of autocorrelation of a spatial measure with similar measures from its spatial surroundings based on a given spatial weight matrix. The Gini index is a measure of statistical dispersion and it is used to measure the inequality in the spatial distribution of measures. The Geographically Weighted Regression (GWR), by Fotheringham et al. (2002), is a regression model in which the coefficients vary spatially. Further details on these techniques are given in the following sections of this chapter where they are applied to the Marche region case study.

31.2.2 Variables Controlling the Energy Potential

This section introduces the most relevant variables commonly used to represent the natural potential of LTGE. The variables to estimate the economic aspects are introduced in the next section. Assuming that the LTGE plant is optimally designed and built, the exploitable potential of LTGE resource depends of the thermal properties of the rocks in the ground surrounding the plant. These properties are concentrated in a single parameter: the *specific Heat Extraction* (sHE) (W/m), which for a given rock type expresses the power produced with 1 meter probe. The sHE generally varies in the range of 40–70 W/m. For each type of rock the sHE has a typical range of values (Verein Deutscher Ingenieure, 2001; Ondreka et al., 2007) though influenced by other factors such as the presence of natural groundwater and local microclimate (Signorelli, 2004). The site of the plant is characterized with the average value of sHE calculated up to the depth reached by the probe. A geospatial model of this resource can be obtained by calculating this variable over a geographical grid (Gemelli et al., 2011). In the following text, the term sHE always refers to average sHE.

31.2.3 Variables Controlling Economic Potential

In a given site the cost of LTGE is the ratio between the cost to build the plant and the extracted energy; it can be expressed in Euro/kWh. The *energy demand* of a residential site determines the potential investment in LTGE on that site. The total investment to supply heating by LTGE over a wide region constitutes the *economic potential* of LTGE. The *economic potential* corresponds to the total cost of replacing the conventional energy-based heating systems over the region with LTGE plants. The economic convenience of investments is the savings calculated as the difference between the cost of LTGE and the cost of an equivalent energy production from conventional sources. Economic models are strongly influenced by the geographical scale and the level of aggregation in the data (Fotheringham et al., 2002). In Dusek (2004) it has been discussed about the *modifiable areal unit problem (MAUP)*, for which the extracted models are affected by two unavoidable approximations: (1) a regional model is arbitrarily separated from the global context and (2) the region is divided into subregions whose shape, size, and number are chosen arbitrarily. Both these approximations condition significantly the econometrics calculations. Along with the results of the analysis it is necessary to provide information on geospatial segmentation adopted, conforming to standards such as NUTS[1] (nomenclature des unités territoriales statistiques) (ISTAT 2008).

The economic potential of a region is proportional to the energy demand which depends on the number of existing houses, their energetic efficiency, and the climate conditions. The energy demand can be estimated from ancillary variables correlated to current consumption or historical data of consumption (Swan and Ugursal, 2009). It is a common practice to estimate the energy demand based on the concept of *degree-days*, a conventional measure that incorporates the effect of climatic conditions (Boyd, 1984). From the *degree-days* indicator it is possible to estimate the energy demand of a house if it is known its energetic efficiency (Santini et al., 2009). In Gemelli et al. (2011), a house having an energy efficiency class "E" and an area of 100 m^2 has been taken as standard consumer. Then the cost of an LTGE plant satisfying the energy demand of a standard house is calculated in each cell of a geographic grid, using the local *degree-days* and the average sHE of the site. This procedure has led to a model of LTGE plant cost across the region.

31.2.4 Hybrid Model

The construction of a model of economic potential is a complex process that involves the following aspects: (a) data fusion; (b) translating decision-making objectives to computational goals; (c) selecting the appropriate computational techniques; and (d) choosing the software. Due to the heterogeneity of the knowledge domain, the model is built on data both physical and socioeconomical; we will refer to it as Hybrid Model. In Voivontas et al. (1998) an informative system leading to the construction of the potential model has been presented. It covers the following aspects: model of the natural potential of the resource; decision making of the authorities; decision making of investors; benefit–cost

models of investment; estimated effects on rates and market. Due to stochastic nature of the geospatial variables, statistics have a fundamental role in supporting the calculations throughout the system. In this chapter, experimentation is conducted based on statistical software R.[2] R is a free software environment for statistical computing and graphics, distributed under the GNU GPL license. It provides an object-oriented programming language and a library of statistical functions. The R platform is widely used in spatial analysis (Bivand et al., 2008), being provided with *sp module* for handling spatial information. The *sp module* implements classes and methods for computational representation of geographic elements, using spatial points, lines, polygons, and geographical grids. Other R modules used in this work are *Maptools, shapefiles, spdep, spgwr e GeoXp*. The GeoXp module supports the ESDA by implementing the calculation of indices such as *I-Moran* and *Gini* and specialized graphs. Also implemented in R is the *data brushing,* a visual analysis tool by which several graphic windows (statistical charts, geographic map, table) representing different "views" of the data are linked together so that when a data subset is selected in any view it is also highlighted in other views.

31.3 SPATIAL EXPLORATORY ANALYSIS OF GEOTHERMAL ECONOMY

31.3.1 Descriptive Statistical Models and Exploratory Statistics

A case study in which the ESDA is applied to calculate the economic potential of LTGE in an Italian region is proposed in this chapter. The Marche region is located in central Italy (Figure 31.1); it covers 9365.86 km^2 and counts about 1.5 million. The region is situated between the Adriatic Sea to the east, and the Italian Apennines to the west. The hills cover two-thirds of the territory: the remaining part is mountainous and lies toward the Italian Apennine. The plains are limited to a narrow coastal strip and to valleys near the mouths of rivers. The region is administratively divided into five provinces: Ancona, Pesaro-Urbino, Macerata, Fermo, and Ascoli Piceno, which are further divided into 239 communes. The territory is economically divided into several productive districts each being an area internally homogeneous for industrial production and target of investments.

For this region, three georeferenced variables constitute the basis for conducting the exploratory analysis: (1) annual energy demand for housing unit; (2) the local upfront cost of LTGE plant for housing unit; and (3) *sHE*. The data were acquired from a previous work (Gemelli et al., 2011) and updated where needed. As a first step in the data processing, the *median values* of the three variables within each commune have been calculated. This set of random variables, to

[1] http://epp.eurostat.ec.europa.eu/portal/page/portal/nuts_nomenclature/introduction

[2] www.r-project.org/

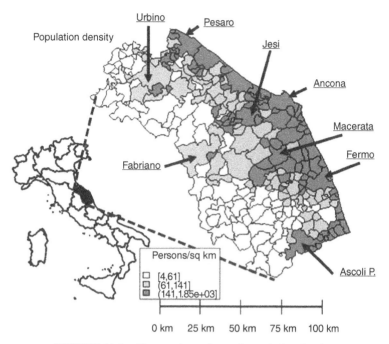

FIGURE 31.1   The marche region and population density.

which we refer in the following text, are named respectively: (1) *specific demand of energy* (kWh/m2 year), (2) *cost of the plant* (Euro), and (3) *sHE* (W/m). Some ancillary data were acquired from other sources, such as the density of the population[3] and the boundaries of provinces and communes in vector format.[4]

Our ESDA begins by examining the statistical distribution of the median values of the variables. The reference model of LTGE natural potential is represented by sHE and is shown in Figure 31.2 upper-left, the distribution is divided into tertiles. Figure 31.2 lower-left shows the specific demand in each commune. Figure 31.2 lower-right shows the cost of plant. In Figure 31.2 upper-right, the range of sHE has been divided into 11 classes of potential and the height of the histogram represents the number of communes falling in each class.

The total demand of energy and the total cost of the plants have been calculated in each commune. To calculate the total demand (Figure 31.3 left) the specific demand is multiplied by the total *inhabited housing surface* in the commune. The inhabited housing surface is calculated by multiplying the population density by the commune surface and by the average living space per capita (which is assumed equals to 36.79 m^2 across the region).

Inhabited surface = population density × commune surface × 36.79

commune demand = specific demand × inhabited surface

In Figure 31.3 right, the total cost of the plants in a commune is calculated by multiplying the cost of the plant by the number of houses in the commune:

Houses = pop.density × commune extension/36.79

commune cost = cost of plant × houses

In Figure 31.4 the data brushing tool is used to analyze the plant cost. The height of the histograms represents the probability of each cost class. Being the map synchronized to the histogram graph, by manually selecting, for example, the histograms corresponding to the first three classes, causes the centroids of the corresponding communes automatically highlighted in gray in the linked map. The remaining centroids are colored in black. The linked map evidences the low cost are located in the mountainous areas of the region and in a narrow coastal strip. In the Figure 31.5 (upper) the brushing is reversed: the centroids within the province of Ancona were manually selected and their contribution to the statistics is automatically highlighted on the linked histograms. To facilitate the selection of the centroids, the boundaries of the provinces have been superimposed on the map. By comparing the statistical table of the region (not shown) with the province of Ancona, it appears that the cost of the plant in this province is above the regional average.

In Figure 31.5 (lower) the centroids of communes falling within the productive district of Jesi, which is also included in the province of Ancona, were selected. On the map the boundaries of production districts have been overlaid. It can be noted in this productive district the cost of plants is lower than in the province that contains it.

[3] ISTAT. 14th Census of population and houses, 2001
[4] ISTAT. Administrative borders updated at January 1, 2011.

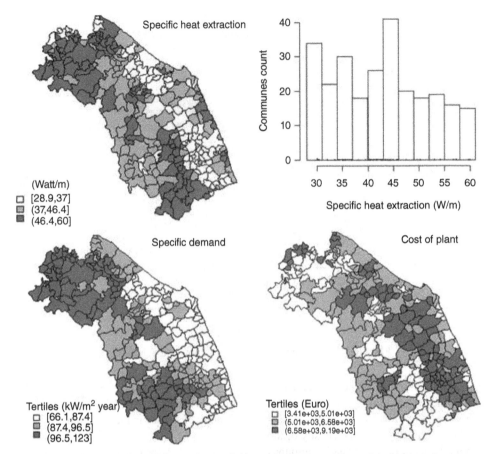

FIGURE 31.2 Upper-left: the natural potential in each commune. Upper-right: the cost frequency. Lower-left: specific demand of energy. Lower-right: cost of the LTGE plant.

The data brushing can be used to identify outliers in the cost of the plants. In Figure 31.6 the cost range has been divided into a large number of classes and the histogram block at the lower end of the scale has been manually selected. In the linked map these outliers appear highlighted and marked with arrows.

The scattergram is a graphical summary of a bivariate dataset aimed to explore visually the relationships between two random variables. In Figure 31.7 (upper) the y-axis of the scattergram is the cost of the plant; the x-axis represents the specific energy demand. In the diagram there is a positive correlation between these two variables. In this diagram

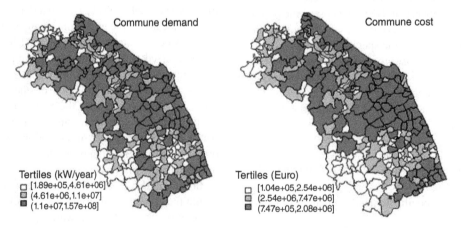

FIGURE 31.3 Left: yearly total energy demand of communes. Right: total commune cost to convert consumes to LTGE.

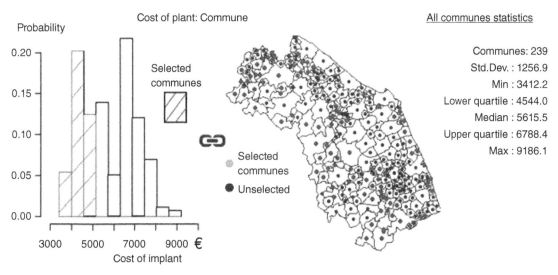

FIGURE 31.4   Spatial distribution and statistics of the cost of LTGE plant across the region.

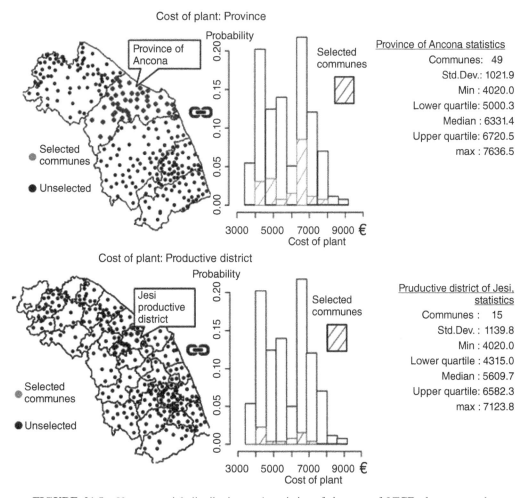

FIGURE 31.5   Upper: spatial distribution and statistics of the cost of LTGE plant across the provinces. Lower: spatial distribution and statistics of the cost of LTGE plant across the productive districts.

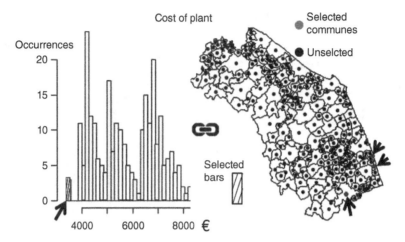

FIGURE 31.6   Identification of outliers in the cost of the LTGE plant.

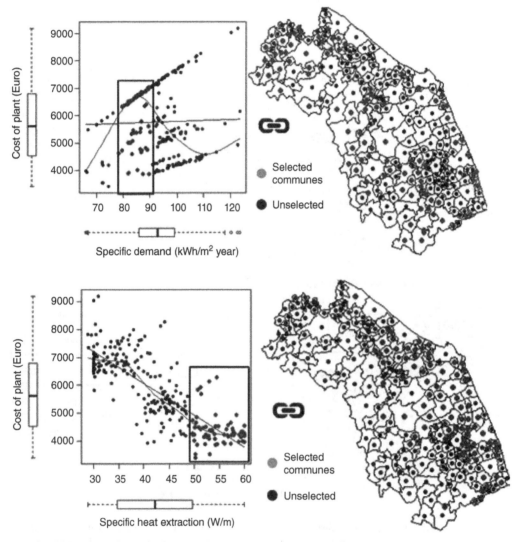

FIGURE 31.7   Upper: scattergram between the cost of the plant and the annual specific demand. Lower: scattergram between the cost of the plant and the sHE.

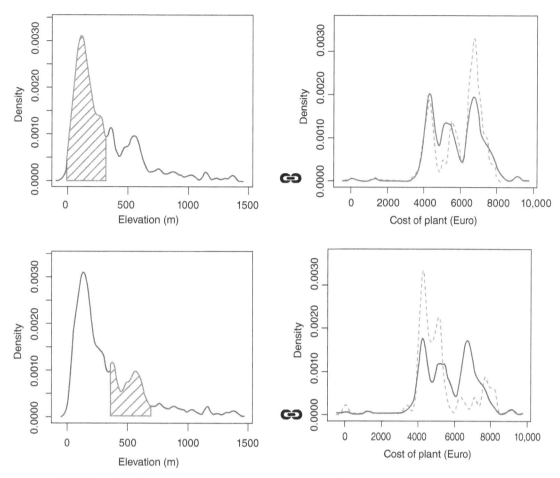

FIGURE 31.8   First example: the upper-left is the probability density function of elevation; the upper-right is the density function of plant cost. Second example: the lower-left: is the density function of elevation; the lower-right is the density function of plant cost.

the brushing is applied by selecting a rectangular area and observing the corresponding centroids automatically highlighted in the map. Notice the aligned points correspond to sites with similar sHE. In Figure 31.7 (lower) the cost of the plant and sHE are compared. The two variables have negative correlation. By selecting the area of greatest cost and highest sHE of the diagram, the coastal and hilly areas of the region are highlighted.

The compared analysis of a resource and the physical aspects of the territory is a typical work of spatial exploratory analysis. In Figure 31.8 the cost of the plant and the elevation are analyzed jointly. In Figure 31.8 upper-left, the probability density function of elevations is drawn. In Figure 31.8 upper-right, the probability density function of the plant cost is shown with the solid line. Between the two diagrams there is an active link, so selecting manually an interval of elevations (the shaded area beneath the curve, on the left figure), the corresponding subdensity function of cost is drawn on right figure (the dashed line curve). The Figure 31.8 lower shows similar experiment but it has been selected at higher

interval of elevations. The two examples differentiate well the mountainous and the hilly part of Marche region from the point of view of cost of LTGE plant.

Focusing our attention to multivariate datasets, the *GWR regression model* has been calculated to represent the data. The dataset input to GWR consists of a dependent variable y, a set X of independent variables, and a set T of points spatially distributed within the region. In the GWR method a linear regression model is calculated at each point $t \in T$, using a subset of measures spatially close and weighted according to their distances from each t stored in the matrix W of spatial weights. The GWR model has a general form:

$$y = X\beta(t) + \varepsilon$$

where, β is the set of computed coefficients.

To evaluate the accuracy of GWR the results are compared with the linear regression technique. The GWR method has been applied to predict the cost of the LTGE plant in two different experiments, respectively, with one and two

TABLE 31.1 R^2 of regression models.

	GWR one predictor	GWR two predictors	Linear regression
	R^2	R^2	R^2
Min	−0.371	0.518	0.001247
Lower tertile	0.341	0.705	
Upper tertile	0.457	0.795	
Max	0.792	0.847	

TABLE 31.2 Residuals of regression models.

	GWR one predictor	GWR two predictors	Linear regression
	Residuals	Residuals	Residuals
Min	−4227.0	−4354.0	−5654.0
Lower tertile	−426.3	−265.2	−696.6
Upper tertile	292.0	207.4	869.2
Max	2209.0	1578.0	2424.0

predictor variables. In the former case the predictive variable is the specific demand of energy while in the latter case the sHE is also used as predictive variable. In Table 31.1 the coefficients R^2 are listed. There is only one R^2value for linear regression, while for the GWR minimum, maximum and tertiles of the R^2 set are listed. It can be observed the simple linear regression R^2 is very small compared to GWR. The highest R^2 is in the GWR with two predictor variables which shows a smaller difference between the tertiles and presumably a higher accuracy. In Table 31.2 the residues of the models are listed.

For the GWR with two predictive variables, the cartograms of coefficients and intercept are shown in Figure 31.9. These models are useful in decision support as they directly represent a location-based assessment of the cost of the plant. It should be emphasized that these methods provide only rough estimates, while the design of a plant requires an accurate assessment that can only come from *in situ* measurements.

In Figure 31.10 a comparative analysis of three variables is done using the method of conditional choropleth. It consists of a graphical representation that highlights the conditional probability of variables. The panels are identified with a pair of indices (i,j). The index i represents the tertile of the specific energy demand and the index j represents the tertile of sHE. The levels of gray represent the value of the third variable namely the cost of plant. For example, the panel (1,3) represents the cost of plant in communes with high sHE and low specific demand, while the panel (3,1) represents the cost of plant in communes with low sHE and high specific demand. However, the overall distribution of communes across the panels suggests the existence of more complex relationships than a simple variability between extreme conditions. For example, the panel (1,1) highlights that there are communes with high cost of plants although there is low specific demand.

31.3.2 Advanced Decisional Modeling

In the previous section we have discussed methods to represent the spatial variables, and the tools to explore a spatial distribution. In this section we focus on building models that emphasize economic differences within the territory to support investment decision-making problems. The first step is to measure the degree of uniformity in the spatial distribution of the resource. The uniformity of LTGE potential distribution can be measured with the *Gini coefficient*. This coefficient is mathematically defined on the *Lorenz curve*. In Figure 31.11 upper, the Lorenz curve for variable sHE is drawn on Cartesian plane. On x-axis the cumulative frequencies (f) of sHE, on y-axis the cumulative amount (g) of

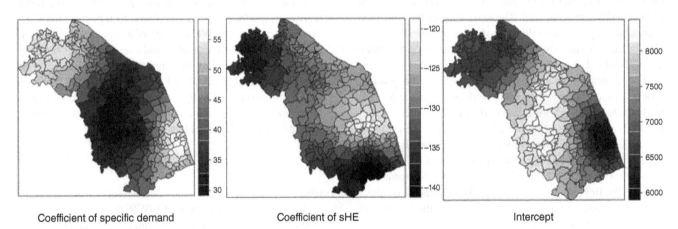

Coefficient of specific demand

Coefficient of sHE

Intercept

FIGURE 31.9 Regression coefficients of specific demand (Left), sHE (Center), and intercept (Right).

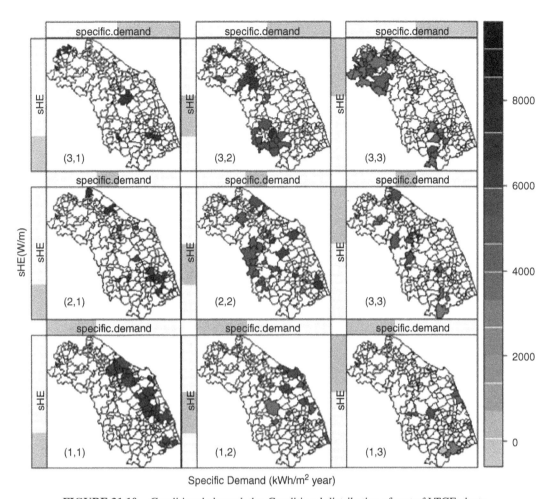

FIGURE 31.10   Conditional choropleths. Conditional distribution of cost of LTGE plants.

sHE. The *Lorenz curve* is more convex when the inequality in the statistic distribution of the variable is greater. The 45° line represents a perfectly uniform distribution. The Gini coefficient is the ratio of the area between perfect uniform distribution and the Lorenz curve, and the total area beneath the perfect uniform distribution line. The Gini coefficient varies between 0 and 1, respectively, the maximum and the minimum possible uniformity in the spatial distribution. In this case the coefficient is 0.121. The *diagram* in diagram in Figure 31.11 is active, thus manually selecting $f = 0.5$ the corresponding cumulative amount $g = 0.4$ is found: these values mean that 50% of the communes owns only 40% of the total potential energy. The data brushing is also active and the communes corresponding to the 40% potential are highlighted on the map. The Figure 31.11 lower shows the Lorenz curve for the cost of the plants. The curve highlights that 90% of the communes requires only 45% of the investment while the remaining 55% of the investment would be focused in the communes with the highest population.

For a quantitative comparison of the potential variations between the provinces, some statistical parameters of the variable sHE and its correlation with some socioeconomic variables are reported in Table 31.3. We notice that in the communes within the province of Ancona the average sHE is relatively low, only 39.5 W/m. The Gini index reveals there is a significant disparity in the distribution of resources among the communes in this province with the LTGE potential that tends to be concentrated in a few communes. The sHE is positively correlated with the specific demand: this is a positive aspect because it indicates the resource is high by chance where there is also a high energy demand due to climatic reasons. However the low correlation (0.08) of sHE with the population density indicates that the population is not concentrated in municipalities with high potential and therefore cannot optimally exploit the LTGE potential. The sHE has the highest value in the province of Pesaro-Urbino and, based on Gini index, it appears more uniformly distributed in the region.

In Table 31.4, five productive districts included in the Province of Pesaro-Urbino have been compared. It is noticed the most favorable conditions for the exploitation of the resource are in the productive district of Cagli because the

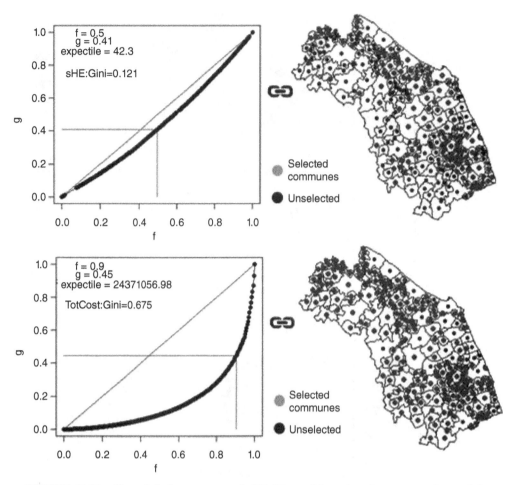

FIGURE 31.11   Upper-left: Lorenz curve of sHE. Upper-right: selected communes. Lower-left: Lorenz curve of commune cost. Lower-right: selected communes.

TABLE 31.3   **Comparison between the distribution of SHE and other variables, by provinces.**

Province	Communes	sHE (W/m)	sHE(W/m) S.D.	sHE Gini	Corr.indx consumption	Corr.Indx dens pop
Pesaro-Urbino	74	46.5	9.34	0.37	0.10	−0.12
Ancona	49	39.5	7.13	0.47	0.49	0.08
Macerata	59	43.18	7.42	0.40	0.22	−0.05
Fermo	42	34.35	8.08	0.50	0.13	0.15
Ascoli Piceno	35	41.2	10.20	0.53	0.26	−0.15

TABLE 31.4   **Comparison between the distribution of sHE and other variables in the productive districts of the Pesaro-Urbino province.**

Productive district	No.	sHE (W/m)	sHE(W/m) S.D.	sHE Gini	Corr.Indx consumption	Corr.indx dens pop
Sassocorvaro	17	44.12	11.55	0.24	0.07	0.23
Fano	16	44.38	7.95	0.39	0.37	0.13
Pergola	10	43.64	4.33	0.23	0.76	−0.36
Pesaro	9	51.44	8.86	0.46	0.49	−0.04
Cagli	5	49.7	5.46	0.25	0.44	−0.02

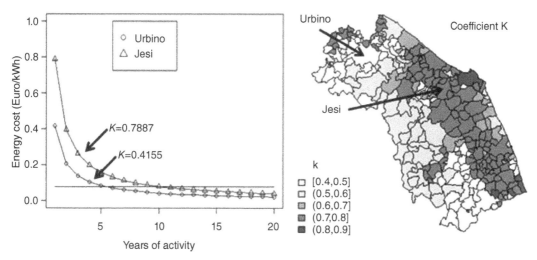

FIGURE 31.12   Left: The characteristic curve of the relationship between the energy produced and the cost of the plant. Right: spatial distribution of the coefficients k.

sHE in this district is relatively high, it has a strong uniformity in the distribution and it is positively correlated with the specific energy needs.

Now we seek a spatial indicator of the effectiveness of the investment. The cost of the LTGE plant is mostly an upfront cost which is amortized over the years by the savings on traditional energy costs. On a given number of years since the construction of the plant, the *specific cost of energy* is equal to the incurred total costs divided by the energy produced. The curve of the specific cost of energy for the communes of Urbino and Jesi is drawn in Figure 31.12 left. On the *x*-axis the years of activity. Comparing the curves we observe the costs in Urbino remains lower than Jesi, this means the former commune offers a more profitable investment because the cost of specific LTGE is lower. The shape of the curve, which is an equilateral hyperbola, is mathematically represented by the coefficient k.

$$\text{Energy specific cost} = \frac{\text{Cost of plant}}{\text{Yearly energy demand} \times \text{years of activity}}$$

$$K = \text{Years of activity} \times \text{energy specific cost} =$$

$$\frac{\text{Years of activity} \times \text{cost of plant}}{\text{Yearly energy demand} \times \text{years of activity}} = \frac{\text{Cost of plant}}{\text{Yearly energy demand}}$$

Furthermore, the coefficient k determines the period of amortization which in the diagram is identified by the intersection between the hyperbola and the horizontal straight line $y = 0.77$ Euro/kWh, which is the cost of energy with a methane plant. The coefficient k is an indicator of the cost-effectiveness of investing in LTGE in a commune and it can be mapped (Figure 31.12 right).

After seeing the economic characterization of the areas we look for a method capable to divide the region into subregions, each internally homogeneous for the LTGE cost-effectiveness. This type of zoning can be made on the basis

of the indicator k by studying its spatial autocorrelation. The *Moran index I* which expresses the degree of spatial autocorrelation is calculated as follows:

$$I = \frac{\sum\limits_i \sum\limits_j W_{ij}(x_i - \bar{x})(x_j - \bar{x})}{\sum\limits_i (x_i - \bar{x})^2}$$

where, x_i is the variable under study in the point i, $\bar{x}$ is the sample mean, W_{ij} is the weight matrix of values inversely proportional to the distance between point i and the point j. In Figure 31.13 the index I has been calculated for the variable k. I corresponds to the slope of the regression line between random variable k, on the x-axes, and the spatially lagged k, based on W, on the y-axes. The dashed lines, corresponding to the medians, separate the plane into quadrants that represent communes with distinct characteristics of spatial autocorrelation. The H–H quadrant distinguishes communes with high value of k having a neighborhood of communes also with relatively high values of k. In the map these points are colored in gray and evidently tend to cluster and highlight large homogeneous subregions. Even the communes of the L–L quadrant tend to cluster in the geographic space. While the H–L communes are scattered in the region and constitutes subregions having small spatial autocorrelation.

Now analyze a typical investment problem in spatial context: *investment siting*. Figure 31.14 depicts the curve of energy production cost. The x-axis represents the total energy production and is obtained by accumulating the demand starting from the communes with low k values. The y-axis represents the corresponding cumulative investment. The variables are normalized in the range 0–1. The convexity of cost curve is due to the fact that in communes with low k the energy production costs less. The indicator k can be used to highlight which municipalities offer higher return on LTGE

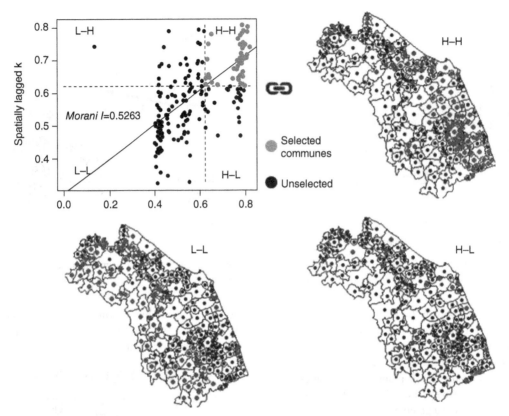

FIGURE 31.13   Upper-left: Moran scattergram. Autocorrelation zones H-H, L-L, H-L.

investments on a given period, hence where to siting the investment.

We apply these observations in the next experiment. In Figure 31.15 upper-left, it was supposed to invest 3,000,000 Euros in the communes to fund the conversion to LTGE; it is wanted to distribute the grants among the communes until exhaustion of the available funds, while giving priority to those communes where the investment is made more profitable due to local favorable conditions. To do this we use the coefficient k as a measure of the favorable condition. In a first scenario, each commune, beginning with those with lowest k, has been funded for 20% of the cost required to achieve a complete LTGE conversion. The communes colored in white are those excluded by this financing policy, because of funds run out to satisfy low k sites. The communes granted are gray shaded in the figure, and were further divided into three groups with different grant median value. The LTGE produced thanks to funds is shown in Figure 31.15 upper center. In the communes where the LTGE plant is more productive there is a higher production and economic saving. In Figure 31.15 upper-right, the *Return On Investment (ROI)* is calculated for each commune by subtracting the grant from the monetary value of energy produced. We notice with this grant policy the return is positive in most municipalities. By contrary, the Figure 31.15 lower panels correspond to an opposite scenario in which it is invested primarily on the communes with high k. Between the two scenarios, the difference in terms of profit and power savings is clear, since in the second scenario after 10 years the ROI is still negative in most part of the communes.

31.4   CONCLUSIONS AND FUTURE DIRECTIONS OF REGIONAL MODELING OF RENEWABLE RESOURCES

The LTGE market has great potential in the Marche and Italy in general, because this resource has been scarcely

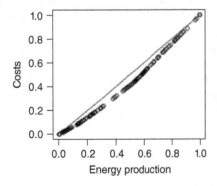

FIGURE 31.14   Cost curve of energy production.

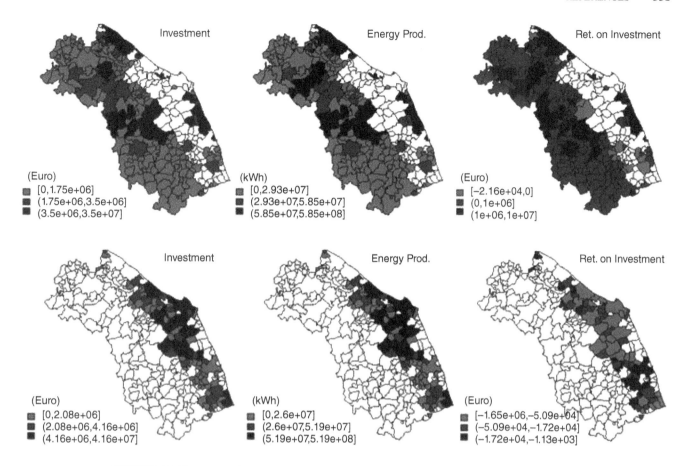

FIGURE 31.15   Upper: first scenario. Investment, production efficiency, and return on investment in the most convenient scenario. Lower: second scenario. Investment, production efficiency, and return on investment in the less convenient scenario.

exploited until now and because the development of all renewable sources will soon undergo rapid increase dictated by incoming changes in the market energy. In this chapter, the ESDA has been applied to analyze the economic potential of LTGE. This case study provides a new approach to the analysis of LTGE, and in general to modeling the potential of renewable energy and distributed energy production in which the spatial location of the resource is crucial to evaluate its effective usefulness. The ESDA produces models and indicators that are tools supporting the decision-making activities by describing the energy resource spatial distribution qualitatively and quantitatively. The ESDA provides a methodical approach of proven efficiency in spatial analysis. The exploratory approach to the analysis of multivariate spatial datasets, composed by physical and socioeconomic data related to an energy resource, highlights opportunities that would otherwise remain overlooked in sector studies. The economic potential of LTGE benefits and conditions locally other economically significant activities such as building renewal market and the use of ground as a storage reservoir of thermal energy. It is likely that the LTGE exploitation

planning in the future will be supported by informative systems able to manage information updated in real time on economic activities in the region, particularly those related to the building activity and their energy efficiency, and will be connected with the databases of cadastral offices. We believe that the object of further studies can be the comparative study of the LTGE potential and detailed characteristics of constructions existing on the territory, and that these study experiences can be applied elsewhere in the country and be exported to foreign emerging economies.

REFERENCES

Anselin, L. (1998). Exploratory spatial data analysis in a geocomputational environment. In: *Geocomputation, A Primer*, Longley, P. A., Brooks, S. M., McDonnell, R. A., Macmillan, B. (editors). John Wiley & Sons, New York.

Biberacher, M., Gadocha, S., Zocher, D. (2008). GIS based model to optimize possible self-sustaining regions in the context of a renewable energy supply. In Proc. of iEMSs 2008: International

Congress on Environmental Modelling and Software Integrating Sciences and Information Technology for Environmental Assessment and Decision Making.

Bivand, R., Pebesma, E. J., Gomez-Rubio, V. (2008). *Applied Spatial Data Analysis with R*. Springer.

Boyd, D. W. (1984) Degree days : the different types, Building Research Note, 138.

Carr, D. B., White, D., MacEachren, A. M. (2005). Conditioned choropleth maps and hypothesis generation, *Annals of the Association of American Geographers*, 95(1):32–53.

De Smith, M., Goodchild, M. F., Longley, P. A. (2007). *Geospatial Analysis—A Comprehensive Guide to Principles, Techniques and Software Tools*. Troubador Publishing Ltd.

Dusek, T. (2004). Spatially aggregated data and variables in empirical analysis and model building for economics, *Cybergeo: Revue européenne de géographie*, 285(9):1278–3366.

Fotheringham, A. S., Brundsdon, C., Charlton, M. (2002). *Geographically Weighted Regression: The Analysis of Spatially Varying Relationships*, John Wiley & Sons.

Gemelli, A., Mancini, A., Longhi, S. (2011). GIS-based energy-economic model of low temperature geothermal resources: a case study in the Italian Marche region, *Renewable Energy*, 36:2474–2483.

Goodchild, M. F., Egenhofer, M. J., Kemp, K. K., Mark, D. M., Sheppard, E. S. (1999). Introduction to the Varenius project, *International Journal of Geographical Information Science*, 13(8):731–745.

Herath, G., Prato, T. (2006). *Using Multi-criteria Decision Analysis in Natural Resource Management*. Ashgate Publishing Limited.

Hiremath, R. B., Shikha, S., Ravindranath, N. H. (2007). Hiremath-decentralized energy planning- modeling and application -a review, *Renewable and Sustainable Energy Reviews*, 11(5):729–752.

Jebaraj, S., Iniyan, S. (2006). A review of energy models, *Renewable and Sustainable Energy Reviews*, 10(4):281–311.

Lund, J., Sanner, B., Rybach, L., Curtis, G., Hellstrom, G. (2004). Heat pumps—a world overview, GHC Bulletin.

Ondreka, J., Rusgen, M. I., Stoberb, I., Czurda, K. (2007). GIS-supported mapping of shallow geothermal potential of representative areas in south-western Germany—possibilities and limitations, *Renewable Energy*, 32(13):2186–2200.

Ramachandra, T. V. (2009). RIEP Regional integrated energy plan, *Renewable and Sustainable Energy Reviews*, 13(2):285–317.

Ramachandra, T. V., Shruthi, B. V. (2007). Spatial mapping of renewable energy potential, *Renewable and Sustainable Energy Reviews*, 11(7):1460–1480.

Renzulli, A., Piscaglia, F., Menichetti, M., Puerini, M., Blasi, A. (2009). Geotermia: dalla produzione di energia elettrica agli utilizzi diretti del calore. In: *Advanced Manufacturing Systems for Geothermal Energy*. Energy Resources, pp. 17–37.

Santini, E., Elia, S., Fasano, G. (2009). *Caratterizzazione dei consumi energetici nazionali delle strutture ad uso ufficio*. ENEA.

Signorelli, S. (2004). Geoscientific investigations for the use of shallow low enthalpy systems, Swiss Federal Institute of Technology Zurich.

Swan, L. G., Ugursal, V. I. (2009). Modeling of end-use energy consumption in the residential sector: a review of modeling techniques, *Renewable and Sustainable Energy Reviews*, 13(8):1819–1835.

Verein Deutscher Ingenieure. (2001). *Thermische Nutzung des Untergrundes–Blatt 2:erdgekoppelte Warmepumpenanlagen*. Beuth Verlag.

Voivontas, D., Assimacopoulos, D., Mourelatos, A., Corominas, J. (1998). Evaluation of renewable energy potential using a GIS decision support system, *Renewable Energy*, 13(3):333–344.

32

DRY COOLING TOWERS FOR GEOTHERMAL POWER PLANTS

Zhiqiang Guan, Kamel Hooman and Hal Gurgenci

All thermal power plants produce waste heat as a byproduct and this waste heat must be continuously dissipated by cooling system to make the plants operate efficiently. The cooling system of a power plant can be either wet or dry cooling. Since geothermal power plants (especially for enhanced geothermal system), in general, are most likely located in arid area where water maybe more valuable than the power produced, the dry cooling towers will be the only cost effective method of heat rejection.

32.1 SELECTION OF COOLING METHODS FOR GEOTHERMAL POWER PLANTS

Geothermal power plants can be broadly classified as dry steam, flash steam, and binary cycle power plants (Ronald, 2005; DiPippo, 2008) depending on the state of the fluid (steam or water) and its temperature. For low-temperature geothermal resources, binary cycle power plants are most likely used in practice. In a binary power plant, the geothermal heat is transferred to a secondary fluid before being converted to power, as shown in Figure 32.1.

Since waste heat from turbine exhaust steam must be continuously rejected to make the plants operate, cooling tower is an integrated part of any geothermal power plant. In general, the heat is dumped to either water or air, because these are the only options that are in relative abundance and at relatively low cost. According to the heat dump choice, the cooling system can be classified as wet cooling and dry cooling (Kroger, 2004).

In a wet cooling tower, the water and air are in direct contact, which results in cooling primarily by evaporation.

As shown in Figure 32.2, hot turbine exhaust steam is cooled and condensed in the condenser by transferring the heat to the cold water pumped from the bottom of the cooling tower. As this happens, the cooling water itself becomes hot. This hot water is then piped to the cooling tower and is distributed onto fill material. Air is introduced by either mechanical draft (fans) or natural draft (tall tower) to move it across the fill material making direct contact with the hot water and carry the heat away. The cooled water is then collected in a cold water basin below the fill from which it is pumped back to the condenser. The fill material increases both the water/air contact area and contact time for better cooling performance.

Wet cooling towers transfer the heat from the water stream to the air stream, which raises the air temperature and its relative humidity to 100%, and this air is discharged to the atmosphere. Since the performance of the wet cooling is depended on the wet bulb temperature of the ambient air, it has higher cooling efficiency than dry cooling tower.

In a wet cooling system, the heat transfer is mainly by latent heat transfer through water evaporation and only partially by sensible heat transfer. This causes large quantities of water to evaporate into the moving air stream and discharged to the atmosphere on top of the tower. This water evaporation can be clearly seen as a plume emanating from the wet cooling towers used in thermal power plant as shown in Figure 32.3. The water lost to evaporation must be continuously replaced. In addition, because evaporation concentrates impurities in the water, some of the circulating water is deliberately drawn off to prevent extensive scale formation. This is called "blow-down". The moving air also carries away of some water droplets, which are called drift. The evaporation loss, blow down water, and drift water

Alternative Energy and Shale Gas Encyclopedia, First Edition. Edited by Jay H. Lehr and Jack Keeley.
© 2016 John Wiley & Sons, Inc. Published 2016 by John Wiley & Sons, Inc.

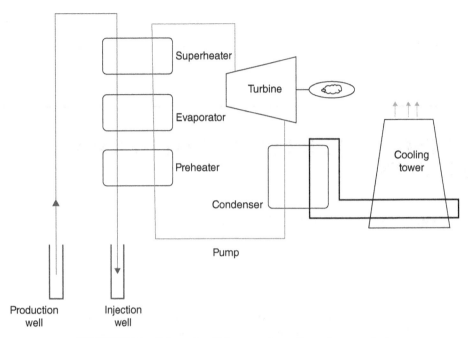

FIGURE 32.1   Schematic of binary cycle geothermal power plant.

correspond to the total water loss in a wet cooling system. Williams and Rasul (2008) reported the water evaporation rate for a power plant of 350 MW capacity in Queensland of Australia was around 1.8 L of water per kWh of power generated. This is about 630 kilo L per hour or 5500 mL per year for a 350 MW coal-fired power plant. NETL (2009) reported that a total makeup water of 5188 gpm was consumed for a 520 MW power plant in the United states, among which, the evaporation water was 3891 gpm and the blown down water was 1297 gpm. This is about 2.27 L of water per kWh of

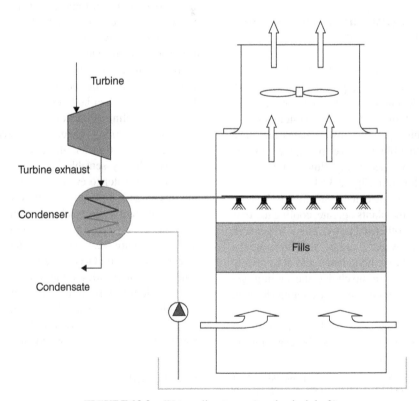

FIGURE 32.2   Wet cooling tower (mechanical draft).

FIGURE 32.3 Photo of wet cooling tower used in a thermal power plant.

power generated. Gurgenci (2010) predicted that total water consumption rate for a geothermal power plant is about 0.4 kg/s per MW of heat rejected. According to this prediction, a 25 MW power plant running at 15% thermal efficiency in Enhanced Geothermal Systems (EGS) will need to dispose of 142 MW of heat. The water consumption for this plant would be 56.8 kg/s or 1.8 million tons per year if wet cooling is used.

Except for the large amount water consumption, wet cooling tower requires frequent water treatment to minimize scaling and fouling to prevent growths of bacteria, fungi and algae, and to eliminate the growth of Legionella that cause Legionnaires' disease. Wet cooling towers also cause other environmental impact such as fog producing.

Based on the amount of water required and cost associated with the water treatment, it will be difficult or expensive to use wet cooling for EGS power plants which are normally located in arid areas. Therefore, dry cooling may be the only alternative for these geothermal power plants.

Dry cooling towers conduct heat transfer through air-cooled heat exchanger that separates the working fluid (steam or water) from the cooling air. Since the working fluid does not contact with the ambient air directly, there is no water evaporation lost in this system. Dry cooling towers have two basic types – direct system and indirect system.

In a direct dry cooling tower, the turbine exhaust steam flows directly into the inside tube of heat exchanger through a large pipe as indicated in Figure 32.4. The large-diameter piping is used to accommodate the relatively low steam densities compared to water and reduce the pressure drop through the pipe. The exhaust steam inside the tube is cooled and condensed by transferring the heat to the flowing air outside the surface of the heat exchanger. The air flows through the outside surface of the heat exchanger either by fans or by natural draft.

With indirect dry cooling tower, as shown in Figure 32.5, the turbine exhaust steam is cooled and condensed in a surface heat exchanger called condenser. In the condenser, the waste heat is transferred to the cold water pumped from the air-cooled heat exchanger located inside the cooling tower. The cooling water gains heat in the condenser and is pumped back to the heat exchanger inside the cooling tower. Air is

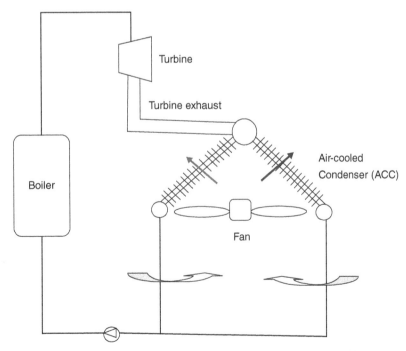

FIGURE 32.4 Direct dry cooling tower (commonly known ACC).

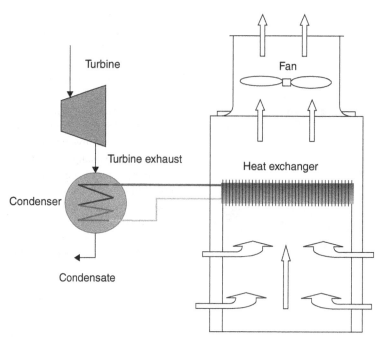

FIGURE 32.5   Schematic of indirect dry cooling tower.

introduced to flow across the outside of the heat exchanger to cool the hot water inside the tube of heat exchanger. The cooled water inside the tube of the heat exchanger is then pumped back to the condenser. Since the turbine exhaust steam is cooled by the circulating water rather by the ambient air directly, it gets the name of indirect cooling.

In a dry cooling tower, air can be introduced by either mechanical draft (fans) or by natural draft (tall tower) to move the air across the air-cooled heat exchangers. Increasing volume flow rate of air through heat exchangers improves the performance of power plants. Since the performance of dry cooling is depended on the air dry-bulb temperature, power plants experience low efficiency at high ambient air temperatures.

While dry cooling system can save significant water resource and is more environment friendly compared with wet cooling system, these water savings can often come at a price, generally in the form of more costly heat exchangers, and reduced plant efficiencies at high ambient temperatures. Therefore, hybrid cooling towers were introduced and studied (Barker, 2007).

Hybrid cooling systems combine both wet and dry cooling towers in one system. In a hybrid cooling system, a moderate amount of water is used during the hot periods of the year to mitigate the large losses in steam cycle capacity and plant efficiency associated with all dry operation. Various hybrid cooling systems have been proposed: wet and dry cooling system in parallel; in series; and inlet air cooling (Barigozzi et al., 2011). Figure 32.6 shows a hybrid cooling system with wet and dry cooling system in parallel (Barker, 2007). The

wet and dry components can be used separately or simultaneously according to the ambient condition.

In the hybrid cooling system shown in Figure 32.6, the dry cooling tower section may be sized for working at the optimum ambient temperature, which is determined by detailed economic analysis. At the peak ambient temperature (higher than the optimum design temperature), the difference between the required total heat rejection and the heat rejected by the dry section at this temperature is carried by the wet section.

For a practical power plant design, the choice between wet and dry cooling system involves a number of trade-offs including the availability and cost of water, environmental effects and the cost of electric power (EPRI, 2003). Since geothermal power plants, in general, are located in arid areas, water conservation is the major factor for selecting dry cooling towers.

32.2   AIR-COOLED HEAT EXCHANGERS

Air-cooled heat exchangers are the most expensive and critical components in a dry cooling system. With heat exchanger employed, the hot fluid flows inside the heat exchanger tubes while the ambient air flows outside the tubes. Since the heat transfer coefficient of the air outside the tubes is much lower than that of the hot fluid inside the tubes, the surface area of the heat exchanger has to be increased by adding fins onto the bare tubes to achieve the required heat transfer rate. This type of heat exchanger is called finned tube heat exchanger.

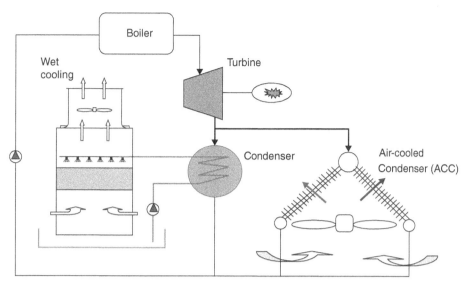

FIGURE 32.6   Hybrid cooling system with separate wet and dry components.

The selection of the finned tube heat exchanger parameters and the material are based on the fluid temperature, environment (corrosion) and cost (Kroger, 2004). For geothermal–thermal power plant, the fluid temperatures are relatively low and several types of finned tube heat exchanger can be used. These finned tube heat exchangers have been constructed with different shapes of tubes and fins, and with different materials. Round, elliptical, and flattened tubes with circular, square, or rectangular fins are commonly used for dry cooling thermal power plants. The materials used for tubes and fins are mostly aluminium and steel.

The finned tube heat exchanger with round tube is formed by winding a strip of the fin material around the round tube under tension as shown in Figure 32.7. In this configuration, the round tube is normally made of steel and the fins are made of aluminium or steel. Several individual finned tube

heat exchangers are then arranged with different rows and columns to form a heat exchanger bundle (Figure 32.8). Tens or even hundreds of heat exchanger bundles are required in a moderate capacity power plant if dry cooling tower is used.

In another configuration of finned tube heat exchanger with round tube, instead of winding the fin on single tube, a number of tubes are arranged into array with several rows (up to six rows) and columns. Continuous large rectangular plate fins are then mechanically bound to the tubes array as shown in Figure 32.9 to form heat exchanger bundle.

Various techniques are employed in attaching the fins onto the round tube to improve the thermal contact and reduce the production cost.

Compared to round tubes, elliptical tubes have larger hydraulic radii and offer lower resistance to the fluid flow.

FIGURE 32.7   Formation of the finned tube heat exchanger with round tube.

FIGURE 32.8   Finned tube heat exchanger bundle with round tube.

FIGURE 32.9   Round tube heat exchanger with rectangular fins.

This feature is especially useful for direct dry cooling system, because lower turbine exit pressure drop significantly improves the power plant performance. The construction of elliptical tube heat exchanger is made with elliptical steel tubes and rectangular steel-plate fins. These fins are made with collar and spacer for the purpose of separating the adjacent fins (Figure 32.10). Fins are attached to the tube by mechanical means. After the assembly of the fins,

FIGURE 32.10   Finned tube heat exchanger with elliptical tube.

FIGURE 32.11   Manufacturing the flattened tube from steel sheet.

galvanize processing is used to ensure the good thermal contact between the tubes and fins.

Elliptical tube is made of larger diameter than the round tube. The bundles of elliptical tube heat exchangers are usually made of single or double rows of finned tubes. This type of heat exchanger is mostly used in air-cooled condenser (ACC) where the turbine exhaust steam is condensed by the air directly.

Flattened tube heat exchangers have been developed for application in direct dry cooling towers and showed good performance characteristics (Heyns, 2008; Yang et al., 2012). In this type of heat exchanger, the flattened tube is formed from steel sheet, which is aluminized on one side, by a special designed machine (Figure 32.11). The aluminium side of the sheet is kept at outside after forming the tube for the fin attachment. The aluminium fins are made in a wavy shape separately as shown in Figure 32.12. The fins are then bounded to the surface of the flattened tube to form the heat exchanger (Figure 32.13).

Single row of tubes is used to form the heat exchanger bundle for direct dry cooling system. This type of heat exchanger has least resistance to the hot fluid flow and has found increasing application in modern power plants as ACC components.

Models predicting thermal characteristics and pressure drop of air-cooled heat exchange have been developed by various researchers for sizing the heat exchanger and cooling tower (Kroger, 2004; Kakac and Liu, 2002). These two parameters are also provided by the manufacturers of heat exchangers, usually by testing data. The following summarizes the steps used in deriving thermal characteristics and pressure drop of heat exchangers.

The heat exchanger rate of an air-cooled heat exchanger can be expressed in the form of (Kroger, 2004):

$$Q = UAF_T\Delta T_{lm} \qquad (32.1)$$

FIGURE 32.12   Wavy fins made of aluminium.

where U – the overall heat transfer coefficient referred to the area A, W/(m² °C);

A – any convenient transfer area referred by U, m²;

F_T – correction factor;

ΔT_{lm} – logarithmic mean temperature difference, °C; for counterflow arrangement, ΔT_{lm} is calculated by the following form:

$$\Delta T_{lm} = \frac{(T_{h2} - T_{c2}) - (T_{h1} - T_{c1})}{\ln[(T_{h2} - T_{c2})/(T_{h1} - T_{c1})]} = \frac{\Delta T_2 - \Delta T_1}{\ln(\Delta T_2 / \Delta T_1)}$$

$$(32.2)$$

FIGURE 32.13   Flattened tube heat exchanger.

where T_{h2} – hot fluid outlet temperature;

T_{h1} – hot fluid inlet temperature;

T_{c1} – cold fluid (air) outlet temperature;

T_{c2} – cold fluid inlet temperature.

The product of the overall heat transfer coefficient and the transfer area (UA) should always be used. When different areas are used, the result of U will be different but the product of U and A will be always the same.

With finned tube heat exchanger, the total air side area A_o includes both the fin area and the exposed tube root area. When calculating this total area, fin efficiency must be taken into account. In this case, the outside area A_o is calculated:

$$A_o = A_r + \eta_f A_f$$

$$(32.3)$$

where, A_r is the exposed root area of the tube, A_f is the fin area and η_f is the fin efficiency. The fin efficiency is an indication of the effectiveness of the fin.

Based on air side area A_o of a finned tube heat exchanger, the overall heat transfer coefficient U can be calculated using the following equations:

$$U = \left(\frac{A_o}{h_i A_i} + \frac{A_o \ln(r_o/r_i)}{2\pi L k} + R_{fo} + \frac{1}{h_o} \right)^{-1}$$

$$(32.4)$$

where, A_i is the tube inside area; h_i and h_o are the convection heat transfer coefficient of the fluid inside and outside; r_o, r_i and L are the outside radius, inside radius and length of the tube; k is the thermal conductivity of the tube material; R_{fo} – resistance caused by fouling, corrosion, etc.

U can be either obtained from the heat exchanger manufacturers or derived by equation 32.4. Since convection heat transfer coefficients h_i (inside fluid) and h_o (outside air) are velocity dependent, various empirical models have been developed to predict these two parameters and they are expressed in the form defined as Nusselt number (Churchill, 1997; Hausen, 1974; Kroger, 2004).

When quoting the value of overall heat transfer coefficient U given by the manufacturers, it is important to identify the reference area associated with the value.

If equation 32.1 is used to calculate the heat transfer of a heat exchanger, one needs to know all four terminal temperatures to obtain the logarithmic mean temperature (equation 32.2). In practical design of the geothermal power plant, outlet temperatures of the hot fluid and the outgoing temperature of the cooling air are not known and need to be found by iterative procedure based on the following heat balance equations.

Heat rejected from the hot fluid inside heat exchanger:

$$Q_w = m_w c_{pw} (T_{h1} - T_{h2})$$

$$(32.5)$$

Heat transferred by air outside heat exchanger:

$$Q_a = m_a c_{pa} \left(T_{c1} - T_{c2} \right) \qquad (32.6)$$

Heat transfer through heat exchanger can be calculated by equation 32.1.

With the heat balance, $Q = Q_w = Q_a$

The inlet temperatures of the hot fluid (T_{h1}) and ambient air (T_{c2}) are known for the geothermal power plant, which leaves three unknowns—the two exit temperatures (T_{h2}, T_{c1}) and the heat transfer rate (Q). These unknowns can be determined from the above three equations (32.1, 32.5, 32.6) through iterative procedure.

If one wants to avoid using the above iterative procedure for the problem solving, the effectiveness-NTU method can be used in this case without applying the iterative processing. The effectiveness of a heat exchanger, e, is defined as the ratio of the actual rate of heat transfer to the maximum possible rate of heat transfer. The maximum possible rate of heat transfer can be expressed as equations 32.7 and 32.8 depending on which of the heat capacity rates (product of the mass flow rate and specific heat—$m_w c_{pw}$ and $m_a c_{pa}$) is smaller.

$$Q_{max} = C_{min} \left(T_{h1} - T_{c2} \right) \qquad (32.7)$$

where, T_{h1} is inlet hot fluid temperature; T_{c2} is ambient air temperature; and C_{min} is the smaller heat capacity rate.

The actual heat transfer of the heat exchanger is:

$$Q = e Q_{max} \qquad (32.8)$$

Calculation of the effectiveness e can be found in various literatures (Kays and London, 1984; Kroger, 2004). The equations for calculation of a counterflow, indirect dry cooling as well as direct cooling are given below:

For counter-flow, indirect dry cooling system,

$$e = \frac{1 - \exp\left[-N\left(1 - C\right)\right]}{1 - C \exp\left[-N\left(1 - C\right)\right]} \qquad (32.9)$$

For ACC (direct cooling),

$$e = 1 - \exp\left(-N\right) \qquad (32.10)$$

where, $N = NTU = UA/C_{min}$; $C = C_{min}/C_{max}$; C_{min} is the smaller one between $m_a c_{pa}$ and $m_w c_{pw}$ and C_{max} is the larger one between them.

The first step is to calculate e by using equation 32.9 or 32.10. The heat transfer can be obtained by equation 32.8. Once Q has been calculated, the two outlet temperatures (T_{h2}, T_{c1}) can be derived using equations 32.5 and 32.6 without iterative processing.

The air side pressure drop of finned tube heat exchanger is also velocity dependent. The friction and contraction causes the pressure drop when air flows over the heat exchange

surface. Air-side fouling has a significant impact on air-side pressure drop (Ahn et al., 2003, Lankinenn et al., 2003). The pressure drop data can be either obtained from the heat exchanger manufacturers or derived from empirical models reported in various literatures.

32.3 MECHANICAL DRAFT DRY COOLING TOWER

Figure 32.14 shows a mechanical draft dry cooling system (GEA), which is commonly called ACC. In this cooling system, the exhaust turbine steam flows directly into the inside tube of heat exchanger through large pipe. The finned tube heat exchangers used are either elliptical or flattened tubes and heat exchange bundles are normally arranged in the form of an A-frame or delta to reduce the land area required. Fans are installed under the A-frame to force air flow through the heat exchanger. The exhaust steam condenses in the heat exchanger due to the cooling of air flowing through the outside surface of the heat exchanger. The amount of the air flow is controlled by fans.

Axial flow fans are normally used in mechanical draft dry cooling tower for geothermal power plants. Large low-noise fans have been developed for modern ACCs. Tens of fans are used to provide the required mass flow rate. Figure 32.15 shows the cooling tower with four fans used for the 1 MW pilot geothermal power plant by Geodynamics of Australia (Guan et al., 2011).

If mechanical draft is used in an indirect dry cooling tower, the turbine exhaust steam is directed and condensed in a condenser, where the cooling water is also directed to the same condenser to take heat away from the exhaust steam. This heat is carried by the water and is rejected in the cooling tower through air-cooled heat exchanger. In this case, the hot fluid flowing inside the heat exchanger is the cooling water flowing through the condenser. The air flow through the outside surface of the heat exchanger is by fans.

Design of mechanical draft dry cooling tower requires heat transfer and pressure drop calculations. The former is needed for sizing the heat exchangers and the latter is used for the fan selections.

The heat transfer of the ACC shown in Figure 32.14 can be expressed as,

$$Q = m_a c_{pa}(T_{ao} - T_{ai}) = m_c i_{fg} = e m_a c_{pa}(T_s - T_{ai}) \qquad (32.11)$$

$$e = 1 - \exp[UA/m_a c_{pa}] \qquad (32.12)$$

where, m_a is the air mass flow rate; T_{ao} is the air temperature exiting the ACC; T_{ai} is the air inlet temperature; $m_c\, i_{fg}$, and T_c are the mass flow rate, latent heat and the temperature of condensate, respectively.

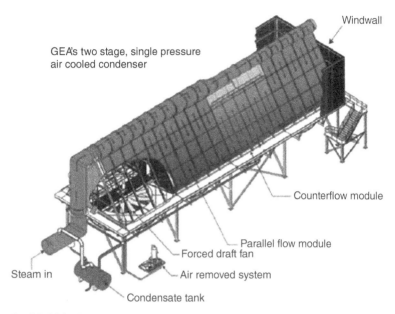

FIGURE 32.14   Mechanical draft, Air-cooled condenser (GEA heat exchanger).

FIGURE 32.15   Mechanical draft, dry cooling tower used in a 1 MW pilot geothermal power plant by Geodynamics of Australia.

For fan selection, draft equation is used and it has the following form:

$$\sum \text{pressure drop (tower + heat exchanger)}$$
$$= \text{Fan pressure (draft force)} \qquad (32.13)$$

The fans can be selected based on the air volume flow rate m_a/ρ and pressure drop to achieve the maximum fan efficiency.

When using mechanical draft dry cooling for geothermal power plants, a relatively larger proportion of power consumption will be required by the fans. It is estimated that about 5–8% of the net output will be needed for mechanical draft dry cooling system in low temperature geothermal power plants.

32.4 NATURAL DRAFT DRY COOLING TOWER

In a natural draft dry cooling tower, no fans are required. The flow of air through the bundles of heat exchangers is by means of buoyancy effects. Buoyancy occurs due to a difference in air density between the inside and outside of the tower resulting from the temperature difference. The greater the temperature difference and the height of the tower structure, the greater the buoyancy force. Therefore, the volume flow rate of air across the heat exchanger bundle is directly proportional to the inside–outside temperature difference and the height of the cooling tower. Large natural draft cooling towers up to 200 m tall have been built for thermal power plant (Busch et al., 2002).

Figure 32.16 shows natural draft cooling towers for both wet and dry cooling in a coal-fired power plant in China. The tower structure has parabolic shape and is made of reinforced concrete (RC). Even though the majorities of the natural draft cooling tower structure were made of RC in the past, the cooling tower was also made of wood (Becher, 2006), steel (Kollar, 1985), rope (Erdmann, 1991) and fiberglass (Wang, 2011). However, large RC tower is more economical for large power stations.

The heat exchangers in a natural draft dry cooling tower can be arranged vertically around the inlet of the tower (Figure 32.17) or horizontally inside the tower at the height of the tower inlet (Figure 32.18). Horizontal arrangement is preferred in geothermal power plants since it reduces the negative effects caused by the soil and plant pollen dusts.

Using natural draft dry cooling tower in a geothermal power plant saves the cost of fans, the power consumed by the fans, and the maintenance associated with the fans. However, the capital cost of constructing the tower structure is relatively high.

The capital cost of tower structure includes materials cost, labor cost (up to 4000 people were employed at peak times) and the equipment cost such as crane hiring. Since most

of natural draft cooling towers previously built were made of RC, the erection of RC tower shell must be with high accuracy in the shape and wall thickness to have designed bending and buckling strength. This requires advanced construction methods and skilled work force. The RC towers are heavy which increases the costs of material transportation, especially to the remote geothermal power sites.

Computational fluid dynamics (CFD) simulation and finite element modeling (FEM) have been used in the tower structure design to optimize the tower structure (Busch et al., 2002, Al-Waked and Behnia, 2004, Arora and Saha, 2011).

The performance of a natural draft dry cooling tower is influenced by the characteristics of heat exchanger, the tower geometry and ambient conditions. Natural draft dry cooling tower must meet the heat balance and draft equation at the specified ambient conditions.

For an indirect dry cooling tower, the heat transfers by air, water, and heat exchanger are expressed in the following equations:

$$Q = m_a c_{pa}(T_{ao} - T_{ai}) = m_w c_{pw}(T_{wi} - T_{wo}) = UAF_T \Delta T_{lm} \qquad (32.14)$$

$$\Delta T_{lm} = \frac{(T_{wo} - T_{ai}) - (T_{wi} - T_{ao})}{\ln[(T_{wo} - T_{ai})/(T_{wi} - T_{ao})]} \qquad (32.15)$$

where, T_{wi} and T_{wo} are the inlet and outlet temperatures of the water.

The draft equation for a natural draft cooling tower takes the following form:

$$\sum \textit{flow resistance} \, (\textit{tower} + \textit{heat exchanger})$$
$$= \textit{natural draft force of the tower}$$

$$\cong (\rho_{ao} - \rho_{ai}) gH \qquad (32.16)$$

In addition to the three unknowns of T_{ao}, T_{wo}, and Q need to be solved by the equations, there is an extra unknown (m_a) also need to be determined. Therefore, iterative processing has to be applied in natural draft dry cooling tower design.

32.5 DEVELOPMENTS OF DRY COOLING TECHNOLOGIES FOR GEOTHERMAL POWER PLANTS

Ambient condition has significant effect on the performance of dry cooling towers. Most geothermal power plants, especially the geothermal power plants using EGS technology, have unique ambient conditions and applications. The following technologies are specially developed for geothermal

FIGURE 32.16   Natural draft cooling tower used for both dry and wet cooling in China.

FIGURE 32.17 Natural draft cooling tower with vertical arranged heat exchangers.

FIGURE 32.18 Natural draft cooling tower with horizontal arranged heat exchangers.

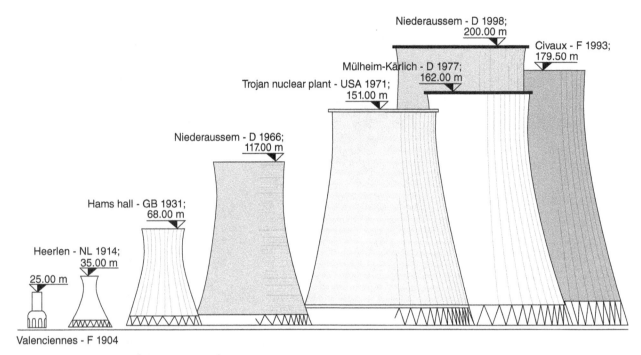

Niederaussem - D 1998;
200.00 m

Civaux - F 1993;
179.50 m

Mülheim-Kärlich - D 1977;
162.00 m

Trojan nuclear plant - USA 1971;
151.00 m

Niederaussem - D 1966;
117.00 m

Hams hall - GB 1931;
68.00 m

Heerlen - NL 1914;
35.00 m

25.00 m

Valenciennes - F 1904

FIGURE 32.19   Natural draft cooling tower height in historical perspective (Gould, 2012).

power plants by the Queensland Geothermal Energy Centre of Excellence (QGECE).

32.5.1   Development of Small Natural Draft Dry Cooling Towers

Building small EGS power plants for Australian remote community is probably more cost effective since no expensive national power transmission lines are required. Therefore, it is important to develop high performance, but small enough natural draft dry cooling towers (NDDCTs) for these plants located in remote areas. Researches on natural draft cooling towers in recent decades are all focused on large towers with tower height of more than 100 m. The building of new towers are historically becoming bigger and bigger (Gould, 2012) as shown in Figure 32.19. Building larger towers needs skilled work force and is very expensive.

Design methods for large NDDCTs do not include crosswind effect, which plays a negative role in the performance of large cooling towers (Gao and Sun, 2007; Wei et al., 1995). This negative effect of crosswind will be much more significant on the performance of small NDDCTs and may even prevent small towers from functioning.

Queensland Geothermal Energy Centre of Excellence (QGECE) development of small NDDCTs are divided into two steps: the first step is to derive smallest sizes of NDD-CTs based on the design methods of large NDDCTs; the second step is to identify the effect of the crosswind on the performance of small tower with 3D CFD simulation, and improve performance with crosswind by introducing windbreak walls.

The 3D simulation shows that the performance of small NDDCT is significantly reduced by the crosswind which especially affects the inlet air flow as shown in Figure 32.20. The heat rejecting capacity of the tower reduces from about 580 kW at no crosswind to 200 kW at the crosswind speed of 10 m/s – 65% reduction as indicated in Figure 32.21. This demonstrates that the design method without taking crosswind into account for small NDDCT is not acceptable.

Three simple but very effective windbreak walls have been introduced to improve the performance of small NDDCTs under crosswind. These walls are used to divert crosswind flow through the heat exchangers to improve the performance of the tower. The walls are installed in the bottom of the tower with 120 degree separation. All three walls start at the center of the tower and end at the outside perimeter of the tower with the same height as the bottom level of the heat exchangers. When there is no crosswind, the cooling air enters into the tower freely without any obstruction from the walls. If crosswind exists, the walls stop the crosswind flowing across the bottom, change the direction of the crosswind, and force its flow through heat exchanger. Since there is more air flow through the heat exchanger, it improves the performance of the tower.

After introducing three windbreak walls, new 3D simulation was carried out for the same small tower with these windbreak walls. The results show that the performance of the

FIGURE 32.20   Crosswind effect on air flow (8 m/s crosswind).

same size of tower with windbreak walls has been increased from 580 kW at no crosswind to about 800 kW with the crosswind speed of 10 m/s – nearly 40% increase as shown in Figure 32.22.

32.5.2   Solar Enhanced Natural Draft Dry Cooling Tower

The design concept of solar enhanced natural draft dry cooling tower is shown in Figure 32.23. The solar enhanced cooling system includes a natural draft tower, solar collectors, and heat exchangers. The solar collectors are arranged radically at the base of the tower, and the heat exchangers are placed vertically at the outside edge of the solar collectors. As the Sun heats the sun roof, the air under the sun roof is heated. Warm air naturally rises through the tower and fresh air is sucked in, thus providing a cooling air flow through the heat

exchanger bundles. By this arrangement, it would enhance the performance of a natural draft dry cooling tower due to the added solar energy to the air stream after the heat exchanger. The extra heat from the solar collectors increases the buoyancy of the air inside the tower and helps to drive more air through the heat exchangers.

The system exploits the solar energy, which is abundant in geothermal site, during the hottest periods at which the conventional dry cooling tower would suffer the lowest performance.

Modeling has been done to predict the performance of NDDCT with solar enhanced (Zou et al., 2012). The first performance indicator is measured by the height of tower required for both solar hybrid system and the system without solar assistance when rejecting the same amount of heat with the same heat exchanger area. A shorter tower height

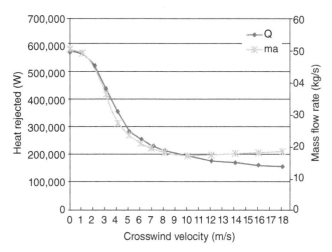

FIGURE 32.21   Heat and mass transfer without windbreak wall.

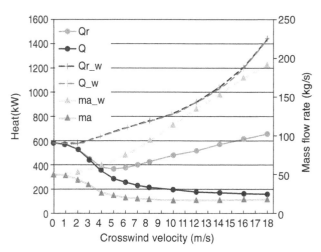

FIGURE 32.22   Heat and mass transfer with windbreak walls (ma_w, Qr_w, Q_w).

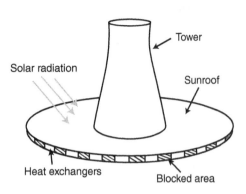

FIGURE 32.23 Concepts of solar enhanced natural draft dry cooling tower.

indicates better performance of the system. Figure 32.24 shows a comparison of tower height required between the solar enhanced system with three different dimensions of solar collector and the conventional natural NDDCT at different heat rejection rates.

Figure 32.24 indicates that, under the same number of heat exchanger bundles installed in the system, the height of tower required in solar hybrid design is lower than its counterpart in NDDCT design. The difference becomes larger as heat rejecting rate increases. If the same height of cooling tower is used, solar enhanced natural draft cooling tower can reject more heat than a conventional natural draft cooling tower, which means more net power can be generated.

The second performance indicator to show the effectiveness of solar enhanced natural draft cooling tower is to compare the heat exchanger areas required while keeping the same tower height. Less heat exchanger area indicates better performance of the cooling system. Zou et al. (2012) reported that the proposed solar enhanced system requires less finned tube bundles than the cooling system without solar enhanced. More than 10% saving of finned tube bundles can be achieved with the solar enhanced system.

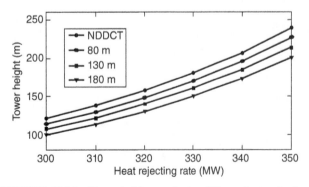

FIGURE 32.24 Tower height required at different heat rejecting rates.

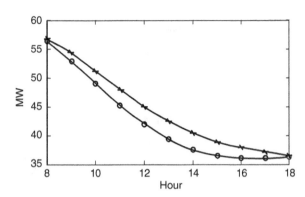

FIGURE 32.25 Net power generation comparison between the solar enhanced system (+) and the NDDCT (o) at a day time.

If the same tower height and heat exchanger area are used for both the solar enhanced natural draft cooling tower and the conventional natural draft cooling tower, solar enhanced cooling system is able to generate more power as shown in Figure 32.25.

32.5.3 Development of Steel Cooling Tower

The erection of RC tower shell requires advanced construction methods and skilled work force and it is also time-consuming process. The heavy RC structure increases the costs of material transportation, especially to the remote geothermal power sites. There is a need to develop a natural draft cooling tower with simplified field erection suitable for the remote sites. Steel cooling tower offers the opportunities. The benefits of using steel cooling tower include the less material transportation, short construction time, and less skilled worked required. It is expected that individual member of the steel cooling tower can be factory made and installed in site quickly.

QGECE develops a steel cooling tower structure for a geothermal power plant with heat rejection of 150 MW. The structure frame of the tower in this design is shown in Figure 32.26. The frame is made of tubular members and each individual member is joined together by welding. The tower has a cylindrical shape at the top half and a truncated cone shape at the bottom half. Twenty four longitude tubular members are connected by 11 latitude tubular members. In the outside of the frame, aluminium cladding (not showing in the figure) is attached to form a closed space. The total weight of the steel tower is about 1/15 of the equivalent RC tower.

32.5.4 Inlet Air Pre-cooling Natural Draft Dry Cooling Tower

This technology uses small amount water to enhance the cooling performance of NDDCTs during periods of high

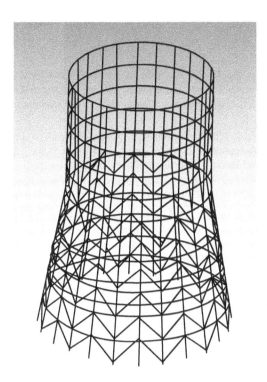

FIGURE 32.26   Steel cooling tower for geothermal power plants.

ambient temperatures. In this system, water is introduced into the inlet air stream of a dry cooling tower. The water evaporates and reduces the entering air dry-bulb temperature theoretically to its wet bulb temperature. The cooler air then cools the dry system more efficiently. Inlet air cooling has been effectively used in gas turbine power plants to increase power plant output (Ibrahim et al., 2011). Power plant applications with inlet air cooling have been studied by EPRI (EPRI reports 2003).

Previous researchers have found that inlet air cooling can provide cost-effective alternatives for power plants. Based on the test results obtained from both the laboratory and field, the EPRI (2003) concluded that the allocation of cooling water in modest quantities increased the power output by 50% or more during the hottest period. Ashwood and Bharathan (2011) conducted modeling analysis using ASPEN Plus to evaluate various inlet air cooling methods for low-temperature geothermal power production and reported that the net power generated during the hottest hours of the day with water spray can be significantly higher than that without water-assistance.

The disadvantages of using inlet air pre-cooling are associated with the service life and maintenance cost of heat exchanger. Frequent use of water spray in a natural draft dry cooling tower can cause corrosion, scaling, and fouling on the heat exchanger bundles, if water droplets are carried by the air stream to the heat exchanger bundles. To avoid this, a system is required to evaporate all water sprayed in the air stream to prevent water droplet contact with heat exchanger

surface. Therefore, special spray nozzles may be required to generate small size of water droplet. Water droplet sizes play an important role in achieving water full evaporation. Small droplet size can evaporate quickly but normally needs special nozzles for water atomization at higher cost. Water quality affects the performance of the nozzles and its maintenance cost. Optimizing the selection of nozzle based on the droplet size and water quality offers greatest challenge for cost-effective design of this system.

The study of QGECE on inlet air pre-cooling is to identify spray nozzles to achieve full evaporation and optimize a spray cooling system which can provide a cost-effective solution for full scale renewable power plants using natural draft dry cooling systems. To achieve this goal, intensive experimental studies have been carried out with the wind tunnel shown in Figure 32.27.

Spray systems with various flow rates and droplet sizes have been tested in this wind tunnel at the controlled condition to characterize their evaporation rate and cooling effect. Wind tunnel provides controlled air temperature, humidity, and uniform speed for the spray system.

The wind tunnel is an open circuit tunnel. The tunnel can operate at the maximum wind speed up to 75 m/sec with the fan motor power of 75 kW. The wind tunnel consists of a centrifugal fan (Blower) to draw air, a diffuser section (with 3 screens) after the fan, a setting chamber (with a honeycomb and 4 screens), contractions (three sizes for various air speeds), a transparent working section, an exhaust air scrubber (exit diffuser), and an exhaust fan. The exhaust scrubber and the exhaust fan are located outside the shed while the rests of the tunnel are under the cover of the shed.

A Phase Doppler Particle Analyzers (PDPA) is used for the spray nozzle characterization. The other transducers include those to monitor fan speed, air temperature, humidity, velocity, spray liquid temperature, pressure, and flow rate.

A natural draft dry cooling tower (20 m height) testing facility has been built by QGECE in the Gatton campus of the University of Queensland, Australia for developing cooling technologies for renewable power plants as shown in

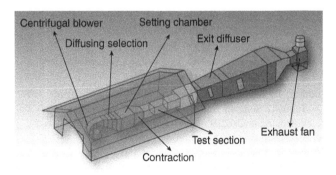

FIGURE 32.27   Wind tunnel in Gatton campus of the University Queensland.

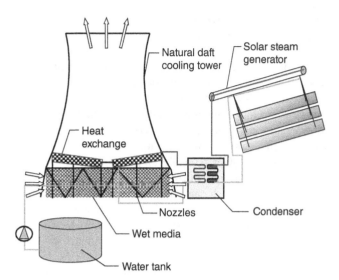

FIGURE 32.28   QGECE cooling research station.

Figure 32.28. The results from the study will be directly used for the design of renewable power plants using inlet air pre-cooling natural draft dry cooling tower.

REFERENCES

Ronald D. (2005). *Geothermal Power Plants: Principles, Applications, Case Studies The Boulevard, Langford Lane, Kidlington,* Oxford OX5 1GB, UK, Elsevier Advanced Technology.

DiPippo R. (2008). *Geothermal Power Plants, Principles, Applications, Case Studies and Environmental Impact,* 2nd edition, Elsevier Ltd.

Kroger D. G. (2004). *Air-Cooled Heat Exchangers and Cooling Towers,* Tulsa, OK, Pennwell Corp.

Williams C. R., Rasul M. G. (2008). Feasibility of a Hybrid Cooling System in a Thermal Power Plant, 3rd IASME/WSEAS International Conference on Energy & Environment, University of Cambridge, UK, February 23–25.

NETL, (2009) Estimating Freshwater Needs to Meet Future Thermoelectric Generation Requirements, Update. http://www.netl. doe.gov/energyanalyses/pubs/2009%20Water%20Needs%20 Analysis%20-%20Final%20(9-30-2009).pdf

Gurgenci H. (2010). AGEC Paper on New Options for Dry Cooling Towers. http://www.uq.edu.au/geothermal/index.html? page=151582. 12 November

Barker B. (2007). Running dry at the power plant, EPRT Journal, Summer. http://www.circleofblue.org/waternews/wp-content/uploads/2010/08/EPRI_Running-Dry-at-the-Power-Plant.pdf.

Barigozzi G., Perdichizzi A., Ravelli S. (2011). Wet and dry cooling systems optimization applied to a modern waste-to-energy cogeneration heat and power plant, *Applied Energy,* 88:1366–1376.

Electric Power Research Institute (EPRI). (2003). Spray Enhancement of Air Cooled Condensers, September 2003.

Yang L., Tan H., Du X., Yang Y. (2012). Thermal-flow characteristics of the new wave-finned flat tube bundles in air-cooled condensers, *International Journal of Thermal Sciences,* 53:166–174.

Heyns J. A. (2008). Performance characteristics of an air-cooled steam condenser incorporating a hybrid (dry/wet) dephlegmator, Master thesis, University of Stellenbosch.

Kakac S., Liu H. (2002). *Heat Exchangers Selecting, Rating, and Thermal Design.* CRC Press.

Churchill S. W. (1977). Comprehensive correlating equations for heat, mass and momentum transfer in fully developed flow in smooth tubes, *Industrial & Engineering Chemistry Fundamentals,* 16(1):109–116.

Hausen H. (1974). Extended equation for heat transfer in tubes at turbulent flow, *Waerme- und Stoffuebertragung,* 7(4):222–225.

Kays W. M., London A. L. (1984). *Compact Heat Exchangers,* New York, McGraw Hill Book Co.

Ahn Y. C., Cho J. M., Dhin H. S., Hwang Y. J., Lee C. G., Lee J. K., Lee H. U., Kang T. W. (2003). An experimental study of the air-side particulate fouling in fin-and-tube heat exchangers of air conditioners, *The Korean Journal of Chemical Engineering,* 20(5):873–877.

Lankinenn R., Suihkonen J., Sarkomaa P. (2003). The effect of air side fouling on thermal-hydraulic characteristics of a compact heat exchanger, *International Journal of Energy Research,* 27:349–361.

GEA, Heat Exchangers | GEA Aircooled Systems. http://www.gea-energytechnology.com/opencms/opencms/gas/en/products/Direct_Air-Cooled_Condensers.html

Guan Z., Pluschke A., Roe T., James K., Gurgenci H. (2011). Dust monitoring for air-cooled heat exchangers in Innamincka geothermal power site. Proceeding of the 15th IAHR Cooling Tower and Air-cooled Heat Exchanger conference, Beijing, China, (pp. 328–334). 23–26 October.

Busch D., Harte R., Kratzig W. B., Montag U. (2002). New natural draft cooling tower of 200 m of height, *Engineering Structures,* 24:1509–1521.

Becher (2006). *Cooling Towers,* The MIT Press, London.

Kollar L. (1985). Large reticulated steel cooling towers, *Engineering Structures,* 7:263–267.

Erdmann W. (1991). Experiences with the first dry cooling tower for the THTR power plant at Schmehausen, FRG, *Construction & Building Materials,* 5(2):68–74.

Wang S. (2011). Steel Frame Mounting Panel Hyperbolic Natural Ventilation Application of Cooling Tower. Proceeding of the 15th IAHR Cooling Tower and Air-cooled Heat Exchanger conference, Beijing, China, (pp. 262–266). 23–26 October 2011.

Al-Waked R., Behnia M. (2004). The performance of natural draft dry cooling towers under crosswind: CFD study. *International Journal of Energy Research,* 28(2):147–161.

Arora P., Saha A. K., (2011). Three-dimensional numerical study of flow and species transport in an elevated jet in crossflow. *International journal of heat and mass transfer*, 54(1–3):92–105.

Gould P. L. (2012). The influence of R&D on the design, construction and damage assessment of large cooling towers, 6th International symposium on cooling towers, Cologne.

Gao M., Sun F. Z. (2007). *Research on the Effect of Cross-wind to Temperature Difference and Efficiency of Natural Draft Counter flow Wet Cooling Tower*, Springer Berlin Heidelberg: Berlin, Heidelberg, pp. 513–517.

Wei Q. D., Zhang B. Y., Liu K. Q., Du X. D., Meng X. Z. (1995). A study of unfavourable effects of wind on the cooling efficiency of dry cooling-towers. *Journal of wind engineering and industrial aerodynamics*, 54:633–643.

Zou Z., Guan Z., Gurgenci H., Lu Y. (2012). Solar Enhanced Natural Draft Dry Cooling Tower for Geothermal Power Applications, Solar Energy, DOI:10.1016/j.solener.2012.06.003.

Ibrahim T. K., Rahman M. M., Abdalla A. N. (2011). Improvement of gas turbine performance based on inlet air cooling systems: A technical review, *International Journal of Physical Sciences*, 6(4):620–627.

Ashwood A., Bharathan D. (2011). Hybrid Cooling Systems for Low-Temperature Geothermal Power Production, NREL technical report.

33

THERMAL STORAGE

Marc A. Rosen

33.1 THERMAL ENERGY STORAGE

Thermal energy storage (TES) or, more commonly, thermal storage, is the storage of thermal energy (heat or cold) for a period of time using a storage material. Thermal energy may be stored by elevating or lowering the temperature of a substance (i.e., altering its sensible heat), by changing the phase of a substance (i.e., altering its latent heat), and by causing a substance to undergo endothermic and exothermic chemical reactions. Consequently, there exist three main types of thermal storage: sensible, latent, or thermochemical. Energy is stored or discharged by changing the temperature of the storage material in sensible storage, changing the phase of the storage material in latent storage, and changing the chemical form of the storage material in thermochemical storage. Typical storage materials for sensible storages include water, rock, and soil; for latent storage include water/ice and salt hydrates; and for thermochemical storage include various reacting pairs of chemicals, with one sometimes being water. Aspects of TES in aquifers, boreholes, phase change materials (PCMs), and thermochemical reactions are covered in many sources (Dincer and Rosen, 2011; Paksoy, 2007).

Thermal storage has many practical applications. Thermal storage helps offset the mismatch between periods when thermal energy (heat or cold) is available and in demand. TES offers the possibility of storing thermal energy for later use in its original form or in conversion to electricity or other energy products. TES has a wide variety of uses, most of which involve heating and cooling applications. Examples of TES are the storage of solar energy for overnight heating, of summer heat for winter use, of winter ice for space cooling in summer, and of the heat or cool generated electrically during off-peak hours for use during subsequent peak

demand hours. Space heating using electric TES has been used extensively. District heating and cooling systems often also incorporate TES and can benefit from its careful integration into the overall system.

A thermal storage system generally consists of a storage medium, a container, and equipment for injecting and recovering thermal energy. The container retains the storage material and prevents losses of thermal energy.

The selection of a TES system mainly depends on such factors as the application, the storage period required, economics, and operating conditions. In some instances, it is advantageous to use sensible storage and in others latent storage. Thermochemical energy storage is relatively new and in development, and thus at present does not have many applications.

Significant research has been undertaken on the TES in the last couple of decades. Extensive research on TES has been undertaken through the Energy Conservation through Energy Storage Implementing Agreement (ECESIA) of the International Energy Agency (www.iea-eces.org) as well as via other research programs. Many investigations and advances in thermal storage are covered at relevant points throughout this chapter.

33.2 SENSIBLE THERMAL STORAGE

Sensible TES systems undergo changes in sensible heat, which are associated with temperature change. Some of the desirable characteristics of a sensible heat storage medium are high specific heat capacity, long-term stability under charging/discharging cycles, compatibility with its containment, and, often most importantly, low cost. Sensible heat

Alternative Energy and Shale Gas Encyclopedia, First Edition. Edited by Jay H. Lehr and Jack Keeley.
© 2016 John Wiley & Sons, Inc. Published 2016 by John Wiley & Sons, Inc.

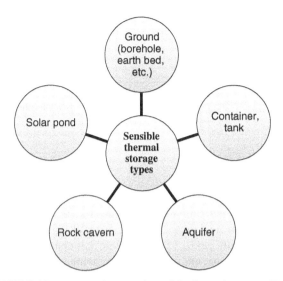

FIGURE 33.1   Several types of sensible thermal storage. Variations of each type exist.

storage may be classified on the basis of the heat storage media as liquid (water, oil, etc.) and solid (rocks, metals, etc.).

Some of the more common types of sensible thermal storage are summarized in Figure 33.1 and are described below (Ataer, 2006; Beckman and Gilli, 1984; Dincer and Rosen, 2011; Dinter et al., 1991; IEA, 2010):

- *Ground*: Heat or cold can be transferred into underground soil for storage and subsequent recovery. Some ground-based storages are shallow (e.g., Earth beds) while others are deep (e.g., borehole systems). The latter consist of a network of tubes inserted into boreholes drilled into the ground, allowing heat or cold to be transferred into underground soil and rock for storage and subsequent recovery. The use of earth as a TES medium is often restricted to new construction, since the application requires installations in the ground, beneath a structure making retrofit work difficult.

- *Container/tank*: Containers and tanks filled with a heat storage medium, such as water or rock, can act as a TES. Such tanks are often made of steel or concrete because of their physical characteristics, cost, availability, and easy processing. Ceramic bricks can also act as a good heat storage medium, especially for uses in new and old buildings, where they are advantageous due to their modular sizes, ease of installation, and high heat-retention abilities.

- *Aquifer*: An aquifer is a groundwater reservoir, in which the water is located in impermeable materials such as clay or rock and moves very slowly. An aquifer TES is typically a permeable, water-bearing rock formation. Aquifers often have large volumes, exceeding millions

of cubic meters, and consist of about 25% water. In aquifer TES, water from the aquifer is extracted and heated or cooled. It is then reinjected at another point in the aquifer for storage and subsequent recovery. With an aquifer system, therefore, two well fields are often tapped: one for cold storage and the other for heat. Aquifer stores are most suited to high-capacity systems. External thermal energy is stored in some aquifer TES systems, while the natural groundwater temperatures are used in others.

- *Rock cavern*: A rock cavern can be filled with a storage medium and used as a TES. The storage medium in such systems depends on the ability of the cavern to hold it. Such TES systems are usually large.

- *Solar pond*: A salinity gradient solar pond is an integrated collection and storage device of solar energy. In an ordinary pond, the sun's rays heat the water which, being less dense, rises to the surface and loses heat to the atmosphere. A solar pond inhibits this phenomenon by dissolving salt into the bottom layer of the pond, making it too dense to rise to the surface, even when hot. The salt concentration increases with depth, forming a salinity gradient. Sunlight which reaches the bottom of the pond remains trapped there as thermal energy. Useful thermal energy is recovered as hot brine.

Not all of the above types of TES are in-ground systems. Tank-based TES can be located in or above ground. Note also that both hot TES and CTES can utilize the above TES categories.

Selected TES system characteristics are compared in Table 33.1 for several types of TES. The sizes range to as large as 500,000 m^3, while the efficiencies vary, for short-term applications, from approximately 60–90%. The specific thermal capacity is the storage capacity per unit volume and per unit temperature rise and allows comparisons of storage capacity for TESs of the same size and same temperature increase.

Sensible heat storage systems commonly use rocks or water as the storage medium due to their availability, cost, ease of use, and thermal characteristics. The high heat capacity of water often makes water tanks a logical choice for TES systems that operate in a temperature range needed for building heating or cooling. A typical application of sensible heat storage is in the solar water heater storage tank, where solar energy is stored in the water during periods of sunshine and can be used during nights or subsequent mornings. The relatively low heat capacity of rocks and ceramics is somewhat offset by the large temperature changes possible with these materials and their relatively high density. Hasnain et al. (1996) identify cast iron as a significant solid material for sensible heat storage, noting that its energy density exceeds that for water storage. However, cast iron is more expensive

TABLE 33.1   Selected characteristics for several types of sensible thermal storage.

	Tank				Ground	
Characteristic	Steel	Concrete	Rock cavern	Aquifer	Earth bed	Drilled boreholes
Specific thermal capacity (kJ/m^3 K)	4180	4180	4180	2700	2520	2270
Volume (m^3)^a	0–100,000	0–100,000	50,000–300,000	50,000–500,000	0–100,000	50,000–400,000
Energy efficiency (%)^b	90	90	80	75	60	70

Data sources: SFEO (1990) and Piette (1990).
^{a}Volumes shown are given in terms of typical ranges of sizes.
^{b}Energy efficiencies shown are typical values for short-term storage applications.

than stone or brick and has a relatively longer payback period. Therefore, pebble beds or rock piles are preferred choice for sensible storage. Sensible thermal storage materials have been investigated in terms of thermophysical and economic characteristics (Fernandez et al., 2010). Packed bed TES integrated with solar air heaters was reviewed by Singh et al. (2010). They report considerable progress on design, void fraction, storage material, heat transfer enhancement, air flow pattern, and pressure drop in packed bed TES. The materials used in most investigations were rocks and/or pebbles. In most studies small- to medium-size packed beds were considered although in a few studies large packed bed thermal storages were examined.

Sensible TES can be used for a wide range of storage durations. Short-term (daily to weekly) systems often use crushed rock beds, earth beds, water tanks, ceramic brick, and building mass. Long-term (as long as seasonal) TESs often utilize rock beds, earth beds, and water tanks. Aquifer TES can be used for storage periods ranging from daily to seasonal.

Some sensible TES systems are designed to take advantage of thermal stratification, the existence of a temperature gradient across the storage. Thermal stratification is desirable as it avoids loss of temperature due to mixing and is simpler to achieve in solid storage media than in fluids.

33.3   LATENT THERMAL STORAGE

Latent heat changes are the heat interactions associated with a phase change of a material, usually from a solid to a liquid and back. Latent heat TES systems store thermal energy as a material changes phase at constant temperature. The latent heat change during phase change is usually much higher than the sensible heat change for a given medium. Thus, latent TES has a high energy storage density. Latent heat TES systems also incorporate a storage containment and heat exchange capability for transferring thermal energy into and out of the storage.

Latent heat TES systems utilize a variety of materials as storage media. The storage medium is usually selected so that it undergoes a phase transition within the desired

operating temperature range. Latent storage media often include PCMs, including inorganic materials, organic materials, large fatty acids, and aromatics. Salt compounds that absorb a large amount of heat during melting (e.g., eutectic salts, salt hydrates, Glauber's salt) are useful for latent TES. Paraffin waxes are another common phase change material since they exhibit high stability and little degradation over repeated cycles of latent TES operation.

The phase change material in a latent heat TES can be contained in a single large vessel or in rods or plastic envelopes. The small modules in the latter case, and the small number of modules required for storage, make latent TES especially convenient.

Latent heat TES can be used to store thermal energy at hot or cold conditions. CTES often utilizes latent heat changes, using such storage media as water/ice, eutectic salts, glycol, brine, and ice slurry.

In a review article on latent TES, Jegadheeswaran and Pohekar (2009) summarized various techniques suggested in the literature to improve the heat transfer rates in PCMs, so as to facilitate faster charging and discharging of thermal energy. Some of the suggested measures include use of extended surfaces, multiple PCMs, thermal conductivity improvements, and micro-encapsulation PCMs. Also, they concluded that the use of exergy methods may be an important area of future research for investigating PCMs.

Recent work has shown that PCMs provide much greater energy storage with smaller temperature fluctuations compared to sensible heat storage and offer a potentially superior alternative in solar walls (Farid et al., 2004; Farid and Kong, 2001; Zalba et al., 2004). Khalifa and Abbas (2009) investigated thermal storage walls with three PCMs (concrete, hydrate salt ($CaCl_2 \cdot 6H_2O$), paraffin wax (*N*-eicosane)) and their ability to maintain comfortable temperature in a living space. They found wall thickness and heat storage media to be significant parameters for controlling temperature. It was concluded that a storage wall of hydrate salt (thickness 0.08 m) can maintain the temperature within 18–22°C.

Investigations are ongoing to enhance PCM thermal conductivity, including efforts to develop appropriate material combinations, use of metal matrixes and high-conductivity particles in materials, improved fin configurations, bubble agitation, and micro-encapsulation (Agyenim et al., 2010; Alkan and Sari, 2008; Bayés-García et al., 2010; Cabeza et al., 2007; Fan and Khodadadi, 2011; Koca et al., 2008; Lai et al., 2010; Sari and Karaipekli, 2009; Sharma et al., 2009; Shilei et al., 2007). Carbon fiber chips and carbon brushes have been shown to have potential as additives to improve PCM heat transfer rates (Hamada et al., 2003), as has the use of paraffin with lessing rings (Velraj et al., 1999). To avoid the reduced heat transfer rate from PCM solidification around the heat exchanger tubes, a coated shell and tube heat exchanger using a specific salt mixture composition was found to be effective (Mathur et al., 2010).

In solar combi-systems for space heating and cooling, large solar factions normally require larger water volumes. Alternative solar thermal storage technologies such as PCMs are promising alternatives to large water volumes. Using dynamic system simulation, Heinz and Schranzhofer (2010) analyzed the potential of PCM relative to water storage and concluded that (1) for a small tank volume with a low solar fraction ratio, PCMs are competitive with sensible storage and (2) for seasonal storage, PCMs may be viable and even superior to traditional heat storage methods.

33.4 THERMOCHEMICAL THERMAL STORAGE

Thermochemical storage utilizes chemical reactions to store and release heat. They are generally much more compact than latent and sensible storage units, which are especially beneficial where space is limited. Thermochemical TES systems are not yet commercial. They have been considered for several decades, with Wettermark (1989) having comprehensively reviewed thermochemical energy storage, its main concepts, and criteria for choosing appropriate storage media and candidate reaction pairs. But more scientific research and development is required to better understand and design thermochemical storage technologies and to resolve other practical aspects before commercial implementation

can occur (Bales, 2006; Hauer and Lavemann, 2007; IEA, 2008; Masruroh et al., 2006; Zondag et al., 2008). In particular, appropriate material pairs are required. The design of thermochemical energy storage systems is complex and requires appropriate consideration of many factors. This category of thermal storage includes sorption and thermochemical reactions. In thermochemical energy storage, energy is stored after a dissociation reaction and then recovered in a chemically reverse reaction. Thermochemical storage is particularly appropriate for long-term storage applications, for example, seasonal storage of solar heat, because the process involves little energy loss during the storing period, which is usually at ambient temperature. In sorption systems (adsorption and absorption), adsorption occurs when an adsorptive accumulates on the surface of an adsorbent and forms a molecular or atomic layer. The adsorptive can be a liquid or gas while the adsorbent can be a solid or liquid. Absorption is a process that occurs when a substance is distributed into a liquid or solid and forms a solution.

Thermochemical TES is based on a chemical reaction that can be reversed:

$$C + heat \leftrightarrows A + B \tag{33.1}$$

Thermochemical material (C) absorbs energy and is converted chemically into two components (A and B), which can be stored separately. The reverse reaction occurs when A and B are combined together and C is formed, releasing the thermal energy that is recovered from the TES. The storage capacity of this system is the heat of reaction when C is formed. Substance C is the thermochemical material for the reaction and can be a hydroxide, hydrate, carbonate, and ammoniate, while A and B are reactants, which can be water, CO, ammonia, and hydrogen. Usually C is a solid or a liquid and A and B can be any phase.

Several thermochemical storage material candidates are listed in Table 33.2. They are based on water and carbon dioxide working fluids. Also listed in Table 33.2 are energy

TABLE 33.2   Selected potential storage materials for thermochemical energy storage.

Materials			Thermal characteristics	
Thermochemical storage material (C)	Solid reactant (A)	Working fluid (B)	Energy storage density (GJ/m³)	Charging reaction temperature (°C)
Water-based reactions				
$MgSO_4 \cdot 7H_2O$	$MgSO_4$	H_2O	2.8	122
$Ca(OH)_2$	CaO	H_2O	1.9	479
$Fe(OH)_2$	FeO	H_2O	2.2	150
$CaSO_4 \cdot 2H_2O$	$CaSO_4$	H_2O	1.4	89
CO_2-based reactions				
$FeCO_3$	FeO	CO_2	2.6	180
$CaCO_3$	CaO	CO_2	3.3	837

Sources: Visscher and Veldhuid (2005) and Wettermark (1989).

TABLE 33.3 Main processes in thermochemical energy storage.

Process	Reaction	Thermal nature of reaction
Charging	$C + heat \rightarrow A + B$	Endothermic
Discharging	$A + B \rightarrow C + heat$	Exothermic

densities and reaction temperatures for the thermochemical materials. These are important thermal factors that strongly affect their application in thermochemical TES systems.

The three main processes (charging, storing, discharging) of a general TES cycle are illustrated for thermochemical energy storage in Table 33.3 and described individually below:

- *Charging*: During the endothermic charging process (equation 33.1 in forward direction), thermal energy is absorbed from an energy resource, causing the dissociation of the thermochemical material (C) into two materials (A and B).
- *Storing*: Substances A and B are stored separately, usually at ambient temperatures. Material degradation can lead to some energy loss, but there is little energy loss except during the initial cooling of components A and B after charging.
- *Discharging*: Substances A and B combine exothermically during discharging (equation 33.1 in backward direction), allowing the stored energy to be recovered and C to be regenerated for reuse in the cycle.

Several investigations have recently been reported of thermochemical TES. Exergy analysis has recently been applied to thermochemical storage (Haji Abedin and Rosen, 2010a, 2010b). An examination of the potentials of $MgCl_2 \cdot 6H_2O$, $CaCl_2 \cdot 2H_2O$, $MgSO_4 \cdot 7H_2O$, and $Al_2(SO_4)_3 \cdot 18H_2O$ as salt hydrates for thermochemical storage showed that each can be dehydrated below 150°C; the hydration and dehydration behavior of various $MgCl_2 \cdot 6H_2O$ was the most promising (van Essen et al., 2010). For a direct floor heating system using flat plate solar collectors, the relations have been

determined between power levels and heating storage capacities for a thermochemical storage using bromide strontium as the reactant and water as the working fluid (Lahmidi et al., 2006). The working pair $MgSO_4$/zeolite has been shown to have reasonable energy densities and thermal properties for seasonal thermochemical heat storage (Hongois et al., 2010). Long-term heat storage for a closed sorption system using the working pair NaOH and water has been contrasted with conventional storage (Weber and Dorer, 2008). The working pair bromide strontium and water has been analyzed experimentally for heating and cooling (Mauran et al., 2008). The thermochemical storage reaction pair ammonia and water has been examined for solar thermal applications, including the characteristics of the materials and their dissociation (Kreetz and Lovergrove, 1999; Lovergrove et al., 1999a, 1999b). The cycling behavior of the storage medium magnesium sulfate ($MgSO_4$) and the dehydration temperature of the reactant have been investigated (Zondag et al., 2007). The reversibility and efficiency of thermochemical storage materials $Ca(OH)_2$ and CaO have been investigated (Azpiazu et al., 2003).

33.5 COMPARISON OF THERMAL STORAGE TYPES

In Table 33.4, some properties of several sensible and latent TES materials are listed and compared. The latent storage materials are PCMs.

For the thermal storage materials listed in Table 33.4, the size of storage required to store a fixed quantity of thermal energy is listed in Table 33.5 in terms of mass and volume. A comparison of the storage masses and volumes required demonstrates that there is significant variability in storage quantities, with the storage having the lowest mass density in Table 33.5 requiring a volume 15 times greater than the storage having the greatest mass density. Similarly, the storage having the lowest volumetric density in Table 33.5 requires a volume 11 times greater than the storage having the greatest volumetric density.

Sensible, latent, and thermochemical thermal storage types are compared in Table 33.6, considering a range of

TABLE 33.4 Selected properties of some typical sensible and latent thermal storage materials.

Property	Thermal storage type			
	Sensible		Latent	
	Rock	Water	Organic PCM	Inorganic PCM
Latent heat of fusion (kJ/kg)	–	–	190	230
Specific heat (kJ/kg)	1.0	4.2	2	2
Density (kg/m³)	2340	1000	800	1600

Sources: Adapted from Hasnan (1998) and Dincer and Rosen (2011).

TABLE 33.5 Size of storage required storing 1 GJ of thermal energy using typical sensible and latent thermal storage materials.

| | Thermal storage type | | | |
| | Sensible | | Latent | |
Property	Rock	Water	Organic PCM	Inorganic PCM
Storage mass (kg)	67,000	16,000	5,300	4,400
Storage volume (m³)	30	16	6.6	2.7

Sources: Data drawn from several sources (Dincer and Rosen, 2011; Hasnan, 1998).

TABLE 33.6 Comparison of selected performance factors for the main types of thermal storage.

| | Type of thermal energy storage | | |
Performance parameter	Sensible	Latent	Thermochemical
Temperature range (°C)	30–400	20–80	20–200
Storage density (GJ/m³)	0.03–0.2	0.3–0.5	0.4–3
Lifetime	20 years	100–1000 storage cycles	Unknown
Status	Commercial	Commercial (for some systems)	Undergoing research and development

Sources: Adapted from several sources, including http://www.preheat.org/technology, Wettermark (1989), Dincer and Rosen (2011), van Helden (2009), and Bakker et al. (2009).

performance factors. The principal advantages and disadvantages of these three storage types are listed and compared in Table 33.7.

33.6 DESIGN, OPERATION, AND ECONOMICS OF THERMAL STORAGE

33.6.1 Thermal Storage Design and Selection

Energy demands in the commercial, industrial, utility, and residential sectors, for such tasks as space and water heating, cooling, and air conditioning, vary on daily, weekly, and seasonal bases. A TES is designed or selected to match the application. Numerous other criteria also affect the selection of TES systems, many of which are summarized in Table 33.8. These criteria include technical factors (e.g., required storage

capacity, storage duration, physical size, space availability, efficiency, installation limitations, reliability, safety, impact on performance of the overall application), environmental factors, and economics (e.g., system cost, lifetime, payback period). Appropriate trade-offs are often made among competing criteria. The Air Conditioning Contractors of America (ACCA) has developed a guide to explain TES using HVAC terminology, aimed at providing designers and contractors a step-by-step approach. The guide aims to increase the understanding of TES technologies and their integration with applications and to elevate comfort levels so that contractors can consider TES solutions more readily (ACCA, 2005; ASHRAE, 2007).

Fernandez et al. (2010) have developed a methodology for the selection of suitable sensible heat storage materials in the temperature range 150–200°C. In this approach the material selection was based on volumetric energy density

TABLE 33.7 Main advantages and disadvantages for the main types of thermal storage.

| | Type of thermal energy storage | | |
Attribute	Sensible	Latent	Thermochemical
Advantages	Relatively low cost	Medium storage density	High storage density/compact
	High reliability	Small volume	Low heat loss
	Ease of application		Long storage period
	Abundant materials		Long distance transport
Disadvantages	Significant heat loss over time	Low heat conductivity	Relatively high cost
	Large volume	Corrosive materials	Technical complexity
		Significant heat loss over time	Not commercial

Sources: Adapted from several sources, including http://www.preheat.org/technology, Wettermark (1989), Dincer and Rosen (2011), van Helden (2009), and Bakker et al. (2009).

TABLE 33.8 Criteria and factors affecting design or selection of a thermal storage for an application.

Requirements	Performance	Limitations	Economics	Environmental impact
Storage capacity	Energy efficiency	Physical size	System capital cost	Resource use in thermal storage installation and operation
Storage duration	Exergy efficiency	Available space	Operating costs	Emissions associated with thermal storage
Rate of charging and discharging	Reliability	Installation limitations	Payback period	Change in environmental impact of application
Variability in thermal demands	Impact on application safety		Lifetime	

as well as cost and thermal conductivity (preferably greater than 1 W/m K). The approach was reported to be helpful in the selection of suitable TESs, based on costs, availability, and environmental benefits.

The enhancement of thermal conductivity is the leading area of research in the PCMs. In recent years considerable efforts have been made to develop appropriate techniques and/or combinations of materials to increase PCM thermal conductivity (Agyenim et al., 2010; Cabeza et al., 2007; Sari and Karaipekli, 2009). Some of the recommended measures include finned tubes of various configurations, bubble agitation, insertion of a metal matrix into the PCM, using PCMs dispersed with high-conductivity particles, microencapsulation, and multi-tube configurations. Hamada et al. (2003) used carbon fiber chips and carbon brushes as additives and observed the carbon fiber chips to be effective in improving the heat transfer rate in PCMs. However, the higher thermal resistance near heat transfer surface was observed to lower the overall heat transfer rate in the fiber chips relative to carbon brushes, indicating that the carbon brushes are superior to fiber chips. Other studies have also reported the enhancements in heat transfer rate for innovative measures (Alkan and Sari, 2008; Bayés-García et al., 2010; Fan and Khodadadi, 2011; Lai et al., 2010). One significant development was reported by Velraj et al. (1999) who investigated paraffin and lessing rings and found that this combination improved the thermal conductivity about 10 times compared to paraffin (i.e., to 2 W/m K from 0.2 W/m K).

A constraint preventing the large-scale commercial use of PCMs for energy storage is salt solidification around the heat exchanger tubes, which slows the discharging of the latent heat stored in PCMs. Terrafore has developed an innovative approach that helps to increase the heat transfer rate by forced convection (Mathur et al. 2010). In this technique a coated shell and tube heat exchanger is used with a specific salt mixture composition (i.e., a dilute eutectic). This heat exchanger with a unique coating of salt mixture has been observed to improve the heat transfer coefficient, and work is continuing to further develop it.

An alternative method of numerical modeling for designing TES systems with PCMs has been investigated by Bruno et al. (2010). The method, based on the effectiveness-number of transfer units (ε-NTU) approach, has been tested experimentally on a cylindrical tank filled with PCM. The results show that the ε-NTU technique provides a useful design tool for sizing and optimizing a TES unit with PCMs.

33.6.2 Thermal Storage Operation

Thermal storage systems are generally designed to operate on a cyclical basis (usually daily, weekly, or seasonally). In considering TES operation, it is useful to characterize TES systems according to storage duration:

- *Short term*: Short-term (or diurnal) TES addresses peak loads lasting a few hours to a day in order to reduce the sizing of systems and/or to take advantage of energy-tariff daily structures or to allow intermittent energy sources to be used throughout the day. The use of diurnal TES for electrical load management in buildings is increasing. TES allows electricity consumption costs to be reduced by shifting electrical heating and cooling loads to periods when electricity prices are lower, usually during the night. Load shifting can also reduce demand charges, which can represent a significant proportion of total electricity costs for commercial buildings.

- Medium-term TES operate on weekly cycles and exhibit many of the characteristics of short-term systems.

- Long-term TES operates on annual or seasonal cycles and usually take advantage of seasonal climatic variations. Seasonal TES systems have a much greater capacity than daily TES, often by two orders of magnitude. Thermal losses are more significant for long-term storage, so more effort is made to prevent thermal losses in seasonal rather than daily TES. While diurnal systems can generally be installed within a building, seasonal storage requires such large storage volumes that special care is required in locating the storage and separate locations are often required. The significance of long-term TES applications is growing in many parts of the world. Recently, for instance, the Canadian federal government with several industry partners completed the Drake Landing Solar Community (DLSC) in the city of Okotoks, Alberta (Dincer and Rosen, 2011; SAIC, 2010). The project has successfully integrated the long-term thermal storage technologies with solar energy. The corresponding district heating system is designed to store solar energy underground in the summer months and supply it to 52 homes of DLSC community in winter months. This project has met almost 90% space heating requirement of the community.

In CTES applications, several strategies are available for charging and discharging so as to meet cooling demand during peak hours. The main strategies are full storage and partial storage. A full-storage strategy shifts the entire peak cooling load to off-peak hours, a strategy that is most attractive when peak demand charges are high or the peak period is short. With partial storage, the chiller operates to meet part of the peak-period cooling load, and the rest is met by drawing from storage. Partial-storage systems are therefore load leveling and demand limiting. A large cold thermal energy system (CTES) is currently being constructed in Chicago to supply two-thirds of the peak-load air conditioning demands of the downtown area. In another large development in Shanghai, China, where the electrical peak load is a serious problem and air conditioning is responsible for a large part of the summer peak load, Hangchow City adopted an incentive rate system for buildings that adopt certain CTES options (Saito, 2002).

33.6.3 Thermal Energy Quality and Exergy

Energy quality as measured by the temperatures of the materials entering, leaving, and stored within a storage is an important consideration in TES. For example, 1 kWh can be stored by heating 1000 kg of water at 0.86°C or by heating 10 kg of water at 86°C. The latter case is more attractive in terms of energy quality as a wider range of tasks can be accomplished with the higher temperature medium upon discharging the storage. But the costs of TES systems that retain thermal energy quality at higher temperatures are often higher than for those that do not. A particularly useful measure of energy quality is "exergy," which can be usefully applied to thermal storage systems analysis and improvement. In fact, one of the performance factors listed in Table 33.8 is exergy efficiency, which is like the more conventional energy efficiency but different (Dincer and Rosen, 2007; Rosen et al., 2004).

Exergy is loosely a measure of energy quality or usefulness or value. It is defined as the maximum amount of work which can be produced by a stream or system as it is brought into equilibrium with a reference environment and can be thought of as a measure of the usefulness or quality

of energy. Exergy is consumed during real processes due to irreversibilities and conserved during ideal processes. Exergy quantities are evaluated with respect to a reference environment. Exergy analysis, which is a tool based on exergy, has been applied to a wide range of processes and systems (Dincer and Rosen, 2007), including thermal storage (Dincer and Rosen, 2011; Rosen et al., 2004). Exergy analysis includes the evaluation of exergy efficiencies, which are efficiencies that always provide a measure of how nearly the operation of a system approaches the ideal or theoretical upper limit. This characteristic is not in general true for energy efficiencies. Exergy inefficiencies are inherently areas in a system having the largest theoretical margins for efficiency improvement.

Exergy analysis is useful to apply to thermal storage systems as it allows meaningful efficiencies to be evaluated and sources of thermodynamic losses to be clearly identified and quantified. Exergy analysis has been applied to various types of thermal storage, including those for storing heat and cold (Dincer and Rosen, 2011).

It is useful at this point to provide an illustration of how exergy efficiencies provide rational measures of performance, in terms of measures of the approach of the performance of a system to the ideal. Consider a perfectly insulated thermal storage containing 1000 kg of water, initially at 40°C. The ambient temperature is 20°C, and the specific heat of water is taken to be constant at 4.2 kJ/kg K. A quantity of 4200 kJ of heat is transferred to the storage through a heat exchanger from an external body of 100 kg of water cooling from 100°C to 90°C. This heat addition raises the storage temperature 1.0°C, to 41°C. After a period of storage, 4200 kJ of heat is recovered from the storage through a heat exchanger which delivers it to an external body of 100 kg of water, raising the temperature of that water from 20°C to 30°C. The storage is returned to its initial state at 40°C. For this storage cycle the energy efficiency, the ratio of the heat recovered from the storage to the heat injected, is 4200 kJ/4200 kJ = 1, or 100%. But the recovered heat is at only 30°C, and of little use, having been degraded even though the storage energy efficiency was 100%. The exergy recovered in this example can be shown to be 70 kJ, and the exergy supplied 856 kJ. Thus the exergy efficiency, the ratio of the thermal exergy recovered from storage to that injected, is 70/856 = 0.082, or 8.2%, a much more meaningful expression of the achieved performance of the TES and one that shows a significant margin for efficiency improvement. Consequently, a device which appears to be ideal on an energy basis is correctly shown to be far from ideal on an exergy basis, clearly demonstrating the benefits of using exergy analysis for evaluating thermal storages.

Haller et al. (2009) have suggested methods for characterizing thermal stratification efficiency in TES and report stratification efficiency to be a vital TES parameter. Existing methods for calculating stratification efficiencies are applied to

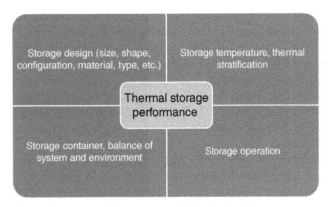

FIGURE 33.2 Factors affecting the performance of a thermal storage.

storage processes and compared with the rate of entropy production caused by mixing. In the analysis, the stratification efficiency of fully mixed tank is taken to be 0% and that for a perfectly stratified tank 100%, and stratification efficiencies are applied to charging, discharging, and storing processes. Exergy analysis has been shown to be effective for assessing thermal stratification and its benefits (Rosen et al., 2004).

33.6.4 Factors Affecting Thermal Storage Performance

TES performance is affected by many factors, as illustrated in Figure 33.2. These factors include design, temperature, storage operation container and balance of system, and the setting in which the storage operates.

33.6.5 Economics of Thermal Storage

The economic justification for TES systems normally requires that the annualized capital and operating costs be less than those required for primary generating equipment supplying the same service loads and periods. In general, TES systems accrue fuel cost savings relative to primary generating equipment, but often at the expense of higher initial capital costs. TES cost-effectiveness is often evaluated considering the distribution of heating, cooling, and electrical loads over time; TES system characteristics (e.g., size) and performance (e.g., control methods, efficiency); and economic factors (e.g., electricity demand charges, time-of-use costs, TES costs, financial incentives). Economic evaluation and comparison criteria often include simple payback period, return on investment, and comparisons of the annualized investment cost with annual energy cost savings. Financial analyses of TES systems are important since they sometimes have relatively high initial capital costs which are justified through savings over time.

TES systems can be cost-effective in both residential and commercial buildings, with projects often having payback periods of less than 3 years. Payback periods for CTES systems vary significantly with application, from less than 1 year to over 10 years in some instances. The economic feasibility of TES can be influenced by incentive programs (e.g., rebates, special electricity rates, tax incentives) from governments, electrical utilities, and other agencies. Heating TES systems can be justified economically for most facilities which have significant space heating needs and are billed under time-of-use electric rate schedules that have large differentials between peak and off-peak electric consumption. Cooling TES systems are generally advantageous for new facilities which have cooling loads that are large in the day and small at night. For retrofit situations, cooling TES is sometimes difficult to justify unless the cooling system is being replaced because of old age or inadequate capacity.

TES systems can shift nearly all energy use for space heating to off-peak hours, whereas only about 50% of the energy for heating is used during off-peak periods in conventional systems. The Electric Power Research Institute indicates that overall HVAC operating costs can be lowered by 20–60% by combining TES with cold-air distribution. TES systems for cooling capacity have been most successful in larger buildings, although the attractiveness of applications in smaller units has been increasing. For TES systems designed to reduce electricity costs, TES allows consumer electricity costs to be reduced by shifting electrical loads to periods of lower electricity prices when time-of-use tariffs exist. Shifting or spreading of the load can also reduce demand charges significantly. Benefits from TES also accrue to the electricity utilities, since shifting electrical loads to off-peak periods reduces peak demands and required generation capacity and reduces the utilization of more expensive and often more polluting generating stations. Hence utilities often offer rebates for TES.

In the future, TES is expected to be able to reduce the capacity of heating and cooling systems by 50–70% and to yield energy savings due to better system efficiencies.

Various factors affecting the economics of TES systems are summarized in Figure 33.3.

33.6.6 Thermal Storage Testing

Several approaches for testing of TES devices have been proposed in the past (Henze et al., 2003; Hill et al., 1977; Scalat

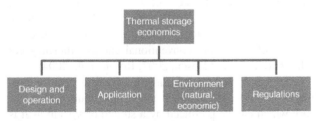

FIGURE 33.3 Factors affecting the economics of thermal storage.

et al., 1996). The development of reliable test methodology which would help in predicting the performance of TES units has been challenging, due to the complexity of the energy storage processes involved. ASHRAE provides testing standards for TES systems, for example, ASHRAE TES standard 94-77: Methods of Testing Thermal Storage Devices Based on Thermal Performance. Also, ASHRAE Standard 943 (Method of Testing Active Sensible TES Devices Based on Thermal Performance) and ANSI/ASHRAE standard 94.1-1985 (Method of Testing Active Latent Heat Storage Devices Based on Thermal Performance) (ASHRAE 2000) are halves of a revision of ASHRAE's first TES standard (94-77). Further, the Canadian Standards Association has published information on the design and installation of underground TES systems for commercial and institutional buildings (CSA 2002).

There are no national and international standards available for the testing of TES in PCMs (Agyenim et al., 2010), even though much research is reported on developing PCMs for various applications. The lack of a unified international standard for testing and performance comparison is considered a hindrance to future growth in the technology and efforts are being made to address it.

33.7 BENEFITS OF THERMAL STORAGE

Thermal storage systems can yield significant benefits, which vary by application (ASHRAE, 2007; Ataer, 2006; Beckman and Gilli, 1984; Dincer and Rosen 2011). Many of the benefits are summarized in Figure 33.4, and are described below:

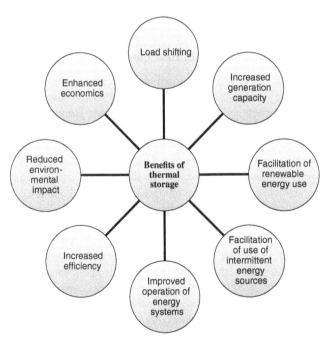

FIGURE 33.4   Several benefits of thermal storage.

- *Load shifting and increased generation capacity*: TES permits energy consumers to shift energy use from high- to low- demand periods. Also, since demands for heating, cooling, or electricity are seldom constant over time, the excess generation capacity available during low-demand periods can be used to charge a TES in order to increase the effective generation capacity during high-demand periods. This benefit allows smaller production units to be installed or increased capacity to be attained without purchasing additional units.

- *Facilitation of use of renewable energy resources and other intermittent energy sources*: TES can facilitate the use of energy sources which are not available continuously, by storing energy between periods of availability and demand. Intermittent energy sources include renewable energies like solar and wind, as well as waste heat. TES thereby allows intermittent energy sources to meet a greater fraction of the loads for which they are used. In these ways, TES can facilitate substitutions of energy resources, on small to large scales.

- *Improved operation of systems in which thermal storage is applied*: The incorporation of thermal storage into a system (e.g., heat pumps, power plants, cogeneration plants) can improve the operation of the system. For instance, TES facilitates improved operation of thermal equipment, in that it can allow such equipment to operate more effectively and flexibly. Also, TES systems can complement heat pumps for heating or cooling by providing hot or cold reservoirs, thereby improving their efficiencies and performances. Also, a cogeneration plant with TES need not follow a thermal load and can be operated more advantageously. This helps overcome a weakness of cogeneration plants, in that they are generally operated to meet the demands of the connected thermal load, which often results in excess electric generation during periods of low electricity demand.

- *Increased efficiency*: By storing heat (e.g., waste heat, solar energy) or cold so that it can be used when needed, with temperature enhancement where necessary via heat pumps or other technologies, the efficiencies of heating and cooling operations can be increased.

- *Reduced environmental impact*: By increasing the efficiency of systems which utilize TES and facilitating the use of renewable energy sources and waste energy, TES systems help reduce emissions of pollutants and environmental impacts. Climate change mitigation with thermal storage has been examined (Paksoy, 2007).

- *Enhanced economics*: Many of the above benefits allow TES systems to provide significant financial gains over their lifetimes. For instance, by facilitating shifting of energy use to low-demand periods, TES allows energy consumers subject to time-of-day pricing to shift energy purchases from high- to low-cost periods. Also, TES can allow thermal equipment to operate more economically.

- *Improved operation of thermal equipment*: TES can allow thermal equipment to operate more effectively and flexibly.

An investigation of applications of TES for cooling capacity (Dincer and Rosen, 2001) that assessed TES with exergy methods in hopes of attaining more realistic determinations of energy saving, emissions reductions, and economics found that the appropriate type of TES provides energy savings of up to 50% and environmental benefits in terms of greenhouse gas emissions reductions up to 40%. This investigation helped confirm the benefits of cold thermal storage in energy systems.

33.8 APPLICATIONS OF THERMAL STORAGE

33.8.1 Direct Applications of Thermal Storage

Various applications are possible for thermal storage. The most common parts of the economy in which thermal storage is employed are the building, industry, and utility (e.g., power generation) sectors. Various applications of thermal storage systems are shown in Figure 33.5.

The integration of TES with solar energy is a promising area, as is CTES, which has found many applications, such as food processing and building air conditioning. For instance, applications of TES systems for cooling capacity have been studied by Dincer and Rosen (2001). The focus of that investigation was to assess TES with exergy methods to obtain more realistic determinations of energy saving, greenhouse gas emissions reductions, and economic savings.

The integration of TES with solar drying has been reviewed by Bal et al. (2010) for the continuous drying of

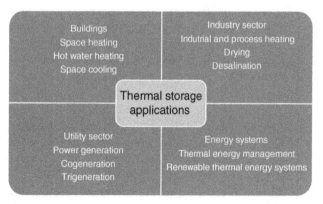

FIGURE 33.5 Selected applications of thermal storage in various application areas.

agricultural food products at moderate temperatures (40–75°C). Numerous solar dryers with TES have been developed but experienced limited commercial success. The integration of TES with solar dryers was identified to extend the drying period, reduce product waste, improve overall drying quality, and improve solar drying cost-effectiveness. For this application, PCMs are suggested to be advantageous over sensible heat storage, as the drying air temperature is maintained close to constant value. Also, more work was reported to be needed to develop a viable and economic TES media for solar drying.

Significant research is presented on using natural and renewable energy resources with thermal storage in heating and cooling applications, and site investigations for underground TES applications and borehole thermal energy storage (BTES) design using earth energy design software (Paksoy, 2007).

The potential of solar thermal walls has been recognized for several decades as useful for passive applications of solar energy. For instance, building materials with PCMs can be integrated into buildings. Some building materials such as gypsum wallboards are suitable for PCM containment.

Solar combi-systems (i.e., systems that provide both space heating and cooling) offer another beneficial application for thermal storage.

Thermal storage is a critical component in efforts to develop net-zero energy buildings and communities, that is, buildings and communities that achieve zero average annual energy consumption at both the building and community levels. Such buildings and communities will likely utilize short-term and seasonal TES in conjunction with building-integrated solar systems, high-performance windows with active control of solar gains, heat pumps, combined heat and power technologies, and smart controls.

33.8.2 Use of TES with Heat Pumps

The heat pump is an important component of many energy efficiency and conservation strategies. Thermal storage can be used beneficially in conjunction with heat pump technology, for heating and cooling applications. This is evidenced in part by the fact that several organizations are actively promoting these technologies. The IEA Heat Pump Centre (www.heatpumpcentre.org) is an international information service center for heat pumping technologies, applications, and markets. The goal of the center is to accelerate the implementation of heat pumps and related heat pumping technologies, including air conditioning and refrigeration. The Centre publishes the IEA Heat Pump Centre Newsletter, which often has articles on TES, for example, Van de Ven (1999). The CANMET Energy Technology Centre (CETC) (www.nrcan.gc.ca/es/etb/cetc) is a Canadian federal government science and technology organization with a mandate to develop and demonstrate energy-efficient, alternative, and renewable energy technologies and processes, including heat pump and TES (CANMET, 2005). The Canadian GeoExchange Coalition is an industrial association representing hundreds of members that promotes the ground-based heat pump industry actively (Canadian GeoExchange Coalition, 2007).

Heat pumps and thermal storage can be combined in various ways, and such a combination can be beneficial as follows:

- Heating with heat pumps and thermal storage can be accomplished using ground-based storage, where the heat source to the heat pump is the ground.
- Heat pumps can also be combined with latent TES. For example, a heat pump can use a latent TES to enable rapid room temperature increases and defrosting. In one such system, the latent TES uses polyethylene glycol as a phase change material, which surrounds a rotary compressor of the air conditioner/heat pump for a room. Heat released from the compressor is transferred to the TES through a finned-tube heat exchanger and recovered for use during start-up and for defrosting. During start-up, the TES halves the time to reach a 45°C discharge air temperature. The integration improves heat capacity by about 10% and coefficient of performance by 5%, and requires the same installation space as a conventional heat pump/air conditioner.
- Heat from solar collectors, which is often used directly for space heating, can instead be used as a heat source for a heat pump. In such a solar-augmented heat pump system, the solar collector outlet temperatures can be lower than with direct heating, increasing energy efficiency and reducing the cost of the solar collector. The higher source temperature also increases the coefficient of performance of the heat pump, reducing its electricity consumption. Such a system can operate in various modes. With small TES, the solar collector improves heat pump efficiency mainly during sunny periods. With larger TES, the solar energy provides a warm storage for heat pump operations during cloudy periods and night. Alternatively, the overall system can be designed so that the heat pump operates only during off-peak hours. Such an approach requires two TESs, one to store solar energy and one to store the heat pump output for space heating at all times.

Heat pumps can be beneficially integrated with both latent and sensible TES for cooling. In the former case, heat pumps can be combined with latent TES using ice and ice slurry. An example of the latter case is a system combining water-based CTES combined with heat pumps, which can make heat pump operation more economic and help level electrical loads. In conventional air conditioning using heat pumps, the heat pumps operate during the day when cooling

demand exists. This operation contributes to electricity daytime demand, which is significant since cooling demand is sometimes responsible for more than one-third of peak electrical demand. In a typical water-based CTES system, half of the daily cooling load can be met by night operation of heat pumps. Combining of a heat pump and a CTES can provide the following benefits:

- *Efficient operation of heat pumps*: Although a heat pump typically achieves maximum efficiency at a certain operating condition, heat pump operation normally cannot be maintained at the most efficient condition in commercial and residential applications because cooling and heating loads vary temporally. This variation reduces the seasonal efficiency of heat pumps. Because heat pumps integrated with CTES can operate independently of the cooling and heating load of buildings, CTES helps avoid this problem by permitting operation of heat pumps at the most efficient operating condition.
- *Load leveling of air conditioning electricity demand*: CTES can shift peak air conditioning loads to the night, increasing significantly the annual dependence on nighttime electricity. A typical CTES, for instance, shifts half the air conditioning load to the night on the peak day and permits the annual dependence on nighttime electricity to reach up to 70% for cooling and 90% for heating. Also, a CTES system can improve the annual load factor of electricity generation facilities and save money for consumers when electric power companies provide discount rates for nighttime electricity.
- *Reduced heat pump size*: For a fixed air conditioning load, the longer operating hours of a heat pump integrated with CTES allow a smaller capacity heat pump to be used, reducing electrical demand peaks and decreasing initial and operating costs.

Combined heating and cooling operations can also be advantageously accomplished by integrating heat pumps with thermal storage, particularly for multi-season applications. For example, with a ground-coupled heat pump with seasonal TES, the ground or groundwater is cooled during heating, as heat from the TES is supplied to the building. After the heating season, the stored cold is used for direct cooling, via cold groundwater from the injection well or cold brine from earth heat exchangers. After some time, the ground temperature may be too high for direct cooling. The system then can be operated as a conventional heat pump, cooling the building space and storing heat in the ground until the next heating season. Also, a heat pump and an aquifer TES allows cold water to be extracted from a cold well during summer and warmed by cooling a building, and then returned to a warm well in the aquifer. A heat pump can cool the cold water further, if necessary. The warmed water increases the temperature of the aquifer near the warm well. The operation is reversed during winter, with warm water extracted from the warm well and boosted in temperature by the heat pump if necessary.

33.8.3 Market Penetration of Thermal Storage

TES applications have achieved various levels of market penetration, depending on country. Diurnal heat storage has achieved a large market share in many countries. Diurnal cold storage in air conditioned buildings for demand-side management is growing. Short-term cool TES for air conditioning is often cost-effective, with numerous applications in the United States, Canada, Japan, and Europe. Individual houses or commercial buildings can use diurnal TES either for heating or cooling applications. Seasonal TES has achieved the greatest success in northern countries, sometimes in conjunction with district heating. Large seasonal TES has often been installed with solar collection systems in large buildings or in association with district heating.

33.9 EXAMPLES

33.9.1 A Solar Energy Community Using Thermal Storage

In 2006, the DLSC was completed in Okotoks, 25 km south of Calgary, Alberta, Canada, consisting of 52 low-rise detached homes, located on two streets running east–west (McClenahan et al., 2006; Sibbitt et al., 2007; Wong et al., 2006). The DLSC energy system (see Figure 33.6) demonstrates the feasibility of replacing substantial residential conventional fuel

energy use with solar energy, collected during the summer and utilized for space heating during the following winter, in conjunction with seasonal TES. The DLSC homes, of a typical size of 138–151 m^2 based on gross floor area, are of high efficiency (meeting Natural Resources Canada's R-2000 standard). Each home has a detached garage at the back facing a lane, and the garages are joined by a roofed breezeway that extends the length of each of the four laneways and incorporates the 800 solar collectors of nearly 2300 m^2 area. The DLSC energy system has numerous key components, including two thermal storage types (short and long term).

For DLSC homes, 90% of heating and 60% of hot water needs are designed to be met using solar energy. Annually, each home uses approximately 110.8 GJ less energy and emits about 5.65 tonnes fewer GHGs than a conventional Canadian home (see Table 33.9), avoiding about 260 tonnes of GHG emissions annually. Each DLSC home is about 30% more efficient than conventionally built houses and is expected to use 65–70% less natural gas to heat water than a conventional new home. Note that large seasonal thermal storage systems require a significant time to charge since the storage medium must be heated up to a minimum temperature before any heat can be extracted, so it will not reach a 90% solar fraction before 5 years of operation.

The DLSC energy system has five main components: (1) solar thermal collectors, which can generate 1.5 MW of thermal power on a typical summer day; (2) the Energy Center, which connects the various DLSC parts and contains short-term thermal storages; (3) the district heating system, which has a capacity of 4.5 MW so as to meet the annual heat demand of 50 GJ per house; (4) the BTES; and (5) the solar domestic hot water system, which uses separate rooftop solar panels and natural gas-based hot water units.

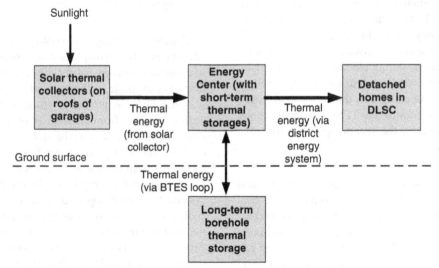

FIGURE 33.6   The main components of the energy system of the Drake Landing Solar Community and the principal energy flows.

TABLE 33.9 Annual energy use for space and domestic hot water heating and greenhouse gas emissions for DLSC and conventional homes.

| Quantity | Home type | | DLSC home savings relative to conventional home |
	Conventional	DLSC	
Energy use for space and water heating (GJ)	126	86	40[a]
Natural gas	126	15	111[b]
Solar energy	0	71	–
GHG emissions for space and water heating (t)	6.4	0.8	5.7
Space heating	5.1	0.3	4.8
Water heating	1.3	0.5	0.9

[a]Savings of energy use for space and water heating reflect savings due to conservation measures in the DLSC homes.
[b]Natural gas savings are reduced natural gas use for DLSC homes relative to conventional homes.

During operation, a glycol heat transport fluid conveys thermal energy from the solar collectors through insulated pipes to the Energy Center. At the Energy Center, a heat exchanger transfers heat from the solar collector loop to two 125,000-L tanks containing 240 m³ of water. Hot water from the tanks can be transferred via a district heating network to DLSC homes. During warmer months, heat is stored in the ground by circulating hot water from the tanks to a BTES beside the Energy Center and used in the subsequent winter. The BTES has 144 boreholes that are 35 m deep and linked in 24 parallel circuits, each with six boreholes in series. The six boreholes in series are arranged in a radial pattern to maintain the highest temperature near the center; water flows from the center to the outer edge when charging the BTES with heat, and from the edge toward the center when recovering heat. Heat transferred to the surrounding earth raises its temperature to up to approximately 80°C. During cold months, water at approximately 40–50°C from the boreholes is transferred by the heat exchanger to the district energy network for circulation to homes. The Energy Center distribution network has a natural gas-fired boiler for peaking requirements during winter and a separate cooling system for the solar collector loop. Pumps for the collector and district heating loops use variable speed drives to reduce electrical power consumption while managing varied thermal power levels.

The short-term thermal storages in the Energy Center act as a buffer between the collector loop, the district energy loop, and the BTES field, receiving and discharging thermal energy as necessary. The short-term storage tanks support the system operation by being able to receive and discharge heat at a much greater rate than the BTES storage, which has a much higher thermal storage capacity. This use of short- and long-term storage is advantageous because the BTES cannot receive energy as quickly as it can be collected during high insolation periods, so it is stored in the short-term TES tanks before transfer to the BTES at night. Also, when heat cannot be discharged from the BTES sufficiently quickly to meet winter peak heating demands, heat is continually removed from the BTES and stored in the short-term storage tanks if not immediately needed.

33.9.2 A University Energy System Using Thermal Storage

A BTES system is utilized in Oshawa, Ontario, Canada, at the University of Ontario Institute of Technology (UOIT), which opened in 2003 and has about 7500 students (Dincer and Rosen, 2011). Most of its seven buildings are designed to be heated and cooled using ground-source heat pumps in conjunction with the BTES and its 384 boreholes, each 213 m deep (Figure 33.7). Thermal energy is upgraded by ground-source heat pumps for building heating. Alternatively

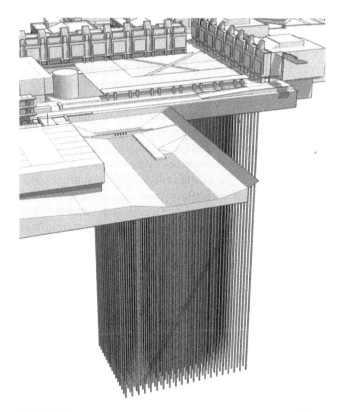

FIGURE 33.7 Borehole thermal energy storage system at UOIT, showing boreholes below the university quadrangle field and adjacent to university buildings.

the ground can absorb energy and be increased in temperature using the heat pump in its cooling (reverse) mode. During winter, fluid circulating through tubing extended into the wells collects heat from the earth and carries it into the buildings. In summer, the system extracts heat from the building and places it in the ground. Thus the BTES provides for both heating and cooling on a seasonal basis.

Chillers pump energy from the buildings into the thermal storage. The heat pump modules assist in this cooling. Chilled water is supplied from two chillers, each having seven 90-ton modules, and two sets of heat pumps with seven 50-ton modules each. The condensing water passes to the BTES field, which stores it for use in the winter (when the heat pumps reverse) and provides low-temperature (53°C) hot water for the campus. Each building is hydronically isolated with a heat exchanger and has an internal distribution system. Supplemental heating is also provided by condensing boilers. In autumn, energy is reclaimed from the BTES field, and the return water is sufficiently hot for "free heating" (heating without using the heat pumps). Technical specifications of the chiller and two heat pumps, each having seven modules, are listed in Table 33.10.

The total cooling load of the campus buildings was anticipated to be about 7000 kW. Thermal conductivity test results (Beatty and Thompson, 2004), which showed the thermal conductivity for the geologic media in a test well to be 1.9 W/m K, determined the requirements for the borehole field in terms of numbers and depths of boreholes to meet the energy service needs. Steel casing was installed in the upper 58 m of each borehole to seal out groundwater in the shallow formations. Water-filled borehole heat exchangers (BHEs) were used improve the system efficiency and extend the borehole lives. The hydrogeologic setting at the vicinity of the site has over 40 m of unconsolidated overburden deposits overlying shale bedrock (Figure 33.8), and the background temperature of the geologic formations at the site is 10°C. Groundwater resources in the Oshawa area are limited to isolated, thin sand deposits (Beatty and Thompson, 2004), and the homogeneous, non-fractured rock is well suited for thermal storage since little groundwater flow exists to transport thermal energy from the site. The BTES field has a volume of approximately 1,400,000 m^3 and contains 1,700,000 tonnes

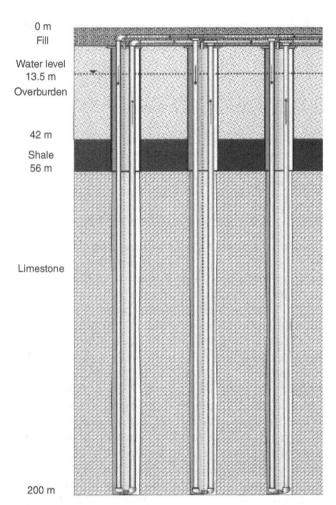

FIGURE 33.8 Illustration of the ground composition and geology for the UOIT BTES system, and several boreholes and the U-shaped tubes within them.

of rock and 600,000 tonnes of overburden. The BHEs are located on a 4.5 m grid and the total field is about 7000 m^2 in area.

Numerous underground TES applications exist (IEA, 2009), but the UOIT borehole TES is the largest and deepest in Canada, and the geothermal well field is one of the largest in North America. The simple payback period when the system was designed was 7.5 years for the geothermal well field and 3–5 years for the high-efficiency HVAC equipment. Annual energy costs are reduced using the BTES system by 40% for heating and 16% for cooling, and the system provides indirect financial benefits (reduced boiler plant costs, reduced annual use of potable water and treatment chemicals, avoided costs of roof cooling towers).

33.10 CLOSING REMARKS

Some thermal storage technology (sensible and some latent) has reached a high level of maturity, while other types

TABLE 33.10 Design values for heat pumps accompanying the UOIT borehole thermal storage system.

Parameter	Heating	Cooling
Energy load (kW)	1386	1236
Load water temperatures: entering/leaving (°C)	41.3/52	14.4/5.5
Source water temperatures: entering/leaving (°C)	9.3/5.6	29.4/35
Design COP	2.8	4.9

(thermochemical) are still in development. Although many types of thermal storage, for heat or cold, have been demonstrated and are commercially available, advances to the technology are continually being made. In recent years, phase change and thermochemical materials have received increased attention due to their greater volumetric TES capacity at relatively constant temperature, for various potential applications. The selection or design of a storage for a given application is normally based on its comparative characteristics, and successful utilization of this technology is possible when it can be developed with acceptable capital and operating costs. The two examples presented of major thermal storage installations illustrate this need. Thermal storage is expected to be increasingly applied with new energy technologies and to contribute to solving the problems societies face regarding energy supply, environmental impact, and overall sustainability. In particular, thermal storage is expected to play a major role in facilitating the use of intermittent sources of thermal energy like solar energy.

ACKNOWLEDGMENTS

The author gratefully acknowledges the financial support provided by the Natural Sciences and Engineering Research Council of Canada.

REFERENCES

ACCA. (2005). *Thermal Energy Storage: A Guide for Commercial HVAC Contractors.* Air Conditioning Contractors of America, Arlington, VA.

Agyenim, F., Hewitt, N., Eames, P., Smyth, M. (2010). A review of materials, heat transfer and phase change problem formulation for latent heat thermal energy storage systems (LHTESS). *Renewable and Sustainable Energy Reviews*, 14:615–628.

Alkan, C., Sari, A. (2008). Fatty acid/poly (methyl methacrylate) (PMMA) blends as form-stable phase change materials for latent heat thermal energy storage. *Solar Energy*, 82:118-124.

ASHRAE. (2007). Thermal storage. In: *ASHRAE Handbook: HVAC Applications.*Chapter 34. American Society of Heating, Refrigerating and Air-Conditioning Engineers, Atlanta, GA.

ASHRAE. (2000). *Standard 943: Method of Testing Active Sensible TES Devices Based on Thermal Performance.* American Society of Heating, Refrigerating and Air-Conditioning Engineers, Atlanta, GA.

Ataer, O. E. (2006). Storage of thermal energy. In: *Energy Storage Systems*, Gogus, Y. A. (editor). In *Encyclopedia of Life Support Systems*, EOLSS Publishers, Oxford, UK.

Azpiazu, M. N., Morquillas, J. M., Vazquez, A. (2003). Heat recovery from a thermal energy storage based on the Ca(OH)$_2$/CaO cycle. *Applied Thermal Engineering*, 23:733–741.

Bakker, M., van Helden, W. G. J., Hauer, A. (2009). Materials for compact thermal energy storage: a new IEA Joint SHC/ECES Task. *Proceeding of the ASME 2009 3rd International Conference of Energy Sustainability*, July 19–23, San Francisco.

Bal, L. M., Satya, S., Naik, S. N. (2010). Solar dryer with thermal energy storage systems for drying agricultural food products: a review. *Renewable and Sustainable Energy Reviews*, 14:2298–2314.

Bales, C. (2006). Chemical and sorption heat storage. *Proceedings of the DANVAK Seminar, DANVAK Seminar (Solar Heating Systems – Combisystems – Heat Storage)*, November, Lyngby, Denmark.

Bayés-García, L., Ventolà, L., Cordobilla, R., Benages, R.Calvet, T., Cuevas-Diarte, M. A. (2010). Phase change materials (PCM) microcapsules with different shell compositions: preparation, characterization and thermal stability. *Solar Energy Materials and Solar Cells*, 94:1235–1240.

Beatty, B., Thompson, J. (2004). 75 km of drilling for thermal energy storage. *Geo-Engineering for the Society and its Environment: Proc. 57th Canadian Geotechnical Conference*, Quebec City, Session 8B, pp. 38–43.

Beckman, G., Gilli, P. V. (1984). *Thermal Energy Storage.* Springer-Verlag, New York.

Bruno, F., Belusko, M., Tay, N. H. S. (2010). Design of PCM thermal storage system using the effectiveness-NTU Method. *Proc. EuroSun 2010 Conf.*, September 28–October 1, Graz, Austria, paper 200.

Cabeza, L. F., Castellón, C., Nogués, M., Medrano, M., Leppers, R., Zubillaga, O. (2007). Use of microencapsulated PCM in concrete walls for energy savings. *Energy and Buildings*, 39:113–119.

Canadian GeoExchange Coalition (2007). Survey of Canadian geoexchange industry: 2004–2006. *GeoConneXion Magazine*, December:10–13.

CANMET. (2005). Ground-source heat pump project analysis. In *Clean Energy Project Analysis.* CANMET Energy Technology Centre – Varennes, Minister of Natural Resource Canada, Ottawa.

CSA. (2002). *Design and Installation of Underground Thermal Energy Storage Systems for Commercial and Institutional Buildings.* Standard CAN/CSA-C448-02. Canadian Standards Association, Mississauga, ON, Canada.

Dincer, I., Rosen, M. A. (2001). Energetic, environmental and economic aspects of thermal energy storage systems for cooling capacity. *Applied Thermal Engineering*, 21:1105–1117.

Dincer, I., Rosen, M. A. (2007). *Exergy: Energy, Environment and Sustainable Development.* Elsevier, Oxford, UK.

Dincer, I., Rosen, M. A. (2011). *Thermal Energy Storage Systems and Applications*, 2nd edition. Wiley, London.

Dinter, F., Ger, M., Tamme, R. (1991). *Thermal Energy Storage for Commercial Applications.* Springer-Verlag, Berlin.

Fan, L., Khodadadi, J. M. (2011). Thermal conductivity enhancement of phase change materials for thermal energy storage: a review. *Renewable and Sustainable Energy Reviews*, 15:24–46.

Farid, M. M., Khudhair, A. M., Razack, S. A. K., Al-Hallaj, S. (2004). A review on phase change energy storage: materials and applications. *Energy Conversion and Management*, 4:1597–1615.

Farid, M., Kong, W. J. (2001). Underfloor heating with latent heat storage. *Proceedings of the Institution of Mechanical Engineers, Part A: Journal of Power and Energy*, 215:601–609.

Fernandez, A. I., Martínez, M., Segarra, M., Martorell, I., Cabeza, L. F. (2010). Selection of materials with potential in sensible thermal energy storage. *Solar Energy Materials and Solar Cells*, 94:1723–1729.

Haji Abedin, A., Rosen, M. A. (2010a). Energy and exergy analyses of a closed thermochemical energy storage system: methodology and illustrative application. *Proc. 23rd Conference on Efficiency, Cost, Optimization, and Environmental Impact of Energy Systems*, 14–17 June, Lausanne, Switzerland, Vol. III, pp. 107–114.

Haji Abedin, A., Rosen, M. A. (2010b). Energy and exergy analysis of a closed thermochemical energy storage system. *Proc. 5th International Green Energy Conference (IGEC-V)*, 1–3 June, Waterloo, Ontario, Canada, pp. 1–17.

Haller, M. Y., Cruickshank, C. A., Streicher, W., Harrison, S. J., Andersen, E.Furbo, S. (2009). Methods to determine stratification efficiency of thermal energy storage processes: review and theoretical comparison. *Solar Energy*, 83:1847–1860.

Hamada, Y., Ohtsu, W., Fukai, J. (2003). Thermal response in thermal energy storage material around heat transfer tubes: effect of additives on heat transfer rates. *Solar Energy*, 75:317–328.

Hasnain, S. M. (1998). Review of thermal energy storage. Part I: heat storage materials and techniques. *Energy Conversion and Management*, 39:1127–1138.

Hasnain, S. M, Smiai, M., Al-Saedi, Y., Al-Khaldi, M. (1996). Energy Research Institute-Internal Report. KACST, Riyadh, Saudi Arabia.

Hauer, A., Lavemann, E. (2007). Open absorption systems for air conditioning and thermal energy consumption, Part VI. In: *Thermal Energy Storage for Sustainable Energy Consumption*, Paskoy, H. O. (editor). Springer, Netherlands, pp. 429–444.

Heinz, A.Schranzhofer, H. (2010). Thermal energy storage with phase change materials: a promising solution? *Proc. EuroSun 2010 Conf.*, September 28–October 1, Graz, Austria, paper 206.

Henze, G. P., Krarti, M., Brandemuehl, M. J. (2003). Guidelines for improved performance of ice storage systems. *Energy and Buildings*, 35:111–127.

Hill, J. E., Kelly, G. E., Peavy, B. A. (1977). A method of testing for rating thermal storage devices based on thermal performance. *Solar Energy*, 19:721–732.

Hongois, S., Kuznik, F., Stevens, P., Roux, J. J., Radulescu, M., Beuarepaire, E. (2010). Thermochemical storage using composite materials: from the material to the system. *Proceedings of Eurosun 2010, The 3rd International Conference on Solar Heating, Cooling and Buildings*, September 28–October 1, Graz, Austria.

International Energy Agency (IEA). (2008). Compact thermal energy storage: material development and system integration. Technical Report (draft), Task 42, Annex 28, Solar Heating and Cooling Programme.

International Energy Agency (IEA). (2009). Energy conservation through energy storage. Available at: http://www.iea-eces.org (Accessed August 14, 2009).

International Energy Agency (IEA). (2010). Energy Technology Perspectives 2010: Scenarios and Strategies to 2050. Report. International Energy Agency, Paris, France.

Jegadheeswaran, S., Pohekar, S. D. (2009). Performance enhancement in latent heat thermal storage system: a review. *Renewable and Sustainable Energy Reviews*, 13:2225–2244.

Khalifa, A. J. N., Abbas, E. F. (2009). A comparative performance study of some thermal storage materials used for solar space heating. *Energy and Buildings*, 41:407–415.

Koca, A., Oztop, H., Koyun, T., Varol, Y. (2008). Energy and exergy analysis of a latent heat storage system with phase change material for a solar collector. *Renewable Energy*, 33:567–574.

Kreetz, H., Lovergrove, K. (1999). Theoretical analysis and experimental results of a 1 kW_{chem} ammonia synthesis reactor for a solar thermochemical energy storage system. *Solar Energy*, 67:287–296.

Lahmidi, H., Mauran, S., Goetz, V. (2006). Definition, test and simulation of a thermochemical storage process adapted to solar thermal systems. *Solar Energy*, 80:883–893.

Lai, C., Chen, R. H., Lin, C. (2010). Heat transfer and thermal storage behaviour of gypsum boards incorporating micro-encapsulated PCM. *Energy and Buildings*, 42:1259–1266.

Lovergrove, K., Luzzi, A., Kreetz, H. (1999a). A solar-driven ammonia-based thermochemical energy storage system. *Solar Energy*, 67:309-316.

Lovergrove, K., Luzzi, A., McCann, M., Freitag, O. (1999b) Exergy analysis of ammonia-based solar thermochemical power systems. *Solar Energy*, 66:103–115.

Masruroh, N. A., Li, B., Klemes, J. (2006). Life cycle analysis of a solar thermal system with thermochemical storage process. *Renewable Energy*, 31:537–548.

Mathur, A., Kasetty, R., Hardin, C. (2010). A practical phase change thermal energy storage for concentrating solar power plants. *Proc. EuroSun 2010 Conf.*, September 28–October 1, Graz, Austria, paper 217.

Mauran, S., Lahmidi, H., Goetz, V. (2008). Solar heating and cooling by a thermochemical process: first experiments of a prototype storing 60 kWh by a solid/gas reaction. *Solar Energy*, 82:623–636.

McClenahan, D., Gusdorf, J., Kokko, J., Thornton, J., Wong, B. (2006). Okotoks: seasonal storage of solar energy for space heat in a new community. *ACEEE, 2006 Summer Study on Energy Efficiency in Buildings*, Pacific Grove, CA.

Paksoy, H. O. (editor) (2007). *Thermal Energy Storage for Sustainable Energy Consumption: Fundamentals, Case Studies and Design*. NATO Science Series II: Mathematics, Physics and Chemistry, Vol. 234. Springer, New York.

Piette, M. A. (1990). *Learning from Experience with Diurnal Thermal Energy Storage Managing Electric Loads in Buildings*. Analysis Support Unit, Centre for Analysis and Dissemination of Demonstrated Energy Technologies (CADDET), International Energy Agency.

Rosen, M. A., Tang, R., Dincer, I. (2004). Effect of stratification on energy and exergy capacities in thermal storage systems. *International Journal of Energy Research*, 28:177–193.

SAIC. (2010). *Drake Landing Solar Community Energy: Annual Report for 2008–2009*. Report CM002171, Science Applications International Corporation (SAIC Canada), Ottawa, February 3.

Saito, A. (2002). Recent advances in research on cold thermal energy storage. *International Journal of Refrigeration*, 25:177–189.

Sari, A., Karaipekli, A. (2009). Preparation, thermal properties and thermal reliability of palmitic acid/expanded graphite composite as form-stable PCM for thermal energy storage. *Solar Energy Materials and Solar Cells*, 93:571–576.

Scalat, S., Banu, D., Hawes, D., Parish, J., Haghighata, F., Feldman, D. (1996). Full scale thermal testing of latent heat storage in wallboard. *Solar Energy Materials and Solar Cells*, 44:49–61.

SFEO. (1990). *Guide to Seasonal Storage*. Swiss Federal Energy Office.

Sharma, A., Tyagi, V. V., Chen, C. R., Buddhi, D. (2009). Review on thermal energy storage with phase change materials and applications. *Renewable and Sustainable Energy Reviews*, 13:318–345.

Shilei, L., Guohui, F., Neng, Z., Li, D. (2007). Experimental study and evaluation of latent heat storage in phase change materials wallboards. *Energy and Buildings*, 39:1088–1091.

Sibbitt, B., Onno, T., McClenahan, D., Thornton, J., Brunger, A., Kokko, J., Wong, B. (2007). The Drake Landing Solar Community project: early results. *Proc. Canadian Solar Buildings Conference*, Calgary, June 10–14.

Singh, H., Saini, R. P., Saini, J. S. (2010). A review on packed bed solar energy storage systems. *Renewable and Sustainable Energy Reviews*, 14:1059–1069.

Van de Ven, H. (1999). Status and trends of the European heat pump market. *IEA Heat Pump Center Newsletter*, 17/1:10–12.

van Essen, V. M., Bleijendaal, L. P. J., Kikkert, B. W. J., Zondag, H. A., Bakker, M., Bach, P. W. (2010). Development of a compact heat storage system based on salt hydrates. *Proceedings of Eurosun 2010, The 3rd International Conference on Solar Heating, Cooling and Buildings*, September 28–October 1, Graz, Austria.

van Helden, W. G. J. (2009). Compact thermal energy storage. Available at: http://www.leonardo-energy.org/drupal/node/3916 (Accessed August 4, 2011).

Velraj, R., Seeniraj, R. V., Hafner, B., Faber, C., Schwarzer, K. (1999). Heat transfer enhancement in a latent heat storage system. *Solar Energy*, 65:171–180.

Visscher, K., Veldhuid, J. B. J. (2005). Comparison of candidate material for seasonal storage of solar heat through dynamic simulation of building and renewable energy system. *Proc. Ninth International IBPSA Conference*, Montreal, Canada, pp. 1285–1292.

Weber, R., Dorer, V. (2008). Long-term heat storage with NaOH. *Vacuum*, 82:708–716.

Wettermark, G. (1989). Thermochemical energy storage. *Proceedings of the NATO Advanced Study Institute on Energy Storage Systems*, 167:673–681.

Wong, W. P., McClung, J. L., Snijders, A. L., Kokko, J. P., McClenahan, D., Thornton, J. (2006). First large-scale solar seasonal borehole thermal energy storage in Canada. *Proc. Ecostock 2006 Conference*, Stockton, NJ.

Zalba, B., Marin, J. M., Cabeza, L. F., Mehling, H. (2004). Free-cooling of buildings with phase change materials. *International Journal of Refrigeration*, 27:839–849.

Zondag, H. A., Kalbasenka, A., van Essen, M., Bleijendaal, L., van Helden, W., Krosse, L. (2008). First studies in reactor concepts for thermochemical storage. *Proceedings of Eurosun 2008, 1st International Conference on Solar Heating, Cooling and Buildings*, 7–10 October, Lisbon, Portugal.

Zondag, H., van Essen, V. M., He, Z., Schuitema, R., van Helden, W. G. J. (2007). Characterization of $MgSO_4$ for thermochemical storage. *Proc. Second International Renewable Energy Storage Conference (IRES II)*, Bonn, Germany.

34

SHALLOW GEOTHERMAL SYSTEMS: COMPUTATIONAL CHALLENGES AND POSSIBILITIES

Rafid Al-Khoury

34.1 INTRODUCTION

Geothermal energy comprises an enormous heat generated in the core of the earth, about 6000 km below the surface. As a result of radiogenic decay of naturally occurring isotopes; particularly those of potassium, uranium, and thorium; temperatures hotter than the surface of the sun are continuously produced inside the earth. The temperature at the core of the earth reaches up to 5000°C. Far from the core and toward the surface of the earth, this temperature decreases gradually to reach around 10°C at the surface. This gradient of temperature drives a continuous flux of heat from the core of the earth to the surface, forming a wide range of energy resources.

Geothermal energy resources range from shallow depths, of the order of few tens to few hundreds of meters; to intermediate depths, of the order of 1 or 2 km; to even deeper depths with extremely high temperatures of molten rocks. Energy systems depending on the first type of energy resources are commonly referred to as shallow geothermal systems, whereas those depending on the other two types of resources are commonly referred to as deep geothermal systems. They offer a number of advantages over conventional fossil-fuel energy resources. Mainly, the geothermal heat resources are renewable and economic and their environmental impact in terms of CO_2 emission is significantly lower.

Shallow geothermal systems constitute important resources of geothermal energy and are suitable for heating and cooling individual buildings and compounds. The continuous thermal interaction between the air and the earth makes the first 100 m sustainable and hence, suitable for supply and storage of thermal energy, though at a relatively low temperature (low-enthalpy energy). The seasonal temperature variation at the ground surface is reduced to a nearly constant temperature of around 12°C at about 20 m below the surface. Under this depth, the temperature is known to increase with an average gradient of 3°C per 100 m depth.

Currently, the technology for building shallow geothermal systems is fairly advanced as drilling equipment, heat pumps, and materials are available, and the installation is relatively cheap to invest. Heat transfer theory of shallow geothermal systems is also well established. However, the computational modeling is still lacking behind. So far, and even though this research field is relatively new, a large number of computational models have been introduced. Yet, they are limited to academic level, though some engineering design tools are in use. This might be attributed to the relative difficulty in modeling two main aspects: the geometry and the physical processes. Modeling these two aspects is computationally challenging. The geometry of a typical shallow geothermal system involves very slender borehole heat exchangers (BHEs) embedded in a vast soil mass. This disproportionate geometrical scale makes the computational procedure awkward and costly. The physical processes in a shallow geothermal system, on the other hand, involve conduction and convection. The governing partial differential equations in such a system are quasi-hyperbolic functions, the solution of which leads most likely to spurious oscillations. As a consequence of these two attributes, most current models are made either extremely simplified or computationally very demanding.

Simplified models are essentially formulated based on the analytical solutions of Carslaw and Jaeger in 1947 (Carslaw and Jaeger, 1959), who seem to be the first to introduce a

Alternative Energy and Shale Gas Encyclopedia, First Edition. Edited by Jay H. Lehr and Jack Keeley.

comprehensive treatment of heat conduction in solids subjected to different combinations of initial and boundary conditions. Heat flow in finite, semi-infinite, and infinite domains subjected to point, line, plane, cylindrical, and spherical heat sources was studied in their works between 1947 and 1959. Based on Carslaw and Jaeger works, Ingersoll and Plass (1948) and Ingersoll et al. (1954) made a significant contribution to the field of heat conduction in solids and provided a practical framework for modeling geothermal systems.

Models, which are computationally demanding, are essentially formulated on the basis of numerical methods, such as the finite difference, the finite volume, or the finite element. Such models, in spite of their versatility, are usually not feasible for utilization in engineering practice, since they normally require large computer memories and CPU time.

To circumvent these problems and acquire computationally feasible tools, which are versatile and efficient, a good combination between a mathematical model and a computational procedure is essential. With such a combination, important aspects of heat transfer mechanisms can be captured, and significant computational processes and time can be gained. In this chapter, we discuss such a combination.

This chapter provides a framework for developing versatile and efficient computational tools for shallow geothermal systems. It describes the challenges encountering the modeling of shallow geothermal systems and the possibilities to tackle them. We provide an introduction to analytical, semi-analytical, and numerical solution techniques suitable for solving heat flow in shallow geothermal systems. Focus is placed on modeling ground-source heat pump (GSHP) systems, in particular, the vertical BHEs and their thermal interactions with the soil mass. Yet, the computational models are generic and can be extended to model other types of geothermal systems.

This chapter is structured into four main parts: introduction to shallow geothermal systems, governing fluid and heat flow equations in a soil mass and BHEs, analytical and semi-analytical modeling, and numerical modeling. Essential parts of the text and Figures 34.2, 34.3, 34.4, and 34.5 are taken (with permission) from Al-Khoury (2012), where an in-depth treatment of this topic is given.

34.2 SHALLOW GEOTHERMAL SYSTEMS

Commonly, there are two systems that make use of shallow geothermal energy: *ground-source heat pump* and *underground thermal energy storage*. Hereafter, an overview of these two systems is given.

34.2.1 Ground-Source Heat Pump

Ground-source heat pump (GSHP), also known as geothermal heat pump, or downhole heat exchanger, is an important

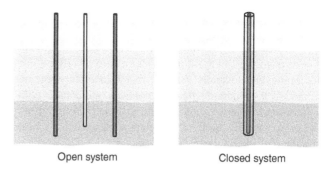

Open system   Closed system

FIGURE 34.1   Open and closed geothermal systems.

renewable energy system based on shallow geothermal resources. In practice, there are mainly two kinds of GSHP: *open* and *closed* (Figure 34.1). The open system, also known as doublet, extracts heat from a wet aquifer using at least two separate wells: one for extraction and another for reinjection. The basic advantage of this system, as compared to the closed, is that it is more efficient, the degree of which depends on the permeability of the aquifer formation and the quality of the groundwater.

The closed system, on the other hand, extracts heat from the earth using heat exchangers embedded in dry or wet soil layers. The heat is carried to a heat pump by a fluid circulating inside the heat exchanger pipes. The heat pump raises the fluid temperature to the required level and then sends it to the heating system inside the building. The main advantage of this system, as compared to the open, is that there is no need for an aquifer, and, in principle, only one borehole is necessary to extract the thermal energy. Furthermore, the circulating fluid is usually clean, and hence the system requires less maintenance.

Closed systems are installed in mainly two ways: horizontal and vertical. The horizontal systems, also known as ground heat collectors or horizontal loops, are relatively easy to install, as less complicated excavation equipment is needed. The main source of heat of this kind of systems is the surface temperature, which arises from direct exposure to the sun. Depending on the area available, the horizontal pipes are laid in different ways, including series, parallel, or spiral (Mands and Sanner, 2005).

The vertical system is widely used, especially in areas where the land is scarce. The heat in such a system is extracted by a BHE, also known as vertical ground heat exchanger or downhole heat exchanger, which consists of a plastic pipe, mainly polyethylene or polypropylene, installed in a borehole in a U-shape (or coaxial) and fixed by filling the borehole with grout. The U-tube carries a circulating fluid, referred to in the literature as refrigerant or working fluid. The U-tube (or the coaxial tube) effectively forms two pipes. One pipe receives the circulating fluid from the heat pump and conveys it downward. We denote this pipe as *pipe-in*. The other pipe collects the circulating fluid at the bottom of pipe-in and brings it out to the surface, to enter the heat pump. We denote this pipe as *pipe-out*. The heat pump, usually located inside the building, extracts designed amount of heat from the fluid and pumps it back to the BHE. The circulating fluid, usually water with 20–25% anti-freezing coolant such as monoethylene glycol, gets into contact with the surrounding soil via the U-tube material and the grout. The grout, usually Bentonite–cement mix, exchanges heat with the soil and the BHE inner pipes.

BHEs are slender heat pipes with dimensions of the order of 30 mm in diameter for the inner pipes, 150 mm in diameter for the borehole, and 100 m in length for the borehole and the inner pipes. In practice, there are different types of BHE. They differ in their configurations, but in practice, there are two main types:

- *U-tubes*: This BHE configuration consists of one (or more) inner pipe, which is inserted in the borehole from its middle to form a U-shape. Two main configurations are in use: *single U-tube* and *double U-tube*. Figure 34.2 shows a sketch of these two types of BHE. Single U-tube BHE consists of pipe-in, where the working fluid enters; pipe-out, where the working fluid leaves; and grout, where the contact with the surrounding soil mass takes place. The double U-tube consists of two pipes-in, two pipes-out, and grout.

- *Coaxial*: This BHE configuration consists of concentric pipes. In practice, there are mainly two coaxial configurations: *annular* (CXA) and *centered* (CXC) (Figure 34.3). In CXA, pipe-out is configured inside

pipe-in, forming an annular inlet and a centered outlet. Heat exchange with the grout occurs along the surface area of pipe-in. In CXC, pipe-in is configured inside pipe-out, forming a centered inlet and an annular outlet. Here, heat exchange with the grout occurs via pipe-out.

The heat pump is an important part of the shallow geothermal system. It is a mechanical device, which transfers thermal energy from a region of a lower temperature to a region of a higher temperature. In the heating mode, the heat pump exerts work on the refrigerant to increase its temperature and conveys the extracted heat to the heating system. In the cooling mode, the system is reversed, and heat from the building is extracted and injected into the ground. The effectiveness of the heat pump is described in terms of the *coefficient of performance* (COP). COP is defined as the ratio of the heat transferred into the system to the work required to transfer that heat, that is,

$$\mathrm{COP} \equiv \frac{\text{heat transferred}}{\text{work done by pump}} = \frac{Q}{W} \qquad (34.1)$$

If the work needed to raise 1 kWh of heat is 0.25 kWh, the COP of the system is 4. That is the heat transferred to the building is four times greater than the work done by the motor of the heat pump. Shallow geothermal systems using geothermal heat pumps typically have COP ranging from 3.5 to 4 at the beginning of the heating season. As the ground temperature decreases due to heat extraction, the COP drops.

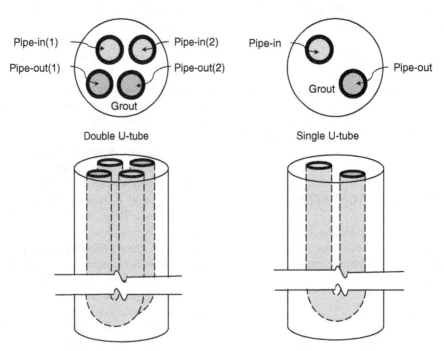

FIGURE 34.2 Single and double U-tubes BHE (see Al-Khoury, 2012).

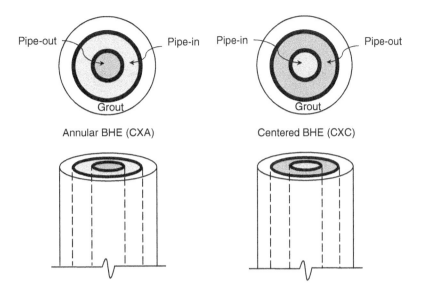

FIGURE 34.3   Annular (CXA) and centered (CXC) coaxial tubes (see Al-Khoury, 2012).

34.2.2   Underground Thermal Energy Storage

Storing energy in shallow underground depths is another renewable energy system, as cold and hot water stored in the ground can be utilized for heating and cooling of buildings all over the year. As for geothermal heat pumps, two systems are in use: open, known as *aquifer thermal storage*, and closed, known as *borehole storage* or *duct thermal storage* (Dincer and Rosen, 2011).

34.2.2.1   *Aquifer Thermal Storage.*   The aquifer thermal storage (ATES) is an open system, where the heat is carried by the groundwater. It is usually built in porous aquifers with medium to high permeability. Typically, two wells are used: one for the warm water and another for the cold water. In winter, the warm water is utilized for heating, and the resulting cold water is stored in the cold reservoir. In summer, the process is reversed. The advantage of this system is that it is environmentally safe. The extracted groundwater does not mix with other types of waters and thus cannot be contaminated. Moreover, the amount of the extracted water is equal to the amount of reinjected water, and hence there is no net loss of water from the ground. The main problem is that this system works only on areas with permeable aquifers.

34.2.2.2   *Duct Thermal Storage.*   The duct thermal storage (DTES) is a closed system and somewhat more complicated than the ATES. In this system, boreholes of typically 50–200 m in depth are drilled and within them are inserted plastic heat exchangers carrying fluid. The circulating fluid carries the thermal energy from the earth to the surface and vice versa. The advantage of this system, as compared to the ATES, is that the pipe network and fittings are much less exposed to scaling and clogging. The efficiency of this system, however, depends on the geology, the temperature of the ground, the groundwater conditions, and the thermal properties of the ground.

Computational modeling of underground thermal storage systems is similar to that of the ground-source geothermal systems. Hence, in this chapter, we focus on modeling ground-source geothermal systems, as modeling of heat transfer in the underground thermal storage systems and other types of shallow geothermal systems can be conducted by fine-tuning and tailoring.

34.3   HEAT EQUATIONS IN SHALLOW GEOTHERMAL SYSTEMS

Temperature distribution of a shallow geothermal system can be computed by a superposition of two thermally interacting entities: the soil mass and the BHE. The governing heat and fluid flow equations of these two entities are given hereafter.

34.3.1   Heat and Fluid Flow Equations of a Soil Mass

Natural soil is a porous medium constituting particles (phases) of different thermodynamic properties. Typical soil mass consists of three phases: solid, water, and air. Within the context of shallow geothermal systems, the soil mass can be considered as fully saturated two-phase porous medium constituting a solid phase, s, and a water phase, w. The macroscopic energy field equations of such a medium can be described as

$$
\begin{aligned}
(\rho c_{p})_{s} \frac{D_{s} T_{s}}{Dt} &= -\operatorname{div} \mathbf{q}_{s} + \rho_{s} Q_{s} \\
(\rho c_{p})_{w} \frac{D_{w} T_{w}}{Dt} &= -\operatorname{div} \mathbf{q}_{w} + \rho_{w} Q_{w}
\end{aligned}
\tag{34.2}
$$

where $(\rho c_{p})_{\pi}$ is the volume heat capacity of π (s or w) phase, ρ_{π} is its mass density, Q_{π} is the intrinsic heat source, and

$$
\mathbf{q}_{\pi} = -\lambda_{\pi} \operatorname{grad} T_{\pi}
\tag{34.3}
$$

is the heat flux (Fourier's law) with λ_{π} the thermal conductivity of π phase and T_{π} its temperature. The operator D/Dt in equation (34.2) is the material time derivate, defined as

$$
\frac{D}{Dt} = \frac{\partial}{\partial t} + \operatorname{grad} \cdot \mathbf{v}
\tag{34.4}
$$

with $\mathbf{v}$ the field velocity vector.

Thermal parameters of the soil constituents are significantly different from each other. For instance, the thermal conductivity of sand reaches up to 4 W/m K, whereas for water it is 0.56 W/m K. Similarly, the specific heat of sand is on the order of 1 kJ/kg K, but for water it is 4.18 kJ/kg K. It is therefore expected that at the microscopic level there is a temperature gradient between the solid particles and the water particles. The significance of this gradient depends on time and on whether there is a heat sink or a source in one of the phases. For a highly transient case, or when there is a heat generation in one of the phases, the material is in *local thermal nonequilibrium*. However, for an ordinary transient conduction or conduction–convection case, the microscopic temperature gradient is generally much smaller than that at the macroscopic level, and hence can be ignored. In this case, the material is in *local thermal equilibrium*. The mechanism of heat transfer in shallow geothermal systems is dominated by the second equilibrium type.

In shallow geothermal systems, the temperature gradient at the microscopic (pore) level is less than that at the macroscopic level, and both are much less than that occurring at the megascopic level, the physical system. That is,

$$
\Delta T_{micro} < \Delta T_{macro} \ll \Delta T_{mega}
\tag{34.5}
$$

This indicates that the local temperature gradient between the phases is negligible, allowing for assuming that the solid and the fluid phases within a representative elementary volume (REV) are in local thermal equilibrium. This means that the average temperatures of the two phases are equal, such that

$$
T_{s} = T_{w} = T
\tag{34.6}
$$

Applying this condition to the energy field equations, equation (34.2), and using equation (34.4) give

$$
(\rho c_{p})_{s} \frac{\partial T}{\partial t} + (\rho c_{p})_{s} \mathbf{v}_{s} \cdot \nabla T + \nabla \cdot \lambda_{s} \cdot \nabla T = 0
\tag{34.7}
$$

$$
(\rho c_{p})_{w} \frac{\partial T}{\partial t} + (\rho c_{p})_{w} \mathbf{v}_{w} \cdot \nabla T + \nabla \cdot \lambda_{w} \cdot \nabla T = 0
\tag{34.8}
$$

Adding equation (34.7) to equation (34.8) and omitting the convective term of the solid, $(\rho c_{p})_{s} \mathbf{v}_{s} \cdot \nabla T$, the system macroscopic field equation reads

$$
(\rho c_{p})_{eff} \frac{\partial T}{\partial t} + (\rho c_{p})_{w} \mathbf{v}_{w} \cdot \nabla T + \nabla \cdot \lambda_{eff} \cdot \nabla T = 0
\tag{34.9}
$$

in which $(\rho c_{p})_{eff}$ and λ_{eff} are the local averaged effective heat capacity and thermal conductivity of the porous medium, respectively. These parameters are usually obtained by a simple volume averaging, such that

$$
(\rho c_{p})_{eff} = n (\rho c_{p})_{w} + (1 - n) (\rho c_{p})_{s}
\tag{34.10}
$$

$$
\lambda_{eff} = n \lambda_{w} + (1 - n) \lambda_{s}
\tag{34.11}
$$

where n is the porosity. In what follows, c will be utilized instead of c_{p}.

Equation (34.9) contains two unknowns, the temperature and the fluid Darcy velocity. To solve this indeterminate equation, it is necessary to introduce the momentum conservation (balance) equation describing the fluid flow. The fluid flow balance equation can be expressed as

$$
S \frac{\partial \varphi}{\partial t} = -\nabla \cdot (\mathbf{k} \nabla \varphi) + Q_{p}(x, y, z)
\tag{34.12}
$$

in which S is the specific storage coefficient, representing the amount of water released per unit area per head gradient, Q_{P} is a water sink or source, and φ is the total head, described as

$$
\varphi = \frac{P}{\rho g} + z
\tag{34.13}
$$

where $P/\rho g$ is the pressure head and z is the elevation head. The Darcy's velocity can be expressed as

$$
\mathbf{v} = -\mathbf{k} \nabla \varphi
\tag{34.14}
$$

in which $\mathbf{k} = \bar{\mathbf{k}} \rho g / \mu$ is the hydraulic conductivity tensor, commonly termed permeability (m/s), with $\bar{\mathbf{k}}$ the intrinsic permeability tensor, μ the dynamic viscosity, and g the gravity. For three-dimensional problems, the permeability is described as anisotropic in the principal axes as

$$
\mathbf{k} = \operatorname{diag}(k_{x}, k_{y}, k_{z})
\tag{34.15}
$$

For an isotropic material $k_{x} = k_{y} = k_{z} = k$.

34.3.2 Heat Equations of Borehole Heat Exchangers

Modeling heat transfer in a BHE is challenging as it involves conductive–convective processes occurring in a thermally interacting multiple component medium. To derive heat equations of a BHE, we need to consider the particularity of its geometry: slenderness and multiple channels. Typical BHEs are quite slender with length to diameter ratio of the order of 700 or more. This entails that the main stream of heat flow is along its axial axis. The radial heat flow is negligible. Therefore, it is reasonably accurate to utilize a one-dimensional heat equation to describe heat flow in the involved BHE components.

A typical BHE consists of one or more plastic U-tubes, surrounded by a bentonite–cement grout. The U-tubes carry a circulating (working) fluid, usually water with 20–25% anti-freezing coolant. Since these materials have different thermal properties, the mechanisms of heat transfer in each component differ significantly from the others and, as a result, affect heat transfer in the whole BHE. Each BHE component acts as a channel for transferring heat along its axial dimension and exchanging heat along its contact surface area with other components.

Following these mechanisms and assumptions, the energy balance equation for each individual pipe component, α, can be described as

$$\rho c_\alpha \frac{\partial T_\alpha}{\partial t} + \frac{\partial q_\alpha}{\partial z} = Q_\alpha, \qquad \alpha = 1, 2, \ldots, n \quad (34.16)$$

where $c_\alpha = (c_p)_\alpha$ is the specific heat of α component (pipe-in, pipe-out, or grout) and Q_α is a heat source, which can be described by a linear function of temperature differences between pipe components in thermal contact as

$$Q_\alpha = \sum_{\beta=1}^{n} b_{\alpha\beta} (T_{\alpha\alpha} - T_{\alpha\beta}) \quad (34.17)$$

where $b_{\alpha\beta}$ is a thermal coefficient describing the thermal interaction between pipe component α and pipe component β. For the whole borehole, the summation of heat sources due to thermal interactions between the involved components must vanish (Aifantis, 1979), that is,

$$\sum_{\alpha=1}^{n} Q_\alpha = 0 \quad (34.18)$$

Based on equations (34.16) and (34.17), it is possible to formulate heat equations of all kinds of BHEs (Al-Khoury, 2012). Here, we formulate four commonly used BHEs, namely, single U-tube (1U), double U-tube (2U), coaxial pipe with annular (CXA), and coaxial pipe with centered inlet (CXC).

34.3.2.1 Heat Equations of a Single U-Tube Borehole Heat Exchanger (1U).

Heat equations of a single U-tube BHE, consisting of pipe-in (denoted as i), pipe-out (denoted as o), and grout (denoted as g), can be expressed as follows:

Pipe-in

$$(\rho c)_{\text{ref}} \frac{\partial T_i}{\partial t} - \lambda_{\text{ref}} \frac{\partial^2 T_i}{\partial z^2} + (\rho c u)_{\text{ref}} \frac{\partial T_i}{\partial z} = b_{ig} (T_i - T_g) \quad (34.19)$$

Pipe-out

$$(\rho c)_{\text{ref}} \frac{\partial T_o}{\partial t} - \lambda_{\text{ref}} \frac{\partial^2 T_o}{\partial z^2} - (\rho c u)_{\text{ref}} \frac{\partial T_o}{\partial z} = b_{og} (T_o - T_g) \quad (34.20)$$

Grout

$$(\rho c)_g \frac{\partial T_g}{\partial t} - \lambda_g \frac{\partial^2 T_g}{\partial z^2} = b_{ig} (T_g - T_i) + b_{og} (T_g - T_o) \quad (34.21)$$

where the subscript ref indicates the refrigerant, u the fluid velocity, and b_{ig} the thermal coefficient describing the thermal interaction between pipe-in and grout, and so on.

34.3.2.2 Heat Equations of a Double U-Tube Borehole Heat Exchanger (2U).

Heat equations of a double U-tube BHE, consisting of two pipes-in (denoted as i1 and i2), two pipes-out (denoted as o1 and o2), and grout (denoted as g), can be described as follows:

Pipes-in

$$(\rho c)_{\text{ref}} \frac{\partial T_{i1}}{\partial t} - \lambda_{\text{ref}} \frac{\partial^2 T_{i1}}{\partial z^2} + (\rho c u)_{\text{ref}} \frac{\partial T_{i1}}{\partial z} = b_{ig1} (T_{i1} - T_g)$$

$$(\rho c)_{\text{ref}} \frac{\partial T_{i2}}{\partial t} - \lambda_{\text{ref}} \frac{\partial^2 T_{i2}}{\partial z^2} + (\rho c u)_{\text{ref}} \frac{\partial T_{i2}}{\partial z} = b_{ig2} (T_{i2} - T_g) \quad (34.22)$$

Pipes-out

$$(\rho c)_{\text{ref}} \frac{\partial T_{o1}}{\partial t} - \lambda_{\text{ref}} \frac{\partial^2 T_{o1}}{\partial z^2} - (\rho c u)_{\text{ref}} \frac{\partial T_{o1}}{\partial z} = b_{og1} (T_{o1} - T_g)$$

$$(\rho c)_{\text{ref}} \frac{\partial T_{o2}}{\partial t} - \lambda_{\text{ref}} \frac{\partial^2 T_{o2}}{\partial z^2} - (\rho c u)_{\text{ref}} \frac{\partial T_{o2}}{\partial z} = b_{og2} (T_{o2} - T_g) \quad (34.23)$$

Grout

$$(\rho c)_g \frac{\partial T_g}{\partial t} - \lambda_g \frac{\partial^2 T_g}{\partial z^2} = b_{ig1}(T_g - T_{i1}) + b_{ig2}(T_g - T_{i2})$$
$$+ b_{og1}(T_g - T_{o1}) + b_{og2}(T_g - T_{o2})$$

$$(34.24)$$

34.3.2.3 Heat Equations of a Coaxial Borehole Heat Exchanger with Annular (CXA). Heat equations of a CXA BHE, consisting of pipe-in, pipe-out, and grout, can be described as follows:

Pipe-in

$$(\rho c)_{ref} \frac{\partial T_i}{\partial t} - \lambda_{ref} \frac{\partial^2 T_i}{\partial z^2} + (\rho c u)_{ref} \frac{\partial T_i}{\partial z}$$
$$= b_{ig}(T_i - T_g) + b_{io}(T_i - T_o) \qquad (34.25)$$

Pipe-out

$$(\rho c)_{ref} \frac{\partial T_o}{\partial t} - \lambda_{ref} \frac{\partial^2 T_o}{\partial z^2} - (\rho c u)_{ref} \frac{\partial T_o}{\partial z} = b_{io}(T_o - T_i)$$

$$(34.26)$$

Grout

$$(\rho c)_g \frac{\partial T_g}{\partial t} - \lambda_g \frac{\partial^2 T_g}{\partial z^2} = b_{ig}(T_g - T_i) \qquad (34.27)$$

34.3.2.4 Heat Equations of a Coaxial Borehole Heat Exchanger with Centered Inlet (CXC). Heat equations of a CXC BHE, consisting of pipe-in, pipe-out, and grout, can be described as follows:

Pipe-in

$$(\rho c)_{ref} \frac{\partial T_i}{\partial t} - \lambda_{ref} \frac{\partial^2 T_i}{\partial z^2} + (\rho c u)_{ref} \frac{\partial T_i}{\partial z} = b_{io}(T_i - T_o)$$

$$(34.28)$$

Pipe-out

$$(\rho c)_{ref} \frac{\partial T_o}{\partial t} - \lambda_{ref} \frac{\partial^2 T_o}{\partial z^2} - (\rho c u)_{ref} \frac{\partial T_o}{\partial z}$$
$$= b_{io}(T_o - T_i) + b_{og}(T_o - T_g) \qquad (34.29)$$

Grout

$$(\rho c)_g \frac{\partial T_g}{\partial t} - \lambda_g \frac{\partial^2 T_g}{\partial z^2} = b_{og}(T_g - T_o) \qquad (34.30)$$

34.4 ANALYTICAL AND SEMI-ANALYTICAL MODELING OF SHALLOW GEOTHERMAL SYSTEMS

In geothermics, there are several models for heat conduction in soil masses subjected to different heat sources that can be classified into two distinct, but overlapping, categories: analytical and semi-analytical. The analytical models are mainly based on the idea of heat conduction in an infinite domain subjected to a point source introduced by Kelvin in 1880, when he used Fourier's theory of heat conduction to study the transmission of electric signals through submarine cables. Based on Kelvin's work, Carslaw and Jaeger introduced in 1947 a comprehensive treatment of heat conduction in solids, including finite, semi-infinite, and infinite domains subjected to different initial and boundary conditions. Between 1948 and 1954, Ingersoll and his coworkers provided practical applications of point and infinite line source solutions and established the framework for more elaborate modeling of shallow geothermal systems (Ingersoll and Plass, 1948; Ingersoll et al., 1954).

The semi-analytical models are more versatile, as the domain can be multidimensional and the boundary conditions can be more complicated. So far, there are few semi-analytical models for shallow geothermal systems. One type of model is based on coupling between the Laplace transforms and the finite difference method, and another is based on eigenfunction expansions and Fourier transforms.

34.4.1 Analytical Models

Heat conduction in a soil mass is commonly simulated using three analytical models: the infinite line source model (ILS), the finite line source model (FLS), and the infinite cylindrical source model (ICS). These models calculate heat conduction in a one-dimensional (mainly radial) soil mass, embedded in which is a BHE emitting a constant heat flow rate per unit length. On the other hand, heat flow in a BHE is commonly modeled using Eskilson and Claesson model (Eskilson, 1987; Eskilson and Claesson, 1988). This model calculates steady-state convective heat flow. In the following, a brief description of these models is presented.

34.4.1.1 Analytical Soil Models.

Infinite Line Source Model (ILS).   ILS model describes heat flow in an infinite soil mass subjected to a constant heat flux emitted from an infinite line source, representing a BHE, embedded along the vertical axis. This arrangement entails that heat flow in the soil mass occurs only in the radial direction, and the thermal interaction between the soil and the borehole takes place along the centerline of the borehole, not along its surface area. The governing initial and boundary value problem of the ILS model is described in the cylindrical coordinate system in terms of the temperature difference, $u(r,t) = T(r,t) - T_0$, with T_0 the initial soil temperature, as

$$\frac{1}{\alpha}\frac{\partial u}{\partial t} = \frac{\partial^2 u}{\partial r^2} + \frac{1}{r}\frac{\partial u}{\partial r}$$

$$u(r,0) = 0$$
$$u(r \to \infty, t) = 0 \qquad\qquad (34.31)$$

$$-\lambda\frac{\partial u}{\partial r}\cdot 2\pi r\bigg|_{r\to 0} = q_0$$

in which $\alpha = \lambda/\rho c$ is the thermal diffusivity and q_0 is the BHE heat flux. Solving this set of equations is usually done using the Laplace transform, which yields

$$T(r,t) - T_0 = \frac{q_0}{4\pi\lambda}\mathrm{Ei}\left(\frac{r^2}{4\alpha t}\right)$$
$$= \frac{q_0}{2\pi\lambda}\mathrm{Ei}\left(\frac{r}{2\sqrt{\alpha t}}\right) \qquad (34.32)$$

where $\mathrm{Ei}(x)$ is the exponential integral function, defined as

$$\mathrm{Ei}(x) = \int_x^\infty \frac{e^{-v}}{v}\,dv \qquad (34.33)$$

in which $v = r^2/4\alpha t$. Ingersoll and Plass (1948) provided tabulated values of the exponential integral function. For $\sqrt{v} < 0.2$, the integral in equation (34.33) can be approximated by

$$I(v) = \ln\frac{1}{v} + \frac{v^2}{2} - \frac{v^4}{8} - 0.2886 \qquad (34.34)$$

Ingersoll and Plass recommended using the ILS model only for applications with Fourier's number, $\alpha t/r_b^2 > 20$. For smaller values, the solution gets distorted in the shorter time scale because the effect of the actual finite length of the line source (BHE) becomes significant.

From equation (34.32) we can readily notice that as $t \to \infty$, $\mathrm{Ei}(0) = \infty$, and thus there exists no steady-state solution to this model. However, for large t, Carslaw and Jaeger have shown that the temperature can be approximated as

$$T(r,t) - T_0 \simeq \frac{q_0}{4\pi\alpha}\left(\ln\frac{4\alpha t}{r^2} - \gamma\right) \qquad (34.35)$$

in which $\gamma = 0.5772...$ is Euler's constant.

For an instantaneous line heat source, where a heat flux is released suddenly at $t = 0$ with a strength q_i (J/m), the

solution of the heat equation leads to (Yener and Kakac, 2008)

$$T(r,t) - T_0 = \frac{q_i}{4\pi\lambda}\frac{1}{t}\exp(-r^2/4\alpha t) \qquad (34.36)$$

Infinite Cylindrical Source Model (ICS). Similar to ILS model, the infinite cylindrical source model simulates heat conduction in a soil mass subjected to a constant heat flow rate. The difference, however, is that the contact area between the BHE and the soil is along the surface area of the borehole, that is, at $r = r_b$, the borehole radius. The initial and boundary value problem of the ICS model is described as

$$\frac{1}{\alpha}\frac{\partial u}{\partial t} = \frac{\partial^2 u}{\partial r^2} + \frac{1}{r}\frac{\partial u}{\partial r}, \quad r > r_b$$

$$u(r,0) = 0$$
$$u(r \to \infty, t) = 0 \qquad\qquad (34.37)$$

$$-\lambda\frac{\partial u}{\partial r}\cdot 2\pi r\bigg|_{r=r_b} = q_0$$

Using Laplace transforms, the solution can be expressed as (Carslaw and Jaeger, 1959)

$$T(r,t) - T_0 = -\frac{2Q_0}{\pi\lambda}\int_0^\infty (1 - e^{-\alpha v^2 t})$$
$$\times\frac{J_0(vr)Y_1(vr_b) - Y_0(vr)J_1(vr_b)}{v^2\left[J_1^2(vr_b) + Y_1^2(vr_b)\right]}dv \qquad (34.38)$$

in which T_0 is the initial soil temperature and $Q_0 = q_0/2\pi r_b$, at $r = r_b$. Carslaw and Jaeger provided approximate solutions for short and long time scales. For a short time scale, the solution gives

$$T(r,t) - T_0 = \frac{2Q_0}{\lambda}\sqrt{\frac{\alpha r_b t}{r}}$$

$$\times\left(\mathrm{ierfc}\frac{r - r_b}{2\sqrt{\alpha t}} - \frac{(3r + r_b)\sqrt{\alpha t}}{4r r_b}\mathrm{i}^2\mathrm{erfc}\frac{r - r_b}{2\sqrt{\alpha t}} + \cdots\right)$$
$$(34.39)$$

where $\mathrm{i}^n\mathrm{erfc}$ is the iterated integrals of the complementary error function, defined as

$$\mathrm{i}^n\mathrm{erf}(z) = \int_z^\infty \mathrm{i}^{n-1}\mathrm{erf}(\zeta)\,d\zeta \qquad (34.40)$$

For a long time scale, the solution can be approximated by

$$T(r,t) - T_0 = \frac{Q_0\, r_b}{2\lambda} \left(\ln \frac{4\alpha t}{Cr^2} + \frac{r_b^2}{2\alpha t} \ln \frac{4\alpha t}{Cr^2} \right.$$

$$\left. + \frac{1}{4\alpha t} \left(r_b^2 + r^2 - 2\, r_b^2 \ln \frac{r_b}{r} \right) + \cdots \right) \quad (34.41)$$

in which $\ln C = 0.5772\ldots$ is Euler's constant.

Finite Line Source Model (FLS). FLS approximates the BHE by a finite line constituting a series of point sources. The temperature at point r in an infinite domain subjected to a point heat source is

$$T(r,t) - T_0 = \frac{q_0}{4\pi r\lambda}\, \mathrm{erfc}\,(r/2\sqrt{\alpha t}) \quad (34.42)$$

in which T_0 is the initial soil temperature. Following this, the temperature at point $P(r,z)$ in a soil mass subjected to a continuous point source representing a line heat source (BHE) can be described as

$$T(r,z,t) - T_0 = \frac{q_0}{4\pi\lambda} \int_0^L \frac{1}{r_1}\, \mathrm{erfc}\,(r_1/2\sqrt{\alpha t})\, d\zeta \quad (34.43)$$

where $r_1 = \sqrt{r^2 + (z - \zeta)^2}$, with ζ being any point along the borehole (see Figure 34.4).

Based on equation (34.43) and the method of images (for taking the effect of ground surface into consideration),

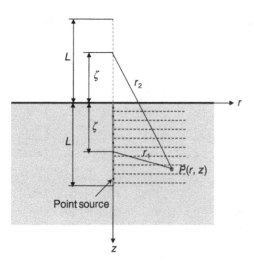

FIGURE 34.4   A schematic representation of the FLS model (see Al-Khoury, 2012).

Eskilson (1987) introduced the FLS model, which reads

$$T(r,z,t) - T_0 = \frac{q_0}{4\pi\lambda} \int_0^L$$

$$\times \left[\frac{1}{r_1}\mathrm{erfc}\left(\frac{r_1}{2\sqrt{\alpha t}} \right) - \frac{1}{r_2}\mathrm{erfc}\left(\frac{r_2}{2\sqrt{\alpha t}} \right) \right] d\zeta \quad (34.44)$$

in which

$$r_1 = \sqrt{r^2 + (z - \zeta)^2}, \quad r_2 = \sqrt{r^2 + (z + \zeta)^2} \quad (34.45)$$

In contrast to ISL, the FSL model enables the calculation of the steady-state temperature as

$$T(r,z) - T_0 = -\frac{q_0}{4\pi\lambda} \int_0^L \left(\frac{1}{r_1} - \frac{1}{r_2} \right) d\zeta \quad (34.46)$$

Solving equation (34.46) gives (Philippe et al., 2009)

$$T(r,z) - T_0 = -\frac{q_0}{4\pi\lambda}$$

$$\times \ln \left(\frac{\sqrt{r^2 + (z - L)^2} - (z - L)}{\sqrt{r^2 + (z + L)^2} + (z + L)} \cdot \frac{\sqrt{r^2 + z^2} + z}{\sqrt{r^2 + z^2} - z} \right)$$

$$(34.47)$$

34.4.1.2 Analytical Borehole Heat Exchanger Models.
Most of the current analytical models for heat flow in a BHE follow the Eskilson and Claesson model. They approximated the BHE by two interacting channels conveying a circulating fluid. The model assumes a steady-state convective heat flow, described as

$$\rho c\, Q\, \frac{dT_i}{dz} = \frac{T_b - T_i}{R_i} - \frac{T_i - T_o}{R_{io}}$$

$$-\rho c\, Q\, \frac{dT_o}{dz} = \frac{T_b - T_o}{R_o} + \frac{T_i - T_o}{R_{oi}} \quad (34.48)$$

in which Q is the flow rate, T_i is the pipe-in temperature, T_o is the pipe-out temperature, T_b is the temperature at the borehole wall, R_i is the thermal resistance in pipe-in, R_o is the thermal resistance in pipe-out, and $R_{io} = R_{oi}$ is the thermal resistance between pipe-in and pipe-out. The relevant boundary conditions are

$$T_i(0,t) = T_{in}(t)$$

$$T_i(L,t) = T_o(L,t) \quad (34.49)$$

in which T_{in} is the incoming temperature at the inlet of pipe-in and L is the length of the pipe. Using Laplace transforms,

Eskilson and Claesson solved this boundary value problem in the region $0 \leq z \leq L$ (Eskilson and Claesson, 1988) as

$$
\begin{aligned}
T_i(z,t) &= T_i(0,t)f_1(z) + T_o(0,t)f_2(z) \\
&\quad + \int_0^z T_b(\zeta,t)f_4(z-\zeta)\,d\zeta \\
T_o(z,t) &= -T_i(0,t)f_2(z) + T_o(0,t)f_3(z) \\
&\quad - \int_0^z T_b(\zeta,t)f_5(z-\zeta)\,d\zeta
\end{aligned}
\tag{34.50}
$$

where

$$
\begin{aligned}
f_1(z) &= e^{\beta z}[\cosh\gamma z - \delta\sinh\gamma z] \\
f_2(z) &= e^{\beta z}\frac{\beta_{12}}{\gamma}\sinh\gamma z \\
f_3(z) &= e^{\beta z}[\cosh\gamma z + \delta\sinh\gamma z] \\
f_4(z) &= e^{\beta z}\left[\beta_1\cosh\gamma z - \left(\delta\beta_1 + \frac{\beta_2\beta_{12}}{\gamma}\right)\sinh\gamma z\right] \\
f_5(z) &= e^{\beta z}\left[\beta_2\cosh\gamma z + \left(\delta\beta_2 + \frac{\beta_1\beta_{12}}{\gamma}\right)\sinh\gamma z\right]
\end{aligned}
\tag{34.51}
$$

in which

$$
\beta_1 = \frac{1}{R_i\rho c\, Q}, \quad \beta_2 = \frac{1}{R_o\rho c\, Q}, \quad \beta_{12} = \frac{1}{R_{io}\rho c\, Q}
$$

$$
\beta = \frac{\beta_2 - \beta_1}{2}, \quad \gamma = \sqrt{\frac{(\beta_1+\beta_2)^2}{4} + \beta_{12}(\beta_1+\beta_2)}
\tag{34.52}
$$

$$
\delta = \frac{1}{\gamma}\left(\beta_{12} + \frac{\beta_1+\beta_2}{2}\right)
$$

At the outlet, and by use of equation (34.49), they showed that the temperature is expressed as

$$
\begin{aligned}
T_{out}(t) &= \frac{f_1(L) + f_2(L)}{f_3(L) - f_2(L)}T_{in}(t) \\
&\quad + \int_0^L \frac{T_b(\zeta,t)[f_4(L-\zeta) + f_5(L-\zeta)]}{f_3(L) - f_2(L)}d\zeta
\end{aligned}
\tag{34.53}
$$

34.4.2 Semi-analytical Models

This kind of models solves an initial and boundary value problem by a combination of fundamental analytical methods and discrete methods. Fundamental analytical methods are based on standard mathematical procedures for solving differential and integral equations. The discrete methods are based on eigenfunction expansions and Fourier series (and possibly numerical techniques) for solving the governing equations.

Eigenfunction expansions are important techniques for solving boundary value problems since they yield, out of a rather complicated function, a series of algebraic terms. Nearly all functions representing physical processes can be decomposed into a set of basis (characteristic) functions. Fourier transforms are one of the most powerful and commonly used eigenfunction expansions. Prior to Fourier's work on *The analytical theory of heat* in 1822, the heat equation could only be solved for simple boundary conditions, mainly for sinusoidal heat source conditions. Fourier showed that any arbitrary function can be expressed by summing over its basis functions. By this, he revolutionized the solution of initial and boundary value problems arising from Sturm–Liouville problems of heat conduction, fluid flow, transport, and so on.

In practice, functions representing a physical system constitute spatial and temporal derivatives, and are subjected to boundary conditions made normally of discrete entries. Solution of such functions can be conducted using the spectral analysis method, where the temporal domain is treated using the fast Fourier transform and the spatial domain is treated using eigenfunction expansions. This technique is accurate, highly efficient, and, compared to the analytical solutions, versatile.

The spectral representation of the time derivative of temperature can be described as

$$
\frac{\partial T}{\partial t} = \frac{\partial}{\partial t}\sum \hat{T}_n e^{i\omega_n t} = \sum i\,\omega_n\,\hat{T}_n e^{i\omega_n t}
\tag{34.54}
$$

usually written as

$$
\frac{\partial T}{\partial t} \rightarrow i\,\omega\,\hat{T}
\tag{34.55}
$$

Similarly, the spectral representation of the spatial derivative of temperature can be described as

$$
\frac{\partial^m T}{\partial z^m} = \frac{\partial^m}{\partial z^m}\sum \hat{T}_n e^{i\omega_n t} = \sum \frac{\partial^m \hat{T}_n}{\partial z^m}e^{i\omega_n t}
\tag{34.56}
$$

usually represented as

$$
\frac{\partial^m T}{\partial z^m} \rightarrow \frac{\partial^m \hat{T}}{\partial z^m}
\tag{34.57}
$$

The transform is represented by $T \Leftrightarrow \hat{T}$.

34.4.2.1 Spectral Soil Model.
Temperature in a soil mass with an initial steady-state temperature and subjected to transient air temperature and BHE temperature variations can be described by the superposition of temperatures of two thermally interacting subsystems as

$$
T_{soil}(r,z,t) = T_z(z,t) + T_{rz}(r,z,t)
\tag{34.58}
$$

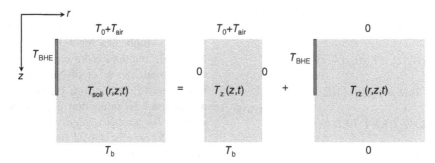

FIGURE 34.5   A schematic representation of the soil temperature (see Al-Khoury, 2012).

where $T_z(z, t)$ represents the soil temperature due to steady-state and transient conditions in the vertical direction, and $T_{rz}(r, z, t)$ represents the soil temperature due to thermal interaction with the BHE. Figure 34.5 shows a scheme of these two subsystems.

Solving T_z (z,t).   This temperature distribution can be described as a superposition of an initial steady-state temperature, T_{st}, and a transient temperature, T_t, such that

$$T_z(t, z) = T_{st}(z) + T_t(t, z) \qquad (34.59)$$

The steady-state temperature arises due to the temperature gradient between the soil surface temperature, T_0, and some deeper layer temperature, T_b. The transient temperature develops as a result of a short-term temperature change in the air, $T_a(t)$. In a typical shallow geothermal system, these temperature gradients occur in the vertical direction.

Assuming that the thermal conductivity is homogeneous, that is, $\lambda_r = \lambda_z = \lambda$, the initial and boundary value problem for a one-dimensional domain can be described as

$$\frac{1}{\alpha}\frac{\partial T_z}{\partial t} = \frac{\partial^2 T_z}{\partial z^2}$$
$$T_z(0, 0) = T_0, \quad T_z(h, 0) = T_b \qquad (34.60)$$
$$T_z(0, t) = T_a(t)$$

in which $\alpha = \lambda/\rho c$ is the thermal diffusivity and h is some depth deeper than the bottom of the BHE.

Substituting equation (34.59) into the first equation of equation (34.60) yields

$$\frac{1}{\alpha}\frac{\partial}{\partial t}[T_{st}(z) + T_t(z, t)] = \frac{\partial^2}{\partial z^2}[T_{st}(z) + T_t(z, t)] \qquad (34.61)$$

Ignoring the independent variables between brackets for simplicity of notation, equation (34.61) can be written in terms of two equations

$$\frac{d^2 T_{st}}{dz^2} = 0 \qquad (34.62)$$

$$\frac{1}{\alpha}\frac{\partial T_t}{\partial t} = \frac{\partial^2 T_t}{\partial z^2} \qquad (34.63)$$

The initial and boundary conditions, second and third equations of equation (34.60), can then be modified to

$$T_t(z, 0) = 0 \qquad (34.64)$$
$$T_t(0, t) = T_a(t) - T_0$$
$$\qquad\qquad = \Delta T \qquad (34.65)$$
$$T_t(h, t) = 0$$

The second boundary condition in equation (34.65) indicates that at the bottom, the temperature is constant.

Introducing the boundary conditions, equation (34.65), into equation (34.62), and solving lead to

$$T_{st}(z) = T_0\left(1 - \frac{z}{h}\right) + T_b\frac{z}{h} \qquad (34.66)$$

Using equations (34.55) and (34.57), Fourier transform of equation (34.63) yields

$$\frac{d^2 \hat{T}_t}{dz^2} - \kappa^2 \hat{T}_t = 0 \qquad (34.67)$$

in which

$$\kappa = \left(\frac{i\omega\rho c}{\lambda}\right)^{1/2} \qquad (34.68)$$

represents the *spectrum relationship* of a typical one-dimensional heat flow. Due to the presence of the imaginary component in the spectrum relationship, the temperature in the soil exhibits strong attenuation. Also, since the relationship in equation (34.68) is nonlinear in frequency, the propagated temperature changes its shape.

Fourier transform of the corresponding boundary conditions, equation (34.65), is

$$\hat{T}_t(0, \omega) = \hat{T}_a(\omega) - T_0\,\delta(\omega) = \Delta\hat{T}(\omega)$$
$$\hat{T}_t(h, \omega) = 0 \qquad (34.69)$$

where δ is the Dirac delta function.

Solving equation (34.67) in terms of equation (34.69) gives

$$T_t(z, \omega_n) = -\frac{\Delta \hat{T}(\omega_n)}{1 - e^{-2\kappa_n h}} e^{\kappa_n(z - 2h)} + \frac{\Delta \hat{T}(\omega_n)}{1 - e^{-2\kappa_n h}} e^{-\kappa_n z}$$

(34.70)

in which $n = 0, 1, 2, \ldots, N - 1$ with N denoting the total number of discrete samples.

Solving $T_{rz}(r,z,t)$. During system operation, the BHE exerts heat flux to the surrounding soil mass. In an axial-symmetric system, the heat flux propagates in the z-direction (along the fluid flow direction) and diffuses away in the radial direction. In such a system, the initial and boundary value problems can be described as

$$\frac{1}{\alpha} \frac{\partial T_{rz}}{\partial t} - \frac{\partial^2 T_{rz}}{\partial r^2} - \frac{1}{r} \frac{\partial T_{rz}}{\partial r} - \frac{\partial^2 T_{rz}}{\partial z^2} = 0 \quad (34.71)$$

$$-\lambda_s \frac{\partial T_{rz}(0, z, t)}{\partial z} = b_{gs}(T_{rz} - T_g) \quad (34.72)$$

$$T_{rz}(R, z, t) = 0 \quad (34.73)$$

In the first boundary condition, equation (34.72), we postulate that the centerline of the soil overlaps with the centerline of the BHE, though the physical contact surface areas are taken into consideration. In the second boundary condition, equation (34.73), we assume that at some distance, R, far from BHE, the heat flux vanishes.

Fourier transform of equations (34.71), (34.72), and (34.73) gives

$$\frac{i\omega}{\alpha} \hat{T}_{rz} - \frac{\partial^2 \hat{T}_{rz}}{\partial r^2} - \frac{1}{r} \frac{\partial \hat{T}_{rz}}{\partial r} - \frac{\partial^2 \hat{T}_{rz}}{\partial z^2} = 0 \quad (34.74)$$

$$-\lambda_s \frac{\partial \hat{T}_{rz}(0, z)}{\partial z} = b_{gs}[\hat{T}_g - \hat{T}_{rz}(0, z)] \quad (34.75)$$

$$\hat{T}_{rz}(R, z) = 0 \quad (34.76)$$

Solution of equation (34.74) can be carried out using the method of separation of variables. Assume

$$\hat{T}_{rz}(r, z) = \hat{R}(r) \hat{Z}(z) \quad (34.77)$$

in which $\hat{R}(r)$ is a function of r only and $\hat{Z}(z)$ is a function of z only. Substituting equation (34.77) into equation (34.74), dividing through by $\hat{R}(r) \hat{Z}(z)$, and equating both sides of the equation by some arbitrary constant, say $-\xi^2$, $\hat{Z}(z)$ may take the form

$$\hat{Z}(z) = A_2 e^{i\zeta z}; \quad \zeta = \left(-\frac{i\omega}{\alpha} - \xi^2\right)^{1/2} \quad (34.78)$$

and $\hat{R}(r)$ may take the form

$$\hat{R}_m(r) = J_0(\xi_m r) = J_0\left(\frac{\beta_m}{R} r\right) \quad (34.79)$$

in which β_m is the infinitely many positive roots of the Bessel function J_0, obtained from solving for the second boundary condition, equation (34.76) (Al-Khoury, 2012). Each m in equation (34.79) corresponds to the mth eigenvalue of the system. It can be noticed that the imposition of the homogeneous boundary condition at $r = R$ has inevitably led to a priori determined discrete set of eigenvalues. This property renders this approach computationally very efficient, as compared to those which involve infinite or semi-infinite spatial domains. In the latter, ξ_m has to be evaluated by solving the involved contour integrals for each calculation point. Such evaluation is computationally cumbersome due to the infinite upper limit of the involved integrands and the oscillatory nature of the Bessel function. These problems intensify at points relatively far from the source, causing spurious oscillations.

The general solution of the heat equation in the spatial domain can then be obtained by summing over all significant eigenvalues as

$$\hat{T}_{rz}(r, z) = \sum_m A_m e^{i\zeta_m z} J_0(\xi_m r) \quad (34.80)$$

Substituting equation (34.80) into the boundary condition, equation (34.75), results in

$$\sum_m A_m e^{i\zeta_m z} = \tilde{T}(z) \quad (34.81)$$

in which

$$\tilde{T}(z) = \frac{b_{gs} \hat{T}_g(z)}{b_{gs} - i\zeta_m \lambda_s} \quad (34.82)$$

Equation (34.81) is a typical complex Fourier series with its coefficient expressed as

$$A_m = \frac{1}{L} \int_0^L \tilde{T}(z) e^{-i\zeta_m z} dz \quad (34.83)$$

in which L is the BHE length.

General Solution of the Soil Heat Equation, $T_{soil}(r, z, t)$. Collecting the above solutions, and applying the inverse Fourier transform, the general solution of the soil heat equations in the time domain can then be expressed as

$$T_{soil}(r, z, t) = T_{st}(z) + T_t(z, t) + T_{rz}(r, z, t) \quad (34.84)$$

in which

$$T_{st}(z) = T_0 \left(1 - \frac{z}{h} \right) + T_b \frac{z}{h} \tag{34.85}$$

$$T_t(z,t) = \sum_n \left(-\frac{\Delta \hat{T}(\omega)}{1 - e^{-2\kappa_n h}} e^{\kappa_n(z-2h)} \right.$$
$$\left. + \frac{\Delta \hat{T}(\omega)}{1 - e^{-2\kappa_n h}} e^{-\kappa_n z} \right) e^{i\omega_n t} \tag{34.86}$$

and

$$T_{rz}(r,z,t) = \sum_n \sum_m A_m e^{i\zeta_m z} J_0(\xi_m r) e^{i\omega_n t} \tag{34.87}$$

where $\kappa_n, \zeta_m, \xi_m, \tilde{T}(z)$, and A_m are given in equations (34.68), (34.78), (34.79), (34.82), and (34.83), respectively.

34.4.2.2 Spectral Borehole Heat Exchanger Model.
As described above, the initial and boundary value problem of a single U-tube BHE consisting of pipe-in, pipe-out, and grout can be described as

$$\rho c \frac{\partial T_i}{\partial t} dV_i - \lambda \frac{\partial^2 T_i}{\partial z^2} dV_i + \rho c u \frac{\partial T_i}{\partial z} dV_i$$
$$- b_{ig}(T_i - T_g) dS_{ig} = 0$$

$$\rho c \frac{\partial T_o}{\partial t} dV_o - \lambda \frac{\partial^2 T_o}{\partial z^2} dV_o - \rho c u \frac{\partial T_o}{\partial z} dV_o \tag{34.88}$$
$$- b_{og}(T_o - T_g) dS_{og} = 0$$

$$\rho c_g \frac{\partial T_g}{\partial t} dV_g - \lambda_g \frac{\partial^2 T_g}{\partial z^2} dV_g - b_{ig}(T_g - T_i) dS_{ig}$$
$$- b_{og}(T_g - T_o) dS_{og} = 0$$

$$T_i(z,0) = T_o(z,0) = T_g(z,0) = T_{st}(z,0) \tag{34.89}$$

$$T_i(0,t) = T_{in}(t) \tag{34.90}$$

$$-\lambda_g \frac{\partial T_g}{\partial z} = b_{gs}[T_g - T_t(z,t)] \tag{34.91}$$

$$T_i(L,t) = T_o(L,t) \tag{34.92}$$

in which T_{st} is any initial soil temperature, $T_t(z,t)$ is the one-dimensional transient temperature of the soil immediately surrounding the BHE, L denotes the BHE length, and b_{gs} is the thermal interaction coefficient between the grout and the soil.

Fourier transform of equation (34.88) yields

$$i\omega \rho c \hat{T}_i dV_i - \lambda \frac{d^2 \hat{T}_i}{dz^2} dV_i + \rho c u \frac{d\hat{T}_i}{dz} dV_i$$
$$- b_{ig}(\hat{T}_i - \hat{T}_g) dS_{ig} = 0$$

$$i\omega \rho c \hat{T}_o dV_o - \lambda \frac{d^2 \hat{T}_o}{dz^2} dV_o - \rho c u \frac{d\hat{T}_o}{dz} dV_o \tag{34.93}$$
$$- b_{og}(\hat{T}_o - \hat{T}_g) dS_{og} = 0$$

$$i\omega \rho c_g \hat{T}_g dV_g - \lambda_g \frac{d^2 \hat{T}_g}{dz^2} dV_g - b_{ig}(\hat{T}_g - \hat{T}_i) dS_{ig}$$
$$- b_{og}(\hat{T}_g - \hat{T}_o) dS_{og} = 0$$

This type of equations can be solved using eigenfunction expansions. Since equation (34.93) has constant coefficients, the solution might be of the form

$$\hat{T}_i = A e^{-ikz}$$
$$\hat{T}_g = \bar{A} e^{-ikz} \tag{34.94}$$
$$\hat{T}_o = \bar{\bar{A}} e^{ikz}$$

in which $A, \bar{A}, \bar{\bar{A}}$ are integration constants and k denotes the eigenvalues of the system that needs to be determined. Note that equation (34.94) states that heat flow in pipe-in and grout is in $z > 0$ direction, and heat flow in pipe-out is in the opposite direction. Substituting equation (34.94) into equation (34.93) and putting the results in a matrix form give

$$\begin{pmatrix} \lambda k^2 dV_i - \rho c u i k dV_i \\ + i\omega \rho c dV_i - b_{ig} dS_{ig} & 0 & b_{ig} dS_{ig} \\[2ex] b_{ig} dS_{ig} & b_{og} e^{2ikz} dS_{og} & \begin{matrix} \lambda_g k^2 dV_g + i\omega \rho c_g dV_g \\ - b_{ig} dS_{ig} - b_{og} dS_{og} \end{matrix} \\[2ex] 0 & \begin{matrix} \lambda k^2 dV_o - \rho c u i k dV_o \\ + i\omega \rho c dV_o - b_{og} dS_{og} \end{matrix} & b_{og} e^{-2ikz} dS_{og} \end{pmatrix}$$
$$\times \begin{pmatrix} A \\ \bar{\bar{A}} \\ \bar{A} \end{pmatrix} = 0 \tag{34.95}$$

The eigenvalues, k, can be determined by letting the determinant of this matrix equal to zero. Six eigenvalues in three complex conjugates are obtained, representing three basic modes, one for each BHE component. Accordingly, solution of the temperature distribution in the three BHE components can be written as

$$\hat{T}_i = A e^{-ik_1 z} + B e^{-ik_2 z} + C e^{-ik_3 z}$$
$$\hat{T}_g = \bar{A} e^{-ik_1 z} + \bar{B} e^{-ik_2 z} + \bar{C} e^{-ik_3 z} \tag{34.96}$$
$$\hat{T}_o = \bar{\bar{A}} e^{ik_1 z} + \bar{\bar{B}} e^{ik_2 z} + \bar{\bar{C}} e^{ik_3 z}$$

Since T_i, T_g, and T_0 are coupled, the constants, $A, B, \ldots, \bar{\bar{C}}$, must be related to each other. The relationships between these constants can be determined from equation (34.95).

The spectral representation of the boundary conditions, equations (34.90), (34.91), and (34.92), can be expressed as

$$\hat{T}_i(0, \omega) = \hat{T}_{in}(\omega) \tag{34.97}$$

$$-\lambda_g \frac{\partial \hat{T}_g(z, \omega)}{\partial z} = b_{gs}[\hat{T}_g(z, \omega) - \hat{T}_t(z, \omega)] \tag{34.98}$$

$$\hat{T}_i(L, \omega) = \hat{T}_0(L, \omega) \tag{34.99}$$

The imposition of the third boundary condition, equation (34.99), indicates that heat flow in pipe-in is continuing in pipe-out, but in the opposite direction. This entails that the system is, in effect, consisting of two components, pipe-in and grout, that can be described basically by two eigenvalues (eigenmodes). The third mode, which is related to pipe-out, would eventually diminish, that is, $C = \bar{C} = \bar{\bar{C}} = 0$. Therefore, the system can be divided into two subsystems: pipe-in–grout and pipe-out–grout. The two subsystems are coupled at the point where pipe-in and pipe-out meet, that is, $z = L$. Accordingly, equation (34.96) can be modified to account for the two subsystems as follows:

Pipe-in–grout

$$\begin{aligned}\hat{T}_i &= A\,e^{-ik_1 z} + B\,e^{-ik_2 z}\\ \hat{T}_{gi} &= \bar{A}\,e^{-ik_1 z} + \bar{B}\,e^{-ik_2 z}\end{aligned} \tag{34.100}$$

Pipe-out–grout

$$\begin{aligned}\hat{T}_0 &= \bar{\bar{A}}\,e^{ik_1 z} + \bar{\bar{B}}\,e^{ik_2 z}\\ \hat{T}_{go} &= \bar{A}\,e^{-ik_1 z} + \bar{B}\,e^{-ik_2 z}\end{aligned} \tag{34.101}$$

Solving equations. (34.100) and (34.101) for As and Bs, and assuming that the soil temperature is constant along the BHE length: $\hat{T}_t(z, \omega) = \hat{T}_t(\omega)$, leads to

Pipe-in–grout

$$A_i = \frac{\left(\left(ik_2 + \frac{b_{gs}}{\lambda_g}\right)\bar{Y}_2 e^{-ik_2 z}\,\hat{T}_{in}(\omega) - \frac{b_{gs}}{\lambda_g}\hat{T}_t(\omega)\right)}{\left(ik_2 + \frac{b_{gs}}{\lambda_g}\right)\bar{Y}_2 e^{-ik_2 z} - \left(ik_1 + \frac{b_{gs}}{\lambda_g}\right)\bar{Y}_1 e^{-ik_1 z}} \tag{34.102}$$

$$B_i = \frac{\left(-\left(ik_1 + \frac{b_{gs}}{\lambda_g}\right)\bar{Y}_2 e^{-ik_1 z}\,\hat{T}_{in}(\omega) + \frac{b_{gs}}{\lambda_g}\hat{T}_t(\omega)\right)}{\left(ik_2 + \frac{b_{gs}}{\lambda_g}\right)\bar{Y}_2 e^{-ik_2 z} - \left(ik_1 + \frac{b_{gs}}{\lambda_g}\right)\bar{Y}_1 e^{-ik_1 z}} \tag{34.103}$$

Pipe-out–grout

$$A_0 = \frac{\left(\left(ik_2 + \frac{b_{gs}}{\lambda_g}\right)\bar{\bar{Y}}_2 e^{-ik_2 z}\,\hat{T}_i(L, \omega) - \frac{b_{gs}}{\lambda_g}\hat{T}_t(\omega)\right)}{\left(ik_2 + \frac{b_{gs}}{\lambda_g}\right)\bar{Y}_2 e^{-ik_2 z} - \left(ik_1 + \frac{b_{gs}}{\lambda_g}\right)\bar{Y}_1 e^{-ik_1 z}}\,e^{-ik_1 L} \tag{34.104}$$

$$B_0 = \frac{\left(-\left(ik_1 + \frac{b_{gs}}{\lambda_g}\right)\bar{\bar{Y}}_2 e^{-ik_1 z}\,\hat{T}_i(L, \omega) + \frac{b_{gs}}{\lambda_g}\hat{T}_t(\omega)\right)}{\left(ik_2 + \frac{b_{gs}}{\lambda_g}\right)\bar{Y}_2 e^{-ik_2 z} - \left(ik_1 + \frac{b_{gs}}{\lambda_g}\right)\bar{Y}_1 e^{-ik_1 z}}\,e^{-ik_2 L} \tag{34.105}$$

where $\bar{Y} = \bar{A}/A$ and $\bar{\bar{Y}} = \bar{A}/\bar{\bar{A}}$, with $\bar{Y}_1$ and $\bar{\bar{Y}}_1$ associated with the wavenumber, k_1, and $\bar{Y}_2$ and $\bar{\bar{Y}}_2$ associated with the wavenumber, k_2.

34.4.2.3 *General Solution of the BHE Heat Equations.*
Having determined the eigenvalues and the integration constants, the general solution of the BHE system of equations can then be obtained by summing over all eigenfunctions (corresponding to k_1 and k_2) and frequencies as

$$\begin{aligned} T_i(z, t) &= \sum_n (A\,e^{-ik_1 z} + B\,e^{-ik_2 z})e^{i\omega_n t}\\ T_0(z, t) &= \sum_n (\bar{\bar{A}}\,e^{ik_1 z} + \bar{\bar{B}}\,e^{ik_2 z})\,e^{i\omega_n t}\\ T_g(z, t) &= \frac{1}{2}\Big[\sum_n \big[(\bar{Y}_1 A + \bar{\bar{Y}}_1 \bar{\bar{A}})e^{-ik_1 z}\\ &\quad + (\bar{Y}_2 B + \bar{\bar{Y}}_2 \bar{\bar{B}})e^{-ik_2 z}\big]\,e^{i\omega_n t}\Big]\end{aligned} \tag{34.106}$$

where the reconstruction of the time domain is obtained using the inverse FFT algorithm.

34.5 NUMERICAL MODELING OF SHALLOW GEOTHERMAL SYSTEMS

In spite of the elegance of the analytical and the semi-analytical solutions, we have seen from the previous section that in order to obtain a feasible solution to the governing equations, several simplifications are necessary to model the involved geometry and initial and boundary conditions. Only limited level of complexities is possible. These limitations make the analytical and the semi-analytical models fall short of the versatility of the numerical models. The numerical models are quite effective in describing complicated geometry and initial and boundary conditions.

Several numerical methods have been employed to model shallow geothermal systems. Commonly, the finite difference, the finite element, and, to a lesser extent, the finite volume methods are utilized for this purpose. The finite difference method seems to dominate this research field, as it is traditionally utilized to solve heat and fluid flow problems. However, it is commonly limited to equidistant and structured grids. Thus, here, we focus on the use of the finite element and the finite volume methods to model shallow geothermal systems, particularly the vertical GSHP.

34.5.1 Finite Element Modeling

The finite element method (FEM) is a versatile numerical technique utilized for solving wide range of partial differential equations. It is particularly suitable for solving initial and boundary value problems occurring in engineering applications, constituting complicated geometry and initial and boundary conditions. The fundamental property of the FEM is that it partitions a continuum of space and time into a set of discrete components, represented by elements and nodes.

During 1960s and early 1970s the FEM was utilized to solve engineering problems related mainly to solid mechanics. The finite difference method was then the tool for solving heat and fluid flow problems. However, starting early 1970s, the FEM was utilized for solving transient and steady-state problems dealing with advective–diffusive transport problems. Since then, it became an important tool for modeling wide range of thermo-hydro mechanical problems in engineering and geosciences.

34.5.1.1 3D Soil Finite Element.
The mechanism of heat flow in a soil mass constituting a single phase is dominated by conduction, but that constituting a multiphase, with groundwater flow, is dominated by conduction–convection. In typical shallow geothermal systems, it is realistic to assume that the groundwater flow is in steady state and occurs in confined fully saturated soil layers. Furthermore, the groundwater flow is independent of temperature, but the temperature of the soil is dependent on the groundwater flow.

Recall the balance equations governing fluid and heat flow in a fully saturated isothermal porous medium:

$$S \frac{\partial \varphi}{\partial t} - \nabla \cdot (\mathbf{k} \nabla \varphi) + Q_p = 0$$

$$\mathbf{v} = -\mathbf{k} \nabla \varphi \qquad (34.107)$$

$$\rho c \frac{\partial T_s}{\partial t} + \rho c_w \mathbf{v} \cdot \nabla T_s - \nabla \cdot (\lambda \nabla T_s) + Q_T = 0$$

The relevant initial and boundary conditions are

$$\begin{aligned} \varphi(x, y, z, 0) &= \varphi_0(x, y, z), \quad t = 0 \\ T_s(x, y, z, 0) &= T_0(x, y, z), \quad t = 0 \end{aligned} \qquad (34.108)$$

$$\begin{aligned} \varphi(x, y, z) &= \varphi_1, \quad \text{on } x = 0 \text{ surface,} && t > 0 \\ \varphi(x, y, z) &= \varphi_2, \quad \text{on } x = b \text{ surface,} && t > 0 \\ \mathbf{k} \nabla \varphi \cdot \mathbf{n} &= J, \quad \begin{array}{l}\text{on any of the}\\ \text{boundary surfaces,}\end{array} && t > 0 \\ \lambda \nabla T \cdot \mathbf{n} + b_{as}(T_s - T_a) &= 0, \quad \begin{array}{l}\text{on the surface in}\\ \text{contact with the air,}\end{array} && t > 0 \\ \lambda_z \frac{\partial T}{\partial n} + b_{gs}(T_s - T_g) &= 0, \quad \begin{array}{l}\text{on the surface in}\\ \text{contact with a BHE,}\end{array} && t > 0 \end{aligned}$$

$$(34.109)$$

in which b is the length (along the x-axis) between two known hydraulic heads, J is a fluid flux, T_a is the air temperature, b_{as} is the convective heat transfer coefficient at the surface in contact with air, T_g is the pipe (grout) temperature, and b_{gs} is the reciprocal of the thermal resistance between the soil and the grout (borehole).

As the fluid velocity influences the temperature distribution, and the temperature does not influence the fluid velocity, the solution is uncoupled and can be conducted in two steps. First, the fluid velocity, $\mathbf{v}$, is computed using the first and second equations of equation (34.107). Then it is utilized as an input to the heat equation, the third equation of equation (34.107).

Using the weighted residual method to discretize the governing equations, the fluid and heat flow can be described as follows:

Fluid flow

$$\int_\Omega \mathbf{W}^T \left[S \frac{\partial \varphi}{\partial t} - \frac{\partial}{\partial x}\left(k_x \frac{\partial \varphi}{\partial x}\right) + \frac{\partial}{\partial y}\left(k_y \frac{\partial \varphi}{\partial y}\right) \right.$$
$$\left. + \frac{\partial}{\partial z}\left(k_z \frac{\partial \varphi}{\partial x}\right) + Q_p \right] d\Omega = 0 \qquad (34.110)$$

Heat flow

$$\int_\Omega \mathbf{W}^T \left[\begin{array}{l} \rho c \frac{\partial T}{\partial t} - \frac{\partial}{\partial x}\left(\lambda_x \frac{\partial T}{\partial x}\right) + \frac{\partial}{\partial y}\left(\lambda_y \frac{\partial T}{\partial y}\right) \\ + \frac{\partial}{\partial z}\left(\lambda_z \frac{\partial T}{\partial x}\right) + \rho c_w \left(v_x \frac{\partial T}{\partial x} + v_y \frac{\partial T}{\partial y} + v_z \frac{\partial T}{\partial z}\right) \\ + Q_T(x, y, z) \end{array} \right] d\Omega = 0$$

$$(34.111)$$

Using Galerkin's FEM, the unknown variables φ and T can be approximated by

$$\begin{aligned} \varphi &= \mathbf{N}_p \bar{\boldsymbol{\Phi}} \\ T_s &= \mathbf{N}_T \bar{\mathbf{T}}_s \end{aligned} \qquad (34.112)$$

where $\bar{\boldsymbol{\phi}}$ and $\bar{\mathbf{T}}_s$ are the total head and temperature nodal vectors, respectively, and $\mathbf{N}_p$ and $\mathbf{N}_T$ are the corresponding finite element shape functions. In practice the two shape functions are made equal, that is, $\mathbf{N}_p = \mathbf{N}_T = \mathbf{N}$. Substituting these approximations into equations (34.110) and (34.111), using integration by parts and Green's theorem, and applying the boundary conditions, equation (34.109), yield

Fluid flow

$$\int_\Omega \mathbf{W}^T S \mathbf{N} \frac{\partial \bar{\boldsymbol{\phi}}}{\partial t} \, d\Omega - \int_\Omega \left(k_x \frac{\partial \mathbf{W}^T}{\partial x} \frac{\partial \mathbf{N}}{\partial x} \right.$$
$$\left. + k_y \frac{\partial \mathbf{W}^T}{\partial y} \frac{\partial \mathbf{N}}{\partial y} + k_z \frac{\partial \mathbf{W}^T}{\partial z} \frac{\partial \mathbf{N}}{\partial z} \right) \bar{\boldsymbol{\phi}} \, d\Omega \quad (34.113)$$
$$- \int_\Gamma \mathbf{W}^T q_p \, d\Gamma - \int_\Omega \mathbf{W}^T Q_p \, d\Omega = 0$$

Heat flow

$$\int_\Omega \mathbf{W}^T \rho c \, \mathbf{N} \frac{\partial \bar{\mathbf{T}}_s}{\partial t} \, d\Omega$$
$$- \int_\Omega \left(\begin{array}{l} \lambda_x \frac{\partial \mathbf{W}^T}{\partial x} \frac{\partial \mathbf{N}}{\partial x} + \lambda_y \frac{\partial \mathbf{W}^T}{\partial y} \frac{\partial \mathbf{N}}{\partial y} + \lambda_z \frac{\partial \mathbf{W}^T}{\partial z} \frac{\partial \mathbf{N}}{\partial z} \\ + \rho_w c_w \mathbf{W}^T \left(v_x \frac{\partial \mathbf{N}}{\partial x} + v_y \frac{\partial \mathbf{N}}{\partial y} + v_z \frac{\partial \mathbf{N}}{\partial z} \right) \end{array} \right) \bar{\mathbf{T}}_s \, d\Omega$$
$$- \int_\Gamma \mathbf{W}^T q_T \, d\Gamma + \int_\Omega \mathbf{W}^T Q_T \, d\Omega = 0 \quad (34.114)$$

The finite element discretization of the boundary conditions is as follows:

Soil–air contact

$$\int_\Gamma \mathbf{W}^T b_{as} \left(\mathbf{T}_s - T_a \right) d\Gamma = \int_\Gamma \mathbf{W}^T b_{as} \mathbf{N} \, d\Gamma \, \bar{\mathbf{T}}_s$$
$$- \int_\Gamma \mathbf{W}^T b_{as} T_a \, d\Gamma \quad (34.115)$$

Soil–BHE contact

$$\int_\Gamma \mathbf{W}^T b_{gs} \left(\mathbf{T}_s - \mathbf{T}_g \right) d\Gamma = \int_\Gamma \mathbf{W}^T b_{gs} \mathbf{N} \, d\Gamma \, \bar{\mathbf{T}}_s$$
$$- \int_\Gamma \mathbf{W}^T b_{gs} d\Gamma \bar{\mathbf{T}}_g \quad (34.116)$$

For a typically slow groundwater flow, the standard Galerkin method is practically adequate for formulating the convective–conductive heat flow in a porous medium, that

is, $\mathbf{W} = \mathbf{N}$. The final finite element relationships can then be expressed as follows:

Fluid flow

$$\mathbf{M}_p \frac{\partial \bar{\boldsymbol{\phi}}}{\partial t} + \mathbf{K}_p \bar{\boldsymbol{\phi}} = \mathbf{F}_p \quad (34.117)$$

where

$$\mathbf{M}_p = \int_\Omega \mathbf{N}^T S \mathbf{N} \, d\Omega$$
$$\mathbf{K}_p = \int_\Omega \mathbf{B}^T \mathbf{k} \mathbf{B} \, d\Omega \quad (34.118)$$
$$\mathbf{F}_p = \int_\Gamma \mathbf{W}^T J \, d\Gamma - \int_\Omega \mathbf{W}^T Q_p \, d\Omega$$

Heat flow

$$\mathbf{M}_T \frac{\partial \bar{\mathbf{T}}_s}{\partial t} + \mathbf{K}_T \bar{\mathbf{T}}_s = \mathbf{F}_T \quad (34.119)$$

where

$$\mathbf{M}_T = \int_\Omega \mathbf{N}^T \rho c \, \mathbf{N} \, d\Omega$$
$$\mathbf{K}_T = \int_\Omega (\mathbf{B}^T \boldsymbol{\lambda} \mathbf{B} + \rho_w c_w \mathbf{N}^T \mathbf{v} \mathbf{B}) d\Omega$$
$$+ \int_\Gamma \mathbf{N}^T b_{as} \mathbf{N} \, d\Gamma + \int_\Gamma \mathbf{N}^T b_{gs} \mathbf{N} \, d\Gamma \quad (34.120)$$
$$\mathbf{F}_T = \int_\Omega Q_t \mathbf{N}^T \, d\Omega + \int_\Gamma \mathbf{N}^T b_{gs} \, d\Gamma \bar{\mathbf{T}}_g + \int_\Gamma \mathbf{N}^T b_{as} T_a \, d\Gamma$$

in which $\mathbf{B} = d\mathbf{N}/dx$.

Equations (34.117) and (34.119) are first-order differential equations in time, referred to as semi-discrete equations. They can be solved mainly by two types of time integration schemes: finite difference or finite element. The theta-method is one of the most utilized finite difference time integration scheme. In this method, the dependent variables can be defined as

$$\varphi_{n+\theta} = \theta \varphi_{n+1} + (1 - \theta)\varphi_n$$
$$T_{n+\theta} = \theta T_{n+1} + (1 - \theta)T_n \quad (34.121)$$

Substituting equation (34.121) into equations (34.117) and (34.119) gives

$$(\mathbf{M}_p + \theta \, \Delta t \, \mathbf{K}_p) \, \bar{\boldsymbol{\phi}}_{n+1} = [\mathbf{M}_p - (1 - \theta) \, \Delta t \, \mathbf{K}_p] \, \bar{\boldsymbol{\phi}}_n$$
$$+ \Delta t \, [\theta \, \mathbf{F}_{p(n+1)} + (1 - \theta) \, \mathbf{F}_{p(n)}]$$
$$(\mathbf{M}_T + \theta \, \Delta t \, \mathbf{K}_T) \, \bar{\mathbf{T}}_{s(n+1)} = [\mathbf{M}_T - (1 - \theta) \, \Delta t \, \mathbf{K}_T] \, \bar{\mathbf{T}}_{s(n)}$$
$$+ \Delta t \, [\theta \, \mathbf{F}_{T(n+1)} + (1 - \theta) \, \mathbf{F}_{T(n)}]$$
$$(34.122)$$

where $0 \leq \theta \leq 1$ and the subscript n represents a calculation time step.

34.5.1.2 Borehole Heat Exchanger Finite Element.
Recall the initial and boundary value problem of a single U-tube BHE:

Pipe-in

$$\rho c \frac{\partial T_i}{\partial t} - \lambda \frac{\partial^2 T_i}{\partial z^2} + \rho c u \frac{\partial T_i}{\partial z} = b_{ig} (T_i - T_g) \quad (34.123)$$

Pipe-out

$$\rho c_f \frac{\partial T_o}{\partial t} - \lambda_f \frac{\partial^2 T_o}{\partial z^2} - \rho c_f u \frac{\partial T_o}{\partial z} = b_{og} (T_o - T_g) \quad (34.124)$$

Grout

$$\rho c_g \frac{\partial T_g}{\partial t} - \lambda_g \frac{\partial^2 T_g}{\partial z^2} = b_{ig} (T_g - T_i) + b_{og} (T_g - T_o) \quad (34.125)$$

where we ignored the subscript ref for simplicity of notation. The initial and boundary conditions are

$$T_i(z, 0) = T_o(z, 0) = T_g(z, 0) = T_s(z, 0), \quad t = 0 \quad (34.126)$$

$$T_i(0, t) = T_{in}(t)$$
$$-\lambda_g \frac{\partial T_g}{\partial n} = b_{gs} (T_g - T_s) \quad (34.127)$$

in which T_s is the soil temperature immediately around the borehole and b_{gs} is the reciprocal of the thermal resistance between the soil and the grout.

The weighted residuals of this problem can be expressed as follows:

Pipe-in

$$\int_t \int_V \left(W_i^T \rho c \frac{\partial T_i}{\partial t} - W_i^T \lambda \frac{\partial^2 T_i}{\partial z^2} + W_i^T \rho c u \frac{\partial T_i}{\partial z} \right) dV \, dt$$
$$- \int_t \int_S W_i^T b_{ig} (T_i - T_g) \, dS \, dt = 0 \quad (34.128)$$

Pipe-out

$$\int_t \int_V \left(W_o^T \rho c \frac{\partial T_o}{\partial t} - W_o^T \lambda \frac{\partial^2 T_o}{\partial z^2} - W_o^T \rho c u \frac{\partial T_o}{\partial z} \right) dV \, dt$$
$$- \int_t \int_S W_o^T b_{og} (T_o - T_g) \, dS \, dt = 0 \quad (34.129)$$

Grout

$$\int_t \int_V \left(W_g^T \rho c_g \frac{\partial T_g}{\partial t} - W_g^T \lambda_g \frac{\partial^2 T_g}{\partial z^2} \right) dV \, dt$$
$$- \int_t \int_S \left[b_{ig} W_g^T (T_g - T_i) + b_{og} W_g^T (T_g - T_o) \right] dS \, dt = 0 \quad (34.130)$$

Using integration by parts and Green's theorem, we obtain

Pipe-in

$$\int_t \int_V W_i^T \rho c \frac{\partial T_i}{\partial t} \, dV \, dt + \int_t \int_V \lambda \frac{dW_i^T}{dz} \frac{\partial T_i}{\partial z} \, dV \, dt$$
$$+ \int_t \int_V W_i^T \rho c u \frac{\partial T_i}{\partial z} \, dV \, dt$$
$$- \int_t \int_S W_i^T b_{ig} (T_i - T_g) \, dS \, dt = 0 \quad (34.131)$$

Pipe-out

$$\int_t \int_V W_o^T \rho c \frac{\partial T_o}{\partial t} \, dV \, dt + \int_t \int_V \lambda \frac{dW_o^T}{dz} \frac{\partial T_o}{\partial z} \, dV \, dt$$
$$- \int_t \int_V W_o^T \rho c u \frac{\partial T_o}{\partial z} \, dV \, dt$$
$$- \int_t \int_S W_o^T b_{og} (T_o - T_g) \, dS \, dt = 0 \quad (34.132)$$

Grout

$$\int_t \int_V W_g^T \rho c \frac{\partial T_g}{\partial t} \, dV \, dt + \int_t \int_V \lambda_g \frac{dW_g^T}{dz} \frac{\partial T_g}{\partial z} \, dV \, dt$$
$$- \int_t \int_S \lambda_g W_g^T \frac{\partial T_g}{\partial z} n_z \, dS \, dt$$
$$- \int_t \int_S \left[W_g^T b_{ig} (T_g - T_i) + W_g^T b_{og} (T_g - T_o) \right] dS \, dt = 0 \quad (34.133)$$

There are many ways to approximate the solution of these equations. In what follows, we use the Galerkin method and the Petrov–Galerkin method.

Galerkin Formulation. Using the standard Galerkin method, the nodal weighting functions and the temperature are described as

$$W(z) = N(z)$$
$$T(z,t) \equiv N(z)\bar{T}(t) \tag{34.134}$$

in which $\bar{T}(t)$ is the nodal temperature. The weighted residual discretization can then be expressed as

Pipe-in

$$\int_V N^T \rho c\, N\, dV \frac{d\bar{T}_i}{dt}$$

$$+ \int_V \left(\frac{dN^T}{dz} \lambda \frac{dN}{dz} dV + N^T \rho c\, u \frac{dN}{dz} \right) dV\, \bar{T}_i$$

$$- \int_S N^T b_{ig}\, N\, dS\, \bar{T}_i + \int_S N^T b_{ig}\, N\, dS\, \bar{T}_g = 0 \quad (34.135)$$

Pipe-out

$$\int_V N^T \rho c\, N\, dV \frac{d\bar{T}_o}{dt}$$

$$+ \int_V \left(\frac{dN^T}{dz} \lambda \frac{dN}{dz} dV - N^T \rho c\, u \frac{dN}{dz} \right) dV\, \bar{T}_o$$

$$- \int_S N^T b_{og}\, N\, dS\, \bar{T}_o + \int_S N^T b_{og}\, N\, dS\, \bar{T}_g = 0 \quad (34.136)$$

Grout

$$\int_V N^T \rho c_g\, N\, dV \frac{d\bar{T}_g}{dt} + \int_V \frac{dN^T}{dz} \lambda_g \frac{dN}{dz} dV\, \bar{T}_g$$

$$- \int_S N^T b_{ig}\, N\, dS\, \bar{T}_g + \int_S N^T b_{ig}\, N\, dS\, \bar{T}_i$$

$$- \int_S N^T b_{og}\, N\, dS\, \bar{T}_g + \int_S N^T b_{og}\, N\, dS\, \bar{T}_o$$

$$- \int_S N^T b_{gs}\, N\, dS\, \bar{T}_g = - \int_S N^T b_{gs}\, dS\, \bar{T}_s \quad (34.137)$$

Putting these equations in a matrix format gives a first-order semi-discrete differential equation of the form

$$\begin{pmatrix} M_{ii} & 0 & 0 \\ 0 & M_{gg} & 0 \\ 0 & 0 & M_{oo} \end{pmatrix} \begin{pmatrix} \dot{\bar{T}}_i \\ \dot{\bar{T}}_g \\ \dot{\bar{T}}_o \end{pmatrix} + \begin{pmatrix} K_{ii} & K_{ig} & 0 \\ K_{gi} & K_{gg} & K_{go} \\ 0 & K_{og} & K_{oo} \end{pmatrix} \begin{pmatrix} \bar{T}_i \\ \bar{T}_g \\ \bar{T}_o \end{pmatrix}$$

$$= \begin{pmatrix} q_{in} \\ F_{gs} \\ 0 \end{pmatrix} \tag{34.138}$$

where

$$M_{ii} = \int_V N^T \rho c\, N\, dV$$

$$M_{oo} = \int_V N^T \rho c\, N\, dV$$

$$M_{gg} = \int_V N^T \rho c_g\, N\, dV$$

$$K_{ii} = \int_V (B^T \lambda_i\, B + N^T \rho c\, u\, B)\, dV - \int_S N^T b_{ig}\, N\, dS$$

$$K_{ig} = \int_S N^T b_{ig}\, N\, dS$$

$$K_{oo} = \int_V (B^T \lambda_o\, B - N^T \rho c\, u\, B)\, dV - \int_S N^T b_{og}\, N\, dS$$

$$K_{og} = \int_S N^T b_{og}\, N\, dS$$

$$K_{gg} = \int_V B^T \lambda_g\, B\, dV - \int_S N^T b_{ig}\, N\, dS \tag{34.139}$$

$$- \int_S N^T b_{og}\, N\, dS - \int_S N^T b_{gs}\, N\, dS$$

$$K_{gi} = \int_S N^T b_{ig}\, N\, dS$$

$$K_{go} = \int_S N^T b_{og}\, N\, dS$$

$$F_{gs} = - \int_S N^T b_{gs}\, dS\, \bar{T}_s$$

with $\dot{T} = d\bar{T}/dt$ and $B = dN/dz$.

The semi-discrete equation, equation (34.138), can be written in a compact form as

$$M\dot{\bar{T}} + K\bar{T} = F \tag{34.140}$$

This equation can be solved in time using the finite element time integration scheme. In this case, the temperature in the time domain can be approximated as

$$T(t) = \sum_j N_j(t)\, T_j \tag{34.141}$$

in which $N_j(t)$ is the temporal shape function. Using a two-node linear element in time $N(t) \in (t_n, t_{n+1}]$ gives

$$T(t) = N_n\, T_n + N_{n+1}\, T_{n+1} \tag{34.142}$$

where

$$N_n = 1 - \frac{\tau}{\Delta t}, \quad N_{n+1} = \frac{\tau}{\Delta t}$$
$$\tau = t - t_n, \qquad \Delta t = t_{n+1} - t_n \tag{34.143}$$

Substituting equation (34.142) into equation (34.140) gives

$$
\left(\frac{\mathbf{M}}{\Delta t} + \theta\,\mathbf{K}\right)\mathbf{T}_{n+1} = \left(\frac{\mathbf{M}}{\Delta t} - (1-\theta)\,\mathbf{K}\right)\mathbf{T}_n \\
+ (1-\theta)\,\mathbf{F}_n + \theta\,\mathbf{F}_{n+1} \quad (34.144)
$$

where the algorithmic parameter θ is expressed as

$$
\theta = \frac{1}{\Delta t}\frac{\int_{t_n}^{t_{n+1}}\mathbf{W}(t)\,\tau\,dt}{\int_{t_n}^{t_{n+1}}\mathbf{W}(t)\,dt} \quad (34.145)
$$

in which $\mathbf{W}(t)$ is the temporal weighting function.

Petrov–Galerkin Formulation. One important approach to obtain an improved algorithm for solving highly transient convective problems is the Petrov–Galerkin FEM. A space–time Petrov–Galerkin element is one of the important techniques in this field. The weighting function of this type of elements is formulated by perturbing the involved shape functions with their derivatives as (Heinrich and Pepper, 1999)

Pipe-in

$$
\mathbf{W}_i(z,t) = \mathbf{N}(z,t) + \frac{\alpha\,h}{2}\frac{d\mathbf{N}(z,t)}{dz} + \frac{\beta\,h\,\Delta t}{4}\frac{d^2\mathbf{N}(z,t)}{dz\,dt} \\ (34.146)
$$

Pipe-out

$$
\mathbf{W}_o(z,t) = \mathbf{N}(z,t) - \frac{\alpha\,h}{2}\frac{d\mathbf{N}(z,t)}{dz} + \frac{\beta\,h\,\Delta t}{4}\frac{d^2\mathbf{N}(z,t)}{dz\,dt} \\ (34.147)
$$

For the grout, the standard Galerkin finite element weighting function can be utilized as

$$
\mathbf{W}_g(z,t) = \mathbf{N}(z,t) \quad (34.148)
$$

Using these weighting functions, the finite element discretization of equations (34.123), (34.124), and (34.125) yields

Pipe-in

$$
\int_t\int_V \mathbf{W}_i\,\rho\,c\,\mathbf{N}\frac{\partial\bar{\mathbf{T}}_i}{\partial t}\,dV\,dt \\
+ \int_t\int_V \left(\mathbf{P}_i^T\,\lambda\,\mathbf{B} + \mathbf{W}_i^T\,\rho\,c\,u\,\mathbf{B} + \mathbf{W}_i^T\,\rho\,c\,\frac{\partial\mathbf{N}}{\partial t}\right)dV\,dt\,\bar{\mathbf{T}}_i \\
- \int_t\int_S \mathbf{W}_i^T\,b_{ig}\,\mathbf{N}\,dS\,dt\,\bar{\mathbf{T}}_i + \int_t\int_S \mathbf{W}_i^T\,b_{ig}\,\mathbf{N}\,dS\,dt\,\bar{\mathbf{T}}_g = 0 \\ (34.149)
$$

Pipe-out

$$
\int_t\int_V \mathbf{W}_o^T\,\rho\,c\,\mathbf{N}\frac{\partial\bar{\mathbf{T}}_o}{\partial t}\,dV\,dt \\
+ \int_t\int_V \left(\mathbf{P}_o^T\,\lambda\,\mathbf{B} - \mathbf{W}_o^T\,\rho\,c\,u\,\mathbf{B} + \mathbf{W}_o^T\,\rho\,c\,\frac{\partial\mathbf{N}}{\partial t}\right)dV\,dt\,\bar{\mathbf{T}}_o \\
- \int_t\int_S \mathbf{W}_o^T\,b_{og}\,\mathbf{N}\,dS\,dt\,\bar{\mathbf{T}}_o + \int_t\int_S \mathbf{W}_o^T\,b_{og}\,\mathbf{N}\,dS\,dt\,\bar{\mathbf{T}}_g = 0 \\ (34.150)
$$

Grout

$$
\int_t\int_V \mathbf{N}\,\rho\,c_g\,\mathbf{N}\frac{\partial\bar{\mathbf{T}}_g}{\partial t}\,dV\,dt \\
+ \int_t\int_V \left(\mathbf{B}^T\,\lambda_g\,\mathbf{B} + \mathbf{N}^T\,\rho\,c_g\,\frac{\partial\mathbf{N}}{\partial t}\right)dV\,dt\,\bar{\mathbf{T}}_g \\
- \int_t\int_S \left(\mathbf{N}^T\,b_{ig}\,\mathbf{N} + \mathbf{N}^T\,b_{og}\,\mathbf{N}\right)dS\,dt\,\bar{\mathbf{T}}_g \\
+ \int_t\int_S \mathbf{N}^T\,b_{ig}\,\mathbf{N}\,dS\,dt\,\bar{\mathbf{T}}_i + \int_t\int_S \mathbf{N}^T\,b_{og}\,\mathbf{N}\,dS\,dt\,\bar{\mathbf{T}}_o \\
- \int_t\int_S \mathbf{N}^T\,b_{gs}\,\mathbf{N}\,dS\,\bar{\mathbf{T}}_g = - \int_t\int_S \mathbf{N}^T\,b_{gs}\,dS\,\bar{\mathbf{T}}_s \quad (34.151)
$$

in which $\mathbf{P}_i = \partial\mathbf{W}_i/\partial z$, and so on. The temperature derivative with space is carried out as

$$
\frac{\partial\mathbf{T}(z,t)}{\partial z} = \frac{\partial}{\partial z}[\mathbf{N}(z,t)\,\bar{\mathbf{T}}(t)] \\
= \frac{\partial\mathbf{N}(z,t)}{\partial z}\bar{\mathbf{T}}(t) \quad (34.152)
$$

and the temperature derivative with time is carried out as

$$
\frac{\partial\mathbf{T}(z,t)}{\partial t} = \frac{\partial}{\partial t}[\mathbf{N}(z,t)\,\bar{\mathbf{T}}(t)] \\
= \frac{\partial\mathbf{N}(z,t)}{\partial t}\bar{\mathbf{T}}(t) + \mathbf{N}(z,t)\frac{\partial\bar{\mathbf{T}}(t)}{\partial t} \quad (34.153)
$$

The finite element equation can then be formulated as

$$
\begin{bmatrix} \mathbf{M}_{ii} & 0 & 0 \\ 0 & \mathbf{M}_{gg} & 0 \\ 0 & 0 & \mathbf{M}_{oo} \end{bmatrix}
\begin{bmatrix} \dot{\bar{\mathbf{T}}}_i \\ \dot{\bar{\mathbf{T}}}_g \\ \dot{\bar{\mathbf{T}}}_o \end{bmatrix}
+ \begin{bmatrix} \mathbf{K}_{ii} & \mathbf{K}_{ig} & 0 \\ \mathbf{K}_{gi} & \mathbf{K}_{gg} & \mathbf{K}_{go} \\ 0 & \mathbf{K}_{og} & \mathbf{K}_{oo} \end{bmatrix}
\begin{bmatrix} \bar{\mathbf{T}}_i \\ \bar{\mathbf{T}}_g \\ \bar{\mathbf{T}}_o \end{bmatrix}
= \begin{bmatrix} q_{in} \\ \mathbf{F}_{gs} \\ 0 \end{bmatrix} \\ (34.154)
$$

in which

$$\mathbf{M}_{ii} = \int_t \int_V \mathbf{W}_i^T \rho c \mathbf{N} \, dV \, dt$$

$$\mathbf{M}_{gg} = \int_t \int_V \mathbf{N}^T \rho c_g \mathbf{N} \, dV \, dt$$

$$\mathbf{M}_{oo} = \int_t \int_V \mathbf{W}_o^T \rho c \mathbf{N} \, dV \, dt$$

$$\mathbf{K}_{ii} = \int_t \int_V \left(\mathbf{P}_i^T \lambda \mathbf{B} + \mathbf{W}_i^T \rho c u \mathbf{B} + \mathbf{W}_i^T \rho c \frac{\partial \mathbf{N}}{\partial t} \right) dV \, dt$$
$$- \int_t \int_S \mathbf{W}_i^T b_{ig} \mathbf{N} \, dS \, dt \qquad (34.155)$$

$$\mathbf{K}_{oo} = \int_t \int_V \left(\mathbf{P}_o^T \lambda \mathbf{B} - \mathbf{W}_o^T \rho c u \mathbf{B} + \mathbf{W}_o^T \rho c \frac{\partial \mathbf{N}}{\partial t} \right) dV \, dt$$
$$- \int_t \int_S \mathbf{W}_o^T b_{og} \mathbf{N} \, dS \, dt$$

$$\mathbf{K}_{gg} = \int_t \int_V \left(\mathbf{B}^T \lambda_g \mathbf{B} + \mathbf{N}^T \rho c_g \frac{\partial \mathbf{N}}{\partial t} \right) dV \, dt$$
$$- \int_t \int_S \left(\mathbf{N}^T b_{ig} \mathbf{N} + \mathbf{N}^T b_{og} \mathbf{N} + \mathbf{N}^T b_{gs} \mathbf{N} \right) dS \, dt$$

$$\mathbf{K}_{ig} = \int_t \int_S \mathbf{W}_i^T b_{ig} \mathbf{N} \, dS \, dt$$

$$\mathbf{K}_{gi} = \int_t \int_S \mathbf{N}^T b_{ig} \mathbf{N} \, dS \, dt$$

$$\mathbf{K}_{go} = \int_t \int_S \mathbf{N}^T b_{og} \mathbf{N} \, dS \, dt \qquad (34.156)$$

$$\mathbf{K}_{og} = \int_t \int_S \mathbf{W}_o^T b_{og} \mathbf{N} \, dS \, dt$$

$$\mathbf{F}_{gs} = - \int_t \int_S \mathbf{N}^T b_{gs} \, dS \, dt \, \bar{\mathbf{T}}_s$$

Using the theta-method, the semi-discrete equation, equation (34.154), can be written as

$$(\mathbf{M} + \theta \, \Delta t \, \mathbf{K}) \bar{\mathbf{T}}_{n+1} = [\mathbf{M} - (1 - \theta) \, \Delta t \, \mathbf{K}] \bar{\mathbf{T}}_n$$
$$+ \Delta t \, [\theta \, \mathbf{F}_{n+1} + (1 - \theta) \mathbf{F}_n] \quad (34.157)$$

where the subscript n represents a calculation time step. Using the time–space Petrov–Galerkin method, we are able to perform only implicit algorithms. This means that we should utilize $1/2 \le \theta \le 1$.

It is important to note that the inclusion of the thermal interaction terms between the BHE components alleviates the need to spatially discretize the individual components, offering thus the possibility to model the BHE by a line element. Such a possibility reduces the number of finite elements dramatically and makes the computational process highly efficient, rendering the use of the FEM feasible for engineering practice. This modeling technique is also applicable to the finite volume model, described hereafter.

34.5.2 Finite Volume Modeling

The finite volume method is a numerical technique used for solving partial differential equations, mainly those related to transport phenomena. One of the important features of the finite volume method, compared to the finite difference method, is that it can be implemented in a structured and unstructured mesh. Also, unlike the FEM, the boundary conditions in the finite volume method are not invasive, giving more stability to the numerical processes.

Using the finite volume method, the volume integrals of the involved partial differential equations are transformed into surface integrals by means of the divergence theorem. These terms are evaluated as fluxes at the surfaces of each finite volume. As the flux entering a given volume is identical to that leaving, the resulting system of equations is locally conservative.

Here, derivatives of the convection terms are discretized using a weighted cell-centered-upwind second-order scheme, and derivatives of the diffusion terms are discretized using the conventional cell-centered scheme. The temperature and the velocity are discretized on a staggered grid, entailing that the temperature is located at the center of the cell and the velocities are located at its surfaces (Figure 34.6). Figure 34.6 shows upstream sides, denoted as w and s, and their corresponding downstream sides, denoted as e and n. The upstream and the downstream in the z-direction, not shown in the figure, are denoted as t and b, respectively.

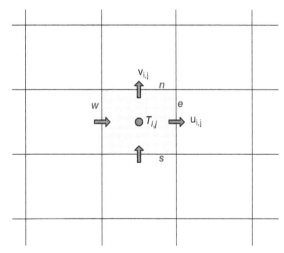

FIGURE 34.6 Control volume.

34.5.2.1   3D Soil Finite Volume

Heat flow

The soil heat equation, equation (34.9), can be written in an integral form as

$$\int_V \left(\rho c \frac{\partial T}{\partial t} + \rho_w c_w (\nabla \cdot \mathbf{v} T) - \nabla.(\lambda \nabla T) - H \right) dV = 0$$

$$(34.158)$$

where the continuity condition $\nabla \cdot \mathbf{v} = 0$ is applied. Using the divergence theorem, the volume integral of the convection term is evaluated, in the x-axis, as

$$\int_V \frac{\partial u T}{\partial x} dV = \int_S u T n_x dS = (u_e T_e - u_w T_w) \Delta y \Delta z$$

$$(34.159)$$

In a cell with a grid point (i, j, k) at the center, the convection term can be evaluated by weighted first-order upwind and central difference, such that

$$\frac{\partial u T}{\partial x_i} = \alpha \left. \frac{\partial u T}{\partial x_i} \right|_{upwind} + (1 - \alpha) \left. \frac{\partial u T}{\partial x_i} \right|_{cen} \quad (34.160)$$

in which α is a weighting factor. In the e–w direction, this gives

$$(u_e T_e - u_w T_w)_{upwind} = \frac{1}{2} [(u_e + |u_e|) T_{i+1,j,k} + (u_e - |u_e|) T_{i,j,k}$$
$$- (u_w + |u_w|) T_{i-1,j,k} - (u_w - |u_w|) T_{i,j,k}]$$

$$(u_e T_e - u_w T_w)_{cen} = \frac{1}{2} [(T_{i,j,k} + T_{i+1,j,k}) u_e - (T_{i,j,k} + T_{i-1,j,k}) u_w]$$

$$(34.161)$$

where $|u_e|$ represents the absolute value of u_e, and so on.

The conduction terms can be discretized using the central difference, giving (in the x-axis)

$$\int_V \lambda_x \frac{\partial^2 T}{\partial x^2} dV = \lambda_x \frac{T_{i+1,j,k} - 2T_{i,j,k} + T_{i-1,j,k}}{\Delta x} \Delta y \Delta z$$

$$(34.162)$$

The volume integral of the temporal derivative in equation (34.158) can be evaluated as

$$\int_V \frac{\partial T}{\partial t} dV = \frac{1}{\Delta t} (T^{n+1} - T^n) \Delta x \Delta y \Delta z \qquad (34.163)$$

in which T^n is the value of T at time t and T^{n+1} is the value at time $t + \Delta t$, with Δt the time step. The temperature and the heat source are evaluated in time using the theta-method as

$$T^{n+\theta} = \theta T^{n+1} + (1 - \theta) T^n$$
$$H^{n+\theta} = \theta H^{n+1} + (1 - \theta) H^n$$

$$(34.164)$$

Putting equations (34.160), (34.161), (34.162), and (34.163) together, using equation (34.164), and dividing by $\Delta V = \Delta x \Delta y \Delta z$ give

$$\rho c \frac{T^{n+1} - T^n}{\Delta t} + \rho_w c_w \left[\alpha \left(\frac{\partial u T}{\partial x} + \frac{\partial v T}{\partial y} + \frac{\partial w T}{\partial z} \right)^{n+\theta}_{upwind} \right.$$

$$+ (1 - \alpha) \left(\frac{\partial u T}{\partial x} + \frac{\partial v T}{\partial y} + \frac{\partial w T}{\partial z} \right)^{n+\theta}_{cen} \Bigg]$$

$$- \left[\lambda_x \frac{T_{i+1,j,k} - 2T_{i,j,k} + T_{i-1,j,k}}{(\Delta x)^2} \right.$$

$$+ \lambda_y \frac{T_{i,j+1,k} - 2T_{i,j,k} + T_{i,j-1,k}}{(\Delta y)^2}$$

$$+ \lambda_z \frac{T_{i,j,k+1} - 2T_{i,j,k} + T_{i,j,k-1}}{(\Delta z)^2} \Bigg]^{n+\theta} + H^{n+\theta}_{i,j,k} = 0$$

$$(34.165)$$

where u, v, and w are the Darcy velocity components along x, y, and z axes, respectively, and the derivative terms are described in equations (34.159), (34.160), and (34.161). $\theta = 0$ gives explicit scheme, $\theta = 1$ gives fully implicit, and $\theta = 1/2$ yields the Crank–Nicholson.

Groundwater flow

Evaluating the total head φ at the center of the cell, and using the discretization procedure outlined above, the finite volume equation of the groundwater flow, equation (34.12), can be described as

$$S \frac{\varphi^{n+1}_{i,j,k} - \varphi^n_{i,j,k}}{\Delta t} + \frac{k_x}{\Delta x^2} (\varphi_{i+1,j,k} - 2\varphi_{i,j,k} + \varphi_{i-1,j,k})^{n+\theta}$$

$$+ \frac{k_y}{\Delta y^2} (\varphi_{i,j+1,k} - 2\varphi_{i,j,k} + \varphi_{i,j-1,k})^{n+\theta}$$

$$+ \frac{k_z}{\Delta z^2} (\varphi_{i,j,k+1} - 2\varphi_{i,j,k} + \varphi_{i,j,k-1})^{n+\theta} + Q^{n+\theta}_{i,j,k} = 0$$

$$(34.166)$$

The Darcy velocities are calculated as

$$u_{i,j,k} = -k_x \frac{\varphi_{i+1,j,k} - \varphi_{i,j,k}}{\Delta x}$$

$$v_{i,j,k} = -k_y \frac{\varphi_{i,j+1,k} - \varphi_{i,j,k}}{\Delta y} \qquad (34.167)$$

$$w_{i,j,k} = -k_z \frac{\varphi_{i,j,k+1} - \varphi_{i,j,k}}{\Delta z}$$

34.5.2.2 *Borehole Heat Exchanger Finite Volume.* Using the discretization procedure for the convection term, equation (34.159), the diffusion term, equation (34.162), and the theta-method for the time integration, the finite volume equations of the BHE give

Pipe-in

$$\rho_r c_r \frac{T_k^{n+1} - T_k^n}{\Delta t}\bigg|_i + \rho_r c_r \left(\alpha \frac{\partial uT}{\partial z}\bigg|_{upwind} + (1-\alpha) \frac{\partial uT}{\partial z}\bigg|_{cen} \right)_i^{n+\theta}$$
$$- \lambda_r \left(\frac{T_{k+1} - 2T_k + T_{k-1}}{(\Delta z)^2} \right)_i^{n+\theta} = b_{ig}(T_k|_g - T_k|_i)^{n+\theta}$$

$$(34.168)$$

Pipe-out

$$\rho_r c_r \frac{T_k^{n+1} - T_k^n}{\Delta t}\bigg|_o - \rho_r c_r \left(\alpha \frac{\partial uT}{\partial z}\bigg|_{upwind} + (1-\alpha) \frac{\partial uT}{\partial z}\bigg|_{cen} \right)_o^{n+\theta}$$
$$- \lambda_r \left(\frac{T_{k+1} - 2T_k + T_{k-1}}{(\Delta z)^2} \right)_o^{n+\theta} = b_{og}(T_k|_g - T_k|_o)^{n+\theta}$$

$$(34.169)$$

Grout

$$\rho_r c_r \frac{T_k^{n+1} - T_k^n}{\Delta t}\bigg|_g - \lambda_r \left(\frac{T_{k+1} - 2T_k + T_{k-1}}{(\Delta z)^2} \right)_g^{n+\theta}$$
$$= b_{ig}(T_k|_i - T_k|_g)^{n+\theta} + b_{og}(T_k|_o - T_k|_g)^{n+\theta}$$
$$+ b_{sg}(T_s - T_k|_g)^{n+\theta}$$

$$(34.170)$$

in which T_s denotes the soil temperature, and

$$\frac{\partial uT}{\partial z}\bigg|_{upwind} = \frac{u_t T_t - u_b T_b}{\Delta z} = \frac{1}{\Delta z}\left[\frac{u_t + |u_t|}{2} T_{k+1} \right.$$
$$\left. + \frac{u_t - |u_t|}{2} T_k - \frac{u_b + |u_b|}{2} T_{k-1} - \frac{u_b - |u_b|}{2} T_k \right]$$

$$\frac{\partial uT}{\partial z}\bigg|_{cen} = \frac{u_t T_t - u_b T_b}{\Delta z} = \frac{1}{\Delta z}\left[u_t \frac{T_k + T_{k+1}}{2} - u_b \frac{T_k + T_{k-1}}{2} \right]$$

$$(34.171)$$

34.6 CONCLUDING REMARKS

Computational modeling of shallow geothermal systems is challenging, stimulating, and inspiring. This chapter addresses this subject briefly and provides researchers and developers in computational mechanics, geosciences, geology, and geothermal engineering with the basic knowledge to develop computational tools capable of modeling the complicated nature of heat flow in shallow geothermal systems. Coupled conduction–convection models for heat flow in BHEs and the surrounding soil mass are formulated and solved using analytical, semi-analytical, and numerical methods. We show that a good combination between a mathematical model and a computational procedure potentially leads to a versatile and highly efficient computational tool.

REFERENCES

Aifantis, E. C. (1979). A new interpretation of diffusion in high-diffusivity paths: a continuum approach. *Acta Metallurgica*, 27:683–691.

Al-Khoury, R. (2012). *Computational Modeling of Shallow Geothermal Systems*. CRC press, Leiden, The Netherlands.

Carslaw, H. S., Jaeger, J. C. (1959). *Conduction of Heat in Solids*. 2nd edition. Oxford University Press.

Dincer, I., Rosen, M. A. (2011). *Thermal Energy Storage, Systems and Applications*. Wiley, Chichester, UK.

Eskilson, P. (1987). Thermal analysis of heat extraction boreholes. Ph.D. Thesis, University of Lund, Sweden.

Eskilson, P., Claesson, J. (1988). Simulation model for thermally interacting heat extraction boreholes. *Numerical Heat Transfer*, 13:149–165.

Heinrich, J. C., Pepper, D. W. (1999). *Intermediate Finite Element Method, Fluid Flow and Heat Transfer Applications*. Taylor & Francis.

Ingersoll, L. R., Plass, H. J. (1948). Theory of the ground pipe heat sources for the heat pump. *Heating, Piping and Air conditioning*, 20:119–122.

Ingersoll, L. R., Zobel O. J., Ingersoll A. C. (1954). *Heat Conduction with Engineering, Geological, and Other Applications*. McGraw-Hill, New York.

Mands, E., Sanner, B., (2005). Shallow geothermal energy. Available at: www.ubeg.de/Downloads/ShallowGeothEngl.pdf.

Philippe, M., Bernier, M., Marchio, D. (2009). Validity ranges of three analytical solutions to the transfer in the vicinity of single boreholes. *Geothermics*, 38:407–413.

Yener, Y., Kakac, S. (2008). *Heat Conduction*. 4th edition. Taylor and Francis Group, New York.

35

GEOTHERMAL BASICS—WHAT IS GEOTHERMAL ENERGY?

35.1 GEOTHERMAL BASICS

35.1.1 What Is Geothermal Energy?

The word geothermal comes from the Greek words *geo* (earth) and *therme* (heat). So, geothermal energy is heat from within the Earth. We can recover this heat as steam or hot water and use it to heat buildings or generate electricity.

Geothermal energy is a renewable energy source because the heat is continuously produced inside the Earth.

35.1.2 Geothermal Energy Is Generated Deep Inside the Earth

Geothermal energy is generated in the Earth's core. Temperatures hotter than the sun's surface are continuously produced inside the Earth by the slow decay of radioactive particles, a process that happens in all rocks. The Earth has a number of different layers:

- The core itself has two layers: a solid iron core and an outer core made of very hot melted rock, called magma.
- The mantle surrounds the core and is about 1800 miles thick. It is made up of magma and rock.
- The crust is the outermost layer of the Earth, the land that forms the continents and ocean floors. It can be 3–5 miles thick under the oceans and 15–35 miles thick on the continents.

The Earth's crust is broken into pieces called plates. Magma comes close to the Earth's surface near the edges of these plates. This is where volcanoes occur. The lava that erupts from volcanoes is partly magma. Deep underground, the rocks and water absorb the heat from this magma. The temperature of the rocks and water gets hotter and hotter as you go deeper underground.

People around the world use geothermal energy to heat their homes and to produce electricity by digging deep wells and pumping the heated underground water or steam to the surface. We can also make use of the stable temperatures near the surface of the Earth to heat and cool buildings.

35.2 WHERE GEOTHERMAL ENERGY IS FOUND

Naturally occurring large areas of hydrothermal resources are called **geothermal reservoirs**. Most geothermal reservoirs are deep underground with no visible clues showing above ground. But geothermal energy sometimes finds its way to the surface in the form of

- Volcanoes and fumaroles (holes where volcanic gases are released)
- Hot springs
- Geysers

35.2.1 Most Geothermal Resources Are Near Plate Boundaries

The most active geothermal resources are usually found along major plate boundaries where earthquakes and volcanoes are concentrated. Most of the geothermal activity in the world occurs in an area called the *ring of fire*. This area encircles the Pacific Ocean.

Alternative Energy and Shale Gas Encyclopedia, First Edition. Edited by Jay H. Lehr and Jack Keeley.
© 2016 John Wiley & Sons, Inc. Published 2016 by John Wiley & Sons, Inc.

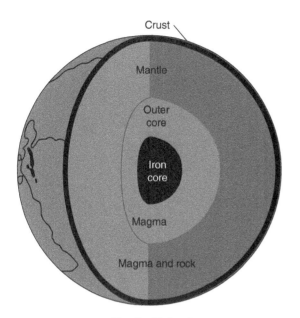

The Earth's interior

FIGURE 35.1   *Source*: Adapted from a National Energy Education Development Project graphic (Public Domain).

When magma comes close to the surface, it heats groundwater found trapped in porous rock or water running along fractured rock surfaces and faults. These features are called hydrothermal. They have two common ingredients: water (hydro) and heat (thermal).

Geologists use various methods to look for geothermal reservoirs. Drilling a well and testing the temperature deep underground is the most reliable method for finding a geothermal reservoir.

35.2.2   US Geothermal Power Plants Are Located in the West

Most of the geothermal power plants in the United States are located in the western states and Hawaii, where geothermal energy resources are close to the surface. California generates the most electricity from geothermal energy. "The Geysers" dry steam reservoir in northern California is the largest known dry steam field in the world and has been producing electricity since 1960.

35.3   USE OF GEOTHERMAL ENERGY

Some applications of geothermal energy use the Earth's temperatures near the surface, while others require drilling miles into the Earth. The three main uses of geothermal energy are

- *Direct use and district heating systems* use hot water from springs or reservoirs near the surface.
- *Electricity generation power plants* require water or steam at very high temperature (300–700°F). Geothermal power plants are generally built where geothermal reservoirs are located within a mile or two of the surface.

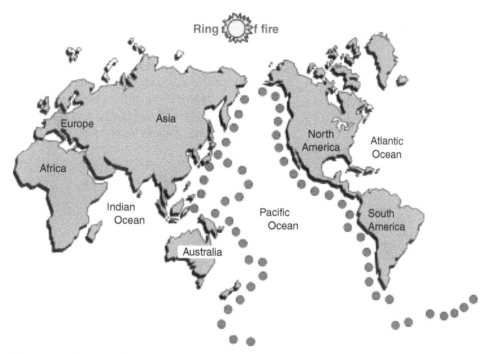

FIGURE 35.2   The Ring of Fire goes around the edges of the Pacific. The map shows that volcanic activity occurs around the Pacific Rim. *Source*: National Energy Education Development Project (Public Domain).

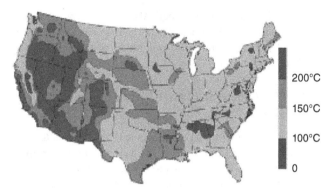

FIGURE 35.3 US Geothermal Resource Map. *Source*: US Department of Energy, Energy Efficiency & Renewable Energy.

- *Geothermal heat pumps* use stable ground or water temperatures near the Earth's surface to control building temperatures above ground.

35.3.1 Direct Use of Geothermal Energy

There have been *direct uses* of hot water as an energy source since ancient times. Ancient Romans, Chinese, and Native American cultures used hot mineral springs for bathing, cooking, and heating. Today, many hot springs are still used for bathing, and many people believe the hot, mineral-rich waters have natural healing powers.

After bathing, the most common direct use of geothermal energy is for heating buildings through *district heating systems*. Hot water near the Earth's surface can be piped directly into buildings and industries for heat. A district heating system provides heat for 95% of the buildings in Reykjavik, Iceland.

Industrial applications of geothermal energy include food dehydration, gold mining, and milk pasteurizing. Dehydration, or the drying of vegetable and fruit products, is the most common industrial use of geothermal energy.

FIGURE 35.4 A geothermic power station. *Source*: Stock photography (copyrighted).

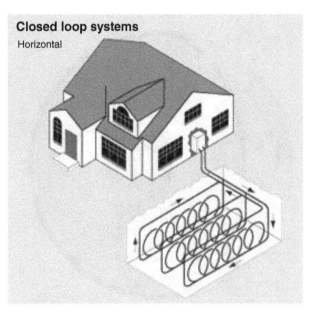

FIGURE 35.5 A type of geothermal heat pump system. *Source*: US Department of Energy, Energy Efficiency & Renewable Energy (Public Domain).

35.3.2 The United States Is the Leader in Geothermal Power Generation

The United States leads the world in electricity generation with geothermal power. In 2012, US geothermal power plants produced about 17 billion kWh, or 0.4% of total US electricity generation. In 2012, six states had geothermal power plants:

- California had 36 geothermal power plants, which produced about 80% of US geothermal electricity.
- Nevada had 21 geothermal power plants, which produced about 16% of US geothermal electricity.
- Utah had two plants, and Hawaii, Idaho, and Oregon each had one geothermal plant.

FIGURE 35.6 Grand Prismatic Spring, Yellowstone National Park, Wyoming. *Source*: Stock photography (copyrighted).

35.3.3 Geothermal Contributes Significant Share of Power Generation in Several Countries

Twenty four countries including the United States had geothermal power plants in 2010, which generated a total of about 63.9 billion kWh. The Philippines was the second largest geothermal power producer after the United States at 9.4 billion kWh, which equaled about 16% of the country's total power generation. Iceland, the seventh largest producer at 4.3 billion KWh, produced 26% of its total electricity using geothermal energy.

35.4 GEOTHERMAL POWER PLANTS

Geothermal power plants use *hydrothermal* resources that have two common ingredients: water (hydro) and heat (thermal). Geothermal plants require high temperature (300–700°F) hydrothermal resources that may come from either dry steam wells or hot water wells. We can use these resources by drilling wells into the Earth and piping the steam or hot water to the surface. Geothermal wells are 1–2 miles deep.

35.4.1 Types of Geothermal Plants

There are three basic types of geothermal power plants:

- *Dry steam plants* use steam piped directly from a geothermal reservoir to turn the generator turbines. The first geothermal power plant was built in 1904 in Tuscany, Italy, where natural steam erupted from the Earth.
- *Flash steam plants* take high-pressure hot water from deep inside the Earth and convert it to steam to drive the generator turbines. When the steam cools, it condenses to water and is injected back into the ground to be used over and over again. Most geothermal power plants are flash steam plants.
- *Binary cycle power plants* transfer the heat from geothermal hot water to another liquid. The heat causes the second liquid to turn to steam which is used to drive a generator turbine.

35.5 GEOTHERMAL HEAT PUMPS

35.5.1 Using the Earth's Constant Temperatures for Heating and Cooling

While temperatures above ground change a lot from day to day and season to season, temperatures 10 ft below the Earth's surface hold nearly constant between 50° and 60°F. For most areas, this means that soil temperatures are usually warmer than the air in winter and cooler than the air in summer. Geothermal heat pumps use the Earth's constant temperatures to heat and cool buildings. They transfer heat from the ground (or water) into buildings in winter and reverse the process in the summer.

35.5.2 Geothermal Heat Pumps Are Energy Efficient and Cost-Effective

According to the US Environmental Protection Agency (EPA), geothermal heat pumps are the most energy-efficient, environmentally clean, and cost-effective systems for temperature control. Although most homes still use traditional furnaces and air conditioners, geothermal heat pumps are becoming more popular. In recent years, the US Department of Energy and the EPA have partnered with industry to promote the use of geothermal heat pumps.

35.6 GEOTHERMAL ENERGY AND THE ENVIRONMENT

The environmental impact of geothermal energy depends on how it is being used. Direct use and heating applications have almost no negative impact on the environment.

35.6.1 Geothermal Power Plants Have Low Emission Levels

Geothermal power plants do not burn fuel to generate electricity, so their emission levels are very low. They release less than 1% of the carbon dioxide emissions of a fossil fuel plant. Geothermal plants use scrubber systems to clean the air of hydrogen sulfide that is naturally found in steam and hot water.

Geothermal plants emit 97% less acid rain-causing sulfur compounds than are emitted by fossil fuel plants. After the steam and water from a geothermal reservoir have been used, they are injected back into the Earth.

35.6.2 Many Geothermal Features Are National Treasures

Geothermal features in national parks, such as geysers and fumaroles in Yellowstone National Park, are protected by law, to prevent them from being disturbed.

36

GEOTHERMAL—CHRONOLOGICAL DEVELOPMENT

1904	The first dry steam geothermal power plant was built in Larderello in Tuscany, Italy. The Larderello plant today provides power to about 1 million households.
1960	The first commercial-scale development tools were placed in California at The Geysers, a 10 MW unit owned by Pacific Gas & Electric.
1970	Reinjection of spent geothermal water back into the production reservoir was introduced as a way to dispose of wastewater and to extend reservoir life.
1972	Deep well drilling technology improvements led to deeper reservoir drilling and access to more resources.
1974	Scientists began to develop the first hot dry rock (HDR) reservoir at Fenton Hill, New Mexico. An HDR power facility was tested at the site in 1978 and started to generate electricity 2 years later.
1978	US Department of Energy (DOE) funding for geothermal research and development was increased substantially. The Public Utility Regulatory Policies Act (PURPA) of 1978 was enacted to promote greater use of renewable energy, cogeneration, and small power projects.
1980	The first commercial-scale binary plant in the United States began operation in Southern California's Imperial Valley.
1980s	California's Standard Offer Contract system for PURPA-qualifying facilities provided renewable electric energy systems a relatively firm, stable market for output, allowing the financing of capital-intensive technologies like geothermal energy facilities.
1982	Geothermal (hydrothermal) electric generating capacity reached a new high of 1000 MW.
1984	Utah's first commercial geothermal power plant began operating at Roosevelt Hot Springs with a 20 MW capacity. Nevada's first geothermal binary power production plant began operating at Wabuska Hot Springs.
1989	DOE and the Electric Power Research Institute operated a 1 MW, geopressured power demonstration plant in Texas, extracting methane and heat from brine liquids.
1990	DOE funding for geothermal energy research and development declined throughout the 1980s and reached a low of $15 million.
1991	The world's first magma exploratory well was drilled in the Sierra Nevada Mountains to a depth of 7588 ft.
1992	The Puna field of Hawaii began electrical generation at a 25 MW geothermal plant.
1994	California Energy became the world's largest geothermal company through its acquisition of Magma Power.
1995	Worldwide geothermal capacity reached 6000 MW. At Empire Nevada, a food-dehydration facility processed 15 million pounds of dried onions and garlic a year, using geothermal resources. A DOE low-temperature resource assessment of 10 Western States identified nearly 9000 thermal wells and springs and 271 communities with a geothermal resource greater than 50°C.
1999	California's geothermal power plants provided 54.9% of the State's electricity.
2000	The DOE and industry worked together on the Geothermal Resource Exploration and Definition Program. It was a cooperative effort to find, evaluate, and define additional geothermal resources throughout the western United States.

(continued)

Alternative Energy and Shale Gas Encyclopedia, First Edition. Edited by Jay H. Lehr and Jack Keeley.
© 2016 John Wiley & Sons, Inc. Published 2016 by John Wiley & Sons, Inc.

2004	Geothermal energy costs dropped from $0.10–0.16 per kilowatt hour to $0.5–0.8 per kilowatt hour.
2006	The US geothermal industry became a $1.5 billion a year business that involved electricity generation and thermal energy in direct use such as indoor heating, greenhouses, food drying, and aquaculture. Alaska installed a 200 kW power plant that used low-temperature (74°C) geothermal water along with cooling water (4°C).
2008	Idaho's first commercial geothermal power plant began operating.

Last revised: July 2008.

Sources: US Department of Energy, Office of Energy Efficiency and Renewable Energy, *A History of Geothermal in the United States*, January 2008; California Energy Commission, *Overview of Geothermal Energy in California*, January 2008; California Energy Commission, *California Geothermal Resources*, January 2008; California Energy Commission, *Geothermal Technologies Program*, January 2008; US Department of Energy, Office of Energy Efficiency and Renewable Energy, *Geothermal Research*, January 2008; Oregon Institute of Technology, JHC Bulletin, *Development and Utilization of Geothermal Resources,* June 2007.

PART IV

HYDROPOWER

37

SUSTAINABILITY OF HYDROPOWER

Joerg Hartmann

Hydropower projects and their impacts are specific for each site, design, and operational regime. The main constraints to a continued expansion of hydropower are not technical or economic, but related to environmental and social (E&S) impacts which have created conflicts over existing and proposed new projects [1]. Many approaches have been developed and are still being developed to support sustainable development of hydropower. This article introduces the main dimensions of sustainability and reviews sustainability approaches available both internationally and within countries. It should be read as an umbrella article, in conjunction with the more detailed exploration of E&S issues in other articles.

37.1 DIMENSIONS OF SUSTAINABILITY

There is widespread agreement that the electricity sector and individual power projects should be "sustainable." There is some uncertainty over the meaning of the term, however, and some overlap with other concepts such as corporate social responsibility and E&S management.

37.1.1 Definition

In 1987, the Brundtland Commission of the United Nations proposed that "sustainable development is development that meets the needs of the present without compromising the ability of future generations to meet their own needs" [2]. The first step toward operationalizing this definition is to recognize that this requires simultaneously meeting economic, social, and environmental demands (the "triple bottom line").

The second step is to effectively balance the three dimensions of sustainability.

In practice, achieving sustainable development requires negotiations in which "workable compromises are found that address the environmental, economic, and human development objectives of competing interest groups" [3]. One such negotiation was the Hydropower Sustainability Assessment Forum (HSAF), a global multi-stakeholder effort from 2008 to 2010 that produced a benchmarking tool (the "Hydropower Sustainability Assessment Protocol" [Protocol]). Besides reiterating support for the sustainability definition of the Brundtland Commission, the HSAF established the following shared principles [4]:

- "Sustainable development embodies reducing poverty, respecting human rights, changing unsustainable patterns of production and consumption, long-term economic viability, protecting and managing the natural resource base, and responsible environmental management.

- Sustainable development calls for considering synergies and trade-offs amongst economic, social and environmental values. This balance should be achieved and ensured in a transparent and accountable manner, taking advantage of expanding knowledge, multiple perspectives, and innovation.

- Social responsibility, transparency, and accountability are core sustainability principles.

- Hydropower, developed and managed sustainably, can provide national, regional, and local benefits, and has the potential to play an important role in enabling communities to meet sustainable development objectives."

Alternative Energy and Shale Gas Encyclopedia, First Edition. Edited by Jay H. Lehr and Jack Keeley.
© 2016 John Wiley & Sons, Inc. Published 2016 by John Wiley & Sons, Inc.

37.1.2 Technical Sustainability

The original meaning of sustainability is the ability to endure over long periods of time. Any attempt to meet the triple bottom line must also be technically feasible. In this sense, hydropower is a unique source of electricity, with many power stations built in the early 20th century still in operation today. In developed countries where most of the hydropower potential has already been developed, the management of a fleet of aging assets is the main challenge. This also includes keeping infrastructure safe, according to current technical understanding and expectations.

There are few absolute technical constraints to extending the lifetime of hydropower stations, though sedimentation of reservoirs over time can present a major operational issue. Experiences with decommissioning hydropower stations are limited and are often linked more to changing societal expectations about environmental performance (e.g., passage of salmon in the US northwest) than to technical problems.

37.1.3 Economic Sustainability

Hydropower projects are characterized by high, and highly site-specific, upfront capital costs. Once operational, hydropower projects have low operating costs and can follow loads and deliver high-value electricity and ancillary services, for long periods of time. Investment in hydropower often requires long-term financing and is subject to a different risk profile than for other sources of electricity.

Geotechnical risks are most relevant during construction and commissioning, while hydrological risks are expected to increase with climate change. Both categories of risk have large influences on investment costs and cash flows. Societal expectations can change radically over the lifetime of a power station and, in some cases, require costly changes to operations. Prices for power services are difficult to predict in the long term. The eligibility of hydropower projects for public funding, subsidies, tax breaks, access to carbon markets, and other financial support is often determined by sustainability perceptions.

37.1.4 Social Sustainability

The social impacts of hydropower projects can be complex: negative and positive, short and long term, local and regional. Equity considerations call, in particular, for the protection of vulnerable groups and individuals. Obtaining stakeholder support or a "social license to operate" requires

- developing open, respectful, and ongoing relationships with stakeholder groups and communities,
- avoiding, minimizing, and compensating for negative social impacts, and

- identifying and pursuing potential positive social impacts, to share the benefits from hydropower development.

37.1.5 Environmental Sustainability

The environmental impacts of hydropower projects can be equally complex as the social ones. The most important positive impact is the direct generation of low-carbon electricity, and the indirect support for the expansion of other renewable sources of electricity, such as wind and solar, which are intermittent and require backup through dispatchable hydropower. Because of this, hydropower has often been seen as a clean source of electricity. However, there are also very significant negative impacts on terrestrial and aquatic biodiversity and on the ecosystem services that rivers provide to people.

Sustainability from an environmental point of view generally requires that "duty of care" is exercised in avoiding, minimizing, and compensating for negative environmental impacts. Increasingly, ambitious objectives such as "zero (net) harm" or "positive net benefit" are established, which call for offsetting any remaining negative impacts [5].

37.2 NATIONAL REGULATIONS

It is primarily the responsibility and prerogative of national and, in some cases, subnational governments to establish policies and regulations for hydropower development, to issue permits and licenses, and to determine eligibility for subsidies. Most governments endorse the principles of sustainability and aim to translate them into their regulatory regime. What is accepted as sustainable depends on difficult judgments on the appropriate balance between the economic, social, and environmental implications of projects, under specific national circumstances. Such judgments ultimately need to be legitimized by governments. This may involve different perspectives from government agencies with responsibilities for different policy areas that can only be resolved by top-level executive, legislative, or judicial authorities.

37.2.1 Environmental and Social Impact Assessments

The assessment of the impact of individual projects was first introduced in the United States through the National Environmental Policy Act 1970 and has since spread to practically all countries. From early assessments, which were focused on describing the impacts, regulatory requirements have continued to evolve. State-of-the-art assessments nowadays provide a platform to influence project choice and design and to define management plans for dealing with any identified impacts. They have also broadened to include social impacts, either integrated with environmental impact assessments (recognizing that many livelihood impacts are due to physical environmental changes, and that social impacts tend to have knock-on environmental effects) or as separate exercises.

It is also widely recognized that the best way to avoid project-level conflicts is to establish regional- or sector-wide plans which resolve trade-offs at higher levels and rank projects for implementation according to their overall costs and benefits. Such planning is generally undertaken by governmental or quasi-governmental organizations, not by project proponents. Part of such exercises should be Strategic Environmental Assessments (SEAs), which address the environmental effects of programs, plans, and policies. The number of countries requiring or otherwise undertaking SEAs is on the rise. The most current evolution of impact assessments is sustainability assessments, which explicitly deal with the multiple dimensions of impacts, and are not yet standardized and widely required [6, 7].

37.2.2 Licensing and Relicensing

The development of new projects as well as the continued operation of existing ones is generally subject to government-issued licenses. Different jurisdictions have developed different licensing regimes. Licenses can be for a project as such or for a specific project stage. For example, the construction permit may be separate from the operating permit, which is only issued when the project is ready for commissioning. One unified or multiple licenses may be required (e.g., for water rights and dam safety). Only a few jurisdictions have time-bound licenses, where licensees have to periodically apply for extensions and projects are subjected to a review of their performance. The main benefit of a relicensing scheme, such as under the US Federal Energy Regulatory Commission (FERC), is that shifts in societal expectations and increases in E&S knowledge can be incorporated into updated license conditions.

37.3 INTERNATIONAL GUIDELINES

Advances in the understanding of sustainability issues frequently first get codified through international guidelines, before they are integrated into national regulations and become binding for individual projects. International guidelines are developed through a variety of bodies, including intergovernmental organizations, professional and business associations, and multi-stakeholder processes. The status of such guidelines ranges from purely voluntary to international law.

Examples for such bodies and the sustainability guidance that they have issued include

- the United Nations and its subsidiary organizations, which have concluded various conventions and issued a number of declarations; recent examples are the Declaration on the Rights of Indigenous People (2007), which establishes their right to free, prior, and informed consent for any project that affects their lands, and the Framework Convention on Climate Change, which requires hydropower projects to demonstrate their contribution to sustainable development in order to be accepted into the Clean Development Mechanism (CDM);
- the International Energy Agency (IEA), which has undertaken various initiatives to promote good practices in hydropower development;
- the World Commission on Dams, whose final report [8] in 2000 is still an important reference on sustainability issues;
- the International Hydropower Association (IHA), which "is dedicated to advancing the role sustainable hydropower has in meeting the world's water and energy needs by championing continuous improvement and sustainable practices... " [9] and has developed sustainability guidelines (2004) and early versions of the Protocol;
- the Hydropower Sustainability Assessment Council (HSAC), which was established to promote and oversee the broad application of the Protocol;
- the International Organization for Standardization (ISO), which has developed various standards on environmental management and corporate social responsibility; and
- the International Commission on Large Dams (ICOLD), which focuses on technical guidelines and dam safety issues.

International guidelines are sometimes outdated or contradictory with each other, so that commitments that a project will "comply with international standards and regulations" are not sufficiently clear. They also often diverge from national laws and regulations, and not all governments accept that an international standard is followed instead of the national one. They nevertheless serve an important function in fostering international debate on sustainability issues. They also provide references for regulators who do not have significant specific hydropower experience; for developers who volunteer to go beyond what national regulations require, and are seeking guidance for what practices are internationally acceptable; and for all other stakeholders.

37.4 BANK SAFEGUARDS

A special group of guidelines is those issued by banks. Since hydropower projects are highly capital-intensive, access to loans from commercial and development banks is a major consideration and banks can exercise influence over project choice and design. Banks protect themselves from material and reputational risks and ensure that projects they fund are in line with their sustainability policies, through lending guidelines or "safeguards." These are usually generic across sectors, though in some cases banks have issued specific guidelines, such as the World Bank on dam safety and HSBC on freshwater infrastructure lending. Sustainability commitments by banks reference either their own standards or other international guidance.

37.4.1 Development Banks

The national development or "policy" banks from countries with major hydropower sectors such as China Development Bank, China Exim Bank, and the Brazilian Banco Nacional de Desenvolvimento Economico e Social (BNDES) are currently among the most important financiers of hydropower projects. Multilateral development banks are also providing funding and have some additional influence through funding of preparatory studies, arranging project finance and guarantees and technical assistance grants, and the drafting of safeguard policies which are then often adopted by other banks. Export credit agencies and development banks from developed countries also provide some financing and have broadly adopted consistent safeguard approaches. Since these banks are owned by individual or multiple governments, their safeguards are additional regulatory instruments and reflect evolving political attitudes toward sustainability.

37.4.2 Commercial Banks

Commercial banks provide loans and other financial services, such as arranging bond placements, for hydropower utilities and contractors. Such services are covered by the banks' individual CSR and sustainability mechanisms and increasingly by common policies such as the Equator Principles. The Equator Principles have been adopted by close to 80 banks and cover about one-half of the global project finance market by loan volumes [10]. They require compliance with the Performance Standards of the International Finance Corporation (IFC).

37.5 CORPORATE RESPONSIBILITY INSTRUMENTS

Hydropower projects often involve multiple partners including developers, contractors, suppliers, consultants, and power off-takers. Many of these partners will have their own sustainability policies. Electric utilities are publicly owned in many countries; in some countries such as China, most of the other partners in a project will also be owned by the state.

In these cases, corporate policies are another expression of the government's perspective on sustainability.

37.5.1 Benchmarking, Measurement, and Certification

An important element of sustainability management is establishing appropriate, measurable objectives and indicators. Companies compare their sustainability performance internally between different projects and externally with their peers in the industry. They benchmark their performance against preestablished objectives and establish business processes to maintain a consistent performance. Some companies decide to seek external verification and recognition of their performance. The Dow Jones Sustainability Indexes and other composite rating schemes track corporate performance for larger, publicly listed companies. At the corporate or facility level, environmental management systems can be certified against ISO 14001. More hydropower-specific tools with a focus on environmental criteria are used by national certification schemes such as the Low Impact Hydropower Institute (LIHI) in the United States and the "naturemade" label in Switzerland. The first comprehensive, globally applicable assessment mechanism is the Protocol, which enables the benchmarking of projects against international "proven best practices" and "basic good practices" on a broad range of sustainability topics.

37.5.2 Stakeholder Engagement, Public Commitments, and Reporting

Internal business processes and their validation by external experts can only go so far in satisfying the expectations of the public. As stated above, given that hydropower projects can substantially affect the interests of many stakeholder groups, obtaining their acceptance and support requires developing relationships. Stakeholders will want to see that the developer has made a public commitment to sustainability, is prepared to establish two-way communications, take stakeholder knowledge and opinions into account when formulating project objectives, and offer a mechanism to resolve grievances. Public commitments can be expressed by own policies or by the endorsement of a code of conduct such as the United Nations "Global Compact."

Transparency in communications with stakeholders will support sustainability objectives. This includes publishing relevant materials such as environmental impact and sustainability assessments and publicly reporting on corporate sustainability performance. Such information should be easily accessible, which may require translation into local languages and dissemination through culturally appropriate processes. Standardized formats will allow the public to understand how projects and corporations compare over time and with each other. Corporate sustainability reporting, for example, should follow the established guidelines of the Global Reporting Initiative (GRI).

REFERENCES

[1] Kumar, A., T. Schei, A. Ahenkorah, R. Caceres Rodriguez, J.-M. Devernay, M. Freitas, D. Hall, Å. Killingtveit, Z. Liu (2011). Hydropower. In *IPCC Special Report on Renewable Energy Sources and Climate Change Mitigation*. Cambridge University Press, Cambridge, UK.

[2] United Nations World Commission on Environment and Development. (1987). *Our Common Future*. Oxford University Press, Oxford, UK.

[3] Robert, K. W., Parris, T. M., Leiserowitz, A. A. (2005). What is sustainable development? *Environment: Science and Policy for Sustainable Development*, 47(3):8–21.

[4] International Hydropower Association. (2011). Hydropower Sustainability Assessment Protocol, IHA, London, UK.

[5] Business and Biodiversity Offsets Programme. (2012). Standard on Biodiversity Offsets, Forest Trends/BBOP, Washington, D.C.

[6] Tetlow, M., Hanusch, M. (2012). Strategic environmental assessment: the state of the art. *Impact Assessment and Project Appraisal*, 30:15–24.

[7] Bond, A., Morrison-Saunders, A., Pope, J. (2012). Sustainability assessment: the state of the art. *Impact Assessment and Project Appraisal*, 30:53–62.

[8] WCD. (2000). Dams and Development: A New Framework for Decision-Making, Earthscan, London, UK.

[9] www.hydropower.org

[10] http://www.mrcmekong.org/assets/Events/Mekong2Rio/5.3b-The-role-of-private-sector-Le-Clerc.pdf

38

ENVIRONMENTAL ISSUES RELATED TO CONVENTIONAL HYDROPOWER

Zhiqun Daniel Deng, Alison H. Colotelo, Richard S. Brown, and Thomas J. Carlson

38.1 ENVIRONMENTAL ISSUES

38.1.1 Fish Passage

Hydropower facilities impact the ability of aquatic organisms to move upstream and downstream within a river system. A common example of this interruption is salmonid migration, where both upstream and downstream seaward migrations are affected. For example, the construction of Grand Coulee and Chief Joseph dams on the Columbia River in the 1930s and 1940s eliminated the upstream migration routes of Pacific salmon and steelhead past the dam structures. In addition, the presence of hydroelectric dams can be a barrier to movement of less migratory species, limiting home ranges. This may lead to population fragmentation among species such as white sturgeon. In addition, this fragmentation can reduce the availability of fish access to critical habitats such as spawning and nursery areas, which may contribute to population decline.

38.1.1.1 Downstream. The route of passage available to fish and other aquatic organisms passing downstream through hydroelectric dams varies depending on the project. In the Columbia River Basin (CRB) where juvenile salmonid migration is a conservation issue, large facilities have three basic routes of passage: over the spillway, through a juvenile bypass facility, or through turbines. At some facilities, spillways and juvenile bypass facilities do not exist, and all fish must pass through turbines. In some cases, management practices have elected to capture and transport juvenile fish via truck or barge downstream past dams to avoid entrainment. In addition, fish screens may be employed to guide fish away from turbines and into bypass or transport facilities.

Survival rates of fish passing through hydroelectric projects are a focus for dam regulators and conservation agencies (Cada, 2001). Researchers use a variety of tools and techniques to better understand the mechanisms of injury and mortality. Overall, juvenile bypass facilities and spillways are generally believed to be the most benign routes of passage, and fish are commonly diverted away from turbines. For fish that do pass through turbines, mechanisms of fish injury include blade strike, shear stress and turbulence, rapid decreases in pressure, cavitation, and grinding (Coutant and Whitney, 2000).

Research during the past several years has focused on understanding how variables including dam-operating condition and design influence the biological consequences of passage. The development of devices such as the autonomous Sensor Fish (Deng et al., 2007a) and acoustic telemetry systems has given researchers and managers a better understanding of the physical conditions experienced by fish as they pass through hydroturbines, over spillways, or through other dam passage routes. This information has led to a better understanding of the mechanisms of injury and mortality for fish passing dams. For example, injuries associated with exposure to shear forces during turbine passage include bruising, eye and gill/opercular damage, and descaling, which can contribute to mortality (Deng et al., 2005). In contrast, barotraumas, resulting from exposure of juvenile salmonids to rapid decreases in pressure, are typically observed as rupture of the swim bladder, exopthalmia, emboli in the fins and gills, and hemorrhaging (Brown et al., 2012).

Survival of undesirable introduced species downstream through dams can also be a problem in some systems. Many reservoirs created by large dams become popular recreation

Alternative Energy and Shale Gas Encyclopedia, First Edition. Edited by Jay H. Lehr and Jack Keeley.
© 2016 John Wiley & Sons, Inc. Published 2016 by John Wiley & Sons, Inc.

sites, particularly for angling, where popular game fish species (e.g., centrarchids, bullheads) are stocked. These non-migratory introduced species may be entrained through dams and outcompete native species. For example, Flaming Gorge Reservoir (in Wyoming and Utah, on the Green River) serves as a popular angling site for species such as kokanee, lake trout, rainbow trout, brown trout, smallmouth bass, and burbot. Successful passage of these nonnative species through the dam has resulted in competition for habitat and other resources with the native and Endangered Species Act (ESA)-listed razorback sucker, Colorado pikeminnow, humpback chub, and bonytail chub.

38.1.1.2   *Upstream.*
Fishways (fish ladders) have been utilized at many hydroelectric facilities to aid upstream passage of adult fish. A large focus of this work has been salmonids as they move upstream from the ocean to their spawning grounds. There are many designs types, depending on species of interest, river characteristics, and hydropower facility configuration. These ladders typically consist of high-flow areas with pools after each step, where fish can rest and regain energy before their next strenuous movement. Despite the need for these structures to allow for upstream migration past hydropower dams and their proven benefit for fish such as salmonids, other migratory species such as lamprey have very different needs for fish passage. In response to recent concern, eel ladders have been installed at hydroelectric facilities to aid in migration of upstream-moving juvenile eels. These ladders provide a rough substrate against which eels can push as they move over dams, necessary for their biology. Fish lifts (or elevators) have also been used to transport fish upstream and are more successful for some species (e.g., American shad) when compared to traditional fishways.

Much of the research devoted to fish passage issues in relation to hydropower has focused on salmonids and other migratory species (e.g., lamprey, herring); however, passage issues also exist for nonmigratory species (e.g., sturgeon). Fish passage enhancements employed to aid in salmonid migration may not be effective for other types of fish, due to morphological and behavioral differences. This obstruction can lead to population isolation, limit habitat access, and hinder upstream migration. For example, the only population of white sturgeon in the Columbia River that is considered healthy is downstream of Bonneville Dam, where fish have access to freshwater, estuarine, and marine environments. The inability of sturgeon to use fish ladders designed for Pacific salmonids has led to population fragmentation and isolation.

38.1.2   Water Quality

Hydroelectric facilities undoubtedly alter the flow and quality of water passing through. Water passing through a hydroelectric facility undergoes significant forces, whether pushing a turbine blade to generate electricity or passing through a spillway, plunging hundreds of feet into the river downstream. Water can also be drawn from different depths within reservoirs, impacting water quality parameters including temperature and dissolved gas.

38.1.2.1   *Total Dissolved Gas.*
An important water quality parameter related to hydropower is the amount of dissolved gas found in the water column. As water passes over spillways, it entrains gases, which can lead to elevated levels of total dissolved gas (TDG) in the water column (Figure 38.1). Typically, TDG levels in nonimpacted river systems are approximately 100%. However, in rivers heavily impacted by hydropower such as in the CRB, TDG levels as high as 140% have been measured. Fish exposed to elevated TDG levels can develop gas bubble disease (GBD) because of the supersaturation of water. Symptoms of GBD include presence of emboli in the fins, gills, and eyes, and can influence the fish survival rates in rivers, reservoirs, and net pens housed in rivers. In contrast, the hydropower facility tailraces, which release hypolimnetic water, can have low dissolved oxygen levels, which can negatively impact fish and aquatic macroinvertebrates.

38.1.2.2   *Water Temperature.*
Water used to generate electricity at hydroelectric facilities is generally stored in reservoirs upstream of the dam. These reservoirs typically have slower flow rates than do unregulated river systems. This slow movement of water can lead to high water temperatures and stratification of the water column, with the cooler water settling under the warmer water. Generally, deeper, colder water is released through the powerhouse to generate electricity, while warmer shallow water is released over the spillway. The strategic release of cooler water can reduce the water temperature downstream, particularly during warm summer months. The use of different temperatures of water to generate power or release water from the dam can drastically influence the water temperature downstream, which can influence fish and other aquatic organisms. The physiological processes of fish are temperature regulated, and as such, behaviors such as migration and spawning are dictated by changes in water temperature (e.g., male smallmouth bass begin making nests at 15°C). Altered river temperatures downstream of hydroelectric facilities due to discharge may influence the physiology and behavior of fish. They may also allow for economically viable fisheries to exist in waters that historically were too warm to support them (e.g., salmonid fisheries).

During winter months in colder climates, the release of hypolimnetic water can alter the environment downstream of hydroelectric facilities. This warmer water (4°C) will gradually cool down to 0°C as it moves downstream, leading to the formation of frazil and anchor ice. Both frazil and anchor ice

FIGURE 38.1   Bonneville Dam spillway on the Columbia River.

can influence the available habitat in tailrace waters, forcing fish out of natural feeding and resting areas.

Warmer water downstream can also reduce the amount of stationary ice cover, which can increase the feeding and activity of fish but also increase predation rates on fish by mammals such as otters and mink and allow the persistence of angling.

38.1.3   Habitat Impacts

Hydroelectric dams can influence the habitats and life stages of fish and other aquatic animals in several ways. Dams typically reduce the extent of flood waters or river ice runs. While this can be beneficial to humans, flooding and extensive runs of broken river ice are important for maintaining quality of substrate used by fish and aquatic insects. This can lead to higher sedimentation of substrate, reducing the potential for spawning and reducing spaces within the substrate that are often used as shelter by small fish.

The presence of dams also transforms areas of rapid flow and relatively shallow water to slow-moving reservoirs. This leads to a loss of spawning and rearing habitats and habitats that are needed for aquatic invertebrates, thus modifying the food base for fish. This can lead to changes in the species present in riverine systems.

38.1.3.1   Flow Rates/Hydropeaking.   Just as water temperatures are affected by discharges, water flow rates are also influenced by the discharge from hydropower facilities. Flow rates are important for the upstream migration of fish, as migratory species are attracted to areas of high flow. In addition, increased flow rates may be required to facilitate downstream fish migration, as is the case for juvenile salmonids in the CRB. Flow rates from hydro dams can also be important for nonmigratory species.

Hydropeaking is caused by the fluctuation of water levels downstream of dams due to variable energy demands. When energy requirements are high, larger volumes of water are passed through the turbines to generate electricity. When energy is in lower demand, water discharge through turbines is decreased, lowering the water levels downstream of the dam. These sudden and variable changes in water levels can lead to stranding of fish and mortality as water levels are reduced. Movements of fish as water level fluctuates downstream of the dam may also lead to increased energy demands among fish.

Hydroelectric facilities in many large river systems now also serve as sources of flood control. Due to the presence of dams, buildings have been constructed in areas that previously may have been impacted by increased river flows in the spring. In years of high flow rates and increased volumes of water during the spring freshet, dams can serve as important regulators of the amount of water moving downstream.

38.1.3.2 Habitat Conditions. Specific flow rates may be necessary for fish to access habitats required for different life stages. Changes to flow rates due to the presence of a dam may be detrimental for native species or may create favorable conditions for invasive species. For example, the disappearance of fast-moving rapids upstream of Bonneville Dam in the CRB has reduced the available spawning habitat for white sturgeon. Recently, the changes in habitat conditions in reservoirs due to changing discharge rates have drawn the attention of managers and regulators. The water levels in reservoirs fluctuate based on the volume of water being released downstream. These fluctuations can dramatically influence the riparian habitat in the reservoir, as aquatic vegetation may die as a result of increased or decreased water levels. The decay of these plants has been shown to increase the production of methane, a gas that could possibly contribute to climate change.

38.2 REGULATORY CRITERIA FOR SURVIVAL AND WATER QUALITY

Hydroelectric facilities are regulated by government agencies that put restrictions on a number of performance metrics. In the United States, the Federal Energy Regulatory Commission (FERC) is the regulatory body that monitors the performance of dams and determines the fate of these structures. In the United States, privately owned hydroelectric facilities undergo relicensing every 50 years and, for successful completion of this process, facility managers must demonstrate several factors such as fish passage survival rates and water quality metrics (FERC, 2015).

38.2.1 Survival Rates for Fish Passage

Attaining high rates of survival for fish passing through hydroelectric dams is a high priority for managers and regulators. Migratory species have been of key concern, specifically salmonids in the Pacific Northwest of the United States and in other areas. For example, 57% (28 of 49) of West Coast salmon and steelhead stocks are currently listed as threatened or endangered under the ESA. These listings occurred over the past 17 years, and regulators have responded by implementing requirements for downstream survival of salmonids at lower Columbia River dams. Several research techniques are employed to investigate routes of passage used by fish

passing downstream and to determine their subsequent survival; these include telemetry (acoustic and radio), tagging with passive integrated transponders (PIT tags), and fish mark-recapture studies (Mathur et al., 1996). In systems without migratory species, entrainment rates of fish may be low, so determination of passage and survival rates does not warrant such attention.

38.2.2 Water Flow

The rate of water discharge through dams is monitored closely. Regulations typically call for minimum flow rates to support downstream habitat requirements. For example, in the Green River, discharge rates from Flaming Gorge Dam (located near the border of Utah, Colorado, and Wyoming) are set to between 4600 and 8600 cfs to meet target flows required by ESA-listed species to access riparian zones. Minimum flow requirements may also be necessary to facilitate migration patterns, as is the case for juvenile salmonids moving downstream in the CRB. Flow rates are also regulated to reduce stranding of fish or to allow fish access to spawning areas.

38.2.3 Water Temperature

Water temperature influences the rates of physiological processes of fish and macroinvertebrates. Therefore, the temperature of water discharged from hydropower facilities can greatly influence the health of downstream populations. Required water temperatures are dependent on biota requirements and downstream river temperatures. For example, in rivers with cold-water fish such as salmonids, temperatures must not exceed a specific threshold due to species-specific requirements.

38.2.4 Dissolved Oxygen Levels

Aquatic organisms require adequate supplies of dissolved oxygen for basic physiological processes. Generally, dissolved oxygen levels below 5 mg/L can be a stressor for aquatic organisms. Required levels are dependent on the species that inhabit a system and their sensitivity to low oxygen concentrations. To alleviate low dissolved oxygen levels, aeration techniques such as using aerating turbines and increasing spill flow are commonly used in systems with deep reservoirs and high water temperatures where low dissolved oxygen in the hypolimnion is more likely to occur.

38.3 METHODS AND EFFORTS TO ADDRESS ENVIRONMENTAL ISSUES

Because of the negative environmental impacts of hydropower, dam operators and design engineers are altering hydroturbine design and operation to make hydropower dams more environmental-friendly, especially safer for fish. To better design environmental-friendly hydropower systems, a better understanding is required of the physical conditions fish experience and the mechanics of the injury process during passage through hydroturbines, over spillways, and through other dam bypass alternatives. Several technologies have been developed and are being applied to provide critical design criteria information.

Fish and other aquatic animals can be injured when exposed to rapidly moving water. The eyes, operculum, or other parts of fish may be damaged. Shear occurs when fish traveling in relatively slow water are exposed to areas of rapidly moving water. Conversely, fish traveling in very high velocity water can sustain injuries when suddenly exposed to areas of low-velocity water. Specialized flumes have been constructed to evaluate fish exposure to shear. These flumes have powerful pumps that can replicate the high-velocity environments at hydropower facilities. Fish can be released from slow- to fast-moving water or propelled from fast water into areas of relatively slow-moving water. After their exposure to these conditions, fish are closely examined to relate injury to flow conditions. Video imaging can be used to examine how flow conditions influence fish movement and the cause of injury. These specialized facilities can also be used to determine how tagging of fish may influence them when passing hydro facilities. Testing can determine if tags influence the movements of fish within the flow, how well the tags are retained, or if injuries occur due to the presence of tags.

Managers are able to better understand the conditions that fish experience during dam passage due to the relatively recent development of an autonomous device called the Sensor Fish. This small device has several different kinds of sensors that measure the pressure, three-dimensional acceleration, three-dimensional rotational velocity, pressure, and temperature that fish experience during passage. Researchers place the device in the water flowing through turbines or through spillways, bypass, or pump storage facilities. The conditions in these routes of passage can then be related to injuries that fish may attain when exposed to dam passage. For example, the device can measure the pressures fish are exposed to during turbine passage. These data can be compared to results of research relating pressure changes to fish injury. In this manner, managers can modify operations or facilities to improve fish survival.

Fish passing hydroturbines can be damaged when they strike the turbine blades. Many factors determine how likely it is for a fish to strike the blade. Some turbines contain a large number of blades (such as a Francis turbine), while others have relatively few blades (such as a Kaplan turbine). Blade design or turbine operation conditions can also influence the likelihood of a fish striking a blade. Testing to determine the likelihood of strike can be done in several ways. Large flumes can entrain fish in high-velocity water and propel them toward a blade, or prototype scaled-down turbines can be developed. Cost-effective numerical blade strike models have been developed to estimate the probability of injury resulting from direct contact between fish and runner blades (Deng et al., 2007b). These blade strike models assume that a fish passing through a turbine will be struck by a runner blade if it does not pass the plane of the leading edge of the blade within sweeps of two successive blades. Testing can also be conducted using computational fluid dynamics (CFD), a computer modeling technique in which computers simulate water flow through a turbine. Then they simulate the presence of fish within the flowing water and estimate the likelihood that it strikes a blade or other part of the turbine. These types of research can be used by engineers to design turbines to improve survival of turbine-passed fish.

Investigations into the injury and mortality of fish exposed to pressure changes associated with dam passage have become more common over the last 15 years. Fish can be injured when they are exposed to rapid decreases of pressure when passing hydroturbines, spillways, or pump storage facilities. Rapid decompression leads to expansion of gas existing within the fish either as bubbles or in the swim bladder. Decompression can also lead to gas coming out of suspension in the blood or tissues.

To determine how pressure changes influence fish, researchers use hyper/hypobaric chamber systems. These systems can both increase (hyper) or decrease (hypo) the pressure in the fish and the water that surrounds them. Researchers simulate the speed and extent of pressure changes during passage through turbines, bypass systems, pump storage facilities, or spillways. Information gathered from this research can be used to design safer passage routes for fish, reducing the environmental impact of hydroelectric facilities.

38.4 ENVIRONMENTAL-FRIENDLY HYDROPOWER SYSTEMS

The US Department of Energy (DOE) established the Advanced Hydropower Turbine Systems (AHTS) program in 1994 in collaboration with the hydropower industry and other agencies to develop environmental-friendly turbines. Partners include the US Army Corps of Engineers (USACE), the National Hydropower Association, Bonneville Power Administration, Electric Power Research Institute (EPRI), the US Geological Survey (USGS), public utility districts, turbine manufacturers, and DOE national laboratories (Brookshier et al., 1995). The USACE also initiated the Turbine Survival Program (TSP) in 1995 to address the biological opinion measures established by the US National Marine Fisheries Service (NMFS) to improve the survival rates of adult and juvenile salmonids within the Columbia and Snake River basins (USACE, 2004). Hydropower dams within these two basins contribute approximately half of the total hydro-electric energy production in the United States. TSP members included design engineers and fisheries biologists from the USACE Portland and Walla Walla districts, the Hydroelectric Design Center, and the Engineer Research and Development Center, with technical support from research organizations such as the Pacific Northwest National Laboratory (PNNL), NMFS, and USGS. Since the beginning of the two programs, there has been a close working relationship between DOE and USACE including sharing technologies and costs for managing their respective programs. In 2010, DOE Secretary Steven Chu, US Department of Interior Secretary Ken Salazar, and USACE representatives signed a memorandum of understanding to provide close collaboration among these federal agencies for the development of environmentally sustainable hydropower. Several major breakthroughs in the environmental-friendly turbine design, including the minimum gap runner (MGR) turbine and Alden turbine, have been made from the collaborative efforts among the two programs, the turbine industry, and various research organizations.

The MGR turbine was designed and manufactured by Voith Hydro and incorporated many of the AHTS program's design features to improve fish survival through turbines while increasing operation efficiency. Over a 14-year period concluding in July 2010, 10 MGR Kaplan turbines were installed at the USACE Bonneville Dam First Powerhouse. The survival rates of juvenile salmonids passing through the powerhouse increased from 97% to 98%. In 2005, an MGR Kaplan turbine also was installed at Wanapum Dam, operated by Public Utility District No. 2 of Grant County (Grant PUD) for testing as part of Grant PUD's relicensing effort to replace the 10 existing aging turbines and provide better fish passage conditions through the turbines. Operating efficiency improved, and power generation increased from 895 to 1118 MW. Grant PUD and DOE funded several studies to evaluate the biological performance of the new MGR turbine. Results indicated the MGR turbine had a 48-hour survival rate of 97% (using balloon tag technology) and provided a better pressure and rate of pressure change environment for fish passage (using Sensor Fish technology), even though the MGR turbine has six blades and the old turbine had five blades (Deng et al., 2010). Based on the test results, the FERC approved the Grant PUD relicensing application for Wanapum Dam, and installation of the remaining nine turbine units have been completed by 2012.

An Alden turbine is a Francis-type turbine designed by Alden Research Laboratory with the support of the DOE AHTS program and EPRI (Hecker and Cook, 2005). Its main features include a three-blade design to reduce blade strike and slow rotation to reduce shear and rapid pressure changes, all of which are major causes of fish injuries. Pilot-scale biological testing showed very high survival rates of 96% for many species when scaled to full-size units. Alden Research Laboratory, EPRI, and Voith Hydro are now working together to continue engineering development and are planning full-scale demonstrations of the Alden turbine.

REFERENCES

Brookshier, P. A., Flynn, J. V., Loose, R. R. (1995). 21st century advanced hydropower turbine system. Waterpower '95 - Proceedings of the International Conference on Hydropower. American Society of Civil Engineers, New York, NY, pp. 2003–2008.

Brown, R. S., Pflugrath, B. D., Colotelo, A. H., Brauner, C. J., Carlson, T. J., Deng, Z. D., Seaburg, A. G. (2012). Pathways of barotrauma in juvenile salmonids exposed to simulated hydroturbine passage: Boyle's law vs. Henry's law. *Fisheries Research*, 121: 43–50.

Cada, G. F. (2001). The development of advanced hydroelectric turbines to improve fish passage survival. *Fisheries*, 26: 14–23.

Coutant, C. C., Whitney, R. R. (2000). Fish behavior in relation to passage through hydropower turbines: a review. *Transactions of the American Fisheries Society*, 129(2):351–380.

Deng, Z., Carlson, T., Duncan, J. P., Richmond, M. C., Dauble, D. D. (2010). Use of an autonomous sensor to evaluate the biological performance of the advanced turbine at Wanapum Dam. *Journal of Renewable and Sustainable Energy*, 2(5):53–104.

Deng, Z., Carlson, T. J., Duncan, J. P., Richmond, M. C. (2007a). Six-degree-of-freedom Sensor Fish design and instrumentation. *Sensors*, 7:3399–3415.

Deng, Z., Guensch, G. R., McKinstry, C. A., Mueller, R. P., Dauble, D. D., Richmond, M. C. (2005). Evaluation of fish-injury mechanisms during exposure to turbulent shear flow. *Canadian Journal of Fisheries and Aquatic Sciences*, 62(7):1513–1522.

Deng, Z., Carlson, T. J., Ploskey, G. R., Richmond, M. C., Dauble, D. D. (2007b). Evaluation of blade-strike models for estimating the biological performance of Kaplan turbines. *Ecological Modelling*, 208:165–176, doi:10.1016/j.ecolmodel.2007.05.019.

Federal Energy Regulatory Commission. (2015). Hydropower. www.ferc.gov/industries/hydropower.asp.

Hecker, G. E., Cook, T. C. (2005). Development and evaluation of a new helical fish-friendly hydroturbine. *Journal of Hydraulic Engineering*, 131(10):835–844.

Mathur, D., Heisey, P. G., Euston, E. T., Skalski, J. R., Hays, S. (1996). Turbine passage survival estimation for Chinook salmon smolts *(Oncorhynchus tshawytscha)* at a large dam on the Columbia River. *Canadian Journal of Fisheries and Aquatic Sciences*, 53(3):542–549.

USACE. (2004). Turbine survival program phase 1 report, 1997-2002. Technical report. U.S. Army Corps of Engineers, Portland, Oregon.

39

SOCIAL ISSUES RELATED TO HYDROPOWER

Joerg Hartmann

39.1 SOCIAL ISSUES

All hydropower projects are site-specific and their impacts differ substantially from site to site. More than any other source of electricity, they can have multiple social impacts, both positive and negative. Larger projects, or multiple smaller projects, can transform entire regions. The systematic assessment of potential social impacts and opportunities for their management are key to the successful development of new hydropower plants. Even in long operating projects, relations to social groups need to be maintained and often improved. The "social license to operate" can be as relevant as the regulatory license. This chapter categorizes social issues and highlights some social groups affected by hydropower, and provides an overview of social issues management approaches.

39.1.1 Rights, Risks, and Responsibilities

Benefits and costs of projects are diverse and unequally distributed within society. They range from gaining access to more reliable energy and employment opportunities, to losing access to burial sites and farmlands. A systematic approach to reducing negative impacts and enhancing positive impacts is required. A comprehensive framework was provided by the World Commission on Dams (2000) [1] for the recognition of rights and assessment of risks, based on the universal values of equity, efficiency, participatory decision-making, sustainability, and accountability. Stakeholders are those whose rights (such as human rights, property rights, or the right to development) may be affected or whose commercial interests, livelihoods, cultural traditions, or other interests may be at risk. It is important to distinguish

between voluntary risk-takers, such as the project developer, and involuntary risk-takers such as a village that may be affected. Once rights and risks are identified, negotiations between stakeholders can lead to mutually acceptable outcomes. It is the primary responsibility of those proposing or operating a project—the voluntary risk-takers—to resolve social issues.

39.1.2 Impacts in the Vicinity of the Hydropower Project

The areas directly affected by hydropower projects include the areas inundated by reservoirs, dams, and weirs, diversion reaches of the river, headrace and tailrace channels and tunnels, penstocks and powerhouses, quarry and spoil areas, and transmission and road access corridors. Impacts on people living within or close to these areas are often quite disruptive; for vulnerable groups they can be traumatic. However, those directly impacted are most easily identified, are often best placed to protest against what they perceive as unfair treatment, and are often the first to receive compensation. In some cases, where local people are explicitly recognized and singled out as owners of land and resources, they have benefited significantly from hydropower projects.

39.1.3 Impacts in the Larger Society

Projects can have a larger indirect social footprint than is often realized. If people need to be resettled, their new host communities will also experience social change. Access roads and employment opportunities will induce migration into formerly remote regions. Access to electricity will enable industrial development in other parts of the country. Where

Alternative Energy and Shale Gas Encyclopedia, First Edition. Edited by Jay H. Lehr and Jack Keeley.

large reservoirs change water quality, flow regimes, and sediment transport, impacts on fishing and riverbank agriculture may reach hundreds of kilometers downstream. Many of these indirect impacts are typically seen as outside the project's responsibility and are not subject to compensation arrangements.

39.2 SOCIAL GROUPS

Three social groups often receive specific attention in hydropower projects, primarily because their rights and risks are more significant and easily distinguished from those of other groups.

39.2.1 People Affected by Displacement and Resettlement

People displaced by hydropower projects number in the tens of millions. Displacement can be either physical (having to move out of a project area) or economic (losing access to resources, such as land, in the project area). People are either resettled as complete communities, resettled as smaller groups or families into host communities, or leave the area, with or without cash compensation, to live elsewhere. Project planning should aim at avoiding displacement as much as possible and mitigating the effects of any remaining displacement. Most countries encourage prior negotiations to achieve consent with people to be displaced, but also have—as a last resort—legislation covering involuntary expropriation and compensation arrangements. Specific resettlement safeguard arrangements also exist for projects financed by most development banks and major commercial project finance banks.[1] Nevertheless, livelihoods for resettled people are often difficult to restore, and for people losing access to resources over which they may have no formal rights, displacement often results in loss of livelihoods without compensation.

39.2.2 Indigenous Groups

As more remote sites are being developed for hydropower, projects frequently impact upon lands owned or used by indigenous peoples. The appreciation of the special cultural contributions, vulnerabilities, and rights of indigenous groups has evolved rapidly. The 2007 United Nations Declaration on the Rights of Indigenous Peoples,[2] supported by almost all countries, says in Article 10 that "No relocation shall take place without the free, prior and informed consent of the indigenous peoples concerned and after agreement on

just and fair compensation ..." and in Article 32 that "States shall consult and cooperate in good faith with the indigenous peoples concerned through their own representative institutions in order to obtain their free and informed consent prior to the approval of any project affecting their lands or territories and other resources, particularly in connection with the development, utilization or exploitation of mineral, water or other resources." In practice, treatment of indigenous groups still differs widely, from forced assimilation or relocation at one end of the spectrum to recognition as equal partners in project joint ventures at the other end.

39.2.3 Employees

As some of the largest infrastructure projects, large dams often involve thousands of workers during the construction phase, many of whom are unskilled migrant workers. Conditions in work camps and interactions with local populations need to be well managed, often indirectly through contractors and subcontractors. Public health and law and order issues may arise. After commissioning, many projects can be operated with relatively small numbers of specialists. In fact, many smaller sites in developed countries are now remotely controlled and dispatched. Energy utilities often pay above average salaries and care strongly about workplace conditions, including safety.

39.3 SOCIAL MANAGEMENT

Social impact assessment leads directly into the management of identified impacts. Applying simple metrics such as people displaced/MW allows project proponents to select low-impact sites [2]. Ideally, social management should start during the preparation and design phase of a project, through an iterative process that avoids, minimizes, mitigates, compensates, and offsets negative impacts, and enhances positive impacts.[3] In the following, a number of key social management concepts are highlighted.

39.3.1 Community Consultation and Grievance Mechanisms

The basis for any functional relationship with directly affected communities is open two-way communication. Communities can help to identify impacts, design solutions, and monitor ongoing impacts even if there is no consensus over the project as such. Credible grievance mechanisms, where people receive feedback on whether and how their concerns are being addressed, are crucial for building relations.

---

[1] For example, World Bank OP 4.12—Involuntary Resettlement, last revised February 2011.

[2] Resolution adopted by the General Assembly in the 107th plenary meeting, September 13, 2007.

[3] A broadly accepted framework for systematically assessing the quality of this process, and the sustainability of its outcomes, is International Hydropower Association's Hydropower Sustainability Assessment Protocol (2011).

39.3.2 HSE (Health, Safety, and Environment) Management

HSE issues, for which a project developer has direct regulatory responsibility and liability, differ from country to country, but they generally receive the most focused attention. In many cases, particularly internationally operating developers go beyond regulatory requirements, establish their own HSE commitments, or obtain certification against international standards. HSE programs generally have a narrow perspective. For example, they will address the risk from spills of hazardous materials in the powerhouse, but not the risk from normal operations of the power plant, releasing flows which may be detrimental to social conditions downstream.

39.3.3 Corporate Social Responsibility

The broader notion of corporate social responsibility has gained significant traction since the 1990s.

Moving on from "corporate charity," many larger companies are now supporting local, regional, and national social development objectives, have signed on to the UN "Global Compact" [3], or are publicly reporting on their sustainability performance according to the Global Reporting Initiative (GRI) [4]. As an indication of how much CSR has become accepted, in 2010 the International Standards Organization (ISO) published its ISO 26000 "Guidance on social responsibility," emphasizing the following rationale for its use [5]: "The perception and reality of an organization's performance on social responsibility can influence, among other things: Competitive advantage; Reputation; Ability to attract and retain workers or members, customers, clients or users; Maintenance of employees' morale, commitment and productivity; View of investors, owners, donors, sponsors and the financial community; Relationship with companies, governments, the media, suppliers, peers, customers and the community in which it operates."

39.3.4 Benefit Sharing

Hydropower projects generate significant benefits over the long term. Some of these are shared by following national rules on compensating for adverse impacts and paying required taxes and royalties. Many developers are exploring how they can go beyond this and can become more active "corporate citizens." Governments are also increasingly attentive to the need for reducing social conflicts and the opportunity to leverage hydropower projects as regional development poles.

Benefit sharing can take many forms and usually involves some public–private partnership mechanism.

39.3.5 Multiple-Purpose Management of Dams

One form of benefit sharing is to use reservoirs, which have often been financed on the basis of expected hydropower revenue, for multiple purposes. Many reservoirs simultaneously provide flood control, storage for water supply and irrigation, recreation, fishing, and aquaculture, environmental flows, navigation, and other benefits. In fact, in some reservoirs hydropower is only a minor concern and the powerhouse is only operated when water is released for other purposes. This illustrates that different operational objectives are often contradictory, if not mutually exclusive, and that the balancing or optimization of operations for the benefit of different social groups can be a challenging and divisive process.

39.3.6 Downstream Flows

The pattern of releases of water from a reservoir, as a result of operational decisions of the dam operator, impacts both the natural characteristics of the river downstream and the social conditions along the river. It has been estimated that "472 million river-dependent people [live] downstream of large dams along impacted river reaches" [6]. In recent years more understanding has been achieved on these impacts, and methodologies have been developed to quantify "environmental flows," defined as the "quantity, timing, and quality of water flows required to sustain freshwater and estuarine ecosystems and the human livelihoods and well-being that depend on these ecosystems" [7].

REFERENCES

[1] World Commission on Dams. (2000) *Dams and Development: A New Framework for Decision-Making.* London: Earthscan.

[2] Ledec, G., Quintero, J. D. (2003). Good dams and bad dams: environmental criteria for site selection of hydroelectric projects. Latin America and Caribbean Region Sustainable Development Working Paper 16. World Bank, Washington, DC.

[3] http://www.unglobalcompact.org/

[4] http://www.globalreporting.org/

[5] ISO. (2010). *Discovering ISO 26000 on Social Responsibility.* ISO, Geneva.

[6] Richter, B. D., Postel, S., Revenga, C., Scudder, T., Lehner, B., Churchill, A., Chow, M. (2010). Lost in development's shadow: the downstream human consequences of dams. *Water Alternatives*, 3(2):14–42.

[7] The Brisbane Declaration. (2007). Environmental Flows are Essential for Freshwater Ecosystem Health and Human Well-Being. 10th International River Symposium, Brisbane, Australia, 3–6 September 2007.

40

SAFETY IN HYDROPOWER DEVELOPMENT AND OPERATION

Urban Kjellén

40.1 INTRODUCTION

Many developing countries have ambitious goals to provide electricity for all, and investments in the construction of power generation and distribution infrastructure grow fast and often exceed the rate of growth of economy as a whole during an early phase of development (ILO, 2001; OECD/IEA, 2011). The construction industry is generally characterized by a high risk of accidents. The fatal accident risk is, according to ILO estimates, five times higher for construction as compared to the general average among employees worldwide (Murie, 2007). Data on fatalities from hydropower projects in developing countries are notoriously bad, but statistics from Norway may illustrate the developments in the accident risk. There were 23 fatalities in the Tokke 1 GW hydropower project between 1956 and 1972 (Skjold, 2006). The Førre 2 GW hydropower plant was constructed between 1974 and 1984 with three fatalities, and the Tyin 182 MW project in 2001–2004 with zero fatalities, illustrating the consequences of improved technology and safety management. There is evidence that some developing countries are operating with a similar risk level as that of Norway 50 years ago, as illustrated by the 16 fatalities in the construction of the 14-km headrace tunnel for the Teesta project in India (Khali and Bisaria, 2008).

This chapter deals with safety in the different phases of development and operation of hydropower projects. It covers safety for hydropower company personnel and contractors as well as the general public. Where published data are lacking, the chapter is based on statistics and experiences from an international hydropower company's operations in five different countries in Asia and South America. This chapter has been written from the perspective of a hydropower company and outlines safety management approaches and measures available to company management to control the risk of accidents.

40.2 ACCIDENT RISKS IN HYDROPOWER DEVELOPMENT AND OPERATION

40.2.1 Overview

Table 40.1 gives an overview of typical accident risks in hydropower development and operation. The overview focuses on hazards that reflect the specific characteristics of hydropower plants and their location and the consequences of these to the safety in the different phases of a hydropower project, including activities before construction (mainly site investigation), during construction, and during operation and maintenance of the plant. The exposure estimate in man-hours applies to company and contractor personnel working in the field, at the construction sites or in the plant. It does not include office work, transportation outside the site, or third-party exposure to accident risks. The estimate is based on a typical medium-sized hydropower project (150 MW) and assuming a construction period of 4 years with a work force of about 2000 construction workers.

40.2.2 Major Accidents with Multiple Fatalities

For conventional hydropower plants with reservoirs, dam failure, although rare, represents the single largest potential of multiple fatalities, because of the exposure of the population downstream of the dam to flooding (ICOLD,

Alternative Energy and Shale Gas Encyclopedia, First Edition. Edited by Jay H. Lehr and Jack Keeley.
© 2016 John Wiley & Sons, Inc. Published 2016 by John Wiley & Sons, Inc.

TABLE 40.1   Typical accident risks in different stages of hydropower development and operation.

	Site investigations and collection of field data	Construction	Operation (incl. maintenance)
Rough exposure estimate	~15,000 man-hours	~16 million man-hours	~30,000 man-hours/year
Major accident risks	• Transportation accident	• Fire in tunnel • Tunnel collapse • Transportation accident • Natural hazards (rockfall, landslide, avalanche, flooding)	• Dam failure • Flooding of powerhouse • Fire, explosion in subsurface powerhouse
Occupational accident risks	• Transportation accident • Accident due to natural hazards • Falling during movement and manual transportation of equipment • Getting squeezed during rigging	• Natural hazards • Falling from height • Hit by load (material handling) • Rockfall, rock burst in tunneling • Fire in tunnel • Transportation accident (rollover, hit by vehicle, driving off road)	• Drowning • Electrocution • Accident with moving machinery parts and equipment
Third-party accident risks	Not applicable	• Drowning • Transportation accident (hit by project vehicle)	• Drowning due to dam break or penstock rupture • Falling into water-conveying canal or dam and drowning

2005). One example is the failure of the Banqiao dam in China in 1975, causing flooding and the loss of 26,000 lives. The potential for dam failure with catastrophic consequences is dependent on the size of the dam, the extent of the inundation area, the size of the population at risk in exposed area downstream of the dam, and the amount of warning time available. Hence large reservoir projects (> 500 MW) in populated areas represent the highest potential for catastrophic dam failure. Failure of dams is an unlikely event. Statistics from ICOLD collected at the beginning of 1990 shows that less than 0.5% of the dams built after 1950 failed in the course of their lifetime (ICOLD, 1995). The statistics has likely improved further due to improved design and monitoring of dams.

Also penstock rupture may have severe consequences as illustrated by the Bieudron HPP in Switzerland, where three lives were lost in 2000 due to this type of event (Bissel et al., 2011).

Flooding of the powerhouse is another type of event with major accident potential. One example is the accident at the Sayano–Shushenskaya hydropower plant in 2009, where the cover of one of the turbines ruptured, resulting in flooding of the machine hall and 75 people being killed mainly through drowning (International Water Power & Dam Construction, 2010).

Explosion of transformers in subsurface powerhouses is a third type of major accident risks in hydropower operation.

The explosion is caused by an arc igniting the transformer oil (Hansen et al., 2002). Depending on the design of the powerhouse, the explosion may escalate outside the transformer cavern and trap the personnel inside the powerhouse. In Tonstad, Norway, three people were killed in 1973 as a result of an explosion in a transformer in a hydropower plant, located 750 m inside the mountain.

During the construction period, up to several thousand workers may be present at the construction site. *Natural hazards* such as landslides, rockfalls, avalanches, and flash floods may have catastrophic consequences (Figure 40.1).

There are three specific hazards related to *tunnel work* that have major accident potential: misfire of explosives, tunnel fire, and tunnel collapse. Depending on the technology in use in tunnel work, up to 30 people may be present in a tunnel simultaneously. Any of these events can cause multiple fatalities among the tunnel crew. Tunnel fire usually starts in a diesel-powered machine or vehicle, resulting in extensive smoke development and trapping of personnel inside the tunnel (Ingason et al., 2010). Tunnel collapse results in a sudden, uncontrolled loss of the cross section of a tunnel or fall of material and is caused by unpredicted geological conditions and human errors during planning and execution (Sidenfuss, 2006).

Transportation of personnel has the potential of a major accident, for example, when a bus drives off a road in a mountainous area.

FIGURE 40.1 Construction road in the Indian Himalayas blocked by falling rock.

40.2.3 Occupational Accidents

Site investigations take place in various project phases before start of construction, with the majority of the work being carried out during the feasibility assessment (Figure 40.2). The purpose is to get a general impression of the suitability of the site and to collect area-specific data on hydrology, meteorology, topography, geology, and biology. Site investigations also include communication with local communities in connection with the environmental and social impact assessment.

The primary hazard in these activities is related to transportation. The sites for hydropower development are often located in remote areas high up in the mountains, where the differences in height are utilized to generate power. Infrastructure such as roads and access to emergency care may be poorly developed at this stage of the project. Accidents due to driving off the road or being hit by rockfall or landslide often have fatal consequences.

At the site, the crew is exposed to natural hazards such as rockfall, landslide, flooding, and avalanche. Due to the nature of data collection, site investigations cannot completely be restricted to times of the year, when the natural conditions are most benign.

Data collection also involves walking and manual transportation of equipment in areas with difficult access due to a rough terrain with steep slopes. Some activities such as core-drilling and hydrologic measurement take place on the river bed or in the river. Consequently, most of the field investigation activities involve hazards with potential fatal consequences due to falling from elevation and drowning. Unpublished data from India indicate that the fatal accident frequency in field investigations is many times higher than in construction, but the actual number is lower due to a considerably lower exposure in man-hours per project.

Construction of a hydropower plant is the phase with the highest exposure to accident risks due to the large number of workers at the construction site. The construction industry is generally characterized by a high risk of accidents. According to ILO estimates, 100,000 workers are killed annually on construction sites globally, and the fatality rate in construction is five times the general average among employees worldwide (Murie, 2007). Hydropower construction like any infrastructure project involves heavy civil works such as earth movement, excavation of tunnels and caverns, and construction of large structures (powerhouse, dams, and waterways). This is followed by the installation of heavy electrical and mechanical equipment. All of these activities are associated with the handling of large energies with the potential of severe accident.

Distribution of fatalities (*N*-14)

Distribution of total recordable injuries (*N*=273)

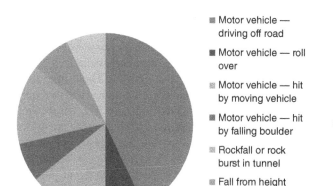
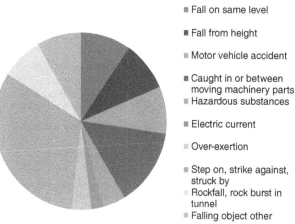

FIGURE 40.2 Accident statistics from hydropower construction projects. *N* = number of fatalities/injuries in the sample.

Figure 40.2 shows statistics from a hydropower company's construction projects in Asia and South America and covers about 17 construction-years for 5 medium-sized hydropower projects. The fatality statistics is dominated by one project in the Himalayas (10 of 14 fatalities).

Transportation of personnel and materials at the construction site dominates the risk of fatal accidents. Typically these accidents are due to loss of control of the vehicle and driving off the road in a mountainous area, loss of control and rollover of the vehicle, and hitting and squeezing a person during vehicle movement. Also geophysical hazards like falling rock or landslide may have fatal consequences when a vehicle is hit.

A second significant cause of fatal accident in construction is falling rock or rock burst in tunnel excavation. Fatal fall from height occurred in transmission tower erection.

The largest category of ordinary occupational accidents resulting in recordable injury is "stepping on, strike against and truck by." This is what could be expected from any construction site. "Caught in or between moving machinery parts" is the second largest group. Here we also find accident with severe, nonfatal consequences like crushing and amputation of limbs.

Plant-specific hazards with high potential (fatality) in hydropower *operation and maintenance* are associated with work in high-voltage systems (electrocution), work in waterways (drowning), and work in the vicinity of rotating equipment. Figure 40.3 shows statistics from an international hydropower company's 16 small- and medium-sized plants in Asia and South America with a total installed capacity of 1500 MW. The statistics is based on 60 plant-operation years. It shows that falls from height dominate the picture with motor vehicle accidents as the second largest group. It also shows that there have been no occupational accidents in work in high-voltage systems or in waterways, indicating the risk of these types of events is adequately managed.

40.2.4 Third-Party Accidents

Drowning in dams and canals belonging to the hydropower plant and transportation accidents in construction dominate the picture of third-party accident risks. In the materials from the international hydropower company's 16 operating plants, there have been 3 cases of drowning in conveying canals or dams.

40.2.5 Influencing Factors

The risk of accidents in hydropower development and operation is influenced by internal as well as external factors. External influencing factors are the frame conditions over which management does not have direct control.

A typical external factor is the standard of the *national regulations and enforcement* in the area of safety. Countries undergoing a transformation into a modern economy will make large investments in infrastructure including power generation as part of this development. Whereas developing countries may have comprehensive safety legislation, enforcement is often weak due to lack of resources (Kjellén, 2012). Legislation may also lag behind in critical areas for hydropower development and operation such as safety in tunnel construction and dam safety.

The *physical environments of the project and site layout differ* considerably between hydropower projects, but are usually located in mountainous areas to exploit the drop in the river. An example is a project in the Indian Himalayas, located between elevations about 2000 and 3000 m above sea level (Kjellén, 2012). It includes 30 km of internal access roads to two separate intakes. The access roads have been carved out of steep mountains and there are significant risks of driving off the roads as well as risks of road erosion, falling boulders, and landslides during the monsoon period. The roads are exposed to winter conditions with ice and snow and the area is avalanche prone.

Dams are often located in narrow canyons to generate head and the physical environment is characterized by steep slopes with risks of falling and landslide. Dams may also be located on the plains and large water volumes will be required to achieve enough hydropower capacity. The consequences of a dam break will be extensive if large areas are affected and the areas downstream the dam are densely populated.

The *standard of the equipment and the production methods* used in construction will have a significant impact on the accident risk (Kjellén, 2012). The use of labor-intensive methods of work rather than modern, mechanized work methods with few persons exposed to the hazards at the sharp end will increase the risk of accidents. Tunnel construction is a typical example, where traditional tunneling methods will

Distribution of total recordable injuries (*N*=36)

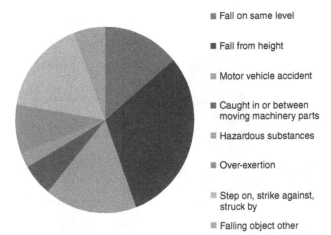

- Fall on same level
- Fall from height
- Motor vehicle accident
- Caught in or between moving machinery parts
- Hazardous substances
- Over-exertion
- Step on, strike against, struck by
- Falling object other

FIGURE 40.3 Accident statistics from operation and maintenance of hydropower plants. *N* = number of injuries in the sample.

require about 25 workers at each tunnel face, as compared to 3 in modern tunneling. The workers operating modern tunneling equipment will also be protected inside an operator cabin for most of the time at the face. These differences have implications for the risk of accidents. Modern and manpower-efficient tunneling methods will reduce the number of workers at the risk-exposed sharp end and thus reduce the risk of accidents (Manu et al., 2010). This is especially evident in tunneling, where the tunnel front constitutes a hazardous zone during certain parts of the excavation cycle.

The maintenance standard of equipment has an impact on the risk of accidents. Poor maintenance will result in frequent breakdowns leading to increased needs of improvisations along with an increased risk of accidents (Saari, 1976).

40.3 MANAGEMENT OF ACCIDENT RISKS

Many energy companies utilize staged decision models to achieve predictability and control of investments (Figure 40.4). These models typically integrate all functions including the management of safety into one process where an investment project is developed from a business opportunity through different phases into an operation plant (Kjellén, 2000). The staged model is used here to illustrate the principles for the management of safety in hydropower development and operation. It is a general principle that the most robust and cost-efficient solutions are found by identifying and mitigating the accident risks in the phase along the value chain before the phase where they are actually substantiated.

The five phases are shortly described here, together with typical safety activities in a comprehensive approach to manage accident risks:

1. *Business development*: screening and investigation of business opportunities including the pre-feasibility study. This is primarily a desk study of a development opportunity, complemented with some collection of field data (hydrology, meteorology, topography, geology, biology, and communication with local communities). The first screening of significant safety issues such as the safety standard of roads and the risk of natural hazards is made at this stage.

2. *Feasibility assessment*: demonstrating at a more precise level whether the proposed investment is profitable and meets the company's policies and strategic objectives. Significant accident risks in construction and operation will be identified and assessed in order to implement mitigation in the cost estimates.

3. *Engineering and tendering*: preparation for construction, including the development of detail design, tendering for the main equipment and construction contracts, obtaining statutory approvals, land acquisition, and so on. Safety requirements will be implemented in specification for design and construction, and safety will be part of the qualification criteria for contractors.

4. *Construction*: detailed planning and mobilization at site, construction, commissioning, and handover to operation. This phase represents the highest risk of accidents due to the large exposure of contractor employees to construction-related hazards. Contractors will be responsible for safety and will be monitored by the owner.

5. *Operation*: operation and maintenance of the plant, refurbishments and modifications also being part of this phase. The owner will implement safety management systems to ensure an adequate level of safety.

As shown in the previous section, the physical hazards representing significant accident risks manifest themselves primarily in field data collection in the pre-feasibility and feasibility assessments, in construction, and in operation. The management of safety in hydropower development and operation falls into three main areas: (1) design; (2) contracting; and (3) operation and maintenance.

The *safety management of design* will address the constructability of the plant during execution as well as the operability and maintainability of the plant during operation. Examples are as follows:

- *Constructability consideration in design*: arrangement of access roads for safe transportation of personnel and materials; tunnel dimensions and layout that take into account needs for ventilation and emergency evacuation; general layout of the construction site considering natural hazards such as flooding and landslides

- *Operability and maintainability considerations in design*: general plant layout (including access roads) considering natural hazards, a layout that promotes safe access and maintenance handling of heavy equipment; fire partitions and firefighting systems; use of noncombustible materials; noise limits in the different areas of the plant; design of escape and evacuation routes; dam safety considerations

Safety management of design starts already in business development, when the hydropower potentials in the area are

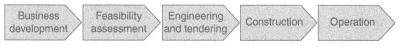

FIGURE 40.4 Typical phases in the value chain of an energy project.

assessed together with the natural conditions and possible development scenarios. Any hydropower concept must be adequately robust against prevailing natural hazards. Consequences of dam failure will be addressed at a coarse level already at this stage. The infrastructure in the area, and especially the road safety standard, is critical to a safe development and will be addressed at this stage as well.

The design will be further matured during the feasibility assessment and engineering until it has reached an acceptable level for tendering and finally for construction.

Safety management of design does not differ in principle from the management of any quality aspect of design and follows the principles of quality management standards such as ISO 9001. The owner's (customer's according to ISO 9001) safety requirements are laid down in internal design specifications as well in national regulations and safety requirements of financial institutions (International Finance Corporation, Asian Development Bank). There are also voluntary international standards and guidelines that may be adapted by the owner that address specific safety aspect of the design of hydropower plants, such as

- BS 6164, Code of practice for safety in tunneling in the construction industry
- NFPA 851, Recommended Practice for Fire Protection for Hydroelectric Generating Plants
- IHA Hydropower Sustainability Assessment Protocol (IHA, 2004)

In the course of design, risk assessments are used to define specific requirements to design and to determine the acceptability of design solutions. Examples of risk assessment methods used in hydropower development (Singh, 1996; Kjellén, 2000) are as follows:

- Hazard Identification (HAZID) is a checklist method that gives an overview of the hazards of a hydropower development by type, causes, consequences, and risk.
- Hazard and Operability (HAZOP) Analysis may be used to analyze the waterways in a hydropower plant and identifies deviations and potential consequences such as flooding of power plant.
- Dam break analyses are used for analysis of the downstream consequences of dam failure and for designing dam safety features including emergency spillways.
- Job Safety Analysis is used to identify and control hazards in the performance of a job. It is used in all phases of a hydropower development in the analysis of critical job with potential for severe accidents.

Management of contracting: Contractors are responsible for the large majority of work carried out at the hydropower development site. This applies to field data collection during

the pre-feasibility and feasibility studies and, most significantly, during construction of the plant. Contractors usually play an important role in maintenance and modifications of existing plants as well. It is essential for the owner to be able to manage the contracting process in an adequate way to ensure an acceptable safety standard in the execution of work.

A systematic approach to safety management in all phases of the contracting process (Figure 40.5) will increase the likelihood of satisfactory safety results. This is accomplished through

- an adequate understanding of the scope of the construction work and the associated hazards and implementation of the necessary mitigating measures as safety requirements in the tender documents and contract;
- ensuring qualified contractors with the ability for safe execution of contract work;
- a joint understanding of the necessity of a good safety culture at the site in order to achieve satisfactory overall and safety results;
- timely implementation of the necessary safety systems and practices at site establishment and start-up;
- systematic monitoring of safety performance during execution of work and timely implementation of corrective and compensatory measures; and
- continuous improvements through experience analysis and follow-up.

Practice often deviates from these ideal premises as illustrated by the following examples:

- The conditions at the site may involve hazards that are not under the control of the project, for example, transportation on public roads and exposure to natural hazards such as landslides and flooding.
- There may be an insufficient competition for contracts involving work at remote locations with significant safety and/or security challenges, resulting in the selection of contractors with less-than-adequate qualifications for safe execution of work.
- The owner may experience that his action alternatives in case of violations by contractors of the safety provisions of the contract are limited. Strong reactions such as termination of contract may have unacceptable cost and schedule consequences, especially in a market with limited competition.

These conditions represent significant threats to the possibilities of a safe execution of contract work and may also result in significant delays and cost overruns. They will call for compensating measures by the owner. There will be a

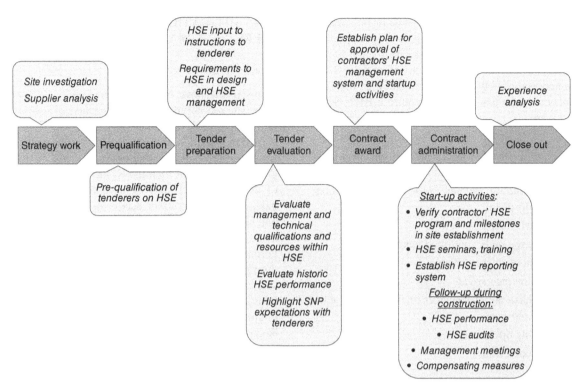

FIGURE 40.5 The contracting process and associated safety activities in a construction project (adapted from Kjellén, 2012). Similar principles apply to the contracting of field investigation work and maintenance work in an operating plant.

balance between the owner's need of control to ensure safe execution of contractor's work and the risk of taking over the responsibility for safety in construction work from the contractor.

Management of operations and maintenance: Generally, safety management of a hydropower plant follows the same basic principles as the management of safety in any industry with potential for major loss, and industry standards for occupational health and safety systems (such as OHSAS 18001, British Standards Institute, 2007) are widely used. Safety management of operation and maintenance is based on defined safety and operational standards, combined with systematic analysis, follow-up work, and correction of deviations from the standard.

Below are typical safety issues listed that need to be addressed in the plant's safety standards and procedures:

- Personal protective equipment (PPE)
- Driving and transportation safety
- Hazardous and toxic substances, procurement, handling, storage and disposal, MSDS
- Personal protective equipment (PPE)

- Cranes and lifting
- Security, perimeter guarding, visitors
- Safety induction and training
- Safety meetings including toolbox meetings
- Working environment control
- Cranes and lifting

- Driving and transportation safety
- Hazardous and toxic substances, procurement, handling, storage and disposal, MSDS
- Incident and unwanted occurrence reporting
- Safety inspections and audits
- Permit to work (PTW), hot work, work in confined space, lockout/tag-out (LO/TO) procedure
- Disciplinary action in case of violation of safety rules
- Right to stop dangerous work
- Electrical safety, substation safety, LO/TO
- Access to tunnels
- Fire safety
- Work at height

- Security, perimeter guarding, visitors
- Safety induction and training
- Safety meetings including toolbox meetings
- Working environment control
- Signage
- Location of first aid and firefighting equipment, escape ways (plot plan)
- Control of communicable diseases
- Health and hygiene
- Prohibitions against smoking, alcohol, drugs
- Instructions for critical jobs (refer Job Safety Analysis)
- Contractors' safety
- Handling of deviations

A critical element is the plant's *permit-to-work system*, which is an administrative system to secure that the necessary barriers have been implemented before start of work on critical systems. A PTW is a written proof of permission to carry out specified work in a specified area or on specified equipment for a limited time period. It is essential in preventing severe accidents in nonroutine tasks. It is often used in combination with a risk assessment (Job Safety Analysis) to define the necessary safety barriers and other measures.

Examples of tasks that normally require a PTW before start of work are

- Work on pressurized process system or system containing hazardous substances
- Hot work (grinding, welding, etc.) in a potentially combustible atmosphere
- Work on high-voltage systems
- Working on remotely controlled machinery
- Deactivation of safety systems
- Entry into confined space
- Work with hazardous substances (including radioactive isotopes, explosives)
- Critical lifting operations
- Work over open water

A PTW procedure includes steps to ensure that the necessary precautions are taken before the job starts and that the workplace is brought back to normal operation after finalizing the task.

Lockout/tag-out (LO/TO) is part of the PTW procedure. It shall ensure that potentially hazardous energy sources (rotating machinery, high-voltage systems, water under pressure, etc.) are properly de-energized or secured behind barriers prior to the start of maintenance work. LO/TO works in conjunction with a physical lock that prevents the energy to be activated and a tag to be affixed to the locked device indicating that it should not be turned on (Figure 40.6).

FIGURE 40.6   Lockout and tag-out of fire protection pipe work.

40.4 MITIGATION OF ACCIDENT RISKS IN SELECTED AREAS

40.4.1 Transportation

As discussed above, transportation of personnel and materials at the site represents a dominating cause of severe accidents in hydropower development projects.

A transportation risk assessment will be needed in order to address the critical safety issues in a systematic way. One critical issue is the quality of the internal site roads and of the public road network used by the project for internal transportation. Road access needs to be addressed as a constructability issue during the feasibility assessment to avoid transportation on dangerous roads. Alternative transportation means such as use of cableways and transportation through project tunnels need to be evaluated. Road quality standards need to be established based on national regulations and international best practice. This includes specifying minimum width, maximum slope, required reinforcement for heavy transportation, roadside barriers at critical sections of the road, and securing of unstable slopes.

The quality of drivers and vehicles needs also to be ensured through acceptance tests when arriving at site and regular inspections and audits and preventive maintenance of vehicles. There will be needs of local traffic rules such as speed limits, driving and rest periods to manage fatigue, and restrictions in driving during extreme weather conditions.

40.4.2 Underground Work

Most hydropower development projects involve the construction of tunnels and caverns. Unless the hazards are adequately mitigated, these are high-risk operations.

Underground works involve many types of hazards. The most significant immediate risks are falling rock and rock bursts, water inflow, misfire of explosives, electrical hazards due to a wet environment, moving vehicles and equipment, fire and smoke, and toxic and explosive gases. All hazards are not directly obvious to the layman. Falling rock and rock burst, for example, are caused by such conditions as high rock stress, expanding clay, squeezing ground, and inadequate rock support.

The following are some examples of measures to ensure an adequate level of safety in underground work (ITA, 2008):

- Risks of falling rock and rock burst are prevented through proper scaling and application of rock support. Rock support must be based on engineering geological mapping and assessment considering rock conditions, the size of the tunnel, available equipment and materials, and the qualifications of the personnel. Figure 40.7 illustrates provisions for rock support through bolting and application of shotcrete.

FIGURE 40.7   Installation of rock support behind the face.

- Air quality and smoke and dust control are essential to secure workers' health and an adequate progress as well. This has to be considered in the design of the tunnels and ventilation systems, in the design and implementation of regular air quality monitoring, and in the availability and use of gas and dust masks.
- Fires, especially hydrocarbon fires, are very dangerous and must be controlled through the selection of tunneling equipment with good fire safety characteristics, use of flame-proof equipment, and adequate firefighting routines.
- Emergency escape ways must be provided in case of fire or other accidental events. Provisions for safe havens (rescue chambers) are an alternative if proper escape through long tunnels is impossible due to heavy smoke development. Counting routines are essential in order to have control of the number of people in the tunnel at any time (Figure 40.8).

FIGURE 40.8   A system to ensure overview of the number of people in the tunnel at any time.

- Use of proper PPE shall ensure adequate visibility (highly visible clothing with reflective bands), protection against falling rock, poor air quality, and so on.
- Handling of explosives can only be done by certified personnel and under strict safety procedures.

40.4.3   Work on High-Voltage Systems

High-voltage electricity is an energy source that is present at several places in a power plant, and it is a source that most likely will kill or severely injure a person that gets exposed to it.

High voltage is normally defined as voltages above 1000 V alternating current or 1500 V direct current. Voltages below this level (low voltage and extra-low voltage) may also be hazardous energy sources, but safety management is simpler.

High-voltage installations must be fenced and locked. A company operating high-voltage installations needs to have a system where only qualified personnel have keys and are allowed access to these areas without guidance. High voltage may cause harm by arc (without physical contact), and safety distances have to be established by qualified personnel. Transmission towers need to be protected from unauthorized access by the general public. Depending on local conditions and policy, some towers are made less accessible by fences or barbwire.

Safety management of maintenance work on high-voltage systems is standardized around the world and is based on the use of trained and qualified personnel and the establishment of several barriers between the high-voltage system and the personnel. In mature organizations, a breach to one barrier is a very serious occurrence even if two barriers remain. The work involves several procedures, routines, and checklists according to the principles of the PTW system. The procedures are strictly formal and require signatures, checkouts, and log registrations.

40.4.4   Work in Proximity of Water under Pressure

Waterways represent a safety hazard in all power plants. The water flow from a ruptured penstock and subsequent landslide is likely to cause severe damages to humans and material. The same applies to a high-velocity water beam from a seal leakage in a system with water under pressure.

The most common accident is unintentional flooding of the powerhouse causing severe material damages and production losses, and in some cases injuries or fatalities. There is a long list of examples in the history of hydropower. Except for plants with Pelton turbines, powerhouses are submerged with several meters by design and have many similarities with a ship. Pump and drain systems are designed to cope only with small and medium leakages.

These accidents are prevented through proper design and professionally executed operation and maintenance.

Securing of maintenance work on high-pressure water systems has many similarities with work on high-voltage systems. Not only is this based on knowledge and caution, but it requires strictly formal PTW procedures with control sequences with signatures, checkouts, lock and tag, and log registrations. This is especially critical since prevention is not as standardized and formalized by authorities and the industry as for work on high-voltage systems.

40.4.5 Public Safety Related to Dams and Waterways

Dams and artificial canals make up part of the infrastructure for hydropower plants and represent a hazard to the general public.

- Dam breaks will cause flooding and inundation of areas downstream of the dam and expose the population in this area to risks of catastrophic events through drowning and destruction of livelihood.
- People may fall into a dam or canal and rescue is often made difficult due to the steep sides and lack of natural footsteps and suction from intakes.
- Changes in water level may occur rapidly and without warning and trapping people before they have time to escape.

Planning of remedial actions needs to be based on risk assessments, identifying and evaluating the specific third-party risks of the facilities. Examples of measures are as follows:

- Dimensioning of the dams' safety features including spillways based on possible flooding scenarios and consequences of dam failure
- Fencing of dams/waterways where required to prevent people from accidentally falling into the water
- Signage to inform the general public about the hazards and safety devices to be used in an emergency
- Emergency response plans to allow for rapid detection of potential flooding scenarios, notification of the general public, and evacuation.

REFERENCES

Bissel, C., Vullioud, G., Weiss, E., Heimann, A., Chène, O., Dayer, J.-D., Nicolet, C. (2011). Recommissioning of the Bieudron powerplant. *Int. J. Hydropower Dams*, 18(3):82–85.

British Standards Institution (2007). Occupational health and safety management systems – requirements. BSI Standard OHSAS 18001:2007. BSI, London.

Hansen, O. R., Wiik, A., Wilkins, B. (2002). Suppression of secondary explosions in transformer rooms, *J. Phys. IV France* 12(7):385–392.

ICOLD (1995). *Dam Failures Statistical Analysis*. Bulletin 99. International Commission on Large Dams, Paris.

ICOLD (2005). *Risk Assessment in Dam Safety Management. A Reconnaissance of Benefits - Methods and Current Applications*. Bulletin 130. International Commission on Large Dams, Paris.

IHA (2004). *Sustainability Guidelines*. International Hydropower Association, London.

ILO (2001). The construction industry in the twenty-first century: its image, employment prospects and skills requirements. Report TMCIT/2001. International Labour Organization, Geneva.

Ingason, H., Lönnermark, A., Frantzich, H., Kumm, M. (2010). Fire incidents during construction work of tunnels. SP Report 2010:83. SP Technical Research Institute of Sweden, Borås, Sweden.

International Water Power & Dam Construction (2010). Sayano Shushenskaya accident – presenting a possible direct cause (http://www.waterpowermagazine.com/story.asp?storyCode= 2058518)

ITA (2008). Guidelines for good occupational health and safety practice in tunnel construction. ITA Report No. 001. ITA, Lausanne.

Khali, R. K., Bisaria, M. S. (2008). Construction of Teesta head race tunnel lot TT 3 - innovative approach of Gammon for this mega project. Proceedings of the World Tunnel Congress 2008, New Delhi.

Kjellén, U. (2000). *Prevention of Accidents through Experience Feedback*. Taylor & Francis, London.

Kjellén, U. (2012). Managing safety in hydropower projects in emerging markets – experiences in developing from a reactive to a proactive approach. *Saf. Sci.*, 50:1941–1951.

Manu, P., Ankrah, N., Proverbs, D., Suresh, S. (2010). An approach for determining the extent of contribution of construction project features to accident causation. *Saf Sci*, 48:687–692.

Murie, F. (2007). Building safety – an international perspective. *Int. J. Occup. Environ. Health*, 13:5–11.

OECD/EIA (2011). *Energy for All – Financing Access for the Poor*. OECD/EIA, Paris.

Saari, J. (1976/77). Characteristics of tasks associated with the occurrence of accidents. *J Occup Accid*, 1:273–279.

Sidenfuss, T. (2006). *Collapses in Tunneling*. Master Thesis, Stuttgart University of Applied Science, Stuttgart.

Singh, V. P. (1996). *Dam Breach Modeling Technology*. Kluwer Academic.

Skjold, D. O. (2006). *Statens kraft*. Universitetsforlaget, Oslo.

41

PUMPED HYDROELECTRIC STORAGE

JOHN P. DEANE AND BRIAN O'GALLACHOIR

41.1 INTRODUCTION

The US Army Corps of Engineers distinguishes between two types of pumped hydroelectric storage (PHES), namely, pure PHES and pump-back PHES (Figure 41.1). Pure PHES plants rely entirely on water that has been pumped to an upper reservoir from a lower reservoir, a river, or the sea. Pure PHES is also known as "closed loop" or "off-stream." Pump-back PHES uses a combination of pumped water and natural inflow to produce power/energy similar to a conventional hydro-electric power plant. The Goldisthal pumped hydro storage plant in Germany is a good example of a closed loop system, whereas the Grand Coulee Dam in the United States is an example of a "pump-back" facility. There has been a renewed commercial and technical interest in PHES recently with the advent of increased variable renewable energy generation and the development of liberalized electricity markets. Currently over 20 GW of PHES is in construction worldwide with many other projects in planning stages.

41.1.1 Technology

PHES is a mature and reliable technology with some of the earliest plants being built in the Alpine regions of Switzerland and Austria in the early 1900s, regions that have a rich hydro resource and a natural complimentary topography for PHES. The world's largest plant is the 2710 MW plant at Bath County, Virginia, United States. PHES is similar in many aspects to conventional hydropower; however, the primary difference is that it has pumping capability through a dedicated pump or through a pump-turbine unit. PHES is a resource-driven facility which requires very specific site conditions to make a project viable. These requirements are

a high head, favorable topography, good geotechnical conditions, access to the electricity transmission networks, and water availability. The most essential of these criteria is availability of locations with a difference in elevation and access to water.

The response time of PHES plant is several minutes, which makes it easy to provide power conditioning for electrical power systems. Its inherent flexibility can also provide many benefits to the power system operation. Its flexible generation can provide both up- and down-regulation in the power system while its quick start capabilities make it suitable for black starts and provision of spinning and standing reserve. It also has the ability to store electricity in the form of hydraulic potential in pumping mode. Most modern PHES plants have round trip efficiencies (where efficiency is measured as the ratio of energy output to energy input) of between 75% and 80%. However, the efficiency is dependent on unit size, length of waterway, and how the unit is operated. This means that a PHES unit is a net consumer of electricity in the power system; however, because of its ability to store electricity generated during off-peak times for use in peak times and allow baseload plant to operate more efficiently, there is a net economic benefit to its operation.

Most PHES plants use radial single-stage pump-turbines as this represents the most economical equipment for plant. This type of machine can be used at sites with available heads from 60 up to 800 m or more with unit capacities ranging from 50 up to 500 MW. The pump-turbine uses a turbo impeller and the rotational speed and pumping head determine the input of the pump-turbine. Therefore, when operating at a certain pumping head, a single-speed pump-turbine driven by a synchronous motor cannot vary the input. However, by using variable speed machines, the

Alternative Energy and Shale Gas Encyclopedia, First Edition. Edited by Jay H. Lehr and Jack Keeley.
© 2016 John Wiley & Sons, Inc. Published 2016 by John Wiley & Sons, Inc.

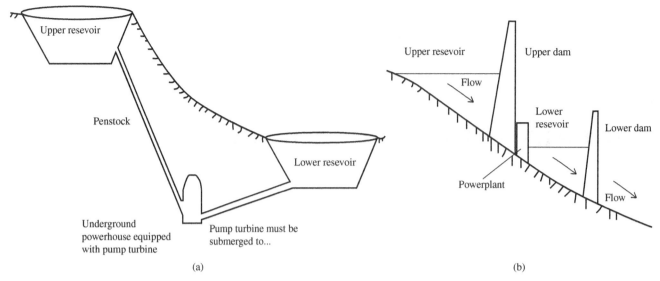

FIGURE 41.1   Pure PHES (a) and pump-back (b).

operating flexibility and plant efficiency can be further improved by covering a wider operation range. A number of existing PHES plants use variable speed pump-turbine units. The advantage of variable speed units is that the plants use asynchronous motor generators that allow the pump-turbine rotation speed to be adjusted. This technology allows regulation of the amount of energy absorbed in pumping mode. This facilitates energy storage when power levels available on the network are low and can help regulate the network frequency or voltage in pumping mode. This technology also allows turbines to operate closer to their optimal efficiency point and can be useful for the absorption of fluctuation renewable energy such as wind and solar.

41.2   DEVELOPMENT OF PUMPED HYDRO ENERGY STORAGE

PHES plants were traditionally built by state-owned utilities as a system tool to supply energy in times of high demand and allow baseload power plants to operate at high efficiencies in periods of low demand. PHES also provided management tasks for such power systems as balancing, frequency stability, and black starts. PHES plants have been built in many countries such as the United States and Japan to act as fast-response peaking plant to complement nuclear power plants, but more recently there has been a renewed interest in the technology as an integrator for variable wind power.

The chronological development of PHES in many countries shows the majority of plants were built from 1960s to the late 1980s. This was in part due to a rush for energy security and nuclear energy after the oil crises in the early 1970s. Fewer facilities were developed post 2000, due to a

natural saturation of the best available (most cost-effective) locations and a decline in growth in nuclear development.

However, since then a number of large PHES plants have come online in Europe, such as Goldisthal in Germany with a capacity of 1060 MW and Kopswerk II with a capacity of 450 MW in Austria. On the worldwide scale the United States and Japan have the highest installed capacities of PHES. The United States has an installed capacity of 21,886 MW. Like the United States, Japan developed PHES to compliment nuclear power facilities, providing peak power in the evenings and pumping when demand is low. Currently Japan has 34 PHES major plants with a total capacity of 24,575 MW.

In the European context the majority of PHES facilities are concentrated in the Alpine regions of France, Switzerland, and Austria; however, Germany has the largest number of PHES plants with 23 operational plants ranging in capacity from 62.5 to 1060 MW. The largest PHES plant in the EU is the 1800 MW EDF-owned "Grand'Maison" facility in the French Alps opened in 1987.

41.3   FUTURE DEVELOPMENT OF PUMPED HYDRO ENERGY STORAGE

Current trends for new PHES development generally show that developers operating in liberalized markets are tending to repower, enhance projects, or build "pump-back" PHES rather than traditional "pure pumped storage." Capital costs per kilowatt for proposed PHES are estimated to be between 2700 and 3300 USD/kW. However, costs are highly site and project specific. An emergence has also been observed in recent PHES developments of the use of variable speed technology. This technology, while incurring slightly higher capital costs, offers a greater operational range.

One of the greatest barriers to new PHES development is the availability of economical and environmentally suitable sites. This has led to the investigation of coastal and underground locations as potential PHES sites. Japan's J-POWER became the first company in the world in 1999 to build and operate a 30 MW seawater PHES plant at Okinawa with a head of 136 m. Research for the plant development started in 1981 and focused on corrosion-preventive measures. Seawater PHES may have several advantages over conventional PHES such as lower civil construction cost and greater site availability. Seawater PHES technology has yet to develop a commercial track record however, with only one completed plant worldwide.

Underground PHES plants have been conceived to be more practical facilities for the future; however, no sites currently exist. Underground pumped hydro systems of two types are possible: freshwater underground PHES that may use artificial underground tunnels as the lower reservoir and a natural lake as the upper reservoir; and seawater underground pumped hydro systems that utilize artificial underground tunnels as the lower reservoir and the ocean as an upper reservoir.

42

GREENHOUSE GAS EMISSIONS FROM HYDROELECTRIC DAMS IN TROPICAL FORESTS

PHILIP M. FEARNSIDE

42.1 INTRODUCTION

Although hydroelectric dams are often presented as "green" energy, meaning an energy source without greenhouse gas emissions, dams do, in fact, emit substantial amounts of gases (e.g., Fearnside, 2007; 2009a; 2009b; Gunkel, 2009). The amounts emitted vary greatly depending on the geographical location, age of the reservoir, external inputs of carbon and nutrients, and characteristics of the reservoir such as water flow, turnover time, area, depth, water level fluctuations, and the positioning of the turbines and spillways. Dams in tropical areas emit more methane than do those in temperate or boreal areas (Barros et al., 2011; Matthews et al., 2005). Bastviken et al. (2011) estimated that reservoirs cover 500,000 km^2 worldwide and emit 20 million tons of methane (CH_4) annually. This is equivalent to 185 million tons of CO_2-equivalent carbon if calculated using the Intergovernmental Panel on Climate Change (IPCC) Fifth Assessment Report global warming potential (GWP) for methane of 34 for 100 years, or 1.7 billion tons if the 20-year GWP of 86 is used (Myhre et al., 2013, p. 714). However, these numbers only include emissions from the surfaces of the reservoirs through ebullition (bubbling) and diffusion (emanation)—not the emissions that occur as methane-rich water emerges (under pressure) from deep in the water column through the turbines and spillways, which can more than double the total (e.g., Abril et al., 2005; Fearnside, 2009a; 2009b; Kemenes et al., 2008). However, the amount of information needed for reliable estimates of these emissions on a dam-by-dam basis makes a global estimate difficult at present.

The factors mentioned above—omission of major emissions sources such as turbines, much higher methane emission from tropical dams as compared to other regions, and ignoring or downplaying the importance of time—explain the conclusion of the IPCC *Special Report on Renewable Energy Sources and Climate Change Mitigation* that hydropower has half or less impact per kilowatt hour of electricity generated as compared to any other source, including wind and solar (Moomaw et al., 2012, p. 982). In the IPCC review, none of the 11 sources used from all climatic zones appears to concern tropical dams (Moomaw et al., 2012, p. 986). However, it is in tropical areas such as Amazonia that much of the world's hydroelectric development is expected in the coming decades.

The review that follows focuses on dams in tropical forest areas in South America (Figure 42.1). Much of the information is applicable to other tropical areas and, to a certain extent, to subtropical and other areas. The rapid expansion of dams planned in Amazonia makes advances in the measurement and modeling of hydroelectric emissions an urgent priority. Brazil's 2013–2022 Decennial Plan for Energy Expansion calls for 18 new large dams in the country's Legal Amazon Region (Brazil, MME, 2013).

42.2 TYPES OF EMISSION

42.2.1 Carbon Dioxide (CO_2)

Hydroelectric dams emit greenhouse gases in various ways throughout the lives of these projects. First, there are emissions from the construction of the dam from the cement, steel, and fuel that are used. These emissions are greater than those for an equivalent facility for generating the same amount

Alternative Energy and Shale Gas Encyclopedia, First Edition. Edited by Jay H. Lehr and Jack Keeley.
© 2016 John Wiley & Sons, Inc. Published 2016 by John Wiley & Sons, Inc.

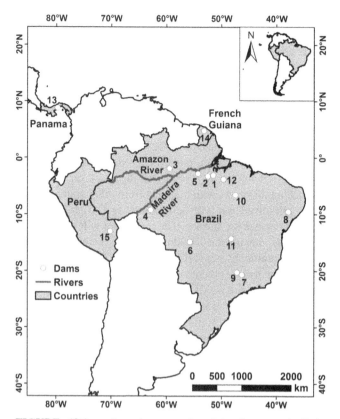

FIGURE 42.1   Locations mentioned in the text: 1, Belo Monte Dam; 2, Babaquara (Altamira) Dam; 3, Balbina Dam; 4, Samuel Dam; 5, Curuá-Una Dam; 6, Manso Dam; 7, Furnas Dam; 8, Xingó Dam; 9, Peixoto Dam; 10, Estreito Dam; 11, Serra da Mesa Dam; 12, Tucuruí Dam; 13, Fortuna Dam; 14, Petit Saut Dam; 15, Inambari Dam.

of electricity from fossil fuels or from alternative sources such as wind and solar. The emissions from dam construction also occur several years before generation of electricity begins—longer than the lead time for other sources. Because time has value for global warming impacts, this time difference adds to the impact of hydropower relative to most other sources (Fearnside, 1997). The construction emissions are estimated at 0.98 million tons of CO_2-equivalent carbon for Brazil's planned Belo Monte Dam and 0.78 million tons for the Babaquara/Altamira Dam if calculated without weighting for time (Fearnside, 2009a; 2009b).

When a landscape is flooded by a reservoir, the emissions and uptakes of the pre-dam landscape must be deducted from the corresponding gas fluxes from the reservoir in order to assess the net impact of the dam. In tropical forest areas the carbon balance of the forest is a critical factor. In the 1990s many believed the Amazon forest to be a major sink for atmospheric carbon, thereby increasing the net impact on global warming of converting forest to other uses, including reservoirs. However, correction for a series of problems in the measurement techniques has subsequently reduced the

estimates of forest uptake by more than fivefold, and the forest is no longer thought to be a major carbon sink on average (e.g., Araújo et al., 2002; Fearnside, 2000; Kruijt et al., 2004).

The amount of carbon uptake by Amazonian forest varies substantially with location (Ometto et al., 2005). The greatest uptake rates were estimated from tree-growth measurements in Peru and Ecuador (Phillips et al., 1998; 2004); unfortunately, there are no towers at these sites for comparable eddy correlation measurements. Uptake rates decline from the Andes to the Atlantic Ocean, a pattern that has been attributed to a corresponding gradient in soil fertility (Malhi et al., 2006). In 2010, Brazil signed an agreement with Peru to allow the Brazilian government electricity company (ELETROBRÁS) to build the first six of over a dozen planned dams in the Amazonian portion of Peru. One of these highly controversial dams (Inambari) is currently suspended by presidential decree.

Deforestation emissions can be substantial as a result of population displacement and stimulation of clearing in the areas surrounding new dams and their access roads, as occurred at Brazil's Tucuruí Dam (Fearnside, 2001). Displaced emissions can occur not only from lost land use, but also from lost water use, for example, to replace fish that were formerly produced in the undammed river. This is a concern for dams under construction on the Madeira River in Brazil (Fearnside, 2014a).

Another major source of emission is the carbon released from above-water decay of the trees that are killed by flooding (Abril et al., 2013). The trees are generally left standing in the reservoir, where they project above the water and rot in the presence of oxygen, releasing their carbon as CO_2. Additional trees are killed in unflooded forest near the shoreline, including forest on islands in the reservoir, due to the rise in the water table. This addition is greatest in reservoirs with convoluted shorelines and many islands, such as Brazil's Balbina Dam (Feitosa et al., 2007). The release of carbon from tree death begins when the reservoir is first filled (well before any generation of electricity), and the bulk of the emission occurs within the first few years of reservoir life. The value of time therefore makes this up-front impact a substantial count against hydropower as compared to generation from fossil fuels, which release the great majority of their CO_2 at the same time that the electricity is produced (e.g., Fearnside 1997). For 1990 (the standard year for the initial greenhouse inventories under the Climate Convention), the annual emission from above-water decay of flooded trees (not counting shoreline mortality) was estimated at 6.4 million tons of carbon for Balbina (Fearnside, 1995), 1.1 million tons for Samuel (Fearnside, 2005a), and 2.5 million tons for Tucuruí (Fearnside, 2002). The Babaquara/Altamira dam, "unofficially" planned for construction upstream of Belo Monte, would, in conjunction with Belo Monte, be likely to become the all-time "champion" for these emissions, with an average in the first 10 years estimated at 9.6 million tons of carbon emission annually from above-water decay of flooded trees plus 0.07 million tons from shoreline emissions (Fearnside, 2009a; 2009b).

The water in the reservoir also emits carbon dioxide, either through bubbling or diffusion through the reservoir surface or from the water being released through the turbines and spillways. This CO_2 comes from various sources, and it is important to avoid double-counting of the carbon. Some is from underwater decay of the trees initially present in the reservoir, either as CO_2 being produced directly if the tree biomass decays in the surface layer of water that contains oxygen, or indirectly if the biomass decays in the deep layers where there is little or no oxygen and the carbon is released as methane, some of which is subsequently converted to CO_2 by bacteria in the surface layers. This pathway, from tree biomass to dissolved methane to dissolved CO_2, is believed to be the major source of CO_2 released from water at Balbina (Kemenes et al., 2011).

Carbon dioxide is also released from soil carbon in the flooded land. Like the trees, this is a fixed source that will eventually be depleted. Similarly, the emission is greatest in the first years. Researchers at the Petit Saut Dam in French Guiana believe soil carbon to be the major source for both CO_2 and methane produced in the initial pulse of emission after flooding (Tremblay et al., nd [C. 2005]).

CO_2 emission from the water includes the carbon released from renewable sources, in addition to those from fixed sources such as trees and soil carbon. Carbon also enters the reservoir as dissolved organic carbon (from leaching) and as sediments coming from soil erosion throughout the hydrographic basin upstream of the reservoir. This carbon is continually being removed from the atmosphere by photosynthesis in the standing forest and converted to soil organic carbon and to direct exports of biomass carbon through the deposition of litter on the forest floor. Substantial amounts of the still-undecomposed litter are washed into the streams during torrential rains (Monteiro, 2005). Some of this carbon is stored in the sediments at the bottom of the reservoir. This storage in sediments has been claimed to be a carbon benefit of dams (e.g., Gagnon, 2002). However, a full accounting would require deducting the portion of the carbon that otherwise would have been carried down the river and deposited in ocean sediments. Some would have been released from the water in the downstream river, the water in the Amazon River being known as a significant emitter of CO_2 (Richey et al., 2002).

Other renewable sources of carbon include photosynthesis in the reservoir itself from phytoplankton, algae, and water weeds (macrophytes). There is also a renewable source from the herbaceous plants that grow in the drawdown zone. This zone is the mudflat that is exposed around the edge of the reservoir each time the water level is lowered for power generation in the dry season. Soft herbaceous plants, such as weeds and grasses, grow quickly in this zone as soon as the water level goes down. The drawdown area can be vast: 659.6 km^2 at Balbina (Feitosa et al., 2007) and 3580 km^2 at the "unofficially" planned Babaquara/Altamira reservoir (Fearnside, 2009a; 2009b). When the water rises again, the plants are killed and then decay quickly because they are soft (in contrast to wood, which contains lignin and decays very slowly underwater). When oxygen is present in the water this carbon will be released as CO_2, but because the weeds are rooted to the bottom, much of the decay will be in the oxygenless water at the bottom of the reservoir and will produce methane. As with methane from other sources, part of this dissolved gas will be oxidized to CO_2 by bacteria before it reaches the surface. The remainder will be released as methane, making the drawdown zone a "methane factory" that will continually convert atmospheric CO_2 into methane, which is much more potent per ton of gas in provoking global warming (Fearnside, 2008a; 2008b).

The CO_2 in the water that has come from renewable sources such as forest litter, phytoplankton, algae, water weeds, and the drawdown zone vegetation must be distinguished from CO_2 coming from initial fixed sources such as flooded trees and soil carbon. The portion from fixed sources represents a net contribution to global warming, taking care not to double-count any of the carbon. The portion coming from renewable sources, however, does not represent a contribution to global warming because the same amount of CO_2 that has been removed from the atmosphere by photosynthesis is simply being returned to the atmosphere in the same form (CO_2) after a period of months or years. If all of the dead tree biomass is counted as an emission from "deforestation," or by difference in biomass stocks between forest and "wetland," as in the case of the IPCC methodology (Duchemin et al., 2006; IPCC, 1997) used in Brazil's first and second inventories under the Climate Convention (Brazil, MCT, 2004; 2010), then some of the same carbon is being counted twice. Calculations of reservoir impact that count all of this CO_2 as a global warming impact (e.g., dos Santos et al., 2008; Kemenes et al., 2011; Saint Louis et al., 2002) therefore err on the high side for this portion of the emission. Research to better quantify the carbon sources from which the reservoir's CO_2 emission is derived should be a high priority. In the meantime, this author has opted to count only methane emissions from the reservoir surface and from the water passing through the turbines and spillways—not CO_2 from these sources (e.g., Fearnside, 2002; 2005b; 2009a; 2009b). Carbon dioxide is only counted for above-water decay of dead trees.

42.2.2 Nitrous Oxide (N_2O)

Nitrous oxide (N_2O) is another greenhouse gas with a contribution from reservoirs. Amazonian reservoir surfaces emit an average of 7.6 kg N_2O km^{-2} day^{-1} (Lima et al., 2002), or

$27.6 \text{ kg ha}^{-1} \text{ year}^{-1}$. Unflooded forest soil emits 8.7 kg ha^{-1} year^{-1} (Verchot et al., 1999, p. 37). The reservoirs therefore emit more than three times as much as the forests they replace. Considering the most recent GWP for nitrous oxide from the IPCC, each ton of N_2O has an impact equivalent to 298 or 264 tons of CO_2 gas over a 100-year or 20-year period, respectively (Myhre et al., 2013, p. 714). Amazonian reservoirs therefore emit 2.26 or $2.00 \text{ Mg ha}^{-1} \text{ year}^{-1}$ of CO_2-equivalent carbon, versus 0.71 or 0.63 for the forest, leaving a net emission of 1.55 or $1.37 \text{ Mg ha}^{-1} \text{ year}^{-1}$ of CO_2-equivalent carbon. For a 3000-km^2 reservoir like Brazil's Balbina Dam this represents 465,000 or 412,000 tons of carbon equivalent per year. Measurements of N_2O emissions at the Petit Saut reservoir in French Guiana and the Fortuna reservoir in Panamá indicate emissions around twice those of tropical forest soils (Guérin et al., 2008). Emissions from forest soils vary considerably among locations, indicating the importance of site-specific measurements for estimating pre-dam emissions. Unlike CO_2 and CH_4, almost all of the N_2O emission from dams occurs through the reservoir surface rather than from downstream degassing (Guérin et al., 2008). The range of emission is large: considering only emissions from the reservoir surface, the share of the global warming impact from N_2O ranges from 29% to 31% of the surface emission total considering CO_2, CH_4, and N_2O in four reservoirs in tropical forest areas: Tucuruí, Samuel, Petit Saut, and Fortuna (Guérin et al., 2008). In reservoirs that are not in tropical forest areas the emissions of N_2O are much lower.

42.2.3 Methane (CH_4)

Methane emission is a major contribution of hydroelectric dams to global warming. Methane (CH_4) is formed when organic matter decays without oxygen being present, for example, at the bottom of a reservoir. The water in a reservoir stratifies into two layers: a surface layer (the epilimnion) where the water is warmer and is in contact with the air, and a bottom layer (the hypolimnion) that lies below a separation known as the "thermocline," because the water below this point is much colder. If expressed in terms of dissolved oxygen content, the separation, which occurs at approximately the same depth of 2–10 m, is known as the "oxycline." The water below the thermocline or oxycline does not mix with the surface water, except for occasional events where the stratification breaks down and bottom water rises to the surface, killing many fish. In Amazonia this occurs during cold snaps (*friagens*), which are a climatic feature in the western but not in the eastern part of Amazonia. Balbina lies approximately at the eastern limit of this phenomenon and has experienced several fish dieoffs from overturning water during cold snaps. Under normal conditions, with the cold water at the bottom staying separated below the thermocline,

the dissolved oxygen in the bottom water is quickly depleted in oxidizing some of the leaves and other organic matter on the bottom of the reservoir, and thereafter essentially all decay must end in CH_4 rather than CO_2. Higher concentrations of gases can be dissolved in water at the bottom of the reservoir because the water is cold and under high pressure.

Natural lakes and wetlands, including the *várzea* and the *pantanal*, are significant global sources of methane (Devol et al., 1990; Hamilton et al., 1995; Melack et al., 2004; Wassmann and Martius, 1997). A hydroelectric reservoir, however, is a substantially greater source of CH_4 per area of water because of one crucial difference: the water leaving the reservoir is drawn from the bottom instead of the surface. Both natural lakes and reservoirs will emit CH_4 through bubbles and diffusion at the surface, but in the case of the reservoir there is an additional source of CH_4 from water passing through the turbines and spillways. These take water from below the thermocline, where it is saturated with methane. The reservoir is like a bathtub, where one pulls the plug and the water drains out of the bottom rather than overflowing from the top like a lake. Because the water emerging from the turbines is under high pressure, the sudden drop in pressure as it emerges downstream will cause most of the methane to form bubbles and be released to the atmosphere. Over a longer time, the warming of the water as it flows downstream below the dam will result in further reduction in solubility and increase in release of gas (Le Chatalier's principle).

For gas in the water flowing downstream below a dam, release to the atmosphere is sufficiently fast for most of the CH_4 to escape being converted to CO_2 by bacteria in the water. In fact, the major release is immediately below the turbines and even inside the turbines themselves. This is the reason why gas flux measurements from the water surface in the river below a dam are not sufficient to measure the impact of emissions from water passing through the turbines—much of the emission is escaping measurement. This is the main explanation, for example, for why the research group mounted by FURNAS (a power company that supplies 40% of Brazil's electricity) was able to claim that hydroelectric dams were "100 times" better than fossil fuels in terms of global warming (Garcia, 2007). Such low values for emissions are in part because the dams studied were in the *cerrado* (central Brazilian savanna) rather than tropical forest and because the estimates omit emissions from degassing at the turbines and spillways (Ometto et al., 2011; 2013). In fact, the flux measurements began at distances below the dam ranging from 50 m at the Furnas, Estreito, and Peixoto dams (dos Santos et al., 2009, p. 835) to 500 m at the Serra da Mesa and Xingó dams (da Silva et al., 2007). They also ignored emissions more than 1 km below the dams (Ometto et al., 2011). The only way to estimate the release without such major biases is to base it on the difference in concentration of CH_4 in the water above and below the dam (e.g., Fearnside, 2002; Kemenes et al., 2007).

Estimates of the impact of Amazonian dams on global warming have varied by many fold. Most people hearing about the different estimates through the press have no information about how the underlying measurements were made and what is included or omitted from the estimates. Examining the original studies on all sides of the debate is essential. Both sides of the extensive debate over greenhouse gas emissions are available in the "Amazon Controversies" section of the website http://philip.inpa.gov.br.

A brief review of reasons for the very disparate results is in order. First, omission of the emissions from the water passing through the turbines and spillways is one that should be obvious. This omission has been a longstanding feature of official Brazilian estimates, as was highlighted during the memorable debate on this topic in the journal *Climatic Change* (see Rosa et al., 2004; 2006; Fearnside, 2004; 2006a). The same omission applies to the greenhouse gas emissions estimate for dams in Brazil's first national communication under the Climate Convention (Brazil, MCT, 2004; Rosa et al., 2002), with results more than 10 times lower than those of this author for dams such as Tucuruí and Samuel (Fearnside, 2002; 2005a). Omission of the turbines and spillways was the major explanation. The major role played by emissions from water released by the turbines is clear from direct measurements above and below the dams at Petit Saut in French Guiana (Abril et al., 2005; Delmas et al., 2004; Galy-Lacaux et al., 1997; 1999; Guérin, 2006) and at Balbina in Brazil (Kemenes et al., 2007; 2008; 2011).

In Brazil's first inventory of greenhouse gases, hydropower emissions were calculated for nine dams, but the results were confined to a box on the side and not included in the tally of the country's emissions (Brazil, MCT, 2004, pp. 152–153). In the second national inventory (Brazil, MCT, 2010), hydroelectric emissions were omitted altogether. However, although the impact of CO_2 release from the trees killed by the reservoir is a major omission of many discussions of the role of dams in global warming, in the case of Brazil's second national inventory the CO_2 release from biomass loss in converting forest to "wetlands" is included as a form of land-use change.

Exaggeration of the pre-dam emission is another way that the net emissions of dams can be underestimated. As already mentioned, natural wetlands are significant sources of methane, and this has been used to argue that the landscape flooded by a dam would have been emitting large amounts of methane anyhow if the dam had not been built. For example, the International Hydropower Association (IHA) considered hydroelectric emissions to be a "zero-sum" issue because they would not exceed pre-dam emissions (Gagnon, 2002). In the Environmental Impact Study (EIA) for the Belo Monte Dam, the area to be flooded was assumed to be emitting 48 mg CH_4 m^{-2} day^{-1} prior to creation of the reservoir, based on two sets of measurements of emission from the river surface and soil at sites near the edge of the river

(Brazil, ELETROBRÁS, 2009, Appendix 7.1.3-1; see Fearnside, 2011). Most of the soil emission measurements in the wet season were in waterlogged areas that had recently been exposed by the falling water level (Brazil, ELETROBRÁS, 2009, Appendix 7.1.3-1, p. 72), resulting in their high CH_4 emission heavily influencing the mean used for all of the land area to be inundated by Belo Monte. However, hydroelectric dams are normally built in places with well-drained soils, sites with rapids and waterfalls being chosen rather than flat wetlands. This is because the steep topography results in greater generation of power. The seasonally flooded soils along the river cannot be generalized to a reservoir area, which in Amazonia is usually unflooded upland (*terra firme*) forest. The soil under *terra firme* forest is generally considered to be a methane sink, rather than a source (Keller et al., 1991; Potter et al., 1996). An unrealistically high estimate of pre-dam emission leads to an underestimate of the net impact. In the case of the Belo Monte EIA, the 48 mg CH_4 m^{-2} day^{-1} is subtracted from the EIA's estimate of 70.7 mg CH_4 m^{-2} day^{-1} for emission in the reservoir (an underestimate for various reasons, including using as half of the estimate a set of measurements at the Xingó Dam in Brazil's semiarid northeast region where emissions would be lower than at an Amazonian dam), leaving only 70.7 − 48.0 = 22.7 mg CH_4 m^{-2} day^{-1} as the net emission.

Another source of lower estimates of hydropower emissions in Brazil is a mathematically erroneous power-law correction that has been repeatedly applied in calculating bubbling and diffusion emissions from reservoir surfaces. This stems from a doctoral thesis (dos Santos, 2000), which is the basis of an ELETROBRÁS report (Brazil, ELETROBRÁS, 2000). The report calculates and tabulates the emissions for all 223 large dams in Brazil at that time, with a total water surface of 32,975 km^2—an area larger than Belgium. The correction continues to be applied (e.g., dos Santos et al., 2008). These ELETROBRÁS adjustments reduce the emission estimates for surfaces by 76% as compared to the simple mean of their measured values in the data from the same study (see Pueyo and Fearnside, 2011a; 2011b). The problem is that bubbles from the reservoir surface normally occur in sporadic episodes with intense bubbling for a short period, followed by long periods with few bubbles. Because the number of samples is inevitably insufficient to represent these relatively infrequent events, a power-law correction can be applied to the measurement data. However, the rare but high-impact events raise rather than lower the real mean emissions. In fact there are at least five major mathematical errors in the ELETROBRÁS calculation, including a reversal of the sign from positive to negative. Note, however, that the underestimate from the errors in application of the power-law correction applies not only to methane but also to CO_2 bubbling, not all of which is a net contribution to global warming. The correct application of the power law results in estimates of surface emissions of methane that are 345% higher than the

ELETROBRÁS estimates (see Pueyo and Fearnside, 2011a; 2011b).

Inappropriate sampling methodology is another way that can lead to emissions several times lower than they should be (Fearnside and Pueyo, 2012). As already mentioned, attempting to estimate the turbine and spillway emissions by relying only on surface-flux measurements below a dam is fated to miss much of the emission, resulting in gross underestimates of the total impact. This is a major factor in low estimates by FURNAS and ELETROBRÁS. Even concentration-based estimates (including the author's) have underestimated emissions because of the sampling methodology used to obtain water from near the bottom of the reservoir. The almost universal method is the Ruttner bottle, which is a tube with "doors" that open at each end. The tube is lowered on a cord with both doors open, then the doors close and the bottle is pulled up to the surface. Water is then removed for chemical analysis. The problem is that gases dissolved in the water under pressure will form bubbles inside the Ruttner bottle as it is pulled to the surface. The gas leaks out around the doors (which are not airtight), and in any case would be lost when the water is removed at the surface (with a syringe) for a head-space determination of gas volume and for chemical analysis. This problem has recently been addressed by Kemenes et al. (2011). Alexandre Kemenes invented a "Kemenes bottle," which collects the water in a syringe that is lowered to the required depth. The syringe has a spring mechanism that draws in the water for the sample, and the gas bubbles that emerge as the sample is raised to the surface are captured and measured. Comparison of the two sampling methods indicates that the average methane concentration for a sample taken at 30 m depth is 116% higher if measured with the Kemenes bottle, thereby more than doubling the amount of methane estimated to pass through the turbines at Balbina. The difference would be even greater for reservoirs with deeper turbines, as at Tucuruí.

Another important factor affecting the calculated impact of hydroelectric dams is the GWP of methane. This is the conversion factor for translating tons of methane into tons of CO_2-equivalents. The values for this conversion have increased in successive estimates by the IPCC. The conversions are based on the 100-year time horizon adopted by the Kyoto Protocol. The IPCC's 1994 interim report estimated a value of 11 for the GWP of methane, that is, the release of 1 ton of methane would have the same impact on global warming as releasing 11 tons of CO_2 (Albritton et al., 1995). This was raised to 21 in the 1995 Second Assessment Report used by the Kyoto Protocol (Schimel et al., 1996). In 2001 it was raised to 23 in the Third Assessment Report (Ramaswamy et al., 2001) and then to 25 in the 2007 Fourth Assessment Report (Forster et al., 2007). The Fifth Assessment Report (AR5) increased this to 28 if calculated in the same way (100-year time horizon and no climate-carbon feedbacks in response to CH_4 emissions), but

also reports a value of 34 when these feedbacks are included (Myhre et al., 2013, p. 714). The uncertainty range for this estimate extends to a value of over 40 (Shindell et al., 2009). The AR5 also reports a value of 86 for the GWP of methane if the time horizon is shortened to 20 years (Myhre et al., 2013, p. 714). This shorter time horizon is much more relevant to establishing policies on mitigating global warming, since it is emissions over this period that will determine whether global mean temperature surpasses the limit now agreed as "dangerous": 2°C increase above the preindustrial mean. As compared to the value of 21 adopted by the Kyoto Protocol for the 2008–2012 First Commitment Period, the value of 34 represents an increase of 62%, whereas the value of 86 effectively quadruples the impact of hydropower. For hydroelectric dams, methane emission represents most of the impact, whereas for fossil fuels almost all of the emission is in the form of CO_2.

42.2.4 Recovery of Methane

Proposals have been made to recover and use some of the methane that is being produced in hydroelectric dams. This would both reduce the amount of methane released to the atmosphere and generate additional electricity without adding to global emissions (Bambace et al., 2007; Lima et al., 2008). One design calls for pumping methane-rich water from below the thermocline (Ramos et al., 2009) while another would capture methane that is degassed immediately below the turbines (Kemenes and Forsberg, 2008). So far, no methane-capture system has been implemented in practice.

42.2.5 Comparisons of Dams with Fossil Fuels

The value of time is crucial in comparing the global warming impact of hydropower and fossil fuels or other energy sources. One difference is the gases emitted. A ton of methane has a very high instantaneous impact in outgoing infrared radiation (increasing surface temperatures), but each molecule only remains in the atmosphere for an average of 12.4 years (Myhre et al., 2013, p. 714). A ton of CO_2 blocks much less infrared radiation than a ton of CH_4 on an instantaneous basis, but the average CO_2 molecule remains in the atmosphere for approximately 10 times longer than the average CH_4 molecule. This is reflected in the much higher value for methane's GWP on a 20-year basis as compared to a 100-year basis like the GWP that was used by the Kyoto Protocol. Any strategy capable of preventing mean global temperature from surpassing the 2°C increase limit defining "dangerous" climate change must include reduction of methane emissions within this time period (Shindell et al., 2012).

Hydropower has a tremendous emission in the initial years from the death of trees, the underwater decay of soil carbon and of leaves and from the original forest, and the explosion of water weeds in the first years due to the higher fertility of the water. In subsequent years this emission will decline to a lower level that will be maintained indefinitely from renewable sources such as the annual flooding of the soft vegetation in the drawdown zone. The huge peak of emissions in the early years creates a "debt" that will be slowly paid off as power generation from the dam substitutes for fossil fuel generation over the succeeding years. The time elapsed can be substantial. For example, in the case of Belo Monte plus the first upstream dam (Babaquara/Altamira), the time needed to pay off the initial emission debt is estimated at 41 years (Fearnside, 2009a; 2009b). This is even with the true impact being underestimated by using the Kyoto Protocol value of 21 as the GWP of methane and by using methane concentrations measured with the traditional Ruttner bottles. A period of 41 years has tremendous importance for Amazonia, where the forest itself is under threat from climate changes projected on this time scale (e.g., Fearnside, 2009c). An energy source that takes 41 or more years just to break even in terms of global warming can hardly be considered as "green" energy.

Dams have many other impacts in addition to greenhouse gas emission, including displacement of human populations and loss of livelihoods (e.g., from fishing) for riverside residents both upstream and downstream of a reservoir (e.g., WCD, 2000). Reservoirs also destroy biodiversity and agricultural and urban land uses. They also provoke methylation of mercury that is present in the soil—a process which occurs in the anoxic conditions at the bottom of reservoirs—leading to accumulation of this toxic form of mercury in fish and in the humans who consume them. Dams also disrupt sediment flows and fish migrations, among other impacts (see reviews for individual dams in Fearnside, 1989; 1999; 2001; 2005a; 2006b; 2013a; 2014a; 2014b). While other energy sources also have impacts, the social and environmental destruction wrought by dams place this option in a class by itself. In addition, the inordinate concentration of hydropower's impacts on local peoples who happen to live in the path of this form of development represents a social cost that is more pronounced in the case of dams than for other energy options, and that makes the impact of dams even greater than if viewed as a hypothetical "average" spread over society as a whole. The contribution of dams to global warming makes a widely unappreciated addition to these impacts.

Controlling global warming will require an accurate accounting of net emissions throughout the world: any emission that is left out or underestimated implies that mitigation agreements designed to contain temperature rise within a specified limit (such as the 2°C limit currently agreed under the Climate Convention) will simply fail to prevent temperatures from continuing to rise. Amazonia is one of the places expected to suffer the most severe consequences if we fail in this responsibility.

42.2.6 Carbon Credit for Hydropower

Carbon credit that is currently granted to hydropower projects through the Clean Development Mechanism (CDM) is one of the most controversial aspects of efforts to mitigate global warming under the United Nations Framework Convention on Climate Change (UNFCCC). Hydroelectric dams are an increasingly important form of mitigation under the CDM, representing 28% of the expected issuance of credits from projects in the "pipeline" for funding as of 1 July 2014, with an expected annual global total to be granted of 342.8 million certified emissions reductions (CERs), meaning carbon credit expressed as tons of CO_2 equivalent (UNEP Risø Centre, 2014). This amount of CO_2 equivalent is equal to 93.5 million tons of carbon per year, approximately equal to Brazil's annual emission from fossil fuels. CDM regulations currently allow hydroelectric projects to claim that they produce little or no emissions (see Fearnside, 2013b; 2013c). This represents a significant loophole, especially since much of the future expansion of hydropower is expected to occur in the tropics where dams have the highest emissions. Even more important is the fact that countries throughout the world build dams as part of national development programs that have nothing to do with concerns for global warming. The willingness of governments and companies to invest vast sums in dams long before any carbon credit is approved also indicates that the dams would be built regardless of any additional income from sale of CERs. The financial calculations included in the carbon projects submitted to the CDM to substantiate claims that the dams would only be built because of the carbon income (i.e., that they are "additional") are at variance with the behavior of the governments and companies building the dams, indicating shortcomings in the CDM's current methodologies for determining the "additionality" of hydropower projects (Fearnside, 2013b; 2013c). When credit is granted to projects that would be built anyway, the countries that purchase the credit subsequently emit this amount of CO_2 without the emission actually having been offset, thus further increasing global warming.

42.3 CONCLUSIONS

Tropical hydroelectric dams emit substantial amounts of greenhouse gases. The amounts emitted vary greatly among dams, but the reported emissions vary even more due to frequent omissions in the emissions reported, such as methane release from water passing through the turbines and spillways. Hydroelectric emissions occur in a large pulse in the first few years after a reservoir is created, followed by a lower but indefinitely sustained emission. Comparison with the emissions impact of power generation from fossil fuels therefore depends heavily on the time horizon and any weighting for time preference used in the comparison. Even without any weighting for time preference, Amazonian dams can take four or more decades to "break even" in terms of their greenhouse impact, making them anything but "green" energy that can be presented as mitigating global warming. Dams also contribute to global warming through carbon credit issued to dams for which emissions are underestimated or ignored and by the effect of credit being granted to dams that would be built regardless of any extra income from sale of the credits.

ACKNOWLEDGMENTS

The author's research is supported exclusively by academic sources: Conselho Nacional do Desenvolvimento Científico e Tecnológico (CNPq: Proc. 305880/2007-1; 304020/2010-9; 573810/2008-7; 575853/2008-5), Fundação de Amparo à Pesquisa do Estado do Amazonas – FAPEAM (Proc. 708565), and Instituto Nacional de Pesquisas da Amazônia (INPA: PRJ15.125).

REFERENCES

Abril, G., Guérin, F., Richard, S., Delmas, R., Galy-Lacaux, C., Gosse, P., Tremblay, A., Varfalvy, L., dos Santos, M. A., Matvienko, B. (2005). Carbon dioxide and methane emissions and the carbon budget of a 10-years old tropical reservoir (Petit-Saut, French Guiana). *Global Biogeochemical Cycles*, 19:GB 4007. doi: 10.1029/2005GB002457.

Abril, G., Parize, M., Pérez, M. A. P., Filizola, N. (2013). Wood decomposition in Amazonian hydropower reservoirs: an additional source of greenhouse gases. *Journal of South American Earth Sciences*, 44:104–107. doi: 10.1016/j.jsames.2012.11.007.

Albritton, D. L., Derwent, R. G., Isaksen, I. S. A., Lal, M., Wuebbles, D. J. (1995). Trace gas radiative forcing indices. In: *Climate Change 1994: Radiative Forcing of Climate Change and an Evaluation of the IPCC IS92 Emission Scenarios*, Houghton, J. T.Meira Filho, L. G.Bruce, J.Lee, H.Callander, B. A.Haites, E.Harris, N.,Maskell, K. (editors.). Cambridge University Press, Cambridge, UK, pp. 205–231; 339 pp.

Araújo, A.C., Nobre, A. D., Kruijt, B. Culf, A. D., Stefani, P., Elbers, J. Dallarosa, R., Randow, C., Manzi, A. O., Valentini, R., Gash, J. H. C., Kabat, P. (2002). Dual tower longterm study of carbon dioxide fluxes for a central Amazônian rain forest: the Manaus LBA site. *Journal of Geophysical Research*, 107(D20):8090.

Bambace, L. A. W., Ramos, F. M., Lima, I. B. T., Rosa, R. R. (2007). Mitigation and recovery of methane emissions from tropical hydroelectric dams. *Energy*, 32:1038–1046.

Barros, N., Cole, J. J., Tranvik, L. J., Prairie, Y. T., Bastviken, D., Huszar, V. L. M., del Giorgio, P., Roland, F. (2011). Carbon emission from hydroelectric reservoirs linked to reservoir age and latitude. *Nature Geoscience*, 4:593–596. doi: 10.1038/NGEO1211.

Bastviken, D., Tranvik, L. J., Downing, J. A., Crill, P. M., Enrich-Prast, A. (2011). Freshwater methane emissions offset the continental carbon sink. *Science*, 331:50.

Brazil, ELETROBRÁS. (2009). *Aproveitamento Hidrelétrico Belo Monte: Estudo de Impacto Ambiental*. Fevereiro de 2009. Centrais Elétricas Brasileiras (ELETROBRÁS). Rio de Janeiro, RJ, Brazil. 36 vols.

Brazil, ELETROBRÁS (Centrais Elétricas Brasileiras S/A). (2000). Emissões de dióxido de carbono e de metano pelos reservatórios hidrelétricos brasileiros: Relatório final. Relatório Técnico. ELETROBRÁS, dea, deea, Rio de Janeiro, RJ, Brazil. 176 pp. Available at : http://wwwq2.eletrobras.com/elb/services/eletrobras/ContentManagementPlus/FileDownload.ThrSvc.asp?DocumentID=%7BCAFECBF7-6137-43BC-AAA2-35181AAC0C64%7D&ServiceInstUID=%7B3CF510BA-805E-4235-B078-E9983E86E5E9%7D.

Brazil, MCT (Ministério da Ciência e Tecnologia). (2004). *Comunicação Nacional Inicial do Brasil à Convenção-Quadro das Nações Unidas sobre Mudança do Clima*. MCT, Brasília, DF, Brazil, 276 pp. Available at: http://www.mct.gov.br/upd_blob/0005/5586.pdf

Brazil, MCT (Ministério da Ciência e Tecnologia). (2010). *Segunda Comunicação Nacional do Brasil à Convenção-Quadro das Nações Unidas sobre Mudança do Clima*. MCT, Brasília, DF, Brazil, 2 vols. 520 pp.

Brazil, MME (Ministério das Minas e Energia). (2013). *Plano Decenal de Expansão de Energia 2022*. MME, Empresa de Pesquisa Energética (EPE), Brasília, DF, Brazil. 409 pp. Available at: http://www.epe.gov.br/PDEE/24102013_2.pdf

da Silva, M., Matvienko, B., dos Santos, M. A., Sikar, E., Rosa, L. P., dos Santos, E., Rocha, C. (2007). Does methane from hydro-reservoirs fiz out from the water upon turbine discharge? *SIL – 2007-XXX Congress of the International Association of Theoretical and Applied Limnology*, Montreal, Québec, Canada. Available at: http://www.egmmedia.net/sil2007/abstract.php?id=1839.

Delmas, R., Richard, S., Guérin, F., Abril, G., Galy-Lacaux, C., Delon, C., Grégoire, A. (2004). Long term greenhouse gas emissions from the hydroelectric reservoir of Petit Saut (French Guiana) and potential impacts. In: Tremblay, A.Varfalvy, L.Roehm, C. and Garneau, M. (editors.) *Greenhouse Gas Emissions: Fluxes and Processes. Hydroelectric Reservoirs and Natural Environments*. Springer-Verlag, New York, NY, pp. 293–312; 732 pp.

Devol, A. H., Richey, J. E., Forsberg, B. R., Martinelli, L. A. (1990). Seasonal dynamics in methane emissions from the Amazon River floodplain to the troposphere. *Journal of Geophysical Research*, 95:16,417–16,426.

dos Santos, M. A. (2000). Inventário de emissões de gases de efeito estufa derivadas de hidrelétricas. Ph.D. thesis in energy planning. Universidade Federal do Rio de Janeiro, Rio de Janeiro, RJ, Brazil. 148 pp. Available at: http://www.ppe.ufrj.br/ppe/production/tesis/masantos.pdf.

dos Santos, M. A., Rosa, L. P., Matvienko, B.dos Santos, E. O., D'Almeida Rocha, C. H. E., Sikar, E., Silva, M. B., Ayr Júnior, M. P. B. (2008). Emissões de gases de efeito estufa por reservatórios de hidrelétricas. *Oecologia Brasiliensis*, 12(1):116–129.

dos Santos M. A., Rosa, L. P., Matvienko, B., dos Santos, E. O., D'Almeida Rocha, C. H. E., Sikar, E., Silva, M. B., Bentes

Júnior, A. M. P., (2009). Estimate of degassing greenhouse gas emissions of the turbined water at tropical hydroelectric reservoirs. *Verhandlungen Internationale Vereinigung für Theoretische und Angewandte Limnologie*, 30(Part 6):834–837.

Duchemin, É., Huttunen, J. T., Tremblay, A., Delmas, R., Menezes, C. F. S. (2006). Appendix 3. CH_4 emissions from flooded land: basis for future methodological development. In: Volume 4: *Agriculture, Forestry and Other Land Use*, Eggleson, S.Buendia, L.Miwa, K.Ngara, T.Tanabe, K. (editors.). Intergovernmental Panel on Climate Change (IPCC) Guidelines for National Greenhouse Gas Inventories, IPCC National Greenhouse Gas Inventories Programme Technical Support Unit, Institute for Global Environmental Strategies, Hayama, Kanagawa, Japan, Irregular pagination, pp. Ap3.1–Ap3.8.

Fearnside, P. M. (1989). Brazil's Balbina Dam: environment versus the legacy of the pharaohs in Amazonia. *Environmental Management*, 13(4):401–423.

Fearnside, P. M. (1995). Hydroelectric dams in the Brazilian Amazon as sources of 'greenhouse' gases. *Environmental Conservation*, 22(1):7–19.

Fearnside, P. M. (1997). Greenhouse-gas emissions from Amazonian hydroelectric reservoirs: the example of Brazil's Tucuruí Dam as compared to fossil fuel alternatives. *Environmental Conservation*, 24(1):64–75.

Fearnside, P. M. (1999). Social impacts of Brazil's Tucuruí Dam. *Environmental Management*, 24(4):485–495.

Fearnside, P. M. (2000). Global warming and tropical land-use change: greenhouse gas emissions from biomass burning, decomposition and soils in forest conversion, shifting cultivation and secondary vegetation. *Climatic Change*, 46(1-2):115–158.

Fearnside, P. M. (2001). Environmental impacts of Brazil's Tucuruí Dam: unlearned lessons for hydroelectric development in Amazonia. *Environmental Management*, 27(3):377–396.

Fearnside, P. M. (2002). Greenhouse gas emissions from a hydroelectric reservoir (Brazil's Tucuruí Dam) and the energy policy implications. *Water, Air and Soil Pollution*, 133(1-4):69–96.

Fearnside, P. M. (2004). Greenhouse gas emissions from hydroelectric dams: controversies provide a springboard for rethinking a supposedly "clean" energy source. *Climatic Change*, 66(2-1): 1–8.

Fearnside, P. M. (2005a). Brazil's Samuel Dam: lessons for hydroelectric development policy and the environment in Amazonia. *Environmental Management*, 35(1):1–19.

Fearnside, P. M. (2005b). Do hydroelectric dams mitigate global warming? The case of Brazil's Curuá-Una Dam. *Mitigation and Adaptation Strategies for Global Change*, 10(4):675–691.

Fearnside, P. M. (2006a). Greenhouse gas emissions from hydroelectric dams: reply to Rosa *et al. Climatic Change*, 75(1-2):103–109.

Fearnside, P. M. (2006b). Dams in the Amazon: Belo Monte and Brazil's hydroelectric development of the Xingu River Basin. *Environmental Management*, 38(1):16–27.

Fearnside, P. M. (2007). *Why Hydropower Is Not Clean Energy*. Scitizen, Paris, France (peer-reviewed website). Available at: http://www.scitizen.com/screens/blogPage/viewBlog/sw_view Blog.php?idTheme=14&idContribution=298

Fearnside, P. M. (2008a). Hidrelétricas como "fábricas de metano": O papel dos reservatórios em áreas de floresta tropical na emissão de gases de efeito estufa. *Oecologia Brasiliensis*, 12(1):100–115.

Fearnside, P. M. (2008b). A framework for estimating greenhouse gas emissions from Brazil's Amazonian hydroelectric dams. [English translation of Fearnside, 2008a]. Available at: http://philip.inpa.gov.br/publ_livres/mss%20and%20in%20press/Fearnside%20Hydro%20GHG%20framework.pdf

Fearnside, P. M. (2009a). As hidrelétricas de Belo Monte e Altamira (Babaquara) como fontes de gases de efeito estufa. *Novos Cadernos NAEA*, 12(2):5–56.

Fearnside, P. M. (2009b). Hydroelectric dams planned on Brazil's Xingu River as sources of greenhouse gases: Belo Monte (Kararaô) and Altamira (Babaquara). [English translation of Fearnside, 2009a]. Available at: http://philip.inpa.gov.br/publ_livres/mss%20and%20in%20press/Belo%20Monte%20emissions-Engl.pdf

Fearnside, P. M. (2009c). A vulnerabilidade da floresta amazônica perante as mudanças climáticas. *Oecologia Brasiliensis*, 13(4):609–618.

Fearnside, P. M. (2011). Gases de efeito estufa no EIA-RIMA da hidrelétrica de Belo Monte. *Novos Cadernos NAEA*, 14(1):5–19.

Fearnside, P. M. (2013a). Decision-making on Amazon dams: politics trumps uncertainty in the Madeira River sediments controversy. *Water Alternatives*, 6(2):313–325.

Fearnside, P. M. (2013b). Carbon credit for hydroelectric dams as a source of greenhouse-gas emissions: the example of Brazil's Teles Pires Dam. *Mitigation and Adaptation Strategies for Global Change*, 18(5):691–699. doi: 10.1007/s11027-012-9382-6

Fearnside, P. M. (2013c). Credit for climate mitigation by Amazonian dams: loopholes and impacts illustrated by Brazil's Jirau Hydroelectric Project. *Carbon Management*, 4(6):681–696. doi: 10.4155/CMT.13.57

Fearnside, P. M. (2014a). Impacts of Brazil's Madeira River dams: unlearned lessons for hydroelectric development in Amazonia. *Environmental Science & Policy*, 38:164–172. doi: 10.1016/j.envsci.2013.11.004.

Fearnside, P. M. (2014b). Brazil's Madeira River dams: a setback for environmental policy in Amazonian development. *Water Alternatives*, 7(1):156–169.

Fearnside, P. M., Pueyo, S. (2012). Underestimating greenhouse-gas emissions from tropical dams. *Nature Climate Change*, 2(6):382–384. doi:10.1038/nclimate1540

Feitosa, G. S., Graça, P. M. L. A., Fearnside, P. M. (2007). Estimativa da zona de deplecionamento da hidrelétrica de Balbina por técnica de sensoriamento remoto. In: *Anais XIII Simpósio Brasileiro de Sensoriamento Remoto, Florianópolis, Brasil 21-26 abril 2007*, Epiphanio, J. C. N. Galvão, L. S. and Fonseca, L. M. G. (editors). Instituto Nacional de Pesquisas Espaciais (INPE), São José dos Campos, SP, Brazil, pp. 6713–6720. Available at: http://marte.dpi.inpe.br/col/dpi.inpe.br/sbsr@80/2006/11.13.15.55/doc/6713-6720.pdf

Forster, P. and 50 others. (2007). Changes in atmospheric constituents and radiative forcing. In: *Climate Change 2007: The Physical Science Basis. Contribution of Working Group to* the Fourth Assessment Report of the Intergovernmental Panel on Climate Change, Solomon, S.Qin, D.Manning, M.,Chen, Z.,Marquis, M.,Averyt, K. B.,Tignor, M.,Miller, H. L. (editors.). Cambridge University Press, Cambridge, UK, pp. 129–234; 996 pp.

Gagnon, L. (2002). The International Rivers Network Statement on GHG Emissions from Reservoirs, a Case of Misleading Science. International Hydropower Association (IHA), Sutton, Surrey, UK, 9 pp.

Galy-Lacaux, C., Delmas, R., Jambert, C., Dumestre, J.-F., Labroue, L.,Richard, S., Gosse, P. (1997). Gaseous emissions and oxygen consumption in hydroelectric dams: a case study in French Guyana. Global Biogeochemical Cycles, 11(4):471–483.

Galy-Lacaux, C., Delmas, R., Kouadio, J., Richard, S., Gosse, P. (1999). Long-term greenhouse gas emissions from hydroelectric reservoirs in tropical forest regions. Global Biogeochemical Cycles, 13(2):503–517.

Garcia R. (2007). Estudo apóia tese de hidrelétrica "limpa": Análise em usinas no cerrado indica que termelétricas emitem até cem vezes mais gases causadores do efeito estufa. Folha de São Paulo, 1 May 2007, p. A-16.

Guérin, F., Abril, G., Richard, S., Burban, B., Reynouard, C., Seyler, P., Delmas, R. (2006). Methane and carbon dioxide emissions from tropical reservoirs: significance of downstream rivers. Geophysical Research Letters 33:L21407. doi: 10.1029/2006GL027929.

Guérin, F., Abril, G., Tremblay, A., Delmas, R. (2008). Nitrous oxide emissions from tropical hydroelectric reservoirs. Geophysical Research Letters, 35:L06404. doi: 10.1029/2007GL033057.

Gunkel, G. (2009). Hydropower – a green energy? Tropical reservoirs and greenhouse gas emissions. CLEAN – Soil, Air, Water, 37(9):726–734.

Hamilton, S. K., Sippel, S. J., Melack, J. M. (1995). Oxygen depletion, carbon dioxide and methane production in waters of Pantanal wetland of Brazil. Biogeochemistry, 30:115–141.

IPCC (Intergovernmental Panel on Climate Change). (1997). Revised 1996 Intergovernmental Panel on Climate Change Guidelines for National Greenhouse Gas Inventories. IPCC, Bracknell, UK, 3 vols.

Keller, M., Jacob, D. J., Wofsy, S. C., Harriss, R. C. (1991). Effects of tropical deforestation on global and regional atmospheric chemistry. Climatic Change, 19(1-2):139–158.

Kemenes, A., Forsberg, B. R., (2008). Potencial ampliado: gerado nos reservatórios, gás de efeito estufa pode ser aproveitado para produção de energia em termoelétricas. Scientific American Brasil, Especial Amazônia, no. 2:18–23.

Kemenes, A., Forsberg, B. R., Melack, J. M. (2007). Methane release below a tropical hydroelectric dam. Geophysical Research Letters, 34:L12809. doi: 10.1029/2007GL029479. 55.

Kemenes, A., Forsberg, B. R., Melack, J. M. (2008). As hidrelétricas e o aquecimento global. Ciência Hoje 41(145):20–25.

Kemenes, A., Forsberg, B. R., Melack, J. M. (2011). CO_2 emissions from a tropical hydroelectric reservoir (Balbina, Brazil). Journal of Geophysical Research, 116:G03004. doi: 10.1029/2010JG001465

Kruijt, B., Elbers, J. A., von Randow, C., Araujo, A. C., Oliveira, P. J., Culf, A., Manzi, A. O., Nobre, A. D., Kabat, P., Moors, E. J. (2004). The robustness of eddy correlation fluxes for Amazon rain forest conditions. Ecological Applications, 14:S101–S113.

Lima, I. B. T., Ramos, F. M., Bambace, L. A. W., Rosa, R. R. (2008). Methane emissions from large dams as renewable energy sources: a developing nation perspective. Mitigation and Adaptation Strategies for Global Change, 13(2):193–206.

Lima, I. B. T., Victoria, R. L., Novo, E. M. L. M., Feigl, B. J., Ballester, M. V. R., Ometto, J. M. (2002). Methane, carbon dioxide and nitrous oxide emissions from two Amazonian reservoirs during high water table. Verhandlungen International Vereinigung für Limnologie, 28(1):438–442.

Malhi, Y., Wood, D., Baker, T. R., Wright, J., Phillips, O. L., Cochrane, T., Meir, P., Chave, J., Almeida, S.Arroyo, L., Higuchi, N., Killeen, T., Laurance, S. G., Laurance, W. F., Lewis, S. L., Monteagudo, A., Neill, D. A., Vargas, P. N., Pitman, N. C. A., Quesada, C. A., Salomão, R., Silva, J. N. M., Lezama, A. T., Terborgh, J., Martínez, R. V., Vinceti, B. (2006). The regional variation of aboveground live biomass in old-growth Amazonian forests. Global Change Biology, 12:1107–1138.

Matthews, C. J. D., Joyce, E. M., St. Louis, V. L., Schiff, S. L., Vankiteswaran, J. J., Hall, B. D., Bodaly, R. A., Beaty, K. G. (2005). Carbon dioxide and methane production in small reservoirs flooding upland boreal forest. Ecosystems, 8:267–285.

Melack, J. M., Hess, L. L., Gastil, M., Forsberg, B. R., Hamilton, S. K., Lima, I. B. T., Novo, E. M. L. M. (2004). Regionalization of methane emission in the Amazon Basin with microwave 645 remote sensing. Global Change Biology, 10:530–544.

Monteiro, M. T. F. (2005). Interações na Dinâmica do Carbono e Nutrientes da Liteira entre a Floresta de Terra Firme e o Igarapé de Drenagem na Amazônia Central. Masters dissertation in Tropical Forest Science, Instituto Nacional de Pesquisas da Amazônia (INPA) & Fundação Universidade do Amazonas (FUA), Manaus, Amazonas, Brazil, pp. 93.

Moomaw, W., Burgherr, P., Heath, G., Lenzen, M., Nyboer, J., Verbruggen, A. (2012). Annex II: Methodology. In: IPCC Special Report on Renewable Energy Sources and Climate Change Mitigation, Edenhofer, O.,Pichs-Madruga, R.,Sokona, Y.,Seyboth, K.,Matschoss, P.,Kadner, S.,Zwickel, T.,Eickemeier, P.,Hansen, G.,Schlomer, S.,von Stechow, C. (editors). Cambridge University Press, Cambridge, UK, pp. 973–1000. Available at: http://www.ipcc.ch/pdf/special-reports/srren/srren_full_report.pdf

Myhre, G. and 37 others. (2013). Anthropogenic and natural radiative forcing. In: Climate Change 2013: The Physical Science Basis. Working Group I Contribution to the IPCC Fifth Assessment Report, Stocker, T. F.,Qin, D.,Plattner, G.-K.,Tignor, M.,Allen, S. K.,Boschung, J.,Nauels, A.,Xia, Y.,Bex, V.,Midgley, P. M. (editors). Cambridge University Press, Cambridge, UK, pp. 661–740. Available at: http://www.ipcc.ch/report/ar5/wg1/

Ometto, J. P., Cimbleris, A. C. P., dos Santos, M. A., Rosa, L. P., Abe, D., Tundisi, J. G., Stech, J. L., Barros, N., Roland, F. (2013). Carbon emission as a function of energy generation in hydroelectric reservoirs in Brazilian dry tropical biome. Energy Policy, 58:109–116. doi: 10.1016/j.enpol.2013.02.041

Ometto, J. P., Nobre, A. D., Rocha, H., Artaxo, P., Martinelli, L. (2005). Amazonia and the modern carbon cycle: lessons learned. *Oecologia*, 143(4):483–500.

Ometto, J. P., Pacheco, F. S., Cimbleris, A. C. P., Stech, J. L., Lorenzzetti, J. A., Assireu, A., Santos, M. A., Matvienko, B., Rosa, L. P., Galli, C. S., Abe, D. S., Tundisi, J. G., Barros, N. O., Mendonça, R. F., Roland, F. (2011). Carbon dynamic and emissions in Brazilian hydropower reservoirs. In: *Energy Resources: Development, Distribution, and Exploitation*, de Alcantara, E. H. (editor). Nova Science Publishers, Hauppauge, NY, pp. 155–188.

Phillips, O. L., Malhi, Y., Higuchi, N., Laurance, W. F., Núñez, P. V., Vásquez, R. M., Laurance, S. G., Ferreira, L. V., Stern, M., Brown, S., Grace, J. (1998). Changes in the carbon balance of tropical forests: evidence from long-term plots. *Science*, 282:439–442.

Phillips, O. L., Baker, T. R., Arroyo, L., Higuchi, N., Killeen, T. J., Laurance, W. F., Lewis, S. L., Lloyd, J., Malhi, Y., Monteagudo, A., Neill, D. A., Vargas, P. N., Silva, J. N. M., Terborgh, J., Martínez, R. V., Alexiades, M., Almeida, S., Almeida, S., Brown, S., Chave, J., Comiskey, J. A., Czimczik, C. I., Di Fiore, A., Erwin, T., Kuebler, C., Laurance, S. G., Nascimento, H. E. M., Olivier, J., Palacios, W., Patiño, S., Pitman, N. C. A., Quesada, C. A., Saldias, M., Lezama, A. T., Vinceti, B. (2004). Pattern and process in Amazon tree turnover, 1976-2001. *Philosophical Transactions of the Royal Society of London* B, 359:381–407.

Potter, C. S., Davidson, E. A., Verchot, L. V. (1996). Estimation of global biogeochemical controls and seasonality on soil methane consumption. *Chemosphere*, 32:2219–2246.

Pueyo, S., Fearnside, P. M. (2011a). Emissões de gases de efeito estufa dos reservatórios de hidrelétricas: implicações de uma lei de potência. *Oecologia Australis*, 15(2):114–127. doi: 10.4257/oeco.2011.1502.02

Pueyo, S., Fearnside, P. M. (2011b). Emissions of greenhouse gases from the reservoirs of hydroelectric dams: implications of a power law. [English translation of Pueyo and Fearnside, 2011a]. Available at: http://philip.inpa.gov.br/publ_livres/mss%20and%20in%20press/Pueyo%20&%20Fearnside-GHGs%20FROM%20%20RESERVOIRS–engl.pdf.

Ramaswamy, V. and 40 others. (2001). Radiative forcing of climate change. In: *Climate Change 2001: The Scientific Basis,* Houghton, J. T., Ding, Y., Griggs, D. G., Noguer, M., Van der Linden, R. J., Xiausu, D. (editors). Cambridge University Press, Cambridge, UK, pp. 349–416; 881 pp.

Ramos, F. M., Bambace, L. A. W., Lima, I. B. T., Rosa, R. R., Mazzi, E. A., Fearnside, P. M. (2009). Methane stocks in tropical hydropower reservoirs as a potential energy source: an editorial essay. *Climatic Change*, 93(1):1–13.

Richey, J. E., Melack, J. M., Aufdenkampe, K., Ballester, V. M., Hess, L. L. (2002). Outgassing from Amazonian rivers and wetlands as a large tropical source at atmospheric CO_2. *Nature*, 416:617–620.

Rosa, L. P., dos Santos, M. A., Matvienko, B., dos Santos, E. O., Sikar, E. (2004). Greenhouse gases emissions by hydroelectric reservoirs in tropical regions. *Climatic Change*, 66(1-2):9–21.

Rosa L. P., dos Santos, M. A., Matvienko, B., Sikar, E., dos Santos, E. O. (2006). Scientific errors in the Fearnside comments on greenhouse gas emissions (GHG) from hydroelectric dams and response to his political claiming. *Climatic Change*, 75(1-2):91–102.

Rosa, L. P., Sikar, B. M., dos Santos, M. A., Sikar, E. M. (2002). *Emissões de dióxido de carbono e de metano pelos reservatórios hidrelétricos brasileiros. Primeiro Inventário Brasileiro de Emissões Antrópicos de Gases de Efeito Estufa. Relatórios de Referência.* Instituto Alberto Luiz Coimbra de Pós-Graduação e Pesquisa em Engenharia (COPPE), Ministério da Ciência e Tecnologia (MCT), Brasília, DF, Brazil, 119 pp. Available at: http://www.mct.gov.br/clima/comunic_old/pdf/metano_p.pdf

Saint Louis, V. C., Kelly, C., Duchemin, E., Rudd, J. W. M., Rosenberg, D. M. (2002). Reservoir surface as sources of greenhouse gases to the atmosphere: a global estimate. *Bioscience*, 20:766–775.

Schimel, D. and 75 others. (1996). Radiative forcing of climate change. In: *Climate Change 1995: The Science of Climate Change,* Houghton, J. T., Meira Filho, L. G., Callander, B. A., Harris, N., Kattenberg, A., Maskell, K. (editors). Cambridge University Press, Cambridge, UK, pp. 65–131; 572 pp.

Shindell, D. T., Faluvegi, G., Koch, D. M., Schmidt, G. A., Unger, N., Bauer, S. E. (2009). Improved attribution of climate forcing to emissions. *Science*, 326:716–718.

Shindell D. T. and 24 others. (2012). Simultaneously mitigating near-term climate change and improving human health and food security. *Science*, 335:183–189. doi: 10.1126/science.1210026

Tremblay, A., Varfalvy, L., Roehm, C., Garneau, M. nd [C.] (2005). The issue of greenhouse gases from hydroelectric reservoirs: from boreal to tropical regions. Unpublished Hydro-Québec manuscript, 11 pp. Available at: http://www.un.org/esa/sustdev/sdissues/energy/op/hydro_tremblaypaper.pdf

UNEP (United Nations Environment Programme) Risø Centre. (2014). *Risoe CDM/JI Pipeline Analysis and Database.* UNEP Risø Centre, Risø, Denmark. Available at: http://www.cdmpipeline.org/

Verchot, L. V., Davidson, E. A., Cattânio, J. H., Akerman, I. L, Erickson, H. E., Keller, M. (1999). Land use change and biogeochemical controls of nitrogen oxide emissions from soils in eastern Amazonia. *Global Biogeochemical Cycles*, 13(1):31–46.

Wassmann, R., Martius, C. (1997). Methane emissions from the Amazon floodplain. In: *The Central Amazon Floodplain – Ecology of a Pulsing System,* Junk, W. J. (editor). Springer-Verlag, Heidelberg, Germany, pp. 137–143; 525 pp.

WCD (World Commission on Dams). (2000). *Dams and Development: A New Framework for Decision Making.* Earthscan, London, 404 pp. Available at: http://www.dams.org/index.php?option=com_content&view=article&id=49&Itemid=29

43

PHYSICAL AND MULTIDIMENSIONAL NUMERIC HYDRAULIC MODELING OF HYDROPOWER SYSTEMS AND RIVERS

Timothy C. Sassaman and Daniel Gessler

43.1 INTRODUCTION

Simulating or modeling hydraulic behaviors, whether using a computer model or a physical laboratory, is the industry-accepted method to determine how a proposed hydropower facility will interact with its environment and may be recommended in situations where other design methods are not applicable or do not provide adequate information. In this era of two-dimensional and three-dimensional computer models, one may wonder why a large, scaled physical model is necessary for a hydropower project and what developers gain from such a study. Computational fluid dynamics (CFD) is being used more than ever in the hydropower industry. Despite increased sophistication, however, there are still some physical phenomena that cannot be accurately predicted through computation. There is an additional trust factor in that physical modeling methods are mature, have been used to develop many of the key hydraulic design elements currently employed, and are tangible and visible. Recent projects show that the current trend is toward relying on a combination of computational and physical modeling (hybrid modeling), rather than on one to the exclusion of the other. This chapter will seek to give the reader an understanding of recent usage of physical and numeric modeling in the hydropower industry and why developers find value in model studies. The conventional wisdom says that modeling anything with a computer is faster and cheaper than performing a physical simulation. In many cases, including many hydropower applications, this is true. In fact, certain types of computer simulations are used routinely in design.

There are many cases, however, particularly those involving complex geometry, where the use of physical models may have a lower cost and shorter schedule. On the other hand, some modifications to a physical model may require breaking concrete and remolding geometry, which may be more quickly done in a CFD model. CFD models allow visualization and recording of key phenomena including streamlines, and also can easily provide flow data throughout the region of interest.

If CFD is less expensive for the majority of cases, then why is so much physical modeling still performed? The primary reason is the level of trust within the industry. While CFD capabilities have made enormous strides in the last 20 years, the hydropower industry's CFD experience does not match the more than 100 years of history in using physical modeling to ensure desired behavior of waterways impacted by a hydropower project. Therefore, hydraulic engineers recommend physical modeling in scenarios where CFD modeling is not perceived as being properly validated or applicable. These scenarios include irregular or nonstandard site-specific conditions, very complex hydraulic conditions, or the use of a nonstandard design in order to improve project performance, constructability, or economics. Even when CFD has been properly validated for the physics involved in a particular project, physical modeling may be required based on the perceived risk involved or simply on the comfort level of the project owner or reviewing agency. Cost, schedule, and the availability of a practitioner who has the appropriate experience in the various modeling methods will also play a role in any decision to use physical modeling as opposed to CFD.

Alternative Energy and Shale Gas Encyclopedia, First Edition. Edited by Jay H. Lehr and Jack Keeley.
© 2016 John Wiley & Sons, Inc. Published 2016 by John Wiley & Sons, Inc.

Beyond validation, historical usage, and trust of CFD results, there are some modeling requirements that CFD cannot meet at its present stage of development. One of these is navigational simulations where a barge can alter the flow field it is moving through, requiring two-way coupling of the motion of the fluid with the motion of the barge. While this type of simulation is theoretically possible, real-time coupling of barge motion and pilot input with the flow field computations is beyond the limitations of computers at this time.

Some further introductory information about the specific methods of physical hydraulic and CFD modeling is warranted.

43.1.1 Physical Hydraulic Modeling

43.1.1.1 Model Similitude. In a fluid with a free surface influenced by gravity, the ratio of inertial to buoyancy forces can be estimated by U^2/gL, where U is a characteristic velocity, g is the gravitational acceleration, and L is a characteristic length. The flow depth is often used for characteristic length L. Customarily, in the United States, the square root of this expression is defined as the Froude number, $Fr = U/\sqrt{gL}$. The flow patterns for large-scale free surface flows, driven by gravity, are dominated by gravitational and inertial forces. Hence, physical models of free surface flows are scaled and operated using Froude number similarity. That is, the Froude numbers of prototype and model must be equal $Fr_\mathrm{m} = Fr_\mathrm{p}$.

Secondary forces affecting flow patterns are viscous forces and surface tension. The significance of these forces relative to the inertial forces can be characterized by the Reynolds number ($Re = UL/\nu$) and Weber number ($We = \rho L U^2/\sigma$), respectively, where ν represents the fluid kinematic viscosity, ρ the fluid density, and σ the surface tension, with L and U as defined previously.

These secondary forces typically have negligible impact on the flow patterns in the prototype and will also have negligible impact on flow patterns in the model as long as the model Reynolds and Weber numbers are sufficiently high (corresponding to a large enough geometric scale of the model). For open channel Froude-scaled models the Reynolds number should be larger than 3000 (using the hydraulic radius as a length scale) to provide fully turbulent flow, and the Weber number should be larger than 120 (Jain et al., 1978).

Water surface profiles and flow patterns are also affected by resistance to flow from channel roughness, which is a function of roughness size and Reynolds number. Flow resistance does not scale correctly in a pure Froude scale model as the relative thickness of the boundary layer to the water depth becomes too large in the physical model. The head loss or resistance to flow can be given by the Darcy–Weisbach

equation, which relates head loss to the Darcy–Weisbach friction factor:

$$h_r = f_r \frac{L_r}{(4R_\mathrm{H})_r} \frac{U_r^2}{(2g)_r} \tag{43.1}$$

where f is the Darcy–Weisbach friction factor and R_H is the hydraulic radius. The subscript r denotes the ratio between model and prototype, L_r is the length scale ratio between the model and the prototype, U_r is the velocity scale ratio between the model and the prototype, and f_r is the Darcy–Weisbach friction factor ratio between the model and the prototype. For a Froude-scaled model, the velocity ratio is related to the length scale ratio as $U_r^2 = L_r$. Therefore, it is necessary to closely match the Darcy–Weisbach friction factor in the model to that in the prototype:

$$f_r = 1 \tag{43.2}$$

such that the head loss scales with the length scale $h_r = L_r$. A model bottom surface roughness is selected to satisfy the condition of equation (43.2) for the model Reynolds number. This can be accomplished with some knowledge of the roughness characteristics in the prototype and the Moody diagram.

43.1.1.2 Physical Model Construction and Instrumentation. Physical hydraulic models of river beds and banks are generally constructed by creating plywood or sheet metal templates (sections) that will reproduce the shape of the topography on selected planes separated by 1–2 ft model scale. The space between these plates is filled by either gravel or screening and then covered by a skim coat of concrete. Models of the relevant man-made structures are built from PVC plastic, acrylic, or plywood, depending upon tolerance requirements.

Flow through the overall model and through individual closed conduits is measured using venturi or orifice meters. Local velocities are measured using Pitot tubes, propeller meters, a camera recording the motion of lighted floats, or fine-scale particle image velocimetry (PIV), depending on the water velocity, the level of detail and accuracy required, and the project budget. At pump or turbine intakes, rotating swirl meters within the closed conduit near the impeller location can be used to measure swirl angles. Pressures and water surface elevations are measured with pressure taps linked to pressure transducers. Qualitative observations may include identification of free surface vortex activity and hydraulic jump locations, the use of dye traces to observe subsurface vortex activity and gross flow patterns, and scaled particles to evaluate sediment transport and deposition patterns.

43.1.2 Computational Fluid Dynamics Modeling

The foundation of CFD modeling involves the conversion of the notoriously challenging partial differential equations that govern continuum fluid flow into algebraic equations, which are amenable to solution by numerical methods. Most tactics of doing this entail breaking up the volume of interest into small "cells." By making certain assumptions about how the solution variables (velocity, pressure, etc.) change within the cells, the differential equation set can be converted to algebraic equations. Commercial tools have been developed to solve flow fields using this methodology, and what is left to the user is generating a computational mesh (all of the cells), setting fluid properties and boundary conditions for the model of interest, and selecting the appropriate models for simulating parametrized effects such as turbulence.

Most flows of interest to the hydropower and water industries are turbulent, meaning the local flow velocities and pressures change rapidly with time due to very small (sub-grid scale) unsteady eddies. The spatial and temporal resolution required to resolve these eddies would be impractical for the scale of solutions that are required in industry. Therefore, steady-state solutions are generally sought for the average flow. Even when unsteady flow is simulated, the time scales of the small eddies are not of interest, so the time-dependent solution can be considered as a time series of local averages. Still, the effects of the eddies on the average flow must be included, and this is generally done through "turbulence modeling." Additional equations are solved that parameterize the creation, transport, and dissipation of turbulent eddies, usually as an enhanced viscosity.

For hydraulics problems, the motion of a free surface may also be of critical interest. Typically, this is handled with a volume of fluid (VOF) model, in which the fraction of water within any given cell is tracked. Very sophisticated methods for refining the shape of the surface and reducing free surface diffusion have been developed in commercially available tools.

In the last 10–20 years, CFD has been validated for a wide range of hydraulics applications and has proven extremely useful for simulation and analysis of spillways, turbine runners, draft tubes, penstocks, and other hydraulic structures. However, one of the more difficult applications for CFD remains predicting unsteady free surface and subsurface vortex motion. While simulation of these phenomena would be theoretically feasible with CFD, the challenge is following the unsteady motion of a vortex with very high velocity gradients. These vortices tend to be artificially diffused in CFD, particularly when the computational grid is not sufficiently refined to resolve the velocity gradients. This challenge is usually tackled by using higher order spatial discretization (schemes that enhance the accuracy of gradient calculations) and very fine computational meshes locally where the gradients are high. When the region of high gradients changes over time (as it does with unsteady vortices), this becomes increasingly difficult. While adaptive meshing (moving the high-density grid based on flow solution values) is becoming more sophisticated, the computational expense is high. The resulting study for a practical problem would be an order of magnitude greater than the cost of running a well-designed physical model.

43.1.3 Computational Sediment Transport Modeling

A specialized subset of the CFD modeling discussed in the previous section is sediment transport modeling. When developing hydroelectric projects, understanding the dynamics of sediment transported with the water is critical. Several one-, two-, and three-dimensional models exist which can model the movement of sediment upstream and downstream of projects, making it possible to predict long-term channel evolution. Application of the models is instrumental in predicting potential changes in downstream channel morphology and dredging considerations. They are also instrumental in predicting reservoir life, the time before a reservoir becomes functionally obsolete due to sediment accumulation. Many hydropower projects require sediment transport modeling.

Sediment transport simulations are typically run for extended periods of time ranging from one flood hydrograph to several decades. Therefore, the models solve a simplified version of the system of equations typically solved for clear water CFD simulations to reduce computational time. The models include a system of equations for predicting the scour, transport, and deposition of sediment. Most sediment transport functions are empirical relationships based on a physical understanding of the forces that erode and transport sediment.

Modeling of hydropower projects is typically performed in order to ensure performance before going to the large expense of building a project. The large capital cost of hydropower projects and benefits that are realized by optimizing performance, avoiding impacts to navigation and other competing uses, and addressing environmental impacts provide justification for the use of these techniques to address key design concerns. The performance metrics addressed most often are the efficiency of a powerhouse, safety, and environmental concerns, such as sedimentation or fish passage. The uses of the methods to optimize all of the above concerns, with the exception of fish passage, are outlined in three short case studies. The first two examples involve the proposed addition of a powerhouse at an existing lock and dam facility on the Ohio River. The third example addresses updated probable maximum flood (PMF) calculations that required confirmation of spillway safety. Some physical modeling was done in each of these projects, often in conjunction with CFD. How the modeling was used to meet the needs of the project owners is outlined in each case.

43.2 SELECTED CASE STUDIES

43.2.1 Low Head Intake and Approach Channel

Hydraulic modeling of the Smithland Hydroelectric Project intake and approach is an example of the coordinated use of CFD and physical modeling to resolve performance, constructability, and cost issues in the design of a low head project. CFD modeling was used to develop and evaluate alternatives and physical modeling was used to confirm performance of the preferred CFD alternative, as well as to evaluate criteria not reliably simulated in the CFD model.

American Municipal Power ("AMP") is developing a 72 MW three-unit bulb turbine powerhouse at the US Army Corps of Engineers' (USACE) Smithland lock and dam on the Ohio River. The powerhouse location, on the left descending riverbank, was selected for foundation, constructability, and environmental considerations. The challenge is to provide practical approach channel geometry that distributes flow to the units uniformly, oriented longitudinally with the unit and with no significant vortex formation. The ideal approach, a long straight channel, is only achievable in rare circumstances or at great cost. At Smithland, a 1500 ft long converging approach channel, which makes an approximately 60° bend, delivers flow to the intake (see Figure 43.1). The intake geometry was developed and agreed upon by the project engineer (MWH Americas) and the turbine generator supplier (Voith Hydro), and was designed to provide economical powerhouse construction as well as acceptable turbine performance. In the designer's experience, the approach is short relative to other intakes for similar projects and required a relatively large convergence angle from the trash rack to the bulb. Hydraulic challenges that were identified from similar plants are flow direction changes at the entrance from the river to the approach channel and at the bend immediately upstream of the powerhouse. The channel geometry causes the flow approaching the intake to separate from the inside (right) embankment while helicoidal flow is induced by the outside (left) embankment of the bend. Hydraulic model studies were undertaken to refine the approach channel geometry. Design concerns also included conformance with Voith Hydro guidelines for intake performance as well as balancing efficient energy generation with construction cost.

The design approach involved both CFD and physical models. The CFD models, validated by the physical model, were developed to compare alternatives while final verification of the selected geometry was completed in the physical model. The CFD model, developed using the FLUENT code from ANSYS, Inc., simulated 7000 ft of the river length upstream of the dam including the entire river width and the dam. The 1:60 Froude scale physical model included 3200 ft of bathymetry upstream of the intake and included a portion of the main river channel width. The spillway bays were not included in the physical model because the powerhouse was located about 2000 ft away from the spillway separated by an overflow weir. The CFD model showed that flow patterns approaching the powerhouse were minimally affected by the spillway operation. Upstream velocity profiles in the physical model were created by baffled headboxes at the upstream and right sides of the model and were checked against measured velocity profiles in a comprehensive 1:120 scale physical model described in Section 43.2.2.

FIGURE 43.1 A physical model of the proposed powerhouse intake channel at the Smithland lock and dam looking downstream. The gravel portion was designed to have a variable geometry in order to optimize the approach flow patterns, for the purpose of avoiding free surface vortices and providing a uniform flow to the turbine intakes.

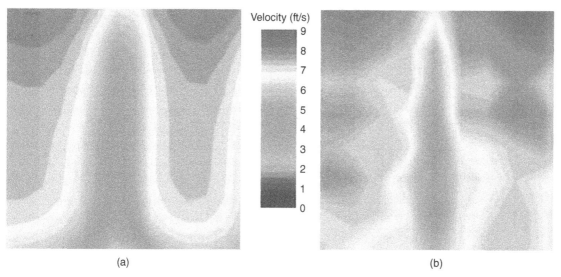

Velocity (ft/s)

9
8
7
6
5
4
3
2
1
0

(a)   (b)

FIGURE 43.2   Intake velocity magnitude profiles at the turbine intake computed by CFD (a) and measured in the physical model (b). The agreement between the two models was deemed sufficient to allow the use of CFD to investigate the effect of varying the approach channel geometry.

Key intake performance parameters are velocity distribution uniformity, flow angle, and vortex strength at the front of the powerhouse. CFD models can accurately simulate velocity distribution and angle but, as described above, detailed vortex simulation is not possible and results are subject to interpretation. Physical models can accurately portray velocities, flow angle, *and* vortex strength, when the scale is kept sufficiently large to avoid viscosity and surface tension effects (Reynolds number greater than 3,000 and Weber number greater than 120, respectively). Velocity distributions computed in the CFD model at the intake bulkhead slot were compared with those measured in the physical model in the same location. Comparisons were made using graduated contours (Figure 43.2) and in the classic form proposed by Fisher and Franke (1987) (see Figure 43.3). The CFD model also provided the ability to visualize the velocity vectors in the intake between the trash rack and the bulb. The agreement between the results was considered sufficient to justify use of the CFD model for alternative design development.

The velocity distributions for the initial design were very close to the Voith Hydro guidelines. Deviations resulted primarily from the vertical velocity distribution, which was influenced by the short intake and resulted in a flow concentration near the top of the intake.

Four alternative geometries were developed and tested in the CFD model to improve the velocity distribution and evaluate potential cost-saving modifications. The most promising of the alternatives resulted in a construction cost savings and improved performance free of adverse hydraulic conditions. Flow visualization provided by the CFD modeling improved the understanding of the intake performance and aided

alternative development. The physical hydraulic model was modified to incorporate the selected alternative and results of the physical and CFD modeling were compared. The physical model indicated an improvement in the performance of the end units and a decline in performance of the middle unit, whereas the CFD model indicated improvements for all three units. Overall, the physical model indicated an improvement in performance and the results of both CFD and physical modeling were similar.

While the CFD modeling served a valuable function in this study, it did not diminish the need for physical modeling. The CFD modeling was able to reduce the necessary scope and budget of the physical model by reducing the number of model geometries considered in the physical model. Significant intake vortex formation should be avoided in order to prevent air entrainment and maintain flow uniformity to optimize turbine performance and life. The Smithland physical model contributed to the development and evaluation of an approach channel geometry where vortices are not a concern.

43.2.2   Impacts of a Hydropower Plant on Navigation

The Ohio River is a major thoroughfare of commercial navigation. Appalachian coal and other bulk commodities are shipped up and down the river in barge trains that may reach 1100 ft in length and 105 ft in width. Lockages at the Smithland facility, about 60 miles upstream of the Ohio and Mississippi River confluence, average 21 per day and the hydro plant described in Section 43.2.1 has the potential to impact that traffic. The project may affect the operation of the existing lock and dam through changes in the upstream

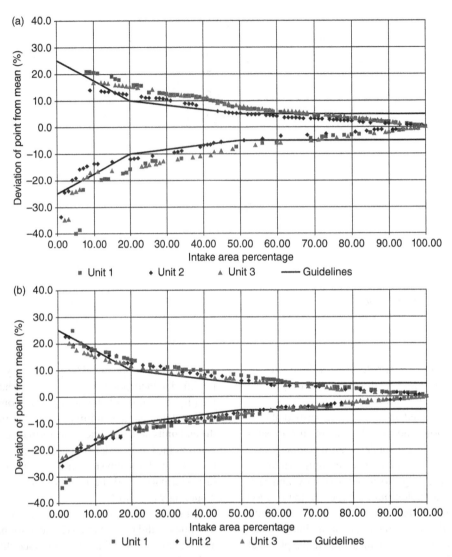

FIGURE 43.3   Velocity deviation from the mean predicted at the Smithland powerhouse intake, plotted in Fisher–Franke format for (a) the CFD model and (b) the physical model.

and downstream flow patterns. During normal river conditions the hydro plant will use the first 55,000 cubic feet per second (cfs) of flow that previously passed over the spillway. The plant outfall is a 2000 ft long tailrace oriented at approximately 45° to the river channel and directed across the river toward the lower approach to the navigation lock on the opposite bank. Possible project impacts on flow conditions warrant investigation of potential effects on the commercial navigation.

The key mission of the USACE Ohio River Division is to provide safe and efficient passage for the navigation traffic. As such, they provide standards and participate in physical model studies to evaluate potential project impacts while developing mitigation measures. The primary USACE concerns are (a) impacts on barge trains entering and leaving the lock, (b) changes in bed sediment movement

impacting the navigation channel, and (c) impacts on dam stability. Computational methods are being developed that may eventually assist or replace physical models, but they are not currently adequate. CFD models are capable of simulating project impacts on flow distribution but the interface with the moving barge is not sufficiently developed.

A comprehensive Froude scale model of the river reach potentially affected by the project was constructed (see Figure 43.4). Acceptable scales range from 1:100 to 1:150. The USACE maintains a fleet of towboats at these scales for use in physical models. The model extent is based on actual site conditions and generally extends from 1 to 1-1/2 miles upstream of the dam to a similar distance downstream. The Smithland model reach included 2 miles upstream of the dam and almost 3 miles downstream. The downstream reach included the confluence of the Cumberland River with the

FIGURE 43.4   Multi-frame photo of Smithland lock and dam navigation model. Each remote-controlled scaled barge and tug combination in the photo represents one frame taken from a test during operation of the physical model. The combined photo helps the US Army Corps of Engineers to evaluate differences in motion between the baseline and "with powerhouse" conditions.

Ohio River, several large islands, and a bend, all of which added complexity. The model was constructed to simulate the existing baseline operation as well as future conditions including the project as proposed. The fixed bed physical model was calibrated/validated with recorded river velocity distributions obtained with an acoustic Doppler current profiler as well as with historic water surface profiles by modifying channel roughness. The physical model program was developed to evaluate potential project impacts over the entire range of project operations, river flows from approximately 7000 cfs up to approximately 250,000 cfs. The model had the capability to handle higher river flows and floods up to approximately 700,000 cfs. Seven flow conditions, ranging from a single unit operating to the maximum navigable river flow, were selected for testing. These conditions were evaluated for two possible conditions of the Cumberland River as its flow is controlled by a separate hydro plant.

The USACE visited the physical model at the completion of validation to evaluate navigation under the existing conditions. The USACE team included the lockmaster, the lead hydraulic engineer from the Louisville District, and navigation experts from the USACE Engineering Research and Development Center (ERDC). They tested the barge movement into and out of the upper and lower lock approaches and recorded the barge travel path with "stop action"

digital photography (Figure 43.4). The lockmaster consulted with the ERDC barge operator regarding the appropriate navigation line and expected behavior. The barge operator developed a "feel" for the barge's interaction with the river. The ship tracks were then compiled to provide a visual record of the effects of river currents on the barges and the operator made extensive notes to document his findings for comparison to later "with project" testing. Detailed records of river velocities and currents were prepared by tracking the paths of lighted floats with digital imaging equipment and special software.

The model was then modified to include the project facilities (powerhouse, tailrace, and approach channels) and a testing program was undertaken to develop the optimum tailrace geometry. Emphasis is placed on the tailrace geometry because of the higher downstream velocities resulting from the shallower depth. The project design engineer worked with the testing laboratory (Alden) to develop a geometry that considers impacts on energy generation (head loss) and construction cost while providing acceptable navigation conditions. As the USACE was not present at these development tests, the impacts on navigation were estimated based on certain key flow conditions. Geometry of the approach flow channel was shown to have minimal impact on the flow patterns at the upper lock entrance.

Development of the tailrace geometry at the Smith-land project resulted in significant improvements for project economics and the ability of barges to enter the lock. The tail-race must be cut through a large island of sediment which has deposited downstream of the fixed weir portion of the dam in the last 20 years. The physical model was used to develop a tailrace that minimized its length without impacting nav-igation. The USACE revisited the model after the tailrace design had been developed and retested the same flow con-ditions but with the plant operating. The tailrace design was modified slightly to improve navigation considerations and project constructability. The physical model modifications were made quickly with available materials and the barge operator was able to assess their impact on the same or fol-lowing day.

Other USACE concerns were also evaluated in the phys-ical model. Studies of "tracer" materials representing sedi-ment were used to simulate patterns of bed load movement under baseline and "with project" conditions. Waves result-ing from powerhouse load rejections were created to deter-mine the impacts on barges in phases of entering the lock. The model was also used to evaluate alternatives to miti-gate potential impacts to an endangered mussel species that inhabits an area near the project tailrace.

The Ohio River at Smithland lock and dam is a sand bed river with a mean grain size of about 1 mm that has exhib-ited depositional characteristics. A comprehensive compu-tational two-dimensional sediment transport model of the river reach downstream of the dam was used to evaluate project impacts on navigation channel dredging. Results of the two-dimensional sediment model were also used to deter-mine the flow distribution around the major downstream island.

43.2.3 Spillway Modeling

In reviewing the hydraulic performance of the spillways of an existing dam, the PMF (Probable Maximum Flood) was recomputed in accordance with present-day requirements. The newly computed pool elevation during the PMF was found to be about 10 ft higher than the design level and about 26 ft above the spillway crest. The spillways were originally designed and tested using a 1:60 scale model at Alden in 1960. Based on the physical model results, for combined flows of up to 50,000 cfs, the spillways can operate for prolonged periods without sustaining significant wear or deterioration in performance. At flow rates greater than 50,000 cfs, the performance characteristics of the spillway were unknown and there was concern that the impact zone on the catch chute may change.

Figure 43.5 provides a detailed view of one of the spill-ways with key features identified:

- The ogee refers to the curved portion of the spillway, integral with the dam and under the bridge which spans the spillway opening.
- The point of separation refers to the point at which the flow separates from the downstream end of the ogee and free fall is initiated.
- The catch chute is where the free falling water impacts and is conveyed to the river.
- The point of impact is the most upstream location where the free falling water impacts the catch chute.

For this investigation, CFD was used for the study in conjunction with the validation data that were available from the 1960 study. The CFD model was validated by simulating

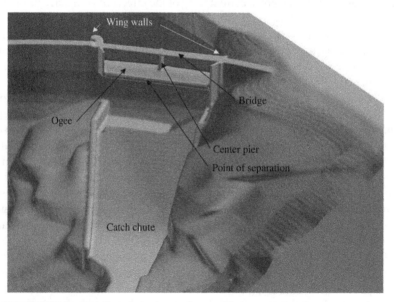

FIGURE 43.5 Detailed view of one of two spillways with key features identified.

the same flow rates as were used in the physical model and comparing results. After favorable validation, the CFD model was run at higher flow rates, for which the performance of the spillways is unknown.

Each simulation was initiated with water up to the spillway crest elevation and with all velocities set to zero. The initial computational grid contained 378,000 cells, with cells concentrated around the two spillways. At the upstream end of the model, a uniform inflow velocity was applied to obtain a target flow rate. The model was run until the spillway discharge was approximately equal to the inflow. Grid refinement was then undertaken, during which the number of computational cells was increased from the low resolution grid with 378,000 cells to the medium resolution grid with 5,168,000 cells. Grid resolution was increased in all parts of the model, with particular emphasis on the area around the spillways. Horizontal grid resolution around the spillways was approximately 1 ft (each spillway is 100 ft wide), sufficient to accurately resolve the shape of the wing walls, center pier, and bridge. Vertical resolution around the spillways varied from about 6 in. to 1 ft. The medium resolution grid was designed to resolve the water velocities and flow patterns in the lake approaching the spillways and over the ogee to where the water initiates free fall.

The trajectory of the free falling water is exclusively a function of the water velocity at the separation point from the ogee. Since air friction and breakup of the jet were not included in the simulation, the flow followed a ballistic trajectory to impact with the catch chute. To determine the full footprint of the impact zone as accurately as possible, the model was further refined and a sub-model was created for each spillway chute. Flow into the sub-model boundaries

was given from the global model. The sub-models for two spillways had between 3 million and 4.5 million grid cells.

Model validation was conducted in two steps. The first was to validate the model physics against a well-documented test case; the second was to validate the model for the specific application.

The CFD tool used for this case was FLOW-3D®, developed and distributed by Flow Science. Model physics had been validated by Flow Science using the laboratory data set collected by Rajaratnam and Chamani (1995). Within the accuracy of the laboratory data, it was demonstrated that for a free discharging weir, FLOW-3D® accurately predicted the upstream water surface elevation, the trajectory of the free falling discharge, and the angle of impact.

Site-specific model validation was accomplished by comparing the CFD model results with the physical model study results performed at Alden in 1960. The physical model had been used to develop the stage versus discharge rating curve for the spillways. The numerical model was used to compute the same rating curve for flows between 30,000 and 50,000 cfs. Matching the observed water level in the physical model confirmed that the losses through the spillway and water velocity on the ogee should be in agreement with the physical model and, consequently, the real world. Since the free falling trajectory is largely the result of the water velocity at separation, this is an important point of validation.

The pool elevation was plotted with the rating curve obtained in the laboratory (Figure 43.6). Physical models are known to underpredict the flow through the spillway, or overpredict the water level in the lake for a given discharge. Alden engineers in 1960 estimated that the 1:60 scale model underpredicted the discharge for a given pool elevation by

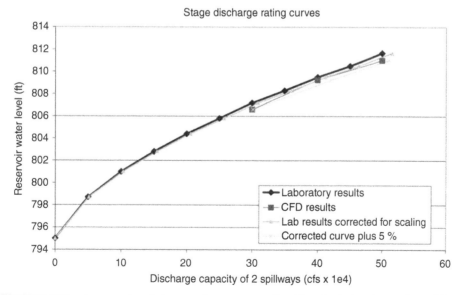

FIGURE 43.6   A comparison between rating curves predicted by physical and numeric models of the spillways.

about 3%. Therefore, Figure 43.6 also shows a corrected rating curve, with the discharge increased by 3%. The following observations about the figure can be made:

- The model was well behaved with consistent and reproducible results under a range of flow conditions.
- A difference of 5% is well within the accuracy of the physical model. The CFD model showed variations in the water surface elevation throughout the reservoir of ±0.5 ft. Therefore, it is reasonable to suggest that there was some subjectivity in establishing what constitutes the reservoir water level and that the exact point of measurement was critical.
- Finally, lower lake levels equate to a smaller wetted cross-sectional area when the flow separates from the ogee. Therefore, the predicted water velocities at separation were systematically higher than those of the physical model, resulting in longer flight trajectories and a lower impact elevation on the catch chute. The objective of the study was identification of excessively long trajectories; therefore the predictions can be considered conservative.

A critical aspect of every CFD study is model grid resolution; generally the finer the grid (high grid resolution), the more accurate the solution. Limitations in computer speed and memory restrict the number of computational cells available for a simulation. Consequently, for large models, grid resolution is a significant practical consideration. With a low resolution grid, model results may become dependent on the grid cell size.

Three levels of grid resolution were used for this study. The water trajectory after separation with the ogee was compared for the three grid resolutions, in order to assess the grid dependency. The higher resolution grids consistently predicted a shorter trajectory than the lower resolution grid, impacting higher on the catch chute. The trend suggests that the actual point of impact with the catch chute should be equal to or higher than that predicted by CFD.

Photo documentation is available from the 1960 physical model study to compare the physical model-predicted performance with that predicted by CFD. Figure 43.7 shows a visual comparison of the numeric and physical models. The general flow patterns look very similar in the two models and the point of impact on the catch chute appears to be about the same. However, it is important to note that this comparison is only qualitative in nature, and the actual point of impact is extremely difficult to determine from a frontal view.

After the above qualitative and quantitative validation, simulations of the newly calculated PMF were run following the same procedure developed for the lower flows. The computational grid used for modeling the PMF pool level was the same as that used at the lower flow rates. An increase in flow

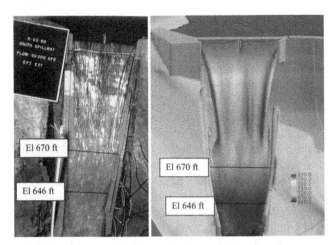

FIGURE 43.7 Spillway at design flow, showing the physical model (left) and the numeric model (right), contours colored by elevation.

rate on the same computational grid is analogous to a small increase in the grid resolution because the cell size relative to the jet size decreases.

Model results from the PMF simulations included predictions of the pool level, a discharge rating curve, and the water trajectory after separation from the ogee. In addition, flow visualization was used to show the approach flow conditions in the lake and the flow across the spillways. The PMF exceeded the capacity of the two spillways, resulting in flow over the crest of the dam.

Results generally show that the increase in discharge does not significantly change the trajectory of the flow or the initial point of impact on the catch chute. To compare the trajectory of the free falling flow at 50,000 cfs and during the PMF, cross-sections were cut through the two spillways at three locations on each. Figure 43.8 shows a typical section from the south spillway at 50,000 cfs and during the PMF.

In each cross-section, the initial point of contact during the PMF occurred downstream and at a lower elevation than during the 50,000 cfs design flow. In all cases, the initial point of contact moved less than about 25 vertical feet down the catch chute, and remained at least 20 ft above the elevation at which the catch chute slope begins to flatten.

The use of CFD to evaluate spillways at greater than design flows is increasing. When performed properly and validated with physical model data or prototype data, CFD can be a very reliable tool for evaluating spillway performance.

43.3 SUMMARY

Physical and computational hydraulic models are useful tools and both are actively used to address performance, navigation, and safety of hydropower systems. The last

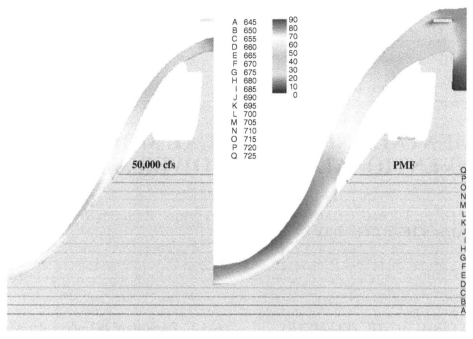

FIGURE 43.8   A section through one of the spillways at design flow (left) and for the PMF (right), contours colored by velocity magnitude.

20 years have shown a dramatic increase in the use of computational modeling augmenting, and in some cases replacing physical modeling in a wide variety of industries, hydropower being no exception. Computer speed, reduced memory costs, and software sophistication have made the increasing use of CFD a cost-effective way to optimize performance of expensive hydropower projects prior to construction and manufacturing. Trust in these methods is slowly growing among regulatory agencies and project owners. There is every reason to believe that these developments will continue. At the same time, however, there will be physical phenomena which are expected to remain beyond the reach of the computational tools, requiring the use of physical modeling for many more years. As the computational tools acquire new capabilities, the physical models will be needed to validate them. The

case studies covered in this chapter have shown how hybrid modeling, which combines usage of the two methods, can provide the most efficient modeling solution.

REFERENCES

Jain, A. K., Raju, K. G. R., Garde, R. J. (1978). Vortex formation at vertical pipe intakes. *Journal of the Hydraulics Division*, 104(HY10):1429.

Fisher, R. K., Franke, G. (1987). The impact of inlet flow characteristics on low head hydro projects. *Proceedings of Water Power '87*, American Society of Civil Engineers, Portland.

Rajaratnam, N., Chamani, M. R. (1995). Energy loss at drops. *Journal of Hydraulic Research*, 33(3):373–384.

44

EXPERIMENTAL AND NUMERICAL MODELING TOOLS FOR CONVENTIONAL HYDROPOWER SYSTEMS

Zhiqun Daniel Deng, Thomas J. Carlson, Gene R. Ploskey, Richard S. Brown, Gary E. Johnson, and Alison H. A. Colotelo

44.1 UNDERWATER ACOUSTIC TELEMETRY

44.1.1 Introduction

Telemetry is the process of making measurements or sourcing a signal containing information at a distance and transmitting it to a receiver. In the case of underwater acoustic telemetry (AT), the communication channel is water. AT is only one of many applications of sound in water. Underwater sound consists of alternating regions of compression (increasing pressure) and refraction (decreasing pressure) that propagate through water. Sound is a preferred means to explore bodies of water and for communication through water because it propagates long distances whereas visible light and other electromagnetic energy only transmit very short distances in water. Therefore where radar (*ra*dio *d*etection *a*nd *r*anging) is a preferred means to detect and track objects in the air, sonar (*so*und *n*avigation *a*nd *r*anging) is preferred to perform very similar tasks such as the location of the bottom or fish relative to a vessel in water (Urick, 1983).

As terrestrial animals have adapted to use sound in air to communicate and sense their surrounding so have aquatic animals. The effectiveness of underwater sensing and communication demonstrated by many aquatic animals has been known for centuries and, over the last few decades, has been studied to extend the range of uses of underwater sound and to improve the performance of man-made underwater sound instruments (Au and Hastings, 2008).

44.1.2 Elements of an Underwater Acoustic Telemetry System

An underwater AT system typically consists of a sound source capable of transmitting a signal containing information and a receiver capable of receiving the signal and decoding it to recover the transmitted information. AT is used to accomplish a wide range of tasks. These range from sensors located underwater to measure water temperature or any of a large number of attributes of water, to communications between a diver and surface support personnel, through transmitters located in fish or other aquatic animals that identify the animal and can be used to detect its presence and estimate its location. Whatever the measurement or communication process being enabled the data flow is the same: encoding of an acoustic transmission at a source, transmittal of the pulse through water, and reception by a receiver where it is decoded, displayed, and perhaps stored.

The necessary elements of a sound source for an AT system that is being used to measure a quantity are a sensor and electronics to handle the sensor output and to translate it into a coded signal and condition it to drive a transducer that can change the encoded electrical signal into a mechanical pressure pulse that can propagate through water. A number of encoding strategies are used to meet specific information transmission needs. These encoding strategies typically manipulate one or more of the amplitude, frequency, and phase of individual pulses in addition to the time between

Alternative Energy and Shale Gas Encyclopedia, First Edition. Edited by Jay H. Lehr and Jack Keeley.
© 2016 John Wiley & Sons, Inc. Published 2016 by John Wiley & Sons, Inc.

pulses. Typically the encoding of pulses is designed to transmit a digital signal where elements of a pulse code signify either 1 or 0 (Kilfoyle and Baggeroer, 2000). AT applications that may not include a sensor, such as simple acoustic transmitters that are attached to or implanted in fish, require the same elements with the exception of a sensor.

44.1.3   Water as a Communication Channel

While radio and light may be used for communications and telemetry application in the air they are of little use under water because both are strongly attenuated in water, particularly so in saline ocean water. The differences between communications in air and water are very large. In air electromagnetic signals can propagate at the speed of light, 3×10^8 m/s, while in water sound propagates at about 1.5×10^3 m/s. In addition, as a pulse of sound propagates its characteristics are modified by frequency-dependent attenuation, where higher frequencies are attenuated more rapidly than lower frequency signals. Also, particularly in shallow water or for sources located near water boundaries, the transmitted pulse can reach the receiver by many paths and may arrive at the receiver at the same time causing distortion in the transmitted pulse, which can compromise detection and decoding of the transmitted signal. These and other factors, such as thermal and salinity gradients and the general anisotropic nature of the oceans in particular, make water a challenging communication channel for AT (Chitre et al., 2008).

44.1.4   Examples of Underwater Acoustic Telemetry Systems

Relatively simple forms of AT would be transmission of encoded signals carrying the output of a single sensor from a location in water to a receiver both of which may be in fixed locations that do not change over their operational life. An example would be a pressure sensor located on a well head submerged some distance from a stationary receiver located on an ocean buoy with the capability to transmit the received measurements by satellite to a location where they are utilized.

AT applications with greater complexity would include transmitters located aboard an unmanned underwater vehicle or implanted in the body of a fish. The transmitters would send information with the identity of the carrier perhaps with information to aid estimation of its locations, such as pressure. The transmitted information would be detected by an array of several hydrophones, and features of the timing of arrival of the transmitted signals would be used with other transmitted information to locate and perhaps track the time-varying position of the transmitters (Deng et al., 2011a; Xiao, 2010).

One of the more complex and challenging AT applications is verbal communication between divers and between divers and support staff located aboard vessels. Verbal communications depend upon retention or recovery of critical temporal and frequency characteristics of speech to make the transmitted sounds understandable to a human receiver. The difficulty of transmitting speech through water is affected not only by phenomenon such as attenuation but also by the bandwidth limitations of the slow speed of propagation of sound in water (Istepanian and Stojanovic, 2002).

44.2   FIXED-ASPECT HYDROACOUSTICS

44.2.1   Introduction

The fixed-aspect hydroacoustic technique uses active sonar with transducers fixed in position to detect fish in water moving through stationary acoustic sampling volumes. This technique is commonly applied to estimate fish entrainment rates at hydropower turbine intakes (Ploskey and Carlson, 1999; Ransom et al., 1996) and cooling water intakes for thermal power plants (Ross et al., 1993), escapement rates of upstream-migrating adult salmonids (Enzenhofer et al., 1998; Maxwell, 2007; Reynolds et al., 2007), and passage rates of downstream-migrating juvenile salmonids at hydropower dams (Johnson et al., 1994; Ransom and Steig, 1994). We focus on fixed-aspect sampling of salmonids to estimate downstream salmonid passage at dams because the breadth of published literature on all fixed-aspect hydroacoustic sampling of fish is too wide to effectively cover in a single article. Nevertheless the basic principles of the sampling technique are similar, although some details like the selection of acoustic beam angles, aiming orientation, and settings may vary for other applications.

Passage data from fixed-aspect hydroacoustic studies, usually rates (number of fish per unit time) and efficiencies (number of fish passing through a selected route relative to the whole dam), are used by fisheries managers and dam operators to develop long-term structural and operational solutions that provide safe passage for juvenile salmonids at dams. In many instances, such studies are necessary because salmonid populations are listed as threatened or endangered under the Endangered Species Act (Thorne and Johnson, 1993). Fixed-aspect hydroacoustic data can be used to estimate horizontal and diel distributions, run timing, and fish passage efficiency of non-turbine routes relative to all routes at the dam. Passage routes can include turbine intakes, sluiceways, surface flow outlets, or spillways, and within each there can be numerous passage locations, such as individual turbine intakes or spill bays. Managers use passage data early in the design process to develop long-term passage solutions and later to evaluate the performance of installed devices or operations. The performance results are integrated with other complementary sources of data to adaptively manage the effort to protect juvenile salmon migrating downstream through dams.

The first reported fixed-aspect hydroacoustic study for juvenile salmonids was on the Columbia River by Carlson et al. (1981). Since then, fixed-aspect hydroacoustic studies have been conducted over multiple years at all 13 main stem dams on the Columbia and Snake rivers that have downstream migrant salmonid populations, and other sites nationwide. For more information, see reviews of fixed-aspect hydroacoustic studies from Bonneville Dam (Ploskey et al., 2007a, 2007b), the Dalles Dam (Johnson et al., 2007; Ploskey et al., 2001), John Day Dam (Anglea et al., 2001), Lower Granite Dam (Johnson et al., 2005a), and Wells Dam (Johnson et al., 1992; Skalski et al., 1996).

Like any monitoring technique, fixed-aspect hydroacoustics has strengths and weaknesses. The most important strength is that it is non-obtrusive. Fish do not have to be collected or handled, nor are they affected by the sampling device, which could bias results. Fixed-aspect hydroacoustics has high-resolution sampling capability over time and space, depending on study objectives. The main limitation is lack of species identification. This can be overcome, though, using concurrent direct capture sampling or by sampling when the fish of interest predominate the targets passing through the ensonified volumes. Statistically significant associations between concurrent hydroacoustic and net catch estimates of fish passage have been used to validate the fixed-aspect technique (Ploskey and Carlson, 1999; Ransom et al., 1996). Care must be taken to ensure passage estimates from fixed-aspect hydroacoustics, as with any hydroacoustic technique, are not biased by detection of nontarget species. For this reason, fixed-aspect hydroacoustic studies are most successful when a few species of fish dominate fish passage through a dam (e.g., juvenile salmonid runs in the Pacific Northwest) or when concurrent net sampling provides robust species composition information to apportion trace counts among species. Fisheries managers and dam operators choose fixed-aspect hydroacoustics when they want non-obtrusive passage data with high temporal and spatial resolution for the run-at-large (i.e., the collective population of fish passing downstream through a dam).

The purpose of this section is to explain the fixed-aspect hydroacoustic technique in enough detail for the reader to understand the basic approach and principles. We cover fixed-aspect hydroacoustic systems, the acoustic screen model, data collection, and data analysis. The citations in the References provide further information, as do *Fisheries Acoustics* by Simmonds and MacLennan (2005) and *Principles of Underwater Sound* by Urick (1983).

44.3 FIXED-ASPECT HYDROACOUSTIC SYSTEMS

A basic fixed-aspect hydroacoustic system consists of a scientific quality echo sounder, cables, transducers, and a

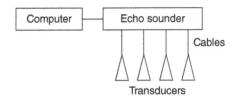

FIGURE 44.1   Basic fixed-aspect hydroacoustic system.

computer (Figure 44.1). Scientific quality means the system permits precise control over transmitted and received pulses, is electrically stable over long periods of time, and can be accurately calibrated. Besides being stationary, a major difference between fixed-aspect and mobile hydroacoustic systems is the array of two to eight transducers typically connected to one echo sounder in a fixed system. Frequencies of fixed-aspect hydroacoustic system usually are ≥ 200 kHz, because these frequencies permit cost-effective transducer designs, smaller transducer sizes, narrower beam widths, and shorter pulse widths than lower frequencies (Horne and Clay, 1998). Higher frequencies over 400 kHz often work best when sampling turbines and spill bays bounded on three or more sides by concrete and steel where sound persistence can be a problem. Although sound attenuation in water increases with increasing acoustic frequency, this is not usually an issue for fixed-aspect applications because most are in freshwater.

Single- or split-beam hydroacoustic systems are typical for most applications. In a single-beam system, the transducer has a single receiving element that permits determination of the range of the target but not its position relative to the axis of the acoustic beam. In a split-beam system, the transducer has four receiving quadrants. The transmission pulse activates the entire transducer but echoes received at the four quadrants are processed separately. The four signals are combined in pairs to produce the phase differences used to calculate the along- and athwart-ship angles necessary to determine the target's position relative to the axis of the beam. Split-beam systems are more powerful than single-beam systems because split beam allows precise determination of position in the acoustic sampling volume and, knowing this, estimation of fish target strength (see sonar equation later).

Most systems deployed at dams have a narrow bandwidth ranging from 5 to 20 kHz and allow pulse durations of 200–400 µs (range resolution = 14.6–29.1 cm) and high pulse repetition rates (15–35 pings/s) that are required to detect fish passing rapidly through acoustic beams. Systems with wider bandwidths of 10–20 kHz have been used to accommodate even shorter pulse durations (0.8–100 µs) to get the range resolution of targets down to 5.8–7.3 cm to sample fish passing into surface flow outlets at dams (Johnson et al., 2005b; Ploskey et al., 2006). Broadband systems greatly increase the bandwidth of the transmission frequency and sample entire echo pulses to provide substantially improved range resolution (~1 cm) and information about each echo to aid in target discrimination (Jung et al., 2003; Simmonds et al., 1996; Stanton et al., 2010). However, broadband systems have yet to be used to sample fish passing through hydropower projects, perhaps because many such specialized systems would be required to sample all passage routes.

Fixed-aspect hydroacoustic systems work like other fisheries hydroacoustic systems. The echo sounder (transceiver) generates an electrical pulse or transmission that passes through a cable to an underwater transducer where the electrical energy in the pulse is converted to mechanical energy. The transducer produces sound pulses that emanate through water and reflect off regions where there is a density difference with the ambient water, such as the swim bladder in a fish. The echo is in turn received by the transducer which reconverts the sound energy to electrical energy. The signal then passes back to the echo sounder where it is amplified by a time-varied gain and other gains. A 40 $\log_{10}$(range) time-varied gain is used because this corresponds exactly to two-way signal strength loss due to beam spreading. Other gains serve to equalize the output from different transducers for a given sized target. The received signal is sent from the echo sounder to a computer. Depending on the system and manufacturer, the electrical signal is processed (digitized) at the transducer, the echo sounder, or the computer. The raw output from a fixed-aspect hydroacoustic system is digitized echoes as a function of range from the transducer and elapsed time (pings) that are recorded in computer files.

Fixed-aspect hydroacoustic systems must be calibrated to produce reliable data. In this case, a "system" is a particular echo sounder–cable–transducer combination. The objective of calibration is to accurately measure three characteristics: system source levels for various transmit powers, system receiving sensitivities, and transducer beam patterns and beam pattern factors. Source level is the sound pressure level produced by the system when it is transmitting. Receiving sensitivity is the signal output level of the system as a function of sound pressure level at the face of the transducer. Beam pattern is a measure of transmit and receive sensitivities as a function of angle off the main axis of the acoustic beam in two planes of rotation. Calibration typically occurs in a tank using a standard transducer of known receiving and transmitting characteristics set at a known distance from and aimed directly at the face of the transducer to be calibrated. The standard transducer is used once in receiving mode to measure source levels of the system and again in transmit mode to measure receiving sensitivities of the system. Beam pattern is estimated by aiming the standard and test transducers directly at each other and then precisely rotating the test transducer in increments of 1° or less through a horizontal plane while recording transmit and receive levels of the test system (Figure 44.2) and then repeating the process for the vertical plane. The two-way amplitude response of a test

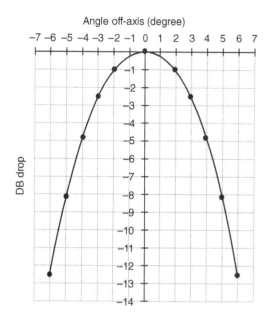

FIGURE 44.2   Example of a beam pattern for one plane of a 6° split-beam transducer.

transducer is the sum of transmit and receive responses and is referred to as the two-way beam pattern factor (2BPF); it is zero for on-axis target signals and progressively negative for signals further off axis. The transducer angle associated with one-half sound intensity (2BPF = –3 dB) is referred to as the nominal beam angle.

The sonar equation is fundamental to the fixed-aspect hydroacoustic technique. For example, given a known source level and other system characteristics, the sonar equation is used to calculate the receiving sensitivity during calibration. The sonar equation uses parameters expressed in decibels (dB re: μPa @ 1 m). The sonar equation is

$$V_{out} = SL - 40 \log R - 2\alpha R + 2BPF + TS + TVG$$
$$+ 2L + G_1 + G_0$$

where

V_{out} = detected output voltage for a single target;
SL = source level, that is, sound pressure level at a distance of 1 m from the transducer;
$40 \log R$ = two-way spreading loss for sound traveling to and from a target at range (R);
$2\alpha R$ = two-way absorption loss, where α is the absorption coefficient;
2BPF = two-way beam pattern factor;
TS = target strength of the fish;
TVG = time-varied gain function in the echo sounder receiver;
$2L$ = two-way cable loss;
G_1 = system receiving sensitivity; and
G_0 = constant receiver gain.

44.4 ACOUSTIC SCREEN MODEL

The fixed-aspect hydroacoustic technique is based on the acoustic screen model (Johnson, 2000). The acoustic screen model is an echo counting technique (Simmonds and MacLennan, 2005) that relies on single echoes from individual fish passing through the region ensonified by successive transmissions of sound energy from an underwater transducer, referred to as the acoustic sampling volume (Foote, 1991). This volume has user-defined beginning and end ranges. The technical edge of the beam is the angle off-axis where the sound intensity has dropped by half; sound intensity drops dramatically as the angle widens. Most importantly, the effective beam angle for detecting a fish is defined by the echo detection threshold and the echo level, which depends on target strength, target position in the beam, and other factors described in the next paragraph. Echoes from fish in the sampling volume are combined into fish tracks using spatial correlation algorithms and the fact that the sampling volume is ensonified successively in fixed-aspect hydroacoustics (Kieser and Mulligan, 1984). This conversion from echoes to tracks essentially converts the sampling volume to a sampling plane, or "acoustic screen."

Detectability is the probability of obtaining some minimum number of echoes from fish passing through an acoustic sample volume, and it is expressed as effective beam angle as a function of range from a transducer. Detectability depends on characteristics of the sampling system such as ping rate, beam geometry, echo or target strength threshold, and the minimum detection criterion (e.g., four consecutive echoes or four echoes in five pings). In general and within limits, detectability increases with ping rate, nominal beam angle, and increasing sampling range. Of course it is possible to ping too fast or deploy a beam that is too wide or to choose a poor minimum detection criterion. Detectability also depends upon fish target strength, range, speed, and trajectory through the acoustic sampling volume. It follows that large fish moving slowly through the beam at long range will have higher detectability than fish moving quickly across the beam at short range regardless of trajectory. The orientation of fish to the axis of the acoustic beam (fish aspect) greatly affects target strength and detectability. Detectability modeling requires special software, but is essential to estimate effective beam angles of acoustic screens. The best detectability models are stochastic and estimate effective beam angle based on simulations of 0.5–1.0 million targets of various sizes passing through an acoustic beam. Effective beam angle, along with the start and end ranges, define the acoustic screen (Figure 44.3).

The acoustic screen model can be used to make relative or absolute smolt passage rate estimates. The principle of equivalency is applied to make relative estimates. It says that the acoustic screen model is valid when elements, such as detectability, are equivalent from location to

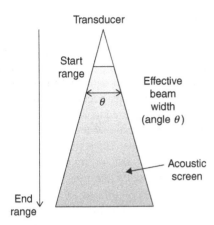

FIGURE 44.3 Depiction of the acoustic screen (gray-shaded area).

location. For example, the model is valid for relative estimates if detectability is 80% of nominal at both spillway and powerhouse sample locations. Absolute estimates of passage rates can be obtained when detectability is well understood. Whether relative or absolute estimates are made, the assumptions of the acoustic screen model must be assessed each time the model is applied.

The assumptions of the acoustic screen model fall into three categories: detection, identification, and weighting. Detection, or the ability of the fixed-aspect hydroacoustic system to accurately acquire fish echo data, assumes that sound energy does not affect fish behavior; all targets of interest in the acoustic beam are actually detected; detectability by range and sampling location is known accurately; and system performance is consistent and stable during a study. Identification is the process mentioned above to convert fish echoes into fish tracks. This process assumes targets do not overlap; there are no false targets; fish tracked at a particular passage route in the dam actually pass there; a given fish is counted only once; and the identification process is consistent across sampling volumes. Weighting is the analysis step where individual fish tracks are extrapolated spatially to the full width (vertically aimed acoustic beams) or height (horizontally aimed acoustic beams) of the passage location being sampled. (This step is explained in detail in Section 44.4.2.) The assumptions for weighting are that the effective beam angle for each sampling location is accurately known; target strength characteristics are the same among sample locations or can be accurately estimated for fish traces; and the horizontal distribution of fish within the sampling location is uniform.

44.4.1 Data Collection

Data collection involves setting hydroacoustic system parameters, deploying the transducers, designing the sampling

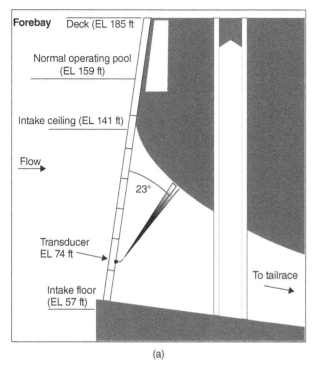
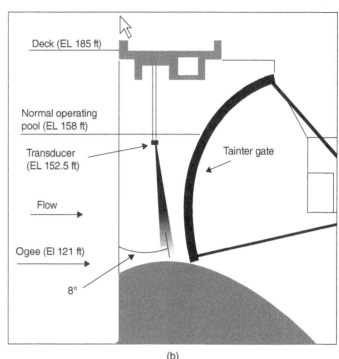

(a)   (b)

FIGURE 44.4   Example transducer deployments for a turbine intake (a) and a spill bay (b). From Johnson et al. (2005b).

scheme, dealing with noise, and processing echoes. Data collection produces computer files of echoes from fish and other sources. Ploskey et al. (2000) provide recommendations for standards and guidelines for the conduct of fixed-aspect hydroacoustic studies.

With the calibration data in hand, the hydroacoustic system parameters are set in the field. The parameters include transmit power, pulse width, bandwidth, ping rate, trigger source, receiving threshold, receiver gain, time-varied gain, start and end ranges, sound velocity, and others. These parameters are system specific. Transmit power is set, within the constraints of the machine, reasonably high to place a reasonable amount of sound energy in the water to detect fish. Pulse width and bandwidth are interrelated; the lower the pulse width, the higher the bandwidth. As noted above, the time-varied gain is $40 \log R$ for echo counting in fixed-aspect hydroacoustics. The maximum ping rate depends on the maximum range of interest and the sound velocity in study site water. In freshwater, 1490 m/s is a sensible sound velocity. The next ping in a repetition cannot begin until the echoes from the previous one have been received.

The system's receiving threshold is set to represent the voltage output for the smallest target strength of interest as far off-axis as is reasonable given the typical transducer beam pattern. Love (1977, 1978) showed the relationship between target strength and fish length. Once the threshold voltage is set, a receiving gain is calculated for each transducer by using the sonar equation to equalize for varying source levels and receiving sensitivities for the sound–cable–transducer combinations in the system. When the system parameter settings are ready, they are documented and a thorough system check is conducted for quality control.

In fixed-aspect hydroacoustics, transducers are located and aimed to have the acoustic sampling volume in a region of interest of flowing water where the route for passage downstream is known and highly probable. Detection of fish that are not committed to passing through a sampled route introduces the possibility that a fish may be counted more than once and should be avoided whenever possible. For example, at hydropower turbine intakes, transducers would be installed downstream of trash racks; at spill bays, this would be as near the spill gate controlling flow as possible (Figure 44.4). In either case, deploying transducers in the forebay upstream could led to biased results because fish detected there might not actually pass into the dam at that location. Thus, the location to be sampled affects the choice of transducer beam width, ping rates, and other factors. Transducer beam angles must be narrow enough to preclude echoes from side lobes reverberating off side walls in confined locations such as turbine intakes. Beam angles of 6° are commonly used inside turbine intakes, whereas wider circular beams (10°–12°) or elliptical beams may be used to sample spill bays. The horizontal position of a transducer at a given location is often randomized to help meet randomization requirements of the sampling design.

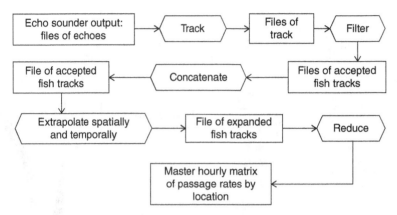

FIGURE 44.5   Flow chart for data analysis.

For a given fixed-aspect hydroacoustic system with multiple transducers, sampling design usually involves collection of systematic samples. Systematic sampling falls under the scope of finite sampling theory (Cochran, 1977). Transducers are sequentially sampled in a prescribed order and, when they all have been interrogated, the cycle starts over. Sometimes two transducers are sampled simultaneously, a process called "fast-multiplexing." Care must be taken so that one transducer does not interfere with the other. Systematic sampling with a sequence chosen at random is used because it is practical to implement, although stratified random sampling has better precision (Skalski et al., 1993). The interrogation time for a given transducer is minimized (~1 minute) within the logistical constraints of the hydroacoustic system to maximize sample size and minimize variance. The number of samples per transducer per hour is dependent on the interrogation time and the number of transducers in the system. The same systematic sample occurs from hour to hour, day to day in a study. Johnson et al. (1992) identified trade-offs in precision for hydroacoustic passage rate estimates and sampling intensity.

Noise is the bane of hydroacoustics. Noise caused by electrical interference from various cables and machines at a dam can be eliminated by shielding or otherwise isolating the hydroacoustic equipment from other electrical sources. Noise from echoes off dam structures can be avoided by proper transducer beam widths, aiming angles, and ping rates. Noise generated from turbulent water can be minimized by placing acoustic sampling volumes away from sources of turbulence such as dam piers. Wind-generated turbulence is difficult to contend with and can result in missing data that have to be interpolated because fish cannot be detected through the noise (see missing data algorithms in Section 44.4.2).

Echo signal processing is the digitization of electrical signals after conversion from sound energy by the transducer. Signal-processing parameters include pulse duration relative to the duration of the transmitted pulse (e.g., range = 0.5–3 times the transmitted pulse duration) and amplitude relative to the threshold (often about −56 dB relative to 1 μPa at 1 m). The output from data collection for fixed-aspect hydroacoustics is files of echoes as a function of target range and time, which are downloaded, archived, and transported to offices for data analysis.

44.4.2 Data Analysis

The analysis process for fixed-aspect hydroacoustic data begins with the digitized echoes from the hydroacoustic system and progresses through a series of steps to a master hourly matrix of passage rates by location (Figure 44.5).

Tracking echo files is the first step in the analysis of fixed-aspect hydroacoustic data. In the 1980s, this work was done using a hard copy chart recording, but today echo files are processed using computer programs. Tracking is the extraction of statistics on linear or collinear series of echoes using the spatial correlation algorithms. There are manual tracking programs that write out trace statistics to a file as a single fish after users select a series of echoes using pointing device. There are also automated tracking programs that identify a series of echoes meeting specific criteria based on common tracking algorithms such as an alpha–beta tracker or Kalman filter. Criteria vary among users but usually require some minimum number of echoes within a specific number of transmissions from a transducer. For example, a user might require four echoes in five pings as a minimum. Beyond that, criteria can range widely and include many other criteria such as mean range, maximum ping gap, threshold, mean target strength (split beam) or echo strength (single beam), echo count, slope, linearity, azimuth direction of travel, noise index, or proximity to noise. Fish traces are usually pretty obvious (Figure 44.6), but the presence of many echoes from turbulence or vortices can make the process difficult or even impossible. For automated tracking, users are careful to set the tracking parameters to include all possible fish traces and

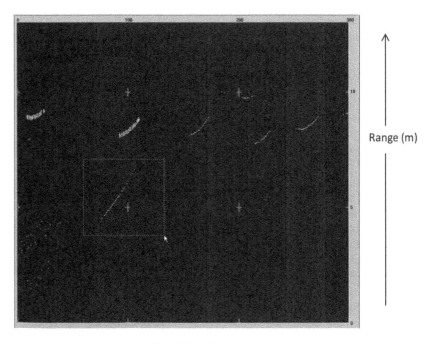

Range (m)

Time (pings)

FIGURE 44.6 Echogram showing two echo traces that have already been selected as fish (echoes have white vertical bars through them), one echo trace (inside the dashed box) that is about to be selected, and four other echo series that should be selected based on a minimum criterion of four echoes in five pings.

rely on subsequent filtering steps to remove questionable tracks.

Data from an automated fish tracking system must be filtered to eliminate echo series at constant range that are from known structure or chance echo alignments embedded in turbulent flow (noise events). Filter parameters include fish track characteristics such as range, echo count, linearity, slope, speed, noise indices, pulse duration, or average echo or target strength. Filters must often be customized for each system, without biasing the data. For quality assurance, the data are manually checked to ensure that valid fish tracks remain after filtering. Often, manual counts by technicians are regressed on automated tracker counts to assure that there is a reasonable correspondence between the two estimates and that the automated tracker is not accepting echo traces that a manual tracker would not accept. After such quality control checks, tracked fish files are finalized and likely contain $< 1\%$ of all the echoes originally detected by the acoustic system.

Classical finite sampling theory (Cochran, 1977) assuming simple random sampling is used to estimate hourly fish passage rates and associated sampling error. Using the acoustic screen model, each accepted fish track is expanded spatially by the ratio of the route width (vertical acoustic beams) or height (horizontal beams) to the diameter of the beam at the range of detection. For example, counts of fish through

a turbine or spill bay sampled by a vertically oriented beam would be expanded as follows:

$$\text{ExpCount} = \frac{\text{RW}}{2R\text{Tan}\left(\frac{\theta}{2}\right)}$$

where ExpCount is the spatially expanded count; RW the route width; R the range from the transducer; Tan the tangent; and θ the beam angle. From least to most desirable, beam angle may be the nominal beam angle of the transducer or an angle calculated from echo strength, target strength, or detectability modeling. Detectability modeling can account for many factors such as route-specific target strength statistics, beam pattern, threshold, ping rate, the minimum detection criterion (e.g., four hits in five pings), and target range, speed, and trajectory through an acoustic beam. The sum of spatially expanded counts of fish is then extrapolated temporally to a whole hour by the hourly sampling fraction (60/total minutes sampled per location). Missing data can be estimated using interpolation from data before and after the missing hour(s) for a given location; extrapolation into noisy regions using clean data; or regression estimators of relationships with adjacent locations.

Hourly variances for each transducer are computed as using the finite samples within that hour. Based on finite sampling theory, variances can be summed across hours and locations. Ninety-five percent confidence intervals (CI) are calculated as follows:

$$CI = \pm 1.96\sqrt{\text{Variance}}$$

The end product is a matrix of hourly passage rate estimates with associated variances for each sampling location. In cases where a fraction of the passage locations for a given route is sampled, such as sampling four of six spill bays at a spillway, passage at the unsampled bays is estimated by extrapolating passage from the sampled bays. The passage rate data are then used to estimate various performance metrics, including fish passage efficiency, spillway efficiency and effectiveness, sluiceway efficiency and effectiveness, and sluiceway passage. For the many details on spatial and temporal expansions, detectability modeling, and calculations of fish passage metrics, we refer readers to technical papers by Johnson (2000), Johnson et al. (2005a), Ploskey et al. (2006), and Skalski et al. (1996).

44.5 PRESSURE TESTING

Fish can be exposed to changes in pressure at several stages of hydro system passage and at several different types of hydro structures. As fish pass by the blade during hydroturbine passage, there is a rapid decrease in pressure which can result in barotraumas (injury due to changes in barometric pressure). These barotraumas occur primarily due to the expansion of gases in the swim bladder and include swim bladder rupture, exopthalmia, hemorrhaging, and emboli in the blood and tissues. The expansion of gases existing within the fish (such as the swim bladder) is proportional to the ratio of pressure change (governed by Boyle's law). For example, in a fish passing through a hydroturbine in which the pressure is decreased by half, the preexisting gas in the body will double in volume. It is very difficult to isolate the sources of barotraumas in field settings since pressures and water quality parameters can vary and are difficult to control. Therefore, research has been directed toward using precise and accurate laboratory equipment which can replicate the pressure profiles experienced by fish passing through hydroturbines and other hydro structures.

Hypo/hyperbaric chambers (such as those in the Mobile Aquatic Barotrauma Laboratory (MABL); Figure 44.7; Stephenson et al., 2010) have been used to gain a better understanding of the mechanisms of injury and mortality for fish exposed to pressures associated with passing hydroturbines and other hydro facilities. The system used by Stephenson et al. (2010) can replicate specific pressure profiles using a motor-controlled piston. The system can also control flow rate of water passing through the chamber and water temperature and total dissolved gas levels using a computer.

FIGURE 44.7 One of the hypo/hyperbaric chambers of the MABL system.

Following exposure, fish can be immediately euthanized and examined for commonly observed barotraumas. In contrast, fish could also be held post-exposure to monitor delayed mortality. In addition, the depth at which fish can achieve neutral buoyancy can be investigated using these hypo/hyperbaric chambers (Pflugrath et al., 2012). This is important since for fish, maintaining neutral buoyancy may be less energetically costly than other states. Also, determining limits of neutral buoyancy can give researchers a better idea of what depths within the water column fish are occupying and how much gas is within fish before they pass hydro projects (Pflugrath et al., 2012).

Research has shown that as the ratio of acclimation to exposure pressure increases, the probability of mortality and injury also increases for juvenile salmonids exposed to rapid decompression characteristic of turbine passage (Brown et al., 2012a, 2012c). For this reason, it is important to precisely and accurately control the pressures that fish are acclimated and exposed to during simulated turbine pressure. In addition, water quality parameters such as flow, temperature, and total dissolved gas levels should also be maintained throughout the acclimation period to simulate conditions representative of a field location. Acclimation requires the researcher to handle the fish in such a way that the amount of gas in its swim bladder would be similar to a fish that would be approaching a hydro facility. This may require fish to be held until they are neutrally buoyant at a relevant approach depth, as is the case for juvenile salmonids. Since some species can only attain neutral buoyancy by gulping air at the water surface (physostomous fish such as salmonids), testing facilities must be able to have a bubble available at the surface of the chamber. This bubble should be removed prior to exposure as air is very compressible and would make the pressure change difficult to achieve. Studies have shown that fish that are negatively buoyant prior to rapid decompression are less likely to acquire barotraumas; thus testing pressure change scenarios using negatively buoyant fish may not be representative of naturally moving or migrating fish (Stephenson et al., 2010).

Exposure pressures are the lowest pressure experienced during hydroturbine passage, typically on the suction side of the runner blade. These pressures depend on a number of different factors including turbine operating conditions, head of the dam, and submergence depth of the turbine. Autonomous Sensor Fish can be used to measure the pressures inside hydroturbines and these pressure profiles can be applied in hyper/hypobaric chambers. Since changes in pressure are experienced by all fish passing through hydroturbines, it is an important factor in understanding how fish are affected by downstream passage at hydroelectric facilities.

CASE STUDY

Influence of telemetry tags on the survival of tagged juvenile Chinook salmon exposed to rapid decompression

The route of passage and survival of fish passing downstream at hydroelectric facilities is often evaluated using telemetry. Acoustic or radio transmitters are commonly surgically implanted into the coelom of the fish, with the assumption that tagged individuals behave in the same manner as non-tagged individuals. Most transmitters are negatively buoyant and when implanted add an excess mass (added weight in water) to the body of the fish, causing the fish to uptake more molecules of gas into their swim bladder to achieve neutral buoyancy in the water column. Until recently, it was unknown how the presence of these negatively buoyant transmitters influenced the survival of juvenile Chinook salmon passing through hydroelectric facilities, particularly through the turbines. To understand the relationship between the excess mass of telemetry tags and the rate of mortality and injury, researchers conducted experiments in hyper/hypobaric pressure chambers exposing tagged juvenile Chinook salmon to rapid decompression associated with simulated turbine passage (Carlson et al., 2012). The results of this study show that as tag burden (mass of the telemetry tags relative to the mass of the fish) increased, the probability of mortality or injury also increased. This suggests that there is a negative bias in

previous survival estimates (estimates are too high) of juvenile salmon passing through hydroturbines, due to the presence of the telemetry tag. These results suggest that future studies investigating survival associated with hydroturbine passage use transmitters which minimize the excess mass added to a fish. In response to these conclusions, a neutrally buoyant externally attached acoustic transmitter has been designed and evaluated in the laboratory and shows promise for field testing (Brown et al., 2012b; Deng et al., 2012).

44.6 BIOTELEMETRY

Biotelemetry is commonly used to assess the behavior and movement patterns of fish in natural settings, such as river environments. It allows researchers to collect information about individuals from remote locations. In freshwater systems, where conventional hydropower facilities exist, passive and active telemetry tags can be used. Tags can be surgically implanted in the coelom, gastrically inserted, or externally attached and there are three basic types of tags. The fundamentals of the AT were described in Section 44.1.1.

Passive tags (such as passive integrated transponder (PIT) tags) contain no internal battery but instead are powered to transmit a unique code when they pass close to a system reader or antenna. Passive tag systems are used in adult fish ladders and juvenile bypass systems to identify individuals passing through specific routes. Since they are externally powered, PIT tags are small compared to other technologies (approximately 0.1 g) and can be used for multiple years (e.g., can be implanted in juvenile salmon and used to identify returning adults years after tagging). The size of the tag allows them to be injected into the coelom of the fish, precluding the need for surgery. The main disadvantage of these tags is that detection ranges are short; between a few centimeters to 1 m depending on antenna strength.

Radio transmitters are often used in freshwater systems. These tags emit radio waves on a unique frequency that can be picked up by a receiver through an antenna. Although transmitters can be implanted into the body cavity of fish, they possess an antenna which is generally exterior to the fish to transmit the signal. Radio telemetry has been used for decades to monitor the routes of passage and survival of fish passing by hydroelectric facilities, particularly juvenile salmonids passing downstream at hydroelectric facilities in the Columbia River basin (CRB). Extensive radio antenna arrays across the face of dams allow researchers to monitor the routes of passage and, in conjunction with downstream antenna systems, survival can be estimated for each route of passage. Unlike passive technologies, radio transmitters do not have to be activated by the antenna system, and therefore detection ranges are larger. However, radio transmitters typically become fully attenuated when passing through 10 m of freshwater or more. Thus they are not typically used in deep water environments or in areas of brackish or saltwater. The tag life of radio transmitters is correlated to the size of the battery, with increasing tag life with increasing battery size.

CASE STUDY

Use of electromyogram telemetry to assess swimming activity of adult Chinook salmon migrating past a Columbia River Dam

In addition to monitoring behavior and survival of fish, radio transmitters can be outfitted with instruments which determine and relay information such as the temperature, depth (pressure), and swimming activity of tagged fish to researchers. Electromyogram (EMG) radio telemetry tags have electrodes which are surgically implanted into the muscle (usually the red muscle which is used for non-burst swimming) of the fish. Following surgical implantation of the tags, the relationship between the EMG signal and swim speeds can be defined by making fish swim at prescribed speeds in a respirometer (a device designed to swim fish at different water velocities). In order to understand how adult migrating spring Chinook salmon behaved as they passed by a large dam in the Columbia River, researchers surgically implanted EMG transmitters into fish and monitored their passage with a combination of above-water and underwater antennas (Brown et al., 2006). Results of the study showed that adult spring Chinook salmon spent most of their time in the tailrace of the dam, when compared with the time spent in the fishways and forebays. In addition, the swim speeds recorded for fish in the tailrace were higher than when they were moving through other parts of the dam. This can be energetically costly for fish and this information directs research to find ways to reduce the time that adult Chinook salmon are spending in the tailraces by improving guidance to the fishways and providing areas where fish can rest as they move upstream through the tailrace.

44.7 SENSOR FISH

44.7.1 Introduction

The Sensor Fish is an autonomous device developed to better understand the physical conditions in the water passageways of hydroturbines and other dam water passageway such as spill. Information about water passageway conditions provides important diagnostics for assessment of the function of hydroturbines and also for the suitability of the passageway as a route for fish to pass a dam. Sensor Fish devices have been extensively used at dams on the Columbia and Snake rivers in the Pacific Northwest. The data acquired by Sensor Fish have been used to identify structural and operational modifications to hydroturbines and spillways that improve their safety for fish passage (Deng et al., 2010).

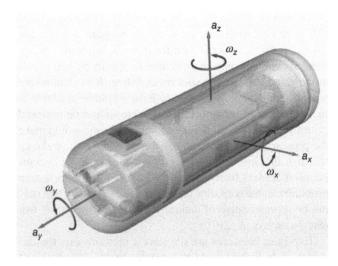

FIGURE 44.8 Drawing of a Sensor Fish showing the axis of rotation for acceleration and rotation measured by the sensors aboard the device.

44.7.2 Sensor Fish Design and Operation

The Sensor Fish is an autonomous 6 degrees of freedom motion sensor that also measures pressure and temperature. The Sensor Fish contains three rotation sensors, a tri-axial accelerometer, a pressure sensor, and a temperature sensor. The analog outputs from the sensors are sampled at a frequency of 2000 Hz. The Sensor Fish housing, which is a 90 mm long and 24.5 mm diameter cylinder with slightly rounded ends, is constructed of clear polycarbonate plastic. The sensor size and density is similar to that of a yearling salmon smolt. A drawing of the Sensor Fish showing its shape and axes of rotation is shown in Figure 44.8.

The Sensor Fish is deployed at dams using an injection system consisting of a system of flexible and rigid pipes that

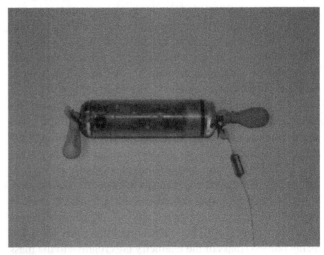

FIGURE 44.9 Photo of a Sensor Fish in pre-deployment condition with balloons and radio transmitter attached.

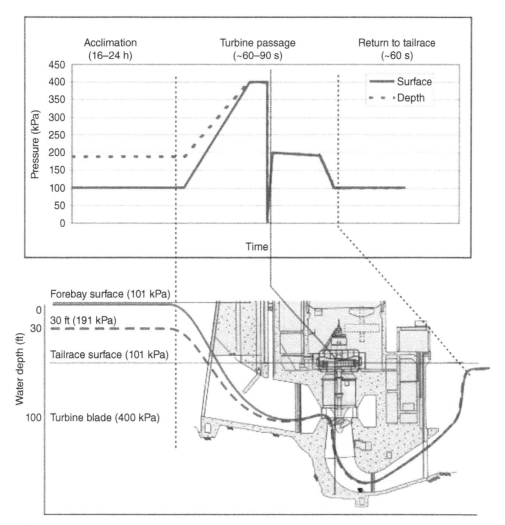

FIGURE 44.10   The top panel shows the pattern in absolute pressure with time that a fish would experience during movement through the forebay of a dam, entry into the turbine environment at passage through the turbine intake trash rack followed by passage through the turbine and exit into the powerhouse tailrace. The bottom panel shows a cross section through the dam showing the features mentioned previously. Lines connect the panels showing where distinct features in the pressure time history occur during dam passage.

extends from the intake deck of a dam into the water passageway of interest. The sensor is placed into the upper end of the system and is flushed through the piping with a flow of water. In preparation for use balloons containing a chemical that produces gas when mixed with water are attached to the Sensor Fish along with a micro radio transmitter (Figure 44.9). Immediately prior to placement in the injection system a small amount of water is injected into the balloons. The injected water begins to erode a capsule containing the gas-producing chemical and within 2–3 minutes, sufficient gas is produced to bring the Sensor Fish to the surface where it is recovered. The Sensor Fish is located in the dam tailrace by a crew that uses a radio receiver and directional antenna to locate the sensor. After recovery the data acquired by the sensor are downloaded to a computer using a wireless

infrared link. The design and operation of the first generation Sensor Fish is more extensively described in Deng et al. (2007a). A new generation Sensor Fish was developed with funding from U.S. Department of Energy and Electric Power Research Institute. The new Sensor Fish includes a internal radio tag and a recovery module that enables the Sensor Fish to float to the surface for recovery after a pre-programmed time (Deng et al., 2014).

44.7.3   Sensor Fish Data

A schematic showing the pressure–time history for fish entering and passing through a hydroturbine is shown in Figure 44.10. The fish passes through the dam reservoir at a depth of its choosing and enters the turbine environment by passing

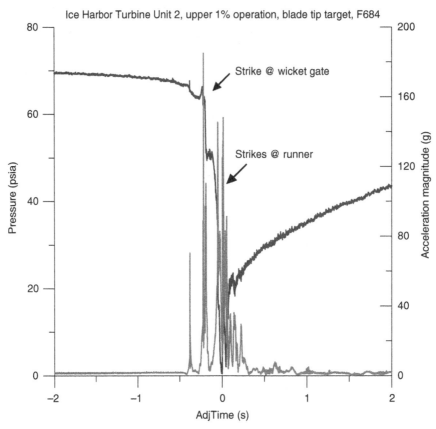

FIGURE 44.11 Pressure and acceleration time histories acquired by a Sensor Fish during passage through an operating Kaplan turbine. The time period covered includes approach to the turbine wicket gates, passage through the turbine runners, and entry into the turbine draft tube. The data show contact by the sensor of the structure during passage through the wicket gates and again during passage through the turbine runner.

through the turbine intake trash rack, then through the turbine exiting through the turbine draft tubes into the powerhouse tailrace. In the figure, two approach paths are shown, one near the surface and the other at a depth of 30 ft. Sensor Fish devices are deployed directly into the turbine environment. Depending upon the portion of the turbine environment to be sampled, a Sensor Fish device is typically injected at any location between the trash racks and the turbine wicket gates. Changes in pressure, acceleration, and rotation are typically small during passage through the turbine intake. During passage through the turbine runner changes in pressure are sudden and large and sensor acceleration may show high levels of turbulence and the occurrence of other events such as collision with turbine structure. Figure 44.11 shows the pressure and acceleration time histories for a Sensor Fish passing through an operating Kaplan turbine where the sensor data show contact with a turbine wicket gate and a turbine runner blade during turbine passage. The large decrease in pressure occurs when the sensor passes from the pressure to the suction side of the turbine runner assembly. The spikes in the acceleration time history occur when the sensor has

contacted the turbine structure or encountered high turbulence. This figure shows passage events that would have probably injured a fish that experienced them (Richmond et al., 2009).

Sensor Fish is also routinely used to evaluate conditions in flow through spillways. Spillways are often used as a means of passing fish past dams to avoid having the fish travel through turbines. Figure 44.12 shows a cross section through a spillway typical of those at dams in the Pacific Northwest. A large radial gate that is lifted to permit water in the dam forebay to flow directly into the dam's tailrace controls the amount of water that flows through the spillway. Samples of pressure and acceleration time histories acquired by Sensor Fish show that passage under the spillway control gate are characterized by a very rapid change in pressure from that at the depth of the sensor in the forebay to atmospheric pressure on the spillway chute. Depending upon several factors, travel down the spillway chute may be very smooth with little turbulence to that shown in Figure 44.12, which is quite turbulent with contact of the sensor on the surface of the spillway chute. At the end of the spillway chute the

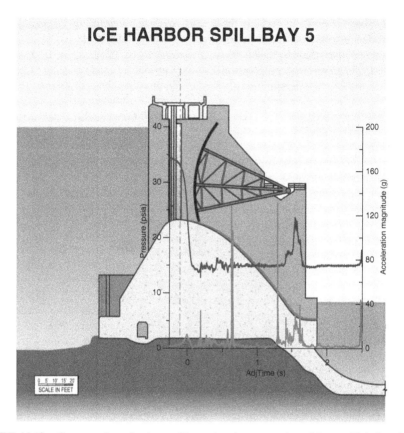

FIGURE 44.12 Cross section of a dam spillway showing an overlay of Sensor Fish data showing the location of pressure and acceleration time history features characteristic of sensor passage through a spillway. The pressure–time history in blue shows the rapid decrease in pressure at passage under the spillway control gate, turbulence during passage down the spillway chute, and finally an increase in pressure when the water is turned by the spillway flow deflector prior to entry into the spillway tailrace. The acceleration–time history shows characteristic features that complement those of pressure plus additional features showing contact of the sensor on the spillway chute and deflector.

water is turned by a deflector, which causes an increase in pressure followed by turbulence as the spillway flow enters the spillway stilling basin.

44.8 BLADE-STRIKE MODELING

Blade strike is the direct contact between a fish and the leading edge of a turbine blade. It is one of the major mechanisms that injure or kill fish. Blade-strike modeling is an important and cost-effective approach to evaluate the biological performance of turbine design and operations. The first deterministic model for predicting probability of strike was proposed by Von Raben in 1957. Other researchers (Ferguson et al., 2008; Turnpenny et al., 1998, 2000) expanded the model and investigated fish passage through turbine runners and associated injuries.

The major assumption behind the model is that a fish will be struck by the blade if it does not pass through the plane of the leading edges of the runner blades within sweeps of two successive blades. The probability of strike is expressed as (see Deng et al., 2007b, 2011b for additional details):

$$P = \frac{t}{t_{cr}} = \frac{l \cos \theta \times n \times \dfrac{N}{60}}{V_{axial}}$$

$$= \frac{l \cos \theta \times n \times \dfrac{N}{60} \times \pi \left(R_{tip}^2 - R_{hub}^2 \right)}{Q}$$

where t_{cr} is the critical passage time (the time between sweeps of two successive blades), Q the turbine discharge, l the fish length in meters, V_{axial} the axial velocity, n the number of blades, N the runner speed in revolutions per minute (rpm), R_{tip} the radius of circle formed by the runner blade tip, R_{hub} the radius of circles formed by the runner hub, and θ the angle between the absolute velocity and the axial velocity at a given fish passage location.

Von Raben (1957) reported that not all fish struck by the blades were injured and introduced mutilation ratio as the percentage of injured fish among the fish struck by the blades. Turnpenny et al. (2000) empirically developed a regression equation of mutilation ratio for different fish lengths:

$$MR = 0.15533 \ \ln{(l)} + 0.0125$$

where MR is the mutilation ratio, ln is the natural logarithm, and l is the fish length (in centimeters), and injury rate is the product of MR and blade-strike probability.

There are two numerical blade-strike models depending on the assumption for the input parameters: deterministic and stochastic models. The deterministic model assumed that fish were rigid bodies oriented perpendicular to the leading edge of the blade and predicted a single unique estimate for each combination of input values. The stochastic model analysis is usually performed using multiple realizations by assigning input parameters distributions of possible values. For example, Deng et al. (2011b) assigned discharge a uniform distribution ranging from the minimum to the maximum observed for each turbine at each turbine discharge treatment, fish passage radius uniform distribution between the runner hub and discharge ring radii, and fish length relative to the leading edge of the runner blade uniform and normal distribution. Out of these two models, the stochastic model will usually have a better agreement with experimental results and provides a more realistic prediction for blade-strike rate and injuries because it incorporates how fish approach the leading edges of turbine runner blades.

REFERENCES

Anglea, S., Poe, T., Giorgi, A. (2001). Synthesis of radio telemetry, hydroacoustic, and survival studies of juvenile salmon at John Day Dam. Report prepared for the U.S. Army Corps of Engineers, Portland District, by Battelle Pacific Northwest Division, Richland, WA.

Au, W. W. L., and Hastings, M. C. (2008). *Principles of Marine Bioacoustics.* Springer, New York, NY.

Brown, R. S., Carlson, T. J., Gingerich, A. G., Stephenson, J. R., Pflugrath, B. D., Welch, A. E., Langeslay, M. J., Ahmann, M. L., Johnson, R. L., Skalski, J. R., Seaburg, A. G., and Townsend, R. L. (2012a). Quantifying mortal injury of juvenile Chinook salmon exposed to simulated hydroturbine passage. *Trans. Am. Fish. Soc.,* 141:147–157.

Brown, R. S., Geist, D. R., and Mesa, M. G. (2006). Use of electromyogram telemetry to assess swimming activity of adult spring Chinook salmon migrating past a Columbia River Dam. *Trans. Am. Fish. Soc.,* 135:281–287.

Brown, R. S., Pflugrath, B. D., Carlson, T. J., and Deng, Z. D. (2012b). The effect of an externally attached neutrally buoyant transmitter on mortal injury during simulated hydroturbine passage. *Journal of Renewable and Sustainable Energy,* 4:013107; doi:10.1063/1.3682062.

Brown, R. S., Pflugrath, B. D., Colotelo, A. H., Brauner, C. J., Carlson, T. J., and Deng, Z. D. (2012c). Pathways of barotrauma in juvenile salmonids exposed to simulated hydroturbines passage: Boyles law vs. Henry's law. *Fish. Res.,* 121–122:43–50.

Carlson, T. J., Acker, W. C., and Gaudet, D. M. (1981). *Hydroacoustic Assessment of Downstream Migrant Salmon and Steelhead at Priest Rapids Dam in 1980.* APL-UW 8016. Applied Physics Laboratory, University of Washington, Seattle, Washington.

Carlson, T. J., Brown, R. S., Stephenson, J. R., Pflugrath, B. D., Colotelo, A. H., Gingerich, A. J., Benjamin, P. L., Langeslay, M. J., Ahmann, M. L., Johnson, R. L., Skalski, J. R., Seaburg, A. G., and Townsend, R. L. (2012). The influence of tag presence on the mortality of juvenile Chinook salmon exposed to simulated hydroturbine passage: Implications for survival estimates and management of hydroelectric facilities. *N. Amer. J. Fish. Man.,* 32(2):249–261.

Chitre, M., Shahabodeen, S., and Stojanovic, M. (2008). Underwater acoustic communications and networking: Recent advances and future challenges, *Mar. Technol. Soc. J.,* 42(1):103–116.

Cochran, W. G. (1977). *Sampling Techniques.* John Wiley & Sons, New York.

Deng, Z., Carlson, T. J., Dauble, D. D., and Ploskey, G. R. (2011b). Fish passage assessment of an advanced hydropower turbine and conventional turbine using blade-strike modeling. *Energies,* 4(1):57–67.

Deng, Z., Carlson, T. J., Duncan, J. P., and Richmond, M. C. (2007a). Six-degree-of-freedom Sensor Fish design and instrumentation. *Sensors,* 7:3399–3415.

Deng, Z., Carlson, T. J., Duncan, J. P., Richmond, M. C., and Dauble, D. D. (2010). Use of an autonomous sensor to evaluate the biological performance of the advanced turbine at Wanapum Dam. *Journal of Renewable and Sustainable Energy,* 2(5):053104.

Deng, Z., Carlson, T. J., Ploskey, G. R., Richmond, M. C., and Dauble, D. D. (2007b). Evaluation of blade-strike models for estimating the biological performance of Kaplan turbines. *Ecol. Model.,* 208:165–176; doi:10.1016/j.ecolmodel.2007.05.019.

Deng Z. D., J. Lu, M. J. Myjak, J. J. Martinez, C. Tian, S. J. Morris, T. J. Carlson, D. Zhou, and H. Hou. (2014). Design and Implementation of a new Autonomous Sensor Fish to Support Advanced Hydropower Development. *Review of Scientific Instruments* 85(11):115001; doi:10.1063/1.4900543.

Deng, Z. D., Martinez, J. J., Colotelo, A. H., Abel, T. K., LeBarge, A. P., Brown, R. S., Pflugrath, B. D., Mueller R. P., Carlson, T. J., Seaburg, A. G., Johnson, R. L., and Ahmann, M. L. (2012). Development of external and neutrally buoyant acoustic transmitters for juvenile salmon turbine passage evaluation. *Fish. Res.,* 113:94–105.

Deng, Z., Weiland, M. A., Fu, T., Seim, T. A., Lamarche, B. L., Choi, E. Y., Carlson, T. J., and Eppard, M. B. (2011a). A cabled acoustic telemetry system for detecting and tracking juvenile salmon: Part 2. Three-dimensional tracking and passage outcomes. *Sensors,* 11(6):5661–5676.

Enzenhofer, H. J., Olsen, N., and Mulligan, T. J. (1998). Fixed-location riverine hydroacoustics as a method of enumerating migrating adult Pacific salmon: Comparison of split-beam acoustics vs. visual counting. *Aquat. Living Resour.*, 11(2):61–74.

Ferguson, J. W., Ploskey, G. R., Leonardsson, K., Zabel, R. W., and Lundquist, H. (2008). Combining turbine blade-strike and life cycle models to assess mitigation strategies for fish passing dams. *Can. J. Fish. Aquat. Sci.*, 65:1568–1585.

Foote, K. G. (1991). Acoustic sampling volume. *J. Acoust. Soc. Am.*, 90:959–964.

Horne, J. K., and Clay, C. S. (1998). Sonar systems and aquatic organisms: Matching equipment and model parameters. *Can. J. Fish. Aquat. Sci.*, 55:1296–1306.

Istepanian, R. S. H., and Stojanovic, M. (2002). *Underwater Acoustic Digital Signal Processing and Communication Systems.* Kluwer Academic Publishers, Boston, MA.

Johnson, G., Beeman, J., Duran, I., and Puls, A. (2007). Synthesis of juvenile salmonid passage studies at the Dalles Dam, volume II: 2001-2005. PNNL-16443. Final report submitted to the U.S. Army Corps of Engineers, Portland District, by the Pacific Northwest National Laboratory, Richland, WA, and the U.S. Geological Survey, Cook, WA.

Johnson, G. E. (2000). Assessment of the acoustic screen model to estimate smolt passage rates at dams: Case study at the Dalles Dam in 1999. Appendix B in Ploskey et al. (2001). Hydroacoustic evaluation of juvenile salmon passage at the Dalles Dam: 1999. U.S. Army Corps of Engineers Research and Development Center Report TR-01-11, Vicksburg, MS.

Johnson, G. E., Anglea, S. M., Adams, N. S., and Wik, T. O. (2005a). Evaluation of the prototype surface flow bypass for juvenile salmon and steelhead at the powerhouse of Lower Granite Dam, Snake River, Washington, 1996-2000. *N. Amer. J. Fish. Man.*, 25:138–151.

Johnson, G. E., Hanks, M. E., Khan, F., Hedgepeth, J. B., Mueller, R. P., Rakowski, C. L., Richmond, M. C., Sargeant, S. L., Serkowski, J. A., and Skalski, J. R. (2005b). Hydroacoustic evaluation of juvenile salmonid passage at the Dalles Dam in 2004. PNNL-15180. Final report submitted to the U.S. Army Corps of Engineers, Portland District, by the Pacific Northwest National Laboratory, Richland, WA.

Johnson, G. E., Skalski, J. R., and Degan, D. J. (1994). Statistical precision of hydroacoustic sampling of fish entrainment at hydroelectric facilities. *N. Amer. J. Fish. Man.*, 14(2): 323–333.

Johnson, G. E., Sullivan, C. M., and Erho, M. W. (1992). Hydroacoustic studies for developing a smolt bypass system at Wells Dam. *Fish Res*, 14:221–237.

Jung, J.-B., Jacobs, J. H., Dowding, G. A., and Simpson, P. K. (2003). Initial species discrimination experiments with riverine salmonids. Proceedings of the International Joint Conference on Neural Networks, Portland, OR, July 20–24, 2003, vol. 2, pp. 1295–1300.

Kieser, R., and Mulligan, T. J. (1984). Analysis of echo counting data: A model. *Can. J. Fish. Aquat. Sci.*, 41:451–458.

Kilfoyle, D. B., and Baggeroer, A. B. (2000). The state of the art in underwater acoustic telemetry. *IEEE J. Oceanic Eng.*, 25(1):3–27.

Love, R. H. (1977). Target strength of an individual fish at any aspect. *J. Acoust. Soc. Am.*, 62:1397–1403.

Love, R. H. (1978). Resonant acoustic scattering by swimbladder-bearing fish. *J. Acoust. Soc. Am.*, 64:571–580.

Maxwell, S. L. (2007). Hydroacoustics: Rivers. In: Johnson, D. H., Shrier, B. M., O'Neal, J. S., Knutzen, J. A., Augerot, X., O'Neil, T. A., and Pearsons, T. N. (editors). *Salmonid Field Protocols Handbook: Techniques for Assessing Status and Trends in Salmon and Trout Populations.* American Fisheries Society, Bethesda, MD, pp. 133–152.

Pflugrath, B. D., Brown, R. S., and Carlson, T. J. (2012). Maximum acclimation depth of juvenile Chinook salmon: Implications for survival during hydroturbine passage. *Trans. Am. Fish. Soc.*, 141(2):520–525.

Ploskey, G., Poe, T., Giorgi, A., and Johnson, G. (2001). Synthesis of radio telemetry, hydroacoustic, and survival studies of juvenile salmon at the Dalles Dam (1982-2000). PNWD-3131. Final report submitted to U.S. Army Corps of Engineers, Portland District, by Battelle Pacific Northwest Division, Richland, WA.

Ploskey, G. P., Carlson, T. J. (1999). Comparison of hydroacoustic and net estimates of fish guidance efficiency of an extended length bar screen at John Day Dam. *N. Am. J. Fish. Man.*, 19:1066–1079.

Ploskey, G. R., Johnson, G. E., Giorgi, A. E., Johnson, R. L., Stevenson, J. R., Schilt, C. R., Johnson, P. N., and Patterson, D. S. (2007a). Synthesis of biological reports on juvenile fish passage and survival at Bonneville Dam, 1939-2005. PNNL-15041. Final report to the U.S. Army Corps of Engineers, Portland District, by Pacific Northwest National Laboratory, Richland, WA.

Ploskey, G. R., Patterson, D. S., Schilt, C. R., and Hanks, M. E. (2000). Workshop on standardizing hydroacoustic methods of estimating fish passage for lower Columbia River dams. ERDC/EL SR-00-01. Final report submitted to U.S. Army Corps of Engineers, Portland District, by U.S. Army Engineer Research and Development Center, Vicksburg, MS.

Ploskey, G. R., Weiland, M. A., and Kim, J. (2007b). Evaluating tradeoffs for efficiently passing juvenile salmonids through Bonneville Dam on the Lower Columbia River. Fourth International Reservoir Fisheries Symposium, Atlanta, GA, American Fisheries Society, Bethesda, MD.

Ploskey, G. R., Weiland, M. A., Zimmerman, S. A., Hughes, J. S., Bouchard, K., Fischer, E. S., Schilt, C. R., Hanks, M. E., Kim, J., Skalski, J. R., Hedgepeth, J. B., and Nagy, W. T. (2006). Hydroacoustic evaluation of fish passage through Bonneville Dam in 2005. PNNL-15944. Final report to the U.S. Army Corps of Engineers, Portland District, by Pacific Northwest National Laboratory, Richland, WA.

Ransom, B. H., and Steig, T. W. (1994). Using hydroacoustics to monitor fish at hydropower dams. *Lake Reserv. Manage.*, 9(1):163–169.

Ransom, B. H., Steig, T. W., and Nealson, P. A. (1996). Comparison of hydroacoustic and net catch estimates of Pacific salmon smolt (Oncorhynchus spp.) passage at hydropower dams in the Columbia River basin, USA. *ICES. J. Mar. Sci.*, 53:477–481.

Reynolds, J. H., Woody, C. A., Gove, N. E., and Fair, L. F. (2007). Efficiently estimating salmon escapement uncertainty using systematically sampled data. In: Woody, C. A. (editor). *Sockeye Salmon Evolution, Ecology, and Management*. Symposium 54, American Fisheries Society, Bethesda, MD.

Richmond, M. C., Deng, Z., McKinstry, C. A., Mueller, R. P., Carlson, T. J., and Dauble, D. D. (2009). Response relationship between juvenile salmon and an autonomous sensor in turbulent flows. *Fish Res*, 97(1–2):134–139; doi:10.1016/j.fishres.2009.01.011.

Ross, Q. E., Ross, D. J., Dunning, R. E., Thorne, J. K., Menezes, G. W., and Tiller, J. K. (1993) Watson Response of alewives to high-frequency sound at a power plant intake on Lake Ontario. *N. Am. J. Fish. Manage.*, 13:766–774.

Simmonds, E. J., Armstrong, F., and Copland, P. J. (1996). Species identification using wideband backscatter with neural network and discriminant analysis. *ICES J. Mar. Sci.*, 53:189–195.

Simmonds, J., MacLennan, D. (2005). *Fisheries Acoustics*. Blackwell Science, Oxford, UK.

Skalski, J. R., Hoffmann, A.,Ransom, B. H., and Steig, T. W. (1993). Fixed-location hydroacoustic monitoring designs for estimating fish passage using stratified random and systematic sampling. *Can. J. Fish. Aquat. Sci.*, 50:1208–1221.

Skalski, J. R., Johnson, G. E., Sullivan, C. M., Kudera, E. A., and Erho, M. W. (1996). Statistical evaluation of turbine bypass efficiency at Wells Dam on the Columbia River, Washington. *Can. J. Fish. Aquat. Sci.*, 53:2188–2198.

Stanton, T. K., Chu, D., Jech, J. M., and Irish, J. D. (2010). New broadband methods for resonance classification and high-resolution imagery of fish with swimbladders using a modified commercial broadband echosounder. *ICES J. Mar. Sci.*, 67:365–378.

Stephenson, J. R., Gingerich, A. J., Brown, R. S., Pflugrath, B. D., Deng, Z., Carlson, T. J., Langeslay, M. J., Ahmann, M. L., Johnson, R. L., and Seaburg, A. G. (2010). Assessing barotrauma in neutrally and negatively buoyant juvenile salmonids exposed to simulated hydro-turbine passage using a mobile aquatic barotrauma laboratory. *Fish. Res.*, 106:271–278.

Thorne, R., and Johnson, G. E. (1993). A review of hydroacoustic studies for estimation of salmonid downriver migration past hydroelectric facilities on the Columbia and Snake rivers in the 1980s. *Rev. Fish. Sci.*, 1:27–56.

Turnpenny, A. W. H. (1998). Mechanisms of fish damage in low-head turbines: An experimental appraisal. In: Jungwirth, M., Schmutz, S., Weiss, S. (editors). *Fish Migration and Fish Bypasses*. Blackwell Publishing, Oxford, UK, pp. 300–314.

Turnpenny, A. W. H., Clough, S., Hanson, K. P., Ramsay, R., and McEwan, D. (2000). Risk assessment for fish passage through small, low-head turbines. ETSUH/06/00054/REP. Technical report for Energy Technology Support Unit, Harwell, UK.

Urick, R. J. (1983). *Principles of Underwater Sound*. 3rd edition. McGraw-Hill, New York, NY.

Von Raben, K. (1957). Regarding the problem of mutilations of fishes by hydraulic turbines. *Die Wasserwirtschaft*, 4:97–100.

Xiao, Y. (2010). *Underwater Acoustic Sensor Networks*. Auerbach Publishers, Boston, MA.

45

THE STATE OF ART ON LARGE CAVERN DESIGN FOR UNDERGROUND POWERHOUSES AND SOME LONG-TERM ISSUES

Ömer Aydan

45.1 INTRODUCTION

Underground space has long been used through the history for the purposes of accommodation, religious ceremony, defense, and food storage by humankind (Photo 45.1). For example, in the Cappadocia Region of Turkey, there exist significant historical underground settlements, such as rock cut dwellings, cities, churches, and semi-underground cliff settlements, and modern man-made cavities used for multipurposes. The use of underground space in civil engineering projects other than mining purposes has become widespread all over the world. It seems that large caverns for various purposes will be needed in places where open space has been already scarce. Large underground caverns have been generally used for powerhouses, storage of crude oil and natural gas, shelters, and sport halls (Photos 45.2 and 45.3). There are also examples of the utilization of underground space for factories manufacturing high-precision equipments.

Underground powerhouses are constructed to house turbines and transformers in conjunction with hydroelectric power schemes and pumped-water energy storage schemes (Figures 45.1 and 45.2). They are generally 20–25 m wide, 40–50 m high, and 100–200 m long. There are also some discussions to house nuclear power reactors in underground caverns.

In the design of underground powerhouses, the anticipated form of instability is of great importance. The instabilities around underground openings can be categorized as global instability and local instability, which are defined in the following sense (Photo 45.4):

Global instability: When the excavated space cannot be kept open and the failure of surrounding mass continues to take place indefinitely unless any supportive measure is undertaken. The global instability would be as a result of exceeding the strength of surrounding rocks due to redistribution of initial ground stresses.

Local instability: After clearance of the failed zone and without taking any supportive measure, if the remaining space can be kept open, the form of instability is termed as the local instability. The main cause of failure will be the dead weight of rock in a particular zone about the cavity, defined by the geometry of caverns and the spatial distribution of discontinuities.

In this section, the outcomes of a survey on the design of large underground caverns for powerhouses are described and some current issues and their possible solutions are presented. The literature survey indicated that many large underground powerhouses are constructed in highly competent rocks and the competency factor is a reliable parameter to evaluate the global stability of underground powerhouses. It seems that the support design of large caverns is generally based on the forms of local instability. However, there are long-term issues of large underground caverns related to cyclic variations of temperature and groundwater table and ground shaking caused by earthquakes and the operation of turbines. Furthermore, the long-term performance of support members such as rockbolts and shotcrete may be a great issue for the underground caverns.

Alternative Energy and Shale Gas Encyclopedia, First Edition. Edited by Jay H. Lehr and Jack Keeley.
© 2016 John Wiley & Sons, Inc. Published 2016 by John Wiley & Sons, Inc.

Photo 45.1   Natural and man-made antique underground caverns.

Photo 45.2   Man-made underground caverns and examples of their utilization.

Photo 45.3 Examples of underground powerhouses for hydroelectric and pumped-water energy storage schemes

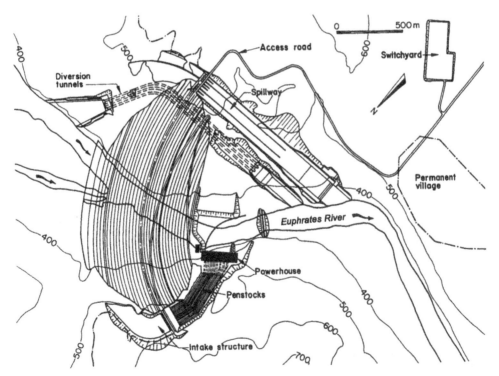

FIGURE 45.1 Layout of Atatürk hydroelectric scheme and location of powerhouse. Courtesy of Devlet Su İşleri (DSİ) of Turkey.

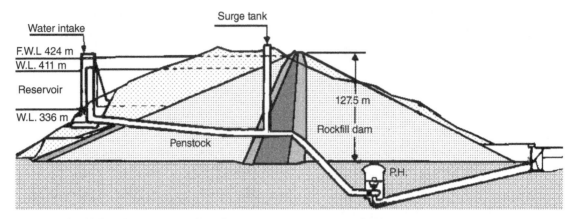

FIGURE 45.2   A cross section of Mazegawa pumped-water energy storage scheme (from Aydan et al. 2012).

45.2 A SURVEY ON LARGE UNDERGROUND CAVERNS

A survey on large underground caverns excavated all over the world has been carried out with an emphasis on those in Japan and a database called "UGCAVERN" has been developed. Underground caverns involve powerhouses, storage caverns, shelters, and sport halls. For this purpose, a datasheet was used in the surveying of underground caverns. The items of the data-sheet include the geometry of the caverns,

Photo 45.4   Some examples of global and local failure modes of underground openings.

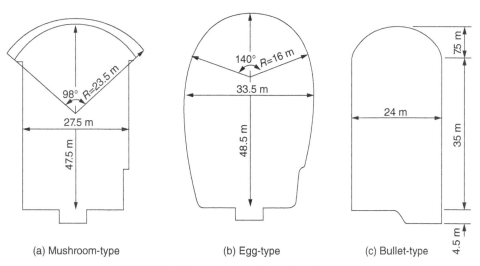

FIGURE 45.3   Typical configuration of caverns.

overburden, geology, initial stress state, some mechanical properties of rock, and measured deformations. The outcomes of this survey are described in the following subsections.

45.2.1 Geometry of Underground Caverns

The geometrical shapes of underground openings are generally mushroom-like, ovaloid (egg-shape), or bullet-like (Figure 45.3). Mushroom-shaped cavern is generally preferred for powerhouses as the rails of cranes can be easily seated on the haunches (abutments). To specify the geometry of the caverns, various parameters are defined as illustrated in Figure 45.4. Figure 45.5 shows a plot of the maximum width and maximum height of the caverns. It seems that the general trend for average height–width ratio is about 1.5:1 to 2:1. The radius–span ratios, which may be called as the arch parameter, are plotted in Figure 45.6. Most of the plotted data are above 0.3 and a common value is about 0.7 particularly for caverns in Japan.

45.2.2 Positioning of the Long-Axis Caverns

The positioning of the long axis of caverns is generally governed by construction constraints, initial stress state, intensity, and orientations of geological structural defects (faults, joints, etc.). The constructional constraints suggest that the long axis of the cavern should be perpendicular or sub-perpendicular to the flow direction of water.

The long axis of caverns is generally chosen to be parallel to the largest in situ principal stress direction in the United States, Japan, and Germany while it is generally chosen perpendicular to that in Scandinavian countries. The initial stress is the most important parameter in the positioning of the

longitudinal axis of the caverns of many underground powerhouses in Japan. Figure 45.7 shows the measured stress state in Japanese archipelago and the alignment of the longitudinal axis of the caverns. Figure 45.8 shows the variation of vertical stress and the ratio of mean horizontal stress to vertical stress in various parts of the earth together with those for Japan (Aydan and Kawamoto, 1997; Aydan, 2014). The ratio of the mean horizontal stress to the vertical stress ranges between 0.5 and 5 depending upon the location. Furthermore, the plotted data for the ratio of the mean horizontal stress to vertical

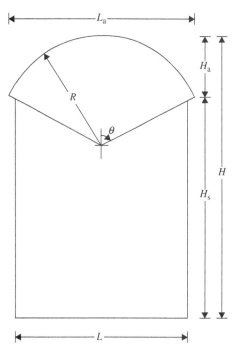

FIGURE 45.4   Geometrical parameters to define cavern shape.

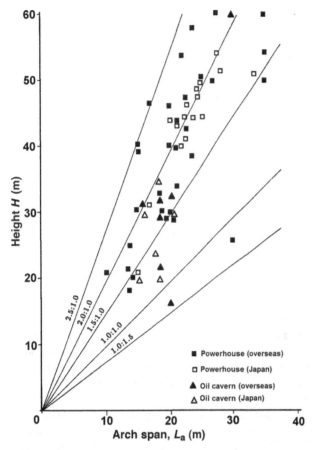

FIGURE 45.5 Relation between height and width of caverns.

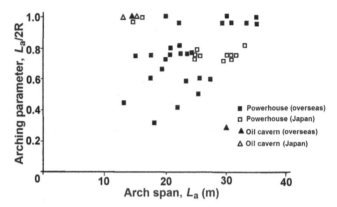

FIGURE 45.6 Arch parameter for various caverns.

stress are bounded by the equations purposed by Terzaghi and Richart (1952) and Aydan and Paşamehmetoğlu (1994), given as

Lower bound (Terzaghi and Richart, 1952)

$$\frac{\sigma_{Hm}}{\sigma_v} = \frac{\nu}{1 - \nu} \qquad (45.1)$$

Upper bound (Aydan and Paşamehmetoğlu, 1994)

$$\frac{\sigma_{Hm}}{\sigma_v} = 1 + \frac{2000}{h} \qquad (45.2)$$

where $\sigma_{Hm}, \sigma_v, \nu$, and h are the mean horizontal stress, vertical stress, Poisson's ratio, and depth, respectively.

The third important item is the geological structures such as faults, shear zones, and persistent joint sets. The long axis of the cavern is generally chosen to be perpendicular to the strike of major geological structures.

45.2.3 Mechanical Properties of Rock and Rock Mass

The caverns are generally located in relatively hard and solid rocks. The ratios of the elastic modulus of intact rock to that

of rock mass are generally in the range of 1–5 as seen in Figure 45.9. Aydan et al. (2014) have recently proposed a new rock classification named as Rock Mass Quality Rating (RMQR). The classification incorporates six basic parameters of rock masses, which are the discontinuity set number (DSN), discontinuity spacing (DS), discontinuity condition (DC), degradation degree (DD), ground seepage condition (GWSC), and groundwater absorption condition (GWAC). This rock classification quantifies the state of rock mass, and possible geo-mechanical properties of rock masses can be estimated using this classification system together with intrinsic geo-mechanical properties of intact rock. Aydan et al. (2014) associated RMQR with six fundamental properties of rock masses normalized by corresponding properties measured in laboratory on the basis of laboratory and in situ tests performed in major rock engineering projects in Japan in the following form:

$$\alpha = \alpha_0 - (\alpha_0 - \alpha_{100})\frac{RMQR}{RMQR + \beta(100 - RMQR)} \qquad (45.3)$$

where α_0 and α_{100} are the values of the function at RMQR = 0 and RMQR = 100 of the property α, and β is a constant to be determined by using a minimization procedure for experimental values of given physical or mechanical properties. Some values for these empirical constants with the consideration of in situ experiments carried out in Japan are given in Table 45.1. These values can be used for assessing the performance and stability of rock engineering structures when rock mass is treated as continuum.

The uniaxial compressive strength (σ_c) of intact rocks surrounding caverns is generally more than 10 MPa. Defining the competency factor as the ratio of σ_c to overburden pressure ($\sigma_v = \gamma H$), the results are plotted for the range of minimum and maximum values in Figure 45.10. Symbols with filled black ink are the caverns where severe instability problems are encountered. Although it is very difficult to find any report clearly stating the instability problems for some official or other reasons, it can be said that some kind of

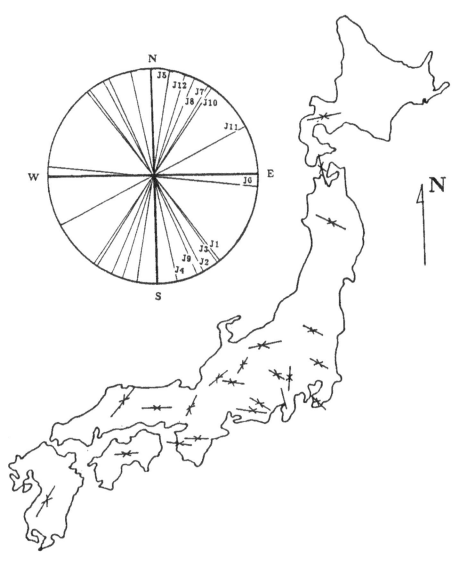

FIGURE 45.7   Measured maximum horizontal stress directions in Japan.

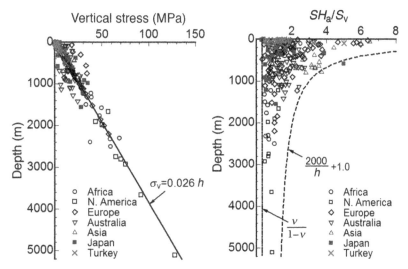

FIGURE 45.8   Measured variation of vertical stress and the ratio of mean horizontal stress to vertical stresses with depth (from Aydan and Kawamoto, 1997; Aydan, 2014).

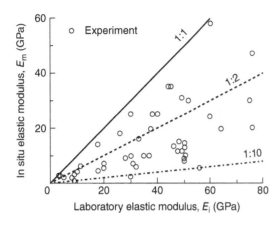

FIGURE 45.9 The relation between laboratory and in situ deformation moduli.

TABLE 45.1 Values of α_0, α_{100}, and β for various properties of rock mass.

Property (α)	α_0	α_{100}	β
Deformation modulus	0.0	1.0	6.0
Poisson's ratio	2.5	1.0	1.0
UCS	0.0	1.0	6.0
Tensile strength	0.0	1.0	6.0
Cohesion	0.0	1.0	6.0
Friction angle	0.3	1.0	1.0

severe instability problems are encountered when the competency factor is less than 4.

Let us define wall and roof displacements as

$$\varepsilon_r = \frac{u_r}{L_u} \times 100 \qquad (45.4)$$

$$\varepsilon_w = \frac{u_w}{H_t} \times 100 \qquad (45.5)$$

The observed results for many caverns are shown in Figure 45.11. As noted from the figure, when the strain exceeds 0.2% it is likely to encounter some instability problems.

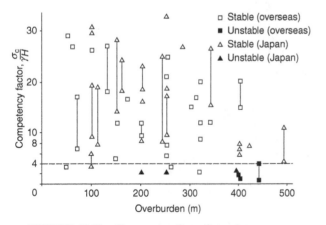

FIGURE 45.10 Competency factor for various caverns.

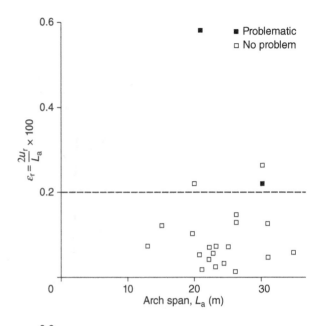

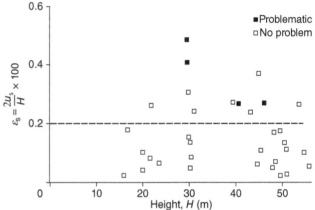

FIGURE 45.11 Measured roof and wall strains.

45.2.4 Excavation and Support Procedure

The excavation of an underground powerhouse is done using drilling and blasting technique (Photo 45.5). Figure 45.12 shows a general layout of excavation of large underground caverns. First an access drift, which is later used as a main service gallery, is excavated at the base of the powerhouse to ease the transportation of debris from the subsequent excavation steps of the powerhouse. Then a central drift in the arch section is excavated, which is followed by two side-drifts. A shaft is excavated to connect the arch to the access drift. And then, the remaining parts between the central and side-drifts are excavated and it is widened to the final configuration of the top arch section. The rest of the powerhouse is excavated in lifts with a height ranging between 4 and 6 m.

Support systems of caverns are usually based on the experiences of each country. The present conventional practice in Japan is to use bolts and shotcrete for the arch as a temporary support and to construct a concrete arch lining with a

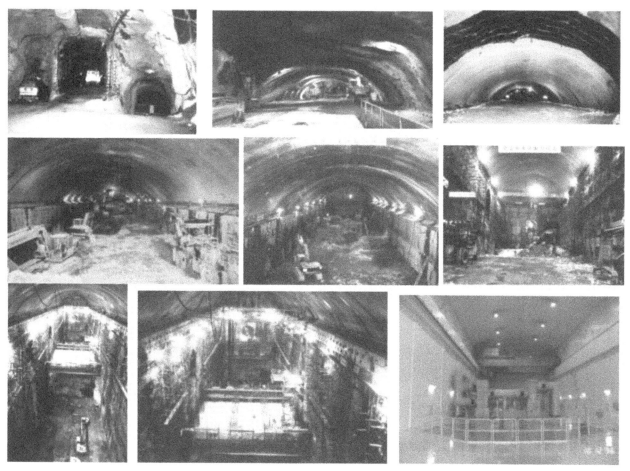

Photo 45.5   Views of different stages of excavation of Okumino Powerhouse. Courtesy of Chubu Electric Company.

thickness of 1000–1200 mm. As for the sidewalls, rock anchors 10–15 m long with a 400–600 kN capacity and rockbolts and shotcrete are used. In some countries, the concrete arch lining has completely disappeared and rock anchors are used instead. Figure 45.13 shows the present tendency in selecting the anchor and bolt lengths. Table 45.2 gives empirical relations between rock mass quality rate (RMQR) and the dimensions of support members normalized by arch span or sidewall height on the basis of data from large caverns built in Japan. It should be noted that these empirical relations can be used with confidence for RMQR greater than 45. For caverns with an arch span of 20–25 m and height of 40–45 m, Table 45.3 may serve for the empirical design of support in competent rock. It may also be used for primary support design when surrounding rock mass is subjected to even stress-induced yielding.

45.3   DESIGN PHILOSOPHY OF CAVERNS

The design of caverns is generally done using empirical methods such as rock classifications and/or with the help

of some theoretical methods and model tests before the computers have become available to geo-engineers. The present tendency is also to use numerical techniques such as finite element and boundary element methods for design purposes. As the rock mass always contains numerous discontinuities, there is no unified method of design of caverns. Nevertheless, the present philosophy of underground powerhouse design may be outlined as follows (Kawamoto et al., 1991a; Kawamoto and Aydan, 1999; Aydan and Kawamoto, 2001) (Figure 45.14):

Stage I—Geological and geophysical investigations and testing. This involves the investigations of geological structure of the site and geological structural defects such as faults, shear zones, and joints. Geophysical explorations are done for the purpose of characterizing the rock mass. The mechanical properties of intact rocks, rock mass, faults, joints, and seismic wave velocities are measured by means of laboratory and in situ tests. In situ stress state is evaluated through measurement and/or inference techniques.

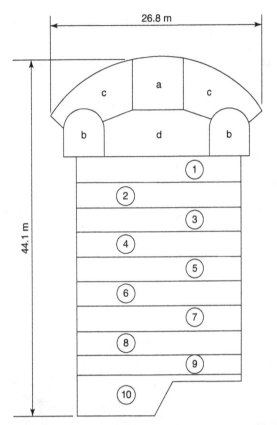

FIGURE 45.12 Excavation stages of Okumino Powerhouse (from Tsuchiyama et al., 1993). *Note:* For order of excavation procedure: a: central adit; b: side drifts; c: upper core; d: lower core 1,2,3,4... 9 stands for the excavation lifts from top to bottom.

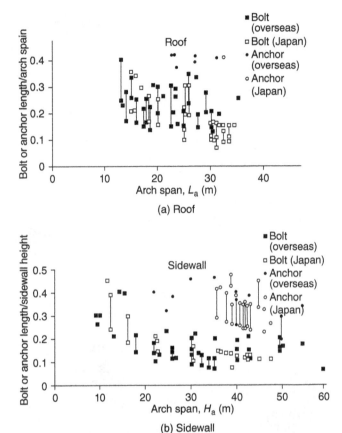

FIGURE 45.13 Ratios of anchor and bolt length to cavern dimensions.

Stage II—General evaluation. A general evaluation of properties and structure of rock mass is carried out on the basis of geological and geophysical investigations, experiments, and in situ testing with the consideration of past case histories and experiences.

Stage III—Global and local stability assessment. Possibility of global and local instability is checked on the bases of rock mass classifications, kinematic stability models, and simplified numerical analysis (i.e., Cording, 1972; Barton et al., 1974; Aydan, 1989; Bieniawski, 1989; Kawamoto et al., 1991b; Aydan et al., 2014;

TABLE 45.2 Empirical relations between rock mass quality rate (RMQR) and the dimensions of support members normalized by arch span or sidewall height.

Support member		Roof arch	Sidewall
RockAnchors	Length	$\dfrac{L}{L_a} = 0.8 - \dfrac{RMQR}{200}$	$\dfrac{L}{H_s} = 0.7 - \dfrac{RMQR}{200}$
	Spacing	$e_{av} = 2 + 0.02RMQR$	$e_{av} = 2 + 0.03RMQR$
Rockbolts	Length	$\dfrac{L_b}{L_a} = 0.35 - \dfrac{RMQR}{500}$	$\dfrac{L_b}{H_s} = 0.30 - \dfrac{RMQR}{500}$
	Spacing	$e_{bv} = 1 + 0.015RMQR$	$e_{bv} = 1 + 0.015RMQR$
Shotcrete	Thickness	$\dfrac{t^{sr}}{L_a} = 0.0125 - \dfrac{RMQR}{10,000}$	$\dfrac{t^{sw}}{H_s} = 0.0075 - \dfrac{RMQR}{18,000}$
Concrete liner	Thickness	1000	None

TABLE 45.3   Support systems for roof in competent discontinuous rock mass ($S = 20$–25 m; $H = 40$–45 m).

	Bolts				Anchors				Shotcrete	
	Roof		Sidewall		Roof		Sidewall		Roof	Sidewall
RMQR range	L (m)	e (m)	L (m)	e (m)	L (m)	e (m)	L (m)	e (m)	t (mm)	t (mm)
$100 \geq$ RMQR > 95	–	–	–	–	–	–	–	–	–	–
$95 \geq$ RMQR > 80	3	2.5	5	3.0	8	4.0	10	5.0	100	80
$80 \geq$ RMQR > 60	4	2.2	6	2.7	10	3.7	12	4.3	150	120
$60 \geq$ RMQR > 40	5	1.9	7	2.3	12	3.3	15	3.6	200	150

L, length; e, spacing; t, thickness; bolt, 200 kN; anchor, 400 kN; UCS of shotcrete, 10 MPa; S, span (width); H, height).

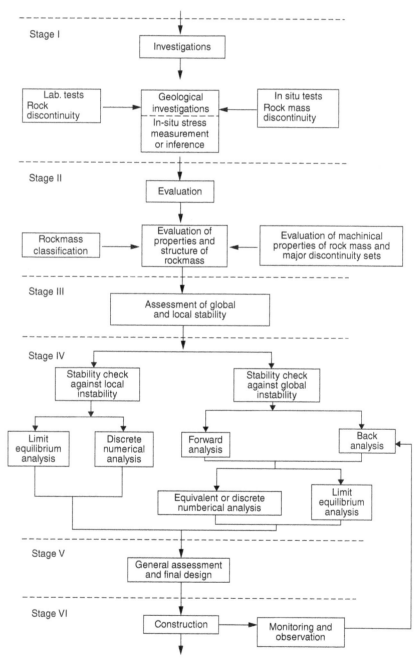

FIGURE 45.14   Flow chart of design and construction of an underground powerhouse (modified and translated from Kawamoto et al., 1991a).

Aydan and Ulusay, 2014). Preliminary support system is designed. The design of support systems is not well established and there are various approaches:

a. The first approach is based on restraining the development of plastic zone or assuming that it acts as a dead weight on the support members and they should be carried by the members or transferred to the elastic zone. The plastic zone is calculated by the closed-form solutions or numerical analysis.

b. The second approach is based on defining the maximum size of potentially unstable zone on the basis of discontinuity surveying and making the support members either to suspend that zone to the stable zone in the roof or to prevent its sliding or toppling into the opening. The block theory also belongs to this group.

c. Third approach is based on the consideration and modeling of support members in numerical or theoretical analysis. Nevertheless, this type of analysis is rarely applied.

Stage IV—Detailed stability analysis against global and local instability modes. First, rock mass is generally modeled by three models for both global and local instability modes:

a. *Equivalent mass approach.* The rock mass is assumed to have mechanical properties which are a fraction of intact rocks (Singh, 1973; Aydan and Kawamoto, 2000, 2001; Aydan et al., 2014). This reduction in the properties is done by either measuring the mechanical properties directly or using rock mass classifications or squared ratio of the elastic wave velocity of the rock mass to that of intact rock.

b. *Semi-explicit continuum models.* The effect of discontinuities of finite length is considered a damage in the body and modeled by some tensorial methods (i.e., fabric, damage tensors) to consider the effect of discontinuities within the frame-work of continuum mechanics (Kawamoto et al., 1988; Oda et al., 1993).

c. *Explicit approaches.* Intact rocks and discontinuities are modeled individually (Goodman et al., 1968; Cundall, 1971; Shi, 1988; Aydan et al., 1996).

Then the methods of analysis of the stability of underground caverns given below are chosen:

(a) *Closed-form methods.* Although closed-form solutions are appropriate for some simple geometry and material behaviors, they are often used in the preliminary design of structures since the error in modeling the rock mass is much larger than the error caused by the difference between the theoretical model and actual structure (Aydan, 1989; Aydan and Genis, 2010).

(b) *Numerical methods.* Once the mechanical model for rock mass is chosen and the constitutive relations are established, it is a simple matter of calculations. The general tendency is to use the elasto-plastic type constitutive relations in analyses (Kawamoto and Aydan, 1999). Nevertheless, the general tendency is to restrict the calculations to elastic case only as the nonlinear models cost more. The numerical models based on the explicit models of rock mass are less used because of the difficulties in presenting the discontinuous nature of rock mass and the huge number of input data efforts.

(c) *Limiting equilibrium approaches.* The limiting equilibrium approaches are generally restricted to investigate the local instability of underground openings. Block theory and other methods are available for this purpose (Shi, 1988; Kawamoto et al., 1991b; Aydan and Kawamoto, 2001; Aydan and Tokashiki, 2011).

With the advance of computers, the design procedures for caverns also utilize computational techniques such as FEM or BEM in recent decades. Nevertheless, the specification of the material properties for the surrounding mass is the most difficult aspect of such designs as the existence of discontinuities in rock mass presents considerable restrictions on the mechanical models used for the rock mass. Most of the techniques model the mass as an equivalent continuum with reduced mechanical properties in order to check the global stability of caverns. Because of the limitation of the equivalent continuum modeling, additional stability analyses are carried out to check the stability of caverns against some modes of instabilities of local kind, which are due to discontinuities.

Stage V—General assessment and final design. On the basis of past experiences, rock classifications, and preliminary and detailed analyses of the stability of the cavern and support system, a general assessment is performed and the final design is decided. Furthermore, a detailed planning of excavation, construction, and monitoring procedures is laid out. In addition, some criteria for monitored response of surrounding rock mass in conjunction with excavation stages are established.

Stage VI—Construction and back analyses. The construction of the cavern is carried out according to the final design. However, the safety and efficiency of the construction are checked from time to time in view of monitoring results and overall advance rate of excavation. If monitored results are somewhat different from the anticipated results, back analyses are carried out and the final design is checked or modified if necessary.

The construction procedure of underground powerhouses is mainly carried out according to the procedure described in

Section 45.2.4. It is almost a common procedure to monitor the behavior and response of surrounding rock mass during excavation. The monitoring schemes generally involve the continuous measurement of displacement of surrounding rock mass and acoustic emissions. Viewing boreholes next to boreholes in which extensometers are installed was recently carried out (Uchida et al. 1993). Additionally, seismic tomography, borehole deformability tests, and permeability test are performed on surrounding rock mass to assess its material property changes. Back-analysis technique proposed by Sakurai and his coworkers (Sakurai, 1993) has also become a common procedure to assess the response of surrounding rock mass to excavation steps.

45.4 CAVERN DESIGN LOADS AND DIMENSIONING OF SUPPORT MEMBERS

Initially, Cording et al. (1972), Barton et al. (1974), Bieniawski (1989), and later the author and his coworkers (Kawamoto et al., 1991a,b; Aydan and Kawamoto, 1992) compiled a number of case histories of cavern excavations all over the world and plotted the results as shown in Figures 45.15 and 45.16 for roof and sidewalls. In the same figures, the cavern excavations compiled by the authors are also plotted. In plotting the data of caverns in Japan, the support

effect of the arch concrete lining is excluded. In estimating the support pressure, the following formulas were used:
Rock anchors

$$p_i^a = \frac{T_y^a}{e_t \times e_l} \tag{45.6}$$

where T_y^a is the yield load of anchor and e_t and e_l are the traverse and longitudinal spacings, respectively.
Rockbolts

$$p_i^b = \frac{T_y^b}{e_t \times e_l} \tag{45.7}$$

Shotcrete

$$p_i^{sr} = \sigma_c^s \frac{t^s}{R} \text{ for roof} \tag{45.8}$$

$$p_i^{sw} = \sigma_c^s \frac{t^s}{H_l} \text{ for sidewall} \tag{45.9}$$

where σ_c^s is the uniaxial compressive strength of shotcrete; and t^s, R, and H_l are the thickness, radius of arch, and height of the sidewall, respectively. Total lengths of rockbolts and anchors are generally obtained by assuming such that they are anchored into the stable zones.

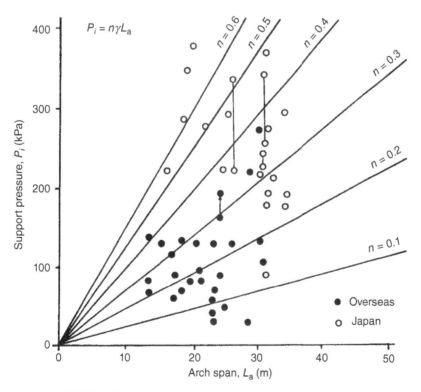

FIGURE 45.15 Support pressures used for arches of caverns.

45.5   METHODS FOR CALCULATING THE DESIGN SUPPORT LOADS AND COMPARISONS AGAINST LOCAL INSTABILITY MODES

Local instabilities about underground openings in a layered or blocky rock mass generally involve block falls, block slides, flexural or block toppling, and combined block toppling and slides. For the sake of simplicity, falls in the roof, sliding in sidewalls (Figure 45.16), and flexural toppling in the roof and sidewalls are only considered and procedures to determine rock loads to be carried by support members in relation to idealized discontinuity patterns are described. More complicated cases can be found in Aydan (1989) and Kawamoto et al. (1991a,b). Though discontinuity patterns in rock mass are closely associated with rock type, the patterns in a blocky rock mass can be generally classified (Aydan et al., 1989) as cross-continuous pattern and intermittent pattern. The intermittent pattern actually represents the most likely pattern in all kinds of rock. If an intermittency parameter introduced by Aydan et al. (1989) is used, it is seen that the cross-continuous pattern is a special case of the intermittent pattern.

First, how to determine the boundaries of the potentially unstable region in relation with the spatial distribution of discontinuity sets and the opening geometry is described. Though the geometry differs from a circular shape to a horseshoe shape, the shape of the opening is assumed to have vertical sidewalls with a circular-arched roof as shown in Figure 45.17a. Next, we assume the problem to be a two-dimensional one and calculate the maximum size of potentially unstable regions, although there are block theories which take into account the three-dimensional nature of the problem. This assumption is justified as far as the safety and the speed of calculations are concerned. Nevertheless, it may sometimes be necessary to carry out more detailed calculations when the cost of support elements is found to be quite high.

45.5.1   Estimation of Suspension Loads

Suspension loads arise when any frictional resistance on the critical bounding planes cannot be mobilized during movements of unstable region toward the opening. Practically when the apex angle of the unstable region is greater than 90° and there is no possibility of this angle becoming less than 90° during movements because asperities may exist on the critical bounding planes, the load due to dead weight of the unstable region is called as the suspension load.

Let us consider that the unstable region is defined by two planes α_1 and α_2 as shown in Figure 45.16a. The area of the potentially unstable region can be obtained using the geometry as

$$A^{\mathrm{r}} = \frac{L_{\mathrm{a}}}{2}\left[L_{\mathrm{a}}\frac{\tan\alpha_1\tan\alpha_2^*}{\tan\alpha_1+\tan\alpha_2^*} - R\left\{2\theta\frac{R}{L_{\mathrm{a}}} - \cos\theta\right\}\right]$$

(45.10)

where L_{a} is the width and R, θ are the radius and angle of arch. $\alpha_2^* = \alpha_2 - \xi$.

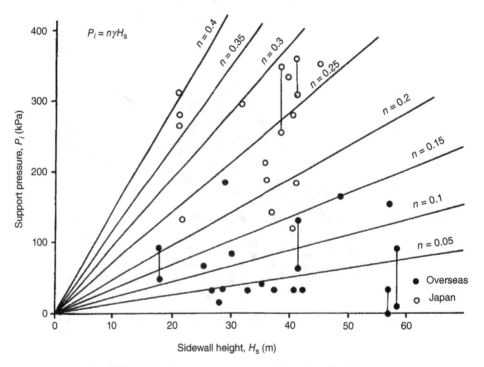

FIGURE 45.16   Support pressures used for sidewalls of caverns.

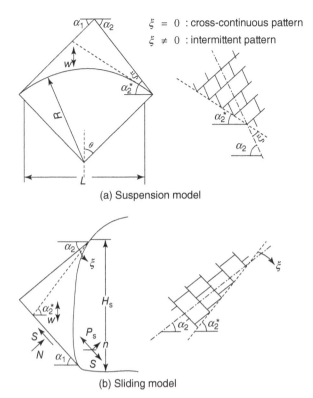

(a) Suspension model

(b) Sliding model

FIGURE 45.17 Computational models for support pressures for suspension and sliding-induced loads.

For a given thickness t, the suspension load can be written in the following form:

$$F_{\text{sus}} = \gamma A^{\text{r}} t \qquad (45.11)$$

Figure 45.18 shows the plotted results for the required support pressure against rock falls for various intersection angles of discontinuity sets. Since the natural discontinuity intersection angle lies between $60°$ and $120°$, the presently utilized arch support pressures provided by rockbolts and anchors for the roof caverns fall into the calculated range. In other words, if the supports of roofs of caverns are designed against the suspension load it will be sufficient and safe, provided that intact rock about the cavern does not yield.

45.5.2 Sliding Loads

Sliding failure is possible when the frictional resistance of the critically orientated discontinuity set satisfies the following condition:

$$\tan \alpha_i < \tan \phi_i \qquad (45.12)$$

where subscript i represents the discontinuity set on which sliding is likely. The sliding load can be determined

from the limiting equilibrium approach in the following form:

$$F_{\text{slid}} = \gamma S_{\text{slid}} t \frac{\sin(\alpha_i - \varphi_i)}{\cos \varphi_i} \qquad (45.13)$$

The area S_{slid} of the sliding region can be easily determined from the geometry of the region prone to sliding in relation to the geometry of opening. For example, the area of sliding body shown in Figure 45.17b can be specifically obtained as follows:

$$S_{\text{slid}} = \frac{H_s}{2} \left[H_s \frac{\tan \alpha_1 \tan \alpha_2^*}{\tan \alpha_1 + \tan \alpha_2^*} \right] \qquad (45.14)$$

Figure 45.19 shows the plotted results for the required support pressure against sliding for various intersection angles of discontinuity sets. Since the natural discontinuity intersection angle lies between $60°$ and $120°$, the presently used support pressures provided by rockbolts and/or anchors for the sidewall of caverns fall into the calculated range. In other words, if the support design of sidewalls of caverns is designed against the sliding load it will be sufficient and safe, provided that intact rock about the cavern does not become plastic.

The above formulations are based on the static loads. If dynamic loads resulting from earthquakes or turbines are present, for example, the method suggested by Aydan et al. (2012) can be utilized.

45.5.3 Loads due to Flexural Toppling

When the rock mass is thinly layered, the layers may not be strong enough to resist the tensile stresses due to bending under gravitational forces. Seismic forces may also induce additional loads easing flexural toppling failure. In such cases, tensile stresses in layers should be reduced below their tensile strength, if the stability is required. If the roof consists of n layers (Figure 45.20), the outer fiber stresses for layers in the sidewall and in the roof take the following forms (Aydan, 1989; Aydan and Kawamoto, 1992):

$$\sigma_t^i = \pm \frac{N_i}{A_i} + \frac{6t_i}{I_i} \left[P_{i+1} \eta h_i - T_{i+1} \frac{t_i}{2} - P_{i-1} \eta h_{i-1} \right. $$
$$\left. - T_{i-1} \frac{t_i}{2} + S_i \right] \qquad (45.15)$$

where $N_i = W_i \cos \alpha - E_i \sin(\alpha + \beta) S_i = W_i \sin \alpha + E_i \cos (\alpha + \beta)$; $W_i = \gamma_i t_i b(h_{i-1} + h_i)/2 A_i = t_i b$; γ_i is the unit weight of layer; h_i and h_{i+1} are the side lengths of the interfaces between layers $i-1, i$ and layers $i, i+1$, respectively; P_{i-1}, P_{i+1} and T_{i-1}, T_{i+1} are the normal and shear forces acting on the interfaces between layers $i-1, i$ and layers $i, i+1$, respectively; t_i is the thickness of layer i; b is the width; α is the layer inclination; and η is the coefficient of

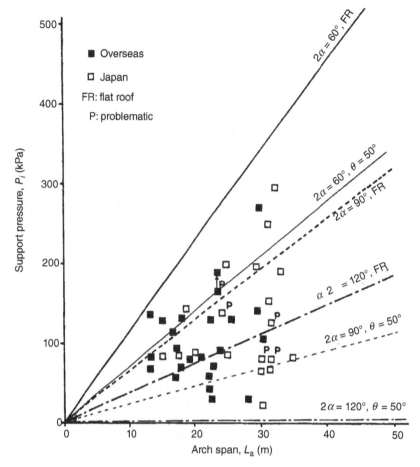

FIGURE 45.18   Support pressures for roofs and comparison with actual examples.

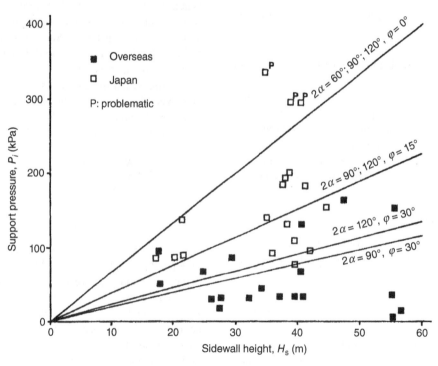

FIGURE 45.19   Support pressures for sidewalls and comparison with actual examples.

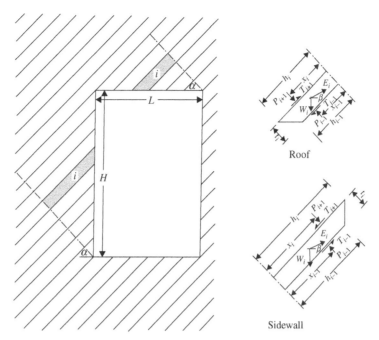

Roof

Sidewall

FIGURE 45.20 Model for limiting equilibrium analysis of flexural toppling of an underground opening (from Aydan and Kawamoto, 1992).

load action location. Sign (+) stands for layers in roof and (−) for layers in sidewalls.

Introducing the yield condition such that the outer fiber stress of the layer is equal to the tensile strength σ_T of the rock with a factor of safety SF as

$$\sigma_t^i \le \frac{\sigma_T}{\text{SF}} \tag{45.16}$$

and assuming the normal and shear forces acting on the interfaces of layers through the frictional yielding condition (ϕ is the friction angle)

$$T_{i+1} = P_{i+1}\mu; \quad T_{i-1} = P_{i-1}\mu; \quad \mu = \tan\phi \tag{45.17}$$

the normal forces acting on layer $i-1$ can be easily obtained as

$$P_{i-1} = \frac{P_{i+1}\left(\eta h_i - \mu\frac{t_i}{2}\right) + S_i\frac{h_i}{2} - \frac{2I_i}{t_i}\left(\frac{\sigma_T}{\text{SF}} \pm \frac{N_i}{A_i}\right)}{\left(\eta h_{i-1} + \mu\frac{t_i}{2}\right)} \tag{45.18}$$

The equation above is solved by "step-by-step method" and rock load P_s is obtained from the following criterion:

$$P_0 > 0 \text{ and } P_0 = P_s \tag{45.19}$$

If $P_0 > 0$, it is interpreted that some support measures are necessary.

The stabilization of underground openings against flexural toppling failure requires reduction in the magnitude of the moment and an increase in the compressive normal forces acting on interfaces. There are a number of ways to provide such an effect through the use of artificial support. Prestressed cables and/or fully grouted rockbolts are effective solutions. When the prestressed cables are used, they should be anchored beyond the basal plane, otherwise prestress forces may cause much higher bending stresses in layers. The alternative is to use fully grouted rockbolts or "dowels." The fully grouted rockbolts would be more economical than rock anchors as they are shorter and do not need to be prestressed. In this subsection, a procedure proposed by Aydan (Aydan, 1989; Aydan and Kawamoto, 1992) is introduced as to how to consider the reinforcement effect of the fully grouted rockbolts against flexural toppling failure.

Fully grouted rockbolts contribute to the shear resistance of discontinuities directly through the shear resistance of steel bar itself and indirectly through the shear and frictional components of the axial force in the bar. The reinforcing effect of n rockbolts on the shear resistance of a discontinuity plane can be expressed in the following form (Aydan, 1989) (Figure 45.21):

$$T_T = nA_b\sigma_b\left(1 + \frac{1}{2}\sin 2\theta \tan\varphi\right) \tag{45.20}$$

where σ_b is the axial stress; A_b is the cross section of the bar; and θ is the angle between the bar axis and discontinuity (installation angle).

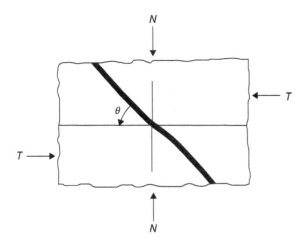

FIGURE 45.21   A model for rockbolt crossing a discontinuity (modified from Aydan and Kawamoto, 1992).

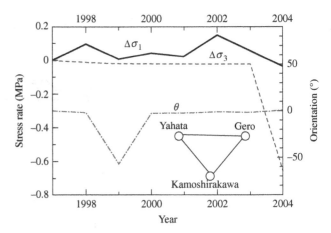

FIGURE 45.22   Stress changes estimated from crustal deformations obtained from GPS in the close vicinity of Mazegawa underground powerhouse (inlet shows the positions of GPS stations with the location of Mazegawa underground powerhouse).

The results of laboratory and field studies and numerical modeling show that the stress in the bar quickly develops and becomes almost equal to its yielding strength at very small displacements in the order of 0.1–3 mm. It is reasonable, therefore, to assume that the stresses in bars are same and equal to their yield strength. The yielding strength of a bar subjected to tension or compression and shearing in equation (13) may be given as

$$\sigma_b = \frac{\sigma_{tb}}{\sqrt{\cos^2\theta + 3\sin^2\theta}} \tag{45.21}$$

where σ_{tb} is the tensile strength of rockbolt.

Considering the contribution of bolts as an addition to the shear resistance across interfaces of layers, one obtains the following:

$$P_{i-1} = $$
$$\frac{P_{i+1}\left(\eta h_i - \mu\frac{t_i}{2}\right) + \left(T_T^{i+1} + T_T^{i-1}\right)\frac{t_i}{2} + S_i\frac{h_i}{2} - \frac{2I_i}{t_i}\left(\frac{\sigma_T}{SF} \pm \frac{N_i}{A_i}\right)}{\left(\eta h_{i-1} + \mu\frac{t_i}{2}\right)}$$
$$\tag{45.22}$$

where T_T^{i-1} and T_T^{i+1} are the resistances provided by rockbolts at the interfaces between layers $i-1, i$ and layers $i, i+1$, respectively.

45.6   SOME LONG-TERM ISSUES

45.6.1   In Situ Stress Variations

The far-field in situ stress state in rock mass around the underground powerhouses may change with time. The earthquakes and crustal deformations may reflect these changes.

However, the most difficult problem is how to assess such changes with time. One possible option to determine the stress field changes could be the use of the GPS technique (Aydan, 2000a, 2004). The other option may be the stress inferences from the focal plane solutions using some methods such as Aydan's method (Aydan, 2000b, 2013). Such methods were applied to an underground powerhouse in Japan (Aydan et al., 2008, 2012). The increments of principal stresses were estimated from the GPS technique in the order of 0.01–0.6 MPa/year (Figure 45.22). The focal plane solutions implied that the lateral stress coefficient might change from 1.0 to 1.5 in the cross section of the cavern (Table 45.4).

Aydan et al. (2012) showed that the changes in in situ stress state induced changes in the sidewall deformation of the cavern and the stresses around the cavern. Furthermore, they noted that the axial stresses of rockbolts and rock anchors would be expected to change when the lateral stress coefficient changes.

45.6.2   Degradation and Creep Properties of Surrounding Rock Mass

It is expected that the deformability and strength of every geomaterial have some time-dependent characteristics. This time dependency may be recoverable or irrecoverable and they may be called as viscoelastic or viscoelasto-plastic (i.e., Aydan et al., 1995; Aydan and Kawamoto, 1997; Aydan and Nawrocki, 1998). When rock mass exhibits creep behavior, deformations would become larger and plastic zones may develop in rock mass around the cavern. In such circumstances, elasto-visco-plastic numerical analyses would be necessary.

Aydan et al. (2012) utilized an elasto-visco-plastic finite element technique for analyzing the performance of

TABLE 45.4   Measured and inferred in situ stresses (from Aydan et al., 2012).

Stress inference	Measurement	Horizontal stress (MPa)	Vertical stress (MPa)	Lateral stress coefficient
Overcoring method	Type 1	0.981	0.981	1.000
	Type 2	1.638	1.638	1.000
Focal mechanism of 1992 Eq.M4.2		1.818	1.275	1.426
Focal mechanism of 2002 Eq.M4.2		2.539	1.275	1.991

$H = 49\,\text{m}; \gamma = 26\,\text{kN/m}^3$.

underground powerhouse in the long term. In their numerical analyses, the parameters of the constitutive law were a function of plastic straining and/or time. The cohesion (c) of rock mass was assumed to degrade with time while the friction angle was kept constant by considering experimental results on the igneous rocks (Aydan and Nawrocki, 1998). They found that the deformation of the cavern became larger and a plastic zone developed in the vicinity of the sidewall, which was not observed during the initial excavation step in short-term analyses.

45.6.3   Degradation and Creep Properties of Support Members

The degradation of support system causes the increase of the displacement field around the cavern (Aydan et al., 2012). However, it should be noted that these types of continuum analyses are valid for global stability of the caverns. The effect of preexisting discontinuities must be taken into account for the cavern stability even though the continuum-type analyses may not imply any stability problems.

The stability analyses for potentially unstable blocks in the close vicinity of the cavern imply that the axial forces may increase during their service life (Figure 45.23) when forces result from vibrations in the vicinity of cavern and/or earthquakes from time to time (Aydan et al., 2008, 2012). These increases may also lead to the rupture of rock anchors in the long term besides the reduction of cross sections of rock anchors due to corrosion.

45.6.4   Effect of Cyclic Water Head and Temperature Variations

Aydan et al. (2008, 2012) carried out a series of coupled finite element analyses to investigate the effect of the underground water on groundwater table variation in the vicinity of rock mass around an underground powerhouse located 50 m below ground surface and adjacent to a rockfill dam. The underground powerhouse was associated with a pumped-water energy storage scheme. When the lining of underground powerhouse was impermeable the water head variation reached steady state after about 300 days and the variation of the

water head in rock mass due to the variation of water level of reservoirs was limited to 60–70 m from the intakes/outlets. There was almost no change on the water head in the vicinity of the powerhouse. When the cavern was unlined, the water head variations are much smaller and they are limited to 60–70 m from the intakes/outlets. However, the steady state was achieved after about 400 days.

The computational results by Aydan et al. (2012) showed that the rock mass below the upper reservoir rebounded while the rock mass below the lower reservoir subsided. However, when the upper reservoir was refilled, it tended to return to its original state while the rock mass below the lower reservoir rebounded. This process repeated itself as the water levels of reservoirs are cyclically varied. The large variations

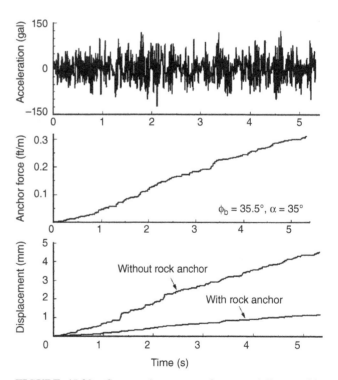

FIGURE 45.23 Computed response of a potentially unstable block in a sidewall and associated axial force development in a rockbolt induced by vibration of turbines (from Aydan et al., 2012).

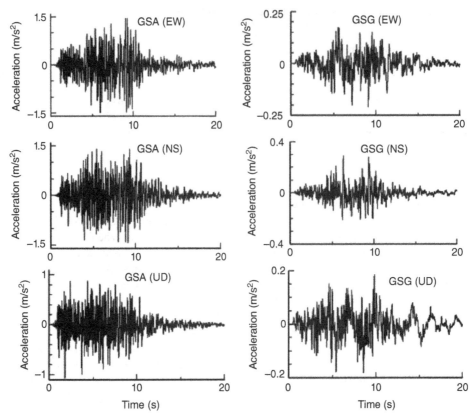

FIGURE 45.24 Acceleration records at GSA and GSG strong motion stations (from Aydan et al., 2009, strong motion records from Italian Strong Motion Network, RAN). *Note:* UD Stands for up and down motion and it is a conventional notation in earthquake engineering.

of water heads in rock mass occurred in the close vicinity of the reservoirs and their effect was quite negligible around the cavern. These results further imply that there was no need to take into account the effective stress changes around the caverns of the pumped storage schemes in the long term.

Aydan et al. (2008, 2012) performed the temperature and humidity measurements in the underground powerhouse at its entrance for several years. The temperatures were different in the cavern and the yearlong variation was within 8–10°. The daily variation ranged between 2° and 3°. However, the variation at the entrance was quite large and the yearlong variation was about 30–32° while the daily variation was around 8–12°. The flow of water through the turbines of the powerhouse also had some effects on the temperature field in the close vicinity of the turbines. Aydan et al. (2008, 2012) performed two-dimensional coupled thermomechanical analyses on an underground powerhouse. It was found from the analyses that variations of temperature in the cavern induced cyclic thermal stresses with a given amplitude in the rock mass adjacent to the cavern and the amplitude of these cycles disappeared when the distance from the cavern wall became three times the cavern radius. These thermal stresses may induce fatigue-type loading in rock mass in the long term.

45.6.5 Effect of Earthquakes

It is well known that the ground motions are generally smaller than those at ground surface. Figure 45.24 shows the acceleration records measured on the ground surface (GSA) and underground (GSG) during the 2009 Mw 6.3 L'Aquila earthquake (Aydan et al., 2009). The GSA station is at Assergi and the GSG station is located in an underground gallery of Gran Sasso Underground Physics Laboratory of Italy. Both stations are founded on Eocene limestone with a shear wave velocity of 1 km/s. Although the epicentral distances and ground conditions are almost the same, the acceleration at ground surface is amplified almost 6.4 times of that in the underground gallery.

It is well known that the underground structures such as tunnels and powerhouses are generally resistant against earthquake-induced motions. However, they may be damaged when permanent ground movements occur in/along the underground structures. The reports on damage to underground powerhouses so far are almost none. The damage by

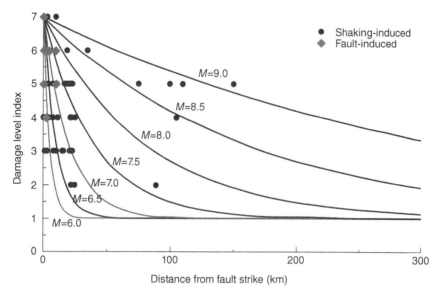

FIGURE 45.25   Relation between distance (Rf) from surface trace of the fault and damage level index (DLI) (from Aydan and Genis, 2012).

the 1999 Chi-chi earthquake in Taiwan to powerhouses in the hanging wall side of the wall was almost none. A slight damage to the Singkarak powerhouse in Sumatra Island of Indonesia by the M6.4 Singkarak earthquake was reported.

Aydan et al. (2010a, 2010b) compiled case histories and developed databases for three different categories of damages, namely, faulting induced (18 cases histories), shaking induced (105 case histories), and slope failure induced (50 cases histories). They also proposed a new classification of damage for underground openings, in which seven levels of damage are defined (Figure 45.25). As the ground motions are generally small at ordinary depths of underground caverns for powerhouses, the shaking-induced damage is thought to be quite limited and it may only induce some local instability modes. However, the effect of permanent ground deformation due to faulting may affect the underground powerhouses if it happens to be crossing the cavern. As the caverns are generally sited in rock mass with the consideration of skipping major faults and fracture zones during design stage, it is quite unlikely to have the underground powerhouses to be influenced by the permanent ground deformation of such major geological structural defects.

Owada and Aydan (2005) and Owada et al. (2004) carried out some model experiments on the development of axial forces in rock anchors and grouted rockbolts stabilizing the potentially unstable blocks in sidewalls of the underground openings using shaking tables. They measured displacement responses of a potentially unstable block in a sidewall and axial force development in a model rock anchor in relation to the applied base acceleration. The inclination of the discontinuity plane was 58° and the friction angle of the discontinuity plane involved in sliding was 24–26°. They experimentally

showed that the anchor force increased after each slip event in a step-like fashion and becomes asymptotic to a certain value thereafter (Figure 45.26). Furthermore, the block did not return to its original position. This implied that the axial force becomes higher than the applied prestress level following each slip event.

Aydan et al. (2012) applied the method proposed by Owada and Aydan (2005) to an underground powerhouse and their computational results also implied that the axial forces of rock anchors and rockbolts would increase during their service life when forces resulted from vibrations in the vicinity of cavern and/or earthquakes from time to time. They pointed out that the increases may lead to the rupture of rock anchors in the long term besides the reduction of cross sections of rock anchors due to corrosion.

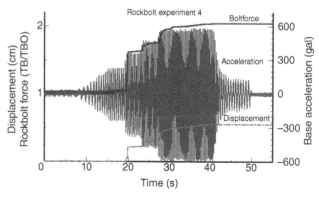

FIGURE 45.26   Responses of the relative displacement of block and the axial force of rock anchor normalized by its initial prestress value in relation to applied base acceleration in a model test.

ACKNOWLEDGMENTS

The author sincerely thanks Emeritus Professor T. Kawamoto of Nagoya University for encouragement and guidance and Chubu Electric Company (Nagoya) for very valuable and detailed information on the underground powerhouses of Mazegawa and Okumino Pumped Water Storage Schemes.

REFERENCES

Aydan, Ö. (1989). The stabilisation of rock engineering structures by rockbolts. Doctorate Thesis, Nagoya University, 204 pages.

Aydan, Ö. (1990). The arch formation effect of rockbolts (in Turkish). *Madencilik*, 28(3):33–40.

Aydan, Ö. (2000a). A stress inference method based on GPS measurements for the directions and rate of stresses in the earth' crust and their variation with time. *Yerbilimleri*, 22:21–32.

Aydan, Ö. (2000b). A stress inference method based on structural geological features for the full-stress components in the earth' crust. *Yerbilimleri*, 22:223–236.

Aydan, Ö. (2004). Implications of GPS-derived displacement, strain and stress rates on the 2003 Miyagi-Hokubu earthquakes. *Yerbilimleri*, No. 30:91–102.

Aydan, Ö. (2013). Inference of contemporary crustal stresses from recent large earthquakes and its comparison with other direct and indirect methods. *6th International Symposium on In Situ Rock Stress (RS2013)*, Sendai, Paper No. 1051, 8 p.

Aydan, Ö. (2014). Methods for In-situ Stress Measurements and its Applications to Turkey (in Turkish)., *Yer Mühendisliği*, No. 2, 46–51.

Aydan, Ö., Geniş, M. (2010). A unified analytical solution for stress and strain fields about a radially symmetric openings in elasto-plastic rock with the consideration of support system and long-term properties of surrounding rock. *International Journal of Mining and Mineral Processing*, 1(1):1–32.

Aydan, Ö., Geniş, M. (2012). Effects of earthquakes on underground structures including subsea tunnels. *Prof. Dr. Mahir Vardar Special Sessions on Geomechanics, Tunnelling, Design of Rock Constructions*, Istanbul, pp. 29–56.

Aydan, Ö., Kawamoto, T. (1987). Toppling failure of discontinuous rock slopes and their stabilization (in Japanese). *Journal of the Mining and Metallurgical Institute of Japan*, 103:763–770.

Aydan, Ö., Kawamoto, T. (1992). The flexural toppling failures in slopes and underground openings and their stabilisation. *Rock Mechanics and Rock Engineering*, 25(3):143–165.

Aydan, Ö., Kawamoto, T. (1997). The general characteristics of the stress state in various parts of the earth's crust. *Int. Symp. on Rock Stress*, Kumamoto, pp. 369–373.

Aydan, Ö., Kawamoto, T. (2000). The assessment of mechanical properties of rock masses through RMR rock classification system. *GeoEng2000, UW0926*, Melbourne.

Aydan, Ö., Kawamoto, T. (2001). The stability assessment of a large underground opening at great depth. *17th Int. Min. Congress and Exhibition of Turkey, IMCET 2001*, Ankara, Vol. 1, pp. 277–288.

Aydan, Ö., Nawrocki, P. (1998). Rate-dependent deformability and strength characteristics of rocks. *Int. Symp. on the Geotechnics of Hard Soils-Soft Rocks*, Napoli, Vol. 1, pp. 403–411.

Aydan, Ö., Paşamehmetoğlu, G. (1994). In-situ stress measurements and lateral stress coefficients in various parts of the earth-crust (in Turkish). *Bulletin of Rock Mechanics, Turkish National Group for Rock Mechanics, ISRM*, Ankara, 10:1–17.

Aydan, Ö., Tokashiki, N. (2011). A comparative study on the applicability of analytical stability assessment methods with numerical methods for shallow natural underground openings. *The 13th International Conference of the International Association for Computer Methods and Advances in Geomechanics*, Melbourne, Australia, pp. 964–969.

Aydan, Ö., Ulusay, R. (2014). Rock Mass Quality Rating (RMQR) System: Its Application to Estimation of Geomechanical Characteristics of Rock Masses and to Rock Support Selection for Underground Caverns and Tunnels. Proc. of the 8th Asian Rock Mechanics Symposium, Sapporo, 2075–2084.

Aydan, Ö., Shimizu, Y., Ichikawa, Y. (1989). The effective failure modes and stability of slopes in rock mass with two discontinuity sets. *Rock Mechanics and Rock Engineering*, 22:163–188.

Aydan, Ö., Tokashiki, N., Seiki, T., Ito, F. (1992). Deformability and strength of discontinuous rock masses. *Int. Conf. Fractured and Jointed Rock Masses*, Lake Tahoe, 1992, pp. 256–263.

Aydan, Ö., Akagi, T., Ito, T., Ito, J., Sato, J. (1995) Prediction of deformation behaviour of a tunnel in squeezing rock with time-dependent characteristics. *Numerical Models in Geomechanics, NUMOG V*, pp. 463–469.

Aydan, Ö., Mamaghani, I. H. P., Kawamoto, T. (1996). Application of discrete finite element method (DFEM) to rock engineering structures. *North American Rock Mechanics Symp.* Montreal, vol. 2, pp. :2039–2046.

Aydan, Ö., Tsuchiyama, S., Kinbara, T., Uehara, F., Tokashiki, N., Kawamoto, T. (2008). A numerical analysis of non-destructive tests for the maintenance and assessment of corrosion of rockbolts and rock anchors. *The 12th International Conference of International Association for Computer Methods and Advances in Geomechanics (IACMAG)*, Goa, India, pp. 40–45.

Aydan, Ö., Kumsar, H., Toprak, S., Barla, G. (2009). Characteristics of 2009 l'Aquila earthquake with an emphasis on earthquake prediction and geotechnical damage. *Journal Marine Science and Technology, Tokai University*, 9(3):23–51.

Aydan, Ö., Ohta, Y., Geniş, M., Tokashiki, N., Ohkubo, K. (2010a). Response and earthquake induced damage of underground structures in rock mass. *Journal of Rock Mechanics and Tunnelling Technology*, 16(1):19–45.

Aydan, Ö., Ohta, Y., Geniş, M., Tokashiki, N., Ohkubo, K. (2010b). Response and stability of underground structures in rock mass during earthquakes. *Rock Mechanics and Rock Engineering*, 43(6):857–875.

Aydan, Ö., Uehara, F., Kawamoto, T. (2012). Numerical study of the long-term performance of an underground powerhouse subjected to varying initial stress states, cyclic water heads, and temperature variations. *International Journal of Geomechanics, ASCE*, 12(1):14–26.

Barton, N. R., Lien, R., Lunde, J. (1974). Engineering classification of rock masses for the design of tunnel support. *Rock Mechanics and Rock Engineering*, 6(4):189–239.

Bieniawski, Z. T. (1989). *Engineering Rock Mass Classifications.* Wiley, New York.

Chubu Electric Power Company (1979). *Construction reports of First and Second Mazegawa Power Stations* (in Japanese), 657 pages, Chubu Electric Power Company, Nagoya.

Chubu Electric Power Company (1999). *Construction Reports of Okumino Power Houses* (in Japanese), 2 Volumes. Chubu Electric Power Company, Nagoya.

Cording, E. J., Hendron, A. J., Deere, D. U. (1972). Rock engineering for underground caverns. *ASCE Symposium on Underground Chambers*, pp. 567–600.

Cundall, P. A. (1971). A computer model for simulating progressive, large-scale movements in blocky rock systems. *Proc. Int. Symp. Rock Fracture*, Paper 11-8, Nancy, France.

Goodman, R. E., Taylor, R., Brekke, T. L. (1968). A model for the mechanics of jointed rock. *Journal of the Soil Mechanics and Foundation Engineering Division ASCE*, 94:637–659.

Kawamoto, T., Aydan, Ö. (1999). A review of numerical analysis of tunnels in discontinuous rock masses. *International Journal of Numerical and Analytical Methods in Geomechanics*, 23:1377–1391

Kawamoto, T., Ichikawa, Y., Kyoya, T. (1988). Deformation and fracturing behaviour of discontinuous rock mass and damage mechanics theory. *International Journal for Numerical and Analytical Methods in Geomechanics*, 12(1):1–30 .

Kawamoto, T., Aydan, Ö., Tsuchiyama, S. (1991a). Analysis and review of design for large underground cavern (in Japanese). *Tunnels and Underground*, 23(3):31–37.

Kawamoto, T., Aydan, Ö., Tsuchiyama, S. (1991b). A consideration on the local instability of large underground openings. *Int. Conf. GEOMECHANICS'91*, Hradec, pp. 33–41.

Oda, M., Yamabe, T., Ishizuka, Y., Kumasaka, H., Tada, H., Kimura, K. (1993). Elastic stress and strain in jointed rock masses by means of crack tensor analysis. *Rock Mechanics and Rock Engineering*, 26(2):89–112 .

Owada, Y., Aydan, Ö. (2005). Dynamic response of fully grouted rockbolts during shaking (in Japanese). *The School of Marine Science and Technology, Tokai University*, 3(9):18.

Owada, Y., Daido, M., Aydan, Ö. (2004). Mechanical response of rock anchors and rockbolts during earthquakes (in Japanese). *34th Japan Rock Mechanics Symposium*, pp. 519–524.

Sakurai, S. (1993). Back analysis in rock engineering. *Comprehensive Rock Engineering*, 4:543–569.

Shi, G. H. (1988). Discontinuous deformation analysis: a new numerical model for the statics and dynamics of block systems. Ph.D. Thesis, Department of Civil Engineering, University of California, Berkeley, 1988, 378p.

Singh, B. (1973). Continuum characterization of jointed rock masses. Part I: the constitutive equations, *International Journal of Rock Mechanics and Mining Science*, 10:311–335.

Terzaghi, K., Richart, F. E. (1952). Stresses in rock about cavities. *Geotechnique*, 3:57–90.

Tsuchiyama, S., Aydan, Ö., Ichikawa, Y. (1993). Deformational behaviour of a large underground opening and its back analysis. *International Symposium on Assessment and Prevention of Failure Phenomena in Rock Engineering*, Istanbul, pp. 865–870.

Uchida, Y., Harada, T., Urayama, M. (1993). Behavior of discontinuous rock during large underground cavern excavation, Proceeding of the International Symposium on Assessment and Prevention of Failure Phenomena in Rock Engineering, Istanbul, Turkey, 807–816.

46

HYDROELECTRIC POWER FOR THE NATION

PUBLIC DOMAIN

Hydroelectric power must be one of the oldest methods of producing power. No doubt, Jack the Caveman stuck some sturdy leaves on a pole and put it in a moving stream. The water would spin the pole that crushed grain to make their delicious, low-fat prehistoric bran muffins. People have used moving water to help them in their work throughout history, and modern people make great use of moving water to produce electricity.

46.1 HYDROELECTRIC POWER FOR THE NATION

Although most energy in the United States is produced by fossil fuel and nuclear power plants, hydroelectricity is still important to the nation, as about 7% of the total power is produced by hydroelectric plants. Nowadays, huge power generators are placed inside dams. Water flowing through the dams spin turbine blades (made out of metal instead of leaves), which are connected to generators. Power is produced and is sent to homes and businesses (Figure 46.1).

46.2 WORLD DISTRIBUTION OF HYDROPOWER

- Hydropower is the most important and widely used renewable source of energy.
- Hydropower represents 19% of total electricity production.

- China is the largest producer of hydroelectricity, followed by Canada, Brazil, and the United States (*Source*: Energy Information Administration).
- Approximately two-thirds of the economically feasible potential remains to be developed. Untapped hydro resources are still abundant in Latin America, Central Africa, India, and China.

Producing electricity using hydroelectric power has some advantages over other power-producing methods. Let us do a quick comparison.

46.2.1 Advantages of Hydroelectric Power

- Fuel is not burned, so there is minimal pollution.
- Water to run the power plant is provided free by nature.
- Hydropower plays a major role in reducing greenhouse gas emissions.
- It involves relatively low operations and maintenance costs.
- The technology is reliable and proven over time.
- It is renewable—rainfall renews the water in the reservoir, so the fuel is almost always there.

Read an expanded list of advantages of hydroelectric power from the Top World Conference on Sustainable Development, Johannesburg, South Africa (2002).

Alternative Energy and Shale Gas Encyclopedia, First Edition. Edited by Jay H. Lehr and Jack Keeley.
© 2016 John Wiley & Sons, Inc. Published 2016 by John Wiley & Sons, Inc.

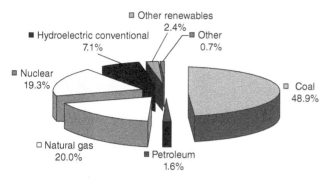

FIGURE 46.1 Sources of electricity in the United States, 2006. Data from http://www.eia.doe.gov/cneaf/electricity/epa/epat1p1.html.

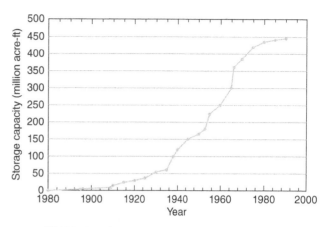

FIGURE 46.2 Construction of surface reservoirs.

46.2.2 Disadvantages of Power Plants That Use Coal, Oil, and Gas Fuel

- They use up valuable and limited natural resources.
- They can produce a lot of pollution.
- Companies have to dig up the earth or drill wells to get the coal, oil, and gas.
- For nuclear power plants there are waste-disposal problems.

46.2.3 Hydroelectric Power Is Not Perfect, Though, and Does Have Some Disadvantages

- High investment costs
- Hydrology dependent (precipitation)
- In some cases, inundation of land and wildlife habitat
- In some cases, loss or modification of fish habitat
- Fish entrainment or passage restriction
- In some cases, changes in reservoir and stream water quality
- In some cases, displacement of local populations

46.3 HYDROPOWER AND THE ENVIRONMENT

46.3.1 Hydropower Is Nonpolluting, But Does Have Environmental Impacts

Hydropower does not pollute the water or the air. However, hydropower facilities can have large environmental impacts by changing the environment and affecting land use, homes, and natural habitats in the dam area.

Most hydroelectric power plants have a dam and a reservoir. These structures may obstruct fish migration and affect their populations. Operating a hydroelectric power plant may also change the water temperature and the river's flow. These changes may harm native plants and animals in the river

and on land. Reservoirs may cover people's homes, important natural areas, agricultural land, and archeological sites. So building dams can require relocating people. Methane, a strong greenhouse gas, may also form in some reservoirs and be emitted to the atmosphere (EPA Energy Kids).

46.3.2 Reservoir Construction Is "Drying Up" in the United States

Gosh, hydroelectric power sounds great—so why do we not use it to produce all of our power? Mainly because you need lots of water and a lot of land where you can build a dam and reservoir, which all takes a LOT of money, time, and construction. In fact, most of the good spots to locate hydro plants have already been taken. In the early part of the century hydroelectric plants supplied a bit less than one-half of the nation's power, but the number is down to about 10% today. The trend for the future will probably be to build small-scale hydro plants that can generate electricity for a single community.

As Figure 46.2 shows, the construction of surface reservoirs has slowed considerably in recent years. In the middle of the 20th century, when urbanization was occurring at a rapid rate, many reservoirs were constructed to serve people's rising demand for water and power. Since about 1980, the rate of reservoir construction has slowed considerably.

46.3.3 Typical Hydroelectric Power Plant

Hydroelectric energy is produced by the force of falling water. The capacity to produce this energy is dependent on both the available flow and the height from which it falls. Building up behind a high dam, water accumulates potential energy. This is transformed into mechanical energy when the water rushes down the sluice and

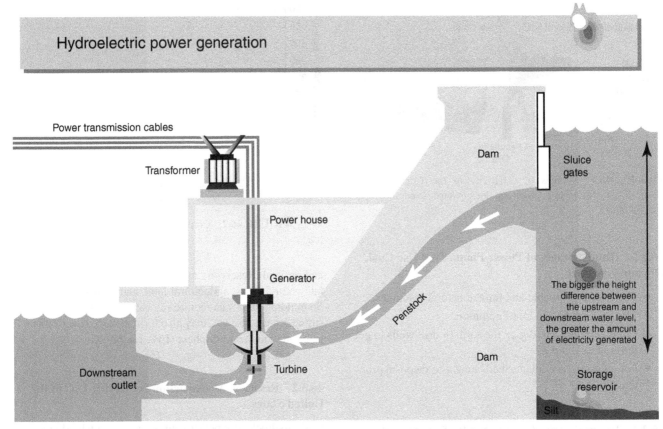

FIGURE 46.3   Hydroelectric power generation. *Source*: Environment Canada.

strikes the rotary blades of the turbine. The turbine's rotation spins electromagnets which generate current in stationary coils of wire. Finally, the current is put through a transformer where the voltage is increased for long-distance transmission over power lines (*Source*: Environment Canada; Figure 46.3).

46.3.4   Hydroelectric Power Production in the United States and the World

As Figure 46.4 shows, in the United States, most states make some use of hydroelectric power, although, as you can expect, states with low topographical relief, such as Florida and Kansas, produce very little hydroelectric power. But some

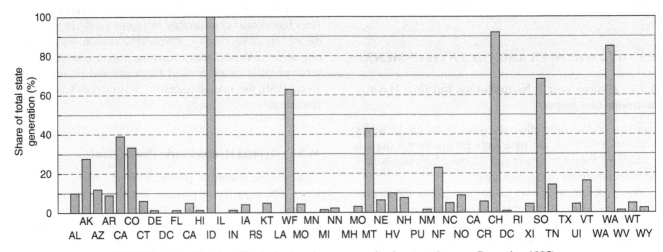

FIGURE 46.4   Electric utility hydroelectric net generation by state (January–December 1995).

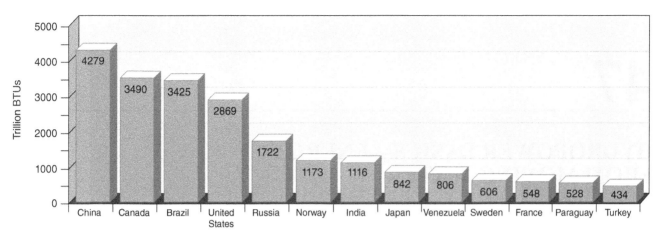

FIGURE 46.5   World net hydroelectric power generation (trillion BTU), 2006. *Source*: Energy Information Administration (EIA). http://www.eia.doe.gov/emeu/international/electricitygeneration .html.

states such as Idaho, Washington, and Oregon use hydro-electricity as their main power source. In 1995, all of Idaho's power came from hydroelectric plants.

Figure 46.5 shows hydroelectric power generation in 2006 for the leading hydroelectric-generating countries in the world. China has developed large hydroelectric facilities in the last decade and now leads the world in hydroelectricity usage. But, from north to south and from east to west, coun-tries all over the world make use of hydroelectricity—the main ingredients are a large river and a drop in elevation (along with money, of course).

47

HYDROPOWER BASICS—ENERGY FROM MOVING WATER

PUBLIC DOMAIN

47.1 HYDROPOWER GENERATES ELECTRICITY

Hydropower is the renewable energy source that produces the most electricity in the United States. It accounted for about 7% of the total US electricity generation and 56% of generation from all renewables in 2012.

47.2 HYDROPOWER RELIES ON THE WATER CYCLE

Understanding the water cycle is important to understand hydropower. In the water cycle (Figure 47.1):

- Solar energy heats water on the surface, causing it to evaporate.
- This water vapor condenses into clouds and falls back onto the surface as precipitation (rain, snow, etc.).
- The water flows through rivers back into the oceans, where it can evaporate and begin the cycle over again.

47.3 MECHANICAL ENERGY IS HARNESSED FROM MOVING WATER

The amount of available energy in moving water is determined by its flow or fall. Swiftly flowing water in a big river, like the Columbia River that forms the border between Oregon and Washington, carries a great deal of energy in its flow. Water descending rapidly from a very high point, like Niagara Falls in New York, also has lots of energy in its flow.

In either instance, the water flows through a pipe, or *penstock*, then pushes against and turns blades in a turbine to spin a generator to produce electricity (Figure 47.2). In a run-of-the-river system, the force of the current applies the needed pressure, while in a storage system, water is accumulated in reservoirs created by dams, then released as needed to generate electricity. Watch a video about hydropower on the Bonneville Power Administration website.

47.4 HISTORY OF HYDROPOWER

Hydropower is one of the oldest sources of energy. It was used thousands of years ago to turn a paddle wheel for purposes such as grinding grain. Our nation's first industrial use of hydropower to generate electricity occurred in 1880, when 16 brush-arc lamps were powered using a water turbine at the Wolverine Chair Factory in Grand Rapids, Michigan.

The first US hydroelectric power plant opened on the Fox River near Appleton, Wisconsin, on September 30, 1882.

Because the source of hydroelectric power is water, hydroelectric power plants must be located on a water source. Therefore, it was not until the technology to transmit electricity over long distances was developed that hydropower became widely used.

For more information about hydropower, see Hoover Dam, a hydroelectric facility completed in 1936 on the Colorado River between Arizona and Nevada. This dam created Lake Mead, a 110-mile-long national recreational area that offers water sports and fishing in a desert setting.

Alternative Energy and Shale Gas Encyclopedia, First Edition. Edited by Jay H. Lehr and Jack Keeley.
© 2016 John Wiley & Sons, Inc. Published 2016 by John Wiley & Sons, Inc.

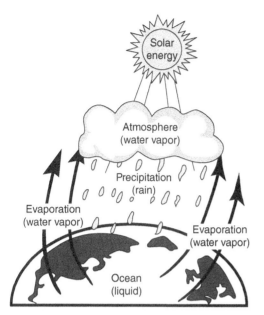

FIGURE 47.1   The water cycle. *Source*: National Energy Education Development Project (public domain).

FIGURE 47.3   Top hydropower-producing states, 2011. Source: U.S. Energy Information Administration, *Electric Power Monthly*, Table 1.13B (February 2012).

47.5   WHERE HYDROPOWER IS GENERATED

47.5.1   Most US Hydropower Is in the West

Over half of the US hydroelectric capacity for electricity generation is concentrated in three states: Washington, Oregon, and California (Figure 47.3). Approximately 29% of the total US hydropower was generated in Washington in 2011,

the location of the nation's largest hydroelectric facility—the Grand Coulee Dam.

Most hydropower is produced at large facilities built by the federal government, such as the Grand Coulee Dam. The west has most of the largest dams, but there are numerous smaller facilities operating around the country.

47.5.2   Most Dams Were Not Built for Power

Only a small percentage of all dams in the United States produce electricity. Most dams were constructed solely to provide irrigation and flood control.

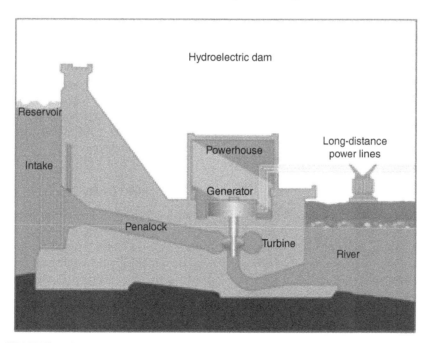

FIGURE 47.2   Hydroelectric dam. *Source*: Tennessee Valley Authority (public domain).

47.6 HYDROPOWER AND THE ENVIRONMENT

47.6.1 Hydropower Generators Produce Clean Electricity, But Hydropower Does Have Environmental Impacts

Most dams in the United States were built mainly for flood control and supply of water for cities and irrigation. A small number of dams were built specifically for hydropower generation. While hydropower (hydroelectric) generators do not directly produce emissions of air pollutants, hydropower dams, reservoirs, and the operation of generators can have environmental impacts.

Fish Ladder at the Bonneville Dam on the Columbia River Separating Washington and Oregon

A dam to create a reservoir may obstruct migration of fish to their upstream spawning areas. A reservoir and operation of the dam can also change the natural water temperatures, chemistry, flow characteristics, and silt loads, all of which can lead to significant changes in the ecology (living organisms and the environment) and rocks and land forms of the river upstream and downstream. These changes may have negative impacts on native plants and animals in and next to the river, and in the deltas that form where rivers empty into the ocean. Reservoirs may cover important natural areas, agricultural land, and archeological sites and cause the relocation of people.

Greenhouse gases, carbon dioxide and methane, may also form in reservoirs and be emitted to the atmosphere. The exact amount of greenhouse gases produced from hydropower plant reservoirs is uncertain. The emissions from reservoirs in tropical and temperate regions, including the United States, may be equal to or greater than the greenhouse effect of the carbon dioxide emissions from an equivalent amount of electricity generated with fossil fuels.

47.6.2 Fish Ladders Help Salmon Reach Their Spawning Grounds

Hydro turbines kill and injure some of the fish that pass through the turbine. The US Department of Energy has sponsored research and development of turbines that could reduce fish deaths to less than 2%, in comparison to fish kills of 5–10% for the best existing turbines.

In the Columbia River, along the border of Oregon and Washington, salmon must swim upstream to their spawning grounds to reproduce, but the series of dams along the river gets in their way. Different approaches to fixing this problem have been used, including the construction of "fish ladders" that help the salmon "step up" and around the dam to the spawning grounds upstream.

47.7 TIDAL POWER

Tides are caused by the gravitational pull of the moon and sun, and the rotation of the earth. Near shore, water levels can vary up to 40 ft due to tides.

Dam of the tidal power plant on the estuary of the Rance River, Bretagne, France

Tidal power is more predictable than wind energy and solar power. A large enough tidal range—10 ft—is needed to produce tidal energy economically.

47.7.1 Tidal Barrages

A simple generation system for tidal plants involves a dam, known as a barrage, across an inlet. Sluice gates (gates commonly used to control water levels and flow rates) on the barrage allow the tidal basin to fill on the incoming high tides and to empty through the turbine system on the outgoing tide, also known as the ebb tide. There are two-way systems that generate electricity on both the incoming and outgoing tides.

A potential disadvantage of tidal power is the effect a tidal station can have on plants and animals in the estuaries. Tidal barrages can change the tidal level in the basin and increase turbidity (the amount of matter in suspension in the water). They can also affect navigation and recreation.

There are currently six tidal power barrages operating in the world. The largest is the Sihwa Lake Tidal Power Station in South Korea with a total power output capacity of 254 MW. The second largest and oldest is in La Rance, France, with 240 MW capacity. The next largest is in Annapolis Royal, Nova Scotia, Canada, at 20 MW, followed by the Jiangxia Tidal Power Station in China at 3.7 MW, a 1.7 MW kilowatt tidal barrage in Kislaya Guba, Russia, and the 1 MW Uldolmok Tidal Power Station in South Korea.

The United States has no tidal plants and only a few sites where tidal energy could be produced economically. France, England, Canada, and Russia have much more potential to use this type of energy.

47.7.2 Tidal Fences

Tidal fences can also harness the energy of tides. A tidal fence has vertical axis turbines mounted in a fence. All the water that passes is forced through the turbines. Tidal fences can be used in areas such as channels between two landmasses. Tidal fences are cheaper to install than tidal barrages and have less impact on the environment than tidal barrages, although they can disrupt the movement of large marine animals.

A tidal fence is planned for the San Bernardino Strait in the Philippines.

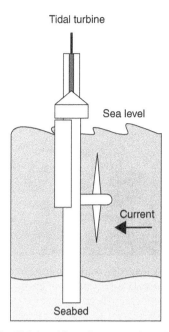

FIGURE 47.4 Tidal turbine. *Source*: Adapted from National Energy Education Development Project (public domain).

47.7.3 Tidal Turbines

Tidal turbines are basically wind turbines in the water that can be located anywhere there is strong tidal flow. Because water is about 800 times denser than air, tidal turbines have to be much sturdier than wind turbines. Tidal turbines are heavier and more expensive to build but capture more energy. There is an operating 1.5 MW tidal turbine in Strangford Lough, Scotland (Figure 47.4).

47.8 WAVE POWER

47.8.1 Waves Have Lots of Energy

Waves are caused by the wind blowing over the surface of the ocean. There is tremendous energy in the ocean waves. It is estimated that the total potential off the coasts of the United States is 252 billion kWh a year, about 6% of the United States' electricity generation in 2012. The west coasts of the United States and Europe and the coasts of Japan and New Zealand are good sites for harnessing wave energy (Figures 47.5, 47.6, and 47.7).

47.8.2 Different Ways to Channel the Power of Waves

One way to harness wave energy is to bend or focus the waves into a narrow channel, increasing their power and size. The waves can then be channeled into a catch basin or used directly to spin turbines.

FIGURE 47.5 Pelamis wave power device in use in Portugal. *Source*: Marine and Hydrokinetic Technologies Program, U.S. Department of Energy, Energy Efficiency and Renewable Energy (public domain).

FIGURE 47.6 Wave energy site. *Source*: Adapted from National Energy Education Development Project (NEED).

FIGURE 47.7 CETO underwater wave energy device. *Source*: Tuscanit, Wikimedia Commons author (GNU free documentation license) (public domain).

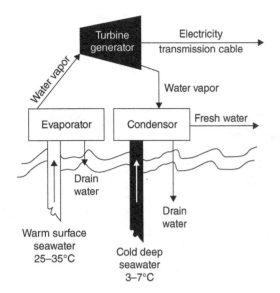

FIGURE 47.8   Ocean thermal energy conversion system.

FIGURE 47.9   An experimental OTEC plant on the Kona Coast of Hawaii, USA. *Source*: U.S. Department of Energy (public domain).

Many more ways to capture wave energy are currently under development. Some of these devices being developed are placed underwater, anchored to the ocean floor, while others ride on top of the waves. The world's first commercial wave farm using one such technology opened in 2008 at the Aguçadora Wave Park in Portugal.

See all the technologies under development at the U.S. Department of Energy's Marine and Hydrokinetic Technology Database.

47.9   OCEAN THERMAL

The energy from the sun heats the surface water of the ocean. In tropical regions, the surface water can be much warmer than the deep water. This temperature difference can be used to produce electricity. The ocean thermal energy conversion

(OTEC) system must have a large temperature difference of at least 77°F to operate, limiting its use to tropical regions (Figure 47.8).

Hawaii has experimented with OTEC since the 1970s (Figure 47.9). There is no large-scale operation of OTEC today, mainly because there are many challenges. The OTEC systems are not very energy efficient. Pumping water is a major engineering challenge.

Electricity generated by the system must be transported to land. It will probably be 10–20 years before the technology is available to produce and transmit electricity economically from OTEC systems.

EIA does not forecast the commercialization of OTEC systems in its most recent *Annual Energy Outlook*. The US Department of Energy's Office of Energy Efficiency and Renewable Energy has supported OTEC technology research and development.

48

HYDROPOWER—CHRONOLOGICAL DEVELOPMENT

PUBLIC DOMAIN

BC	• Hydropower was used by the Greeks to turn water wheels for grinding grains more than 2000 years ago.
Mid-1770s	• French hydraulic and military engineer Bernard Forest de Belidor wrote *Architecture Hydraulique,* a four-volume work describing vertical- and horizontal-axis machines.
1880–1895	• Hydropower was beginning to be used for electricity. The first hydroelectric plants were direct current (DC) stations used to power nearby arc and incandescent lighting.
1880	• Michigan's Grand Rapids Electric Light and Power Company generated DC electricity, using hydropower at the Wolverine Chair Factory. A dynamo belted to a water turbine at the factory generated electricity to light 16 brush-arc lamps in the storefront.
1881	• Street lamps in the city of Niagara Falls were powered by hydropower (DC).
1882	• The world's first central DC hydroelectric station provided power for a paper mill in Appleton, WI.
1886	• Between 40 and 50 hydroelectric plants were operating in the United States and in Canada.
1888	• About 200 electric companies relied on hydropower for at least part of their generation.
1889	• The nation's first AC hydroelectric plant, Willamette Falls Station, began operation in Oregon City, OR.
1893	• The Austin Dam, near Austin, TX, was completed. It was the first dam specifically designed for generating hydropower.
1895–1896	• The Niagara Falls hydropower station opened. It originally provided electricity to the local area. One year later, when a new AC power line was opened, electric power from Niagara Falls was sent to customers over 20 miles away in Buffalo, NY.
1899	• The Rivers and first Federal Water Power Act required special permission for a hydroelectric plant to be built and operated on any stream large enough for boat traffic.
1901	• The first Federal Water Power Act required special permission for a hydroelectric plant to be built and operated on any stream large enough for boat traffic.
1902	• The Reclamation Act of 1902 created the United States Reclamation Service, later renamed the US Bureau of Reclamation. The Reclamation Service was formed to manage water resources and was given the authority to build hydropower plants at dams.
1905	• The Reclamation Service installed a hydropower plant at the Arizona construction site of the Theodore Roosevelt Dam. The power plant was originally built to provide electricity for constructing the dam, but sales of extra electricity helped pay for the project and improved life in the local community.
1920	• Federal Power Act established the Federal Power Commission (later replaced by the Federal Energy Regulatory Commission) to issue licenses for hydropower development on public lands.

(continued)

Alternative Energy and Shale Gas Encyclopedia, First Edition. Edited by Jay H. Lehr and Jack Keeley.
© 2016 John Wiley & Sons, Inc. Published 2016 by John Wiley & Sons, Inc.

1933	• The Tennessee Valley Authority (TVA) was established to take charge of the hydroelectric potential of the Mississippi River in the Tennessee Valley. • Construction of the Grand Coulee Dam began on the Columbia River. Originally built to meet irrigation needs, it had more electric generating capacity than any other dam in North America.
1935	• Federal Power Commission authority was extended to all hydroelectric projects built by utilities engaged in interstate commerce.
1936	• Boulder Dam (later renamed the Hoover Dam) began operating on the Colorado River. The hydropower plant produced up to 130,000 kW of electricity. (Read about Energy Ant's visit to the Hoover Dam.)
1937	• The US Army Corp of Engineers finished the Bonneville Dam, on the Columbia River in Oregon and Washington. • The Bonneville Power Administration (BPA) was established.
1941	• Grand Coulee, the nation's largest hydroelectric dam, began operation.
1949	• Almost one-third of the nation's electricity came from hydropower.
1961	• The Columbia River Treaty was signed between the United States and Canada. Under the treaty, Canada built two dams for storage and one dam for generation. This resulted in greater power and flood control, which benefited US facilities downstream.
1977	• The Federal Power Commission was disbanded by Congress. A new agency was created, the Federal Energy Regulatory Commission (FERC), to regulate energy production and transmission.
1978	• Congress passed the Public Utility Regulatory Policies Act (PURPA) of 1978. The act required utilities to purchase electricity from qualified independent power producers. Portions of the act stimulated growth of small-scale hydro plants to help meet the nation's energy needs.
1980	• Conventional hydropower plant capacity nearly tripled in the United States since 1940. • Poor salmon runs in the Columbia River system prompted Congress to pass the Pacific Northwest Power Planning and Conservation Act of 1980. This act established the Northwest Power Planning Council, responsible for the protection and recovery of salmon runs in the Columbia River system. These laws resulted in a more complex, expensive process to obtain a license for a hydroelectric facility.
1986	• Congress amended the Federal Power Act to increase the environmental review of hydropower projects.
1988	• The Northwest Power Planning Council designated 44,000 miles of Pacific Northwest streams as *protected areas* because of their importance as critical fish and wildlife habitats.
1994	• Court ruled that the 1993 Biological Opinion, which guided coordinated use of the Columbia River System, failed to meet legal standards associated with the Endangered Species Act.
2006	• The United States ranked among the top four countries in the world for hydroelectric generation, along with China, Canada, and Brazil. These countries generated 44% of the world's electricity from hydropower.
Today	• Between 6% and 10% of US electricity comes from hydropower, depending on water supply and annual rainfall. In total, the United States has about 80,000 MW of conventional capacity and 18,000 MW of pumped storage capacity.

Last Revised: January 2009. Sources: Energy Information Administration, *The Changing Structure of the U.S. Electric Power Industry 2000*, Appendix A: History of US Electric Power Industry: 1882–1991, 2000; U.S. Department of Interior, Bureau of Reclamation, *The History of Hydropower Development in the United States*, November 2008; U.S. Department of Energy, Office of Energy Efficiency and Renewable Fuels, *History of Hydropower*, October 2004; U.S. Library of Congress, *America's Story - Gilded Age (1878–1889)*, January 2009; The National Energy Education Development Project, *Secondary Energy Infobook*, 2008; The Foundation for Water and Energy Education, *Timeline of Hydroelectricity and the Northwest*, August 2008.

PART V

BATTERIES AND FUEL CELLS

49

FUEL CELL CONTROL

Winston Garcia-Gabin and Darine Zambrano

49.1 INTRODUCTION

Fuel cells are electrochemical membrane reactors that are able to convert chemically stored energy directly to electrical energy at high thermodynamic efficiencies by combining a gaseous fuel and oxidizer. They provide an environmentally friendly power source without Carnot's limitation of efficiency.

There is a wide range of possible applications, from small systems like portable and mobile systems or medical devices like as implantable enzymatic fuel cells, to high power systems like propulsion of vehicles, buses, and submarines, or combined heat and high power generation. In these practical uses, fuel cells have a wide power range from milliwatt to tens of megawatt. Their number of applications continues to increase at a significant rate, because they are seen as feasible technology that can help to solve the nowadays environmental challenges, such as global warming, or the harmful levels of local pollutants produced by vehicles in cities.

The automotive market is highly regulated, particularly in terms of its environmental impact. Fuel cell is considered one of the most promising technologies for application in hybrid vehicles due to low operating temperature, zero pollution, and its high power density required to meet the space constraints in a light-duty vehicle. To be practical in automotive traction applications, fuel cell systems must provide power output performance that rival with internal combustion engines. As consequence, the transient behavior is one of the keys for success of fuel cell systems in vehicles. The fuel cell performance can be significantly improved with suitable control strategies, because the control of the fuel cell dynamic response is important for vehicular applications due to their frequent fluctuations in power demand. Thus, the control system must operate the fuel cell at the optimal operating point designed by the manufacturer. Therefore, it is important to design a robust control technique to satisfy the aforementioned power fluctuation requirements.

Although the fuel cell behavior is well understood, it requires a control strategy for robust performance operation even if it is applied to stationary applications, for example, fuel cell systems producing electrical energy for office buildings. Without regulation, fuel cell power output will drift over time, even if reactant flow rates are held steady. Fuel cell systems are complex and must be controlled to reliably achieve desired power levels. This is a requirement that is independent of fuel cell system power rating.

49.2 CONTROL OBJECTIVES

Fuel cells are nonlinear processes with a lot of variables involved in their behavior, with the control objective changing according to the application. On the one hand, a stationary use with a constant electrical load, for example, a fuel cell system producing electrical energy in a power plant, does not require a very complex control strategy. On the other hand, as in the case of fuel cell for transportation (trucks, buses, cars) it works in different operating conditions due to large load variations most of the time embedded in a hybrid system with multiple energy sources. In other words, a fuel cell that is used at only a few operating points will not require a complex control system as another that experiences unpredictable and widely varying changes in power demand. Thus,

Alternative Energy and Shale Gas Encyclopedia, First Edition. Edited by Jay H. Lehr and Jack Keeley.
© 2016 John Wiley & Sons, Inc. Published 2016 by John Wiley & Sons, Inc.

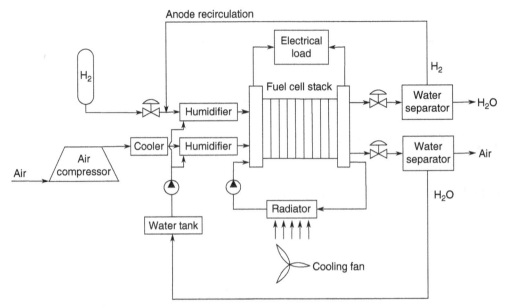

FIGURE 49.1   Fuel cell diagram.

control complexity needs to be increased with the performance demand.

Figure 49.1 shows a general diagram of a fuel cell. The main elements are a compressor used to supply pressurized air and avoid the starvation of air in the fuel cell and regulate the optimum stoichiometric amount of oxygen; a heat exchanger or intercooler to remove heat from the air; a humidification system for the air stream in order to operate the fuel cell properly, because humidity has an important effect on conductivity; and a back-pressure valve to control the system pressure. The anode and cathode back-pressure valves must be adjusted to obtain the optimal pressure for each volume according to the operating conditions. This adjustment produces an improvement in the system transient response. A cooling system is used to control the stack temperature. A tight temperature control is not usual, but keeping the stack temperature in a particular range according to the operating point will optimize the fuel cell behavior and extend the useful life time. The anode recirculation line provides a return of gas from the anode output to the anode input of the fuel cell stack. It is used to reduce hydrogen waste, maintaining the pressure difference between anode and cathode to a minimum. An anode purging is included for preventing anode flooding.

To sum up, the optimal operation of controlled variables of the fuel cell is achieved by manipulating mainly the compressor motor, humidifier pump, back-pressure valve, cooling pump, and anode recirculation. The control objective is to track trajectories for the best fuel cell power efficiency, while avoiding water flooding of the membrane, oxygen starvation, and rejection of disturbances.

49.3   CONTROL PROBLEM FORMULATION

Fuel cells are complex reactors that need to be controlled during their operation mainly when they have high and frequent load changes in the normal operating conditions. The general control scheme is shown in Figure 49.2. As it is illustrated, the methods used to control the fuel cells are feedforward control and feedback control. They can be used to regulate the ideal amounts of reactants, cell humidity, stack temperature, and output power. Many different control techniques can be used to accomplish these goals, while their complexity increases with the performance demands placed on the fuel cell.

Traditional control strategies based on simplifying the system by linearizing about an operating point and controlling only a part of the system can be used when large excursions from loads are not expected. This condition simplifies the control solution, allowing linear control to be used. A first level of control strategy is to avoid the irreversible damage in the fuel cell as those produced by starvation. Stack starvation is produced when there is a lack of suitable stoichiometric relationship between oxygen and hydrogen to deliver the load current. It causes a voltage drop in the stack and results in a high nonuniform distribution of current density that damages the fuel cell membrane and catalyst layers due to hot spots [1].

The best way to prevent oxidant starvation and to allow for a dynamic operation of the fuel cell is to rapidly adjust the excess ratio of oxygen by increasing the mass flow into the cathode. This increase is limited by the inertia of the actuators. Particularly at fast load changes, the risk of starvation is

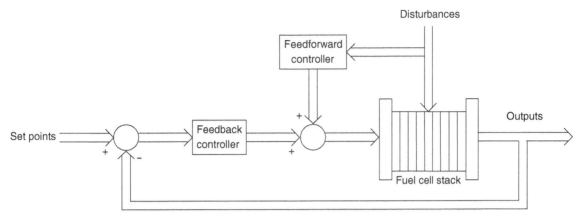

FIGURE 49.2   Feedback–feedforward control scheme.

high [2]. As consequence, the regulation of λ_{O_2} is therefore a crucial issue. λ_{O_2} is defined as the ratio between the oxygen entering the cathode ($W_{O_2,in}$) and the oxygen reacting in the fuel cell stack ($W_{O_2,reacted}$):

$$\lambda_{O_2} = \frac{W_{O_2,in}}{W_{O_2,reacted}} \qquad (49.1)$$

The value for this variable to provide high efficiency is a function of the fuel cell itself and the operating objective. Also, it can be used as a set point [3]. Notice that positive deviations of λ_{O_2} above the reference imply lower efficiency, since excess oxygen replenishment into the cathode will cause power waste, and negative deviations increase the probability of the starvation phenomena. This stoichiometric imbalance is produced by changes in the load. Because the current demand can be measured easily it is evident to use this information to avoid starvation, replenishing rapidly the air supply by the compressor.

49.3.1   Feedforward

Feedforward is a well-known control technique used for eliminating the disturbance effect when the source of disturbance can be measured. In this case, the demanded stack current is used to determine the required air flow for the reaction, and the compressor speed is adjusted accordingly via the compressor motor control signal [4, 5]. The feedforward controller can be implemented as a simple static feedforward [6] or a bidimensional lookup table [7]; as shown in equation (49.2) the compressor feedforward control signal $u_{ff}(t)$ is calculated based on the desired λ_{O_2} level and the stack current value I_{stack}:

$$u_{ff}(t) = f(I_{stack}(t), \lambda_{O_2}) \qquad (49.2)$$

Whenever a good model of the fuel cell is available, these points of the lookup table can be obtained analytically. However, it is also feasible to obtain this relationship using a direct test on the real fuel cell for several loads.

The advantage of the feedforward controller is that it is easy to implement, it considers fuel cell system nonlinearity, and it is a function of the optimized value of λ_{O_2}. In Ref. [8], an optimum profile for this variable is proposed that gives the highest efficiency for each value of the load.

49.3.2   Feedback–Feedforward

Feedforward is not able to handle all disturbances that affect the controlled variable, that is why the configuration shown in Figure 49.2 is required. Feedback controller produces the additional control signal required to compensate the modeling errors when the feedforward was designed. Figure 49.3 presents the λ_{O_2} control of a 1.2 kW polymer electrolyte membrane (PEM) fuel cell with a feedback–feedforward configuration. Notice that the main contribution in the control signal is given by the feedforward component (u_{ff}); however, feedback compensation is required to track the reference trajectory.

The feedback–feedforward control scheme (Figure 49.2) can incorporate different measured outputs and disturbances to design more complex control structures. Obtaining the optimal power response requires to maintain at specific optimal values the controlled variables, according to each operating condition for the partial pressure of the reactants across the membrane, membrane hydration, stack temperature, and anode recirculation.

49.4   CONTROL STRUCTURES

Fuel cells have a highly coupled nonlinear behavior and strong disturbances given by the load changes, and their

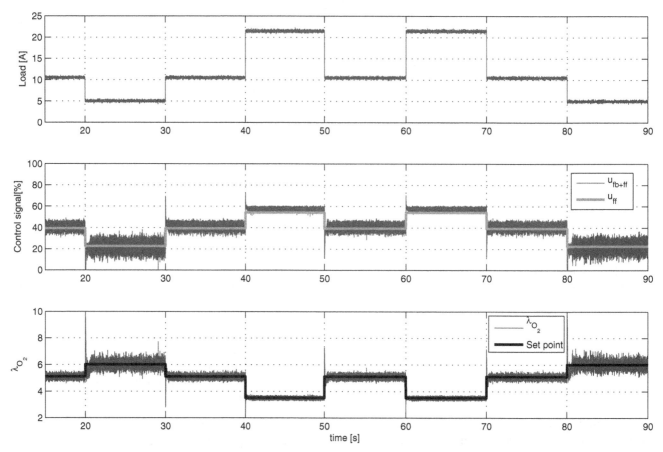

FIGURE 49.3 Feedback–feedforward control signals and controlled variable (λ_{O_2}) versus electrical load changes.

parameters vary with time and temperature. As a consequence, advanced control techniques are necessary to achieve efficient performance, prolong the fuel cell's life, guarantee safety, and achieve low-cost operation. In automotive traction applications fuel cell systems must provide different power output levels according to changing load requirements. That is why operation in a wider range and transient behavior are keys for success of fuel cell systems in vehicles. Linearization of the system and the design of linear control approach give an adequate performance only in the vicinity of the point of linearization. Fuel cells working in different operation points require nonlinear control techniques to handle their nonlinear behavior. For such conditions, some approaches have been proposed to allow operation in a wider range.

A nonlinear control scheme is proposed for fuel cells in Ref. [9]. The controller design is based on the concept of passivity. The state–space feedback law and the observer are combined to an output-feedback controller, which guarantees stability of the closed-loop system over a wide range of operation conditions. A rule-based, output-feedback multivariable control was implemented with fuzzy logic with triangular membership functions and coupled with a nonlinear

feedforward approach in Ref. [10]. The air and fuel supply subsystems are controlled to achieve an optimum pressurization of air flow for maximum system efficiency during load transients. A fuel cell was modeled as a two-input two-output multivariable system and a nominal plant was identified for the controller design [11]. The robust PID controller was designed for the fuel cell by considering the frequency and time-domain specifications. The air pump and the hydrogen flow were the manipulated variables, and current and voltage of the load were the measured output signals of the system.

Sliding mode control is a robust control technique which has been successfully implemented in simulations [12] and laboratory tests [7, 13]. Satisfactory performance was obtained to solve the air supply control problem in PEM fuel cells. The simulations and experimental tests showed excellent results regarding dynamic oxygen stoichiometry regulation and robust performance against uncertainty and load changes.

Model predictive control (MPC) is, after PID, the most popular control methodologies in industry. The basic idea of MPC is to calculate a sequence of future control signals in such a way that it minimizes the multistage cost function (equation (49.3)) defined over a prediction horizon. The typically used cost function is given by

$$
\min_{\Delta u} \left\{ J = \sum_{j=N_1}^{N_p} \| \delta_y \left(w(k+j) - \hat{y}(k+j|k) \right) \|^2 \\
+ \sum_{j=1}^{N_u} \delta_{\Delta u} (k+j|k) \|^2 \right\}
$$

(49.3)

subject to:

$$
\begin{aligned}
u_{\min} &\le u_k \le u_{\max} \\
\Delta u_{\min} &\le \Delta u_k \le \Delta u_{\max} \\
y_{\min} &\le y_k \le y_{\max}
\end{aligned}
$$

where N_1 and N_p define the beginning and end of the prediction horizon; N_u is the control horizon; $w(k+j)$ is the reference trajectory; $\hat{y}(k+j|k)$ is the j-step prediction of the output on data up to instant k; δ_y and $\delta_{\Delta u}$ are the weights on the error and control action increments, respectively; Δu is the input increment; $y_{\min/\max}$, $u_{\min/\max}$, and $\Delta u_{\min/\max}$ are the minimum and maximum constraints imposed on the output, input, and input increments, respectively. The performance index to be optimized is the expectation of a quadratic function measuring the distance between the predictive system output and a predictive reference sequence over the horizon, plus a quadratic function measuring the control effort. In order to implement an MPC, a model of the plant is used to predict the future outputs. This prediction is based on past and current values of the input and the output of the plant. Feedforward compensation and time delay are inherently considered in the prediction model of MPC. An experimental validation of a constrained MPC strategy on a real fuel cell is presented in Ref. [14]. In Ref. [2], an MPC is used to prevent fuel cell starvation by guaranteeing a minimum value of the excess ratio of oxygen and a minimum value of the fuel cell pressure. An MPC satisfies the power demands while optimizing the fuel efficiency of the entire system. The use of suitable constraints results in significant improvements in fuel efficiency [15]. A nonlinear MPC based on the moving horizon state estimation is designed for the solid oxide fuel cell in Ref. [16]. The current density, fuel, and air molar flow rates are chosen as manipulated variables to control the output power, fuel utilization, and temperature. A nonlinear MPC using a reliable and efficient dynamic optimization approach is implemented in a PEM fuel cell system [17]. The optimization is performed using a direct transcription method that handles the optimal control problem as a nonlinear programming problem. The MPC ensures optimum power generation following a variable load demand with

acceptable response time while avoiding oxygen starvation and minimizing hydrogen consumption.

MPC is a flexible control technique that may incorporate new developments easily. It can handle hybrid systems that allow the integration of continuous and discrete variables. In Ref. [3], a hierarchical predictive control strategy to optimize both power utilization and oxygen control simultaneously for a hybrid proton exchange membrane fuel cell/ultracapacitor system is applied. The control employs fuzzy clustering-based modeling, constrained MPC, and adaptive switching among multiple models. The proposed approach can provide a globally optimized solution for power distribution and oxygen control, avoiding oxygen starvation by trading off transient current demand from the fuel cell to the ultracapacitor.

Neural networks is a class of modeling technique that results in a nonlinear model based on experimental data. It is a black-box model organized in sequential layers containing neurons. The network output is obtained as a weighted sum of input through the hidden layers. The summation weights are found by means of training process which tries to minimize the error between desired and network output. Neural networks have been used in the context of fuel cells [18, 19]. Fuzzy logic control takes the input variables and maps them into fuzzy variables by sets of membership functions. Each input variable has a determined value's degree of membership in a fuzzy set. The set of rules are built generally based on expert knowledge. The signal is processed applying the rules and generating a result for each, then combining the results of these rules. Finally, the fuzzy controller output is obtained via defuzzification combining the result back into a specific control output value. In Ref. [20], a fuzzy logic controller is presented to regulate a fuel cell. The experimental results showed that the fuzzy controller consumed less bio-hydrogen than a PID controller. A nonlinear temperature predictive control algorithm based on an improved Takagi–Sugeno fuzzy model is presented in Ref. [21]. How the proposed controller applied to a solid oxide fuel cell stack keeps the temperature within a specific range in order to maintain good performance of the stack is shown. In Ref. [22], the improvement of behavior and performance of a fuel cell using fuzzy PID control to regulate the airflow rate is shown. A full control of a fuel cell requires the handling of a lot of variables. As consequence, the complexity of neural networks and fuzzy logic controllers increases when large numbers of variables and rules are considered. However, they can be used in applications where classic control methods fail.

H_∞ is a control theory to synthesize controllers achieving robust performance, ensuring stabilization. It is easily applicable to process involving multivariable systems with cross-coupling between variables. The general idea behind H_∞ control design is to find a feedback gain K for $u(t) = Kx(t)$ such that transfer function matrix of the system is minimized according to the H_∞ norm. In Ref. [23], a multivariable H_∞

control is applied to a proton exchange membrane fuel cell, where the inputs are air and hydrogen flow rates and the outputs are cell voltage and current. The H_∞ controller was designed to provide robust performance and to reduce the hydrogen consumption of this system. From the experimental results, multivariable robust control is shown to provide steady output responses and significantly reduce hydrogen consumption. In Ref. [24], a nonlinear multivariable robust control strategy is designed for a proton exchange membrane fuel cell. The state feedback exact linearization approach can achieve the global exact linearization via the nonlinear coordinate transformation and the dynamic extension algorithm, such that H_∞ robust control strategy can be directly utilized to guarantee the robustness of the system.

The feedback gain matrix K for $u(t) = Kx(t)$ can be obtained using linear quadratic regulator (LQR) optimal design. This control technique is concerned with operating a dynamic system at minimum cost that results in some balance between system errors and control effort. The cost function is as follows:

$$\min_{\Delta u} \left\{ J = \int_0^\infty (x^T Q x + u^T R u)\mathrm{d}t \right\} \qquad (49.5)$$

where matrices Q and R are the weighting factors, which can be used to adjust the importance of control parameters.

The effectiveness of the LQR controller applied to fuel cell is illustrated in Refs. [4, 6, 10, 19, 25, 26].

49.5 DYNAMIC MODELS FOR CONTROL

Designing a control approach requires models of the fuel cell, which can be obtained using experimental data [27]. In Refs. [7, 11], the transfer function models are obtained setting the fuel cell in different operating conditions of load, then the data are stored and used to identify and validate the model. The same idea but using nonlinear black-box model is used in Refs. [18, 19] where neural networks are trained using input–output data.

In case of nonlinear controller, a nonlinear model must be used to design the controller. Fuel cells are complex systems which involve liquid, vapor, and gas mixed flow transportation and electrochemical reactions. The power of a fuel cell also strongly depends on operating conditions such as flow rates, relative humidity, and temperature of the gases. A fundamental fuel cell model consists of five principles of conservation: mass, momentum, species, charge, and thermal energy. The variables of the physical parameters change spatially in the volume of the fuel cell. In real-time control applications, where computation time to obtain each control signal is limited, models assume that parameters are not spatially varying. That is, their values do not change along any direction inside the cell.

A dynamic model that includes the flow and inertia dynamics of the compressor, the manifold filling dynamics (both anode and cathode), reactant partial pressures, and membrane humidity is developed in Ref. [28]. The fuel cell stack temperature is treated as a parameter rather than a state variable of this model because of its long time constant. In Ref. [5], a fuel processor system model is developed, which represents the conversion of hydrocarbon fuel to a hydrogen-rich mixture that is directly fed to the proton exchange membrane fuel cell stack. These models could be enhanced by considering transient dynamics of stack cooling system, anode recirculation pump, air cooler, and reactant humidification systems.

In summary, models for control must have an excellent trade-off between accuracy and computation time cost. For this reason, there is a continuous effort to improve and validate mathematical models based on first principles for fuel cells [29–36]. Control models must ignore parameters that do not have a significant effect on the dynamic behavior. Dynamic models of mass flow rates, pressures, temperatures, and voltages must have lumping spatially varying parameters. In order to simplify, the ideal gas law must be used for both pure and mixed gases; mixtures with water vapor are considered to be single phase; and volumes are assumed to be under isothermal conditions. Also all cells are lumped into one equivalent cell and the stack voltage and power can be determined from the number of cells in the stack. In general, slow dynamics are considered with constant values, as a consequence they are neglected; and fast dynamics are modeled by gains or input–output maps. A control model must include the dynamic states of current, voltage, relative humidity, stoichiometry of air and hydrogen, cathode and anode pressures, cathode and anode mass flow rates, and stack temperatures. To build a complete model around a stack, it must comprise compressor, injection pump, humidifier, cooler, inlet and outlet manifolds, and stack cooling. Also, anode recirculation must be included with the injection pump, as well as anode purging, for preventing anode flooding.

49.6 CONCLUSIONS

Fuel cell is a promising and growing technology that will be more and more present in practical applications. From portable electric devices to electrical power units generating tens of megawatt the success of this technology is based on obtaining a high efficiency. The control approach is the key to achieve this goal. Automotive sector, where high load changes must be operated frequently and two or more energy sources must be combined, is a clear example of the challenge that the control engineer has to handle. They are complex systems with multiple final control elements, which must operate the fuel cell at the optimal operating point of the controlled variables to achieve a high efficiency.

Feedforward control approach increases considerably the dynamic behavior of the controlled variable; in this manner load changes and any other measured disturbance must be incorporated into the control scheme. MPC is considered as a powerful control technique to deal with complex control requirements. It can handle continuous and discrete variables and nonlinear hybrid models, including constraints in its design to obtain optimal control sequences minimizing a cost function that can be changed according to the operating conditions. Finally, this chapter gives a panorama of the most important control approaches applied to fuel cells; however, it must be kept in mind that selecting the proper control approach is a difficult task. The criterion should always be to choose the simplest controller to do the job.

REFERENCES

[1] Varigonda, S., Kamat, M. (2006). Control of stationary and transportation fuel cell systems: progress and opportunities. *Computers and Chemical Engineering*, 30(10–12):1735–1748.

[2] Danzer, M. A., Wittmann, S. J., Hofer, E. P. (2009). Prevention of fuel cell starvation by model predictive control of pressure, excess ratio, and current. *Journal of Power Sources*, 190(1):86–91.

[3] Chen, Q., Gao, L., Dougal, R. A., Quan, S. (2009). Multiple model predictive control for a hybrid proton exchange membrane fuel cell system. *Journal of Power Sources*, 191(2):473–482.

[4] Pukrushpan, J., Stefanopoulou, A., Peng, H. (2004). *Control of Fuel Cell Power Systems: Principles, Modeling and Analysis and Feedback Design*. Advances in Industrial Control Series. Springer.

[5] Pukrushpan, J., Stefanopoulou, A., Varigonda, S., Eborn, J., Haug-stetter, C. (2006). Control-oriented model of fuel processor for hydrogen generation in fuel cell applications. *Control Engineering Practice*, 14(3):277–293.

[6] Kim, Y. B., Kang, S. J. (2010). Time delay control for fuel cells with bidirectional DC/DC converter and battery. *International Journal of Hydrogen Energy*, 35:8792–8803.

[7] Garcia-Gabin, W., Dorado, F., Bordons, C. (2010). Real-time implementation of a sliding mode controller for air supply on a PEM fuel cell. *Journal of Process Control*, 20(3):325–336.

[8] Bordons, C., Arce, A., Del Real, A. J. (2006). Constrained predictive control strategies for PEM fuel cells. *American Control Conference*, Minneapolis, USA, pp. 2486–2491.

[9] Mangold, M., Buck, A., Hanke-Rauschenbach, R. (2010). Passivity based control of a distributed PEM fuel cell model. *Journal of Process Control*, 20:292–313.

[10] Al-Durra, S., Yurkovich, A., Guezennec, Y. (2010). Study of nonlinear control schemes for an automotive traction PEM fuel cell system. *International Journal of Hydrogen Energy*, 35:11291–11307.

[11] Wang, F. C., Ko, C. C. (2010). Multivariable robust PID control for a PEMFC system. *International Journal of Hydrogen Energy*, 35:10437–10445.

[12] Park, G. (2014). A simple sliding mode controller of a fifth-order nonlinear PEM fuel cell model. *IEEE Transactions on Energy conversion*, 29(1):65–71.

[13] Kunusch, C., Puleston, P. F., Mayosky, M. A., Fridman, L. (2013). Experimental results applying second order sliding mode control to a PEM fuel cell based system. *Control Engineering Practice*, 21:719–726.

[14] Gruber, J., Doll, M., Bordons, C. (2009). Design and experimental validation of a constrained MPC for the air feed of a fuel cell. *Control Engineering Practice*, 17(8):874–885.

[15] Golbert, J., Lewin, D. R. (2007). Model-based control of fuel cells (2): optimal efficiency. *Journal of Power Sources*, 173(1):298–309.

[16] Zhang, X. W., Chan, S. H., Ho, H. K., Li, J., Li, G., Feng, Z. (2008). Nonlinear model predictive control based on the moving horizon state estimation for the solid oxide fuel cell. *International Journal of Hydrogen Energy*, 33(9):2355–2366.

[17] Ziogoua, C., Papadopouloua, S., Georgiadis, M. C., Voutetakis, S. (2013). On-line nonlinear model predictive control of a PEM fuel cell system. *Journal of Process Control*, 23:483–492.

[18] Almeida, P. E. M., Simoes, M. G. (2005). Neural optimal control of PEM fuel cells with parametric CMAC networks. *IEEE Transactions on Industry Applications*, 41(1):237–245.

[19] Hasikos, J., Sarimveis, H., Zervas, P. L., Markatos, N. C. (2009). Operational optimization and real-time control of fuel-cell systems. *Journal of Power Sources*, 193(1):258–268.

[20] Huang, S. R., Lin, C. Y., Wu, C. C., Yang, S. J. (2008). The application of fuzzy controller for fuel cell generating studies. *International Journal of Hydrogen Energy*, 33:5205–5217.

[21] Yang, J., Li, X., Mou, H. G., Jian, L. (2009). Predictive control of solid oxide fuel cell based on an improved Takagi-Sugeno fuzzy model. *Journal of Power Sources*, 193(2):699–705.

[22] Tekin, M., Hissel, D., Pera, M. C., Kauffmann, J. M. (2006). Energy consumption reduction of a PEM fuel cell motor-compressor group thanks to efficient control laws. *Journal of Power Sources*, 156(1):57–63.

[23] Wang, F. C., Chen, H. T., Yang, Y. P., Yen, J. Y. (2008). Multivariable robust control of a proton exchange membrane fuel cell system. *Journal of Power Sources*, 177(2):393–403.

[24] Li, Q., Chen, W., Wang, Y., Jia, J., Han, M. (2009). Nonlinear robust control of proton exchange membrane fuel cell by state feedback exact linearization. *Journal of Power Sources*, 194(1):338–348.

[25] Niknezhadi, A., Allu-Fantova, M., Kunusch, C., Ocampo-Martinez, C. (2011). Design and implementation of LQR/LQG strategies for oxygen stoichiometry control in PEM fuel cells based systems. *Journal of Power Sources*, 196:4277–4282.

[26] Mueller, F., Jabbari, F., Brouwer, J., Junker, S. T., Ghezel-Ayagh, H. (2009). Linear quadratic regulator for a bottoming solid oxide fuel cell gas turbine hybrid system. *Journal of*

Dynamic Systems, Measurement and Control, Transactions of the ASME, 131(5):1–9.

[27] Tirnovan, R., Giurgea, S., Miraoui, A., Cirrincione, M. (2008). Surrogate model for proton exchange membrane fuel cell (PEMFC). *Journal of Power Sources*, 175(2):773–778.

[28] Pukrushpan, J., Huei, P., Stefanopoulou, A. (2004). Control-oriented modeling and analysis for automotive fuel cell systems. *Journal of Dynamic Systems, Measurement, and Control*, 126:14–25.

[29] Haraldsson, K., Wipke, K. (2004). Evaluating PEM fuel cell system models. *Journal of Power Sources*, 126(1–2):88–97.

[30] Biyikoglu, A. (2005). Review of proton exchange membrane fuel cell models. *International Journal of Hydrogen Energy*, 30(11):1181–1212.

[31] Meyer, R. T., Yao, B. (2006). Modelling and simulation of a modern PEM fuel cell system. *The 4th International Conference on Fuel Cell Science, Engineering and Technology*, Irvine, USA, June 19–21, 2006.

[32] Philipps, S. P., Ziegler, C. (2008). Computationally efficient modeling of the dynamic behavior of a portable PEM fuel cell stack. *Journal of Power Sources*, 180(1):309–321.

[33] Brown, T. M., Brouwer, J., Samuelsen, G. S., Holcomb, F. H., King, J. (2008). Dynamic first principles model of a complete reversible fuel cell system. *Journal of Power Sources*, 182(1):240–253.

[34] Siegel, C. (2008). Review of computational heat and mass transfer modeling in polymer-electrolyte-membrane (PEM) fuel cells. *Energy*, 33(9):1331–1352.

[35] Sharifi, S. M., Rowshanzamir, S., Eikani, M. H. (2010). Modelling and simulation of the steady-state and dynamic behaviour of a PEM fuel cell. *Energy*, 35(4):1633–1646.

[36] Wang, Y., Chen, K. S., Mishler, J., Cho, S. C., Cordobes-Adroher, X. (2011). A review of polymer electrolyte membrane fuel cells: technology, applications, and needs on fundamental research. *Applied Energy*, 88(4):981–1007.

50

RECENT TRENDS IN THE DEVELOPMENT OF PROTON EXCHANGE MEMBRANE FUEL CELL SYSTEMS

Amornchai Arpornwichanop and Suthida Authayanun

50.1 INTRODUCTION

Because of the need for clean energy and an efficient method for its production, substantial effort has been expended toward the development of fuel cells for electricity generation. Typically, a fuel cell generates electricity via the electrochemical reaction of hydrogen and oxygen and produces only water and heat as by-products. Among the various types of fuel cells, the proton exchange membrane fuel cell (PEMFC) offers the highest energy density and shortest start-up time because of its low operating temperature (60–80°C). In addition, PEMFCs provide a rapid response to changes in power demand and have low weights and volumes.

In general, pure hydrogen or hydrogen-rich gas from reforming processes can be used as a fuel for PEMFCs. Until suitable hydrogen infrastructure and storage are readily available, a fuel cell system integrated with a fuel processor that allows hydrogen generation from hydrocarbon fuels represents an effective solution. However, to avoid catalyst poisoning during PEMFC operation, the reformate gas must be highly purified to reduce the amount of carbon monoxide (CO) to less than 10 ppm. This purification requirement necessitates the use of a sophisticated CO clean-up unit [1].

Currently, most hydrogen is produced from natural gas, which contains methane as the major component; however, other fuels, such as liquid petroleum gas (LPG), diesel fuel, gasoline, and methanol, are also used to produce hydrogen for PEMFCs. Because of increased power demand and environmental concerns, the development of a new sustainable feedstock for hydrogen production is necessary. In the long term, renewable energy sources, such as biomass, biogas, and bio-ethanol, will become the most important fuel source [2].

Because the use of biodiesel as a petroleum-diesel substitute is growing, glycerol generated from the biodiesel production process is also a potential fuel to be used for the production of hydrogen.

Typically, a fuel processing setup for PEMFCs consists of a sulfur removal unit, a reformer, a water–gas shift (WGS) reactor, and additional CO removal units. The reforming of fuels to a reformate gas with a high hydrogen content is a main step in the processing of fuels. In general, fuel reforming processes can be classified into three main reactions: steam reforming (SR), partial oxidation (POX), and autothermal reforming (ATR), which combines the SR and the POX processes. Each reforming process has its own advantages and limitations; thus, the selection of the reformer type depends on the system design and intended application of the PEMFC. Because of a CO poisoning effect in PEMFCs, the WGS reactor and CO removal processes are added to the fuel processors for PEMFC systems. An amount of CO in the reformate gas can be reduced to an acceptable level for PEMFC operation through a variety of approaches, such as the preferential oxidation (PROX) process, pressure swing adsorption, the CO methanation process, or membrane separation. In addition, the application of a membrane reactor to produce pure hydrogen has been continuously developed to improve the quality of the reformate gas and to reduce the complexity of the fuel processor. More details on the fuel types and the fuel processor for the production of hydrogen-rich gas for PEMFCs will be discussed in this chapter.

Apart from the design issues of the fuel processor, several problems associated with PEMFC operation still exist that need resolution, including issues of water management and CO poisoning. These difficulties and their solutions for

Alternative Energy and Shale Gas Encyclopedia, First Edition. Edited by Jay H. Lehr and Jack Keeley.
© 2016 John Wiley & Sons, Inc. Published 2016 by John Wiley & Sons, Inc.

PEMFC systems that use reformate gas obtained from a fuel processor will be discussed. The effect of contaminants and key operating conditions with respect to the fuel cell performance and the complexity of the PEMFC-integrated system are reviewed. Because of limited hydrogen storage and transportation capabilities, the use of a reformate gas, which is derived from fuel processors, as an alternative hydrogen source for PEMFCs is unavoidable. Various aspects related to the development of PEMFCs that run on reformate gas are discussed, including issues that concern oxygen bleeding, catalysts with high CO tolerance, and the high-temperature proton exchange membrane fuel cell (HT-PEMFC). Finally, the prospect of using a PEMFC-integrated system for stationary and automotive applications is discussed.

50.2 FUEL SOURCES FOR HYDROGEN PRODUCTION

Hydrogen production is one of the key processes in the chemical and petrochemical industries because hydrogen is used as a reactant to produce various valuable products. In addition, it is considered an important, clean energy carrier and can be used in fuel cells to generate electricity through electrochemical reactions. Because hydrogen is not readily available, it is necessary to produce it from other sources. Many studies on hydrogen production from various nonrenewable and renewable fuel sources, such as natural gas, ethanol, methanol, and biogas, have been conducted.

50.2.1 Nonrenewable Fuels

To date, many nonrenewable fuels have been utilized to produce hydrogen for PEMFCs. Fossil-derived fuel is among the most widely used nonrenewable energy sources for PEMFC applications because of its wide availability and because of the well-established infrastructure for its distribution. However, petroleum fuel is limited and insufficient for an increased power demand. In addition, the use of this fuel type detrimentally affects the environment through the generation of greenhouse gases, such as carbon dioxide. Currently, natural gas is the most common fuel that is employed for hydrogen production because of its cost effectiveness for industrial hydrogen. The production of hydrogen from natural gas is currently used in the production of ammonia and alcohol [3]. Natural gas consists of methane, as the major component, as well as nitrogen, CO_2, ethane, propane, butane, pentane, and traces of other components. Because of the high hydrogen-to-carbon ratio of methane, a product with a high hydrogen concentration is obtained from the reforming of natural gas. The SR of methane has been reported to yield the highest hydrogen content of 75–78 vol.% [4]. When synthesis gas with a higher hydrogen concentration is fed to fuel cells, the efficiency of the fuel cell system is improved because the hydrogen fraction directly affects the performance of the fuel cell. However, the obtained system efficiency also relies on other factors, such as the system design and the operating conditions of each unit in the fuel cell system.

In addition to natural gas, LPG is also a nonrenewable fuel with a strong potential to be a hydrogen carrier for PEMFCs [5, 6]. This potential is due to its ease of storage, the presence of existing infrastructure, and its safety. Moreover, like natural gas, LPG is easier to decompose in the presence of water. The maximum thermal efficiency of a LPG reforming system that uses membrane reactor technology is 82.4%, as calculated based on the high heating value (HHV) of the fuel; thus, a PEMFC system coupled with the LPG reformer would have a system efficiency of 36.1% [5].

Diesel, gasoline, and methanol are among the most common nonrenewable liquid fuels used to produce hydrogen for PEMFCs. They are liquids at room temperature and thus can be easily stored and transported. Infrastructure for the distribution of diesel and gasoline already exists, whereas no infrastructure currently exists for the distribution of methanol. Therefore, the associated start-up cost of fuel cell vehicles fueled with diesel would be significantly less than a methanol-fueled fuel cell system [7]. However, the large molecules of diesel and gasoline represent an obstacle in the reforming process. The reforming of diesel and gasoline fuels is complicated and requires high temperatures [8]. Furthermore, because of the low H/C ratio in diesel and gasoline molecules, the reforming process of these fuels results in a reformate with a low hydrogen concentration. However, methanol molecules are smaller than those of diesel and gasoline, and methanol is thus easily converted into hydrogen. In addition, it has a potentially high production capacity [9]. The high H/C ratio and the absence of carbon–carbon bonds in methanol allow it to be reformed into hydrogen at low operating temperatures. These conditions favor the WGS reaction and the consequent reduction of the CO content in the reformate gas, which significantly influences the activity of the Pt catalyst in PEMFCs. In addition, a sulfur removal unit is not required for methanol but would be necessary for any sulfur-containing petroleum-based fuels. As a result, the methanol processor is less complicated than that for other nonrenewable energy sources.

For stationary applications, natural gas and LPG are good candidates for use in the hydrogen production process for PEMFCs because of their high obtained efficiency [10, 11]. However, for automotive applications, liquid fuels, such as gasoline, diesel, and kerosene, are preferable because they are easily transported and their refueling stations are well established in many parts of the world [12–14]. For portable devices, methanol is a suitable fuel for the production of hydrogen because of its lower operating temperature and the lower CO content in the reformate gas compared to the operating temperature and CO content in the reformate from petroleum-based hydrocarbons. These factors make methanol an attractive fuel for on-board hydrogen production [15–17].

50.2.2 Renewable Resources

Because of increased energy demand and environmental awareness, it is necessary to find new potential feedstocks that are environmentally friendly and readily available for the production of hydrogen. In the long term, renewable energy sources, such as biomass, biogas, and bio-ethanol, will become the most important feedstocks for the production of hydrogen because these fuels are derived from agricultural products and plants, which consume CO_2 in their life cycles and offer a nearly closed carbon cycle [8, 9].

Liquid biofuels, specifically ethanol and glycerol, have received significantly more attention than other fuels over the past few years as alternative fuel sources for PEMFC systems [18–20]. Ethanol is also considered a good candidate for hydrogen production because of its renewability, low toxicity, and high hydrogen content. It can be produced from several biomass sources, including energy plants, waste materials from agro-industrial wastes, and forestry residue materials [21]. Ethanol appears to be an appropriate fuel for the production of hydrogen for PEMFCs because, similar to the reforming process for methanol, the process for ethanol can be operated at low temperatures. However, the use of ethanol is preferable to the use of methanol because of methanol's toxicity.

To date, glycerol is also regarded as an alternative fuel for hydrogen production. A significant amount of glycerol is generated in response to increased biodiesel demand because glycerol has been found to be a by-product of the typical transesterification process of vegetable oils and alcohol for biodiesel production. In general, crude glycerol always contains impurities, whereas the purification process for crude glycerol requires high operational costs and is not economical [22]. Therefore, the utilization of glycerol for hydrogen production is an interesting choice [23–25]. Apart from glycerol, biodiesel itself is also used to produce hydrogen for PEMFCs in automotive applications because of its renewability and ease of transport [26]. However, its large molecules require high operating temperature and a significant amount of steam to produce hydrogen via the SR process; thus, its usefulness in hydrogen production is limited. In this regard, biodiesel is similar to other large-molecule hydrocarbons, such as diesel and gasoline.

Biomass is another attractive renewable fuel for use in the production of hydrogen for PEMFCs. Most biomass sources are solids and can be converted into hydrogen through a gasification process. However, the hydrogen-rich gas produced by such processes contains some impurities, such as particulates, tars, alkali compounds, sulfur compounds, and halogen compounds. These compounds are often harmful to

fuel cells, especially PEMFCs; therefore, a complex purification process is needed for its use in this application. Consequently, the use of biomass for hydrogen production is not suitable for use in automobiles because of the sophisticated system required for its use and the difficulties related to its transportation. In contrast, a biomass-fueled PEMFC based on a microcogeneration system for stationary applications is an interesting topic [27]. Only a few studies related to the integration of a biomass gasification system with a PEMFC system have been published.

Biogas is a renewable fuel gas with properties similar to those of natural gas; however, biogas exhibits a higher carbon dioxide content and the biogas-fed PEMFC system provides a lower power density. Depending on the raw materials and production technologies used, biogas sometimes contains various components, such as sulfur and ammonia, that are harmful to the catalysts used for hydrogen production [28]. In general, biogas produced via the anaerobic digestion of animal and human wastes, agricultural residues, aquatic weeds, and other organic matter consists primarily of methane (CH_4) and carbon dioxide and may contain as much as 60 vol.% CH_4 [29]. The composition of biogas derived from various feedstocks is dependent upon the nature of biomass feed and the digestion conditions. Currently, biogas-based hydrogen production systems have been widely used in high-temperature fuel cells, such as a molten carbonate fuel cells (MCFCs) and solid oxide fuel cells (SOFCs). However, limited work on a PEMFC system that runs on biogas has been published [30].

50.3 FUEL PROCESSING FOR PEMFCS

A fuel processor converts fuels into hydrogen-rich gas, which is supplied to fuel cells for electricity generation. In the fuel processor, the reformation of fuels is among the main steps. There are several thermochemical methods to produce hydrogen from gas and liquid hydrocarbons, such as SR, POX, and ATR. The hydrogen-rich gas derived from the reforming process is generally composed of H_2, CO, CO_2, and CH_4. However, PEMFCs, which are operated at low temperatures, require high-purity hydrogen fuel with a low CO content to avoid catalyst poisoning; the CO content in the hydrogen feed for PEMFCs must be less than 10 ppm. However, the level of CO contamination in the hydrogen-rich reformate gas normally exceeds this concentration, and the reformate gas fed to the PEMFC must therefore be further treated. Consequently, a number of hydrogen production and purification processes for PEMFC applications have been proposed.

50.3.1 Conventional Fuel Processing

Typically, the fuel processor for PEMFCs consists of a sulfur removal unit, a reformer, WGS reactors, and other CO

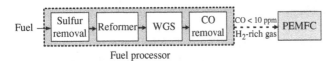

FIGURE 50.1   A fuel processor in conventional PEMFCs.

removal units, such as PROX processors or pressure swing adsorption, CO methanation, or membrane separation systems, to reduce the content of CO in the reformate gas to an acceptable level. An overall system for the conventional fuel processing for a PEMFC is illustrated in Figure 50.1. The sulfur removal unit is installed in the fuel reforming process to eliminate sulfur from a fuel before it is fed to the reformer. In general, fuel reforming processes for hydrogen production can be classified into one of three common methods: SR, POX, and ATR, which combines SR and POX in a single unit. The main reactions of these three processes are shown in Table 50.1. SR results in a higher hydrogen yield of 20% compared to ATR because in ATR a portion of the fuel feed is consumed by the combustion reaction to supply heat to the system. However, the SR process requires high external energy input to maintain the reformer operation at the desired temperature [31]. Therefore, ATR provides the highest reforming efficiency [32]. The required heat input due to the endothermicity of the SR reaction is considered a major drawback of the SR process.

In addition to the energy consumption of each reforming process, catalyst deactivation is another important issue because it also affects the reforming efficiency. Rabenstein and Hacker [33] have investigated catalyst deactivation due to coke formation during the reforming of ethanol. They found that the possibility of coke formation in each reforming process decreases in the order POX > SR > ATR. However, an analysis of the net system efficiency of a fuel-processor-integrated PEMFC system when methane was used as the fuel to produce hydrogen showed that the PEMFC system with the SR of methane provides the highest system efficiency [34]. This result was observed because the heat recovery from the anode off-gas can be used to sustain the SR operation; thus, the supply of the external heat for SR is reduced. Ahmed and Krumpelt [35] explained that SR is well suited for a long period of steady-state operation, whereas the POX and ATR

processes are more attractive for automotive applications, which require a fast start and a dynamic response.

During the operation of a reformer, other products in addition to hydrogen, for example, CO, CO_2, and CH_4, are produced. Thus, a posttreatment unit for the reformate gas is necessary to produce high-purity hydrogen. In general, a WGS reactor is used to reduce CO contamination and enhance the hydrogen content. Nonetheless, the reformate gas at the outlet of the WGS reactor still contains CO in concentrations that exceed the acceptable level for PEMFCs. A CO purification unit is further required to reduce the CO concentration. Normally, a PROX reactor is preferably used to purify hydrogen through the elimination of CO [36,37]. Oxygen is fed to the PROX reactor to convert CO into CO_2 (equation 50.1). However, a loss of hydrogen is unavoidable due to the hydrogen oxidation reaction (HOR) (equation 50.2). The difficulty in the design of this process is that, if a large amount of oxygen is fed into the PROX reactor, CO conversion will be enhanced, but hydrogen loss will increase as well. However, when a low oxygen-to-CO ratio is applied to the PROX reactor to avoid the loss of hydrogen, CO cannot be reduced to a concentration less than 10 ppm using a single PROX reactor. In this situation, multiple stages of the PROX reactor, as shown in Figure 50.2, might be required to reduce the CO concentration to an acceptable level [37]. Therefore, the oxygen/air feeding ratio should be carefully selected because it could affect the complexity of the system. Currently, catalysts with high selectivity for CO oxidation have been developed, and the selectivity of CO oxidation in the PROX process is approximately 0.4–0.9, which is dependent on the catalyst and operating parameters used [38–40]. In addition, when the effluent gas of the WGS process contains more CO, a greater amount of oxygen must be fed to the PROX process, which will result in a greater hydrogen loss. This result implies that the quality of the reformate gas obtained from the PROX process depends on the amount of CO and hydrogen derived from the WGS process:

$$CO + 0.5O_2 \leftrightarrow CO_2 \qquad (50.1)$$
$$H_2 + 0.5O_2 \leftrightarrow H_2O \qquad (50.2)$$

In addition to a PROX reactor for CO removal, a CO methanation process is also used to reduce the CO content in

TABLE 50.1   **Main reactions that occur in three common methods of hydrogen production.**

Processes	Main reaction	External energy
SR	$C_mH_n + mH_2O \rightarrow mCO + \left(m + \frac{n}{2}\right)H_2$	Required
POX	$C_mH_n + \frac{1}{2}mO_2 \rightarrow mCO + \frac{1}{2}nH_2$	Not required
ATR	$C_mH_n + \frac{1}{2}mH_2O \rightarrow mCO + \left(\frac{m}{2} + \frac{n}{2}\right)H_2$	Not required

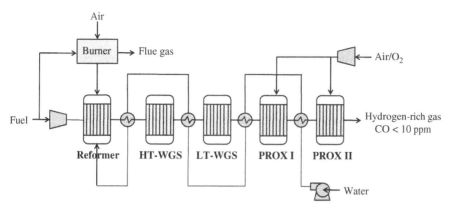

FIGURE 50.2 Fuel processor with two stages of PROX reactors.

the reformate gas for PEMFCs. In this process, no additional reactant is fed to the reactor; CO will react with hydrogen in the presence of selective catalysts to produce methane and water (equation 50.3). However, the methanation process is not usually applied in the hydrogen production process because a significant amount of the hydrogen product is consumed during the elimination of CO compared to that consumed in the PROX reactor. Notably, the control system for the methanation process is simpler than that for the PROX reactor, where the oxygen supply rate and the reaction temperature must be carefully monitored [41]. This monitoring is necessary because the methanation reaction is less exothermic than the CO and hydrogen oxidation reactions. Unlike the PROX process, a compressor is unnecessary for the methanation process, and, thus, no parasitic losses from the pressurized unit are incurred:

$$CO + 3H_2 \leftrightarrow CH_4 + H_2O \qquad (50.3)$$

In general, the PROX and CO methanation reactors provide only a partial CO clean-up; however, a pressure swing absorption (PSA) system and a membrane purification system can completely eliminate CO in the reformate gas. PSA is the absorption system used to separate unwanted components, specifically CO and CO_2, from hydrogen-rich gas and can produce hydrogen with a purity of up to 95–99.99 mol% [42]. However, a large-size reactor and a regeneration process are required for PSA systems. The schematic of the operation of a PSA system is shown in Figure 50.3. Contaminates such as CO and CO_2 are adsorbed onto the surface of the adsorbent, and hydrogen is separated with high purity, whereas a portion of the produced hydrogen is back-purged to the absorber during the regeneration step [43].

Membrane separation is a hydrogen purification method that not only removes CO but also eliminates other inert gases, such CO_2 and N_2, as well as other trace impurities. This purification causes an increase in the H_2 partial pressure in the anode feed of fuel cells. However, this method

of purifying hydrogen is limited by the cost of the noble metal membrane and by the membrane's durability. Moreover, the requirement of high differential pressure across the membrane results in a high parasitic loss. Salemme et al. [32] concluded that the use of a membrane separation unit placed between the WGS reactor and the PEMFC system is not energetically convenient because of the high-pressure operation required to maximize the efficiency of the system. Notably, the PSA and membrane separation methods are not suitable for small-scale applications because an additional compressor is required to increase the pressure. A summary of each CO removal process is presented in Table 50.2.

50.3.2 Innovative Fuel Processing

One of the most serious obstacles to the operation of conventional fuel processors is caused by the thermodynamic equilibrium of reversible reforming reactions. For conventional reformers, high reaction temperatures are required to achieve complete fuel conversion; however, such operating conditions favor the formation of carbon, which leads to the deactivation of the reforming catalysts [44]. Several approaches have been proposed to improve the performance and reduce

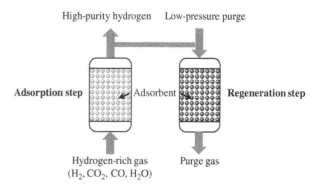

FIGURE 50.3 Schematic of the PSA process.

TABLE 50.2   Summary of CO removal process.

CO removal process	Methods to remove CO	Purity of hydrogen
Preferential oxidation	CO is removed from reformate gas by reacting with oxygen via the CO oxidation reaction	Reformate gas (CO < 10 ppm)
CO methanation	CO is removed from reformate gas by reaction with hydrogen via the methanation reaction	Reformate gas (CO < 10 ppm)
Pressure swing absorption	Hydrogen is separated from CO and other gases in the reformate gas by an absorption technique	Almost pure hydrogen
Membrane separation	Hydrogen is separated from CO and other gases in the reformate gas by a membrane technique	Almost pure hydrogen

the complexity of the fuel processor. The coupling of reaction and separation systems can improve reactant conversion or product selectivity. This coupling results in a lower operating temperature compared to that required for a conventional reforming process, and the lower temperature reduces both the possibility of catalyst fouling and the amount of energy consumed by the process. The selection of different components in the fuel processor significantly affects the efficiency and the cost of the fuel processor.

A membrane reactor is an attractive technology for hydrogen purification. During the reforming of fuel for hydrogen-rich gas production, hydrogen is continually removed from the reaction zone; consequently, the equilibrium-limited reforming reaction is shifted forward to the product side [45]. Typically, the membranes used for purifying hydrogen for PEMFCs are dense metallic membranes made of palladium alloys because such membranes provide good performance and selectivity and can provide almost pure hydrogen for the PEMFC [46, 47]. Actually, a phase change of the material occurs when a pure palladium membrane is employed; therefore, palladium is usually alloyed with other elements. In general, silver (Ag) is commonly used as an alloying element, and the performance of the silver–palladium membrane is better than that of the pure palladium membrane [48].

In the operation of a Pd-based membrane reactor, fuel and steam (in the case of SR) are fed to the reformer, and the reforming reaction occurs at the surface of the catalysts. Hydrogen is produced and diffuses through the membrane simultaneously because of the difference in hydrogen partial pressures between the two sides of the membrane. Other product components, such as CO_2, CO, CH_4, and the remaining steam, which cannot cross the membrane, are emitted from the membrane reactor at the retentate side. In practice, a sweep gas is used to decrease the hydrogen partial pressure at the permeated side, which can enhance the hydrogen permeation. Steam is typically used as a sweep gas for membrane reactors for PEMFC systems because the hydrogen feed stream of PEMFCs needs to be humidified to prevent the membrane from drying out. The schematic

diagram of the Pd-based membrane reactor is shown in Figure 50.4.

Numerous previous studies have shown that a membrane-based fuel processor integrated with a PEMFC system provides higher system efficiency than a conventional fuel processor [49, 50]. Lattner and Harold [7] reported that the total volume of a conventional fuel processor that consisted of a reformer, a WGS reactor, and a PROX system for a PEMFC was 20% greater than that of a fuel processor based on a membrane reactor. However, because of the relatively low hydrogen permeability of the Pd-based membrane, a low hydrogen recovery of the reforming process was observed. In addition, another serious problem of the reforming membrane reactor is its low membrane stability at high reforming temperatures [51].

Because a WGS reaction is thermodynamically limited, as is the reforming reaction, the integration of Pd-based membranes with WGS reactors is also a promising technology to reduce the system complexity and to provide CO-free fuel for PEMFC systems [52, 53]. A schematic of a fuel processor based on a WGS membrane reactor is presented in Figure 50.5. Compared to the reforming membrane reactor, the WGS membrane reactor exhibits greater membrane stability at low WGS temperatures and requires a lower membrane area due to the high hydrogen driving force [51]. Therefore, the WGS membrane reactor has been investigated by numerous researchers [54–56]. In the WGS membrane reactor, the hydrogen content of the reformate gas increases because

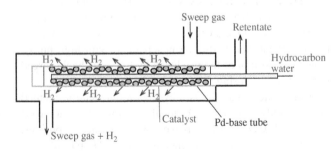

FIGURE 50.4   Schematic of a Pd-based membrane reactor.

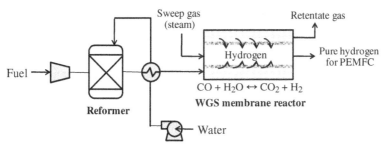

FIGURE 50.5   A WGS membrane fuel processor.

of the WGS reaction. The continuous removal of hydrogen enhances the conversion of CO over its thermodynamic equilibrium limitation. Notably, an optimal membrane reactor design is required to minimize the required membrane area for a given hydrogen recovery.

A comparison of the various aspects of conventional and innovative fuel processing methods for PEMFC applications is summarized in Table 50.3. Because a membrane reactor requires a lower operating temperature than does a conventional process to achieve the desired conversion of fuel, its energy consumption for hydrogen production is reduced. In addition, the use of the membrane reactor can reduce the complexity of the fuel processor and improve the quality of the obtained product gas. As previously mentioned, a conventional fuel processor integrates a reformer, a WGS reactor, and CO removal units to produce hydrogen-rich gas for PEMFCs; in contrast, only one (e.g., a membrane-based reformer) or two reactors (e.g., a reformer integrated with a WGS membrane reactor) are needed in innovative fuel processing. High-purity hydrogen is also obtained through the use of a membrane-based reformer, whereas the conventional fuel processor produces hydrogen fuel gas mixed with CO_2, CH_4, and traces of CO (<10 ppm). Notably, in addition to the CO poisoning effect, the concentration of hydrogen also affects the performance of a PEMFC: a higher hydrogen concentration gives better PEMFC performance. The need for high-pressure operation of membrane-based reformers causes high parasitic losses from auxiliary units, such as a compressor and a pump.

TABLE 50.3   Comparison of conventional and innovative fuel processing.

	Conventional fuel processor	Innovative fuel-processing-based membrane reactor
Reformer temperature	High	Low
Complexity	High	Low
Hydrogen purity	Low	High
Parasitic loss from auxiliary unit	Low	High
Cost	Low	High

50.4   PEMFC SYSTEM AT REFORMATE GAS OPERATION

There are several difficulties associated with the operation of a PEMFC that still need to be resolved. The intrinsic problems of PEMFCs are mainly attributed to water management and a CO poisoning effect. Either pure hydrogen or hydrogen-rich gas (reformate gas) from a reforming process can be used as fuel in PEMFC anodes. The water management is the critical issue for PEMFCs. Because of the proton conductivity mechanism of the membrane electrolyte in PEMFCs, which depends on the water content, the fuel stream must be humidified to avoid the loss of PEMFC performance due to membrane dehydration. The liquid water generated at the cathode catalyst layer due to the electrochemical reaction and the proton conductivity mechanism (electro osmotic drag) causes a water flooding problem. The liquid water blocks the reactant gas (oxygen) to reach active sites, which results in a decreased voltage. To solve the water flooding problem, numerous researchers have investigated the water management issues within the cell [57,58]. Several solutions to this critical issue have been proposed, including the addition of a hydrophobic material to the gas diffusion layer, the addition of a microporous layer (MPL), and the design of an effective flow field.

The CO poisoning problem occurs when reformate gas is used as fuel in a PEMFC. At present, the reformate gas derived from fuel processors is the preferred fuel for PEMFCs until methods for hydrogen transport and storage become readily available. However, contaminants in the reformate gas, especially CO, poison the anode's platinum electro-catalyst, which reduces the catalyst's activity and, consequently, the fuel cell's power output. Various aspects of the performance and reliability of PEMFC systems run on reformate gas will be reviewed and compared with the performance of PEMFC systems run on hydrogen fuel.

50.4.1   Effects of Contaminants

The reformate gases obtained from a fuel processing method typically contain traces of CO, which can strongly adsorb onto the surface of Pt and occupy HOR sites. The adsorption of CO and H_2 onto the Pt surface can be described by Langmuir adsorption isotherms, as shown in equations (50.4) and (50.5). The CO molecule adsorbs associatively onto Pt at temperatures less than 500 K, whereas H_2 is dissociatively adsorbed [59]. Trace CO dramatically reduces the activity of Pt or Pt alloys in the anode, which results in the deterioration of PEMFC performance:

$$CO(g) + Pt \rightarrow Pt - CO \qquad (50.4)$$

$$H_2(g) + 2Pt \rightarrow 2Pt - H \qquad (50.5)$$

The performance of PEMFCs fed by reformate gas typically relies on the fraction of CO in the fuel feed stream, the temperature, and the current density. Das et al. [60] have reported that the CO poisoning effect is diminished at high temperatures and at low current density. However, only trace quantities of CO can cause a significant voltage loss in the fuel cell, and the CO content in the hydrogen feed typically must be less than 10 ppm.

In addition to the CO poisoning effect, the influence of other contaminants in the reformate gas, such as CO_2, on PEMFC performance has also been studied by numerous research groups [61–65]. Typically, the carbon dioxide content in the reformate gas is approximately 20–25%, depending on the type of reactants and operating conditions used [62]. Hedstrom et al. [65] have found that the dilution of fuel by CO_2 has a minimal effect on PEMFC performance. Based on the Nernst equation, the thermodynamic loss due to the dilution effect is approximately 17 mV at a hydrogen concentration of 30% and a temperature of 60°C. However, CO_2 has a more significant effect on cell performance than do other inert gases, such as nitrogen. In addition, numerous researchers have shown that the presence of carbon dioxide in the fuel can lead to a significant degradation of the stack performance. This degradation is mainly due to the reverse WGS reaction (equation 50.6) [66]. As a result, the content of CO in the fuel cell stack increases, and the activity of the Pt catalyst decreases. Notably, the CO_2 poisoning problem is more pronounced at low CO concentrations in the fuel feed stream and at low current density [67]. In addition, Giddey et al. [68] studied the effects of methane, carbon dioxide, and water in the hydrogen fuel with respect to the cell performance. The presence of methane at concentrations up to 10% was found to have a negligible effect on the cell voltage:

$$CO_2 + H_2 \rightarrow CO + H_2O \qquad (50.6)$$

50.4.2 Operating Conditions

Operating conditions, for example, temperature, relative humidity (RH), pressure, and flow rate, significantly affect the fuel cell performance. From some points of view, the influence of operating conditions on the performance of PEMFCs operated on reformate gas differs from that of PEMFCs operated on pure hydrogen. Indeed, an increase in PEMFC operating temperatures causes a reduction of the theoretical cell voltage based on the Nernst equation. However, the actual cell voltage is typically enhanced with an increase in temperature because voltage losses tend to decrease when the fuel cell temperature is increased. For PEMFCs operated on pure hydrogen, their performance is improved with an increase in the temperature from 65°C to 75°C, is unchanged between 75°C and 80°C, and begins to decrease at 85°C [69]. The increased cell performance in the temperature range of 65–75°C is because gas diffusivity is enhanced and, thus, the voltage losses caused by a mass transfer limitation are diminished. However, a continued increase in the cell temperature causes a low RH of the fuel gas, which results in the dehydration of membrane and, consequently, poor performance of the PEMFC.

In the case of a PEMFC run on reformate gas, high-temperature operation has been reported to improve the performance of the PEMFC [70]. From a thermodynamic point of view, the adsorption of CO onto the Pt surface can be diminished by increasing the temperature and/or decreasing the CO concentration [71]. As a result of the high operating temperature, the CO tolerance of PEMFCs is enhanced. In general, CO coverage on Pt is decreased at higher temperatures. Because the adsorption of CO is an exothermic process, the number of Pt-active sites for hydrogen adsorption increases. This increase implies that the possibility of hydrogen adsorption, which is less exothermic than CO adsorption, increases at higher temperatures. Li et al. [72] have also investigated the relative activity of Pt catalysts for hydrogen oxidation as a function of temperature at different CO concentrations and found that the oxidation of hydrogen can be promoted through high-temperature operation.

Although the operation of PEMFCs at high temperatures can overcome the CO poisoning problem, high-temperature operation leads to dehydration of the membrane and to loss of the membrane's ionic conductivity. The efficiency of conventional PEMFCs relies on the presence of water in the membrane, and its performance is degraded when the water content is low. Figure 50.6 shows the saturated vapor pressure at different temperatures; the saturated vapor pressure increases with temperature. This figure indicates that, at elevated temperatures and at atmospheric pressure, the PEMFC must be operated at low RH or at low hydrogen and oxygen concentrations. Hydrogen and oxygen directly influence cell performance, especially during high-current-density operation. To address this difficulty, the PEMFC should be operated at high pressure to maintain high RH and high fuel concentration during high-temperature operation. However, the operating pressure for conventional PEMFCs is usually limited to less than 4 atm; consequently, supplying reformate gas with a high RH is difficult [73–75]. Therefore, a newly

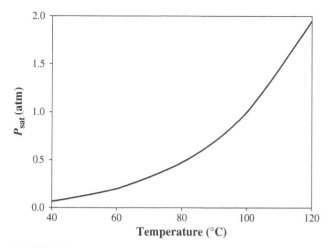

FIGURE 50.6 Saturated vapor pressure at different temperatures (REF.).

modified membrane that can be operated at low humidity has been developed [76, 77].

50.4.3 System Complexity

Typically, the design of a PEMFC system depends on the available or chosen fuels. As previously mentioned, either pure hydrogen or hydrogen-rich gas from reforming processes can be used as a fuel in PEMFC anodes. A PEMFC system operated on hydrogen-rich gas would be more complicated than a system operated on pure hydrogen because fuel processing, which consists of hydrogen production and hydrogen purification units, would be added to the system to produce hydrogen-rich gas for the PEMFC. For operation of a PEMFC with pure hydrogen, the hydrogen can be produced elsewhere and then stored as a part of the system. In such cases, the PEMFC system consists only of a storage tank, a fuel cell, and some auxiliary units. CO poisoning is an important problem with PEMFCs operated on reformate gas. A humidifier unit is necessary for a PEMFC operated on pure hydrogen; however, a humidifier is not necessary at the anode side of a PEMFC when reformate gas from the SR process is used because it is usually saturated with water.

With respect to the heat integration of a PEMFC and the reforming processes, the energy required for the reforming process can be supplied by waste heat from the fuel cell to preheat the fuel or to supply the reforming processes with heat. Furthermore, because the fuel processor consumes steam and the fuel cell simultaneously produces some steam, the balancing of steam recovered from the fuel cell exhaust gas with that required for the reforming process is possible. Consequently, the fuel processor needs to be fully integrated with the fuel cell system. The fuel processor not only provides fuel to the fuel cell but also uses heat from the fuel cell

exhaust gases. In addition, the fuel processor and fuel cell share air, water, coolant, and control subsystems.

50.4.4 Technology Development for PEMFCS Operated With Reformate Gas

To support the use of reformate gas as a fuel for PEMFCs, several technologies, including oxygen bleeding, catalysts with a high CO tolerance, and HT-PEMFCs, have been developed. The objectives for the development of such systems include improving the PEMFC system efficiency.

50.4.4.1 HT-PEMFCs. As previously mentioned, the operation of PEMFCs at high temperatures can mitigate the CO poisoning problem. Therefore, a high-temperature PEMFC (HT-PEMFC) operated at 100–200°C has been developed. Under high-temperature operation, the amount of CO that adsorbs onto the Pt catalyst used in HT-PEMFCs is reduced, which results in a high CO tolerance. However, the operation of PEMFCs at high temperatures leads to dehydration of the membrane and to the loss of the membrane's ionic conductivity. Therefore, numerous researchers have attempted to develop new membranes that can be operated at temperatures greater than 100°C and that exhibit high conductivity under low-humidity conditions. A polybenzimidazole (PBI) was used in HT-PEMFCs because it can be operated at low RH. However, the proton conductivity of PBI is lower than that of Nafion; thus, PBI was doped with phosphoric acid or other dopants to increase its proton conductivity [76, 78]. This membrane can be operated under dry conditions, and the humidifier can therefore be eliminated from the fuel cell system.

Table 50.4 summarizes the CO tolerance properties of HT-PEMFCs reported in the literature. Li et al. [72] demonstrated that a HT-PEMFC can tolerate CO at concentrations up to 3% at a temperature of 200°C and can generate electricity at 0.8 A/cm [2] with voltage losses of less than 10 mV. Das et al. [60] reported that the CO poisoning problem of a PEMFC operated at high temperatures was diminished compared to that of a PEMFC operated at low temperatures. When the PEMFC was operated at a temperature of 180°C or greater, a reforming gas with 2–5% CO was used without a significant loss of cell performance. Later, Mamlouk and Scott [79]

TABLE 50.4 Summary of CO tolerance.

Literatures	Temperature (°C)	% CO tolerance
Li et al. [72]	200	3
Korsgaard et al. [80]	160	2
Das et al. [60]	180	5
Mamlouk and Scott [79]	175	10
Jiao et al. [81]	190	10

proposed that a HT-PEMFC can be operated at CO concentrations as high as 10% by volume, whereas Korsgaard et al. [80] reported that a HT-PEMFC operated at 160°C can tolerate CO at a concentration of 2%. Recently, Jiao et al. [81] developed a non-isothermal model for a HT-PEMFC by considering the effect of CO on the performance of the fuel cell. They found that CO drastically affects the performance of a HT-PEMFC at 190°C when the CO fraction in the hydrogen feed is greater than 10%. They also reported that the reformate gas from the methanol reformer (CO ≈ 1–2%) can be fed directly to a HT-PEMFC operated at 160°C. In addition to the fuel cell operating temperature, the catalyst used also affects the %CO tolerance of HT-PEMFCs.

Furthermore, according to the literature, the reverse WGS reaction ceases under the HT-PEMFC operating conditions due to the higher cell temperature and the higher acceptable concentration of CO in the anode feed stream. In addition to the dilution effect, CO_2 in the reformate gas does not have any effect on cell performance [81]. However, some studies have shown that water in the hydrogen feed can diminish the CO poisoning effect and enhance cell performance [82]. This effect occurs because a large amount of water favors the WGS reaction (equation 50.7), and the hydrogen produced from this reaction can be used to generate protons and elec-

tricity through the HOR represented in equation (50.8). The overall reaction (CO oxidation reaction) can be described by equation (50.9).

Water–gas shift reaction:

$$CO + H_2O \Leftrightarrow CO_2 + H_2 \qquad (50.7)$$

Hydrogen oxidation reaction:

$$H_2 \rightarrow 2H^+ + 2e^- \qquad (50.8)$$

CO oxidation reaction:

$$CO + H_2O \rightarrow CO_2 + 2H^+ + 2e^- \qquad (50.9)$$

Because of the high CO tolerance of HT-PEMFCs, reformate gas can be used directly as a fuel for HT-PEMFCs without a complicated purification process, which could make the design of the fuel processor for HT-PEMFCs simpler than that for conventional PEMFCs. The CO removal unit, such as a PROX unit, is unnecessary for this type of fuel cell. The reformate gas can be fed directly to the HT-PEMFC. Alternatively, a WGS reactor can be installed between the reformer and the HT-PEMFC to increase the hydrogen content and reduce the CO content of the fuel. One possible design for a HT-PEMFC system is illustrated in Figure 50.7.

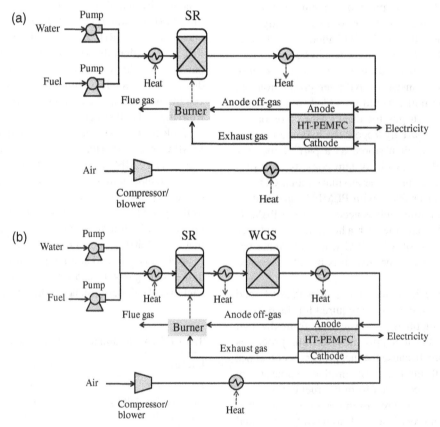

FIGURE 50.7   The possible designs for HT-PEMFC systems: (a) a HT-PEMFC with only a steam reformer and (b) a HT-PEMFC with a reformer and a water–gas shift reactor.

The higher operating temperature of PEMFCs also allows the efficient utilization of waste heat from the fuel cell to preheat the fuel feed stream or to supply heat to the reforming process. Jespersen et al. [83] have claimed that a HT-PEMFC is more suitable than a LT-PEMFC for the integration of an on-board fuel processing unit. Furthermore, the high-temperature operation of PEMFCs increases the electrochemical reaction rate at the anode and cathode and simplifies water management within PEMFCs. When PEMFCs are operated at temperatures greater than 100°C, water is present only in the vapor phase. For this reason, the flooding problem is solved and the transport of water is easily balanced.

50.4.4.2   CO-Tolerant Catalysts.

The development of CO-tolerant catalysts is one approach to alleviate the CO poisoning problem in PEMFCs. The new generation of CO-tolerant catalysts can reduce anode activation losses caused by the presence of CO in the fuel feed, which decreases the reformate clean-up load and simplifies the design of a PEMFC system integrated with a hydrocarbon fuel processing unit for hydrogen production. Numerous research groups have modified Pt by alloying it with other metals, including Ru, Sn, Mo, Co, Cr, Fe, Ni, Pd, Os, and Mn, and have demonstrated the improved CO tolerance properties of these alloyed catalysts [84–87]. Among the various types of catalysts, Pt–Ru is one of the most interesting catalysts for PEMFCs that run on reformate gas [88–90]. Compared to Pt, Pt–Ru alloys show lower overpotentials and greater poisoning tolerance [91]. Pt–Ru has been reported to provide four times greater activity than pure Pt in the presence of 100 ppm CO/H$_2$ anode fuel [85]. Three possible mechanisms have been proposed to explain the increased CO tolerance of the Pt–Ru catalysts compared to that of pure Pt.

Bifunctional mechanism. In this mechanism, ruthenium was suggested to provide oxygen-containing species at more negative potentials than those of platinum [92]. Ruthenium is more efficient in the dissociation of water and provides OH groups to oxidize the CO adsorbed at Pt sites to CO$_2$ [93,94].

Ligand-effect mechanism. Under the ligand-effect mechanism, the combination of platinum and ruthenium causes a decrease in the Pt–CO bond strength, and the rate of CO oxidation at the surface is consequently increased [95].

Detoxification mechanism. This mechanism presumes that the presence of ruthenium leads to a weakening of the Pt–CO bonds and that a lower equilibrium CO surface coverage exists at the Pt–Ru surface. Consequently, free sites for hydrogen oxidation are more available [96,97].

However, PEMFCs based on Pt–Ru catalysts can tolerate CO only to a concentration of 100 ppm [98,99]. A number of investigations to synthesize PEMFC catalysts that can tolerate high CO content in the hydrogen feed stream have been reported. These include the development of ternary catalysts, such as PtRuW, PtRuSn, PtRuIr, and PtRuMo [100–104].

50.4.4.3   O$_2$/Air Bleeding.

O$_2$/air bleeding is a common method to mitigate the CO poisoning problem at the anode of low-temperature PEMFCs. With this approach, a small amount of O$_2$ or air is added to the H$_2$-rich gas feed to reduce the fraction of CO that is permanently adsorbed onto the Pt catalyst by reacting with CO to form CO$_2$. The reduction of CO in the hydrogen feed increases the active site for the HOR. The reaction pathway is shown in equations (50.10), (50.11), and (50.12), and the overall reaction is given in equation (50.13).

$$Pt + O_2 \rightarrow 2(Pt - O) \qquad (50.10)$$

$$Pt + CO \rightarrow Pt - CO \qquad (50.11)$$

$$Pt - CO + Pt - O \rightarrow 2Pt + CO_2 \qquad (50.12)$$

$$O_2 + 2CO \rightarrow 2CO_2 \qquad (50.13)$$

Gottesfeld and Pafford [105] found that the cell performance could be completely recovered when 4.5% O$_2$ was injected in the presence of 100 ppm CO in the hydrogen feed. Sung et al. [106] also suggested that the CO poisoning effect can be reduced by injecting a small amount of air into the anode feed stream. An injection of 5% air (1% O$_2$) into the PEMFC anode resulted in the best cell performance with 53 ppm CO in the anode fuel. Through durability testing, they observed that the cell voltage degradation was less than 3% with 25 ppm CO after more than 3000 hours of testing.

Recently, Chung et al. [107] reported that the air bleeding technique can improve the CO tolerance of the cell and recover its performance, irrespective of the timing of air addition and the type of catalyst used (pure Pt vs. Pt–Ru) at a CO level of 52.7 ppm. Bhatia and Wang [108] reported that O$_2$ bleeding does not solve the CO poisoning problem when the CO concentration in the anode feed stream is high (greater than 100 ppm). When Pt catalysts are used at the anode, the maximum CO concentration that can be treated by air bleeding is approximately 100 ppm. In addition, Murthy et al. [109] concluded that the performance of a PEMFC operated with 500 ppm CO in the anode feed can be recovered after being air bled with 5 vol% air when a CO-tolerant Pt alloy is used. However, a portion of the hydrogen in the fuel feed also reacts with O$_2$ to form hydrogen peroxide (H$_2$O$_2$), which severely affects the proton exchange membrane during long-term operation [110]. In addition, the O$_2$ bleeding causes a hot spot and is not a perfect solution to the CO poisoning problem because it degrades the fuel cell performance in the

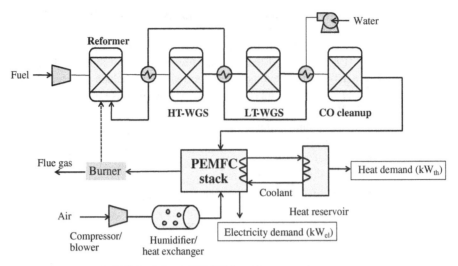

FIGURE 50.8   A PEMFC-based cogeneration system.

long term [106]. In addition, the limitation of O_2/H_2 mixtures to 5% O_2 in H_2 should be observed for safety [111].

50.5   PROSPECT OF USING AN INTEGRATED PEMFC SYSTEM FOR STATIONARY AND AUTOMOTIVE APPLICATIONS

Until a suitable infrastructure for the storage and transport of is available, the use of reformate gas from fuel processors is a promising alternative hydrogen source for PEMFCs. Therefore, integrated systems of a fuel processor and a PEMFC have been continuously studied [112–114]. As previously noted, PEMFCs have some limitations and specifications with respect to their operation. Therefore, many studies on integrated PEMFC systems have focused on the design and selection of an efficient reformer, on the CO removal process, on the types of fuel used, and on the management of heat and water between the fuel processor and the fuel cell to improve the system efficiency. Notably, although integrated fuel processor and PEMFC systems have been widely investigated, the process design for the PEMFC system for real applications, such as stationary and automotive applications, is still limited. In general, the complexity of the integrated system depends on the fuel used and the specific requirements for each fuel cell application [115]. For automotive applications, weight and size of the overall system are critical. A fast start-up and a rapid response to load changes are also preferred in this application; thus, the ATR of fuels is more suitable for this application than is SR [116]. Research on the use of PEMFCs in this application has concentrated on pure hydrogen operation [117, 118].

In contrast, stationary applications require high overall efficiency without specific weight or size restrictions [119]. Heat and power cogeneration is favorably suited to stationary applications as an effective way to improve the overall system efficiency. For low-temperature fuel cells, a high system efficiency can be achieved when the low-quality heat released from cell is utilized for household use [120]. The PEMFC cogeneration system is considered to be an effective system that meets both the electrical and thermal demands of residential applications [121]. A schematic of a PEMFC cogeneration system is shown in Figure 50.8.

Based on literature reviews, reformed natural gas is widely used to fuel PEMFCs in stationary applications. Mathiak et al. [122] developed a 2.5 kW steam reformer with a 1 kW$_{el}$ PEMFC for a one-family household. They reported that the recirculation of the anode off-gas to the reformer burner improved the overall efficiency. Automatic control was required to reduce the load-change time of the fuel cell system in a residential application. Furtado et al. [123] found that a fuel cell integrated with a natural gas reformer in a cogeneration system produced electricity at a cost 20% lower than that of the pure electrical generation case. They claimed that based on the hydrogen obtained from the natural gas reformer, the system operated under excellent electrothermal stability conditions and, thus, the energy conversion efficiency and the economics of the cogeneration power plant improved. Oh et al. [120] developed a 1 kW PEMFC system operated on reformed natural gas for residential applications in Korea. They also studied the combined heat and power of this system and found that 20% of the operational cost can be saved, irrespective of the installation cost.

50.6   CONCLUSIONS

PEMFCs are a promising clean and effective technology, and a PEMFC-integrated system is considered to be a potential electricity generation solution for small-scale stationary and automotive applications. In this chapter, the thermal-chemical processes and fuel resources for hydrogen production for PEMFCs were reviewed. For environmental and availability reasons, renewable resources, such as biomass, biogas, bio-ethanol, and glycerol, are promising fuels for a PEMFC-integrated system. In addition, liquid fuels, such as ethanol, methanol, diesel, and gasoline, are preferred because they are easily transported and their refueling infrastructure is well established. The low-temperature operation and low CO content in hydrogen-rich gas produced from methanol reforming is a suitable hydrogen production process for low-temperature fuel cells such as the PEMFC. The design of a hydrogen purification process is one of the key issues in improving the overall efficiency and compactness of an integrated PEMFC system. The trade-off between the performance, size, and efficiency of a PEMFC system should also be considered. The oxygen bleeding technique, CO-tolerant catalysts, and the HT-PEMFC are among the newly developed technologies to solve the CO poisoning problem in PEMFCs. Research on an integrated system of a fuel processor and a PEMFC should be performed to improve the reliability of PEMFC systems for real applications.

ACKNOWLEDGMENTS

Support from the Thailand Research Fund is gratefully acknowledged.

REFERENCES

[1] Semelsberger, T. A., Borup, R. L. (2005). Fuel effects on start-up energy and efficiency for automotive PEM fuel cell systems. *International Journal of Hydrogen Energy*, 30:425–435.

[2] Authayanun, S., Mamlouk, M., Arpornwichanop, A. (2012). Maximizing the efficiency of a HT-PEMFC system integrated with glycerol reformer. *International Journal of Hydrogen Energy*, 37:6808–6817.

[3] Farrauto, R. J. (2005). Introduction to solid polymer membrane fuel cells and reforming natural gas for production of hydrogen. *Applied Catalysis B: Environmental*, 56:3–7.

[4] Heinzel, A., Vogel, B., Hubner, P. (2002). Reforming of natural gas-hydrogen generation for small scale stationary fuel cell system. *Journal of Power Source*, 105:202–207.

[5] Perna, A., Cicconardi, S. P., Cozzolino, R. (2001). Performance evaluation of a fuel processing system based on membrane reactors technology integrated with a PEMFC stack. *International Journal of Hydrogen energy*, 36:9906–9915.

[6] Laosiripojana, N., Assabumrungrat, S. (2006). Hydrogen production from steam and autothermal reforming of LPG over high surface area ceria. *Journal of Power Sources*, 158:1348–1357.

[7] Lattner, J. R., Harold, M. P. (2004). Comparison of conventional and membrane reactor fuel processors for hydrocarbon-based PEM fuel cell systems. *International Journal of Hydrogen energy*, 29:393–417.

[8] Amphlett, J. C., Mann, R. F., Peppley, B. A., Roberge, P. R., Rodrigues, A., Salvador, J. P. (1998). Simulation of a 250 kW diesel fuel processor/PEM fuel cell system. *Journal of Power Sources*, 71:179–184.

[9] Agrell, J., Birgersson, H., Boutonnet, M. (2002). Steam reforming of methanol over a Cu/ZnO/Al$_2$O$_3$ catalyst: a kinetic analysis and strategies for suppression of CO formation. *Journal of Power Sources*, 106:249–257.

[10] Adachi, H., Ahmed, S., Lee, S. H. D., Papadias, D., Ahluwalia, R. K., Bendert, J. C., Kanner, S. A., Yamazaki, Y. (2009). A natural-gas fuel processor for a residential fuel cell system. *Journal of Power Sources*, 188:244–255.

[11] Chipiti, F., Recupero, V., Pino, L., Vita, A., Lagana, M. (2006). Experimental analysis of a 2 kWe LPG-based fuel processor for polymer electrolyte fuel cells. *Journal of Power Sources*, 157:914–920.

[12] Severin, C., Pischinger, S., Ogrzelwalla, J. (2005). Compact gasoline fuel processor for passenger vehicle APU. *Journal of Power Sources*, 145:675–682.

[13] Castaldi, M. J., Barrai, F. (2007). An investigation into water and thermal balance for a liquid fueled fuel processor. *Catalysis Today*, 129:397–406.

[14] Danial, D. E., Kumar, R., Ahluwalia, R. K., Krumpelt, M. (2001). Fuel processors for automotive fuel cell systems: a parametric analysis. *Journal of Power Sources*, 102:1–15.

[15] Faungnawakij, K., Kikuchi, R., Eguchi, K. (2006). Thermodynamic evaluation of methanol steam reforming for hydrogen production. *Journal of Power Sources*, 161:87–94.

[16] Chein, R. Y., Chen, Y. C., Lin, Y. S., Chung, J. N. (2011). Experimental study on the hydrogen production of integrated methanol-steam reforming reactors for PEM fuel cells. *International Journal of Thermal Sciences*, 50:1253–1262.

[17] Lattner, J. R., Harold, M. P. (2005). Comparison of methanol-based fuel processors for PEM fuel cell systems. *Applied Catalysis B: Environmental*, 56:149–169.

[18] Oliva, D. G., Francesconi, J. A., Mussati, M. C., Aguirre, P. A. (2010). Energy efficiency analysis of an integrated glycerin processor for PEM fuel cells: comparison with an ethanol-based system. *International Journal of Hydrogen Energy*, 35:709–724.

[19] Lopes, D. G., da Silva, E. P., Pinto, C. S., Neves, N. P., Camargo, J. C., Ferreira, P. F. P., Furlan, A. L., Lopes, D. G. (2012). Technical and economic analysis of a power supply system based on ethanol reforming and PEMFC. *Renewable Energy*, 45:205–212.

[20] Rossetti, I., Biffi, C., Tantardini, G. F., Raimondi, M., Vitto, E., Alberti, D. (2012). 5 kW$_e$ + 5 kW$_t$ reformer-PEMFC energy generator from bioethanol first data on the fuel processor from a demonstrative project. *International Journal of Hydrogen Energy*, 37:8499–8504.

[21] Lima, S. M., Colman, R. C., Jacobs, G., Davis, B. H., Souza, K. R., de Lima, A. F. F., Appel, L. G., Mattos, L. V., Noronha,

F. B. (2009). An integrated fuel processor comprising ethanol steam reforming and preferential oxidation of CO. *Catalysis Today*, 146:110–123.

[22] Demirbas, A. (2009). Biodiesel from waste cooking oil via basecatalytic and supercritical methanol transesterification. *Energy Conversion Management*, 50:923–927.

[23] Byrd, A. J., Pant, K. K., Gupta, R. B. (2008). Hydrogen production from glycerol by reforming in supercritical water over Ru/Al$_2$O$_3$ catalyst. *Fuel*, 87:2956–2960.

[24] Pompeo, F., Santori, G., Nichio, N. N. (2010). Hydrogen and/or syngas from steam reforming of glycerol. Study of platinum catalysts. *International Journal of Hydrogen Energy*, 35:8912–8920.

[25] Chen, H., Ding, Y., Cong, N. T., Dou, B., Dupont, V., Ghadiri, M., Williams, P. T. (2011). A comparative study on hydrogen production from steam-glycerol reforming: thermodynamics and experimental. *Renewable Energy*, 36:779–788.

[26] Martin, S., Worner, A. (2011). On-board reforming of biodiesel and bioethanol for high temperature PEM fuel cells: comparison of autothermal reforming and steam reforming. *Journal of Power Sources*, 196:3163–3171.

[27] Toonssen, R., Woudstra, N., Verkooijen, A. H. M. (2009). Decentralized generation of electricity from biomass with proton exchange membrane fuel cell. *Journal of Power Sources*, 194:456–466.

[28] Schmersahl, R., Scholz, V. (2005). Testing a PEM fuel cell system with biogas fuel. *Agricultural Engineering International: the CIGR Ejournal*, 7:1–12.

[29] Pandya, J. D., Ghosh, K. K., Rastogi, S. K. (1988). A phosphoric acid fuel cell coupled with biogas. *Energy*, 13:383–388.

[30] Hedstrom, L., Tingelof, T., Alvfors, P., Lindbergh, G. (2009). Experimental results from a 5 kW PEM fuel cell stack operated on simulated reformate from highly diluted hydrocarbon fuels: efficiency, dilution, fuel utilisation, CO poisoning and design criteria. *International Journal of Hydrogen Energy*, 34:1508–1514.

[31] Vagia, E. C., Lemonidou, A. A. (2008). Thermodynamic analysis of hydrogen production via autothermal steam reforming of selected components of aqueous bio-oil fraction. *International Journal of Hydrogen Energy*, 33:2489–2500.

[32] Salemme, L., Menna, L., Simeone, M. (2009). Analysis of the energy efficiency of innovative ATR-based PEM fuel cell system with hydrogen membrane separation. *International Journal of Hydrogen Energy*, 34:6384–6392.

[33] Rabenstein, G., Hacker, V. (2008). Hydrogen for fuel cells from ethanol by steam-reforming, partial-oxidation and combined auto-thermal reforming: a thermodynamic analysis. *Journal of Power Sources*, 185:1293–1304.

[34] Ersoz, A., Olgun, H., Ozdogan, S. (2006). Reforming options for hydrogen production from fossil fuels for PEM fuel cells. *Journal of Power Sources*, 154:67–73.

[35] Ahmed, S., Krumpelt, M. (2001). Hydrogen from hydrocarbon for fuel cells. *International Journal of Hydrogen Energy*, 26:291–301.

[36] Hu, J. E., Pearlman, J. B., Jackson, G. S., Tesluk, C. J. (2010). Evaluating the impact of enhanced anode CO tolerance on performance of proton-exchange-membrane fuel cell systems fueled by liquid hydrocarbons. *Journal of Power Sources*, 195:1926–1935.

[37] Giunta, P., Mosquera, C., Amadeo, N., Laborde, M. (2007). Simulation of a hydrogen production and purification system for a PEM fuel-cell using bioethanol as raw material. *Journal of Power Sources*, 164:336–343.

[38] Ahluwalia, R. K., Zhang, Q., Chmielewski, D. J., Lauzze, K. C.Inbody, M. A. (2005). Performance of CO preferential oxidation reactor with noble-metal catalyst coated on ceramic monolith for on-board fuel processing applications. *Catalysis Today*, 99:271–283.

[39] Cipiti, F., Recupero, V. (2009). Design of a CO preferential oxidation reactor for PEFC systems: a modelling approach. *Chemical Engineering Journal*, 146:128–135.

[40] Marino, F., Descorme, C., Duprez, D. (2005). Supported base metal catalysts for the preferential oxidation of carbon monoxide in the presence of excess hydrogen (PROX). *Applied Catalysis B: Environmental*, 58:175–183.

[41] Galletti, C., Specchia, S., Saracco, G., Specchia, V. (2010). CO-selective methanation over Ru–γAl$_2$O$_3$ catalysts in H$_2$-rich gas for PEMFC applications. *Chemical Engineering Science*, 65:590–596.

[42] Vasiliev, L. L., Kanonchik, L. E., Kulakov, A. G., Mishkinis, D. A., Safonova, A. M., Luneva, N. K. (2007). New sorbent materials for the hydrogen storage and transportation. *International Journal of Hydrogen Energy*, 32:5015–5025.

[43] Majlan, E. H., Daud, W. R. W., Iyuke, S. E., Mohamad, A. B., A. Kadhum A. H., Mohammad, A. W., Takriff, M. S., Bahaman, N. (2009). Hydrogen purification using compact pressure swing adsorption system for fuel cell. *International Journal of Hydrogen Energy*, 34:2771–2777.

[44] Trimm, D. L. (1997). Coke formation and minimization during steam reforming reactions. *Catalysis Today*, 37:233–238.

[45] Tong, J., Matsumura, Y. (2006). Pure hydrogen production by methane steam reforming with hydrogen-permeable membrane reactor. *Catalysis Today*, 111:147–152.

[46] Rakib, M. A., Grace, J. R., Lim, C. J, Elnashaie, S. S. E. H., Ghiasi, B. (2010). Steam reforming of propane in a fluidized bed membrane reactor for hydrogen production. *International Journal of Hydrogen Energy*, 35:6276–6290.

[47] Yu, C. Y., Lee, D. W., Park, S. J., Lee, K. Y., Lee, K. H. (2009). Study on a catalytic membrane reactor for hydrogen production from ethanol steam reforming. *International Journal of Hydrogen Energy*, 34:2947–2954.

[48] Shu, J., Grandjean, B. P. A., Van Neste, A., Kaliaguine, S. (1991). Catalytic palladium-based membrane reactors: a review. *Canadian Journal of Chemical Engineering*, 69:1036–1060.

[49] Roses, L., Gallucci, F., Manzolini, G., Campanari, S., Annaland, M. S. (2011). Comparison between fixed bed and fluidized bed membrane reactor configurations for PEM based micro-cogeneration systems. *Chemical Engineering Journal*, 171:1415–1427.

[50] Campanari, S., Macchi, E., Manzolini, G. (2008). Innovative membrane reformer for hydrogen production applied to PEM micro-cogeneration: simulation model and thermodynamic analysis. *International Journal of Hydrogen Energy*, 33:1361–1373.

[51] Abdollahi, M., Yu, J., Liu, P. K. T., Ciora, R., Sahimi, M., Tsotsis, T. T. (2012). Ultra-pure hydrogen production from reformate mixtures using a palladium membrane reactor system. *Journal of Membrane Science*, 390–391:32–42.

[52] Koc, R., Kazantzis, N. K, Ma, Y. H. (2011). Process safety aspects in water-gas-shift (WGS) membrane reactors used for pure hydrogen production. *Journal of Loss Prevention in the Process Industries*, 24:852–869.

[53] Babita, K., Sridhar, S., Raghavan, K. V. (2011). Membrane reactors for fuel cell quality hydrogen through WGSR-review of their status, challenges and opportunities. *International Journal of Hydrogen Energy*, 36:6671–6688.

[54] Mendes, D., Chibante, V., Zheng, J. M., Tosti, S., Borgognoni, F., Mende, A., Madeira, L. M. (2010). Enhancing the production of hydrogen via water gas shift reaction using Pd-based membrane reactors. *International Journal of Hydrogen Energy*, 35:12596–12608.

[55] Iyoha, O., Enick, R., Killmeyer, R., Howard, B., Morreale, B., Ciocco, M. (2007). Wallcatalyzed water-gas shift reaction in multi-tubular Pd and 80 wt% Pd–20 wt% Cu membrane reactors at 1173 K. *Journal of Membrane Science*, 298:14–23.

[56] Bi Y., Xu, H., Li, W., Goldbach, A. (2009). Water–gas shift reaction in a Pd membrane reactor over $Pt/Ce_{0.6}Zr_{0.4}O_2$ catalyst. *International Journal of Hydrogen Energy*, 34:2965–2971.

[57] Meng, H. (2009). Multi-dimensional liquid water transport in the cathode of a PEM fuel cell with consideration of the micro-porous layer (MPL). *International Journal of Hydrogen Energy*, 34:5488–5497.

[58] Li, H., Tang, Y., Wang, Z., Shi, Z., Wu, S., Song, D., Zhang, J., Fatih, K., Zhang, J., Wang, H., Liu, Z., Abouatallah, R., Mazza, A. (2008). Review of water flooding issues in proton exchange membrane fuel cell. *Journal of Power Sources*, 178:103–117.

[59] Zhang, J., Xie, Z., Zhang, J., Tang, Y., Song, C., Navessin, T., Shi, Z., Song, D., Wang, H., Wilkinson, D. P., Liu, Z., Holdcroft, S. (2006). High temperature PEM fuel cells. *Journal of Power Sources*, 160:872–891.

[60] Das, S. K., Reis, A., Berry, K. J. (2009). Experimental evaluation of CO poisoning on the performance of a high temperature proton exchange membrane fuel cell. *Journal of Power Sources*, 193:691–698.

[61] Janssen, G. J. M. (2004). Modelling study of CO_2 poisoning on PEMFC anodes. *Journal of Power Sources*, 136:45–54.

[62] Smolinka, T., Heinen, M., Chen, Y. X., Jusys, Z., Lehnert, W., Behm, R. J. (2005). CO_2 reduction on Pt electrocatalysts and its impact on H_2 oxidation in CO_2 containing fuel cell feed gas – a combined in situ infrared spectroscopy, mass spectrometry and fuel cell performance study. *Electrochimica Acta*, 50:5189–5199.

[63] Tingelof, T., Hedstrom, L., Holmstrom, N., Alvfors, P., Lindbergh, G. (2008). The influence of CO_2, CO and air bleed on the current distribution of a polymer electrolyte fuel cell. *International Journal of Hydrogen Energy*, 33:2064–2072.

[64] Zamel, N., Li, X. (2011). Effect of contaminants on polymer electrolyte membrane fuel cells. *Progress in Energy and Combustion Science*, 37:292–329.

[65] Hedstrom, L., Tingelof, T., Alvfors, P., Lindbergh, G. (2009). Experimental results from a 5 kW PEM fuel cell stack operated on simulated reformate from highly diluted hydrocarbon fuels: efficiency, dilution, fuel utilisation, CO poisoning and design criteria. *International Journal of Hydrogen Energy*, 34:1508–1514.

[66] Bruijn, F. A., Papageorgopoulos, D. C., Sitters, E. F., Janssen, G. J. M. (2002). The influence of carbon dioxide on PEM fuel cell anodes. *Journal of Power Sources*, 110:117–124.

[67] Yan, W., Chu, H., Lu, M., Weng, F., Jung, G., Lee, C. (2009). Degradation of proton exchange membrane fuel cells due to CO and CO_2 poisoning. *Journal of Power Sources*, 188:141–147.

[68] Giddey, S., Ciacchi, F. T., Badwal, S. P. S. (2005). Fuel quality and operational issues for polymer electrolyte membrane (PEM) fuel cells. *Ionics*, 11:1–10.

[69] Yan, Q., Toghiani, H., Causey, H. (2006). Steady state and dynamic performance of proton exchange membrane fuel cells (PEMFCs) under various operating conditions and load changes. *Journal of Power Sources*, 161:492–502.

[70] Xu, H., Song, Y., Kunz, H. R., Fenton, J. M. (2006). Operation of PEM fuel cells at 120–150 °C to improve CO tolerance. *Journal of Power Sources*, 159:979–986.

[71] Bellows, R. J, Marucchi-Soos, E. P, Buckley, D. T. (1996). Analysis of reaction kinetics for carbon monoxide and carbon dioxide on polycrystalline platinum relative to fuel cell operation. *Industrial & Engineering Chemistry Research*, 35:1235–1242.

[72] Li, Q., He, R., Jensen, J. O., Bjerrum, N. J. (2003). Approaches and recent development of polymer electrolyte membranes for fuel cells operating above 100 °C. *Chemistry of Material*, 15:4896–4915.

[73] Zhang, J., Xie, Z., Zhang, J., Tang, Y., Song, C., Navessin, T., Shi, Z., Song, D., Wang, H., Wilkinson, D. P, Liu, Z., Holdcroft, S. (2006). High temperature PEM fuel cells. *Journal of Power Sources*, 160:872–891.

[74] Zhang, J., Tang, Y., Song, C., Xia, Z., Li, H., Wang, H., Zhang, J. (2008). PEM fuel cell relative humidity (RH) and its effect on performance at high temperatures. *Electrochimica Acta*, 53:5315–5321.

[75] Yang, C., Costamagna, P., Srinivasan, S., Benziger, J., Bocarsly, A. B. (2001). Approaches and technical challenges to high temperature operation of proton exchange membrane fuel cells. *Journal of Power Sources*, 103:1–9.

[76] Asensio, J. A., Gomez-Romero, P. (2005). Recent developments on proton conducting poly(2,5-benzimidazole) (ABPBI) membranes for high temperature polymer electrolyte membrane fuel cell. *Fuel Cells*, 5:336–343.

[77] He, R., Li, Q., Xiao, G., Bjerrum, N. J. (2003). Proton conductivity of phosphoric acid doped polybenzimidazole and its composites with inorganic proton conductors. *Journal of Membrane Science*, 226:169–184.

[78] Li, Q., Jensen, J. O., Savinell, R. F., Bjerrum, N. J. (2009). High temperature proton exchange membranes based on polybenzimidazoles for fuel cells. *Progress in Polymer Science*, 34:449–477.

[79] Mamlouk, M., Scott, K. (2010). The effect of electrode parameters on performance of a phosphoric acid-doped PBI membrane fuel cell. *International Journal of Hydrogen Energy*, 35:784–793.

[80] Korsgaard, A. R., Nielsen, M. P., Kær, S. K. (2008). Part one: a novel model of HTPEM-based micro-combined heat and power fuel cell system. *International Journal of Hydrogen Energy*, 33:1909–1920.

[81] Jiao, K., Alaefour, I. E., Li, X. (2011). Three-dimensional non-isothermal modeling of carbon monoxide poisoning in high temperature proton exchange membrane fuel cells with phosphoric acid doped polybenzimidazole membranes. *Fuel*, 90:568–582.

[82] Modestov, A. D., Tarasevich, M. R., Filimonov, V. Y., Davydova, E. S. (2010). CO tolerance and CO oxidation at Pt and Pt–Ru anode catalysts in fuel cell with polybenzimidazole–H_3PO_4 membrane. *Electrochimica Acta*, 55:6073–6080.

[83] Jespersen, J. L., Schaltz, E., Kaer, S. K. (2009). Electrochemical characterization of a polybenzimidazole-based high temperature proton exchange membrane unit cell. *Journal of Power Sources*, 191:289–296.

[84] Cecilia, B., Valerio, I., Sergio, D. R., Leonardo, G. (2000). Anodic catalysts for polymer electrolyte fuel cells: the catalytic activity of Pt/C, Ru/C and PtRu/C in oxidation of CO by O_2. *Catalyst Today*, 55:45–49.

[85] Pozio, A., Giorgi, L., Antolini, E., Passalacqua, E. (2000). Electroxidation of H_2 on Pt/C, PtRu/C and PtMo/C anodes for polymer electrolyte fuel cell. *Electrochimica Acta*, 46:555–561.

[86] Massong, H., Wang, H. S., Samjeske, G., Baltruschat, H. (2000). The cocatalytic effect of Sn, Ru and Mo decorating steps of Pt(111) vicinal electrode surfaces on the oxidation of CO. *Electrochimica Acta*, 46:701–707.

[87] Alcaide, F., Alvarez, G., Cabot, P. L., Miguel, O., Querejet, A. (2010). Performance of carbon-supported PtPd as catalyst for hydrogen oxidation in the anodes of proton exchange membrane fuel cells. *International Journal of Hydrogen Energy*, 35:11634–11641.

[88] Roth, C., Papworth, A. J, Hussain, I., Nichols, R. J, Schiffrin, D. J (2005). A Pt/Ru nanoparticulate system to study the bifunctional mechanism of electrocatalysis. *Journal of Electroanalytical Chemistry*, 581:79–85.

[89] Desai, S., Neurock, M. (2003). A first principles analysis of CO oxidation over Pt and Pt66.7%Ru33.3% (111) surfaces. *Electrochimica Acta*, 48:3759–3773.

[90] Liao, M. S., Cabrera, C. R., Ishikawa, Y. (2000). A theoretical study of CO adsorption on Pt, Ru and Pt–M (M = Ru, Sn, Ge) clusters. *Surface Science*, 445:267–282.

[91] Pitois, A., Pilenga, A., Tsotridis, G. (2010). CO desorption kinetics at concentrations and temperatures relevant to PEM fuel cells operating with reformate gas and PtRu/C anodes. *Applied Catalysis A: General*, 374:95–102.

[92] Chun, H. J., Kim, D. B., Lim, D. H., Lee, W. D., Lee, H. I. (2010). A synthesis of CO tolerant Nb_2O_5-promoted Pt/C catalyst for direct methanol fuel cell; its physical and electrochemical characterization. *International Journal of Hydrogen Energy*, 35:6399–6408.

[93] Gasteiger, H. A., Markovic, N., Ross, P. N. (1995). H_2 and CO electrooxidation on well-characterized Pt, Ru, and Pt–Ru. 2. Rotating disk electrode studies of CO/H_2 mixtures at 62 degree °C. *Journal of Physical Chemistry*, 99:16757–16767.

[94] Davies, J. C., Hayden, B. E., Pegg, D. J. (1998). The electrooxidation of carbon monoxide on ruthenium modified Pt(110). *Electrochimica Acta*, 44:1181–1190.

[95] Christoffersen, E., Liu, P., Ruban, A., Skriver, H. L., Norskov, J. K. (2001). Anode materials for low temperature fuel cells. *Journal of Catalysis*, 199:123–131.

[96] Igarashi, H., Fujino, T., Zhu, Y. M., Uchida, H., Watanabe, M. (2001). CO tolerance of Pt alloy electrocatalysts for polymer electrolyte fuel cells and the detoxification mechanism. *Physical Chemistry Chemical Physics*, 3:306–314.

[97] Davies, J. C., Tsotridis, G. (2008). Temperature-dependent kinetic study of CO desorption from Pt PEM fuel cell anodes. *Journal of Physical Chemistry C*, 112:3392–3397.

[98] Gu, T., Lee, W. K., Zee, J. W. V., Murthy, M. (2004). The effect of reformate components on PEMFC performance: dilution and reverse water gas shift reaction. *Journal of the Electrochemical Society*, 151:A2100–A2105.

[99] Enback, S., Lindbergh, G. (2005). Experimentally validated model for CO oxidation on PtRu/C in a porous PEFC electrode. *Journal of the Electrochemical Society*, 152:A23–A31.

[100] Wee, J. H., Lee, K. Y. (2006). Overview of the development of CO-tolerant anode electrocatalysts for proton-exchange membrane Fuel Cells. *Journal of Power Sources*, 157:128–135.

[101] Yoo, J. S., Kim, H. T., Joh, H. I., Kim, H., Moon, S. H. (2011). Preparation of a CO-tolerant $PtRu_xSn_y/C$ electrocatalyst with an optimal Ru/Sn ratio by selective Sn-deposition on the surfaces of Pt and Ru. *International Journal of Hydrogen Energy*, 36:1930–1938.

[102] Umeda, M., Ojima, H., Mohamedi, M., Uchida, I. (2004). Methanol electrooxidation at Pt-Ru-W sputter deposited on Au substrate. *Journal of Power Sources*, 136:10–15.

[103] Liao, S., Holmes, K. A., Tsaprailis, H., Birss, V. I. (2006). High performance PtRuIr catalysts supported on carbon nanotubes for the anodic oxidation of methanol. *Journal of the American Chemical Society*, 128:3504–3505.

[104] Lu, G., Cooper, J. S., McGinn, P. J. (2006). SECM characterization of Pt-Ru-WC and Pt-Ru-Co ternary thin film combinatorial libraries as anode electrocatalysts for PEMFC. *Journal of Power Sources*, 161:106–114.

[105] Gottesfeld, S., Pafford, J. (1988). A new approach to the problem of carbon monoxide poisoning in fuel cells operating at low temperatures. *Journal of Electrochemical Society*, 135:2651–1652.

[106] Sung, L. Y., Hwang, B. J., Hsueh, K. L., Tsau, F. H. (2010). Effects of anode air bleeding on the performance of CO-poisoned proton-exchange membrane fuel cells. *Journal of Power Sources*, 195:1630–1639.

[107] Chung, C. C., Chen, C. H., Weng, D. Z. (2009). Development of an air bleeding technique and specific duration to improve the CO tolerance of proton-exchange membrane fuel cells. *Applied Thermal Engineering*, 29:2518–2526.

[108] Bhatia, K. K, Wang, C. Y. (2004). Transient carbon monoxide poisoning of a polymer electrolyte fuel cell operating on diluted hydrogen feed. *Electrochimica Acta*, 34:2333–2341.

[109] Murthy, M., Esayian, M., Hobson, A., MacKenzie, S., Lee, W. K., Van Zee, J. W. (2001). The performance of a PEM fuel cell exposed to transient CO concentrations. *Journal of Electrochemical Society*, 148:A1141–A1147.

[110] Inaba, M., Sugishita, M., Wada, J., Matsuzawa, K., Yamada, H., Tasaka, A. (2008). Impacts of air bleeding on membrane degradation in polymer electrolyte fuel cells. *Journal of Power Sources*, 178:699–705.

[111] Gottesfeld, S., Zawodzinski, T. A. (1997). Polymer electrolyte fuel cells. In: *Advances in, Electrochemical Science and Engineering*, Alkire, R. C., Gerischer, H., Kolb, D. M., Tobias, C. W. (editors). Wiley-VCH, New York, pp. 219–225.

[112] Aicher, T., Full, J., Schaadt, A. (2009). A portable fuel processor for hydrogen production from ethanol in a 250 W_{el} fuel cell system. *International Journal of Hydrogen Energy*, 34:8006–8015.

[113] Jannelli, E., Minutillo, M., Galloni, E. (2007). Performance of a polymer electrolyte membrane fuel cell system fueled with hydrogen generated by a fuel processor. *Journal of Fuel Cell Science and Technology*, 4:435–440.

[114] Wu, W., Pai, C. (2009). Control of a heat-integrated proton exchange membrane fuel cell system with methanol reforming. *Journal of Power Sources*, 194:920–930.

[115] Lindstrom, B., Karlsson, J. A. J., Ekdunge, P., De, Verdier, L., Haggendal, B., Dawody, J., Nilsson, M., Pettersson, L. J. (2009). Diesel fuel reformer for automotive fuel cell applications. *International Journal of Hydrogen Energy*, 34:3367–3381.

[116] Sopena, D., Melgara, A., Briceno, Y., Navarrob, R. M., Alvarez-Galvan, M. C., Rosa, F. (2007). Diesel fuel processor for hydrogen production for 5 kW fuel cell application. *International Journal of Hydrogen Energy*, 32:1429–1436.

[117] Corbo, P., Migliardini, F., Veneri, O. (2008). Experimental analysis of a 20 kWe PEM fuel cell system in dynamic conditions representative of automotive applications. *Energy Conversion and Management*, 49:2688–2697.

[118] Erdinc, O., Uzunoglu, M. (2010). Recent trends in PEM fuel cell-powered hybrid systems: investigation of application areas, design architectures and energy management approaches. *Renewable and Sustainable Energy Reviews*, 14:2874–2884.

[119] Barbir, F. (2006). PEM fuel cells: theory and practice. In: *Fuel Cell Technology Reaching Towards Commercialization*, Sammes, N. (editor). Springer Academic Press, Germany, pp. 27–52.

[120] Oh, S. D., Kim, K. Y., Oh, S. B., Kwak, H. Y. (2012). Optimal operation of a 1-kW PEMFC-based CHP system for residential applications. *Applied Energy*, 95:93–101.

[121] Barelli, L., Bidini, G., Gallorini, F., Ottaviano, A. (2011). An energetic–exergetic analysis of a residential CHP system based on PEM fuel cell. *Applied Energy*, 88:4334–4342.

[122] Mathiak, J., Heinzel, A., Roes, J., Kalk, T., Kraus, H., Brandt, H. (2004). Coupling of 2.5 kW steam reformer with a 1 kW_{el} PEM fuel cell. *Journal of Power Sources*, 131:112–119.

[123] Furtado, J. G. M., Gatti, G. C., Serra, E. T., Almeida, S. C. A. (2010). Performance analysis of a 5 kW PEMFC with a natural gas reformer. *International Journal of Hydrogen Energy*, 35:9990–9995.

51

INTEGRATED SOLID OXIDE FUEL CELL SYSTEMS FOR ELECTRICAL POWER GENERATION—A REVIEW

Suttichai Assabumrungrat, Amornchai Arpornwichanop, Vorachatra Sukwattanajaroon, and Dang Saebea

51.1 INTRODUCTION

Solid Oxide Fuel Cell (SOFC) is a promising technology for direct conversion of chemical energy in a fuel to electrical power. It offers several benefits such as high efficiency, fuel flexibility, low noise and environmental impacts, less electrolyte management problem, variety of electrode catalysts, size flexibility, and possibility for combined heat and power (CHP) operation [1]. Although it has a wide variety of applications ranging from small-scale auxiliary power units (APUs) to large-scale stationary power generation, SOFC operation still suffers from many limitations such as high cost, short lifetime due to pulse demands and impurities of feed, poor thermal shock, long start-up time, and risk for carbon deposition [2]. A number of multidisciplinary research and development efforts are required to overcome these problems in order to make SOFC a marketable technology. The research and development can be mainly classified into improvement of material and component properties and integration of SOFC with various technologies and functions [3]. Many materials for the cell components (anode, cathode, and electrolyte) have been tested to achieve better cell performance (e.g., high power density, stability, compatibility, and durability) and lower cost [4, 5]. An anode-supported SOFC is considered to be the best candidate in terms of polarization standpoint when compared with a conventional electrolyte-supported SOFC and a cathode-supported SOFC. A number of research efforts have focused on the development of intermediate-temperature SOFC (IT-SOFC) (500–750°C) which offers several important advantages including lower cost of construction materials, an enhanced

range of applications, and improved durability and lifetime [6]. Another key development of SOFC is based on the integration of SOFC with other technologies and/or functions. Various concepts/strategies for SOFC-based integration system have been proposed and several benefits have been reported [2]. The following sections will outline the fundamental of SOFC and concept of process integration as well as review the development in SOFC-based hybrid systems.

51.2 FUNDAMENTAL OF SOFC SYSTEM

SOFC generates electricity from the electrochemical reaction of a fuel and an oxidant. Although high operating temperature of SOFC allows direct use of a variety of fuels such as hydrocarbons and alcohols, hydrogen and/or synthesis gas is a major fuel for practical SOFC in order to extend the lifetime of the cells. The major components of fuel cell are anode, cathode, and electrolyte. There are two types of electrolyte: oxygen ion and proton-conducting electrolytes. For SOFC, the former is most commonly used due to its availability. The electrochemical reactions at the electrodes are as follows:

$$\text{Anode: } H_2 + O^{2-} \rightarrow H_2O + 2e^-$$
$$\text{Cathode: } {}^1/_2 O_2 + 2e^- \rightarrow O^{2-}$$

A fuel processor is typically installed in an SOFC system to convert a fuel (e.g., natural gas, biogas, hydrocarbons, and biomass) via different reactions, such as steam reforming (SR), partial oxidation (PO), and autothermal reforming (ATR), to a hydrogen-rich gas. It is then fed to the anode

Alternative Energy and Shale Gas Encyclopedia, First Edition. Edited by Jay H. Lehr and Jack Keeley.
© 2016 John Wiley & Sons, Inc. Published 2016 by John Wiley & Sons, Inc.

together with air at the cathode of an SOFC stack where the electrochemical reaction takes place to generate electricity at high efficiency. The presence of fuel remaining in the anode exhaust gas and heat generated from irreversibility of the electrochemical reaction allows the SOFC to be integrated with various technologies to achieve high energy efficiency and other benefits. The next section will provide fundamental concepts of SOFC-integrated processes.

51.3 SOFC-INTEGRATED PROCESSES

As mentioned earlier, SOFC has been known as a high-efficiency power generator which offers more than 50% of electrical efficiency [7]. To extend the capability of SOFC for use with various fuel supplies apart from the traditional hydrogen supply, a supplementary processing system is integrated as components in the balance of plant (BoP) of the SOFC system to facilitate the use of such fuels. High performance and energy requirement of the fuel conversion system are important issues to be considered for improving the performance of the SOFC system. In addition, energy loss or waste heat generated within the SOFC system is another issue which significantly obstructs the system in achieving high efficiency. Technical strategies such as heat recovery processes and waste energy minimization methods are applied to obtain suitable solutions for such a problem. To achieve a higher efficiency of SOFC system, the use of process analysis tools as performance assessment indicators is considered together with a principle of exergy analysis to indicate the content of available energy and to identify energy loss (exergy destruction) distribution in the overall system. These approaches could help find causes of the problem in particular subsystems or the whole system, and then guidelines for the improvement are obtained. In the BoP of SOFC system, there are two major sections that impact the system efficiency: (i) a fuel processing section (hydrogen production system) and (ii) a heat recovery section. Concept and strategy reviews for both the sections are provided next.

51.3.1 Process Integration Concept

Efficient energy usage in chemical process industries has been realized as an important issue, driven by the energy crisis in the 1970s. This topic is then taken into consideration in a process engineering design approach. Several process design methodologies have been investigated and developed from academic and industrial sectors in order to target energy conservation and waste minimization in chemical processes. Similar to other chemical process improvements, SOFC systems for power generation, containing integrated systems, for example, fuel preconditioning, fuel processor, and heat and mass recovery systems, need to be analyzed through process configurations and process integration techniques to find a

feasibility of the system efficiency development. The sequential progress of process integration development started with the energy perspective in the earlier stage. Heat integration concept was initiated by Hohmann [8] who investigated an optimum network for heat exchanges but his work was not recognized at that time until Linnhoff and Flower [9] brought Hohmann's work to further develop a pinch technology. The aim of this method is to identify energy-saving opportunities in order to design an optimal heat exchanger network (HEN) to meet the minimum utility targets, such that the overall process not only has minimum energy requirement (MER) but also owns a maximum energy recovery. The procedure of this method starts with classifying a list of hot and cold streams. Overall heat calculated from each hot stream is considered the heat sources that need to be removed to reduce their temperatures, while all cold streams are calculated to gain overall heat requirement defined as the heat sinks to increase their temperatures. This stage can initially identify the net MER. In the step of pinch analysis, several methods are introduced to minimize the energy requirement, for example, temperature-interval method and graphical method using composite heating and cooling curves. The minimum utility loads (heat input and heat removal loads) are carried out from these calculation methods. Based on a thermodynamic principle, thermal energy is transferred from a hot stream with a higher temperature to a cold stream having a lower temperature. Therefore, heat transfer mechanism can occur if the hot and cold streams have a temperature difference high enough to create a thermal driving force. For the overall system, heat recovery can be maximized if the difference temperature of the hot and cold streams is close to a minimum difference temperature $\Delta T_{\min}$. As illustrated with the temperature–enthalpy diagram in Figure 51.1, the heat composite curve is located above the cooling composite curve. Thermal energy is input to the system above pinch temperature (heat sink), while heat is removed from the system below pinch temperature (heat source). $\Delta T_{\min}$ has a specific interval range, typically 10°F or 5°C. In case when the value of $\Delta T_{\min}$ is too high, the system requires more minimum heating ($Q_{H,\min}$) and cooling ($Q_{C,\min}$) loads. The results lead

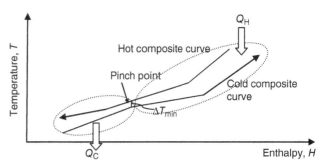

FIGURE 51.1 Temperature–enthalpy diagram with pinch point dividing by heat source and heat sink regions.

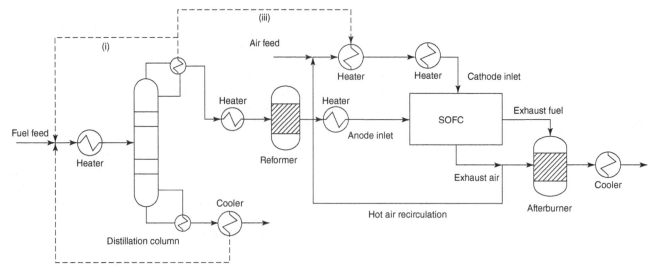

FIGURE 51.2   HEN design options diagram of the SOFC system integrated with distillation column (modified from Reference 11, p. 195).

to a lower heat recovery capacity of the system. If the interval ΔT_{min} is too low, heat cannot be transferred in each hot and cold stream, especially when ΔT_{min} is set to be zero; this case is called pinch point/temperature.

The information from pinch analysis is used to design HENs and is further applied to design the minimum number of heat exchangers. Applications of heat integration through HENs for the fuel cell system have been reported in scientific literatures. For example, the optimization of a proton exchange membrane fuel cell (PEMFC) system using process integration techniques and simulation model was studied by Godat and Marechal [10]. The system consists of a methane steam reformer, water gas shift reactor, preferential oxidation reactor, PEMFC, and combustor unit. The influence of major parameters (e.g., steam-to-carbon ratio, SR temperature, fuel cell temperature, and fuel utilization) was investigated. In case when excess air is employed to control the water balance of the cathode, it was shown that the combustion temperature is reduced to be close to the SR temperature and then the high pinch point occurs at the reformer temperature. However, with optimal operating conditions and HEN configuration, the system efficiency increases from 35% to 49%. A bioethanol-fueled SOFC system integrated with distillation column was improved in terms of the electrical efficiency by using the MER network [11]. Different hot streams and cold streams within the system were matched, leading to various cases of process configurations as illustrated in Figure 51.2. Heat exchange between the condenser (heat source) and bioethanol feed stream (heat sink) and using hot air cathode recirculation stream are the suitable options. This process configuration was applied with MER network configuration and the results showed that it can reduce the minimum cooling loads from 73.4 to 55.9 kW and offer the overall

efficiency at 40.8%. Basically, it can be stated that the use of heat integration concept for SOFC system applications aims to utilize its internal heat sources (post-combustion unit) and useful residual heat accumulated in the exhaust streams (hot residual air and fuel and hot stream product) for supplying to other heat-demanding units. This approach leads to an efficient energy management.

In case of high pinch temperature, especially for the SOFC system integrated with an endothermic SR as the fuel processor for hydrogen production, the classical energy integration theory based on a thermal pinch analysis is generally not enough to deal with this problem [12]. According to the rule "no hot utility below pinch point," heat released from units below the pinch temperature needs to be removed from the system. To recover this useful heat, new techniques with the integration of fuel and mechanical power (turbine engine) under the concept of a CHP have to be considered [13]. Townsend and Linnhoff [14, 15] developed the concept of heat engine combined with heat integration system (i.e., HEN). They observed that the heat engine should be placed above or below the pinch but not across the pinch point because no reduction in heating or cooling loads is achieved from this arrangement. The role of heat engine is to transform thermal energy in hot streams into work but it still discharges residual heat due to the incomplete conversion explained by a second law of thermodynamics. As shown in Figure 51.3, the heat engine is located below the pinch point of the HEN system to recover heat in case of high pinch temperature as mentioned earlier. Surplus heat below pinch point normally removed uselessly by cooling loads is fed into the heat engine. This device offers the benefit of not only removing heat below the pinch but also partially converting heat into useful work. On the contrary, if excessive heat is supplied

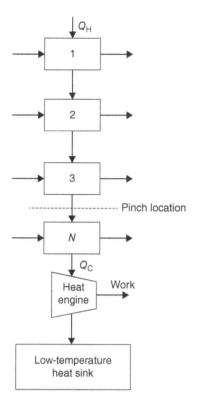

FIGURE 51.3 Placement of heat engine below the pinch point (modified from Reference 16, p. 269).

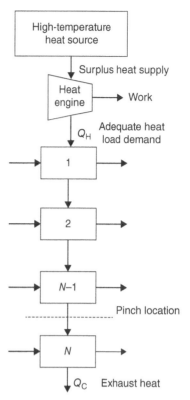

FIGURE 51.4 Placement of heat engine above the pinch point (modified from Reference 16, p. 268).

to the above pinch region more than the minimum heating load demand, the heat engine is installed to partly convert the surplus heat to work and the remaining heat is used for heating utility served as a heat sink of heat engine (Figure 51.4). Further applications of heat engine to the SOFC system are provided in the section of SOFC system integrated with gas turbine (GT).

In addition to the process integration from the energy conservation point of view, the concept of mass integration is another important issue involving waste utilization and reduction. The new process design methodologies which take this principle into account have been developed, including mass pinch analysis and other related methods used for the design of waste recovery process [17, 18]. Mass exchange network (MEN) synthesis based on mass pinch technology, a similar concept as the thermal pinch technology, originated from the mass integration concept. In general, the MEN principle is applied with separating agents to build recovery or purification units such as absorption column, adsorber, and liquid–liquid extraction. The properties of mass separating agent (MSA), for example, solubility limit, thermodynamic constraints, and absorption/adsorption capability used to recover the target component, are essential for the performance of MEN design. In physical terms, direct contact and counter currents are preferential configurations for

mass exchange system design. However, operating and capital costs are crucial to be considered before selecting these recovery systems. For SOFC systems, the concept of mass integration can be employed to design a recovery system of useful residual exhaust gases from the anode and cathode of a fuel cell stack.

Hartono et al [19]. applied the mass integration concept to high-temperature fuel cell plants (i.e., SOFC and MCFC), leading to a simplified system design. Previous studies by Fellows [20] and Lisbona et al [21]. indicated that increase of anode recycle ratio shows negative effects on the system efficiency. In their work, the utilization of anode exhaust gas of SOFC power plant was studied by two approaches: (i) basic stoichiometric calculation and (ii) detailed modeling of overall system, to find a proper minimal anode recirculation ratio. As seen from Figure 51.5, it should be noted that increasing anode exhaust gas flow rate causes a higher power consumption of the blower. The optimal anode recycle ratio depends on the types of fuel. Among the studied fuel types (methane, ethanol, wood, methanol, and anthracite), methanol requires the lowest recycle ratio, whereas anthracite needs the highest one. The two approaches used to analyze the SOFC system achieve similar results; however, the stoichiometric method is simpler and can predict the minimal recycle ratio and steam-to-carbon ratio to prevent carbon formation.

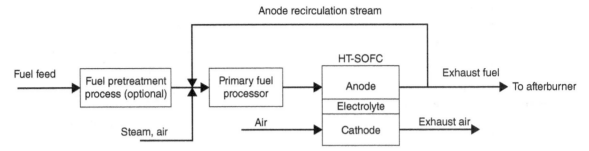

FIGURE 51.5   Schematic diagram of the mass-integrated high-temperature fuel cell plant (modified from Reference 19, p. 7241).

In general, there are several system analysis tools that can be used for a mass integration-based design methodology such as a stream mapping diagram used to find an opportunity for direct or mixed recycle, a path diagram used to analyze the process configuration of specific required species represented in the form of mass flow, an interception technology used to analyze a proper purification/pretreatment approach, and a reaction pathway modification used to convert undesired species to useful/desired products [22]. The reaction pathway modification method can be further applied for MEN synthesis, leading to reactive mass exchange networks (REAMENs). A recent progress in the process integration development is focused on the combination of heat and mass integration concepts. Several process design methodologies which follow this combined concept have been developed [22], for example, heat-induced separation networks (HISENs), energy-induced separation networks (EISENs), membrane separation networks (MSNs), heat-induced waste minimization networks (HIWAMINs), energy-induced waste minimization networks (EIWAMINs), and waste interception and allocation networks (WINs). For the SOFC system, a fuel cell combined with a fuel processor is a distinct representation of the concept that uses a combined heat and mass integration; efficient processing units in part of the BoP are selected and combined for supporting the SOFC module to achieve better performance under an effective energy and mass management, that is, a self-sufficient condition ($Q_{net} = 0$) and low level of exhaust fuel residues.

The design of SOFC-integrated systems requires some useful analysis tools to help focus or identify some deficits and evaluate the performance of newly proposed SOFC system configurations. Exergy analysis is an effective tool that is capable to indicate possible improvement together with process configurations to ensure the optimal operating conditions and integrated process design. Exergy is destroyed by irreversible process conditions leading to the loss of work. Optimization of exergy efficiency deals with minimization of lost work that is equivalent to the concept of minimization of the entropy production followed by Gouy–Stodola theorem [23]. Exergy can be transferred in and out through the open system (such as SOFC system) by three forms:

heat, work, and mass with the same principle of energy balance.

Exergy calculation depends on states of the system and reference environment (restricted dead state). The exergy destruction is mainly caused by high-temperature processing units such as post-combusting unit and water vaporizer [24,25]. For a plug-flow reformer, the entropy/exergy loss is due to a heat transfer, frictional flow, and chemical reactions [26]. To solve this problem, equal distributions of entropy production and forces along the length of reactor are suggested. This solution approach is known as "equipartition of entropy production [22] and equipartition of forces [27]." Increasing the reactor length to increase the residence time and using optimal control theory [28] were demonstrated to reduce the entropy production [29].

51.4   SOFC WITH HYDROGEN PRODUCTION SYSTEM

51.4.1   Hydrogen Production Processes

Typically, hydrogen is a desired fuel to be introduced to the anode side of SOFC module for power generation, regardless of internal reforming within the fuel cell. Hydrogen is considered a clean and sustainable energy carrier in terms of chemistry theory point of view because it is widely available in the form of other substances such as water, gas/liquid/solid hydrocarbons, and other hydrogen-compounded substances. However, it does not have any natural conditions that can offer pure hydrogen readily in use. Complex infrastructures, for example, treatment, production, distribution and storage handling, and economic competitiveness need to be tackled to support this promising energy. As a result of these factors, an efficient fuel processor for producing hydrogen from other fuel sources is a key consideration for SOFC system. The selection of appropriate fuel for hydrogen production is another essential topic because it affects the complex degree of a fuel processor design and in turn the whole SOFC system. General criteria such as renewability, cost, storage, and availability are considered for selecting a proper fuel. In addition, the selected fuel should have a well-developed technology support for converting it into hydrogen with low complexity, fewer requirements of equipment, and low energy consumption. Basically, all hydrocarbon fuels in different forms, that is, solid, liquid, and gas phase, can be employed as raw material for hydrogen production. In general, the overall reforming reaction of hydrocarbon fuels can be written as

$$C_aH_bO_c + dO_2 + 2\left(a - d - \frac{c}{2}\right)H_2O \rightarrow$$
$$aCO_2 + 2\left(a - d - \frac{c}{2} + \frac{b}{4}\right)H_2 \quad (51.1)$$

This reaction simply shows the ideal reforming of hydrocarbon fuels into only hydrogen and carbon dioxide without considering reactant residues, steam, carbon monoxide, and other low-carbon compounds. Typically, the hydrogen production process from hydrocarbon and alcohol fuels is performed via three common reforming reactions: steam reforming (SR), partial oxidation (PO), and autothermal reforming (ATR). In each reaction, there are individual characteristics as follows:

Steam reforming: This reaction is commonly used in industrial processes because of long-period operation stability and high yield of the hydrogen product. Steam reacts with fuel to produce hydrogen and carbon dioxide via strongly endothermic reaction (equation (51.2)) at temperatures above 500°C. Heat transfer is an essential consideration for this reactor design. High steam-to-carbon ratio, high temperature, and low pressure are suitable conditions for preventing coke formation and gaining high hydrogen yield.

$$C_aH_bO_c + 2\left(a - \frac{c}{2}\right)H_2O \rightarrow aCO_2 + \left(2a + \frac{b}{2} - c\right)H_2$$
$$(51.2)$$

Partial oxidation: Oxygen is used instead of steam to partially react with fuels to generate hydrogen and carbon monoxide (equation (51.3)). This reaction consumes less energy than the other two reactions (SR and ATR) but requires good catalysts to avoid the complete oxidation that leads to the presence of carbon dioxide and water. Furthermore, control of reaction conditions is important to prevent a runaway reaction due to its highly exothermic nature and carbon formation on catalyst.

$$C_aH_bO_c + \frac{1}{2}(a - c)O_2 \rightarrow aCO + \frac{b}{2}H_2 \quad (51.3)$$

Autothermal reforming: An SR and a PO take place simultaneously in the ATR. Heat released from the PO is absorbed by the endothermic SR. This thermal balance in the autothermal process leads to a lower-temperature operation. However, the ATR needs to be controlled to operate under thermal

neutral or little exothermic condition. Selection of appropriate ratio of fuel, air, and steam and use of proper catalysts can control the heat transfer between the PO and the SR.

Apart from these three basic reactions, a number of hydrogen production technologies have been investigated and reported in literatures, for example, super critical water reforming (SCWR) for methanol [30] and biomass [31] and catalytic dehydrogenation (CDH) of methane [32]. The CDH of methane is a single-step reaction which produces high purity of hydrogen and also a valuable carbon nanotube byproduct. However, economic and practical issues for a large-scale production are still the main barriers for these novel processes to be developed further.

Chemical-looping reforming (CLR) technology has been used for hydrogen production and carbon dioxide capture. The evolution of this technique was initiated by Conrad Arnold [33] from the Standard Oil Development Company, who first proposed the chemical-looping combustion (CLC) process for carbon monoxide and hydrogen production and was granted a patent since the early 1950s. Thereafter, the CLR has been developed and applied to many applications: chemical-looping steam reforming (CLSR), chemical-looping CO_2 acceptor reforming (CLCAR), chemical-looping partial oxidation (CLPO), and autothermal reforming (CLATR) as shown in Figure 51.6. All CLR types are based on the same principle that consists of coupled processes: cyclic oxidation and reduction of metallic oxide carriers (e.g., CLSR, CLPO, and CLATR) or CaO as CO_2 sorbent (e.g., CLCAR) in two separate reactors. Considering their performances, both the CLPO and CLATR need to keep at a low ratio of air and fuel to produce syngas and prevent

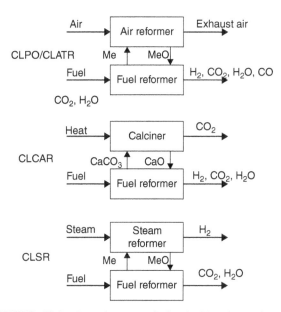

FIGURE 51.6 Several types of chemical-looping reforming (modified from Reference 34, p. 20).

the complete combustion of fuel to CO_2 and water, likewise the CLC [35]. CLSR is an interesting process among the other types. Similar to ordinary SR characteristics, fuels can be completely oxidized in the CLSR fuel reactor, and also metallic oxide carriers are entirely reduced to metal form before moving into the steam reactor where metal particles are reacted with steam to produce pure hydrogen and solid metal oxide particles for looping return. In the effluent stream of the fuel reactor, there are steam and carbon dioxide apart from metal particles. Carbon dioxide can be easily separated by condensing steam to water. The advantage of CLSR is that there is no need to install any further processing units such as a water gas shift reactor or a pressure swing adsorption [36].

A number of metals have been investigated to select a suitable metal oxide for an oxygen carrier. Iron is considered an interesting metal because it is environmentally benign and relatively inexpensive [37]. In addition, after the oxygen carrier is used in a number of cycles, it can be recycled to a steel industry [38]. The CLR process that uses iron oxide is known as a steam-iron process developed by Messerschmitt [39] and Lane [40]. Fossil fuels such as natural gas and the synthesis gas derived from coal/lignite gasification are used in the CLSR. Based on natural gas (methane) feed, the CLSR process has the following coupled reactions:

$$Fe_3O_4 + CH_4 \rightarrow 3Fe + CO_2 + 2H_2O \qquad (51.4)$$
$$3Fe + 4H_2O \rightarrow Fe_3O_4 + 4H_2 \qquad (51.5)$$

Equation (51.4) is an endothermic reduction of fuel carried out in the fuel reactor, whereas equation (51.5) is an exothermic oxidation of iron oxide and steam occurring in the steam reformer. Ideally, both the reactions can transfer thermally neutral heat to each other; however, thermal balance does not absolutely occur in the overall CLSR process. Additional heat needs to be supplied to maintain its smooth operation,

optionally retrieving heat residues from other neighboring processing parts (e.g., the water gas shift reactor) within the system or relying on an external heat source. The CLSR process needs to be further developed for solving some problems such as recirculation of metal oxide particles between two fluidized bed reactors, low conversion efficiency, and slow reaction rate of oxygen carrier particles.

In conclusion, the selection of hydrogen production process relies not only on a proper type of fuel but also on the reforming reaction. Furthermore, the hydrogen production process should be compatible with SOFC system applications. For example, when the SOFC system has a problem of inadequate thermal energy supply, the ATR may be an interesting choice. If high yield of hydrogen is a main target of SOFC systems to maximize their performance, the SR can serve their requirement. By-products obtained from the hydrogen production process should also be considered as they affect the difficulty level of a hydrogen purification process.

51.4.2 Integration of SOFC System with Hydrogen Production Processes

A fuel processor is an essential unit used to convert various fuels to hydrogen-rich gas for SOFC. Based on the heat and mass integration concepts, an integrated fuel cell (SOFC) and fuel processor (hydrogen production unit) system is considered to enhance its performance. Lee et al [41]. studied the recirculation of anode outlet stream from the SOFC to the fuel processor. Due to the internal reforming property and CO tolerance material of SOFC, only a primary reformer without having an additional gas clean-up unit (CO removal) is used for the fuel processor. The hydrogen pretreating unit is normally used for a low-temperature fuel cell system (e.g., PEMFC) as illustrated in Figure 51.7. In a conventional fuel

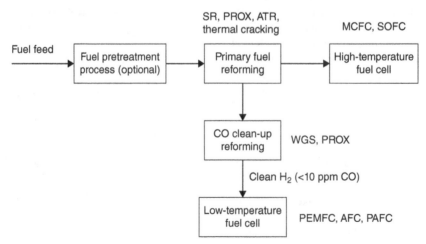

FIGURE 51.7 Basic process flow diagram in comparison with PEMFC system and SOFC system (modified from Reference 42, p. 3555).

cell power system, the system volume of 33% is occupied by the fuel processor [43]. As a result of this, the development of a fuel processor can take significant effects on the SOFC system efficiency. Over the past two decades, many research efforts have been devoted to develop and improve a fuel processor for hydrogen production. Mass and energy management strategies, novel structural designs of reformer and SOFC module, process intensification, and fuel processing-assisted units have been proposed.

51.4.2.1 Structural Design of SOFC.

Due to the high-temperature operation of SOFC (800–1000°C), reforming of fuel can occur within the fuel cell. Unconverted hydrocarbons and CO from the external reformer are further converted to hydrogen at the anode. It is noted that the net efficiency of the SOFC system will be decreased when more steps of hydrogen production are added to the fuel processor. The internal reforming characteristics of the SOFC result in the compactness of the system, faster loading response, cost reduction, and increased system efficiency [44]. Furthermore, the endothermic internal reforming reaction can consume the excessive heat released from the exothermic hydrogen oxidation within the fuel cell. In general, the internal reforming can be classified into two types: (i) direct internal reforming and (ii) indirect internal reforming. A porous ceramic anode electrode containing reforming catalysts, for example, Pt, Rh, and Ru, is used for coupling the reforming and electrochemical reactions in the direct internal reforming [45]. For the indirect internal reforming operation of SOFC, the reforming section is separated but located close to the anode electrode. Anode catalyst deactivation by the formation of carbon is a major drawback of internal reforming. Regarding an SOFC structural modification, since solid electrolyte material is used, the SOFC can be fabricated in various shape designs, that is, tubular, planar, and monolithic [46]. An integrated planar design is also proposed, which combines tubular and planar geometries, leading to a good thermal expansion compliance, low-cost component fabrication, short current path [47], and a higher ratio of power per volume [48]. The aim of these structural designs is to optimize the performance of fuel cell. Fuel cell-integrated concept is another aspect to be considered for process modification. High fuel utilization and high operating voltage are desirable to achieve high efficiency of the fuel cell. Practically, increasing fuel utilization leads to a decrease in the Nernst potential which limits the cell operating voltage. The ways to solve this problem are to operate the fuel cell at a high fuel recycle ratio [49] and to apply a multistage oxidation concept. Selimovic and Palsson [50] applied multistage oxidation concept by networking SOFC stacks to improve the SOFC/GT system (Figure 51.8). The benefits of this solution approach are improved thermal balance, higher total fuel utilization, and increased power efficiency. From their study on the SOFC stack arrangement as illustrated in Figure 51.9, it was found that the SOFC stack arranged in

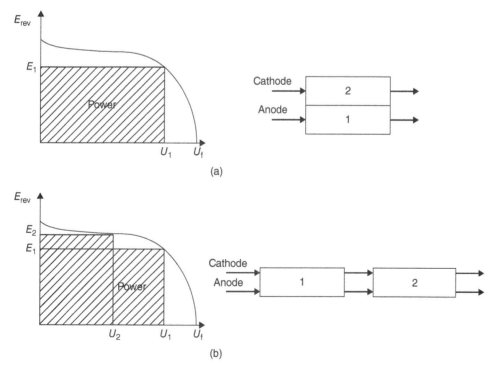

FIGURE 51.8 Basic concept of fuel cell stacks arrangement with fuel flow configurations: (a) single and (b) two-stage stacks (modified from Reference 50, p. 2).

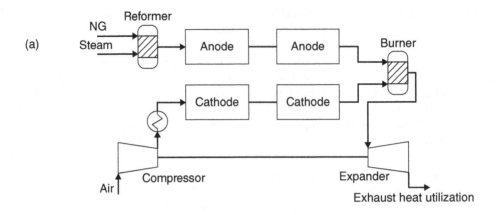

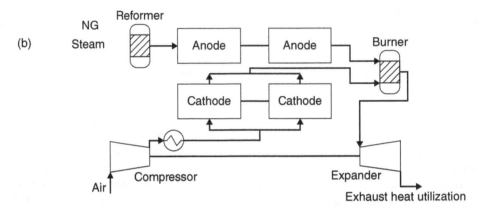

FIGURE 51.9   Networked SOFC–GT system with (a) both reactant streams in series and (b) only fuel stream in series and parallel air streams (modified from Reference 50, pp. 80–81).

series shows an increase in system efficiency. However, this stack networking was found not to be worthwhile in large-scale SOFC system.

Shin'ya Obara [51] studied a combined SOFC–PEMFC system for supplying energy to 30 houses in Sapporo, Japan (Figure 51.10). It is known that the exhaust heat of SOFC

is not released consistently due to its nonuniform system load. In the proposed system, flexible time shift operation plan was employed to manage the inconsistent SOFC high-temperature exhaust heat for the generation of the reformed gas from natural gas, which is used for electricity production in the low-temperature PEMFC. As schematically shown in

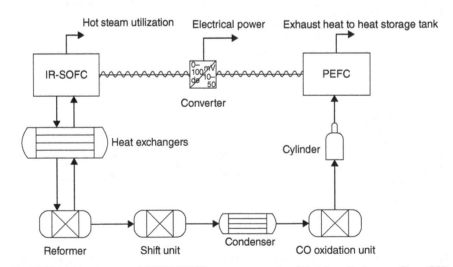

FIGURE 51.10   Integrated SOFC–PEFC power system (modified from Reference 51, p. 758).

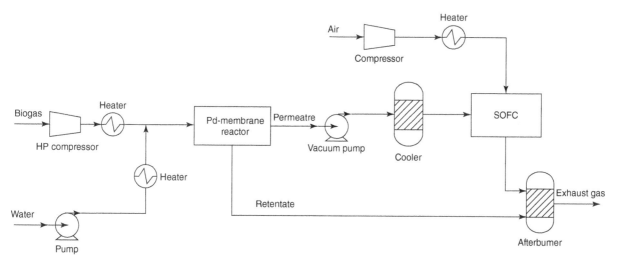

FIGURE 51.11 SOFC system integrated with palladium membrane reactor (modified from Reference 58, p. 3897).

Figure 51.11, the SOFC is operated with internal reforming. Heat released from the SOFC is supplied to a steam reformer (R/M) where hydrogen is produced. Water is removed by an air-cooling condenser (C/S) and CO content is reduced by a CO oxidation unit (C/O). Finally, the reformed gas with high hydrogen content is stored in the cylinder tank before being fed to the PEMFC. With the target of effective exhaust heat utilization to obtain a higher total power generation efficiency, operation methods with different power load patterns related to the amount of heat storage and power efficiency were studied. The results indicated that it can maintain average power generation efficiency near 48% in winter and summer seasons.

51.4.2.2 Structural Design of Reformer.
Generally, a structural design of reformer is proposed with the attempt to achieve a high surface-to-volume ratio and efficient heat transfer to allocate/utilize thermal energy. Energy consumption in the reformer depends on the type of reforming reactions and fuels. Monolithic structure is a type of shape-fabricated design which can increase the active surface area; however, it has the limitation in temperature control [52]. Membrane separation technique is applied to assist reforming mechanisms and enhance heat integration in the reformer. With appropriate membrane type to remove only a preferential substance (product) from the reformer, reaction rates can be increased and by-products are separated at the same time. For example, a hydrogen transport membrane (HTM) reactor was used to concentrate produced hydrogen and promote the reaction selectivity at low temperatures [53]. Membrane material should tolerate a high-temperature operation in the reformer. Generally, polymeric-made membranes cannot be used at temperatures above 100–150°C [54]. Hence, a metallic-made or inorganic membrane is a promising choice

used in the reformer. In addition, it can also be fabricated in various shapes to improve a high surface area. Palladium and its alloys (e.g., Pd–Ag, Pd–Cu) have been used in dense membrane reactors for maximizing hydrogen production [55–57]. Integration of SOFC and Pd membrane reactor fed by biogas/methane was subjected to technical and economic analysis [58]. It was found that the proposed system was not a good choice from the economic viewpoint as it required a large high-pressure compressor for fuel gas feeding to a membrane reactor.

Heat integration and catalyst-assisted techniques were employed in structural designs of the reformer. A heat exchanger with wave-shaped plates design was proposed by Isogawa et al [59]. to enhance heat integration of the combustion and endothermic reforming reactions. Heat exchange steam reformer was developed by Ida Tech [60]. Heat from the combustion of unreacted fuel or raffinate was supplied to a pre-reforming feed entering the catalyst bed in the reforming section. It can reduce heat loss and increase the thermal efficiency to the reformer. Catalyst was applied together with a thermal integration in the reformer design. Lesieur and Corrigan [61] proposed a catalytic wall reactor to enhance heat integration by transferring the thermal energy through a thin wall. Noble metal catalysts were coated on one side of the wall for endothermic reforming and on the other side of the wall for fuel combustion. This means that the burner, reformer, and heat exchanger are integrated into a single unit making the system compact and with higher efficiency.

A pressure swing reformer (PSR) was developed by ExxonMobil [62]. It has low-pressure combustion cycles, with the benefit to heat high-pressure SR within the catalyst bed. Division of multiple reaction zones in the reformer, that is, impurities removal zones and reforming zones such as

HT/LT-WGS, PROX, and SR, is another strategy of structural reformer design to utilize thermal energy effectively and perform a waste heat reduction [63, 64]. However, temperature control in each zone and difficulty in scale-up of this kind of reformer are problems that need to be further improved.

51.4.2.3 Fuel Types for SOFC System and Additional Fuel Processing-Assisted Units.

As mentioned above, the degree of complexity in the fuel processor SOFC-integrated system design depends on a selection of fuel and reforming technology. Factors that need to be considered when choosing fuel and performing process system design are the availability of technological supports at the desired scale production, capital and operation costs, catalyst technology supports, extent of additional fuel processing-assisted units (fuel pretreatment/purification), and especially efficiency of the system gained from those design configurations. Summarized here are the details of different fuel types.

Gaseous fossil fuels. This fuel type normally refers to natural gas which is considered initially in converting to hydrogen-rich gas that is used to feed in the multistage fuel cell system [65] and SOFC stationary system [66] because of its existing infrastructure supports, availability, and high conversion efficiency. Methane is a main component in natural gas (70–90%), meaning that a plenty of reactions used for methane conversion can be used with natural gas. Sulfur species such as H_2S (0–5%) or sulfur-containing odorants in natural gas can affect catalyst degradation, especially for nickel catalyst [67]. Therefore, a desulfurizer unit such as a zinc oxide polisher and a hydrodesulfurizer is required to pretreat natural gas by reducing the sulfur species to an acceptable level (e.g., 1 ppm sulfur odorants and < 1 ppm H_2S) for fuel cell applications [7]. Liquefied petroleum gas (LPG), another interesting fuel containing mainly propane, was used for hydrogen generation [68, 69] and was also demonstrated as fuel supply for an SOFC [70]. A prototype LPG reformer was designed with the reactions: SR, water gas shift, and methanation, for 5 kW IR-SOFC system. Selection of a suitable catalyst was studied and its operating regime to avoid catalyst deactivation due to carbonaceous depositions was given.

Liquid fossil fuels. Liquid fossil fuel has similar benefits as gaseous fossil fuels but is more secure and convenient in terms of fuel handling and storage. However, in the BoP design, a vaporizer is required for liquid fuels. Generally, they need high temperatures in the vaporization and fuel conversion for hydrogen generation. Gasoline requires temperature higher than 650°C for its catalytic conversion and over 1000°C in case of no catalyst. However, gasoline has twice the power density, compared to methanol [71]. Diesel requires vaporization temperature higher than 400°C. Although use of this fuel is interesting to apply for SOFC

APU, a large amount of sulfur and carbon formations in the fuel processor still needs to be tackled. Basically, the active metal sites are deteriorated because sulfur will react with metal catalysts to form stable sulfide complexes during the reforming process [72–74]. Therefore, catalysts with high sulfur tolerance and high activity of aromatics are required for the reforming of diesel [73, 75, 76]. Noble metals, that is, Pt and Rh, with gadolinium-doped ceria support showed good catalyst performance [77–79].

Solid fossil fuels. Solid fossil fuels such as coal is a common source for syngas production [80]. Gasification methods, operating conditions, and raw material components of coal can influence the composition of the produced syngas [81, 82]. The power efficiency of the coal-based SOFC system mainly depends on the fuel content and impurities in syngas. Basic coal gasification systems can be classified into three types: (i) moving bed, (ii) fluidized bed, and (iii) entrained bed [7]. Steam and air are used to partially burn coal into syngas. The contaminants produced from the gasifier consist of COS, H_2S, HCN, NH_3, tars, phenol, and particulates that need to be removed. Impurity removal/gas clean-up units such as cyclone for particulate removal, H_2S scrubber, COS hydrolysis reactor, and ammonia scrubber are employed. These units increase the complexity of BoP, leading to higher cost and imbalance of the thermal efficiency in the overall SOFC system. Power cycles are introduced to improve the power system performance [83]. A 500 MW coal-fueled SOFC system with integrated gasification combined cycles (IGCC) was studied [84]. This integrated power system targets to increase the overall efficiency with low emission. Regarding environmental aspect, the coal gasification combined with carbon dioxide capturing technology for hydrogen production was studied using a ChemCAD package [85]. However, it was indicated that using a coal fuel with the SOFC system requires a number of pretreatment and fuel processing units including carbon capture and storage (CCS) units in case of environmental concerns. Economic evaluation should be taken into account by comparing the prices of coal raw materials, efficiency of system (product outputs), and total investment cost.

These solid fossil-derived fuels are still regarded as nonrenewable and high pollution resources even though all of them can be effectively used in SOFC system. However, the environmental aspect rises up renewable resources outstandingly as a clean fuel for the SOFC system.

Biomass. Biomass is a promising alternative energy source because it is available from various agricultural materials. Typically, biomass is converted into a synthesis gas by high-temperature gasification/pyrolysis and catalytic reforming, respectively [86]. However, high thermal energy is required to devolatile biomass to gaseous product. A novel allothermal biomass steam gasification and a CHP integrated with

SOFC system were investigated [87] with an attempt to max-imize/allocate heat utilization within the system. The results reported that 36% electrical efficiency and 14% thermal efficiency were obtained. For other alternatives, the aqueous phase reforming and supercritical water oxidation at medium temperatures offer the promising choice for solid reforming process to get high yield of hydrogen without a large amount of heat requirement [88]. Biomass fermentation is another interesting option, relying on the advancement of biochemical technologies to produce biofuels, that is, bioethanol, biomethanol, biogas, bioglycerol, and biodiesel [52]. Performance analysis of the integrated SOFC system with three different renewable fuels (i.e., biogas, ethanol, and glycerol) was investigated [89]. The results indicated that the SOFC system running on ethanol offered the highest electrical and thermal efficiencies. Considering the use of bioethanol, it is inappropriate to directly feed into the steam reformer because of its low ethanol concentration (5–12 wt%) [90]. Excess water needs to be removed to obtain a suitable proportion of ethanol and water for ethanol SR. Several methods such as distillation [11], pervaporation [91], and hybrid pervaporation–vapor permeation [92] were proposed to precondition bioethanol for use in the SOFC system.

51.5 SOFC WITH GAS TURBINE

Due to its high-temperature operation, SOFC generates a high-quality exhaust gas that can be used for preheating fuel and air before they are fed to the SOFC stack. This causes the energy efficiency of the system to increase. A number of researchers have concentrated on heat management of the SOFC system. In particular, the hybrid system of SOFC and GT or steam turbine (ST) has received much attention. Regarding the integration of SOFC and GT, the exhaust gas from SOFC can be directly or indirectly fed to GT. The SOFC–GT hybrid system can theoretically have an overall electrical efficiency of up to 70% [93].

51.5.1 Development of SOFC–GT Hybrid Systems

The enhancement of SOFC systems with integration of GT is a promising technology for electricity power plants. At the beginning, pressurized SOFC–GT hybrid was proposed by the National Fuel Cell Research Center in Irvine, California. This system was designed and built by Siemens Westinghouse Power Corporation (SWPC). The overall power of this system is 220 kW; the SOFC generated power of 200 kW while the micro turbine generator produced 20 kW. After that, the SOFC hybrid system received much attention. A number of research groups have focused on SOFC–GT hybrid systems in different issues. The National Energy Technology Laboratory (NETL) tested pressurized SOFC integrated with two turbines based on Siemens Westinghouse

technology to scale up the power plant to 20 MW. Furthermore, the NETL takes part in the project of the Hybrid Power Generation Systems Division of General Electric. The aim of this project is to demonstrate the distributed power generation by development of SOFC using a thin film electrolyte technology fabricated by the tape calendaring method and thin-foil metallic interconnects, which is expected to be a low cost, high-performance, compact planar SOFC. In 1999, the US Department of Energy (DOE) started the solid-state energy conversion alliance (SECA) program devoted to develop a hybrid system of SOFC and GT so as to achieve a high-effective and low-cost power system leading to commercial level. Under this program, a number of industrial partners were involved such as Cummins Power Generation, Accumetrics, Siemens Westinghouse, Fuel Cell Energy, General Electric, and Delphi. Korea Institute of Energy Research [94] designed and constructed an atmospheric and pressurized 5 kW class SOFC power generation system with a pre-reformer for the fuel cell/GT hybrid system. In the hybrid mode of operation at 3.5 atm, the SOFC stack combined with an LNG pre-reformer and a micro-GT produces 5.1 kW based on the fuel utilization of 33.2%. The results confirmed the success of their design and fabrication technologies for a pressurized anode-supported planar SOFC system combined with a micro-GT.

Apart from interests from governmental and industrial sections, many academic research teams have also studied and developed SOFC–GT hybrid systems. The thermochemical power group at the University of Genoa investigated the cycle layout and the part-load dynamics of SOFC hybrid system [95–98]. A detailed dynamic model of the internal reforming SOFC integrated with GT was developed and its reliability was validated by comparing the model prediction with data obtained from the SOFC–GT system at the National Fuel Cell Research Center [99–102]. Furthermore, the model of SOFC based on planar technology integrated with GT was proposed by the research group at the University of Lund [103–106]. A 200 kW power plant of pressurized SOFC–GT hybrid system by a combined cycle system with 75 kW was studied at Toyohashi University of Technology [107].

Presently, the research goal is to improve and implement the SOFC–GT hybrid system to an industrial level. The challenging issues involve development of SOFC materials, appropriate management of energy, effective design of the SOFC system and determination of optimal conditions for SOFC and GT operations.

51.6 SOFC–GT HYBRID SYSTEMS

A main principle of a hybrid system is that one system needs to meet the requirement of another system. For an SOFC–GT hybrid system, the exhaust heat of SOFC can be recovered for

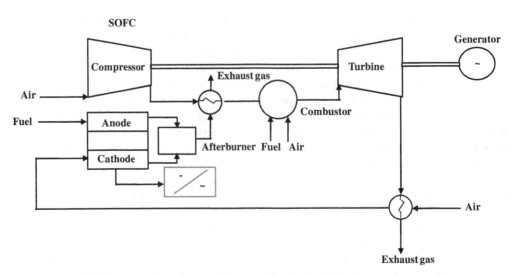

FIGURE 51.12   Indirect integrated system of a solid oxide fuel cell and a gas turbine.

use in other heat-requiring units of the SOFC system via heat and power cogeneration. There are two configurations considered for SOFC-based system: Brayton (gas) regenerative cycle and Rankine (steam) cycle. The Rankine cycle is essentially a heat engine with vapor power and is combined with the SOFC using a direct thermal coupling scheme. The commonly used working fluid is water and the system operates in the liquid–vapor mode. Thus, the Rankine cycle cannot be combined with SOFC using direct thermal coupling scheme. Due to the use of a gas-based working fluid, the Brayton cycle is a favorable candidate for SOFC integration [2].

51.6.1   Basic Configuration

For the SOFC and Brayton cycle integrated system, two major system configurations are possible depending on the operating pressure of the SOFC, namely, non-pressurized and pressurized hybrid cycles. Figure 51.12 shows the non-pressurized SOFC–GT hybrid system. Air flowing out of the compressor is heated by the fuel cell exhaust through a heat exchanger before being fed to the combustor of the

GT, and the SOFC can be operated under atmospheric conditions. The system efficiency for this hybrid configuration is expected in the range of 50–60% [108]. The advantage of this system is that GT operation does not depend directly on the SOFC and it is a simple cycle system. Moreover, it reduces the sealant requirement in the SOFC stack. Nevertheless, the heat exchanger has to be operated at very high temperatures and pressure differences, and consequently the requirement of high effective materials is the main problem of this atmospheric SOFC–GT hybrid system.

Pressurized SOFC–GT hybrid system involves the direct integration of an SOFC and a GT system, as seen in Figure 51.13. The combustion chamber of the GT engine is replaced by the SOFC and the afterburner. The pressurized air from the compressor is fed into the SOFC. The exhaust gas from the SOFC goes to the afterburner and the resulting high-temperature and pressure exhaust gas enters the turbine. Compressed air is preheated by the turbine exhaust gas through the recuperative heat exchanger. In this case, the SOFC is operated at high pressure, which further improves its performance. Heat exchangers are installed after the turbine

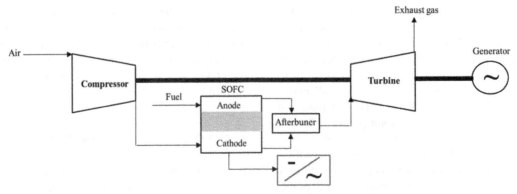

FIGURE 51.13   Direct integrated system of a solid oxide fuel cell and a gas turbine.

to further utilize the waste heat for preheating the streams entering the SOFC stack. This increases the thermal efficiency of the SOFC system as well.

In comparison with the SOFC operated under atmospheric pressure, the pressurized SOFC–GT hybrid system can achieve higher efficiency (up to 10%) and specific work values [109]. This is because the ambient pressure SOFC system cannot be run when the GT is operated at a high pressure ratio; the required turbine inlet temperature is too low [110]. Nonetheless, the pressurized SOFC–GT hybrid system has some limitations. The operation of the SOFC has direct influence on the GT. High interaction between SOFC and GT leads to difficulty in system operation. The pressurized SOFC–GT hybrid system should be modified with the addition of an external combustor so as to simplify the control system of the GT as it will be independent of the SOFC operation.

51.6.2 Configuration of SOFC–GT Hybrid System Integrated with a Heat Recovery Steam Generator

An SOFC–GT hybrid system integrated with a heat recovery steam generator (HRSG) is another system design to improve its efficiency (Figure 51.14). Due to its high-quality heat source, the exit gas from a recuperative heat exchanger can be used for steam generation via the HRSG consisting of economizer and evaporator parts. Water is pressurized at the desired value before it is vaporized by using the HRSG unit. It

is found that the total exergy output of the system is increased because the second product, saturated steam, is produced without the requirement of more fuel supplied to the system. The steam generated by the HRSG is employed to drive an ST cycle for additional power generation [111, 112]. As a result, the SOFC–GT hybrid system integrated with an ST can achieve a higher system efficiency at off-design. Chan et al [113]. studied the integration of SOFC, GT, and HRSG. The HRSG was applied to produce steam for an SR of fuel. The results showed that the SOFC–GT hybrid system has net electrical efficiency of higher than 60%. When the waste heat recovery for steam generation is considered, a system efficiency of 80% can be achieved.

Motahar and Alemrajabi [114] presented a comparison of the conventional SOFC–GT hybrid system with the retrofitted SOFC system in which steam produced by the HRSG using a hot GT exhaust gas is injected into a GT. From the detailed exergy analysis, it was found that the steam injection decreases the wasted exergy from the system and boosts the exergetic efficiency by 12.11%. Park et al [115]. examined two different SOFC configurations: a pressurized SOFC system and a non-pressurized SOFC system. The effect of steam injection on the recovered heat from the exhaust gas and the system performance was considered. The results showed that the pressurized system hardly takes advantage of the steam injection in terms of the system efficiency. On the other hand, the steam injection contributes to the efficiency improvement of the non-pressurized SOFC system in some design

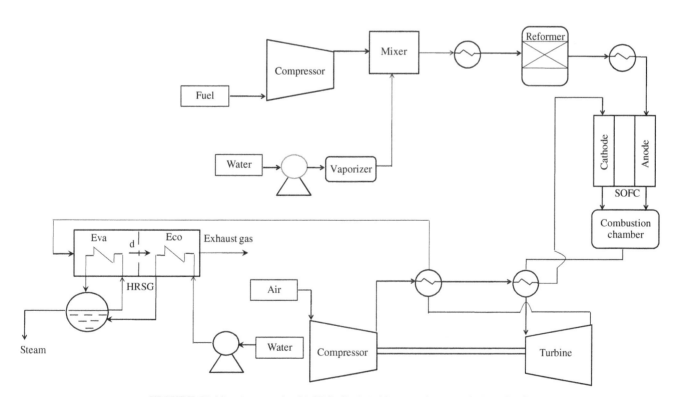

FIGURE 51.14 A pressurized SOFC–GT hybrid system integrated with HRSG.

conditions. In particular, a higher pressure ratio provides a better chance of efficiency increase due to the steam injection.

51.7 FUEL PROCESSOR COMBINED WITH SOFC–GT HYBRID SYSTEM

Hydrogen is a major fuel for electricity generation in fuel cells; however, its uses are still facing several issues such as its economical production, storage, and distribution. In general, hydrogen can be derived from primary fuels such as natural gas, methanol, gasoline, and coal via a fuel processor. To date, natural gas is widely used to produce hydrogen for fuel cell applications. Although natural gas is a cost-effective feedstock, it is also a limited and non-renewable resource. To date a number of studies have been conducted to explore the means to produce hydrogen from alternative renewable resources to support the future use of fuel cells. In particular, biomass, biogas, and bioethanol have been considered as attractive and potential feedstocks for SOFC systems.

The use of biomass for hydrogen production seems to be more suitable for supplying to SOFCs than other fuel cells because the SOFC has a higher tolerance toward contaminants in the hydrogen-rich gas [116–118]. In this case, an SOFC-based power plant involves gasification and cleaning processes to convert biomass to hydrogen fuel. There are several works investigating the performance of a combined biomass gasification and SOFC–GT hybrid plant [119–123]. As the design of gasification and gas cleaning processes

has an effect on the efficiency of the SOFC–GT power system, Toonssen et al [124]. studied the use of different gasification technologies (such as an atmospheric indirect steam gasification and a pressurized direct air gasification) and gas cleaning technologies (such as low-temperature and high-temperature gas cleaning processes) in the SOFC–GT hybrid system fueled by biomass. Their results indicated that the SOFC system based on pressurized direct air gasification and high-temperature gas cleaning process shows the highest electrical exergy efficiency of 49.9%. Apart from biomass, the use of liquid fuels in SOFC–GT power plant is promising, especially in remote areas [125]. Santin et al [125]. studied the SOFC–GT hybrid system run on two liquid fuels, methanol and kerosene, and showed that the methanol-fueled SOFC hybrid system shows better performance from the thermodynamic and economic points of view.

Another design option of the SOFC hybrid system is based on a fuel reforming technology. In general, hydrocarbon fuels can be internally reformed into a hydrogen-rich gas within SOFC stack which is known as an internal reforming SOFC because the operating temperature of the SOFC is in the same range as that of the reforming reactions. However, this would cause carbon formation leading to the degradation of anode catalyst and thus loss of fuel cell performance. To cope with this difficulty, the application of an external reforming process to produce hydrogen for fuel cells is a potentially better option. Since the product gas from the reformer can be further purified to achieve higher purity of hydrogen, the SOFC performance should be enhanced. Figure 51.15 shows a

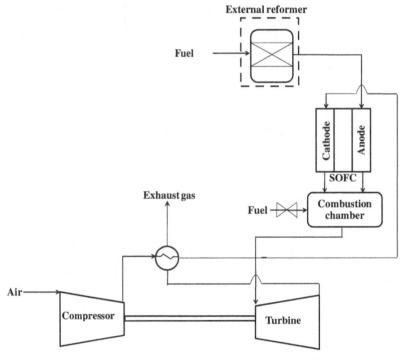

FIGURE 51.15 A conventional SOFC–GT system combined with a fuel processor.

typical diagram of a combined reforming process and SOFC–GT system. Any gaseous or liquid fuels can be fed to the reformer to convert into the synthesis gas with high hydrogen content. Cocco and Tola [126] investigated SOFC–MGT plants fueled by methanol and dimethyl ether (DME). External reformer was employed to convert methanol and DME to a synthesis gas. It was found that the reforming temperature of methanol and DME (200–350°C) is lower than that of natural gas (700–900°C) and the SOFC system efficiency can be enhanced by improving the heat recovery of the exhaust gas. The optimum operating temperature of the methanol reforming process is approximately 240°C and the system efficiency is about 67–68%, whereas in the case of using DME as a fuel for the SOFC system, the optimal temperature is 280°C and the efficiency is 65%. Yang et al [127]. studied the SOFC–GT system using internal and external reforming processes. The effect of SOFC temperature and GT inlet temperature on the system performance was analyzed. The SOFC system with internal reforming operation provides better efficiency and power capacity for all design conditions.

51.8 COMBINED COOLING, HEATING, AND POWER GENERATION FOR SOFC SYSTEM

In general, SOFC is practically operated at a moderate fuel utilization and thus the outlet gas from a cell stack contains valuable residual fuels. Furthermore, due to the high-temperature operation (up to 1000°C) and the irreversible electrochemical process of SOFC, a waste heat is also produced in the fuel cell power section [128]. As a result, the utilization of the high-quality heat and the remaining fuel exiting SOFC can further improve the

efficiency of the SOFC system. Following this approach, the waste heat is utilized as a heat source for the system through a series of recuperative heat exchangers. The cogeneration of heat with electricity (a CHP system) can increasingly enhance the system thermal efficiency. In addition, the waste heat can be employed to produce cooling water for air conditioning via an absorption chiller in which a refrigerant is another working fluid cycle. A combined cooling, heating, and power (CCHP) generation or tri-generation becomes an alternative method for energy management in the SOFC power system that is potential to solve energy-related problems such as energy security, energy shortage, emission control, economy, and conservation of energy. This method is related to the production of cooling, heating, and power simultaneously from a single fuel source with the principle of energy cascade utilization [129, 130].

For operation of the absorption chiller, the refrigerant is evaporated at low pressure. Then, it is absorbed by an absorbent and becomes a solution at high pressure. Two widely used refrigerants (or absorbents) are lithium bromide and ammonia aqueous solutions. The selection of the refrigerant depends on the system application. Lithium bromide is suitable for air conditioning system that requires temperature above 0°C, whereas ammonia solution is applied to system with a lower cooling temperature (below 0°C) [2]. In general, the absorption chiller requires external heat for separating the refrigerant from the aqueous solution. Therefore, the combined system of SOFC and absorption chiller can lead to an efficient energy usage by using an exhaust gas from the SOFC system to drive the absorption chiller. The conventional integrated process of SOFC and absorption chiller is illustrated in Figure 51.16 [131]. The heat from exhaust gas is used to generate a cooling water via a

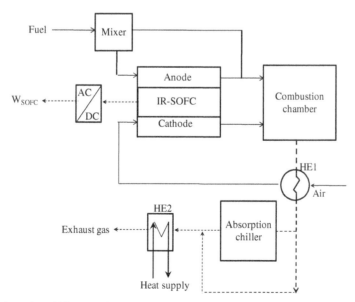

FIGURE 51.16   SOFC and adsorption chiller combined system.

refrigeration cycle using the adsorption chiller. Additionally, a portion of the exhaust gas can be applied for hot water production. The amount of the produced cooling water and hot water can be controlled by adjusting a ratio of the exhaust gas. Yu et al [131]. studied an integrated tri-generation system incorporating an internal reforming SOFC, a HRSG, and a double-effect water–lithium bromide absorption chiller. The simulation results showed that total efficiency of the studied tri-generation system of higher than 84% can be achieved. The increased fuel flow results in an increase in the cooling, heating, and electrical power output.

A new CCHP system consisting of SOFC, GT system, and ammonia–water thermodynamic cycle was proposed by Ma et al [132]. It was found that although the SOFC–GT hybrid system shows good performance, the high-quality exhaust gas still remains. To improve the overall energy conversion efficiency, the waste heat from the exhaust gas of the GT was recovered by using the ammonia–water cycle. The thermal efficiency of the proposed CCHP system is 80%. Operating parameters of the absorption chiller such as ammonia concentration and operating pressure affect the CCHP system in terms of electrical efficiency.

51.9   CONCLUSION

Integrated SOFC systems for electrical power generation are reviewed with particular focus on integration of SOFC with hydrogen production process and GT system. Heat generated from the SOFC can be efficiently utilized for hydrogen production, the complexity of the HEN depending significantly on the type of fuel and fuel processing technology. SOFC–GT hybrid systems are promising integration strategies for improving the overall electrical efficiency by utilizing waste heat for further generation of electrical power. Depending on applications, the SOFC system can be integrated with other units such as steam generator and absorption chiller for combined cooling, heating, and electrical power generation.

ACKNOWLEDGMENT

The support from Thailand Research Fund is gratefully acknowledged.

REFERENCES

[1] Mekhilef, S., Saidur, R., Safari, A. (2012). Comparative study of different fuel cell technologies. *Renewable and Sustainable Energy Reviews*, 16:981–989.

[2] Zhang, X., Chan, S. H., Li, G., Ho, H. K., Li, J., Feng, Z. (2010). A review of integration strategies for solid oxide fuel cells. *Journal of Power Sources*, 195:685–702.

[3] Hemmes, K., Barbieri, G., Lee, Y. M., Drioli, E., De Wit J. H. W. (2012). Process intensification and fuel cells using a multi-source multi-product approach chemical engineering and processing. *Process Intensification*, 51:88–108.

[4] Sun, C., Hui, R., Roller, J. (2010). Cathode materials for solid oxide fuel cells: a review. *Journal of Solid State Electrochemistry*, 14:1125–1144.

[5] Liu, M., Lynch, M. E., Blinn, K., Alamgir, F. M., Choi, Y. M. (2011). Rational SOFC material design: new advances and tools. *Materials Today*, 14:534–546.

[6] Brett, D. J. L., Atkinson, A., Brandon, N. P., Skinner S. J. (2008). Intermediate temperature solid oxide fuel cells. *Chemical Society Reviews*, 37:1568–1578.

[7] EG&G Technical Services, Parsons Inc. (2004). *Fuel Cell Handbook*, 7th ed., US Department of Energy, Office of fossil Energy, Federal Energy Technology Center, Morgantown, p. 427.

[8] Hohmann, E. C. (1971). Optimum Networks for Heat Exchange. PhD Thesis, University of Southern California, Los Angeles.

[9] Linnhoff, B., Flower, J. R. (1978). Synthesis of heat exchanger networks: I. Systematic generation of energy optimal networks. *AIChE Journal*, 24(4):633–642.

[10] Godat, J., Marechal, F. (2003). Optimization of a fuel cell system using process integration techniques. *Journal of Power Sources*, 118:411–423.

[11] Jamsak, W., Assabumrungrat, S., Douglas, P. L., Croiset, E., Laosiripojana, N., Suwanwarangkul, R., Charojrochkul, S. (2007). Thermodynamic assessment of solid oxide fuel cell system integrated with bioethanol purification unit. *Journal of Power Sources*, 174:191–198.

[12] Linnhoff, B., Townsend, D. W., Boland, D., Hewitt, G. F., Thomas, B. E. A., Guy, A. R., Marsland, R. H. (1994). *A User Guide on Process Integration for the Efficient Use of Energy*, IChemE, Rugby, p. 247.

[13] Kalitventzeff, B., Marechal, F., Closon, H. (2001). Better solutions for process sustainability through better insight in process energy integration. *Applied Thermal Engineering*, 21:1349–1368.

[14] Townsend, D. W., Linnhoff, B. (1983). Heat and power networks in process design. I. Criteria for placement of heat engines and heat pumps in process networks. *AIChE Journal*, 29:742–747.

[15] Townsend, D. W., Linnhoff, B. (1983). Heat and power networks in process design. II. Design procedure for equipment selection and process matching. *AIChE Journal*, 29:747–754.

[16] El-Halwagi, M. M. (2006). *Process Systems Engineering Volume 7: Process Integration*, 1st ed., Academic Press, San Diego, pp. 268–269.

[17] El-Halwagi, M. M. (1997). *Pollution Prevention through Process Integration: Systematic Design Tools*, Academic Press, San Diego, p. 334.

[18] El-Halwagi, M. M., Manousiouthakis, V. (1989). Design and analysis of multicomponent mass-exchange networks. In: *AIChE Annual Meeting*, San Francisco.

[19] Hartono, B., Heidebrecht, P., Sundmacher, K. (2011). A mass integration concept for high temperature fuel cell plants. *International Journal of Hydrogen Energy*, 36:7240–7250.

[20] Fellows, R. (1998). A novel configuration for direct internal reforming stacks. *Journal of Power Sources*, 71:281–287.

[21] Lisbona, P., Corradetti, A., Bove, R., Lunghi, P. (2007). Analysis of a solid oxide fuel cell system for combined heat and power applications under non-nominal conditions. *Electrochimica Acta*, 53:1920–1930.

[22] Dunn, R. F., Bush, G. E. (2001). Using process integration technology for CLEANER production. *Journal of Cleaner Production*, 9:1–23.

[23] Kjelstrup, S., Sauar, E., Hansen, E. M., Lien, K. M., Hafskjold, B. (1995). Analysis of entropy production rates for design of distillation columns. *Industrial & Engineering Chemistry Research*, 34:3001.

[24] Douvartzides, S. L., Couteliens, F., Tsiakaras, P. (2004). Exergy analysis of a solid oxide fuel cell power plant fed by either ethanol or methane. *Journal of Power Sources*, 131:224–230.

[25] Chan, S. H., Low, C. F., Ding, O. L. (2002). Energy and exergy analysis of simple solid oxide fuel cell power systems. *Journal of Power Sources*, 103:188–200.

[26] Kjelstrup, S., Bedeaux, D., Johannessen, E. (2006). *Elements of Irreversible Thermodynamics for Engineers*, Tapir Academic Press, Trondheim, p. 135.

[27] Sauar, E., Kjelstrup, S., Lien, K. M. (1996). Equipartition of forces: a new principle for process design and optimization. *Industrial & Engineering Chemistry Research*, 35:4147.

[28] Johannessen, E., Kjelstrup, S. (2004). Minimum entropy production in plug flow reactors: an optimal control problem solved for SO_2 oxidation. *Energy*, 29:2403.

[29] Van der Ham, L. V., Gross, J., Verkooilen, A., Kjelstrup, S. (2009). Efficient conversion of thermal energy into hydrogen: comparing two methods to reduce exergy losses in a sulfuric acid decomposition reactor. *Industrial & Engineering Chemistry Research*, 48:8500–8507.

[30] Boukis, N., Diem, V., Habicht, W., Dinjus, E. (2003). Methanol reforming in supercritical water. *Industrial & Engineering Chemistry Research*, 42:728–735.

[31] Byrd, A. J., Pant, K. K., Gupta, R. B. (2007). Hydrogen production from glucose using Ru/Al_2O_3 catalyst in supercritical water. *Industrial & Engineering Chemistry Research*, 46:3574–3579.

[32] Shah, N., Panjala, D., Huffman, G. P. (2001). Hydrogen production by catalytic decomposition of methane. *Energy & Fuels*, 15:1528–1534.

[33] Arnold, C. (1950). Process for Production of Carbon Monoxide and Hydrogen. Patent specification 636206, The Patent Office, London, UK.

[34] Moghtaderi, B. (2012). Hydrogen enrichment of fuels using a novel miniaturized chemical looping steam reformer. *Chemical Engineering Research and Design*, 90:19–25.

[35] Ortiz, M., Abad, A., de Diego, L. F., Garcia-Labiano, F., Gayan, P. (2011). Optimization of hydrogen production by chemical-looping auto-thermal reforming working with Ni-based oxygen-carriers. *International Journal of Hydrogen Energy*, 36:9663–9672.

[36] Go, K. S., Son, S. R., Kim, S. D., Kang, K. S., Park, C. S. (2009). Hydrogen production from two-step steam methane reforming in a fluidized bed reactor. *International Journal of Hydrogen Energy*, 34:1301–1309.

[37] Abad, A., Adanez, J., Garcia-Labiano, F., de Diego, L. F., Gayan, P., Celaya, J. (2007). Mapping of the range of operational conditions for Cu-, Fe-, and Ni-based oxygen carriers in chemical-looping combustion. *Chemical Engineering Science*, 62:533–549.

[38] Cormos, C. (2011). Hydrogen production from fossil fuels with carbon capture and storage based on chemical looping systems. *International Journal of Hydrogen Energy*, 36:5960–5971.

[39] Messerschmitt, A. (1910). Process for Producing Hydrogen. US patent 971206.

[40] Lane, H. (1913). Process for the Production of Hydrogen. US patent 1078686.

[41] Lee, T. S., Chung, J. N., Chen, Y. C. (2011). Design and optimization of a combined fuel reforming and solid oxide fuel cell system with anode off-gas recycling. *Energy Conversion and Management*, 52:3214–3226.

[42] van den Oosterkamp, P. F. (2006). Critical issues in heat transfer for fuel cell systems. *Energy Conversion and Management*, 47:3552–3561.

[43] Qi, A., Peppley, B., Karan, K. (2007). Integrated fuel processors for fuel cell application. A review. *Fuel Processing Technology*, 88:3–22.

[44] Samms, S. R., Savinell, R. F. (2002). Kinetics of methanol-steam reformation in an internal reforming fuel cell. *Journal of Power Sources*, 112:13.

[45] Fasano, B. V., Prettyman, K. M. (2000). Solid Oxide Fuel Cell Having a Catalytic Anode. US patent 6051329A.

[46] Hirschenhofer, J. H., Stauffer, D. B., Engleman, R. R., Klett, M. G. (1928). *Fuel Cell Handbook*, 4th ed., US Department of Energy, Office of Fossil Energy, Federal Energy Technology Center, Morgantown, p. 268.

[47] Gardner, F. J., Day, M. J., Brandon, N. P., Pashley, M. N., Cassidy, M. (2000). SOFC technology development at Rolls-Royce. *Journal of Power Sources*, 86:122–129.

[48] Pramuanjaroenkij, A., Kakac, S., Zhou, X. Y. (2008). Mathematical analysis of planar solid oxide fuel cells. *International Journal of Hydrogen Energy*, 33:2547–2565.

[49] EG&G Technical Services, Parsons Inc. (2000). *Fuel Cell Handbook*, 5th ed., US Department of Energy, Office of Fossil Energy, Federal Energy Technology Center, Morgantown, p. 691.

[50] Selimovic, A., Palsson, J. (2002). Networked solid oxide fuel cell stacks combined with a gas turbine cycle. *Journal of Power Sources*, 106:76–82.

[51] Obara, S. (2010). Power generation efficiency of an SOFC-PEFC combined system with time shift utilization of SOFC

exhaust heat. *International Journal of Hydrogen Energy*, 35:757–767.

[52] Xuan, J., Leung, M. K. H., Leung, D. Y. C, Ni, M. (2009). A review of biomass-derived fuel processors for fuel cell systems. *Renewable and Sustainable Energy Reviews*, 13:1301–1313.

[53] Gallucci, F., Paturzo, L., Basile, A. (2004). A simulation study of the steam reforming of methane in a dense tubular membrane reactor. *International Journal of Hydrogen Energy*, 29:611–617.

[54] Nunes, S. P., Peinemann, K. V. (2001). *Membrane Technology in the Chemical Industry*, Wiley-VCH Verlag GmbH, Geesthacht, p. 319.

[55] Paturzo, L., Basile, A., Drioli, E. (2002). High temperature membrane reactors and integrated membrane operations. *Reviews in Chemical Engineering*, 18:511–551.

[56] Wieland, S., Melin, T., Lamm, A. (2002). Membrane reactors for hydrogen production. *Chemical Engineering Science*, 57:1571–1576.

[57] Tosti, S., Basile, A., Bettinali, L., Borgognoni, F., Chiaravalloti, F., Gallucci, F. (2006). Long-term tests of Pd-Ag thin wall permeator tube. *Journal of Membrane Science*, 284:393–397.

[58] Piroonlerkgul, P., Kiatkittipong, W., Arpornwichanop, A., Soottitantawat, A., Wiyaratn, W., Laosiripojana, N., Adesina, A. A., Assabumrungrat, S. (2009). Integration of solid oxide fuel cell and palladium membrane reactor: technical and economic analysis. *International Journal of Hydrogen Energy*, 34:3894–3907.

[59] Isogawa, R., Nobata, Y., Kondo, M., Ogino, S., Kimura, K., Taki, M., Nishimura, A., Yamaoka, M., Negishi, Y., Taguchi, H., Saito, N., Takumi, A. Fuel Reformer For Mounting on a Vehicle (2000). US patent 6390030B1.

[60] Loffler, D. G., Taylor, K., Mason, D. (2003). A light hydrocarbon fuel processor producing high-purity hydrogen. *Journal of Power Sources*, 117:84–91.

[61] Lesieur, R. R., Corrigan, T. J. Compact fuel gas reformer assemblage (1999). US patent 6203587B1.

[62] Berlowitz, P. J., Hershkowitz, F. (2004). *Fuel Cell Seminar*, Mira Digital Publishing, San Antonio.

[63] Ahmed, S., Ahluwalia, R. K., Lee, S. H., Lottes, S. (2004). *Fuel Cell Seminar*, Mira Digital Publishing, San Antonio.

[64] Bentley, J. M., Mitchell, W. L., Clawson, L. G., Cross, J. C. Reactor for producing hydrogen from hydrocarbon fuels (2001). US patent 6783742B2.

[65] Overview of 11 MW fuel cell power plant. (1989). Non-published information from Tokyo Electric Power Company, September.

[66] Wilkinson, D. P., Steck, A. E. (1997). General progress in the research of solid polymer fuel cell technology at Ballard. In: *Proceedings of the Second International Symposium on New Materials for Fuel Cells and Modern Battery Systems*, Montreal.

[67] Twigg, M. V. (1989). *Catalyst Handbook*, Wolfe Publishing Ltd., Frome, p. 608.

[68] Rampe, T., Heinzel, A., Vogel, B. (2000). Hydrogen generation from biogenic and fossil fuels by autothermal reforming. *Journal of Power Sources*, 86:536–541.

[69] Aartun, I., Gjervan, T., Venvik, H., Gorke, O., Pfeifer, P., Fathi, M., Holmen, A., Schubert, K. (2004). Catalytic conversion of propane to hydrogen in microstructured reactors. *Chemical Engineering Journal*, 101:93–99.

[70] Ahmed, K., Gamman, J., Foger, K. (2002). Demonstration of LPG-fueled solid oxide fuel cell systems. *Solid State Ionics*, 152–153:485–492.

[71] Villegas, L., Guilhaume, N., Provendier, H., Daniel, C., Masset, F., Mirodatos, C. (2005). A combined thermodynamic/experimental study for the optimization of hydrogen production by catalytic reforming of isooctane. *Applied Catalysis A: General*, 281:75–83.

[72] Chen, I., Shiue, D. W. (1988). Resistivity to sulfur poisoning of nickel-alumina catalysts. *Industrial & Engineering Chemistry Research*, 27:1391–1396.

[73] Liu, D. J., Krumpelt, M. (2005). Activity and structure of perovskites as diesel-reforming catalysts for solid oxide fuel cell International. *Journal of Applied Ceramic Technology*, 2:301–307.

[74] Ferrandon, M., Mawdsley, J., Krause, T. (2008). Effect of temperature, steam-to-carbon ratio, and alkali metal additives on improving the sulfur tolerance of a Rh/La-Al_2O_3 catalyst. *Applied Catalysis A: General*, 342: 67–77.

[75] Flytzani-Stephanopoulos, M., Voecks, G. E. (1983). Autothermal reforming of aliphatic and aromatic hydrocarbon liquids. *International Journal of Hydrogen Energy*, 8:539–548.

[76] Palm, C., Cremer, P., Peters, R., Stolten, D. (2002). Small-scale testing of a precious metal catalyst in the autothermal reforming of various hydrocarbon feeds. *Journal of Power Sources*, 106:231–237.

[77] Krumpelt, M., Krause, T. R., Carter, J. D., Kopasz, J. P., Ahmed, S. (2002). Fuel processing for fuel cell systems in transportation and portable power applications. *Catalyst Today*, 77:3–16.

[78] Hennings, U., Reimert, R. (2007). Noble metal catalysts supported on gadolinium doped ceria used for natural gas reforming in fuel cell applications. *Applied Catalysis B: Environmental*, 70:498–508.

[79] Yoon, S., Kang, I., Bae, J. (2009). Suppression of ethylene-induced carbon deposition in diesel autothermal reforming. *International of Journal of Hydrogen Energy*, 34:1844–1851.

[80] Yamashita, K., Barreto, L. (2005). Energyplexes for the 21st century: coal gasification for co-producing hydrogen, electricity and liquid fuels. *Energy*, 30:2453–2473.

[81] Liu, K., Chunshan, S., Subramani, V. (2010). *Hydrogen and Syngas Production and Purification Technologies*, John Wiley & Sons, Inc., Hoboken, p. 545.

[82] Souza-Santos, M. (2010). *Solid Fuels Combustion and Gasification: Modeling, Simulation and Equipment Operation*, Taylor & Francis, Boca Raton, p. 486.

[83] Horlock, J. H. (1995). Combined power plants: past, present and future. *Transactions of ASME, Journal of Engineering Gas Turbines and Power*, 117:608–616.

[84] George, T. J., James, R., Lyons, K. D. (1998). Multi-Staged Fuel cell Power Plant (Targeting 80% Lower Heating Value Efficiency), *Power Generation International 1998 Conference*, Florida.

[85] Cormos, C., Starr, F., Tzimas, E., Peteves, S. (2008). Innovative concepts for hydrogen production processes based on coal gasification with CO_2 capture. *International Journal of Hydrogen Energy*, 33:1286–1294.

[86] Evans, R., Czernik, S., Magrini-Bair, K. (2004). DOE Hydrogen Program, FY2004 Progress Report, p. 65.

[87] Panopoulos, K. D., Fryda, L. E., Karl, J., Poulou, S., Kakaras, E. (2006). High temperature solid oxide fuel cell integrated with novel allothermal biomass gasification. Part I: modeling and feasibility study. *Journal of Power Sources*, 159:570–585.

[88] Huber, G. W., Shabaker, J. W., Dumesic, J. A. (2003). Raney Ni-Sn catalyst for hydrogen from biomass-derived hydrocarbons. *Science*, 300:2075–2077.

[89] Saebea, D., Authayanun, S., Patcharavorachot, Y., Paengjuntuek, W., Arpornwichanop, A. (2012). Performance analysis of SOFC systems integrated with steam reforming of different renewable fuels. In: *International Conference on Renewable Energies and Power Quality (ICREPQ'12)*, Santiago de Compostela, Spain.

[90] Huang, H. J., Ramaswamy, S., Tschirner, U. W., Ramarao, B. V. (2008). A review of separation technologies in current and future biorefineries. *Separation and Purification Technology*, 62:1–21.

[91] Choedkiatsakul, I., Charojrochkul, S., Kiatkittipong, W., Wiyaratn, W., Soottitantawat, A., Arpornwichanop, A., Laosiripojana, N., Assabumrungrat, S. (2011). Performance improvement of bioethanol-fuelled solid oxide fuel cell system by using pervaporation. *International Journal of Hydrogen Energy*, 36:5067–5075.

[92] Sukwattanajaroon, V., Charojrochkul, S., Kiatkittipong, W., Arpornwichanop, A., Assabumrungrat, S. (2011). Performance of membrane-assisted solid oxide fuel cell system fuelled by bioethanol. *Engineering Journal*, 15:53–66.

[93] Kandepu, R., Imsland, L., Foss, B. A., Stiller, C., Thorud, B., Bolland, O. (2007). Modeling and control of a SOFC-GT-based autonomous power system. *Energy*, 32:406–417.

[94] Lim, T. H., Song, R. H., Shin, D. R., Yang, D. R., Jung, J., Jung, H., Vinke, I. C., Yang, S. S. (2007). Operating characteristics of a 5 kW class anode-supported planar SOFC stack for a fuel cell/gas turbine hybrid system. *International Journal of Hydrogen Energy*, 33:1076–1083.

[95] Costamagna, P., Magistri, L., Massardo, A. F. (2001). Design and part-load performance of a hybrid system based on a solid oxide fuel cell reactor and a micro gas turbine. *Journal of Power Sources*, 96:352–368.

[96] Franzoni, A., Magistri, L., Traverso, A., Massardo, A. F. (2008). Thermoeconomic analysis of pressurized hybrid SOFC systems with CO_2 separation. *Energy*, 33:311–320.

[97] Ferrari, M. L., Traverso, A., Magistri, L., Massaro, A. F. (2005). Influence of the anodic recirculation transient behavior on the SOFC hybrid system performance. *Journal of Power Sources*, 149:22–32.

[98] Ferrari, M. L., Sorce, A., Pascenti, M., Massardo, A. F. (2011). Recuperator dynamic performance: experimental investigation with a microgas turbine test rig. *Applied Energy*, 88:5090–5096.

[99] Robert, R., Brouwer, J., Jabbari, F., Junker, T., Ghezel-Ayagh, H. (2006). Control design of an atmospheric solid oxide fuel cell/gas turbine hybrid system: variable versus fixed speed gas turbine operation. *Journal of Power Sources*, 161:484–491.

[100] Kaneko, T., Brouwer, J., Samuelsen, G. S. (2006). Power and temperature control of fluctuating biomass gas fueled solid oxide fuel cell and micro gas turbine hybrid system. *Journal of Power Sources*, 160:316–325.

[101] Mueller, F., Gaynor, R., Auld, A. E., Brouwer, J. (2008). Synergistic integration of a gas turbine and solid oxide fuel cell for improved transient capability. *Journal of Power Sources*, 176:229–239.

[102] Mueller, F., Jabbari, F., Brouwer, J. (2009). On the intrinsic transient capability and limitations of solid oxide fuel cell systems. *Journal of Power Sources*, 187:452–460.

[103] Palsson, J., Selimovic, A., Sjunnesson, L. (2000). Combined solid oxide fuel cell and gas turbine systems for efficient power and heat generation. *Journal of Power Sources*, 86:442–448.

[104] Selimovic, A., Palsson, J. (2002). Networked solid oxide fuel cell stacks combined with a gas turbine cycle. *Journal of Power Sources*, 106:76–82.

[105] Moller, B. F., Arriagada, J., Assadi, M., Potts, I. (2004). Optimisation of an SOFC-GT system with CO_2-capture. *Journal of Power Sources*, 131:320–326.

[106] Azra, S., Kemm, M., Torisson, T., Assadi, M. (2005). Steady state and transient thermal stress analysis in planar solid oxide fuel cells. *Journal of Power Sources*, 145:463–469.

[107] Inui, Y., Matsumae, T., Koga, H., Nishiura, K. (2005). High performance SOFC-GT combined power generation system with CO_2 recovery by oxygen combustion method. *Energy Conversion and Management*, 46:1837–1847.

[108] Zhao, Y., Sadhukhan, J., Lanzini, A., Brandon, N., Shah, N. (2011). Optimal integration strategies for a syngas fuelled SOFC and gas turbine hybrid. *Journal of Power Sources*, 196:9516–9527.

[109] Traverso, A., Magistri, L., Massardo, A. F. (2010). Turbomachinery for the air management and energy recovery in fuel cell gas turbine hybrid system. *Energy*, 35:764–777.

[110] Park, S. K., Kim, T. S. (2006). Comparison between pressurized design and ambient pressure design of hybrid solid oxide fuel cell–gas turbine systems. *Journal of Power Sources*, 163:490–499.

[111] Akkaya, A. V., Sahin, B., Erdern, H. H. (2008). An analysis of SOFC/GT CHP system based on exergetic performance criteria. *International Journal of Hydrogen Energy*, 33:2566–2577.

[112] Arsalis, A. (2008). Thermoeconomic modeling and parametric study of hybrid SOFC–gas turbine–steam turbine power plants ranging from 1.5 to 10 MWe. *Journal of Power Sources*, 181:313–326.

[113] Chan, S. H., Ho, H. K., Tain, Y. (2003). Multi-level modeling of SOFC–gas turbine hybrid system. *International Journal of Hydrogen Energy*, 28:889–900.

[114] Motahar, S., Alemrajabi, A. A. (2009). Exergy based performance analysis of a solid oxide fuel cell and steam injected gas turbine hybrid power system. *International Journal of Hydrogen Energy*, 34:2396–2407.

[115] Park, S. K., Kim, T. S., Sohn, J. L. (2009). Influence of steam injection through exhaust heat recovery on the design performance of solid oxide fuel cell - gas turbine hybrid systems. *Journal of Mechanical Science and Technology*, 23:550–558.

[116] Omosun, A. O., Bauen, A., Brandon, N. P., Adjiman, C. S., Hart, D. (2004). Modelling system efficiencies and costs of two biomass-fuelled SOFC systems. *Journal of Power Sources*, 131:96–106.

[117] Singh, D., Hernandez-Pacheco, E., Hutton, P. N., Patel, N., Mann, M. D. (2005). Carbon deposition in an SOFC fueled by tar-laden biomass gas: a thermodynamic analysis. *Journal of Power Sources*, 142:194–199.

[118] Athanasiou, C., Coutelieris, F., Vakouftsi, E., Skoulou, V., Antonakou, E., Marnellos, G., Zabaniotou, A. (2007). From biomass to electricity through integrated gasification/SOFC system-optimization and energy balance. *International Journal of Hydrogen Energy*, 32:337–342.

[119] Aravind, P. V., Woudstra, T., Woudstra, N., Spliethoff, H. (2009). Thermodynamic evaluation of small-scale systems with biomass gasifiers, solid oxide fuel cells with Ni/GDC anodes and gas turbines. *Journal of Power Source*, 190:461–475.

[120] Abuadala, A., Dincer, I. (2011). Exergoeconomic analysis of a hybrid system based on steam biomass gasification products for hydrogen production. *International Journal of Hydrogen Energy*, 36:12780–12793.

[121] Merida, W., Maness, P. C., Brown, R. C., Levin, D. B. (2004). Enhanced hydrogen production from indirectly heated, gasified biomass, and removal of carbon gas emissions using a novel biological gas reformer. *International Journal of Hydrogen Energy*, 29:283–290.

[122] Fryda, L., Panopoulos, K. D., Kakara, E. (2008). Integrated CHP with autothermal biomass gasification and SOFC–MGT. *Energy Conversion and Management*, 49:281–290.

[123] Sucipta, M., Kimijima, S., Suzuki, K. (2007). Performance analysis of the SOFC–MGT hybrid system with gasified biomass fuel. *Journal of Power Sources*, 174:124–135.

[124] Toonssen, R., Sollai, S., Aravind, P. V., Woudstra, N., Verkooijen, A. H. M. (2011). Alternative system designs of biomass gasification SOFC-GT hybrid systems. *International Journal of Hydrogen Energy*, 35:10414–10425.

[125] Santin, M., Traverso, A., Magistri, L. (2009). Liquid fuel utilization in SOFC hybrid systems. *Applied Energy*, 86:2204–2212.

[126] Cocco, D., Tola, V. (2009). Externally reformed solid oxide fuel cell–micro-gas turbine (SOFC–MGT) hybrid systems fueled by methanol and di-methyl-ether (DME). *Energy*, 34:2124–2130.

[127] Yang, W. J., Park, S. K., Kim, T. S., Kim, J. H., Sohn, J. L., Ro, S. T. (2006). Design performance analysis of pressurized solid oxide fuel cell/gas turbine hybrid systems considering temperature constraints. *Journal of Power Sources*, 160:462–473.

[128] Zink, F., Lu, Y., Schaefer, L. (2007). A solid oxide fuel cell system for buildings. *Energy Conversion and Management*, 48:809–818.

[129] Wu, D. W., Wang, R. Z. (2006). Combined cooling, heating and power: a review. *Progress in Energy and Combustion Science*, 32:459–495.

[130] Yu, Z., Han, J., Cao, X. (2011). Investigation on performance of an integrated solid oxide fuel cell and absorption chiller tri-generation system. *International Journal of Hydrogen Energy*, 36:1256–12573.

[131] Yu, Z., Han, J., Cao, X., Chen, W., Zhang, B. (2010). Analysis of total energy system based on solid oxide fuel cell for combined cooling and power applications. *International Journal of Hydrogen Energy*, 35:2703–2707.

[132] Ma, S., Wang, J., Yan, Z., Dai, Y., Lu, B. (2011). Thermodynamic analysis of a new combined cooling, heat and power system driven by solid oxide fuel cell based on ammonia–water mixture. *Journal of Power Sources*, 196:8463–8471.

52

POLYMER ELECTROLYTES FOR LITHIUM SECONDARY BATTERIES

FIONA M. GRAY AND MICHAEL J. SMITH

52.1 INTRODUCTION

The efficient conversion and storage of energy is undoubtedly one of the greatest challenges that scientists and engineers must meet in the twenty-first century. The energy requirements of a modern society dependent on portable electronic devices must be supplied by lightweight power sources with a low environmental impact. On the other hand, the concept of zero-emission, commercially viable high power, and high-energy-density storage systems to power electric vehicles (EVs) is very attractive, particularly in a world of upwardly spiraling fuel costs. Lithium batteries are generally considered to be the most promising energy sources for both these market sub-domains because of their potentially high volumetric and gravimetric energy densities and superior power capability. These characteristics, together with the development of thin film lithium battery engineering, became the enabling technology for the proliferation of portable electronic devices, with enormous commercial impact in notebook computers and mobile phone applications.

To satisfy the energy demands of commercial devices, cells are usually connected in series or parallel combinations in batteries. Over the last 150 years a wide range of different battery technologies have been developed to satisfy consumer demand and naturally the technical performance of these battery systems are determined by their chemical components. The characteristics of different battery systems are normally described, classified, and compared through a number of operational parameters:

- Specific (or gravimetric) energy density—the amount of electrical energy stored per unit weight (Wh kg^{-1}).

- Volumetric energy density—the amount of electrical energy stored per unit volume (Wh dm^{-3}).

- Capacity—the amount of electrical charge a cell can store. A small-volume cell generally has less capacity than a large one (Ah). The practical capacity is evaluated by multiplying the value of the discharge current by the time required to reach a terminal reference voltage.

- Specific capacity—the capacity given in terms of mass (Ah kg^{-1}).

- Rate capacity—the amount of electrical charge delivered on discharge, under a specified discharge rate (Ah). A battery rated 200 Ah for a 10-hour rate will deliver 20 A current for 10 hours.

- C-rate—a parameter that allows the discharge (or charge) rate to be expressed in terms of the nominal cell capacity; the higher the C-rate the more current drawn from the battery and the shorter the period of discharge (or charge). For example, a battery with a rating of 200 Ah may be discharged at a $C/10$ rate. At this rate the current can be calculated from the following relationship, $C/10$ rate (A) = 200 Ah/10 h = 20 A.

- Power density (specific power)—ratio of the power available from a battery to its volume (W dm^{-3}). Specific power generally refers to the ratio of power to mass (W kg^{-1}).

Most practical applications require the interconnection of individual cells to form an assembly with the desired voltage or capacity. The voltage is determined by the number of cells in series and the capacity by the number in parallel.

Alternative Energy and Shale Gas Encyclopedia, First Edition. Edited by Jay H. Lehr and Jack Keeley.
© 2016 John Wiley & Sons, Inc. Published 2016 by John Wiley & Sons, Inc.

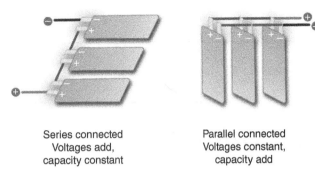

Series connected
Voltages add,
capacity constant

Parallel connected
Voltages constant,
capacity add

FIGURE 52.1 Multi-cell lithium polymer (LiPo) battery configurations. Many applications require both higher voltage and higher current. LiPo packs can be assembled with a combination of both series and parallel configurations.

Figure 52.1 demonstrates how the voltage and capacity of a thin film battery pack can be altered.

Individual cells comprise a cathode and an anode, separated by an electrolyte, the medium which enables charge transfer between the two electrodes. In order to attain high gravimetric and volumetric energy densities, the electrode materials should provide the most efficient and complete conversion of their chemical energy and have the highest cell voltage and cell capacity with respect to their weight and volume. For a given cell size, larger values of volumetric energy density (Wh dm^{-3}) and gravimetric energy density (Wh kg^{-1}) translate into smaller and lighter cells. Figure 52.2 shows a comparison of different battery technologies in terms of volumetric and gravimetric energy density.

The preference for a battery technology based on a lithium anode, therefore, is not surprising; lithium is the lightest and most electropositive of metals (3.04 V vs. SHE). Lithium-based cells are competitive as power sources because they

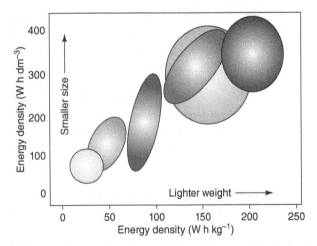

FIGURE 52.2 A comparison of different battery technologies in terms of volumetric and gravimetric energy density. Reprinted by permission from Macmillan Publishers Ltd: *Nature*, 414:359–367, copyright 2001, http://www.nature.com/

exhibit very favorable gravimetric and volumetric energy densities. The specific capacity of lithium metal is 3800 Ah kg^{-1}, which compares extremely favorably with more traditional materials such as lead (260 Ah kg^{-1}) and cadmium (480 Ah kg^{-1}). The highly desirable characteristics of lithium as a battery material have been exploited in the lithium primary (non-rechargeable) cell market for several decades in applications such as power sources for watches, calculators, and medical implants. The development of secondary, lithium metal-based cells has, however, been much more problematic. Lithium is highly electropositive, is thermodynamically unstable in contact with many reducible materials, and reacts with components of most electrolytes to form a passivating layer at the electrode/electrolyte interface. Such films ensure long shelf life for primary lithium cells but in secondary cells, although the interfacial passivating film prevents reaction beyond the lithium surface, it leads to uneven lithium plating as the cell is charged and discharged over a number of cycles. The growth of dendrites ultimately results in total cell failure through short-circuiting as well as serious safety issues due to local overheating.

A major breakthrough in the development of rechargeable lithium batteries came about as a result of developments in the area of inorganic intercalation compounds. These compounds are able to reversibly insert alkali metal ions from an electrolyte into the host structure, with a concomitant redox reaction occurring within the host matrix. The potential of such layered or framework structured compounds to be used as electrode materials was recognized in the 1970s but their initial development was restricted to applications as cathodic materials [1–5]. The difficulties associated with the metallic lithium anode prompted a widening of the scope of research into lithium insertion compounds to include those that could substitute metallic lithium as the anode. The choice of intercalation compounds for both anode and cathode materials was demonstrated by the end of the 1970s [6, 7] and led to the development of the lithium-ion (Li-ion, sometimes called rocking-chair) cell a decade later. The intercalation anodic material of choice has, to date, been carbon. In the charged state of the carbonaceous anode, lithium exists in its ionic rather than metallic form, thus eliminating any possibility of dendrite formation. Other advantages are highlighted in the low cost of carbon and the high lithium-ion activity in the intercalation compound; the latter renders an anode potential close to that of lithium metal and minimizes any energetic penalty. The Sony Corporation launched the first commercial Li-ion cell in 1991 [8] based on a carbon (petroleum coke) anode [9] and LiCoO$_2$ cathode. This type of Li-ion cell is used in many high-performance portable electronic devices. These batteries are light, compact, and work with a voltage of the order of 3.6–4.0 V, with a gravimetric energy density range of 100–150 Wh kg^{-1} (two to three times that of comparable Ni–Cd batteries). Figure 52.3 shows a schematic representation of a common Li-ion battery.

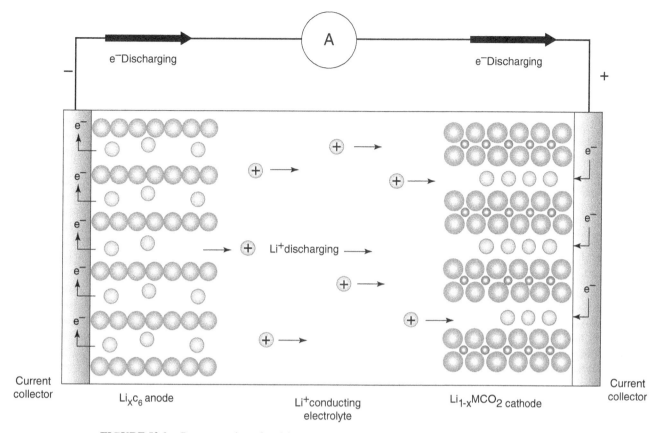

FIGURE 52.3   Representation of a Li-ion battery consisting of a graphite anode, a cathode formed from a lithium metal oxide ($LiMO_2$, e.g., $LiCoO_2$), and an electrolyte containing a lithium salt.

With the commercial success of Li-ion technology secured, it soon became apparent that substituting the liquid electrolyte with a polymeric electrolyte could result in improved device performance. This change simplifies the cell structure by eliminating the separator and reducing the need for efficient seals. Removal of architectural or form factor restrictions imposed on cells by the use of conventional electrolytes allows manufacturers to meet growing demands for smaller, thinner, and more flexible batteries for installation in electronic devices. As an added advantage, part of the assembly technology for these devices was already available in low-cost, high production capacity, within the paper and coating industries.

Solid polymer electrolytes (SPEs) have attracted enthusiastic attention from academic and industrial researchers since the early 1970s [10]. First-generation so-called "dry" polymer electrolytes or SPEs contain no organic solvent and are formed by the direct dissolution of a lithium salt into a host polymer matrix. They suffered from serious practical disadvantages, but subsequent electrolyte formulations show significantly improved electrochemical and mechanical characteristics. The principal limitation of dry polymer electrolytes has been the poor low-temperature conductivity.

Industrial investment has continued but R&D is presently directed toward lithium metal–polymer (LMP) battery technology with a particular focus on large-capacity batteries that are applied in EV traction and back-up power, which operates above ambient temperature [11–15]. A clear change in the direction of development was required in order to achieve a viable commercial product for the portable electronics market within a realistic timescale. That direction was toward the hybrid polymer electrolyte, often called a gel polymer electrolyte (GPE). This system comprises a polymer matrix swollen with a lithium-salt-containing liquid electrolyte. In 1996, Bellcore [16] announced the first reliable and practical plastic lithium-ion polymer (PLiON) rechargeable battery, heralding the development of what would become better known as lithium-ion polymer (LiPo or LiPoly) batteries. The LiPo battery may be considered as an evolved version of the Li-ion battery, sharing much of the processing technology. One of the most important consequences of using a polymer gel electrolyte is that there are no liquid components and therefore the traditional metal canister can be substituted by a lightweight pouch laminate. This permits a further weight reduction of the cell with a consequential increase in gravimetric energy density and allows for more flexible packaging design and higher packing efficiency. The past decade has seen LiPo batteries establish themselves as a major market subdivision.

52.2 POLYMER ELECTROLYTES

Polymer electrolytes are ion-conducting materials that have attracted interest as components of batteries, fuel cells, capacitors, sensors, display devices, and solar cells since the 1970s. The first reported ion-conducting polymer was high molecular weight poly(ethylene oxide) (PEO). This host polymer is capable of dissolving guest alkali metal salts to form SPEs with moderately high ionic conductivities ($\geq 10^{-4}$ S cm^{-1} above 60°C). The potential of the resulting free-standing films as the electrolyte component of an all-solid-state thin film cell was recognized at an early stage [10, 17–20]. For 30 years both the academic community and industrial sector have pursued research in this area attracted by the potential technological applications of these materials. In addition to numerous studies dealing with ion transport, thermal, and structural characterization, various theoretical approaches have been proposed to describe the mechanism of ion transport in SPEs as well as the physical and chemical processes occurring at the electrode/electrolyte interfaces [10, 17–22].

The most important performance characteristics for technologically viable lithium-ion-conducting polymer electrolytes include

- high ionic conductivity at ambient and sub-ambient temperatures
- good mechanical stability (good molding characteristics at the electrode interface)
- high lithium-ion transference number
- thermal stability
- chemical compatibility with electrodes and electrochemical stability
- safety under cell abuse conditions

In addition to these characteristics and looking beyond cell performance achievements, cost, sustainability, and end-of-life disposal issues also need to be factored into any assessment of commercial viability. Polymer electrolytes have developed through successive generations and evolved in different directions in order to improve performance characteristics. The versatility of these materials with respect to their electrochemical properties can largely be attributed to the diversity of ion-conducting materials along with the relatively large variations of polymer-salt solutions which can be created within each defined category. This is highlighted in the classification system for ion-conducting polymers given below, which acknowledges both well-established polymer

electrolyte materials together with the more innovative materials developed specifically to achieve practical performance targets [18, 23]:

- Liquid-solvent-free ("dry") macromolecular systems that can dissolve appropriate salts to form an ionically conducting phase. The system may be solid or liquid, depending on the polymer used [10, 18–20].
- Gel electrolytes that are formed by dissolving a salt in a polar liquid. Acceptable mechanical stability is achieved by adding an inert immobilizing polymeric material [24].
- Plasticized polymer electrolytes are prepared by the addition of high dielectric constant solvent [19] or ionic liquid (IL) [25, 26] to other classes of polymer electrolyte to enhance the ionic conductivity.
- Ionic rubber (polymer-in-salt) materials are formed by adding small amounts of high molecular weight polymer to a low-temperature liquid ionic mixture [27].
- Single-ion-conducting polyelectrolytes that consist of polymer chains that have ionic functional groups chemically bonded to the backbone [19].
- Hybrid inorganic–organic, multi-phase polymer electrolytes
  - Nanocomposite electrolytes consisting of classical polymer electrolytes into which oxide nanopowders have been introduced to promote ionic conductivity and improve chemical, mechanical, and electrochemical stability [28].
  - Organically modified ceramics (ORMOCERS) as polymer electrolytes consist of inorganic and organic nano-domains formed via a sol-gel reaction [29].
- Hybrid inorganic–organic, mono-phase polymer electrolytes
  - Aluminate or borate three-dimensional inorganic–organic ionically conducting structures with bridging anionic species fixed to the polymer backbone to give single ion (cation) conductors [30].
  - Aluminate or borate complex structures prepared by introducing metallic or nonmetallic atoms with weak Lewis acid characteristics into the host skeleton [31]
  - Zeolitic inorganic–organic polymer electrolytes contain inorganic clusters formed by the aggregation of inorganic coordination complexes that can be positively or negatively charged and are bridged by organic macromolecules [32]
  - Hybrid gel electrolytes prepared by a sol-gel reaction of inorganic or organometallic precursors bearing an ion-conducting cation and low molecular weight or macromolecular liquids [28].

The reality of commercial lithium polymer batteries has been approached from two distinct directions: dry, all-solid polymer electrolytes and hybrid organic solvent-containing gel electrolytes (GPEs), and consequently the following discussions will consider these subclasses separately.

52.2.1 Dry Polymer Electrolytes (SPE)

Since Wright [33] in the 1970s reported ionic conductivity in solid-state solutions formed by the addition of various salts to high molecular weight PEO, and Armand et al. [34] highlighted their potential as practical battery components, the worldwide research and development effort in this area has been enormous. These so-called "dry" SPEs represented an elegant solution to several problems which had plagued the lithium cell market for many years. Polyethers have strong coordinating atoms along the chain and can dissolve a wide variety of salts. In low molecular weight solvents, solvation of the cation depends mainly on the number of molecules that can fit around the cation. In high molecular weight polymers, the chain must accommodate the cation without excessive strain. PEO provides just the right spacing for maximum solvation. PEO is not only an ideal solvent for many salts but also offers excellent electrochemical stability as the C–O bond in the ether linkages is of relatively low reactivity. The ability to transfer cations between coordinating sites on the polymer host relies to a greater extent on local macromolecular chain flexing. Segmental motion of the polymer chains continuously creates free volume into which the ions migrate, and this process allows them to progress through the electrolyte [28]. The necessity for polymer chain flexibility and constant short-range restructuring explains why ion transport in SPEs is many orders of magnitude higher in amorphous polymer-salt complexes at temperatures above their glass transition temperature (T_g). As a result, the general focus of research has been on the synthesis of materials with low T_g and hence high levels of segmental motion in order to enhance the ionic conductivity. Recently, it has been established that conductivity is not just confined to the amorphous phase. In the 1970s, Armand suggested the possibility of ion transport along ordered PEO helical coils [34], and indeed, a mechanism of ion hopping along fixed pathways has been demonstrated [35]. Under the conditions normally applied to film preparation this ion motion would contribute little to the overall conductivity as the crystalline domains would form in the plane of the film with random orientation. For practical purposes the polymer chains need to be unidirectional and oriented perpendicular to the plane of the electrolyte film. One approach to altering the structure of a PEO-based polymer electrolyte has been to cast LiI-based SPEs under an applied magnetic field. This enhances both intra- and inter-chain ion mobility by about one order of magnitude at 65°C. The magnetic field effect is more pronounced

in polymer electrolytes which incorporate diamagnetic and ferrimagnetic nanofillers [36,37].

52.2.1.1 Ionic Conductivity.
For applications in all-solid-state electrochemical devices, the polymer electrolyte should possess an ionic conductivity greater than 10^{-4} S cm^{-1} at room temperature to achieve a performance level close to that of liquid electrolyte-based devices. Solid PEO-based electrolytes display a marked tendency to crystallize below about 70°C, and therefore support very poor ambient temperature conductivities. A very considerable effort has been invested into finding methods of enhancing the conductivity without compromising the many attributes of PEO that make it such an appealing electrolyte host. Although useful ionic conductivities are obtained at temperatures above ambient, this does restrict the market choice for their application. Various strategies for enhancing ambient temperature conductivities in PEO have been reported. One of the earliest approaches was to modify PEO polymer architecture to reduce crystallinity. These modifications included synthesizing random copolymers, block copolymers, crosslinked polymer networks, and comb-shaped polymers with short oligo-oxyethylene chains attached to the polymer backbone [18,19,38,39]. Another approach has been to incorporate low levels of plasticizers, such as organic liquids or low molecular weight ethylene glycols, in the polymer host [18,19]. Unfortunately, although conductivities of the order of 10^{-4}–10^{-5} S cm^{-1} can be achieved in all these modified PEO electrolytes and, in particular, the plasticized systems, this improvement is accompanied by a loss of mechanical integrity and of chemical compatibility with the lithium metal electrode, one of the most important intrinsic features of the PEO-based SPE.

Room temperature conductivities as high as 10^{-2} S cm^{-1} have been reported for rubbery "polymer-in-salt" systems in which lithium salts are mixed with small quantities of PEO and poly(propylene oxide) (PPO), for example, the AlCl$_3$–LiBr–LiClO$_4$–PPO system [27]. Despite their very promising conductivities, these guest salts tend to be corrosive and crystallize at lower temperatures. This in turn compromises the electrochemical stability of the electrolytes.

52.2.1.2 Lithium Salts.
In view of the difficulties arising from the use of organic liquid plasticizers, much attention has focused on synthesizing nonvolatile lithium salts with bulky anions. The motivation for using low lattice energy salts with large stable anions containing strong electron-withdrawing substituents is that they increase amorphicity through their plasticizing action on PEO even in the absence of any liquid component. Anion symmetry or chain flexibility has been shown to introduce a plasticizing effect which lowers the glass transition temperature (T_g) of the polymer electrolyte. The reduced electrostatic anion-to-cation attraction also favors a high degree of ion dissociation and cation mobility. Examples of these plasticizing salts include [39–41]

- Lithium bis(trifluoromethane sulfonyl) imide—LiN(SO$_2$CF$_3$)$_2$ (LiTFSI)
- Lithium tris(trifluoromethane sulfonyl) methide—LiC(SO$_2$CF$_3$)$_3$ (LiTriTFSM)
- Lithium pentafluoro sulfur difluoromethylene sulfonate—LiSO$_3$CF$_2$SF$_5$
- Lithium bis(trifluoromethane sulfonyl) methane—LiCH(SO$_2$CF$_3$)$_2$
- Lithium 4,5-dicyano-1,2,3-trizolate—Li (LiDCTA)
- Lithium bis(oxalate)borate—LiB[(OCO)$_2$]$_2$ (LiBOB)

Phase diagrams reveal the extent of crystallinity in PEO-LiX complexes at different polymer-salt compositions (normally represented as the EO:Li ratio) and temperatures and confirm that some PEO-LiX complexes are more effective at suppressing the melting temperature of the eutectic than others [42]. The ability to lower the temperature at which the electrolyte components crystallize increases the attractiveness of PEO-based SPE films. The structure of PEO$_3$LiTFSI as determined from X-ray powder diffraction data confirms that only one -SO$_2$ group of each imide ion is involved in Li$^+$ coordination. The uncoordinated fragment of the anion projects into the inter-chain volume, disturbing polymer chain alignment, and thereby inhibiting crystallization. The partial phase diagram of PEO-LiTFSI shown in Figure 52.4 lends support to this explanation, demonstrating that the polymer complex remains completely amorphous at temperatures of 50°C and lower, over the salt concentration range 6 < O:Li <12.

LiTFSI in particular has become the salt of choice for PEO-LiX electrolytes. Conductivities of 10^{-4} S cm^{-1} at 30°C and 10^{-3} S cm^{-1} at 80°C can be achieved.

52.2.1.3 Composite Polymer Electrolytes. An alternative approach to enhancing the ion transport properties in the PEO-based SPEs has been through the dispersion of nanoparticle-sized ceramic fillers such as TiO$_2$, Al$_2$O$_3$, and SiO$_2$ within the polymer matrix [43, 44]. The idea of adding electrochemically inert particulate fillers to PEO-based polymer electrolytes dates back to the early 1980s when this strategy was originally employed to increase the mechanical stability of the polymer. The formation of such polymer–ceramic composites was also found to enhance the ionic conductivities of the electrolytes [44]. The ceramic additive, due to its large surface area, can act as a solid plasticizer for PEO by inhibiting chain crystallization after annealing the composite in the amorphous state. This inhibition leads to stabilization of the amorphous phase at lower temperatures and thus to an increase in the useful range of electrolyte conductivity. Figure 52.4 shows the conductivity Arrhenius plot of a representative example of nanocomposite polymer electrolytes [44].

Figure 52.5 reveals that conductivity enhancement also occurs at temperatures where the composite materials are amorphous in nature. The role of the ceramic, therefore, cannot be exclusively to prevent crystallization of the polymer

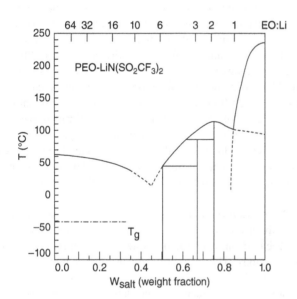

FIGURE 52.4 Phase diagram for PEO-LiN(SO$_2$CF$_3$)$_2$ showing the eutectic equilibrium between PEO (mw = 1 × 10^6) and the 6:1 (salt wt. fraction 0.52) intermediate compound. Adapted with permission from C. Labrèche, I. Lévesque and J. Prud'homme, Macromolecules, Vol. 29, 7795, Copyright (1994) American Chemical Society, and from S. Lascaud, M. Perrier, A. Vallée, S. Bresner, J. Prud'homme and M. Armand, Macromolecules, Vol. 27, 7469, Copyright (1994) American Chemical Society.

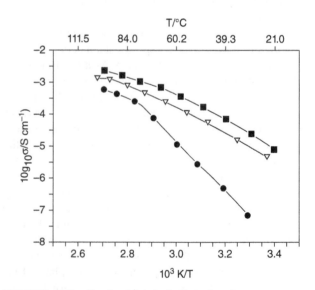

FIGURE 52.5 Conductivity Arrhenius plots for nanocomposite PEO-based polymer electrolytes and a ceramic-free system. From Ref. [39]. http://onlinelibrary.wiley.com/advanced/search/results?start = 1&resultsPerPage = 20. ▽, composite containing Al$_2$O$_3$; ■, composite containing TiO$_2$; •, ceramic free.

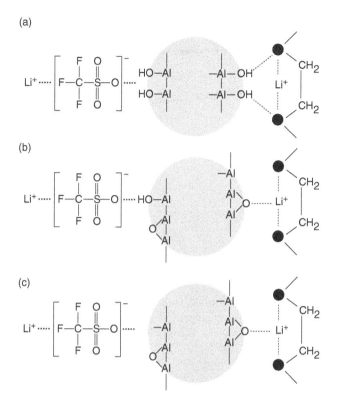

FIGURE 52.6 Schematic model of the surface interactions between ceramic particles and the polymer chain, and with the salt anion in PEO-LiX nanocomposite polymer electrolytes under (a) acidic, (b) neutral, and (c) basic conditions. From Ref. [39]. http://onlinelibrary.wiley.com/advanced/search/results?start=1&resultsPerPage=20

chains. Lewis acid groups on the ceramic nanoparticles (e.g., the -OH groups on the Al_2O_3 surface) may well complex with both the anion of the guest salt and the coordinating ether oxygen atoms on the PEO host as depicted in Figure 52.6. These interactions may reduce the ability of the PEO to reorganize and hence introduce structural modifications of the polymer chains promoting lithium-ion-conduction pathways at the ceramic surface. In addition, Lewis acid–base interactions with the ionic species might also be expected to reduce ionic coupling, promoting dissociation through an ion–ceramic complex formation. Such a scenario would increase the number of mobile lithium cations accounting for the observed enhancement of conductivity at all temperatures. Further evidence to support this model is found in the lithium-ion transference number measurements which will be discussed in the next section.

52.2.1.4 Lithium-Ion Mobility.

Another important practical issue that has delayed the commercial development of dry, PEO-based polymer electrolytes is the low lithium-ion transference number. The mobility of lithium ions in cells is a factor that has a crucial impact in terms of fast and/or

deep discharge, energy and power density, and cycle number, and low lithium transference gives rise to concentration polarization of the cell as the intercalation process which drives lithium batteries is based exclusively on the transfer of lithium ions.

Various techniques have been used to measure transference numbers on a wide range of dry polymer electrolytes. All these techniques provide values of the lithium-ion transference number of less than 0.5. The most reliable data have been obtained from diffusion coefficient measurements on poly(ethylene oxide)-$LiN(CF_3SO_2)_2$ (PEO-LiTFSI) systems and in this case a cationic transference number of only 0.25–0.3 was reported, suggesting that the cation is indeed very much the minor current carrier. Several methods of improving cationic conductivity have been applied: introducing large and heavy anions of low mobility, immobilizing anions on the polymer chain, or lowering the mobility of the anions by complexation with Lewis acid additives ($AlBr_3$, $SnCl_4$, or boron compounds). These methods have usually led to a significant drop in ionic conductivity of the electrolyte, although this effect may be attenuated through the use of calixarenes or calixpyrroles as anion receptors [45–47]. This strategy has led to cationic transference numbers close to unity, accompanied by a lowering of the overall conductivity by less than one order of magnitude [47].

An alternative but effective approach to achieving the goals of high ionic conductivity, mechanical stability, and good electrode/electrolyte interfacial properties involves the introduction of nanomeric inorganic fillers into the polymer electrolyte matrix. As described in the previous section, one of the key contributions of the filler originates in its ability to act as a solid-state plasticizer, kinetically inhibiting the PEO chain crystallization. The scheme presented in Figure 52.6 also demonstrates how the ceramic filler can facilitate salt dissociation and decrease anion mobility. The degree of success depends upon the choice of ceramic filler and, in particular, the nature of its surface states. Aluminum or boron derivatives and functionalized zirconia have been identified as a very promising group of compounds, as a result of their favorable Lewis acid properties [48]. Sulfate-promoted superacid zirconia ($S-ZrO_2$) has an acid strength more than twice that of H_2SO_4, associated with the Zr^{4+} cations. A high density of both Lewis and Brønsted-type acidic sites are therefore present on the surface of the oxide. Dispersion of $S-ZrO_2$ in PEO-$LiBF_4$ gives a composite polymer electrolyte with a lithium transference number of 0.81 ± 0.05, compared to 0.42 ± 0.05 for the ceramic-free electrolyte. This increase in transference number is accompanied by an enhancement of ionic conductivity at all temperatures, as demonstrated in Figure 52.7. The nanocomposite approach appears to be the most effective in the suppression of anion mobility while simultaneously increasing the overall ionic conductivity through enhanced lithium-ion mobility [47–49]. There is clear evidence that nanocomposite polymer electrolytes also present

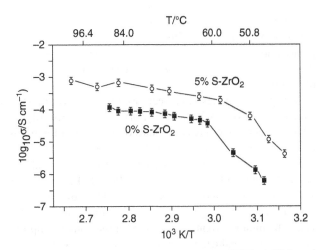

FIGURE 52.7 Conductivity Arrhenius plots of $PEO_{20}LiClO_4$ and a S-ZrO_2 composite with $PEO_{20}LiClO_4$. Reprinted from Croce, F., Sacchetti, S., Scrosati, B. (2006). Advanced high-performance composite polymer electrolytes for lithium batteries. *Journal of Power Sources*, 162:685–689, copyright 2006 with permission from Elsevier.

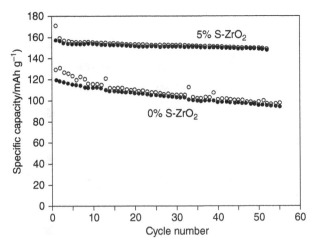

FIGURE 52.8 Capacity versus charge–discharge cycle number for a Li/$PEO_{20}LiClO_4$ + 5% S-ZrO_2/LiFePO$_4$ cell (upper curve) and a Li/$PEO_{20}LiClO_4$ /LiFePO$_4$ cell (lower curve). Reprinted from, Croce, F., Sacchetti, S., Scrosati, B. (2006). Advanced high-performance composite polymer electrolytes for lithium batteries. *Journal of Power Sources*, 162:685–689, copyright 2006 with permission from Elsevier.

a less reactive lithium–electrolyte interface, an aspect that will be discussed in the next section.

52.2.1.5 *Electrochemical Stability.*

Passivation is a process whereby products from the reaction at the electrode/electrolyte interface form a protective film that covers the pristine surface of the electrode and limits sustained reaction. The occurrence of passivation is the foundation on which many high-energy-density battery chemistries, including lithium, are built. Electrolyte components are sacrificed to form the protective film and therefore increase the chemical stability of the new electrode surface. Unfortunately, this passivated interface also acts as a barrier to the facile transfer of species between the electrolyte and electrode and, more often than not, may impose practical limitations on battery performance.

Lithium metal is stable in nonaqueous electrolytes as a result of surface passivation. Once formed, the film is not completely removed during stripping/deposition of lithium. The film acts as a solid electrolyte interface and therefore the rate determining step for a redox process at the electrode is the diffusion of lithium ions through this film.

The great attraction of dry polymer electrolytes is, of course, the possibility of using lithium metal as the anode. For secondary lithium batteries based on liquid electrolytes, the limitations caused by the formation of a passivating layer at the electrode/electrolyte interface and the consequential growth of lithium dendrites are too severe: cycle life and safety of the battery are compromised. PEO-based electrolytes also form a passivating layer at the electrode/electrolyte interface. The nature of this layer, as

for all electrolytes, depends on the electrolyte salt [50]. Under appropriate conditions, however, passivation in PEO-based electrochemical cells can be suppressed, opening up real possibilities for lithium metal anode-based secondary batteries.

The nanocomposite approach once again appears to be an effective route to enhancing the SPE performance. Depending on the type of filler and particle size, enhancement of both cycle efficiency and the chemical stability of the polymer electrolyte in contact with the lithium anode is often observed [49, 51]. The performance of a lithium-ion battery incorporating a lithium metal anode and a $PEO_{20}LiClO_4$ electrolyte with and without S-ZrO_2 nanofiller is shown in Figure 52.8. A higher charge–discharge cycle number, a lower capacity decay upon cycling, and, in particular, a more stable charge–discharge efficiency (and therefore cell lifetime) are observed for an electrolyte containing 5% nanofiller.

By the mid-1990s, significant improvements had been achieved in many of the properties of solvent-free polymer electrolyte materials but even so they were still not able to fulfill the stringent operational requirements of commercial rechargeable batteries in terms of transport properties. As a result, the focus of polymer electrolyte development switched to gel systems formed by polymers and nonaqueous solutions of lithium salts or plasticized polymer electrolytes in which the ionic transport simultaneously occurs in the liquid and polymeric phases. With lithium-ion intercalating anode technology already well developed, this was certainly the obvious direction for the major thrust of polymer electrolyte R&D.

52.2.2 Gel Polymer Electrolytes

The restricted temperature range over which dry SPEs exhibit adequate ionic conductivities has prevented them from realizing their otherwise promising potential in ambient temperature battery applications. In contrast, liquid electrolyte-based lithium-ion technology provides better performance over a much wider temperature range, although electrolyte leakage and safety issues remain a constant concern. Midway between the SPE and the liquid electrolyte is the hybrid polymer electrolyte concept. Suitable plasticizers can be chosen from the same polar solvents as used in liquid electrolytes, including propylene carbonate (PC), ethylene carbonate (EC), dimethyl carbonate (DMC), diethyl carbonate (DEC), dimethoxyethane (DME), γ-butyrolactone, or short-chain polyethylene glycol ethers [19, 24]. Figure 52.9 compares the temperature variation of the ionic conductivity for a sample of SPEs, liquid electrolytes, and GPEs.

The addition of moderate amounts of plasticizer (10–25% additive) can improve the conductivity by an order of magnitude and this method has been applied to PEO-based polymer electrolytes as a valid approach to improving conductivity. Interestingly, if the polymer host is a high molecular weight polyether, the interaction takes place through the molecular chains as in dry SPE systems, rather than the carbonate solvents as these are less able to donate electron pairs.

A GPE can take one of two forms. On the one hand it can constitute a dry polymer electrolyte plasticized with

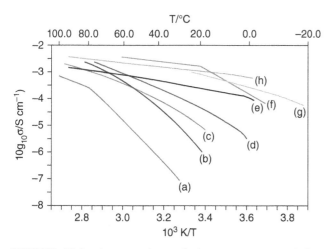

FIGURE 52.9 A comparison of the temperature variation of the ionic conductivity for representative SPEs, GPEs, and liquid electrolytes. (a) PEO-LiClO$_4$; (b) PEO-Li((CF$_3$SO$_2$)$_2$)N; (c) PEO-LiClO$_4$-10 wt% TiO$_2$ ceramic; (d) Plasticized high mw PEO-based electrolyte: PEO-Li((CF$_3$SO$_2$)$_2$)N-25% w/w (polyethylene glycol dimethylether mw 250); (e) gel electrolyte: crosslinked PEO dimethacrylate-Li((CF$_3$SO$_2$)$_2$)N-70% PC; (f) gel electrolyte: P(VDF-HFP)-40 wt% TiO$_2$ ceramic-EC/DMC-LiPF$_6$; (g) gel electrolyte: P(VDF-HFP)-EC/DMC-LiPF$_6$; (h) liquid electrolyte: PC/DME-LiCF$_3$SO$_3$.

an organic solvent. Ionic transport occurs simultaneously in both the liquid and polymeric phases. Alternatively, a GPE can be formed by incorporating a nonaqueous solution of lithium salt into an inert porous polymer. Ion transport in this instance only occurs in the liquid phase. Porous GPE films typically contain 60–95% liquid electrolyte trapped in pores of the polymer network. They are only two to five times less conductive than their liquid counterparts [52]. The GPE approach to lithium-ion battery technology combines the positive attributes of both the solid and liquid systems; these materials show the cohesive properties of solids and the diffusive properties of liquids. GPEs are less flammable than their liquid electrolyte counterparts and, in common with dry SPEs, they are obtained as flexible, self-supporting films. The all-solid-state assembly confers an attractive tolerance to shock, vibration, leakage, and deformation.

The variety of gel-based polymer electrolytes and production strategies that have been researched and exploited at a commercial level, together with the number of patents which have been registered, are an elegant testimony to the difficulties encountered in achieving a viable solution to the problems associated with reliable mass production of this type of high-performance cell. Polymers which have been used as components in the preparation of gel electrolytes include PEO and PEO-based crosslinked networks, polyacrylonitrile (PAN), polymethylmethacrylate (PMMA), polyvinylidenefluoride (PVdF), and PVdF-hexafluoropropylene (PVdF-HFP) [24]. Just as in dry polymer electrolyte research, the objective for gel electrolytes has been to maximize the ionic conductivity, electrochemical stability, and transference number while preserving adequate mechanical stability. Many blend formulations with a variety of lithium salts, plasticizing salts, crosslinked networks, crystalline/amorphous matrices, and low molecular weight plasticizers have been studied. As with PEO-based SPEs, the plasticizing salt LiTFSI is used extensively in GPE formulations [53]. As well as promoting high lithium-ion mobility, and being extremely resistant to oxidation, its safety and lack of toxicity make it an ideal candidate for all polymer electrolyte materials.

Although plasticizers may alter the overall properties of gel electrolytes in a significant manner, the role of the polymer host is generally complex and dominant. The choice of individual components of the gel membrane and optimization of their concentrations are crucial; their chemistry and compatibility ultimately determine both the mechanical and electrochemical properties of the final material. PAN-based gel electrolytes, where the salt and plasticizer are dispersed at a molecular level, have shown promising properties; they offer room temperature ionic conductivities in the region of 10^{-3} S cm^{-1}, lithium-ion transference numbers of 0.5–0.7, and an electrochemical stability window of 4.5 V. Unlike other polymer gel electrolytes, the mechanical stability is good at plasticizer concentrations which afford viable ionic conductivities. The inclusion of ceramic fillers within

PEO-based electrolytes to improve conductivity, mechanical stability, and interfacial stability has also been successfully extended to GPEs [54–59]. PAN-based gel networks containing a lithium salt and Al_2O_3, for example, are found to support high ionic conductivity and to display a wide electrochemical stability and also high chemical integrity, even at temperatures above ambient. Remarkably, the compatibility of a lithium metal anode with its gel-type electrolyte is noticeably improved by incorporating ceramic nanoparticles. These also appear to inhibit solvent evaporation, improving cyclability.

PVdF has been presented as another attractive polymer host by virtue of its strong electron-withdrawing functional group and high dielectric constant, conferring good anodic stability and the capacity to support a significant concentration of dissolved charge carriers, respectively. The conductivity of PVdF-based gel electrolytes is largely dominated by the uptake of electrolyte solution. Porous membranes of PVdF are widely used in a variety of biomedical applications and the methodology for preparing such membranes has been applied to PVdF gel electrolytes to great effect. Ionic conductivities are enhanced by several orders of magnitude but without serious loss of mechanical stability. PVdF itself is highly crystalline but when copolymerized with HFP, the degree of crystallinity is greatly reduced. This structural change modifies the mechanical properties of the resulting copolymer, and as the material can be prepared in a microporous form, it has effectively become an electrolyte matrix of choice for lithium-ion polymer battery R&D.

The interest in PVdF-HFP systems arises not only as a result of their favorable electrochemical performance and thermal stability, but also from the superior processing capabilities of these gel electrolytes. The copolymer is made up of an amorphous phase which traps large amounts of solvent containing the dissolved lithium salt and a much smaller crystalline fraction which imparts mechanical stability to the membrane without the need to crosslink the final copolymer. Mechanically stable, dry membranes are first prepared by solvent casting and the cells preassembled before plasticizer is added at the final stage to activate the system. This "phase inversion" technique is strategically advantageous in mass production as components are assembled in the solid state and thus the need for critical moisture control is limited to the final cell-activation step [56].

52.2.2.1 Polymer Electrolytes Containing Room Temperature Ionic Liquids.
Recently, research has focused on replacing conventional flammable and volatile organic alkyl carbonate electrolytes with highly stable IL-based solutions [60]. ILs are room temperature molten salts typically consisting of bulky asymmetric organic cations and inorganic anions. The interest in these materials arises from desirable properties such as nonvolatility, nonflammability, high ionic conductivity, and high thermal stability.

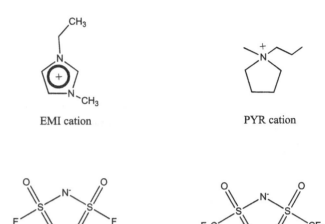

FIGURE 52.10 The structure of commonly studied ionic liquids 1-ethyl-3-methyl imidazolium (EMI), *N*-methyl-*N*-propylpyrrolidinium (PYR), bis(trifluoromethanesulfonyl)imide (TFSI), and bis(fluorosulfonyl)imide (FSI).

Along with their favorable electrochemical characteristics they are also expected to greatly reduce risks of thermal runaway and fire hazards and contribute to an enhanced safety margin of all lithium-based devices, particularly lithium polymer batteries. In addition, they have been shown to improve the behavior of the electrolyte/lithium interface, providing further encouragement for the development of the lithium metal-based Li-ion and LiPo secondary batteries.

The most studied ILs are those formed by the combination of imidazolium-based cations, such as 1-ethyl-3-methyl imidazolium (EMI) and *N*-methyl-*N*-propylpyrrolidinium (PYR), and delocalized-charge anions, including bis(trifluoromethanesulfonyl)imide (TFSI) and more recently, bis(fluorosulfonyl)imide (FSI) (Figure 52.10). These systems, like most ILs, do not contain mobile Li^+ ions and therefore a lithium salt is also incorporated, most commonly LiTFSI.

As a result of the addition of the lithium salt, the electrolyte conductivity decreases due to an increase in viscosity and due to the interactions between Li^+ cations and anions. This is not a critical drawback, however, since the conductivity of the lithium salt IL solution is still sufficiently high for battery applications. A conductivity of $1.06 \times 10^{-2} \, S \, cm^{-1}$ at 303 K has, for instance, been reported for LiTFSI-EMI-TFSI [61], a value of the same order of magnitude as found with solvent-based electrolytes.

Just as the ambient temperature conductivity of PEO-based polymer electrolytes can be enhanced by the incorporation of organic liquids or plasticizing lithium salts, so ILs offer a promising approach to achieving more liquid-like conductivities in a polymeric matrix [62–67].

The lithium-ion conductivity in PEO-based polymer electrolytes is found to be greatly enhanced by the presence of certain ILs [68]. On the other hand the contribution of the lithium ion to the overall ionic transport has been found to be quite low in IL-based GPEs which may limit their use in electrochemical cells. A variety of ILs have been studied, including families of imidazolium-, pyrrolidinium-, morpholinium-, and guanidinium-based ILs. The incorporation of several N-alkyl-N-methylpyrrolidinium bis(trifluoromethanesulfonyl)imide (PYR-TFSI) ILs into PEO electrolytes increases the ionic conductivity to values greater than 10^{-4} S cm^{-1} at 20°C [64].

PEO and poly(vinylidene fluoride-co-hexafluoropropylene) copolymer (PVdF-HFP) have been compared as polymer hosts for the hydrophobic IL 1-n-propyl-2,3-dimethylimidazolium tetrafluoroborate (MMPIBF$_4$) and hydrophilic IL 1-n-propyl-2,3-dimethylimidazolium hexafluorophosphate (MMPIPF$_6$). On the basis of ionic conductivity and electrochemical behavior, the PVdF-HFP gels perform better than the PEO gels. The superior behavior of the nonpolar PVdF-HFP is justified by the supposition that the membrane gives structural integrity while establishing ionically conductive channels in which the lithium-containing IL resides. The mobile ions move through these channels without restrictions caused by interactions with the polymeric structure. The choice of a hydrophobic-type plasticizer is also preferable [69]. Polymer electrolytes with IL N-ethyl-N-methyl morpholinium bis(trifluoromethanesulfonyl) amide and PVdF-HFP were successfully prepared and characterized [70].

One of the few remaining concerns regarding the application of ILs is related to their electrochemical stability, specifically related to the value of the reduction potential of the cation, which is generally too positive to allow lithium deposition. For instance, the cathodic limit of the EMI cation is about 1.0 V versus Li/Li$^+$ [71] and is associated with the tendency of the imidazolium-based cations to be reduced by electrochemical deprotonation. This prevents the use of IL-based solutions with common low-voltage anode materials such as lithium metal, graphite, or Li$_4$Ti$_5$O$_{12}$. Many authors have attempted to improve the electrochemical stability by alteration of the IL structure. Small quantities of film-forming additives can be included in the electrolyte formulation to avoid cathodic decomposition and improve the interfacial compatibility toward a lithium metal anode [72]. Cationic ILs, more resistant to reduction than members of the imidazolium family, would be expected to show improved characteristics and quaternary ammonium cations which have no acidic protons are a good example. Pyrrolidinium-based ILs show a much wider electrochemical stability window and better compatibility toward lithium [62, 64]. The result of adding N-methyl-N-propyl pyrrolidinium-TFSI to P(EO)$_{20}$LiTFSI polymer electrolytes and using these membranes as electrolytes in solid-state Li/V$_2$O$_5$ batteries has been investigated [63]. The Li/V$_2$O$_5$ cells showed excellent reversible cyclability with a capacity fade of only 0.04% per cycle over several hundred cycles at 60°C. An emerging class of ILs for battery applications has arisen from the replacement of the TFSI anion with an FSI anion. It is found that with the same IL cations, changing from a TFSI to an FSI anion vastly improves the electrochemical properties of the system [73]. In particular, this alteration drastically increases the cycling reversibility of the graphitized carbon electrode [74].

One significant drawback to the use of ILs is their prohibitively high production cost. With industrial scale-up and increasing demand, however, the current cost of ILs is expected to fall.

52.3 BATTERY TECHNOLOGY

52.3.1 Electrodes for Lithium Polymer Batteries

The choice of electrode material depends on whether the cell is a rechargeable Li-metal or Li-ion type. If metallic lithium is used as the negative electrode, the positive electrode does not need to be lithiated before cell assembly. Manufacturers of metallic lithium-based SPE cells therefore normally use V$_2$O$_5$ or its derivatives as the positive electrode. In contrast, for Li-ion batteries, a carbonaceous anode is used and because the carbon is not lithiated, the positive electrode must supply the initial charge of lithium, thus requiring use of air-stable Li-based intercalation compounds to facilitate the cell assembly. LiCoO$_2$ has been the most popular choice of cathode material for commercial Li-ion batteries because of the high energy density available from this material. However, the long-term commercial availability of cobalt is not certain and the relatively high cost has led manufacturers to seek alternative cathodic materials.

Current interest in positive electrode materials is focused on manganese-based compounds, including LiMn$_2$O$_4$, and olivine lithium metal phosphates, LiMPO$_4$, where M = Mn, Co, or Fe. Operational issues have delayed the commercial development of the former group [75] but olivine phosphates, and in particular lithium iron phosphate, LiFePO$_4$, have reached commercial development. Their high capacity (170 Ah kg^{-1} for LiFePO$_4$), safety (due to the strength of the P–O covalent bond), low cost, and low environmental impact make them particularly attractive electrode materials [76]. LiFePO$_4$ unfortunately suffers from a very high intrinsic resistance due to its structure and was largely ignored as a potential cathodic material. By reducing the LiFePO$_4$ particle size and coating these particles with electronically conducting carbon, a significant increase in conductivity has been obtained and full capacity under prolonged cycling and high rate delivery has been reported.

To date, carbons have been used as the anode material of choice in commercial rechargeable lithium batteries, although new materials such as lithium titanium oxide ($Li_4Ti_5O_{12}$, LTO) are entering the portable energy market [75,77]. The ability of the carbonaceous material to accommodate lithium is greatly influenced by its crystallinity, microstructure, and micro-morphology. Carbonaceous materials suitable for lithium intercalation are commercially available in many forms and qualities. These include

- Graphite—naturally occurring, highly ordered, crystalline, composed of graphene planes packed in parallel configuration. This material can accommodate lithium up to a stoichiometry of LiC_6.
- Hard carbon—made from organic precursors that char as they pyrolyze (non-graphitizable). There are no neatly stacked graphene layers, the material is noncrystalline, it is macroscopically anisotropic, and because of this structure there is almost no volume change on intercalation and thus better cyclability.
- Soft carbon—made from organic precursors which melt before they pyrolyze (graphitizable). The graphene layers are neatly stacked but have less long-range order than in graphite. The inter-plane spacing is greater than for graphite, but variable.
- Single- and multiwall carbon nanotubes.

The graphitic materials may be applied in the form of flakes, fibers, chopped fibers, or beads [78]. Carbon nanotubes have, not surprisingly, been investigated and recent reports suggest that cells using functionalized multiwalled carbon nanotubes as an anodic material can deliver about 10 times the power of conventional Li-ion cells [79]. The cost of synthesizing these materials, however, leaves their short-term viability open to question. The future use of nanomaterials will probably, in the long term, lead to benefits in terms of capacity, power, cost, and materials sustainability [80].

In the search for alternatives to carbon anodes, lithium alloys containing aluminum, antimony, tin, and silicon have been explored because they offer much larger specific capacities. Indeed, a fully lithiated lithium–silicon alloy, $Li_{4.4}Si$, has a theoretical specific capacity of 4200 Ah kg^{-1} exceeding that of metallic lithium. Unfortunately the favorable electrochemical characteristics of these alloys are counterbalanced by very large charge/discharge volume changes; 100–300% variations are common. In comparison, the volume change of lithiated graphite is only 10%. These volume changes cause enormous internal strains in the metal grains resulting in disintegration of the electrode with drastic consequences for the cell's cyclability. Intermetallic alloys such as Cu_6Sn_5, InSb, and Cu_2Sb retain a strong structural relationship to their lithiated products, Li_2CuSn and Li_3Sb and the application of these materials may provide a solution to the problem of volume expansion. In spite of the novelty and elegance of

the design of these intermetallic electrodes, they still suffer from poor cyclability, particularly during the initial cycle [11]. Nanotechnology, however, may once again provide the answer to achieving desirable material characteristics. Metal–carbon nanocomposites such as Sn–C, for example, are capable of buffering the large volume changes and therefore providing long cycle life combined with high specific capacity [75]. Other suitable modifications include the formation of nanowires and three-dimensional porous particles. As a result of these developments, ternary lithium metal Sn–Co–C alloys have reached the marketplace as anode materials for commercial batteries [79].

LTO has an operating voltage of 1.5 V versus Li and a theoretical capacity of 175 Ah kg^{-1}. Although this capacity is lower than that of graphite (372 Ah kg^{-1}) and the voltage level of the electrode is higher, leading to lower specific energy, it is still an attractive anodic material. The less advantageous electrochemical performance may be offset by other, more favorable, properties:

- No electrolyte decomposition and therefore no SEI layer formation; the reversible capacity is consequently high.
- Less than 0.2% volume change from fully discharged ($Li_4Ti_5O_{12}$) to fully charged ($Li_7Ti_5O_{12}$,).
- Good thermal stability and safety performance.
- Long cycle life.
- Excellent power and low-temperature behavior.

Confirmation of these advantages is demonstrated by the recent incorporation of LTO in batteries for plug-in hybrid electric vehicles (PHEVs) [77].

The voltages versus capacities of many positive and negative electrode materials that have been studied as possible candidates for rechargeable Li-based cells are shown in Figure 52.11 [11]. The large difference in capacity between lithium metal and other anodic materials is evident and explains why the continued efforts to establish this as an anodic material for rechargeable lithium-based batteries are worth pursuing.

52.3.2   Commercial Lithium Polymer Cells

Liquid electrolyte-based Li-ion batteries have been established on the market since the early 1990 s. Li-ion batteries in spite of many years of development still present some serious safety issues related to the use of a liquid electrolyte and as a result there has been significant attention devoted to the development of polymer electrolytes with adequate practical performance. The use of polymeric electrolytes for Li-ion batteries had several false starts prior to 1999, when finally, following the demonstration of Bellcore's PLiON rechargeable battery in 1996 [16], the first batteries were commercialized, for example, the Ultralife Polymer™ battery [81].

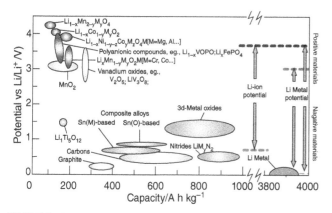

FIGURE 52.11 Voltage versus capacity for positive and negative electrode materials relevant to rechargeable Li-based cells. The output voltage values for Li-ion or Li-metal cells are represented. Note the very large difference in capacity between Li-metal and other negative electrodes. Adapted from Ref. [11]. Reprinted by permission from Macmillan Publishers Ltd: *Nature*, 414:359–367, copyright 2001, http://www.nature.com/

Cells with polymeric electrolytes can offer some significant advantages over the conventional cell design, for example, no leakage, flexibility, and very low profile formats. The adoption of the polymer electrolyte cell configuration has led to the substitution of the traditional cylindrical metal casing of Li-ion cells by a lightweight laminated aluminum bag. The packing efficiency of this new cell configuration, generally designated as the pouch cell, is about 90–95%. The electrical contacts are simply conducting foil tabs which pass through the walls of the thermally sealed pouch, as depicted in Figure 52.1. This novel cell presentation brings a significant reduction in weight, packaging costs, and simplification of the manufacturing process. The simplified architecture also facilitates interconnection between modular cells to provide series and parallel combinations with adaptable voltage and current characteristics suitable for a wide range of applications.

The so-called GPEs are particularly well-suited to commercial products for consumer electronics applications, with their performance characteristics approaching those of liquid electrolyte cells. There are two techniques which may be applied to processing GPEs: in the first, liquid electrolyte is added to the polymer film and is immobilized within the pores of the matrix; in the second, polymer and electrolyte form a homogeneous phase. This latter condition may be achieved either by mixing polymer powder and electrolyte in a common solvent (which is then evaporated) or by adding a polymer precursor to the electrolyte and promoting polymerization in situ. The former approach, using the separately formed gel electrolyte, is applied by Sony in commercial cells. Porous copolymer PVdF-HFP is used with $LiPF_6$ in PC-EC. The graphite anode is coated with amorphous carbon, in order to avoid decomposition in the presence of PC.

By optimizing the salt concentration, PC/EC and VDF/HFP ratios, and copolymer molecular weight, a conductivity of 9 $\times 10^{-3}$ S cm^{-1} which is comparable to a liquid electrolyte can be achieved at 25°C [82]. The latter in situ technique has been exploited by Sanyo in their commercial Li-ion polymer cells. The cell is preassembled in accordance with the conventional procedure for liquid electrolyte-based devices and, in the final step, a crosslinkable PEO precursor is added and heat applied thereby trapping the electrolyte, LiTFSI in EC-DEC, within the polymerized host [82].

The important contribution that the pouch cell has made to the improvement of cell safety is widely acknowledged. Both Li-ion and LiPo cells are designed and assembled to operate with a variety of safeguards. Naturally, as with any other primary or secondary cell, there are general handling precautions which must be observed. In common with the Li-ion cell, the LiPo cell assembly is particularly sensitive to overcharging, discharging beyond certain well-defined voltage limits, and overheating. In order to achieve and maintain high levels of performance, manufacturers have developed specific charging equipment incorporating electronic circuitry to avoid short-circuit, overvoltage, undervoltage, and thermal runaway. When a high level of performance is required, a cell discharge control module can monitor the voltage of each cell in series- and parallel-connected battery packs so that no cell is permitted to discharge below the safe limit of 2.5 V. Similarly, charge control modules balance cell status and restrict deviation of the nominal pack voltage to less than 10 mV (0.01 V), resulting in a permanently matched cell pack. Although not mandatory, charge devices may also permit automatic cell cycling to maintain optimum safety, power and capacity performance, and longevity.

The advantage of sharing electrode components with the Li-ion cell is that many safeguards originally developed for this technology can be transferred to the polymer gel version. Chemically, the formulations used by manufacturers of polymer gel electrolytes contain significantly lower quantities of volatile liquids than those used in comparable Li-ion cells. In exchanging the traditional hard steel casing for a flexible foil pouch, the result of heat generation during abuse is swelling, rather than violent rupture with dangerous projection of casing fragments. Traditional penetration, crush, and heat exposure tests carried out with LiPo cells have confirmed the superior performance of this class of cells. While many performance parameters are similar to those of current Li-ion cells, the features which provide major advantages for the LiPo cells are those of low profile and volume adaptability. Market analysts have identified these features as crucial to the ultimate commercial success of the latter cell.

52.3.2.1 *Commercial LiPo Cell Performance.* Due to the extremely demanding nature of the market for high energy and power density secondary cells and to the rapid transfer of technical solutions from prototype to commercial cells, it is

TABLE 52.1   Comparison of three competing secondary battery technologies.

	Ni-MH	Li-ion prismatic	Li-ion polymer
Gravimetric energy density (Wh kg^{-1})	60–120	110–200^a	150–220
Volumetric energy density (Wh dm^{-3})	200–300	240–400^a	150–380
Cycle life (to 80% of initial capacity)	300–500	500–600	300–550
Duration of fast charge (h), C-rate	0.5–4, 2 °C	1–3, 1 °C	0.3–3, 1–3 °C
Overcharge tolerance	Low	Very low	Low
Self-discharge per month	≈20%	≈10%^{b}	≈10%^{b}
Cell voltage (V)	1.2	3.7	3.7
Discharge performance (C)			
Peak	1–5	2–5	2–30
Best result	0.5	0.2–1	0.2–0.5
Typical operating temperature (°C)	–20 to 60	–20 to 60	–20 to 60^c
Maintenance required	60–90 days	Not required	Not required

The data are presented as representative of typical commercial devices (prismatic, 950–1000 mAh) at the time of publication.
^{a}Energy density is significantly influenced by the battery form factor. In an attempt to provide a reasonable basis for comparison between battery chemistries the data have been collected from manufacturers with prismatic cells of similar dimensions and capacities.
^{b}The protection circuits used with these cells may consume up to 3% of the cell energy content per month.
^{c}The lower operational temperature limits vary between manufacturers but –10 to –20 are typical specifications. Charge at less than 0 or greater than 45 °C is not recommended.

difficult to obtain precise details of the chemical constitution of current devices. Some examples of performance data for different battery technologies are provided in Table 52.1. The inclusion of comparable data for a Li-ion cell allows direct comparison of cells with a nominal capacity of 1000 mAh (3600 Coulombs).

Cost differences between Li-ion and LiPo have decreased markedly in the last few years. As a result of an improved understanding of the LiPo cell chemistry, high discharge rate behavior has improved dramatically with an increase in continuous discharge rates from 3-5C to 20C and short pulse discharge up to 40C. This advance in LiPo cell technology has coincided with price erosion of Li-ion cells and the result is an alteration in the commercial balance between these competing systems.

52.3.3   Lithium Metal–Polymer (LMP) Batteries

The development of batteries based on the Li-metal anode and polymeric electrolytes has taken much longer than their intercalation electrode/GPE-based counterparts, but there are good prospects for significant developments. A quarter of a century of R&D by Hydro-Quebec (Canada), and various strategic partners, has led to the development of a unique LMP battery technology. PEO-based electrolytes incorporating lithium salts such as LiTFSI have proved themselves in lithium secondary batteries operating at 40–80°C. In 2001, Avestor, a Hydro-Quebec and Kerr-McGee (USA) partnership, set out to commercialize this battery for the hybrid and electric car market, where it is easy to regulate the temperature above ambient, and for load leveling and emergency energy supply of remote telecommunications relays, where

temperatures above 40°C can vastly shorten the lifespan of ordinary electrochemical systems [13].

Modules are assembled from a three-component laminate composed of a dry polymer electrolyte of an ethylene oxide-based copolymer in which LiTFSI salt is dissolved, a lithium metal anode, and vanadium oxide (LiV_3O_8) cathode. This latter component is mixed with conductive carbon and the polymer electrolyte to form a composite that binds the particles while increasing ion conduction pathways. The laminate is less than 100 μm in thickness. In comparison to an equivalent lead-acid battery, the LMP battery is one-third the size, is one-fifth the weight, and has two to four times the service life expectancy. A solid electrolyte interfacial (SEI) passivating film forms on the lithium metal surface. The film restricts the extent of anode reaction with the electrolyte but supports exchange of lithium ions between metal and electrolyte. Although dendrite growth is far less serious with a dry polymer electrolyte than with liquid electrolytes, metal deposits have been detected in PEO-based cells. Consequently, in order to guarantee trouble-free lithium plating, the cells are structured to ensure strong adhesion to the metal surface and a uniform distribution of pressure throughout.

Since 2004, Avestor's LMP technology batteries have been deployed in volume by various North American telecommunications operators and LMP batteries were successfully tested in numerous prototype cars. In 2001, the Bolloré Group (France) set up a subsidiary BatScap in which EDF (Électricité de France) has a 5% stake. BatScap's objective was to take LMP batteries and supercapacitors which Bolloré had been developing since 1998 to the marketplace. In 2007, the Bolloré Group acquired the assets of Avestor, and the LMP battery technology has continued to develop under

TABLE 52.2   A comparison of LMP characteristics with those for other EV modules.

Technology	Specific energy (Wh kg^{-1})	Energy density (Wh dm^{-3})	Specific power (Wh kg^{-1})	Cycle life (cycles)
BatScap LMP	110	110	320	–
Avestor prototype LMP	120	140	240	300
Avestor-projected LMP	120	160	–	800
Li-ion	140	210	430	550
Ni-MH	65	150	200	800
Ni-Cd	50	90	120	800
Lead-acid	35	85	180	600

Reprinted from Pistoia, G. (2007). Nonaqueous batteries used in industrial applications. In: *Industrial Applications of Batteries: From Cars to Aerospace and Energy Storage*, Broussely, M., Pistoia, G. (editors). pp. 1–52, Copyright 2007 with permission from Elsevier.

the responsibility of BatScap. A comparison of the performance of different chemistries in EV modules is made in Table 52.2. The currently used Ni-MH battery is challenged by LMP and Li-ion batteries.

BatScap's LMP batteries are currently undergoing road tests in the Bluecar®, previously known as B0 or B Zero. The Bluecar® is a product of the partnership between Bolloré and Pininfarina; it is a production car, not a concept car, and went into production in late 2009. The LMP battery is supported by a supercapacitor, allowing greater acceleration, increased range, and a longer lifespan for the battery. Some physical and electrical battery characteristics are highlighted in Table 52.3. A range of 250 km, a top speed of 130 km/h, and a 0–60 km/h acceleration time of 6.3 seconds [15] have been announced for the vehicle.

52.4   LOOKING TO THE FUTURE

An analysis of the scientific literature relating to polymer electrolytes shows that interest in this field has grown continuously and exponentially over the last three decades

TABLE 52.3   Physical, thermal, and electrical characteristics of the Bluecar® LMP battery.

Volume	300 dm^3
Mass	300 kg
Lifespan	approx. 200,000 km
Range between charges[a]	250 km
End of life	100% reusable or recyclable
Operating temperature	−20°C to 60°C
Internal temperature	60–80°C
Nominal voltage	410 V
Power rating	30 kWh
Peak power output	45 kW (30 s)
Capacity at *C*/4	75 Ah
Specific energy density	100 Wh kg^{-1}
Energy density	100 Wh dm^{-3}

Adapted from Ref. [15].
[a]For the Bluecar® powered by BatScap's LMP battery and supercapacitor.

[23]. Today, the variety of applications for which polymer electrolytes are being considered has gone beyond just rechargeable batteries. Examples include

- Devices for the conversion and storage of energy, including batteries, fuel cells, supercapacitors, and ultracapacitors;
- Sensors and actuators;
- Photo-electrochemical displays;
- Microelectronic components; and
- Devices for applications in biotechnology.

Figure 52.12 gives an indication of the growth of interest in different fields through annual publications and reviews.

The next question raised of course is "which direction will lithium battery technology take in the future?" The best-performing combination of electrode and electrolyte can only be achieved through selective use of existing and new anodic and cathodic materials and by choosing the most

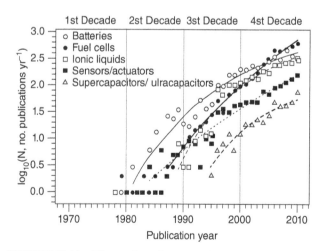

FIGURE 52.12   The number of papers and reviews per year relating directly to applications of polymer electrolytes. Reprinted from Di Noto, V., Lavina, S., Giffin, G. A., Negro, E., Scrosati, B. (2011). Polymer electrolytes: present, past and future. *Electrochimica Acta*, 57:4–13, copyright 2011 with permission from Elsevier.

appropriate electrolyte–electrode combinations, in order to minimize detrimental reactions associated with the electrolyte–electrode interface. Lithium-ion batteries are attracting a very large research investment, for small-capacity portable applications and also large-capacity plug-in hybrid (PHEV) and full electrical vehicle (EV) usage [83]. At the present time, the rate of improvement of specific energy density and power attainable using the lithium-ion technology applied in consumer electronics has slowed, with limits of around 200–250 Wh kg^{-1} and 1 kW kg^{-1} based on current expectations of electrode material performance. Upgrading one or both electrode components would be an important step in the right direction but battery performance and safety may still be compromised by the liquid organic solvent. Electrolytes currently used in commercial LiPo cells are similar in chemistry and operation to their liquid electrolyte Li-ion counterparts. The characteristics and performance of Li-ion and LiPo systems are now very similar but the all-solid-state LiPo battery offers several unique advantages. Whatever advances are made with respect to electrode materials for Li-ion cells, a solid, liquid-free, flexible, battery may well still offer the ideal compromise for commercial acceptability.

For PHEV and EV applications, a new generation of materials is required to provide even higher energy density, lower cost, and safety, with a focus on environmental sustainability. A cathode material previously discussed, modified nanostructured LiFePO4, may provide access to appropriate properties [84]. An alternative contender may be the fluorosulfate family, LiMSO$_4$F, M = Fe, Ni, Co. For non-lithium metal anodes, nanostructured intercalation materials such as nano-sized titanate electrodes are considered to be capable of providing an increase in energy densities by 20–30%. Both the academic and industrial communities are committed to the development of new materials to bring vehicle-compatible Li-ion type batteries to the marketplace. Scaled-up, 100 Ah capacity, LiPo batteries are commercially available [85] and PHEVs, powered by lithium-ion batteries, have already been announced by leading car manufacturers [86].

An alternative technology path toward higher energy densities may include lithium-air batteries. Metal–air technologies have already been exploited, the most successful example being the zinc–air battery. By reacting lithium directly with oxygen from air, according to the reaction 2Li $+$O$_2$ $\rightleftharpoons$ Li$_2$O$_2$, a very high capacity of 1200 Ah kg^{-1} may be obtained, much greater than that achievable by any insertion cathode. Practical considerations, however, still prevent the development of this battery; even the trace amounts of water produced introduce a safety risk in the presence of lithium metal, and selective, high-cost catalysts are required to deal with detrimental reaction mechanisms in the pores of the oxygen cathode. However, the use of hydrophobic ILs as an electrolyte component has the potential to solve the water issue, and expensive catalysts may eventually be substituted by a nanostructural design for the cathode [12, 87].

In spite of the enormous advances, diversification, and growth of interest in polymer–salt materials since polymer electrolytes were first proposed as ideal candidates for lithium rechargeable batteries in the late 1970s, the original concept of solid-state Li-ion batteries based on a dry polymer electrolyte is still held as an ideal. Even after more than 30 years of research, no ambient temperature solvent-free ion-conducting polymer capable of satisfying all operational and commercial criteria has been designed and developed. The use of nanostructural inorganic fillers has gone some way to meeting these expectations but not far enough. Control over the alignment of the crystalline phase could open new prospects for further development by enhancing ionic conductivity below 40°C. Recent investigations of nanoscale PEO-based polymer electrolyte thin films based on the self-assembly of AB and ABA block copolymers have potential merit. The phase separation inherent to these systems not only results in good mechanical properties, but also offers a way of increasing dissociation by partitioning anions and cations in the two microphases. The creation of directional diffusion channels for lithium ions in the flexible polyether phase could lead to enhancement of the conductivity and the incorporation of a hard, high shear modulus component could potentially benefit these membranes by suppressing dendrite growth on a lithium metal surface [88–90].

Whatever the direction of future battery development and technology, polymer electrolytes, with their potential for satisfying so many performance criteria, have undeniably played a major role in steering R&D to its current level of success. This is unlikely to change; the literature clearly indicates a healthy global, multidisciplinary, research activity which will undoubtedly leave its mark on the next generation of high-performance batteries.

REFERENCES

[1] Steele, B. C. H. (1973). Chemical diffusion. *Fast Ion Transport in Solids*, Van Gool, W., editor. North-Holland, Amsterdam, pp. 103–109.

[2] Armand, M. B. (1973). New electrode material. *Fast Ion Transport in Solids*, Van Gool, W., editor. North-Holland, Amsterdam, pp. 665–673.

[3] Murphy, D. W., Christian, P.A. (1979). Solid state electrodes for high energy batteries. *Science*, 205:651–656.

[4] Whittingham, M. S. (1976). The role of ternary phases in cathode reactions. *Journal of Electrochemical Society*, 123:315–320.

[5] Whittingham, M.S. (1979). Intercalation chemistry and energy storage. *Journal of Solid State Chemistry*, 29:303–310.

[6] Murphy, D. W., DiSalvo, F. J., Carides, J. N., Waszczak, J. V. (1978). Topochemical reactions of rutile related structures with lithium. *Materials Research Bulletin*, 13:1395–1402.

[7] Lazzari, M., Scrosati, B. (1980). A cyclable lithium organic electrolyte cell based on two intercalation electrodes. *Journal of the Electrochemical Society*, 127:773–774.

[8] Nagaura, T., Tozawa, K. (1990). Lithium ion rechargeable battery. *Progress in Batteries and Solar Cells*, 9:209.

[9] Mohri, M., Yanagisawa, N., Tajima, Y., Tanaka, H., Mitate, T., Nakajima, S., Yoshida, M., Yoshimoto, Y., Suzuki, T., Wada H. (1989). Rechargeable lithium battery based on pyrolytic carbon as a negative electrode. *Journal of Power Sources*, 26:545–551.

[10] Gray, F. M. (1991). *Solid Polymer Electrolytes: Fundamentals and Technological Applications*. VCH Publishers, New York.

[11] Tarascon, J. M., Armand, M. (2001). Issues and challenges facing rechargeable lithium batteries. *Nature*, 414:359–367.

[12] Armand, M, Tarascon, J. M. (2008). Building better batteries. *Nature*, 451:652–657.

[13] Dorval, V., St-Pierre, C., Vallée, A. (2004). The impact of lithium-metal-polymer battery characteristics on telecom power system design. *Lithium-Metal-Polymer Batteries: From the Electrochemical Cell to the Integrated Energy Storage System*. Available at: http://www.battcon.com/papersfinal2004/valleepaper2004.pdf (accessed January 2012).

[14] AVESTOR Battery Deployed Throughout North America. *Infobatt 2005 Presentation*. Available at: www.infobat.com (accessed January 2012).

[15] http://www.bluecar.fr/en/pages-innovation/batterie-lmp.aspx (accessed January 2012).

[16] Tarascon, J. M., Gozdz, A. S., Schmutz, C., Shokoohi, F., Warren, P. C. (1996). Performance of Bellcore's plastic rechargeable Li-ion batteries. *Solid State Ionics*, 86–88:49–54.

[17] Scrosati, B. (1993). *Applications of Electroactive Polymers*. Chapman Hall, London.

[18] Gray, F. M. (1997). *Polymer Electrolytes*. RSC Materials Monographs, The Royal Society of Chemistry, Cambridge.

[19] MacCallum, J. R., Vincent, C. A. (1987). *Polymer Electrolyte Reviews –I*. Elsevier, London.

[20] MacCallum, J. R., Vincent, C. A. (1989). *Polymer Electrolyte Reviews –II*. Elsevier, London.

[21] Alamgir, M., Abraham, K. M. (1994). Room temperature polymer electrolytes. *Lithium Batteries: New Materials, Developments and Perspectives*, Pistoia, G. editor, Elsevier, Amsterdam, pp. 93–136.

[22] Meyer, W. H. (1998). Polymer electrolytes for lithium-ion batteries. *Advanced Materials*, 10:438–439.

[23] Di Noto, V., Lavina, S., Giffin, G. A., Negro, E., Scrosati, B. (2011). Polymer electrolytes: present, past and future. *Electrochimica Acta*, 57:4–13.

[24] Stephan, A. M. (2006). Review on gel polymer electrolytes for lithium batteries. *European Polymer Journal*, 42:21–42.

[25] Armand, M., Endres, F., MacFarlane, D. R., Ohno, H., Scrosati, B. (2009). Ionic liquid materials for electrochemical challenges of the future. *Nature Materials*, 8:621–629.

[26] Le Bideau, J., Viau, L., Vioux, A. (2011). Ionogels, ionic liquid based hybrid materials. *Chemical Society Reviews*, 907–925.

[27] Angell, C. A., Liu, C., Sanchez, E. (1993). Rubbery solid electrolytes with dominant cationic transport and high ambient conductivity. *Nature*, 362:137.

[28] Gray, F. M., Armand, M. (2000). Polymer electrolytes. *Energy Storage Systems in Electronics*, Osaka, T., Datta, M. (editors). Gordon & Breach, Amsterdam, pp. 351–406.

[29] Popall, M., Andrei, M., Kappel, J., Kron, J., Olma K., Olsowski, B. (1998). ORMOCERs as inorganic-organic electrolytes for new solid state lithium batteries and supercapacitors. *Electrochimica Acta*, 43:1155–1161.

[30] Xu, W., Williams, M. D., Angell, C. A. (2002). Novel polyanionic solid electrolytes with weak coulomb traps and controllable caps and spacers. *Chemistry of Materials*, 14:401–409.

[31] Sun, X., Angell, C. A. (2001). "Acid-in-chain" versus "base-in-chain" anionic polymer electrolytes for electrochemical devices. *Electrochimica Acta*, 46:1467–1473.

[32] Di Noto, V., Zago, V., Pace, G., Fauri, M. (2004). Inorganic-organic polymer electrolytes based on PEG400 and Al[OCH(CH$_3$)$_2$]$_3$: i. Synthesis and vibrational characterizations. *Journal of Electrochemical Society*, 151(2):A224–A231.

[33] Wright, P. V. (1975). Electrical conductivity in ionic complexes of poly(ethylene oxide). *British Polymer Journal*, 7(5):319–327.

[34] Armand, M. B., Chabango, J. M., Duclot, M. J. (1979). Polyethers as solid electrolytes. In: *Fast Ion Transport in Solids*, Vashishta, P., Mundy, J. N., Shenoy, G. K. (editors). North-Holland, Amsterdam, pp. 131–136.

[35] Gadjourova, Z., Andreev, Y. G., Tunstall, D. P., Bruce, P. G. (2001). Ionic conductivity in crystalline polymer electrolytes. *Nature*, 412:520–523.

[36] Golodnitsky, D., Livshits, E., Kovarsky, R., Peled, E., Chung, S. H., Suarez, S., Greenbaum, S. G. (2004). New generation of ordered polymer electrolytes for lithium batteries. *Electrochemical and Solid-State Letters*, 7(11):A412–A415.

[37] Livshits, E., Kovarsky, R., Lavie, N., Hayashi, Y., Golodnitsky, D., Peled, E. (2005). New insights into structural and electrochemical properties of anisotropic electrolytes. *Electrochimica Acta*, 50:3805–3814.

[38] Dias, F. B., Plomp, L., Veldhuis, J. B. J. (2000). Trends in polymer electrolytes for secondary lithium batteries. *Journal of Power Sources*, 88:169–191.

[39] Bellon-Maurel, V., Calmon-Decriaud, A., Chandrasekhar, V., Hadjichristidis, N., Mays, J. W., Pispas, S., Pitsikalis, M., Silvestre, F. (editors) (1998). *Block Copolymers, Polyelectrolytes, Biodegradation*. Advances in Polymer Science, No. 135, Springer-Verlag, Berlin.

[40] Egashira, B. S., Armand, M., Beranger S., Michlot, C. (2003). Lithium dicyanotriazolate as a lithium salt for poly(ethylene oxide) based polymer electrolytes. *Electrochemical and Solid State Letters*, 6:A71.

[41] Appetecchi, G. B., Zane, D., Scrosati, B. (2004). PEO-based electrolyte membranes based on LiBC4O$_8$ salt. *Journal of Electrochemical Society*, 151(9):A1369–A1374.

[42] Gray, F. M., Armand, M. (1999). Polymer electrolytes. In: *Handbook of Battery Materials*, Besenhard, J. O. (editor). Wiley-VCH, Weinheim, pp. 499–523.

[43] Croce, F., Appetecchi, G. B., Persi, L., Scrosati, B. (1998). Nanocomposite polymer electrolytes for lithium batteries. *Nature*, 394:456–458.

[44] Croce, F., Scrosati, B. (2003). Nanocomposite lithium ion conducting membranes. *Annals of New York Academy of Sciences*, 984:194–207, doi: 10.1111/j.1749–6632.2003.tb06000.x.

[45] Plewa, A., Chylinski, F., Kalita, M., Bukat, M., Parzuchowski, P., Borkowska, R., Siekierski, M., Zukowska, G. Z., Wieczorek, W. (2006). Influence of macromolecular additives on transport properties of lithium organic electrolytes. *Journal of Power Sources*, 159:431–437.

[46] Blazejczyk, A., Szczupak, M., Wieczorek, W., Cmoch, P., Appetecchi, G. B., Scrosati, B., Kovarsky, R., Golodnitsky, D., Peled, E. (2005). Anion-binding calixarene receptors: synthesis, microstructure, and effect on properties of polyether electrolytes. *Chemistry of Materials*, 17:1535–1547.

[47] Kalita, M., Bukat, M., Ciosek, M., Siekierski, M., Chung, S. H., Rodrıguez, T., Greenbaum, S. G., Kovarsky, R., Golodnitsky, D., Peled, E., Zane, D., Scrosati, B., Wieczorek, W. (2005). Effect of calixpyrrole in PEO-LiBF$_4$ polymer electrolytes. *Electrochimica Acta*, 50:3922–3927.

[48] Croce, F., Sacchetti, S., Scrosati, B. (2006). Advanced, lithium batteries based on high-performance composite polymer electrolytes. *Journal of Power Sources*, 162:685–689.

[49] Agrawal, R. C., Pandey, G. P. (2008). Solid polymer electrolytes: materials designing and all-solid-state battery applications: an overview. *Journal of Physics D: Applied Physics*, 41:223001–223018.

[50] Xu, K. (2004). Nonaqueous liquid electrolytes for lithium-based rechargeable batteries. *Chemical Reviews*, 104:4303–4417.

[51] Croce, F., Sacchetti, S., Scrosati, B. (2006). Advanced, high-performance composite polymer electrolytes for lithium batteries. *Journal of Power Sources*, 161:560–564.

[52] Stallworth, P.E., Fontanella, J. J., Wintersgill, M. C., Scheidler, C. D., Immel, J. J., Greenbaum, S. G., Gozdz, A. S. (1999). NMR, DSC and high pressure electrical conductivity studies of liquid and hybrid electrolytes. *Journal of Power Sources*, 81–82:739–747.

[53] Choe, H. S., Giaccai, J., Alamgir, M., Abraham, K. M. (1995). Preparation and characterization of poly(vinyl sulfone)-and poly(vinylidene fluoride)-based electrolytes. *Electrochimica Acta*, 40:2289–2293.

[54] Appetecchi, G. B., Romagnoli, P., Scrosati, B. (2001). Composite gel membranes: a new class of improved polymer electrolytes for lithium batteries. *Electrochemistry Communications*, 3:281–284.

[55] Cho, B. W., Kim, D. H., Lee, H. W., Na, B. K. (2007). Electrochemical properties of gel polymer electrolyte based on poly(acrylonitrile)-poly(ethylene glycol diacrylate) blend. *Korean Journal of Chemical Engineering*, 24(6): 1037–1041.

[56] Arora, P., Zhang Z. (2004). Battery separators. *Chemical Reviews*, 104:4419–4462.

[57] Rajendran, S., Babu, R., Sivakumar, P. (2008). Investigations on PVC/PAN composite polymer electrolytes. *Journal of Membrane Science*, 315:67–73.

[58] Moreno, M., Santa Ana, M. A., Gonzalez, G., Benavente, E. (2010). Poly(acrylonitrile)-montmorillonite nanocomposites: effects of the intercalation of the filler on the conductivity of composite polymer electrolytes. *Electrochimica Acta*, 55:1323–1327.

[59] Rahman, M. Y. A., Ahmad, A., Lee, T. K., Farina, Y., Dahlan, H. M. (2011). LiClO$_4$ salt concentration effect on the properties of PVC-modified low molecular weight LENR50-based solid polymer electrolyte. *Journal of Applied Polymer Science*, 124:2227–2233. doi: 10.1002/app.35255.

[60] Armand, M., Endres, F., MacFarlane, D. R., Ohno, H., Scrosati, B. (2009). Ionic-liquid materials for the electrochemical challenges of the future. *Nature Materials*, 8:621–629.

[61] Seki, S., Kobayashi, Y., Miyashiro, H., Ohno, Y., Usami, A., Mita, Y., Kihira, N., Watanabe, M., Terada, N. (2006). Lithium secondary batteries using modified-imidazolium room-temperature ionic liquid. *Journal of Physical Chemistry B*, 110:10228–10230.

[62] Shin, J. H., Henderson, W. A., Passerini, S. (2003). Ionic liquids to the rescue? Overcoming the ionic conductivity of polymer electrolytes. *Electrochemistry Communications*, 5:1016–1020.

[63] Shin, J. H., Henderson, W. A., Appetecchi, G. B., Alessandrini, F., Passerini, S. (2005). Recent developments in the ENEA lithium metal battery project. *Electrochimica Acta*, 50:3859–3865.

[64] Kim, G. T., Appetecchi, G. B., Alessandrini, F., Passerini, S. (2007). Solvent-free PYR 1A TFSI ionic liquid-based ternary polymer electrolyte systems: I. Electrochemical characterization. *Journal of Power Sources*, 171:861–869.

[65] Yun, Y. S., Song, S. W., Lee, S. Y., Kim, S. H., Kim, D. W. (2010). Lithium metal polymer cells assembled with gel polymer electrolytes containing ionic liquid. *Current Applied Physics*, 10:e97–e100.

[66] Li, M., Yang, L., Fang, S., Dong, S., Jin, Y., Hirano, S., Tachibana, K. (2011). Li/LiFePO$_4$ batteries with gel polymer electrolytes incorporating a guaidinium-based ionic liquid cycled at room temperature and 50 C. *Journal of Power Sources*, 196:6502–6506.

[67] Chew, S.Y., Sun, J., Wang, J., Liu, H., Forsyth, M., MacFarlane, D. R. (2008). Lithium-polymer battery based on an ionic liquid-polymer electrolyte composite for room temperature applications. *Electrochimica Acta*, 53:6460–6463.

[68] Pandey, G. P., Kumar, Y., Hashmi, S. A. (2010). Ionic liquid incorporated polymer electrolytes for supercapacitor application. *Indian Journal of Chemistry*, 49A:743–751.

[69] Sutto, T. E. (2007). Hydrophobic and hydrophilic interactions of ionic liquids and polymers in solid polymer gel electrolytes. *Journal of Electrochemical Society*, 154:P101–P107.

[70] Kim, K. S., Park, S. Y., Choi, S., Lee, H. (2006). Ionic liquid-polymer gel electrolytes based on morpholinium salt and PVdF (HFP) copolymer. *Journal of Power Sources*, 155:385–390.

[71] Holzapfel, M., Jost, C., Prodi-Schwab, A., Krumeich, F., Würsig, A., Buqa, H., Novák, P. (2005). Stabilisation of lithiated graphite in an electrolyte based on ionic liquids: an electrochemical and scanning electron microscopy study. *Carbon*, 43:1488–1498.

[72] Koch, V. R., Nanjundiah, C., Appetecchi, G. B., Scrosati, B. (1995). The interfacial stability of Li with two new solvent-free ionic liquids: 1,2-dimethyl-3-propylimidazolium imide and methide. *Journal of the Electrochemical Society*, 142: L116–L118.

[73] Guerfi, A., Dontigny, M., Kobayashi, Y., Vijh, A., Zaghib, K. (2009). Investigations on some electrochemical aspects of lithium-ion ionic liquid/gel polymer battery. *Journal of Solid State Electrochemistry*, 13:1003–1014.

[74] Guerfi, A., Duchesne, S., Kobayashi, Y., Vijh, A., Zaghib, K. (2008). LiFePO$_4$ and graphite electrodes with ionic liquids based on bis(fluorosulfonyl) imide (FSI)- for Li-ion batteries. *Journal of Power Sources*, 175:866–873.

[75] Scrosati, B., Garche, J. (2010). Lithium batteries: status, prospects and future. *Journal of Power Sources*, 195:2419–2430.

[76] Patil, A., Patil, V., Shin, D. W., Choi, J. W., Paik, D. S., Yoon, S. J. (2008). Issue and challenges facing rechargeable thin film lithium batteries. *Materials Research Bulletin*, 43:1913–1942.

[77] www.altairnano.com; www.ener1.com (accessed January 2012).

[78] van Schalkwijk, W. A., Scrosati, B. (editors) (2002). *Advances in Lithium-Ion Batteries*, Kluwer Academic/Plenum Publishers, New York.

[79] Lee, S. W., Yabuuchi, N., Gallant, B. M., Chen, S., Kim, B. S., Hammond, P. T., Shao-Horn, Y. (2010). High-power lithium batteries from functionalized carbon-nanotube electrodes. *Nature Nanotechnology*, 5:531–537.

[80] Arico, A. S., Bruce, P., Scrosati, B., Tarascon, J. M., van Schalkwijk, W. (2005). Nanostructured materials for advanced energy conversion and storage devices. *Nature Materials*, 4:366–377.

[81] Cuellar, E. A., Manna, M. E., Wise, R. D., Gavrilov, A. B., Bastian, M. J., Brey, R. M., DeMatteis, J. (2001). Ultralife's polymer electrolyte rechargeable lithium-ion batteries for use in mobile electronics industry. *Journal of Power Sources*, 96:184–198.

[82] Pistoia, G. (2009). Battery Operated Devices and Systems, Elsevier B.V., Amsterdam, Oxford.

[83] Hassoun, J., Panero, S., Reale, P., Scrosati, B. (2009). A new, safe, high-rate and high-energy polymer lithium-ion battery. *Advanced Materials*, 21(47):4807–4810.

[84] Recham, N., Dupont, L., Courty, M., Djellab, K., Larcher, D., Armand, M., Tarascon, J.-M. (2009). Ionothermal synthesis of tailor-made LiFePO$_4$ powders for Li-ion battery applications. *Chemistry of Materials*, 21(6):1096–1107.

[85] Salaneh, Z. M., Kim, B. G. (2009). *IEEE Power & Energy General Meeting*. doi: 10.1109/PES.2009.5275404; Available at: http://ieeexplore.ieee.org/xpls/abs_all.jsp?arnumber = 5275404&tag = 1 (accessed January 2012).

[86] Tellefson, J (2008). Charging up the future. *Nature*, 456:436–440.

[87] Tarascon, J. M. (2010). Key challenges in future Li-battery research. *Philosophical Transactions of the Royal Society A*, 368:3227–3241.

[88] Sadoway, D. R., Huang, B., Trapa, P. E., Soo, P. P., Bannerjee, P., Mayes, A. M. (2001). Self-doped block copolymer electrolytes for solid-state, rechargeable lithium batteries. *Journal of Power Sources*, 97–98, 621–623.

[89] Ghosh, A., Kofinas, P. (2008). Nanostructured block copolymer dry electrolyte. *Journal of the Electrochemical Society*, 155(6):A428–A431.

[90] Mullin, S. A., Stone, G. M., Panday, A., Balsara, N. P. (2011). Salt diffusion coefficients in block copolymer electrolytes. *Journal of the Electrochemical Society*, 158(6):A619–A627.

53

RECYCLING AND DISPOSAL OF BATTERY MATERIALS

Michael J. Smith and Fiona M. Gray

53.1 INTRODUCTION

Portable electronic devices that depend on batteries as a source of power have rapidly evolved to become essential accessories of everyday life and whether the device in question is a remote-control command, a child's toy, a smart phone/organizer, or a laptop computer, it relies on its battery pack to provide a source of continuous electrical power. In many cases the efficiency of these batteries also largely determines the commercial success of the device. While the batteries applied in these equipment safely contain their chemical components during normal use, inappropriate disposal at the end of their working lifetime can result in the release of toxic chemicals to the environment. The consequences of disposal of a wide variety of battery types in landfill sites or municipal incinerators instead of responsible chemical neutralization and recovery of active components have led battery producers and vendors, legislators, and the consumer population to recognize the need to introduce improved procedures, and battery recycling has become a priority on the environmental and political agenda.

Global estimates of annual battery consumption are close to 300 million units at a cost of about 50 billion USD. The end-of-life treatment that batteries receive in different European Community member states, North America, and Asia differs very significantly and the percentage of household cells recycled ranges from almost zero in some countries to a modest high of about 50% of annual sales in the best case scenario. With current estimates of global *per capita* and *per annum* battery consumption of between 300 and 500 g, respectively, even this best case figure leaves a massive residue of untreated battery waste to enter municipal landfill sites.

Over the last few years, legislation aimed at regulating the use of toxic chemicals in batteries, and creating a practical framework for the obligatory recycling of spent batteries, has been enacted in various countries. This legislation will have a widespread impact on future practice, both in the battery manufacturing industry and on domestic consumer used-battery disposal habits, and hence it is timely to look at key issues, such as environmental consequences, public awareness and acceptance, current good practice, challenges and practicalities, and the consequences of legislation, that are being addressed within Europe, North America, and Asia.

In this chapter the topic of battery recycling is discussed and a summary of the key concepts, useful to a wide scientific and technological audience, is presented. The legislation that is driving a change of approach to spent battery recycling within Europe, North America, and Asia and the best collection, sorting, and processing strategies are outlined to provide entry points to useful sources of information for the nonspecialist in battery recycling.

53.2 CHEMISTRY OF BATTERIES

Electrochemical power sources or batteries are devices that convert energy stored in chemicals into electrical energy. A battery is made up of an assembly of two or more cells connected in a series or parallel configuration [1–6]. Over the last few decades, however, the terms *cell* and *battery* have become synonymous. Although credit for the original invention that demonstrated the viability of the concept is generally attributed to Alessandro Volta (1800) [6], the need to improve the efficiency of energy storage and conversion, in response to the requirements of commercial devices, resulted

Alternative Energy and Shale Gas Encyclopedia, First Edition. Edited by Jay H. Lehr and Jack Keeley.
© 2016 John Wiley & Sons, Inc. Published 2016 by John Wiley & Sons, Inc.

in a sustained effort to develop higher performance power sources. Ever since the pioneering days of battery science, the search for better performance has been driven by the introduction of new devices in the electronics sector.

Batteries are conveniently classified as primary (single use) or secondary (rechargeable), with further subdivision into categories of household (for consumer goods such as smart telephones, flashlights, radios, watches, cameras, or computers), SLI (for starting, lighting, ignition of internal combustion motor-driven vehicles), traction batteries (a market segment currently under rapid expansion), and industrial applications (for reserve network power or local energy backup). The principal commercial battery chemistries are listed in Table 53.1, together with examples of typical applications.

The energy densities indicated in Table 53.1 are included to provide an indication of battery merit. These indicated values can only be considered to be estimates as they vary (often considerably) with manufacturer, operating temperature, cell size, and discharge regime. In general, commercial success is a consequence of high electrochemical performance, a good safety record, competitive cost per energy unit, and low component toxicity or ready component recycling. Further details of the operational characteristics of these cells may be obtained from [1–5].

The efficiency of a specific battery chemistry depends on the chemical reactions taking place at the electrodes, the nature of the electrolyte present, and the state of charge of the cell. In addition to the active components, commercial cells also contain inactive components that perform essential support functions and ensure correct cell operation. These include the casings, often made of steel, and separators, seals, or labels, typically fabricated from polymers, paperboard, or paper. The active battery chemicals that are currently of greatest environmental concern are those based on mercury, cadmium, and lead, and to a lesser degree nickel, cobalt, copper, lithium, zinc, and silver [7]. The primary objective of recycling many commercial products is to reduce the costs associated with manufacture by recovery of useful ingredients (glass, paper, aluminum, and certain polymers). In the case of batteries, however, the secondary objective is of comparable importance and that is to avoid transfer of toxic components to the environment. Often inactive components may also make a useful contribution to the value of marketable resources recovered from specific classes of spent cells (see Table 53.2).

The global environmental impact of any substance depends not only on the component toxicity but also on the amount of substance contained in each refuse item, the number of items present in the waste stream [8–11], and the total transfer of the components to the environment that occurs during the "cradle-to-grave" lifetime of the device. In a life cycle impact analysis [8, 12] of battery systems there are opportunities for toxic component transfer to the environment during the extraction and preparation of raw active materials, battery manufacture, commercial distribution, practical application of the device by the end user, collection and transport at the conclusion of service, and recycling or waste management. The high-risk stages of this lifetime in terms of the environment are during the manufacturing and recycling/elimination steps.

53.3 ENVIRONMENTAL ISSUES RELATING TO SPENT BATTERY DISPOSAL

Under all normal conditions of use the manufacturer's design of case and seal assembly is able to ensure that the chemicals present in batteries present no threat to human health or to the environment. However, when subjected to careless disposal within household or workplace waste, or to extreme abuse, inevitable damage and degradation of the battery case changes this situation. The environmental impact of batteries in landfills [10–14] depends on the battery chemistry, the residual capacity of the battery, the local conditions of temperature, moisture, and oxygen content, the design and maintenance of the landfill, and the proximity of surface or groundwater.

Cells classified as household cells are mainly of the cylindrical zinc–carbon and alkaline manganese types and button or coin format cells, generally zinc–air, zinc–silver oxide, and lithium chemistries. This group of primary cells continues to make up the great majority of cells purchased by consumers, accounting for about 90% of the market of cells applied in portable devices [5, 10–14]. The commercial success of aqueous electrolyte-based cells (zinc–carbon, alkaline manganese, zinc–air, and zinc–silver oxide) is due to low material costs, ease of manufacture, and performance characteristics that suit a wide range of electronic devices with modest energy and power requirements. Although these cells are based on some of the oldest chemistries, they have been subjected to continuous improvement and have now attained a fairly stable commercial market share. It is noteworthy that a significant percentage of the alkaline manganese, zinc–air, and zinc–silver oxide miniature cells (button or coin format) currently in use, hoarded by consumers, or entering the waste stream may still contain small quantities of mercury. Hoarding, in this context, is the well-documented tendency of many consumers to accumulate cells, frequently installed in obsolete electronic devices, in their home instead of delivering them to collection centers or municipal waste collection [15]. The practical consequence of hoarding is that toxic components, including mercury, that are no longer present in modern cells, persist in waste streams and therefore the treatment of collected cells must be capable of dealing with this contaminant. Mercury was added to these cells as a corrosion-suppressing additive for the anode, to inhibit gas formation. In Europe, for example, the marketing of any

TABLE 53.1 **Principal cell and battery chemistries present in household, traction, and industrial markets.**

Primary cell chemistries present in household waste stream						
	Battery components				Energy density	
Designation	Anode	Electrolyte	Cathode	Typical applications	(Wh/dm³)	(Wh/kg)
Zinc–carbon	Zn	NH_4Cl	MnO_2	Used in a diminishing number of portable electronic devices. Moderate energy and power characteristics partially compensated by low cost	80–150	35–50
Zinc–carbon, heavy duty	Zn	$ZnCl_2$	MnO_2	Improved version of the zinc–carbon cell with better, but still rather modest performance characteristics	120–190	75–90
Alkaline manganese	Zn	KOH/NaOH	MnO_2	The highest performance version of the Zn–C cell with alterations to the cell architecture, cathode and electrolyte formulations	300–380	120–150
Zinc–mercuric oxide	Zn	KOH/NaOH	HgO	This battery system has been discontinued because of environmental concern. Typical applications were in cameras, hearing aids, and small electronic devices	400–500	123–140
Zinc–silver oxide	Zn	KOH/NaOH	Ag_2O and/or AgO	Typical commercial application in watches and calculators. The cell provides a flat discharge curve; however, Ag content increases cell production cost	350–650	130–150
Zinc–air	Zn	KOH	O_2	Principal niche market in hearing aids, high cell performance with nominal 1.4 V but significant self-discharge rate	1300–1500	325–450
Lithium	Li	Various	Various	Various cells with different electrolyte and cathode systems are exploited commercially. Good electrochemical performance is shared by these systems and applications are found in many portable high-end market devices	500–1100	250–600
Secondary cell chemistries present in household, traction, and industrial waste stream						
	Battery components				Energy density	
Designation	Anode (negative electrode)	Electrolyte	Cathode (positive electrode)	Typical applications	(Wh/dm³)	(Wh/kg)
Lead–acid	Pb	H_2SO_4	PbO_2	Widespread use in SLI applications, reserve power and storage batteries. High component toxicity. Nominal 2 V cell potential. Easy to recycle	65–85	30–50
NiCad	Cd	KOH	NiO(OH)	Still present but in diminishing number of portable electronic devices. High cycle life, but suffers from memory effect. Nominal 1.2 V cell potential. Cd anode is highly toxic	140–200	45–80

(continued)

TABLE 53.1   (*Continued*)

Designation	Battery components			Typical applications	Energy density	
	Anode (negative electrode)	Electrolyte	Cathode (positive electrode)		(Wh/dm^3)	(Wh/kg)
Ni–metal hydride	AB$_5$ or AB$_2$ intermetallic compounds	KOH	NiO(OH)	Substitute for NiCd cell. Improved in both electrochemical and environmental aspects. Nominal 1.2 V cell potential	200–330	50–120
Li-poly or LiPo	C.Li	Organic solvent + Li salt	Li$_{(1-x)}$M$_n$O$_n$	Proposed as substitute for Li ion, probably will evolve as a cheaper and safer battery with comparable performance. Nominal 3.7 V cell potential	300–350	130–225
Lithium ion	C.Li	Organic solvent + Li salt	Li$_{(1-x)}$M$_n$O$_n$	High-performance cell widely used in a wide range of portable electronic equipment. Low environmental impact. Nominal 3.7 V cell potential	300–550	125–275

class of cell containing more than 0.0005% of mercury and specifically button cells containing more than 2% of mercury by weight has been prohibited since January 2000. In view of the significant contribution [16] made by cells installed in imported electronic devices, particularly toys, this legislation certainly does not completely eliminate the risk of mercury contamination. Zinc–silver oxide, zinc–air, and alkaline manganese button cells which contain 5 ppm ≤ mercury ≤ 2% per cell must also be labeled as not suitable for disposal in household waste. These restrictions have obliged manufacturers of various button cell systems [17] to introduce structural changes including the use of zinc alloy powder

TABLE 53.2   Components of environmental concern and recoverable materials present in cells and batteries.

Cell/battery designation	Chemical components of concern	Recoverable materials
Zinc–carbon[a]	Zn, NH$_4$Cl, carbon black, and MnO$_2$	Steel, zinc, inorganic salts, and ferromanganese
Zinc–carbon, heavy duty[a]	Zn, ZnCl$_2$, carbon black, and MnO$_2$	Steel, zinc, inorganic salts, and ferromanganese
Alkaline manganese[a,b]	Zn, KOH or NaOH, carbon black, and MnO$_2$	Steel, zinc, inorganic salts, and ferromanganese
Zinc–mercuric oxide	Zn, KOH or NaOH, carbon black, and HgO	Steel, zinc, inorganic salts, and mercury
Zinc–silver oxide[c]	Zn, KOH or NaOH, carbon black, and Ag$_2$O or AgO	Steel, zinc, inorganic salts, and silver
Zinc–air[d]	Zn, KOH or NaOH, and mercury	Steel, zinc, inorganic salts, and mercury
Lithium primaries	Li, various lithium salts, and various organic solvents	Steel, lithium, and inorganic salts
Lead–acid	Pb, Pb salts, and sulfuric acid	Lead, acid, and plastic casing components
Nickel–cadmium	Cd, NiO(OH), and KOH	Steel, cadmium, nickel, and inorganic salts
Nickel–metal hydride	MH alloy, NiO(OH), and KOH	Steel, nickel, cobalt, and inorganic salts
Lithium-ion polymer secondaries	Metal oxides, carbon black, lithium salts, and various organic solvents	Lithium salts, metal oxide cathodes, and carbon black
Lithium-ion secondaries	Metal oxides, carbon black, lithium salts, and various organic solvents	Lithium salts, metal oxide cathodes, steel, and carbon black

[a]Since about 1993 these cylinder-form cells have been mercury- and lead-free.
[b]Mercury-free button and coin cells have been manufactured since about 2009.
[c]Various manufacturers introduced mercury-free versions from about 2004.
[d]Mercury-free from about 2001.

anodes, the development of new corrosion suppressors, gas-absorbing components, and modified cathode formulations, in order to maintain pre-legislation performance. An important advance in environmental protection was achieved with the introduction of mercury-free anodes for button cells based on zinc–silver oxide (2004) and alkaline manganese (2009) chemistries. These two subclasses of primary cells represent a significant proportion of the commercial button cell market.

The lithium nonaqueous primary cell market sector has also progressed significantly since the early 1970s [18, 19]. Although substantial commercial market expansion has been observed, the greater unit cost of lithium-based primaries is only justified in specific applications where higher cell performance is essential. From the environmental point of view, the subsector of button and coin cells based on lithium is effectively mercury-free.

The high specific density and moderate operating potential of the lead–acid battery result in low gravimetric and volumetric energy densities. However, as commercial cells and batteries can be assembled with large surface areas and low internal resistances, the discharge current available can be quite high, providing relatively high power densities. This feature makes this system quite attractive as a basis for batteries for SLI applications in vehicles where they must provide high engine cranking currents but generally undergo quite shallow discharge. Structurally different versions of the lead–acid battery are also produced for deeper cycle applications, for example, as associated energy storage for photovoltaic systems, traction batteries in electric vehicles, and uninterruptible power supplies. In these applications batteries are designed to provide a good deep-cycle response but the chemical reactions of the cell are essentially the same. More details regarding the electrochemistry and design of the various classes of commercial cells are readily available from a variety of sources [1–4].

The lead–acid battery is indisputably the best example of successful recycling with substantially higher recovery yields than glass, paper, or aluminum. The principal driving factors that have led to this satisfactory situation are the toxicity of lead, the inherent value of the scrap metal, the effective spent-battery replacement and collection procedure applied in this market segment, and the relatively simple structure of the cell and straightforward nature of lead recovery through the smelting process.

The NiCd secondary battery has been commercially available since 1950 and effectively dominated the household secondary cell market until about 1990. It is still produced in large-capacity format and contests certain subsectors of the commercial market previously occupied by lead–acid batteries. NiCd cells have rather modest energy/power densities and high component toxicity and are therefore a diminishing presence in the portable electronics sector of the market. The highly toxic cadmium anode, along with the nickel hydroxide cathode and concentrated lithium/sodium hydroxide electrolyte, presents an environmental dilemma.

In 1990, NiMH cells with their improved electrochemical performance became available commercially and also presented a less aggressive environmental impact. While the electrolyte and cathode compositions are similar to those of a NiCd cell, a hydrogen storage anode of nickel–cobalt–rare-earth metal alloy replaces the toxic cadmium electrode.

NiMH technology continues to be viewed as a stop-gap, to be superseded by lithium-based battery technology. There has been significant electrochemical development in the secondary lithium sector, first with the launch of the lithium-ion cell and more recently with the lithium polymer (Li-poly) cell. A move to lithium-based batteries (both primary and secondary) represents an important advance in terms of environmental impact. Although the anodic materials are nontoxic, lithium-ion cells do contain flammable electrolytes and may also contain moderately toxic composite cathodes. Lithium polymer cells contain similar anode and cathode constituents to the Li-ion cells but incorporate a polymeric gel electrolyte. In this subclass of cells current collectors, electrodes, electrolyte, and separators are wound together and encapsulated in a laminated Al foil/polymer pouch that isolates the cell content from the environment. The many advantages of this new cell format, including high electrochemical and better safety performance, high packaging efficiency, and a thin prismatic cell profile that allows manufacturers to adapt cells to fit available space in new devices, are expected to lead to significant growth in this battery market subsector. It will also require alterations in disposal strategies. An unexpected consequence of the Li-poly cell roll format is that the inactive component content is very low, contributing to their high energy and power density, but also reducing the recyclable content.

53.4 DISPOSAL OPTIONS

Many batteries still end up in landfill or are incinerated because of incorrect disposal by consumers or inefficient national battery collection and recycling schemes. This is highly undesirable because of the risk of hazardous chemicals contributing to leachate or runoff from landfill (a 25 g NiCd phone battery can contaminate 750,000 L of groundwater to the maximum acceptable concentration limit [20]) or to gaseous emissions and ash residues from incineration plants. Airborne contaminants can enter the food chain through plant adsorption and bio-accumulate causing serious health disorders in mammals [21]. The quantities of hazardous emissions produced by incineration depend on furnace temperature, the volatility of the elements, and the efficiency of local treatments applied to the furnace gaseous emissions and plant treatment of ash residue and process water. The majority of cell components are not substantially changed by the conditions in municipal incinerators and therefore heavy elements are concentrated in the furnace slag and require specific and expensive secondary treatment.

In the rare cases where disposal is the only end-of-life option, it is possible to treat heavy metals by stabilization and inertization to avoid leaching. These processes reduce the toxicity, make insoluble or immobilize the hazardous waste, and involve chemical reactions between constituents in the waste or reactions with species in the solid encapsulation matrix added to the residue. Inertization is generally considered as financially nonviable. It requires a specific battery collection scheme, the cost and inconvenience of which are frequently cited as being a major factor leading to inappropriate disposal and, unlike recycling, the inertized materials have no residual commercial value.

53.5 THE CHALLENGES OF RECYCLING

As governments around the world face a mounting crisis caused by the need to dispose of household and industrial waste, the public in general has compounded the problem by demanding that all environmentally harmful material be separated and dealt with in accordance with the best known practices. The focus is therefore on two distinct issues: recycling components that may contribute to reduced energy consumption associated with extraction of components from raw materials, and minimizing the release of toxic chemicals to landfill or incineration. The solution to the specific problem of treatment of household and industrial batteries is complex and may only be achieved by cooperation between consumers, battery manufacturers, scrap recyclers, and governmental agencies. The difficulties that arise are conveniently dealt with separately in the following sections that address issues of battery collection, cell class sorting, and legislation.

53.5.1 Battery Collection

The concept of total manufacturer responsibility at the entry and exit points of the commercial circuit has been applied consensually to many consumer goods and accepted practice often involves direct participation of the manufacturer in both new-product sales and used-product return. The reverse logistics of this strategy generally make good sense from various points of view. The original equipment manufacturer has most accurate knowledge of product content and specifications and is best equipped for manipulation of sensitive materials. From both energy and material standpoints industry has much to gain from using recycled components rather than processing new raw material. The end user also has an easily identifiable point of return, often with part-exchange of old/new components.

In the battery industry attributing producer responsibility is met by a uniform, rather unenthusiastic response from manufacturers' associations [22–27]. The difficulty in motivating the public to participate in battery return, the cost of publicity campaigns, the expense of collection from end user-convenient, but dispersed, locations, all contribute to

the industrial sector's counterproposal that responsibility for battery collection should be shared with the state or partially subsidized by an eco-tax. In many countries funding of battery recycling is based on an additional consumer point-of-sale eco-tax that includes a contribution to cover the cost of collection, storage, and transportation of spent batteries. The sale of the recycled components to battery manufacturers, metal foundries, or the materials industry covers outstanding processing/treatment costs and provides sufficient profit to guarantee commercial viability of the cycle. While certain segments of the global battery market benefit from specific collection routines (e.g., the lead–acid batteries discussed above, or decommissioning programs of large-capacity energy storage batteries) the most challenging market segment is certainly that of household cells. The challenges in this segment are collecting (with a low environmental impact) the geographically dispersed batteries from the population, creating conditions for safe temporary storage of cells and batteries in various stages of discharge, developing suitable procedures to preselect a broad variety of battery chemistries, and handling the large portion of the global cell market that this commercial segment represents. Efficient collection of household batteries is critically dependent on legislation and the ability to inform and motivate the consumer population to participate in the return of spent cells. Recent studies [28, 29] confirm that high recycling rates, measured as a percentage of the mass of recycled batteries to the mass of batteries sold for any given financial year, can be achieved. In Belgium [23], for example, the collection rate per person is the highest in the world. This result was achieved by investing in an intense and continuous public awareness campaign to inform the public about national laws, leading to widespread participation in collection programs and a sustained alteration in battery disposal habits. The Belgian program involves schools, public and private services, civic associations, point-of-sale outlets (supermarkets, jewelers, photographic shops, pharmacies, toy stores), and municipal eco-yards.

53.5.2 Sorting Strategies

Battery collection programs accept all types of household and industrial batteries and at the preliminary collection stage only a very limited preselection, on the basis of cell reaction chemistry, is implemented. Generally industrial batteries are simply identified and side-streamed for disassembly and reintegration into the processing stream at a future point. The crucial task of separation by chemical content takes place only on arrival at the recycling installation. Most component recovery procedures are very sensitive to battery-type purity and therefore separation by chemical content constitutes the most critical processing stage.

Over the last decade various automatic separating systems have been developed based on magnetic/electromagnetic separation [30, 31], photographic identification [32, 33], UV label detection [34], barcode detection, and X-ray radiographic fingerprinting [35, 36]. Improvements in the detection step technology now permit fast identification and selection of spent cells; sophisticated high-rate sensors enable sorting at rates from 5 to about 50 cells per second [37]. The average sorting rate at large-scale plants treating thousands of tonnes of household waste every year has risen to about 15 cells per second. The sorting process also includes mechanical steps that supply a continuous feed to the selection sensors and air-jet cell ejection or transfer to fraction-collecting bins. The rates of these crucial steps are improving but are not yet able to sustain the very rapid speeds possible in cell-chemistry identification. Cell recognition must not only be fast, but also efficient, in view of the sensitivity to contamination of the treatment procedure, and current sorting techniques have already achieved class purity of about 95–99%.

Although progress in reaching targets set by the legislation has been relatively slow, and recycling rates vary notably between countries participating in specific programs, significant changes have been reported. Difficulties in achieving high-level consumer involvement and commitment to recycling are unlikely to be overcome in the near future. In the same way as previous campaigns normalized the reuse of paper, glass, and aluminum, it is likely that continued encouragement to collaborate in battery collection and recycling will eventually lead to the high levels stipulated by current legislation. In view of the rapid progress registered in recent years, the sorting phase of battery treatment no longer represents the limiting step of the recycling process.

53.5.3 Legislation

The current world market value for all primary and secondary batteries is estimated at about 50 billion USD with a primary/secondary division of between 85% and 90% primary and between 10% and 15% secondary. The projected overall annual growth rate is generally given as about 3–5%. The preparation of statistical information and the prediction of market growth in this specific industrial subsector are greatly complicated by the inconsistencies in the presentation of industrial production data. Information may be presented as tonnage of manufactured cells, as market values of the commercial sector (normally expressed in USD) or even as the number of units manufactured (without specifying whether the numbers indicate multiple-cell battery packs or single cells). The battery industry is a commercial domain subject to rapid change [38–42] and the combined effects of new portable electronic devices, technical evolution in cell design, or changes in production volume altering manufacturers' pricing can significantly influence annual market estimates.

In spite of rapid changes inherent in the portable energy marketplace, one conclusion is unavoidable: the sheer tonnage of cells and batteries that are disposed off throughout the world is massive. An appropriate response to the impact of these cells on the Earth's environment is therefore essential.

Although there are many differences in the ways that governments approach health and environmental issues, the legislative content is similar in respect to the intention to reduce battery residue and increase recovery and recycling. In the earliest stages of battery regulation in Europe [43–45], Asia [46, 47], and North America [48–52], initial objectives included the prohibition of heavy metal components (specifically Hg, Cd, and Pb) in the manufacture of household cells and batteries. This preliminary legislation also provided economic instruments to encourage separate collection and recycling of spent cells and batteries.

To facilitate implementation of these measures, the directives required that cells, batteries, and appliances were correctly labeled to inform consumers of battery content. Specific collection targets were defined for recycling household and industrial cells and batteries. These targets will become progressively more demanding as consumer habits change, with increased acceptance of battery recycling. Legislation has developed as a consequence of accumulated experience in the recycling industry and working committees that were created to enable consultation between legislators and battery producers. This consultation allowed consensual positions to be established and resulted in the foundation of associations of manufacturers and importers [23–26] that assumed responsibility for coordinating battery elimination and recycling. In spite of the desire for harmonization, some differences still remain between working groups. Japan, for example, continues to landfill batteries considered hazard-free by the Battery Association of Japan. In contrast, some European countries collect all cells for recycling, regardless of their chemical constitution. Although legislation is in place to regulate battery disposal and recycling, poor public awareness, lack of enforcement, and insufficient budget allocation to environmental issues are significant barriers in many countries to its effective implementation [53, 54]. Achieving targets set by the legislation has been relatively slow, and recycling rates vary between countries participating in specific programs, but despite this, significant changes have been made and it is already clear that acceptable consumer, manufacturer, and state agency cooperation will eventually be achieved.

53.6 RECYCLING AND RECOVERY METHODS

The diversity of battery chemistries has naturally led to a correspondingly wide range of recycling treatments [55–57]. Regardless of the specific method of treatment applied, the preliminary processing stage involves removal of labels, opening of cell casings, and destroying seals and separators by procedures based on mechanical cutting/chopping/pounding, vacuum milling, or primary pyrolysis (Figures 53.1 and 53.2). The secondary stages of recycling are broadly classified as hydrometallurgic or pyrometallurgic.

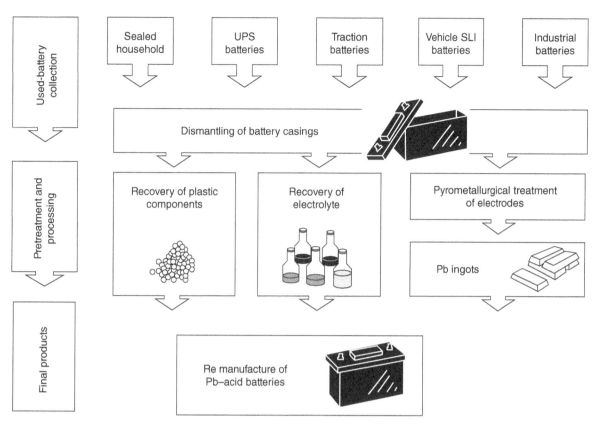

FIGURE 53.1 Treatment and recycling pathway for lead–lead oxide cells and batteries.

Hydrometallurgical techniques applied to the cell fragments include acid, alkaline, solvent, or supercritical carbon dioxide solvent extraction [58–62]. These extraction procedures yield metal solutions that are subsequently subjected to pH-controlled precipitation, selective reactions, electrolysis, electro-winning, or electrodialysis [63–66] to isolate the purified materials. The application of hydrometallurgical processes has several clear advantages, the most important being the high yields of metal recovery from feedstock and the wide variety of metals that respond to this treatment. In general the products isolated are of high purity, the process itself is very energy efficient, and emissions released by the plant to the atmosphere are relatively innocuous. The principal disadvantage of the hydrometallurgical recovery of metal components is associated with the large volume of process solutions that are necessary and the need for careful treatment of plant effluents.

Pyrometallurgical procedures [62, 67–69], using high temperatures to separate metals, may be subdivided by the product "final destination." On the one hand, the processing treatments are aimed at extracting metals specifically for steel production; on the other, specialized processes are designed to yield purified elements for reentry into a variety of industrial feedstocks. While the nickel, chromium, and manganese residues from recycled batteries are acceptable components in steel production, the levels of cadmium, copper, and zinc impurity must be carefully monitored to avoid deterioration

of the properties of the steel-foundry product. At the extremely high furnace temperatures used in steel production, any residual zinc, cadmium (and mercury, should it be present) will evaporate, oxidize, and be emitted from fume stacks as fly-ash loaded with hazardous dust. Although useful, this strategy for battery waste treatment has certain limitations. Various companies specialize in the production of purified zinc, cadmium, lead, mercury, and nickel from battery feedstock. The pure elements are supplied to other metallurgical companies as raw material, and the separated slag or bottom-ash containing unwanted residues is used in road or building foundations. The treatments described in this section are represented in Figure 53.2 and encompass a wide variety of waste battery types.

Procedures for recycling lithium battery feedstocks, also represented in Figure 53.2, have been developed by various companies. In the Toxco (hydrometallurgical) treatment [69], lithium is ultimately recovered as the metal or lithium hydroxide. Initial processing of battery feedstock involves cryoshredding and reacting with water to produce hydrogen that can be burnt off above the reaction liquid. In pyrometallurgical procedures, component recovery is limited to cobalt and steel-making residues. Other treatments involve a combined pyro-hydrometallurgical process where punctured cells are incinerated and cobalt is subsequently recovered from metallic waste through the application of standard hydrometallurgical procedures [70–74]. With alternative, less

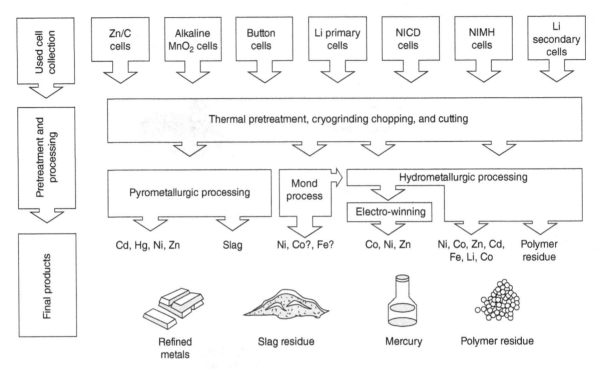

FIGURE 53.2   Household primary and secondary cell treatment and recycling pathway.

vigorous, purely hydrometallurgical procedures [75, 76], components from electrolyte and electrode material may also be recovered from the disassembled cells. This latter option is more attractive and, even with fluctuations in the market value of recycled materials, the fundamental profitability of the process is supported by the sale of products. This is clearly a better solution than the alternative of using charges levied on battery end users.

53.7 DEVELOPING CELL TECHNOLOGIES AND IMPLICATIONS FOR RECYCLING

Information provided by manufacturers and recycling agencies confirms that treatment of spent cell and battery residues has arrived at a critical moment when old responsibilities are being addressed with new strategies. More than ever before politicians and consumers are being made aware of their duty to adopt a socially and scientifically correct response to used-battery disposal in order to preserve the quality of our environment.

An ever-increasing number of equipment manufacturers are using high-performance lithium-based secondary cells in their products. Such cells are increasingly of the Li-poly class and this poses an interesting conundrum. With laminated aluminum-bag containers substituting the traditional steel casing, the cells have relatively low recyclable content and combine competitive electrochemical performance with negligible environmental impact. Future versions of Li-poly

secondary cells may represent a truly ecological choice of power source in which the toxic chemical content is so low that they can safely be disposed of as municipal solid waste.

Very significant advances are also being made in fuel cell technology with many new companies investing in the design and manufacture of high-performance fuel cells adapted to the portable electronics market, emerging applications in backup energy, and new generations of electric vehicles [77–81]. These hydrogen or methanol-fuelled devices draw their chemical energy from a quick-fill reservoir outside the cell (or stack) structure. As the source of chemical energy is not part of the cell, the task of recycling these units is much simpler than that of conventional or advanced batteries. The use of precious metal catalysts in the composite electrode component of fuel cells also provides a strong economic motivation for their end-of-life collection and recycling treatment. The impact of these devices on the market share for batteries in portable devices, industrial, or transport energy applications is difficult to predict at this time.

In many applications where energy is supplied from batteries, electrical circuits may require peaks of high power for periods of several seconds. Recently supercapacitors or electrochemical capacitors [82–84] have been proposed as a practical solution for short duration and intense energy injection. Prototypes already achieve quite high values of energy storage capacity, good cycling efficiency, high specific power density, low toxic substance content, and excellent cycle life. Devices with these characteristics may contribute to an increase in the performance of batteries in certain applications and therefore have a positive impact on the battery market. Again, at this stage, it is difficult to predict how this emerging technology may alter the current status of the portable energy market.

With these two examples it is clear that even before the routines for end-of-life processing of current primary and secondary cells have become well established, and before widespread collection strategies have been implemented at a local level, there are indications that new technological developments may compete with or complement conventional battery-based power sources. Naturally these innovations have been duly registered by the electronics industry and it is clear that the portable electronics industry is prepared to welcome any innovation that offers an advantage to future consumers. It is unclear, however, how the emerging technologies will affect both short-term and long-term investment in conventional battery recycling and treatment strategies.

REFERENCES

[1] Vincent, C. A., Scrosati, B. (1997). *Modern Batteries: An Introduction to Electrochemical Power Sources*, 2nd edition. Arnold, London.

[2] Crompton, T. R. (2000). *Battery Reference Book*, 3rd edition. Elsevier Science, London.

[3] Dell, R. M., Rand, D. A. J. (2001). *Understanding Batteries*. RSC Publishing, London.

[4] Linden, D., Reddy, T. B. (2010). *Handbook of Batteries*, 4th edition. McGraw-Hill, New York.

[5] Pistoia, G. (2005). Spent battery collection and recycling, *Batteries for Portable Devices*. Elsevier Science, London.

[6] Heise, G. W., Cahoon, C. N. (1960). *Primary Batteries*, Vol. 1. John Wiley & Sons, Inc., New York.

[7] Lange, R. (1988). *The Sigma Aldrich Library of Safety Data*, 2nd edition. Sigma-Aldrich Corporation, Milwaukee, WI.

[8] Pistoia, G. (2005). World battery market. In: *Batteries for Portable Devices*, Pistoia, G. (editor). Elsevier, Amsterdam, The Netherlands.

[9] Directive of the European Parliament and of the Council on Batteries and Accumulators and Spent Batteries and Accumulators. (2003). Commission Staff Working Paper, Brussels (November 24, 2003). Available at: www.aeanet.org/GovernmentAffairs/gajgEU_batteries_impactassessment.asp (accessed November 2011).

[10] Broussely, M. (2007). Spent battery collection and recycling. In: *Industrial Applications of Batteries: From Cars to Aerospace and Energy Storage*, Pistoia, G. (editor). Elsevier Science, London.

[11] Hurd, D. J., Muchnik, D. M., Schedler, T. M. (1993). *Recycling of Consumer Dry Cell Batteries-Pollution Technology Review*, no. 213. Notes Data Corporation, New Jersey.

[12] Lund, H. F. (2001). *The McGraw-Hill Recycling Handbook*. McGraw-Hill Professional, New York.

[13] Pistoia, G., Wiaux, J. P., Wolsky, S. P. (2001). *Used Battery Collection and Recycling*. Elsevier, Amsterdam, The Netherlands.

[14] Pistoia, G. (2008). *Battery Operated Devices and Systems*. Elsevier Science, Amsterdam, The Netherlands.

[15] Wiaux, J.-P. (2001). Portable rechargeable batteries in Europe: sales, uses, hoarding, collecting and recycling. In: *Used Battery Collection and Recycling*, Pistoia, G., Wiaux, J.-P., Wolsky, S. P. (editors). Elsevier Science, Amsterdam, The Netherlands.

[16] Hahn, D. (2008). Through the open door: searching for deadly toys in China's Pearl River Delta. *Harper's Magazine*, p. 49.

[17] United Nations Environmental Program (UNEP). (2008). Report on the major mercury-containing products and processes, their substitutes and experience in switching to mercury-free products and processes, July 14, 2008, p. 39.

[18] Vincent, C. A. (2000). Lithium batteries: a 50-year perspective, *Solid State Ionics*, 134:159–167.

[19] Tamura, K., Horiba, T. (1999). Large-scale development of lithium batteries for electric vehicles and energy power storage applications. *Journal of Power Sources*, 81–82:156–161.

[20] National primary drinking water regulations. (2011). Environmental Protection Agency, Washington, DC. Available at: http://www.epa.gov/gateway/science/water.html.

[21] Pacyna, E. G., Pacyna, J. M., Steenhuisen, F., Wilson, S. (2006). Global anthropogenic mercury emission inventory for 2000. *Atmospheric Environment*, 40:4048.

[22] Gemeinsames Rücknahmesystem Batterien (GRS Batterien), stiftung gemeinsames rucknahmesystem batterien, Hamburg, Germany. Annual reports, available from 2011. Available at: http://www.grs-batterien.de/facts_and_figures.html (accessed August 2015).

[23] Bebat, Fonds Ophaling Batterijen – VZW, St-Stevens-Woluwe, Belgium. Battery recycling report and current status, available for each year. Available at: http://www.bebat.be/ (accessed August 2015).

[24] Stibat, Stichting Batterijen, Den Haag, The Netherlands. Battery recycling report and current status available for each year. Available at: http://www.stibat.nl/ (accessed August 2015).

[25] Rechargeable Battery Recycling Corporation (RBRC), Atlanta, GA. call2recycle, Annual report and current updates available annually. Available at: http://www.rbrc.org/call2recycle/ (accessed August 2015).

[26] Portable Rechargeable Battery Association (PRBA), Washington, DC. Association site with current information under permanent renewal. Available at: http://www.prba.org/ (accessed August 2015).

[27] Vassart, A., Recycling of used portable batteries in Europe (2002), European Battery Recycling Association, Bruxelles, Belgium Available at: http://www.ebrarecycling.org (accessed August 2015).

[28] Recycling around Europe. European Portable Battery Association, Bruxelles, Belgium. Information available at a permanently-updated site. Available at: http://www.epba-europe.org (accessed August 2015).

[29] Hansmann, R., Loukopoulos, P., Scholz, R. W. (2009). Characteristics of effective battery recycling slogans: a Swiss field study. *Resources, Conservation and Recycling*, 53:218.

[30] Fricke, J. L. (2008) *Disposal of Battery Sets (Entsorgung verbrauchter geratebatterien)*, March 2008. GRS Batterien, Hamburg, Germany. Available at: http://www.grs-batterien.de/ (accessed November 2011).

[31] Bernardes, A. M., Espinosa, D. C. R., Tenório, J. A. S. (2004). Recycling of batteries: a review of current processes and technologies. *Journal of Power Sources*, 130:291–298.

[32] Sattler, P. H. (1998). See the label, know the type. In: Proceedings of the 4th International Battery Recycling Congress, Hamburg, Germany.

[33] Watson, N. (2001), Advances in sorting technology. In: Proceedings of the International Congress for Battery Recycling 2001, ICM AG, Birrwil, Switzerland.

[34] Rausch, S. (2001).Sorbarec X-ray technology for sorting of spent batteries. In: Proceedings of the International Congress for Battery Recycling 2001, ICM AG, Birrwil, Switzerland.

[35] Fricke, J. L., Luhrsen, G. (2008). *Annual Performance Review*. Documentation 2007, GRS Batterien, Hamburg, Germany. Available at: http://www.grs-batterien.de/ (accessed November 2011).

[36] Rausch, S., Molchim, T., Nowak, K., Timm, G. (2003). Proceedings of the 8th International Congress for Battery Recycling, Lugano, Switzerland, June.

[37] Mitsubishi Battery Sorter, Mitsubishi Heavy Industries, Ltd., Tokyo, Japan. CTM application, 7th July 1999.

[38] Buchmann, I. (2011). *Batteries a Portable World*, 3rd edition. Cadex Electronics Inc., Richmond, BC, Canada. Available at: http://batteryuniversity.com/learn/article/battery_statistics (accessed November 2011).

[39] Batterien-Montage-Zentrum GmbH, Karlstein, Germany. Available at: www.bmz-gmbh.de. (accessed August 2015).

[40] Frost & Sullivan, Rockville Centre, NY. Annual reports on progress in battery industry. Available at: www.batteries.frost.com (accessed August 2015).

[41] Freedonia Group, Cleveland, OH. Industrial market research reports published with annual updates. Available at: www.freedoniagroup.com.

[42] Battery Association of Japan, Tokyo, Japan. Annual data from Japanese manufacturers association. Available at: http://www.baj.or.jp/e/statistics/01.html.

[43] Directive 2006/66/EC, 6 September 2006, the battery directive, accumulators and waste batteries disposal. (2006). *Official Journal of the European Union* (September 29, 2006). Available at: http://ec.europa.eu/environment/waste/batteries/index.htm (accessed November 2011).

[44] Summary of EU legislation, "Disposal of spent batteries and accumulators". Site with results from EC Commission on various aspects of batteries and environment, under permanent update. Available at: http://europa.eu/legislation_summaries/environment/waste_management/l21202_en.htm (accessed August 2015).

[45] Council Directive 91/157/EEC, 18 March 1991. (1991). Available at: http://eur-lex.europa.eu/smartapi/cgi/sga_doc? smartapi!celexplus!prod!CELEXnumdoc&numdoc=391L0157&lgamp;=EN (accessed November 2011).

[46] Battery Association of Japan, Tokyo, Japan. Available at: www.baj.or.jp/; Nippon Recycle Center Corp., Osaka, Japan. Annual statistical data from Japanese manufacturers including production, sales and disposal. Available at: http://www.recycle21.co.jp/recycle-e/index.html (accessed August 2015).

[47] Ministry of Economy, Trade and Industry, Japan. (2010). Information pamphlets, "Towards a 3R-orientated sustainable society: legislation and trends", August 2010. Available at: http://www.meti.go.jp/policy/recycle/main/english/pamphlets/index.html (accessed November 2011).

[48] Environmental Protection Agency, USA. (1996). Universal waste rule, hazardous waste management system modification of the hazardous waste recycling regulatory program, Federal Register, May 11, 1995, and EPA530-F-95-025, February 1996. Available at: www.epa.gov/epawaste/hazard/wastetypes/universal/batteries.htm (accessed November 2011).

[49] Environmental Protection Agency, USA. (1996). The Battery Act, EPA 300-N-02-002. Public Law 104-142, May 1996. Available at: http://www.epa.gov/wastes/laws-regs/state/policy/p1104.pdf (accessed November 2011).

[50] EPA-USA. (1997). Implementation of the Mercury-Containing and Rechargeable Battery Management Act, EPA530-K-97-009, November 1997. Available at: http://nlquery.epa.gov/epasearch/epasearch (accessed November 2011).

[51] Environmental Protection Agency, USA. (2002). Envirosense: AF Center for Environmental Excellence—Fact sheet on batteries disposal and The Battery Act, EPA Enforcement Alert, Vol. 5, no. 2, March 2002. Available at: www.ehso.com/ehshome/batteries.php (accessed November 2011).

[52] Common batteries collection scheme. Gemeinsames Rücknahmesystem Batterien–(GRS Batterien), Stiftung Gemeinsames Rucknahmesystem Batterien, Hamburg, Germany. Success Monitor. Hamburg March 2007. Available at: www.grs-batterien.de/ (accessed August 2015).

[53] Rogulski, Z., Czerwinski, A. (2006). Used batteries collection and recycling in Poland. *Journal of Power Sources*, 159:454–458.

[54] Bernardes, A. M., Espinosa, D. C. R., Tenorio, J. A. S. (2003). Collection and recycling of portable batteries: a worldwide overview compared to the Brazilian situation. *Journal of Power Sources*, 124:586–592.

[55] Ellis, T. W., Mirza, A. H. (2006). Battery recycling: defining the market and identifying the technology required to keep high value materials in the economy and out of the waste dump. Available at: http://www.nist.gov/tip/wp/pswp/publicly-submitted-white-papers-energy.cfm.

[56] Cheret, D. (2007). Battery collection and recycling. In: *Industrial Applications of Batteries: From Cars to Aerospace and Energy*, Broussely, M., Pistoia, G. (editors), Chapter 14. Elsevier Science, Amsterdam, The Netherlands.

[57] Pistoia, G. (2001). Spent battery collection and recycling. In: *Batteries for Portable Devices*, Pistoia, G. (editor), Chapter 9. Elsevier Science, Amsterdam, The Netherlands.

[58] Zhang, P. W., Yokoyama, T., Itabashi O., Wakui, Y., Susuki, T. M., Inoue, K. (1998). Hydrometallurgical process for recovery of metal values from spent nickel-metal hydride secondary batteries. *Hydrometallurgy*, 50:61.

[59] Kleinsorgen, K., Kohler, U., Bouvier, A., Folzer, A. (2000). Process for the recovery of metals from used nickel/metal/rare earth hydride storage batteries. US Patent no. 6110433.

[60] Watson, N. (2001). Primary battery recycling in Europe. In: *Used Battery Collection And Recycling*, Pistoia, G., Wiaux, J.-P., Wolsky, S. P. (editors), Chapter 7. Elsevier Science, Amsterdam, The Netherlands.

[61] Kuzuya, T., Naito, T., Sano, H., Fujisawa, T. (2003). Hydrometallurgical process for recycling of spent nickel-metal hydride batteries. In: *Metallurgical and Materials Processing: Principles and Techniques*, Kongoli, F. (editor). Yazawa International Symposium, San Diego, CA, March 2–6, 2003, Vol. 3, pp. 365–372. TMS, Warrendale, PA.

[62] Nan J., Han, D., Yang, M., Cui, M. (2006). Dismantling, recovery and re-use of spent nickel-metal hydride batteries. *Journal of Electrochemical Society*, 153:A101–A105.

[63] Froelich, S., Sewing, D. (1995). Simultaneous recovery of zinc and manganese dioxide from household alkaline batteries through hydrometallurgical processing. *Journal of Power Sources*, 57:27–30.

[64] Binsfield, M., Pesic, B., Storhok, V. (1996). Recovery of zinc and manganese from Bunker Hill water treatment plant sludge, part II: electrowinning process development. In: Proceedings of the 2nd International Symposium on Extraction and Processing for the Minimization of Waste, Ramachandran, V., Nesbit, C., (editors). TMS, Warrendale, PA.

[65] Lupi, C., Pilone, D., (2002). Battery recycling: defining the market and identifying the technology required to keep high value materials in the economy and out of the waste dump. *Waste Management Resources*, 22:871.

[66] David, J. (2001). Nickel-cadmium and nickel-metal hydride battery treatments. In: *Used Battery Collection and Recycling*, Pistoia, G., Wiaux, J.-P., Wolsky, S. P. (editors), Chapter 6. Elsevier Science, Amsterdam, The Netherlands.

[67] Pescetelli, A., Paolucci, E., Tinè, A. (2001). Lead-acid batteries. In: *Used Battery Collection and Recycling*, Pistoia, G., Wiaux, J.-P., Wolsky, S. P. (editors), Chapter 8. Elsevier Science, Amsterdam, The Netherlands.

[68] The Unicore process: recycling of Li-ion and NiMH batteries via a unique industrial closed loop, April 2010. (2010). Umicore Battery Recycling, Olen, Belgium. Available at: www-recyclingsolutions.umicore.com.

[69] Toxco Inc., Anaheim, CA. Treatment implemented by industrial company. Available at: http://www.toxco.com/ (accessed November 2011).

[70] Miller, D. G., McLaughlin, B. (2001). Recycling the lithium battery. In: *Used Battery Collection and Recycling*, Pistoia, G., Wiaux, J.-P., Wolsky, S. P. (editors), Chapter 9. Elsevier Science, Amsterdam, The Netherlands.

[71] Smith, D. (1998). Recycling of Lithium Cells and Batteries, Proceedings of 10th International Seminar on Battery Waste Management, Florida.

[72] Fricke, J. L., Knudsen, N., The disposal of portable batteries (2003) In: *Battery Technology*, Kiehne, H.A., Chapter 19. CRC Press, Taylor&Francis Group, 6000 Broken Sound Parkway, NW Suite 300, Boca Raton FL 33487, USA. Available at: www.grs-batterien.de.

[73] Sohn, J. S. (2003). Collection and Recycling of Spent Batteries in Korea, Proceedings of 8th International Congress for Battery Recycling, Lugano, Switzerland, June.

[74] Sohn, J. S., Shin, S. M., Yang, D. H., Kim, S. K., Lee, C. K. (2006). Hydrometallurgical approaches for selecting the effective recycle process of spent lithium ion batteries. In: *Advanced Processing: of Metals and Materials*, Kongoli, F., Reddy, R. G. (editors). Sohn International Symposium Proceedings, San Diego, CA, August 27–31, Vol. 6, pp. 135–143. TMS, Warrendale, PA.

[75] Lain, M. (2001). Recycling of lithium cells and batteries. *Journal of Power Sources*, 97–98:736–738.

[76] Contestabile, M., Panero, S., Scrosati, B. (2001). A laboratory-scale lithium-ion battery recycling process. *Journal of Power Sources*, 92:65–69.

[77] Cook, B. *An Introduction to Fuel Cells and Hydrogen Technology*. (2001). Heliocentris, Vancouver, BC, Canada. Available at: http://www.ogniwa-paliwowe.info/download/introduction_to_fuel_cells_and_hydrogen_technology.pdf (accessed August 2015).

[78] Hydrogen & our energy future (2005), eere.energy.gov/hydrogenandfuelcells/pdfs/hydrogenenergyfuture_web.pdf. (accessed August 2015).

[79] UPP Hydrogen Fuel Cell Power, product specification sheet (2015), Intelligent Energy, Charnwood Building, Holywell Park, Ashby Road, Loughborough, LE11 3GB, UK. www.beupp.com (accessed August 2015).

[80] Ballard delivers first prototypes of third generation long-life fuel cell for residential cogeneration. (2006). Ballard Power Systems Inc., Burnaby, BC, Canada. Available at: www.ballard.com (accessed November 2011).

[81] Honda Fuel Cell Power FCV Concept. (2015). Press information 2015.01, http://world.honda.com/news/2015/4150113FCV-Concept-North-American-Debut/. (accessed August 2015).

[82] Graphene supercapacitor breaks storage record. Available at: http://physicsworld.com/cws/article/ news/44477 (accessed November 2011).

[83] Vehicle technologies program – Multi-Year Program Plan. (2011–2015). U.S. Department of Energy, Energy efficiency and renewable energy, http://www1.eere.energy.gov/vehiclesandfuels/pdfs/program/vt_mypp_2011-2015. (accessed August 2015).

[84] Ultracapacitors for stationary, industrial, consumer and transport energy storage - an industry, technology and market analysis, February 2010, report code ET-111. (2010). iRAP Innovative Research and Products Inc., Stamford, CT. Available at: http://www.innoresearch.net/report_summary.aspx?id=71&pg=171&rcd=ET-111&pd=2/1/2010 (accessed November 2011).

54

AC OR DC

M. ARAM AZADPOUR

AC/DC. No! This is not the Scottish transplanted Australian rock-n-roll band's name I am referring to . . . this is about alternating current (AC) and direct current (DC) forms of transmitting electricity.

The name is well known only to those in the electromagnetic fields, as the unit of measuring magnetic field strength is named after Nikola Tesla. Nikola was a Serbian who found employment at one of Thomas Edison's companies in Paris, France, but was later transferred to work in the New York office in 1885 to be close to Edison. Due to his ingenuity and innovation, Tesla was promised 50,000 dollars if he could redesign Edison's motors and generators to be more efficient. However, that monetary promise was a poor joke on part of Edison. After Tesla successfully completed the tasks, Edison refused to pay him the bounce saying that the offer was a joke. Instead, he offered to raise Tesla's salary from 18 to 28 dollars a week. Not being amused by the poor joke, Tesla resigned to collaborate with another of Edison's foes, George Westinghouse.

Thomas Edison was of the opinion that DC was the way to go (mainly for safety), while Nikola Tesla was of the opinion that AC was the way to go (mainly for the ability to produce and to transmit massive amounts of electricity over long distances). While Tesla did not have a lot of business savvy, he was a driven person due to his engineering training and his exceptional genius. Tesla, in collaboration with George Westinghouse, designed, built, and put in place world's first hydro turbines which produced AC power at the Niagara Falls Power Company in 1895 and built transmission lines to turn on the lights in Buffalo, NY. That demonstration showed that AC can be transmitted over long distances (over 20 miles for that demonstration) whereas DC would lose voltage when transmitted over a mile. The voltage step-down transformer provides the means for AC to be better suited for long-distance transmission at high voltage and then reduced in voltage to suit the household consumption. The same step down of voltage would result in too much of a power loss using DC, hence, making it unsuitable for long-distance transmission.

Edison, to battle Tesla on the notion that DC is safer than AC power, commissioned others to build an electrocution chair and killed off animals in his poor choice demonstration of safety. Later on, the electric chair was adopted for the execution of condemned persons by various state correctional authorities throughout the United States. Tesla was able to show that AC transmission was suitable for having a source of power many tens of miles away from the destination of consumption and developed a power line network which remains the model of present-day electric grid systems.

As it is not possible to store AC power, DC remains the optimal form of electricity consumption for small applications such as a battery-powered hand-held electronic device. However, for the purpose of power delivery over long distances, AC remains the choice form of electricity generation and transmission.

Although both men were raised in abject poverty, Edison died a rich, publically well known, and revered innovator. Tesla, on the other hand, died a broken, destitute, and lonely man. Sadly over time, while the rest of us enjoy and prosper in a better life due to the products of Tesla's amazing career and mind, only those in the engineering and scientific communities remember Tesla's name and appreciate his genius.

Alternative Energy and Shale Gas Encyclopedia, First Edition. Edited by Jay H. Lehr and Jack Keeley.
© 2016 John Wiley & Sons, Inc. Published 2016 by John Wiley & Sons, Inc.

PART VI

RENEWABLE ENERGY CONCEPTS

55

WILL RENEWABLES CUT CARBON DIOXIDE EMISSIONS SUBSTANTIALLY?

HERBERT INHABER

55.1 INTRODUCTION

The French economist Bastiat[1] developed the concept of "that which is and is not seen." He used the story of a boy breaking a window. This is what is seen, and it appears to generate economic activity as the glazier is paid to repair the window. But what is not seen is what the glazier would do if the window were not broken, perhaps installing glass in new houses.

What does this have to do with whether renewables reduce carbon dioxide emissions? Magazines and newspapers often show wind turbines, solar collectors, and other renewables—that which is seen. What is not seen or depicted is the backup—usually natural gas single-cycle turbines—for when the wind does not blow and the sun does not shine.

Most renewables (with the exception of ethanol) do not generate CO_2 in energy production. A relatively small amount is originated in the materials—steel, copper, aluminum, and so on—used in construction. However, when gas turbines are turned on and off to smooth out the variability of "natural" energy sources, considerable CO_2 is emitted. Studies in a number of countries indicate that as electrical grid penetration of renewables increases, the savings of CO_2 diminish rapidly. By the time that penetration reaches approximately 20%, CO_2 savings are small, of the order of a few percent.[2]

This result has policy implications, as the Kyoto Protocol and suggested follow-up treaties concentrate only on the reduction of CO_2, although other greenhouse gases are discussed.

55.2 INTERMITTENCY, STORAGE, AND BACKUP

Some renewable sources, like geothermal and hydroelectricity, need no backup because they produce power steadily. They have an energy store: in the case of geothermal, heat below the earth's surface, and for hydro, potential energy behind a dam. Solar thermal, photovoltaics, and wind are highly intermittent (the former two less so in certain desert areas). Since regulatory bodies require all electrical generation to have a high degree of reliability, outages of the order of no more than a few hours a year, some form of backup or storage is required for solar and wind.

55.2.1 Storage

If renewable energy could be somehow stored for use when it is not available, it would become more competitive with traditional (mostly fossil fuel) systems. In some countries, one attempt (applicable to both renewable and fossil fuel sources) is pumped storage. In the United States, there are 17 pumped storage systems, in which water is pumped from a lower reservoir to an upper one.[3] Water flowing down from the upper one runs a turbine for a generator.

[1]Bastiat, F. (1995). *Selected Essays on Political Economy.* The Foundation for Economic Education, Inc., Irvington-on-Hudson, NY.
[2]Inhaber, H. (2011). Why wind power does not deliver the expected emissions reductions. *Renewable and Sustainable Energy Systems*, 15:2557–2562.

[3]Anonymous. (2010). Pumped-storage plants in the USA. Data courtesy of Robert Margolis and Kelly Taylor of Florida Power and Light. Available at: http://www.industcards.com/ps-usa.htm

Alternative Energy and Shale Gas Encyclopedia, First Edition. Edited by Jay H. Lehr and Jack Keeley.
© 2016 John Wiley & Sons, Inc. Published 2016 by John Wiley & Sons, Inc.

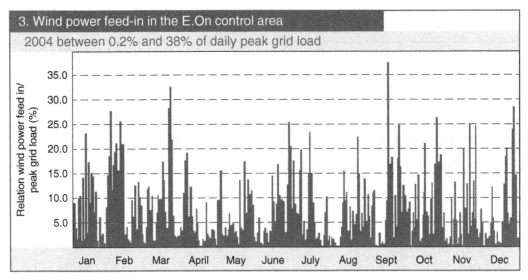

FIGURE 55.1   Wind power production in the E.On Netz system covering much of Germany, 2004.[4]

However, pumped storage is highly dependent on local terrain and geography, and as a result the 17 pumped storage systems generate only a tiny fraction of US electrical capacity.

Another storage system is batteries, used in all-electric vehicles like the Chevrolet Volt and Nissan Leaf. These lithium-ion batteries are highly expensive for the energy they store (the cheapest battery per unit energy stored is still the lead-acid one developed by Planté over a century ago). The cost to store energy for an entire city would be prohibitive.

One method of self-storage for renewables that has been suggested is averaging over a wide geographical area. In principle, if the wind is not blowing in one location, it might be available in a nearby area. Transmission lines might obviate the need for storage or backup.

However, experience in Germany suggests otherwise. Figure 55.1[4] shows 2004 wind variability in Germany. These "wind turbines are spread out over all of Germany, from Bavaria in the South to offshore in the German Bight in the North."[5]

55.2.2   Backup

It is then evident from the above brief discussion that storage is either physically unfeasible or financially unattainable. Coal and nuclear are generally used as baseload electricity sources. They cannot respond quickly to the rapid changes in renewables output.

This leaves natural gas plants, which can react reasonably quickly to intermittency. Backup from gas turbines is generally a closed cycle, that is, once-through. These are inherently less energy efficient than combined cycle turbines which use the exhaust as an energy source for another heat engine.

Thus renewables, due to their intermittency, will inevitably be accompanied by CO_2 emissions from natural gas backup. It is clear that when renewables form a small part of the grid, backup will be rarely if ever used, and thus CO_2 savings will be close to 100%. The question then is, how much CO_2 will be emitted from the renewable–gas turbine combination as grid penetration increases.

55.3   AUTOMOBILE FUEL EFFICIENCY

Cycling (increasing and decreasing power output from gas-fired turbines) increases CO_2 production. An analogy may be drawn with auto fuel efficiency. New automobiles have two values of efficiency posted on their windshields. The first, higher than the second, shows highway efficiency, with few stops or starts. The second, labeled city driving, takes into account many cycles, similar to the gas turbines noted above. The first value (in miles per gallon [mpg] or kilometers per liter) is usually substantially higher than the second.

Values for the Toyota Camry, the best-selling auto in the United States for many years, are shown in Figure 55.2. Highway driving yields 32 mpg; city, 22. The same principle applies to the cycling of natural gas plants caused when intermittent sources of power enter the electrical grid—efficiency decreases and the emissions (CO_2 and others) per mile or kilometer increases.

[4]E.On Netz GmbH (2005). *Wind Report 2005.* E.On Netz GmbH, Bayreuth, Germany.
[5]de Groot, K., le Pair, C. Hidden fuel costs of wind generated electricity. Available at: http://www.wind-watch.org/documents/hidden-fuel-costs-of-windgenerated-electricity/ and http://www.clepair.net/windefficiency.html

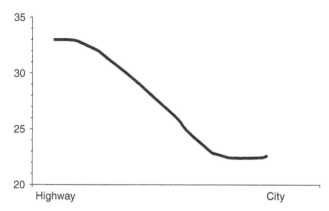

FIGURE 55.2 Schematic graph of miles per gallon for the 2011 Toyota Camry, the best-selling automobile in the United States. Mileage on the highway is about 32, and for city driving, about 22. While only two points on the curve are known, the intermediate values likely follow the curve shown.

55.3.1 Wind CO_2 Emissions Reductions in Real Time

Now that wind and some other renewables have achieved a substantial portion of total electricity production in certain locations, it is possible to compile data and make approximate calculations on how those reductions vary with grid penetration. In the interests of brevity, the analyses will be referred to briefly.

Baker et al.[6] note

> For this analysis, one-to-one backup is required when electricity production capacity from PVs [photovoltaics] is 20 percent of total capacity in any region.

This result will form the right-hand side of Figure 55.3, that is, at around 20% penetration CO_2 savings are close to zero. While the hour-by-hour variability of PVs is not exactly the same as wind, the main focus of the present analysis, the order of magnitude of overall intermittency between various renewables is similar.

Gross et al.[7] developed one of the most comprehensive studies on CO_2 savings due to renewables use. The study estimated greenhouse gas savings of 0–48%, indicating that a wide variety of assumptions went into the estimates of the papers surveyed. Many of the papers are theoretical and statistical constructs, not based on real-life data. Gross notes,

> Almost all of the literature deals with the impacts of intermittency using a statistical representation of the main factors, or through simulation models based upon statistical principles.

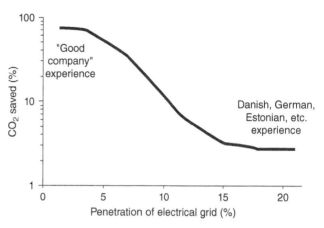

FIGURE 55.3 A schematic graph of CO_2 reductions as a function of wind (or other intermittent renewables) penetration into an electrical grid. Penetration is defined as the average fraction of energy contributed by wind to overall energy consumption. As shown in Figure 55.1, dealing with a large German system, that fraction varied from about 35% to almost zero. While a small penetration should yield close to 100% CO_2 savings, the graph starts at about 90%, based on CO_2 being generated in producing steel, aluminum, rare earths, and other materials.[18]

One of the aims of this chapter is to evaluate the small number of empirical data gathered to date.

55.3.2 Germany

E.on Netz[8] of Germany (in 2005 having the largest wind-power capacity in the world) concluded,

> ... traditional power stations with capacities equal to 90% of the installed wind power capacity must be permanently online in order to guarantee power supply at all times.

This implies that any carbon dioxide reduction would be at most 10%. However, this will depend on the fractional contribution of different fossil and other fuel sources, such as natural gas, coal, and nuclear.

55.3.3 Denmark and Bentek

A report from Denmark, which has perhaps the highest fraction of wind generation in the world, says that it has saved 2.4 million metric tons of carbon dioxide per year.[9] Using data from the US Energy Information Administration,[10] this corresponds to about 4% CO_2 savings.

[6]Baker, E., Chon, H., Keisler, J. (2007). Advanced solar R&D: applying expert elicitations to inform climate policy. February 28, 2007. p. 10. Available at: http://www.internationalenergyworkshop.org/pappdf/Baker.pdf
[7]Gross, R., et al. (2006). *The Costs and Impacts of Intermittency*. Imperial College, London.

[8]E.On Netz GmbH (2005). *Wind Report 2005*. E.On Netz GmbH, Bayreuth, Germany.
[9]Center for Politiske Studier (2009). Wind Energy: The Case of Denmark, Copenhagen; September 2009. Part 1.
[10] Inhaber, ibid.

55.3.3.1 Bentek—Colorado and Texas. Bentek[11] estimates 2% savings of CO_2 production at the J. T. Deeley plant in Texas due to cycling, a reversal of sign compared to Colorado. It states,

> CO_2 emissions were higher in 2009 by between 0.8 and 1.1 thousand tons in 2009 and ranged from a very small savings to 0.6 thousand tons incremental emissions in 2008. The range amounts to less than 1% of total CO_2 emissions in either year.

However, due to Bentek's inability to determine hourly wind production in these two states, it was not possible to estimate CO_2 reduction as a function of wind energy penetration.

55.3.4 US National Academy of Sciences

Droz quotes a committee of the NAS as saying

> the committee estimates that wind energy development probably will contribute to offsets of approximately 4.5 percent in U.S. emissions of CO_2 from electricity generation by other electricity-generation sources by the year 2020.[12]

The document does not estimate this CO_2 reduction as a function of wind grid penetration.

55.3.5 Denmark Again

White[13] notes

> There is no CO_2 saving in Danish exchange with Norway and Sweden because wind power only displaces CO_2-free generated power . . . Hence spinning reserve is essential, although it leads to a minimal CO_2 saving on the system. Innogy made the same observation about the operation of the UK system.[14]

55.3.6 Ireland

A wind capacity of 400 MW (about one-tenth) of Ireland's power capacity at the time of writing would reduce CO_2 emissions by 5.4%.[15]

55.3.7 Estonia

The Tallinn Technical University of Estonia stated[16]

> . . . In reality, only keeping the necessary additional reserve capacity will increase the fuel consumption (emissions) by up to 8.1% . . . The calculations were repeated for several values of power system load and the results showed at least 8–10% increase of fuel consumption and emissions compared with the steady operation of thermal stations under constant mean power of wind turbines.

55.4 CONCLUSIONS

It can be seen from the scattered data quoted above that compiling a curve relating windpower (and other renewables) grid penetration to CO_2 savings is not simple. However, an approximate schematic diagram relating these two quantities is shown in Figure 55.3. Note that the vertical scale is logarithmic, indicating substantial decreases in CO_2 savings as penetration of windpower into the grid increases. The right-hand side of the curve shows the Danish, German, and Estonian experiences, with relatively little CO_2 savings due to cycling of backup. The curve may be approximated by the equation

$$Q = 200/(1 + e^{cx})$$

where Q is the CO_2 reduction in percent, x is the wind or intermittent renewable penetration of the grid in percent, and c is a constant, of the order of 0.2 in Figure 55.3.

There are considerably uncertainties in this curve, too numerous to list here. Inhaber[17,18] mentions at least nine major ones, such as the mix of fossil fuels used as backup and the variable quality of the data on CO_2 savings. However, it is clear from both experimental and theoretical calculations that CO_2 savings drop off rapidly with renewable grid penetration.

Results have implications for future energy policy. The reduction of CO_2 and other greenhouse gases is the focus of the Kyoto Protocol and proposed follow-on treaties, in turn designed to mitigate global warming. If CO_2 reductions due to widespread use of wind and other renewables are less than anticipated, the effect on avoiding global warming due to the implementation of renewables could be slight.

[11] Bentek Energy LLC (2010). *How Less Became More: Wind, Power and Unintended Consequences in the Colorado Energy Market.* Bentek Energy LLC, Evergreen, CO.

[12] Committee on Environmental Impacts of Wind-Energy Projects (2007). *Environmental Impacts of Wind Energy Projects.* National Research Council, National Academies Press, Washington, DC, pp. 5–6.

[13] White, D. J. (2004). Danish wind: too good to be true? *The Utilities Journal*, July:37–39.

[14] Tolley, D. (2003). *Paper presented to the Institute of Mechanical Engineers.*

[15] Byrne, S. E. (2004). *ESB National Grid, Impact of Wind Power Generation in Ireland on the Operation of Conventional Plant and the Economic Implications.* EirGrid-WindImpact-Main.pdf.

[16] Liik, O., Oidram, R., Keel, M. (2003). *Estimation of Real Emissions Reduction Caused by Wind Generators.* Tallinn Technical University, Tallinn, Estonia. Presented at International Energy Workshop at Laxenburg, Austria. Quoted in Stelling, E. (2007). *Calculating the Real Cost of Industrial Wind Power*, November, Bruce County, Ontario.

[17] Inhaber, ibid., p. 2662.

[18] Greenspon, J. (2010). Here comes the sun. David (Las Vegas); July 30–31; p. 48.

56

THE CONCEPT OF BASE-LOAD POWER

Mark Diesendorf

KEY POINTS

- Matching base-load power stations to base-load demand is useful in electricity supply based predominantly on coal or nuclear power.

- However, in systems with high penetrations of renewable energy, base-load power stations are unnecessary.

- The key requirement for the reliability of a renewable electricity system is a mix of "flexible" and "inflexible" technologies.

- Reliability of electricity supply is a property of the whole supply–demand system, not of any single power station within that system.

- "Smart grids," in which demand and supply are treated together, will further assist the integration of large amounts of renewable energy.

56.1 BALANCING SUPPLY AND DEMAND IN CONVENTIONAL SYSTEMS

An electricity supply system is actually a delicate balancing act between supply and demand. As weather, business, industrial output, and residential activities vary, demand varies with time of day, day of the week, and season. Because the storage of electricity on a large scale is still very expensive, supply in a conventional system has to track demand with a high degree of precision. Even small deviations between the two variables produce changes in the frequency of the alternating current that can damage electronic equipment. Large deviations can produce blackouts.

The most challenging task for an electricity supply system is to follow the large and rapid demand variations over a 24-hour time period while maintaining reliability. Figure 56.1 shows the typical diurnal demand variations, from midnight to midnight, in summer (left-hand side) and winter (right-hand side) in an industrialized country. In summer there is a single large peak in the afternoon due to air conditioning. The two largest peaks in winter are due mainly to residential space heating during the periods around breakfast and dinner.

The trough in demand between midnight and dawn defines *base-load* level of demand; the big peaks are called *peak-load* demand; and between them is *intermediate-load*.

In the past it was appropriate to match each of these demand categories with a specific type of power station. Coal-fired and nuclear power stations are inflexible in terms of varying their outputs. They can take a day or more to bring from cold to full power. Once they are running, they cannot follow the diurnal variations in demand without incurring great expenses in operation and maintenance. They have high capital costs and, provided they are operated with approximately constant outputs, low fuel and other operating costs. So it is preferable that they are operated continuously at rated power. Thus it is natural to match them with base-load demand and label them as *base-load power stations*.

To meet the peaks in demand and to help fill the gap in supply when a base-load power station breaks down unexpectedly, *peak-load power stations* are used. These are flexible, because they can be started up within minutes and their outputs can be varied rapidly. There are two principal types of peak-load power station: gas turbines and hydroelectric stations based on storage dams. The gas turbines used for electricity generation are similar to jet engines on aircraft. They can be brought from cold to full power in about 10 minutes.

Alternative Energy and Shale Gas Encyclopedia, First Edition. Edited by Jay H. Lehr and Jack Keeley.
© 2016 John Wiley & Sons, Inc. Published 2016 by John Wiley & Sons, Inc.

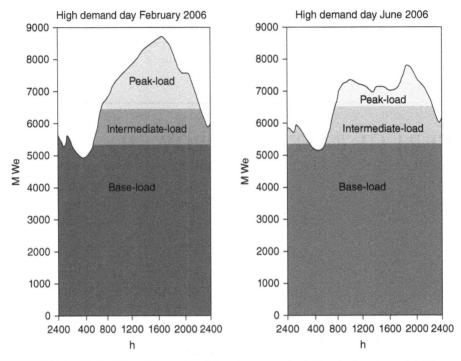

FIGURE 56.1   Typical diurnal demand variations in summer (on left) and winter (on right). Diagram reproduced from Diesendorf (2014) with the permission of UNSW Press.

They have low capital costs in dollars per kilowatt ($/kW) of rated generating capacity, but high running costs, because the fuels they burn—natural gas, oil, or renewable equivalents—are much more expensive than the coal or uranium fuelling base-load power stations. Thus they are operated as little as possible. They may be considered to be a type of reliability insurance.

Storing water is the principal conventional way of (indirectly) storing electricity on a large scale. Although rainfall varies often substantially by season and year, a large dam can smooth out these natural fluctuations. In the few countries with very large hydroelectric capacity (e.g., Norway, Iceland, and Brazil), hydro can be operated as base-load power. However, in most countries, storage capacity is limited and hydro is peak-load or at best intermediate-load. The output of hydro stations with large dams can be varied even more rapidly than that of gas turbines. However, unlike the latter, hydro stations generally have high capital costs. Although their "fuel," the water they store, appears to be free, for storages that are not gigantic this limited resource has to be used sparingly, as if it had a high fuel price.

Intermediate-load power stations, as the name suggests, are intermediate to base and peak in several characteristics. Their output can be varied more quickly than that of base-load stations, but more slowly than peak-load. Their fuel costs generally lie between those of base-load coal and peak-load gas. Their capital costs in dollars per kilowatt of rated

generating capacity fall between those of base-load and peak-load stations. They are usually combined-cycle gas turbine stations or old, small, coal-fired power stations whose capital costs have been paid off.

In an electricity supply system that is based predominantly on fossil and/or nuclear fuels, there is a combination or mix of base-, intermediate-, and peak-load power stations that gives the minimum annual cost. This is known as the "optimal mix." If there is too much base-load plant, costs are high, because of the high annual interest and repayments on the capital costs. If there is too much peak-load gas turbine use, costs are high because of the high use of the expensive fuel of the peak-load plant. Striking the correct balance between the different types of conventional power station to achieve the optimal economic mix is difficult, especially if average annual demand is changing significantly from year to year. It involves projection and prediction, which can be very inaccurate. The calculation is even more complicated in supply systems with significant fraction of hydroelectricity.

56.2   ENTER THE NEW RENEWABLE ENERGY SOURCES

A conventional electricity supply system is designed to handle two sources of random variation: fluctuations in demand and fluctuations in supply caused by the unexpected breakdown of power stations. It does this mainly by having sufficient peak-load plant and reserve base-load plant, some of which is kept hot. When non-hydro storage renewable energy is introduced into the grid, it is a tiny fraction of national electricity generation. Then the fluctuations in output from wind and solar photovoltaic (PV) power are generally smaller than the fluctuations in demand and so can be handled easily by the existing intermediate-load, peak-load, and reserve base-load stations. Small penetrations into the grid of wind and PV can be treated simply as negative demand. Among electrical engineers a conventional view was that, although these small fluctuations were not a problem, large contributions from renewable energy, especially wind and PV, would introduce new sources of random variation. Therefore, large penetrations of wind and PV would have to await the development and commercialization of base-load renewable electricity technologies and/or low-cost electrical storage. Some proponents of renewable energy responded by arguing that soon the growth in geothermal power, both conventional and hot rock, would provide the "necessary" base-load renewable technology.

However, the late twentieth and early twenty-first centuries saw the rapid growth of non-hydro renewable energy technologies from a small base to substantial levels in several countries. The two most rapidly growing technologies were (and are) wind power and solar PV power. In 2014, wind supplied 39% of the annual electricity generation in Denmark, 33% in South Australia, and about 22% each of Spain and Portugal. In several regions during windy periods, the major proportion of electricity generation came from wind. For instance, on September 5, 2012 wind power was supplying 75% of South Australia's electricity for a few hours. Solar PV too was beginning to make an impact on the operation of electricity systems: on a sunny day, May 25, 2012, solar PV power generated 22 GW of power in Germany, equivalent to half the country's electricity demand.

While there are challenges in handling large penetrations of renewable energy into the electricity grid, they are not as difficult or expensive to handle as previously thought. Practical experience, together with computer simulations of such systems, has led to changes in the conceptual framework of an electricity supply system. In particular, detailed hour-by-hour computer simulations of supply and demand by the US National Renewable Energy Laboratory (NREL) have shown that

renewable electricity generation from technologies that are commercially available today, in combination with a more flexible electric system, is more than adequate to supply 80% of total electricity generation in 2050 while meeting electricity demand on an hourly basis in every region of the United States … [furthermore] increased electric system flexibility, needed to enable electricity supply-demand balance with

high levels of renewable generation, can come from a portfolio of supply- and demand-side options, including flexible conventional generation, grid storage, new transmission, more responsive loads, and changes in power system operations. (Hand et al., 2012).

In the 80% renewable electricity scenario by NREL, half the electricity generation comes from fluctuating wind and solar PV. Scenarios for 100% renewable electricity, also based on hourly simulations—using real data on demand, wind, and solar energy—have been published for several countries and regions, including Germany and Australia (summarized in Diesendorf 2014, chapter 3).

56.3 FLEXIBLE AND VARIABLE RENEWABLE ELECTRICITY SOURCES

An important result of these and other renewable electricity simulations is that there is no need to supply base-load demand by means of base-load supply. Indeed, in a predominantly renewable energy system there is no need for any kind of base-load supply. The important constraint is to maintain the reliability of the whole supply–demand system at the required level. This can be done with a combination of renewable energy sources as shown in Figure 56.2 for a summer day. Instead of attempting to classify renewable supply technologies into base-, intermediate-, and peak-load, they are better described as either "variable/inflexible" or "flexible." Among commercially available technologies the variable power stations include wind, solar PV without storage, and run-of-river hydro. Actually, even they have a limited degree of flexibility, because their outputs can be easily reduced, or diverted to other uses, when they would otherwise push supply above demand. However, they are inflexible to the extent that their outputs cannot be increased if there is insufficient wind, sunshine, or river flow, respectively.

Flexible renewable energy power stations are needed to balance the fluctuations in variable ones and so maintain the reliability of the generating system. Among commercially available technologies they include concentrated solar thermal (CST) power with thermal storage, hydro with water storage, and gas turbines fueled on liquid or gaseous biofuels. The degree of flexibility of these technologies depends on the amount of storage they have.

The mix of renewable energy technologies in systems with high penetrations will depend on the particular region concerned as well as the economics of the various technologies. For instance, in Northern Europe and the southern coastline of Australia the dominant source is likely to be wind, while in the Middle-East, north Africa, south-west United States, and inland Australia the dominant sources could be CST and PV.

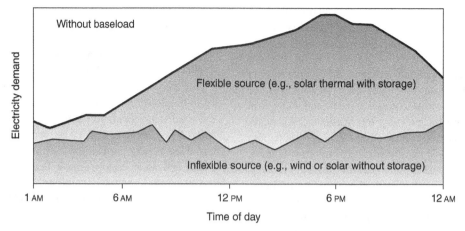

FIGURE 56.2   The new conceptual framework for electricity supply with no base-load. Diagram reproduced from Diesendorf (2014) with the permission of UNSW Press.

56.4   DEMAND REDUCTION AND "SMART" GRIDS

The traditional approach to planning electricity generation is to take demand (and demand growth) as given and to build a supply system that can follow variations in demand. However, even in this approach demand is not completely fixed: a limited amount of demand management is included. For instance, to reduce the difference between base-load and peak-load demand, some electricity utilities boost base-load demand by offering cheap electricity between midnight and dawn to heat off-peak electric hot water systems. To avoid blackouts when a large power station breaks down, utilities sometimes off-load a few very large electricity consumers, such as aluminum smelters, for periods of up to an hour, without damaging the industrial process. The electricity prices paid by the smelters are very low, in part reflecting the need for occasional requests by utilities to off-load the smelters.

The advent of new technologies, such as "smart" meters and other "smart" devices, offers the opportunity to greatly expand demand management by both electricity customers and suppliers. This would be particularly helpful in supply systems with high penetrations of variable renewable energy technologies. Spot pricing of electricity, that reflects the changing values of supply and demand, would play an important role. For example, during periods of low wind or sunshine, electricity prices would rise and residential refrigerators could be off-loaded for short periods by means of a smart device programmed by the householder to follow the changing price of electricity.

As batteries gradually become less expensive, the growth of the electric car industry would provide a large amount of potential electrical storage that could be used to buffer the fluctuations in grid electricity that is supplied predominantly by renewable sources. Flows of electricity in both directions between vehicle and the grid are now technically feasible and could be encouraged by appropriate pricing systems, with the motorist becoming both a buyer and seller of electricity. If batteries become less expensive and spot pricing of electricity becomes widespread for residential consumers, then householders in sunny places are likely to install batteries together with solar PV, so that they can sell electricity to the grid on evenings when the price is highest. If and when batteries become very cheap, many households could disconnect from the grid.

56.5   CONCLUSION

The assumption that base-load power has to supply base-load demand has been useful for conventional electricity generation systems based on fossil and nuclear fuels. However, with the rise of renewable energy and scenarios in which several countries, and possibly the whole world (Delucchi and Jacobson, 2011; Hand et al., 2012; Jacobson and Delucchi, 2011; WBGU, 2011), make the transition to a renewable electricity system, the old concept is now of limited usefulness. Renewable electricity supply systems need a portfolio or mix of flexible and variable technologies. Even without any base-load power stations they can be as reliable as conventional supply systems. Indeed, conventional base-load power stations are too inflexible in operation to be good partners for very large-scale renewable electricity.

REFERENCES

Delucchi, M. A., Jacobson, M. Z. (2011). Providing all global energy with wind, water, and solar power, Part II: reliability, system and transmission costs, and policies. *Energy Policy*, 39:1170–1190.

Diesendorf, M. (2014) *Sustainable Energy Solutions for Climate Change.* UNSW Press, Sydney, Australia, and Routledge-Earthscan, London, UK.

Hand, M. M., Baldwin, S., DeMeo, E., Reilly, J. M., Mai, T., Arent, D., Porro, G., Meshek, M., Sandor, D. (2012) *Renewable Electricity Futures Study.* NREL/TP-6A20-52409.National Renewable Energy laboratory, Golden, CO. Available at: http://www.nrel.gov/analysis/re_futures.

Jacobson, M. Z.Delucchi, M. A. (2011). Providing all global energy with wind, water, and solar power, Part I: technologies, energy resources, quantities and areas of infrastructure, and materials. *Energy Policy*, 39:1154–1169.

WBGU (German Advisory Council on Global Change) (2011). *World in Transition: A social contract for sustainability.* Available at: http://www.wbgu.de.

57

TIDAL POWER HARNESSING

Roger H. Charlier

Providing incentives for energy-efficiency and clean energy
are the right thing to do for our future,
because the nation that leads the clean energy economy
will be the nation that leads the global economy.

Barack Obama, President U.S.A.
State of the Union, January 28, 2010

57.1 INTRODUCTION

Tidal energy has been used for centuries with both tidal current and rise and fall of tides put to work. They provided power for flour mills, saw mills, breweries, and so on. Tide mills dotted not only several geographic regions of Europe from the Netherlands to Spain and from Wales to England, but also coastal areas of the United States and Canada (Figure 57.1). They may well be considered the forerunners of the power-generating tidal power stations. These are not numerous, except for the mini plants in China. Besides the Rance River plant in Brittany (France), facilities were built in Russia and Canada. Tides provided the dismal Russian North with the electricity needed to develop a rather desolate region.[1] The Canadian plant, in Nova Scotia, is more a trial run than a badly needed plant. Originally geographically limited to coasts with large tidal ranges, the development of very small head turbines permits today the implantation of tidal power plants[2] in many more locations. The development of the tidal power plant went hand-in-hand with or at least was boosted by that of the bulb turbine (France, Russia)[3] and later of the Straflo turbine (Canada).[4]

57.2 THE ANCESTORS OF THE TIDAL POWER PLANT

Tide mills are of course not different from run-of-the-river mills, except that they include an impounding basin where the water brought in by the incoming (flood) tide is stored: at ebb tide the water is released but has to pass through a channel wherein the mill wheel is set. Some more sophisticated mills even captured power from both ebb and flood tides. And still others captured the energy of the horizontal movement of water from tides. The tide mills' demise in man's industrial arsenal was slow but their numbers declined rapidly and abruptly, as newer technology unfolded.

The mill itself may be built in the middle or at either end of the dike, the exact place being usually determined by the topography and/or hydrology. Industrial mills, however, are usually built at one end of the causeway as this offers easier access.

From the mid-nineteenth century, industrial mills could have three, four, and even five storeys. The floor plan can be up to three times larger than that of the traditional mills they often replaced (e.g., Rochegoude mill, from 8.30 m × 10 m to 13 m × 20.50 m). These traditional mills show considerable variation in size. Most of them are of a more or less pronounced rectangular shape (e.g., in Galicia,

---

[1]Bernshtein, L. B., Usachev, I. N. (1957). Utilization of tidal power in Russia in overcoming the global and ecological crisis. *La Houille Blanche–Revue Internationale de l'Eau*, 52(3):96–102.
[2]Henceforth referred to by the acronym TPP.

[3]Charlier, R.H., 1982, *op.cit.* fn. 4
[4]Charlier, R. H., Justus, J. R. (1993). *Ocean Energies*. Elsevier, Amsterdam-New York, pp. 316–320.

Alternative Energy and Shale Gas Encyclopedia, First Edition. Edited by Jay H. Lehr and Jack Keeley.
© 2016 John Wiley & Sons, Inc. Published 2016 by John Wiley & Sons, Inc.

FIGURE 57.1 Tide mill at Suliac.

9 m × 13.70 m for the De Cura mill, 6.20 m × 20.50 m for the Acea da Ma mill). Although most are one or two storeys buildings, in the South a third floor may be added for the miller's dwelling. From the nineteenth century, in Brittany, the miller's dwelling tends to be built away from the mill proper (Figure 57.2).

In mills equipped with horizontal internal wheels, the cavity under the building is reserved for these and for the mechanisms transmitting power to the grinding stones in all cases. When wheels are vertical, this cavity is virtually empty (e.g., Rance estuary).

57.3 THE EXISTING TIDAL POWER PLANTS

Though power stations have thus been constructed during the second half of the twentieth century in France (Rance River estuary) and Canada (Bay of Fundy, Hog Island), on Kislaya Guba, and "mini" stations in China, interest in building more plants waned, as did most other schemes to tap the energies of the ocean to generate electricity (Figure 57.3). However, utilizing the tidal processes has received considerable attention during the last few years as oil reserves, and especially prices, and climate changes cause increasing concern. China is considering building a plant and has received feasibility study funds from the World Bank; on Lake Shiwha, Korea (R.O.K.) has undertaken the construction of the largest ever tidal power station.

Government funding, particularly in Europe, is currently more common as several new approaches are getting refined.

The first utilization of tidal power in the United Kingdom dates back to a Bristol 1931 scheme, on the heels of the Aber

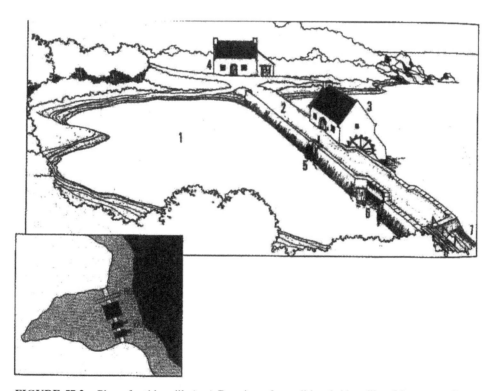

FIGURE 57.2 Plan of a tide mill. (top) Drawing of a traditional tide mill and its surroundings. Legend: 1, pond; 2, causeway; 3, mill; 4, miller's dwelling; 5, exit sluice and hydraulic wheel; 6, entrance sluice gate; 7, bridge (Drawing by A. de La Vernhe. In Boithias and de La Vernhe, 1988). (left) Simplified schematic presentation of a traditional tide mill. Roue, wheel; vannes, sluice gates (*Source*: EDF, France).

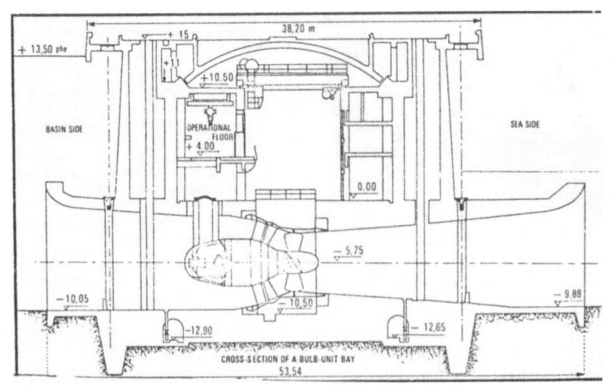

FIGURE 57.3   Rance River plant aerial view.

W'rach aborted attempt (Brittany) and just before the United States' Passamaquoddy project got the go-ahead signal. Work came rapidly to a halt at "Quoddy" under pressure from traditional electrical power providers. In fact it took another half century before, as the anecdote goes, French President Charles de Gaulle, irritated by the dilly-dallying, ordered the construction of the Rance River TPP (Figure 57.4). As the French dug and built, the British looking at the Severn Estuary kept talking, as they still do. By 1968 French power was tapped in the English Channel....

Russians in Kislaya Bay and the Canadians in Fundy Bay decided to fall in step with the French and constructed pilot plants.[5],[6] The expected resurgence of interest in TPPs (Wilson, 1983) after completion of the Annapolis-Royal (Nova Scotia) plant did not materialize, yet such mega-schemes as the Chausey Islands Plant (France) found its way in print (Banal 1982; Bonnefille 1982) (Figure 57.5).

Various try-outs, studies, plans, and projects are on-going; in China the technology of the LonWorks Fieldbus digital communication network has been applied in the 70 kW

tidal current electricity generating system. In Singapore, four sites—Lighthouse, Pasir Panjang, Sembawang, and Tanjong Pagar—have been considered for the implant of a tidal power plant, locations that are rather remote; thus these facilities would provide small quantities of power, ranging from 1 to 10 MW. The limited information available about Chinese TPPs hampers the view of China; the Kianghsia facility, however, was briefly described in the [French] *Revue de l'Énergie*, and the recent call for specialists for a pre-feasibility study for a TPP in Zhijiang Province bears witness of continued interest.[7] The project is buttressed by funding from the World Bank.[8]

Even though 2006 dated information pertaining to tidal energy use in Russia indicates a current disinterest, Usachev et al. published in 2004 both a review of its large-scale use and the observations made over 35 years of operation of

[5]Chaineux, M.-C. P., Charlier, R. H. (2008). Women's tidal power plant: Kislaya Guba's 40 candles. *Renewable & Sustainable Energy Review*, 12(à8):2508–2517.

[6]De Lory, R. P. (1987). Prototype power plant achieves 99% availability. *Sulzer Technical Review*, 1:3–8.

[7]Anonymous (1981). China looks at tidal power. *Electrical Review*, 3:208.; Zhenzia, L., Donking, X., Mingzuo, F. (1989). Evaluation on sited of two tidal power stations. *Coastal Zone '89*, pp. 2203–2209; Anonymous (1998). *Experience in the Topographical Survey for a Tidal Power Plant.* Design Institute of the Yao-Ning Province Administration of Water Management and Power Construction, Beijing [a Russian translation of the Chinese version].

[8]Ch'iu Hou-Ts'ung (1958). The building of the Shamen TPP. *Tien Chi-Ju Tung-Hsin*, 9:52–56.

FIGURE 57.4   Rance River TPP: (a) exterior view, (b) view of cofferdams used to build in the dry, and (c) side view of a bulb turbine.

the Kislaya Bay TPP (Figure 57.6). The TPP has operated successfully, with moving parts underwater, under extreme climatic conditions. The plant has been used as a biological test site in a nearly closed-off basin, including fisheries and mariculture observations. A variety of units were tested among which were cathodic behavior, sorption, and electrolytic. The reporters confirmed that a TPP can be integrated into a power system, in basic and peak periods of the load diagram, allowing for the specific conditions of tidal energy generation.

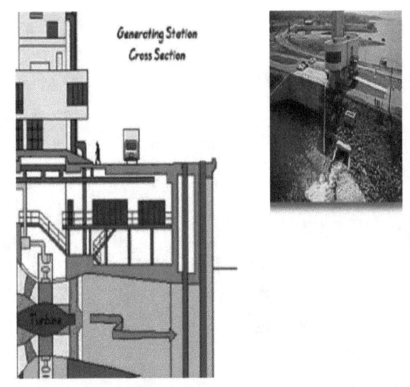

FIGURE 57.5 Canada, Annapolis-Royal TPP views.

The World Energy Council sees considerable opportunities between the present and the end of the next decade for renewable energy sources. A rapid turnover is excluded because of investments in traditional facilities. The choices that are made today, however, will determine whether the cortege of ill consequences of climate warm-up will be blocked by the development and implementation of better technologies or will proceed with its scientific uncertainties, an Absalom sword.

57.4 ENVIRONMENTAL IMPACT

The Rance River TPP's construction phase proved particularly environment damaging, yet a different and diversified flora and fauna took over once the plant began operating. Migratory organisms are able to pass through sluices and turbines, and even a female seal passed through; the animal was removed repeatedly from the retaining basin and released to the open sea, but stubbornly returned, well decided to establish its habitat in the basin.

This author does not share the viewpoint that environmental concerns are the reason for the small number of TPPs—rather the situation should be ascribed predominantly to economic motives—but shares the view that EIAs are useful tools to identify impacts and that applications of modern appropriate technologies might help abate the objectionable effects of a tidal power plant.

Already in 1993 additional research had been suggested pertaining to the interface of the TPP output with national grids, that a sound assessment of such plants, their economic interest, design, and implementation according to the site, and environmental effects was both desirable and urgent (Frau, 1993).

Environmental effects have been again reexamined by Van Walsum in 2003. Interestingly that study addressed two-basin schemes, an approach rarely still considered today, though such scheme had been consistently considered for the Bay of Fundy and nearby sites such as Quoddy.

An unusual reaction was triggered by the release of the UN Environmental Panel on Climate Change report that held—for the second time—that human activity strongly impacts the climate, with the combustion of fossil fuels playing a major part (2000). The International Energy Agency predicts, if no measures are taken to reduce emissions from transportation and power sectors, a still increasing release into the atmosphere of greenhouse gases. It holds that population growth and GNP are proportional to such increase. At the 2002 meeting of the IEEE Power Engineering Society, a diametrically opposed view was presented by a group of scientists who, through the paper authored by Meisen, dispute these findings.

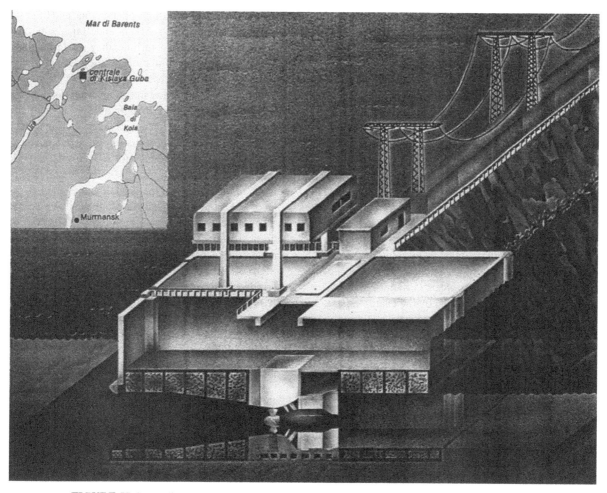

FIGURE 57.6   Stylized view of Kislaya, Russia plant with location map. Fowey, Sangrin; Sergey, Fan; Myers and Bajah.

To reduce emissions not only the UN Development Program, but also the Royal Dutch Shell Company long-range planners suggest the use of renewable energies. The price of these, however, must be made competitive with that of conventional fuels. It must be mentioned here that each time the cost of oil rises, automatically the price gap narrows. Another condition is that means have to be found to secure systems' reliability, as intermittent resources, such as tidal power among others, expand into the daily load mix. And there is a psychological factor as well, linked to the final recognition by governments—particularly the United States—to wit acceptance by the public at large and the governments' willingness to address the problem of climate change.

While Gibrat proposed a formula to rate the suitability of sites for location of a TPP, Sergey (2003) established a semi-quantitative rating of environmental disturbance brought about by power-generating schemes. The thermal power industry is, according to that scale, the most damaging (coefficient 74) and the wave energy capturing

industry the most benign (coefficient 31). Second to best is tidal power with a score of 42, just one point "better" than that of solar power (43).

57.5   MODUS OPERANDI MODIFICATIONS AT THE RANCE

Rance River TPP operators are assisted since the 1990s by a sophisticated computer enabling to carry out simultaneously automatic control and human shift reduction. Automatization insures higher reliability and a more sensitive response to external conditions. Alternators were rebuilt in 1997, the bulb turbines renovated and underwent design changes.

The AGRA program was put in use at the Rance TPP in 1970. The special features of the AGRA new operational model for the Rance, introduced in 1982, encompass close optimization by dynamic programming, modular program structure, and developed graphical outputs (Merlin 1982; Sandrin 1980).

As for other French projects, the Rance Plant has followed the trend for automation (Ferdinand).

57.6   NEW TURBINES

In operating tidal power plants bulb (Rance, Kislaya) or Straflo (Hog Island) turbines have been installed. Low heads being a limiting factor for tidal power plants, Darrieus turbines might offer a solution. Shiono et al. (2000) developed a Darrieus turbine specifically for use in tidal current energy harnessing. Darrieus turbines, originally designed for windmills and adapted for use in water, can always revolve in a same direction without being influenced by tidal flow direction. Difficult to start, these turbines, now equipped with spiral blades, have overcome that problem. Efficiency of tidal current power generation is influenced by the characteristics of the turbine, and Darrieus turbines are affected by the solidity and the number of blades. The experimental turbine was tested in ad hoc channels to ascertain the best values for rotor solidity and the number of blades.

Tide variations in the Amazon River basin provided an opportunity to study turbine rotor blades with simple curvature for use in small hydropower plants with maximal generation of 10 kW. Application in tidal power plants seemed promising and although rotor yield proved to be low, decentralized power generation would be viable (Mauad 2000).

57.7   TIDE CURRENT [TIDAL STREAM]

Over 20,000 Gwh per year could be generated on the West Coast of Canada by harnessing tidal current energy. Tidal stream [current] has also been considered in Ireland. Tidal power has been investigated at Bull's Mouth: Achill, the largest Irish island, could be electrically self-sufficient using winds and tides as energy sources, provided the network be upgraded (Ottewell 2003).

Besides assessing the resource, estimated to have a mean output of 130 MW in several sites, Bryans studies (2004, 2005) looked at methods of deployment and control inclusive of downrating of the generator in regard of turbine size and operational output reduction to reduce capital costs, capacity factor increase, and reduction of impact on the grid. Two tidal energy devices involved horizontal axis turbines and one had an undulating wing design.

Relatively little research has been conducted to determine the characteristics of turbines running in water to convert kinetic energy. The fundamental issues likely to play important roles in MCT (Marine Current Turbines[TM]) systems implementation include particularly harsh marine environments—though already examined in the case of

Kislaya Bay TPP—cavitations phenomena, and high stresses (Bajah and Myers, 2003). Myers and Bajah submitted the design of a horizontal axis turbine; they used tidal data from Alderney Race to run simulations to show a potential annual energy output of 1340 Gwh.[9] Such "races" exist also near Scotland's west coast. Land mass constrictions in these geographical locations are at the origin of high current velocities at depths suitable for the placement of turbines. In a preceding paper (2004) the same authors had emphasized that the power density—for marine currents—for a horizontal axis turbine is similar, in form, to that of a wind turbine and depends on the cube of the speed and the fluid density (water ds = 1000 × air ds). A tidal current turbine can thus be smaller than the wind turbine. Energy yields exceeding 7.4 TWh could—theoretically—be realized, which would satisfy 2% of the demand of the United Kingdom (2000).

Marine Current Ltd used a single-turbine variant of the one originally developed by IT Power. Two axial flow rotors with 17–23 m approximately in diameter drive a generator via a gearbox, mounted on a tubular steel monopile about 2.7–3.3 m in diameter, set into a hole drilled in the seabed (Sanford, 2003).

Gorlov of Boston's Northeastern University has been active for decades in developing systems for capturing tidalenergy.[10] A few decades ago he proposed a "removable barrage" scheme and energy storage with pneumatic chambers. He recently (Gorlov, 2003) introduced a helical turbine that could be used in harnessing tidal as well as wave energy. The scheme dispenses with dams and can operate in free or ultra-low head water currents. The cross-flow unidirectional rotation machine is usable in reversible tidal streams and estuaries, for instance.

In line with the philosophy that even small quantities of power are worth harnessing, Bryden et al. (2000) proposed to link tidal current turbines with energy storage mechanisms even if small. They furthermore (Bryden et al., 2004a, 2004b) developed a simple model of tidal current energy extraction and showed that flow in a simple channel is altered. Ten percent of the flux ought to be the limit to assess the extractable amount. Admittedly the limitation may be less stringent where sea lochs are concerned.

Using a two-dimensional tide-driven hydrodynamic model, Blunden and Bahaj (2006) assessed the Portland Bill (Dorset, GB) proposed site for a tidal current [stream] TPP. Their results could eventually be used if a TPP was indeed implanted there, a site close to population centers and thus economically attractive.

[9]Myers, L., Bahaj, A. S. (2005). Simulated electrical power potential harnessed by marine current turbine arrays in the Alderney Race. *Renewable Energy*, 30(11):1713–1731.
[10]Gorlov, A., A.M., 1982, Hydropneumatic approach to harnessing tidal power. *New Approaches Tidal Power Conf. Bedford Inst. Oceanog. Proc.*, pp. 8–16.

The capability of tidal currents to provide firm power was defended recently, stressing concomitantly the finite nature of wind power expansion (Clarke et al., 2006).

Tidal current tapping must take into consideration that increasing the number of turbines will not only have an unfavorable environmental effect but also block the flow and thereby reduce the power generated. The latter is substantially less than the average kinetic energy flux observed. According to Garrett (2004, 2005) the maximum average power lies between 20% and 24% of the peak tidal pressure head (from one end of the channel to the other) in the undisturbed channel. Garrett also showed that an array of turbines placed at the entrance of a bay, gulf, or inlet is quite effective if uniformly distributed across the entrance.[11] The continuous operation in a bay's entrance of current turbines is not less productive than accumulating the high tide waters in a retaining basin for release at outgoing tide, and power generation is compatible with flushing and purposes other than power generation.

Tide-generated coastal currents are part of the nearshore ocean resources that governments are eyeing to provide relief from global warming (Jones, 2005). Studies of tidal current harnessing were launched as collaborative public/private project early in 2005. The EPRI (Electric Power Research Institute) plays a leading role in the project (Siddiqui and Bedard, 2005). The promised technical-economic results for tidal current power plants and several sites have not yet been disclosed. A keen interest has also been manifested by San Francisco for a hydro-venturi approach which, contrary to the Rance River one, would involve no moving parts placed underwater; the value of Blue Energy or Verdant Power vertical or horizontal axis-type propellers might be determinant in any implementation here. Conventional air-driven "turbines are to be placed on land and generate the electricity from the venturi suction."[12] Indeed, this system causes the least disturbance for the fauna of the bay area (Hammons, 2005, O'Donnell, 2005, Proceedings of IEEE Conference 2005). Arlington (Virginia) based Verdant Power is a major cog in the project to place tidal current turbines in New York City's East River (Grad, 2004) (Figure 57.7).

San Francisco city officials passed a resolution committing the city to the development of a 1 MW TPP during the 2003–2006 span. But 2006 is long gone, and no further news concerning this proposal has been received.

Conversion of marine currents' kinetic energy, different from tidal currents', has been proposed for many decades, witness being the gigantic Coriolis Project. The technology is similar to that for wind turbines but we are still quite far from

FIGURE 57.7   East River, New York: artist's view.

a cost-effective large-scale system. Marine Currents Turbine Ltd [London, UK] has been involved in the development of such system for some time (Fraenkel, 2002). Marine Current Turbines Ltd [MCT] (Basingstoke, UK) participated in the Seaflow Project whose aim was the development of a tidal power plant based on the windmill concept. Such a scheme was established off the Devon Coast in the English Channel (Lang, 2003). A 130 Mt unit was cemented to the seafloor at about 1.1 km from the shore, protruding a few meters above sea level. "Tidal currents turn the eleven-meter-long rotor, but as they reverse direction, the rotor's blades can be pitched to accept flow from the opposite direction. [...] The rotor turns slowly in water, [but] at 17 rpm the speed is sufficient, with appropriate gearing, to harness the tide's energy and drive a turbine." Rotor speed varying, a power-conditioning system (with AC-DC-AC conversion) allows to get a current output of 50 Hz (grid frequency). Mean power reaches 100 kW with a peak of 300 kW, fulfilling the needs of about 500 homes.

Japan actively searches ways to extract energy from the ocean: researchers there examine the possibility of adapting reciprocating flow turbines developed for tapping wave energy to tidal energy harnessing (Takenouchi et al., 2006). This would evidently widen the field of tidal power use with extra-low head and time-varying energy characteristics. They concluded the output of the plant depends [evidently] on the tidal [range] and a pond inundation area. Others developed a new type of two-way diffuser suitable for a fluid flow energy conversion. Power from such fluid flow is proportional to the cube of the flow's velocity, thence by increasing velocity results in more power. Based on experience with wind turbines, a system was developed for tidal streams (Kaneko, 2004).

---

[11] The maximum power available from a quadratic drag law occurs when the tidal range inside the bay is reduced to 0.74 of the original amplitude.

[12] In a venturi tube the velocity is increased and the pressure lowered for a fluid that is conveyed through it, by a constricted passage. Named after the Italian G. B. Venturi [1746–1822].

FIGURE 57.8 Lake Shiwha TPP site, Korea.

57.8 NEW APPROACHES AND NEW TECHNOLOGIES

A new approach has been developed by Venturigen in its tidal power-generating concept.[13] Tidal power installations have been traditionally conceived as coastal project. Offshore tidal power has now been suggested by Ullman (2002). The installation would involve a self-contained impoundment structure. Likely sites include southwest Alaska, Puerto Mont in Chile, and Swansea Bay in the United Kingdom (Figures 57.8 and 57.9).

The specific materials requirements for a tidal power plant have been discussed by Gooch (2000) with a view on capital costs reduction, harsh environment, longevity prolongation, and generation efficiency increase.

In its effort to remain at the forefront of wave and tidal power development, the United Kingdom decided to support facilities that aim to switch from research to implementation. These centers are at Blyth and in Scotland's Western Isles. The Severn River is probably the tidal river that has been most thoroughly studied, and the most often as well, with an eye on building a TPP. It has also been shown that had a plant been built after WW II, it would have quite long ago paid for itself. Because of environmental effects of a barrage, and no doubt the economic aspects of any project, no TPP has ever been even started. Fry (2005) has described alternatives to a TPP-with-barrage. Among these, tidal lagoons schemes have been proposed and so has the modest utilization of the tidal stream (tidal current).

The same concern about environmental effects of a barrage led to the Hastings (2005) proposal. The difference in time of tides between the two sides of the Cotentin

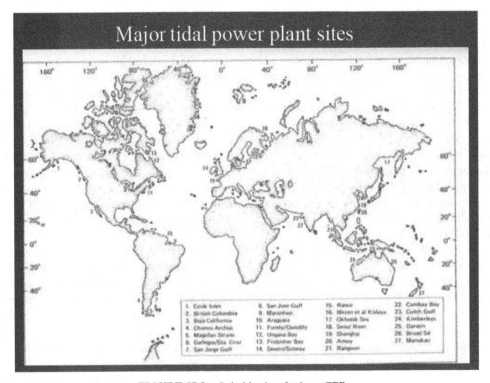

FIGURE 57.9 Suitable sites for large TPPs.

[13]Venturigen (2006). Tidal flow energy system. *Lewis Engineering Technology* 5(10):19–20.

Peninsula (France) has retained the attention of geographers and engineers alike. The tidal phase difference, or tidal delay, could be put to use, and the landform constitutes a natural "barrage." The Australian company Woodshed Technologies proposes to utilize the phenomenon to capture tidal energy. Still another scheme to dispense with the TPP barrage has been put forth by Appleyard (2005). The hydroelectricity generator of Hi-Spec Research (Fowey, UK) is an offshore system that has a series of four channels created from two outer walls and three rows of chambers housing turbines.

A new perspective is perhaps added to utilization of tidal stream (current) by Mueller and Baker (2005) in their analysis of conversion to electricity of energy produced by marine energy converters, using direct drive electrical power take-off, without the use of complex pneumatic, hydraulic, or mechanical linkages. To this end, the linear vernier hybrid permanent magnet machine and the air-cored tubular permanent machine were examined. Taking into account the electrical properties of the topologies of these machines, problems of sealing, lubrication, and corrosion were studied.

An attempt at defining the viability of tidal energy-generated power was made as recently as 2005 (Wood, 2005); however, the rise in oil prices has not been fully taken into account and a revised schedule will be timely.

The European Commission funded CAOC,[14] and put in place a platform enabling devices developers, wave and tidal energy researchers, and standard agencies to share information and knowledge in order to facilitate transition of wave and tidal energy from an energy research technology to one approaching commercial competitiveness with a medium-term time frame (Johnstone, 2006)

REFERENCES

Appleyard, O. (2005). High tide (tidal power). *International Water Power & Dam Construction*, 57(4):36–37.

Baker, A. C. (1991). *Tidal Power*. Peregrinns, London.

Banal, M. (1997a). History of tidal power in France. *La Houille Blanche – Journal International de l'Eau*, 52(3):14–15.

Banal, M. (1997b). The technical origins of the tidal power station. *La Houille Blanche – Journal International de l'Eau*, 52(3):16–17.

Blunden, B. S., Bahaj, A. S. (2006). Initial evaluating of tidal power resource in Portland Bill, UK. *Renewable Energy*, 31(2):121–132.

Bonnefille, R. (1976). Les réalisations de l'Electricité de France concernant l'énergie marémotrice. *La Houille Blanche*, 31(2):87–149.

Bryden, I. G., Melville, G. T. (2004). Choosing and evaluating sites for tidal power development. *Proceedings of the Institute of Mechanical Engineers*, 219(A-3):235–247.

Charlier, R. H., Finkl, C. W. (2009). *Ocean Energy. Tides and Tidal Power*, Springer Verlag, Heidelberg.

Clarke, J. A., Connor, G., Grant, A. D., Johnstone, C. M. (2006). Regulating the output characteristics of tidal power current stations to facilitate better base load matching over the lunar cycle. *Renewable Energy*, 31(2):173–180.

Frau, J. P. (1993). Tidal energy—promising projects. La Rance, a successful industrial-scale experiment. *IEEE Transactions on Energy Conversion*, 8:552–558.

Fry, C. (2005). All at sea (tidal power). *Power Engineering*, 19(5):24–27.

Garrett, C., Cummins, F. (2004). Generating power from tidal currents. *Journal of Waterway, Port, Coastal and Ocean Engineering*, 130(3):114–118.

Gibrat, R. (1976). The current revival of tidal power. *ASTED*, Montreal.

Gibrat, R., Auroy, F. (1956). Problèmes posés par l'utilisation de l'énergie des marées. *World Power Conference*, Vienna: 12, 111, H/22, pp. 4299–4328 [wave and tidal power generation].

Gooch, D. J. (2000). Materials issues in renewable power generation. *International Materials Reviews*, 45(I):1–14.

Gorlov, A. M. (2000). Tidal power. *Encyclopedia of Ocean Sciences*. p. 2955.

Gorlov, A. M. (2003). The helical turbine and its application for tidal and wave power. *Oceans 2003*, 4:1996–1997.

Grad, P. (2004). Changing tide of power generation. *Engineers Australia* 76(11):51–52.

Hammons, T. J. (2005). Energy potential of the oceans: tidal, wave, currents and OTEC. *Proceedings of the 40th International Universities Power Conference*, 1, 2, pp. 1047–1057.

Hastings, S. (2005). Working with nature: tidal power. *International Power Generation*, 28(3):27–28.

O'Donnell, P. (2005). Update '05: ocean wave and tidal power generation in San Francisco. *Proceedings of the IEEE Power Engineering Society General Meeting*, No. 2, pp. 1990–2003.

Ottewell, S. (2003). Ireland's renewable island. *Power Engineering*, 17(3):10–11.

Sanford, L. (2003). Winning the tidal race. *Modern Power Systems*, 23(7):11–12.

Shiono, M., Suzuki, K., Kiho, S. (2000). An experimental study of the characteristics of a Darrieus turbine for tidal power generation. *Electrical Engineering in Japan*, 132(3):38–47.

Siddiqui, O., Bedard, R. (2005). Feasibility assessment of offshore wave and tidal power production: a collaborative public/private partnership. *Proceedings of the IEEE Power Engineering Society General Meeting*, 2, pp. 2004–2010.

Takenouchi, K., Okuma, K., Furukawa, A., Setoguchi, T. (2006). On applicability of reciprocating flow turbines developed for wave power to tidal power conversion. *Renewable Energy*, 31(2):209–223.

---

[14]Co-ordinated Action on Ocean Energy.

Ullman, P. W. (2002). Offshore tidal power—beyond the barrage. *Modern Power System,* 22(6):38–39. See also *International Water Power & Dam Construction* 54(9):24–27.

Usachev, I. N., Shpolyanskii, Y. B., Istorik, B. P. (2004). Performance control of a marine power plant in the Russian Arctic coast and prospects for the wide-scale use of tidal energy. *Power Technology and Engineering*, 37(4):201–206.

Van Walsum, E. (2003). Barriers against tidal power. *International Water Power & Dam Construction,* 55(9):34–41.

Wilson, E. M. (1977). Tidal energy and system planning. *Consulting Engineer London,* 41(4):25.

Wood, J. (2005). Marine renewables face paperwork barrier. *IEE Review,* 51(4):26–27.

58

THE LOADING OF WATER CURRENT TURBINES: THE BETZ LIMIT AND DUCTED TURBINES

D. P. Georgiou and N. G. Theodoropoulos

58.1 INTRODUCTION

Turbines constitute an alternative for renewable energy concepts exploiting the potential provided by river streams or open sea currents [1]. The former refers to rivers on flat earth [2], where we are not able to build a dam. The latter originates on the tidal action of the moon and refers to situations where (tidal) dam building is too costly. In both these cases, the relevant power plants extract part of the stream kinetic energy in a manner similar to that of the conventional wind turbines. As a result, the loading of the water current turbines may be assessed in a way similar to that of the wind turbines. This means the application of the Betz limit for unducted concepts or the ducting of the power-generating turbines.

However, the nature of the water and the corresponding currents introduce a number of additional parameters on the design of such plants. These include

(i) the higher Reynolds number for the same speed;
(ii) the cavitation limit;
(iii) the existence of the free water surface, acting as a hydrodynamic image wall on the performance of the turbine; and
(iv) the interaction of the surface waves with the rotating rotor.

Hence, the analysis here will study these interactions against the reference conditions of a turbine operating on an infinitive incompressible fluid.

58.2 THE BETZ ANALYSIS AND THE CORRESPONDING LIMIT

The model studied by Betz [3] applies to the water current turbines as well. The full analysis has been discussed in the chapter 3 on wind turbines. Here only the main points will be illustrated.

58.2.1 The Current Turbine Efficiency

The power transferred by any turbine results from the volumetric flux through it and the total pressure difference across the stages of the turbomachine. This may be improved by increasing any one of the above two parameters. In water current turbine applications, this improvement is correlated against

(i) the water volumetric flux intercepted by the turbine rotor and
(ii) the dynamic head of the wind blowing far upstream.

The combined effectiveness of a given wind turbine concept in enhancing these two parameters is usually expressed through the "power factor" (C_P), where

$$C_P = \frac{\dot{W}}{\frac{1}{2}\rho A_R V_0^3} \qquad (58.1)$$

Here $\dot{W} \rho, A_R$, and V_0 represent the actual power generated by the rotor, the air density, the rotor cross-sectional area, and wind far upstream velocity, respectively, and

$$\dot{W} = \eta_R (A_R V_R)(\Delta P_{0R}) \qquad (58.2)$$

Alternative Energy and Shale Gas Encyclopedia, First Edition. Edited by Jay H. Lehr and Jack Keeley.
© 2016 John Wiley & Sons, Inc. Published 2016 by John Wiley & Sons, Inc.

where η_R, V_R, and ΔP_{0R} represent the rotor efficiency, the wind velocity, and the corresponding total pressure rise across the rotor, respectively. For an incompressible flow, the velocity remains (nearby) constant across the turbine, so that $\Delta P_{0R} = \Delta P_R$. The Betz analysis assumes that $\eta_R \approx 1.0$. Equation (58.1) may take the form

$$C_P = C_T Y \tag{58.3}$$

where

$$C_T = \frac{\Delta P_R}{\frac{1}{2}\rho V_0^2} = \frac{T}{\frac{1}{2}\rho A_R V_0^2} \tag{58.4}$$

is called the "loading factor." T represents the thrust applied on the rotor by the blowing wind, while

$$Y = V_R/V_0 \tag{58.5}$$

represents the wind "acceleration factor" from far upstream up to the rotor plane. The loading and the acceleration parameters interact among themselves, depending on the plant configuration.

58.2.2 The Betz Limit

Synoptically, Betz [3] employed the one-dimensional mass, momentum, and energy conservation equations of an incompressible flow through an actuator disc to relate the incoming and exhaust flows of the streamtube enclosing the disc. The control volume employed in the analysis is illustrated in Figure 58.1 and corresponds to the area $C_1FGB_1C_1$.

As shown in the chapter 3 on wind turbines for the induction factor (α) employed by Betz

$$Y = 1 - \alpha \tag{58.6}$$

and

$$X = 1 - 2\alpha \tag{58.7}$$

Then

$$C_P = \frac{\Delta P_R A_R V_R}{\frac{1}{2}\rho V_0^3 A_R} = C_T Y = 4\alpha (1 - \alpha)^2 \tag{58.8}$$

while

$$C_T = 4\alpha (1 - \alpha) \tag{58.9}$$

Equation (58.9) implies that

$$0 \leq \alpha \leq 0.5 \tag{58.10}$$

The optimum value for the C_P coefficient appears when

$$\partial C_P / \partial \alpha = 0 \tag{58.11}$$

This leads to the conclusion that $\alpha = 1/3$ and $C_{Pmax} = 16/27 = 0.592$.

The last conclusion is known as the "Betz limit." In other words, any ideal water current turbine may absorb no more than 59.2% of the kinetic energy of the water current intercepted by the turbine rotor, while only 70% of this flux actually passes through the rotor.

The predicted $C_P(\alpha)$ and $C_T(\alpha)$ relationships (for $0 < \alpha < 0.5$) are illustrated in Figure 58.2.

58.2.3 Experimental Verification for the Betz Analysis

No experiment has been conducted and published that tests the validity of the Betz predictions. However, an indirect

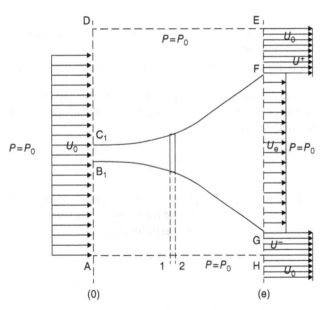

FIGURE 58.1 The control volume employed by Betz in the development of his model.

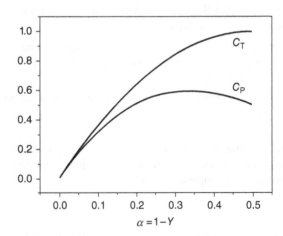

FIGURE 58.2 The relationships $C_P(\alpha)$ and $C_T(\alpha)$ as predicted by Betz.

verification may be established by analyzing experimental data from measurements where both the power and the thrust coefficients were recorded for the same rotor at the same tip to free stream velocity ratio, as the data provided by Bahaj et al. [4] on small-scale models of sea current turbines. The measurements involved various pitch angles for the turbine blades, so as to introduce the turbine efficiency into the analysis. Hence, equation (58.3) becomes

$$C_P = \eta_T Y C_T \tag{58.12}$$

Instead of the induction factor (α), one may introduce the "extended induction factor" ($\alpha*$). Then

$$\alpha^* = 1 - \eta_T Y = 1 - \frac{C_P}{C_T} \tag{58.13}$$

For well-designed and operated turbines, $\eta_T \simeq 0.85 - 0.9$. Hence, the parameter α^* is fairly close to the Betz parameter α. Figure 58.3 was assembled from the C_P and C_T data provided by Bahaj et al. [4]. Each measurement point was deduced from $C_P(\lambda)$ and $C_T(\lambda)$ data, where λ represents the blade tip to free stream velocity ratio.

The results agree fairly well with the Betz analysis, although the efficiency of the turbine appears to introduce a third parameter influence that is quite significant under certain conditions.

58.3 CAVITATION AND MULTISTAGED TURBINES

In sea current turbine applications cavitation introduces a further complication, since it limits the minimum pressure that may appear on the blade hydrofoils. The problem is

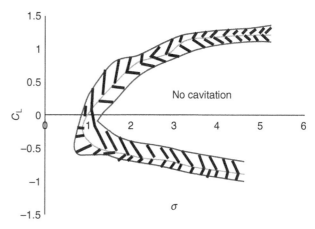

FIGURE 58.4 The loading of marine current turbine hydrofoils, as deduced from the data in [5].

similar to that of the ship propellers. The main criterion is the "cavitation number"

$$\sigma = \frac{P_{at} + \rho g \Delta z - P_v}{\frac{1}{2}\rho W^2} \tag{58.14}$$

where P_{at} and P_v represent the atmospheric and water vapor pressures, ρ is the water density, $g = 9.81$ m s^{-2} is the gravitational acceleration, Δz represents the depth of the blade tip nearest the sea (river) surface, and W is the relative velocity of the water toward the tip hydrofoils. Batten et al. [5] have tested various water current hydrofoils and based on these data Figure 58.4 was drawn to act as a rough guide for the permitted hydrofoil lift coefficient (C_L) against the operating cavitation number.

The need to reduce the hydrofoil loading has brought forth the concept of multistage turbine rotors with two (or more)

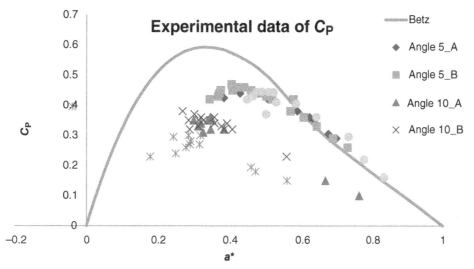

FIGURE 58.3 The $C_P(\alpha*)$ relationship for sea current turbines based upon the experimental data provided by [4].

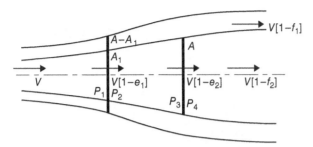

FIGURE 58.5   Double actuator disc [6].

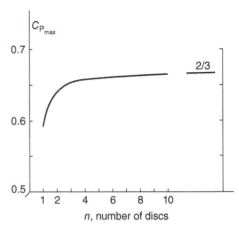

FIGURE 58.7   The maximum power coefficient as a function of the number of discs [6].

turbine rotors on the same shaft. Newman [6] has shown that under ideal flow conditions the limit for a coupled rotor turbine increases to 16/25 or 64% for two actuator discs in line. The streamtube geography employed for such a plant ($n = 2$) is illustrated in Figure 58.5.

According to Newman's analysis, the maximum power factor (C_P) depends on the number of coupled rotors (n). For the case $n = 2$, this leads to

$$C_{P\,\text{max}_{\text{double rotor discs}}} = \frac{8n\,(n+1)}{(2n+1)^2} = \frac{16}{25} = 64\% \quad (58.15)$$

The analysis was extended to a greater number of actuator discs (n of them), as illustrated in Figure 58.6. It was shown that the maximum power coefficient for a very large number of discs reaches the limit of 2/3 (or 66.7%), which is 13% larger than the maximum for a single disc [6]. The influence of the stage number (n) is established by equation (58.1), by computing the proper power coefficient as a function of n. Figure 58.7 shows the variation of $C_{P\text{max}}$ against the number of discs n. The limiting value for large n is 2/3 and is reached almost completely when $n = 4$.

58.4   FREE SURFACE IMPLICATIONS

In many applications the blade tips of water current turbine plants have to operate near the water surface or the sea/river bottom. Under such conditions the boundary surfaces act as

image walls and may affect the loading significantly. Bahaj et al. [4] measured C_P reductions of the order of percent and C_T reductions of the order of percent when the rotor tip was at a depth at the nearest point to the water surface. In their publication they provide an estimate of the influence of these walls on the turbine performance based on actuator disc analysis. In general, wave effects will influence the wake evolution and hence the turbine performance, although the expected depth-based Froude numbers will be small for actual applications. The problem has also been studied by Barltrop et al. [7], with similar conclusions.

58.5   DUCTED MARINE TURBINE CONCEPTS

The need to reduce the diameter of the marine current turbines (MCT) has led to an extensive investigation of the ducted turbine concept. Similar considerations limit the acceleration permitted for ducted turbine concepts in sea/river currents. The ducted turbine concept has been analyzed recently by Georgiou and Theodoropoulos [8], while experimental studies were conducted earlier (on wind turbines) by Gilbert and Foreman [9], who reported a power augmentation factor of

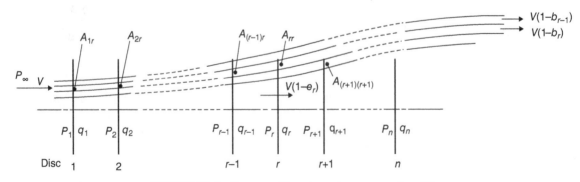

FIGURE 58.6   The streamlines through n actuator discs [6].

4.25 for a turbine with a diffuser. They used slots to draw in high-energy flow from outside the diffuser for boundary layer control and were able to achieve this result with a short, wide-angle diffuser. Riegler [10] showed that the theoretical maximum power coefficient for a diffuser-augmented turbine based on turbine area is 3.3 times higher than the Betz limit. Similar conclusions are reached in the study by Shives and Crawford [11].

REFERENCES

[1] Güney, M. S., Kaygusuz, K. (2010). Hydrokinetic energy conversion systems: a technology status review. *Renewable and Sustainable Energy Reviews*, 14(9):2996–3004.

[2] Khan, M. J., Iqbal, M. T., Quaicoe, J. E. (2008). River current energy conversion systems: progress, prospects and challenges. *Renewable and Sustainable Energy Reviews*, 12(8):2177–32193.

[3] Betz, A. (1928). Windmills in the light of modern research. NACA TN-474.

[4] Bahaj, A. S., Mollard, A. F., Chaplin, J. R., Batten, W. M. J. (2007). Power and thrust measurements of marine current turbines under various hydrodynamic flow conditions in a cavitation tunnel and a towing tank. *Renewable Energy*, 32:407–426.

[5] Batten, W. M. J., Bahaj, A. S., Mollard, A. F., Chaplin, J. R. (2008). The prediction of the hydrodynamic performance of marine current turbines. *Renewable Energy*, 33:1085–1096.

[6] Newman, B. G. (1986). Multiple actuator-disc theory for wind turbine. *Journal of Wind Engineering and Industrial Aerodynamics*, 24:215–225.

[7] Barltrop, N., Varyani, K. S., Grant, A., Clelland, D., Pham, X. P. (2007). Investigation into wave–current interactions in marine current turbines. *Proceedings of the Institution of Mechanical Engineers, Part A: Journal of Power and Energy*, 221:233–242.

[8] Georgiou, D. P., Theodoropoulos, N. G. (2010). Grounding and the influence of the total pressure losses in ducted wind turbines. *Wind Energy*, 13(6):517–527.

[9] Gilbert, B. L., Foreman, K. M. (1979). Experimental demonstration of the diffuser- augmented wind turbine concept. *Journal of Energy*, 3(4):235–240.

[10] Riegler, G. (1983). Principles of energy extraction from a free stream by means of wind turbines. *Wind Engineering*, 7(2):115–126.

[11] Shives, M., Crawford, C. (2010). Overall efficiency of ducted tidal current turbines. IEEE Conference Publications, OCEANS 2010, pp. 1–6.

59

BOTTLED GAS AS HOUSEHOLD ENERGY

Masami Kojima

59.1 PATTERNS OF HOUSEHOLD ENERGY USE

Household energy use is a strong function of income. In developed countries, virtually all households use electricity for lighting and electric appliances, and many households use electricity also for cooking and heating. Other households use a gaseous fuel—natural gas, or, if natural gas is not available, liquefied petroleum gas (LPG)—for cooking, and those who do not use electricity for heating use either a gaseous fuel or heating oil.

These patterns of household energy use stand in stark contrast to those in developing countries, where nearly three billion people are estimated to be using some form of a solid fuel for cooking and heating, and in some cases, even for lighting. These solid fuels include woodfuels (firewood and charcoal), agricultural residues, dung, and coal. When burned in traditional stoves, the simplest of which is a three-stone fire, these solid fuels emit pollutants that are extremely damaging to health, most notably fine particulate matter. The World Health Organization estimates that indoor air pollution from solid fuel use kills four people every minute (UNDP and WHO 2009). Advanced-combustion stoves for solid fuels can substantially reduce these harmful emissions, but they are more expensive, and solid evidence from field experience that demonstrates sustained durability of the stoves' emissions performance is still largely absent.

Damage to health is only one of several costs of using solid fuels. Pots, pans, and even light bulbs could become black with soot. Particularly in rural areas, many households collect biomass. The time spent on fuel collection can be considerable. Children are taken out of school, and parents, all too often mothers, spend time that might otherwise be spent on childcare and alternative productive activities on

fuel collection instead. Cooking also takes longer, especially in rainy seasons when getting a fire going with damp wood could be challenging. For all these reasons, more than half the world does not use any solid fuel for cooking.

Environmental costs of solid fuel use can be high. Where biomass is not harvested sustainably, its use can lead to forest degradation or loss of tree resources. The main driver of deforestation globally remains conversion of forest to agricultural land, but concentrated consumption of woodfuels—typically in urban areas by residential and industrial users—can lead to loss of forest cover. Of the wood removed from productive forests in the last decade, woodfuel accounted for half. While the pace of wood removals *fell* at an annual rate of 0.5 percent between 1990 and 2000, it *increased* at 1.1 percent annually between 2000 and 2010 (FAO 2010). Where tree resources are diminishing, households may have to travel greater distances to collect biomass fuel. In urban areas, the scarcity of tree resources in some Sub-Saharan African cities and elsewhere has driven up woodfuel prices.

A 2011 editorial in *Energy for Sustainable Development* observes that the richer half of the world uses just two types of energy for cooking: gas and electricity. Citing rural women in India who asked why improved biomass stoves are compared to traditional biomass stoves when the focus should be how closely improved stoves can mimic "the stove everyone wants, gas," the editorial suggests a gas stove is the "aspirational appliance that every woman knows and would prefer, for convenience, controllability, time savings, and modernity," setting a gold standard against which to measure improved biomass stoves (Smith and Dutta 2011).

Early studies on the use of fuels for the basic needs of lighting, cooking, and heating centered on the concept of

Alternative Energy and Shale Gas Encyclopedia, First Edition. Edited by Jay H. Lehr and Jack Keeley.

an energy "ladder." Later studies, however, suggested that a portfolio, or fuel-stacking, approach is more realistic. One study describes the differences between the two approaches as follows (ESMAP 2003a, pp. 11–12):

> The energy ladder model envisions a three-stage fuel switching process. The first stage is marked by universal reliance on biomass. In the second stage households move to "transition" fuels such as kerosene, coal and charcoal in response to higher incomes and factors such as deforestation and urbanization. In the third phase households switch to LPG, natural gas, or electricity... Yet the ladder image is perhaps unfortunate because it appears to imply that a move up to a new fuel is simultaneously a move away from fuels used hitherto... Evidence from a growing number of countries is showing multiple fuel use to be fairly common... The new perspective on household energy choice sees it as a portfolio choice more than as a ladder.

A study conducted in rural Mexico provides a striking example of multiple fuel use. The study found households moving to multiple-fuel cooking with rising income in a process in which separate, coexisting factors simultaneously pushed households away from biomass and pulled them back. Even when households had been using LPG for many years, they rarely abandoned firewood use. Firewood savings from using LPG ranged from 35 percent on average in one village to as little as 0 percent. Expenditures on firewood were higher in some cases in households using both firewood and LPG than in firewood-only households (Masera, Saatkamp, and Kammen 2000).

Analogous to lack of access to electricity, the use of solid fuels for cooking and heating is a form of energy poverty. A household is energy poor if it does not use electricity regularly for lighting and other essential needs normally met using electricity—because the household is not connected to electricity, or is connected but electricity is not available during peak demand due to frequent and long-lasting power outages, or does not have the financial means to pay for electricity—or is burning solid fuels in a traditional stove as its primary fuel for cooking, heating, or both. Because the use of advanced-combustion stoves is not yet widespread, most of the three billion people using solid fuels are energy poor by this definition. About half of them also do not have access to reliable electricity.

Noting, amongst others, widespread energy poverty in the world, the United Nations General Assembly in December 2010 passed a resolution declaring 2012 the "International Year of Sustainable Energy for All." In 2011, following this declaration, the United Nations launched a global initiative, "Sustainable Energy for All." The first of the three goals set for 2030 under the initiative is universal access to modern energy services: electricity, and clean cooking and heating solutions. The *World Energy Outlook 2011* (IEA 2011a) suggests that, if no new policies are adopted and implemented beyond those proposed by mid-2011, the number of people without access to clean cooking solutions will actually increase slightly between 2009 and 2030 because of the increase in population. To achieve universal access to modern energy services for cooking, an estimated 1.25 billion people, who would otherwise have continued to rely on traditional use of solid biomass, will use advanced biomass cookstoves; 1.2 billion people will use LPG; and 350 million people will use biogas.

Among different fuel options, gaseous fuels are the cleanest. There are three types of gaseous fuels: biogas, natural gas, and LPG. Biogas, made from organic wastes, is the only renewable fuel among the gaseous fuels and therefore ranks high in environmental sustainability. It is typically made from livestock manure in developing countries, limiting its adoption to rural areas with farm animals. Biogas digesters' costs are high; 2010 estimates range from about US$440 in India to US$924 in Sub-Saharan Africa (IEA 2011a). Natural gas burns more cleanly than LPG and, absent subsidies, is typically much cheaper per unit of energy, making it the fuel of choice in areas where both natural gas and LPG are available. For energy-intensive activities such as space heating, natural gas is also typically cheaper than electricity. But laying down a natural gas pipeline network is capital-intensive with very large economies of scale, and developing a gas supply infrastructure takes time. Even when the supply infrastructure is fully developed, as in some high-income countries, natural gas is available only to urban and some peri-urban households where scale economy can be exploited. In such countries, residential LPG consumers are predominantly in rural areas, which cannot be reached by natural gas pipelines.

In developing countries, the presence of an extensive natural gas pipeline network connecting households is not common even in urban areas, and this, combined with the need for a relatively high concentration of consumers who are able to pay for regular use of LPG to establish a commercially viable LPG market, makes urban LPG consumers far more numerous than those in rural areas. Over the very long run, as income rises to the level that creates sufficient demand for natural gas in cities, and households become capable of paying for natural gas connection and use, many urban households in developing countries, too, will shift largely to natural gas, electricity, or both for cooking and heating. As such, LPG is a transition fuel for many, if not most, urban households over the very long run. For the foreseeable future, however, LPG will be the most widely available of three gaseous fuels in developing countries. LPG is marketed in both urban and rural areas, although in developing countries it tends to be more widely available in urban areas where its supply costs are lower.

Two most important determinants of household fuel choice are income and relative fuel prices. However, as the above Mexican study illustrates, a household does not automatically switch to a cleaner commercial fuel as soon as it becomes affordable, nor does it necessarily abandon, or even decrease, traditional use of solid fuels with rising income. Many add new sources of energy to the existing ones to suit their budgets and preferences, for example, using LPG or electricity for making tea, dung for simmering, and firewood for the rest of cooking. The benefits of using cleaner forms of energy are diminished correspondingly with continuing use of solid fuels. An examination of data from national household surveys in 63 developing countries found that more than 80 percent of the households in the top (richest) income quintile named biomass as the primary cooking fuel in 10 countries, and the corresponding number rose to 23 countries for the fourth quintile (Kojima, Bacon, and Zhou 2011). Adequate income is therefore a necessary, but not a sufficient, condition for shifting households away from solid fuel use. If there are frequent LPG shortages, if LPG delivery is not reliable, if unsafe handling of cylinders has led to publicized fires, or if non-household members are dealing with biomass (e.g., a housekeeper), the incentive to consider LPG is weakened.

This chapter covers household use of LPG. Autogas (automotive use of LPG) is also covered to the extent that it competes with household LPG use in some countries. Available data from 110 developing countries on household energy for cooking illustrate the scale of LPG adoption (Kojima, Bacon, and Zhou 2011). More than half of all households named LPG as their primary cooking fuel in 30 percent of the countries and solid fuels in 49 percent. In more than a dozen countries, less than one-half of the households used solid fuels and widely used forms of energy other than LPG as cleaner-energy alternatives: kerosene in three countries and natural gas and electricity in eight countries each, mostly in the former Soviet Union. Predictably, the share of households using LPG in most cases is higher in urban than in rural areas, except in seven countries where urban households had access to natural gas. In the above analysis of household survey data from 63 developing countries, the percentage of households using LPG rose with quintile in 51 countries. These observations suggest that LPG is indeed likely to play an increasingly important role in the coming years with rising income as well as growing awareness about the convenience and health benefits of LPG.

59.2 LPG SUPPLY AND ECONOMICS

59.2.1 Supply

LPG consists primarily of propane and butane. LPG is used as a gas but sold as a liquid because gaseous propane and butane take up about 250 times as much space as in the liquid state. To conserve space, LPG is pressurized in metal containers at ambient temperature or else refrigerated to transport and store

as a liquid. The need to keep LPG pressurized or refrigerated and associated metal management add considerably to the supply cost of LPG—for example, specialized steels are used to manufacture tanks for LPG ships—and result in large economies of scale. LPG is more costly to distribute than liquid fuels, but less so than natural gas, which cannot be compressed into a liquid and which requires a much lower temperature to refrigerate than LPG.

About two-fifths of LPG supply comes from crude oil refining, one-quarter from natural gas associated with crude oil production, and most of the balance from unassociated natural gas. Slightly more than half of global LPG demand is for cooking and heating by residential and commercial consumers. China is the world's largest residential–commercial consumer of LPG. Residential and commercial demand accounts for about 95 percent of the total in the Indian subcontinent and Africa (consumption in Africa occurs primarily in North Africa) and three-quarters in Latin America and the Caribbean (Hart, Gist, and Otto 2011). Among developing countries, Turkey, the Russian Federation, Mexico, and Thailand were among the top 10 autogas consumers in 2008 (WLPGA 2010).

Commercial participants in the LPG supply chain include the following actors:

- *Producers* sell LPG at the refinery or natural gas processing plant gate.
- *Traders* and *marketers* buy LPG in bulk from producers or from overseas markets, store it in large primary terminals, and sell it to other marketers, distributors, retailers, and final consumers.
- *Transporters* and *distributors* truck, rail, or pipe bulk LPG to their regional depots where it is stored in large pressure vessels, and then supply LPG to bulk customers by small road tankers. LPG is bottled in cylinders and distributed to retailers.
- *Retailers* sell LPG to small customers, including households. The retail outlets may be retail branches or commission agents of a marketer, or independent resellers who purchase and resell LPG in marketer-owned and branded cylinders. Autogas is sold at filling stations.

Equipment and service industries supporting the supply chain include cylinder manufacturing, testing, repair, and recertification; LPG appliances and equipment such as stoves, valves, hoses, and cylinder regulators; and bulk tank manufacturing and installation services. For autogas, associated equipment and services include fuel tanks, valves, hoses, fillers, and conversion of light-duty gasoline vehicles to run on LPG. Autogas is covered in this report because its taxation policy can have a significant effect on the availability and price levels of LPG for household use.

59.2.2 Economics

LPG, natural gas, and kerosene are three modern commercial fuels that can potentially be used for cooking and heating for household energy in developing countries. Kerosene can be burned in either wick or high-pressure stoves. If not properly maintained, wick stoves can be low in fuel efficiency and polluting. Pressure stoves vaporize kerosene first, allowing it to burn cleanly, but they are more costly and noisy than wick stoves. Natural gas is the cleanest of the three fuels.

The steep rise in oil prices since 2004 has affected the global LPG market. The price increase for LPG has been nearly as large as that for oil: between 2003 and 2011, LPG prices rose by a factor of 2.9 and oil prices by a factor of 3.6. In real terms, the LPG price increase corresponded to 140 percent, or an average annual increase of 10 percent, far above the increase in the income of most households.

One way of comparing the economics of the three fuels is to normalize prices based on the energy content of each fuel. Figure 59.1 shows (net-of-tax) monthly international average prices per million British thermal units (Btu), a unit of energy, since January 2003. Prices are for free on board (FOB) for LPG and kerosene and landed costs for natural gas. The LPG price chosen is the Saudi Aramco contract price, the benchmark price for LPG in Asia. To be consistent with the basis for LPG, other fuel prices are also taken from the Asian market: kerosene in the region's largest refining center for export, Singapore, and natural gas in the world's largest importer of liquefied natural gas, Japan. Because the price of natural gas in Asia in recent years has been higher than those in other major gas markets, European gas prices are also shown.

The figure should be interpreted with caution because the prices do not include internal distribution costs or, in the case of kerosene and LPG, freight costs for importing the fuel. The costs of distributing kerosene are lower than for LPG because kerosene does not incur bottling and other cylinder management costs. On the other hand, the efficiency of kerosene stoves tends to be lower than that of LPG stoves, requiring more kerosene per unit of cooking and hence pushing up the effective price of kerosene. The upfront cost of laying down a pipeline network for natural gas is very large, and that cost is also not reflected in the figure.

These limitations notwithstanding, Figure 59.1 illustrates the trend observed globally. Where natural gas is available, it is the fuel of choice for households: excluding the connection cost, natural gas is cheaper than LPG, in addition to being potentially cleaner and safer. The Saudi Aramco contract prices for propane and butane were higher than the landed cost of natural gas in Japan, the highest of three major natural gas markets (Asia, Europe, and North America), in 104 out of 110 months between January 2003 and February 2012. The difference between LPG and natural gas prices is markedly greater in the United States, where spot prices of natural gas in 2011 averaged only US$4.00 per million Btu. In contrast, the Saudi Aramco contract prices of propane and butane were lower than the FOB Singapore price of kerosene in 96 months during the same period. This price difference has been even greater in North America where propane prices have been lower than the Saudi Aramco contract prices in recent years. Where there are no price subsidies in favor of kerosene, households prefer LPG to kerosene because LPG is generally cheaper.

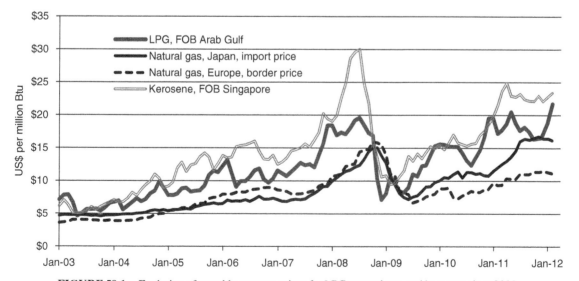

FIGURE 59.1 Evolution of monthly average prices for LPG, natural gas, and kerosene since 2003. *Sources:* Reuters for propane and butane, World Bank's Development Research Group for natural gas, various issues of the Platts Price Oilgram for kerosene, and World Bank staff calculations. *Note:* The price of LPG is the average of Saudi Aramco contract prices for propane and butane.

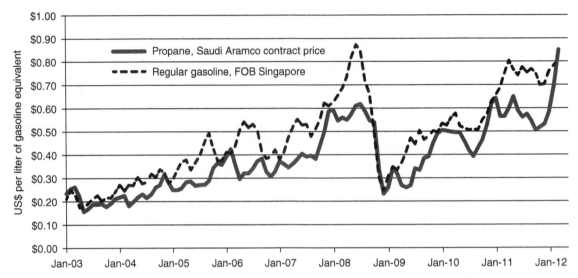

FIGURE 59.2   Monthly average prices for autogas and gasoline in Asia since 2003. *Sources:* Reuters for propane, Platts Price Oilgram for unleaded gasoline with a research octane number of 92, and World Bank staff calculations.

For autogas, the relevant comparison is that between the price of gasoline and the gasoline-equivalent price of propane quoted by Saudi Aramco.[1] These prices are shown in Figure 59.2. Between January 2003 and February 2012, propane was cheaper than regular gasoline in 102 of the 110 months. Although converting a gasoline car to run on LPG incurs an initial upfront cost, this additional expenditure can be quickly recovered if LPG is taxed much less, which is the case in most large autogas markets. For example, despite a rising excise tax on LPG in recent years, the excise tax on regular gasoline in the Republic of Korea during the second quarter of 2011 was 74 percent of the net-of-tax fuel price, but the corresponding figure for autogas was 29 percent. The retail price of autogas was 24 percent lower per liter of gasoline equivalent. In earlier years, this difference in taxation was even larger; taking 2003 as an illustration, despite nearly comparable FOB prices of gasoline and LPG, the retail price of autogas was 40 percent lower per liter of gasoline equivalent (IEA 2011b).

59.3   FACTORS AFFECTING HOUSEHOLD CHOICE OF LPG

Many governments conduct national household surveys to collect data on household characteristics and all expenditures, including those on energy sources. The data are used to assess the level of poverty, employment patterns, education attained, and expenditure patterns. Specific questions vary from survey to survey, but those with detailed questions about energy ask primary sources of energy for cooking and lighting, expenditures spent and quantities purchased of each form of energy, and possession of appliances and equipment using energy.

A recent econometric analysis of data from household surveys in Albania, Brazil, Guatemala, India, Indonesia, Kenya, Mexico, Peru, Pakistan, and Sri Lank examined the determinants of household choice of LPG (Kojima, Bacon, and Zhou 2011). The surveys, carried out between 2004 and 2009, contained sufficiently detailed data on energy for this purpose. Annual per capita expenditure, expressed in 2005 U.S. dollars at purchasing power parity, during the data collection period ranged from US$704 in India to US$4566 in Brazil. More than half of all households lived in urban areas in five out of ten countries. In each of these five countries, annual per capita expenditure exceeded US$3000.

For descriptive statistics, households were divided into quintiles based on per capita expenditure, with each quintile containing the same number of people. The quintiles were further divided into rural and urban. Because poverty tends to be concentrated in rural areas, in most countries, the rural population made up more than half of the total in the bottom (i.e., poorest) quintile, and conversely the urban population made up more than half in the top quintile.

59.3.1   Use of LPG in Survey Countries

Eight surveys out of ten selected for econometric analysis asked households to name their primary cooking fuel. The results for LPG and biomass are presented in Figure 59.3 by expenditure quintile. Biomass dominated the bottom four

---

[1] Both LPG and gasoline run on vehicles using spark ignition, making fuel switching from gasoline to LPG more straightforward than that from diesel (which is used for vehicles with compression ignition) to LPG.

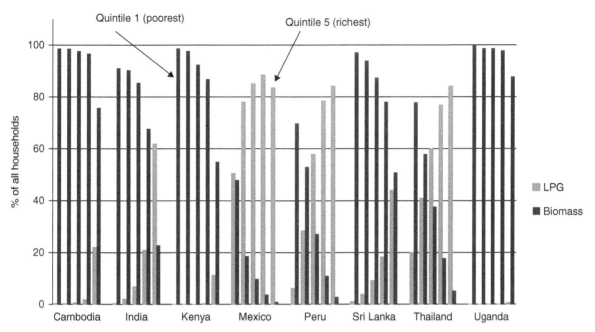

FIGURE 59.3   LPG and biomass as the primary cooking fuel as a function of expenditure quintile.
Sources: Bacon, Bhattacharya, and Kojima 2010; Kojima, Bacon, and Zhou (2011)

quintiles in Cambodia, India, Kenya, Sri Lanka, and Uganda, and was more important than LPG even in the top quintile in Cambodia, Kenya, Sri Lanka, and Uganda. These patterns are in sharp contrast to the dominance of LPG for all quintiles in Mexico and for the top three quintiles in Peru and Thailand.

Table 59.1 provides summary statistics of LPG use by location. With the exception of Mexico, the users in the table are inclusive of all households that reported purchasing or consuming LPG, and not only those that cited LPG as their primary cooking fuel. A greater share of households use LPG in urban areas than in rural, except in Pakistan where two-thirds of urban households had access to natural gas; among LPG users, however, urban households in Pakistan consumed 60 percent more than rural households. More generally, urban user-households consumed markedly more LPG than rural ones in India, Mexico, Pakistan, Peru, and Sri Lanka. Kilograms (kg) of LPG consumed per user-household were markedly higher in Mexico than in any other country in both rural and urban areas. In Albania, Pakistan, and Sri Lanka, urban user-households devoted a greater share

TABLE 59.1   Summary statistics on LPG use.

Country	Share of households using LPG (%)		Quantity per user-household (kg/month)		Expenditure share of user-households (%)	
	Rural	Urban	Rural	Urban	Rural	Urban
Albania[a]	71	72	11	12	2.8	3.3
Brazil[a]	81	89	8.0	8.4	3.5	2.5
Guatemala	24	74	11	12	2.7	2.5
India	12	59	9.1	12	4.8	4.3
Indonesia	2.2	13	11	12	3.4	2.3
Kenya	1.2	13	10	9.6	4.5	2.7
Mexico[a,b]	54	87	25	30	5.6	4.4
Pakistan	8.4	7.0	6.8	11	3.1	4.6
Peru[a]	21	85	7.9	9.6	2.7	2.0
Sri Lanka	21	60	6.7	9.3	1.8	2.3

Source: Kojima, Bacon, and Zhou (2011).
[a]Quantities were not reported and instead estimated by dividing expenditures on LPG by national average prices.
[b]The shares of households using LPG are based on the statistics for those using LPG as the primary cooking fueland exclude those using LPG but not as the primary fuel.

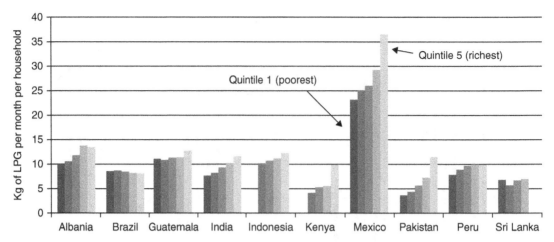

FIGURE 59.4   Monthly household LPG consumption by user-households by quintile. *Source:* World Bank staff calculations using data from national household expenditure surveys in Kojima, Bacon, and Zhou (2011). *Note:* Consumption in Albania, Brazil, Mexico, and Peru is estimated from expenditures and national average LPG prices. The bottom quintile in Indonesia is not recorded because there were only two LPG-using households in the sample.

of their total expenditure to LPG than rural households. The share of total household expenditure spent on LPG varied from 1.8 percent in rural Sri Lanka to 5.6 percent in rural Mexico, and from 2.0 percent in urban Peru to 4.6 percent in urban Pakistan.

Monthly consumption of LPG in kg averaged across all user-households is shown by quintile in Figure 59.4. Consumption followed the expected pattern of rising with quintile in most countries, and particularly in Albania, India, Kenya, Mexico, and Pakistan.

59.3.2 Affordability of LPG

One measure of affordability is the percentage of total household expenditure needed to purchase 1 kg of LPG. The results are shown in Table 59.2. The table also shows FOB prices of LPG averaged during the survey period taken from a geographically relevant market for each country. Comparison of retail and FOB prices indicates that retail LPG prices were subsidized in India, Indonesia, and Mexico.

The combination of subsidies and relatively high income made LPG most "affordable" in Mexico. Whether Brazil or

TABLE 59.2   Affordability of LPG.

Country	Share of households using LPG (%)	Quantity per user-household (kg/month)	Retail price of LPG (US$/kg)[a]	FOB price of LPG (US$/kg)	1 kg of LPG / total household expenditure (%)
Albania[b]	71	12	1.51	0.88	0.2
Brazil[b]	88	8.4	1.39	0.59	0.2
Guatemala	51	12	1.11	0.54	0.2
India	24	11	0.47	0.39	0.6
Indonesia	7.1	12	0.41	0.39	0.4
Kenya	4.0	9.7	1.63	0.48	1.0
Mexico[b,c]	80	29	0.86	0.80	0.1
Pakistan	7.9	8.1	0.71	0.39	0.5
Peru[b]	63	9.4	1.08	0.42	0.2
Sri Lanka	26	7.6	0.75	0.52	0.4

Source: Kojima, Bacon, and Zhou (2011).

Notes: FOB prices are taken from Mont Belvieu propane prices in Texas, United States, for Guatemala, Brazil, Mexico, and Peru; the average of propane and butane prices from the North Sea for Albania; and the average of propane and butane Saudi Aramco contract prices for India, Indonesia, Kenya, Pakistan, and Sri Lanka.

[a]Local retail prices of LPG converted to US dollars using the market exchange rate at the time of the survey.

[b]Quantities are estimated using national average prices.

[c]The share of households using LPG is based on the statistics for those using LPG as the primary cooking fuel and excludes those using LPG but not as the primary fuel.

Mexico had the highest share of households using LPG could not be determined from the survey analysis, because the way the question was asked about expenditures on LPG in Mexico gave responses that missed a large number of households using LPG. The low ratio of price to income seemed to increase the average quantity purchased in Mexico, which was nearly triple the average of the remaining nine countries. The affordability of LPG in Albania, Brazil, Guatemala, and Peru was comparable, but the selection rate ranged from 51 percent to 88 percent, suggesting the potential to increase LPG use further in the countries with low selection rates and help reduce reliance on solid fuels with their attendant problems.

The data collected on LPG consumption in these household surveys enable formulation of simplifying assumptions to calculate a measure of affordability in a larger number of developing countries. First, 3 kg of LPG per person per month is assumed to be a representative amount needed for regular use of LPG in a developing country. This is a notional figure in that there are economies of scale in LPG use and the amount consumed is not directly proportional to the household size, and there are also climatic and other factors that affect LPG use. For example, if LPG is used as the primary fuel for heating, much more than 3 kg per person per month would be required in a cold climate. These shortcomings notwithstanding, this assumption implies 12 kg per month for a family of four, which is broadly in line with the findings reported for most countries in Table 59.1. The analysis also takes data for the household share of gross domestic product (GDP)—for which the most recent data available are from 2010—in each country as a proxy for household income. Retail prices of bottled LPG in December 2010 are multiplied by 3 to arrive at monthly expenditure on LPG per person. The share of household income per capita spent on LPG calculated in this way provides a measure of affordability.

The results are shown in Figure 59.5. This affordability metric varies from nearly 20 percent in Tanzania to 0.1 percent in the Arabic Republic of Egypt, *Republica Bolivariana de Venezuela*, and Ecuador, where LPG is heavily subsidized. The figure also compares the above measure of affordability to residential consumption of LPG available in the International Energy Agency's database. The last year for which the data exist is 2009. For this purpose, countries known for extensive use of natural gas by urban households (e.g., Pakistan) or electricity for cooking and heating

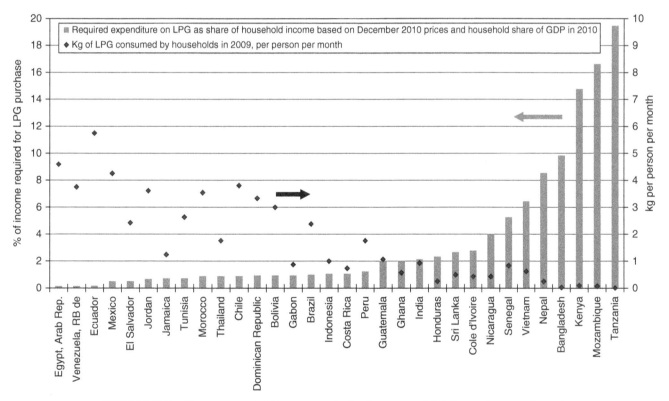

FIGURE 59.5   Share of household income required to be spent on LPG for regular use and actual consumption in 2009. *Sources:* Population, GDP, and household share of GDP from the World Bank's World Development Indicators (World World Bank 2011), LPG consumption from IEA (2011c), and LPG prices from government sources, LPG marketing companies, and local media reports.

(e.g., Albania, South Africa) were not considered. While there is some scatter in household consumption of LPG in countries where LPG was highly affordable, LPG use was minimal in countries where LPG was not affordable by this measure. Affordability and consumption were inversely correlated, with a correlation coefficient of −0.57.

59.3.3 Quantitative Analysis of Factors Affecting Household Use of LPG

The econometric analysis used a two-stage Heckman-type model (Kojima, Bacon, and Zhou 2011) to probe factors potentially influencing households' decision to use LPG (selection) and, for the households that have decided to use it, how much to use (consumption). The surveys in Guatemala, India, Indonesia, Kenya, Pakistan, and Sri Lanka had data to enable full modeling, including quantities of fuels purchased or consumed (from which prices faced by individual households were calculated). The surveys in Albania, Brazil, Mexico, and Peru—where LPG use was more widespread and data were more recent—lacked data on quantities. In these four countries, national average prices were used to estimate quantities of LPG purchased. Although the impact of fuel prices on LPG demand could not be examined in the latter four countries, they were nevertheless investigated to see if their other findings generally supported those from the first six countries.

The results for the six countries with household-level fuel prices show that, in every country, LPG *selection* increased with household expenditure and the highest levels of education attained by female and male household members, and LPG *consumption* increased with household expenditure and decreasing price of LPG. Variables serving as proxies for the level of infrastructure development (electricity connection, urban rather than rural residence) increased selection (except urban residence in Pakistan where natural gas was widely available) and consumption, while engagement in agriculture (broadly associated with biomass availability) reduced selection and consumption. Rising prices of kerosene and firewood, which compete with LPG, also increased LPG selection.

As expected, income and LPG prices had by far the two largest effects. Education also had significant effects on selection. Having a male household member with the highest level of education available in the country would have the same effect as increasing household expenditure by 30 to 80 percent relative to a household with average expenditure and education levels. Women's education had even larger effects; a female household member with the highest level of education available would have the same effect as increasing household expenditure by at least 50 percent.

The gender of the head of household was significant only in Sri Lanka for selection and in India for consumption, and the coefficient was positive in both cases. This may suggest that male-headed households tend to have more assets and better access to credit and employment, providing them with more cash to pay for LPG. Where the age of the head of household was significant, the coefficient was positive. One interpretation is that households headed by older members tend to be more established and have more resources available for LPG purchase.

Because firewood is typically much cheaper than LPG, moving households away from firewood to LPG would be expected to require a very large increase in the prices of firewood. Indeed, for consumption, the price elasticities for LPG were about an order of magnitude larger than those for firewood, suggesting that firewood prices would indeed have to rise steeply before households would start consuming more LPG. In contrast, the magnitudes of elasticity for kerosene were comparable to those for LPG.

The study calculated unconditional marginal effects, which combine the marginal effect on demand among users and the increased probability of use. Income elasticities varied between 0.3 and 0.7. Own price elasticities were close to 1 in India, Pakistan, and Sri Lanka. The price elasticity for kerosene was high in India and Pakistan, and household size generally had a negative effect on LPG consumption.

The results of modeling of the data from the four countries that did not report quantities of fuels consumed were broadly aligned with the foregoing results. The magnitudes of the coefficients for household expenditures (significant in every equation) and agricultural engagement were broadly comparable between the two groups of countries for both selection and consumption. Those for education, the age of the head of household, indigenous head of household (less likely to select LPG), and electricity connection were comparable for selection. The coefficient for the dummy for a male head of household was significant in three equations, and, as with Sri Lanka and India, the sign was again positive in each case.

It has been argued by some that the gender of the head of household—who often controls decisions about expenditures—may influence LPG selection and consumption. Other things being equal, female-headed households might be more willing to pay for LPG, and those using it would purchase more LPG, because LPG use would benefit women much more than men. This study found that, once education was taken into account, the dummy for a male head of household was statistically significant in only five out of 21 equations, and the sign was positive in every case. As mentioned earlier, these results do not necessarily suggest that women are less willing to switch to LPG, but may simply indicate that, at the same total household spending as defined in this study, female-headed households suffer from economic disadvantages—such as having greater difficulties accessing credit, not having title to land, and having fewer employment opportunities—thereby constraining their ability to earn cash and hence being less able to purchase LPG.

59.4   LPG MARKETS IN DEVELOPING COUNTRIES

The previous section points to the importance of relative fuel prices and their ratios to income, as well as the level of awareness about the costs and benefits of LPG (with education levels serving as a proxy), in determining household use of LPG. The way LPG markets are structured and managed affects the availability and prices of LPG, the amount of information available to consumers, the quality of service including reliability of refills and safety, and the extent of commercial malpractice harming consumers and other LPG marketers, all factors that influence the willingness and ability of households to use LPG.

59.4.1   Pricing Policy and Retail Prices of LPG

All governments set taxes for goods, and LPG is no exception. The number of taxes can vary from one—general consumption tax such as value-added tax, although setting it to zero is equivalent to having no tax—to several, such as import duty and an excise tax in addition to general consumption tax. By lowering taxes, governments can decrease retail prices. For example, some countries have reduced taxes repeatedly in the wake of rising world prices, and Kenya has even eliminated them altogether by setting even the value added tax to zero. To the extent that a general consumption tax should be levied on all goods, its reduction or elimination constitutes a form of subsidy.

Governments may levy additional charges on LPG, some unrelated to the LPG sector. Examples include a (small) fee for monitoring and enforcement and a contribution to a social fund unrelated to LPG, such as an unemployment benefit fund.

All governments intervene if they believe that prices are kept artificially high through collusion, although demonstrating collusion is difficult. These suspicions increase in times of high and volatile oil prices. Some industry observers have argued that the relationship between world prices and retail prices appears to exhibit an asymmetric, "rockets and feathers," pattern—when world prices rise, retail prices follow quickly (like a rocket), but when world prices fall, retail prices go down more slowly (like a father). Antitrust laws make it illegal for competitors to engage in any conduct that unreasonably restrains competition, such as colluding to fix prices. But governments are rarely able to demonstrate explicit collusion, because firms are careful not to leave evidence of explicit collusion, and tacit collusion is by nature difficult to prove.

Over and above antitrust concerns, many governments in developing countries intervene to varying degrees to keep retail prices lower than otherwise. A number of governments set price ceilings, especially if the level of competition is considered inadequate. The Ministry of Economy in Mexico, for example, gazettes maximum consumer prices in 145 distribution zones every month on the grounds that the Federal Competition Commission has not yet deemed effective competition to exist in the market, and the Minister of Energy in South Africa similarly issues LPG regulations every month with maximum retail prices for residential consumers in 54 zonal areas. Although these procedures in the two countries appear similar, Mexico subsidizes LPG prices whereas South Africa does not. Price ceilings are typically, but not always, set for retail prices. A counter-example is Pakistan, where the government sets maximum producer prices for LPG from gas fields and domestic refineries, while leaving the prices of imported LPG deregulated. The government of El Salvador sets price ceilings at four points along the supply chain every month. If price levels or ceilings are set at a regular interval, such as on a weekly or monthly basis, price setting usually follows a formula, which may include elements of a subsidy. The size of the subsidy may not be obvious from looking at the pricing formula. For example, both Ghana (NPA 2012) and Thailand (EPPO 2012) publish the price structures of various fuels regularly, but the subsidy at the ex-refinery level, which is large in both countries, is not indicated.

Some countries—Cameroon, Ghana, Morocco, Peru, and Thailand among them—have used a price stabilization fund to keep LPG prices low. In theory, price stabilization funds are intended to smooth prices and be self-financing. Retail prices are meant to be kept relatively high in times of low world prices and the difference is transferred into the fund, and conversely they are kept relatively low in times of high world prices by transferring money out of the fund. But even when prices fluctuate around a reasonably constant mean, the nature of world price movement is such that the fund's cumulative balance can be negative for years at a time, posing a considerable challenge to the financing of the fund. This problem is exacerbated when world prices are on a generally increasing trend, as in recent years (Bacon and Kojima 2008). As a result, and particularly in recent years, these funds have required (large) budgetary transfers.

A number of governments set prices instead of price ceilings, often equalizing them throughout the country. Pan-territorial pricing always involves geographical cross-subsidization because transport costs differ: consumers close to supply sources cross-subsidize those who are located far away from import terminals, gas processing plants, or refineries supplying LPG. Some governments, such as those of the Arab Republic of Egypt and *Republica Bolivariana de Venezuela,* have frozen pan-territorial prices at very low levels for years, making LPG prices in these countries among the lowest in the world.

Figure 59.6 shows retail prices of LPG for residential use (as opposed to the prices of autogas) in January 2012. Aside from the United States, the remaining 46 markets are all in developing countries. Of the top ten most expensive countries, seven are in Sub-Saharan Africa, and only one is

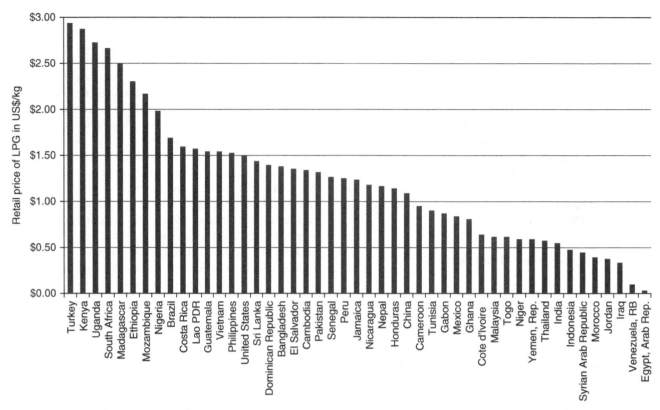

FIGURE 59.6   End-user prices of bottled LPG in January 2012. *Sources:* Government Web sites, oil company Web sites, and local news articles.

landlocked. The prices across the 46 markets vary by two orders of magnitude. A large number of countries provide universal price subsidies, which are highly regressive, as discussed in the next section.

Large-scale use of autogas presents an additional challenge. Autogas is used in vehicles with spark-ignition engines, which are virtually universally fueled by gasoline otherwise. Unless the use of autogas is legally required, it must be cheaper than gasoline before anyone would consider using it, because the cost of converting the gasoline-fueled vehicle to run on autogas or the higher purchase cost of a vehicle made by original equipment manufacturers to run on autogas needs to be recovered through lower fuel costs. In addition, a widespread network of filling stations typically does not exist for autogas and hence, absent a much lower fuel price, it would not be worth the inconvenience of switching to autogas and being constrained in when and where to refuel. These challenges are even greater for compression-ignition engines (fueled by diesel), for which conversion from compression ignition to spark ignition is not nearly as straightforward. Diesel engine technology is robust, and, because the price of diesel has an economy-wide effect, diesel tends to be taxed less than gasoline in many developing countries, resulting in lower end-user prices. For these reasons, substitution of LPG for diesel is much rarer.

Where the price difference is sufficiently large, taxis are often the first to switch—the large kilometers per year taxis are driven reduce the payback period, and refueling is not as significant a challenge as for other vehicle types because taxis usually operate within a confined area and a handful of filling stations can serve them. Once the autogas market surpasses a critical size, it is possible for the number of filling stations to start rising rapidly, making autogas increasingly convenient and attractive.

As Figure 59.2 shows, FOB autogas prices have been generally lower than gasoline prices in recent years. This price difference in favor of autogas is amplified when, as is typically the case, gasoline is taxed more than LPG. Particularly in developing countries, gasoline is often the most heavily taxed petroleum product, because much of its consumption is by private car owners and taxis. Private car owners are among the better-off, as are users of taxis. This makes gasoline taxation generally progressive.

Regarding LPG and gasoline as substitutes and taxing them equally to maintain the progressive taxation of gasoline would reduce the attractiveness of autogas and make LPG for household use expensive. Introducing different tax rates for autogas and bottled LPG for household use to make the latter more affordable would lead to illegal diversion of LPG intended for residential consumers to the automotive sector, often accompanied by illegal and unsafe installation of household cylinders in vehicles as fuel tanks. In terms of the willingness and ability to pay for LPG, taxi and private car drivers are more willing and capable on average than households, especially when drivers are comparing autogas with more expensive gasoline, while households are comparing LPG with cheaper solid-fuel alternatives. Countries with a large autogas market tax autogas at a lower rate than gasoline.

The differences in fuel prices between autogas and gasoline and in the ability to pay have led to a rapid rise in fuel switching from gasoline to autogas in some developing countries. On the one hand, market expansion should enable greater scale economy and more competition, both of which can lower prices. On the other hand, the generally lower tax rates for autogas reduce government revenue. Especially in low-income countries, petroleum product taxation is an important revenue source, because collecting fuel taxes is relatively straightforward and the consumption of fuels as a group is highly income elastic, ensuring buoyant revenue as income rises (Bacon 2001). The loss of revenue would be particularly damaging if autogas is not only taxed less but is actually subsidized.

Where there is suppressed demand due to lack of adequate LPG supplies, effectively rationed subsidies, or both, residential users tend to lose to autogas users. The expansion of the autogas market as well as large price subsidies have been blamed in recent months for acute LPG shortages in Ghana, prompting commercial LPG vehicle drivers to ask the government for subsidy removal so as to ensure reliable supply (All Africa 2011a). Because autogas and LPG for other uses are equally subsidized, the end result is that a sizable fraction of LPG is benefitting private and commercial autogas users, making the subsidy highly regressive. Against the backdrop of rising retail prices of gasoline and static and heavily subsidized LPG prices, the number of autogas filling stations in Thailand is said to have increased dramatically, from 140 recorded in December 2010 to 1030 by mid-2011 *(Platts Oilgram News 2011)*.

In Turkey, which does not subsidize LPG but taxes LPG less than gasoline, autogas consumption surpassed gasoline consumption in 2009, and there are more than 8500 filling stations today, the largest number of autogas stations in the world (Polat 2010). Although lower than that for gasoline, the tax component of LPG in Turkey is still very high. According to information provided by Tupras, taxes and fees amounted to US$1.12 per kg in October 2011, corresponding to 37.5 percent of the retail price (Matthews and Zeissig 2011).

An oil price stabilization fund has historically been used to subsidize LPG in Thailand and Peru. The budgetary transfers for the price stabilization fund in Peru in 2010 alone amounted to S/. 1195 million (US$420 million) and rose to S/. 2479 million (US$900 million) in 2011, although a breakdown by fuel type is not available (BCRP 2012). LPG prices in Thailand have been subsidized for decades. In February 2012, the government's Energy Policy and Planning Office reported that the unit subsidy was 20 baht (US$0.65) per kg, exceeding the retail price for residential consumers of 18.3 baht (US$0.60) per kg (Thai Financial Post 2012). The new government that came into power in July 2011 began reducing subsidies for autogas in 2012 but is holding the price of residential LPG constant until the end of 2012. In Ghana, the ex-refinery price was subsidized by an estimated US$0.42/kg in July 2011, and the cross-subsidy from gasoline provided an additional subsidy of US$0.12/kg, totaling US$0.54/kg (Matthews and Zeissig 2011). A government official reported in August 2011 that the government was spending US$10 million a month on LPG subsidies (All Africa 2011b). Mexico's national oil company, Pemex, posts a graph of estimated annual subsidies since 2003 on its Web site. The under-recoveries of costs by Pemex amounted to US$2.4 billion in 2008, US$0.5 billion in 2009, US$1.9 billion in 2010, and US$1.5 billion during the first seven months of 2011 (PGPB 2011). The most subsidized LPG on a unit basis is in Egypt, where the government's primary strategy for reducing the subsidy bill is to shift households to natural gas rather than to raise the retail price of 2.5 Egyptian pounds for a 12.5 kg cylinder, or a mere US$0.03 cents per kg, this despite the fact that about half of LPG is imported. In February 2012, noting that the official price of LPG sold in 12.5-kg cylinders has not changed since 1991, the petroleum minister reported to the parliament that the annual cost of the LPG subsidy was about 20 billion Egyptian pounds, or more than US$3 billion (Trend News Agency 2012).

59.4.2   Distributional Impact of Universal Price Subsidies

Analysis of the quantities of LPG consumed from national household surveys enables examination of distributional effects of hypothetical universal price subsidies. This illustrative exercise takes data from household surveys in ten developing countries and asks what would have been the impact if, at the time of the survey, the government in each country was providing a universal price subsidy of US$0.20 per kg of LPG. For the purpose of this illustration, the poor are defined as those in the bottom two income quintiles in every country, making up 40 percent of the total population (but a smaller percentage of the total number of households, because household size usually declines with increasing income). The exercise focuses only on LPG consumed by households. The distributional impact assessment poses the following questions, the first three of which are subsidy performance indicators discussed by Komives, Foster, Halpern, and Wodon 2005:

- How well does the subsidy target the poor **(benefit targeting)?** More specifically, what percentage of the total subsidy reaches the poor as opposed to the non-poor?

If more than 40 percent reaches the poor, the subsidy is progressive; conversely, if less that 40 percent reaches the poor, the subsidy is regressive. A benefit-targeting indicator of 100 percent would mean that the subsidized good is completely self-targeting. The gap between 40 percent and the actual percentage (**equity gap**) indicates how far the subsidy scheme falls short of (or exceeds) an equal distribution across all income groups. As an example, if the benefit-targeting indicator is 30 percent, then the equity gap is (40–30) ÷ 40 = 25%.

- How inclusive of the poor is the subsidy (**inclusion**)? That is, what percentage of poor people receives this subsidy?
- How material is this subsidy (**materiality**)? Specifically, what is the total subsidy received by the poor, expressed as a percentage of their total household expenditures?
- What is the total bill to the government of this hypothetical universal price subsidy if only LPG sold to households is considered?

The results are shown in Table 59.3. It is striking that, even after including some lower-middle class households who are not poor, in no country is the subsidy neutral or progressive (zero or negative equity gap). The smallest equity gap is 20 percent in Brazil, rising to virtually 100 percent in Kenya where almost all users of LPG were in the top three quintiles. The hypothetical subsidy hardly covers the poor in Guatemala, India, Indonesia, Kenya, Nepal, Pakistan, and Sri Lanka, all with inclusion of 11 percent or lower. These unfavorable trends would worsen once non-residential users of LPG are included. In terms of materiality, Brazil has the highest indicator at 0.5 percent of the total household expenditure, reflecting the relatively high percentage of households in the bottom two quintiles purchasing LPG. These results suggest that a universal price subsidy for LPG, particularly in low-income and lower-middle-income countries, would not be equitable or cost-effective for social protection.

59.4.3 Other Forms of Subsidy

Some governments have chosen targeted assistance. In Brazil, assistance to enable the poor to use LPG takes the form of vouchers for LPG in *Bolsa Familia*, the government's social welfare program. The government of the Dominican Republic abolished the universal price subsidy for LPG in September 2008 and replaced it with a social protection scheme, whereby 800,000 *bonogas* cards were issued to poor families, entitling them to a monthly reduction of RD$228 (about US$6) for LPG purchase. As of mid-2011, about 750000 families were benefitting from this scheme.

Another approach is to help lower the upfront cost. The start-up cost of LPG consists of paying for the cylinder, cylinder regulator, and the hose in the form of a deposit fee or purchase, and purchasing a stove. Recent prices charged for new cylinders include US$44 for a 14.5-kg cylinder Ghana; US$33 for a 12.5-kg cylinder in Sri Lanka; and US$23 including a 14-percent value-added tax for 9-, 12-, and 14-kg cylinders in South Africa. LPG stoves cost comparable amounts, although small camping stoves are cheaper. For example, a stove that fits on top of a 2-kg cylinder in Sri Lanka costs US$17 in 2011.

Lowering the costs of cylinder and stove acquisition for residential consumers would lower the barrier to switching to LPG and could help a segment of middle-class households who are able to pay for LPG but find it challenging to save enough money to pay for the start-up cost. This segment, however, may be narrow. It is worth noting the difference between electricity and LPG in the effectiveness of lowering the start-up cost. In the power sector, the upfront connection cost may be two orders of magnitude larger than the monthly expenditure on electricity, and most households without electricity are incurring cash expenditure for kerosene purchase for lighting. For cooking with LPG, however, the difference between the start-up cost and monthly fuel cost for regular use of LPG is not nearly as large, and many households are spending little or no cash on fuel acquisition. As a result, a large number of households that cannot pay for an

TABLE 59.3 **Distributional performance of hypothetical universal price subsidies.**

Country	Benefit targeting (%)	Equity gap (%)	Inclusion (%)	Materiality (%)	Total hypothetical subsidy (US$) million per year
Albania	24	40	66	0.03	16
Brazil	32	20	89	0.5	757
Guatemala	9.6	76	11	0.1	38
India	1.6	96	2	0.06	1301
Indonesia	1.0	98	1	0.007	105
Kenya	0.07	100	0	0.0005	6
Nepal	7.0	82	7	0.06	22
Pakistan	8.5	79	5	0.04	30
Peru	13	68	28	0.2	96
Sri Lanka	4.9	88	6	0.05	21

Sources: National household surveys and World Bank staff calculations.

electricity connection can pay for a modest amount of electricity consumption by redirecting cash expenditure on kerosene, but many who are unable to pay for an LPG cylinder and a stove are also unable to pay for regular use of LPG (ESMAP 2003b). For this reason, before asking the government to subsidize the acquisition of the first cylinder, stove, or both, the distributional impact and potential effectiveness of such a subsidy would need to be carefully reviewed.

If the option of lowering the start-up cost is to be pursued, one question is who pays for the reduction in the upfront cost: government, other consumers, or the consumers benefiting from the lower upfront payment by spreading it over time. In some countries, LPG marketing companies cover half or even more of the cost of cylinder manufacture. The required amount for a refundable deposit fee is usually about 130 percent of the replacement cost of the cylinder. If marketing companies are offering a discount on cylinders, as opposed to giving the option of an installment plan for cylinder acquisition, then all consumers in effect subsidize new consumers—through slightly higher LPG prices to enable cost recovery or through higher safety risks if companies compromise cylinder maintenance and renewal to compensate for the loss in revenue.

Setting low cylinder deposit fees or purchase prices is not without risks. If dealer incentives or government subsidies lower deposit fees by a sufficiently large margin, the cylinders may be resold on the black market or smuggled out of the country, and similarly with stoves. Customers obtaining cheap or free LPG cylinders in developing countries have also been known to use them for other purposes, such as metal beating or as flower pots. If LPG marketers are providing the subsidy, they suffer a loss because the business that these subsidies are intended to generate does not materialize. These losses will eventually have to be covered by higher prices elsewhere, and also weaken the incentive to implement full and proper maintenance procedures.

59.4.4 Cylinder Management

A factor that has significant effects on supply costs, the quality of service, and safety is cylinder management, which is unique to LPG.

59.4.4.1 Cylinder Size. The need to keep LPG under pressure and distribute it in cylinders adds significantly to the cost of distributing it safely. There are large economies of scale in cylinder management, and scale economy has two distinct implications.

First, absent subsidies, buying LPG in large cylinders makes it more affordable for an individual household because the cost per kg decreases with increasing cylinder size. Not only is making one 15-kg cylinder cheaper than making five 3-kg cylinders, the costs of cylinder inspection, maintenance, and refilling also rise with an increasing number of cylinders. A large cylinder is also more convenient in that that it lasts longer, reducing the frequency of refills. But requiring a large sum of cash upfront for LPG purchase— for example, at US$1.50 per kg, it would cost US$22.50 to refill a 15-kg cylinder—would make LPG accessible only to the better-off in many developing countries, because LPG is sold only in discrete quantities matching cylinder sizes available on the market. In contrast, kerosene can be purchased in any quantity and the unit price of kerosene is typically the same whether a household buys only half a liter or 20 liters, because there is no equivalent of cylinder management with attendant scale economy for kerosene.

Second, to develop a commercially viable LPG market, a relatively high concentration of regular users of LPG—that is, a high population density with a good share of households willing and able to buy 10 kg or more a month—is preferably required. A high volume turnover may enable LPG bottling plants to be located close to consumption centers, lowering the cost of LPG distribution. Where consumers are spread out over a large area, they tend to be poor and consume little. The cost of transporting numerous cylinders over long distances and delivering only a few cylinders at a time to several villages is very high, and yet these are the very consumers who are least capable of paying high prices and for whom much cheaper alternatives exist.

Small cylinders are easy to carry around so that home delivery by the retailer is not essential, and the cost of each LPG refill is relatively small. For this reason, small cylinders have been suggested as one way of making LPG affordable to the poor. Aside from lower costs per refill, cylinder deposit fees or purchase costs are also lower. LPG marketing companies in some developing countries mount burners on top of small cylinders and market them as a package. But, absent cross-subsidy, the price per kg would be much higher, and the number of cylinders in circulation would be correspondingly large, posing a challenge to enforcement of safety standards. And if LPG delivery is irregular or unreliable—as in many rural areas, because there is not enough demand to justify sending a delivery truck frequently—then the need to refill cylinders frequently can mean that a household must accept running out of LPG for days at a time. Fuel outage can be avoided by keeping two or more filled cylinders, but doing so adds to the initial start-up cost.

International experience with small LPG cylinders has been mixed. In some countries, small cylinders for household use have been tried by the industry in the past but eventually abandoned. For example, the LPG industry in Turkey experimented with 2, 3, 5, 6, and 12 kg, and eventually chose to retain only 2 and 12 kg, with 2 kg reserved primarily for camping.

Use of small cylinders is widespread only when LPG sold in small cylinders is cross-subsidized and where there is a long history of subsidizing LPG. In Senegal, 95 percent of LPG for residential consumers is sold in 2.7- and 6-kg cylinders, which have historically been heavily subsidized and, until recently, by far the lowest price per kg was offered for the smallest cylinder. In Morocco, there are more 3-kg cylinders (again heavily subsidized) than any other size: 20 million 3-kg cylinders in circulation in 2009, followed by 13 million 12-kg cylinders, with the same unit price regardless of the cylinder size. Small cylinders are available in most other markets, although they are used primarily or solely for camping and not for regular household use. In most countries, apart from LPG for camping, the most commonly used cylinders are in the range 10–15 kg even where smaller-size cylinders are available (Matthews and Zeissig 2011). Still, especially with rising international prices of LPG, marketers in some countries are giving renewed consideration to small cylinders.

Households in developed countries, and those in some developing countries, use much larger cylinders or tanks. In Mexico, the trend has shifted from cylinders to stationary tanks, with only half of households purchase LPG in cylinders. According to the energy ministry, 91 percent of LPG cylinders in circulation in June 2009 were split evenly between 20 and 30 kg. The remaining LPG-user-households were supplied by very large stationary tanks on their premises (filled by LPG delivery trucks) or by local pipeline networks fed from similar stationary tanks. According to the statistics collected by the energy ministry, monthly household consumption averaged 23 kg in 2009 (SENER 2010). The cylinder market in Brazil is also migrating to LPG tanks, as many new residential buildings are built with LPG tanks, similar to commercial clients (Petrobras 2011).

59.4.4.2 Cylinder Ownership, Refilling, and Safety.
Whether cylinder ownership can be clearly traced; who assumes the responsibility *in practice* for cylinder maintenance, repair, and replacement; and how customers obtain cylinder refills are all inter-related questions. The responses affect LPG prices as well as the safety of LPG cylinder use.

The cost of cylinder handling depends on the number of cylinders in circulation. In general, for a customer not to run out of LPG, there needs to be almost one cylinder in transit (transport, storage, bottling) for every cylinder in use by the household, although the number of extra cylinders in the system may fall to as low as 0.5 in an extremely efficient system. More specifically, a customer with two cylinders typically requires 2.8–3.0 cylinders in total in the supply chain on average, a customer with one cylinder a total of 1.8–2.0 cylinders, and so on (Matthews and Zeissig 2011). These extra cylinders add to the cost of supply. In practice, even with an extra cylinder in transit, a customer with only one cylinder at home is likely to run out of LPG for half an hour or longer while obtaining a refill when the cylinder runs empty.

Fiberglass composite cylinders, a relatively recent innovation, are much lighter, and hence transporting them is easier,

which is an important consideration if customers assume the responsibility for home delivery, as in Ghana. They are also transparent, so that it is easy to check how much LPG is left. Being able to see the amount of LPG in the cylinder can help curb short-selling, and makes it easier to estimate when the cylinder will run out. However, these cylinders are much more expensive. For example, even in a country as wealthy as Saudi Arabia, they were introduced only in July 2010 at more than double the price of regular cylinders: 350 riyals (US$93) compared to 150 riyals (US$40) (Arab News 2010). As a result, the market penetration of composite cylinders in developing countries has been limited or non-existent.

Unlike natural gas, which rises when leaked, LPG is denser than air and hence sinks to the ground upon release. This means a spark can trigger an explosion near ground level where people are located. Maintaining the physical integrity of LPG cylinders is therefore imperative for public safety. Cylinder ownership influences how an LPG marketer maintains cylinders and whom to hold accountable if there is an accident. If ownership is unclear, there could be far less incentive for individual marketing companies to spend time, money, and effort inspecting cylinders carefully and repairing or scrapping faulty ones. If empty cylinders are exchanged for filled ones, a given cylinder goes through the hands of numerous customers. Where customers go to different marketing companies to refill, no one company may want to assume full responsibility.

There are two ways by which customers obtain cylinder refills:

1. **Centralized system of filling,** whereby empty company-owned or customer-owned cylinders are exchanged for filled ones. If cylinders are company-owned, the company is responsible for ensuring the safety of cylinders. If customer-owned, the customer does not retain the same cylinder. If cross-filling is prohibited, the brand company from which the customer purchases LPG is responsible for cylinder safety. If there is interchangeability of cylinders across companies, incentives for any given company to repair and replace cylinders are considerably weakened without a way of enforcing rigorous maintenance and replacement across all companies.

2. **Decentralized, bulk-supplied system with mini-filling plants,** whereby customers take their own cylinders to mini-filling plants. Each cylinder remains in the possession of the owner through its entire life. This modality poses serious safety risks because the only time LPG suppliers come in contact with cylinders is when they are being refilled. Cylinder safety depends critically on the diligence of plant operators in inspecting each cylinder and refusing to refill those that are defective.

In the centralized system, which is the dominant mode in most countries, if there are different colors, markings, or both to identify cylinders belonging to different companies and repainting or filling cylinders belonging to others is legally prohibited (except where there are mutual arrangements), the incentives for maintenance and replacement are stronger because accidents can be immediately traced to the companies responsible. South Africa and Turkey have clear rules about ownership and cross-filling (prohibited except where mutual agreements exist), which are effectively enforced. In many countries, however, there is a significant failure in the enforcement of cross-filling prohibition. Absent enforcement, there is no guarantee that repaired or replaced cylinders will not be diverted to other companies. The end result is poor maintenance and physical deterioration of the cylinder stock. The key is to set up a system where companies can be assured through effective enforcement that spending money on maintenance and replacement of their cylinder stock will not end up benefiting free-riders.

Globally, the decentralized system is rare. Ghana is an exception. By virtually eliminating commercial transport of cylinders, this system can substantially slash costs for LPG marketing companies, and conversely increase the indirect costs for consumers who are entirely responsible for cylinder transport. In Canada, where both modalities exist, the retail prices in the decentralized system are much lower than in the centralized system (Matthews and Zeissig 2011). However, this approach tends to compromise safety. In Ghana, basic cylinder maintenance—checking for leaks, replacing valves, and repairing leaking cylinder heads—is carried out at some filling plants, but operators often do not check for recertification dates, carry out prescribed visual inspection of cylinder condition, or reject cylinders that are due for testing or that do not pass a visual inspection.

In fact, in both cylinder-filling modalities, inadequate cylinder inspection and testing is all too common in developing countries. In Nepal, LPG dealers reported recently that only 1.2 million of 4 million cylinders in circulation had passed the pressure test and as many as 30 percent of the cylinders in circulation might be at risk of explosion (Kathmandu Post 2012).

59.4.5   Regulatory and Institutional Framework

The nature of legislation governing LPG depends on country circumstances. In every country, to the extent that environment, industrial health and safety, taxation, and legislation governing competition and commercial malpractice apply to all industries, general laws and regulations apply to LPG, even if LPG is not explicitly mentioned. Sector legislation can have more specific details applicable to LPG. In countries with commercially recoverable reserves of hydrocarbons (oil and natural gas), a petroleum law may govern both the upstream and downstream, of which LPG legislation falls

under the latter. Some countries with significant hydrocarbon reserves may have separate upstream and downstream laws. In countries with no commercial reserves of hydrocarbons, downstream petroleum may be covered in an energy law covering both fuels and electricity, or even in a broader law covering multiple sectors including energy.

It would be rare to have a law devoted only to LPG. Brazil has such a law, narrowly focused and mandating transparent weighing of LPG cylinders (Brazil 1995). Although not dedicated only to LPG, Turkish law 5307, entitled "Liquefied Petroleum Gas (LPG) Market Law and the Law Amending the Law on Electricity Market" (Turkey 2005), is among the most comprehensive laws for LPG in developing countries. As a mark of a good law, this law establishes the framework and principles for sector operation, but leaves implementation details largely to regulations. It addresses autogas, cylinder, and bulk LPG suppliers separately. It requires each licensed distributor to hold 20 days of supply in storage; submit annual sales projections and quarterly updates to the Energy Market Regulatory Authority; employ only those who have received training and obtained certificates in their main fields of activities; educate and inform consumers if the license holder is a dealer; and pay a fee to support the activities of the regulatory authority, which is set at 0.1 percent of the net sales up to US$2 million. The law requires LPG marketing companies to handle only cylinders that are marked with the distributor's brand name and logo (except where there are prior agreements with other distributors); inspect cylinders before, during, and after filling for any defects; and not fill cylinders that are rusty, unpainted, or bear marks of damage, including those from fire. These requirements are backed up by effective enforcement.

Technical specifications for LPG and associated equipment are fairly standard and well established globally. Over decades, numerous specialized national and international industry organizations and government entities across the world have developed, and constantly update, hundreds of technical standards and codes of practice for protection of occupational health, industrial safety, and the environment; design, construction, maintenance, and fire protection of buildings, electrical, and other installations; construction materials, transport equipment, machinery and other equipment; appliances such as LPG stoves; and quality-testing methods. Examples include LPG standards and codes issued by the International Organization for Standardization (ISO), the European Union, ASTM International (previously American Society for Testing and Materials), American Petroleum Institute (API), and the (U.S.) National Fire Protection Association (NFPA). Not much can be gained by recreating product specifications and health, safety, and environment standards in each individual developing country. One effective way of taking advantage of the standards that already exist is to adopt them formally by reference, rather than transcribe all or part of the text of an international code in local regulations and standards. By formally adopting international codes, local regulations and standards will be automatically updated when changes are made to the original codes. Codes adopted from outside the country may need to be adapted to local circumstances. An example of a failure to do so is wholesale adoption of Canadian standards by a tropical country without excluding detailed protection against freezing temperatures. The approach adopted by the regulatory agency for LPG in the U.S. state of Texas illustrates good practice in this regard. The agency has formally adopted the most important industry standards together with all other standards referenced therein, but lists the exclusions, changes, and additional requirements and corrections as appropriate for each section of the code.

Countries such as Albania, Ghana, Moldova, Morocco, Pakistan, and Sri Lanka have formally adopted international standards. Many countries, however, have taken, but not formally adopted, the international standards so that their domestic standards, which are the ones that are enforced, are not automatically updated when the standards on which they are based are revised. For example, the hydrocarbon law of the Dominican Republic stipulates that equipment and facilities comply with international technical standards accepted in the petroleum industry if domestic standards do not exist; such wording creates ambiguities.

A well-developed regulatory framework includes requirements for consumer education as well as training and certification of management and personnel of the operators. Consumer education is particularly important for LPG because the cylinders, valves, and cylinder regulators remain under the sole control of the consumers for extended periods in their homes, where accidents can occur. It is critically important that the regulators and the suppliers educate the small distributors and consumers by appropriate means, because what governs safety to a large measure is the knowledge of small operators at the very end of the supply chain and of individual consumers. They need to realize that they are handling a potentially dangerous product. Legislation tends to contain detailed provisions for enforcement and sanctions, but rarely does it require training of the operators and dissemination of information to the public as in, for example, the 1992 Consumer Protection Law of Mexico. Turkey's law 5307 requires licensed distributors to educate LPG vendors and consumers.

Short-selling is always a problem in fuel sale and is particularly problematic with LPG. The tare weight of the cylinder may not be clearly marked and, even if it is, its accuracy is difficult to verify; the customer may not be able to weigh the cylinder before and after refill; and the cylinder may be filled with substances other than LPG. It is no coincidence that Brazil's only law devoted to LPG—Lei N° 9.048 of Brazil—concerns transparency in weighing cylinders, starting with the requirement that the cylinder tare weight must be clearly visible to the customer. The 2009 LPG regulations in Kenya require every retailer to have a properly calibrated weighing instrument.

The retail end of the supply chain is frequently handled by very small operators who deliver cylinders to homes and by consumers who pick them up at local filling stations or retail outlets. In most developing countries, rural retailers are small grocery shops. Regional sub-dealers use small trucks to deliver cylinders to larger villages, from which LPG may be carried by animals and people to smaller villages. Regulating this end of the supply chain poses a considerable challenge.

59.4.6   Fuel Shortages

Aside from serious accidents, nothing destroys consumer confidence like crippling LPG shortages. In addition to the inconvenience of running out of the cooking and heating fuel, acute shortages can, and almost always do, lead to sharp price hikes, doubling or even tripling end-user prices. As one extreme illustration, LPG shortages in Egypt in recent months have caused black-market prices to soar to as high as 55 Egyptian pounds a cylinder from the official price of 2.5 pounds (IHS 2011). The shortage of LPG in Senegal in 2009 doubled the price of LPG and pushed up the price of charcoal as households turned to charcoal instead. Supply shortages and rising prices in turn prompt consumers to stockpile LPG, exacerbating the problem. For business users of LPG, shortages could mean shutting down business altogether, as happened to restaurants in South Africa in October 2011. The immediate cause is typically supply disruption at any point along the supply chain, often coupled with inadequate storage capacity. Increasing storage capacity and requiring and enforcing minimum operational and/or strategic stock could help, especially in countries where the total national storage capacity is of the order of only one week.

Some shortages are caused by exceptional circumstances beyond the control of marketing companies and governments, such as the severe winter of 2011–2012 leading to LPG shortages even in Europe, a strike by LPG transporters in some states in India in January 2012, and the simultaneous shut-down of its four refineries in South Africa for unplanned repairs in October 2011. However, all too often, shortages are a direct result of government policy. Many countries with frequent shortages seem to have one thing in common: universal price subsidies for all consumers, or more typically for LPG sold in relatively small cylinders intended for use by households. Where there is a large price difference between LPG sold in small cylinders—for example, 15 kg or smaller—and LPG for commercial and industrial users, illegal diversion to the latter occurs. Such illegal diversion was blamed for worsening LPG shortages in Egypt in 2011 and 2012 and in some parts of India in the first few months of 2012. India has a system of allocating LPG cylinders to households who can then obtain subsidized LPG. Cylinder allocation is intended to ration LPG to limit the subsidy outlay, but there are frequent reports of LPG shortages in different parts of the country and of raids on hotels and restaurants uncovering illegal possession of 14.2-kg cylinders, intended for household use only.

Large universal price subsidies for petroleum products are fiscally unsustainable, in the worst case leading to an inability to pay for product imports or even crude oil for refining. In Ghana, the only refinery in the country has had to shut down repeatedly for lack of crude oil, in part because of its inability to obtain letters of credit from banks to buy crude oil. Rising demand for autogas and for industrial use of LPG has also been blamed for the LPG shortages. Inadequate storage capacity exacerbates the problem: Ghana's total storage capacity is 6500 tonnes, against weekly consumption in mid-2011 of 4000–5000 tonnes, leaving a buffer of only 1500–2000 tonnes, or about 3 days of consumption (Matthews and Zeissig 2011). Senegal experienced a prolonged LPG shortage in early 2009, caused by the inability to pay for imports due to mounting subsidies; the subsidies to the refinery and for LPG in 2009 are estimated to have amounted to US$70 million (IHS 2010). The Nepal Oil Corporation, the only importer of LPG and refined products into Nepal, has frequently encountered trouble paying its supplier, the Indian Oil Corporation, on time, resulting in delays in fuel shipment and consequently LPG shortages, most recently in the winter of 2011–2012. LPG importers in Egypt have similarly faced trouble paying for the imports.

Frequent LPG shortages can also occur in countries without price subsidies. Kenya's aging refinery, limited port capacity, and inadequate storage capacity have all contributed to frequent acute LPG shortages in the country. Tankers carrying LPG are said to have to wait up to as long as a month in Mombasa to unload, at a cost of hundreds of thousands of dollars (All Africa 2012). These problems affect not only the Kenyan market, but all the neighboring countries that import fuel from Kenya. As a result, Kenya's neighbor Uganda has also experienced frequent LPG shortages followed by price spikes. Similarly, when South African refineries were unexpectedly shut down in October 2011, Mozambique, which imports from South Africa, ran out of LPG altogether, a shortage that continued into December.

59.4.7 Reducing Costs

There are economies of scale in the marine transport of LPG, raising shipping costs for small importing markets. Importers in Turkey contract parcels as large as 20000 tonnes, enabled not only by the country's large LPG market but also in part by joint procurement activities and hospitality (swapping) arrangements for import-terminal storage facilities. This contrasts with parcel sizes of 1200–5000 tonnes in smaller markets in Sub-Saharan Africa and elsewhere (such as Ghana, Kenya, Senegal, and Sri Lanka). While these large parcels help reduce import shipping costs, Turkey suffers from high demurrage expenses on account of port congestion.

Establishing hospitality arrangements for storage facilities and granting third-party access to terminals lower the barrier to entry and reduce costs. The barrier to entry is also lowered if, as in Peru, there is a large bulk supplier who is not engaged as a competitor in the downstream business and a new LPG bottler/distributor can avoid having to invest in a marine terminal by sourcing LPG from the bulk supplier. In Pakistan, private-sector terminal operators are engaged in receiving, storing, and onward shipping LPG for third parties for set fees, potentially avoiding duplication of surplus infrastructure. Having large companies capable of exploiting economies of scale import LPG as much as possible reduces costs. Scale economy may require expanding import terminals to be able to receive larger parcels.

Port demurrage charges can add considerably to LPG prices; port congestion and customs clearance are outside the control of LPG operators and need to be tackled by the government. Many ports in developing countries do not have round-the-clock port staffing. Paying to operate ports at all hours, which benefits all imports and exports, should yield high economic returns.

Having enough companies to enable price competition is important, although the sheer number of marketing companies and retailers alone does not provide an accurate picture of the nature of competition. For example, market conditions that prompt large, creditworthy companies to exist, leaving many small operators who are not capable of raising funds or exploiting scale economy, could result in diminishing investment in the supply infrastructure and may even increase costs.

59.5 CONTRIBUTING TO UNIVERSAL ACCESS TO MODERN ENERGY

International LPG prices have more than doubled since 2000 in real terms, and nearly tripled between 2000 and the first three months of 2012. The average price of food, an even more important expenditure item, has also doubled since 2000 in real terms (FAO 2011), placing considerable pressure on household income. And yet per capita GDP increased by 85 percent or more in real terms between 2000 and 2010, the last year for which GDP statistics are available, in only 11 countries, half of them major oil exporters. This poses a significant challenge to fuel switching to LPG as one pathway toward the UN goal of universal access to modern energy by 2030.

An estimate of the level of income required to start using LPG as the primary cooking fuel can be calculated by making several simplifying assumptions. For the purpose of this illustration, the average of Saudi Aramco's contract prices for propane and butane are taken, and the combined cost of transport, bottling, distribution, retailing, and taxes are assumed to add US$350 per tonne to the contract prices to arrive at the end-user price. These assumptions put the retail price in March 2012 at US$1.56/kg. Based on the findings from household survey data analysis, monthly consumption of LPG is assumed to be 12 kg, sufficient for cooking and water heating in many parts of the developing world, and the share of total household expenditure spent on LPG is assumed to be 4 percent. The results show that monthly expenditure on LPG would have risen, in 2011 U.S. dollars, from about US$7.30 in the first three months of 2002 to US$16.60 in the first three months of 2012. Currency appreciation against the U.S. dollar offsets the rise in global LPG prices, but there are only four developing countries where the local currency appreciated by more than 50 percent between 2000 and 2010, and none where the appreciation was greater than 60 percent. At the opposite end of the spectrum, in 28 developing countries, the local currency *depreciated* against the U.S. dollar by more than 40 percent, exacerbating the challenge of high and volatile LPG prices on the world market. The minimum monthly household expenditure required to start using LPG regularly remained largely below US$250 until mid-2005, but rose above US$450 in March 2012.

These observations suggest that LPG is unlikely to be the fuel of the poor, because the magnitude of financial assistance required to enable the poor to use LPG as the primary source of household energy would not be fiscally sustainable in most developing countries. Although restricting a price subsidy to particular users of LPG, such as low-income households, would lessen the fiscal burden compared to that for universal price subsidies, it would be difficult to control illegal diversion of subsidized LPG to commercial establishments and the automotive sector. For protecting the poor, allowing market forces to govern LPG prices and assisting low-income households through social protection measures to help these households with expenditures on all essential goods, as with Brazil's *Bolsa Familia* program, would be preferable and more efficient.

But even without subsidies, there is still considerable scope for LPG to play an important role in reducing *energy poverty*. First, approximately three billion people globally are not using modern household energy, and a significant portion of the three billion is not *income-poor*. Lack of satisfactory LPG supply services, price levels that are higher than what the market could achieve under fair competition, inefficiency in the supply chain, poor regulation leading to unsafe handling of LPG cylinders and commercial malpractice, frequent supply disruptions leading to acute shortages and price spikes, and unfamiliarity with LPG are among the reasons why more households, who can otherwise afford it, are not using LPG today. Second, relative fuel prices also matter. As prices of competing fuels, such as charcoal, rise—and they are rising sharply in some urban settings—LPG becomes increasingly attractive, although if prices of *all* commercial fuels rise steeply, the poor lacking access to free biomass may have trouble consuming any form of energy.

Steps to facilitate fuel switching to LPG by households may include

- lowering supply costs, thereby reducing prices charged to consumers;
- enhancing safety;
- targeting financial assistance;
- minimizing shortages; and
- raising awareness and involving consumers in improving market conditions.

Table 59.4 gives examples of specific measures under each category. Not all options are applicable to a given market. Country-specific recommendations can draw upon the options listed in the table, tailored to specific country circumstances.

One of the most important and useful roles a government can play is to make information available to the public. Consumers may be interested in learning about trends in international and domestic prices, investors about industry statistics and government policy and regulations, and energy analysts in both. High world oil prices in recent years have increased politicization of fuel pricing in a number of countries. Governments can help foster informed debate by posting detailed information on prices. As long as price information is being collected, it would make sense to retain and post historical prices in addition to current ones. Where LPG prices are kept artificially low, it is informative to list international prices in U.S. dollars and in local currency and show the price build-up starting with ex-refinery or landed cost, so that consumers can see which segments of the supply chain are subsidized. Without such information, it would be extremely difficult, if not impossible, to have a sensible discussion on whether and how to reduce and eventually eliminate universal price subsidies. Consumers would also benefit from information about safety features of LPG, which would help them understand how to handle LPG and allay fears about explosions and other accidents, as well as learn what the government is doing to curb commercial malpractice (such as short-selling) and what steps consumers can take to minimize it.

Given increasing use of the Internet, Web posting is one of the fastest and cheapest ways to disseminate information widely. The ease of navigating a Web site and finding information is important, as is the ease of downloading information, particularly for time-series data—if data have to be downloaded one by one for each time period, temporal comparison can become time-consuming. Where LPG prices are deregulated, how prices vary over time, from location to location, and among marketers may influence household purchasing decisions, although such information can be resource-intensive to collect and compile. The Web sites of the regulatory agencies in Peru, Pakistan, and Turkey post

TABLE 59.4   Examples of options for facilitating household use of LPG.

Goal	Option	Example
Lower costs to consumers	Exploit economies of scale	Hospitality arrangements, third-party access
		Bulk purchase, joint purchase, large import parcels
		Large refineries
	Lower barrier to entry	Hospitality arrangements, third-party access
	Minimize demurrage charges	Rapid customs clearance
		Reduced port congestion
		Round-the-clock staffing by port authorities
		Adequate port receiving capacity
	Minimize short-selling	Clear marking of cylinder tare weight
		Enforcement of scale calibration and date of last scale calibration visible to customers
		Customer's right to check cylinder weight
		Industry association's (voluntary) seal of quality/certification for those shops that have been certified not to short-sell
		Publication of names of companies found short-selling
	Increase price competition	Posting of prices by company, location, and cylinder size on government Web site
		Competition policy
	Improve auxiliary infrastructure	Improved road conditions
		Improved port infrastructure in importing countries
Enhance safety	Establish clear regulations	Formal adoption of international standards by reference
		All regulations posted in one place on the Web in reverse chronological order
		Training of supply personnel legally required
		Education of consumers about safe handling of LPG legally required
	Enforce safety regulations	Where there is a ban on cross-filling, ban effectively enforced
		Small fee levied to finance monitoring and enforcement
		Registry of certified installers
		Clearly marked date of last cylinder recertification
		Registry of certified private inspectors operating under government supervision
		Training workshops organized by LPG industry association
		Publication of names of companies violating safety rules
	Educate consumers	Pictorial guides in local languages, newspaper/radio/TV advertisements, Web posting of safety information
		Neighborhood demonstrations by retailers, industry association, and consumer groups
		In-house demonstration of proper cylinder and stove handling by installers
Target financial assistance	Move away from universal price subsidies	Expansion of social safety net program to help pay for LPG, such as cash transfer or vouchers
	Spread or reduce upfront adoption cost	Dealer incentives for cylinder deposit fee and stove
		Dealer-financed installment plan
		Microfinance scheme
		Small cylinders in niche market
Minimize shortages	Require minimum commercial and/or strategic stockholding in regulations	
	Ensure reasonable returns (through, e.g., removal of universal price subsidies) to efficient operators to build up capital for construction of storage facility	
	Encourage hospitality and third-party access	

(*continued*)

TABLE 59.4 *(Continued)*

Goal	Option
Raise awareness and involve consumers in improving market conditions	Government: Publish price information, industry statistics, frequently asked questions, safety tips, and names of companies violating rules that directly affect consumers on the Web and in reports; establish a simple mechanism for registering complaints
	Industry association: Publish information, frequently asked questions, and safety tips on the Web; publish brochures; take out newspaper/radio/TV advertisements; publicize information on retailer location and contact details; establish quality control and issue seals of quality for companies in compliance; establish a simple mechanism for registering complaints against members
	Companies: Disseminate information on proper handling of LPG cylinders, frequently asked questions, and safety tips; have installers show new customers in their homes how to handle an LPG cylinder and stove properly; establish a simple mechanism for registering complaints

prices by company and location, enabling consumers to compare prices.

LPG marketing companies and industry associations also post information on their Web sites. LPG industry associations, by virtue of their focus, can help disseminate useful information to all parties, from industry actors to residential consumers. The Web site of the LPG Safety Association of Southern Africa has brochures on safety in three languages. The Web site of Liquigas in Brazil has frequently asked questions and safety information. Graphs comparing international and domestic prices and estimated annual and cumulative subsidies since 2003, posted on the Web site of Pemex, visually informs the Web viewers of the scale of subsidies and their increases in recent years in Mexico.

Taken together, the foregoing findings suggest that campaigns and programs to promote household use of LPG to substitute biomass are likely to be more effective if they first focus on areas where biomass is diminishing, the costs of biomass cooking are high, and infrastructure for reliable LPG delivery does not impose an undue cost or time burden on households. The last aspect would include the existence of a tarred road connecting the nearest bottling plant to a shop selling LPG and inexpensive means of getting to the shop by consumers or home delivery at a nominal fee, if any.

In promoting household use, it would make sense to first target households whose income is sufficiently high to start using LPG without subsidies and who already live in areas with LPG marketers, because these households are best placed to switch entirely to LPG and sustain its use. Such a shift will also help ease growing pressure on biomass resources, which will continue to be used by the rural poor for the foreseeable future and the mitigation policy for which must include cleaner-burning, efficient stoves for solid fuels. Other households will start using LPG but will retain solid fuel use. Starting to use LPG is not the same as abandoning solid fuels altogether, but is the first step nevertheless and, with experience, a household will feel increasingly comfortable handling LPG. Increasing use of LPG in the community

could in turn lead others to consider LPG through demonstration effects.

In countries that do not yet have such associations, a helpful development would be the formation of national LPG associations that include marketing companies, cylinder manufacturing and maintenance companies, government representatives, and consumer organizations. Such associations can contribute to regulation and policy formulation, promotion of fair competition, adoption of international safety standards, collection and dissemination of information, and education and training. Governments, LPG marketing companies, and civil society organizations can all contribute especially to consumer education and awareness-raising, as well as watching out for commercial malpractice and helping to create market conditions whereby efficient, law-abiding firms increase their market share to the benefit of all consumers.

REFERENCES

All Africa. (2011a). Drivers want subsidy on LPG removed. August 19.

All Africa. (2011b). Crippling gas crisis – Energy ministers under fire. August 19.

All Africa. (2012). Ship yet to offload gas two weeks after docking. January 10.

Arab News. (2010). Response to fiberglass LPG cylinders lukewarm. July 9.

Bacon, R. (2001). *Petroleum Taxes: Trends in Fuel Taxes (and subsidies) and the Implications*. Viewpoint 240. Washington, DC: World Bank.

Bacon, R., Bhattacharya, S.Kojima, M. (2010). *Expenditure of Low-Income Households on Energy*. Extractive Industries for Development Series #16. Washington, DC: World Bank.

Bacon, R., Kojima, M. (2008). *Coping with Oil Price Volatility*. ESMAP Special Report 005/08. Washington DC: World Bank. Brazil, Government of. 1995. "Lei No 9.048."

BCRP (Banco Central de Reserva del Peru) (2012). Notas de studios del BCRP, No. 6.

EPPO (Energy Policy and Planning Office). (2012). Excel files price structures posted at "Price Structure of Petroleum Products in Bangkok," available at www.eppo.go.th/petro/price/index.html

ESMAP (Energy Sector Management Assistance Program). (2003a). *Household Energy Use in Developing Countries: A Multicountry Study.* Technical Report, October 2003. Washington, DC: World Bank.

ESMAP (Energy Sector Management Assistance Program). (2003b). Chapter 3, Promoting LPG among the rural poor: lessons from the deepam scheme in Andhra Pradesh. In: *India Access of the Poor to Clean Household Fuels.* Report 263/03. Washington DC: World Bank.

FAO (Food and Agriculture Organization). (2010). *Global Forest Resources Assessment 2010: Main Report.* Rome.

FAO (Food and Agriculture Organization). (2011). Food Price Index. Online database. www.fao.org/worldfoodsituation/wfs-home/foodpricesindex/en/

Hart, W., Gist, R., Otto, K. (2011). Stability of LPG markets threatened by world events. *Oil & Gas Journal,* 109(13):98–106.

IEA (International Energy Agency). (2011a). *World Energy Outlook 2011.* Paris.

IEA (International Energy Agency). (2011b). *Energy Prices and Taxes: Third Quarter.* Paris.

IEA (International Energy Agency). (2011c). World Energy Statistics 2011 Edition, online database.

IHS. (2010). Mauritania supplies butane gas to Senegal to alleviate shortages. IHS Global Insight Daily Analysis, October 22.

IHS. (2011). Butane crisis continues in Egypt amidst mounting payment pressure. IHS Global Insight Daily Analysis, December 12.

Kathmandu Post. (2012). Eight-pt pact seeks to smoothen LPG supply. February 22.

Kojima, M., Bacon, R., Zhou, X. (2011). Who uses bottled gas? Evidence from households in developing countries. Policy Research Working Paper 5731. Washington, DC: World Bank.

Komives, K., Foster, V., Halpern, J., Wodon, Q. (2005). *Water, Electricity and the Poor. Who Benefits from Utility Subsidies?* Washington, DC: World Bank.

Masera, O. R., Saatkamp, B. D., Kammen, D. M. (2000). From linear fuel switching to multiple cooking strategies: a critique and alternative to the energy ladder model. *World Development,* 28(12):2083–2103.

Matthews, W., Zeissig, H. (2011). *Residential Market for LPG: A Review of Experience of 20 Developing Countries.* Houston International Business Corp.

NPA (National Petroleum Authority). (2012). Petroleum products price buildup. for various dates posted at www.npa.gov.gh/npa_new/news.php

Petrobras. (2011). Relatorio da Administracao 2010. Available at www.liquigas.com.br/wps/wcm/connect/da74b80046416 56e8d65cddd2947447b/LIQUIGAZ_JN_2737_11.pdf?MOD = AJPERES7

PGPB. (Pemex Gas y Petroqufmica Basica). (2011). Mercado gas LP: subsidio. Available at www.gas.pemex.com/PGPB/Productos+y+servicios/Gas+licuado/Mercado+gas+LP/Subsi dio/

Platts Oilgram News. (2011). Thailand's LPG demand rises 24%, led by subsidies. August 1.

Polat, E. P. (2000). Turkey: Largest LP gas parc in the world. Available at http://worldlpgas.com/gain/autogas-is-best/turkey-largest-lp-gas-parc-in-the-world/

SENER *(Secretaria de Energia)* (2010). Prospectiva del Mercado de Gas Licuado del Petroleo: 2010–2025.

Smith, K., Dutta, K. (2011). Editorial: cooking with gas. *Energy for Sustainable Development,* 15(2):115–16.

Thai Financial Post. (2012). EPPO predicts LPG price will continue rising. February 26.

Trend News Agency. (2012). Cooking gas crisis due to distribution: Petroleum minister. February 6.

Turkey, Government of. (2005). "Syvylabtyrylmyp Petrol Gazlary (LPG) Piyasasy Kanunu ve Elektrik Piyasasy Kanununda DeSipiklik Yapylmasyna Dair Kanunda DeSipiklik Yapylmasyna Ylipkin Kanun."

UNDP. (United Nations Development Program) and WHO (World Health Organization). (2009). *The Energy Access Situation in Developing Countries.* New York.

WLPGA. (World LP Gas Association). (2010). Annual Report 2009.

World Bank. (2011). World Development Indicators, December 2011, online database.

60

EXERGY ANALYSIS: THEORY AND APPLICATIONS

Marc A. Rosen

60.1 INTRODUCTION

Exergy analysis is a thermodynamic tool that can be used for the analysis, design, improvement, and optimization of energy-intensive systems and processes. The tool is based on the quantity exergy, which is a measure of quality or usefulness or value of energy. Exergy analysis is similar to energy analysis, but different in many ways. Exergy analysis is based on the second law of thermodynamics and has several advantages over energy analysis, which is based on the first law of thermodynamics and the principle of conservation of energy. For instance, more meaningful efficiencies are evaluated with exergy analysis since exergy efficiencies are always a measure of the approach to the ideal, and the inefficiencies in a process are better pinpointed with exergy analysis in that the types, causes, and locations of the losses are identified and quantified. For these reasons and others, exergy analysis has been increasingly applied over the last several decades.

In this article, we examine the theory and application of exergy analysis, for purposes of assessing, designing, improving, and optimizing energy-intensive systems. The scope of material covered in this article is as follows. First, we describe exergy, including a statement of its definition, and explanations of the reference environment used in evaluating exergy, exergy quantities, and exergy balances. Then, the tool exergy analysis is presented along with discussions of the benefits it provides, exergy efficiencies, and the relations of exergy analysis to both economics and environmental impact. Finally, applications of exergy analysis are discussed and illustrated, and case studies are presented for a coal-fired power plant—along with a tutorial—and a national energy system.

More information on fundamentals and applications of both exergy and exergy analysis can be found elsewhere (Dincer and Rosen, 2013; Feidt, 2009; Gaggioli, 1998; Kotas, 1995; Moran, 1989; Sato, 2005; Szargut, 2005; Szargut et al., 1988; Wall, 2003; Tsatsaronis, 2008).

60.2 DEFINITION OF EXERGY

Exergy is rigorously defined as the maximum amount of work which can be produced by a stream or system as it is brought into equilibrium with a reference environment (Dincer and Rosen, 2013; Kotas, 1995; Moran, 1989; Szargut, 2005; Tsatsaronis, 2008; Moran et al., 2011).

To provide a more interpretive context, exergy can more loosely be defined or viewed as a measure of the usefulness or quality or value of energy.

60.3 REFERENCE ENVIRONMENT FOR EXERGY

Exergy quantities are evaluated with respect to a reference environment. The intensive properties of the reference environment in part determine the exergy of a stream or system. The exergy of the reference environment is zero. The reference environment is in stable equilibrium, with all parts at rest relative to one another. No chemical reactions can occur between the environmental components. The reference environment acts as an infinite system, and is a sink and source for heat and materials. It experiences only internally reversible processes in which its intensive state remains unaltered (i.e., its temperature T_o, pressure P_o, and the chemical potentials μ_{ioo} for each of the i components present remain constant). Information on reference-environment models can be found

Alternative Energy and Shale Gas Encyclopedia, First Edition. Edited by Jay H. Lehr and Jack Keeley.
© 2016 John Wiley & Sons, Inc. Published 2016 by John Wiley & Sons, Inc.

elsewhere (Dincer and Rosen, 2013; Kotas, 1995; Moran, 1989; Szargut, 2005; Tsatsaronis, 2008). The reference environment is often modeled as the natural environment.

To avoid confusion, several terms related to the reference environment used in exergy analysis, or similar to those, are defined and explained here:

- Reference environment: The reference environment is an idealization of the natural environment which is characterized by a perfect state of equilibrium, that is, absence of any gradients or differences involving pressure, temperature, chemical potential, kinetic energy, and potential energy.
- Dead state: The dead state is the state of a system when it is in thermal, mechanical, and chemical equilibrium with the reference environment.
- Environmental state: The environmental state is the state of a system when it is in thermal and mechanical equilibrium with the reference environment, that is, at the pressure and temperature of the reference environment.

60.4 EXERGY QUANTITIES

Exergy is associated with all material and energy quantities, and exergy flows are associated with all material and energy flows. These exergy quantities and flows are described below, following the presentations by Moran (1989); Kotas (1995) and Dincer and Rosen (2013):

- Exergy of work: The exergy associated with work is equal to the energy.
- Exergy of electricity: The exergy associated with electricity is also equal to the energy.
- Exergy of matter: The exergy Ex of a system can be expressed in terms of physical, chemical, kinetic, and potential components. That is

$$Ex = Ex_{ph} + Ex_o + Ex_{kin} + Ex_{pot} \qquad (60.1)$$

The exergy of a flow of matter Ex_{flow} can also be expressed in terms of physical, chemical, kinetic, and potential components.
- Exergy of thermal energy: The thermal exergy Ex_Q associated with a thermal energy transfer Q can be expressed as

$$Ex_Q = Q(1 - T_o/T_i) \qquad (60.2)$$

where T_i and T_o are the system and reference-environment temperatures, respectively. Thus, thermal exergy is the exergy associated with a heat interaction, that is, the maximum amount of shaft work obtainable

from a given heat interaction using the environment as a thermal energy reservoir.

To clarify and avoid confusion regarding exergy terms, the following definitions and explanations are provided:

- Exergy: Exergy, also known as available energy, is a general term for the maximum work potential of a system, stream of matter or a heat interaction in relation to the reference environment.
- Exergy of matter: The exergy of matter (or of a flow of matter) is the maximum amount of shaft work obtainable when a flow of matter is brought from its initial state to the dead state by means of processes involving interactions only with the reference environment.
- Physical exergy: Physical exergy, also known as thermomechanical exergy, is the maximum amount of shaft work obtainable from a substance when it is brought from its initial state to the environmental state by means of physical processes involving interaction only with the environment.
- Chemical exergy: Chemical exergy is the maximum work obtainable from a substance when it is brought from the environmental state to the dead state by means of processes involving interaction only with the environment.

The above exergy quantities all appear in general in an exergy balance, as described in the next section.

60.5 EXERGY BALANCE

Unlike energy, exergy is consumed during real processes due to irreversibilities and conserved during ideal processes. Thus the balance for exergy is fundamentally different than that for energy.

Like for energy, balances can be written for exergy for a general process or system. Before giving a general exergy balance, it is instructive to consider an energy balance for a system, which may be written as

Energy input − Energy output = Energy accumulation

$$(60.3)$$

Energy input and output refer respectively to energy entering and exiting through system boundaries. Energy accumulation refers to build-up (either positive or negative) of the quantity within the system. Clearly, energy is subject to a conservation law.

By combining the conservation law for energy and non-conservation law for entropy (which states that entropy is created during a process due to irreversibilities), an exergy balance can be obtained:

$$\text{Exergy input} - \text{Exergy output} - \text{Exergy consumption}$$
$$= \text{Exergy accumulation} \qquad (60.4)$$

Exergy is consumed due to irreversibilities. Thus, exergy consumption, which is also known as dissipation and lost work, is the exergy consumed or destroyed during a process due to irreversibilities within the system boundaries. Exergy consumption is proportional to entropy creation. These two equations demonstrate an important main difference between energy and exergy: energy is conserved while exergy, a measure of energy quality or work potential, can be consumed.

60.6 EXERGY ANALYSIS

Exergy analysis involves the application of exergy concepts, balances, and efficiencies to evaluate and improve energy and other systems and processes. Being a thermodynamic analysis technique for systems and processes that is primarily based on the second law of thermodynamics gives exergy analysis several important advantages over energy analysis. Energy analysis is based essentially from the principle of the conservation of energy as it is founded on the first law of thermodynamics.

The performance of exergy analysis involves a straightforward procedure. Exergy analysis is usually performed in parallel with energy analysis, but this is not required. In the procedure outlined here, exergy analysis can be carried out alone or in tandem with energy analysis.

The following procedure can be utilized in applying exergy analysis:

- Subdivide the process under consideration into as many sections as desired, depending on the depth of detail and understanding desired from the analysis.
- Perform conventional mass and energy balances on the process, and determine all basic quantities (e.g., work, heat) and properties (e.g., temperature, pressure).
- Based on the nature of the process, the acceptable degree of analysis complexity and accuracy, and the questions for which answers are sought, select a reference-environment model.
- Evaluate exergy values, relative to the selected reference-environment model. (Evaluate energy values also if a simultaneous energy analysis is being carried out or desired.)
- Perform exergy balances.
- Determine exergy consumptions due to irreversibilities.
- Select exergy-based efficiency definitions, depending on the measures of merit desired, and evaluate the efficiencies. (Select and evaluate energy efficiencies also if a simultaneous energy analysis is being carried out or desired.)
- Interpret the results, and draw appropriate conclusions and recommendations, relating to such issues as design changes, retrofit plant modifications, etc.

60.7 EXERGY EFFICIENCIES

Although numerous definitions for energy and exergy efficiencies exist, energy and exergy efficiencies are for many processes based on the following general definitions:

$$\text{Energy efficiency} = (\text{Energy in products})/(\text{Total energy input}) \qquad (60.5)$$

$$\text{Exergy efficiency} = (\text{Exergy in products})/(\text{Total exergy input}) \qquad (60.6)$$

Exergy efficiencies are measures of how well input exergy is converted to useful exergy products or to useful exergy services. Exergy efficiencies are more meaningful than energy efficiencies since exergy efficiencies are always a measure of the approach to the ideal (at which point they are 100%).

Inversely, the inefficiencies in a process or system are better pinpointed with exergy analysis in that the types, causes and locations of the losses are identified and quantified.

Exergy efficiencies clearly evaluate theoretical limitations to efficiency, by quantifying the maximum efficiency theoretically attainable for a process by the laws of thermodynamics. This limitation must be clearly understood to assess the realistic potential for increased efficiency. Lack of clarity on this issue has in the past often led to confusion, in part because energy efficiencies generally are not measures of how nearly the performance of a process or device approaches the theoretical ideal. The consequences of such confusion can be very significant. For example, extensive resources have at times been directed toward increasing the energy efficiencies of devices that in reality were efficient and had little potential for improvement. Conversely, devices at other times have not been targeted for improved efficiency, even though the difference between the actual and maximum theoretical efficiencies, which represents the potential for improvement, has been large.

Exergy efficiencies are reduced below 100% due to practical factors like friction, heat transfer across finite temperature differences, and unconstrained expansion. We normally do not seek to maximize efficiency, as such aims lead to unrealistic results. Rather, the objective of efficiency-improvement efforts is usually to achieve an optimal trade-off between efficiency and such factors as economics, sustainability, environmental impact, safety, and societal and political acceptability. Exergy analysis can help guide efforts to increase efficiency.

For instance, exergy methods can help identify meaningful qualitative and quantitative incentives and disincentives to foster improved efficiency.

60.8 BENEFITS OF EXERGY ANALYSIS

Many engineers and scientists suggest that devices can be well evaluated and improved using exergy analysis in addition to or in place of energy analysis.

The difficulties inherent in energy analysis are largely attributable to the fact that it considers only quantities of energy and ignores energy quality, which is continually degraded during real processes. Exergy analysis overcomes this and other problems associated with energy analysis.

60.9 EXERGY ANALYSIS AND ECONOMICS

An area related to but different from thermodynamics, where exergy analysis is increasingly applied, is economics. In the analysis and design of energy systems, techniques are often used which combine scientific disciplines (especially thermodynamics) with economics to achieve optimum designs. For energy systems, costs are conventionally based on energy. Many researchers, however, have recommended that costs are better distributed among outputs, based on exergy. Methods of performing exergy-based economic analyses have evolved recently and are referred to by such names as thermoeconomics, second-law costing and exergoeconomics. These analysis techniques recognize that exergy, more than energy, is the commodity of value in a system, and assign costs and/or prices to exergy-related variables. These techniques usually help determine the appropriate allocation of economic resources so as to optimize the design and operation of a system, and/or the economic feasibility and profitability of a system (by obtaining actual costs of products and their appropriate prices). The relation between exergy and economics is discussed elsewhere (Dincer and Rosen, 2013; Rosen, 2011; Kotas, 1995; Moran, 1989; Szargut, 2005; Tsatsaronis, 2008).

60.10 EXERGY ANALYSIS AND ENVIRONMENTAL IMPACT

An area beyond thermodynamics where applications of exergy are increasing is that of environmental impact. Many suggest that the impact of energy resource utilization on the environment and the achievement of increased resource-utilization efficiency should be addressed by considering exergy, among other factors. Although the exergy of an energy form or a substance is a measure of its usefulness, exergy is also a measure of its potential to cause change. The

latter point suggests that exergy may be, or provide the basis for, an effective measure of the potential of a substance or energy form to impact the environment. Relations between exergy and the environment are discussed elsewhere (Dincer and Rosen, 2013; Rosen, 2012).

60.11 APPLICATIONS OF EXERGY ANALYSIS

Exergy analysis has been increasingly applied over the last several decades to a wide range of processes and systems.

60.11.1 Types and Range of Applications of Exergy Analysis

The types of applications of exergy analysis to energy and other systems and processes that have been reported in the past include the following:

- Thermal processes:
  a. Heating systems for industrial and other applications.
  b. Cooling systems for industrial and other applications.
- Energy generation:
  a. Electricity generation, using both conventional devices as well as alternative devices such as fuel cells, from various resources including oil, natural gas, coal, and nuclear power plants.
  b. Energy generation/harvesting from renewable energy sources like solar, wind, geothermal, and biomass energy.
  c. Cogeneration systems for simultaneously producing heat and electricity.
  d. Trigeneration systems for simultaneously producing heat, cool, and electricity.
- Engines and motive force generators:
  a. Engines of various types.
  b. Combustion technologies and systems.
- Chemical processing:
  a. Chemical processes, such as sulfuric acid production,
  b. Separation and distillation processes, including air separation and water desalination.
  c. Petrochemical processing and synthetic fuels production.
  d. Metallurgical processes, such as lead smelting and refining of ores.
- Energy storage:
  a. Batteries and electrochemical energy storage.
  b. Pumped hydro storage.
  c. Thermal and thermochemical energy storage.

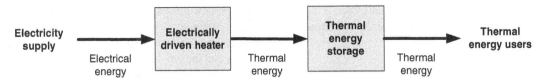

FIGURE 60.1   Integrated system for electrical heating, consisting of an electrically driven heater and a thermal energy storage.

- End-use applications:
  a. Building energy systems, including space heating and cooling as well as hot water heating.
  b. Transportation systems for land, air, and water.

Although the applications of exergy analysis considered here relate to thermodynamics and efficiency assessment and improvement, applications of exergy analysis to other types of processes and systems are described elsewhere (Dincer and Rosen, 2011; 2013). Applications beyond thermodynamics, in areas like environmental impact and economics, are also described elsewhere (Dincer and Rosen, 2013; Kotas, 1995; Moran, 1989; Szargut, 2005; Tsatsaronis, 2008).

Many applications of exergy analysis to energy systems have been reported in the literature, and several extensive bibliographies have been compiled, including one by Göran Wall (see web site http://exergy.se). In addition, a journal devoted to exergy matters and entitled the *International Journal of Exergy* was established in 2004 by Inderscience.

60.11.2   Illustrative Example of Application of Exergy Analysis

An illustrative example of an application of exergy analysis is presented based on electrical space heating integrated with thermal energy storage (TES) (Figure 60.1). The electric space heater converts electricity to heat at a temperature suitable for keeping a room at a comfortable temperature, and the heat is then stored in a TES and discharged at a subsequent time for use. This technology combination can be beneficial in situations such as (1) heating using electricity derived from renewable energy, which is intermittent and often generated at times other than when the heating is needed, and (2) using electrical heating in an environment where time-of-day pricing exists, making it advantageous to convert electricity to heat during off-peak periods for use during peak periods.

The first part of this example considers an electrical space heater. Such a device can be based on electrical resistance heating (Figure 60.2), which has an energy efficiency exceeding 99%. The implication clearly is that the maximum possible energy efficiency for electric resistance heating is 100%, corresponding to the most efficient device possible. This understanding is flawed, however, as energy analysis ignores the fact that in this process high-quality energy (electricity) is used to produce a relatively low-quality product (warm air). Exergy analysis recognizes the difference in energy qualities, and indicates the exergy of the heat delivered to the room to be about 5–25% of the exergy entering the heater. Thus, the exergy efficiency of electric resistance space heating is about 5–25%. Since thermodynamically ideal space heating has an exergy efficiency of 100%, the same space heating can in theory be achieved using as little as 5% of the electricity used in conventional electric resistance space heating. This exergy result can be exploited in practical terms by noting that space heating can be achieved using an electric heat pump with a "coefficient of performance" of 4 (Figure 60.3) using 25% of the electricity required for electric resistance heating.

In the second part of this example, a perfectly insulated TES receives thermal energy from the heater and holds the energy until it is required (Figure 60.4). The conventional energy storage efficiency is an inadequate measure as it is neither rational nor meaningful, but an efficiency based on the ratio of exergy quantities does provide a rational measure of performance, since it indicates the approach of system performance to the ideal. The inadequacy of energy efficiency

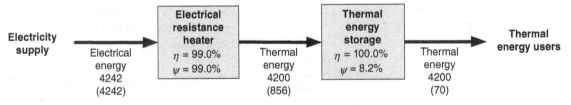

FIGURE 60.2   Energy and exergy flow diagram for the integrated system for electrical heating, consisting of an electrical resistance heater and a thermal energy storage. Flow rates (in MW) are shown of energy (positive values) and exergy (positive values in parentheses). Negative values in parentheses denote exergy consumptions. The overall energy and exergy efficiencies are 99.0% and 1.7%, respectively.

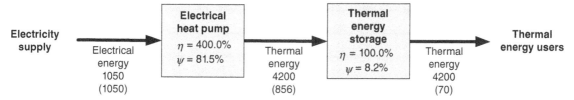

FIGURE 60.3   Energy and exergy flow diagram for the integrated system for electrical heating, consisting of an electrical heat pump and a thermal energy storage. Flow rates (in MW) are shown of energy (positive values) and exergy (positive values in parentheses). Negative values in parentheses denote exergy consumptions. The overall energy and exergy efficiencies are 400.0% (corresponding to a COP of 4.0) and 6.7%, respectively.

as a measure of TES performance is demonstrated here for a TES containing 1000 kg of water, initially at 40°C. The ambient temperature is 20°C, and the specific heat of water is taken to be constant at 4.2 kJ/kg K. A quantity of 4200 kJ of heat is transferred to the storage through a heat exchanger from an external body of 100 kg of water cooling from 100°C to 90°C. This heat addition raises the storage temperature 1.0°C to 41°C. After a period of storage, 4200 kJ of heat is recovered from the storage through a heat exchanger which delivers it to an external body of 100 kg of water, raising the temperature of that water from 20°C to 30°C. The storage is returned to its initial state at 40°C. For this storage cycle the energy efficiency, the ratio of the heat recovered from the storage to the heat injected, is 100% (4200 kJ/4200 kJ). But the recovered heat is at only 30°C, and of little use, having been degraded even though the storage energy efficiency was 100%. The exergy recovered in this example is evaluated as 70 kJ, and the exergy supplied as 856 kJ. Thus the exergy efficiency, the ratio of the thermal exergy recovered from storage to that injected, is 8.2% (70 kJ/856 kJ), a much more meaningful expression of the achieved performance of the TES. Consequently, a TES which appears to be ideal on an energy basis is correctly shown to be far from ideal on an exergy basis, clearly demonstrating the benefits of using exergy analysis for evaluating TESs.

Data for the energy and exergy flows and efficiencies for the overall system and its components are presented for the system using electrical resistance heating in Table 60.1 and for the system using a heat pump for heating in Table 60.2. For the overall system (including the assumption of no heat losses from the TES), the energy and exergy efficiencies are found to be 99.0% and 1.7% when electrical resistance heating is used, and 400.0% (corresponding to a COP of 4) and 6.7% when a heat pump is used for heating. The overall exergy efficiency is increased notably with the heat pump, almost four times, but is still quite low. The example considered here, consisting of electrical heating with TES, is clearly assessed differently on energy and exergy bases. Energy analysis overstates the efficiencies of both parts of the system, while exergy analysis presents a clear and meaningful understanding of the behavior.

60.12   CASE STUDY 1: NATIONAL ENERGY SYSTEM

The energy-utilization efficiency of a large conglomerate of energy systems that cover a region such as a city or country can be assessed beneficially using exergy analysis. This section illustrates how the efficiency of energy utilization in a

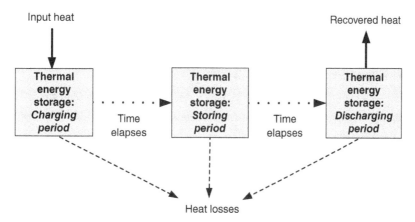

FIGURE 60.4   Three main periods of a thermal energy storage process and the principal energy flows.

TABLE 60.1 Energy and exergy flows and efficiencies for an integrated heating system using electrical resistance heating and thermal energy storage.

Device	Energy flows (kJ)		Exergy flows (kJ)		Efficiency (%)	
	Driving inputs	Product outputs	Driving inputs	Product outputs	Energy	Exergy
Electrical resistance heater	4242	4200	4242	856	99.0	20.2
Thermal energy storage[a]	4200	4200	856	70	100.0	8.2
Overall	4242	4200	4242	70	99.0	1.7

[a] The thermal energy storage is assumed to have no heat losses for simplicity.

TABLE 60.2 Energy and exergy flows and efficiencies for an integrated heating system using a heat pump and thermal energy storage.

Device	Energy flows (kJ)		Exergy flows (kJ)		Efficiency (%)	
	Driving inputs	Product outputs	Driving inputs	Product outputs	Energy	Exergy
Heat pump	1050	4200	1050	856	400.0[a]	81.5
Thermal energy storage[b]	4200	4200	856	70	100.0	8.2
Overall	1050	4200	1050	70	400.0[a]	6.7

[a] An energy of 400.0% corresponds to a coefficient of performance of 4.0.
[b] The thermal energy storage is assumed to have no heat losses for simplicity.

national energy system can be examined using both energy and exergy analyses. The country considered is Canada, and the case study is based on a prior investigation by the author of energy utilization in Canada using energy and exergy analyses (Rosen, 1992).

60.12.1 Energy System Model

The relation between energy resources and sectors of a national energy system can be modeled as in Figure 60.5. The energy resources considered here are: coal, petroleum, natural gas, nuclear energy, and hydraulic energy. Renewable energy resources (e.g., wind, solar, wave, tidal) are neglected since the quantity used in Canada is minor relative to other

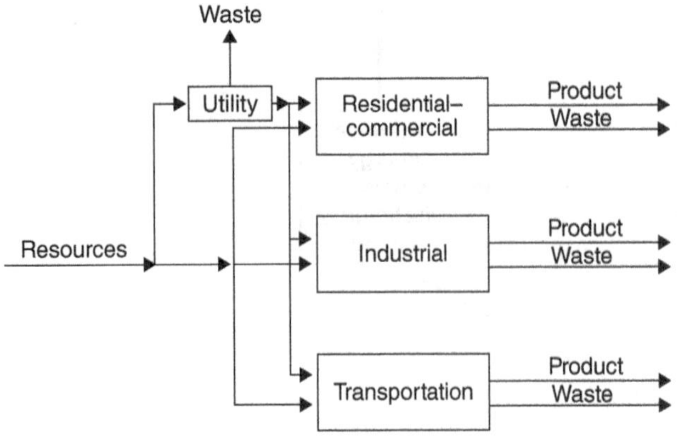

FIGURE 60.5 Illustration of the energy flows in a national energy system.

resources. Four sectors are considered. The residential–commercial sector, which is driven mainly by fossil fuels and electricity, includes residential, public administration, commercial, and other institutional buildings, and involves as principal devices space heaters, water heaters, cooking appliances, and clothes dryers. The transportation sector is primarily driven by fossil fuels and includes rail, marine, air, road and urban transport, pipelines, and retail pump sales. The industrial sector has as its main tasks work production and heating, and includes mines, steam generation, pulp and paper mills, cement, iron and steel mills, petroleum refining, smelting and refining, other manufacturing, agriculture, chemicals, and non-energy uses. The utility sector includes processes for electricity generation from coal, petroleum, natural gas, and nuclear and hydraulic energy sources.

60.12.2 Analysis

The main types of processes encountered, for which energy and exergy efficiencies are evaluated are as follows:

- Heating: Electric and fossil-fuel heating processes are taken to generate product heat at a constant temperature.
- Work production: Electric and fossil-fuel work-production processes produce shaft work.
- Electricity generation: Electricity is generated from fossil fuel, nuclear, and hydraulic sources. The hydraulic source is taken to be the change in kinetic and potential energy of flowing water (even though coal equivalents are often used to represent hydraulic energy inputs). The nuclear source is taken to be nuclear-derived heat

TABLE 60.3   Transportation sector energy and efficiency data.

Transport mode and primary energy form	Breakdown of sector energy use, by mode of transport		Energy efficiencies (%)	
	(PJ)	(%)	Rated load	Estimated operating
Road (gasoline, diesel, propane, electricity)[a]	1257.3	77.8	25	16
Rail (diesel)	75.39	4.6	35	28
Air (jet fuel)	131.2	8.1	35	28
Marine (diesel, gasoline)	74.82	4.6	–	15
Pipelines (natural gas)	86.21	5.3	–	29

[a]Road transport includes retail pump sales for road use (for which energy use is 1093 PJ) and urban transit, and the energy efficiencies are weighted means of the efficiencies for each of these.

TABLE 60.4   Electricity generation data.

Generation mode	Sector served	Electricity generated (PJ)	Energy efficiency (%)	Energy input (PJ)					
				Flowing water	Uranium	Coal	Petroleum	Natural gas	Total
Fossil fuel[a]	Utility and industry	293.0	32	0	0	835	61.3	19.2	915.6
Hydraulic[b]	Utility	1000	85	1176	0	0	0	0	1176
Nuclear	Utility	230.1	30	0	767.0	0	0	0	767.0

[a]28.6% of total fossil fuel energy is input to the utility sector and 71.3% to the industrial sector.
[b]The hydraulic source is taken to be the change in kinetic and potential energy of flowing water, and not a fossil fuel equivalent.

at a temperature so high that the energy and exergy of the nuclear-derived heat are equivalent. The efficiency definitions for nuclear plants used here follow nuclear industry conventions (although these efficiency definitions are inadequate because they are based on the heat released from the nuclear fuel rather than its energy or exergy content).

- Kinetic energy production: Fossil fuel-driven kinetic energy production processes occur in some transportation-sector devices (e.g., turbojet engines) and increase the kinetic energy of matter.

Energy and exergy efficiencies are based in this analysis on the general definitions in Equations (60.5) and (60.6).

The reference-environment model used has a temperature T_o of 25°C, a pressure P_o of 1 atm, and a chemical composition consisting of air saturated with water vapor, and the following condensed phases at 25°C and 1 atm: water (H_2O), gypsum ($CaSO_4 \cdot 2H_2O$) and limestone ($CaCO_3$). Exceptions are noted where relevant.

The sectoral analysis methodologies can be summarized as follows:

- For the transportation sector, weighted means are obtained for the transportation-mode energy efficiencies in Table 60.3, where the weighting factor is the fraction of the total sector energy input which supplies each transportation mode (fourth column in Table 60.3).

- For the utility sector, the mean overall energy and exergy efficiencies for the utility sector are evaluated using data from Table 60.4.

- For the residential–commercial sector, the end-use energy quantities (Table 60.5) and the operating data for its principal devices (Table 60.6) are used to determine the device exergy efficiencies. Weighted mean energy and exergy efficiencies are calculated for the overall residential–commercial sector using a three-step

TABLE 60.5   Residential–commercial sector energy use.

Device	Breakdown (in %) of energy use in the sector, by type		
	Fossil fuel	Electricity	Total
Heaters			
Space	79.5	15.0	66.0
Hot water	15.0	14.0	10.0
Subtotal	94.5	29.0	76.0
Appliances			
Cooking	4.5	5.0	1.7
Clothes drying	1.0	5.0	1.3
Subtotal	5.5	10.0	3.0
Others (fans, pumps, air conditioners, refrigerators, freezers, washers, lights, etc.)	0.0	61.0	21.0
Total	100	100	100

TABLE 60.6   Residential–commercial sector device efficiencies[a].

Device	Product-heat temperature (K)[b]	Heating energy efficiency (%)		Heating exergy efficiency (%)	
		By electricity	By fuel	By electricity	By fuel
Space heaters	328	100.0	65.0	17.1	11.1
Water heaters	350	93.0	62.0	25.4	14.0
Cooking appliances	394	80.0	–	22.5	–
Clothes dryers	350	50.0	50.0	9.6	10.3

[a]Corresponding reference-environment temperatures are 272 K for space heating and 283 K for all other devices.
[b]Product-heat temperatures are the same for fossil fuel and electrical heaters, except water heaters have a fossil fuel product-heat temperature of 374 K.

TABLE 60.7   Industrial sector heating temperatures and energy efficiencies.

Product-heat temperature		Heating energy efficiency (%)	
Range (°C)[a]	Category	By fuel	By electricity
<121	Low	65	100
121–399	Medium	60	90
>399	High	50	70

[a]Temperature ranges are subjectively determined based on many commercial processes.

process. First, weighted means are obtained for the electrical energy and exergy efficiencies for the devices in Table 60.6, where the weighting factor in both cases is the ratio of the electrical energy input to the device to the total electrical energy input to all of the devices (Table 60.5). Second, weighted mean efficiencies for the fossil-fuel-driven devices in Table 60.6 are similarly determined. Third, overall weighted means are obtained for the energy and exergy efficiencies for the electrical and fossil-fuel processes together, where the weighting factor is the ratio of the total energy (either fossil-fuel or electrical) input to the residential–commercial sector to the total energy input to the sector. These means are representative of the sector, since the devices in Table 60.6 account for more than half of the total sector energy consumption.

• For the industrial sector, the five most significant industries are identified, and a single representative commercial process for each is examined. The identified industries with the selected commercial processes are as follows: pulp and paper (integrated paper mill plant), iron and steel (blast furnace and steel mill plant), cement (hydraulic cement factory), petroleum (petroleum refining plant), and chemical [NEC (salt) chemical preparation plant]. These five industries are taken to be representative of the sector since they account for approximately half of the sector energy use. In the sector, only heating and mechanical processes are considered, since these processes account for 81% of the industrial sector energy consumption (66% for heating and 15% for mechanical drives). Following the temperature categories in Table 60.7, process heat data for the five selected commercial processes are listed in Table 60.8, with the percentages of the heat in each range supplied by fossil fuels and electricity. Energy consumptions in the form of electricity and fossil fuels for each of the five main industries are obtained by averaging the percentage of electrical and fuel energy demands for each of the industries. Three steps are used in evaluating industry efficiencies. First, energy and exergy efficiencies are obtained for process heating for each of the temperature range categories in Table 60.8, using data from Tables 60.7 and 60.8. Second, mean heating energy and exergy efficiencies for each of the five industries (see the fifth and sixth columns of Table 60.9) are evaluated

TABLE 60.8   Process heating data for all categories of product-heat temperature for the industrial sector.

Industry	Product-heat temperature category	Mean product-heat temperature (°C)	Breakdown of energy use in product-heat temperature category, by type	
			Electricity (%)	Fuel (%)
Pulp and paper	Low/Medium/High	67/160/732	100/0/0	0/ 83/17
Iron and steel	Low/Medium/High	45/–/983	4/0/96	0/0/100
Cement	Low/Medium/High	31/163/1482	92/0/8	1/9/90
Petroleum	Low/Medium/High	57/227/494	10/9/80	14/23/64
Chemical	Low/Medium/High	42/141/–	63/38/0	0/100/0

TABLE 60.9   Industrial sector energy use and efficiencies.

Industry	Breakdown of sector energy use, by type (PJ)			Energy efficiency (%)		Exergy efficiency (%)	
	Electrical	Fossil fuel	Total	Heating	Total	Heating	Total
Pulp and paper	181.4	169.6	351.0	80.0	82.0	20.4	33.0
Iron and steel[a]	155.7	284.5	440.2	58.0	64.0	43.4	52.0
Cement	5.38	52.29	57.7	56.0	62.3	37.5	47.0
Petroleum	16.45	87.81	104.4	65.0	70.0	29.4	41.0
Chemical	63.00	140.8	203.8	71.0	74.5	18.5	32.0

[a]Iron and steel includes smelting and refining operations.

using a two-step procedure: (i) weighted mean efficiencies for electrical heating and for fuel heating are evaluated for each industry in Table 60.8, using as weighting factors the values listed in the last two columns of Table 60.8; and (ii) weighted mean efficiencies for all heating processes in each industry are evaluated with these values, using as weighting factors the ratio of the industry energy consumption (electrical or fuel) to the total consumption of both electrical and fuel energy (see Table 60.9). Third, weighted mean overall (i.e., heating and mechanical drive) efficiencies for each industry (see the last two columns in Table 60.9) are evaluated using the values in Table 60.9, using as weighting factors the fractions of the total sector energy input for both heating and mechanical drives, which serve each of heating and mechanical drives (as given earlier).

60.12.3   Input Data and Results

Figures 60.6 and 60.7 illustrate, respectively, the flows of energy and exergy in Canada for the year considered. The left sides of the figures consist of the total energy or exergy inputs to all sectors, by fuel type. These quantities are then branched according to sector demand. Fuels such as coal, gas, and petroleum are routed to "Fuel for Electricity," and include fuels used to produce electricity by utilities as well as by industry. Similarly, "Fuel for End Uses" represents the net fuel requirement for all sectors and is routed together with electricity and steam flows to the sectors. The final branching into useful and waste energy is achieved by multiplying the energy for end use (fuel, steam, and electricity) in each sector by the corresponding efficiency (see the values summarized in Figure 60.8). The energy-flow diagram indicates that 51% of the total energy consumed in Canada is converted to

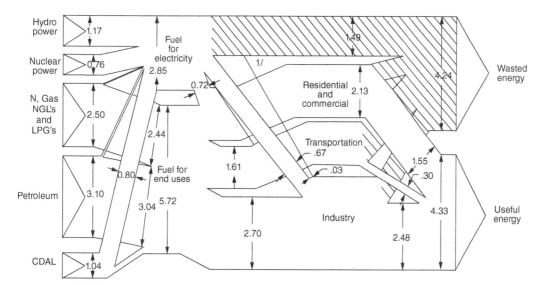

FIGURE 60.6   Flows of energy (in EJ) in Canada. Widths of lines are proportional to the magnitude of the quantity represented. Hydraulic energy is reported in kinetic energy equivalent. The portions of coal, petroleum, and natural gas used for industrial sector electricity generation are routed to the utility sector and then back to the industrial sector in the form of electricity. The flow associated with the symbol 1/ represents steam produced in the utility sector for the end-use sectors.

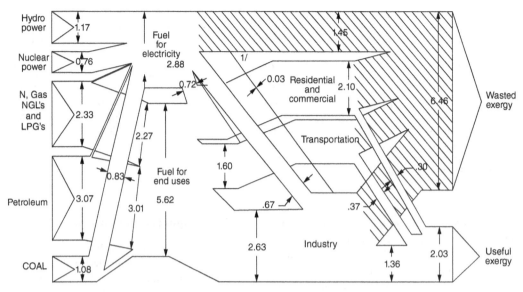

FIGURE 60.7   Flows of exergy (in EJ) in Canada. Details are as for Figure 60.6.

useful energy for end uses, and that the transportation and utility sectors cause the most wastes. The exergy-flow diagram differs markedly from the energy diagram in that only 24% of Canadian exergy consumption is converted to useful exergy for end uses, with the residential–commercial and industrial sectors causing the greatest wastes.

For the transportation sector, a breakdown by mode of transport of sector energy use is presented in the first three columns of Table 60.3, and the corresponding energy efficiencies in the last two columns. Since vehicles generally are not operated at full load, a distinction is made between rated load (full load) and operating (part load) efficiencies. The energy efficiencies in Table 60.3 are equal to the corresponding exergy efficiencies, and the weighted mean overall efficiency for the transportation sector based on energy, which is equal to that based on exergy, is 18%.

For the utility sector, energy efficiencies and the quantities of electricity generated and energy inputs for the different types of electricity generation are listed in Table 60.4. The energy efficiencies in Table 60.4 are equal to the

corresponding exergy efficiencies, and the mean overall energy and exergy efficiencies for the utility sector are found to be equal at 53%.

For the residential–commercial sector, end-use energy quantities are listed in Table 60.5 and operating data for the principal devices are listed in Table 60.6. For the devices in Table 60.6, weighted means for the energy and exergy efficiencies, respectively, are found to be 89% and 20% for electrical devices and 64% and 12% for the fossil-fuel-driven devices. Of the 2129 PJ input to the residential–commercial sector, 34% is electrical energy and 66% fuel energy. Weighted mean energy and exergy efficiencies are found to be 73% and 14%, respectively, for the overall sector.

For the industrial sector, which is composed of many industries having a wide variety of complex plants, we consider only heating and mechanical processes, since these processes account for 81% of the industrial sector energy consumption (66% for heating and 15% for mechanical drives), and the five most significant industries, which account for about 50% of the sector energy use. Process heat data for selected commercial processes are listed in Table 60.8, following the temperature categories in Table 60.7, with the percentages of the heat in each range supplied by fossil fuels and electricity. Canadian energy consumption in the form of electricity and fossil fuels for each of the five main industries is listed in Table 60.9, with a breakdown of the total energy demand. Note that steam (from utilities) is treated as an industrial sector product and so passes through the sector unchanged, electricity generation in the industrial sector is included in the utility sector and mechanical drive energy efficiencies are assumed to be 90%. Weighted means for the weighted mean overall energy and exergy efficiencies (listed in the last two columns of Table 60.9) for the five principal industries in the industrial sector are obtained, using as

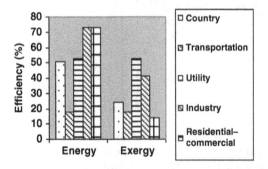

FIGURE 60.8   Energy and exergy efficiencies (in %) for Canada and the four main sectors of its economy.

weighting factors the fraction of the total industrial energy demand supplied to each industry (given in the fourth column of Table 60.9), as 73% and 41%, respectively.

60.12.4 Further Discussion

Generally, exergy efficiencies of the sectors of a national energy system are less than the energy efficiencies, suggesting that energy analysis indicates a more limited margin for efficiency improvement than exergy analysis. Yet only exergy analysis indicates the margin for improvement that actually exists. By using exergy rather than energy analysis, therefore, energy utilization efficiency in the national energy system is more clearly illuminated, and more rational assessments are obtained of allocations of energy-related R&D effort.

The efficiencies for the four sectors and for the overall Canadian economy are compared in Figure 60.8. The energy efficiencies range from 18% to 73% and the exergy efficiencies from 14% to 53%. The most efficient sectors on energy and exergy bases are different. The residential–commercial sector exhibits the greatest differences for energy and exergy evaluations. The most significant differences between energy and exergy efficiencies are attributable to heating processes. For example, the overall exergy efficiency of the residential–commercial sector (14%) is much lower than the overall energy efficiency (73%) because high-quality energy (i.e., high exergy) sources are used for low-quality energy (i.e., low exergy) demands. The higher exergy efficiency of the industrial sector is mainly attributable to the high temperatures required for most industrial heating (≈400°C) compared to residential–commercial heating (≈25°C). The industrial sector thus utilizes relatively more of the quality or work potential of fuels.

The material presented in this section has important implications. In particular, exergy analyses more than energy analyses provide insights into better directions for R&D effort in the sectors of a country, in terms of more promising for significant efficiency gains. By focusing R&D effort on those plant sections or processes with the lowest exergy efficiencies, the effort is being directed to those areas that inherently have the largest margins for efficiency improvement. By focusing on energy efficiencies, on the other hand, one can inadvertently expend R&D effort on areas for which little margins for improvement, even theoretically, exist. Of course, R&D decisions should not be based solely on the results of exergy analyses, and should take into account economic, environmental, safety, social and political factors.

60.13 CASE STUDY 2: COAL-FIRED POWER PLANT

Energy and exergy analyses are applied to a conventional 500 MW coal-fired electrical generating station. This

example clearly illustrates how exergy analysis allows process inefficiencies to be better pinpointed than does an energy analysis, and efficiencies to be more rationally evaluated. Consequently, the example identifies areas where the potential for performance improvement is high, and trends that may aid in the design of future stations.

The analysis presented here is based primarily on an earlier assessment performed by the author of a coal-fired power plant (Rosen, 2001), and draws heavily on that work. The plant primarily involves power generation and internal use of cogeneration. It is pointed out that detailed exergy analyses have been carried out for power plants using coal and other fuels (Dincer and Rosen, 2013; Rosen, 2001; Rosen and Dincer, 2003a; 2003b; Rosen and Horazak, 1995; Rosen and Tang, 2007; 2008; Kwak et al., 2003; Tsatsaronis et al., 1994; Zhang et al., 2007), as well as for cogeneration plants for electrical and thermal energy (Soltani et al., 2014; Dincer and Rosen, 2013; Abusoglu and Kanoglu, 2009; Rosen et al., 2005; Silveira and Tuna, 2004; Vieira et al., 2006) and considering other co-products (Dincer and Rosen, 2013; Tsatsaronis et al., 2008). Many of the articles cited above add exergy-based economic assessments to the exergy analyses.

60.13.1 Description of Coal-Fired Power Plant

The analysis considers the Nanticoke Generating Station, a typical coal-fired power plant (Ontario Hydro, 1996). Located in Ontario, Canada, the plant has been operated since 1981 by the provincial utility, Ontario Power Generation (formerly Ontario Hydro).

Each unit in the coal-fired power plant has a net electrical output of approximately 500 MW. The main process data are listed in Table 60.10 for a plant unit. A process diagram for a single plant unit is shown in Figure 60.9, the main material and energy flows (which are described in detail elsewhere

TABLE 60.10 Main process data for full-load operation of a coal-fired power plant.

Extensive properties (flow rates)		Intensive properties	
Quantity	Value	Quantity	Value
Mass (kg/s)		Temperature (°C)	
Primary steam	454	Boiler feedwater	253
Reheat steam	411	Primary steam	538
Coal	47.9	Reheat steam	538
Cooling water	18,636	Flue gas	120
Electricity (MW)		Cooling water rise	8.3
Gross output	524	Pressure (MPa)	
Internal use	19	Primary steam	16.9
Net output	505	Reheat steam	4.0
		Condenser	0.005

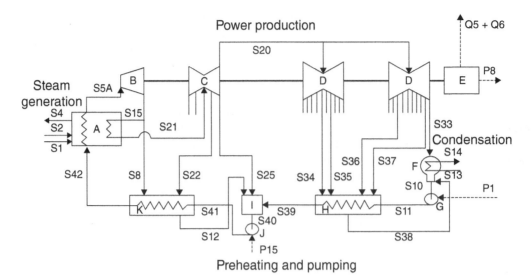

FIGURE 60.9   Single unit of the coal-fired power plant. For the devices, A: steam generator and reheater, B: high-pressure turbine, C: intermediate-pressure turbine, D: low-pressure turbines, E: generator and transformer, F: condenser, G: hot well pump, H: low-pressure heat exchangers, I: open deaerating heat exchanger, J: boiler feed pump, K: high-pressure heat exchangers. Material flows are denoted with a prefix S, thermal flows with a prefix Q and electrical flows with a prefix P. For the material flows, flow S1 denotes coal, S2 air and S4 exhaust combustion gases, and all other flows denote water/steam. Thermal flows Q5 and Q6 denote waste heat from the generator and transformer, respectively, while electrical flows P1 and P15 denote pump work and flow P8 denotes the gross plant output. Lines exiting turbines represent flows of extraction steam.

(Rosen, 2001)). In Figure 60.9, the external inputs for device A are coal and air, and the output is stack gas and solid waste, while the external outputs for device E are electricity and waste heat. Also, electricity is input to devices G and J, and cooling water enters and exits device F. The plant operates, based on the four main sections identified in Figure 60.9, as follows:

- *Preheating* (devices G–K in Figure 60.9). The temperature of the feedwater is increased in several heaters and the pressure is increased in several pumps, to design levels for entering the steam generator.
- *Steam generation* (device A). Eight pulverized-coal-fired steam generators each produce 453.6 kg/s steam at 16.89 MPa and 538°C, and 411.3 kg/s of reheat steam at 4.00 MPa and 538°C. Air is supplied to the furnace by two motor-driven forced draft fans, and regenerative air preheating is used. After treatment in an electrostatic precipitator, the flue gas exits through two multi-flued, 198 m-high stacks.
- *Power production* (devices B–E). Steam passes through a series of turbine generators, attached to a transformer. The turbine generator contains one single-flow high-pressure cylinder, one double-flow intermediate-pressure cylinder, and two double-flow low-pressure cylinders. Steam from the high-pressure cylinder is

reheated in the combustor, and several steam flows for feedwater heating are extracted.
- *Condensation* (device F). The low-pressure turbines exhaust at 5 kPa to the condenser, where the steam is condensed with cooling water from Lake Erie, which is restricted to a specified temperature rise.

60.13.2   Analysis, Assumptions and Data

For the energy and exergy analyses performed, base enthalpy and chemical exergy values reported elsewhere (Rosen and Dincer, 2003a) are used. A common reference-environment model (Gaggioli and Petit, 1977; Rodriguez, 1980) is employed in evaluating energy and exergy, but with a temperature set at the approximate mean for the lake cooling water or 15°C. The reference-environment pressure is 1 atm, and the chemical composition is air saturated with water vapor and the following condensed phases at 15°C and 1 atm: water, gypsum, and limestone. It is noted that realistic variations in reference-environment properties do not significantly affect the energy and exergy results (Rosen and Scott, 1987).

Several simplifications are applied: (1) coal is modeled as graphite and air as 79% nitrogen, and 21% oxygen by volume, (2) the turbine isentropic and mechanical efficiencies are 80% and 95%, respectively, and (3) the generator and the transformer efficiencies are both 99%, with heat losses from their surfaces occurring at 15°C.

Based on Equations (60.5) and (60.6), the overall energy efficiency η is evaluated as

$$\eta = \frac{\text{Net energy output wit h electricity}}{\text{Energy input}} \quad (60.7)$$

and the overall exergy efficiency ψ as

$$\psi = \frac{\text{Net exergy output wit h electricity}}{\text{Exergy input}} \quad (60.8)$$

For most plant components and sections, similar efficiency expressions are applied. Note that a "condenser efficiency" is not straightforwardly defined, as its purpose is not to generate a product, but rather to reject waste heat.

60.13.3 Energy and Exergy Flows and Efficiencies

60.13.3.1 Overall Plant.
Figures 60.10a and 60.10b illustrate the net energy and exergy flows and exergy consumptions for the four main process sections in Figure 60.9. Device exergy consumptions are provided in Table 60.11. Detailed energy and exergy flow rates for material and energy flows are given elsewhere (Rosen, 2001).

Overall energy and exergy efficiencies, where coal provides the only input energy or exergy, can be written using values from Figure 60.10 and Equations (60.7) and (60.8), respectively, as follows:

$$\eta = \frac{(524 - 13)\ \text{MW}}{1368\ \text{MW}}(100\%) = 37\%$$

$$\psi = \frac{(524 - 13)\ \text{MW}}{1368\ \text{MW}}(100\%) = 36\%$$

The difference in efficiencies is minor, and attributable to the specific chemical exergy of coal being slightly greater than its specific base enthalpy.

60.13.3.2 Plant Sections.
Exergy consumptions in the steam generation section are substantial, accounting for 659 MW or 72% of the 916 MW total exergy loss. Of the exergy consumption rate in this section, 444 MW is due to combustion and 215 MW to heat transfer. Energy and exergy efficiencies for the steam generation section, considering the increase in energy or exergy of the water as the product, are 95% and 49%, respectively. The steam generation section is significantly more efficient based on energy than exergy, implying that although 95% of the input energy preheats water, the energy is degraded as it is transferred. Exergy analysis highlights this degradation.

In the condensers, (1) a large quantity of energy enters (775 MW), of which close to the entire amount is rejected; and (2) a small quantity of exergy enters (54 MW), of which about 25% is rejected and 75% is internally consumed. Thus, energy results erroneously suggest that almost all losses in

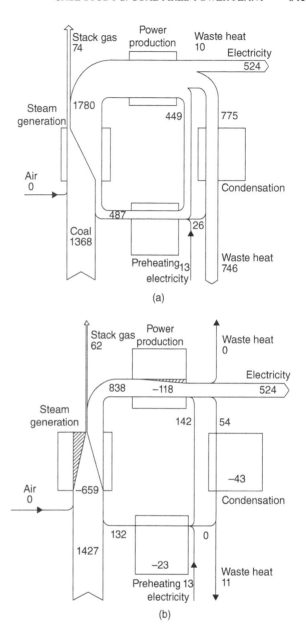

FIGURE 60.10 Flow rates (in MW) through the four main sections of the coal-fired power plant of (a) energy and (b) exergy. Thicknesses of flow lines are proportional to the magnitude of the flow rate of the relevant commodity. Hatched regions and negative values in part (b) denote exergy consumptions.

electricity-generation potential are associated with condenser heat rejection, while exergy analyses demonstrate quantitatively and directly that the condensers are responsible for little of these losses (Figure 60.10). This discrepancy arises because condenser heat is rejected nearly at the environment temperature. Exergy results show that the condenser waste is relatively insignificant for the process. The exergy rejected by the condensers is less than 4% of the net exergy produced by the plant, while the energy rejected exceeds the net energy produced by approximately 50%.

TABLE 60.11 **Exergy consumption rates for coal-fired power plant.**

Section	Device	Exergy consumption rate (MW)
Steam generation	Steam generator (including combustor)	659.0
Power production	Turbines	107.9
	Generator and transformer	10.6
	Section total	118.5
Condensation	Condenser	43.1
Preheating	Heat exchangers	22.2
	Pumps	1.2
	Section total	23.4
Overall		844.0

In the power production section, energy losses are small (less than 10 MW total), and exergy losses are moderately small (around 150 MW). Similarly, energy losses are minor in the preheating section (under 10 MW), as are exergy losses (about 20 MW). Internal consumptions comprise most of the exergy losses in both of these sections.

60.13.3.3 Principal Findings. Several findings about coal-fired power generation have been acquired, the principal ones of which are described here. First, although overall energy and exergy efficiencies for the power plant are similar, exergy analyses – but not energy analyses – identify systematically the locations and causes of inefficiencies. Second, energy losses are mainly associated with emissions (i.e., condenser heat rejection) and exergy losses with consumptions (mainly in combustion and steam generation) and somewhat with cooling water and stack gases. Third, the use of heat rejected by condensers increases the exergy efficiencies little, even though they increase the energy efficiencies notably. Cogeneration can have greater efficiencies than conventional power generation, but the degree of improvement is best assessed with exergy. Relying on energy analyses is problematic since they tend to exaggerate improvements. Fourth, since processes with the largest thermodynamic losses have the greatest margins for efficiency improvement, efforts to increase the efficiencies of coal-fired power generation should focus on the combustor, for example, technologies that generate electricity without combustion like fuel cells or utilize heat at high temperatures could increase efficiencies significantly.

60.14 TUTORIAL ON APPLYING EXERGY ANALYSIS TO PROCESSES RELATING TO POWER GENERATION AND COGENERATION

The tutorial presented in this section attempts to help the reader better understand the application of exergy analysis by illustrating its use and the insights obtained through it, by way of a simplified illustration relating to power generation and cogeneration. A small subsection of a coal-fired steam power plant assessed in the previous section is considered. The objective of the tutorial is to make exergy analysis less confusing to potential users or recipients of the results of exergy analyses, to show that exergy analysis is not excessively complicated and onerous to use, and to highlight how exergy analysis results are meaningful and straightforward to interpret and apply.

60.14.1 Subsystem Considered

The subsystem of the coal-fired power plant considered in this detailed analysis includes the low-pressure turbine (device D in Figure 60.10) and the electrical generator and the transformer (together shown as device E in Figure 60.10). A detailed flow diagram for this subsystem is shown in Figure 60.11. The flows in Figure 60.11 are described in Table 60.12 for material flows and in Table 60.13 for electrical and thermal flows and work interactions.

The analysis covers many aspects of exergy analysis. Various types of energy are involved in this analysis, as is the conversion of energy among different types: thermal to mechanical, mechanical to electrical, electrical to electrical. Note that the choice of subsection considered in this example avoids material flows other than water/steam, allowing chemical exergy to be omitted from the analysis, which helps simplify it greatly. Information on the chemical exergy considerations for the overall plant are included in the previous section.

In the system considered, steam enters the low-pressure turbines and passes to the condenser after expansion. Several steam flows for feedwater heating are extracted at intermediate pressures. The low-pressure turbines each contain a double-flow low-pressure cylinder. Work generated by the turbines drives the generator, which is attached to the transformer. The low-pressure turbines exhaust at 5 kPa to the condenser, where the steam is condensed with cooling water from Lake Erie (with a flow rate of 18,636 kg/s and a temperature rise restricted to 8.3°C).

60.14.2 Analysis

Comprehensive energy and exergy analyses are performed for the subsystem considered of the plant. Base enthalpy and chemical exergy values reported elsewhere (Rosen and Dincer, 2003a) are used.

Several simplifications and assumptions are employed. These include the following, along with relevant data:

- the turbine isentropic efficiency η_{isen} is taken to be 80%.
- the turbine mechanical efficiency η_{mech} is taken to be 95%.

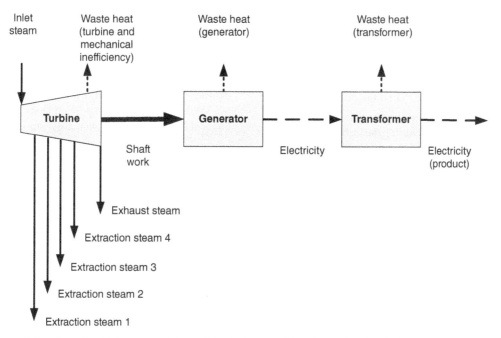

FIGURE 60.11 Subsystem of the coal-fired plant considered, showing the devices included and the flows of material and energy as well as work interactions. Flows of thermal energy are denoted by dashed lines, of electrical energy by dotted lines, of material by thin solid lines, and of work transfers by thick solid lines.

TABLE 60.12 Data for material flows (all H$_2$O) in subsystem of coal-fired power plant in Figure 60.11.

	Intensive flow properties			Flow rates		
Flow	Temperature (°C)	Pressure (MPa)	Vapor fraction[a]	Mass (kg/s)	Energy (MW)	Exergy (MW)
Inlet steam (S20 in Figure 60.9)	360.50	1.03	1.0	367.85	1211.1	411.2
Exhaust steam (S33 in Figure 60.9)	35.63	0.0045	0.93	309.62	774.7	54.1
Extraction steam 4 (S34 in Figure 60.9)	253.22	0.379	1.0	10.47	32.3	9.2
Extraction steam 3 (S35 in Figure 60.9)	209.93	0.241	1.0	23.88	71.7	18.8
Extraction steam 2 (S36 in Figure 60.9)	108.32	0.0689	1.0	12.72	35.8	7.1
Extraction steam 1 (S37 in Figure 60.9)	60.47	0.0345	1.0	11.16	30.4	5.0

[a]Vapor fraction is listed as 0 for liquids and 1 for saturated or superheated vapors or gases.

TABLE 60.13 Data for electrical and thermal flows and work interactions in subsystem of coal-fired power plant in Figure 60.11.

Flow/interaction	Energy flow rate (MW)	Exergy flow rate (MW)
Thermal flows		
Waste heat (turbine and mechanical efficiency)	0.0	8.4
Waste heat (generator)	2.6	0.2
Waste heat (transformer)	2.6	0.2
Electrical flows		
Electricity (exiting generator)	255.2	255.2
Electricity (exiting transformer)	252.6	252.6
Work interactions		
Shaft work (turbine)	257.8	257.8

- the generator efficiency η_{gen} is taken to be 99%, that is, 1% of the shaft work entering the generator is lost as waste heat rather than converted to electricity.

- the transformer efficiency η_{tran} is taken to be 99%, that is, 1% of the electrical energy entering the transformer is lost as waste heat rather than transformed.

- the temperature of the turbine "surface" from which heat losses are emitted (including the turbine rotor shaft) is taken to be the same as the reference-environment temperature T_o, that is, 15°C.

- the temperature of the generator "surface" from which heat losses are emitted is taken to be 40°C.

- the temperature of the transformer "surface" from which heat losses are emitted is taken to be 40°C.

Note that the "surface" temperatures described in the last three bullets are not actual device surface temperatures, but are the temperatures of the control volumes taken to be surrounding each device. These control volumes can be arbitrarily located and can include some of the air surrounding each device.

For the plant components in the system considered, and for the overall system, the energy efficiency η and the exergy efficiency ψ are evaluated as in Equations (60.5) and (60.6), respectively.

As for the analysis of the overall plant (Rosen, 2001), the reference-environment model of Gaggioli and Petit (1977) and Rodriguez (1980) is employed in evaluating energy and exergy quantities, but with a temperature set at the approximate mean for the lake cooling water or 15°C. The reference-environment pressure is 1 atm, and the chemical composition is air saturated with water vapor and the following condensed phases at 15°C and 1 atm: water, gypsum, and limestone.

60.14.3 Energy and Exergy Flows and Efficiencies

An energy/exergy flow diagram for the subsystem of the coal-fired plant considered is shown in Figure 60.12. Flows of thermal energy are denoted by dashed lines, of electrical energy by dotted lines and of material by thin solid lines, while work transfers are denoted by thick solid lines. Indicated are flow rates of energy (values not in parentheses) and exergy (positive values in parentheses) for flows and interactions, and exergy consumption rates (negative values in parentheses) for devices.

Only the portion of the work and electricity outputs due to the low-pressure turbine are shown here, because only the low-pressure turbine is considered in this analysis. Thus, the shaft work exiting the turbine in Figure 60.12 only represents the output of the low-pressure turbine. In reality, more work is produced in the plant since the turbine shaft of the

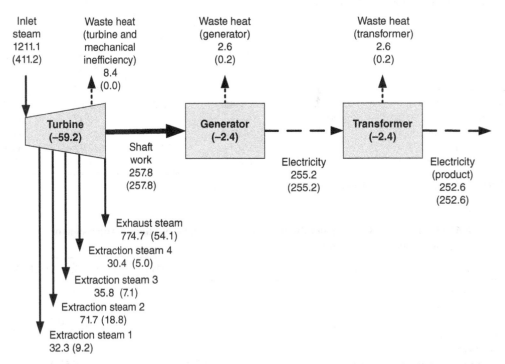

FIGURE 60.12 Energy and exergy flow diagram for the subsystem of the coal-fired plant considered. Flows of thermal energy are denoted by dashed lines, of electrical energy by dotted lines and of material by solid lines. Indicated are flow rates of energy (values not in parentheses) and exergy (positive values in parentheses) for flows and interactions, and exergy consumption rates (negative values in parentheses) for devices. All values are units of MW.

TABLE 60.14   Energy and exergy balances for the system considered.

	Energy		Exergy	
Flow/loss	MW	% of overall input energy	MW	% of overall input exergy
Inputs				
Inlet steam	1211.1	100.0	411.2	100.0
Outputs and losses				
Output products (electricity and extraction steam flows)	422.8	34.9[a]	292.7	71.1[b]
Output losses (exhaust steam and waste heat flows)	788.3	65.1	54.5	13.3
Internal losses (turbine and shaft, generator and transformer)	–	–	64.0	15.6
Total outputs and losses	1211.1	100.0	411.2	100.0

[a]This value is the overall energy efficiency of the subsystem considered.
[b]This value is the overall exergy efficiency of the subsystem considered.

low-pressure turbine is also linked to high- and intermediate-pressure turbines.

It is seen in Figure 60.12 that the energy entering the system via steam is much greater than the exergy entering, and that the energy associated with each of the steam extraction flows and the turbine exhaust steam is much greater than the corresponding exergy. The main exergy loss is associated with the turbine and is in the form of an exergy consumption, which is due to the irreversibilities of the expansion process in the turbine.

It is also observed that the exergy and energy rates are the same for work and electricity, while the energy and exergy rates differ for material and thermal energy flows. Here, the exergy rates of material and thermal flows are less than the energy rates, but this is not the case in general.

Note that if the extraction steam flows are eliminated, the work generated by the turbine would increase. However, the heat input to the overall plant would simultaneously have to increase at a higher proportion, resulting in a reduction in the overall energy and efficiencies of the power plant. On an energy basis, the amount of heat removed by extraction steam flows is large, but these values are all correspondingly much smaller on exergy bases.

60.14.4   Energy and Exergy Balances

Energy and exergy balances for the system considered are presented in Table 60.14, in absolute units and as a percentage of the overall input energy or exergy. Again, it is pointed out that only the portion of the work and electricity outputs due to the low-pressure turbine are shown here. The observations noted via the flow diagram in Figure 60.12 are reinforced by Table 60.14, but in a different manner. In particular, the differences in the energy and exergy losses are highlighted.

Normalized energy and exergy balances, in the form of pie charts, are presented in Figure 60.13. The left half of each pie chart represents inputs (of either energy or exergy),

while the right half represents outputs (and consumptions in the case of exergy). The difference in the relative magnitude of the energy and exergy of the products and of the losses are emphasized. The electrical product is a much greater proportion of the outputs and losses based on exergy rather than energy. Exergy losses include consumptions while energy losses do not. The main exergy losses are associated with consumptions while the main energy losses are associated with waste emissions.

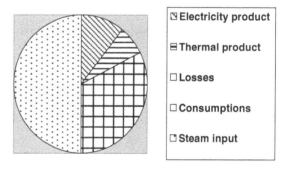

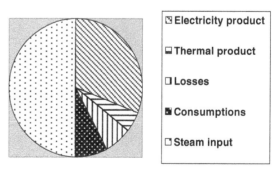

FIGURE 60.13   Normalized balances on the basis of energy (top) and exergy (bottom). The left half of each pie chart represents inputs, while the right half represents outputs (and in the case of exergy consumptions).

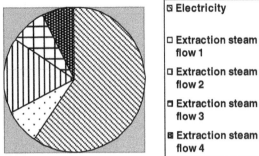

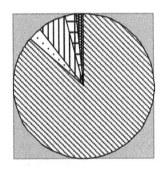

FIGURE 60.14   Breakdown of products (electrical and thermal) on the basis of energy (top) and exergy (bottom).

60.14.5   Breakdown of Energy and Exergy Products

Breakdowns of the products (electrical and thermal) for the system considered are presented in Figure 60.14, on both energy and exergy bases. This figure emphasizes how electricity is much more valuable than heat at finite temperatures, and the exergy results demonstrate this while the energy results do not.

60.14.6   Breakdown of Energy and Exergy Losses

A breakdown is shown in Table 60.15 of the energy and exergy losses for the system, in absolute units and as a percentage of the overall input energy or exergy. The losses are divided into two types: external losses (i.e., waste emissions) and internal losses (i.e., consumptions). External losses of both energy and exergy occur, but internal losses are only possible for exergy and these are the exergy consumptions, which are often referred to as exergy destructions. Internal losses are not possible for energy because energy is conserved.

On energy and exergy bases, the losses appear significantly different. The energy loss is totally due to waste emissions, but for exergy losses the contributions of exergy consumption and waste exergy emission are both significant at 54% and 46% of the total exergy loss, respectively.

The main exergy loss is associated with internal consumptions, predominantly in the turbine, where 50% of the total exergy loss occurs. The second highest exergy loss is associated with the low-pressure turbine exhaust steam flow passing to the condenser (which is responsible for 46% of the total exergy loss).

A significant observation in Table 60.15 is that the exergy loss rate (788.3 MW) is over six times the exergy loss rate (118.5 MW). This is typical of successful industrial systems which tend not to release significant quantities of exergy with waste flow, usually because the usefulness or quality of these flows makes them valuable and worth recovering. There is no corresponding logical behavior observed in general for energy losses, which can be larger or smaller and do not generally reflect quality or value.

60.14.7   Energy and Exergy Efficiencies

Component and overall efficiencies for the subsystem considered are presented in Table 60.16, on energy and exergy bases.

TABLE 60.15   Breakdown of losses for the system considered.

Loss source	Energy loss rate		Exergy loss rate	
	MW	% of overall input energy	MW	% of overall input exergy
External				
Exhaust steam to condenser	774.7	98.3	54.1	45.7
Waste heat from turbine and shaft	8.4	1.1	0.0	0.0
Waste heat from generator	2.6	0.3	0.2	0.2
Waste heat from transformer	2.6	0.3	0.2	0.2
Subtotal	788.3	100.0	54.5	46.0
Internal				
Turbine	–	–	59.2	50.0
Generator	–	–	2.4	2.0
Transformer	–	–	2.4	2.0
Subtotal	–	–	64.0	54.0
Overall total	788.3	100.0	118.5	100.0

TABLE 60.16   Component and overall efficiencies for the system considered.

Device	Product outputs	Inputs	Energy efficiency, η (%)	Exergy efficiency, ψ (%)
Turbine (overall)	Shaft work, extraction steam	Low-pressure steam	35.3	72.4
Turbine (work production only)	Shaft work	Low-pressure steam	21.3	62.7
Generator	Electricity	Shaft work	99.0	99.0
Transformer	Electricity	Electricity	99.0	99.0
Overall	Electricity, extraction steam	Low-pressure steam	34.9	71.1

For clarity, the product outputs and inputs for each device are listed. Note that other efficiencies than those reported here are possible. For example, isentropic and polytropic efficiencies are often cited for turbines.

In Table 60.16, two efficiencies are listed for the turbine: one considering electricity and useful heat as the "product" and one considering only electricity as the product. Because the exergy of the heat is less than its energy, the energy efficiencies vary greatly by an absolute increment of 14% (or the exergy efficiency with heat included as a product is 65.7% greater than without), but the exergy efficiency varies by less (only 9.7% in absolute terms), or the exergy efficiency with heat considered as a product is only 15.5% greater than without. The overall efficiency in Table 60.16 considers electricity and useful heat as a product.

The turbine efficiencies are both much less than those of the generator or transformer. The overall efficiencies are similar to the turbine (overall) efficiency because the efficiencies of the generator and transformer are very high. The exergy efficiency of the turbine is significantly greater than the energy efficiency because the turbine converts a high proportion of the "work potential" (i.e., exergy) in the incoming steam to work, but a much lower proportion of the energy with incoming steam, which is not a measure of its "work potential," to work. Thus exergy analysis indicates that the turbine is much more efficient thermodynamically than does energy analysis, reflecting a meaningful and useful efficiency for the device.

The extraction steam flows represent an internal use of cogeneration in the coal-fired power plant. The efficiencies in Table 60.16 change if this use of cogeneration is expanded to recover the exhaust heat flow from the low-pressure turbine rather than sending it to the condenser where its heat is rejected to the lake. Even though the temperature of the low-pressure turbine exhaust steam is relatively low (approximately 36°C), such a use of expanded cogeneration could find some applications (e.g., aquaculture). If the efficiencies obtained here are reevaluate, it is seen that cogeneration has a small effect on the exergy efficiency, but a large effect on the energy efficiency. This observation suggests that exergy efficiencies realistically assess the benefits of cogeneration, while energy efficiencies tend to exaggerate the benefits.

60.14.8   Further Discussion

Several illuminating insights have been attained about the subsystem of the coal-fired power generation plant considered in the present article:

- Energy losses are mainly in waste emissions (i.e., low-pressure exhaust steam, which passes to the condenser where heat is rejected), and exergy losses are primarily with consumptions (mainly due to expansion in the turbine) and somewhat with waste heat emissions.
- The energy and exergy efficiencies of the subsystem considered and the components within it are in general different, even though they are sometimes similar. Energy analyses do not identify the locations and causes of inefficiencies, while exergy analyses do so systematically. Thus exergy analysis results identify the margin for improvements, unlike energy analysis.
- Since devices with the largest thermodynamic losses have the greatest margins for efficiency improvement, efforts to increase the efficiencies of the subsystem considered here of the coal-fired power plant should focus on the turbine. Significant improvements may be achievable through alternative expanders and technologies that generate shaft work from steam more efficiently or turbine-efficiency improvement measures.
- The use of the heat of the turbine exhaust steam, which is rejected by the condensers, only increases the exergy efficiency of the subsystem considered by a few per cent. Cogeneration systems, which produce heat at useful temperatures at the expense of reduced electrical output, can have greater efficiencies than conventional power generation, but the merit of cogeneration systems must be determined using exergy because energy analyses tend to overstate performance.

It is pointed out that the largest exergy consumption in the overall power plant occurs elsewhere in the plant (rather than the three devices in the subsection considered here), as can be seen in the case study examined in the previous section. Specifically, the largest plant losses are associated with the combustion of fuel and in heat transfer from the combustion gases to the working fluid which occurs across large temperature differences. Both of these processes are highly irreversible, leading to their large exergy losses. The exergy consumption in those processes is an order of magnitude greater than the exergy loss for subsystem including the three devices considered here.

If a use for the low-pressure turbine exhaust steam can be found, it becomes a by-product and the energy efficiency increases significantly (and the energy loss rate decreases to 13.6 MW from 788.3 MW). The exergy efficiency is much less affected, with the exergy loss rate decreasing to 64.4 MW from 118.5 MW. This is analogous to converting the entire plant to cogeneration (noting that thermal energy in extraction flows are already used for productive purposes of preheating, in essence acting as internal cogeneration). But this idea is dependent on the temperature of the low-pressure turbine exhaust steam being large enough to have a use.

The results obtained in this analysis have important economic repercussions. It can be seen in Figure 60.14 that if the costs of products are weighted on the basis of energy, much different results are obtained than if they are weighted on the basis of exergy. Specifically, it is noted that

- Weighting by energy overvalues heat and undervalues electricity.
- Weighting by exergy is more realistic and emulates what is observed in practice more closely (implying an exergy weighting yields more realistic costs or prices).

The results obtained in this analysis also have significant environmental aspects. The waste energy emission is seen in Figure 60.13 to be large, which emphasizes a large energy emission to the environment. But this large magnitude is somewhat misleading in terms of potential for environmental impact of this waste, because the release is in or nearly in equilibrium with the environment, since the heat release is almost at the reference-environment temperature T_o. The exergy reflects this reality, as there is little exergy in the waste heat emissions.

Results similar to those obtained from this analysis of a coal-fired power plant have been observed for nuclear, natural gas, and oil plants, based on previous energy and exergy analyses of such plants by the author (Rosen, 2001; Rosen and Dincer, 2003b; Rosen and Horazak, 1995; Rosen and Tang, 2007; 2008).

Through this tutorial, an attempt has been made to illustrate how exergy analysis improves understanding of the thermodynamic performance of a process and identifies areas having significant potential for improvement. By considering a subsystem of an actual coal-fired power plant consisting the low-pressure turbine, the electrical generator and the electrical transformer, the tutorial quantifies energy and exergy efficiencies and losses—including internal and external exergy losses. It is demonstrated that exergy analysis is straightforward to apply and its results are readily interpretable and utilizable.

60.15 CLOSING REMARKS

Exergy analyses of processes and systems provide useful information, which can have direct implications on designs, improvements, and applications. Exergy analysis results typically suggest improvement efforts should concentrate more on internal rather than external exergy losses, based on thermodynamic considerations, with a higher priority for the processes having larger exergy losses. Note that this statement does not indicate that effort should not be devoted to processes having low exergy losses, as simple and cost-effective ways to increase efficiency by reducing small exergy losses should certainly be considered when identified. But, exergy analyses more than energy analyses provide insights into better directions for R&D effort, where "better" is loosely taken to mean "more promising for significant efficiency gains." There are two main reasons for this statement:

- Exergy losses represent true losses of the potential that exists to generate the desired product from the given driving input. This is not true in general for energy losses. Thus, if the objective is to increase efficiency while accounting for energy degradation, focusing on exergy losses permits R&D to focus on reducing losses that will affect the objective.
- Exergy efficiencies always provide a measure of how nearly the operation of a system approaches the ideal, or theoretical upper limit. This is not in general true for energy efficiencies. By focusing R&D effort on those plant sections or processes with the lowest exergy efficiencies, the effort is being directed to those areas that inherently have the largest margins for efficiency improvement. By focusing on energy efficiencies, on the other hand, one can inadvertently expend R&D effort on areas for which little margins for improvement, even theoretically, exist.

Decisions on designs, improvements, applications, and R&D generally should consider energy and exergy in concert with such other factors as economics, environmental impact, safety, and social and political implications.

NOMENCLATURE

Ex	exergy
Ex_{dest}	exergy destruction
Ex_{flow}	exergy of a matter flow
Ex_{kin}	kinetic exergy

Ex_o	chemical exergy
Ex_{ph}	physical exergy
Ex_{pot}	potential exergy
Ex_Q	thermal exergy
P_o	reference-environment pressure
Q	heat
T	temperature
T_i	temperature of thermal flow i
T_o	reference-environment temperature
η	energy efficiency
η_{isen}	turbine isentropic efficiency
η_{mech}	turbine mechanical efficiency
η_{gen}	generator efficiency
η_{tran}	transformer efficiency
ψ	exergy efficiency

Acknowledgments

The author gratefully acknowledges the financial support provided by the Natural Sciences and Engineering Research Council of Canada.

REFERENCES

Abusoglu, A., Kanoglu, M. (2009). Exergetic and thermoeconomic analyses of diesel engine powered cogeneration. Part 1: Formulations and Part 2: Application. *Applied Thermal Engineering*, 29:234–249.

Dincer, I., Rosen, M. A. (2011). *Thermal Energy Storage: Systems and Applications*, 2nd edition. Wiley, London.

Dincer, I., Rosen, M. A. (2013). *Exergy: Energy, Environment and Sustainable Development*, 2nd edition. Elsevier, Oxford, UK.

Feidt, M. (2009). Optimal thermodynamics—New upperbounds. *Entropy*, 11:529–547.

Gaggioli, R. A. (1998). Available energy and exergy. *International Journal of Applied Thermodynamics*, 1:1–8.

Gaggioli, R. A., Petit, P. J. (1977). Use the second law first. *Chemtech*, 7:496–506.

Kotas, T. J. (1995). *The Exergy Method of Thermal Plant Analysis*, Reprint edition. Krieger, Malabar, Florida.

Kwak, H.-Y., Kim, D.-J., Jeon, J.-S. (2003). Exergetic and thermoeconomic analyses of power plants. *Energy*, 28:343–360.

Moran, M. J. (1989). *Availability Analysis: A Guide to Efficient Energy Use*, Revised edition. American Society of Mechanical Engineers, New York.

Moran, M. J., Shapiro, H. N., Boettner, D. D., Bailey, M. B. (2011). *Fundamentals of Engineering Thermodynamics*, 7th edition. Wiley, New York.

Ontario Hydro. (1996). Nanticoke Generating Station Technical Data. Information document.

Rodriguez, L. S. J. (1980). Calculation of available-energy quantities. In: *Thermodynamics: Second Law Analysis (ACS Symposium Series*, Vol. 122). American Chemical Society, Washington, DC, pp. 39–60.

Rosen, M. A. (1992). Evaluation of energy utilization efficiency in Canada using energy and exergy analyses. *Energy—The International Journal*, 17:339–350.

Rosen, M. A. (2001). Energy- and exergy-based comparison of coal-fired and nuclear steam power plants. *Exergy, An International Journal*, 1:180–192.

Rosen, M. A. (2011). *Economics and Exergy: An Enhanced Approach to Energy Economics*. Nova Science Publishers, Hauppauge, NY.

Rosen, M. A. (2012). *Environment, Ecology and Exergy: Enhanced Approaches to Environmental and Ecological Management*. Nova Science Publishers, Hauppauge, NY.

Rosen, M. A., Dincer, I. (1997). Sectoral energy and exergy modeling Turkey. *ASME Journal of Energy Resources Technology*, 119:200–204.

Rosen, M. A., Dincer, I. (2003a). Thermoeconomic analysis of power plants: an application to a coal-fired electrical generating station. *Energy Conversion and Management*, 44:2743–2761.

Rosen, M. A., Dincer, I. (2003b). Exergoeconomic analysis of power plants operating on various fuels. *Applied Thermal Engineering*, 23:643–658.

Rosen, M. A., Horazak, D. A. (1995). Energy and exergy analyses of PFBC power plants. In: *Pressurized Fluidized Bed Combustion*, Alvarez Cuenca, M., Anthony, E. J. (editors). Chapter 11, Chapman and Hall, London, pp. 419–448.

Rosen, M. A., Le, M. N., Dincer, I. (2005). Efficiency analysis of a cogeneration and district energy system. *Applied Thermal Engineering*, 25:147–159.

Rosen, M. A., Scott, D. S. (1987). On the sensitivities of energy and exergy analyses to variations in dead-state properties. In: *Analysis and Design of Advanced Energy Systems: Fundamentals*, AES-Vol. 3-1, Moran, M. J., Gaggioli, R. A., (editors). American Society of Mechanical Engineers, New York, pp. 23–32.

Rosen, M. A., Tang, R. (2007). Improving electrical generating station efficiency by varying stack-gas temperature. International Journal of Green Energy, 4:589–600.

Rosen, M. A., Tang, R. (2008). Improving steam power plant efficiency through exergy analysis: effects of Altering excess combustion air and stack-gas temperature. *International Journal of Exergy*, 5:31–51.

Sato, N. (2005). *Chemical Energy and Exergy: An Introduction to Chemical Thermodynamics for Engineers*. Elsevier, Oxford, UK.

Silveira, J. L., Tuna, C. E. (2004). Thermoeconomic analysis method for optimization of combined heat and power systems. Part II. *Progress in Energy and Combustion Science*, 30:673–678.

Soltani, R., Mohammadzadeh Keleshtery, P., Vahdati, M., Khoshgoftar Manesh, M. H., Rosen, M. A., Amidpour, M. (2014). Multi-objective optimization of a solar-hybrid cogeneration cycle: application to CGAM problem. *Energy Conversion and Management*, 81:60–71.

Szargut, J. (2005). *Exergy Method: Technical and Ecological Applications*. WIT Press, Southampton, UK.

Szargut, J., Morris, D. R., Steward, F. R. (1988). *Exergy Analysis of Thermal, Chemical and Metallurgical Processes*. Hemisphere, New York.

Tsatsaronis, G. (2008). Recent developments in exergy analysis and exergoeconomics. *International Journal of Exergy*, 5:489–499.

Tsatsaronis, G., Kapanke, K., Marigorta, A. M. B. (2008). Exergoeconomic estimates for a novel zero-emission process generating hydrogen and electric power. *Energy*, 33:321–330.

Tsatsaronis, G., Lin, L., Tawfik, T., Gallaspy, D. T. (1994). Exergoeconomic evaluation of a KRW-based IGCC power plant. *Journal of Engineering for Gas Turbines and Power*, 11:300–306.

Vieira, L. S., Donatelli, J. L., Cruz, M. E. (2006). Mathematical exergoeconomic optimization of a complex cogeneration plant aided by a professional process simulator. *Applied Thermal Engineering*, 26:654–662.

Wall, G. (2003). Exergy tools. *Proceedings of the Institution of Mechanical Engineers, Part A: Journal of Power and Energy*, 217:125–136.

Yantovskii, E. I. (1994). *Energy and Exergy Currents (An introduction to Exergonomics)*. Nova Science Publishers, New York.

Zhang, C., Chen, S., Zhen, C., Lou, X. (2007). Thermoeconomic diagnosis of a coal fired power plant. *Energy Conversion and Management*, 48:405–419.

61

GLOBAL TRANSPORT ENERGY CONSUMPTION

Patrick Moriarty and Damon Honnery

61.1 GLOBAL TRANSPORT ENERGY STATISTICS: PAST AND PRESENT

61.1.1 Introduction: Basic Concepts

All motorized transport, whether for moving passengers or freight, requires the expenditure of energy. Even non-motorized transport requires the expenditure of food energy by humans, or fodder by animals. Today, transport energy uses nearly one-quarter of global primary energy. It is useful to distinguish between two categories of energy: primary and secondary. The International Energy Agency (IEA) [1] defines total primary energy supply (TPES) as follows: "production + imports – exports – international marine bunkers – international aviation bunkers ± stock changes. For the world total, international marine bunkers and international aviation bunkers are not subtracted from TPES". Also, imports and exports are irrelevant at the global level. Primary transport energy consumption thus includes not only the energy of the gasoline in the tank, for example, but also the energy needed to discover, produce, transport, and refine the crude oil, and transport the gasoline to the service station. Following the total transport energy chain from oil wells to the motive power provided by the engine is often called a "well-to-wheels energy" analysis.

However, vehicles do not use crude oil as fuel, but refined oil products such as gasoline, diesel, and aviation turbine fuel. These refined fuels are called secondary energy and their combined use "total final consumption" (TFC), to use the IEA terminology. The IEA defines TFC as follows: "(TFC) is the sum of consumption by the different end-use sectors. Backflows from the petrochemical industry are not included in final consumption". Thus in addition to petroleum-derived fuels, delivered natural gas and electricity are also included

in transport TFC. Energy analyses that just look at vehicle gasoline or diesel consumption are often termed "tank-to-wheels" analyses, as in the familiar "miles per gallon" (mpg) data for cars.

Vehicles engines consume their fuel in order to deliver transport services: passenger-km for passenger transport and tonne-km for freight. Two people in a car who go on a 30-km round trip generate 60 passenger-km, for example. For public transport, travel by drivers is excluded. For freight travel, only tonne-km by the payload is included, unless our interest is solely in engine efficiency. Road vehicles use up fuel to provide these services because they must overcome the "road load". This road load has three components: rolling resistance of the tyres, air resistance (drag), and inertial resistance—the energy needed to increase the speed of a vehicle. Together with auxiliary power consumption of vehicles for lighting, air-conditioning, power steering, and so on, the road load accounts for a vehicle's total fuel use in liters or gallons of refined fuel. More detail is given in Section 60.3.1.

61.1.2 Historical Growth of Transport Energy Demand

Before the advent of steam locomotive, the only transport energy was food and fodder for humans, horses, and oxen, and wind energy for sailing ships—all renewable energy sources. Although both stationary steam engines for pumping water from mines and later for powering cotton mills and flanged steel rail systems were widely used by the second half of the 18th century, these two technologies were first brought together in commercial operation with the opening of the Stockton and Darlington railway in the United Kingdom in 1825 [2]. With the rise of motorized transport for rail, road,

Alternative Energy and Shale Gas Encyclopedia, First Edition. Edited by Jay H. Lehr and Jack Keeley.
© 2016 John Wiley & Sons, Inc. Published 2016 by John Wiley & Sons, Inc.

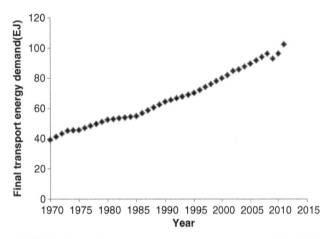

FIGURE 61.1   Global final transport energy demand, 1970–2011 [2, 7].

ocean, and finally air transport, total global transport energy use grew rapidly, and by the year 2011 global final energy demand for transport was 2445 million tonnes of oil equivalent (Mtoe) or 102.4 EJ (EJ = exajoule = 10^{18} joule) [1].

Figure 61.1 shows the growth in final global transport energy demand from 1970 to 2010. Vehicular transport levels—whether passenger or freight—were very low in 1900, but rose rapidly after 1950. For example, world passenger travel was only 121 passenger-km per capita in 1900, reached 1311 in 1950, and 6020 in 2005 [3, 4].

Most of this growth in transport energy has been in countries outside the OECD: Organization for Economic Cooperation and Development. The IEA data [1] show that OECD final transport energy rose from 29.0 EJ in 1973 to 49.9 in 2008, before falling to 45.2 EJ in 2011. For non-OECD countries, final transport energy was only 16.3 EJ in 1973, but rose to 57.2 EJ in 2011 (51.8% of the total), with the rise being especially rapid over the past decade. The global financial crisis saw a fall in world transport energy in both 2009 and 2010.

61.1.3   Fuels Used in Transport

Although coal was the original transport fuel for locomotive-powered trains, and electric motor and steam-powered road vehicles were popular at the turn of the century, oil-derived fuels have long dominated the energy mix for global transport. Table 61.1 shows the breakdown in various energy sources in final global energy demand in the transport sector. The table shows that use of both gas and biofuels has risen since 1973, with coal continuing its steady decline. For the dominant oil-derived fuels, diesel and aviation turbine fuels have both increased their share at the expense of gasoline. Nearly all trucks, buses, and trains (except those using electric traction) run on diesel, and the share of diesel-fueled passenger cars is also increasing, especially outside North

TABLE 61.1   Fuel shares of final world transport energy use in 1973 and 2011 [1].

Fuel	1973 (%)	2011 (%)
Oil	94.30	92.65
Coal	3.05	0.14
Gas	1.64	3.78
Electricity	0.98	1.03
Biofuels	0.03	2.40
Total	100.00	100.00

America. Per liter of fuel, diesel vehicles are more efficient than petrol ones.

61.1.4   Modal Shares of Transport Energy

For the year 2000, the breakdown of the 76.9 EJ of final transport energy demand by mode is given in Table 61.2. As expected, road transport dominates total energy consumption, and passenger transport energy is much larger than freight energy. The percentages for rail, air, and sea include freight as well as passenger transport. For the United States, the US Energy Information Administration (EIA) [6] showed that the total share by road is much higher (78.6%), and that for sea lower (3.7%) than the world average. Total final US transport energy in 2011 was 28.6 EJ.

61.2   GLOBAL TRANSPORT ENERGY PROJECTIONS

61.2.1   Transport Energy Forecasts

A number of organizations, both those with a dominant interest in oil (BP, Shell, the Organization of the Petroleum Exporting Countries (OPEC)), as well as those with a more general interest in energy (EIA, IEA, and the Intergovernmental Panel on Climate Change (IPCC)), have projected either transport energy use and/or the global transport task out for the next few decades. For air travel, Airbus has forecast passenger and freight transport from 2014 to 2033.

TABLE 61.2   Modal shares of final world transport energy use in 2000 [5].

Mode	Share (%)
Road	72.3
Private passenger vehicles	*46.1*
Freight vehicles	*25.0*
Buses	*6.2*
Rail	1.5
Air	11.6
Sea	9.5
All modes	*100.0*

The global oil company, BP [7] in their 'most likely' assessment of the future, project world final energy consumption in transport experiencing steady but slow growth out to 2035, with greater growth in non-OECD countries. The Shell oil company [8] presented two scenarios for future energy demand out to 2060, termed 'Oceans' (roughly corresponding to 'business-as-usual') and 'Mountains', where resource and environmental concerns are more important. For the year 2060, final annual transport energy demand for the world was projected as 143.4–187.3 EJ (a 48%–93% rise over the 2010 value), with the lower values and higher values for the Mountains and Oceans scenarios respectively.

OPEC [9], for the world in 2011, gave a figure of 938 million private transport vehicles together with 189 million commercial vehicles. For the year 2040, OPEC forecast 2159 million cars, and 504 million commercial vehicles. Most cars (64.6% of total) are still owned in OECD countries, but OPEC projected that the annual growth in private vehicles in the OECD will be under one percent over the next two decades. Although vehicle ownership is very unevenly distributed worldwide, this picture is changing, particularly for rapidly industrializing countries. For China, OPEC forecast that car numbers will grow from 73 million in 2010 to 549 million in 2040. Overall, OPEC foresaw oil used for all transport purposes growing from 2010 to 2040, with a decrease in OECD countries more than compensated by non-OECD growth.

Airbus [10] also forecast strong growth in air transport for the next two decades. For passenger travel, revenue passenger-km are expected to rise from 5.5 trillion in 2012 at an annual rate of 4.7% to 14.0 trillion in 2033. As with transport overall, faster growth rates were forecast for non-OECD countries. Air freight was projected to grow at an average annual rate of 4.7% over the next 20 years. Air fuel consumption will rise less rapidly, because further energy efficiency improvements are anticipated. OPEC [9] assumed an average annual growth rate in aviation fuel consumption of 3.4% over the period 2011–2040.

In its Annual Energy Outlook 2014, the EIA [6] saw US transport final energy consumption decreasing out to 2040 at an average annual rate of 0.2% in the base case scenario. Although vehicle-km for passenger and freight vehicles were both projected to grow, passenger vehicles would see declining energy use, but freight vehicles would experience rising energy demand. The EIA also looked at various scenarios involving higher and lower values for economic growth rates compared with those used in the reference scenario. Even for the higher economic growth case, transport energy use in 2040 was projected to be only marginally larger than in 2012. Actual 2008 and 2009 transport energy use in the US was well below that for 2007, because of the global financial crisis. The recent overall forecast for negative growth, however, represents a significant reduction from the more optimistic forecasts of earlier years. Interestingly, the EIA see little growth in alternative transport fuels; oil is assumed to account for over 96% of final transport fuel use in 2040.

61.2.2 Discussion

All the above forecasts are optimistic in the sense that growth in transport energy use is not assumed to be limited by oil availability problems, global climate change, or further international financial turmoil. The Association for the Study of Peak Oil and Gas (ASPO) [11] doubts that the optimistic forecasts for future oil consumption discussed earlier in this section can be realized. Whereas OPEC, the IEA, and the EIA all forecast global oil use for all purposes in 2030 of the order of 100 million barrels per day (about 200 EJ per year), the ASPO forecast is for little more than half this value. Given the almost total reliance of transport on oil, both now and at least for the next two decades, if ASPO is anywhere near correct in their forecast, the consequences for transport energy and transport services are profound.

Other researchers, such as Mitchell [12], and Pelton [13] have also seen future reductions in travel, because of the possibilities for substitution offered by the new information technology (IT). They have argued that not only surface travel but also all passenger travel, both surface and air travel, will decline because of substitution by IT. Massachusetts Institute of Technology (MIT) planner William Mitchell used the term "demobilization" to denote the substitution of work, shopping, and other trips by networked computers. However, the argument that IT will radically reduce travel has now been made for almost three decades. Actual results in the form of travel reductions that can be ascribed to teleworking or teleshopping so far have been disappointing [14–16]. As with the telephone, IT can act to replace physical travel in some circumstances (in the case of teleconferencing, for example), but overall may increase travel by expanding people's social and professional contacts in an increasingly interconnected world.

Even if reductions in travel have not naturally occurred by IT substituting for existing travel, IT has the potential to be one of the ways to cope with reduced travel, should very large reductions in either greenhouse gas emissions and oil consumption be needed in the coming years, given the limited ability of technical solutions like alternative fuels or efficiency gains to deliver very large reductions in energy use in the limited time frame available.

61.3 FUTURE TRANSPORT: ENERGY USE, EFFICIENCY

61.3.1 Reducing Transport Energy Consumption

Transport energy use can be cut in several possible ways:

- Increase the efficiency of the vehicle (vehicle-km per MJ of final energy)
- Increase the efficiency of conversion of primary into secondary energy

- Increase the vehicle loading rates
- Shift from less to more energy efficient transport modes.

The first option has received by far the most attention, but the last, modal shift, is now becoming an important option, especially for urban passenger transport.

61.3.1.1 Efficiency of Vehicular Power Plant and Transmission.
Two approaches are possible for increasing the vehicle-km/MJ (MJ = Megajoule = 10^6 J) for a given vehicle; increase the efficiency of the vehicle's power plant and transmission, or reduce the road load. As shown in Section 60.1.1, the road load has three components—air friction, rolling friction, and inertial load. The advantage of road load reductions is that the improvements can be applied to any vehicle, whereas engine efficiency improvements are often engine-specific; improvements to the combustion efficiency of a conventional car will be of no help in an all-electric vehicle.

According to Cullen et al., [17], large energy savings are possible for cars by reductions in rolling friction (reducing both the coefficient of rolling friction and the vehicular body mass), in wind friction (reducing both the drag coefficient and the vehicle frontal area), and in inertial load (by reducing the vehicular body mass). Truck efficiency improvements can be made by similar methods.

These authors give the practical limit to energy efficiency improvements for each transport mode as follows:

Cars	91%
Trucks	54%
Trains	57%
Air	46%
Ship	63%
Weighted average	*70%*

The figures above are only for reducing the energy *needs* for transport vehicles. Still, further energy savings are possible by engine and transmission improvements. In summary, we can both reduce the amount of fuel needed to produce a given power output, and we can also reduce the average power output needed to move the vehicle and its contents.

However, there are great difficulties in achieving these potential efficiency gains, as they can conflict with other desirable vehicle properties. For cars, high tyre pressures reduce rolling friction—but also ride comfort. Smaller and lighter vehicles may have reduced safety, comfort, durability, and performance. With smaller and lighter cars, the disparity in size and mass with heavy trucks sharing the same road space would further increase, with potential safety problems. Overall (apart from the Corporate Average Fuel Efficiency (CAFE) efficiency improvements in the United States), the gains in car efficiency over recent decades have been disappointing. Increases in engine and transmission efficiency

have been offset by the desire for larger, higher-performance vehicles, and rising power needs for auxiliaries such as air-conditioning and power-steering [18, 19].

61.3.1.2 Efficiency of Conversion of Primary into Secondary Energy.
Although transport efficiency is often measured on a vehicle-km/liter of fuel (or mpg) basis, the adverse effects from vehicular travel—oil depletion, CO_2 emissions, and air pollution—are the result of primary energy consumption. In other words, conversion energy losses cannot be ignored. Because of global depletion of conventional oil reserves, the world will need to rely increasingly on unconventional oil. Although the term is often restricted to heavy oil, shale oil, and bituminous sands, ASPO [11] extends the term to cover oil from deep water and from polar regions, and so unconventional oil is already an important share of total production. The justification is that these sources can also have high monetary and environmental costs.

Even for conventional oil, Murphy and Hall [20] have shown that the ratio of energy output for a given level of input energy (the energy ratio) has fallen steadily over the years. For unconventional sources, the energy ratio will fall still further, and the CO_2 and environmental costs will rise alongside the monetary costs per liter of fuel delivered to the fuel tank. The result will be that any further energy efficiency improvements in vehicular tank-to-wheels efficiency will be at least partly offset by rising well-to-tank energy costs.

61.3.1.3 Vehicle Loading Rates.
Especially for OECD car travel, seat occupancy rates are low, typically about 1.5 persons per vehicle. For a 5-seater car, this gives a mere 30% seat occupancy rate. For public transport travel, occupancy rates vary enormously from country to country, and can be very high, not only in industrializing countries, but also in high-income Asian cities such as Tokyo and Hong Kong [21]. Increasing occupancy rates is attractive because it can be achieved with the existing public transport or private vehicle stock. However, for private transport it is very difficult to implement, especially if household occupancy is falling and car ownership and incomes are both rising. In both the United States and Australia, occupancy rates have fallen from around 2.0 in the 1950s to less than 1.5 today [22]. Raising car occupancy rates has proved difficult when tried because it entails car-sharing with non-family members, and much of the convenience of car travel over public transport, such as ability to choose one's own departure time and travel privacy would be lost.

It can also be difficult for public transport, but if public transport patronage is rising, as is the case in many OECD countries today, higher occupancy rates occur naturally. For many industrializing countries, public transport vehicles are often grossly overcrowded. Although this ensures both high energy and land use efficiency, loadings should be reduced in the interests of comfort and safety. For both public and private transport, occupancy rates vary by time of day and trip purpose. For public transport, loadings, and thus energy efficiency, are highest for the work trip, while for private travel, the opposite is the case, with congestion lowering fleet energy efficiency still further.

For air travel, loadings are already high, with little scope for further improvement [23]. Freight transport typically has large payload to tare weight ratios compared with passenger transport. (According to Cullen et al. [17], transported goods make up about 60% of total truck mass, but for cars only 5–10% of mass is occupants and their luggage.) For specialized transport (bulk ore carriers, oil tankers, etc) back trips are usually made empty, but it is difficult to see how this could be changed.

61.3.1.4 Shifting From Less to More Energy Efficient Transport Modes.
The various transport modes can vary greatly in terms of energy efficiency, even with similar levels of loading. When differences in seat occupancy rates are allowed for, public transport modes are usually much more energy efficient than car travel, as can be seen most clearly by comparing bus and car travel, both of which use petroleum fuels. Comparison of electric public transport (trains and trams) with car travel is complicated because of different efficiencies in primary to secondary energy conversion. Important reasons for the higher efficiency of rail travel compared with cars and trucks are the very low coefficient of rolling friction with steel wheels running on steel rails and the lower air friction per unit load because of the small frontal area to length ratio.

The difference in energy efficiency is far more pronounced for freight modes, with sea freight up to two orders of magnitude more efficient (tonne-km/primary MJ) than air freight. The difficulty is that in terms of tonne-km, most world freight is already sent by energy-efficient water or rail modes. There is some scope for shifting some bulk freight from road to rail—if the necessary rail infrastructure is there—but far less for non-bulk freight or urban deliveries. Air freight accounts for only a tiny share of world tonne-km and is used when delivery speed is at a premium [23].

61.3.2 Discussion and Conclusions

Can both passenger and freight travel continue their rapid growth of the 20th century, as forecast by the various organizations like the IEA and the EIA discussed in Section 60.2.1? Given the huge unmet demand for both vehicle ownership and transport outside the OECD countries, a strong case can be made for a continuation of high growth levels for both the transport task itself and transport energy. After all, when countries are ranked on either air travel trips per capita or car ownership, nearly three orders of magnitude difference exist between the highest and the lowest countries.

On the other hand, the serious problems thrown up by continued growth in travel and accompanying energy use may force the world to limit this growth. These transport challenges include air pollution, especially a serious problem in large Asian cities, traffic casualties, oil depletion, and the risk of climate change from carbon dioxide and other greenhouse gas emissions. Technical solutions have been successful in reducing vehicle pollution emissions in the OECD, through the use of unleaded fuels and 3-way catalytic converters. Although as shown in Section 60.3.1, the potential for vehicular energy efficiency improvements is large, especially for cars, but the record so far is disappointing. Transport energy use has continued to rise worldwide (Figure 61.1), largely because the growth in transport vehicle numbers has swamped efficiency improvements. (For the United States, an important factor was the popularity of sports utility vehicles.) Even for air transport, where aircraft efficiency gains have been remarkable, demand for aviation turbine fuel is still rising. If oil depletion and global climate change force serious reductions in transport energy, travel itself will probably need to be limited, in addition to a major shift from car travel to public transport, and some freight from trucks to water or rail transport.

If such a reduction is needed, a number of ideas have been put forward to reduce travel demand and thus transport energy. For freight transport, LaVaute [24] has suggested a goods "kilometer tax" to discourage excessive haulage of goods long distances when locally produced substitutes are available. There is now considerable interest in "food miles" for various food products; these calculations normally show that the average distance a given foodstuff is transported from producer to consumer is rising over time. For passenger travel, particularly urban travel, the focus needs to shift to ways of reducing travel demand itself [18, 22, 25]. As mentioned in Section 61.2.2, IT could be used to cope with reduced personal travel, should it become necessary. Reductions in transport energy is not just a future possibility: in Japan [26], total final demand for transport energy has been in decline for more than a decade. Japan could be a blueprint, first for other OECD countries, and later for the world overall.

REFERENCES

[1] International Energy Agency (IEA). (2014). *Key world energy statistics 2014*. IEA/OECD, Paris. (Also earlier editions.)

[2] Wikipedia. (2015). History of rail transport. Accessed on 1st July 2015 at: http://en.wikipedia.org/wiki/History_of_rail_transport.

[3] Moriarty, P., Honnery, D. (2004). Forecasting world transport in the year 2050. *Int. J. Vehicle Design*, 35(1/2):151–165.

[4] Schafer, A., Heywood, J. B., Jacoby, H. D., Waitz, I. A. (2009). *Transportation in a Climate-Constrained World*. MIT Press, Cambridge, MA.

[5] Kahn Ribeiro, S., Kobayashi, S., Beuthe, M., et al. (2007). Transport and its infrastructure. In: *Climate change 2007: Mitigation*, Metz, B., Davidson, O. R., Bosch, P. R., et al. (editors), CUP, Cambridge, UK.

[6] Energy Information Administration (EIA). (2014). *Annual Energy Outlook 2014 with Projections to 2040*. Accessed on 20th April 2015 at: www.eia.gov/forecasts/aeo. Also earlier reports.

[7] BP (2011). *BP Energy Outlook 2035*. BP, London, 2014.

[8] Shell International BV. (2014). *New Lens Scenarios: A Shift in Perspective for a World in Transition*. Shell International BV, The Hague, NL.

[9] Organization of the Petroleum Exporting Countries (OPEC). (2014). *World Oil Outlook 2014*. OPEC, Vienna.

[10] Airbus (2014). *Global Market Forecast: Flying on Demand 2014–2033*. Accessed on 20th June 2015 at: www.airbus.com.

[11] Association for the Study of Peak Oil and Gas (ASPO). (2009). ASPO Newsletter No 100 May (Also earlier newsletters). Available at: www.peakoil.net. Accessed 6 May 2011.

[12] Mitchell, W. J. (2003) *Me++: the Cyborg Self and the Networked City*. MIT, Cambridge, MA.

[13] Pelton, J. N. (2004). The rise of telecities: decentralizing the global society. The Futurist, Jan-Feb:28–33.

[14] Zhu, P. (2012). Are telecommuting and personal travel complements or substitutes. *Annals Regional Sci.*, 48:619–639.

[15] Moriarty, P., Honnery, D. (2010). *Rise and Fall of the Carbon Civilisation*. Springer, London, UK.

[16] Moriarty, P., Honnery, D. (2014). Reconnecting technological development with human welfare. *Futures*, 55:32–40.

[17] Cullen, J. M., Allwood, J. M., Borgstein, E. H. (2011). Reducing energy demand: what are the practical limits? *Environ. Sci. Technol.*, 45:1711–1718.

[18] Moriarty, P., Honnery, D. (2013). Greening passenger transport: a review. *J. Clean. Prod.*, 54:14–22.

[19] Moriarty, P., Honnery, D. (2008). The prospects for global green car mobility. *J Clean. Prod*, 16:1717–1726.

[20] Murphy, D. J., Hall, C. A. S. (2010). Year in review—EROI or energy return on (energy) invested. *Ann. N. Y. Acad. Sci.*, 1185:102–118.

[21] Newman, P., Kenworthy, J. (1989). *Cities and Automobile Dependence: An International Sourcebook*. Gower, London.

[22] Moriarty, P., Honnery, D. (2012). Reducing personal mobility for climate change mitigation. Chapter 51. In: *Handbook of CChange Mitigation*, Chen, W.-Y., Seiner, J. M., Suzuki, T., et al. (editors). Springer, New York.

[23] Moriarty, P., Honnery, D. (2011) Reducing transport's impact on climate change. Chapter 4. In: *Environmental Pollution and its Relation to Climate Change*, El Nemr, A. (editor). Nova Science Publishers, Inc., NewYork.

[24] LaVaute, P. (2007). It's time for a business tax. *In Business*, 29(3):20–21.

[25] Moriarty, P., Honnery, D. (2008). Low mobility: the future of transport. *Futures*, 40(10):865–872.

[26] Statistics Bureau Japan (2015). *Japan Statistical Yearbook 2014*. Statistics Bureau, Tokyo. (Also earlier editions). Accessed on 25 June 2015 at: http://www.stat.go.jp/english/data/nenkan/index.htm.

62

BIOMASS: RENEWABLE ENERGY FROM PLANTS AND ANIMALS

PUBLIC DOMAIN

62.1 BIOMASS BASICS

Biomass is organic material made from plants and animals (microorganisms). Biomass contains stored energy from the sun. Plants absorb the sun's energy in a process called photosynthesis. The chemical energy in plants gets passed on to animals and people that eat them.

Biomass is a renewable energy source because we can always grow more trees and crops, and waste will always exist. Some examples of biomass fuels are wood, crops, manure, and some garbage.

When burned, the chemical energy in biomass is released as heat. If you have a fireplace, the wood you burn in it is a biomass fuel. Wood waste or garbage can be burned to produce steam for making electricity, or to provide heat to industries and homes.

62.1.1 Converting Biomass to Other Forms of Energy

Burning biomass is not the only way to release its energy. Biomass can be converted to other useable forms of energy, such as methane gas or transportation fuels, such as ethanol and biodiesel.

Methane gas is the main ingredient of natural gas. Smelly stuff like rotting garbage and agricultural and human waste release methane gas, also called "landfill gas" or "biogas."

Crops like corn and sugar cane can be fermented to produce ethanol. Biodiesel, another transportation fuel, can be produced from left-over food products like vegetable oils and animal fats.

62.1.2 How Much Biomass Is Used for Fuel?

Biomass fuels provided about 5% of the energy used in the United States in 2012. Of this, about 45% was from wood and wood-derived biomass, 44% from biofuels (mainly ethanol), and about 11% from municipal waste. Researchers are trying to develop ways to burn more biomass and less fossil fuels. Using biomass for energy may cut back on waste and greenhouse gas emissions.

62.2 WOOD AND WOOD WASTE

62.2.1 Burning Wood Is Nothing New

The most common form of biomass is wood. For thousands of years people have burned wood for heating and cooking. Wood was the main source of energy in the United States and the rest of the world until the mid-1800s. Wood continues to be a major source of energy in much of the developing world.

In the United States, wood and wood waste (bark, sawdust, wood chips, wood scrap, and paper mill residues) provide about 2% of the energy we use today.

62.2.2 Using Wood and Wood Waste

About 80% of the wood and wood waste fuel used in the United States is consumed by industry, electric power producers, and commercial businesses. The rest, mainly wood, is used in homes for heating and cooking.

Alternative Energy and Shale Gas Encyclopedia, First Edition. Edited by Jay H. Lehr and Jack Keeley.
© 2016 John Wiley & Sons, Inc. Published 2016 by John Wiley & Sons, Inc.

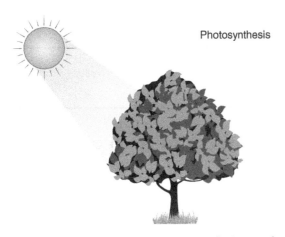

Photosynthesis

In the process of photosynthesis, plants convert radiant energy from the sun into chemical energy in the form of glucose-or sugar.

water + carbon dioxide + sunlight $\rightarrow$ glucose + oxygen

$6\,H_2O$ + $6\,CO_2$ + radiant energy $\rightarrow C_6H_{12}O_6$ + $6\,O_2$

Source: The National Energy Education Project (Public Domain).

Many manufacturing plants in the wood and paper products industry use wood waste to produce their own steam and electricity. This saves these companies money because they do not have to dispose of their waste products and they do not have to buy as much electricity.

Hybrid Poplar Wood Chips Being Unloaded in Crookston, Minnesota *Source:* National Renewable Energy Laboratory, U.S. Department of Energy (Public Domain).

62.3   WASTE-TO-ENERGY

62.3.1   Energy from Garbage

Garbage, often called municipal solid waste (MSW), is used to produce energy in waste-to-energy plants and in landfills in the United States. MSW contains biomass (or biogenic)

Types of biomass

Wood

Crops

Garbage

Landfill gas

Alcohol fuels

Source: The National Energy Education Project (Public Domain).

Total MSW generation (by material), 2011
250 Million Tons (before recycling)

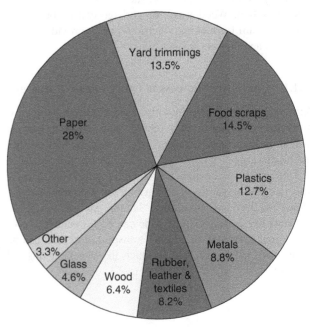

Source: U.S. Environmental Protection Agency, *Municipal Solid Waste in the United States: 2011 Facts and Figures* (May 2013).

Management of MSW in the United States, 2011

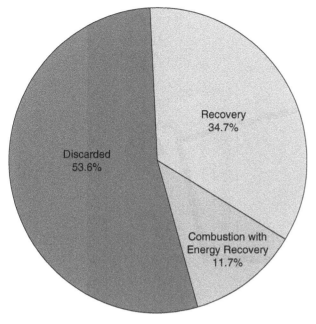

Source: U.S. Environmental Protection Agency, *Municipal Solid Waste in the United States: 2011 Facts and Figures* (May 2013).

materials like paper, cardboard, food scraps, grass clippings, leaves, wood, and leather products, and other non-biomass combustible materials, mainly plastics and other synthetic materials made from petroleum.

In 1960, the average American threw away 2.7 pounds of trash a day. Today, each American throws away about 4.4 pounds of trash every day. Of that, about 1.5 pounds are recycled or composted. What do we do with the rest? One option is to burn it. (Burning is sometimes called combustion.) About 85% of our household trash is material that will burn, and most of that is biogenic, or material that is made from biomass (plant or animal products). About 71% of MSW (by weight) is biogenic.

62.3.2   Waste-to-Energy Plants Make Steam and Electricity

MSW is burned in special waste-to-energy plants that use its heat energy to make steam to heat buildings or to generate electricity. There are about 86 waste-to-energy plants in the United States that generate electricity or produce steam. In 2011, waste-to-energy plants generated 14 billion kilowatt hours of electricity, about the same amount used by 1.3 million US households. The biogenic material in MSW contributed about 51% of the energy of the MSW that was burned in waste-to-energy facilities that generated electricity. Many large landfills also generate electricity with

methane gas that is produced as biomass decomposes in the landfills.

62.3.3   Waste-to-Energy Plants Also Dispose of Waste

Providing electricity is not the major advantage of waste-to-energy plants. It actually costs more to generate electricity at a waste-to-energy plant than it does at a coal, nuclear, or hydropower plant.

The major advantage of burning waste is that it reduces the amount of material that we bury in landfills. Waste-to-energy plants burned about 29 million tonnes of MSW in 2011. Burning MSW reduces the volume of waste by about 87%.

62.4   BIOGAS

62.4.1   Collecting Gas from Landfills

Landfills can be a source of energy. Anaerobic bacteria that live in landfills decompose organic waste to produce a gas called biogas that contains methane.

Methane is the same energy-rich gas that is in natural gas, which is the fuel used for heating, cooking, and producing electricity. Methane is colorless and odorless, and a very strong greenhouse gas. Natural gas utilities add an odorant (bad smell) so people can detect natural gas leaks from pipelines. Landfill biogas can also be dangerous to people or the environment. New rules require landfills to collect methane gas for safety and pollution control.

Some landfills simply burn the methane gas in a controlled way to get rid of it. But the methane can also be used as an energy source. Landfills can collect the methane gas, treat it, and then sell it as a commercial fuel. It can then be burned to generate steam and electricity.

62.4.2   Landfill Gas Energy Projects

As of July 2013, there were 621 operational landfill gas energy projects in the United States. California had the most landfill gas energy projects in operation (77), followed by Pennsylvania (44), and Michigan (41).

62.4.3   Using Animal Waste

Some farmers produce biogas in large tanks called "digesters" where they put manure and bedding material from their barns. Some cover their manure ponds (also called lagoons) to capture biogas. Biogas digesters and manure ponds contain the same anaerobic bacteria in landfills. The biogas can be used to generate electricity or heat for use on the farm, or to sell electricity to an electric utility.

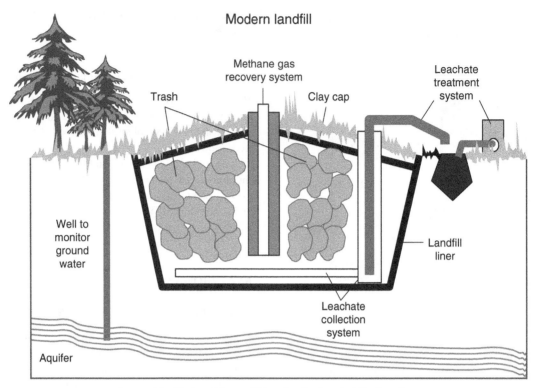

Source: The National Energy Education Project (public domain).

Read about a field trip to a real waste-to-energy plant or learn about the history of MSW.

62.5 BIOMASS AND THE ENVIRONMENT

62.5.1 Using Biomass for Energy Can Have Positive and Negative Impacts

Using biomass for energy can have both positive and negative impacts on the environment. For example, burning biomass may result in more or less air pollution depending on the type of biomass and the types of fuels or energy sources that it replaces. Biomass, such as lumber mill sawdust, paper mill sludge, and yard waste in household trash, used for energy may avoid the use of fossil fuels, coal, petroleum, or natural gas. Burning fossil fuels and biomass releases carbon dioxide (CO_2), greenhouse gas, but when the plants that are the sources of biomass are grown, a nearly equivalent amount of CO_2 is captured through photosynthesis. Sustainable cultivation and harvesting of biomass can result in no net increase in CO_2 emissions. Each of the different forms and uses of biomass impact the environment in a different way.

62.5.2 Burning Wood

Using wood, and charcoal made from wood, for heating and cooking can replace fossil fuels and may result in lower CO_2

emissions. Wood may be harvested from forests or woodlots that have to be thinned or come from urban trees that fall down or have to be cut down anyway. However, wood smoke contains harmful pollutants like carbon monoxide and particulate matter. Burning wood in an open fireplace is very inefficient and produces lots of air pollution. Modern wood stoves and fireplace inserts are designed to release small amounts of particulates. Fireplace inserts are like wood stoves that fit into a fireplace. Wood and charcoal are major cooking and heating fuels in poor countries, and the wood may be harvested faster than trees can grow. This results in deforestation. Planting fast-growing trees for fuel and using fuel-efficient cooking stoves can help slow deforestation and improve the environment.

62.5.3 Burning MSW or Wood Waste

Burning MSW or garbage and wood waste to produce energy means that less of it has to get buried in landfills. Like power plants that burn coal, waste-to-energy plants produce air pollution when the fuel is burned to produce steam or electricity. Burning garbage releases the chemicals and substances found in the waste. Some of these chemicals can be dangerous to people, the environment, or both, if they are not properly controlled.

Incinerators and waste-to-energy power plants must use technology to prevent harmful gases and particles from

Panicum virgatum (switchgrass) being grown *Source:* Wikimedia Commons (public domain).

coming out of their smoke stacks. The Environmental Protection Agency (EPA) applies strict environmental rules to waste-to-energy plants. The EPA requires waste-to-energy plants to use air pollution control devices, including scrubbers, fabric filters, and electrostatic precipitators to capture air pollutants.

Scrubbers clean chemical gas emissions by spraying a liquid into the gas stream to neutralize the acids. Fabric filters and electrostatic precipitators remove particles from the combustion gases. The particles—called fly ash—are then mixed with the ash that is removed from the bottom of the waste-to-energy plant's furnace.

A waste-to-energy furnace burns at such high temperatures (1800–2000°F) that many complex chemicals break down into simpler, less harmful compounds. This chemical change is a kind of built-in antipollution device.

62.5.4   Disposing of Ash from Waste-to-Energy Plants

Another challenge is the disposal of the ash after combustion. Ash can contain high concentrations of various metals that were present in the original waste. Textile dyes, printing inks, and ceramics, for example, contain the metals lead and cadmium.

Separating waste before combustion can solve part of the problem. Because batteries are the largest source of lead and cadmium in municipal waste, they should not be put into regular trash. Florescent light bulbs should also not be put in trash because they contain small amounts of mercury.

The EPA tests ash from waste-to-energy plants to make sure it is not hazardous. The test looks for chemicals and metals that would contaminate ground water by trickling through a landfill. Ash that is considered safe is used in MSW landfills as a daily or final cover layer, or to build roads and make cement blocks.

62.5.5   Collecting Landfill Gas or Biogas

Biogas is a gas composed mainly of methane and CO_2 that form as a result of biological processes in sewage treatment plants, waste landfills, and livestock manure management systems. Methane is one of the greenhouse gases associated with global climate change.[1] Many of these facilities capture and burn the biogas for heat or electricity generation. Burning methane is actually beneficial because methane is a stronger greenhouse gas than CO_2. The electricity generated from biogas is considered "green power" in many states and may be used to meet state renewable portfolio standards (RPS). The electricity may replace electricity produced by burning fossil fuels and result in a net reduction in CO_2 emissions.

62.5.6   Liquid Biofuels: Ethanol and Biodiesel

Ethanol and biodiesel were the fuels used in the first automobile and diesel engines, but lower cost gasoline and diesel fuel made from crude oil became the dominant vehicle fuels. The federal government has promoted ethanol use in vehicles to help reduce oil imports since the mid-1970s. In 2007, the government set a target for biofuels use of 36 billion gallons by the year 2022. As a result, nearly all gasoline now sold in the United States contains some ethanol.

Biofuels may be considered to be carbon-neutral because the plants that are used to make biofuels (such as corn and sugarcane for ethanol, and soybeans and palm oil trees for biodiesel) absorb CO_2 as they grow and may offset the CO_2 produced when biofuels are made and burned.

Growing plants for biofuels is controversial however, as the land, fertilizers, and energy used to grow biofuel crops could be used to grow food crops instead. Also, in some parts of the world, large areas of natural vegetation and forests have been cut down to grow sugar cane for ethanol and soybeans and palm oil trees to make biodiesel. The US government is supporting efforts to develop alternative sources of biomass that do not compete with food crops and use less fertilizer and pesticides than corn and sugar cane, and methods to produce ethanol that require much less energy than conventional fermentation. Ethanol may also be made from waste paper, and biodiesel can be made from waste grease and oils and even algae.

Ethanol and ethanol–gasoline mixtures burn cleaner and have higher octane than pure gasoline, but have higher "evaporative emissions" from fuel tanks and dispensing equipment. These evaporative emissions contribute to the formation of harmful, ground-level ozone and smog. Gasoline requires extra processing to reduce evaporative emissions before it is blended with ethanol. Compared to petroleum diesel, biodiesel combustion produces less sulfur oxides, particulate matter, carbon monoxide, and unburned and other hydrocarbons, but more nitrogen oxide.[1]

[1]U.S. Environmental Protection Agency, Climate Change Science.

63

PLANTING AND MANAGING SWITCHGRASS AS A BIOMASS ENERGY CROP

PUBLIC DOMAIN

63.1 INTRODUCTION

Switchgrass (*Panicum virgatum* L.) is a perennial, native, deep-rooted, warm-season grass that occurs naturally throughout the United States except for the states of Oregon, Washington, and California (U.S. Department of Agriculture Natural Resources Conservation Service (USDA NRCS) 2006) (Figure 63.1). It tolerates a wide range of soil and climatic conditions and is widely acclaimed as a conservation plant for erosion control, pasture and hayland forage, wildlife habitat, and native prairie restoration (Alderson and Sharp 1994). The US Department of Energy has designated switchgrass as one of the principal biofuel species recommended for combustion, gasification, and liquid fuel production (McLaughlin et al. 1996). Some of the characteristics that make switchgrass an ideal renewable energy crop include commercial seed availability of high-yielding cultivars adapted to different geographical regions of adaptation, relative ease of planting and establishment, compatibility with conventional farming equipment for establishment and harvest management, production of large amounts of biomass under a wide range of environmental conditions, and excellent wildlife cover provided (Wright 2007) (Figure 63.2).

Switchgrass also has great potential to protect the environment through carbon sequestration and erosion control (Dewald et al. 1996; Liebig et al. 2008).

63.2 ESTABLISHING SWITCHGRASS

Establishing switchgrass can be a challenge to producers unfamiliar with the cultural aspects of planting a tallgrass

prairie species. Like commodity crops grown for food and fiber, switchgrass has its own set of cultural requirements to plant and manage it as a biomass crop. There are several factors that a producer must consider during the planning stages of a switchgrass biomass production system. Cultural specifications and management considerations for this type of system are described in this chapter.

63.2.1 Field Selection

A primary goal of biomass production is to produce high yields with minimal inputs. When selecting fields for switchgrass biomass production, use soil and field criteria typically used for row crop agriculture. Avoid fields with perennial weed infestations, steep or irregular terrain, and excessively wet areas because of their negative effect on stand establishment and management and on field and harvest operations. Productivity is highest on medium- to high-fertility soils that are moderately fine textured, well- to moderately well drained, and have a pH of 5.5–6.5.

Switchgrass can be produced on a wide range of soil and landscape conditions. Although it is most productive on class I, II, and III lands, switchgrass can also grow on class IV and VI lands as long as the proper growing conditions and crop requirements exist. Highly productive class I and II lands will produce more biomass than class III and IV lands, and usually with more efficient production from inputs of water, nutrients, and energy. Landscapes that are suitable to seeding, application, and harvesting equipment are suitable for switchgrass production. Current marginal land for row crop production can be efficiently used for switchgrass production if the proper management and inputs are performed. Refer to

FIGURE 63.1 Switchgrass area of occurrence.

FIGURE 63.3 Switchgrass 6 weeks after planting in North Dakota. *Source:* USDA NRCS Bismarck, ND.

section II in the local Field Office Technical Guide (FOTG) for specific information on suitable field characteristics for establishing switchgrass.

63.2.2 Preparing the Field

If the field selected for switchgrass biomass production was previously in a row crop, minimal, if any, tillage is needed to convert it from row crop production to switchgrass production. One of the biggest advantages of planting switchgrass in a previously cropped field can be low weed pressure (Figure 63.3). If the field has a history of poor weed control, remedial measures will be needed due to weed seed present in the soil seed bank. Herbicide application from previous

FIGURE 63.2 Cave-in-Rock switchgrass is a high biomass producing cultivar released from the USDA NRCS Plant Materials Center, Elsberry, MO. *Source:* USDA NRCS Elsberry, MO.

cropping system must be considered to avoid potential carry-over effects. Consult manufacturers for herbicide labeling or contact the county extension agent for assistance with herbicide questions. If the field has not been cropped for several years, an additional year of site preparation may be required to control existing weeds. This can be done by chemically or mechanically fallowing the field or growing a row crop such as glyphosate-tolerant soybeans (*Glycine max*) on the field. This system, unlike fallowing, gives the producer flexibility in controlling annual and perennial weeds, maintains a canopy cover for further weed suppression, improves soil quality, and provides the producer with an income from the field for another year.

In northern States, it is important to avoid fields with somewhat poorly drained or variable soil drainage conditions because of the danger of frost heaving, which can be a serious threat to switchgrass stand establishment. Additionally, weed control and early planting are very important to allow sufficient root system and tiller development by the end of the first season to reduce susceptibility to frost heaving.

A soil test should be conducted the year before planting the field to correct any major nutrient deficiencies or soil pH problems. Procedures for taking soil samples and submitting them for analysis are available at the local county extension office. If no specific soil test response curve is available for switchgrass biomass production, treat it as an annual warm-season grass crop (sudex or sudangrass (*Sorghum* spp.)). Nutrient requirements for switchgrass are much lower than for annual row crops, which remove more nutrients with the harvested portion of the crop. Excessive fertilization may cause lodging, leading to harvesting difficulty and possible stand reduction, and may contribute to weed competition (Fike et al. 2006) (see Section 63.4).

63.2.3 Seedbed Preparation and Seeding

Switchgrass seed is smaller than most row crop seed and should be planted no more than a quarter- to half-inch deep. Seed can be no-till planted into stale seedbeds if crop and/or weed residue has been controlled prior to planting. There are several advantages to no-till planting including conserving soil moisture and saving fuel, labor, and time. Another advantage of no-tilling switchgrass seed is less soil disturbance, which reduces soil erosion and may reduce weed population.

Switchgrass can also be planted into a conventionally prepared seedbed. It is critical that the seedbed seed is firm and free of competing vegetation. Correct seedbed firmness is when an adult footprint is only slightly visible on the prepared seedbed prior to the seeding operation (Figure 63.4). Loose, uneven, fluffy, or cloddy seedbed can result in reduced stands or complete failure.

A prepared, tilled seedbed can be firmed, if needed, by pulling a cultipacker or homemade packer or roller over the field.

Regardless of the planting system used, surface crop or weed residue can affect seeding depth and seed-to-soil contact. The allowable amount of residue depends on the seeding equipment to be used. Tillage, fire, and mowing can be used to manage residue prior to seeding. Heavy accumulation of residue should be distributed over a larger area to make seed placement more accurate.

63.2.4 Cultivar Selection

Commercially available switchgrass cultivars with high biomass production potential have been evaluated in field trials by the USDA Agricultural Research Service (ARS) and NRCS, universities, and private research institutes (Lemus et al. 2002; Cassida et al. 2005; Lee and Boe 2005; McLaughlin and Kszos 2005; Adler et al. 2006; USDA NRCS 2007,

FIGURE 63.5 Switchgrass biomass cultivars at the USDA NRCS Plant Materials Centers in Manhattan, KS. *Source:* USDA NRCS Manhattan, KS.

2009) (Figure 63.5). These field trials have provided valuable data on cultivar performance, management, and persistence on various soils and under various climatic conditions in northern, mid, and southern latitudes. In general, these studies have shown that lowland cultivars, such as Alamo and Kanlow, are better adapted to the southern and mid-latitudes, whereas the upland to intermediate cultivars, such as Cave-in-Rock, Sunburst, and Forestburg, are better adapted to mid and northern latitudes (McLaughlin and Kszos 2005) (Figure 63.6).

In addition, other research has shown that upland populations cannot be moved south of their point of origin by more than one plant hardiness zone, and lowland populations should not be moved north of their point of origin by more than one hardiness zone without expecting severe losses in biomass yield and survival (Casler et al. 2004). Adaptation zones for some of the currently recommended biomass switchgrass cultivars are shown in Table 63.1. These

FIGURE 63.4 Slightly visible adult footprint on a firm seedbed for planting switchgrass. *Source:* USDA NRCS Bismarck, ND.

FIGURE 63.6 Dacotah switchgrass (left) matures earlier than Sunburst (right). Sunburst is one of the recommended cultivar for biomass production in the northern Great Plains. *Source:* USDA NRCS Bismarck, ND.

TABLE 63.1 Switchgrass cultivar origin and adaptation expressed in United States plant hardiness zones.

Cultivar	State of origin	Plant hardiness zone	
		Origin	Adaptation
Alamo	Texas	8b	7a, 7b, 8a, 8b
Kanlow	Oklahoma	7a	5a, 5b, 6a, 6b, 7a, 7b, 8a, 8b
Carthage	North Carolina	7a	6a, 6b, 7a, 7b, 8a, 8b
Cave-in-Rock	Illinois	5b	4b, 5a, 5b, 6a, 6b
Forestburg	South Dakota	4b	3b, 4a, 4b
Summer	Nebraska	5a	3b, 4a, 4b, 5a, 5b, 6a,
Sunburst	South Dakota	4b	3a, 3b, 4a, 4b

USDA Plant Hardiness Zone Map (http://www.usna.usda.gov/Hardzone/ushzmap.html).

adaptation zones should only serve as a general guide to the areas where these cultivars can be planted, since selection and breeding can rapidly change the range of potential adaptation of switchgrass populations to latitudinal differences (Casler et al. 2004). In addition to those cultivars listed in Table 63.1, there will soon be other suitable cultivars that have been developed specifically for bioenergy production by universities, the USDA ARS, and private research institutes. A recent example is BoMaster, which was released jointly by USDA ARS and North Carolina State University (Burns et al. 2008). These are in the early stages of commercial increase and will be available for use by producers in the near future.

63.2.5 Pests, Insects, and Diseases

There are some known pests and insects that negatively impact the production of switchgrass. Rust (*Puccinia* sp.) has been reported on switchgrass in South Dakota (Gustafson et al. 2003) and noted on switchgrass at the NRCS Jamie L. Whitten PMC near Coffeeville, MS, and the East Texas PMC near Nacogdoches, TX. A smut fungus (*Tilletia maclaganii*) that caused significant seed and biomass loss was identified on switchgrass at the USDA NRCS Big Flats, NY, PMC and reported on Cave-in-Rock switchgrass in southern Iowa (Gravert et al. 2000). This smut was also prevalent in the switchgrass cultivars Blackwell, Cave-in-Rock, Pathfinder, Shelter, Summer, and accession 9006010 at the NRCS Manhattan, KS, PMC, but did not occur in the cultivar Kanlow at this site (Stuteville et al. 2001).

A switchgrass moth (*Blastobasis repartella*) was reported on young tillers of switchgrass that could potentially impact stand and production in the northern Great Plains (http://nathist.sdstate.edu/BioEnergy/blastobasis.html). The likelihood of additional pests and insects that affect switchgrass production may increase as more acres are planted for biomass production.

63.2.6 Seed Quality and Planting Rate/Date

Switchgrass biofuel stands are expected to have a production life span of more than 10 years (Parrish and Fike 2005; Fike et al. 2006). Planting the best available seeds goes a long way to ensure successful stand establishment and maximum longevity. To ensure the best quality seed is used, plant only certified seed. Certified seed guarantees the best quality seed, true to the cultivar stated on the seed tag that has met rigorous germination and purity analysis and weed seed standards. Using noncertified seed may result in poor stands or a field of weeds that compete with the switchgrass for nutrients, light, and moisture.

Switchgrass seed is purchased and planted on a pure live seed (PLS) basis because germination rates can be poor and dormancy rates can be great. This can occur even when produced under the best growing conditions. A planting rate expressed as pounds PLS per acre is not the same as a planting rate in pounds bulk seed per acre.

To calculate the amount of switchgrass seed to plant in pounds per acre, divide the recommended PLS planting rate by the calculated percent PLS of your purchased seed (expressed as a decimal value) (Figure 63.7). More information on determining percent PLS can be found in the PMC publication *Reading Seed Packaging Labels and Calculating Seed Mixtures*, Aberdeen, ID (http://plant-materials.nrcs.usda.gov/pubs/idpmstn04265.pdf).

It is important to be aware that there can be wide variation among switchgrass seed lots between germinating seed and dormant seed even though the percent PLS is the same. In most cases, a seed lot that has a high percent purity and germination rate with little dormant seed will be the most advantageous to purchase.

It would also be beneficial to know whether seed germination was determined with or without stratification. A standard stratification treatment is to prechill the seed for 2 weeks

% PLS = (% purity × % viability) × 100

% viability = % germination + % dormant seed

FIGURE 63.7 Calculating PLS. *Source:* USDA NRCS Bismarck, ND.

using a cold, moist stratification period (AOSCA 1993). This short prechill period helps break the dormancy of some of the seeds resulting in a higher germination percentage than when not given a prechill treatment. Switchgrass seeds can be prechilled and properly stored until planting. Some seed growers offer stratified (prechill) switchgrass seed as an alternative to nonstratified seed. However, stratified seed must be planted into a warm, moist seedbed to achieve maximum germination. If stratified seeds are planted in a dry seedbed or the seedbed dries out before germination, the seed becomes susceptible to a secondary dormancy and may not germinate until the next spring (Shen et al. 2001).

An alternative method that has been used in colder regions of the United States for breaking seed dormancy in switchgrass is no-till planting the seed in late November or December. The seed overwinter in the soil and the cool, moist soil conditions will result in a natural stratification process that will help break dormancy. However, stand consistency of late fall and early winter plantings vary from year to year. If nonstratified seed is going to be used, seeding rates may need to be increased if the seed lot contains a high percentage of dormant seed. One easy way to ensure that seed dormancy is not excessive is to purchase seed that is at least 1 year old because percent seed dormancy reduces as the seed ages.

Seeding rates and planting dates can vary widely across the area of adaptation for switchgrass cultivars (Table 63.2). Factors that influence seeding rates and planting dates include, but are not limited to the region of the United States, soil moisture and temperature, row spacing, and number of seeds per pound. Consult the local FOTG or local extension service for seeding rates and planting dates for switchgrass cultivars recommended for the State and county.

Switchgrass seed should be planted at a depth between a quarter to half inch when soil moisture and temperature are at levels acceptable for germination. In general, soil moisture conditions for planting are ideal when the soil can be readily formed into a ball in the palm of the hand and break easily when dropped. The optimum planting depth depends on soil texture. In general, switchgrass seed is planted a half-inch deep in coarse-textured soil and a quarter-inch deep in fine-textured soil. The purpose of a deeper planting depth on coarse-textured soils is to provide higher soil moisture levels for germination. A minimum soil temperature of 60°F is needed for switchgrass seed to germinate. Germination and seedling growth will be slow at this temperature, but will increase at higher temperatures.

Because of a narrower planting window in the northern areas of its range, it may be necessary to plant switchgrass seed in the spring before soil temperatures reaches 60°F. Planting seed this early has some advantages. First, the seed is exposed to a natural, light stratification, which may increase its germination rate. Second, adequate soil moisture for germination is usually present. Third, there will be a longer growing season to allow for better root development.

63.2.7   Planting Equipment

Switchgrass seed are small and requires planting equipment that can properly place the seed at a uniform, shallow depth and in a uniform distribution pattern that will allow for good seed-to-soil contact. Cultipacking before and after planting is an option. Grass or small grain drills can place the seed properly as long as they can maintain a shallow and consistent planting depth and are equipped with press wheels to firm the soil around the seed (Figure 63.8). Drills equipped with depth bands on double disk openers that ensure a consistent, proper planting depth for the seed are particularly good for switchgrass planting. Practical experience has shown that if a few seeds are not present on the soil surface regardless of the planter type, suspect that the seeds are being planted too deep (USDA NRCS 2003).

Row planting is preferred over broadcast planting. Research in the southern plains has shown there to be only a minimal increase in biomass yield when row spacing of 7, 14, and 21 inches were used (Muir et al. 2001). In this case, either spacing would be acceptable unless additional research supports another recommendation. In the upper South and Northeast, a 7-inch row spacing is recommended because it provides early canopy closure that helps reduce weed competition. In the Deep South, a 21-inch row planting, in contrast to 7- to 10-inch spacing, was shown to be advantageous for early canopy closure for weed control and maintenance of switchgrass (harvesting, side dressing, and row cultivation, if necessary). In this case, the wider row spacing permitted equipment to travel between rows, resulting in less plant damage from tire traffic and improving the ease of harvest

TABLE 63.2   Examples of the variation in recommended seeding rates and planting dates of switchgrass cultivars in different regions of the United States.

Region	Cultivars	Planting rate (lb PLS/acre)	Approximate planting dates
Northern Great Plains	Summer, Sunburst, Forestburg	3.5–5.0	Mid-May to late June
Midwest	Kanlow, Cave-in-Rock	5.0–6.5	Late April to mid-June
Southern Plains	Alamo, Kanlow	3.0	Mid-April to mid-May
Southeast	Alamo	6–8	Late March to mid-June
Northeast	Cave-in-Rock, Kanlow, Carthage, Sunburst	8	Late April to mid-June

Refer to the eFOTG for specific seeding rates and planting dates for specific areas.

FIGURE 63.8   Drill equipped with depth bands and packer wheels. *Source:* USDA NRCS Bismarck, ND.

operations. This row spacing width may also accommodate the interplanting of winter annual and perennial legumes between rows. These interseeded legumes can provide partial nitrogen requirements of the switchgrass crop (Figure 63.9). Interseeding with cool season perennial legumes is not recommended in the Northeast because they become too competitive with the switchgrass plants.

63.2.8   Stand Evaluation

Stand success is generally evaluated by determining seedling density and distribution of seedlings across the field. This assessment should occur within 6–10 weeks after spring planting. A simple, fast, and reliable method for evaluating stand establishment is using a frequency grid (Vogel and Masters 2001). The frequency grid is 2.5 by 2.5 square feet that contains 25, 6-by-6-inch open cells. The grid is placed on the soil surface and the number of seedlings in each cell is counted, and then the grid is flipped three times, end-over-end, until a total of 100 cells is observed (Figure 63.10). A running total of the number of cells containing one or more seedlings is made. To determine percent stand, divide the total number of cells containing a seedling by 100. The grid should be used to take measurements at several representative locations across the field. If 40% or more of the grids contain seedlings, it is considered an acceptable stand of switchgrass for bioenergy production (Schmer et al. 2005). Another method for determining seedling density is counting the number of seedlings in a 1 square-foot quadrant at several representative locations in the field. Three to six plants per square foot is considered an acceptable switchgrass stand at 6–10 weeks after planting. One to two plants per square foot is considered an adequate stand for biomass production at the end of the second year (Figure 63.11).

FIGURE 63.9   Ball clover interseeded into Alamo switchgrass at the USDA NRCS East Texas Plant Materials Center near Nacogdoches, TX (early spring growth of Alamo and ball clover (top)/late spring growth of Alamo and ball clover (bottom)). *Source:* USDA NRCS Nacogdoches, TX.

63.2.9   Weed Control

Slow germination of switchgrass, slower seedling shoot growth compared to annual weeds, and relatively low tolerance for shade make weed control at planting and during the establishment year important.

There are herbicides labeled for pre- and postemergence control of weeds in switchgrass; however, it is important to consult with extension weed specialists or the county extension agent to determine which ones are registered for use in the State and county. Any information regarding herbicide application is based on current availability and labeling. If herbicides are to be used for weed control in a switchgrass biomass production system, follow current Land Grant University recommendations and adhere to label instructions.

FIGURE 63.10 Frequency grid used to assess stand success of switchgrass. *Source:* USDA NRCS Manhattan, KS.

A nonselective, broad-spectrum herbicide can be applied prior to planting or before switchgrass emergence. Weeds can also be controlled mechanically with a rotary mower. The goal is to reduce the weed canopy so the switchgrass seedlings can grow and to prevent weeds from producing a seed crop. Adjust the rotary mower to a height at or above the top of the switchgrass seedlings to prevent cutting or severely damaging the growing points on the seedlings. Mowing should be performed as often as necessary to control weeds during the first year.

Traditional row crop agriculture is generally depicted as uniform fields of a single crop such as corn (*Zea mays*) growing on level to gentle rolling fields that are weed free. However, switchgrass grown for biomass may not necessarily portray this type of agriculture. Weeds can be common the

FIGURE 63.11 Switchgrass reaches full production 3 years after planting. *Source:* USDA NRCS Booneville, AR.

first few years in switchgrass fields. Do not assume that the planting is a failure if weeds are present in the field and the switchgrass plant population appears sparse. As the switchgrass plants produce tillers and plants increase in size over time, they will begin to shade other plants around them and become more competitive than most of the annual and perennial weeds. Furthermore, in the southern United States, switchgrass plants recover from winter dormancy period better than most introduced warm-season grasses and broadleaf weeds. This means that switchgrass is more competitive than many warm-season weed species early in the growing season. Also, annual weed pressure is usually less the second year, and in subsequent years, due to lack of further soil disturbance.

63.3 POSSIBLE REASONS FOR POOR STAND ESTABLISHMENT

Several factors can cause stand failures in switchgrass including, but not limited to, cultural practices, environmental stresses, and seedling damage. Some of the more common causes are listed below:

- poorly prepared seedbedseedbed too loose for good seed-to-soil contact
- seed planted too deeplyseed germinate but seedlings fail to fully emerge
- high seed dormancyseed did not germinate uniformly in the rows
- excessive weed pressurecompetition with seedlings for moisture and light
- lack of rainfall after plantingseedlings emerge but die due to prolonged drought conditions
- frost heavingseedlings are pushed out of the soil by frost action, exposing the roots to drying
- herbicide damageinjury to young seedlings results in low stand percentage or seedling death
- mowing the seedlings too closely in the establishment yearclipping below the growing point of the young seedlings
- damage by wildlifeslow growth due to reduced leaf area or seedlings killed by grazing too closely

63.4 FERTILIZATION NEEDS IN THE FIRST AND SUBSEQUENT YEARS

63.4.1 Nitrogen—Establishment Year

Because switchgrass is slow to establish, applying nitrogen fertilizer the first year or until a stand of grass is deemed adequate is discouraged (see Section 63.2.8). Applying nitrogen fertilizer to a newly planted field of switchgrass will only encourage weed competition, which hinders switchgrass seedling growth and development.

63.4.2 Nitrogen—Production Years

Nitrogen management for switchgrass biomass production differs markedly from switchgrass managed as livestock forage. For forage production, nitrogen application rates and timing are directed toward producing high nutritional quality forage, whereas nitrogen rates for switchgrass grown for bioenergy seek to maximize biomass production. Nitrogen concentrations in switchgrass forage (harvested at the boot stage) ranged from 1.6% to 2.2% (Ball et al. 2002), while nitrogen concentrations for a switchgrass biomass crop ranged from 0.3% to 0.6% when harvested as a single crop in late fall (Lemus et al. 2009). Less nitrogen has shown to be removed in the biomass in the late-fall harvest compared to a two-cut harvest (midsummer and late fall) because of translocation of nitrogen and other nutrients in late fall to the plant crown and root systems. This indicates that less nitrogen may be required for switchgrass production when the biomass is harvested in late fall (Fike et al. 2006; Lemus et al. 2008).

Studies have shown that switchgrass biomass production is not consistent with different levels of nitrogen fertilization. Differences between studies with regard to cultivars, cutting management, and soil type on which the research was conducted plus climate and precipitation extremes at northern and southern latitudes are some of the factors that contributed to the variation in yield response to nitrogen (Muir et al. 2001; Vogel et al. 2002; McLaughlin and Kszos 2005). For this reason, blanket recommendations on nitrogen fertilization cannot be made without considering the soil and site characteristics.

Sustainable switchgrass production will depend on long-term adjustments to soil fertility. In addition to periodic soil sampling and analyses to monitor soil nutrients, there are some intrinsic physical characteristics that can help guide annual soil fertility management. Soils that provide for deeper rooting depth will allow the switchgrass plant to explore a greater soil volume and utilize nutrients from deep in the profile. Rooting depths greater than 20 inches are considered ample for supplying moisture and nutrients for root growth and plant vigor. Likewise, loam, silty, and clayey soil textures retain soil nutrients and moisture better than coarse-textured soils. Soil with high organic matter levels (>2.5%) can mineralize nitrogen that is then available to the crop. When switchgrass is grown on these more productive soils, apply 50–100 pounds per acre per year nitrogen during the second year (year after establishment) of production to accelerate tillering and plant growth, followed by 50 pounds per acre per year in subsequent production years (Garland 2008; Lemus et al. 2009). Soils with a shallow, eroded topsoil, a coarse profile texture in the root zone, and low soil organic matter have minimum nitrogen supplying power and will need annual nitrogen applications greater than 100 pounds per acre per year. Nitrogen fertilizer needs to be applied when the plant begins active growth or when regrowth reaches 6–8 inches. Split applications are more suitable for rates greater than 100 pounds per acre or for a two-cut harvest system.

63.4.3 Lime, Phosphorus, and Potassium

Establishment year—Prior to planting, adjust P and K levels in the soil to a medium level according to soil test analysis and crop recommendation using a warm-season grass, such as sudangrass, as a soil test guide. If soil pH is less than 5.0, apply sufficient lime to raise pH to 6.0–6.5 to allow for more efficient utilization of soil nutrients.

Production—A single harvest in late fall has produced switchgrass yields in excess of 9–10 tons per acre in the southern United States, 4–5 tons per acre in the Northeastern and Midwestern United States, and more than 6 tons per acre in the Mid-Atlantic States (McLaughlin and Kszos 2005). Removal of P and K from the soil in the harvested biomass must be considered to maintain long-term sustainable switchgrass yields. Removal rates of P and K vary considerably and are dependent on growing season, yield, timing of harvest, precipitation, and soil quality. One ton of switchgrass may remove approximately 4 pounds per acre P and 40 pounds per acre K (Fixen 2007). These nutrients will need to be replaced through proper fertilization following soil test recommendations. A soil test should be taken every 3 years to maintain P and K at a medium soil test level. Once soil test levels of P and/or K drop into the low category, annual soil test monitoring plus fertilizer applications are recommended. Like nitrogen fertilizer, P and K can be broadcast applied in the spring when the plants begin active growth or when regrowth reaches 6–8 inches.

63.5 SOIL CONDITIONS AT HARVEST

As with any crop production system, field and crop conditions at harvest will depend on the weather. Heavy rains and wet soils during harvest will provide poor drying conditions for the harvestable biomass. Biomass spoilage and formation of molds during storage are possible. Wet soils at harvest can reduce field access and cause rutting and damage to plant crowns.

Dry or frozen field conditions will provide support to harvest equipment without damaging the switchgrass plants or soil surface. In parts of the North and Northeast, snow accumulation can effect late fall harvesting.

63.6   HARVEST MANAGEMENT

Weather, soil conditions, farm workload, and the quality of the feedstock to satisfy the various biofuel conversion systems of direct combustion, ethanol production, and thermochemical conversion processes (Lewandowski and Kicherer 1997; Boateng et al. 2006) may dictate frequency of harvest for optimizing biomass production. Harvesting the switchgrass once a year appears to be the most economical harvest system for biomass production. The timing of the harvest should be approximately 1 month after the first killing frost (McLaughlin and Kszos 2005; Parrish and Fike 2005). However, it may take several killing frosts before the plant is totally dormant. In addition, delaying harvest until several weeks after frost allows for translocation of nutrients in the biomass back to the stem bases and root system. It also improves feedstock quality due to low moisture content in the aboveground biomass, as well as a reduction in nitrogen content, other minerals (K, Ca), and ash concentrations that can cause direct combustion systems to fail (Miles et al. 1996; Lewandowski and Kicherer 1997). This strategy conserves nutrients in the soil while reducing the amount of nutrients removed in the harvestable biomass. However, delaying harvest for an extended period after frost may result in significant biomass loss. This is due to winter weather conditions which causes loss of dry matter (leafy tissue) and degradation of the biomass, and complicates harvesting operations and ultimately harvests efficiency (Adler et al. 2006) (Figure 63.12).

Factors such as weather and field conditions in the fall in different regions of the United States may complicate harvesting and baling of the switchgrass (Cherney 2005). A two-cut system may be the only reliable option for harvesting switchgrass biomass on some soils in the wetter, cooler regions of the United States. Generally in a two-cut system, the switchgrass is harvested in mid- to late summer and the

second harvest is made in the fall, as weather permits. One problem is that the nitrogen concentration in the biomass is relatively high for the first harvest. An option for handling switchgrass biomass from a summer harvest is to leave it windrowed in the field until the nutrients and ash concentrations meet the desired levels for a specific conversion process (Cherney 2005). The costs of a second harvest must also be considered when making this management decision (Vogel et al. 2002). Another consideration with a two-cut system is the quality of the feedstock and fertilizer management. Because of the potentially high removal of nitrogen in the harvestable biomass, additional nitrogen fertilizer may need to be applied following the first harvest in a two-cut system.

Regardless of harvest system, switchgrass should be mowed no lower than a 6-inch stubble height to reduce tire damage (Garland 2008), trap snow/retain moisture, reduce soil erosion, and provide some cover for wildlife. Cutting the switchgrass lower than 6 inches will produce stiff, hard stems that may puncture tires on tractors or harvesting equipment. A 6-inch cutting height will also allow switchgrass to maintain nutrients and carbohydrates in the stem bases that are necessary for spring regrowth and result in improved plant stands (Figure 63.13).

A traditional, sidebar-type sickle mower can be used to cut switchgrass the first and possibly the second year harvest after establishment. However, in subsequent years as the switchgrass plants reach full maturity, that type of mower will not be able to cut the switchgrass, especially when making a single harvest in late fall due to the large diameter of the stems and the inability of the sickle to maintain a consistent cutting height of 6 inches. A disc-type mower becomes more effective, but may also present a challenge as the height of the switchgrass may reach 6–8 ft by late July in the Southeast and by the fall in many regions of the United States. Mowers equipped with swinging flail blades may have more

FIGURE 63.12 Delaying harvest of switchgrass until after frost improves feedstock quality. *Source:* Don Tyler, University of Tennessee.

FIGURE 63.13 A 6-inch stubble height may reduce tire damage on harvesting equipment. *Source:* USDA NRCS Manhattan, KS.

merit than a disc-type mower for cutting mature switchgrass. Crushing or conditioning the biomass prior to baling is not recommended.

Moisture content is an important consideration in baling switchgrass biomass. High moisture content of the biomass affects feedstock quality and excess moisture increases transportation costs. Switchgrass is typically baled at moisture contents of 13–15% (McLaughlin et al. 1996). Large round (5 ft by 4 ft) or square (4 ft by 4 ft by 8 ft) bales have been shown to be acceptable methods for transporting and storing switchgrass. Square bales are reported to be easier to handle, store, and transport (Garland 2008). Another potential alternative is to use or modify a silage chopper to harvest the switchgrass as a loose, noncompacted material.

As the biomass industry continues to evolve, development of new equipment and/or modifications of existing equipment will provide the producer with better technology and methods to meet future challenges of harvesting and transporting biomass from the field to the biorefinery.

63.7 WILDLIFE CONSIDERATIONS

Landscape changes in the Great Plains are associated with change in the communities of birds and other animals that rely on grassland habitats. For example, dramatic declines in grassland bird species since the 1950s are attributed to changes in the agricultural landscape of the region (Gerard 1995). Cultivation and management of switchgrass for biomass production has the potential to provide substantial benefits for grassland-associated wildlife. Wildlife use of switchgrass fields during spring and summer is probably most critical. Field selection, establishment, maintenance, and harvest should be given special consideration if managing for wildlife is one of the objectives of the landowner (Tables 63.3–63.5, W. Hohman, personal communication). Field selection should include fields in close proximity to

TABLE 63.3 Field selection for switchgrass biomass production.

Field selection	Production[a]	Wildlife
Field size and width	Varies depending on the farm size and land use	10–20 acres or > 60 ft
Field location/ configuration	Clustered blocks of land	Clustered blocks of land
Adjacent land-use	Varies with crop and rotation system	Perennial herbaceous or annual cropland

[a]Characteristics largely influenced by land ownership and existing land use, but production efficiency would favor clustering of large blocks in close proximity to the biorefinery.

TABLE 63.4 Establishment considerations for switchgrass biomass fields.

Establishment	Production	Wildlife
Seedbed preparation	Varies	Minimal tillage
Planting method	Varies	No-till
Seed mixtures	Monoculture[a]	Multiple species
Seeding rate	Recommended rate	25–30% reduction in recommended seeding rate
Weed control	Pre- or postemergent	Preemergent

[a]Production efficiency for diverse native plant stands may be competitive with monocultures (Tilman et al. 2006).

other switchgrass fields or near pastureland, hayland, and small grain crops. Reduce hard edges by removing mature trees from fence row. Between switchgrass fields and mature forest, establish a feathered edge comprised of herbaceous plants, shrubs, or small trees. Minimize production activities such as fertilizing and weed control during the nesting season until the young are mobile and have become independent. Furthermore, consider time of harvest in relation to wildlife usage. For example, if a dense, overwinter cover is needed for pheasants, then consider partial harvests and retain an unharvested block adjacent to cropland.

Harvesting switchgrass after a killing frost will avoid harmful effects on wildlife, such as grassland nesting songbirds that use fields in spring and summer. Harper and Keyser (2008) suggest that postponing harvest until late winter would provide greater benefits for wildlife over winter.

TABLE 63.5 Maintenance and harvest considerations for switchgrass biomass production.

Maintenance	Production	Wildlife
Fertilizer applications	As needed	Avoid during the nesting season
Pesticide applications	As needed	Consider spot treatment
Season of harvest	Dormant	Dormant
Frequency of harvest	Annual[a]	Annual[a]/rotational harvest[b]
Harvest considerations	Moisture <15%	Daytime only
Height of harvested biomass	4–6 inches	8–10 inches
Pattern	Continuous	Minimize edge

[a]Assumes a one-cut harvest system in late fall or early winter.
[b]If field is less than 40 acres, harvest the entire field every other year. If field is greater than 40 acres, harvest one half per year or retain a minimum of a 20-acre unharvested block (not strip).

FIGURE 63.14 Field borders established between switchgrass fields and/or row crops can provide food and cover for wildlife and pollinators. *Source:* USDA NRCS Elsberry, MO.

Delaying harvest may result in yield loss, but a higher biofuel quality may occur as a result of additional weathering of the biomass (Adler et al. 2005). Field borders and hedge rows established between fields and rights-of-way, with a diverse plant community that includes forbs, flowering trees, and shrubs, are beneficial and highly recommended for pollinators (Figure 63.14). To obtain more information on managing wildlife in switchgrass biomass production, contact the local NRCS wildlife biologist.

63.8 STAND LONGEVITY

If the best adapted cultivar is planted on a moderately fertile soil, soil nutrient levels are maintained through proper soil fertility management and harvest management, it is believed

FIGURE 63.15 Switchgrass stands should remain productive beyond 10 years with proper management. *Source:* USDA NRCS.

that a switchgrass bioenergy production field could potentially remain productive well beyond 10 years (Parrish and Fike 2005; Fike et al. 2006). Regardless of stand age, if there is a steady decline in switchgrass production over consecutive growing seasons and/or yields drop 30% from the normal production levels under proper management and growing conditions, it may be necessary to reestablish the switchgrass stand. An evaluation of the switchgrass cultivar, soil conditions (drainage, fertility, pH), and management techniques should be conducted to determine the cause of stand decline before proceeding with reestablishment (Figure 63.15).

REFERENCES

P. R. Adler, M. A. Sanderson, A. A. Boateng, P. J. Weimer, H.-J. G. Jung. (2006). Biomass yield and biofuel quality of switchgrass harvested in fall or spring. *Agron. J.*, 98:1518–1525.

J. Alderson, W. C. Sharp. (1994). *Grass Varieties in the United States. USDA Agriculture Handbook 170*. US Government Printing Office, Washington, DC.

Association of Official Seed Analysts (AOSA). (1993). Rules for testing seeds. *J. Seed Technol.* 16:1–113.

D. M. Ball, C. S. Hoveland, G. D. Lacefield. (2002). Forage quality parameters for selected forage crops. In: *Southern Forages.* 3rd edition. Potash and Phosphate Inst., Norcross, GA. p. 298.

A. A. Boateng, K. B. Hicks, K. P. Vogel. (2006). Pyroloysis of switchgrass (*Panicum virgatum*) harvested at several stages of maturity. *J. Anal. Appl. Pyrol.*, 75:55–64.

J. C. Burns, E. B. Godshalk, D. H. Timothy. (2008). Registration of 'BoMaster' Switchgrass. *J. Plant Regist.*, 2:31–32.

M. D. Casler, K. P. Vogel, C. M. Taliaferro, R. L. Wynia. (2004). Latitudinal adaptation of switchgrass populations. *Crop Sci.*, 44:293–303.

K. A. Cassida, J. P. Muir, M. A. Hussey, J. C. Read, B. C. Venuto, W. R. Ocumpaugh. (2005). Biomass yield and stand characteristics of switchgrass in south-central U.S. environments. *Crop Sci.*, 45:673–681.

J. H. Cherney (2005). Management of grasses for biofuel. Bioenergy Info. Sheet #4. Cornell Univ. Coop. Ext. Available at: http://grassbioenergy.org/downloads/ Bioenergy_Info_Sheet_4.pdf (Accessed July 31, 2009).

C. Dewald, J. Henry, S. Bruckerhoff, J. Richie, D. Shepard, S. Dabney, J. Douglas, D. Wolfe. (1996). Guidelines for the establishment of warm-season grass hedge for erosion control. *J. Soil and Water Conserv.*, 1:16–20.

J. H. Fike, D. J. Parrish, D. D. Wolfe, J. A. Balasko, J. T. Green, Jr., M. Rasnake, J. H. Reynolds. (2006). Long-term yield potential of switchgrass–for biofuel systems. *Biomass and Bioenergy*, 30:198–206.

P. E. Fixen (2007). Potential biofuel influence on nutrient use and removal in the U.S. *Better Crops.*, 91:12–14.

C. D. Garland (2008). Growing and harvesting switchgrass for ethanol production in Tennessee. Ext. Bull. SP701-A. Available at: http://utextension.tennessee.edu/publications/spfiles/SP701-A.pdf (Accessed July 31, 2009).

P. Gerard (1995). Agricultural practices, farm policy, and the conservation of biological diversity. Biological Science Report 4, U.S. Department of the Interior, National Biological Service. 28 pp.

C. E. Gravert, L. H. Tiffany, G. P. Munkvold. (2000). Outbreak of smut caused by *Tilletia maclaganii* on cultivated switchgrass in Iowa. *Plant Disease*, 84:596.

D. M. Gustafson, A. Boe, Y. Jin. (2003). Genetic variation for *Puccinia emaculata* infection in switchgrass. *Crop Sci.*, 43:755–759.

C. A. Harper, P. D. Keyser. (2008). Potential impacts on wildlife of switchgrass grown for biofuels. Ext. Bull. SP704-A. Available at: http://utextension.tennessee.edu/publications/spfiles/SP704-A.pdf (Accessed July 31, 2009).

D. K. Lee, A. Boe. (2005). Biomass production of switchgrass in central south Dakota. *Crop Sci.*, 45:2583–2590.

R. Lemus, E. C. Brummer, K. J. Moore, N. E. Molstad, C. L. Burras, M. F. Barker. (2002). Biomass yield and quality of 20 switchgrass populations in southern Iowa. *Biomass Bioenergy*, 23:433–442.

R. Lemus, D. J. Parrish, O. Abaye. (2008). Nitrogen-use dynamics in switchgrass grown for biomass. *Bioenerg. Res.*, 1:153–162.

R. Lemus, D. J. Parrish, D. D. Wolfe. (2009). Nutrient uptake by 'Alamo' switchgrass used as an energy crop. Bioenergy Res., (online). DOI 10.1007/ s12155-009-9032-3. Available at: http://www.springerlink.com/content/u6214j86601p437q/fulltext.pdf (Accessed July 31, 2009).

I. Lewandowski, A. Kicherer. (1997). Combustion quality of biomass: practical relevance and experiments to modify the biomass quality of *Miscanthus* × *giganteus*. *Eur. J. Agron.*, 6:163–177.

M. A. Liebig, M. R. Schmer, K. P. Vogel, R. B. Mitchell. (2008). Soil carbon storage by switchgrass grown for bioenergy. *Bioenerg. Res.*, (1):215–222.

S. B. McLaughlin, L. A. Kszos. (2005). Development of switchgrass (*Panicum virgatum*) as a bioenergy feedstock in the United States. *Biomass and Bioenergy*, 28:515–535.

S. B. McLaughlin, R. Samoson, D. Bransby, A. Wiselogel. (1996). Evaluating physical, chemical and energetic properties of perennial grasses and biofuels. *Proc., BIOENERGY '96—The Seventh National Bioenergy Conference: Partnerships to Develop and Apply Biomass Technologies*, September 15–20, 1996, Nashville, TN. Available at: http://bioenergy.ornl.gov/papers/bioen96/mclaugh. html (Accessed July 31, 2009).

T. R. Miles, T. R. Miles, Jr., L. L. Baxter, R. W. Bryers, B. M. Jenkins, L. L. Oden. (1996). Boiler deposits from firing biomass fuels. *Biomass Bioenergy*, 10:125–138.

J. P. Muir, M. A. Sanderson, W. R. Ocumpaugh, R. M. Jones and, R. L. Reed. (2001). Biomass production of 'Alamo' switchgrass in response to nitrogen, phosphorus, and row spacing. *Agron. J.*, 93:896–901.

D. J. Parrish, J. H. Fike. (2005). The biology and agronomy of switchgrass for biofuels. *Critical Reviews in Plant Sciences*, 24:423–459.

M. R. Schmer, K. P. Vogel, R. B. Mitchell, L. E. Moser, K. M. Eskridge, R. K. Perrin. (2005). Establishment stand thresholds for switchgrass grown as a bioenergy crop. *Crop Sci.*, 46:157–161.

Z. X. Shen, D. J. Parrish, D. D. Wolf, G. E. Welbaum. (2001). Stratification in switchgrass seeds is reversed and hastened by drying. *Crop Sci.*, 41:1546–1551.

D. L. Stuteville, R. L. Wynia, J. M. Row. (2001). Reaction of switchgrass cultivars to seed smut caused by *Tilletia maclaganii*. Available at: http://www.apsnet.org/meetings/div/nc01abs.asp (Accessed July 31, 2009).

D. Tilman, J. Hill, C. Lehman. (2006). Carbon-negative biofuels from low-input high diversity grassland biomass. *Science*, 314:1598–1600.

U.S. Department of Agriculture, Natural Resources Conservation Service (USDA NRCS). (1991). *Native Perennial Warm-Season Grasses for Forage in Southeastern United States (Except South Florida)*. Ecological Science and Planning Staff, Fort Worth, TX.

U.S. Department of Agriculture, Natural Resources Conservation Service (USDA NRCS). (2003). *Five Keys to Successful Grass Seeding*. Plant Materials Center, Bismarck, ND. Available at: http://plant-materials.nrcs.usda.gov/pubs/ndpmcbr04959.pdf (Accessed July 31, 2009).

U.S. Department of Agriculture, Natural Resources Conservation Service (USDA NRCS). (2006). *The PLANTS Database*. National Plant Data Center, Baton Rouge, LA. Available at: http://plants.usda.gov (Accessed July 31, 2009).

U.S. Department of Agriculture, Natural Resources Conservation Service (USDA NRCS). (2007). Switchgrass biomass trials in North Dakota, South Dakota, and Minnesota. Available at: http://www.plant-materials.nrcs.usda.gov/pubs/ndpmcpu7093.pdf (Accessed July 31, 2009).

U.S. Department of Agriculture, Natural Resources Conservation Service (USDA NRCS). (2009). Switchgrass for biomass production by variety selection and establishment methods for Missouri, Illinois, and Iowa. Agron. Tech. Note. MO-37, January 2009. Available at: http://efotg.nrcs.usda.gov/references/public/IA/AgronomyTechNote35.pdf (Accessed July 31, 2009).

K. P. Vogel, R. A. Masters. (2001). Frequency grid—a simple tool for measuring grass establishment. *J. Range Manage*, 544:653–655.

K. P. Vogel, J. J. Brejda, D. T. Walters, D. R. Buxton. (2002). Switchgrass biomass production in the Midwest USA. Harvest and nitrogen management. *Agron J.*, 94:413–420.

L. Wright (2007). Historical perspective on how and why switchgrass was selected as a "Model" high-potential energy crop. Oak Ridge National Lab/TM-2007/109. Available at: http://www1.eere.energy.gov/biomass/pdfs/ornl_switchgrass.pdf (Accessed July 31, 2009).

64

MUNICIPAL SOLID WASTE—CHRONOLOGICAL DEVELOPMENT

PUBLIC DOMAIN

Municipal Solid Waste (MSW) (i.e., trash) can be a source of energy by either burning the MSW in waste-to-energy plants or capturing biogas. Over half of MSW is biogenic, or made from renewable materials.

1898	Energy recovery from garbage incineration started in New York City.
1970s	First-generation research was followed by construction of refuse-derived fuel systems and pyrolysis units in the late 1970s. The US Navy, Wheelabrator, and Ogden acquired the European mass burn technologies that would dominate the US industry by the late 1980s.
1976	The Resource Conservation and Recovery Act (RCRA) of 1976 empowered the Environmental Protection Agency (EPA) to regulate residues from solid waste incinerators. Unclear wording made application of the law to MSW power plants uncertain, and the issue was taken to court.
1978	The Public Utility Regulatory Policies Act (PURPA) of 1978 mandated the purchase of electricity from qualifying facilities at a utility's avoided cost of energy and capacity. This legislation was used to require utilities to pay a higher price for power from MSW power plants than the plants had traditionally received.
	The US Supreme Court defined *waste* to be an article of interstate commerce that cannot be discriminated against unless there is some reason, apart from its origin, to treat it differently, or unless Congress specifies otherwise for particular articles of commerce.
1986	The Tax Reform Act of 1986 eliminated the tax-free status of MSW power plants financed with industrial development bonds, reduced accelerated depreciation, and eliminated the 10% tax credit. The Act also reduced State caps on private tax-exempt bonds in 1988, further reducing funding sources and increasing the cost of capital.
1987	Landfill tipping fees doubled, and doubled again every 2 years owing to rising landfill costs resulting from the RCRA. Siting issues became increasingly difficult.
1989	An EPA report on recycling, *The Solid Waste Dilemma: An Agenda for Action*, advocated recycling as a waste management tool.
1990	The EPA recognized MSW power as a renewable fuel that would qualify for up to 30,000 sulfur dioxide emission allowances from a special pool of 300,000 designed to promote conservation and renewable energy. The EPA also required MSW power plants that could process over 250 tonnes per day to use the best available control technology (BACT).
1991	Under Subtitle D of the RCRA, the EPA announced that small, unlined landfills would be required to close by December 31, 1993. Most landfills requested and received extensions. This action spurred the infant recycling industry and increased tipping fees around the country.

(continued)

Alternative Energy and Shale Gas Encyclopedia, First Edition. Edited by Jay H. Lehr and Jack Keeley.
© 2016 John Wiley & Sons, Inc. Published 2016 by John Wiley & Sons, Inc.

1992	An EPA memorandum excluded ash from regulation as a hazardous waste, under Subtitle C of the RCRA, as long as it was not characterized as *toxic*.
	Recycling legislation was adopted by 15 States.
	The US Supreme Court ruled that State-imposed waste import restrictions were illegal. "Economic protectionist" measures that violated the Commerce Clause and were, therefore, unconstitutional.
	President Bush issued Executive Order 12780; it stimulated waste reduction, recycling, and the buying of recycled goods in all Federal agencies.
1994	The US Supreme Court ruled that the exemption of MSW (from a hazardous waste definition) under the RCRA did not extend to ash. MSW ash must be tested and disposed of in hazardous waste landfills, if found to exceed EPA regulations on hazardous wastes under RCRA.
	The term *flow control* is defined as: The official authority of waste managers to direct waste generated in a city or area to a designated landfill, recycling, or waste-to-energy facility (e.g., in another State).
	The US Supreme Court upheld challenges to flow control. As a result, existing flow control contracts could be rendered invalid under specific situations (on a case-by-case basis). Several plants have shut down as a result. The California Supreme Court also ruled against flow control.
	The EPA strengthened air emission standards for MSW combustion plants by requiring maximum achievable control technologies (MACT). It also included plants as small as 40 tonnes per day under regulations.
	President Clinton issued Executive Order 12873; it required Federal agencies to establish waste prevention and recycling programs and to buy and use recycled and environmentally preferable products and services. Clinton created the Office of the Federal Environmental Executive to enforce the order.
1995	The Senate passed a flow control bill that grandfathered in existing flow control contracts to prevent the major risk of MSW bond default in 14 States.
	A total of 208 million tonnes of MSW was generated in 1995. This reflected a decrease of more than 1 million tonnes from 1994, when MSW generation was over 209 million tonnes.
	About 7000 curbside recycling programs and nearly 9000 drop-off centers for recyclables operated in the United States.
	The United States had 112 waste-to-energy combustion facilities.
1996	With a 25% recycling rate achieved in the United States, the EPA set a recycling goal of 35%.
	The Olympic Games in Atlanta introduced a system for voluntary recycling and composting.
2000	The EPA established a link between global climate change and solid waste management, noting that waste reduction and recycling can help reduce greenhouse gas emissions.
2001	EPA policy required its offices to use paper with 100%-recycled content and 50%-postconsumer content.
2003	Recycling and composting diverted more than 72 million tonnes from disposal in landfills.
2004	The EPA reported the amount of toxics released decreased 45% from 1998 to 2004, from 6.7 billion to 3.7 billion pounds.
2005	A total of 245.7 million tonnes of MSW was generated in the United States.
	32% of the MSW was recycled or composted.
2006	Biogenic MSW provided the most energy-from-waste energy at 42%.
	The second largest share, 37%, came from landfill gas.
2007	The EIA decided to split MSW into biogenic and nonbiogenic components for future data releases.

Last revised: June 2008

Sources: Energy Information Administration. (2006). Renewable Energy Trends in Consumption and Electricity. Available at http://www.eia.gov/cneaf/solar.renewables/page/trends/trends.pdf), last accessed July 2008.

Energy Information Administration. (1995). Renewable Energy Annual 1995. Available at http://www.eia.gov/FTPROOT/renewables/060395.pdf), last accessed December 2005.

Energy Information Administration, Renewable Energy Sources: A Consumer's Guide. Available at http://www.eia.gov/neic/brochure/renew05/renewable.html, last accessed December 2005.

US Environmental Protection Agency. Municipal Solid Waste Publications. Available at http://www.epa.gov/epawaste/inforesources/pubs/municipal_sw.htm, last accessed June 2008.

US Census Bureau, Geography and Environment Tables. Available at http://allcountries.org/uscensus/geography_and_environment.html, last accessed June 2008.

65

ETHANOL—CHRONOLOGICAL DEVELOPMENT

1826	Samuel Morey developed an engine that ran on ethanol and turpentine.
1860	German engine inventor Nicholas Otto used ethanol as the fuel in one of his engines. Otto is best known for his development in 1876 of a modern internal combustion engine (referred to as *the Otto Cycle*).
1862	The Union Congress put a $2 per gallon excise tax on ethanol to help pay for the Civil War. Before the Civil War, ethanol was a major illuminating oil in the United States. After the tax was imposed, the cost of ethanol increased too much to be used this way.
1896	Henry Ford built his first automobile, the quadricycle, to run on pure ethanol.
1906	Over 50 years after imposing the tax on ethanol, Congress removed it, making ethanol an alternative to gasoline as a motor fuel.
1908	Henry Ford produced the Model T. As a flexible fuel vehicle, it could run on ethanol, gasoline, or a combination of the two.
1917–1918	During World War I, the need for fuel drove up ethanol demand to 50–60 million gallons per year.
1920s	Gasoline became the motor fuel of choice. Standard Oil began adding ethanol to gasoline to increase octane and to reduce engine knocking.
1930s	Fuel ethanol gained a market in the Midwest. Over 2000 gasoline stations in the Midwest sold gasohol, which was gasoline blended with 6–12% ethanol.
1941–1945	Ethanol production for fuel use increased, owing to a massive wartime increase in demand for fuel, but most of the increased demand for ethanol was for nonfuel wartime uses.
1945–1978	Once World War II ended, with reduced need for war materials and with low price of fuel, ethanol use as a fuel was drastically reduced. From the late 1940s until the late 1970s, virtually no commercial fuel ethanol was available anywhere in the United States.
1974	The first of many legislative actions to promote ethanol as a fuel, the Solar Energy Research, Development, and Demonstration Act of 1974, led to research and development of the conversion of cellulose and other organic materials (including wastes) into useful energy or fuels.
1975	The United States begins to phase out lead in gasoline. Ethanol becomes more attractive as a possible octane booster for gasoline. The Environmental Protection Agency (EPA) issued the initial regulations, requiring reduced levels of lead in gasoline in early 1973. By 1986, no lead was allowed in motor gasoline.
1978	The term *gasohol* was defined, for the first time, in the Energy Tax Act of 1978. Gasohol was defined as a blend of gasoline with at least 10% alcohol by volume, excluding alcohol made from petroleum, natural gas, or coal. For this reason, all ethanol to be blended into gasoline is produced from renewable biomass feedstocks. The Federal excise tax on gasoline at the time was 4 cents per gallon. This law amounted to a 40-cents-per-gallon subsidy for every gallon of ethanol blended into gasoline.

(continued)

Alternative Energy and Shale Gas Encyclopedia, First Edition. Edited by Jay H. Lehr and Jack Keeley.
© 2016 John Wiley & Sons, Inc. Published 2016 by John Wiley & Sons, Inc.

1979	The marketing of commercial alcohol-blended fuels began by the Amoco Oil Company, followed by Ashland, Chevron, Beacon, and Texaco.
	About $1 billion eventually went to biomass-related projects from the Department of the Interior and Related Agencies Appropriation Act.
1980–1984	The first US survey of ethanol production was conducted. The survey found that fewer than 10 ethanol facilities existed, producing about 50 million gallons of ethanol per year. This was a major increase from the late 1950s until the late 1970s, when virtually no fuel ethanol was commercially available.
	Congress enacted a series of tax benefits to ethanol producers and blenders. These benefits encouraged the growth of ethanol production.
	The Energy Security Act of 1980 offered insured loans for small ethanol producers (less than 1 million gallons per year), up to $1 million in loan guarantees for each project that could cover up to 90% of construction costs on an ethanol plant; price guarantees for biomass energy projects; and purchase agreements for biomass energy used by Federal agencies.
	Congress placed an import fee (tariff) on foreign-produced ethanol. Previously, foreign producers, such as Brazil, were able to ship less-expensive ethanol into the United States.
	The Gasohol Competition Act of 1980 banned retaliation against ethanol resellers.
	The Crude Windfall Tax Act of 1980 extended the ethanol–gasoline blend tax credit.
1983	The Surface Transportation Assistance Act of 1982 (signed in early 1983) increased the ethanol subsidy to 50 cents per gallon.
1984	The number of ethanol plants in the United States peaked at 163.
	The Tax Reform Act of 1984 increased the ethanol subsidy to 60 cents per gallon.
1985	Many ethanol producers went out of business, despite the subsidies.
	Only 74 of the 163 commercial ethanol plants (45%) remained operating by the end of 1985, producing 595 million gallons of ethanol for the year.
1988	Ethanol was first used as an oxygenate in gasoline. Denver, CO, mandated oxygenated fuels (i.e., fuels containing oxygen) for winter use to control carbon monoxide emissions.
	Other oxygenates added to gasoline included MTBE (methyl tertiary butyl ether, made from natural gas and petroleum) and ETBE (ethyl tertiary butyl ether, from ethanol and petroleum).
	MTBE dominated the market for oxygenates.
1990	The Omnibus Budget Reconciliation Act of 1990 decreased the ethanol subsidy to 54 cents per gallon of ethanol.
	Ethanol plants began switching from coal to natural gas for power generation and adopting other cost-reducing technologies.
	An expanding market and the high cost of fructose corn syrup encouraged expansion of wet mill plants that produce the syrup as a by-product of the ethanol production process.
1992	The Energy Policy Act of 1992 provided for two additional gasoline blends (7.7% and 5.7% ethanol). The act defined ethanol blends with at least 85% ethanol as *alternative transportation fuels*. It also required specified car fleets to begin purchasing alternative fuel vehicles, such as vehicles capable of operating on E-85 (a blend of 85% ethanol and 15% gasoline). The act also provided tax deductions for purchasing (or converting) a vehicle capable of running on alternative fuel, such as E-85, and for installing equipment to dispense alternative fuels.
	The Clean Air Act Amendments mandated the *wintertime* use of oxygenated fuels in 39 major carbon monoxide nonattainment areas (areas where EPA emissions standards for carbon monoxide had not been met) and required year-round use of oxygenates in 9 severe ozone nonattainment areas in 1995.
	MTBE was still the primary oxygenate used in the United States.
1995	The excise tax exemption and income tax credits were extended to ethanol blenders that produced ETBE.
	The EPA began requiring the use of reformulated gasoline, year round, in metropolitan areas with the most smog.
1995–1996	With a poor corn crop and the doubling of corn prices in the mid-1990s to $5 a bushel, some States passed subsidies to help the ethanol industry.
1997	Major US auto manufacturers began mass production of flexible-fueled vehicle models capable of operating on E-85, gasoline, or both. Despite their ability to use E-85, most of these vehicles used gasoline as their only fuel because of the scarcity of E-85 stations.
1998	The ethanol subsidy was extended through 2007 with a gradual reduction from 54 cents per gallon to 51 cents per gallon in 2005.
1999	Some States began to pass bans on MTBE use in motor gasoline because traces of it were showing up in drinking water sources, presumably from leaking gasoline storage tanks. Because ethanol and ETBE are the main alternatives to MTBE as an oxygenate in gasoline, these bans increased the need for ethanol as they went into effect.
2000	The EPA recommended that MTBE should be phased out nationally.
2001	A 1998 law reduced the ethanol subsidy to 53 cents per gallon, starting January 1, 2001.

2002	US automakers continued to produce large numbers of E-85-capable vehicles to meet Federal regulations that required a certain percentage of fleet vehicles capable of running on alternative fuels. Over 3 million of these vehicles were in use.
	At the same time, several States were encouraging fueling stations to sell E-85.
	With only 169 stations in the United States selling E-85, most E-85-capable vehicles are still operating on gasoline instead of on E-85.
2003	A 1998 law reduced the ethanol subsidy to 52 cents per gallon, starting January 1, 2003.
	As of October 2003, a total of 18 States had passed legislation that would eventually ban MTBE.
	California began switching from MTBE to ethanol to make reformulated gasoline, resulting in a significant increase in ethanol demand by mid-year, even though the California MTBE ban did not officially go into effect until 2004.
2005	The Energy Policy Act of 2005 was responsible for regulations that ensured gasoline sold in the United States contained a minimum volume of renewable fuel, called the *Renewable Fuels Standard.* The regulations aimed to double, by 2012, the use of renewable fuel, mainly ethanol made from corn.
2007	The Energy Independence and Security Act of 2007 expanded the Renewable Fuels Standard to require that 36 billion gallons of ethanol and other fuels be blended into gasoline, diesel, and jet fuel by 2022. In 2007, the United States consumed 6.8 billion gallons of ethanol and 0.5 billions gallons of biodiesel.
	An Argonne National Laboratory study compared data on water, electricity, and total energy usage from 2001 and 2006. During this period, America's ethanol industry achieved improvements in efficiency and resource use while it increased production to nearly 300%.
2008	As of March 2008, US ethanol production capacity was at 7.2 billion gallons, with an additional 6.2 billion gallons of capacity under construction.

Last revised: June 2008.

Sources: Energy Information Administration, Policies to Promote Non-hydro Renewable Energy in the United States and Selected Countries, February 2005; Energy Information Administration, *Renewable Energy Annual 1995*, December 2005; Fuel Testers.com, Ethanol Fuel History (http://www.fuel-testers.com/ethanol_fuel_history.html), May 2008.

66

THERMAL PROPERTIES OF METHANE HYDRATE BY EXPERIMENT AND MODELING AND IMPACTS UPON TECHNOLOGY

Robert P. Warzinski, Isaac K. Gamwo, Eilis J. Rosenbaum, Evgeniy M. Myshakin, Hao Jiang, Kenneth D. Jordan and Niall J. English, and David W. Shaw (Public Domain)

66.1 INTRODUCTION

The National Energy Technology Laboratory (NETL) has been involved in hydrate research since the early 1980s [1]. The current effort involves both experimental and theoretical research that is focused on obtaining pertinent, high-quality information on gas hydrates that will benefit the development of models and methods for predicting the behavior of gas hydrates in their natural environment under production or climate change scenarios. The modeling effort comprises both fundamental and reservoir-scale simulations and economic modeling. NETL has also recently established a new virtual Institute for Advanced Energy Studies (IAES) that includes research on gas hydrates in their natural environment. The IAES involves professors and students from universities in the western Pennsylvania and northern West Virginia regions.

The emphasis of part of the hydrate research at NETL has been on obtaining thermal properties of methane hydrate. These properties are important for hydrate production, seafloor stability, and climate change scenarios. In particular, the emphasis has been on thermal conductivity. The purpose of this chapter is to present the results of recent work in this area by NETL and others and indicate the importance of these findings to understanding the thermal behavior of methane hydrate and how this behavior impacts technology. This chapter will describe recent research using laboratory experiments, molecular level modeling, and reservoir simulation. A brief background will be provided in each of these sections.

66.2 EXPERIMENTAL MEASURMENTS

66.2.1 Thermal Conductivity

Stoll and Bryan were the first to measure and note the unusual thermal conductivity behavior of clathrate hydrates [2]. Unlike ice that has a higher conductivity than water, the conductivity of both methane and propane hydrate was about 30% lower than water. They made a similar observation when ice/sand and hydrate/sand samples were compared. Ross et al. were the first to observe the unusual temperature dependence of clathrate hydrates [3]. Unlike other crystalline, nonmetallic solids at temperatures greater than 100 K, the conductivity of THF hydrate increased with temperature. Other investigations since then have, in general, also observed these unusual properties with various clathrate hydrates [4–20]. Some of this experimental work was performed at temperatures down to 2 K to help elucidate the mechanisms for the anomalous behavior of clathrate hydrates compared to ice. However, no theoretical model currently exists that permits a quantitative description of the thermal conductivity of clathrate hydrates over a wide range of temperatures [12].

With respect to naturally occurring clathrate hydrates, of which methane is generally the predominant guest species, the thermal conductivity of methane hydrate at geologically relevant conditions has only been measured by a small number of investigators owing to the difficulties encountered at working at the pressures required to form and stabilize methane hydrate in a manner suitable for thermal property measurements.

Alternative Energy and Shale Gas Encyclopedia, First Edition. Edited by Jay H. Lehr and Jack Keeley.
© 2016 John Wiley & Sons, Inc. Published 2016 by John Wiley & Sons, Inc.

Stoll and Bryan first measured the thermal conductivity of a porous sample of methane hydrate formed from water and methane using a needle probe based on the technique of von Herzen and Maxwell [2, 21]. Based on the limited information in this paper, they obtained a value similar to that for tests with a somewhat compacted sample of propane hydrate, which was 0.39 W m^{-1} K^{-1} at 275 K. Cook and Leaist, using a guarded hot plate cell, obtained a value of 0.45 W m^{-1} K^{-1} at 216 K on a disc-shaped sample of methane hydrate that may have contained some ice [8]. The sample was prepared in a mold using methane hydrate powder under a mechanical pressure of 100 MPa. Waite et al. [16] synthesized porous methane hydrate around a needle probe from granular ice according to the method of Stern et al. [22, 23] and found the conductivity to range from 0.36 to 0.34 W m^{-1} K^{-1} over a temperature range of 253–278 K, respectively. Waite et al. also reported results for compacted samples of methane hydrate in their device [18, 19]. For a sample prepared in a similar manner and radially compacted around the needle probe at 32 MPa, they obtained values that ranged from 0.456 to 0.452 W m^{-1} K^{-1} over a temperature range from 243 to 268 K [18]. In the same system modified to apply more uniform radial compaction pressure they more recently obtained values of 0.62–0.63 W m^{-1} K^{-1} over a temperature range from 253 to 290 K for a sample compacted at ∼102 MPa [19]. Using a commercially available transient-plane-source (TPS) instrument based on the technique developed by Gustafsson [24, 25], Huang and Fan formed a sample of methane hydrate from methane and water around a relatively flat TPS sensor with the aid of a surfactant (0.971 mol m^{-3} aqueous sodium dodecyl sulfate) [15]. They obtained conductivity results over a temperature range from 263 to 278 K of 0.334–0.381 W m^{-1} K^{-1} for the sample without compaction and 0.564–0.587 W m^{-1} K^{-1} with compaction at 2 MPa.

With respect to all of the above techniques, the needle probe has been the only one used successfully for in situ measurements in geologic samples [21]. Using such a probe in hydrate-cemented sediments requires precision drilling and sealing [26] or otherwise good contact with a solid surface [17]. NETL has recently developed a modified TPS technique that may be more suitable for in situ or field measurements [27]. In this approach, a single-sided TPS technique is used in which the TPS element, two configurations of which are shown in Figure 66.1, is mounted onto an insulating support to permit the device to be used as a contact sensor. The TPS element also requires less sample contact area than a needle probe.

A sensor similar to that shown in Figure 66.1a was mounted on a polyvinylchloride (PVC) base and used to measure the thermal properties of a small, disc-shaped sample of methane hydrate [20]. The NETL device also permitted direct mechanical compaction of the sample to effectively minimize porosity. Thermal conductivity results of 0.68 ±

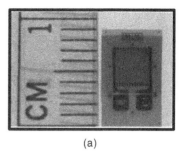 

(a)   (b)

FIGURE 66.1 (a) Rectangular TPS element used in NETL research [20]. (b) Schematic of double-spiral element similar to that used in the work of Huang and Fan [15].

0.01 W m^{-1} K^{-1} over a temperature range of 261.5–277.4 K were obtained after the sample was compacted onto the TPS element at 45 MPa.

Figure 66.2 illustrates the thermal conductivity data for methane hydrate obtained using the NETL device and compares it to the more recent published data or correlations of others discussed above at temperatures typically encountered in hydrate-bearing formations. Also shown in this figure are results for water at 5.5 MPa pressure [28], which was the average gas phase pressure in the NETL experiments, and the results from molecular dynamics simulations that will be discussed below [20, 29]. These experimental and theoretical values for methane hydrate are all above the thermal conductivity of water and have an average value of 0.65 W m^{-1} K^{-1}. The differences in the experimental values may be due to the differences in the techniques used to prepare the samples and to determine the thermal conductivity. The porosity of the sample of Huang and Fan is likely higher than the other

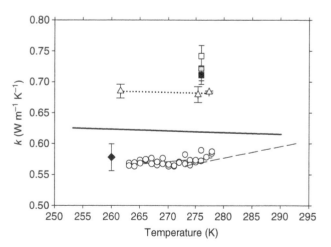

FIGURE 66.2 Thermal conductivity data as a function of temperature. *Experimental*: methane hydrate: Δ, NETL [20]; —, correlation of Waite et al. [19]; ○ Huang and Fan [15]; - - -, water: NIST [28]. *MDS*: methane hydrate: ■, TIP4P-FQ, 100% occupied, ▫, 90%, □, 80% [20]; ◆, COS/G2 100% [29].

samples owing to the low compaction pressures used [15]. The NETL sample is also the smallest, being only about 1 cm^3 [20]; whereas, the sample of Waite et al. was 175 cm^3 [19] and that of Huang and Fan was similar (~200 cm^3) [15]. The compaction of the NETL sample was likely more efficient and is evident in the fact that the slope of the data is the lowest of the three data sets.

The thermal conductivity of a laboratory-formed porous sample of methane hydrate was recently determined by inverse modeling of temperature measurements during thermal cycling of the sample [30]. X-ray computed tomography scans were also taken that permitted assessment of the density of the sample. An arithmetic mixing model was also used by these authors to predict the zero-porosity thermal conductivity of methane hydrate, which was 0.70 ± 0.04 W m^{-1} K^{-1}.

As documented above, the thermal conductivity of low-porosity methane hydrate has now been determined in a manner that precludes removal of the sample for compaction or other handling, thus preventing deterioration of the sample by contamination or decomposition. These measurements were performed using two separate measurement techniques and arrived at similar values that are both higher than previously determined values. The results are also close to values predicted by molecular simulation and an average value that includes both experiment and simulation results is given above. If only the values of Rosenbaum et al. [20] and Waite et al. [19] are averaged, the value is the same, 0.65 W m^{-1} K^{-1}. The NETL experimental and theoretical values for thermal conductivity, 0.68 ± 0.01 and 0.74 ± 0.02 W m^{-1} K^{-1}, respectively, bracket the value of 0.70 ± 0.04 W m^{-1} K^{-1} determined by inverse modeling [30]. Owing to the observed insensitivity to temperature over the range of temperatures of geological interest, no temperature correlations appear to be necessary for most modeling and simulation purposes.

66.2.2 Thermal Diffusivity

Compared to the efforts described above for thermal conductivity, fewer attempts have been made to determine the thermal diffusivity of hydrates. Figure 66.3 contains all of the thermal diffusivity measurements for methane hydrate on both unconsolidated and compacted samples. The first measurements were made by deMartin on a sample of methane hydrate that was compacted in a separate apparatus outside of hydrate equilibrium conditions and appeared to have been contaminated with ice, which began to melt at the highest temperatures investigated [31].

Data for ice, which are about an order of magnitude higher than water, are not shown in Figure 66.3. In the temperature range investigated, the diffusivity of ice is about 1.2 × 10^{-6} m^2 s^{-1} [32].

The data of Waite et al. on a compacted sample [19] and of Kumar et al. on a porous sample [33] are close and fall in the

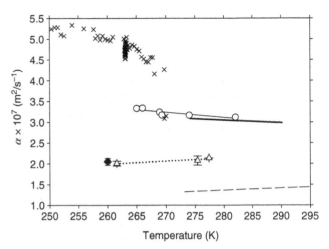

FIGURE 66.3 Thermal diffusivity data as a function of temperature. *Experimental*: methane hydrate: Δ, NETL [20]; —, correlation of Waite et al. [19]; ○, Kumar et al. [33]; × deMartin [31]. Water: —, calculated from NIST data [28]. *MDS*: ♦ COS/G2 [29].

same range as the last two, supposedly ice free, data points of deMartin [31]. This would indicate that porosity has little effect on thermal diffusivity as discussed by Waite et al. [19]. Both the NETL experimental and modeling data are lower. The NETL experimental data also have a slope similar to that of water. As discussed in Rosenbaum et al. [20], finite difference modeling of the single-sided TPS technique with PVC as the support material shows that these data are likely lower than they should be owing to the sample having a higher thermal diffusivity than the PVC. We are continuing to study this issue and expect a resolution of this problem soon. The thermal diffusivity data of Waite et al. [19] are at present the only reliable data set for low-porosity methane hydrate.

66.2.3 Specific Heat Capacity

Only one set of measurements has been made of the specific heat capacity of methane hydrate. Handa determined this value, along with values for ethane and propane hydrate in a heat flow calorimeter over the temperature range of 85–270 K [34]. His values along with those calculated by Waite et al. [19] from their data are shown in Figure 66.4. In this figure, the correlation by Waite et al. [19] was determined only from the data above 273 K; however, it is extended below this temperature for comparison purposes.

66.3 MOLECULAR DYNAMIC SIMULATIONS

MDS has been used at NETL to study the thermal properties of methane hydrate. Rosenbaum et al. reported the use of three rigid potential models to determine the thermal conductivity of methane hydrate at 276 K and at pressures from

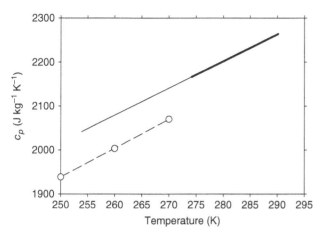

FIGURE 66.4 Specific heat capacity data as a function of temperature for methane hydrate: — correlation of Waite et al. [19]; ∘ Handa [34].

0.1 to 100 MPa [20]. One of the models, TIP4P-FQ, includes the effects of in-plane polarization of water. The simulations were performed with 80%, 90%, and 100% occupancy of the lattice cavities. Simulations were also performed on water and ice, which are shown in Table 66.1 along with experimental values from the literature [20, 29].

The thermal conductivity results obtained with the TIP4P-FQ model at a pressure of 10 MPa are shown in Figure 66.2 along with the previously discussed experimental data. The results are slightly higher than the experimental data as would be expected for a hydrate crystal as compared to the experimental samples that likely contained some structural defects and residual porosity. The removal of methane from the lattice cages appears to result in a slightly higher thermal conductivity, although the trend is almost within the simulation variability.

Recently, more extensive simulations of the thermal properties of methane hydrate have been performed by NETL [29] using the COS/G2 model that allows for both in-plane and out-of-plane polarization of the water molecules [35]. In this work, a range of temperatures was employed from

30 to 270 K. Simulations with ice at 150 K and water at 298 K were also performed and these results are contained in Table 66.1. The lower ice results are likely due to quantum effects [36].

The thermal conductivity of methane hydrate obtained using the COS/G2 model in the temperature range of geologic interest is shown in Figure 66.2. It is close to the experimental values of Huang and Fan [15] and lower than the experimental values of Waite et al. [19] and Rosenbaum et al. [20]. A complete account of this more recent MDS work has been published in Ref. [29].

66.4 RESERVOIR SIMULATION

An accurate knowledge of the thermal properties of methane hydrate is important for predictions involving resource production, greenhouse gas evolution impacting climate change, and seafloor stability. Concise descriptions of the importance of these thermal properties and scenarios have been published [19, 37].

With respect to the technology of gas production from hydrate-containing reservoirs, the inclusion of the hydrate phase has only recently been added to advanced simulation codes [38,39]. A recent assessment also highlights the importance of having accurate thermal properties for methane hydrates, both pure and in porous media [39].

To assess the impact of thermal conductivity on gas production from a hydrate-bearing reservoir, simulations were performed at NETL using the Tough+Hydrate (T+H) simulator [40] with two different values for thermal conductivity, 0.50 and 0.68 W m^{-1} K^{-1}. To accomplish this, we had to use a different expression for composite thermal conductivity, k_θ, than is currently used in this simulator [41], which is shown in the following equation:

$$k_\theta = k_{\mathrm{Rd}} = \left(\sqrt{S_{\mathrm{H}}} + \sqrt{S_{\mathrm{W}}} \right) (k_{\mathrm{Rw}} - k_{\mathrm{Rd}}) + \phi S_I k_I \quad (66.1)$$

In this equation, ϕ represents the porosity of the sediment (rock) phase; S_{H}, S_{W}, and S_I represent the hydrate, water, and ice saturations in the pores, respectively; and k_{Rw}, k_{Rd}, and k_I represent the thermal conductivities of wet rock, dry rock, and ice in the reservoir, respectively. However, equation (66.1) does not specifically account for the thermal conductivity of the hydrate phase. An arithmetic model [41, 42] was used instead and is shown in the equation below:

$$k_\theta = k_{\mathrm{Rd}} + \phi(S_{\mathrm{W}} k_{\mathrm{W}} + S_{\mathrm{H}} k_{\mathrm{H}} + S_I k_I) \quad (66.2)$$

In this equation, k_{W} and k_{H} represent the thermal conductivities of the water and hydrate phases, respectively. Equation (66.1) is currently preferred over equation (66.2) for hydrate reservoirs [41]; however, Moridis et al. acknowledge

TABLE 66.1 **Comparison of the thermal conductivity values obtained by simulation for water and ice with values obtained from the literature (in W m^{-1} K^{-1}).**

	Water	Ice
TIP4P-FQ	0.67 ± 0.03[a]	2.42 ± 0.04[b]
Experiment	0.61[a]	2.2[b]
COS/G2	0.62 ± 0.01[c]	2.2 ± 0.02[d]
Experiment	0.61[c]	4.3[d]

[a]298 K, 100 kPa [20].
[b]273 K, 0 MPa [20].
[c]298 K [29].
[d]150 K [29].

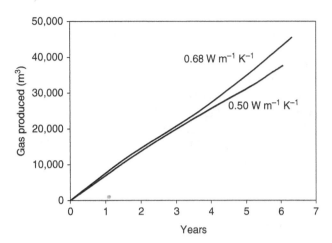

FIGURE 66.5 Simulation of production from a Class 3 hydrate-containing reservoir using two different values for thermal conductivity.

that there is significant room for new relationships for k_θ to be developed [39].

Figure 66.5 depicts the simulation results for gas production from a Class 3 [39] hydrate-bearing formation induced by the depressurization method [39] coupled with thermal stimulation provided by constant temperature boundaries. Equation (66.2) was used to describe the composite thermal conductivity of the hydrate-containing formation. This type of formation does not contain any underlying free gas or water layer; it only consists of the hydrate-containing sediment with overburden and underburden that are impermeable to fluid flow. Thermal conductivities of 0.50 and 0.68 W m^{-1} K^{-1} were used in the simulations.

The simulation results show that initially the variation of thermal conductivities has a negligible effect on the production of methane as the hydrate area affected by the thermal stimulation is relatively small. As years progress, the thermal stimulation affects substantial hydrate area and the effect of thermal conductivity on methane production becomes apparent. As shown in Figure 66.5, after 5 years the production of methane increases by 14% when the thermal conductivity is increased from 0.50 to 0.68 W m^{-1} K^{-1}. It is anticipated that similar observations would be made if higher thermal conductivity values were used in climate change or seafloor stability simulations, that is, the rate of hydrate dissociation would be proportional to the thermal conductivity value used in the simulation.

66.5 CONCLUSIONS

Experimental and theoretical research over the past several years has helped refine the thermal properties of methane hydrate. This is especially true for the thermal conductivity

of methane hydrate. Based on experiment, molecular simulation, and inverse modeling, it is recommended that values in the range of 0.65–0.70 W m^{-1} K^{-1} should be used in simulations of systems that contain low-porosity methane hydrate.

With respect to the other thermal properties, there have been only a few measurements or simulations of thermal diffusivity and only one set of measurements that permitted calculation of the specific heat of methane hydrate. At this time, the values obtained by Waite et al. should be used for simulation purposes [19]. Additional experimental and theoretical simulations are warranted in this area to provide additional validation of these results. Such research is being conducted at NETL.

The TPS technique under development at NETL is also in the process of being adapted to devices for field use. The one-sided approach makes it particularly attractive for this application in natural hydrate-bearing systems.

ACKNOWLEDGMENTS

E. M. M. and N. J. E. performed this work under contract DE-AM26-04NT41817, subtask 41817.660.01.03, and H. J. and K. D. J. performed this work under the same contract, subtask 41817.606.06.03, in support of NETL's Office of Research and Development (ORD). D. W. S. performed this work with support of NETL's ORD through the ORISE Part-Time Faculty Program.

NOMENCLATURE

C_p specific heat capacity
k thermal conductivity
S pore saturation
α thermal diffusivity
ϕ porosity of the rock phase

SUBSCRIPTS

H   hydrate phase
I   ice phase
Rd   dry rock phase
Rw   wet rock phase
W   water phase
Θ composite property

REFERENCES

[1] NETL. (2008). *The National Methane Hydrates R&D Program.* [cited 2008 March 18, 2008]. Available at: http://www.netl.doe.gov/technologies/oil-gas/FutureSupply/Methane Hydrates/rd-program/rd-program.htm.

[2] Stoll, R. D., Bryan, G. M. (1979). Physical properties of sediments containing gas hydrates. *Journal of Geophysical Research*, 84(B4):1629–1634.

[3] Ross, R. G., Andersson, P., Backstrom, G. (1981). Unusual PT dependence of thermal conductivity for clathrate hydrate. *Nature*, 290:322–323.

[4] Krivchikov, A. I., Manzhelii, V. G., Korolyuk, O. A., Gorodilov, B. Y., Romantsova, O. O. (2005). Thermal conductivity of tetrahydrofuran hydrate. *Physical Chemistry Chemical Physics*, 7:728–730.

[5] Ahmad, N., Phillips, W. A. (1987). Thermal conductivity of ice and ice clathrate. *Solid State Communications*, 63(2):167–171.

[6] Andersson, P., Ross, R. G. (1983). Effect of guest molecule size on the thermal conductivity and heat capacity of clathrate hydrates. *Journal of Physical Chemistry: Solid State Physics*, 16:1423–1432.

[7] Ashworth, T., Johnson, L. R., Lai, L.-P. (1985). Thermal conductivity of pure ice and tetrahydrofuran clathrate hydrates. *High Temperatures - High Pressures*, 17(4):413–419.

[8] Cook, J. G., Leaist, D. G. (1983). An exploratory study of the thermal conductivity of methane hydrate. *Geophysical Research Letters*, 10(5):397–399.

[9] Handa, Y. P., Cook, J. G. (1987). Thermal conductivity of xenon hydrate. *The Journal of Physical Chemistry*, 91(25):6327–6328.

[10] Tse, J. S., White, M. A. (1988). Origin of glassy crystalline behavior in the thermal properties of clathrate hydrates: a thermal conductivity study of tetrahydrofuran hydrate. *Journal of Physical Chemistry*, 92(17):5006–5011.

[11] Andersson, O., Suga, H. (1996). Thermal conductivity of normal and deuterated tetrahydrofuran clathrate hydrates. *Journal of Physics and Chemistry of Solids*, 57(1):125–132.

[12] Krivchikov, A. I., Gorodilov, B. Y., Korolyuk, O. A., Manzhelii, V. G., Romantsova, O. O., Conrad, H., Press, W., Tse, J. S., Klug, D. D. (2006). Thermal conductivity of Xe clathrate hydrate at low temperatures. *Physical Review B (Condensed Matter and Materials Physics)*, 73(6):064203–6.

[13] Krivchikov, A. I., Gorodilov, B. Y., Korolyuk, O. A., Manzhelii, V. G., Conrad, H., Press, W. (2005). Thermal conductivity of methane-hydrate. *Journal of Low Temperature Physics*, 139(5):693–702.

[14] Fan, S., Huang, D., Liang, D. (2005). Thermal conductivity of combination gas hydrate and hydrate-sand mixtures. *Fifth International Conference on Gas Hydrates*, Trondheim, Norway.

[15] Huang, D., Fan, S. (2004). Thermal conductivity of methane hydrate formed from sodium dodecyl sulfate solution. *Journal of Chemical Engineering Data*, 49(5):1479–1482.

[16] Waite, W. F., deMartin, B. J., Kirby, S. H., Pinkston, J., Ruppel, C. D. (2002). Thermal conductivity measurements in porous mixtures of methane hydrate and quartz sand. *Geophysical Research Letters*, 29(24):82–1 to 82–4.

[17] Waite, W. F., Gilbert, L. Y., Winters, W. J., Mason, D. H. (2006). Estimating thermal diffusivity and specific heat from needle probe thermal conductivity data. *Review of Scientific Instruments* 77(4):044904.

[18] Waite, W. F., Pinkston, J., Kirby, S. H. (2002). Preliminary laboratory thermal conductivity measurements in pure methane hydrate and methane hydrate-sediment mixtures: a progress report. *Fourth International Conference on Gas Hydrates*, Yokohama, Japan.

[19] Waite, W. F., Stern, L. A., Kirby, S. H., Winters, W. J., Mason, D. H. (2007). Simultaneous determination of thermal conductivity, thermal diffusivity and specific heat in sI methane hydrate. *Geophysical Journal International*, 169:767–774.

[20] Rosenbaum, E. J., English, N. J., Johnson, J. K., Shaw, D. W., Warzinski, R. P. (2007). Thermal conductivity of methane hydrate from experiment and molecular simulation. *Journal of Physical Chemistry B*, 111:13194–13205.

[21] von Herzen, R. P., Maxwell, A. E. (1959). The measurement of thermal conductivity of deep-sea sediments by a needle probe method. *Journal of Geophysical Research*, 84:1629–1634.

[22] Stern, L. A., Kirby, S. H., Durham, W. B. (1996). Peculiarities of methane clathrate hydrate formation and solid-state deformation, including possible superheating of water ice. *Science*, 273(5283):1843–1848.

[23] Stern, L. A., Kirby, S. H., Durham, W. B. (1998). Polycrystalline methane hydrate: synthesis from superheated ice, and low-temperature mechanical properties. *Energy & Fuels*, 12(2):201–211.

[24] Gustafsson, S. E. (1991). Transient plane source techniques for thermal conductivity and thermal diffusivity measurements of solid materials. *Review of Scientific Instruments*, 62(3):797–804.

[25] Gustafsson, S. E. (1991). Device for measuring thermal properties of a test substance–the transient plane source (TPS) method. U.S. Patent 5,044,767 Thermetrol AB (SE).

[26] MacDonald, I. R., Bender, L. C., Vardaro, M., Bernard, B., Brooks, J. M. (2005). Thermal and visual time-series at a seafloor gas hydrate deposit on the Gulf of Mexico slope. *Earth and Planetary Science Letters*, 233(1-2):45–59.

[27] Warzinski, R. P., Lynn, R. J., Shaw, D. W., Rosenbaum, E. (2007). Thermal property measurements of methane hydrate using a transient plane source technique. *AAPG Bulletin*; in press.

[28] NIST. (2007). *NIST Chemistry Webbook*. Available at: http://webbook.nist.gov/chemistry/fluid/ (Accessed May 22, 2007).

[29] Jiang, H., Myshakin, E. M., Jordan, K. D., Warzinski, R. P. (2008). Molecular dynamics simulations of the thermal conductivity of methane hydrate. *Journal of Physical Chemistry B*, 112:10,207–10,216.

[30] Gupta, A., Kneafsey, T. J., Moridis, G. J., Seol, Y., Kowalsky, M. B., Sloan, E. D. (2006). Composite thermal conductivity in a large heterogeneous porous methane hydrate sample. *Journal of Physical Chemistry B*, 110(33):16384–16392.

[31] deMartin, B. J. (2001). Laboratory measurements of the thermal conductivity and thermal diffusivity of methane hydrate

at simulated in-situ conditions. M.S. Thesis, Georgia Institute of Technology, Atlanta, Georgia. p. 135.

[32] Turner, D. J., Kumar, P., Sloan, E. D. (2005). A new technique for hydrate thermal diffusivity measurements. *International Journal of Thermophysics*, 26(6):1681–1691.

[33] Kumar, P., Turner, D., Sloan, E. D. (2004). Thermal diffusivity measurements of porous methane hydrate and hydrate-sediment mixtures. *Journal of Geophysical Research*, 109(B01207):8.

[34] Handa, Y. P. (1986). Compositions, enthalpies of dissociation, and heat capacities in the range 85 to 270 K for clathrate hydrates of methane, ethane, and propane, and enthalpy of dissociation of isobutane hydrate, as determined by a heat-flow calorimeter. *The Journal of Chemical Thermodynamics*, 18(10):915–921.

[35] Yu, H., van Gunsteren, W. F. (2004). Charge-on-spring polarizable water models revisited: from water clusters to liquid water to ice. *The Journal of Chemical Physics*, 121(19):9549–9564.

[36] de la Pena, L. H., Razul, M. S. G., Kusalik, P. G. (2005). Quantum effects in ice Ih. *The Journal of Chemical Physics*, 123(14):144506–144509.

[37] Ruppel, C. (2003). Thermal state of the gas hydrate reservoir. In: *Natural Gas Hydrate in Oceanic and Permafrost Environments*, Max, M. D. (editor). Kluwer Academic, Dordrecht, pp. 29–42.

[38] Boswell, R. (2007). Resource potential of methane hydrate coming into focus. *Journal of Petroleum Science and Engineering*, 56(1-3):9–13.

[39] Moridis, G. J., Collett, T. S., Boswell, R., Kurihara, M., Reagan, M. T., Koh, C., Sloan, E. D. (2008). Toward production from gas hydrates: current status, assessment of resources, and simulation-based evaluation of technology and potential. *SPE 114163*.

[40] Moridis, G. J., Kowalsky, M. B., Pruess, K. (2005). TOUGH-Fx/HYDRATE v1.0 User's Manual: a code for the simulation of system behavior in hydrate-bearing geologic media. Report No. LBNL-58950, Lawrence Berkeley National Laboratory.

[41] Moridis, G. J., Seol, Y., Kneafsey, T. J. (2005). Studies of reaction kinetics of methane hydrate dissociation in porous media. *Fifth International Conference on Gas Hydrates*, Trondheim, Norway.

[42] Bejan, A. (1984). *Convection Heat Transfer*. John Wiley and Sons, Inc., New York.

PART VII

SHALE GAS

67

SHALE GAS WILL ROCK THE WORLD

AMY MYERS JAFFE

Huge discoveries of natural gas promise to shake up the energy markets and geopolitics. And that is just for starters.

There is an energy revolution brewing right under our feet. Over the past decade, a wave of drilling around the world has uncovered giant supplies of natural gas in shale rock. By some estimates, there is 1000 trillion cubic feet recoverable in North America alone—enough to supply the nation's natural-gas needs for the next 45 years. Europe may have nearly 200 trillion cubic feet of its own. We have always known the potential of shale; we just did not have the technology to get to it at a low-enough cost. Now new techniques have driven down the price tag—and set the stage for shale gas to become what will be the game-changing resource of the decade. I have been studying the energy markets for 30 years, and I am convinced that shale gas will revolutionize the industry—and change the world—in the coming decades. It will prevent the rise of any new cartels. It will alter geopolitics. And it will slow the transition to renewable energy. To understand why, you have to consider that even before the shale discoveries natural gas was destined to play a big role in our future. As environmental concerns have grown, nations have leaned more heavily on the fuel, which gives off just half the carbon dioxide of coal. But the rise of gas power seemed likely to doom the world's consumers to a repeat of OPEC, with gas producers like Russia, Iran, and Venezuela coming together in a cartel and dictating terms to the rest of the world. The advent of abundant, low-cost gas will throw all that out the window—so long as the recent drilling catastrophe does not curtail offshore oil and gas activity and push up the price of oil and eventually other forms of energy. Not only will the shale discoveries prevent a cartel from forming, but the petro-states will lose lots of the muscle they now have in world affairs,

as customers over time cut them loose and turn to cheap fuel produced closer to home. The shale boom is also likely to upend the economics of renewable energy. It may be a lot harder to persuade people to adopt green power that needs heavy subsidies when there is a cheap, plentiful fuel out there that is a lot cleaner than coal, even if gas is not as politically popular as wind or solar. But that is not the end of the story: I also believe this offers a tremendous new longer term opportunity for alternative fuels. Since there is no longer an urgent need to make them competitive *immediately* through subsidies, since we can use natural gas now, we can pour that money into R&D—so renewables will be ready to compete without lots of help when shale supplies run low, decades from now.

To be sure, plenty of people (including Russian Prime Minister Vladimir Putin and many Wall Street energy analysts) are not convinced that shale gas has the potential to be such a game changer. Their arguments revolve around two main points: that shale-gas exploration is too expensive and that it carries environmental risks. I would argue they are wrong on both counts.

Take costs first. Over the past decade, new techniques have been developed that drastically cut the price tag of production. The Haynesville shale, which extends from Texas into Louisiana, is seeing costs as low as $3 per million British thermal units, down from $5 or more in the Barnett shale in the 1990s. And more cost-cutting developments are likely on the way as major oil companies get into the game. If they need to do shale for $2, I am willing to bet they can, in the next 5 years.

When it comes to environmental risks, critics do have a point: They say drilling for shale gas runs a risk to

Alternative Energy and Shale Gas Encyclopedia, First Edition. Edited by Jay H. Lehr and Jack Keeley.
© 2016 John Wiley & Sons, Inc. Published 2016 by John Wiley & Sons, Inc.

groundwater, even though shale is generally found thousands of feet below the water table. If a well casing fails, they argue, drilling fluids can seep into aquifers.

They are overplaying the danger of such a failure. For drilling on land, where most shale-gas deposits are, the casings have been around for decades with a good track record. But water pollution *can* occur if drilling fluids are disposed of improperly. So, regulations and enforcement must be tightened to ensure safety. More rules will raise costs—but, given the abundance of supply, producers can likely absorb the hit.

Already, some are moving to nontoxic drilling fluids, even without imposed bans. But the skeptics are not just overstating the obstacles. They are missing two much bigger points. For one thing, they are ignoring history: The reserves and production of new energy resources tend to *increase* over time, not decrease. They are also not taking into account how quickly public opinion can change. The country *can* turn on a dime and embrace a cheaper energy source, casting aside political or environmental reservations.

This has happened before, with the rapid spread of liquefied-natural-gas terminals over the past few years. In short, the skeptics are missing the bigger picture—the picture I think is the much more likely one. Here is a closer look at what I am talking about and how I believe the boom in shale gas will shake up the world.

One of the biggest effects of the shale boom will be to give Western and Chinese consumers fuel supplies close to home—thus scuttling a potential natural-gas cartel. Remember: Prior to the discovery of shale gas, huge declines were expected in domestic production in the United States, Canada, and the North Sea. That meant an increasing reliance on foreign supplies—at a time when natural gas was becoming more important as a source of energy.

Even more troubling, most of those gas supplies were located in unstable regions. Two countries in particular had a stranglehold over supply: Russia and Iran. Before the shale discoveries, these nations were expected to account for more than half the world's known gas resources.

Russia made no secret about its desire to leverage its position and create a cartel of gas producers—a kind of latter-day OPEC. That seemed to set the stage for a repeat of the oil issues that have worried the world over the past 40 years.

As far as I am concerned, you can now forget all that. Shale gas will breed competition among energy companies and exporting countries—which in turn will help economic stability in industrial countries and thwart petro-suppliers that try to empower themselves at our expense. Market competition is the best kryptonite for cartel power.

For one measure of the coming change, consider the prospects for liquefied natural gas, which has been converted to a liquid so it can be carried in a supertanker like oil. It is the easiest way to move natural gas to very long distances, so it gives a good picture of how much countries are relying on foreign supplies.

Before the shale discoveries, experts expected liquefied natural gas, or LNG, to account for half of the international gas trade by 2025, up from 5% in the 1990s. With the shale boom, that share will be more like one-third.

In the United States, the impact of shale gas and deep-water drilling is already apparent. Import terminals for LNG sit virtually empty, and the prospects that the United States will become even more dependent on foreign imports are receding. Also, soaring shale-gas production in the United States has meant that cargoes of LNG from Qatar and elsewhere are going to European buyers, easing their dependence on Russia. So, Russia has had to accept far lower prices from formerly captive customers, slashing prices to Ukraine by 30%, for instance.

But the political fallout from shale gas will do a lot more than stifle natural-gas cartels. It will throw world politics for a loop—putting some longtime troublemakers in their place and possibly bringing some rivals into the Western fold.

Again, remember that as their energy-producing influence grew, nations like Russia, Venezuela, and Iran became more successful in resisting Western interference in their affairs—and exporting their ideologies and strategic agendas through energy-linked deal-making and threats of cutoffs.

In 2006 and 2007, disputes with Ukraine led Russia to cut off supplies, leaving customers in Kiev and Western Europe briefly without fuel in the dead of winter. That cutoff effectively shifted Ukraine's internal politics: The country turned away from the pro-NATO, anti-Moscow candidate and toward a coalition more to Moscow's liking. It looked like the United States and Europe would see their global power eclipse as they kowtowed to their energy suppliers. But shale gas is going to defang the energy diplomacy of petro-nations. Consuming nations throughout Europe and Asia will be able to turn to major US oil companies and their own shale rock for cheap natural gas and tell the Chavezes and Putins of the world where to stick their supplies—back in the ground.

Europe, for instance, receives 25% of its natural-gas supply via pipelines from Russia, with some consumers almost completely dependent on the big supplier.

In the wake of Russia's strong-arming of Ukraine, Europe has been actively diversifying its supply, and shale gas will make that task cheaper and easier. Shale-gas resources are believed to extend into countries such as Poland, Romania, Sweden, Austria, Germany, and Ukraine. Once European shale gas comes, the Kremlin will be hard-pressed to use its energy exports as a political lever. I would also argue that greater shale-gas production in Europe will make it harder for Iran to profit from exporting natural gas. Iran is currently hampered by Western sanctions against investment in its energy sector; so, by the time it can get its natural gas ready for export, the marketing window to Europe will likely be closed by the availability of inexpensive shale gas.

And that may lead Tehran to tone down its nuclear efforts. Look at it this way: If Iran cannot sell its gas in Europe, what options does it have? Piping to the Indian subcontinent is impractical, and LNG markets will be crowded with lower cost, competing supplies. It is admittedly a long shot, but if the regime acts rationally, it will realize it has a chance to win some global goodwill by shifting away from nuclear-power efforts—and using its cheap natural-gas supplies to generate electricity at home.

Overall, the Middle East might get a bit poorer as gas eats into the market for oil. If the drop in revenue is severe enough, it could bring instability. Shale-gas development could also mean big changes for China. The need for energy imports has taken China to problematic nations such as Iran, Sudan, and Burma, making it harder for the West to forge global policies to address the problems those countries create. But with newly accessible natural gas available at home, China could well turn away from imports—and the hot spots that produce them.

The less vulnerable China is to imported oil and gas, the more likely it would be to support sanctions or other measures against petro-states with human-rights problems or aggressive agendas. Moreover, the less Beijing worries about US control of sea lanes, the easier it will be for the United States and China to build trust. So, domestic shale gas for China may help integrate Beijing into a Pax Americana global system.

With natural gas cheap and abundant, the prospects for renewable energy will change just as drastically. I have been a big believer that renewable energy was about to see its time. Prior to the shale-gas revolution, I thought rising hydrocarbon prices would propel renewables and nuclear power into the marketplace easily—albeit with a little shove from a carbon tax or a cap-and-trade system.

But the shale discoveries complicate the issue, making it harder for wind, solar, and biomass energy, as well as nuclear, to compete on economic grounds. Subsidies that made renewables competitive with shale gas would get more expensive, as would loan guarantees and incentives for new nuclear plants. Shale gas also hurts the energy independence argument for renewables: Shale gas is domestic, just like wind and solar, so we would not be shipping those dollars to the Middle East.

But that does not mean we should stop investing in renewables. As large as our shale-gas resources are, they are still exhaustible, and eventually we will still need to transition to energy that is cleaner and more plentiful. So, what should we do?

First, avoid the urge to protect coal states and let cheaper natural gas displace coal, which accounts for about half of all power generated in the United States. Ample natural gas for electricity generation could also make it easier to shift to electric vehicles—once again helping the environment and lessening our dependence on the Middle East.

Then, I think we still need to invest in renewables—but smartly. States with renewable energy potential, such as windy Texas or sunny California, should keep their mandates that a fixed percentage of electricity must be generated by alternative sources. That will give companies incentives and opportunities to bring renewables to market and lower costs over time through experience and innovation. Yes, renewables may seem relatively more expensive in those states as shale gas hits the market. And, yes, that may mean getting more help from government subsidies. But I do not think the cost would be prohibitive, and the long-term benefits are worth it. Still, I do not believe we should set national mandates—which *would get prohibitively expensive in states without abundant renewable resources.* Instead of pouring money into subsidies to make such a plan work, the federal government should invest in R&D to make renewables competitive down the road *without* big subsidies.

In the end, what is important to understand is that shale gas may be the key to solving some of our most pressing short-term crises, a way to bridge the gap to a more-secure energy and economic future. The trade deficit has crippled our economy and shows no signs of abating as long as we remain tethered to imported energy. Why ship dollars abroad where they can destabilize global financial markets—and then hit us back in lost jobs and savings—when we can develop the resources we have here in our own country? Shall we pay Vladimir Putin and Mahmoud Ahmadinejad to develop our natural gas—or the citizens of Pennsylvania and Louisiana?

67.1   CORRECTION AND AMPLIFICATION

Natural gas from a working well flows directly through a pipeline system to a processing facility, and then on to consumers through other pipeline systems.

68

WHAT IS SHALE GAS?

ENERGY INFORMATION ADMINISTRATION (PUBLIC DOMAIN)

68.1 WHAT IS SHALE GAS AND WHY IS IT IMPORTANT?

Shale gas refers to natural gas that is trapped within shale formations. Shales are fine-grained sedimentary rocks that can be rich sources of petroleum and natural gas.

68.2 HORIZONTAL DRILLING AND HYDRAULIC FRACTURING

Over the past decade, the combination of horizontal drilling and hydraulic fracturing has allowed access to large volumes of shale gas that were previously uneconomical to produce. The production of natural gas from shale formations has rejuvenated the natural gas industry in the United States.

68.3 THE UNITED STATES HAS ABUNDANT SHALE GAS RESOURCES

Of the natural gas consumed in the United States in 2009, 87% was produced domestically; thus, the supply of natural gas is not as dependent on foreign producers as is the supply of crude oil, and the delivery system is less subject to interruption. The availability of large quantities of shale gas will further allow the United States to consume a predominantly domestic supply of gas.

According to the EIA Annual Energy Outlook 2011, the United States possesses 2552 trillion cubic feet (Tcf) of potential natural gas resources. Natural gas from shale resources, considered uneconomical just a few years ago, accounts for 827 Tcf of this resource estimate, more than double the estimate published last year.

68.4 ENOUGH FOR 110 YEARS OF USE

At the 2009 rate of US consumption (about 22.8 Tcf per year), 2552 Tcf of natural gas is enough to supply approximately 110 years of use. Shale gas resource and production estimates increased significantly between the 2010 and 2011 Outlook reports and are likely to increase further in the future.

Did You Know? Sedimentary rocks are rocks formed by the accumulation of sediments at the Earth's surface and within bodies of water. Common sedimentary rocks include sandstone, limestone, and shale.

68.5 WHAT IS A SHALE "PLAY"?

Shale gas is found in shale "plays," which are shale formations containing significant accumulations of natural gas and which share similar geologic and geographic properties. A decade of production has come from the Barnett Shale play in Texas.

Experience and information gained from developing the Barnett Shale have improved the efficiency of shale gas development around the country. Other important plays are the Marcellus Shale and Utica Shale in the eastern United States and the Haynesville Shale and Fayetteville Shale in Louisiana and Arkansas. Surveyors and geologists identify suitable well locations in areas with potential for economical gas production by using both surface-level observation techniques and computer-generated maps of the subsurface (Figure 68.1).

68.6 HORIZONTAL DRILLING

Two major drilling techniques are used to produce shale gas. Horizontal drilling is used to provide greater access to the

Alternative Energy and Shale Gas Encyclopedia, First Edition. Edited by Jay H. Lehr and Jack Keeley.
© 2016 John Wiley & Sons, Inc. Published 2016 by John Wiley & Sons, Inc.

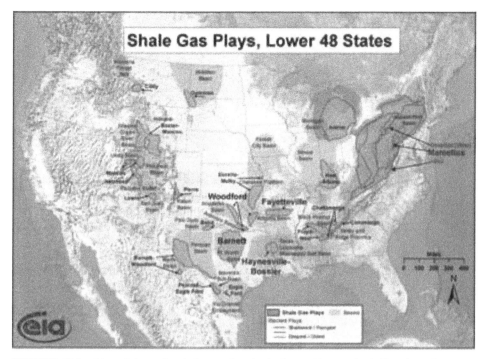

FIGURE 68.1   Map of the major shale gas plays in the lower 48 states including the sedimentary basins which contain them.

gas trapped deep in the producing formation. First, a vertical well is drilled to the targeted rock formation. At the desired depth, the drill bit is turned to bore a well that stretches through the reservoir horizontally, exposing the well to more of the producing shale.

68.7   HYDRAULIC FRACTURING

Hydraulic fracturing (commonly called "fracking" or "hydrofracking") is a technique in which water, chemicals, and sand are pumped into the well to unlock the hydrocarbons

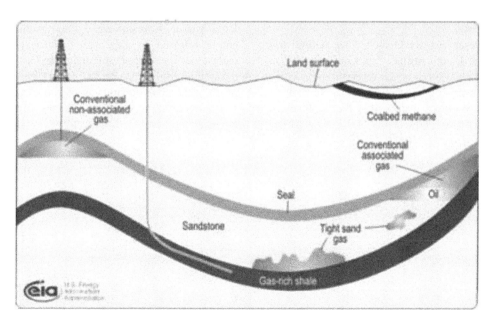

FIGURE 68.2   Diagram showing the geometry of conventional and unconventional natural gas resources. *Source*: Image by EIA.

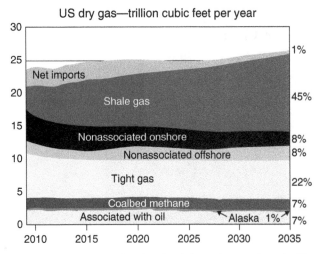

US dry gas—trillion cubic feet per year

Net imports — 1%
Shale gas — 45%
Nonassociated onshore — 8%
Nonassociated offshore — 8%
Tight gas — 22%
Coalbed methane — 7%
Associated with oil   Alaska 1% — 7%

FIGURE 68.3   Chart showing the shale gas forecast from the EIA Annual Energy Outlook 2011. *Source*: Image by EIA.

trapped in shale formations by opening cracks (fractures) in the rock and allowing natural gas to flow from the shale into the well. When used in conjunction with horizontal drilling, hydraulic fracturing enables gas producers to extract shale gas at reasonable cost. Without these techniques, natural gas does not flow to the well rapidly, and commercial quantities cannot be produced from shale (Figure 68.2).

Did You Know? Shale gas in 2009 made up 14% of total US natural gas supply. Production of shale gas is expected to continue to increase and constitute 45% of US total natural gas supply in 2035, as projected in the EIA Annual Energy Outlook 2011 (Figure 68.3).

68.8   SHALE GAS VERSUS CONVENTIONAL GAS

Conventional gas reservoirs are created when natural gas migrates toward the Earth's surface from an organic-rich source formation into highly permeable reservoir rock, where it is trapped by an overlying layer of impermeable rock. In contrast, shale gas resources form within the organic-rich shale source rock. The low permeability of the shale greatly

inhibits the gas from migrating to more permeable reservoir rocks. Without horizontal drilling and hydraulic fracturing, shale gas production would not be economically feasible because the natural gas would not flow from the formation at high enough rates to justify the cost of drilling.

68.9   NATURAL GAS: A CLEAN-BURNING FUEL

Natural gas is cleaner burning than coal or oil. The combustion of natural gas emits significantly lower levels of key pollutants, including carbon dioxide (CO_2), nitrogen oxides, and sulfur dioxide, than does the combustion of coal or oil. When used in efficient combined-cycle power plants, natural gas combustion can emit less than half as much CO_2 as coal combustion, per unit of energy released.

68.10   ENVIRONMENTAL CONCERNS

There are some potential environmental issues that are associated with the production of shale gas. Shale gas drilling has significant water supply issues. The drilling and fracturing of wells require large amounts of water. In some areas of the country, significant use of water for shale gas production may affect the availability of water for other uses and can affect aquatic habitats.

Drilling and fracturing also produce large amounts of wastewater, which may contain dissolved chemicals and other contaminants that require treatment before disposal or reuse. Because of the quantities of water used, and the complexities inherent in treating some of the chemicals used, wastewater treatment and disposal is an important and challenging issue. If mismanaged, the hydraulic fracturing fluid can be released by spills, leaks, or various other exposure pathways. The use of potentially hazardous chemicals in the fracturing fluid means that any release of this fluid can result in the contamination of surrounding areas, including sources of drinking water, and can negatively impact natural habitats.

Source: Republished from a December 2010 "Energy in Brief" by the Energy Information Administration; http://www.eia.gov/energy_in_brief/

69

DIRECTIONAL AND HORIZONTAL DRILLING IN OIL AND GAS WELLS

PUBLIC DOMAIN

Directional and horizontal drilling are the methods used to increase production and hit targets that cannot be reached with a vertical well.

69.1 WHAT IS DIRECTIONAL DRILLING?

Most wells drilled for water, oil, natural gas, information, or other subsurface objectives are vertical wells—drilled straight down into the earth. However, drilling at an angle other than vertical can obtain information, hit targets, and stimulate reservoirs in ways that cannot be achieved with a vertical well. In these cases, an ability to accurately steer the well in directions and angles that depart from the vertical is a valuable ability.

When directional drilling is combined with hydraulic fracturing, some rock units which were unproductive when drilled vertically can become fantastic producers of oil or natural gas. Examples are the Marcellus Shale of the Appalachian Basin and the Bakken Formation of North Dakota.

69.2 WHY DRILL WELLS THAT ARE NON-VERTICAL?

Directional and horizontal drilling have been used to reach targets beneath adjacent lands, reduce the footprint of gas field development, increase the length of the "pay zone" in a well, deliberately intersect fractures, construct relief wells, and install utility service beneath lands where excavation is impossible or extremely expensive. Below is a list of six reasons for drilling non-vertical wells:

A. *Hit targets that cannot be reached by vertical drilling.*

Sometimes a reservoir is located under a city or a park where drilling is impossible or forbidden. This reservoir might still be tapped if the drilling pad is located on the edge of the city or park and the well is drilled at an angle that will intersect the reservoir.

B. *Drain a broad area from a single drilling pad.*

This method has been used to reduce the surface footprint of a drilling operation. In 2010, the University of Texas at Arlington was featured in the news for drilling 22 wells on a single drill pad that will drain natural gas from 1100 acres beneath the campus. Over a 25-year life-time the wells are expected to produce a total of 110 billion cubic feet of gas.

This method significantly reduced the footprint of natural gas development within the campus area.

C. *Increase the length of the "pay zone" within the target rock unit.*

If a rock unit is 50 ft thick, a vertical well drilled through it would have a pay zone that is 50 ft in length. However, if the well is turned and drilled horizontally through the rock unit for 5000 ft then that single well will have a pay zone that is 5000 ft long—this will usually result in a significant productivity increase for the well.

When combined with hydraulic fracturing, horizontal drilling can convert unproductive shales into fantastic reservoir rocks.

D. *Improve the productivity of wells in a fractured reservoir.*

Alternative Energy and Shale Gas Encyclopedia, First Edition. Edited by Jay H. Lehr and Jack Keeley.
© 2016 John Wiley & Sons, Inc. Published 2016 by John Wiley & Sons, Inc.

This is done by drilling in a direction that intersects a maximum number of fractures. The drilling direction will normally be at right angles to the dominant fracture direction. Geothermal fields in granite bedrock usually get nearly all of their water exchange from fractures. Drilling at right angles to the dominant fracture direction will drive the well through a maximum number of fractures.

E. *Seal or relieve pressure in an "out-of-control" well.*

If a well is out of control a "relief well" can be drilled to intersect it. The intersecting well can be used to seal the original well or to relieve pressure in the out-of-control well.

F. *Install underground utilities where excavation is not possible.*

Horizontal drilling has been used to install gas and electric lines that must cross a river, cross a road, or travel under a city.

69.3 ROCK UNITS THAT BENEFIT MOST FROM HORIZONTAL DRILLING

Vertical wells can effectively drain rock units that have a very high permeability. Fluids in those rock units can flow quickly and efficiently into a well over long distances. However, where permeability is very low fluids move very slowly through the rock and do not travel long distances to reach a well bore. Horizontal drilling can increase the productivity in low-permeability rocks by bringing the well bore much closer to the source of the fluid.

69.4 HORIZONTAL DRILLING AND HYDRAULIC FRACTURING IN SHALES

Perhaps the most important role that horizontal drilling has played is in the development of the natural gas shale plays. These low-permeability rock units contain significant amounts of gas and are present beneath very large parts of North America.

The Barnett Shale of Texas, the Fayetteville Shale of Arkansas, the Haynesville Shale of Louisiana and Texas, and the Marcellus Shale of the Appalachian Basin are examples. In these rock units the challenge is not "finding" the reservoir, the challenge is recovering the gas from very tiny pore spaces in a low-permeability rock unit.

To stimulate the productivity of wells in organic-rich shales, companies drill horizontally through the rock unit and then use hydraulic fracturing to produce artificial permeability that is propped open by frac sand. Done together, horizontal drilling and hydraulic fracturing can make a productive well where a vertical well would have produced only a small amount of gas.

69.5 DRILLING METHODOLOGY

Most horizontal wells begin at the surface as a vertical well. Drilling progresses until the drill bit is a few hundred feet above the target rock unit. At that point the pipe is pulled from the well and a hydraulic motor is attached between the drill bit and the drill pipe. The hydraulic motor is powered by a flow of drilling mud down the drill pipe. It can rotate the drill bit without rotating the entire length of drill pipe between the bit and the surface. This allows the bit to drill a path that deviates from the orientation of the drill pipe.

After the motor is installed the bit and pipe are lowered back down the well and the bit drills a path that steers the well bore from vertical to horizontal over a distance of a few hundred feet. Once the well has been steered to the proper angle, straight-ahead drilling resumes and the well follows the target rock unit. Keeping the well in a thin rock unit requires careful navigation. Downhole instruments are used to determine the azimuth and orientation of the drilling. This information is used to steer the drill bit.

Horizontal drilling is expensive. When combined with hydraulic fracturing a well can cost up to three times as much per foot as drilling a vertical well. The extra cost is usually recovered by increased production from the well. These methods can multiply the yield of natural gas or oil from a well. Many profitable wells would be failures without these methods.

69.6 A NEW LEASE AND ROYALTY PHILOSOPHY

In the production of gas from a vertical well the gas is produced beneath a single parcel of property. Most states have long-established mineral rights rules that govern the ownership of gas produced from vertical wells. The gas is often shared by all landowners in a block of land or a radius distance from the producing well.

Horizontal wells introduce a new variable: a single well can penetrate and produce gas from multiple parcels with different owners. How can the royalties from this gas be fairly shared? This question is normally answered prior to drilling through a combination of government rules and private royalty-sharing agreements. How royalties are divided and how "hold-out" landowners are treated can be more complex than with a vertical well.

70

HYDRAULIC FRACTURING OF OIL AND GAS WELLS DRILLED IN SHALE

PUBLIC DOMAIN

70.1 WHAT IS HYDRAULIC FRACTURING?

Hydraulic fracturing is a procedure that can increase the flow of oil or gas from a well. It is done by pumping liquids down a well into subsurface rock units under pressures that are high enough to fracture the rock. The goal is to create a network of interconnected fractures that will serve as pore spaces for the movement of oil and natural gas to the well bore.

Hydraulic fracturing combined with horizontal drilling has turned previously unproductive organic-rich shales into the largest natural gas fields in the world. The Marcellus Shale, Utica Shale, Barnett Shale, Eagle Ford Shale and Bakken Formation are examples of previously unproductive rock units that have been converted into fantastic gas or oil fields by hydraulic fracturing.

70.2 HOW LONG HAS HYDRAULIC FRACTURING BEEN USED?

Hydraulic fracturing was first used to stimulate oil and natural gas wells in the United States over 60 years ago. Haliburton Oil Well Cementing Company was issued a patent for the procedure in 1949. The method successfully increased well production rates and the practice quickly spread. It is now used throughout the world in thousands of wells every year. Our gasoline, heating fuel, natural gas, and other products made from petroleum products would cost a lot more if hydraulic fracturing had not been invented.

70.3 SUCCESSFUL USE OF HYDRAULIC FRACTURING IN SHALE

In the early 1990s, Mitchel Energy began using hydraulic fracturing to stimulate the production of natural gas from wells drilled into the Barnett Shale of Texas. The Barnett Shale contained enormous amounts of natural gas; however, the Barnett rarely produced natural gas in commercial quantities.

Mitchel Energy realized that gas in the Barnett Shale was trapped in tiny pore spaces that were not interconnected. The rock had pore space but lacked permeability. Wells drilled through the Barnett Shale would usually have a show of gas but not enough gas for commercial production. Mitchel Energy solved this problem by hydraulic fracturing the Barnett Shale to create a network of interconnected pore spaces that enabled a flow of natural gas to the well.

Unfortunately many of the fractures produced by the hydraulic fracturing process snapped closed when the pumps were turned off. The Barnett Shale was so deeply buried that confining pressure closed the new fractures. This problem was solved by adding sand to the fracturing fluid. When the rock fractured, the rush of water into the newly opened pore space would carry sand grains deep into the rock unit. When the water pressure was reduced the sand grains "propped" the fracture open and allowed a flow of natural gas through the fractures and into the well bore. Today there are a variety of natural and synthetic products that are sold under the name of "frac sand."

Alternative Energy and Shale Gas Encyclopedia, First Edition. Edited by Jay H. Lehr and Jack Keeley.
© 2016 John Wiley & Sons, Inc. Published 2016 by John Wiley & Sons, Inc.

Mitchel Energy further improved the yield of their wells by drilling them horizontally through the Barnett Shale. Vertical wells were started at the surface, steered to a horizontal orientation and driven through the Barnett Shale for thousands of feet. This multiplied the length of the pay zone in the well. If a rock unit was 100 ft thick, it would be having a pay zone of 100 ft in a vertical well. However, if the well was steered horizontal and stayed horizontal for 5000 ft through the target formation then the length of the pay zone was 50 times longer than the pay zone of a vertical well.

Mitchel Energy used hydraulic fracturing and horizontal drilling to multiply the productivity of Barnett Shale wells. In fact, many of their highly successful wells would have been failures if they were vertical wells without hydraulic fracturing.

70.4 HYDRAULIC FRACTURING IN OTHER SHALE PLAYS

As others learned of Mitchell Energy's success in the Barnett Shale of Texas the methods of horizontal drilling and hydraulic fracturing were tried in other organic-rich shales. These methods quickly succeeded in the Haynesville Shale and Fayetteville Shale of Louisiana, Texas, and Arkansas—then in the Marcellus Shale in the Appalachian Basin. The methods worked in many other shales and are now being used to develop organic-rich shales in many parts of the world (Figure 70.1).

Hydraulic fracturing has also enabled production of natural gas liquids and oil from many wells. Rock units such as the Bakken Shale of North Dakota and the Niobrara Shale of Colorado, Kansas, Nebraska, and Wyoming are now yielding significant amounts of oil from hydraulic fracturing.

70.5 FRACTURING FLUIDS

Water is the driving fluid used in the hydraulic fracturing process. Depending upon the characteristics of the well and the rock being fractured a few million gallons of water can be required to complete a hydraulic fracturing job.

When the water is pumped into the well the entire length of the well is not pressurized. Instead, plugs are inserted to isolate the portion of the well where the fractures are desired. Only this section of the well receives the full force of pumping. As pressure builds up in this portion of the well, water opens fractures and the driving pressure extends the fractures deep into the rock unit. When pumping stops, these fractures quickly snap closed and the water used to open them is pushed back into the borehole, back up the well, and is collected at the surface. The water returned to the surface is a mixture of the water injected and pore water that has been trapped in the rock unit for millions of years. The pore water is usually a brine with significant amounts of dissolved solids.

Chemicals are often added to the water used in hydraulic fracturing. These additives serve a variety of purposes. Some thicken the water into a gel that is more effective at opening fractures and carrying proppants deep into the rock unit. Other chemicals are added to reduce friction, keep rock debris suspended in the liquid, prevent corrosion of equipment, kill bacteria, control pH, and other functions. Most companies have been resistant to revealing the composition of their hydraulic fracturing fluids. They believe that this information should be kept private to protect their competitive research. However, regulators are starting to demand the information and some companies are starting to share the information voluntarily.

70.6 PROPPANTS

A variety of proppants are used in hydraulic fracturing. These are small crush-resistant particles that are carried into the fractures by the hydraulic fracturing fluid. When the pumps are turned off and the fractures collapse these crush-resistant particles hold the fracture open, creating pore space through which natural gas can travel to the well. Frac sand is the proppant most commonly used today but aluminum beads, ceramic beads, sintered bauxite, and other materials have also been used. Over 1 million pounds of proppants can be used while fracturing a single well.

FIGURE 70.1   A satellite view of a Utica Shale drilling site where nine horizontal wells have been constructed and stimulated with hydraulic fracturing.

70.7   ENVIRONMENTAL CONCERNS

There are a number of environmental concerns related to hydraulic fracturing. These include the following:

1. Fractures produced in the well might extend directly into shallow rock units that are used for drinking water supplies. Or, fractures produced in the well might communicate with natural fractures that extend into shallow rock units that are used for drinking water supplies.

2. The casing of a well might fail and allow fluids to escape into shallow rock units used for drinking water supplies.

3. Accidental spills of hydraulic fracturing fluids or fluids expelled during a fracturing job might seep into the ground or contaminate surface water.

70.8   PRODUCTION BENEFITS

Hydraulic fracturing can significantly increase the yield of a well. When it is combined with horizontal drilling, unprofitable rock formations are often converted into productive natural gas fields. The technique is largely responsible for the development of the Barnett Shale, Haynesville Shale, Fayetteville Shale, and Marcellus Shale gas fields. It can also liberate oil from tight rock units as has been done with the Bakken Shale and Niobrara Shale.

The hydraulic fracturing process and the chemicals used with them cause the greatest amount of concern to environmental advocates who watch the natural gas industry. A regulatory environment is needed which will allow these techniques to be employed and provide environmental safeguards to protect water supplies and people who live in the areas where drilling occurs.

71

HYDRAULIC FRACTURING: A GAME-CHANGER FOR ENERGY AND ECONOMIES

Isaac Orr

71.1 INTRODUCTION

Vast reserves of oil and natural gas have been known to exist in shale formations throughout the United States for decades, but extracting these resources was not economically viable until the advent of "smart drilling" technology—the combination of horizontal drilling, hydraulic fracturing techniques, and computer-assisted underground monitoring.[1] This technology, along with confidence that oil prices would remain high, gave producers the incentive to discover and develop shale and other unconventional sources of oil and gas around the nation.[2]

Hydraulic fracturing, more commonly known as "fracking," has transformed the way oil and natural gas are produced in the United States. Before hydraulic fracturing became widely practiced, US energy production forecasts were bleak and threatened to exert downward pressure on the economy as a whole.

For example, in his testimony before the House Energy and Commerce Committee in 2003, the then Federal Reserve Chairman Alan Greenspan warned high natural gas prices were particularly worrisome for industries dependent upon large amounts of natural gas, such as the chemical, fertilizer, and steel and aluminum processing industries. Greenspan

noted, "The perceived tightening of long-term demand–supply balances is beginning to price some industrial demand out of the market."[3] Greenspan suggested the United States increase natural gas imports from abroad in the form of liquefied natural gas (LNG) to satisfy domestic demand and reduce consumer costs.[4]

Greenspan's forecast appeared to be accurate as natural gas prices continued to climb in 2005, reaching $16 per million British thermal units (MMbtu) in some production area spot markets due to growing demand and a drop in production resulting from damage to offshore drilling infrastructure caused by Hurricane Katrina.[5]

Despite these supply problems, 2003, the year Greenspan delivered his pessimistic forecast on the future of natural gas, marked a turning point in the national narrative on oil and gas production when scientists in the Barnett Shale of northern Texas, using the smart drilling technology developed by George Mitchell in the 1990s, demonstrated that a combination of hydraulic fracturing and horizontal drilling could transform previously uneconomic plays into viable drilling

---

[3]Diane Hess, "Greenspan Discusses Natural Gas Woes," *The Street*, July 10, 2003, http://www.thestreet.com/story/10099308/1/greenspan-discusses-natural-gas-woes.html

[4]Dan Thanh Dang, "Greenspan Warns of Natural Gas Supply Gap," *The Baltimore Sun*, June 11, 2003, http://articles.baltimoresun.com/2003-06-11/news/0306110217_1_natural-gas-markets-gas-for-heating-supplies-of-natural

[5]Joseph T. Kelliher, "High Natural Gas Prices: The Basics," Federal Energy Regulatory Commission, December 2005, http://www.ferc.gov/legal/staff-reports/high-gas-prices-1.pdf

---

[1]North Dakota Industrial Commission, "Bakken Formation Reserve Estimates, Executive Summary," n.d., www.nd.gov/ndic/ic-press/bakken-form-06.pdf

[2]*Ibid.*

Alternative Energy and Shale Gas Encyclopedia, First Edition. Edited by Jay H. Lehr and Jack Keeley.
© 2016 John Wiley & Sons, Inc. Published 2016 by John Wiley & Sons, Inc.

options, and in doing so, changed US oil and natural gas supply forecasts from scarcity into abundance.[6,7]

Thirty-five states now participate in what has been christened America's Shale Revolution. This development has resulted in a 34% increase in US natural gas production since 2005, which has made the United States the world's largest producer of natural gas.[8]

The Shale Revolution has also brought US oil production to a 20-year high and created thousands of energy sector jobs, in addition to thousands of jobs outside the energy sector.[9,10] In 2012, US oil production increased by 14% over the previous year, the greatest increase among countries annually producing a million barrels or more. The year 2012 also marked the largest 1-year increase in oil production in US history.[11] In the process, oil imports as a percent of US consumption have fallen from 70% in 2009 to 37% in February 2013, despite policies from Washington that have caused production of oil, natural gas liquids, natural gas, and coal on federal land to fall in quantity and as a percentage of total production.[12, 13] Furthermore, North America is projected to become energy independent by 2020, a development that would have been impossible prior to the invention of smart drilling.[14]

The economic impact of hydraulic fracturing is not limited to the energy sector or the communities near drilling sites. Increased domestic production of natural gas has resulted in lower natural gas prices. According to the Yale Graduates in Energy Study Group, natural gas consumers saved more than $100 billion in 2011; the study suggests the overall benefit derived from recovering shale gas outweighs the costs by a ratio of 400 to 1.[15] Inexpensive natural gas is also driving investment in energy-intensive industries such as steel and aluminum processing, fertilizer production, and manufacturing as energy becomes more affordable due to the switch from coal to cleaner and less-expensive natural gas for electricity generation.

The United States is the world leader in developing hydraulic fracturing technology and development of shale reserves. State and federal regulators are responsible for developing rules and guidelines to protect the public interest. Every society utilizes natural resources, and doing so will have an impact on the environment. People must weigh the costs of developing the resource against the benefits that would be derived from doing so, and develop that resource in the most environmentally friendly way.

Although stories of economic opportunity are a major focus in newsrooms across the nation, environmental groups have raised concerns about the impact this new technique for oil and gas extraction could have on the environment, including concerns about groundwater contamination, water consumption, wastewater disposal, earthquakes, and greenhouse gas emissions. Some environmental groups and policymakers have called for increased regulation of the hydraulic fracturing industry, and others have advocated fracking moratoria, such as the one that has been in place in New York since 2008.

Those raising these fears have taken advantage of the public's limited understanding of the smart drilling process, limited knowledge of geology, and lack of knowledge of current federal and state regulations on oil and gas production. This chapter has been written to explain the advantages and disadvantages of smart drilling and the alternatives so that a better-informed discussion can take place.

In Section 71.2, the author reviews the background and potential of hydraulic fracturing in the United States and then puts that potential in the context of the supply of and demand for oil and gas. Section 71.3 addresses the environmental impacts of hydraulic fracturing, both positive and negative, as well as public safety issues that have been raised by activists, such as potential harm to drinking water supplies. Section 71.4 discusses how oil and gas production is regulated at the state and national levels and discusses the proper interaction of these two levels of government. Section 71.5 offers concluding remarks.

This chapter concludes hydraulic fracturing can be done in a safe and environmentally responsible manner. State governments have done a commendable job working with

[6]Marc Airhart, "The Father of the Barnett Natural Gas Field, George Mitchell," http://geology.com/research/barnett-shale-father.shtml

[7]Institute for Energy Research, "Barnett Shale Fact Sheet," 2012, http://www.instituteforenergyresearch.org/wp-content/uploads/2012/08/Barnett-Shale-Fact-Sheet.pdf

[8]U.S. Energy Information Administration, "U.S. Natural Gas Marketed Production," June 28, 2013, http://www.eia.gov/dnav/ng/hist/n9050us2A.htm

[9]Garry White and Emma Rowley, "Fracking to the Rescue as US Oil Production Hits 20-Year High," *The Telegraph*, March 4, 2013, http://www.telegraph.co.uk/finance/commodities/9905821/Fracking-to-the-rescue-as-US-oil-production-hits-20-year-high.html

[10]Taylor Smith, "The Benefits of Shale Energy on Other Industries," *Research & Commentary*, The Heartland Institute, March 28, 2013, http://heartland.org/policy-documents/research-commentary-benefits-shale-energy-other-industries

[11]Mark Perry, "U.S. Oil Output in 2012: Largest 1-Year Gain in U.S. History," American Enterprise Institute, June 12, 2013, http://www.aei.org/publication/us-oil-output-in-2012-largest-1-year-gain-in-us-history/print/

[12]David Blackmon, "The Texas Shale Oil & Gas Revolution—Leading the Way to Enhanced Energy Security," *Forbes.com*, March 19, 2013, http://www.forbes.com/sites/davidblackmon/2013/03/19/the-texas-shale-oil-gas-revolution-leading-the-way-to-enhanced-energy-security/

[13]United States Energy Information Administration, "Sales of Fossil Fuels Produced on Federal and Indian Lands, FY 2003 Through 2012," May 2013, http://www.eia.gov/analysis/requests/federallands/pdf/eia-federallandsales.pdf

[14]Megan L. O'Sullivan, "'Energy Independence' Alone Won't Boost U.S. Power," *Bloomberg*, February 14, 2013, http://www.bloomberg.com/news/2013-02-14/-energy-independence-alone-won-t- boost-u-s-power.html

[15]Robert M. Ames, Anthony Corridore, Joel Nathan Ephross, Edward Hirs, Paul MacAvoy, and Richard Tavelli, "The Arithmetic of Shale Gas," June 15, 2012, http://ssrn.com/abstract=2085027 or http://dx.doi.org/10.2139/ssrn.2085027

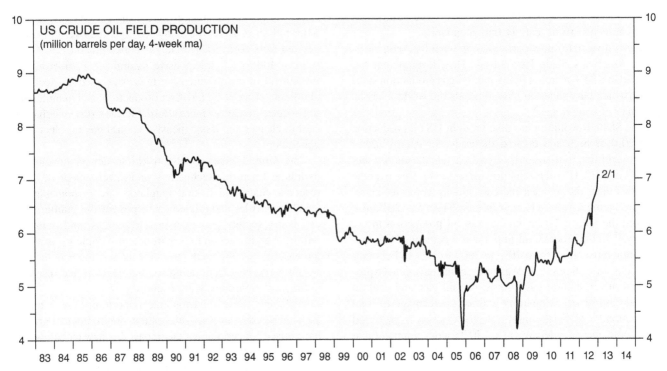

FIGURE 71.1   US oil production 1983–2012. Following nearly two decades of declining oil production, hydraulic fracturing has reversed the US oil production trend. *Source*: US Department of Energy. According to the International Energy Agency, the United States is projected to become the world's largest oil producer by 2017, surpassing Saudi Arabia. Peg Mackey, "U.S. to Overtake Saudi Arabia as Top Oil Producer: IEA," *Reuters*, November 12, 2012, http://www.reuters.com/article/2012/11/12/us-iea-oil-report-idUSBRE8AB0IQ20121112

environmental and industry leaders to craft legislation that protects the environment while permitting oil and gas production to move forward. Federal regulations would be duplicative, resulting in higher costs without significantly increasing environmental protections.

71.2   THE EXTRACTION PROCESS

71.2.1   Conventional Oil and Gas Extraction

Traditionally, oil and gas wells were drilled straight down into permeable, oil- and gas-rich host rocks such as sandstone or limestone, where oil and gas could easily flow through the rock, into the well, and up to the surface.

Think of it as drinking the last of a soft drink from a fast food restaurant. The straw is the well that extends to the bottom of the cup, the beverage is the oil or gas, and the ice is the permeable host rock. The high permeability of the ice allows the soft drink to be sucked out of the cup relatively easily. Although this analogy is imperfect (oil and gas rise through the well due to pressure, whereas the beverage is brought up by a vacuum), it provides a basic understanding of how oil and gas flow through the host rock during the process of conventional drilling.

As production from conventional wells began to decline after 1985 (see Figure 71.1), scientists searched for alternative methods of extracting oil and gas. The result was smart drilling, the combination of hydraulic fracturing and horizontal drilling. Although hydraulic fracturing and horizontal drilling had been used separately to stimulate production at conventional wells since 1947 and 1929, respectively, the combination of these methods has enabled scientists to extract oil and gas trapped in impermeable source rocks such as shale, well-cemented sandstone, and coal bed methane deposits once considered too costly to develop.[16]

Media reports of "fracking" refer to the combination of these two techniques. Throughout this chapter, the author will use the terms hydraulic fracturing, fracking, unconventional energy production, and smart drilling interchangeably to mean the combination of hydraulic fracturing and horizontal drilling, except in the next section entitled "What Is Fracking?" where the author provides a more detailed explanation of these two techniques and their relation to unconventional energy production.

---

[16]U.S. Environmental Protection Agency, "Directional Drilling Technology," http://www.epa.gov/cmop/docs/dir-drilling.pdf

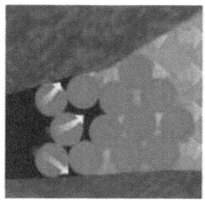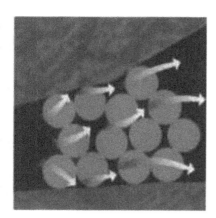

FIGURE 71.2   Resource flow through proppant. Proppant prevents the fissures created during the fracking process from collapsing and allows oil and gas to flow freely to the well. *Source*: Image modified from momentivefracline.com

71.2.2   What Is Fracking?

Hydraulic fracturing is the process of breaking up low-permeability oil- and gas-rich source rocks, enabling the oil and gas to flow freely toward the well. It is accomplished by injecting a mixture of water, sand, and chemical additives at extremely high pressures of 10,000–15,000 pounds per square inch (PSI) into wells drilled in the source rocks.[17] This mixture, commonly referred to as fracking fluid, is composed of 90% water, 9.5% sand, and 0.49% chemical additives.[18]

The sand, typically referred to as frac sand, is generally well rounded (almost spherical), well sorted (all the grains are generally the same size), and durable (able to withstand compressive stresses of 4000–6000 psi). The sand acts as a proppant keeping open the fissures created during the fracking process, increasing the porosity and permeability of the rocks, thus allowing more oil and gas to be recovered (see Figure 71.2).[19]

71.2.2.1   Fracking Fluid.
Although chemical additives constitute a very small percentage of the fracking fluid, as shown in Figure 71.3, they serve a wide variety of important functions such as preventing corrosion in the well, reducing surface tension in liquids, stabilizing clay particles, adjusting pH, and eliminating bacteria.[20]

Many of the chemicals used in fracking fluid are found in common household items. Table 71.1 shows the category of the chemical, the name of the chemical, its purpose in the well, and where it can be found in your bathroom, laundry room, or garage.[21]

Concerns about the safety of some of these chemicals have prompted the oil and gas industry to undertake two major courses of action: one, increasing the disclosure of the chemicals used in the smart drilling process, and two, in the case of one company, making the chemicals used in fracking fluid nontoxic.

Chemical disclosure has occurred through Web sites such as FracFocus.org and its European counterpart, NGS-Facts.org.[22] These Web sites provide the public access to the reported chemicals used in wells in a particular area. FracFocus.org allows individuals to learn which chemicals are utilized at each specific well in its database, an important feature because the chemicals used vary based on local chemistry, causing them to differ from well to well.[23]

Further responding to concerns that the chemicals in fracking fluid could harm the environment, Halliburton has introduced CleanStim, a new nontoxic mixture of fracking fluid additives consisting entirely of ingredients used in the food industry. Colorado Governor John Hickenlooper, a Democrat who holds a master's degree in geology, grabbed headlines after he told the US Senate Committee on Energy

[17]"Fracking," Marcellus Shale, http://www.marcellus-shale.us/fracking.htm

[18]"Fracking Fluids," EnergyFromShale, http://www.energyfromshale.org/hydraulic-fracturing/hydraulic-fracturing-fluid

[19]Mike O'Driscoll, "Frac Sand Frenzy: Focus on Supply & Demand for Hydraulic Fracturing Sand," *Industrial Minerals*, March 2012, http://www.indmin.com/downloads/MODFracSandFrenzySilicaArabia201213312.pdf

[20]Phillip B. Kaufman, Glenn S. Penny, and Javad Parktinat, "Critical Evaluation of Additives Used in Shale Slickwater Fractures," SPE-119900, Society of Petroleum Engineers, November 2008, https://www.onepetro.org/conference-paper/SPE-119900-MS

[21]United States Department of Energy, "Modern Shale Gas Development in the United States: A Primer," April 2009, http://fracfocus.org/node/93

[22]Alessandro Torello, "New OGP Website Identifies Chemicals Used in Shale Gas Operations," International Association of Oil and Gas Producers, June 18, 2013, http://www.ogp.org.uk/news/press-releases/new-ogp-website-identifies-chemicals-used-in-shale-gas-operations/

[23]Lee Lane, "Institutional Choices for Regulating Oil and Gas Wells," Hudson Institute, February 2013, http://www.westernenergyalliance.org/wp-content/uploads/2009/05/Institutional-Choices-for-Regulating-Oil-and-Gas-Wells-Hudson-Institute.pdf

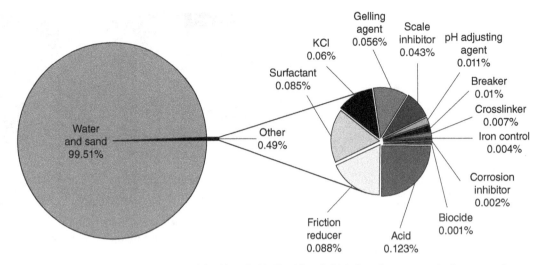

FIGURE 71.3   Composition of fracking fluid. Fracking fluid is largely composed of water and sand (90% water and 9.51% sand). Chemical additives represent just 0.49%. *Source*: United States Department of Energy, "Modern Shale Gas Development in the United States: A Primer," April 2009, p. 62, http://fracfocus.org/node/93

TABLE 71.1   Fracking fluid chemicals and their daily applications.

Additive type	Main compound(s)	Purpose	Common use of main compound
Diluted acid (15%)	Hydrochloric acid or muriatic acid	Helps dissolve minerals and initiate cracks in the rock	Swimming pool chemical and cleaner
Biocide	Glutaraldehyde	Eliminates bacteria in the water that produce corrosive by-products	Disinfectant; sterilize medical and dental equipment
Breaker	Ammonium persulfate	Allows a delayed breakdown of the gel polymer chains	Bleaching agent in detergent and hair cosmetics, manufacture of household plastics
Corrosion inhibitor	*N,N*-dimethyl formamide	Prevents the corrosion of the pipe	Used in pharmaceuticals, acrylic fibers, plastics
Cross-linker	Borate salts	Maintains fluid viscosity as temperature increases	Laundry detergents, hand soaps, and cosmetics
Friction reducer	Polyacrylamide	Minimizes friction between the fluid and the pipe	Water treatment, soil conditioner
	Mineral oil		Makeup remover, laxatives, and candy
Gel	Guar gum or hydroxyethyl cellulose	Thickens the water in order to suspend the sand	Cosmetics, toothpaste, sauces, baked goods, ice cream
Iron control	Citric acid	Prevents precipitation of metal oxides	Food additive, flavoring in food and beverages; lemon juice ~ 7% citric acid
KCl	Potassium chloride	Creates a brine carrier fluid	Low-sodium table salt substitute
Oxygen scavenger	Ammonium bisulfite	Removes oxygen from the water to protect the pipe from corrosion	Cosmetics, food and beverage processing, water treatment
pH-adjusting agent	Sodium or potassium carbonate	Maintains the effectiveness of other components, such as cross-linkers	Washing soda, detergents, soap, water softener, glass, and ceramics
Proppant	Silica, quartz sand	Allows the fractures to remain open so the gas can escape	Drinking water filtration, play sand, concrete, brick mortar
Scale inhibitor	Ethylene glycol	Prevents scale deposits in the pipe	Automotive antifreeze, household cleansers, and de-icing agent
Surfactant	Isopropanol	Used to increase the viscosity of the fracture fluid	Glass cleaner, antiperspirant, and hair color

The specific compounds used in a given fracturing operation will vary depending on company preference, source water quality, and site-specific characteristics of the target formation. The compounds shown above are representative of the major compounds used in hydraulic fracturing of gas shales. *Source*: United States Department of Energy, *Modern Shale Gas Development in the United States: A Primer*, April 2009, p. 63, http://fracfocus.org/node/93

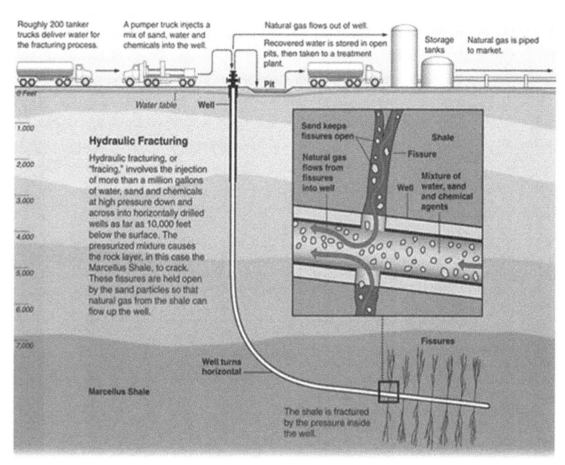

Roughly 200 tanker trucks deliver water for the fracturing process.

A pumper truck injects a mix of sand, water and chemicals into the well.

Natural gas flows out of well.

Recovered water is stored in open pits, then taken to a treatment plant.

Storage tanks

Natural gas is piped to market.

Pit

0 Feet

Water table   Well

1,000

2,000

3,000

4,000

5,000

6,000

7,000

Hydraulic Fracturing

Hydraulic fracturing, or "fracing," involves the injection of more than a million gallons of water, sand and chemicals at high pressure down and across into horizontally drilled wells as far as 10,000 feet below the surface. The pressurized mixture causes the rock layer, in this case the Marcellus Shale, to crack. These fissures are held open by the sand particles so that natural gas from the shale can flow up the well.

Sand keeps fissures open

Natural gas flows from fissures into well

Shale

Fissure

Well

Mixture of water, sand and chemical agents

Well turns horizontal

Marcellus Shale

Fissures

The shale is fractured by the pressure inside the well.

FIGURE 71.4 Cross-section of an unconventional well. *Source*: Graphic by ProPublica/Creative Commons, available at "How Does Hydraulic Fracturing (Fracking) Work?" State Impact: Texas, NPR, http://stateimpact.npr.org/texas/tag/fracking/

and Natural Resources that he drank CleanStim and it was a "benign fluid in every sense."[24] It remains to be seen whether entirely nontoxic chemical additives will become widely used in fracking fluid, but as chemical disclosure becomes more prevalent, products like CleanStim have the potential to change the chemical makeup of fracking fluid at wells across the nation.

71.2.2.2 The Drilling Process. An average unconventional gas well is drilled to a depth of 7500 ft, the equivalent of five Willis Towers (formerly known as the Sears Tower) stacked on top of each other.[25] During the initial drilling phase, wells are drilled vertically to a depth below

the deepest drinking water and irrigation aquifers. Steel surface casing is then inserted down the length of the drilled hole, and cement is pumped the length of the hole to create a barrier of cement and steel between the well, groundwater aquifers, and sensitive geologic formations.[26] Figure 71.4 describes the hydraulic fracturing process in greater detail.

After the well has been cased with steel and cement, operators begin the horizontal drilling phase of the process by angling the well toward the rock layer containing the oil or gas, which is known as the target formation or the "pay zone." Horizontal drilling increases the surface area of the well located in the "pay zone," increasing the amount of oil and gas that can be recovered with one well hole (see Figure 71.5). A single horizontal well can extend up to 2 miles away from the drilling pad and, depending on local factors such as geology and the availability of mineral leases, can

[24]Ben Wolfgang, "I Drank Fracking Fluid, Says Colorado Governor John Hickenlooper," *The Washington Times*, February 12, 2013, http://www .washingtontimes.com/blog/inside-politics/2013/feb/12/colorado-gov-hickenlooper-i-drank-fracking-fluid/

[25]"Fracking Explained with Animation," A2L Consulting, http://www .youtube.com/watch?v=qm7e553S7fg

[26]Institute for Energy Research, "Hydraulic Fracturing—Is It Safe?" May 3, 2011, http://www.instituteforenergyresearch.org/2011/05/03/hydraulic-fracturing-is-it-safe/

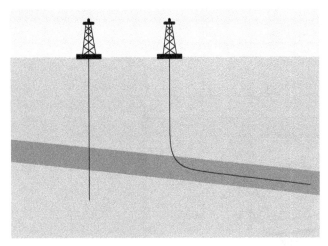

FIGURE 71.5   Vertical drilling compared to horizontal drilling. The dark gray area represents the rock formation containing oil or gas reserves. By using horizontal drilling techniques, producers are able to access more of the target formation and drill fewer holes. *Source*: Geology.com

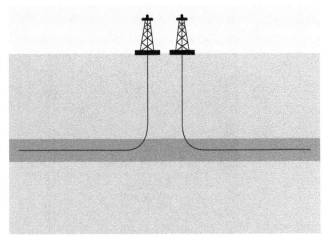

FIGURE 71.6   Multidirectional drilling. Horizontal and directional drilling allows for multiple wells to be drilled in different directions, increasing access to oil and gas resources while decreasing surface disturbance. *Source*: Geology.com

produce 25–30 times more oil or gas on average than a conventional vertical well during the course of its lifetime.[27]

In addition to increasing exposure of the target formation to the well, horizontal drilling makes it possible to drill multiple wells in different directions from the same drilling pad. In 2010, the University of Texas at Arlington drilled 22 wells on a single platform to recover the natural gas from about 1100 acres of area beneath the campus (see Figure 71.6).[28]

A recent study by Cornell University found 83% of the wells drilled in the Marcellus Shale during 2011 were on multi-well pads. The study also found drilling multiple wells on a single pad works in the field, and not just on a university campus, which could further reduce the surface disturbance and associated industrial infrastructure of up to 12 wells, or possibly more, in a single area.[29]

Figure 71.7 shows the layout of a network of wells in Parachute, Colorado. In total, 51 wells were drilled from one well pad, enabling gas producers to access 640 acres of gas reserves from a single 4.6 acre pad.[30] A statement by

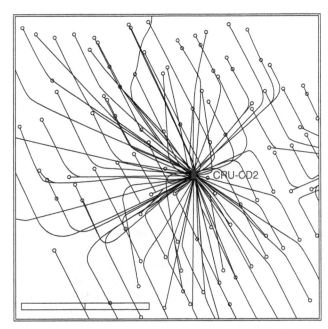

FIGURE 71.7   The "Octopus" A series of 51 wells drilled near Parachute, Colorado, christened "The Octopus" by Brian Hicks. *Source*: *Energy & Capital*, December 13, 2012.

Devon Energy indicates well pads generally take up the same amount of space regardless of the number of wells located on a site, meaning it would have required 51 individual conventional wells disturbing the surface of 243 acres to achieve the same amount of energy production currently achieved by disturbing the surface of only 4.6 acres.[31]

[27] Joe Carroll and Edward Klump, "Shale Drillers Squeeze Costs as Era of Exploration Ends: Energy," *Bloomberg*, June 14, 2013, http://www.bloomberg.com/news/2013-06-13/shale-drillers-squeeze- costs-as-era-of-exploration-ends-energy.html

[28] Hobart King, "Directional Drilling in Oil and Gas Wells," http://geology.com/articles/horizontal-drilling/

[29] Jim Ladee and Jeffery Jaquet, "The Implications of Multi-Well Pads in the Marcellus Shale," Community and Regional Development Institute, Cornell University, Research and Policy Brief Series, No. 43, September 2011, http://cardi.cornell.edu/cals/devsoc/outreach/cardi/publications/loader.cfm?csModule=security/getfile&PageID=1016988

[30] Brian Hicks, "Multi-Well Pad Will Sink OPEC," Energy & Capital, December 13, 2012, http://www.energyandcapital.com/articles/multi-well-pad/2892

[31] Devon Energy, "Multi-Well Pads Are the Norm," http://www.devonenergy.com/featured-stories/multi-well-pads-are-the-norm

Multi-well drill pads increase the profitability of drilling operations by creating economies of scale and minimizing the total number of wastewater ponds, treatment facilities, access roads, and gathering lines required to produce a particular amount of energy, while greatly reducing the amount of land requiring reclamation after the drilling is over.

The Interstate Oil and Gas Compact Commission, a multi-state government agency representing oil- and gas-producing states in Congress, estimates hydraulic fracturing has been used in more than 1 million wells since 1947. Currently, hydraulic fracturing is used to stimulate production in 90% of domestic oil and gas wells (conventional and unconventional), though smart drilling methods utilize fracking to a much greater degree than conventional oil and gas development.[32]

The United States is an energy-hungry nation, and this demand must be met through increased domestic production or reliance on foreign imports. Smart drilling has enabled the United States to increase domestic oil and gas production to meet demand while reducing the surface disturbance and creating thousands of jobs across the nation.

71.2.3 Frack to the Future: The Economic Impacts of Hydraulic Fracturing

With the increased application of smart drilling, the United States has increased domestic natural gas production by 34% since 2005, becoming the world's leading producer.[33] Net imports of natural gas have fallen nearly 50% since 2007, and as a result imports account for only 8% of total US natural gas consumption. Reduced reliance on imports has improved the US trade balance and stimulated the economy as domestic supply has grown to meet demand.

Since smart drilling was found to be economically viable in 2003, unconventional energy production has resulted in the creation of thousands of new jobs across the country. In 2012 the unconventional oil and gas industry supported 360,000 direct jobs and 537,000 indirect supply-industry jobs (equipment manufacturers, frac sand mine operators, steel workers, truckers, etc.), and an estimated 850,000 induced jobs were supported by oil and gas workers spending their paychecks in the general economy at grocery stores, dentist offices, movie theaters, auto dealerships, and so on.[34]

Employment is projected to increase further by 2020, when the unconventional oil and gas industries are estimated to support a total of 600,000 direct jobs, 900,000 indirect jobs, and 1.5 million induced jobs.

Although job creation from a specific industry, especially induced jobs, can be a debatable metric of economic impact, the likelihood is that the new jobs created by unconventional oil production are occurring because domestic oil production is displacing purchases of imports rather than domestically produced substitutes, and it is therefore likely the benefits of increased production are felt by the US economy.

Average wages for drilling and extraction jobs vary from approximately $20/hour for skilled and semiskilled workers (some having only a high school diploma) to $35/hour for professional workers, exceeding the national median by 11% and 15%, respectively.[35]

Nowhere is the economic potential of shale oil and gas production more evident than in towns like Williston, North Dakota. Williams County, which includes the city of Williston, has experienced a 316% growth in jobs, from 8671 in 2000 to 36,107 in the third quarter of 2012.

Since oil production in the Bakken Formation began, North Dakota has climbed from eighth largest oil-producing state in the nation in 2007 to second place today, eclipsing Alaska (see Figure 71.8).[36] The growth in oil production in North Dakota has supported the creation of 70,000 new jobs statewide in the past 5 years. These jobs are a major reason why the North Dakota unemployment rate has remained the lowest in the nation, steadily holding near 3.2%. North Dakota has also experienced the highest population growth rate in the nation, increasing 2.17% in 2012 and 9% since 2000.[37]

Additionally, personal income growth in North Dakota has been the highest in the nation for 5 of the past 6 years.[38] Average weekly wages in the 12 Bakken counties have increased by 49% from 2010 to 2012 as employers have been forced to raise wages to compete with oil production companies for labor. For example, wages at the local McDonald's start at $25,000 annually.[39]

The growth in smart oil and gas production is fueled primarily by the exploration and production industries, which

[32] Railroad Commission of Texas, Testimony Submitted to the House Committee on Energy and Commerce by Victor Carillo, Chairman, Texas Railroad Commission, Representing the Interstate Oil and Gas Compact Commission, February 10, 2005.

[33] U.S. Energy Information Administration, *supra* note 8.

[34] IHS Inc., "America's New Energy Future: The Unconventional Oil and Gas Revolution and the U.S. Economy," October 2012, http://www.energyxxi.org/sites/default/files/pdf/americas_new_energy_future-unconventional_oil_and_gas.pdf

[35] *Ibid.*

[36] U.S. Energy Information Administration, "U.S. Crude Oil Production in First Quarter 2012 Highest in 14 Years," June 8, 2012, http://www.eia.gov/todayinenergy/detail.cfm?id=6610

[37] Dickinson Press Staff "ND Up 100k Jobs Since 2000," *The Dickinson Press*, April 10, 2013, http://www.thedickinsonpress.com/content/nd-100k-jobs-2000

[38] Morgan Brennan, "North Dakota Leads List of America's Fastest Growing States," *Forbes.com*, December 20, 2012, http://www.forbes.com/sites/morganbrennan/2012/12/20/north-dakota-leads-list-of-americas-fastest-growing-states/

[39] Kris Hudson, "Oil Boom By-Product: Unaffordable Housing," *Wall Street Journal*, April 4, 2013, http://online.wsj.com/article/SB10001424127887324883604578396491794558304.html

Crude oil production

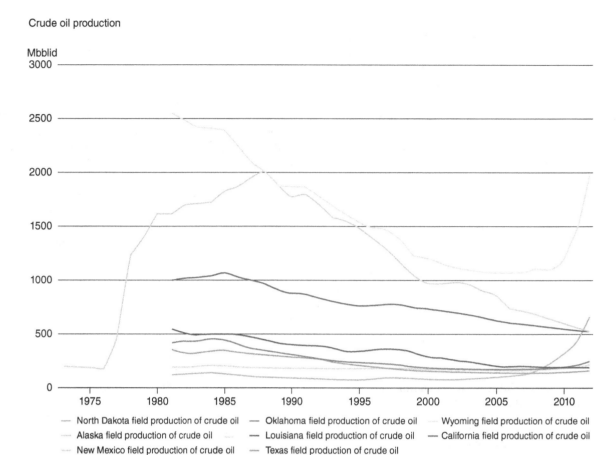

FIGURE 71.8   Crude oil production in the United States. During much of the previous decades, North Dakota was the eighth largest oil-producing state. Since the development of the Bakken oil field, it has become the second largest oil-producing state in the nation. *Source*: US Energy Information Administration.

invested $87 billion in 2012. In addition, the majority of the tools, technology, and knowledge needed to perform hydraulic fracturing are home-grown, resulting in an overwhelming majority of every dollar spent entering the domestic supply chain and supporting domestic jobs, making boomtowns like Williston possible. Annual capital investments are projected to grow throughout the decade, from $87 billion currently to $172.5 billion by the end of the decade.[40]

This investment has spurred the rapid expansion in the nation's shale plays. Figure 71.9 is a map labeling the key oil and gas deposits in the lower 48 states. US shale reserves are the second largest in the world,[41] constituting approximately 2384 trillion cubic feet of natural gas, enough to meet current domestic demand for 98 years. Additionally, it is important to note that reserve estimates are not a static number; they fluctuate based on the ability of oil and gas producers to extract

the resource from the ground economically, meaning reserve estimates could grow with increasingly efficient technology or rising prices.

The shale boom places the United States in the enviable position of having the largest demand for natural gas and among the lowest global prices. Although natural gas prices have been historically volatile, supply stability in the coming years is projected to reduce price fluctuations. Once considered supply-constrained, smart drilling and the resulting increases in gas production have made the US natural gas market demand-constrained. Prices have plummeted from the highs of $16/MMbtu briefly experienced in 2005 to approximately $4.15/MMbtu today. Low natural gas prices have made it a more cost-effective option for electricity generation than coal in many situations; as a result, electricity prices have fallen.

Natural gas accounted for 25% of the fuel burned for US electricity generation in 2011 (see Figure 71.10). According to EIA projections, much of the growth in electricity demand in the coming decades will be met by natural gas. These

[40]IHS Inc., *supra* note 34.
[41]Shale reserve figures were not available for Russia, a major player in the oil and gas industry.

FIGURE 71.9   Shale plays of the lower 48 states. "Shale play" is the term used in the oil and gas industry for a geological formation that has been targeted for exploration. *Source*: US Energy Information Administration; Eagle Ford Shale Blog, "What Is a Shale Play," http://eaglefordshaleblog.com/2010/03/03/what-is-a-shale-gas-play/

projections have been substantiated by the fact that in 2012, the United States produced 30% of its electricity by burning natural gas, a level it was not projected to achieve until 2040.[42] The manufacturing sector will also use more natural gas for energy and as a feedstock for plastics and chemicals.

Low natural gas and electricity prices have given the United States a competitive advantage in energy- and feedstock-intensive industries such as manufacturing, fertilizer production, steel and aluminum processing, petrochemical production, and plastic production, relative to their European and Asian counterparts (see Figure 71.11).[43] As a result, foreign investors such as Voestalpine, an Austrian steel firm, and Japanese oil refiner Idemitsu Kosan and

trading house Mitsui & Co. have opened operations in the United States.[44]

Low natural gas prices are one reason the US manufacturing sector has seen much faster growth of output compared to other advanced nations.[45] Since 2010, the US manufacturing industry has created nearly half a million jobs and the National Association of Manufacturers estimates the shale boom will add 1 million manufacturing jobs to the economy by 2025. Additionally, it has been reported that Boeing, General Electric, and Apple will bring back some of the jobs they moved overseas to cut costs, aided partially by lower natural gas prices.[46]

Unconventional oil and gas production are also generating revenue for state, local, and federal governments. In

[42]U.S. Energy Information Association, AEO2013 Early Release Overview, DOE/EIA-0383ER (2013), December 5, 2012, http://www.eia.gov/forecasts/aeo/er/early_elecgen.cfm
[43]"Shale Boom Sparks U.S. Industrial Revival," *CNBC*, March 26, 2013, http://www.cnbc.com/id/100592605

[44]*Ibid.*
[45]Marilyn Geewax, "Cheap Natural Gas Pumping New Life into U.S. Factories," National Public Radio, March 28, 2013, http://www.npr.org/2013/03/28/175483517/cheap-natural-gas- pumping-new-life-into-u-s-factories
[46]CNBC, *supra* note 43.

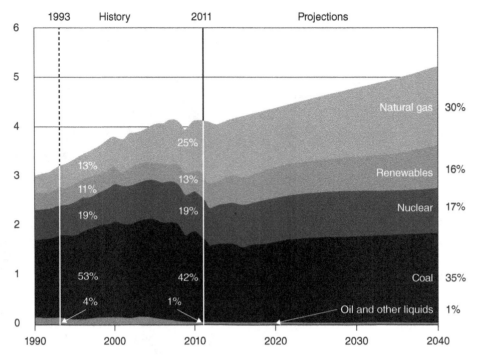

FIGURE 71.10   Electricity generation by fuel 1990–2040. (trillion kilowatt hours per year) Growing demand for electricity will be met largely by using more natural gas for electricity generation as its share of the electricity generation market is predicted to rise to 30% in 2040. *Source*: Energy Information Administration.

2012, total tax revenues from unconventional oil and gas production were approximately $62 billion, with 2020 revenues projected to grow to more than $111 billion. These taxes include corporate income taxes from the production and supply chain of industries and personal income taxes from direct, indirect, and induced employees.[47]

States such as California and Illinois, two states with perennial budget deficits, have noticed this revenue and are looking to the oil and gas industries as a possible way of reducing their budget deficits. A recent study by the University of Southern California estimates hydraulic fracturing in California would yield 2.8 million jobs and $24.6 billion in tax revenue by 2020.[48] The State of Illinois has adopted regulations approved by industry and environmental groups to facilitate the growth of unconventional energy production, due in no small part to the desire for additional income in the state treasury.[49]

Hydraulic fracturing has benefited both the private and the public sectors, but these benefits are not without costs. A study by the Yale Graduates in Energy Study Group provides insights into the cost–benefit analysis of hydraulic fracturing.

The study calculated the benefit derived from hydraulic fracturing to be $100 billion annually in the form of lower

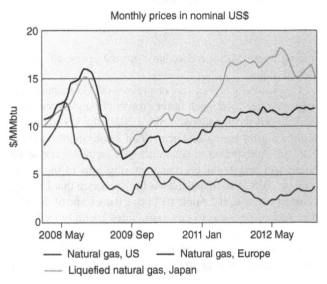

FIGURE 71.11   Comparison of global natural gas prices. The United States currently enjoys a competitive advantage over Europe and Japan as natural gas is three to four times less expensive in America. *Source*: World Bank Commodity Price Data (Pink Sheet), April 2013.

[47] IHS Inc., *supra* note 34.
[48] Alison Vekshin, "California Fracking Fight Has $25 Billion Taxes at Stake," *Bloomberg*, March 13, 2013, http://www.bloomberg.com/news/2013-03-17/california-fracking-fight-has-25-billion-taxes-at-stake.html
[49] Patrick Yeagle, "What to Expect with Fracking," *Illinois Times*, June 27, 2013, http://www.illinoistimes.com/Springfield/article-12566-what-to-expect-with-fracking.html

natural gas prices for consumers. To calculate the cost of hydraulic fracturing, the authors assumed 100 spills for every 10,000 new wells drilled and a cost of approximately $2.5 million if 5000 gallons of wastewater was spilled and it was necessary to remove 5000 cubic yards of contaminated soil. If groundwater wells were contaminated, the cost of providing water to residents would be approximately $5000 per well. In all, the costs associated with 100 accidents every year would be approximately $250 million, making the cost–benefit ratio 400:1.[50]

These projected costs, while vastly outweighed by the benefits, are not insignificant. While policymakers decide on their next course of action, it is important to understand the impacts oil and gas production could have on the environment. The next section focuses on the environmental impact using studies from peer-reviewed publications in an attempt to provide the most accurate and up-to-date information possible.

71.3 ENVIRONMENTAL IMPACT

71.3.1 Environmental Concerns

Benefits from hydraulic fracturing are realized in economic and social terms, whereas the costs are realized in the form of potential environmental impacts. The prospect of large-scale hydraulic fracturing has provoked fears of contaminated water and a new era of manmade earthquakes. These fears led New Jersey,[51] New York, North Carolina, and Vermont to place moratoria on hydraulic fracturing.

Environmental damage is a legitimate concern, yet it must be viewed realistically and in light of cost–benefit analysis and not absolute terms. Among the key areas of concern cited by opponents of unconventional oil and gas production are water consumption, groundwater contamination, wastewater disposal, earthquakes, and greenhouse gas emissions.

This study addresses each of these impacts and determines that reasonable measures short of moratoria and bans can be taken to mitigate environmental damage while allowing for the responsible extraction of oil and gas resources.

71.3.1.1 Water Consumption. Hydraulic fracturing is often portrayed as a water-intensive industry due to the large volumes of water injected into wells to break up shale formations (approximately 2 million to 4 million gallons per well). However, when compared to other uses, such as conventional oil and gas drilling, coal mining, biofuel production, household, and agricultural use, the amount of water consumed for unconventional oil and gas production is relatively small.

Water consumption is determined by calculating the amount of water that is evaporated, contaminated, or stored. In 2011, shale gas production consumed approximately 135 billion gallons of water nationwide. Using the most recent freshwater consumption data from the United States Geological Survey (USGS), the United States consumed 43,800 billion gallons of water in 2005. Using the 2005 water consumption data as a baseline for current use, hydraulic fracturing accounted for just 0.3% of the total water consumed, compared to the 0.5% used to irrigate golf courses annually.[52]

Additionally, shale gas production consumes less water per unit of energy generated than onshore oil production, ethanol production, and washing coal after it has been mined, as demonstrated in Figure 71.12.[53]

Although the scale of the hydraulic fracturing conducted in the Marcellus Shale represents an increase in total water consumption, it rivals the water levels used for irrigation and is dwarfed by other uses such as the public water supply, power generation, and other industrial and mining consumption, as demonstrated in Figure 71.13.[54]

Additionally, although Texas has the highest unconventional gas and oil production in the nation, hydraulic fracturing accounted for just 1% of freshwater withdrawals there, with the water consumed at the Barnett Shale, the largest shale play in the state, equal to about 9% of the annual water consumption of the city of Dallas.[55]

Although hydraulic fracturing utilizes considerable quantities of water, shale gas development has been mischaracterized as an especially water-intensive industry. In arid regions where water supplies are scarce, local authorities are best equipped to make decisions regarding appropriate water use, but when compared to other water uses such as thermal energy production, agriculture, and public consumption, the impact of hydraulic fracturing on freshwater supplies is relatively small.

71.3.1.2 Groundwater Contamination. Public opposition to hydraulic fracturing often stems from fear that methane

[50]Christopher Helman, "The Arithmetic of Shale Gas," *Forbes.com*, June 22, 2012, http://www.forbes.com/sites/christopherhelman/2012/06/22/the-arithmetic-of-shale-gas/

[51]New Jersey enacted a 1-year moratorium that has since expired.

[52]Jesse Jenkins, "Energy Facts: How Much Water Does Fracking for Shale Gas Consume?" *The Energy Collective*, April 6, 2013, http://theenergycollective.com/jessejenkins/205481/friday-energy-facts-how-much-water-does-fracking-shale-gas-consume

[53]Erik Mielke, Laura Diaz Anadon, and Venkatesh Narayanamurti, "Water Consumption of Energy Resource Extraction, Processing, and Conversion," Discussion Paper 2010–15, Energy Technology Innovation Policy Research Group, Belfer Center for Science and International Affairs, Harvard Kennedy School, October 2010, http://belfercenter.ksg.harvard.edu/files/ETIP-DP-2010-15-final-4.pdf

[54]James Brian Mahoney, "The Role of Wisconsin 'Frac' Sand in the U.S. Energy Portfolio," Presentation prepared on behalf of the University of Wisconsin Eau Claire Geology Department, received via email, April 3, 2013.

[55]Jean Philippe Nicot and Bridget Scanlon, "Water Use in Shale Gas Production in Texas, US," *Environmental Science and Technology* 46 (2012), pp. 3580–3586, http://www.beg.utexas.edu/staffinfo/Scanlon_pdf/Nicot+Scanlon_ES&T_12_Water%20Use%20Fracking.pdf

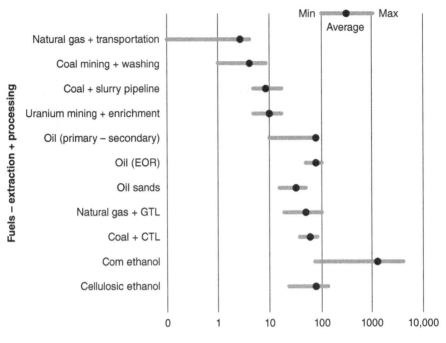

FIGURE 71.12   Water consumption per unit of energy produced (log scale). Shale gas (0.6–1.8 gal/ MMbtu), coal mining and washing (1–8 gal/MMbtu), onshore oil (1–62 gal/btu), and ethanol (more than 1000 gal/MMbtu). *Source*: Erik Mielke, Laura Diaz Anadon, and Venkatesh Narayanamurti, "Water Consumption of Energy Resource Extraction, Processing, and Conversion," Discussion Paper 2010–15, Energy Technology Innovation Policy Research Group, Belfer Center for Science and International Affairs, Harvard Kennedy School, October 2010, Chart 4-1 Water consumption for extraction and processing of fuels (log scale).

(the main hydrocarbon in natural gas) and fracking fluid chemicals will contaminate groundwater aquifers and compromise drinking water supplies.

Perhaps the most powerful image associated with hydraulic fracturing is the scene from the movie *Gasland* where Mike Markham ignites the water running from the faucet of his Colorado home. Later analysis by the Colorado Oil and Gas Conservation Commission (COGCC) determined the methane found in Markham's well was biogenic in origin, naturally occurring, and found in the coal formations present within the aquifer supplying his drinking water, not thermogenic methane, a key component of natural gas. The well did not test positive for chemicals used in the fracking process, providing further evidence that oil and gas production was not the cause of contamination.[56]

A study conducted by Duke University analyzing 68 water wells in the Marcellus Shale found 85% of wells contained methane regardless of gas industry operations. Researchers concluded the methane in these wells was thermogenic in

origin, but they were unable to establish a direct, causal link between hydraulic fracturing and well water contamination, due to a lack of historic data for baseline analysis. Here too, no evidence of fracking fluid was found in water samples.[57]

Instances of groundwater impairment due to activities associated with natural gas and oil production, such as surface spills and leaks from improperly cased wells, have been known to occur in conventional wells in addition to unconventional wells. However, there has been no conclusive evidence provided to support the claim that hydraulic fracturing has caused groundwater contamination.[58] This conclusion was reiterated by David Neslin, director of the COGCC, in a letter providing written answers to follow-up questions by the Senate Committee on Environment and Public Works. In that letter, Neslin testified the COGCC had found

[56]Department of Natural Resources, Colorado Oil and Gas Conservation Commission, "Gasland Document," n.d., http://cogcc.state.co.us/library/ GASLAND%20DOC.pdf

[57]Jeffrey C. King, Jamie Lavergne Bryan, and Meredith Clark, "Factual Causation: The Missing Link in Hydraulic Fracture— Groundwater Contamination Litigation," *Duke Environmental Law & Policy Forum* 22 (Spring 2012), pp. 341–360, http://scholarship.law.duke.edu/ cgi/viewcontent.cgi?article=1234&context=delpf
[58]*Ibid.*

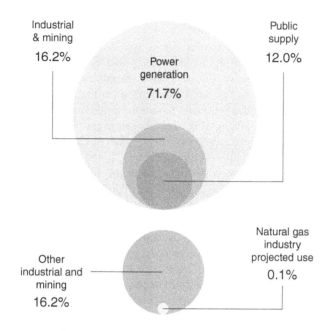

Total water use in the Marcellus area:
3.6 trillion gallons per year

Industrial
& mining
16.2%

Power
generation
71.7%

Public
supply
12.0%

Other
industrial and
mining
16.2%

Natural gas
industry
projected use
0.1%

Other uses: Irrigation 0.1% Livestock: 0.01%

FIGURE 71.13   Water use in the Marcellus Shale. Although shale gas production constitutes a significant increase in water use in the Marcellus Shale, it is comparable to irrigation and dramatically lower than other uses. *Source:* J. Brian Mahoney, "The Role of Wisconsin 'Frac' Sand in the US Energy Portfolio," Presentation prepared on behalf of the University of Wisconsin Eau Claire Geology Department, received via email, April 3, 2013.

no verified incidences of hydraulic fracturing contaminating groundwater.[59]

Current regulations set minimum standards regarding well bore strength, well casing procedures, fluid and gas migration prevention, and other possible sources of contamination, designed to protect the environment while facilitating the responsible extraction of natural resources. Businesses have strong motives to avoid oil- and gas-related activities that compromise drinking water, such as surface spills, faulty casing, machinery malfunction, and operator error. Such accidents increase expenses, interrupt production, create legal liabilities, may violate government regulations, and may require restitution for those affected.

To aid investigators in examining future claims of contamination, baseline data from areas near proposed wells should be obtained prior to drilling, thus providing the information necessary to determine the source of groundwater contamination and hold accountable those whose actions led to contamination.

With no confirmed cases of water well or groundwater contamination directly linked to the process of hydraulic fracturing, calls for moratoria are not supported by science. States should instead seek to create a regulatory framework to mitigate environmental damage, rather than ban hydraulic fracturing.

71.3.1.3 Wastewater. Most of the water used to hydraulically fracture unconventional wells remains underground after the fracking process is complete, but approximately 15–20% is returned to the surface through a steel-cased well bore and temporarily stored in steel tanks or lined pits. This wastewater is referred to as flowback water.[60]

In the February 2013 issue of *Water Resource Research*, Brian Lutz and Aurana Lewis state shale gas production in the Marcellus Shale—the largest shale gas play in the country, producing 10% of US natural gas—produces approximately 65% less wastewater per unit of natural gas recovered than conventional gas drilling.[61] The study quantified natural gas and wastewater production at 2189 shale gas wells in the Marcellus Shale. Lutz and Lewis determined that although shale gas wells produce 10 times more wastewater on average than conventional wells, they produce approximately *30 times* as much natural gas as conventional wells.

Although hydraulically fractured wells in the Marcellus Shale were found to produce 65% less wastewater per unit of gas recovered than conventional wells, the rapid expansion of natural gas production in Pennsylvania has caused related wastewater to increase by 570% since 2004, and it must be disposed of properly.[62] Depending on well location, disposal comes in two primary forms: recycling and injection wells.

Injection wells, also known as disposal wells, are deep wells that pump waste materials or carbon dioxide (CO_2) into isolated formations far below the Earth's surface and are designed to provide multiple layers of protective casing and cement.[63] Injection wells are the most common form

[59]David Neslin, "Written Answers to Follow Up Questions from the Senate Committee on the Environment and Public Works," Colorado Oil and Gas Conservation Commission, submitted May 17, 2011, http://www .savecoloradofromfracking.org/whatgoeswrong/Resources/Neslin%20 Testimony%20EnviroPublicWorksQA.pdf

[60]Jeff Fort, "Exploring the Disposal of Fracking Waste Water—UIC Class II Wells in Ohio," *Oil and Gas Law Report*, April 13, 2013, http://www .oilandgaslawreport.com/2013/04/13/exploring-the-disposal-of-fracking-waste-water-uic-class-ii-wells-in-ohio/

[61]Brian D. Lutz, Aurana N. Lewis, and Martin W. Doyle, "Generation, Transport, and Disposal of Wastewater Associated with Marcellus Shale Gas Development," *Water Resources Research* 49, No. 2 (February 2013), pp. 647–656.

[62]*Ibid.*

[63]U.S. Environmental Protection Agency, "Basic Information about Underground Injection Wells," May 2012, http://water.epa.gov/type/groundwater/ uic/basicinformation.cfm

of wastewater disposal in much of the country because they provide a cost-effective means of disposing of large quantities of water in an environmentally responsible way.

The United States Environmental Protection Agency (EPA) regulates the use of injection wells under the Safe Drinking Water Act (SDWA) and considers them a safe and effective way to protect water and soil resources. Approximately 30,000 class II injection wells are used to dispose of oil and gas waste. Although economical, this method requires transportation of waste fluid to disposal sites, and it has been known to contribute to induced seismic activity, which will be discussed in a later section.

As concerns over the use of injection wells and the availability of freshwater have become more prominent, recycling the wastewater generated at hydraulic fracturing sites has become a more popular option. Gas producers in the Marcellus Shale have formed the Center for Sustainable Shale Development, a consortium dedicated to establishing best practices, which include recycling 90% of their wastewater.[64,65] In Texas, the Railroad Commission recently adopted rules removing regulatory hurdles for recycling flowback water.

As more emphasis is placed on water conservation, recycling flowback water and experimental waterless fracturing techniques will likely become more prominent in oil and gas production.

71.3.1.4 *Earthquakes.*
Some environmental groups claim greater use of hydraulic fracturing will usher in a new era of manmade earthquakes. Although studies have found the process of hydraulic fracturing can cause small earthquakes, there is a greater risk of induced seismicity associated with the use of injection wells for the disposal of wastewater. The first earthquake caused by an injection well in the United States occurred in 1965 at a military complex in Colorado, where a disposal well was used to dispose of military waste, causing what became known as the "Denver Earthquakes."[66]

The risk of earthquakes directly related to hydraulic fracturing is small compared to other human activities, which can cause comparatively large tremors. A study conducted by Dr. Richard Davies, a professor of earth sciences at Durham University, compiled 198 published examples of induced seismic activity (manmade earthquakes) from around the world registering at a magnitude greater than or equal to 1.0 since 1929. The study found hydraulic fracturing was responsible for only three earthquakes large enough to be felt on the surface—one in Canada, one in the United States, and one in the United Kingdom. Of these three earthquakes, the largest occurred in 2011 in the Horn River Basin of Canada, registering at M3.8, a magnitude on the lower end of earthquakes that can be felt by people.[67]

Other human activities can trigger much larger earthquakes. For example, building dams and filling reservoirs have caused earthquakes ranging in magnitude from M2.0 to M7.9, mining can cause earthquakes in the M1.6–M5.6 range, using injection wells for carbon capture and sequestration and disposing of wastewater can trigger M2.0–M5.3 quakes, and using geothermal energy wells has caused earthquakes in the M1.0–M4.6 range.[68]

Earthquakes must generally be in the M5.5–M6 range before slight damage occurs to buildings.[69] Furthermore, it is important to note that an M3.5 earthquake is not half as strong as an M7, because the moment magnitude scale is logarithmic—each whole number on the scale is 10 times as large as the preceding number. Therefore, an M6 is *100 times more intense* than an M4. For earthquakes to be felt on the surface, they must be near M4.0. This reasoning prompted Davies to give the following statement:

> We have concluded that hydraulic fracturing is not a significant mechanism for inducing felt earthquakes. It is extremely unlikely that any of us will ever be able to feel an earthquake caused by fracking.[70]

As noted earlier, it is estimated that hydraulic fracturing has been used to stimulate more than 1 million oil and gas wells in the United States since 1947, making the incidence of earthquakes capable of being felt on the surface caused by the fissuring of deep rock formations, to date, 1 in 1 million.

71.3.1.5 *Injection Wells and Earthquakes.*
The Denver earthquakes were triggered when military wastes were injected into a hole 12,044 ft deep.[71] This series of earthquakes prompted further research investigating the link between injection wells and seismic activity. Although injection wells are an effective way of disposing of waste and sequestering carbon, they pose a larger risk of earthquakes than the hydraulic fracturing process itself.

[64]Susan Brantley and Anna Meyendorff, "The Facts on Fracking," *New York Times*, March 13, 2013, http://www.nytimes.com/2013/03/14/opinion/global/the-facts-on-fracking.html?pagewanted=all&_r=2&
[65]Center for Sustainable Shale Development, "Performance Standards," March 2013, http://037186e.netsolhost.com/site/wp-content/uploads/2013/03/CSSD-Performance-Standards-3-27-GPX.pdf
[66]Richard Davies, Gillian Foulger, Annette Bindley, and Peter Styles, "Induced Seismicity and Hydraulic Fracturing for the Recovery of Hydrocarbons," *Marine and Petroleum Geology*, 2013, https://pangea.stanford.edu/researchgroups/scits/sites/default/files/10679_0.pdf

[67]*Ibid.*
[68]*Ibid.*
[69]Michigan Technological University, UPSeis, Earthquake Magnitude Scale, http://www.geo.mtu.edu/UPSeis/magnitude.html
[70]Richard Davies, quoted in "New Fracking Research Led by Durham University Finds It Is 'Not Significant' in Causing Earthquakes," *Durham University News*, April 10, 2013, http://www.dur.ac.uk/news/newsitem/?itemno=17347
[71]Richard Davies *et al.*, *supra* note 66.

A series of small tremors ranging from M2.4 to M4 were linked to the use of injection wells in Ohio and Arkansas in 2011. There were no reported injuries and only minor property damage occurred. The earthquakes prompted the Ohio Department of Natural Resources to promulgate rules restricting the use of injection wells. The requirements mandate drillers to submit geologic mapping data before drilling; if these data do not exist, companies must hire scientists to obtain them.[72] Wells also must contain state-of-the-art pressure- and volume-monitoring equipment, along with an automatic shutoff system in case pressures exceed the limits the state has set for each well.[73]

Approximately 30,000 disposal wells are used for the disposal of oil and gas waste. According to John Bredehoeft, a geological expert who has held research and management positions at the USGS, the vast majority of these wells are geologically stable, and scientists believe increased data collection and monitoring can help prevent future earthquakes near injection well sites.[74]

As injection wells are increasingly utilized for disposal of oil and gas waste and carbon sequestration around the nation, policies similar to those implemented in Ohio and Arkansas will help reduce the occurrence of manmade earthquakes.

71.3.1.6 Greenhouse Gas Emissions.
As natural gas prices dipped below \$2/MMbtu, power plants began to shift away from coal toward less-expensive natural gas, resulting in lower electricity prices and fewer carbon dioxide and sulfur dioxide emissions into the atmosphere. Burning natural gas emits just 1% as much sulfur dioxide (which at high ambient levels is linked to adverse effects on the respiratory system) as is emitted by coal.[75,76]

Natural gas is the cleanest burning fossil fuel, emitting 25% less carbon dioxide than petroleum and approximately 50% fewer emissions than coal,[77] which is one of the main

reasons natural gas historically has been praised by environmental groups looking for ways to reduce US greenhouse gas emissions (see Figure 71.14).

Natural gas accounted for roughly 30% of the fuel used in electricity generation in 2012, compared to 16% in 2000. Coal has fallen from 52% of the electricity generation market in 2000 to 38% in 2012. Increasing reliance on natural gas and shrinking energy demand due to a struggling economy have been key reasons why US CO_2 emissions have fallen 12% since 2005, to their lowest levels since 1994.[78] The US Department of Energy predicts carbon dioxide emissions will begin to rise again on a year-to-year basis beginning in 2015 but will not reach 2005 levels again through 2040.

Although CO_2 emissions have fallen as natural gas has replaced coal for electricity generation, environmental groups argue unconventional wells emit more "fugitive" methane into the atmosphere, negating the benefits derived from burning natural gas. Although many of these claims are based on discredited data, EPA regulations passed in April 2012 will soon render the argument irrelevant.

These air quality standards will require the use of "green well completion," a technology designed to capture 95% of fugitive methane emissions, by 2015 when the equipment necessary for its implementation will be more widely available. Until then, producers are required to burn excess gas (also known as flaring) to remove volatile hydrocarbons.

Green completion technology is already used in 50% of unconventional gas wells nationwide,[79] and a recent study by the Massachusetts Institute of Technology found green completion technology pays for itself 95% of the time and would still pay for itself 83% of the time if the cost to "green complete" a well were doubled, indicating that in most cases, companies would lose money by not implementing this technology.[80,81] These measures are expected to be so successful that Gina McCarthy, at the time assistant administrator of the EPA Office of Air and Radiation and recently confirmed as EPA administrator, stated the agency does not see a need to take further action on industry methane emissions.

Hydraulic fracturing and the resulting shale gas boom have provided the United States a cost-effective, clean, and

[72]Douglas J. Guth, "Ohio Tries to Avoid Repeat of 2011 Injection Well Quakes," *Midwest Energy News*, April 29, 2013, http://www.midwestenergynews.com/2013/04/29/ohio-tries-to-avoid-repeat-of-2011-fracking-wastewater-injection-well-earthquakes/

[73]John Funk, "Waste-Water Injection Well Caused 12 Earthquakes in Ohio, Investigation Shows," *The Plain Dealer*, March 9, 2012, http://www.cleveland.com/business/index.ssf/2012/03/shale_gas_drilling_caused_smal.html

[74]*Ibid.*

[75]Seamus McGraw, "Is Fracking Safe? The Top 10 Controversial Claims about Natural Gas Drilling," *Popular Mechanics*, n.d., http://www.popularmechanics.com/science/energy/coal-oil-gas/top-10-myths-about-natural-gas-drilling-6386593#slide-3

[76]United States Environmental Protection Agency, "Sulfur Dioxide," updated May 21, 2013, http://www.epa.gov/airquality/sulfurdioxide/

[77]World Nuclear Association, "Comparison of Lifecycle Greenhouse Gas Emissions of Various Electricity Generation Sources," July 2011, http://www.world-nuclear.org/uploadedFiles/org/WNA/Publications/Working_Group_Reports/comparison_of_lifecycle.pdf

[78]Russell Gold, "Rise in U.S. Gas Production Fuels Unexpected Plunge in Emissions," *The Wall Street Journal*, April 18, 2013, http://online.wsj.com/article/SB10001424127887324763404578430751849503848.html

[79]Pam Kasey, "EPA Requires Green Completion, Cuts Pollution from Fracked Wells," *The State Journal*, May 18, 2012, http://www.statejournal.com/story/17526063/epa-requires-green-completion-cuts-pollution-from-fracked-wells

[80]John Hangar, "MIT Blockbuster Study Finds Shale Gas Methane Emissions Greatly Exaggerated: Shale Gas Is Cleaner Than Even EPA Number," John Hangar's Facts of the Day, November 30, 2012, http://johnhanger.blogspot.com/2012/11/mit-blockbuster-study-finds-shale-gas.html

[81]Francis O'Sullivan and Peggy Paltsev, "Shale Gas Production: Potential versus Actual Greenhouse Gas Emissions," *Environmental Research Letters* 7 (2012), http://iopscience.iop.org/1748-9326/7/4/044030/pdf/1748-9326_7_4_044030.pdf

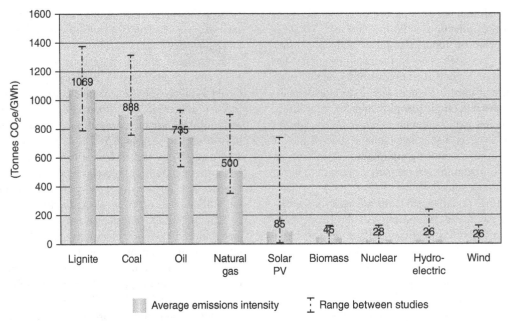

FIGURE 71.14   Fossil fuel emissions of carbon dioxide. Natural gas is the cleanest burning fossil fuel, emitting approximately half of the CO_2 emissions of coal and approximately 25% the CO_2 that is emitted burning oil. *Source*: World Nuclear Association, "Greenhouse gas emissions avoided through use of nuclear energy," 2015.

abundant source of fuel that will stimulate economic growth while reducing greenhouse gas emissions in a practical and significant way.

71.4   OIL AND GAS REGULATION

Regulation of the oil and gas industries has been a shared responsibility between state and federal governments for decades, with a majority of regulations enacted and enforced at the state level. This has resulted in a variance in regulations ranging in scope from laws that facilitate the responsible extraction of natural resources to moratoria banning the use of hydraulic fracturing.

As hydraulic fracturing has become more prevalent, so too have calls for expanded federal oversight. According to a study by Lee Lane of the Hudson Institute, production of oil and natural gas poses a variety of challenges, some better suited for regulation by state governments, others better suited for federal regulation. Which challenges should be handled by which level of government is a point of contention, Lane writes.[82]

Proponents of expanded federal authority, such as Jody Freeman, former counselor for energy and climate change for the Obama administration, argue that although some states will work proactively to create the highest standards, others will lag behind, creating a patchwork of state regulatory systems.[83] Freeman asserts broad public support is necessary to maintain hydraulic fracturing and standardized regulations will more effectively calm the fears of the public and protect the environment.

Although standardized regulations may sound appealing, Lane notes shale oil and gas regulations are not as easily standardized as automobile specifications or air quality regulations. Localized factors such as geology, hydrology, economics, and the difficulties that can arise from local factors in the event of potential problems such as water contamination and earthquakes are better served by state governments and agencies more familiar with the area than federal authorities, especially when these difficulties require immediate action. States will be better able to tailor regulations to their specific needs, whereas standardized regulations create a one-size-fits-all system based on the "average" situation. Regulations appropriate for the average situation are seldom appropriate in specific ones.[84] When problems arise in oil and gas production, specific knowledge of a site is far more helpful than standard protocol.

Unconventional oil and gas production is a dynamic industry with advancements, and accidents, occurring rapidly,

[82]Lee Lane, *supra* note 23.

[83]Julia Bell, "Subtraction Through Addition via the FRAC ACT," Energy In Depth, February 8, 2013, http://energyindepth.org/national/subtraction-through-addition-via-the-frac-act/

[84]Lee Lane, *supra* note 23.

forcing industry to change practices along with them. When earthquakes caused by injection wells shook parts of Ohio and Arkansas, regulators shut down five injection wells and new construction rules were enacted requiring placement of pressure-monitoring equipment in the well, forbidding drilling into basement rock, and requiring companies to provide detailed geologic information to state regulators before wells can be drilled.[85]

Richard Simmers, head of the Ohio Department of Natural Resources Division of Oil and Gas Resources Management, told the House Natural Resources Committee he "unequivocally" believes state regulation of oil and gas production to be the "most effective, efficient and economical." He also testified states can respond more quickly to problems than the federal government.[86] States must be effective at regulating problems because they have the primary responsibility for protecting their environment, the public health, and facilitating economic growth within their borders.

As a result, state regulators must typically be more pragmatic because they do not have the luxury of inefficiency that is afforded to federal regulators. For example, enforcement and promulgation of EPA rules is partially determined by funding. When budget battles rage in Washington, EPA funding is used as a political pawn, as congressional members have learned to use the appropriations process as a potent tool for reining in some of EPA's more costly programs.[87]

The Fracturing Responsibility and Awareness of Chemicals (FRAC) Act would increase the scope of EPA authority regarding hydraulic fracturing and require drilling companies to disclose the chemicals used in fracking fluid, but it has failed to gain any traction in Congress. On the other hand, Colorado, Texas, and Wyoming already require disclosure of fracking fluids, and 10 states—Colorado, Louisiana, Mississippi, Montana, North Dakota, Ohio, Oklahoma, Pennsylvania, Texas, and Utah use the Web site FracFocus for official state chemical disclosure.[88]

These efforts are part of the reason Lisa Jackson, former head of EPA, acknowledged the ability of states to handle oil and gas regulation, stating, "States are doing a good job. It doesn't have to be the EPA that regulates the 10,000 wells

that might go in."[89] State regulators must be quick, efficient, and practical in their actions, addressing problems as they arise. The current culture in Washington does not lend itself to quick, efficient, or practical action.

Finances are another reason states are pragmatic when it comes to oil and gas regulation. As stated previously, total government revenues from unconventional US oil and gas production were more than $62 billion in 2012. This revenue potential has caused states such as California, which has a history of imposing strict environmental standards, to refrain from imposing a moratorium on hydraulic fracturing.[90] Budgetary constraints are one reason the state of Illinois has passed the most comprehensive regulatory bill in the nation, satisfying the interests of both industry and environmental groups.[91]

Illinois' Hydraulic Fracturing Regulatory Act, adopted in June 2013, is one possible model for policymakers in other states.[92] The law establishes numerous technical criteria fracking site operators must satisfy, including a detailed permitting process where applicants must register, disclose their corporate pedigree, prove possession of $5 million in insurance, and disclose any "serious" previous violations. Drilling applications must include the location and depth of the well, angle of the well bore, lowest potential freshwater along the length of the entire well bore, and a detailed description of the geologic formations affected. The law also requires several plans to be submitted with the application: freshwater withdrawal and management plan; a plan for handling storage, transportation, and disposal or reuse of hydraulic fracturing fluid and hydraulic fracturing flowback; well site safety plan; containment plan; casing and cement plan; traffic management plan; and a work plan for water quality monitoring.[93]

The new law also establishes setbacks from schools, residences, and potable wells and prohibits the injection of benzene into freshwater. Applicants must fully disclose the chemicals and proppants to be used in fracking operations at least 21 days prior to the start of fracking operations, and

[85] Aaron Marshall, "New Wave of Injection Wells on the Way in Ohio for Fracking Waste," *The Plain Dealer*, November 23, 2012, http://www.cleveland.com/open/index.ssf/2012/11/new_wave_of_injection_wells_on.html

[86] Sabrina Eaton, "Federal Hydraulic Fracturing Rules Aren't Needed, Ohio Official Tells Congressional Committee" *The Plain Dealer*, April 17, 2013, http://www.cleveland.com/open/index.ssf/2013/04/federal_hydraulic_fracturing_r.html

[87] Richard Lazarus, "Congressional Descent: The Demise of Deliberative Democracy in Environmental Law," *The Georgetown Law Journal* 94 (Winter 2006), pp. 619–681, http://papers.ssrn.com/sol3/papers.cfm?abstract_id=847648

[88] FracFocus, "About Us," http://fracfocus.org/welcome

[89] Nicolas Loris, "Hydraulic Fracturing: Critical for Energy Production, Jobs, and Economic Growth," The Heritage Foundation, August 28, 2012, http://www.heritage.org/research/reports/2012/08/hydraulic-fracturing-critical-for-energy-production-jobs-and-economic-growth

[90] Taylor Smith, "Hydraulic Fracturing in California," *Research & Commentary*, The Heartland Institute, June 17, 2013, http://heartland.org/policy-documents/research-commentary-hydraulic-fracturing-california

[91] Law Offices of Carey S. Rosemarin, P.C., "Fracking in Illinois, The General Assembly Should Pass HB2615, and Reject the Proposed Moratorium," March 19, 2013, http://www.rosemarinlaw.com/Environmental-Law-Blog/2013/March/Fracking-in-Illinois-The-General-Assembly-Should.aspx

[92] Illinois General Assembly, Status of SB 1715, http://www.ilga.gov/legislation/BillStatus.asp?DocNum=1715&GAID=12&DocTypeID=SB&LegID=72606&SessionID=85&SpecSess=&Session=&GA=98

[93] *Ibid.*

the use of nondisclosed chemicals in the fracking fluid is prohibited. The bill also requires notification of the Illinois DNR 48 hours before fracking operations begin, and numerous tests must be satisfied before the event can take place. Air emissions must be minimized and fracking fluids must be stored in aboveground tanks until they are disposed of in class II injection wells.[94]

Fracking operations must immediately be suspended if monitoring data indicate the well has been compromised. The Illinois State Geological Survey will work to develop rules to mitigate induced seismicity from class II injection wells. The law also requires baseline data be taken in water sources within 1500 ft of the well site prior to fracking, with continued monitoring thereafter at 6, 18, and 30 months.

Illinois' new law shows state governments are quite able to promulgate rigorous rules that can satisfy the demands of industry and environmentalists alike, enabling the development of natural resources while protecting the environment. There are nonetheless some areas where federal regulation is more efficient than state laws.

The federal government regulates wastewater injection wells as class II wells under the Safe Drinking Water Act, and Underground Injection Control standards prevent the use of diesel fuel in fracking fluid and drilling mud. Natural gas and oil pipelines crossing state boundaries are regulated by the Federal Energy Regulatory Commission, and the Clean Water Act limits the discharge of pollutants into surface waters. Additionally, EPA recently enacted air quality standards, following the examples set by Wyoming and Colorado, requiring oil and gas drillers by 2015 to capture "volatile compounds" (methane, hydrocarbons, and additional particulates) from wells as they migrate up to the surface during the drilling process. In the meantime, drillers are required to flare, or burn, these gases as they flow out of wells. These factors are best regulated by the federal government because of their interstate nature, whereas additional federal regulation in intrastate matters would be unnecessary, costly, and duplicative.[95]

Illinois' monitoring standards will contribute to the knowledge of how hydraulic fracturing affects the environment. States that have imposed moratoria, by contrast, are not producing data or records of experience that will enable them to better understand the costs and benefits of fracking. New York's initial study, for example, concluded more study must be done, and the preliminary results have not produced the data that will be produced in Illinois.

At the time of this writing, three states have active moratoria on hydraulic fracturing: New York, North Carolina, and Vermont, even though there has never been a confirmed

case of groundwater contamination caused by hydraulic fracturing and the science available does not suggest that moratoria are necessary.[96,97]

Of the three states with moratoria, only New York, which sits on portions of the Marcellus and Utica Shales, has any notable gas reserves. The moratorium has prevented growth in the natural gas industry and the creation of thousands of jobs in New York. One economic analysis projects the incomes of those who live in the 28 New York counties above the Marcellus Shale could grow by as much as 15% over the next 4 years if the moratorium on hydraulic fracturing were lifted.[98]

Although regulation of oil and gas production is important, moratoria and duplicative federal laws are unnecessary and will serve only to create extra costs. As more states enact standards agreed upon by industry and environmental groups, better practices will be developed and the laboratory of the states will identify the most efficient means of extracting oil and gas while protecting the environment.

71.5 CONCLUSION

The combination of hydraulic fracturing and horizontal drilling is a technological breakthrough that has transformed the energy outlook of the United States, turning once uneconomic oil and gas deposits into "America's Shale Revolution." Once facing the prospect of supply shortages, the United States has become the largest producer of natural gas in the world, and the nation is projected to eclipse Saudi Arabia as the top petroleum producer by 2017.[99] These achievements would have been impossible if not for hydraulic fracturing.

The dramatic increase in production led to the creation of thousands of jobs in 2012, with 360,000 jobs supported directly by the oil and gas industries, 537,000 jobs in supply industries, and 850,000 induced jobs supported by oil and gas workers spending their paychecks in the general economy

[94] *Ibid.*
[95] Lee Lane, *supra* note 23.

[96] Senate Committee on Natural Resources and Water, Senate Committee on Environmental Quality, "The Regulation of Hydraulic Fracturing Oil and Gas Production in California," Joint Informational Hearing, February 12, 2013, http://sntr.senate.ca.gov/sites/sntr.senate.ca.gov/files/Hydraulic%20fracturing%20background.pdf
[97] Ben Wolfgang, "Methane Study, EPA Debunk Claims of Water Pollution, Climate Change from Fracking," *The Washington Times*, April 29, 2013, http://www.washingtontimes.com/news/2013/apr/29/pa-environment-agency-debunks-fracking-water-claim/?page=1
[98] Diana Furchtgott-Roth, "The Case for Fracking in New York," *MarketWatch: The Wall Street Journal*, May 9, 2013, http://www.marketwatch.com/story/the-case-for-fracking-in-new-york-2013-05-09
[99] Peg Mackey, "U.S. to Overtake Saudi Arabia as Top Oil Producer: IEA," *Reuters*, November 12, 2012, http://www.reuters.com/article/2012/11/12/us-iea-oil-report-idUSBRE8AB0IQ20121112

at grocery stores, dentist offices, movie theaters, and auto dealerships around the country.

Low energy costs have breathed new life into the American manufacturing sector, which is projected to add 1 million jobs by 2025, thanks to abundant and affordable oil and natural gas.

Despite the misleading theatrics seen in the movie *Gasland*, there has yet to be a confirmed case of hydraulic fracturing contaminating drinking water. There are consequences and risks associated with the production of unconventional oil and natural gas, but the costs are vastly outweighed by the benefits.

Hydraulic fracturing can be done in a safe and environmentally responsible manner. State governments are uniquely qualified to work with environmental and industry leaders to craft legislation that protects the environment while maintaining the vibrancy of the oil and gas industry without duplicative federal regulations that raise costs without significantly increasing environmental protection.

72

ZERO DISCHARGE WATER MANAGEMENT FOR HORIZONTAL SHALE GAS WELL DEVELOPMENT

Technology Status Assessment

WEST VIRGINIA WATER RESEARCH INSTITUTE (PUBLIC DOMAIN)
WEST VIRGINIA UNIVERSITY, MORGANTOWN, WV

72.1 CURRENT STATE OF TECHNOLOGY

72.1.1 Summary of Existing Industry

Hydrofracturing technology teamed with horizontal drilling has facilitated exploitation of huge gas reserves in the Devonian Marcellus Formation of the Appalachian Basin. Existing technology uses fresh water mixed with chemicals and sand injected under high pressure to fracture (frac) the formation. Disposal of the water returned to the surface is then either hauled to an underground injection control (UIC)-certified well for disposal, treated or processed through a municipal sewage treatment plant. Withdrawal or disposal may conflict with aquatic life, recreational, industrial, or domestic uses. Transportation, disposal, or treatment of the frac return water (FRW) can be very expensive depending on well location and water contaminants.

72.1.2 Technologies/Tools Being Used

Treatment at a municipal sewage treatment plant is usually preceded by alum addition to precipitate barium and strontium. However, the sewage treatment process does not address the high sodium and chloride content of the FRW and this practice is being reevaluated. This is a costly alternative that will be less widely used in the future as small towns and cities find that the high salt concentrations are detrimental to their digesters [1]. The number of UIC wells is extremely limited in the region and, consequently, transportation costs are exorbitant. A third option is evaporation/crystallization. One such unit, operated by AOP Clearwater in Fairmont, is near commissioning. Other units are under consideration. Tolls for this facility are reported to be around \$140/1000 gallons while reverse osmosis (RO) costs range from \$6 to \$10/1000 gallons. These are large, capital-intensive, and fixed facilities, so transportation to site is an added cost.

In addition, RO generates substantial volumes of reject disposal (up to 30%) that must be disposed.

A more cost-effective alternative is on-site treatment of the produced water to the degree needed for reuse as frac water. Recycling of produced water lowers transportation costs, environmental conflicts, and the risk of interruption in the well development schedule. In order to be feasible, the returned frac water must be treated to a level where dissolved and suspended solids will not cause scaling in the injection train or clogging of pore space in the formation.

72.1.3 Benefits and Inadequacies of Current Technology

On-site treatment of produced water has been successfully accomplished using RO as the primary treatment technology. However, extension of the RO technology to the treatment of flowback water from hydraulic fracture operations has required pretreatment technologies designed to protect, and thereby extend, the life of the RO unit.

A consortium of oil producers and service companies, led by Texas A&M University, has developed a unique RO

Alternative Energy and Shale Gas Encyclopedia, First Edition. Edited by Jay H. Lehr and Jack Keeley.
© 2016 John Wiley & Sons, Inc. Published 2016 by John Wiley & Sons, Inc.

process that is specifically adapted to oilfield wastewater purification, which was licensed to GeoPure Water Technologies. The RO pilot system was extensively tested in the laboratory and field, and results show that dissolved solid levels up to 50,000 mg/L can be reduced to the level of freshwater. Extension of this technology to treating frac water returns required the RO unit to be protected to prevent early plugging of the membranes. Under a DOE contract, Texas A&M is evaluating pretreatment options that include combinations of liquid–liquid centrifuges, organoclay absorbents, microfiltration, and different oil-resistant membrane materials and membrane types. GeoPure subsequently produced a 200-gpm unit incorporating some of the pretreatment options and used it successfully for about 9 months to treat Barnett Shale frac water of at least 13,400 mg/L total dissolved solids [2]. The unit is not currently in use. These pretreatment technologies add to the cost of the unit, which may discourage producers to adopt on-site treatment of Marcellus Shale flowback using this approach.

ProChemTech invented a sequential precipitation process for treatment of flowback water. The first precipitation removes suspended solids, iron, and barium from the flowback water as a solid, typically nonhazardous, sludge cake. The remaining scale formers (calcium, iron, magnesium, manganese, and strontium) are precipitated to make one solid sludge cake which could be used as an alkaline soil amendment or for abandoned mine spoil reclamation. The process does not materially change the concentration of dissolved solids in the FRW. ProChem concludes that the only known alternative to their process is the GE Thermal Evaporation process which uses a vertical-tube evaporator to convert industrial wastewaters into distilled water for reuse [3].

Other companies have identified water treatment systems for solving the FRW problems, all based on existing technologies. Some prominent examples include a partnership between Devin Energy and Fountain Quail Water Management LLC (mobile vapor distillation @ $3.35 net per bbl), Ecosphere Technologies Inc. (Ozonix™ mobile advanced membrane), and 212 Resources (thermal distillation and evaporation combined with polishing) [4].

72.2   DEVELOPMENT STRATEGIES

72.2.1   Why New Technology and Research Are Required

Research is needed to clean the FRW to a point where it can be economically recycled. State-of-the-art technologies have not yet proven to be economically attractive to the haul and injection option. Development of more efficient pretreatment and filtration technologies can solve these problems.

72.2.2   Problems Being Addressed in This Research Project

The FilterSure technology offers a new approach ideally suited to remove suspended solids associated with the return of the frac water to the surface while promising an order of magnitude improvement in operating efficiency with associated lower costs. The technology is expected to replace all of the alternative pretreatment steps and, in so doing, dramatically reduce the costs and enhance the attractiveness of water reuse. This technology has the potential to anchor a recycle process that will compete with haulage and injection well disposal.

The key to the research is to remove enough of certain problem-causing constituents of the frac flowback water such that they will not interfere with the subsequent frac job. This means that the frac water should have as few suspended solids as reasonably possible. Additionally, certain dissolved solids must be reduced to low concentrations to eliminate the potential for forming "scale" compounds. After removal of multivalent metallic ions, the remaining dissolved solids such as sodium chloride, calcium chloride, magnesium chloride, and other salts may also need to be reduced. Because these chlorides are highly soluble making them less likely to contribute to scale forming, and because they contribute to shale stability, moderate concentrations can be tolerated or even helpful. Water with total chloride content on the order of 25,000 ppm can readily be reused for a frac fluid base (personal communication with BJ Services' Research Group, Houston, TX, 713-462-4239). Higher chloride concentrations can be diluted with fresh makeup water to create an acceptable frac fluid base. If necessary, higher chloride concentrations may be removed by vacuum distillation, salt crystallization, electrocoagulation, or membrane filtration. The FilterSure technology is being tested with other high-potential technologies to develop a viable water cleanup approach.

72.3   FUTURE

72.3.1   Barriers To Be Overcome

The most difficult problem is to develop a cost-effective system to handle the broad spectrum of the suspended and dissolved materials preventing reuse of the FRW. These include heavy metals, natural hydrocarbons, water hardness, biological contaminants, and possibly radioactive contaminants.

FilterSure has already identified appropriate filter media for the system and is currently testing media sequence at the WVU laboratory located in Morgantown, WV. WVU is developing an effective complementary process for increasing the efficiency and effectiveness of the basic filter unit so that the effluent from the system can provide a high-quality fluid for reuse in fracturing nearby gas wells.

72.3.2   Impact on the US Domestic Gas Supply Industry

The successful development of the FilterSure technology for cleanup and reuse of FRW will both improve economics and resolve environmental issues. Improved economics will be achieved by reducing FRW trucking and disposal costs. By reusing the FRW for subsequent frac treatments, the need for new, fresh frac water for future wells will be reduced by 20–40%, depending on the amount of injected water that is returned after the frac. This will save costs by reduced freshwater hauling. Because the mobile unit will operate continuously with little or no need for an attendant, system operating costs will also be minimized.

Significant environmental benefits include the reduction in freshwater needed for future fractures, lowering the demand stress on local streams. Fewer trips with water trucks will cause less damage to local roads, create less fugitive dust emissions, create less engine exhaust, create less mud and muddy water, and create significant "good will" in the local community.

72.3.3   Deliverables (Tools, Methods, Instrumentation, Products)

Water treatment approaches have been identified and research is currently underway to evaluate each alternative for possible use in a mobile unit. Industry has provided water samples for testing with these samples covering the likely range of water that will need to be treated in field operations. Phase I will lead to a decision to proceed with the construction of a mobile unit. In phase II, a 30-gpm mobile system will be constructed for testing at an active field drilling site.

REFERENCES

[1] Pennsylvania Senate Environmental Resources and Energy Committee. (2010). Testimony on Marcellus Shale wastewater treatment issues, January 27, 2010.

[2] Geopure. (2007). News release/abstract, May 2007.

[3] Keister, T. (2008). Marcellus gas well water supply and wastewater disposal, treatment, and recycle technology. ProChem white paper, following USPTO Patent application 61/199,588, Process for Treatment of Gas Well Completion, Fracture, and Production Wastewaters for Recycle, Discharge, and Resource Recovery, filed November 19, 2008.

[4] *Permian Basin Oil & Gas Magazine.* (2008). The future of water recycling, issue no. 2, July 2008.

Prepared for: United States Department of Energy National Energy Technology Laboratory.
Acknowledgment: "This material is based upon work supported by the Department of Energy under Award Number DE-FE0001466."
Disclaimer: "This report was prepared as an account of work sponsored by an agency of the United States Government. Neither the United States Government nor any agency thereof, nor any of their employees, makes any warranty, express or implied, or assumes any legal liability or responsibility for the accuracy, completeness, or usefulness of any information, apparatus, product, or process disclosed, or represents that its use would not infringe privately owned rights. Reference herein to any specific commercial product, process, or service by trade name, trademark, manufacturer, or otherwise does not necessarily constitute or imply its endorsement, recommendation, or favoring by the United States Government or any agency thereof. The views and opinions of authors expressed herein do not necessarily state or reflect those of the United States Government or any agency thereof."

73

ABOUT OIL SHALE—WHAT IS OIL SHALE?

Basic information on oil shale, oil shale resources, and recovery of oil from oil shale.

73.1 WHAT IS OIL SHALE?

The term *oil shale* generally refers to any sedimentary rock that contains solid bituminous materials (called *kerogen*) that are released as petroleum-like liquids when the rock is heated in the chemical process of pyrolysis. Oil shale was formed millions of years ago by deposition of silt and organic debris on lake beds and sea bottoms. Over long periods of time, heat and pressure transformed the materials into oil shale in a process similar to the process that forms oil; however, the heat and pressure were not as great. Oil shale generally contains enough oil that it will burn without any additional processing, and it is known as "the rock that burns."

Oil shale can be mined and processed to generate oil similar to oil pumped from conventional oil wells; however, extracting oil from oil shale is more complex than conventional oil recovery and currently is more expensive. The oil substances in oil shale are solid and cannot be pumped directly out of the ground. The oil shale must first be mined and then heated to a high temperature (a process called *retorting*); the resultant liquid must then be separated and collected. An alternative but currently experimental process referred to as *in situ retorting* involves heating the oil shale while it is still underground, and then pumping the resulting liquid to the surface.

73.2 OIL SHALE RESOURCES

While oil shale is found in many places worldwide, by far the largest deposits in the world are found in the United States in the *Green River Formation*, which covers portions of Colorado, Utah, and Wyoming. Estimates of the oil resource in place within the Green River Formation range from 1.2 to 1.8 trillion barrels. Not all resources in place are recoverable; however, even a moderate estimate of *800 billion barrels* of recoverable oil from oil shale in the Green River Formation is three times greater than the proven oil reserves of Saudi Arabia. Present US demand for petroleum products is about 20 million barrels per day. If oil shale could be used to meet a quarter of that demand, the estimated 800 billion barrels of recoverable oil from the Green River Formation would last for more than 400 years [1].

More than 70% of the total oil shale acreage in the Green River Formation, including the richest and thickest oil shale deposits, is under *federally owned and managed lands*. Thus, the federal government directly controls access to the most commercially attractive portions of the oil shale resource base.

73.3 THE OIL SHALE INDUSTRY

While oil shale has been used as fuel and as a source of oil in small quantities for many years, few countries currently produce oil from oil shale on a significant commercial level. Many countries do not have significant oil shale resources, but in those countries that do have significant oil shale resources, the oil shale industry has not developed because, historically, the cost of oil derived from oil shale has been significantly higher than conventional pumped oil. The lack of commercial viability of oil shale-derived oil has in turn inhibited the development of better technologies that might reduce its cost.

Alternative Energy and Shale Gas Encyclopedia, First Edition. Edited by Jay H. Lehr and Jack Keeley.
© 2016 John Wiley & Sons, Inc. Published 2016 by John Wiley & Sons, Inc.

Relatively high prices for conventional oil in the 1970s and 1980s stimulated interest and some development of better oil shale technology, but oil prices eventually fell, and major research and development activities largely ceased. More recently, prices for crude oil have again risen to levels that may make oil shale-based oil production commercially viable, and both governments and industry are interested in pursuing the development of oil shale as an *alternative to conventional oil*.

73.4 OIL SHALE MINING AND PROCESSING

Oil shale can be mined using one of two methods: *underground mining* using the room-and-pillar method or *surface mining*. After mining, the oil shale is transported to a facility for retorting, a heating process that separates the oil fractions of oil shale from the mineral fraction. The vessel in which retorting takes place is known as a *retort*. After retorting, the oil must be upgraded by further processing before it can be sent to a refinery, and the spent shale must be disposed of. Spent shale may be disposed of in surface impoundments or as fill in graded areas; it may also be disposed of in previously mined areas. Eventually, the mined land is reclaimed. Both mining and processing of oil shale involve a variety of *environmental impacts*, such as global warming and greenhouse gas emissions, disturbance of mined land, disposal of spent shale, use of water resources, and impacts on air and water quality. The development of a commercial oil shale industry in the United States would also have significant *social and economic impacts* on local communities. Other impediments to development of the oil shale industry in the United States include the relatively high cost of producing oil from oil shale (currently greater than $60 per barrel) and the lack of regulations to lease oil shale.

73.5 SURFACE RETORTING

While current technologies are adequate for oil shale mining, the technology for surface retorting has not been successfully applied at a commercially viable level in the United States, although technical viability has been demonstrated. Further development and testing of surface retorting technology is needed before the method is likely to succeed on a commercial scale.

73.6 IN SITU RETORTING

Shell Oil is currently developing an *in situ conversion process (ICP)*. The process involves heating underground oil shale, using electric heaters placed in deep vertical holes drilled through a section of oil shale. The volume of oil shale is heated over a period of 2–3 years, until it reaches 650–700°F, at which point oil is released from the shale. The released product is gathered in collection wells positioned within the heated zone.

Shell's current plan involves using a ground-freezing technology to establish an underground barrier called a *"freeze wall"* around the perimeter of the extraction zone. The freeze wall is created by pumping refrigerated fluid through a series of wells drilled around the extraction zone. The freeze wall prevents groundwater from entering the extraction zone and keeps hydrocarbons and other products generated by the *in situ* retorting from leaving the project perimeter.

Shell's process is currently unproven at a commercial scale, but is regarded by the US Department of Energy as a very promising technology. Confirmation of the technical feasibility of the concept, however, hinges on the resolution of two major technical issues: controlling groundwater during production and preventing subsurface environmental problems, including groundwater impacts [1].

Both mining and processing of oil shale involve a variety of *environmental impacts*, such as global warming and greenhouse gas emissions, disturbance of mined land, impacts on wildlife and air and water quality. The development of a commercial oil shale industry in the United States would also have significant *social and economic impacts* on local communities. Of special concern in the relatively arid western United States is the large amount of water required for oil shale processing; currently, oil shale extraction and processing require *several barrels of water* for each barrel of oil produced, though some of the water can be recycled.

REFERENCES

[1] J. T. Bartis, T. LaTourrette, L. Dixon, D. J. Peterson, and G. Cecchine (2005). *Oil Shale Development in the United States: Prospects and Policy Issues*. MG-414-NETL, RAND Corporation, Santa Monica, CA.

74

NATURAL GAS BASICS—HOW WAS NATURAL GAS FORMED?

PUBLIC DOMAIN

The main ingredient in natural gas is methane, a gas (or compound) composed of one carbon atom and four hydrogen atoms. Millions of years ago, the remains of plants and animals (diatoms) decayed and built up in thick layers. This decayed matter from plants and animals is called organic material—it was once alive. Over time, the sand and silt changed to rock, covered the organic material, and trapped it beneath the rock. Pressure and heat changed some of this organic material into coal, some into oil (petroleum), and some into natural gas—tiny bubbles of odorless gas (Figure 74.1).

In some places, gas escapes from small gaps in the rocks into the air; then, if there is enough activation energy from lightning or a fire, it burns. When people first saw the flames, they experimented with them and learned they could use them for heat and light.

74.1 HOW DO WE GET NATURAL GAS?

The search for natural gas begins with geologists, who study the structure and processes of the Earth. They locate the types of rock that are likely to contain gas and oil deposits.

Today, geologists' tools include seismic surveys that are used to find the right places to drill wells. Seismic surveys use echoes from a vibration source at the Earth's surface (usually a vibrating pad under a truck built for this purpose) to collect information about the rocks beneath. Sometimes it is necessary to use small amounts of dynamite to provide the vibration that is needed.

Scientists and engineers explore a chosen area by studying rock samples from the earth and taking measurements. If the site seems promising, drilling begins. Some of these areas are on land but many are offshore, deep in the ocean. Once the gas is found, it flows up through the well to the surface of the ground and into large pipelines.

Some of the gases that are produced along with methane, such as butane and propane (also known as "by-products"), are separated and cleaned at a gas-processing plant. The by-products, once removed, are used in a number of ways. For example, propane can be used for cooking on gas grills.

Dry natural gas is also known as consumer-grade natural gas. In addition to natural gas production, the US gas supply is augmented by imports, withdrawals from storage, and supplemental gaseous fuels.

Most of the natural gas consumed in the United States is produced in the United States. Some is imported from Canada and shipped to the United States in pipelines. A small amount of natural gas is shipped to the United States as liquefied natural gas (LNG).

We can also use machines called "digesters" that turn today's organic material (plants, animal wastes, etc.) into natural gas. This process replaces waiting for millions of years for the gas to form naturally.

74.2 GETTING NATURAL GAS TO USERS

74.2.1 Natural Gas Is Often Stored Before It Is Delivered

Natural gas is moved by pipelines from the producing fields to consumers. Because natural gas demand is greater in the winter, it is stored along the way in large underground storage

Alternative Energy and Shale Gas Encyclopedia, First Edition. Edited by Jay H. Lehr and Jack Keeley.
© 2016 John Wiley & Sons, Inc. Published 2016 by John Wiley & Sons, Inc.

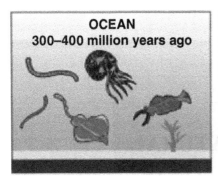
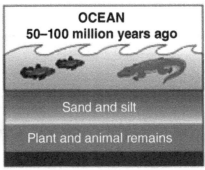
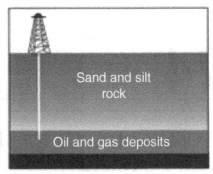

OCEAN
300–400 million years ago

OCEAN
50–100 million years ago

Sand and silt

Plant and animal remains

Sand and silt
rock

Oil and gas deposits

Tiny sea plants and animals died and were buried on the ocean floor. Over time, they were covered by layers of silt and sand.

Over millions of years, the remains were buried deeper and deeper. The enormous heat and pressure turned them into oil and gas.

Today, we drill down through layers of sand, silt, and rock to reach the rock formations that contain oil and gas deposits.

FIGURE 74.1 Petroleum and natural gas formation. *Source*: U.S. Energy Information Administration (public domain).

systems, such as old oil and gas wells or caverns formed in old salt beds. The gas remains there until it is added back into the pipeline when people begin to use more gas, such as in the winter to heat homes.

When the gas gets to the communities where it will be used (usually through large pipelines), it flows into smaller pipelines called "mains." Very small lines, called "services," connect to the mains and go directly to homes or buildings where it will be used (Figure 74.2).

74.2.2 Natural Gas Can Also Be Stored and Transported as a Liquid

When chilled to very cold temperatures, approximately –260°F, natural gas changes into a liquid and can be stored in this form. Because it takes up only 1/600th of the space that it would in its gaseous state, LNG can be loaded onto tankers (large ships with several domed tanks) and moved across the ocean to other countries. When this LNG is received in the United States, it can be shipped by truck to be held in large

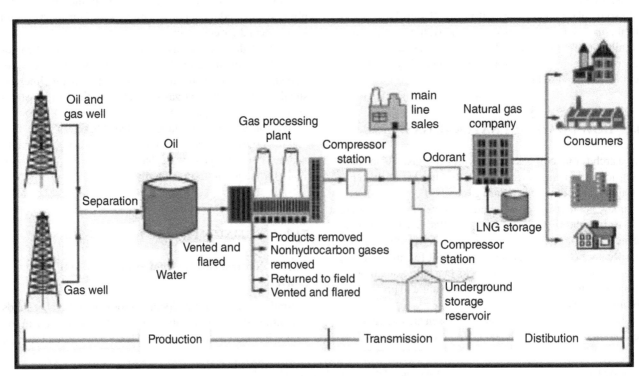

FIGURE 74.2 Natural gas industry.

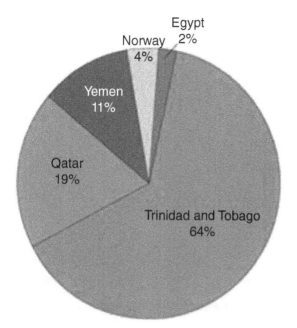

FIGURE 74.3 US liquefied natural gas imports by country (%), 2012. *Source*: U.S. Energy Information Administration, *Natural Gas Monthly*, Table 4 (March 2013).

chilled tanks close to users or turned back into gas when it is ready to put in the pipelines.

74.3 WHAT IS LIQUEFIED NATURAL GAS?

Trinidad and Tobago provide approximately two-thirds of our LNG imports (Figure 74.3).

LNG is natural gas that has been cooled to about −260°F for shipment and/or storage as a liquid. The volume of the liquid is about 600 times smaller than in its gaseous form. In this compact form, natural gas can be shipped in special tankers to receiving terminals in the United States and other importing countries. At these terminals, the LNG is returned to a gaseous form and transported by pipeline to distribution companies, industrial consumers, and power plants.

In 2012 the total US natural gas production was equal to 94% of the US natural gas consumption. Net natural gas imports contributed to the remaining consumption volume.[1]

In 2012, LNG imports contributed about 1% of total natural gas consumption. LNG imports from Egypt, Qatar, Trinidad and Tobago, and Yemen contributed about 96% of total LNG imports.

Even though the United States is primarily an importer of LNG, it is also an exporter. Sometimes LNG originally

imported to the United States is "re-exported" to new destinations where prices are higher.

Since LNG is more energy dense than gaseous natural gas (since natural gas is 600 times smaller in liquid form), there is increasing interest in using LNG in end-use applications, such as for heavy-duty vehicle fuel or other transportation (i.e., marine or rail) applications.

74.4 WHERE OUR NATURAL GAS COMES FROM

74.4.1 Most of the Natural Gas Consumed in the United States Comes from Domestic Production

The US natural gas production and consumption were nearly in balance through 1986. After that, consumption began to outpace production, and imports of natural gas rose to meet the US demand for fuel. Production increased from 2006 through 2011, when it reached the highest recorded annual total since 1973. The increases in production were the result of more efficient, cost-effective drilling techniques, notably in the production of natural gas from shale formations.

Share of 2011 natural gas marketed production (Figure 74.4):

- Texas (29%)
- Wyoming (9%)
- Federal Offshore Gulf of Mexico (8%)
- Louisiana (13%)
- Oklahoma (8%)

In 2011, 90% of net imports came by pipeline, primarily from Canada, and 10% came by LNG tankers carrying gas from five different countries.

74.4.2 Supplemental Gas Supplies

Supplemental gas supplies include blast furnace gas, refinery gas, propane–air mixtures, and synthetic natural gas (gas made from petroleum hydrocarbons or from coal). These supplemental supplies totaled 60 billion cubic feet (Bcf) in 2011. The largest single source of synthetic gas is the Great Plains Synfuels Plant in Beulah, North Dakota, where coal is converted to pipeline-quality gas (Figure 74.5).

74.4.3 Natural Gas Is Stored Underground

There were about 410 active underground storage fields (salt fields, aquifers, or depleted fields) in the United States during 2011. Natural gas is injected into these fields primarily during April through October and withdrawn primarily from November through March during the peak heating season. As of April 2012, the estimated capacity for peak working gas storage was 4239 Bcf.

---

[1] In 2012 the United States imported 3,135,346 MMcf of natural gas and exported 1,618,946 MMcf, resulting in net imports of 1,516,400 Mcf, or 6% of US 2012 natural gas consumption.

FIGURE 74.4 Top natural gas-producing states, 2011. *Source*: U.S. Energy Information Administration, *Natural Gas Monthly*, Table 7 (April 2012).

74.5 NATURAL GAS IMPORTS AND EXPORTS

The United States consumes more natural gas than it produces. While most of the natural gas consumed in the United States is produced domestically, imports from other countries are also an important source of supply. The United States is a "net importer" of natural gas, meaning that it imports more natural gas than it exports.

Reliance on natural gas imports has declined in recent years due to a surge in natural gas production, resulting from

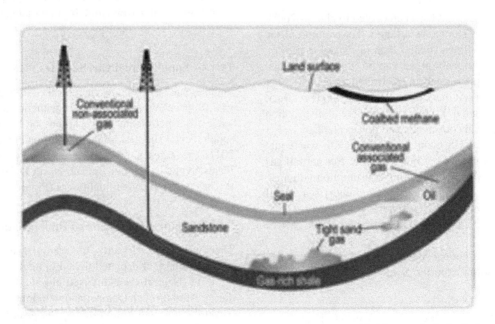

FIGURE 74.5 Schematic geology of natural gas resources. *Source*: Modified from U.S. Geological Survey Fact Sheet 0113-01 (public domain).

more efficient, cost-effective drilling techniques, notably in the production of natural gas from shale formations. Net imports (imports minus exports) of natural gas accounted for 6% of the US natural gas consumption in 2012, compared to the 2007 peak of 16%.

Natural gas can be transported in two ways:

1. Via pipeline—more than 210 natural gas pipeline systems
2. Via ship—in the form of LNG

In 2012, 94% of net imports came by pipeline, primarily from Canada, and 10% came by LNG tankers carrying gas from around the world.

74.5.1 Pipeline Imports of Natural Gas Are Mostly from Canada

In 2012, net pipeline imports totaled 1373 Bcf, or 5% of total natural gas consumption. The United States received almost 99% of its pipeline-imported natural gas from Canada with the remainder from Mexico.

74.5.2 The United States Imports a Small Amount of Liquefied Natural Gas

In 2012, LNG imports totaled 175 Bcf, or about 1% of total natural gas consumption. LNG imports from Egypt, Qatar, Trinidad and Tobago, and Yemen contributed about 96% of total LNG imports.

74.5.3 Most Natural Gas Exports Go to Canada and Mexico

Exports of natural gas peaked in 2012, largely due to expanded pipeline exports to Canada and Mexico. Canada accounted for 61% of pipeline natural gas exports, and Mexico accounted for 39%. The US exports of natural gas also include

- Domestically produced natural gas shipped to Japan as LNG
- LNG originally imported to the United States that is "re-exported" to new destinations where prices are higher

74.6 USES OF NATURAL GAS

74.6.1 Natural Gas Is a Major Energy Source for the United States

About 25% of energy used in the United States came from natural gas in 2012. The United States used 25.46 trillion cubic feet (Tcf) of natural gas in 2012.

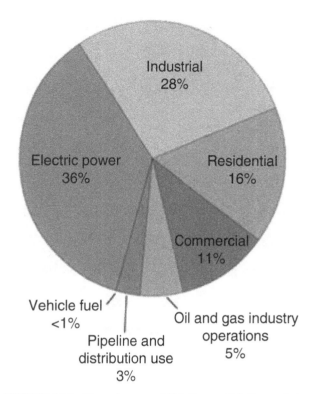

FIGURE 74.6   Natural gas use, 2012. *Source*: U.S. Energy Information Administration, *Natural Gas Monthly* (March 29, 2013).

74.6.2 How Natural Gas Is Used

Natural gas is used to produce steel, glass, paper, clothing, brick, electricity, and as an essential raw material for many common products. Some products that use natural gas as a raw material are paints, fertilizer, plastics, antifreeze, dyes, photographic film, medicines, and explosives (Figure 74.6).

Slightly more than half of the homes in the United States use natural gas as their main heating fuel. Natural gas is also used in homes to fuel stoves, water heaters, clothes dryers, and other household appliances.

The major consumers of natural gas in the United States in 2012 included

- Electric power sector: 9.1 Tcf
- Industrial sector: 7.1 Tcf
- Residential sector: 4.2 Tcf
- Commercial sector: 2.9 Tcf

74.6.3 Where Natural Gas Is Used

Natural gas is used throughout the United States, but the top natural gas-consuming states in 2011 were

- Texas
- California

FIGURE 74.7 Natural gas well drilling operation. *Source*: Bureau of Land Management (public domain).

- Louisiana
- Florida
- New York
- Illinois

74.7   NATURAL GAS AND THE ENVIRONMENT

Natural gas has many qualities that make it an efficient, relatively clean, and economical energy source. There are, however, environmental and safety issues with its production and use. Many of the areas that are now being explored and developed for natural gas production are relatively pristine and/or are wilderness areas, and development of these areas has large impacts on the area's environment, wildlife, and human populations (Figure 74.7).

74.7.1   Natural Gas Is a Relatively Clean Burning Fossil Fuel

Burning natural gas for energy results in much fewer emissions of nearly all types of air pollutants and carbon

dioxide (CO_2) per unit of heat produced than coal or refined petroleum products. About 117 pounds of carbon dioxide are produced per million Btu equivalent of natural gas compared to over 200 pounds of CO_2 per million Btu of coal and over 160 pounds per million Btu of distillate fuel oil. These clean burning properties have contributed to an increase in natural gas use for electricity generation and as a transportation fuel for fleet vehicles in the United States.

74.7.2   Natural Gas Is Mainly Methane, a Strong Greenhouse Gas

Natural gas is made up of mostly methane, which is a potent greenhouse gas. Some natural gas leaks into the atmosphere from oil and gas wells, storage tanks, pipelines, and processing plants. These leaks were the source of about 25% of the total US methane emissions, but only about 3% of the total US greenhouse gas emissions in 2011.[2] The oil and natural gas industry tries to prevent gas leaks, and where natural gas is produced but cannot be transported economically, it is "flared" or burned at well sites. This is considered to be safer and better than releasing methane into the atmosphere because CO_2 is not as potent a greenhouse gas as methane.

74.7.3   Natural Gas Exploration, Drilling, and Production Has Many Environmental Impacts

When geologists explore for natural gas deposits on land, they may have to disturb vegetation and soils with their vehicles. A gas well on land may require a road and clearing and leveling an area to make a drill pad. Well drilling activities produce air pollution and may disturb wildlife. Pipelines are needed to transport the gas from the wells, and this usually requires clearing land to bury the pipe. Natural gas production can also result in the production of large volumes of contaminated water. This water has to be properly handled, stored, and treated so that it does not pollute land and water.

While the natural gas that we use as a fuel is processed so that it is mainly methane, unprocessed gas from a well may contain many other compounds, including hydrogen sulfide, a very toxic gas. Natural gas with high concentrations of hydrogen sulfide is usually flared. Natural gas flaring produces CO_2, carbon monoxide, sulfur dioxide, nitrogen oxides, and many other compounds depending on the chemical composition of the natural gas and how well the gas burns in the flare. Natural gas wells and pipelines often have engines to run equipment and compressors, which produce additional air pollutants and noise.

---

[2]Based on carbon-dioxide equivalents; the most recent year for which the United States Environmental Protection Agency (EPA) has estimates.

74.7.4 Advances in Drilling and Production Technologies Have Positive and Negative Impacts

New drilling and gas recovery technologies have greatly reduced the amount of area that has to be disturbed to produce each cubic foot of natural gas. Horizontal and directional drilling techniques make it possible to produce more gas from a single well than in the past, so fewer wells are needed to develop a gas field. Hydraulic fracturing (commonly called "hydrofracking," or "fracking," or "fracing") of shale rock formations is opening up large reserves of gas that were previously too expensive to develop. Hydrofracking involves pumping liquids under high pressure into a well to fracture the rock and allow gas to escape from tiny pockets in the rock. However, there are some potential environmental concerns that are also associated with the production of shale gas.

- The fracturing of wells requires large amounts of water. In some areas of the country, significant use of water for shale gas production may affect the availability of water for other uses and can affect aquatic habitats.

- If mismanaged, hydraulic fracturing fluid—which may contain potentially hazardous chemicals—can be released by spills, leaks, faulty well construction, or other exposure pathways. Any such releases can contaminate surrounding areas.

- Hydrofracturing also produces large amounts of wastewater, which may contain dissolved chemicals and other contaminants that require treatment before disposal or reuse. Because of the quantities of water used and the complexities inherent in treating some of the wastewater components, treatment and disposal is an important and challenging issue.

- According to the United States Geological Survey, hydraulic fracturing "causes small earthquakes, but they are almost always too small to be a safety concern. In addition to natural gas, fracking fluids and formation waters are returned to the surface. These wastewaters are frequently disposed of by injection into deep wells. The injection of wastewater into the subsurface can cause earthquakes that are large enough to be felt and may cause damage."

- Natural gas may be released to the atmosphere during and after well drilling, the amounts of which are being investigated.

74.7.5 Strict Safety Regulations and Standards Are Required for Natural Gas

Because a natural gas leak can cause an explosion, there are very strict government regulations and industry standards in place to ensure the safe transportation, storing, distribution, and use of natural gas. Because natural gas has no odor, natural gas companies add a strong smelling substance called mercaptan to it so that people will know if there is a leak. If you have a natural gas stove, you may have smelled this "rotten egg" smell of natural gas when the pilot light has gone out.

75

NATURAL GAS—CHRONOLOGICAL DEVELOPMENT

PUBLIC DOMAIN

75.1 NATURAL GAS

Natural gas is mostly a mixture of methane, ethane, and propane, with methane making up 73% to 95% of the total. Often found when drilling for oil, natural gas was once considered mainly a bother. When there were no uses or markets to sell natural gas, it was simply flared (burned off) at the wellhead. Major flaring sites were sometimes the brightest areas visible in nighttime satellite images. Today, however, the gas is mostly injected back into the ground for later use and to encourage greater oil production.

200 B.C.	The Chinese used natural gas to make salt from salt water (brine) in gas-fired evaporators.
1626	French explorers discovered Native Americans burning gases that were seeping into and around Lake Erie.
1816	Natural gas was used in Baltimore to fuel street lamps. During the 19th century, natural gas was used in Europe and in North America as a lighting fuel. Most of the natural gas produced at that time was manufactured from coal and not extracted from the earth, as it is today.
1821	In Fredonia, New York, William Hart dug the first successful well that was intended to produce natural gas. Hart dug a 27-foot well to try and bring a larger flow of gas to the surface. Expanding on Hart's work, the Fredonia Gas Light Company was eventually formed, becoming the first American natural gas company.
1859	Edwin Drake drilled the first commercial well and hit oil and natural gas at 69 feet below the earth's surface. A 2-inch diameter pipeline was built, running $5\frac{1}{2}$ miles from the well to the village of Titusville, Pennsylvania. This milestone may be considered the beginning of the natural gas industry in America.
1885	Robert Bunsen invented what is now known as the Bunsen burner. The Bunsen burner produced a flame that could be safely used for cooking and heating by mixing the right proportion of natural gas and air. The invention of thermostatic devices allowed the flame's temperature to be adjusted and monitored.
1890s	Electricity began to replace natural gas for lighting purposes.
1891	One of the first lengthy pipelines was constructed, which was 120 miles long, and carried natural gas from wells in central Indiana to the city of Chicago. This early pipeline was not very efficient at transporting natural gas.
1925	The first all-welded pipeline, over 200 miles in length, was built—from Louisiana to Texas.
1937	Natural gas distributors began adding mercaptan, with its rotten-egg smell, to the otherwise odorless natural gas—so that leaks can be easily detected.
1906–1970	U.S. residential demand for natural gas grew fifty times.
1940s–1960s	The nation began a massive expansion of its pipeline network, which led to rapid growth of natural gas markets. During the 1950s and 1960s, thousands of miles of pipeline were constructed throughout the United States. Today, the U.S. interstate pipeline network, laid end-to-end, would stretch almost 12 times around the earth.
1959	*Methane Pioneer*, a converted cargo ship, was used to carry liquified natural gas between Lake Charles, Louisiana, and the United Kingdom.
1971	Gas well productivity peaked at 435 thousand cubic feet per well per day.
1973	U.S. natural gas production reached a record-high of 21.7 trillion cubic feet before starting a long period of decline.

(continued)

Alternative Energy and Shale Gas Encyclopedia, First Edition. Edited by Jay H. Lehr and Jack Keeley.
© 2016 John Wiley & Sons, Inc. Published 2016 by John Wiley & Sons, Inc.

1983	The cost of natural gas for residential users set a record high of $10.06 per thousand cubic feet (measured in constant 2004 dollars).
1986–present	Consumption of natural gas began to grow faster than production.
	Net imports, as a share of natural gas consumption, more than tripled. These imports nearly all came by pipeline from Canada. Small shipments were brought by tanker as liquefied natural gas from Algeria and, in recent years, from a few other countries.
	New drilling technology made offshore sites more important. Over the next 20 years, about one-fifth of all U.S. production came from offshore sites.
1990	The New York Mercantile Exchange (NYMEX) issued the first natural gas futures contract. A futures contract is an agreement today on the price of a commodity (or financial instrument) to be paid for and delivered in the future.
	The Clean Air Act Amendments required many changes to fossil fuels to make them pollute less. The use of these cleaner fuels was phased-in during the 1990s. Natural gas was promoted as cleaner burning fuel in power generation and transportation, increasing the use of natural gas.
1998	About 5.1 billion cubic feet of natural gas were reported as being used for vehicles.
2000	Natural gas consumption peaked at 23.3 trillion cubic feet.
2001	The share of natural gas coming from imports peaked at 16.2%.
2003	After years of decline, gas well productivity reached a record low at 124 thousand cubic feet per day. The average natural gas well produced only 29% as much as in 1971.
2004	Over one-fourth of U.S. production came from Texas.
2005	The record-setting hurricane season of 2005 caused massive damage to the U.S. natural gas and petroleum infrastructure. The Gulf of Mexico, one of the nation's largest sources of oil and gas production, was dealt a one-two punch by Hurricanes Katrina and Rita. Many Gulf of Mexico wells, terminals, processing plants, and pipelines went off-line.
	U.S. residential natural gas prices were the highest ever recorded in September, reaching $16.66 per thousand cubic feet.
2006	A record 31,687 natural gas wells were drilled.
2007	U.S. imports of liquefied natural gas (LNG) reached a record level of 771 billion cubic feet.
2010	On April 20, 2010, an explosion and fire occurred on the offshore drilling rig Deepwater Horizon, which had been drilling an exploratory well in the Gulf of Mexico. The accident killed 11 crewmembers and left oil leaking from the unfinished well into the ocean for months.
	On May 27, 2010, Secretary of the Interior Salazar announced a 6-month hold or "moratorium" on deepwater drilling.

Last Revised: June 2010.

Sources: U.S. Energy Information Administration, http://www.eia.gov/kids/energy.cfm?page=tl_naturalgas

76

ENERGY MINERAL DIVISION OF THE AMERICAN ASSOCIATION OF PETROLEUM GEOLOGISTS, SHALE GAS AND LIQUIDS COMMITTEE ANNUAL REPORT, FY 2014

EMD Shale Gas and Liquids Committee

NEIL S. FISHMAN, CHAIR

Hess Corporation, Houston, TX

Vice Chairs:

- Brian Cardott, (Vice Chair, Government), Oklahoma Geological Survey, Norman, OK
- Harris Cander (Vice Chair, Industry), BP, Houston, TX
- Sven Egenhoff, (Vice Chair, University), Colorado State University, Fort Collins, CO

Advisory Committee (in alphabetical order):

- Kent Bowker, Bowker Petroleum, The Woodlands, TX
- Ken Chew, IHS (retired), Perthsire, Scotland
- Thomas Chidsey, Utah Geological Survey, Salt Lake City, UT
- Russell Dubiel, U.S. Geological Survey, Denver, CO
- Catherine Enomoto, U.S. Geological Survey, Reston, VA
- William Harrison, Western Michigan University, Kalamazoo, MI
- Ursula Hammes, Bureau of Economic Geology, Austin, TX
- Shu Jiang, University of Utah, Salt Lake City, UT
- Margaret Keller, U.S. Geological Survey, Menlo Park, CA

- Julie LeFever, North Dakota Geological Survey, Grand Forks, ND
- Peng Li, Arkansas Geological Survey, Little Rock, AR
- Jock McCracken, Egret Consulting, Calgary, AB
- Stephen Nordeng, North Dakota Geological Survey, Grand Forks, ND
- Rich Nyahay, Nyahay Geosciences LLC, Cobleskill, NY
- Stephen Sonnenberg, Colorado School of Mines, Golden, CO
- Michael D. Vanden Berg, Utah Geological Survey, Salt Lake City, UT
- Rachel Walker, Countrymark Energy Resources, LLC, Indianapolis, IN

INTRODUCTION

This report contains information about specific shales across the United States, Canada, Europe, and China from which hydrocarbons are currently being produced or shales that are of interest for hydrocarbon exploitation. Given the intense interest in shales as "unconventional" hydrocarbon reservoirs, this report contains information available at the time of its compilation, and the reader is advised to use links provided herein to remain as up-to-date as possible. The price of

Alternative Energy and Shale Gas Encyclopedia, First Edition. Edited by Jay H. Lehr and Jack Keeley.
© 2016 John Wiley & Sons, Inc. Published 2016 by John Wiley & Sons, Inc.

natural gas, however, has affected gas production in many of the plays in the United States, which is clear from information provided in several of the sections below, including the Antrim, Barnett, Fayetteville, and Haynesville Shales. This report is organized so that the reader can examine contributions from members of the EMD Shale Gas and Liquids Committee on various shales in the United States (presented in alphabetical order by shale name or region), Canada (by province), Europe (by country), and China.

ANTRIM SHALE (DEVONIAN), MICHIGAN BASIN, USA

BY DR. WILLIAM B. HARRISON, III (WESTERN MICHIGAN UNIVERSITY)

The Michigan Basin Antrim Shale play is currently 27 years old, having begun the modern phase of development in 1987. The total number of producing wells drilled in the play through end of October 2013 is approximately 11,550 with about 9672 still online. Total cumulative gas production reached 3.216 trillion cubic feet (Tcf) by the end of October 2013. Michigan Antrim production is reported by project rather than by individual well or lease. Projects may be only a few wells or more than 70 wells. There were 768 separate projects at the end of October 2013. Cumulative production for first 10 months of 2013 was 84,525,378 million cubic feet (Mcf), which was a 1.45% decline from the same period in 2012.

There were 30 operators with production at the end of October 2013. There were 9672 wells online at the end of October 2013. There were 111 new wells drilled in 2009, only 58 in 2010, 13 drilled in 2012, and 5 new wells drilled in 2013. That is, a 48% decrease in wells drilled from 2009 to 2010, a continuing drop of 33% in 2011, a 78% drop in 2012, and 61% in 2013. Overall drilling activity in Michigan was down 2% in 2012 compared to 2011. Most of the production comes from a few operators. The top 10 operators produced 82.5% of the total Antrim gas in 2013.

Although some wells can initially produce up to 500 Mcf/day, generally wells settle at less than 100 Mcf/day. Play wide average production at the end of October 2013 was 28 Mcf/day per well. Many Michigan Antrim wells begin with high water production and begin to increase gas production as the water is pumped off. Water production generally continues throughout the project life, although it usually declines through time. Play wide gas to water production ratio reached almost 3 Mcf/BBL in 1998, in 2004 it was 2.21 Mcf/BBL, 1.56 Mcf/BBL in 2009, 1.57 Mcf/BBL in 2011, and 1.60 Mcf/BBL through October 2013. Play wide water ratios have begun to decrease relative to gas production as old wells are dewatered and very few new wells are being drilled.

CO_2 is also an issue in the produced Antrim gas that is mostly of biogenic origin. Most wells begin with very low amounts of CO_2 in the produced gas; however, the percentage of CO_2 increases through time. Some projects that have a long production history may now exceed 30% CO_2 in the produced gas. The play wide average was just over 12.5% CO_2 in 2012.

Wells are produced from depths as shallow as 350 feet (ft) to just over 3000 ft, although the vast majority of wells are completed from 1000 to 2500 ft deep. They are typically drilled with water and an attempt is made to keep the wells in a balanced or slightly under-balanced. Wells are fraced with water and sand. Some of them are fraced using nitrogen or foam.

Production and well data are available online at the Michigan Public Service Commission at http://www.cis.state.mi.us/mpsc/gas/prodrpts.htm

Various kinds of oil and gas information are also available at the Michigan Office of Geological Survey site at http://www.michigan.gov/deq/0,1607,7-135-3311_4111_4231—,00.html

Cores, samples, and other kinds of data are available at the Michigan Geological Repository for Research and Education at Western Michigan University (http://wst023.west.wmich.edu/MGRRE%20Website/mgrre.html).

Top 10 Operators, Antrim Shale:

Linn Operating, Inc.
Chevron Michigan LLC
Terra Energy Ltd
Breitburn Operating Limited Partnership
Ward Lake Energy
Muskegon Development Co.
Trendwell Energy Corp Jordan Development Co. LLC
Merit Energy Co.
Delta Oil Co. Inc.

Significant Trends – New drilling has almost ceased during 2011 through 2013 due to low gas prices. Production continues to decline as do the total number of active wells. Daily gas production per well declined by 1.45% in the first 10 months of 2013. However, daily water production per well decreased 5.67% in 2013 compared to the same period in 2012. The numbers of horizontal completions still represent less than 5% of total wells.

BAKKEN FORMATION (UPPER DEVONIAN–LOWER MISSISSIPPIAN), WILLISTON BASIN, USA

BY JULIE LEFEVER AND STEPHAN NORDENG (NORTH DAKOTA GEOLOGICAL SURVEY)

In 2008, the United States Geological Survey (USGS) used a standardized assessment regime that concluded that the

Bakken Petroleum System in the entire Williston Basin contains an undiscovered 3.65 BBbls of oil, 1.85 Tcf of natural gas, and 148 million barrels of natural gas liquids that are technically recoverable with current technologies (Pollastro et al., 2008). The North Dakota Department of Mineral Resources (Bohrer et al., 2008) estimates that, within the North Dakota portion of the Williston Basin, the Bakken Formation contains 2.3 BBbls of recoverable oil in place (OIP) and the underlying Three Forks Formation contains an additional 2.1 BBbls (Nordeng and Helms, 2010). A reassessment of the Bakken Petroleum System is currently underway by the USGS with an expected publication date September 2013, although an update is expected to be provided at the AAPG Annual Meeting in Pittsburgh.

Development of Elm Coulee Field in 1996 resulted from the first significant oil production from the middle member of the Bakken Formation. Production from the middle member was established in the Kelly/Prospector #2-33 Albin FLB following an unsuccessful test of the deeper Birdbear (Nisku) Formation. Subsequent porosity mapping outlined a northwest-southeast trending stratigraphic interval containing an unusually thick dolomitized carbonate shoal complex within the middle member. Horizontal wells drilled through this shoal complex in 2000 resulted in the discovery of the giant Elm Coulee Field in eastern Montana. As with the previous Bakken producing fields, production at Elm Coulee depends on fracturing but in this case the productive fractures are found in the middle member of the formation. Since its discovery, more than 600 horizontal wells have been drilled in the 450-square-mile field from which more than 94 MMBbls of oil have been recovered. The productive portions of the reservoir contain between 3% and 9% porosity with an average permeability of md (millidarcies). A pressure gradient in the Bakken of 0.53 psi/ft indicates that the reservoir is overpressured. Laterals are routinely stimulated by a variety of sand-, gel-, and water-fracturing methods. Initial production (IP) from these wells is between 200 and 1900 barrels of oil/day (BOPD) (Sonnenberg and Pramudito, 2009).

The Bakken middle member play moved across the line into North Dakota when Michael Johnson noted that wireline logs of the Bakken Formation along the eastern limb of the Williston Basin in Mountrail County, North Dakota resembled those from Elm Coulee. Even though the kerogen within the Bakken shales appeared immature and thus might not be generating oil, free oil in DSTs (drill stem tests) and some minor Bakken production encouraged Johnson to pursue a Bakken play in Mountrail County (Durham, 2009). In 2005, EOG Resources demonstrated with the #1-24H Nelson-Farms (SESE Sec. 24, T156N, R92W) that horizontal drilling coupled with large-scale hydraulic fracture stimulation of the middle Bakken Formation could successfully tap significant oil reserves along the eastern flank of the Williston Basin. In the following year, EOG Resources drilled the #1-36 Parshall and #2-36 Parshall that resulted in wells with initial produc-

tion rates in excess of 500 BOPD. Well stimulation of the early wells typically involved large single-stage fracs using over 2 million pounds of proppant and over a million gallons of water. More recently, single-stage fracture stimulations have been replaced by multistage stimulations. These fracture the lateral with similar amounts of fluid and proppant but that is distributed over 10 to 40 or more separate stages. In a few instances, different laterals in the same well as well as laterals in adjacent wells are stimulated at the same time. Whiting Oil and Gas Corporation has installed a microseismic array in the Sanish Field in order to better visualize the real-time generation of induced fractures during stimulation process.

Subsequent horizontal drilling coupled with staged fracture stimulation has resulted in wells with IPs in excess of 2000 BOPD. The Parshall field is currently averaging 779 MBbls of oil/month from 233 wells. Sanish Field, next to Parshall, is averaging 1.5 MMBbls of oil/month from 331 wells.

Over 354.5 million bbls of oil have been recovered from the 3157 wells in the 240 middle Bakken producing fields put into service since 2004. The 882 horizontal wells drilled into the Three Forks Formation since 2006 have produced a total of 71.5 million bbls of oil. Currently there are 173 fields with Three Forks production. Sixty-eight wells have been completed in both the Bakken and Three Forks Formations. The majority of these wells were drilled in 2010.

The increase in information from recent drilling has resulted in the definition of a new member of the Bakken Formation called the Pronghorn. Additionally, to conform to adjoining states and provinces the original members have been formalized. New standard subsurface reference sections have also been designated. The formation now consists of four members, including Upper, Middle, Lower, and Pronghorn.

As the play moves into the production phase, multiple wells are now drilled from single pads with a closed mud system to minimize the footprint. Also, there has been an increase in the number of acquisitions. The latest is the purchase of Denbury's Bakken holdings by Exxon-Mobil.

The North Dakota portion of the Williston Basin is extremely active with 186 rigs running. The top 10 producers in the play are

1. EOG Resources (405 wells; up from 267)
2. Whiting Oil and Gas Corporation (360 wells; up from 161 wells)
3. Hess Corporation (473 wells; up from 172 wells)
4. Continental Resources, Inc. (480 wells; up from 210 wells)
5. Marathon Oil Company (280 wells; up from 177 wells)
6. Slawson Exploration Company, Inc. (139 wells; up from 73 wells)
7. Brigham Oil & Gas, L.P. (194 wells)

8. Burlington Resources Oil & Gas Company, LP. (165 wells; up from 92 wells)
9. XTO Energy Inc. (112 wells; up from 83 wells)
10. Petro-Hunt, LLC (188 wells)

Additional Information

The Bakken Source System was the focus of the past year's Williston Basin Petroleum Conference. The materials presented are available at the following link:

http://ndoil.org/?id=279&page=2012+WBPC+ Presentations

North Dakota Geological Survey Website:

https://www.dmr.nd.gov/ndgs/bakken/bakkenthree.asp

Recent Publications

First 60–90 Day Average Bakken Horizontal Production by Well, North Dakota Geological Survey, Geological Investigations 149.

LeFever, J.A., Nordeng, S.H., and Nesheim T., 2012, Core Workshop Booklet, North Dakota Geological Survey, Geological Investigations 155.

LeFever, J.A., and Nordeng, S.H., 2012, Preliminary Report on the Bakken Formation, North-Central ND, North Dakota Geological Survey, Geological Investigations 156.

Nordeng, S.H., and Helms, L.D., 2010, Bakken Source System: Three Forks Formation Assessment, https://www.dmr.nd.gov/ndgs/bakken/bakkenthree.asp

BARNETT SHALE (MISSISSIPPIAN), FORT WORTH BASIN, USA

BY KENT BOWKER (BOWKER PETROLEUM, LLC)

Daily gas production from the primary Barnett field (Newark, East) continues to decline but at a rate less than might have been expected given the decrease in drilling and completion activity over the past 5 years. The current daily gas production is right at 5 billion cubic feet (Bcf) while oil/condensate production is at 9300 bbls.

The total producing-well count in Newark, East field is 17,494; roughly 10,000 of those wells were drilled in the past 5 years. There are approximately 19,200 total Barnett wells in the Fort Worth basin (in several fields) and 14,400 of those are horizontal producers. The number of permits issued for the field has dropped to a rate of 700/year, well below the 4065 permits that were issued in 2008.

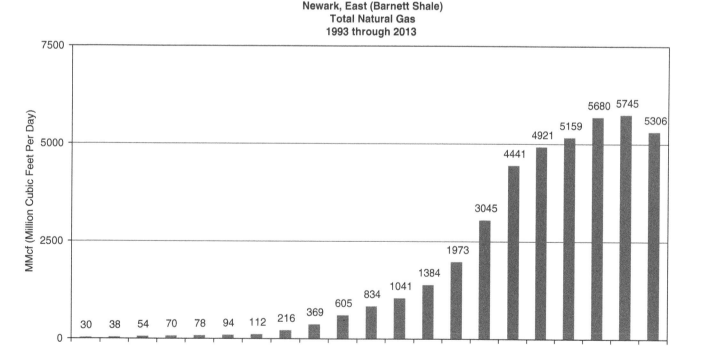

http://www.rrc.state.tx.us/barnettshale/NewarkEastField_1993-2013.pdf

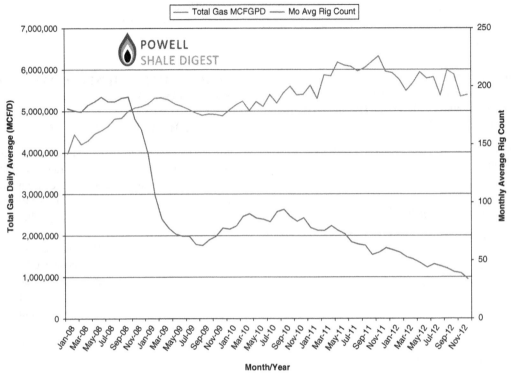

Barnett Shale Daily Avg. Production & Rig Count Jan. 2008 - Dec. 2013

Sources: Railroad Commission of Texas, RigData

Powell Shale Digest, Mar. 18, 2014

(used with permission of Gene Powell)

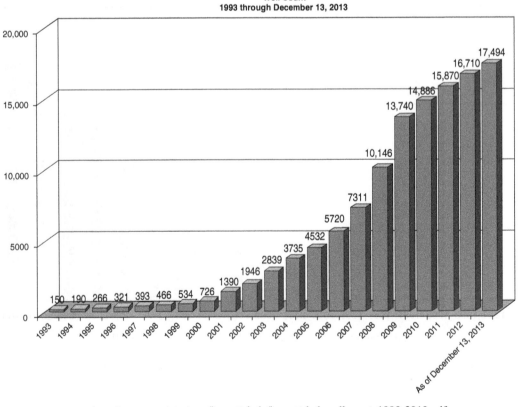

Newark, East (Barnett Shale)
Well Count
1993 through December 13, 2013

http://www.rrc.state.tx.us/barnettshale/barnettshalewellcount_1993-2013.pdf

Barnett Shale Annual Daily Average to January 1, 2014[a]

Year	Total gas per day MCFGPD	Total condensate/oil per day BOPD	Drilling rig count average
2003	861,518	5698	47
2004	1,080,271	6123	61
2005	1,430,783	6518	101
2006	2,039,992	8193	142
2007	3,144,209	8193	175
2008	4,617,514	9565	183
2009	5,088,198	9255	81
2010	5,271,068	11,555	86
2011	5,958,680	33,002	69
2012	5,702,415	25,492	48
2013	5,224,193	21,287	31

[a]Used with permission of Gene Powell.

Abbreviations: MCFGPD, million cubic feet gas/day; BOPD, barrels of oil/day.

The current rig count in the Barnett is 26 in 14 counties, which is a drop of 10 from the count 6 months ago. The bulk of the rigs operating in the Barnett are in the more liquid-rich portions of the play (Montague, Denton, and northern Wise counties), as noted in the Powell Shale Digest (www.shaledigest.com). Devon continues to be the most active operator in the play.

Over the past several months there has been a growing concern about the increase in earthquakes near the small town of Azle, northwest of Fort Worth. The earthquakes have all been less than magnitude 4, but several have been felt by residents in the area. There is a possibility that Class II (oil-field) waste-water injection wells could be the cause of these temblors. In response to these events, the Texas Railroad Commission has begun the process to hire a staff seismologist to help evaluate the possible seismic hazards of Class II injection wells (http://www.rrc.state.tx.us/pressreleases/2014/010714.php), which the Commission regulates. Southern Methodist University, in conjunction with the USGS (http://www.usgs.gov/blogs/features/usgs_top_story/man-made-earthquakes), has also recently announced a scientific investigation into the earthquakes around Azle (http://smu.edu/smunews/earthquakestudy/). A detailed map showing the epicenters of the earthquakes and the locations of Class II injection wells in the area can be found at http://www.shaledigest.com/documents/2014/earthquakes/AZLE_PP_MAPS_012114.pdf As seen in the map, there is some uncertainty as to the actual epicenters and foci of the earthquakes; these uncertainties will be addressed as part of the SMU study. The SMU researchers have cautioned that it could take years to determine the cause of the earthquakes. Interestingly, the USGS has reported only one earthquake in the area (M 2.5) since January 1.

EAGLE FORD SHALE AND TUSCALOOSA MARINE SHALE (CRETACEOUS), GULF COAST BASIN, USA

By Russell Dubiel (US Geological Survey)

The Cretaceous (Cenomanian–Turonian) Eagle Ford Shale of southwest Texas continues to be an important play producing thermogenic gas, oil, and condensate. The Eagle Ford play trends across Texas from the area of the Maverick Basin, northeast into the Karnes Trough, where it is variably a target for dry gas, wet gas/condensate, or oil (Figure 76.1). Completed wells display a steady decline in production similar to those in other shale plays. Recently drilled shale oil wells have shown initial production rates of several hundred to as much as 1000 BOPD. As of July 2014, there were more than 6400 oil wells and more than 3000 gas wells in the Eagle Ford (Texas Railroad Commission http://www.rrc.state.tx.us/eagleford/). The trend occurs at an average depth of 11,000 ft, and it is overpressured.

Similar to the Barnett Shale and Haynesville Formation, the Eagle Ford is a viable target for hydrocarbon exploitation because of advances in the application of horizontal drilling and hydraulic fracturing. Mineralogy of the Eagle Ford is somewhat different from other gas shales, however, in that where it is being explored, the Eagle Ford contains significant marlstone beds that are brittle and enhance the opportunity for induced fractures. Most operators are drilling horizontal well laterals of 3500 to 5000 ft and are stimulating the wells with slick water or acid in at least 10 different fracture stages. For more information on Eagle Ford production, please refer to the Texas Railroad Commission web link at http://www.rrc.state.tx.us/eagleford/.

Activity and success in the Eagle Ford in Texas has generated renewed interest in the laterally equivalent Cenomanian–Turonian marine shale of the Tuscaloosa Formation in the eastern Louisiana and southern Mississippi. Initial exploration in the Tuscaloosa marine shale in the 1970s has been followed by minimal exploration and production in the 1980s, 1990s, and early 2000s. Since 2010, several companies have begun significant leasing in the eastern Louisiana and southern Mississippi. Over the last 4 years, those companies have begun exploration and initial development drilling for the Tuscaloosa marine shale. This activity is based in part on the historical record of hydrocarbon generation and proven, but minimal, production from the unit, the current high price for oil, corresponding low price for natural gas, and the recent success of horizontal drilling in the Eagle Ford in Texas. The Tuscaloosa marine shale trend averages about 12,000 to 15,000 ft in depth and is overpressured. Since 2010, several companies have drilled successful horizontal wells, with about 42 wells currently producing in eastern Louisiana and southern Mississippi (http://dnr.louisiana.gov/; http://www.sonris.com/).

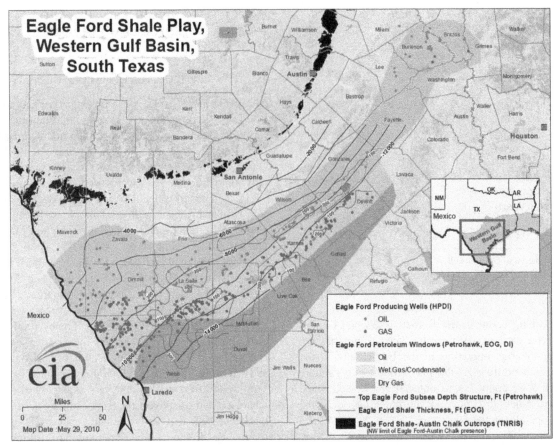

FIGURE 76.1   Map showing extent and hydrocarbon windows of the Eagle Ford Shale (from EIA, 2010).

This production trend is comparable to the approximately 42 wells that were current in the Eagle Ford in early 2009 (Texas Railroad Commission http://www.rrc.state.tx.us/eagleford/). Reported IPs are encouraging, in the neighborhood of several hundred BOPD, but currently only minimal yearly production data are available to evaluate the play's future success.

REFERENCES

EIA. (2010) Eagle Ford Shale Play, Western Gulf Basin, South Texas, http://www.eia.gov/oil_gas/rpd/shaleusa9.pdf, accessed September 4, 2014.

Texas Railroad Commission, (2014) http://www.rrc.state.tx.us/eagleford/, accessed August 10, 2014.

Louisiana Department of Natural Resources. (2014) SONRIS Database Access, http://dnr.louisiana.gov/; http://www.sonris.com/, accessed August 2014.

FAYETTEVILLE SHALE (MISSISSIPPIAN), ARKOMA BASIN, USA

By Peng Li (Arkansas Geological Survey)

The Upper Mississippian Fayetteville Shale play is the current focus of a regional shale-gas exploration and development program within the central and eastern Arkoma Basin of Arkansas. Approximately 2.5 million acres have been leased in the Fayetteville Shale gas play (Figure 76.2). Production of thermogenic gas from the Fayetteville began in 2004 and continues to the present.

U.S. Energy Information Administration (EIA) reports in 2011 that the Fayetteville contains Tcf of technically recoverable gas resource, in which 27.32 Tcf is attributable to the core producing area (aka eastern area) and 4.64 Tcf for the uncore producing area (aka western area). The EIA also reports that the proved gas reserves of the Fayetteville Shale in 2011 is 14.8 Tcf, 2.3 Tcf of increase from the 2010 estimates. Estimated ultimate recovery (EUR) for a typical horizontal Fayetteville gas well increased from 1.8 Bcf in 2008 to 3.2 Bcf in 2011 (OGJ, 2012).

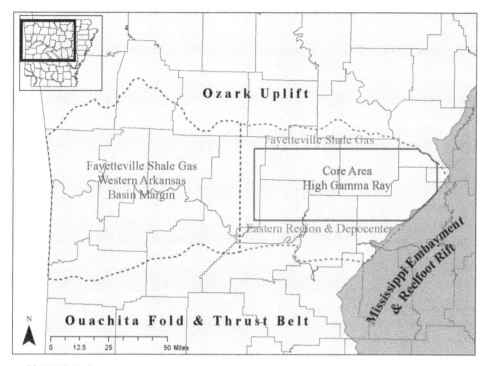

FIGURE 76.2   Primary area of Fayetteville Shale exploration and development in Arkansas.

Estimated cumulative production of gas from the Fayetteville Shale as of November 2013 has totaled 4,569,552,339 Mcf from 5060 wells. For the period of January through November 2013, the Fayetteville Shale zone has yielded 942,951,674 Mcf dry gas from 4843 wells. Annual production of Fayetteville Shale for 2012 is 1,027,711,866 Mcf from 4434 wells. The daily production has amounted to 2.6 Bcf in 2012. Initial production rates of horizontal wells have recently averaged about 3136 Mcf/day. Notably, even in the face of the challenging gas price environment in 2011 and 2012, production from the Fayetteville continued to grow at a fast clip, increasing by over 21% since January 2011. For more Fayetteville Shale production information, please refer to the Arkansas Oil and Gas Commission (AOGC) web link at http://www.aogc.state.ar.us/Fayprodinfo.htm

Like other dry gas plays, the Fayetteville has seen a dramatic decline in its rig count. According to Baker Hughes (BHI), the number of gas-directed rigs active in the play has dropped from 33 rigs in February 2011 to just 9 rigs in February 2014. The continued production growth, in spite of the sharply lower rig count, is explained by the truly remarkable gains in rig productivity and operating efficiencies as the transition towards the full development mode in many areas is beginning to bear fruit. Since 2013, Southwestern Energy has drilled its average well in just 6.5 days, re-entry to re-entry, compared to 11 days in 2010. The comparison is even more impressive given that the average length of the lateral is expected to increase by over 10%.

Fayetteville Shale reports from the AOGC have noted well increases from 24 in 2004, 33 in 2005, 129 in 2006, 428 in 2007, 587 in 2008, 839 in 2009, and 874 in 2010. Since then the numbers of new completed wells declined in three consecutive years, with 829 in 2011, 675 in 2012, and 557 in 2013. As of February 2014, there are a total of 5087 producing gas wells in the Fayetteville Shale play. Most Fayetteville Shale wells are drilled horizontally and have been fracture-stimulated using slickwater or cross-linked gel fluids. Baker Hughes' FracPoint Multi-stage fracturing system has provided most of the hydraulic fracturing completions in the Fayetteville Shale. Completed lateral lengths have increased 82% over the last 4 years while holding total well costs flat at about $2.9 million. Horizontal wells drilled from 2010 to 2012 averaged 5600 ft in lateral length with some wells up to 8000 ft. Fayetteville Shale gas production generally ranges over a vertical depth between 1500 and 6500 ft. The thickness of Fayetteville Shale varies from 50 ft in the western portion of the Arkoma Basin of Arkansas (fairway area) to 550 ft in the central and eastern regions (primary producing area).

Since the play's inception, the Fayetteville Shale play has been dominated by a small number of large players. Three operators – Southwestern Energy, BHP Billiton, and XTO Energy (a subsidiary of Exxon Mobil) – accounts for over 99% of gross operated production from the field. The three companies hold close to 2 million net acres under lease in the play. Southwestern, with 925,000 net acres and more than 3000 producing wells, is by far the largest operator among the three companies and accounts for about two-thirds of the field's total production volume. Exxon and BHP are approximately equal in terms of their acreage and gross operated production. During 2012, Southwestern contributed 724.8 Bcf in Fayetteville gas sales, good for 70.3% of the play's total sales that year. The BHP sold 155.7 Bcf (15.1%) and XTO Energy sold 147.9 Bcf (14.3%). The remaining 0.3% of sales, or 2.4 Bcf, was spread out among nine companies.

The top three operators of the Fayetteville gas shale play as of February 2014 based on numbers of producing wells are as follows (Figure 76.3):

1. SEECO Inc. (an exploration subsidiary of Southwestern Energy) (3264 wells)
2. BHP Billiton Petroleum (965 wells)
3. XTO Energy, Inc. (a subsidiary of ExxonMobil) (817 wells)

Two different maps are available that illustrate the location and types of wells located in the Fayetteville Shale producing area. Web links for the Fayetteville Shale maps and the associated federal and state agencies are listed below:

1. The home page of the Arkansas Geological Survey (AGS) website is http://www.geology.arkansas.gov/ home/index. htm and the AGS Fayetteville Shale

well location maps can be viewed at http://www. geology.arkansas.gov/home/fayetteville_play.htm. The AGS updates these maps and associated well databases online (in Excel® format) approximately every 2 weeks.
2. The home page of the U.S. Energy Information Administration (EIA) website is http://www.eia.doe.gov/ and the EIA Fayetteville Shale map is available at http:// www.eia.doe.gov/pub/oil_gas/natural_gas/analysis_ publications/maps/maps.htm

The AGS has completed two extensive geochemical research projects on the Fayetteville Shale and has provided this information to the oil and gas industry and the public to assist with exploration and development projects. These studies are available at the Arkansas Geological Survey as Information Circular 37 (Ratchford et. al., 2006) and Information Circular 40 (Li et al., 2010) and integrate surface and subsurface geologic information with organic geochemistry and thermal maturity data.

The AGS continues to partner with the petroleum industry to pursue additional Fayetteville Shale-related research. Ongoing AGS research is focused on the chemistry and isotopic character of produced gases, mineralogy of the reservoir, and outcrop to basin modeling.

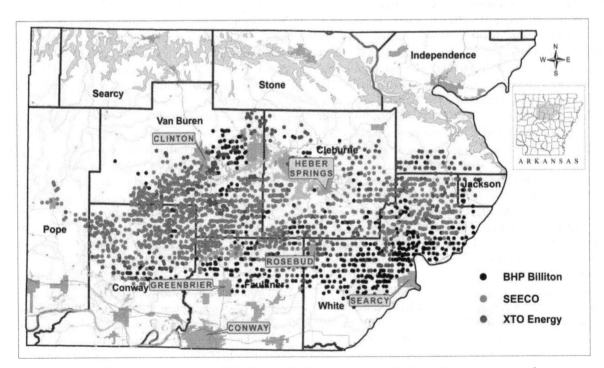

FIGURE 76.3   Location map of the Fayetteville Shale producing wells by top three operators as of February 2014.

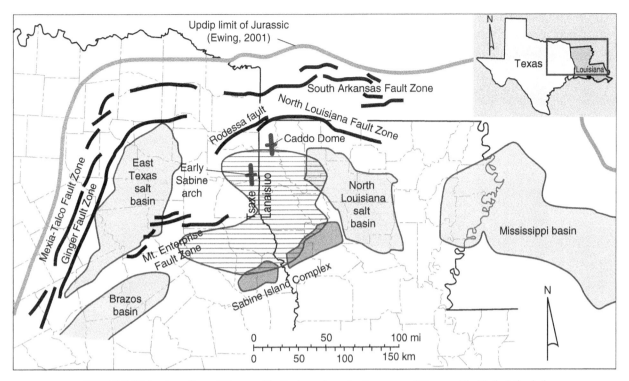

FIGURE 76.4 Location of Haynesville Basin and productive zone of Haynesville Shale (stippled pattern; from Hammes et al., 2011).

HAYNESVILLE/BOSSIER SHALE (JURASSIC), TEXAS AND LOUISIANA, USA

By Ursula Hammes (Bureau of Economic Geology)

The Kimmeridgian Haynesville Shale spans more than 16 counties/parishes along the boundary of eastern Texas and western Louisiana. Basement structures and salt movement influenced carbonate and siliciclastic sedimentation associated with the opening of the Gulf of Mexico forming the Haynesville basin. The Haynesville shale is an organic- and carbonate-rich mudrock that was deposited in a deep, partly euxinic and anoxic basin during Kimmeridgian to early Tithonian time, related to a second-order transgression that deposited organic-rich black shales worldwide. The Haynesville basin was surrounded by carbonate shelves of the Smackover and Haynesville lime Louark sequence in the north and west. Several rivers supplied sand and mud from the northwest, north, and northeast into the basin. Haynesville mudrocks contain a spectrum of facies ranging from more calcareous in the southern part of the productive area to more siliceous and argillaceous rich in the northern and eastern part of the productive area (Figure 76.4; Hammes et al., 2011). Haynesville reservoirs are characterized by overpressuring, porosity averaging 8–12%, Sw (water saturation) of 20–30%, nano-darcy permeabilities, reservoir thickness of 200–300 ft (70–100 m), and initial production ranging from 3

March 14, 2014	All Hayne sville/MBS Wells		
Operator	**LA**	**TX**	**Total**
Chesapeake	8	0	8
Anadarko	2	3	5
Exco Resources	3	2	5
Petrohawk	3	0	3
XTO Energy	0	3	3
J-W Operating	2	0	2
SWEPI LP	1	0	1
Total	**19**	**8**	**27**

Parish/County	**14-Mar**	**7-Mar**	**28-Feb**
DeSoto, LA	10	11	12
Panola, TX	4	4	5
Bossier, LA	3	3	2
Caddo, LA	3	1	1
Sabine, LA	2	2	1
San Augustine, TX	2	2	2
Shelby, TX	2	2	2
Red River, LA	1	1	1
Total	**27**	**26**	**26**

FIGURE 76.5 Rig counts for March 2014 (www.haynesvilleplay.com; accessed March 20, 2014)

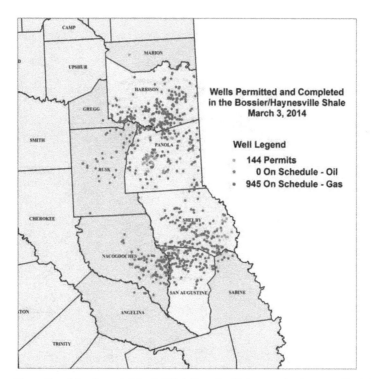

FIGURE 76.6 Texas Haynesville well activity through March 2014. Permits (light gray dots), gas well (dark gray dots)

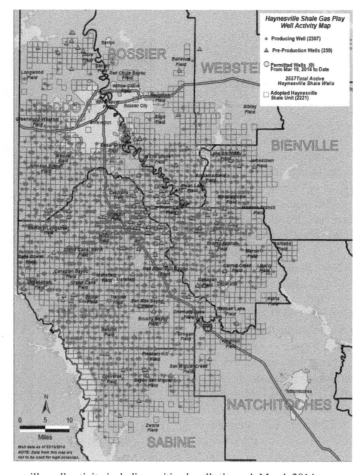

FIGURE 76.7 Louisiana Haynesville well activity including unitized wells through March 2014.

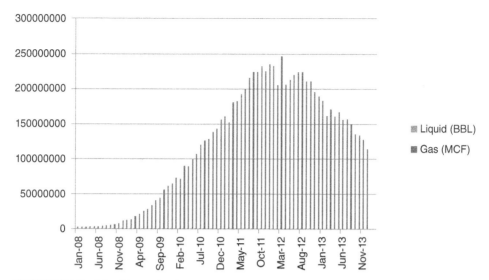

FIGURE 76.8   Monthly production chart (MCF) through December 2013 for Haynesville Shale (data from IHS Enerdeq).

to 30 MMCFD (Wang and Hammes, 2010). Reservoir depth ranges from 9000 to 14,000 ft (3000–4700 m), and lateral drilling distances are 3000–5000 ft (1000–1700 m). Optimal frac stages are 12–16 with an optimum number of 15 with multiple perforation clusters (Wang et al., 2013).

Haynesville drilling permits and rig count showed an increase in Texas over Louisiana (Figures 76.5–76.7). The Haynesville Shale in Texas and Louisiana experienced a slight revival during the last quarter because of higher prices for natural gas and more electricity and export demand

in Louisiana. However, production fell to half from what was produced at its peak in 2011 (Figure 76.8). Cumulative production reached 8.6 Tcf at the end of 2013 (Figure 76.9). Liquids production is still increasing due to additional exploitation in Arkansas and parts of northern Louisiana (Figure 76.10). Additional information on the Haynesville can be found at the Louisiana Oil & Gas association http://www.loga.la/haynesville-shale-news/, accessed March 21, 2014.

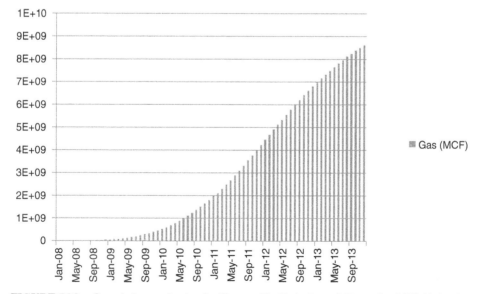

FIGURE 76.9   Cumulative production for Haynesville Shale through December 2013 Shale (data from IHS Enerdeq).

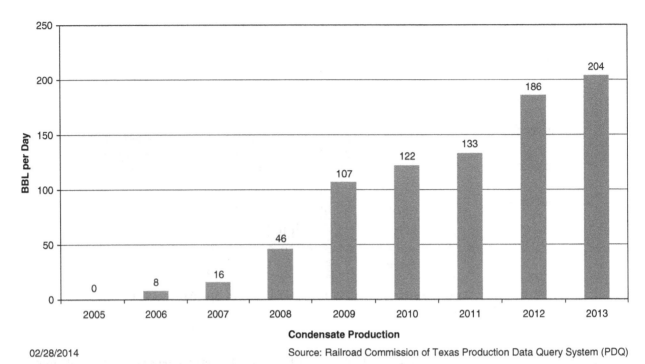

02/28/2014   Source: Railroad Commission of Texas Production Data Query System (PDQ)

FIGURE 76.10 Texas Haynesville Shale condensate production through December 2013 shows steady increase in liquids production.

REFERENCES

Hammes, U., Hamlin, H. S., Ewing, T. E. (2011). Geologic analysis of the Upper Jurassic Haynesville Shale in east Texas and west Louisiana. *AAPG Bull*, 95:1643–1666.

Hammes, U., Frébourg, G. (2012). Haynesville and Bossier mudrocks: a facies and sequence stratigraphic investigation, East Texas and Louisiana, USA. *Mar Petrol Geol*, 31: 8–26, doi: 10.1016/j.marpetgeo.2011.10.001.

Wang, F. P., Hammes, U. (2010). Effects of petrophysical factors on Haynesville fluid flow and production. *World Oil*, D79–D82.

Wang, F. P., Hammes, U. Li, Q. (2013). An overview of Haynesville Shale production. In: Hammes, U., J. Gale, eds. Geology of the Haynesville Gas Shale in East Texas and West Louisiana, USA. *AAPG Memoir*, 105:155–178.

MAQUOKETA AND NEW ALBANY SHALES, ILLINOIS BASIN

BY RACHEL WALKER (COUNTRYMARK ENERGY RESOURCES, LLC)

The 2013/2014 shale activity in the Illinois Basin continues to be limited due to poor gas prices, with some New Albany permitting and drilling taking place, largely within existing fields. No known Maquoketa-related activity has occurred beyond the apparent initial leasing interest in 2011/2012.

Maquoketa Shale

Indiana: No Maquoketa wells have been permitted in Indiana since the last report in November 2013. There is no known production from the Maquoketa in Indiana.

Illinois: No Maquoketa wells have been permitted in Illinois since the last report in November 2013. While there has been speculation in the basin that several deep permitted wells may have been looking at the Maquoketa, there is little publicly available data to support this. There is no known production from the Maquoketa in Illinois.

Kentucky: No Maquoketa wells have been permitted in Kentucky since the last report in November 2013. There is no known production from the Maquoketa in Kentucky.

New Albany Shale

Indiana: In July 2013, Riverside Operating Co. permitted a New Albany well in the Plainville Field in Daviess County (not yet drilled). Riverside continues to operate 26 New Albany Shale gas wells in Daviess County, producing 3500 million cubic feet gas/day (MCFGPD). Although there are several existing NAS fields in Indiana, the Indiana Geological Survey does not have good records of New Albany gas production, so it is difficult to report on volume produced to date. A number of NAS fields were shut in when gas prices dropped and it is unclear when they will be operational again.

Illinois: Other than the Core Minerals New Albany well in the Salem Field of Marion County (drilled July 2013 and temporarily abandoned), there have been no further New Albany specific wells permitted in Illinois.

Kentucky: In May and June 2013, Endeavor Energy Resources permitted three New Albany wells in the Rockvale Consolidated Field in Breckinridge County (not yet drilled). In September 2013, Irvin Anderson Jr. permitted two New Albany wells – one in the Glen Dean Field and one in the McQuady Field, both in Breckinridge County (not yet drilled). In February 2014, GTC Land & Energy also permitted a New Albany well in the McQuady Field (not yet drilled). In March 2014, Trico-Tiger permitted a New Albany well in the Flaherty Field of Meade County (not yet drilled). Trico-Tiger have drilled and completed several New Albany wells in this area, testing between 163 and 360 MCFGPD at a very shallow 500–800 ft depth.

REFERENCES

Indiana Department of Natural Resources, Division of Oil and Gas, Indianapolis, IN.

Illinois Department of Natural Resources, Division of Oil and Gas, Springfield, IL.

Commonwealth of Kentucky Department of Mines and Minerals, Division of Oil and Gas, Frankfort, KY The Scout Check Report, LLC, Evansville, IN.

MARCELLUS SHALE (DEVONIAN) – APPALACHIAN BASIN, USA

By Catherine Enomoto (U.S. Geological Survey)

The Middle Devonian Marcellus Shale of the Appalachian Basin is the most extensive shale play in the United States, covering about 66,600,000 acres (USGS Marcellus Shale Assessment Team, 2011). Extending from Tennessee to New York, the gross thickness of the Marcellus Shale increases to the northeast, with the thickest area located in northeastern Pennsylvania (Milici and Swezey, 2006; Wrightstone, 2009). The organic-rich zone of the Marcellus Shale has a net thickness of 50 to over 250 ft, and exists at drilling depths of about 2000 to 9000 ft (Milici and Swezey, 2006; Wrightstone, 2009). The organic-rich Marcellus Shale has high radioactivity responses, and thus high gamma ray values on well logs, because the organic matter tends to concentrate uranium ions (Harper, 2008). According to the studies during and after the Eastern Gas Shales Project (EGSP), there is a strong relationship between higher-than-normal gamma

ray response and total gas content in the black, organic-rich Marcellus Shale. As reported in Milici and Swezey (2006), Repetski et al. (2008), and Ryder et al. (2013), analyzed samples of the Marcellus Shale had mean random vitrinite reflectance values between 1.0% and 2.5% in the majority of the currently productive area, where most production has been natural gas. However, in southwest Pennsylvania, eastern Ohio, and northern West Virginia reported production included condensate and oil from wells producing from the Marcellus Shale. Published data indicates that the total organic carbon (TOC) content of the Marcellus Shale is as high as 11% (Repetski et al., 2008).

As in other shale plays, horizontal drilling and hydraulic fracturing increase production rates of hydrocarbons, which improves the commerciality of hydrocarbon production from this formation.

The orientation of the horizontal sections of the wells and the design of the staged hydraulic fracturing operations enhance the natural fracture trends in the Marcellus Shale. "Slick-water fracs" have provided the best method for recovering large volumes of natural gas efficiently. This method uses sand as a proppant and large volumes of freshwater that have been treated with a friction reducer such as a gel. The slick-water frac maximizes the length of the induced fractures horizontally while minimizing the vertical fracture height (Harper, 2008). Water supply for large volume fracturing is a concern, as are the potential environmental impacts related to handling and management of produced formation water and used hydraulic fracturing fluid, called "flowback" fluid (Engle and Rowan, 2014; Skalak et al., 2014; Capo et al., 2014). The management of produced formation water and used hydraulic fracturing fluid have been addressed with a variety of approaches including 1) treatment followed by discharge into receiving basins or streams, 2) injection into subsurface disposal wells, or 3) treatment to remove solids and unwanted contaminants followed by reuse.

The U.S. Energy Information Administration (EIA) published a report in February 2014 (U.S. Energy Information Administration, 2014), which contained analyses of drilling data through January 2014, and projected production through March 2014. According to this report, the number of rigs that completed wells in the Marcellus Shale decreased from over 140 per month on January 1, 2012, to about 100 per month, as of January 31, 2014. However, the daily gas production rate of new Marcellus Shale wells increased from less than 3 to over 6 Mcf/day for the same period. Including production declines in legacy wells, the total production from the Marcellus Shale is about 14 Bcf of gas per day and almost 50,000 barrels (bbls) of oil per day, as of January 31, 2014, according to the EIA report (U.S. Energy Information Administration, 2014).

In August 2011, the U.S. Geological Survey (USGS) published Fact Sheet 2011-3092, "Assessment of undiscovered oil and gas resources of the Devonian Marcellus Shale of the Appalachian Basin Province" (Coleman et al., 2011). According to this publication, the USGS estimated a mean undiscovered, technically recoverable natural gas resource of about 84 Tcf and a mean undiscovered, technically recoverable natural gas liquids resource of 3.4 billion barrels in continuous-type accumulations in the Marcellus Shale. The estimate of natural gas resources ranged from 43 to 144 Tcf (95% to 5% probability, respectively), and the estimate of natural gas liquids resources ranged from 1.6 to 6.2 billion bbls (95% to 5% probability, respectively). This re-assessment of the undiscovered continuous resources in the Marcellus Shale updated the previous assessment of undiscovered oil and gas resources in the Appalachian Basin performed by the USGS in 2002 (Milici et al., 2003), which estimated a mean of about 2 Tcf of natural gas and 11.5 million bbls of natural gas liquids in the Marcellus Shale.

The new estimates are for resources that are recoverable using currently available technology and industry practices, regardless of economic considerations or accessibility conditions, such as areas limited by policy and regulations. The Marcellus Shale assessment covered areas in Kentucky, Maryland, New York, Ohio, Pennsylvania, Tennessee, Virginia, and West Virginia. Shown in Figure 76.11 are the extents of three assessment units (AU) defined in this latest assessment. Ninety-six percent of the estimated resources reside in the Interior Marcellus AU.

The increase in undiscovered, technically recoverable resources is due to new geologic information and engineering data. In late 2004, the Marcellus Shale was recognized as a potential reservoir rock, instead of only a regional hydrocarbon source rock. Technological improvements resulted in improved commerciality of gas production from the Marcellus Shale and caused rapid development of this new play in the Appalachian Basin, the oldest producing petroleum province in the United States. According to the Pennsylvania Department of Environmental Protection, the first horizontal wells drilled specifically for the Marcellus Shale were drilled in 2005 (Pennsylvania Department of Environmental Protection, 2014a). Natural gas production was reported from horizontal wells that were completed in the Marcellus Shale in West Virginia as early as 2007 (West Virginia Geological & Economic Survey, 2014).

Following is a summary of activity in each state where the Marcellus Shale is mapped.

Kentucky: As of 2011, there was no reported production from the Marcellus Shale in Kentucky (Kentucky Division of Oil and Gas, 2013).

Maryland: There were no wells drilled in Maryland for the Marcellus Shale between 2004 and 2013. In 2009, four companies submitted applications for permits to drill Marcellus Shale wells. None of these applications were approved, and all of the applications were withdrawn by the companies. There is currently (2014) no reported production from the Marcellus Shale in Maryland. Due to the estimated thermal maturity of the Marcellus Shale in Maryland (Repetski et al., 2008), it is likely that dry gas will be found if wells are drilled and completed in the Marcellus Shale. The permit process to drill and produce natural gas from the Marcellus Shale in Maryland is under review. On June 6, 2011, the Governor of Maryland signed an Executive Order

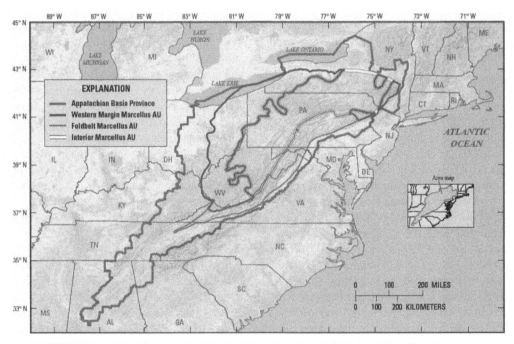

FIGURE 76.11 Map of the Appalachian Basin Province showing the three Marcellus Shale assessment units (AU) (Coleman et al., 2011).

establishing the Marcellus Shale Safe Drilling Initiative. The Order required the Maryland Department of the Environment (MDE) and Department of Natural Resources (DNR) to undertake a study of drilling for and extracting natural gas from shale formations (see Maryland Department of the Environment, 2014a). In December 2011, both the MDE and DNR developed four recommendations regarding revenue and three recommendations regarding standards of liability. Comments were solicited and accepted through September 10, 2013. On September 25, 2013, the results of a survey completed by members of the Marcellus Shale Safe Drilling Initiative Advisory Commission indicating their agreement or disagreement with the recommended best practices were published on the MDE website (Maryland Department of the Environment, 2013). A final report titled "Marcellus Shale Safe Drilling Initiative Study Part II, Interim Final Best Practices" was submitted July 11, 2014, and is available on the MDE website (Maryland Department of the Environment, 2014b).

New York: The Marcellus Shale extends into the northern most part of the Appalachian basin in central New York. The organic-rich thickness of the Marcellus Shale increases from 20 ft in the west to 250 ft in the eastern part of the basin in New York (Smith and Leone, 2010). The depths of the Marcellus Shale range from zero to as much as 7000 ft in the eastern part of basin in south-central New York (Smith and Leone, 2010). In 2012, according to the New York Department of Environmental Conservation (DEC), there were 31 vertical wells with Marcellus Shale listed as the productive formation, but only 15 reported production in 2012 (New York Department of Environmental Conservation, 2014b). Natural gas production from the Marcellus Shale in 2012 was 23.3 Mcf, down from reported production of 25.6 Mcf in 2011, and down from the high of 64 Mcf reported for 2008. There was no reported oil production. According to the New York DEC, there were over 250 Mcf of gas produced from the Marcellus Shale between 2000 and 2012. The New York DEC also reported that between 1967 and 1999, there may have been as much as 543 Mcf of gas produced from the Marcellus Shale (New York Department of Environmental Conservation, 2014b).

A search of the New York DEC wells database returned 184 well permit applications where the objective formation was "Marcellus." Thirty-seven of these wells have been drilled (correction from the EMD Shale Gas and Liquids Committee 2013 Annual Report), and all are vertical wells (correction from the EMD Shale Gas and Liquids Committee 2013 Annual Report). To assess issues unique to horizontal drilling and high-volume hydraulic fracturing, the New York DEC published a Preliminary Revised Draft Supplemental Generic Environmental Impact Statement (SGEIS) in July 2011 (New York Department of Environmental Conservation, 2014a). Additional information was added and a Revised Draft SGEIS was released September 7, 2011.

The public comment period ended January 11, 2012, after which New York DEC was required to refile the draft regulations covering high-volume hydraulic fracturing. New York DEC held public hearings for the SGEIS and for the regulations in November 2012. The public comment period closed on January 11, 2013. While the process of reviewing the SGEIS is ongoing, any company that applies for a drilling permit for horizontal drilling in the Marcellus Shale will be required to undertake an individual, site-specific environmental review (New York Department of Environmental Conservation, 2014a).

Ohio: The Ohio Division of Oil & Gas Resources reported that about 1.23 Bcf of gas and over 17,700 bbls of oil were produced from the Marcellus Shale from 2007 through 2012 (Ohio Division of Oil & Gas Resources, 2013a). There were less than 30 wells that reported production from the Marcellus Shale in 2012. The productive wells were in Athens, Belmont, Jefferson, Meigs, Monroe, Noble, and Washington counties. According to the Ohio Division of Oil & Gas Resources, there were about 188 Mcf of gas and about 770 bbls of oil produced in 2012. This was a decrease from 2011, for which reported production was about 340 Mcf of gas and about 8000 bbls of oil (Ohio Division of Oil & Gas Resources, 2013a). As of July 2014, 44 Marcellus Shale horizontal well permits were issued, 27 horizontal wells have been drilled into the Marcellus Shale, and 11 horizontal wells were producing from the Marcellus Shale. The horizontal Marcellus Shale wells reported as productive are in Belmont, Carroll, Jefferson, and Monroe counties (Ohio Department of Natural Resources, 2014).

The Ohio Geological Survey published a map of the area of potential within the Marcellus Shale (Erenpreiss et al., 2011a), which includes the counties of Ashtabula, Trumbull, Mahoning, Columbiana, Carroll, Jefferson, Harrison, Belmont, Guernsey, Monroe, and Washington. The maximum thickness of the Marcellus Shale in Ohio is 70 ft, and the average thickness is about 40 ft in the prospective area in easternmost Ohio (Erenpreiss et al., 2011b). The Ohio Division of Oil & Gas Resources Management Division developed rules pertaining to horizontal well site construction, which took effect from August 1, 2012 (Ohio Division of Oil & Gas Resources, 2013b).

Pennsylvania: The Marcellus Shale is deepest in north-central Pennsylvania, and the deepest wells to test the Marcellus Shale have been drilled to 8500 ft in Clinton County (Harper and Kostelnik, undated). The areas of greatest activity in the Marcellus Shale are in southwestern and northeastern Pennsylvania. The production of oil and condensate from fields in southwest Pennsylvania made this area attractive to operators. Pennsylvania has continued to be the state with the most drilling into, and production from, Marcellus Shale. According to the Pennsylvania Department of Environmental Protection (DEP), as of March 2014, the counties with the most drilling and production activity in the Marcellus Shale were Greene, Washington, Fayette, Lycoming, Tioga, Bradford, Wyoming, and Susquehanna. According to the Pennsylvania DEP, 2976 permits for unconventional (horizontal) wells were issued in 2013 (Pennsylvania Department of Environmental Protection, 2014b). Susquehanna County was the location of most of the permits for unconventional wells, and Bradford County had the second highest number of permits. In 2013, 1207 unconventional wells and 960 conventional (vertical) wells were drilled in Pennsylvania. The majority of the unconventional wells were drilled in Washington County, followed closely by Susquehanna County (Pennsylvania Department of Environmental Protection, 2014b). According to PA*IRIS/WIS, the Pennsylvania database of oil and gas records, and Pennsylvania DEP, by the end of 2013, over 3100 wells reported production from the Marcellus Shale. About 85% of these productive wells were horizontal wells. According to PA*IRIS/WIS and Pennsylvania DEP (Pennsylvania Department of Environmental Protection, 2014a), about 1800 Bcf of gas, about 1 million bbls of condensate, and about 43,400 bbls of oil were produced from the Marcellus Shale in 2013. During the last 6 months of 2013, Chesapeake Appalachia LLC was the largest producer of natural gas from the Marcellus Shale, having reported over 245 Bcf in the 6-month reporting period. Chesapeake was followed by Cabot Oil & Gas Corporation, Range Resources Appalachia LLC, Anadarko E&P Onshore LLC, and Talisman Energy USA Inc., each with production of over 60 Bcf of natural gas in the last 6 months of 2013 (Pennsylvania Department of Environmental Protection, 2014a).

Tennessee: According to de Witt et al. (1993), the Marcellus Shale is present in the subsurface in northeastern Tennessee. Therefore, the USGS determined that the Foldbelt Marcellus Assessment Unit extends into Tennessee (Figure 76.11). According to the Tennessee Department of Environment and Conservation, Division of Water Resources, Oil and Gas Section, there is no production from the Marcellus Shale in Tennessee (M. Burton, 2014, personal communication).

Virginia: According to the Virginia Division of Gas & Oil (DGO), there were no wells drilled exclusively for the Marcellus Shale in Virginia between 2004 and 2013. It is likely that natural gas was produced from the Marcellus Shale commingled with other zones in vertical wells in Virginia, but the quantity is unknown. A significant fraction of potentially productive acreage in Virginia is on national forest land. In an effort to update the 1993 George Washington National Forest (GWNF) Plan, the U.S. National Forest Service (NFS) issued the Draft Environmental Impact Statement (DEIS) and Draft Revised Land and Resource Management Plan for the GWNF in April, 2011 (U.S. Department of Agriculture, Forest Service, 2014). The NFS preferred alternative forest plan included the restriction that, on lands administratively available for gas and oil leasing within the GWNF, no horizontal drilling will be allowed. The public comment period for the Draft Forest Plan and Draft Environmental

Impact Statement ended on October 17, 2011. Staff of the NFS continue to review and analyze the comments, and as of December 12, 2013, had not released the final Forest Plan and Environmental Impact Statement (U.S. Department of Agriculture, Forest Service, 2014).

West Virginia: Total production from wells completed in the Marcellus Shale from 1979 through 2012 was over 660 Bcf of gas and almost 1.8 million bbls of oil, according to information from the West Virginia Geological and Economic Survey (WVGES) (West Virginia Geological & Economic Survey, 2014). The first production reported from a horizontal well completed in the Marcellus Shale in West Virginia was in 2007. Between 2007 and 2012, about 547 Bcf of gas were produced from horizontal wells completed in the Marcellus Shale, as well as about 1.5 million bbls of oil. The oil volumes reported by operators include natural gas liquids. West Virginia is second to Pennsylvania in cumulative production of hydrocarbons from the Marcellus Shale. According to WVGES (West Virginia Geological & Economic Survey, 2014), over 1400 vertical wells reported production from the Marcellus Shale that was commingled with production from other formations in 2012. There were 612 horizontal wells from which there was reported production from the Marcellus Shale in 2012. In 2012, production of over 324 Bcf of gas and almost 925,000 bbls of oil were reported from both vertical and horizontal wells completed in the Marcellus Shale. In 2012, the counties from which most of the liquids were produced were Wetzel, Ohio, Brooke, Doddridge, Marshall, and Tyler. Most of the completed Marcellus Shale wells that are reported as "deviated", meaning horizontal, were located in Marion, Marshall, Wetzel, Ohio, Taylor, Harrison, Doddridge, and Upshur counties. In most of these counties, the thickness of the Marcellus Shale with high gamma-ray readings is 40 to 100 ft, according to WVGES (West Virginia Geological & Economic Survey, 2014). Based on the volume of gas production from the Marcellus Shale through 2012, the major producers include Chesapeake Appalachia, LLC, Antero Resources Appalachian Corp., EQT Production Company, XTO Energy, Inc., and Stone Energy. The companies with the most oil production in 2012 from the Marcellus Shale were Chesapeake Appalachia LLC, EQT Production Company, and Stone Energy (West Virginia Geological & Economic Survey, 2014).

REFERENCES

Capo, R. C., Stewart, B. W., Rowan, E. L., Kohl, C. A. K., Wall, A. J., Chapman, E. C., Hammack, R. W., Schroeder, K. T. (2014). The strontium isotopic evolution of Marcellus Formation produced waters, southwestern Pennsylvania. *Int J Coal Geol*, 126:57–63.

Coleman, J. L., Milici, R. C., Cook, T. A., Charpentier, R. R., Kirschbaum, M., Klett, T. R., Pollastro, R. M., Schenk, C. J. (2011). Assessment of undiscovered oil and gas resources of the

Devonian Marcellus Shale of the Appalachian Basin Province, 2011. U.S. Geological Survey Fact Sheet FS2011-3092, 2 p., http://pubs.usgs.gov/fs/2011/3092/, accessed October 25, 2011.

de Witt, W., Jr., Roen, J. B., Wallace, L. G. (1993). Stratigraphy of Devonian black shales and associated rocks in the Appalachian basin. In: Roen, J.B., Kepferle, R.C., eds. *Petroleum Geology of the Devonian and Mississippian Black Shale of Eastern North America*. U.S. Geological Survey Bulletin 1909, p. B1–B57.

Engle, M. A., Rowan, E. L. (2014). Geochemical evolution of produced waters from hydraulic fracturing of the Marcellus Shale, northern Appalachian Basin: a multivariate compositional data analysis approach. *Int J Coal Geol*, 126:45–56.

Erenpreiss, M. S., Wickstrom, L. H., Perry, C. J., Riley, R. A., Martin, D. R., et al. (2011a). Areas of Utica and Marcellus potential in Ohio, Ohio Department of Natural Resources, Division of Geological Survey, scale 1 inch equals 27 miles, http://geosurvey.ohiodnr.gov/portals/geosurvey/Energy/Utica/Utica_Marcellus_Ohio_8x11.pdf, accessed July 18, 2014.

Erenpreiss, M. S., Wickstrom, L. H., Perry, C. J., Riley, R. A., Martin, D. R., et al., (2011b). Regional organic-thickness map of the Marcellus Shale with additional organic-rich shale beds in the Hamilton Group included for New York, Pennsylvania, and West Virginia: Ohio Department of Natural Resources, Division of Geological Survey, scale 1 inch equals 52 miles, http://geosurvey.ohiodnr.gov/portals/geosurvey/Energy/Marcellus/Regional_Marcellus%20thickness_cntrlpts_8x11.pdf, accessed May 23, 2014.

Harper, J. A. (2008). The Marcellus Shale – an old "new" gas reservoir in Pennsylvania. *Pennsylvania Geology*, 38(1):2–13, http://www.dcnr.state.pa.us/cs/groups/public/documents/document/dcnr_006811.pdf, accessed July 18, 2014.

Harper, J. A., Kostelnik, J. (undated). The Marcellus Shale play in Pennsylvania, Part 3: The modern Marcellus Shale play: Pennsylvania Department of Conservation and Natural Resources, Bureau of Topographic and Geologic Survey, http://www.dcnr.state.pa.us/cs/groups/public/documents/document/dcnr_007594.pdf, accessed June 6, 2014.

Kentucky Division of Oil and Gas. (2013) http://oilandgas.ky.gov/Pages/ProductionReports.aspx, accessed July 20, 2014.

Maryland Department of the Environment. (2013). Marcellus Shale Safe Drilling Advisory Commission, http://www.mde.state.md.us/programs/Land/mining/marcellus/Documents/Survey_Commissioners_Final.pdf, accessed July 18, 2014.

Maryland Department of the Environment. (2014a). Marcellus Shale Safe Drilling Initiative, http://www.mde.state.md.us/programs/Land/mining/marcellus/Pages/index.aspx, accessed July 18, 2014.

Maryland Department of the Environment. (2014b). Marcellus Shale Safe Drilling Initiative Study, Part II, Interim Final Best Practices, http://www.mde.state.md.us/programs/Land/mining/marcellus/Documents/7.10_Version_Final_BP_Report.pdf, accessed July 18, 2014.

Milici, R. C., Ryder, R. T., Swezey, C. S., Charpentier, R. R., Cook, T. A., Crovelli, R. A., Klett, T. R., Pollastro, R. M., Schenk, C.

J. (2003). Assessment of undiscovered oil and gas resources of the Appalachian basin province. 2002: U.S. Geological Survey Fact Sheet FS-009-03, 2 p., http://pubs.usgs.gov/fs/fs-009-03/FS-009-03-508.pdf, accessed May 23, 2014.

Milici, R. C., Swezey, C. S. (2006). Assessment of Appalachian Basin Oil and Gas Resources: Devonian Shale – Middle and Upper Paleozoic Total Petroleum System: U. S. Geological Survey Open-File Report 2006-1237, 70 p., with additional figures and tables, http://pubs.usgs.gov/of/2006/1237/, accessed March 27, 2012.

New York Department of Environmental Conservation. (2014a). Marcellus Shale, http://www.dec.ny.gov/energy/46288.html, accessed July 18, 2014.

New York Department of Environmental Conservation. (2014b). New York Natural Gas & Oil Production, http://www.dec.ny.gov/energy/1601.html, accessed July 18, 2014.

Ohio Department of Natural Resources (2014). Horizontal Marcellus Shale activity in Ohio: Columbus, scale 1:1,300,000, revised 7/11/2014, http://geosurvey.ohiodnr.gov/Portals/geosurvey/Energy/Marcellus/HorizontalWells_Monthly MarcellusPagesize_07052014.pdf, accessed July 20, 2014.

Ohio Division of Oil & Gas Resources. (2013a). Oil & Gas Well Database, http://oilandgas.ohiodnr.gov/mineral/OHRbdmsOnline/WebReportAccordion.aspx, accessed July 20, 2014.

Ohio Division of Oil & Gas Resources. (2013b) Well construction rule package to JCARR, http://oilandgas.ohiodnr.gov/laws-regulations/well-construction-rule-package-to-jcarr, accessed July 18, 2014.

Pennsylvania Department of Environmental Protection. (2014a). Oil & Gas Reporting Website – Statewide data downloads by reporting period, https://www.paoilandgasreporting.state.pa.us/publicreports/Modules/DataExports/DataExports.aspx, accessed July 18, 2014.

Pennsylvania Department of Environmental Protection. (2014b). Oil and Gas Reports, Permits issued and wells drilled maps, http://files.dep.state.pa.us/OilGas/BOGM/BOGMPortalFiles/OilGasReports/2012/2013Wellspermitted-drilled.pdf, accessed July 18, 2014.

Repetski, J. E., Ryder, R. T., Weary, D. J., Harris, A. G., Trippi, M. H. (2008). Thermal maturity patterns (CAI and %Ro) in Upper Ordovician and Devonian rocks of the Appalachian Basin: a major revision of USGS Map I-917-E using new subsurface collections: U. S. Geological Survey Scientific Investigations Map 3006, http://pubs.usgs.gov/sim/3006/, accessed July 18, 2014.

Ryder, R. T., Hackley, P. C., Alimi, H., Trippi, M. H. (2013). Evaluation of thermal maturity in the low maturity Devonian shales of the northern Appalachian Basin: American Association of Petroleum Geologists Search and Discovery Article #10477 (2013), http://www.searchanddiscovery.com/documents/2013/10477ryder/ndx_ryder.pdf, accessed May 23, 2014.

Skalak, K. J., Engle, M. A., Rowan, E. L., Jolly, G. D., Conko, K. M., Benthem, A. J., Kraemer, T. F. (2014). Surface disposal of produced waters in western and southwestern Pennsylvania: potential for accumulation of alkali-earth elements in sediments. *Int J Coal Geol*. 126:162–170.

Smith, L. B., Leone, J. (2010). Integrated characterization of Utica and Marcellus black shale gas plays, New York State: American Association of Petroleum Geologists, Search and Discovery Article #50289, 36, http:// www.searchanddiscovery.com/ pdfz/ documents/ 2010/ 50289smith/ ndx_smith.pdf.html, accessed May 23, 2014.

U.S. Department of Agriculture, Forest Service. (2014). George Washington & Jefferson National Forests, George Washington Plan Revision, http:// www.fs.usda.gov/ detail/ gwj/ landmanagement/ ?cid=fsbdev3_000397, accessed July 18, 2014.

U.S. Energy Information Administration. (2014). Drilling Productivity Report for key tight oil and shale gas regions, February 2014: U.S. Energy Information Administration, http:// www.eia.gov/ petroleum/ drilling/ pdf/ dpr-full.pdf, accessed May 23, 2014.

USGS Marcellus Shale Assessment Team. (2011). Information relevant to the U.S. Geological Survey assessment of the Middle Devonian Shale of the Appalachian Basin Province, 2011: U.S. Geological Survey Open-File Report 2011-1298, 22 p., http:// pubs.usgs.gov/ of/ 2011/ 1298/, accessed July 18, 2014.

West Virginia Geological & Economic Survey. (2014). Marcellus and other Devonian Shales, http:// www.wvgs.wvnet.edu/ www/ datastat/ devshales.htm, accessed July 18, 2014.

Wrightstone, G. (2009). Marcellus Shale – Geologic controls on production: American Association of Petroleum Geologists Search and Discovery Article #10206 ©AAPG Annual Convention, June 7–10, 2009, 10, http:// www.searchanddiscovery.com/ documents/ 2009/ 10206wrightstone/, accessed May 23, 2014.

NIOBRARA FORMATION (CRETACEOUS), ROCKY MOUNTAIN REGION, USA

By Stephen Sonnenberg (Colorado School of Mines)

The Niobrara is a significant, self-sourced, resource play throughout the Rocky Mountain region. New technology of horizontal drilling and multi-stage, hydraulic-fracture stimulation is unlocking reserves that previously were not obtainable.

Known production comes from both fracture and matrix porosity systems (dual porosity). High matrix porosity is present in the shallow biogenic gas accumulations of eastern Colorado and Western Kansas. The shallow biogenic play is important for natural gas production at burial depths of less than 3500 ft. The deeper Niobrara thermogenic accumulations generally occur at burial depths greater than 7000 ft. Burial diagenesis (chemical and mechanical compaction and cementation) reduces porosities to values less than 10% in the deeper parts of the various basins where the Niobrara is prospective. Mature Niobrara source rocks are located in these areas of low porosity. Natural fractures are important contributors to production in the deeper areas.

The Niobrara Petroleum System contains all aspects of a large resource play (e.g., widespread mature source and reservoir rocks, self-sourced). The Niobrara was deposited in the Western Interior Cretaceous (WIC) Basin and is a widespread unit in the Rocky Mountain region (Figure 76.12). The WIC Basin was broken into numerous smaller basins during the Laramide orogeny. The Niobrara contains reservoir rocks, rich source beds and abundant seals. The various productive lithologies all have low porosity and permeability. The TOC values in shales locally range from 2% to 8% in the eastern WIC area and are reduced to 1–3% because of siliciclastic dilution in the western WIC area. Laramide structural events exert the primary control on fracturing within the Niobrara as well as thermal maturity. Neogene extension fracturing is also thought to be an important component for locating production "sweet spots." Understanding the thermal maturity of the source rocks will assist in predicting the distribution of hydrocarbon accumulations. Hydrocarbon generation may enhance the tectonic fractures

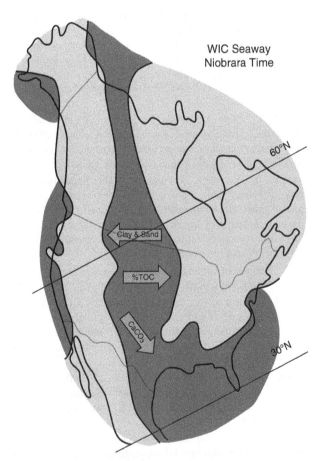

FIGURE 76.12 The Western Interior Cretaceous (WIC) Basin during Niobrara time (modified from Longman et al., 1990). Source area for clastics is dominantly to the west, total organic carbon content in the Niobrara increases to the east, carbonate content generally increases on the eastern side of the WIC seaway and to the southeast.

and may also create new ones as a result of overpressuring associated with this process.

A summary of factors thought to be important for Niobrara production in the Rocky Mountain region are as follows: presence of favorable reservoir facies (brittle chalk) and a diagenetic history that enables open fracture systems to exist; presence of mature source rocks to enable a continuous oil column to exist in the trap; source rocks interbedded with respect to the reservoir limestone (chalk); a favorable tectonic history for fracture formation. Most fracture systems fall into two major categories: structure-related fractures and regional orthogonal fractures.

Resistivity mapping can be used to determine both the presence of a hydrocarbon accumulation and the maturity of source rocks for the Niobrara. The presence of oil in open fracture systems is thought to be the cause of the high resistivity anomalies in chalk beds. A relationship between increasing resistivity of source shales with increasing thermal maturity has also been demonstrated.

Knowledge of the distribution and occurrence of hydrocarbon source and reservoir rocks in the Niobrara interval will greatly aid future exploration.

Regional Setting

The Upper Cretaceous Niobrara (Coniacian–Campanian; ~82 to 89.5 million years ago) was deposited in a foreland basin setting in the WIC Seaway of North America during a time of a major marine transgression (Figure 76.12). This major transgression probably represents the maximum sea-level highstand during the Cretaceous and may contain the best source rocks in the Cretaceous. The present-day basins in the Rocky Mountain region formed during the Late Cretaceous to Early Tertiary Laramide orogeny.

The WIC Basin was an asymmetric foreland basin with the thickest strata being deposited along the western margin of the basin (Figures 76.12, 76.13). The WIC Basin is a complex foreland basin that developed between mid to late Jurassic to Late Cretaceous time. The basin was bordered by mountainous areas to the west (zone of plutonism, volcanism, and thrusting that formed the Cordilleran thrust belt) and a broad stable cratonic zone to the east. The foreland basin subsided in response to thrust and synorogenic sediment loading and pulses of rapid subduction and shallow mantle flow.

During sea-level highstands, coccolith-rich and planktonic foraminifera-rich carbonate sediments (chalks) accumulated on the eastern half of the seaway. Chalky beds extend into Montana and southern Canada (where they are called the White Spec zones) and into the Gulf Coast region (Austin Chalk). Chalk-rich carbonate facies change westward into siliciclastic-rich beds.

Stratigraphy and Depositional Setting

The Niobrara represents one of the two most widespread marine invasions and the last great carbonate-producing

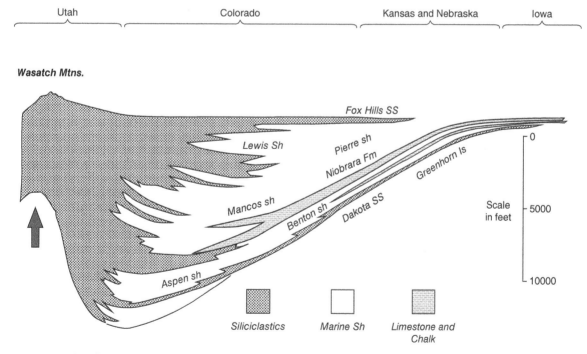

FIGURE 76.13   Generalized cross-section across the Western Interior Cretaceous Basin. The Niobrara is Upper Cretaceous in age. Limestone and chalk beds are present over the eastern two-thirds of the basin (modified from Kauffman, 1977).

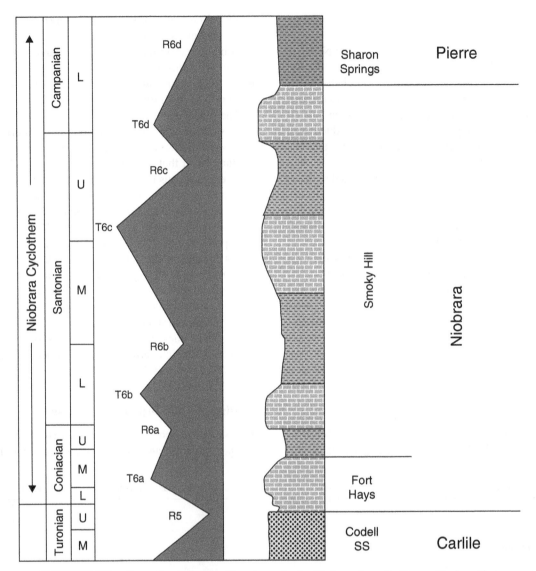

FIGURE 76.14 Generalized stratigraphic column for the Niobrara from the Denver Basin setting. The Niobrara ranges in age from Coniacian to lower Campanian. Several transgressive and regressive cycles are noted for the Niobrara interval. Four chalk-rich intervals were deposited during trangressive events, calcareous shales during regressive events (modified from Longman et al., 1998; Barlow and Kauffman, 1985).

episode of the WIC Basin (the first widespread event is represented by the Greenhorn chalks). The dominant lithologies of the Niobrara Formation are limestones (chalks) and interbedded with marls and calcareous shales (Figures 76.13, 76.14). The chalk-shale cycles are interpreted to represent changes from normal to brackish water salinities possibly related to regional paleo-climatic factors or sea-level fluctuations. The chalk lithologies are thought to represent deposition in normal to near-normal marine salinities having a well-mixed water column and well oxygenated bottom waters. The chalks reflect influx of warm Gulfian currents into the WIC seaway during relatively high sea levels. The interbedded shale/marl cycles are interpreted to be caused by an increase

in fresh water runoff caused by increased rainfall which may be related to climatic warming. The fresh water runoff creates a brackish water cap and salinity stratification. Vertical mixing of the water column is inhibited causing anoxic conditions in the bottom waters. This enhances preservation of organic material and results in organic-rich source rocks. The decrease in water salinities is also suggested by oxygen isotopic values. The shalier intervals may reflect lower sea levels and greater influx of clastic material from the west. The chalks have previously been interpreted to represent higher sea levels during Niobrara time.

Three major facies are present in the Niobrara and equivalents across the Rocky Mountain region (Figures 76.12 and

76.14). On the western side of the area, a sandstone facies is present which changes laterally to the east into a calcareous shale facies, and which, in turn, changes eastward into a limestone and chalk facies. These various lithologies interfinger and the facies changes are very gradational. The Niobrara name is used for chalk and shale units located on the eastern side of the Western Interior Seaway; whereas the term Mancos is generally used for the equivalent shale, and siltstone units in the western part of the area. The equivalent shoreline and non-marine sandstone units further to the west are known by a variety of names. The limestone facies is composed of coccolith-rich fecal pellets probably derived from pelagic copepods, inoceramid and oyster shell fragments, planktonic foraminifer tests, micrite, clay, and quartz silt. The thick siltstone facies was derived from highlands to the west. The shales found in the Mancos/Niobrara are dark-gray to black and generally organic-rich (>1% TOC). The shales are fair to excellent source rocks and also provide seals for the chalky and sandy reservoir facies. The TOC content in the interval increases to the east (Figure 76.12).

The chalks of the Niobrara are rich in organic matter and organic-related material (e.g., pyrite). On the east side of the WIC basin, the Niobrara consists of four chalk beds and three shale intervals (Figure 76.14). The basal chalk bed is known as the Fort Hays limestone member and the unit contains some of the purest chalk in the Western Interior. The Fort Hays is regionally extensive and ranges in thickness from 50 ft in southeast Colorado to 120 ft in New Mexico to less than 10 ft in southeast Wyoming. Carbonate content persists from the Denver Basin to southwest Colorado into the Laramie, North Park, South Park, and Sand Wash basins. The Fort Hays interval is difficult to distinguish from the remainder of the Niobrara north of the Laramie Basin.

The Fort Hays is overlain by the Smoky Hill member. The Smoky Hill consists of organic-rich shales to chalky shale (marls) to massive chalk beds. The interval has been subdivided by various authors into several units. Figure 76.14 illustrates a six-member subdivision.

The Niobrara ranges in thickness from 100 to 300 ft along the eastern side of the WIC basin to over 1500 ft on the west side of the WIC basin. Figure 76.15 illustrates an isopach map of the Niobrara across the northern Rockies region. Thinning occurs is a northeast trend across the map area. This thin trend was related to paleotectonic movement on the Transcontinental arch. Superimposed on the Transcontinental arch are northeast axes of thinning (Figure 76.15). Thinning in the Niobrara is believed to result from differing rates of sedimentation (i.e., convergence or divergence of section) and unconformities at the base, within, and at the top of the formation.

Niobrara deposition in the Western Interior Basin was strongly influenced by the interplay of warm north-flowing currents from the paleo-Gulf of Mexico and cooler southward-flowing currents from the Arctic region along with sea-level fluctuations. Warm waters from the Gulf brought in rich carbonate flora of coccoliths and promoted carbonate production and deposition. Siliciclastic input from the west and cooler Arctic currents inhibited carbonate production and deposition.

Chalks and marls are abundant in the Denver Basin. The section changes to marl and is shalier west of the Front Range and north of the Hartville Uplift. Chalk intervals extend into the Laramie, Hanna, North Park, Sand Wash, and Piceance basins. The section in the Piceance consists of interbedded sandstone, siltstone, and shale. In the San Juan Basin, the Niobrara consists of a mixture of siliciclastic and marl lithologies.

The Niobrara is overlain by the Pierre Shale in the eastern part of the Western Interior Basin and its age equivalent Mancos shale in the western part. The Niobrara overlies the Carlile Formation across much of the Western Interior basin (and its members: Codell Sandstone, Sage Breaks Shale, etc.). The Sharon Springs member of the Pierre shale overlies the Niobrara in most of the eastern Colorado. The Sharon Springs is an excellent source rock with TOCs ranging from 2 to 8 weight percent.

The type locality for the Niobrara Chalk is Knox County in northeastern Nebraska.

Source Rocks

Several workers have discussed the organic-rich nature of the Niobrara Formation and the increased thermal maturity and resistivity with increased burial depth. Vitrinite reflectance and resistivity of the organic-rich shale both increase with increasing thermal maturity. These values can be mapped to show the areas of source rock maturity.

The Niobrara Formation has been analyzed using the Rock-Eval instrument by several workers (Figure 76.16). Organic-rich beds in the formation have TOC values which average 3.2%. A plot of hydrogen index versus oxygen index (modified van Krevelen diagram) illustrates the type and level of maturity of the source rocks for different depths across the Denver basin. The plot also illustrates that the kerogen present in the Niobrara is Type-II or oil-prone (sapropelic).

Reservoir Rocks

The lithology of the Niobrara changes from east to west across the Western Interior Basin (Figure 76.13). In the Denver Basin, the lithology consists of interbedded calcareous shale, shaley limestones, marls, and limestones (Figure 76.14). Westward, the lithology becomes shalier and sandier (Figure 76.12). The carbonates are still present in the western area but clastics begin to dominate.

Most Niobrara reservoir rocks have undergone mechanical and chemical compaction and are low porosity and permeability rocks. Burial depth is the single most important

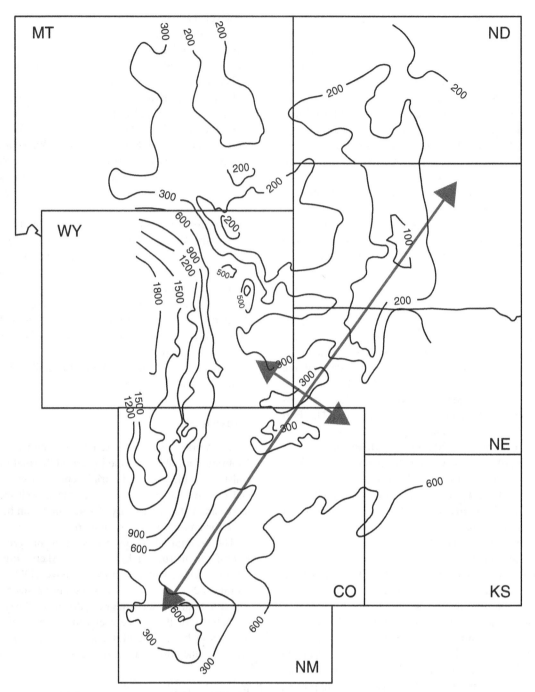

FIGURE 76.15   Isopach of the Niobrara across the northern Rockies (modified from Longman et al., 1998). The Niobrara ranges in thickness from less than 100 ft to over 1800 ft. Thinning occurs in a northeast trend across the area known as the Transcontinental arch (Weimer, 1978).

factor affecting porosity. Chalks have high original porosities (50% or greater). Initial dewatering and mechanical compaction is the first diagenetic phase. Grain and fossil breakage and re-orientation reduce porosity. Initial coccolith grain sizes are 0.2 to 1 micron. Chemical compaction is characterized by calcite dissolution along with wispy dissolution seams, microstylolites, and stylolites. Grain-to-grain dissolution along microstylolites is common and the dissolved calcite is reprecipitated locally.

Hydrocarbon Production

Niobrara production represents some of the oldest established production in the Rocky Mountain region. The oldest field in

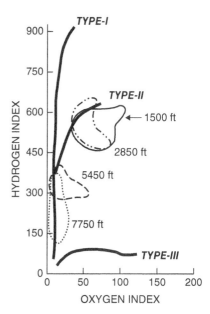

FIGURE 76.16 Niobrara Van Krevelen diagram. Niobrara source rocks are Type-II (oil prone) kerogens based on Rock Eval data. With increasing burial depth and thermal maturity the HI values decrease significantly. Data from Rice (1984); Barlow (1985); Pollastro (1985); after Sonnenberg and Weimer (1993).

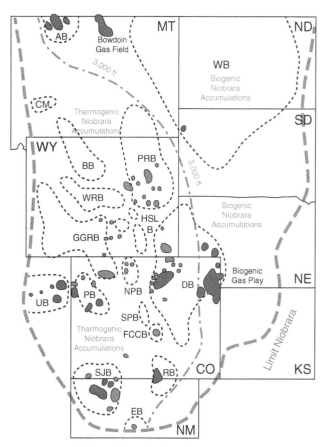

FIGURE 76.17 Niobrara producing areas across the north Rockies (Oil fields: Oil fields: light gray; gas fields dark gray) (modified from Longman et al., 1998). Basin abbreviations are as follows: AB, Alberta Basin; CM, Crazy Mountain; WB, Williston Basin; BB, Bighorn Basin; PRB, Powder River Basin; WRB, Wind River Basin; GGRB, Greater Green River Basin; NPB, North Park Basin; PB, Piceance Basin; UB, Uinta Basin; SPB, South Park Basin; FCCB, Florence-Canon City Basin; SJB, San Juan Basin; RB, Raton Basin; DB, Denver Basin; EB, Estancia Basin. Distribution of sapropelic oil-generation-prone Niobrara source rocks within dashed line (Meissner et al., 1984). Dot-dashed line equals 3000 ft current burial depth; biogenic accumulations east of line; thermogenic accumulations west of line (from Lockridge and Scholle, 1978).

the region is the Florence-Cañon City, which was discovered in 1881 (Figure 76.17). The field produces from the Pierre shale immediately above the Niobrara and is believed to be sourced from the Niobrara and Sharon Springs. The Boulder oil field (western Denver basin) was discovered in 1901 and is also productive from fractured Pierre shale but also sourced from the Niobrara. Fractured Mancos shale production was found in Rangely (northwest Colorado) in 1902. Niobrara production was established in Tow Creek (Sand Wash basin) in 1924. The Berthoud field of the western Denver basin is productive from several horizons including the Niobrara and discovered in 1927. Gas in the Niobrara was discovered in Beecher Island (eastern Colorado) in 1919 (commerciality was not established until 1972, however). The Niobrara interval is productive in the Bowdoin field of Montana, which was discovered in 1913. The reason for these early discoveries is that many of them are associated with surface structures, which were the primary targets of early explorers.

Hydrocarbon production comes from all three major Niobrara lithofacies: 1) microporous and fractured coccolith- and planktonic foraminifer-rich limestone (eastern part of WIC basin); 2) fractured marls and shales (mainly in the central part of the seaway); and 3) fractured sandstone and siltstone-rich facies, mainly in the western and southwestern parts of the seaway. Production occurs in the Laramide-aged Powder River, Denver, North Park, Greater Green River (including Sand Wash), Raton, San Juan, and Piceance basins and in north-central Montana. The widespread distribution of the production along with many wells with hydrocarbon shows

across these basins suggests a large resource play may exist. The majority of recent drilling activity in the Niobrara has been in the Denver Basin, north of Wattenberg field and in southeast Wyoming around the Silo field. Hydrocarbon production from chalk reservoirs occurs along the shallow eastern margin of the Denver basin. Many of the gas accumulations in this area occur in structural traps and reservoirs require hydraulic-fracture stimulation. The gas is biogenic or microbial in origin. Production in the shallow play comes from the upper chalk bench or Beecher Island member of the Niobrara and is mainly from microporosity within the chalks but is enhanced by natural fracturing. Production from the shallow Niobrara from eastern Colorado is 600 billion cubic feet gas (BCFG). Beecher Island Field is one of the largest and first fields discovered in the shallow Niobrara. Commercial production dates back to 1972 (initial discovery in 1919) and the cumulative for the field is 100 BCFG. Three-dimensional seismic data have been used effectively to improve development and exploration success ratios in fields.

Shallow gas production from the Niobrara also occurs in north central Montana. Bowdoin Dome has produced 62 BCFG and 19 MBO from the Niobrara. Additional Niobrara fields are located to the west the Alberta Basin extends into Montana. The largest field to date is the St. Joe Road field, which was discovered in 2001 and has produced 18.2 BCFG.

Deeper in the Denver Basin, the Niobrara is oil productive in a number of fields. The porosity of the chalks in the deeper part of the basin has been dramatically reduced by compaction and burial diagenesis. Production is attributed to the presence of fractures in the chalky intervals. Some attempts have been made to establish production from some of the rich, shaley intervals within the Niobrara. The shale gas and fractured chalk potential of the deep Denver basin area is significant as shown by fields like Wattenberg and Silo. Silo Field was discovered in 1981 and has produced approximately 10.4 MMBO and 8.9 BCFG.

The Niobrara is productive on the Casper Arch of Wyoming at Salt Creek and Teapot fields. Total production has been 1.5 MMBO and 0.2 BCFG. In the deeper, Powder River Basin production has been established in a number of accumulations including Fetter, Hilight, Brooks Draw, and Flat Top. Hilight has produced 411 MBO and 0.8 BCFG to date.

The western portion of the region is productive in a variety of traps and lithologies (mainly siliciclastic) and there is significant potential for hydrocarbon production in many of the western basins. The basal part of the Niobrara equivalent in the west yields oil and gas in the San Juan basin from a sandstone and shale interval (Tocito and Gallup sandstones). Examples of producing fields from the Gallup are Bisti and Verde fields. Bisti Field has produced 41.8 MMBO and 79.2 BCFG. Verde Field has produced 8.1 MMBO and 2.5 BCFG. Examples of fields producing from the Tocito Sandstone are the Blanco South and Chipeta fields. These fields have produced 4.2 MMBO and 18.8 BCFG. Production is from interparticle porosity but is enhanced by fractures. The upper Niobrara equivalent (Smoky Hill member) is productive in the Sand Wash basin from fractured reservoirs (Figure 76.17) and perforated intervals are commonly long. Field examples are Buck Peak and Tow Creek. Buck Peak has produced 4.8 MMBO and 8.5 BCFG. Tow Creek has produced 3 MMBO and 0.3 BCFG. Farther to the west where the Niobrara equivalents are dominantly shale, production is found in the Rangely and Douglas Creek Arch fields. Production form fractured Mancos shale at Rangely represents some of the oldest production in Colorado (1902).

The Mancos at Rangely has produced around 11.9 MMBO and 0.2 BCFG. Neogene age extensional faulting is a key to production at Buck Peak and Rangely. The extensional fracture trend is N60W. The Douglas Creek arch production comes mainly from Cathedral Field. The field has produced 56.5 BCFG and 40.6 MBO from the Mancos (mainly the Mancos B zone).

Other production equivalent to the upper Niobrara zone comes from the Mancos interval in the San Juan basin. Examples of Mancos producing fields are East and West Puerto Chiquito, Rio Puerco; Gavilan, Basin, and Boulder. These fields are interpreted to be fractured reservoirs and producing intervals are hundreds of feet thick. The Puerto Chiquito fields have produced 19.3 MMBO and 55.5 BCFG. Gavilan Field has produced 7.8 MMBO and 111 BCFG. Boulder Field has produced 1.8 MMBO and 1.6 BCFG. Basin Field has produced 120 MBO and 4.1 BCFG. Rio Puerco Field has produced 1.3 MMBO and 1.4 TcfG.

The Mancos is gas productive in the deeper parts of the Uinta basin in several fields including Natural Buttes. The Mancos is also productive in some silty and very fine-grained sandstone zones in the Cathedral field of the Douglas Creek Arch. New Mancos/Niobrara production has been established in several areas of the deeper Piceance Basin (e.g., Mamm Creek field).

Exploration Methods

Methods of exploration for fractured Niobrara reservoirs should incorporate many, if not all, of the following: seismic acquisition, aeromagnetics study, surface lineament analysis, subsurface mapping, isoresistivity mapping, logging technology, and technology to produce the reservoir. The 2D and 3D seismic is extremely important to map structural anomalies. Three-dimensional three-component (compressional and shear wave data) methods have also proved to be effective in analyzing the fractured reservoir. Aeromagnetics is a tool that may identify basement shear zone areas of potential fractures having gradient changes such as narrow zones of steep gradients. Aeromagnetic data examined in the Silo field area illustrates possible northwest-trending shear zones. If basement fracture systems propagate all the way to the surface, then a surface lineament analysis may also be effective. Northwest-trending surface lineament in the Silo area has been mapped by using remote-sensing techniques. Resistivity mapping is important to show areas of oil accumulation. When resistivity mapping is combined with subsurface mapping, the most probable areas of fracturing can be predicted. Logging technologies available to identify fractured reservoirs are geophysical logs such as the FMS, FMI, and CAST. Horizontal drilling and multi-stage hydraulic fracturing offer technologies to economically produce hydrocarbons from the reservoir.

An understanding of the regional stress field is important in most tight oil and gas plays. The direction of maximum horizontal stress (Shmax) is generally the direction of open fractures. Regional horizontal stress maps have been published for North America. The present-day stress field reflects Neogene extensional tectonics and the epeirogenic uplift that has taken place in the western United States.

Summary

Widespread source and reservoir rocks make the Niobrara Formation an attractive target for exploration across the Rocky Mountain region. The Niobrara contains mature source rocks interbedded with brittle limestones (chalks) in the deeper parts of many basins in the Rocky Mountain region. Thermogenic production occurs from the chalk intervals in the eastern part of the region and from siliciclastics and shales in the western and southwestern parts of the Rocky Mountain regions (Uinta and San Juan basins). Biogenic gas production occurs at shallow depths along the eastern Rocky Mountain region in Colorado, Kansas, and Nebraska. Generally production comes from depths less than 3500 ft. Shallow gas production also occurs in several areas of north-central Montana. The shallow gas production generally is structurally controlled.

The Niobrara reservoirs generally have low permeabilites so natural fracturing plays a role in economic production. The limestone (chalk) beds behave in a brittle manner, whereas the adjacent calcareous shales often behave in a ductile manner. Fractures occur for a variety of reasons and several models can be used for exploration. Early created fractures are susceptible to extreme diagenesis and thus generally completely cemented. Late stage structural movement can help re-open old fractures or create new ones. Regional epeirogenic uplift of western North America and subsequent erosion (denudation) may play a role in Niobrara microfractures. The removal of overburden results in lowered effective stress in rocks that may also be overpressured. This mechanism may be important in all tight-reservoir plays in the Rocky Mountain region.

REFERENCES

Barlow, L. K. (1986). An integrated geochemical and paleoecological approach to petroleum source rock evaluation, lower Niobrara Formation (Cretaceous), Lyons, Colorado. *Mount Geol*, 23:107–112.

Clayton, J. L., Swetland, P. J. (1980). Petroleum generation and migration in Denver Basin. *AAPG Bulletin*, 64:1613–1633.

Friedman, M., Wiltschko, D. V. (1992). An approach to exploration for naturally fractured reservoirs, with examples from the Austin Chalk. In: J.W.Schmoker, E.B.Coalson, C.A.Brown, eds.

Geological Studies Relevant to Horizontal Drilling: Examples from Western North America: RMAG Guidebook. p. 143–153.

Landon, S. M., Longman, M. W., Luneau, B. A. (2001). Hydrocarbon source rock potential of the Upper Cretaceous Niobrara Formation, Western Interior seaway of the Rocky Mountain region. *Mount Geol*, 38:1–18.

Lockridge, J. P. (1977). Beecher Island field, Yuma County, Colorado. In: H.K.Veal, ed., *Exploration Frontiers of the Central and Southern Rockies: RMAG Guidebook*, p. 272–279.

Lockridge, J. P., Scholle, P. A. (1978). Niobrara gas in eastern Colorado and northwestern Kansas. In: J.D.Pruit, P.E.Coffin, eds. *Energy Resources of the Denver Basin: RMAG Guidebook*, p. 35–49.

Longman, M. W., Luneau, B. A., Landon, S. M. (1998). Nature and distribution of Niobrara lithologies in the Cretaceous Western Interior Seaway of the Rocky Mountain region. *Mount Geol*, 35:137–170.

Lorenz, J. C., Teufel, L. W., Warpinski, N. R. (1991a). Regional fractures 1: a mechanism for the formation of regional fractures at depth in flat-lying reservoirs. *AAPG Bulletin*, 75:1714–1737.

Mallory, W. W. (1977). Oil and gas from fractured shale reservoirs in Colorado and northwest New Mexico. RMAG Special Publication no. 1, 38 p.

Meissner, F. F., Woodward, J., Clayton, J. L. (1984). Stratigraphic relationships and distribution of source rocks in the greater Rocky Mountain Region. In: Woodward, J., Meissner, F. F., Clayton, J. L., eds, *Hydrocarbon Source Rocks of the Greater Rocky Mountain Region: RMAG Guidebook*, p. 1–34.

Perry, S. L. (1991). A statistical approach for fracture analysis using satellite imagery, Niobrara Formation: Wyoming and Colorado examples (abs.). In: RMAG Short Course Notes: Exploration for Hydrocarbons in the Niobrara Formation, Rocky Mountain Region, p. 17.

Pollastro, R. M. (1992). Natural fractures, composition, cyclicity, and diagenesis of the upper cretaceous Niobrara Formation, Berthoud field, Colorado. In: J.W.Schmoker, E.B.Coalson, C.A.Brown, eds. *Geological Studies Relevant to Horizontal Drilling—Examples From Western North America: RMAG Guidebook*, p. 243–255.

Pollastro, R. M., Martinez, C. J. (1985). Mineral, chemical and textural relationships in rhythmic-bedded, hydrocarbon-productive chalk of the Niobrara Formation, Denver Basin, Colorado. *Mount Geol*, 22:55–63.

Pollastro, R. M., Scholle, P. A. (1986). Exploration and development of hydrocarbons from low-permeability chalks–an example from the Upper Cretaceous Niobrara Formation, Rocky Mountain region. In: C.W.Spencer, R. J.Mast, eds. *Geology of Tight Gas Reservoirs: AAPG Studies in Geology* No. 24, p. 129–141.

Rice, D. D. (1975). Origin of and conditions for shallow accumulations of natural gas. In: F.Exum, G.George, eds. *Geology and Mineral Resources of the Bighorn Basin: Wyoming Geological Association 27th Annual Field Conference Guidebook*, p. 267–271.

Rice, D. D., (1984). Relation of hydrocarbon occurrence to thermal maturity of organic matter in the Upper Cretaceous Niobrara Formation, eastern Denver Basin: evidence of biogenic versus thermogenic origin of hydrocarbons. In: J.Woodward, F. F.Meissner, J. C.Clayton, eds. *Hydrocarbon Source Rocks of the Greater Rocky Mountain Region: RMAG Guidebook*, p. 365–368.

Rice, D. D., Claypool, G. E. (1981). Generation, accumulation, and resource potential of biogenic gas. *AAPG Bull*, 65:5–25.

Smagala, T. M., Brown, C. A., Nydegger, G. L. (1984). Log-derived indicator of thermal maturity, Niobrara Formation, Denver Basin, Colorado, Nebraska, Wyoming. In: J.Woodward, F.F.Meissner, and J.C.Clayton, eds. *Hydrocarbon Source Rocks of the Greater Rocky Mountain region: RMAG Guidebook*, p. 355–363.

Sonnenberg, S. A., Weimer, R. J. (1981). Tectonics, sedimentation and petroleum potential of the northern Denver Basin, Colorado, Wyoming, and Nebraska. *Colorado School of Mines Quarterly*, 7(2)45.

Sonnenberg, S. A., Weimer, R. J. (1993). Oil production from Niobrara Formation, Silo field, Wyoming: fracturing associated with a possible wrench fault system. *Mount Geol*, 30:39–53.

Stearns, D. W., Friedman, M. (1972). Reservoirs in fractured rock: geologic exploration methods. In: H.R.Gould, ed., *Stratigraphic Oil and Gas Fields–Classification, Exploration Methods, and Case Histories: AAPG Memoir*, Vol. 16, pp. 82–106.

Stone, D. S. (1969). Wrench faulting and rocky mountain tectonics. *Mount Geol*, 6:67–79.

Svoboda, J. O. (1995). Is Permian salt dissolution the primary mechanism of fracture genesis at Silo Field, Wyoming? In: R.R.Ray, ed., *High-Definition Seismic: 2-D, 2-D Swath, and 3-D Case Histories: RMAG Guidebook*, p. 79–85.

Vincelette, R. R., Foster, N. H. (1992). Fractured Niobrara of northwestern Colorado. In: J.W.Schmoker, E.B.Coalson, C.A.Brown, eds. *Geological studies relevant to horizontal drilling—Examples from western North America: RMAG Guidebook*, p. 227–239.

Weimer, R. J. (1960). Upper cretaceous stratigraphy, Rocky Mountain Area. *AAPG Bull*, 44:1–20.

Weimer, R. J. (1978). Influence of Transcontinental Arch on Cretaceous marine sedimentation: a preliminary report. In: J.D.Pruit, P.E.Coffin, eds. *Energy Resources of the Denver Basin: RMAG Guidebook*, p. 211–222.

UTAH SHALES, USA

BY THOMAS C. CHIDSEY, JR. AND MICHAEL D. VANDEN BERG (UTAH GEOLOGICAL SURVEY)

The high price of crude oil, coupled with lower natural gas prices, has generated renewed interest in exploration and development of liquid hydrocarbon reserves in Utah. Following on the success of the recent shale gas boom and employing many of the same well completion techniques, petroleum companies are now exploring for liquid petroleum in shale formations in the state. In fact, many shales or low-permeable ("tight") carbonates recently targeted for natural gas include areas in which the zones are more prone to liquid production. Organic-rich shales in the Uinta and Paradox Basins have been the source for significant hydrocarbon generation, with companies traditionally targeting the interbedded sands or porous carbonate buildups for their conventional resource recovery. With the advances in horizontal drilling and hydraulic fracturing techniques, operators in these basins are now starting to explore the petroleum production potential of the shale and interbedded tight units themselves.

Uinta Basin

Overview: The Uinta Basin is the most prolific petroleum province in Utah. It is a major depositional and structural basin that subsided during the early Cenozoic along the southern flank of the Uinta Mountains. Lake deposits filled the basin between the eroding Sevier highlands to the west and the rising Laramide-age Uinta Mountains, Uncompahgre uplift, and San Rafael Swell to the north, east, and south, respectively. The southern Eocene lake, Lake Uinta, formed within Utah's Uinta Basin and Colorado's Piceance Creek Basin.

The Green River Formation consists of as much as 6000 ft of sedimentary strata (Hintze and Kowallis, 2009; Sprinkel, 2009) and contains three major depositional facies associated with Lake Uinta sedimentation: alluvial, marginal lacustrine, and open lacustrine (Fouch, 1975). The marginal lacustrine facies, where most of the hydrocarbon production is found, consists of fluvial-deltaic, interdeltaic, and carbonate flat deposits, including microbial carbonates. The open-lacustrine facies is represented by nearshore and deeper water offshore muds, including the famous Mahogany oil shale zone, which represents Lake Uinta's highest water level.

The Uinta Basin is asymmetrical, paralleling the east-west trending Uinta Mountains. The north flank dips 10–35° southward into the basin and is bounded by a large north-dipping, basement-involved thrust fault. The southern flank gently dips between 4° and 6° north-northwest.

Activity: Recent tight-oil drilling and exploration activities in the Uinta Basin are targeting relatively thin porous carbonate beds of the Uteland Butte Limestone Member of the lower Green River Formation (Figures 76.18 and 76.19), particularly in an area referred to as the "Central Basin region" between Altamont-Bluebell field to the north and Monument Butte field to the south. The Uteland Butte has historically been a secondary oil objective of wells tapping shallower overlying Green River reservoirs and deeper fluvial–lacustrine Colton Formation sandstone units in the western Uinta Basin.

The Uteland Butte records the first major transgression of Eocene Lake Uinta after the deposition of the alluvial Colton Formation, and thus it is relatively widespread in the basin (Figure 76.20). The Uteland Butte ranges in thickness from less than 60 ft to more than 200 ft and consists of limestone,

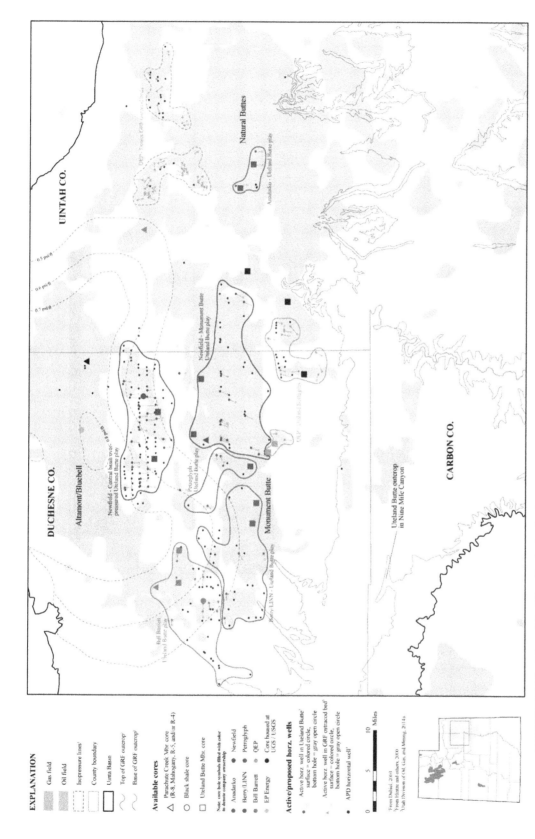

FIGURE 76.18 Map of the Uinta Basin, Utah, showing play areas for the Uteland Butte Limestone Member of the Tertiary Green River Formation, APD horizontal well locations, and active horizontal wells by operators.

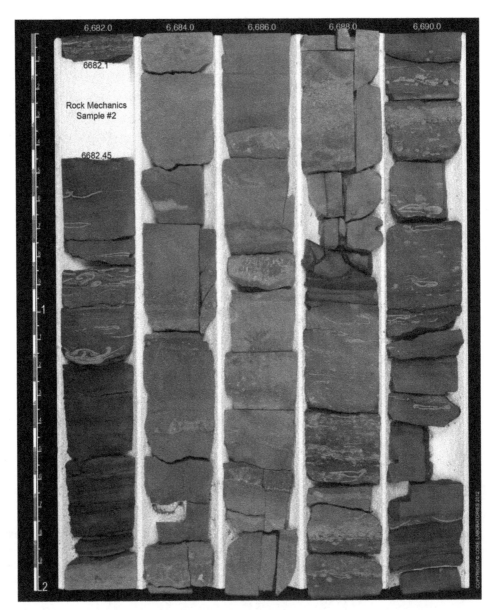

FIGURE 76.19   Uteland Butte core from the Bill Barrett 14-1-46 well. The horizontal drilling target is the roughly 5-ft light brown dolomitic interval (depth 6,684.0 ft–6,689.0 ft). Porosity in this interval ranges from 15% to 30% and permeability averages 0.06 mD. The dolomite is interbedded with organic-rich mudstones and limestones averaging between 1% and 3% TOC. Note the abundant shell fragments indicating deposition in a freshwater lacustrine environment.

dolomite, organic-rich calcareous mudstone, siltstone, and rare sandstone (Figures 76.20–76.22). The dolomite (Figure 76.19), the new horizontal drilling target, often has more than 20% porosity, but is so finely crystalline that the permeability is very low (single mD or less).

Several companies (Berry/LINN, Bill Barrett Corporation, EP Energy, Newfield, QEP Resources, Devon, and Petroglyph) have had recent and continued success targeting the Uteland Butte with horizontal wells in both the central, normally pressured part of the basin near Greater Monument Butte field, and farther north in the overpressured zone in western Altamont field (Figure 76.18). There are over 50 active horizontal wells producing from the Uteland Butte. Production from these wells averages 500–1500 BOE/day from horizontal legs up to 4000 ft in length. However, at the end of 2013, Newfield Production Company announced the completion of three high-volume producers in the Central Basin region with initial rates of 1213–1337 barrels per day (IHS Inc., 2013d). As of January 2014, there were over 200 APDs for horizontal wells targeting the Uteland Butte and

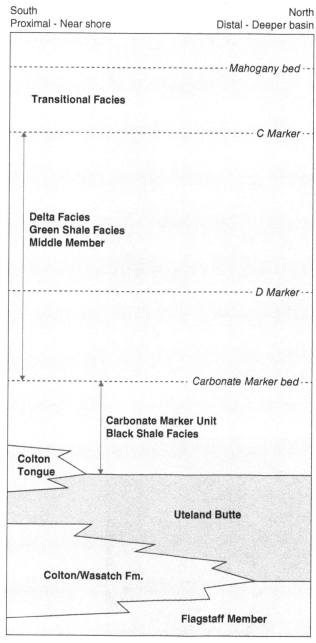

South
Proximal - Near shore

North
Distal - Deeper basin

---------------------------------- *Mahogany bed* --

Transitional Facies

---------------------------------- *C Marker* --

Delta Facies
Green Shale Facies
Middle Member

---------------------------------- *D Marker* --

---------------------------- *Carbonate Marker bed* --

Carbonate Marker Unit
Black Shale Facies

Colton
Tongue

Uteland Butte

Colton/Wasatch Fm.

Flagstaff Member

FIGURE 76.20 General stratigraphy of the lower to middle Green River Formation in the western Uinta Basin (not to scale).

other potential Green River tight-oil zones (Figure 76.18) (Utah Division of Oil, Gas, and Mining, 2014a).

Paradox Basin

Overview: The Paradox Basin is located mainly in southeastern Utah and southwestern Colorado with small portions in northeastern Arizona and the northwestern corner of New Mexico. The Paradox Basin is an elongate,

northwest-southeast-trending, evaporitic basin that predominately developed during the Pennsylvanian, about 330 to 310 Ma. The basin was bounded on the northeast by the Uncompahgre Highlands as part of the Ancestral Rockies. As the highlands rose, an accompanying depression, or foreland basin, formed to the southwest – the Paradox Basin. Rapid basin subsidence, particularly during the Pennsylvanian and continuing into the Permian, accommodated large volumes of evaporitic and marine sediments that intertongue with non-marine arkosic material shed from the highland area to the northeast. Deposition in the basin produced a thick cyclical sequence of carbonates, evaporites, and organic-rich shale of the 500- to 5000-ft-thick Pennsylvanian Paradox Formation (Hintze and Kowallis, 2009). Rasmussen (2010) divided the middle part of the Paradox Formation in the evaporite basin into as many as 35 salt cycles, some of which onlap onto the basin shelf to the west and southwest (Figure 76.23). Each cycle consists of a clastic interval/salt couplet. The clastic intervals are typically interbedded dolomite, dolomitic siltstone, anhydrite, and black, organic-rich shale—the sources of the petroleum in the basin. Clastic intervals typically range in thickness from 10 to 200 ft and are generally overlain by 200 to 400 ft of halite.

The Paradox Basin can generally be divided into three areas: the Paradox fold and fault belt in the north, the Blanding sub-basin in the south-southwest, and the Aneth platform in the southernmost part in Utah. The area now occupied by the Paradox fold and fault belt was the site of greatest Pennsylvanian/Permian subsidence and salt deposition. Folding in the Paradox fold and fault belt began as early as the Late Pennsylvanian as sediments were laid down thinly over, and thickly in areas between, rising salt. Spectacular salt-cored anticlines extend for miles in the northwesterly trending fold and fault belt. Reef-like buildups or mounds of carbonates consisting of algal bafflestone and oolitic/skeletal grainstone fabrics in the Desert Creek and Ismay zones of the Paradox Formation are the main hydrocarbon producers in the Blanding sub-basin and Aneth platform. Oil in these zones is sourced above, below, or within the organic-rich Gothic, Chimney Rock, Hovenweep, and Cane Creek shales (Figure 76.23).

Activity: The Cane Creek shale zone of the Paradox Formation has been a target for tight-oil exploration on and off since the 1960s and produces oil from several small fields (Figure 76.24). The play generated much interest in the early 1990s with the successful use of horizontal drilling. Currently, eight active fields produce from the Cane Creek in the Paradox Basin fold-and-fault belt, with cumulative oil production over 5 million barrels and 4 Bcf of gas (Utah Division of Oil, Gas, and Mining, 2014b). Once again, the Cane Creek and other Paradox zones are being targeted for exploration using horizontal drilling.

The Cane Creek shale zone records an early stage of a transgressive–regressive sequence (cycle 21) in the

FIGURE 76.21 Outcrop of the Uteland Butte Member of the Green River Formation, Nine Mile Canyon, central Utah.

FIGURE 76.22 Fresh road cut exposure of Uteland Butte Member of the Green River Formation consisting of interbedded dolomite and mudstone/limestone, Nine Mile Canyon, central Utah.

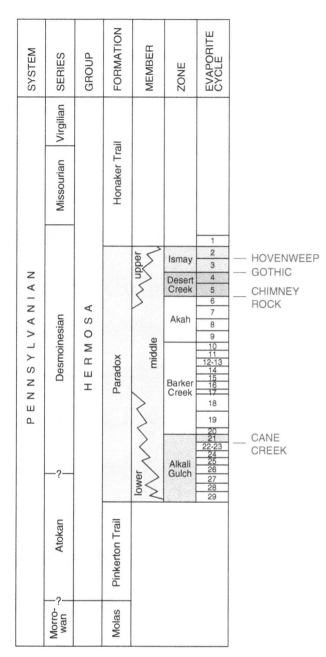

FIGURE 76.23 Pennsylvanian stratigraphic chart for the Paradox Basin, informal organic-rich shale units are highlighted. Modified from Hite (1960), Hite and Cater (1972), and Reid and Berghorn (1981).

Paradox Formation and consists of organic-rich marine shale with interbedded dolomitic siltstone and anhydrite (Figure 76.25). The unit is up to 160 ft thick and areally extensive within the Paradox Basin. It is divided into the A, B, and C zones, with the shale and silty carbonates of the B zone considered both the source rock and reservoir. The A and C zones are anhydrite-rich and provide an upper and lower seal to the B zone. The unit is highly overpressured, with

measurements ranging between 5000 and 6200 psi, which is probably the result of hydrocarbon generation between very impermeable upper and lower anhydrite seals. The B zone is naturally fractured, and oriented cores show that fractures trend northeast-southwest, matching the regional structural trend.

Two new Cane Creek shale horizontal well discoveries were announced in 2013 (Figure 76.18): Fidelity Exploration & Production Company's 17-1 Cane Creek Unit (SWSE section 17, T. 26 S., R. 20 E., Salt Lake Basin Line & Meridian [SLBL&M], Grand County) just east of Park Road field, and Southwestern Energy Production Company's 1-16H SEPCO-State 30-23 (SESE section 16, T. 30 S., R. 23 E., SLBL&M, San Juan County). The 17-1 well was drilled in a south-southeast direction and averaged 524 BOPD and 242 MCFGPD in June 2013. Fidelity completed additional wells in Big Flat and Park Road fields in 2012 and 2013. The 36-1 Cane Creek Unit well (SWSW section 36, T. 25 S., R. 19 E., SLBL&M, San Juan County) in Big Flat field flowed over 1250 BOPD (IHS Inc., 2013b). The company has identified 50 to 75 Cane Creek locations with an estimated recovery of 250,000 to 1 million bbls of oil per well (IHS Inc., 2013a). The 1-16H well was drilled in a northwest direction and in October 2013 was testing oil into onsite production facilities. It is located 5 miles south of Stone Energy Corporation's 2011 29-28 La Sal Unit Cane Creek discovery (SESE section 29, T. 29 S., R. 23 E., SLBL&M, San Juan County). That well initially flowed 248 BOPD and 320 MCFGPD from a northeast-directed horizontal lateral. At the end of 2013, Fidelity staked horizontal wells targeting the Cane Creek, one northeast of Hell Roaring field and two west of Big Flat field.

The U.S. Geological Survey (2012), Whidden et al. (2014), and Anna et al. (2014) re-assessed the undiscovered oil resource in the Cane Creek at 103 million barrels at a 95% confidence level and 198 million barrels at a 50% confidence level. In addition to the Cane Creek, several other organic-rich shale zones are present in the Paradox Formation, creating the potential for significant reserve base additions. The Gothic, Chimney Rock, and Hovenweep shales (Figure 76.23) in the Blanding sub-basin and Aneth platform are estimated to hold an undiscovered oil reserve of 126 million barrels at a 95% confidence level and 238 million barrels at a 50% confidence level (U.S. Geological Survey, 2012; Whidden et al., 2014; Anna et al., 2014). These zones are also actively being evaluated for tight oil potential and several wells were staked in 2013. Anadarko E&P Onshore LLC drilled a horizontal well, the 3424-2-1H Lewis Road-Fee (SWSW section 2, T. 34 S., R. 24 E., SLBL&M, San Juan County), targeting the Gothic shale; no information has been released (IHS Inc., 2013c). Anadarko has three additional horizontal Gothic wells staked 5, 8, and 17 miles to the south, southwest, and southeast, respectively, from the Lewis Road well.

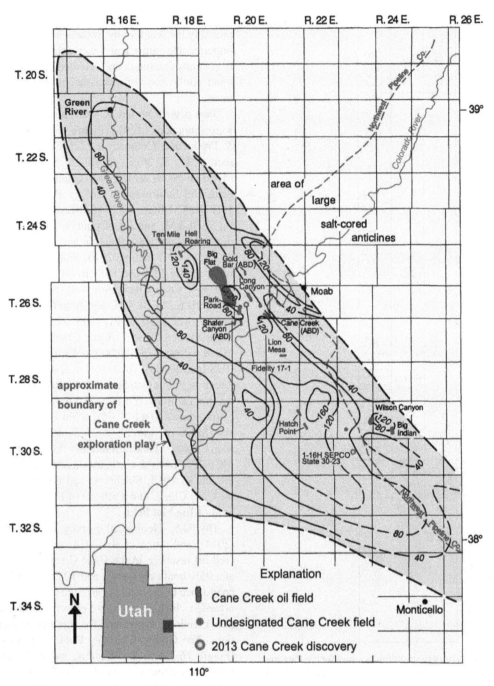

FIGURE 76.24   Thickness map and exploration play area of the Cane Creek shale zone of the Pennsylvanian Paradox Formation, northern Paradox Basin, Utah, showing Cane Creek fields and 2013 discoveries. Contour interval = 40 ft.

New Research: The Utah Geological Survey has been awarded funding from the National Energy Technology Laboratory, U.S. Department of Energy for a three-year project titled "Liquid-Rich Shale Potential of Utah's Uinta and Paradox Basins: Reservoir Characterization and Development." The overall goal of this study is to provide reservoir-specific geological and engineering analyses of the (1) emerging Green River Formation tight oil plays (such as the Uteland Butte Limestone Member, Black Shale facies, deep Mahogany zone, and other deep Parachute Creek member high-organic units) in the Uinta Basin and (2) the established, yet understudied Cane Creek shale (and possibly other shale

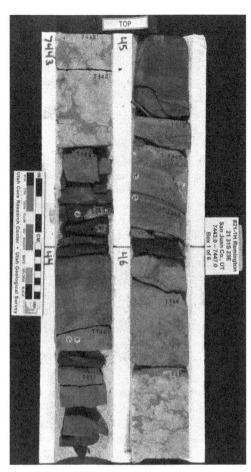

FIGURE 76.25 Cane Creek core from the Union Pacific Resources Remington 21-1H well (section 21, T. 31 S., 23 E., SLBL&M, San Juan County, Utah) displays interbedded medium gray dolomite with organic-rich dark gray/black shale. Also present is mottled light gray to white anhydrite.

units such as the Gothic and Chimney Rock shale zones) of the Paradox Formation in the Paradox Basin. To accomplish this goal, the project will:

- Characterize geologic, geochemical, and petrophysical rock properties of target zones in the two designated basin areas by compiling various sources of data and by analyzing newly acquired and donated core, well logs, and well cuttings.
- Describe outcrop reservoir analogs of Green River Formation plays and compare them to subsurface data (not applicable in the Paradox Basin since the Cane Creek shale is not exposed).
- Map major regional trends for targeted liquid-rich intervals and identify "sweet spots" that have the greatest oil production potential.
- Suggest techniques to reduce exploration costs and drilling risks, especially in environmentally sensitive areas.

- Improve drilling and fracturing effectiveness by determining optimal well completion design.
- Suggest techniques to reduce field development costs, maximize oil recovery, and increase reserves.

The project will therefore develop and make available geologic and engineering analyses, techniques, and methods for exploration and production from the Green River Formation tight oil zones and the Paradox Formation shale zones where operations encounter technical, economic, and environmental challenges.

In addition to a thorough geologic characterization of the target zones, tests will be performed to characterize the geomechanical properties of the zones of interest to inform/guide well completion strategies. The brittle characteristics of the target intervals will be studied in detail using energy-based calculations. This approach acknowledges mechanical properties and *in situ* stress conditions, as well as geometric lithologic constraints and the mineralogy that regulates fracturing. The study will establish a template for more effective well planning and completion designs by integrating the geologic characterization and formation evaluation with state-of-the-art rock mechanical analyses. Results from the study will help companies access oil in known reservoir rock that has been previously technically difficult to recover.

To aid in the identification of hydrocarbon "sweet spots," novel concepts for exploration will be employed such as the use of low-cost, low-environmental impact, epifluorescence analysis of regional core and well cuttings. Epifluorescence microscopy is a technique used to provide information on diagenesis, pore types, and organic matter (including "live" hydrocarbons) within sedimentary rocks. It is a rapid, non-destructive procedure that uses a petrographic microscope equipped with reflected-light capabilities, a mercury-vapor light, and appropriate filtering. Epifluorescent intensities obtained from core and cuttings will be mapped to help identify areas with potential for significant hydrocarbon production. The detailed reservoir characterization and rock mechanics analyses will provide the basis for identification of "sweet spots" and improve well completion strategies for these undeveloped and under-developed reservoirs.

For more information about this ongoing project, including available posters and talks (in pdf), refer to the Utah Geological Survey's project website: http://geology.utah.gov/emp/shale_oil

Recent Presentations

1. Liquid-Rich Shale Potential of Utah's Uinta Basin: Reservoir Characterization and Development, by Michael D. Vanden Berg, January 10, 2013, quarterly Uinta Basin Oil and Gas Collaborative Group Meeting, Vernal, Utah.

2. The Uteland Butte Member of the Eocene Green River Formation: An Emerging Unconventional Carbonate Tight Oil Play in the Uinta Basin, Utah, by Michael D. Vanden Berg, Craig D. Morgan, Thomas C. Chidsey, Jr., and Peter Nielsen, May 19–20, 2013, AAPG Annual Convention, Pittsburgh, Pennsylvania.

3. Current Understanding of the Sedimentology, Stratigraphy, and Liquid-Oil Potential of the Pennsylvanian Cane Creek Shale of the Paradox Formation, Southeastern Utah, by Peter Nielsen, Craig D. Morgan, and Michael D. Vanden Berg, September 22–24, 2013 AAPG Rocky Mountain Section Meeting, Salt Lake City, Utah.

4. Detailed Sedimentology and Stratigraphy of the Remington 21-1H Cane Creek Shale Core, Pennsylvanian Paradox Formation, Southeastern Utah: Implications for Unconventional Hydrocarbon Recovery, by Peter J. Nielsen, Craig D. Morgan, and Michael D. Vanden Berg, September 22–24, 2013 AAPG Rocky Mountain Section Meeting, Salt Lake City, Utah.

5. Reservoir Characterization of the Uteland Butte Formation in the Uinta Basin, by Jason Anderson and John Roesink, September 22–24, 2013 AAPG Rocky Mountain Section Meeting, Salt Lake City, Utah.

6. Temporal and Spatial Variations in Lacustrine Depositional Controls from the Middle to Upper Green River Formation, Central and Western Uinta Basin, Utah," by Leah Toms, Lauren Birgenheier, and Michael D. Vanden Berg, September 22–24, 2013 AAPG Rocky Mountain Section Meeting, Salt Lake City, Utah.

7. Geologic Evaluation of the Cane Creek Shale, Pennsylvanian Paradox Formation, Paradox Basin, Southeastern Utah," by Stephanie M. Carney, Peter J. Nielsen, and Michael D. Vanden Berg, April 9, 2014 AAPG Annual Convention, Houston, Texas.

REFERENCES

Anna, L. O., Whidden, K. J., Lillis, P. G., Pearson, K. M. (2014). Assessment of continuous oil and gas reservoirs, Paradox Basin, Utah, Colorado, New Mexico, and Arizona. *Mount. Geol.*, 51(2):139–160.

Dubiel, R. F. (2003). Geology, Depositional Models, and Oil and Gas Assessment of the Green River Total Petroleum System, Uinta-Piceance Province, Eastern Utah and Western Colorado, in U.S. Geological Survey Uinta-Piceance Assessment Team, compilers, Petroleum systems and geologic assessment of oil and gas in the Uinta-Piceance Province, Utah and Colorado: U.S. Geological Survey Digital Data Series DDS-69-B.

Fouch, T. D. (1975). Lithofacies and related hydrocarbon accumulations in tertiary strata of the Western and Central Uinta Basin, Utah. In: Bolyard, D.W., ed., *Symposium on Deep Drilling Frontiers in the Central Rocky Mountains: Rocky Mountain Association of Geologists Guidebook*, p. 163-173.

Hintze, L. F., Kowallis, B. J. (2009). Geologic history of Utah. Brigham Young University Geology Studies Special Publication 9, 225.

Hintze, L. F., Willis, G. C., Laes, D. Y. M., Sprinkel, D. A., Brown, K. D. (2000). Digital geologic map of Utah. Utah Geological Survey Map 179DM, scale 1:500,000.

Hite, R. J. (1960). Stratigraphy of the saline facies of the Paradox Member of the Hermosa Formation of southeastern Utah and southwestern Colorado, In: Smith, K.G., ed., *Geology of the Paradox Basin Fold and Fault Belt: Four Corners Geological Society, Third Field Conference Guidebook*, pp. 86–89.

Hite, R. J., Cater, F. W. (1972). Pennsylvanian rocks and salt anticlines, Paradox Basin, Utah and Colorado. In: Mallory, W.W., ed. *Geologic Atlas of the Rocky Mountain region: Rocky Mountain Association of Geologists Guidebook*, p. 133–138.

IHS Inc. (2013a). Rocky Mountain regional report. (August 14, 2013) non-paginated.

IHS Inc. (2013b). Rocky Mountain regional report. (November 4, 2013) non-paginated.

IHS Inc. (2013c). Rocky Mountain regional report. (November 19, 2013) non-paginated.

IHS Inc. (2013d). Rocky Mountain regional report. (November 25, 2013) non-paginated.

Rasmussen, D. L. (2010) Halokinesis features related to flowage and dissolution of Pennsylvanian Hermosa salt in the Paradox Basin, Colorado and Utah [abs]: American Association of Petroleum Geologists, Rocky Mountain Section Meeting Program with Abstracts, 59.

Reid, F. S., Berghorn, C. E. (1981). Facies recognition and hydrocarbon potential of the Pennsylvanian paradox formation. In: Wiegand, D.L., ed. *Geology of the Paradox Basin: Rocky Mountain Association of Geologists Guidebook*, p. 111–117.

Sprinkel, D. A. (2009). Interim geologic map of the Seep Ridge 30' x 60' quadrangle, Uintah, Duchesne, and Carbon Counties, Utah, and Rio Blanco and Garfield Counties, Colorado: Utah Geological Survey Open-file Report 549, compact disc, GIS data, 3 plates, scale 1:100,000.

Utah Division of Oil, Gas, and Mining. (2014a). Well-file database: Online, oilgas.ogm.utah.gov, accessed May 2014.

Utah Division of Oil, Gas, and Mining. (2014b). Oil and gas summary production report by field, December 2013: Online, fs.ogm.utah.gov/pub/Oil&Gas/Publications/Reports/Prod/Field/Fld_Dec_2013.pdf, accessed May 2014.

U.S. Geological Survey. (2012). Assessment of undiscovered oil and gas resources in the Paradox Basin province, Utah, Colorado, New Mexico, and Arizona, 2011: U.S. Geological Survey Fact Sheet 2012-3021, March 2012, 4 p., online, http://pubs.usgs.gov/fs/2012/3031/FS12-3031.pdf, accessed July 25, 2012.

Whidden, K. J., Lillis, P. G., Anna, L. O., Pearson, K. M., Dubiel, R. F. (2014). Geology and total petroleum systems of the Paradox Basin, Utah, Colorado, New Mexico, and Arizona. *Mount Geol*,51(2):119–138.

UTICA SHALE (ORDOVICIAN), APPALACHIAN BASIN, USA

BY RICH NYAHAY (NYAHAY GEOSCIENCES, LLC, COBLESKILL, NY)

The Ordovician Utica (Indian Castle), Dolgeville, and Flat Creek are the formations of interest. These shales and interbeded limestones range in TOC from 1% to 5% in the dry gas window. They cover an area from Mohawk Valley south to the New York State boundary line with Pennsylvania and extend west to the beginning of the Finger Lakes region and east to the Catskill Mountain region. These three formations have a total thickness from 700 to 1000 ft.

In Ohio and Pennsylvania, the Utica is underlain by organic-rich Point Pleasant Formation that is in part the lateral equivalent of the upper portion of the Trenton limestone and is in the gradational relationship with the overlying Utica shale which thickens into the Appalachian Basin (Wickstom, 2011). The Utica Point Pleasant interval is up to 300 ft thick in Ohio and over 600 ft thick in southwestern Pennsylvania. The TOC in this interval ranges from 1% to 4% (Harper, 2011). In Ohio, gas-prone areas will be found in the deeper parts of the basin well as appreciable amounts of oil (Ryder, 2008).

In Michigan, the Utica is underlain by the Collingwood Formation in the northern central part of the state. This formation consists of shales that are black to brown and dark gray in color, with a thickness between 25 and 40 ft and TOC range between 2% and 8% (Snowdon, 1984).

Geology

The Late Ordovician Utica shale was deposited in a foreland basin setting adjacent to and on top of the Trenton and Lexington carbonate platforms. Initial deposition of the Trenton and Lexington platform began on a relatively flat Black River passive margin. Early tectonic activity from the Taconic orogeny created the foreland bulge that would become the Trenton and Lexington platforms. Carbonate growth was able to keep up with the overall rise in sea level while areas stayed relatively deeper until increased subsidence in the foreland basin lowered the ramps out of the photic zone and inundated the passive margin with fine grained clastics. (Willan et al. 2012).

The Trenton/Lexington limestone through the Utica Shale comprise the trangressive systems tract (TST) of a large second-order sequence, superimposed with four, smaller scale third-order composite sequences. Third-order sequences are regional correlative, aggradational, and lack lowstand deposits. Sequences are separated by type 3 sequence boundaries that amalgamate with trangressive surfaces and separate underlying highstand system tracts (HSTs) from overlying TSTs (McClain, 2012).

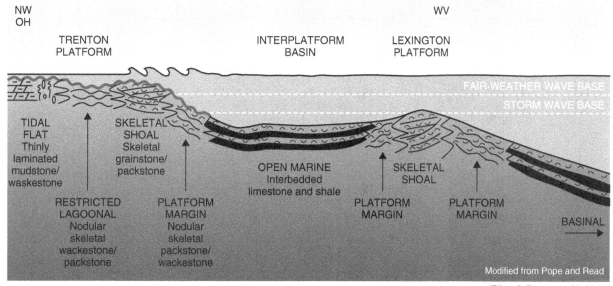

(Riley & Baranowski, 2011)

Smith (2013) proposes that the organic-rich sections were deposited in shallow water to the west and become progressively less organic-rich while approaching what was the deepest part of the basin due to progressively more dilution from clay and silt that are sourced from the highlands to the east, but it may be the longest duration of anoxic conditions occurred in the shallowest water.

Well Activity

With the current regulatory moratorium in place in New York, well activity has been focused in eastern Ohio, western Pennsylvania, and northern Michigan. Ohio's current drilling activity as of March 2014 lists 1148 Utica permits, 770 wells drilled, and 385 producing wells (ODNR, 2014). The three biggest operators in the play are Chesapeake Energy with 584 permits, Gulfport Energy with 109 permits, and Antero Resources with 62 permits (Downing, 2014). Well activity has been concentrated further to the south producing mostly dry gas and companies are moving slowly to the west in search of more oil-rich deposits (see Initial map production figure below).

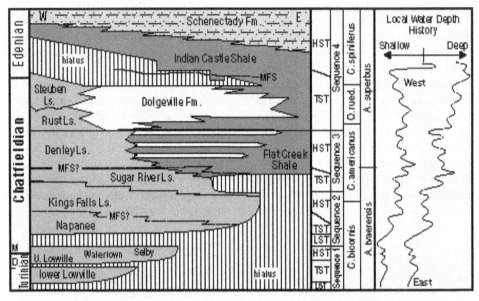

(Joy et al. 2000)

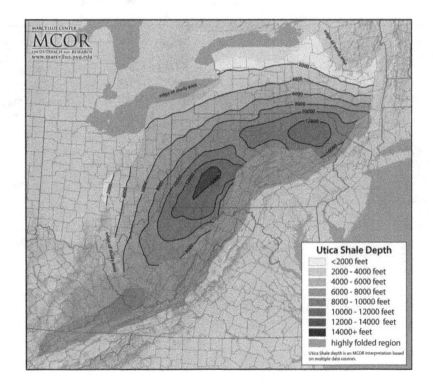

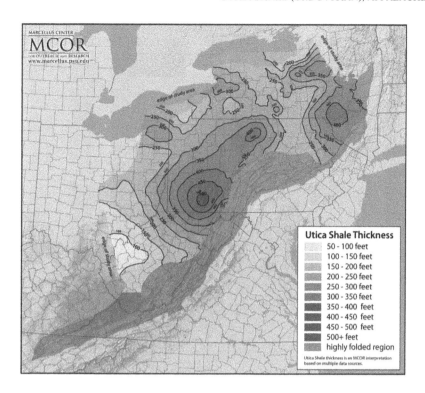

Earlier expectations have been dampened by larger companies selling large acreage parcels, pipeline infrastructure not in place, and construction of gas-processing units. This concern may be changing because the current activity has doubled the number of producing wells again from the latest third quarter production reports of 2013. Three cryogenic natural gas plants have been added in Columbia, Harrison, and Noble counties to separate and purify natural gas (Downing, 2013).

Kinder Morgan Energy Partners LP and Mark West Energy Partners LP and planning a Utica Marcellus Pipeline to Texas. This project has a target date of second quarter 2016. Spectra Energy Corp is also planning a Utica to Gulf Coast pipeline to operational by November 2015 (Knox, 2014).

On average Ohio Utica well takes between 22 and 23 days to drill and costs between 8 and 10 million dollars (Downing, 2014).

Ohio Three wells in Monroe County are permitted to tap both the Marcellus and Utica Shales. The Magnum Resources Stadler No 3UH was drilled to a depth of 10,653 ft and horizontally drilled 5050 ft and completed in 20 stages producing 32.5 MMCFPD of 97% methane (Oil & Gas Journal, 2014).

Pennsylvania Mercer County of Pennsylvania has been gaining interest with well permits by Halcon Resources (7), Hilcorp Energy (5), Shell Western E&P Inc. – a division of Royal Dutch Shell (3), and Cheveron – Appalchia (3). Shell Western E&P Inc. – a division of Royal Dutch Shell also permitted a well in Lawrence County, PA.

Halcon Reources tested wells in Crawford, Mercer, and Venango counties, Pennsylvania and has taken 1844 ft of conventional core.

- The Staab1H was drilled to a total measured depth of 11,166 ft, a lateral length of 4364 ft and completed with17 frac stages in Crawford County.
- The Pilgrim 2-3H was drilled to a total measured depth of 16,185 ft, a lateral length of 7320 ft and completed with 31 frac stages in Mercer County
- The Yoder 2H was drilled to a total measured depth of 10,825 ft, a lateral length of 3810 ft and completed with14 frac stages in Mercer County.
- The Phillips 1H was drilled to a total measured depth of 12,411 ft, a lateral length of 5360 ft and completed with 20 frac stages in Mercer County. It has 304 ft of net pay with an average of 5.1% effective porosity and 72% hydrocarbon saturation.
- The Allam 1H was drilled to a total measured depth of 14,300 ft, a lateral length of 5580 ft and completed with 21 frac stages in Crawford County. It has 266 ft of net pay with an average of 4.4% effective porosity and 75% hydrocarbon saturation.

Both Phillips and Allam wells were located in the gas/condensate window of the play and have BTU contents of 1250 and 1210, respectively (Halcon Resources June 13, 2013 news release).

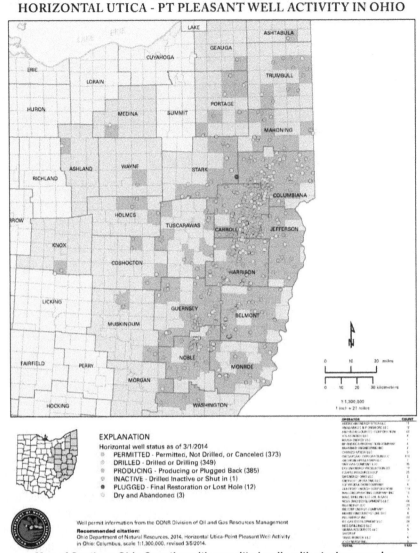

OHIO DEPARTMENT OF NATURAL RESOURCES

HORIZONTAL UTICA - PT PLEASANT WELL ACTIVITY IN OHIO

EXPLANATION

Horizontal well status as of 3/1/2014

PERMITTED - Permitted, Not Drilled, or Canceled (373)
DRILLED - Drilled or Drilling (349)
PRODUCING - Producing or Plugged Back (385)
INACTIVE - Drilled Inactive or Shut in (1)
PLUGGED - Final Restoration or Lost Hole (12)
Dry and Abandoned (3)

Well permit information from the ODNR Division of Oil and Gas Resources Management

Recommended citation:
Ohio Department of Natural Resources, 2014, Horizontal Utica-Point Pleasant Well Activity in Ohio: Columbus, scale 1:1,300,000, revised 3/5/2014.

Map of Southern Ohio Counties with permitted wells with stack reservoirs

Michigan Michigan Department of Environmental Quality said it is proposing new rules for the operation commonly known as fracking and expects to implement them next year (Downing, 2013). This might signal increased exploration activity of the Collingwood.

Completion Techniques A new technique used to test the possible productivity of a new well is to set a permanent plug isolating the final stage or the stage closest to the well head while letting the other stages rest, usually 3 or more months (EID, 10/10/2012).

Gulfport Energy found a 225-foot optimum stage length and is now thinking about 250 ft between laterals.

The basic completion concept is to drill with long laterals, have short stages, and shut in the well for a determined resting period.

The Utica is still in its infancy, everyone is still trying to get the oil and liquids from the Formation. With all the wells scheduled to be drilled in 2013, looking at more geological and petrophysical data that will be generated from each well will further define the future of Utica Play in the Appalachian Basin.

Geochemistry

In May of 2012, a TOC map generated by the Ohio Geological Survey caused a fall out between the State Geologist and critics from southeastern counties of Ohio. The main criticism of the map was the limited amount data points in the southeastern part of the state which may have caused limited interest and lower lease and bonus prices offered to landowners. This newer map does seem to justify extending the play to the southeast.

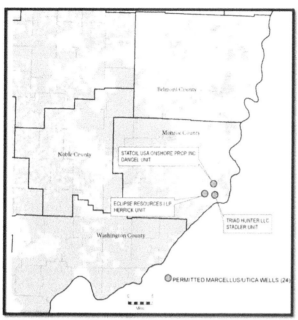

(Wickstrom and Shumway, 2014)

New SEM images from Gulfport Energy well show porosity and permeability is associated with organic matter and organic horizontal fractures might indicate overpressure.

Issues: The Ohio Division of Natural Resources confirmed 11 low magnitude earthquakes that occurred near Hilcorp Energy Corp's well pad operations in Poland Township, Ohio. The series of earthquakes were recorded by both the USGS and Columbia University's Lamont-Doherty Earth Observatory. The USGS confirmed five earthquakes ranging from 2.1 to 3.0 magnitude and Columbia University's Lamont-Doherty Earth Observatory registered six lower magnitude shocks in other places in the region (McParland, 2014)

The first quake occurred at a depth of 1.2 miles and the second quake was recorded at a depth of 3 miles (Obrien, 2014). The Precambrian basement is at 9000 ft and the vertical depth of the Hilcorp Energy well in the area is at a depth of 7900 ft.

These earthquakes have brought attention to whether fracking causes these earthquakes or are they naturally occurring. Ohio records reveal that between 1950 and 2009, there

Map of Initial Production

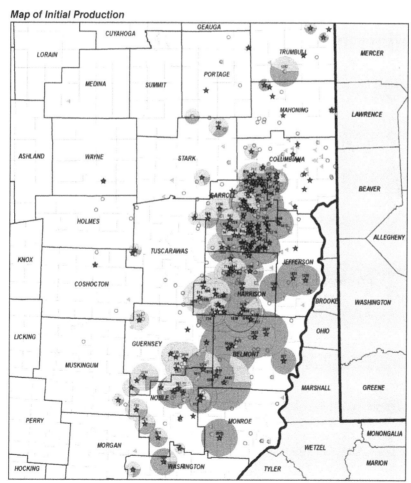

Wickstrom and Shumway, 2014

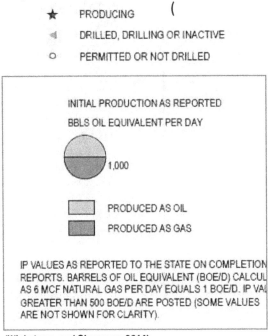

Legend for Initial Production

★ PRODUCING

◁ DRILLED, DRILLING OR INACTIVE

○ PERMITTED OR NOT DRILLED

INITIAL PRODUCTION AS REPORTED

BBLS OIL EQUIVALENT PER DAY

1,000

PRODUCED AS OIL

PRODUCED AS GAS

IP VALUES AS REPORTED TO THE STATE ON COMPLETION REPORTS. BARRELS OF OIL EQUIVALENT (BOE/D) CALCUL AS 6 MCF NATURAL GAS PER DAY EQUALS 1 BOE/D. IP VAL GREATER THAN 500 BOE/D ARE POSTED (SOME VALUES ARE NOT SHOWN FOR CLARITY).

(Wickstrom and Shumway, 2014)

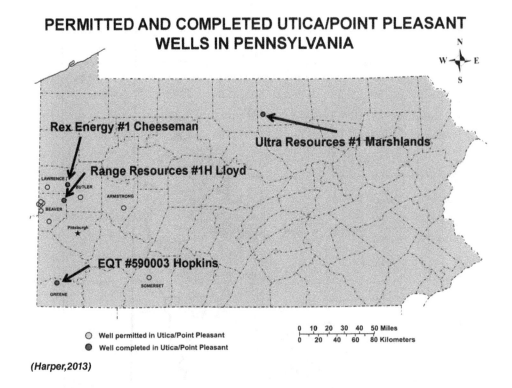

PERMITTED AND COMPLETED UTICA/POINT PLEASANT WELLS IN PENNSYLVANIA

○ Well permitted in Utica/Point Pleasant
● Well completed in Utica/Point Pleasant

0 10 20 30 40 50 Miles
0 20 40 60 80 Kilometers

(Harper, 2013)

was an average of two earthquakes each year that had a magnitude of 2.0 or greater. Between 2010 and 2014, the average rose to nine (Drabold, W., 2014). The Ohio Division of Natural Resources has halted operations and has been petitioned by local residences to set up a seismic network in the area to monitor operations. The area is close to the Youngstown where an injection well was determined to cause earthquakes in 2011.

Northwest PA – Utica/Point Pleasant Potential

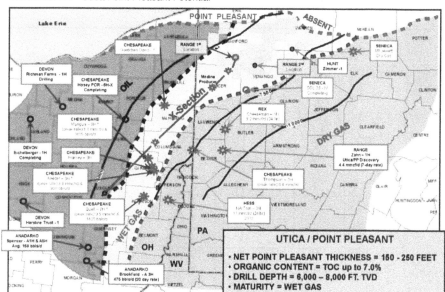

Note: * CHK rates include ethane

Target is Point Pleasant Carbonate Section

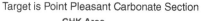

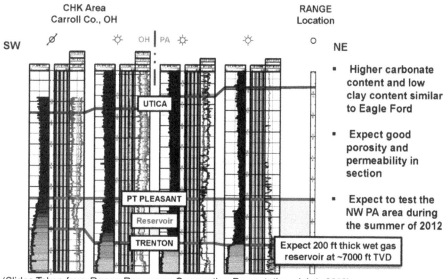

(Slides Taken from Range Resources Corporation Presentation, July1, 2012)

Governor Kasich has proposed a flat tax of 2.75% on producer's gross receipts (Provance, J., 2014).

Sites

http://geosurvey.ohiodnr.gov/major-topics/interactive-maps – This website provides the downloadable data of oil and gas in Ohio as well as information on type logs, cores, and instructions on how to download digital and raster geophysical logs.

http://esogis.nysm.nysed.gov – This website provides information on well logs, formation tops, core, and well samples. In this website, many studies on New York reservoirs sponsored by NYSERDA can be downloaded for free of cost.

http://www.dec.ny.gov/energy/205.html – This website is used to find out information on wells being permitted,

HIGH VOLUME HYDRAULIC FRACTURING
ACTIVE APPLICATIONS AND ISSUED PERMITS - SINCE 2008*
AS OF 03/18/2014

(Michigan Department of Environmental Quality, 2014)

HIGH VOLUME (>100,000 gallons) HYDRAULIC FRACTURING SINCE 2008 - ACTIVE PERMITS

#	Permit #	Company Name	Well Name	Well No	County	Wellhead T R S	Pilot Boring	Comments	Target formation	Well Type	Well Status	Confidential
1	59112	BEACON EXPLORATION AND PRODUCTION CO LLC	SCHULTZ	1-36	SANILAC	12N 15E 36	NA	well completed Feb. 2012	A1 Carbonate	Oil	Shut-in	NO
2	59173	CIMAREX ENERGY CO	SOPER	1-25 HD1	OSCEOLA	17N 10W 25	ACOW	well completed Aug. 2008	Antrim	Gas	Plugging complete	NO
3	59979	ENCANA OIL AND GAS USA INC	PIONEER	1-3 HD1	MISSAUKEE	24N 7W 3	59919	well completed Feb 2010	Utica-Collingwood	Gas	Temporarily abandoned	NO
4	60041	MERIT ENERGY COMPANY	HUBBEL	2-22 HD1	MONTMORENCY	29N 1E 22	NA	well completed June 2010	Niagaran	Oil	Producing	NO
5	60161	ATLAS RESOURCES LLC	STATE NORWICH	1-6 HD1	MISSAUKEE	24N 8W 6	NA	well not hydraulically fractured to date	Utica-Collingwood	Dry Hole	Temporarily abandoned	NO
6	60170	ENCANA OIL AND GAS USA INC	STATE KOEHLER & KENDALL	1-27 HD1	CHEBOYGAN	36N 2W 33	60133	well completed Oct 2010	Utica-Collingwood	Oil	Temporarily abandoned	NO
7	60199	ATLAS RESOURCES LLC	LUCAS	1-13 HD1	KALKASKA	28N 8W 13	60138	well not hydraulically fractured to date	Utica-Collingwood	Not available	Temporarily abandoned	NO
8	60212	COUNTRYMARK RESOURCES INC	KELLY ET AL	1-26 HD1	HILLSDALE	8S 2W 26	NA	well completed Sept. 2011	Black River (Van Wert)	Oil	Producing	NO
9	60389	ENCANA OIL AND GAS USA INC	STATE EXCELSIOR	1-13 HD1	KALKASKA	27N 9W 24	60367	well completed Nov 2011	Utica-Collingwood	Gas	Producing	NO
10	54698	TIGER DEVELOPMENT LLC	STATE GARFIELD & TIGER	1-14	KALKASKA	25N 8W 14	NA	Well completed Oct. 2012	Collingwood	Gas	Temporarily abandoned	NO
11	60380	DEVON ENERGY PRODUCTION COMPANY LP	CRONK	1-24 HD1	GLADWIN	19N 1W 24	60379	well completed April/May 2012	A1 Carbonate	Dry Hole	Plugging complete	NO
12	60398	ENCANA OIL AND GAS USA INC	STATE EXCELSIOR	1-26 HD1	KALKASKA	28N 8W 1	NA	well completed Nov 2011	Utica-Collingwood	Gas	Producing	NO
13	60452	DEVON ENERGY PRODUCTION COMPANY LP	WILEY	1-18 HD1	GLADWIN	18N 2W 18	59451	well completed May/June 2012	A1 Carbonate	Gas	Plugging complete	NO
14	60526	DEVON ENERGY PRODUCTION COMPANY LP	SCHICK	1-7HD1	CLARE	19N 3W 7	60525	well not hydraulically fractured	A1 Carbonate	Other	Plugging complete	NO
15	60537	CONTINENTAL RESOURCES INC	MCNAIR ET AL	1-36 HD1	HILLSDALE	8S 2W 36	59536	well completed August 2012	Black River (Van Wert)	Oil	Producing	NO
16	60545	ENCANA OIL AND GAS USA INC	STATE EXCELSIOR	2-25 HD1	KALKASKA	28N 8W 1	NA	well completed Oct 2012	Utica-Collingwood	Gas	Producing	NO
17	60546	ENCANA OIL AND GAS USA INC	STATE EXCELSIOR	3-25 HD1	KALKASKA	28N 8W 1	NA	well completed Oct 2012	Utica-Collingwood	Gas	Producing	NO
18	60560	DEVON ENERGY PRODUCTION COMPANY LP	STATE RICHFIELD	1-34 HD1	ROSCOMMON	22N 1W 27	60559	well completed Nov 2012	Collingwood	Gas	Plugging complete	NO
19	60575	ALTA ENERGY OPERATING LLC	RILEY	1-22 HD1	OCEANA	15N 18W 22	60874	Well completed Dec. 2012/May 2013	A1 Carbonate	Oil	Well complete	NO
20	60579	ENCANA OIL AND GAS USA INC	STATE GARFIELD	1-25 HD1	KALKASKA	25N 8W 36	NA	Well completed Dec. 2012	Utica-Collingwood	Gas	Well Complete	NO
21	60592	DEVON ENERGY PRODUCTION COMPANY LP	DAVID'S ACRES, LLC	1-19 HD1	OGEMAW	22N 4E 19	60681	well not hydraulically fractured	A1 Carbonate	Other	Plugging complete	NO
22	60597	COUNTRYMARK ENERGY RESOURCES LLC	ARNO	1-26 HD1	HILLSDALE	6S 2W 26	NA	permit for horizontal well	Black River (Van Wert)	Location	Permitted Well	NO
23	60598	COUNTRYMARK ENERGY RESOURCES LLC	ARNO & TIMMONS	1-24 HD1	HILLSDALE	6S 3W 24	NA	permit for horizontal well	Black River (Van Wert)	Location	Permitted Well	NO
24	60601	ENCANA OIL AND GAS USA INC	WESTERMAN	1-32 HD1	KALKASKA	29N 9W 29	60600	No completion records on file	Utica-Collingwood	Location	Drilling complete	NO
25	60606	ENCANA OIL AND GAS USA INC	STATE MENTOR	1-17 HD1	CHEBOYGAN	34N 3W 17	NA	permit for horizontal well (60682)	Utica-Collingwood	Location	Permitted Well	YES
26	60615	ROSETTA RESOURCES OPERATING LP	STATE ORANGE & CHRISTENSEN	1-21 HD1	IONIA	6N 8W 21	60614	Well completed June 2013	A1 Carbonate	Dry Hole	Well complete	NO
27	60617	DEVON ENERGY PRODUCTION COMPANY LP	YOUNGMAN	1-29 HD1	MISSAUKEE	21N 8W 29	00581	well not hydraulically fractured	Utica	NA	Plugging complete	NO
28	60621	ENCANA OIL AND GAS USA INC	STATE BEAVER CREEK	1-23 HD1	CRAWFORD	25N 4W 11	60620	well completed May 2013	Utica-Collingwood	Gas	Well Complete	NO
29	60662	COUNTRYMARK ENERGY RESOURCES LLC	STIVERSON & FRENCH	1-25 HD1	HILLSDALE	6S 2W 24	NA	well not hydraulically fractured	Black River (Van Wert)	Dry Hole	Plugging complete	NO
30	60672	ENCANA OIL AND GAS USA INC	STATE ROSCOMMON	1-7 HD1	ROSCOMMON	21N 4W 17	60670	permit for horizontal well	Utica-Collingwood	Location	Permitted Well	YES
31	60674	MUZYL OIL CORPORATION	BURNS	A1-23 HD1	HILLSDALE	6S 2W 23	NA	permit for horizontal well	Black River (Van Wert)	Location	Permitted Well	NO
32	60688	ENCANA OIL AND GAS USA INC	STATE GARFIELD	1-23 HD1	KALKASKA	25N 8W 26	60665	permit for horizontal well	Utica-Collingwood	Location	Permitted Well	YES
33	60718	JORDAN DEVELOPMENT CO. LLC	STATE JEROME & STAHNKE	15-8 HD1	MIDLAND	6N 1W 8	60717	No completion records on file	Dundee	Location	Drilling complete	NO
34	60746	ENCANA OIL AND GAS USA INC	STATE EXCELSIOR	1-14 HD1	KALKASKA	27N 8W 24	NA	permit for horizontal well	Utica-Collingwood	Location	Permitted Well	YES
35	60747	ENCANA OIL AND GAS USA INC	STATE EXCELSIOR	1-12 HD1	KALKASKA	27N 8W 24	NA	permit for horizontal well	Utica-Collingwood	Location	Permitted Well	YES
36	60748	ENCANA OIL AND GAS USA INC	STATE EXCELSIOR	1-11 HD1	KALKASKA	27N 8W 24	NA	permit for horizontal well	Utica-Collingwood	Location	Permitted Well	YES
37	60749	ENCANA OIL AND GAS USA INC	STATE EXCELSIOR	2-14 HD1	KALKASKA	27N 8W 24	NA	permit for horizontal well	Utica-Collingwood	Location	Permitted Well	YES
38	60750	ENCANA OIL AND GAS USA INC	STATE EXCELSIOR	2-12 HD1	KALKASKA	27N 8W 24	NA	permit for horizontal well	Utica-Collingwood	Location	Permitted Well	YES
39	60755	ROSETTA RESOURCES OPERATING LP	SWANSON TRUST	1-1 HD1	MUSKEGON	9N 14W 1	60754	permit for horizontal well	A1-Carbonate	Location	Permitted Well	YES
40	60785	ENCANA OIL AND GAS USA INC	STATE EXCELSIOR	3-12 HD1	KALKASKA	27N 8W 24	NA	permit for horizontal well	Utica-Collingwood	Location	Permitted Well	YES
41	60786	ENCANA OIL AND GAS USA INC	STATE EXCELSIOR	4-12 HD1	KALKASKA	27N 8W 24	NA	permit for horizontal well	Utica-Collingwood	Location	Permitted Well	YES
42	60787	ENCANA OIL AND GAS USA INC	STATE EXCELSIOR	5-12 HD1	KALKASKA	27N 8W 24	NA	permit for horizontal well	Utica-Collingwood	Location	Permitted Well	YES
43	60789	UNION GAS OPERATING COMPANY	MERTEN	1-24 HD1	OCEANA	15N 17W 24	60787	permit for horizontal well	A1-Carbonate	Location	Permitted Well	NO
44	60803	WHITING OIL AND GAS CORPORATION	WALKER	11-25 HD1	SANILAC	12N 15E 25	60008	permit for horizontal well	A1-Carbonate	Location	Permitted Well	NO
45	60811	GEOSOUTHERN OPERATING LLC	SHERWOOD	1-22 HD1	LIVINGSTON	4N 3E 22	60804	permit for horizontal well	A-1 Carbonate	Location	Permitted Well	YES
46	60818	ENCANA OIL AND GAS USA INC	STATE OLIVER	3-13 HD1	KALKASKA	28N 8W 1	NA	permit for horizontal well	Utica-Collingwood	Location	Permitted Well	YES
47	60819	ENCANA OIL AND GAS USA INC	STATE EXCELSIOR	4-25 HD1	KALKASKA	28N 5W 1	NA	permit for horizontal well	Utica-Collingwood	Location	Permitted Well	YES
48	60820	ENCANA OIL AND GAS USA INC	STATE OLIVER	3-13 HD1	KALKASKA	28N 8W 1	NA	permit for horizontal well	Utica-Collingwood	Location	Permitted Well	YES
49	60821	ENCANA OIL AND GAS USA INC	STATE OLIVER	1-13 HD1	KALKASKA	28N 8W 1	NA	permit for horizontal well	Utica-Collingwood	Location	Permitted Well	YES
50	60822	ENCANA OIL AND GAS USA INC	STATE EXCELSIOR	5-25 HD1	KALKASKA	28N 8W 1	NA	permit for horizontal well	Utica-Collingwood	Location	Permitted Well	YES
51	60828	WHITING OIL AND GAS CORPORATION	STATE WHEATLAND & REINELT	11-7 HD1	SANILAC	13N 14E 7	60825	permit for horizontal well	A1 Carbonate	Location	Permitted Well	YES
52	60848	ENCANA OIL AND GAS USA INC	STATE PIONEER	3-4 HD1	MISSAUKEE	24N 7W 3	NA	permit for horizontal well	Utica-Collingwood	Location	Permitted Well	YES
53	60891	ENCANA OIL AND GAS USA INC	STATE NORWICH	3-12 HD1	KALKASKA	25N 8W 36	NA	permit for horizontal well	Utica-Collingwood	Location	Permitted Well	YES
54	60892	ENCANA OIL AND GAS USA INC	BLACK RIVER CONSERVATION ASSN	1-8 HD1	KALKASKA	27N 5W 26	NA	permit for horizontal well	Utica-Collingwood	Location	Permitted Well	YES
55	60922	O I L ENERGY CORP	STATE CUSTER AND MUNN	D3-8	ANTRIM	29N 7W 8	NA	permit for directional well	Antrim	Location	Permitted Well	NO
56	60930	WHITING OIL AND GAS CORPORATION	RICH ET AL	11-33 HD1	SANILAC	12N 15E 4	60927	permit for horizontal well	A-1 Carbonate	Location	Permitted Well	YES

HIGH VOLUME (>100,000 gallons) HYDRAULIC FRACTURING PROPOSALS - ACTIVE APPLICATIONS

#	App #	Company Name	Well Name	Well No	County	Wellhead T R S	Pilot Boring	target formation	comments
1	A130162	ENCANA OIL AND GAS USA INC	BLACK RIVER CONSERVATION ASSN	6-9 HD1	KALKASKA	27N 5W 26	NA	Utica-Collingwood	application for horizontal well
2	A130164	O I L ENERGY CORP	STATE CUSTER AND BGC	C3-31	ANTRIM	29N 7W 8	NA	Antrim	application for a directional well

(Michigan Department of Environmental Quality, 2014)

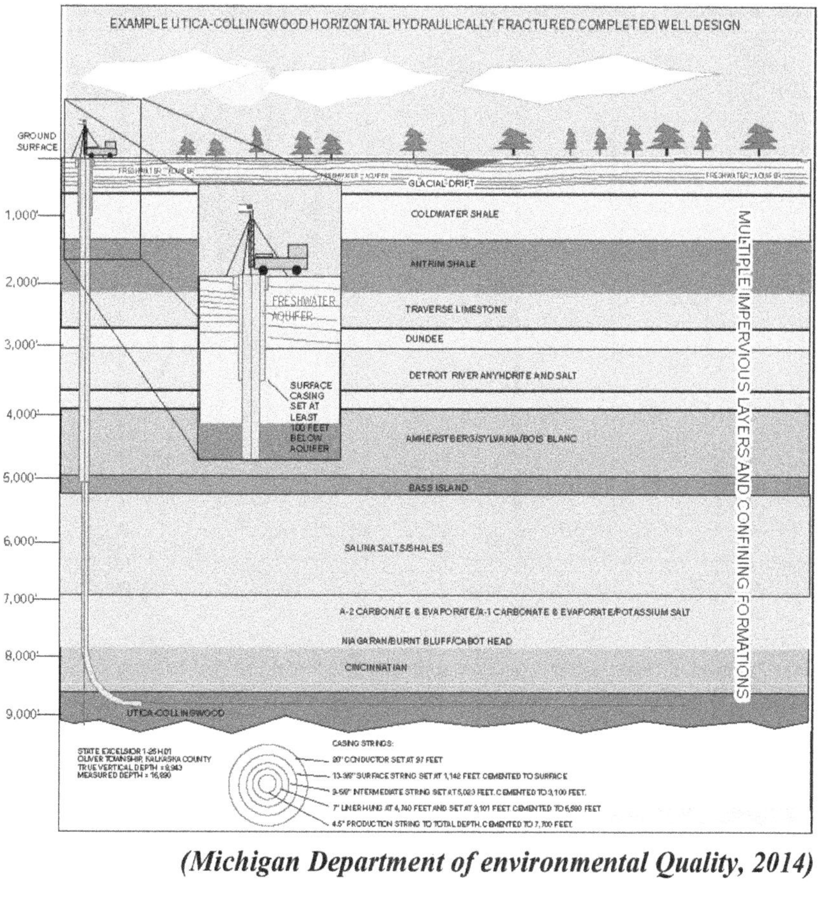

(Michigan Department of environmental Quality, 2014)

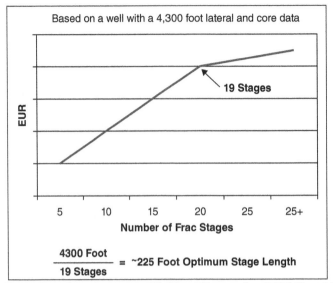

Optimum Stage Length   *(Gulfport Energy Inc DUG East 11-14-2012)*

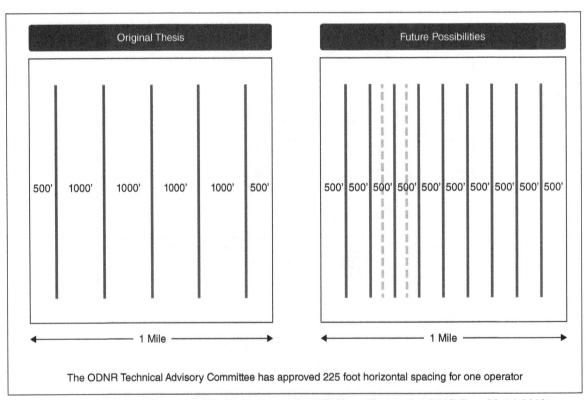

Lateral spacing consideration of Gulfport Energy Inc (Gulfport Energy Inc DUG East 11-14-2012)

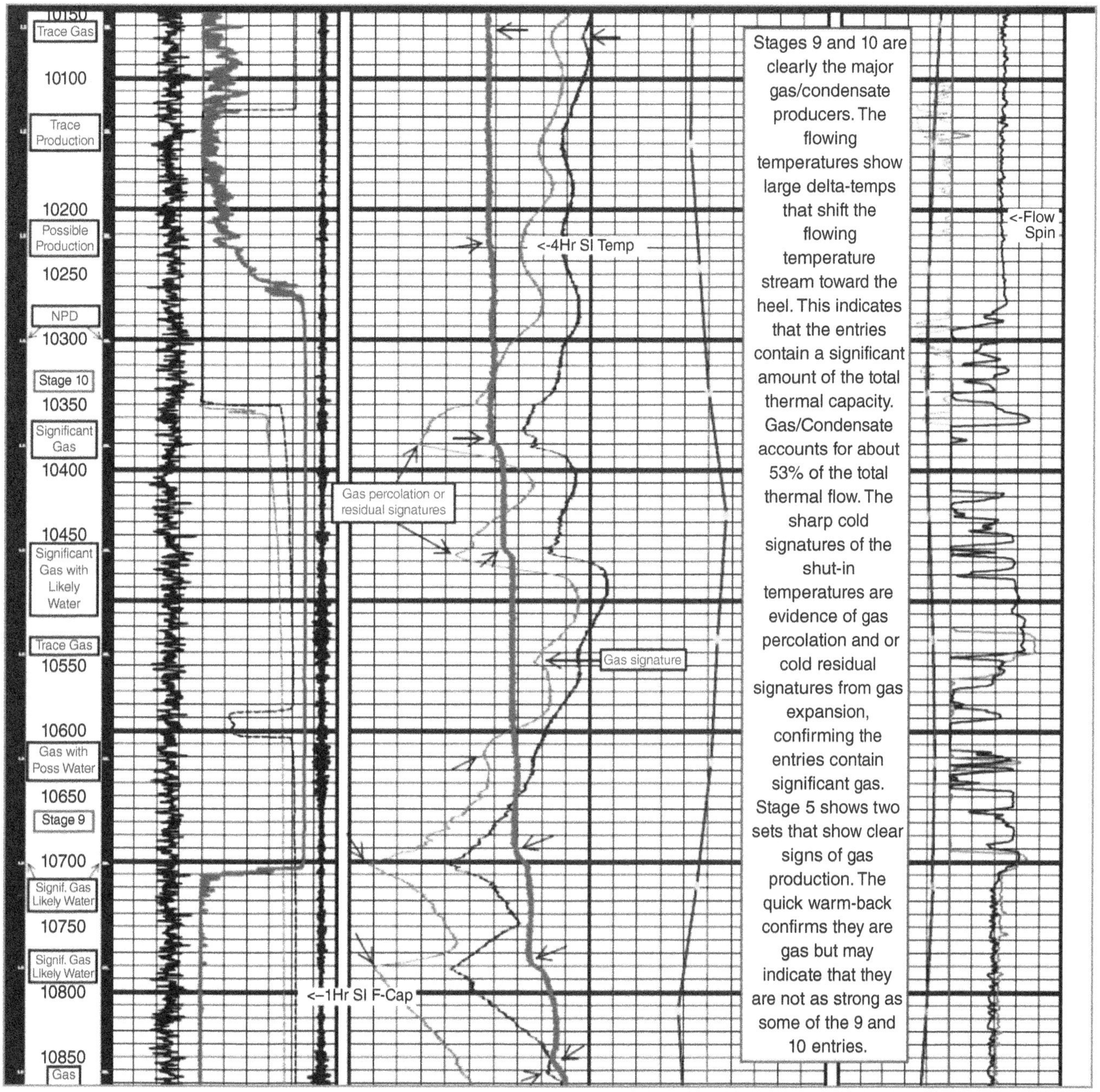

Trace Gas
10100
Trace Production
10200
Possible Production
10250
NPD
10300
Stage 10
10350
Significant Gas
10400
10450
Significant Gas with Likely Water
Trace Gas
10550
10600
Gas with Poss Water
10650
Stage 9
10700
Signif. Gas Likely Water
10750
Signif. Gas Likely Water
10800
10850
Gas
<-4Hr SI Temp
Gas percolation or residual signatures
Gas signature
<-1Hr SI F-Cap
<-Flow Spin
Stages 9 and 10 are clearly the major gas/condensate producers. The flowing temperatures show large delta-temps that shift the flowing temperature stream toward the heel. This indicates that the entries contain a significant amount of the total thermal capacity. Gas/Condensate accounts for about 53% of the total thermal flow. The sharp cold signatures of the shut-in temperatures are evidence of gas percolation and or cold residual signatures from gas expansion, confirming the entries contain significant gas. Stage 5 shows two sets that show clear signs of gas production. The quick warm-back confirms they are gas but may indicate that they are not as strong as some of the 9 and 10 entries.

Production Log from a Gulfport Energy well (Gulfport Energy Inc DUG East 11-14-2012)

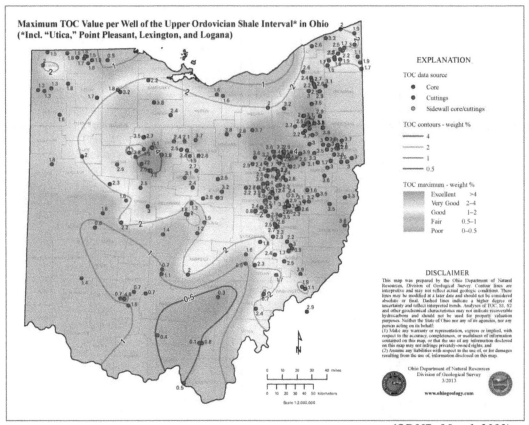

(ODNR, March 2013)

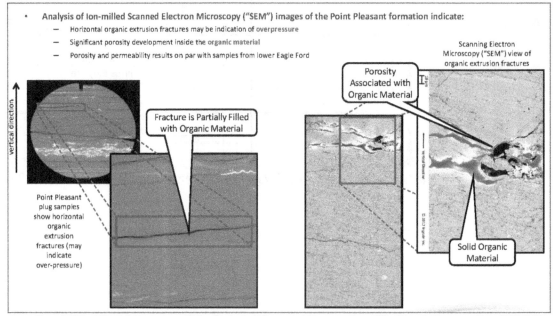

Ion Milled SEM Images of the Point –Pleasant Formation (Gulfport Energy Inc DUG East 11-14-2012)

well spacing, and all state regulations regarding oil and gas well drilling. In this website, 1000 page draft Supplemental Generic Environmental Impact can be downloaded.

http://oilandgas.ohiodnr.gov/shale – This website helps to get weekly activity and yearly production information in Ohio.

REFERENCES

Baird, G. C., Brett, C. E. (2002). Indian Castile: late synorogenic siliciclastic succession in an evolving Middle to Late Ordovician foreland basin, eastern New York State. *Phys Chem Earth*, 27:203–230.

Cole, G. A., others. (1987). Organic geochemistry and oil-source correlations, Paleozoic of Ohio. *AAPG Bull*, 71(7):788–809.

Cross, G. E. (2004). Fault-related mineralization and fracturing in the Mohawk Valley, Eastern New York State. Master's Thesis, SUNY at Buffalo, 251 p.

Drabold, W. (2014) Scientists study Ohio's quakes, fracking, http://www.dispatch.com/content/stories/local/2014/03/16/scientists-study-quakes-fracking.html

Downing, R. (2014). More details from anticipated state report on Utica shale drilling. Posted March 17, 2014, http:/Ohio.com

Energy In Depth, (2012) Update II-Gulfport Has More Exciting New's For Ohio's Utica Shale. Posted October 10, 2012, http:/www.eidohio.org/gulfport-has-more-exciting-news for Ohioans/

Fisher, D. W. (1979). Folding in the Foreland, Middle Ordovician Dolgeville Facies, Mohawk Valley. *Geology*, 7:455–459.

Goldman, D., Mitchell, C. E., Bergstrom, S. M., Delano, J. W., Tice, S. (1994). K-Bentonite and graptolite biostratigaraphy in the Middle Ordovician of New York State and Quebec: a new chronstratigraphic model. *Palaios*, 9:124–143.

Gulfport Energy Incorporated, Proving up the Utica's Liquids Window, Presentation made at DUG in Pittsburgh, PA, November 14, 2013.

Halcon Resources. (2013). Halcon Provides Operational Update: Halcon Resources Press Release, June 10, 2013, http//investors.halconresources.com/releasedetail.cfm?Release ID=770152

Jarvie, D. M. (2012). Geochemical Assessment of the Utica and Point Pleasant Shale, Ohio, USA: Abstracts with Programs Association of Petroleum Geologist, Eastern Section Meeting, Cleveland, Ohio, p. 38.

Jacobi, R. D., Mitchell, C. E. (2002). Geodynamical interpretation of a major unconformity in the Taconic Foredeep: slide scar or onlap unconformity. *Phys Chem Earth*, 27:169–201.

Joy, M. P., Mitchell, C. E., Adhya, S. (2000). Evidence of a tectonically driven sequence succession in the Middle Ordovician Taconic Foredeep. *Geology*, 28:727–730.

Joy, M. P. (1996). Evolution of the Facies Mosaic in the Taconic Foredeep Basin, Central Mohawk Valley, NY: A Graphic Correlation Approach. Unpublished MA Thesis, SUNY Buffalo, Buffalo, New York, 101 p.

Knox, T. (2014). Open season over for Kinder Morgan–Mark West pipeline project. Posted March 03, 2014, http://www.bizjournals.com/columbus/news/2014

Laughrey, C. D., Ruble, T. E., Wickstrom, L. H., Kostelnik, J., Drozd, R. J., Heim, L. R., Alcorn, G. A. (2012). Geochemical Interpretation Pitfalls in the Utica, Point Pleasant, and Trenton Formation (Upper Ordovician) Tight Petroleum Play, Eastern Ohio, USA: Abstracts with Programs Association of Petroleum Geologist, Eastern Section Meeting, Cleveland, Ohio, p. 43.

Martin, J. P., Hill, D. G., Lombardi, T. E., Nyahay, R. (2008). A primer on New York's gas shales. New York State Geological Association Guidebook, p. A1–A32.

McClain, T. G. (2012). Sequence Stratigraphy and Petrophysics of the Utica Shale and Associated Late Ordovician Strata, Eastern Ohio and Western Pennsylvania: Abstracts with Programs Association of Petroleum Geologist, Eastern Section Meeting, Cleveland, Ohio, p. 47.

McPArland, T (2014). Frackfree group expresses concern, http://www.vindy.com/news/2014/mar/13/data-on-poland-quakes-studied/?fracking

Mehrtens, C. J. (1988). Bioclastic turbidites in the Trenton Limestone: Significance and criteria for recognition. In: Keith, B.D., ed. The Trenton Group (Upper Ordovician Series) of eastern North America; deposition, diagenesis, and petroleum. American Association of Petroleum Geologists Studies in Geology, 29, p. 87–112

Michigan Department of Environmental Quality. (2014). http://michigan.gov/deq

Morgan, R. Youngstown residents react to fracking wastewater dump. Timesonline.com., Februray 6, 2013.

Nyahay, R. E., Leone, J., Smith, L., Martin, J., Jarvie, D. (2007). Update on Regional Assessment of Gas Potential in the Devonian Marcellus and Ordovician Utica Shales of New York, Thirty-Sixth Meeting of the Eastern Section, American Association of Petroleum Geologists, Lexington, KY, September 16–18, 2007, Search and Discovery Article #10136.

Obrien, D. (2014). Drilling halted after quakes near Hilcorp wells, http://businessjournaldaily.com/drilling-down/drilling-halted-after-quakes-near-hilcorp-wells-2014-3-11

Ohio Division of Natural Resources. (2014). http://oilandgas.ohiodnr.gov/ shale

Oil and Gas Journal. (2014). Magnum-Hunter find Utica gas, http://www.ogj.com/articles/2014/02/magnum-hunter-finds-utica-gas.html

Passey, Q. R., Creany, S., Kulla, J. B., Moretti, F. J., Stroud, J. D. (1990). A practical model for organic richness from porosity and resistivity logs. *Am Assoc Petrol Geol Bull*, 74(12):1777–1794

Provance, J. (2014). Ohio Funding–Governor proposes tax changes-Kasich eyes boosting cigarette, drilling fees, http://www.toledoblade.com/Politics/2014/03/12/Governor-proposes-tax- changes.html

Reed, J., Brown, S., Zumberge, J. (2012). Hydrocarbon Potential of the Utica/Point Pleasant In Eastern Ohio and Western Pennsylvania: Abstracts with Programs Association of Petroleum Geologist, Eastern Section Meeting, Cleveland, Ohio, p. 54.

Range Resources Corporation. (2012). Infection Report "An event that results in a significant change in the progress of a company": Range Resources Corporation Presentation July 1, 2012.

Riley, R. A. (2010). A Utica-Point Pleasant type log for eastern Ohio: Ohio Department of Natural Resources, Division of Geological Survey, one sheet (PDF), http://www.dnr.state.oh.us/Portals/ 10/Energy/Utica/TuscarawasWellRockAnalyses.pdf

Riley, R. A., Baranowski, M. (2011) Regional Stratigraphic and Facies Relationships of the Ordovician Utica/Point Pleasant Interval in the Appalachian Basin: Taking a Deeper Look at Shales: Geology and Potential of the Upper Ordovician Utica Shale in the Appalachaian Basin, June 21, 2011, New Philadelphia, Ohio.

Ryder, R. T. (2008). Assessment of Appalachian basin oil and gas resources: Utica-Lower Paleozoic Total Petroleum System: U.S. Geological Survey Open-File Report 2008–1287

Ryder, R. T., Burruss, R. C., Hatch, J. R. (1998). Black shale source rock and oil generation in the Cambrian and Ordovician of the central Appalachian Basin, USA. *Am Assoc Petrol Geol Bull*, 82:412–441.

Smith, L. B. (2013). Shallow transgressive onlap model for Ordovician and Devonian organic-rich shales, New York State. SPE Unconventional Resources Technology Conference, 12–14 August, Denver, Colorado, USA

Snowdon, L. R. (1984). A comparison of rockeval pyrolysis and solvent extract from the Collingwood and Kettle Point oil shales, Ontario. *Bull Canadian Petrol Geol*, 32:327–334.

Uniteds States Geological Survey. (2012). Assessment of undiscovered oil and gas resources of the Ordovician Utica Shale of the Appalachian Basin Province, 2012. USGS Fact sheet 2012–3116, 6 p.

Wallace L. G., Roen, J. B. (1989). Petroleum source rock potential of the Upper Ordovician black shale sequence, northern Appalachian basin. U. S. Geological Survey Open-file report 89–488, 66 p.

Willan, C. G., McCallum, S. D., Warner, T. B. (2012). Regional Interpretation of the Late Utica Ordovician Shale Play in the Appalachian Basin: Abstracts with Programs Association of Petroleum Geologist, Eastern Section Meeting, Cleveland, Ohio, p. 62.

Wickstrom, L. H., Gray, J. D., Stieglitz, R. D. (1992). Stratigraphy, structure, and production history of the Trenton Limestone (Ordovician) and adjacent strata in northwestern Ohio. Ohio Division of Geological Survey, Report of Investigations No. 143, 78 p., 1 p l.

Wickstrom, L. H., Erenpreiss, M., Riley, R., Perry, C., Martin, D. (2012). Geology and Activiy Update of the Ohio Utica-Point Pleasant Play, http://www.ohiodnr.com/portals/10/energy/Utica/Utica-PointPleasantPlay.pdf

Wickstrom, L. H., Shumway. (2014). 2013 Ohio Exploration Report, http://ooga.org/wp- content/uploads/2014OOGAWM-FRI-Exploration-Wickstrom-small.pdf

WOODFORD SHALE (LATE DEVONIAN-EARLY MISSISSIPPIAN), ANADARKO, ARKOMA, AND ARDMORE BASINS, USA

BY BRIAN CARDOTT (OKLAHOMA GEOLOGICAL SURVEY)

The Oklahoma Geological Survey has a database of all Oklahoma shale gas and shale oil well completions (http://www.ogs.ou.edu/level3-oilgas.php). The database of 3195 well completions from 1939 to February 2014 contains the following shale formations (in alphabetical order) and number of completions: Arkansas Novaculite (3), Atoka Group shale (1), Barnett Shale (2), Caney Shale or Caney Shale/Woodford Shale (107), Excello Shale/Pennsylvanian (2), Sylvan Shale or Sylvan Shale/Woodford Shale (13), and Woodford Shale (3037). Shale wells commingled with non-shale lithologies are not included. Exceptions include 18 Sycamore Limestone/Woodford Shale, 9 Mississippian/Woodford Shale, and 3 Hunton Group carbonate/Woodford Shale horizontal completions where non-Woodford perforations were minimal. The database was originally restricted to shale-gas wells. Shale-oil wells have been added since 2004. Figure 76.26 illustrates 2996 Oklahoma shale gas and shale oil well completions (1939–2013) on a geologic provinces map of Oklahoma.

Since 2004, the Woodford Shale-only plays of Oklahoma have expanded from primarily one (Arkoma Basin) to four geologic provinces (Anadarko Basin, Ardmore Basin, Arkoma Basin, and Cherokee Platform) and from primarily gas to gas, condensate, and oil wells (Figure 76.27). The recent low price of natural gas has shifted the focus of the plays more toward condensate ("Cana" for western Canadian County or "SCOOP" for "South Central Oklahoma Oil Province" play in the Anadarko Basin and western Arkoma Basin) and oil (northern Ardmore Basin, "Cana", "SCOOP", and north-central Oklahoma) areas. Of the 3017 Woodford-only well completions from 2004 to February 2014, 420 wells are classified as oil wells and 2614 wells are horizontal wells. Vertical depths range from 388 ft (Mayes Co.) to 16,259 ft (Caddo Co.). Initial potential gas rates range from a trace to 12 Mcf/day. Initial potential oil/condensate rates range from a trace to 1059 barrels per day. Reported oil gravities range from 21 to 66 API degrees.

The annual peak of 534 Woodford Shale wells occurred in 2008 (Figure 76.28). Following the drop in natural gas prices in 2008, the emphasis in the Woodford Shale plays has been for oil- and condensate-producing wells. Figure 76.29, showing Woodford Shale wells from 2011–2013, illustrates the expansion of the Woodford Shale condensate play in the

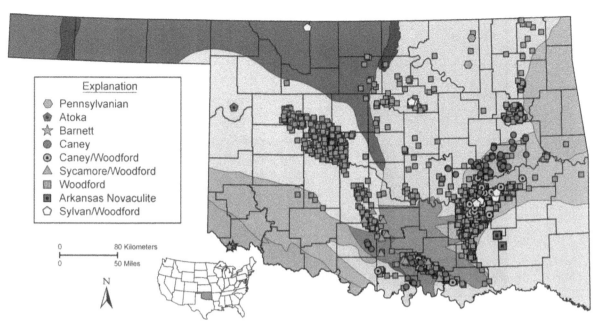

FIGURE 76.26   Map showing Oklahoma shale gas and shale oil well completions (1939–2013) on a geologic provinces map of Oklahoma.

Anadarko Basin which began in Canadian County (Cana) in 2007 and South Central Oklahoma Oil Province (SCOOP) in 2012.

The four Woodford shale plays in Oklahoma are as follows:

1. Western Arkoma Basin in eastern Oklahoma with thermogenic methane production at thermal maturities

from <1% to >3% vitrinite reflectance (VRo) and condensate production up to 1.67% VRo (Figure 76.30);

2. Anadarko Basin ("Cana" and "SCOOP" plays) in western Oklahoma with thermogenic methane production at thermal maturities from 1.1% to >1.6% VRo and condensate production at thermal maturities up to ~1.5% VRo (Figure 76.31);

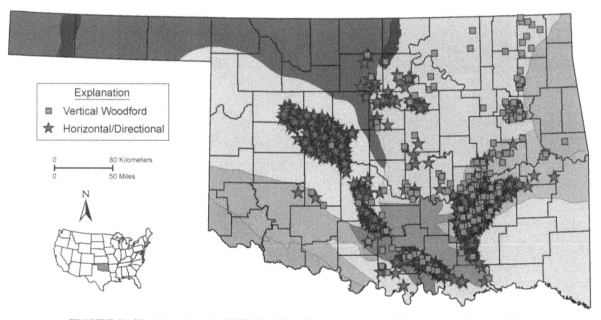

FIGURE 76.27   Map showing 2822 Woodford Shale-only gas and oil well completions (2004–2013) on a geologic provinces map of Oklahoma.

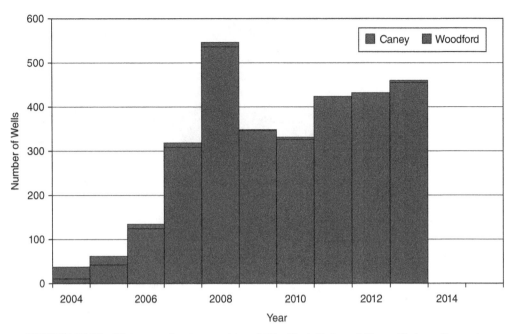

FIGURE 76.28   Histogram showing numbers of Woodford Shale and Caney Shale well completions, 2004–February 2014.

3. Ardmore Basin in southern Oklahoma with oil, condensate, and thermogenic methane production at thermal maturities in the oil window (<1.8% VRo) (Figure 76.26);

4. North-central Oklahoma (Cherokee Platform and Anadarko Shelf) with oil and thermogenic methane production at thermal maturities <1.0% VRo (Figure 76.27).

Of 32 operators active in Oklahoma shales during 2013, the top 10 operators (for number of wells drilled during 2013) are as follows:

1. Devon Energy Production Co. LP (145)
2. XTO Energy (103)
3. Cimarex Energy (41)
4. Continental Resources (37)

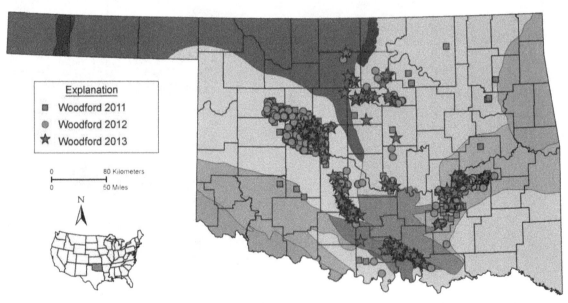

FIGURE 76.29   Map showing Woodford Shale-only gas and oil well completions (2011–2013) on a geologic provinces map of Oklahoma.

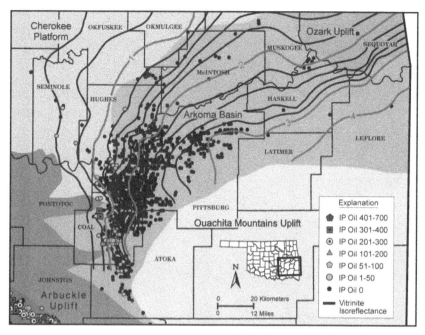

FIGURE 76.30   Map showing initial potential liquid hydrocarbon production of Woodford Shale-only gas and oil well completions (2004–2013) in the Arkoma Basin of eastern Oklahoma.

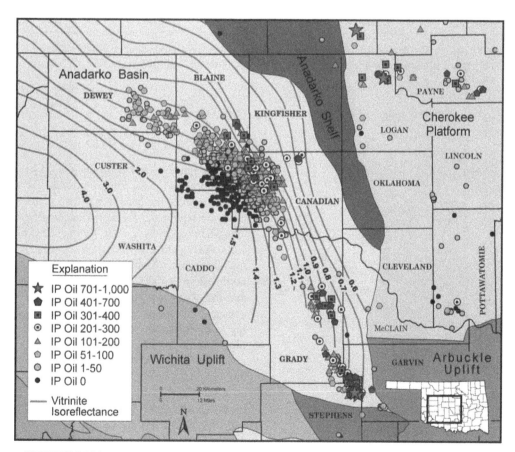

FIGURE 76.31   Map showing initial potential liquid hydrocarbon production of Woodford Shale-only gas and oil well completions (2004–2013) in the Anadarko Basin of western Oklahoma.

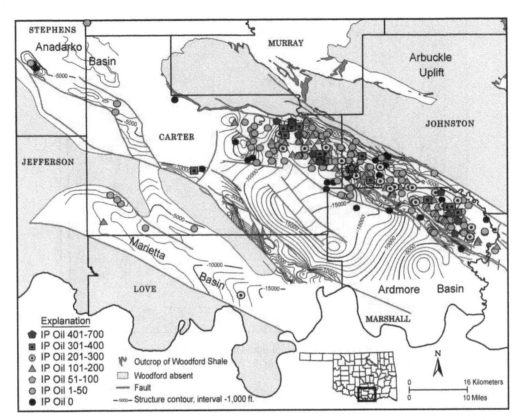

FIGURE 76.32 Map showing initial potential liquid hydrocarbon production of Woodford Shale-only gas and oil well completions (2004–2013) in the Ardmore and Marietta basins of southern Oklahoma.

5. Newfield Exploration Mid-Continent Inc. (29)
6. Red Fork (USA) Investments (18)
7. QEP Energy (14)
8. Petroquest Energy (12)
9. Marathon Oil (9)
10. BP America Production Company (8)

REFERENCES

Cardott, B. J. (2012). Thermal maturity of Woodford Shale gas and oil plays, Oklahoma, USA. *Int J Coal Geol*, 103:109–119.

Cardott, B. J. (2013a). Woodford Shale: From hydrocarbon source rock to reservoir: AAPG Search and Discovery Article 50817, 85.

Cardott, B. J. (2013b). Overview of Oklahoma shale resource plays: Oklahoma Geological Survey, Oklahoma Shale Gas and Oil Workshop presentation, 56 slides, http://www.ogs.ou.edu/MEETINGS/Presentations/2013Shale/2013ShaleCardott.pdf

Cardott, B. J. (2014). Determining the thermal maturity level at which oil can be economically produced in the Woodford Shale: American Business Conferences, Woodford Oil Congress 2014 Presentation, 92 slides, http://www.ogs.ou.edu/fossilfuels/pdf/2014CardottOilCong.pdf

CANADIAN SHALES

BY JOCK MCCRACKEN (EGRET CONSULTING)

Even though Canada has an abundance of conventional oil and natural gas, new unconventional gas, liquids, and oil plays dominate the headlines. Most of these shale opportunities lie within the Western Canadian Sedimentary Basin (WCSB), which is a vast sedimentary basin underlying 1,400,000 sq km (540,000 sq mi) of Western Canada including southwestern Manitoba, southern Saskatchewan, Alberta, northeastern British Columbia, southeast corner of the Yukon and the southwest corner of the Northwest Territories. It consists of a massive wedge of sedimentary rock extending from the Rocky Mountains in the west to the Canadian Shield in the east. This wedge is about 6 km (3.7 mi) thick under the Rocky Mountains, but thins to zero at its eastern margins. The WCSB contains one of the world's largest reserves of petroleum and natural gas and supplies much of the North American market, producing about 658,000 BOPD and 14 MMCFD gas. It also has huge reserves of coal. Of the provinces and territories within the WCSB, Alberta has most of the oil and gas reserves and almost all of the oil sands.

The first announcement of new discoveries in shale occurred in Canada at the beginning of 2008, 6 years ago.

Now, about 25% of Canada's natural gas is coming from unconventional which would include tight sands. The state of development for the shale plays range from speculative to exploratory to emerging to developing and under increasing commercial production. Typically, production numbers from government websites are up to 1 year or more behind. Additional production numbers and exploration statistics for this report are therefore gathered from press releases and presentations from some of the key companies involved with the plays. As a result of the low natural gas prices, operators have been focusing exploration and production into the liquids-rich hydrocarbons. The following plays are under development and increasing the production yearly: Horn River and Montney in NE BC, Duvernay and Alberta Bakken in Alberta, and the Bakken oil play (tight oil play encased in shale) in Saskatchewan and Manitoba.

There have been other shales that have disappointments for technical and regulatory reasons. Significant shale gas wells have been drilled and tested in the St. Lawrence Lowlands of Québec but a government freeze on fracking because of about environmental concerns will slow down any future exploration and production. The positive announcements out of New Brunswick have been tempered by recent disappointing results and low gas prices. To date, there is shale exploration activity in 9 provinces of Canada out of the 10 with Prince Edward Island being the exception. One of the three Territories of Canada, the Northwest Territories, is just now witnessing the drilling of their first wells into a possible oil-bearing shale section. The Yukon is evaluating their shale plays as well.

As a further note, there has been significant public concern in the press about hydraulic fracturing in various locations across Canada which is hindering or slowing down exploration and/or production. These concerns, especially in Provinces where there is limited oil and gas exploration and production, are being with by greater transparency and self-imposed industry guidelines.

http://www.capp.ca/aboutUs/mediaCentre/NewsReleases/Pages/GuidingPrinciplesforHyraulicFracturing.a spx

http://www.capp.ca/getdoc.aspx?DocId=210903&DT=NTV

A number of provincial and territorial governments are reviewing these practices (Quebec, Nova Scotia, New Brunswick and Newfoundland and both the Yukon and North West Territory). Alberta recently updated their regulations. It is hopeful, at the end of this discussion, hydraulic fracturing will be managed such that it will minimize potential risks and allow the public to have a balanced and realistic sense of the costs and benefits.

Northeast British Columbia

Northeast British Columbia contains Cretaceous to Devonian-aged shale deposits that potentially could contain 1200 Tcf of natural gas of which 78 Tcf is estimated to be marketable. This shale gas interest has therefore dominated the sale of petroleum and natural gas (PNG) rights from the province in the last 8 years with the Horn River Basin, the Cordova Embayment and the Montney Play trend generating the most interest. Recently, the Liard Basin or

Beaver River Area has come on to the radar screen with most of the basin almost entirely licensed. British Columbia has developed Natural Gas and Liquefied Natural Gas Strategies considering the immensity of this resource. The Montney play is garnering much interest because of its liquid components and is now producing at more than 1.74 BCFD as of March 2013. Crown land sale bonuses for these northeast (NE) British Columbia (BC) areas accounted for almost $5 billion since the record year in 2008. The total for 2013 was $225 million (see chart below). Capital spending by operators was estimated to be approximately 11 billion in 2011 and 2012. Operators are now conducting evaluation and drilling programs on their accumulated lands. These BC shale regions are summarized below from Adams (2013).

http://www.empr.gov.bc.ca/OG/oilandgas/petroleumgeol ogy/statsactivity/Documents/Summary%20of%20ShalE% 20Gas%20Activity%20in%20NEBC%202012_Oil%l20and% 20Gas%20Reports%202013-1.pdf

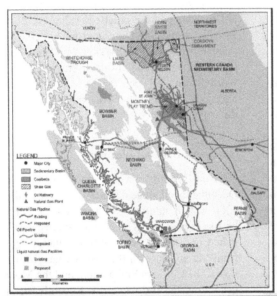

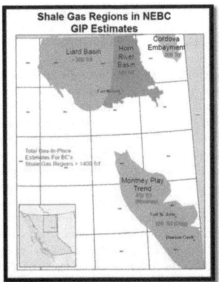

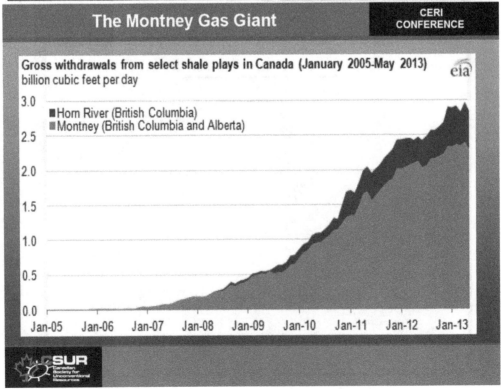

Bonuses Paid for PNG Rights in BC's Shale Gas Regions

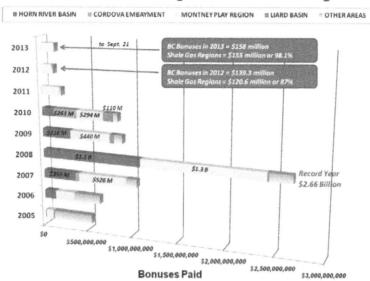

- ➤ Land brokers and producers have focused on purchasing PNG rights in the Montney play trend in 2011, 2012 and 2013.

- ➤ Shale gas regions accounted for 87% of total bonuses in 2012.

- ➤ So far this year, over 98% of the bonus total has resulted from PNG rights sold within the Montney play region.

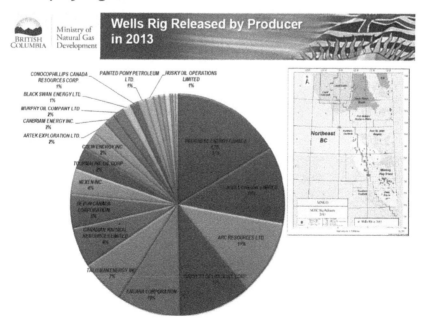

TABLE 1. POTENTIAL SHALE GAS FORMATIONS
IN NORTHEAST BRITISH COLUMBIA

PROSPECTIVE HORIZONS						
	Formations	Description	Depth	Average Thickness	Total Organic Carbon	Gas in Place
LOWER CRETACEOUS	Wilrich and Buckinghorse shales	Potential interbedded sand/ siltstone	800 to 1,200 metres	100 metres	2.3%	60 Bcf per section
JURASSIC	Nordegg and Fernie shales	Recognized source rocks	1,200 to 2,500 metres	Up to 30 m organic rich section	up to 14%	>20 Bcf per section
TRIASSIC	Doig, Doig Phosphate and Montney	Montney turbidites may increase permeability. Phosphate units have high TOC and are excellent source rocks	1,200 to 3,000 metres	300 to 500 metres	0.5 to >10%	10 to 110 Bcf per section
DEVONIAN	Exshaw, Besa River, Fort Simpson Horn River and Muskwa	Exshaw and Muskwa are widely distributed organic shales. Fort Simpson and Besa River are thick basin-filling shales	1,800 to 3,500 metres	Huge thicknesses are common with some high TOC intervals	0.5 to >10%	10 to 100 Bcf per section
GEOLOGIC ANALOGUE						
MISSISSIPPIAN	Barnett Shale (Fort Worth Basin)	Marine-shelf deposit	2,000 to 2,500 metres	100 metres	4.5%	140 Bcf per section

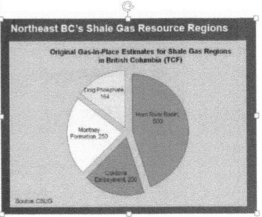

Upper and Middle Devonian, Evie (Klua), Otter Park and Muskwa members of the Horn River Formation Horn River Basin, Cordova Embayment and the Liard Basin

http://www.empr.gov.bc.ca/OG/oilandgas/petroleumgeology/statsactivity/Documents/5th%20NE%20BC%20Nat%20Gas%20Summit.pdf

http://www.empr.gov.bc.ca/OG/oilandgas/petroleumgeology/statsactivity/Documents/Adams.pdf

The gas production for the Horn River and Montney are also shown.

http://www.csur.com/images/CSUG_presentations/Allan_presentation_CERI_2014_Natural_Gas_Confer nce.pdf

Of these very far north basins, the Horn River has the most activity. Recent government reports state that the Horn River production was, March 2013, approximately 480 MMCFD from 121+ producing wells increasing from roughly 80 MMCFD at the end of 2009 and a cumulative gas production of approximately 343 BCF (April 2013).

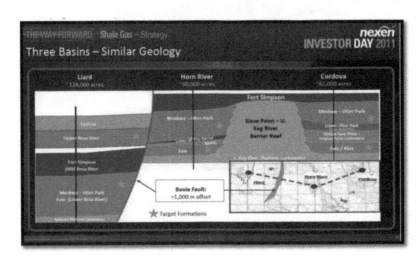

The following 11 operators form the Horn River Producers Group, which is a cooperative endeavor set up to share ideas and reduce the footprint: Apache, ConocoPhillips, Devon, Encana, EOG Resources, Imperial Oil, Nexen, Pengrowth, Suncor, Quicksilver, and Stone Mountain. There are another 18+ companies working in this area including SMR Oil and Gas, Spoke Resources, Husky, Harvest, Taqa North, Storm Gas Resources, Canadian Natural Resources, Ramshorn Canada, Husky Oil, and Delphi. There have been 379 shale gas-directed wells drilled since 2003. The companies with the most acreage are Encana, 260,000 ac; ExxonMobil, 250,000 ac; Apache, 200,000 ac; EOG, 130,000 ac; Quicksilver, 127,000 ac; Devon, 100,000 ac; Nexen, 88,000 ac; and Taqa North, 31,500 ac (Hart E&P May 2011). It is said that the entire Horn River has marketable gas at 78 Tcf (http://www.energy- pedia.com/news/canada/new-150721). The seven companies with the most drilling, as of end of 2011, were Apache, Encana, Nexen, EOG, Imperial Oil, Devon, and Quicksilver. Encana has been the most active, drilling 82 shale wells since 2003 with their production (April 2013) at 144 MMCFD. They lead the way with multiple fracture stimulations of up to 28 per well as well as the longest laterals of up to 3 km long. Apache has been active since 2003. Their production (April 2013) in the Horn River was 45 MMCFD from 72 wells on 7 pads. Their net recoverable resource is 9.2 Tcf. Apache has up to 16 wells on a pad.

Encana and Apache have entered into a partnership sourcing fracing water from the Mississippian Debolt Formation. Encana and Kogas Canada Ltd., a subsidiary of Korea Gas Corporation (KOGAS), have entered into a 3-year exploration and production agreement with the first well pad expecting gas production this summer. The EOG completed 11 wells in 2010 but planning minimal drilling after that to hold the acreage. They drilled three wells testing and producing the Evie member at 16–22 MMCFD. In the first quarter of 2013, they decided to sell their interest in the Kitimat LNG project as well as 28,000 acres of their Horn River acreage to Chevron. They still hold 127,000 acres with an estimated 7 Tcf potential until 2020. Nexen, which was recently purchased by CNOOC Limited of China, is expecting production to be 200 MMCFD by 2013. By the third quarter of 2011, they fractured and completed and brought on stream gas from their 9-well pad. With the completion of this pad, they expect to ramp up production to 50 MMCFD. They have another pad with 18 wells coming on stream in late 2012. They have drilled, as of April 2013, 66 wells and are producing 52 MMCFD. Their previously announced shale gas joint venture with INPEX Gas British Columbia Ltd. (IGBC) closed in August. Devon with 174,000 acres has the potential to produce up to 700 MMCFD based on its good land position and in the thickest part of the Basin. They have drilled 26 wells. Eight horizontal wells are producing.

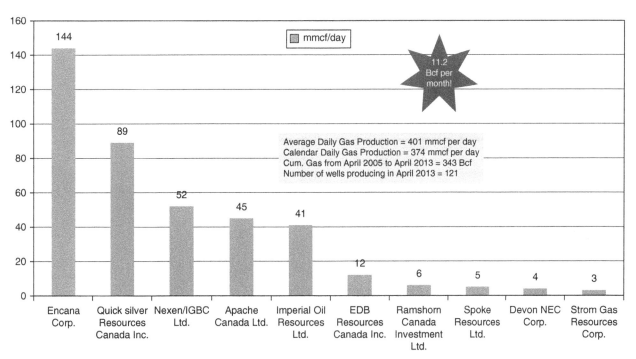

➢ Calendar daily gas production in April 2013 was approximately 374 mmcf per day from 121 producing shale gas wells.

➢ Cumulative production of 343 Bcf from April 2005 to April 2013.

➢ No production volumes reported for wells that remain on confidential status.

ExxonMobil and its 50% partner Imperial Oil have been encouraged by their 10 plus wells over their large 340,000 acre position. They are therefore setting up a multi-well horizontal pad pilot development in one of their areas. Production began on schedule from an eight-horizontal-well pad in August 2012 to assess productivity and improve development costs. Imperial continues to evaluate information from the pilot phase to determine long-term plans for the development.

Quicksilver has drilled 16 wells on their 140,420 acreage and are now producing at an average 50 MMCFD. They were also planning to drill an oil-rich horizontal leg into a section considered to be "Exshaw/Bakken zone" within the Horn River area which was probably uneconomic.

The Laird Basin, straddling the Yukon, North West Territory, and British Columbia, has great potential with 3 million acres and 5 km of section from the Cambrian to the Upper Cretaceous. It remains relatively unexplored with only a few

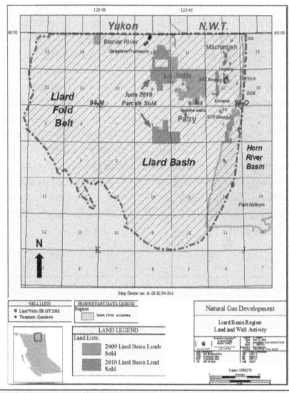

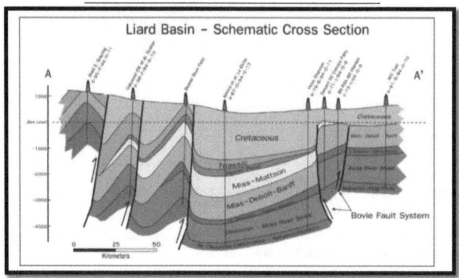

recent shale-targeted wells but Houston-based independent Apache Corp. calls the Lower Besa River black shale "the best unconventional gas reservoir evaluated in North America with excellent vertical and lateral reservoir continuity." http://www.ogj.com/articles/print/vol-110/issue-8b/regular-features/journally-speaking/liard-and-the-besa-river.html http://geoconvention.org/archives/2013abstracts/263_ GC2013_Liard_Basin_Hydrocarbon_Project.pdf From Apache presentation, June 2012.

http://www.google.com/url?sa=t&rct=j&q=&esrc= s&source=web&cd=6&cad=rja&uact=8&ved=0CEwQFj AF&url=http%3A%2F%2Ffiles.shareholder.com%2Fdown loads%2FAPA%2F2148221537x0x 577564%2Fd3ace688-47ba-48e4-8179-83b15adb6881%2FApache_Investor-Day_ 20120614.pdf&ei=NJg4U53sA-66yAGXsoDACQ&usg= AFQjCNHNYyC2VdJ9Etmrv3Uxkf8KzrfBRw&bvm=bv. 63808443,d.aWc

In 2012, Apache, with 430,000 acres, reported that one of their wells (Apache HZ Patry d-34-K/94-O-5) recorded a 30-day initial production rate of 21.3 MMCFD on a six-stage fracturing operation (3.6 MMCFD per hydraulic fracture). The well was drilled in 2010 to a vertical depth of 3843 m with a horizontal leg of 885 m and has an estimated ultimate recovery (EUR) of 17.9 BCF. It is considered to

be one of the best shale gas resource tests in any of North America's unconventional reservoirs (Apache Canada Ltd., 2012).

Apache is targeting the Upper Devonian Lower Besa River Black Shale and estimates that its Liard Basin lands carry a net gas in place (GIP) of 201 Tcf, which could yield net sales gas of 48 Tcf. The shale is 400–1000 ft thick lying at depths of 9500–15,000 ft. Porosity range is 3–8% and water saturation is 15–20%. Total organic carbon values are 3–6 wt. %. Apache showed a development model that would involve recovery of 54 Tcf of raw gas using 731 well locations on 61 pads with two drilling rigs per pad

The Apache's vertical well, C-86-F, was drilled to 15,000 ft and had a 30-day initial production of 9.9 MMCFD. A second vertical well, D-28-B, was drilled to 13,200 ft and flowed at an initial rate of 4.6 MMCFD. The two vertical wells had only a single frac apiece. Net pay thickness is 1024 ft at C-86-F and 708 ft at D-28-B. In its development model, Apache envisions drilling horizontal wells with 7050–8040-ft laterals with 18 fracs per lateral. The company estimates 400 ft spacing between fracs and 600 m between wells. Drilling time is 110–120 days/well. The company plans to drill tenure wells in this year's second half followed by more concept wells in 2013.

LIARD BASIN

BEST UNCONVENTIONAL GAS RESERVOIR IN NORTH AMERICA

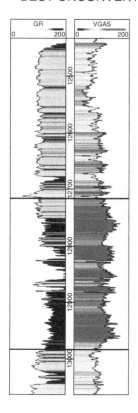

Reservoir	Units	Liard	NE-PA Marcellus	Haynesville
Depth	(ft)	9500 - 15000	7000 - 11000	10000 - 13000
Thickness	(ft)	400 - 1000	150 - 400	100 - 300
Porosity	(%)	3 - 8	6 - 12	4 - 7
Water Saturation	(%)	15 - 20	15 - 45	20 - 40
OGIP / Sec	(BCF)	170 - 500	30 - 200	50 - 100
Thermal Maturity	(VRo)	>1.5	>1.6	>1.7
Pressure	(Psi/ft)	0.85 - 0.92	0.5 - .65	~0.85
GOR		Dry Gas	Dry Gas	Dry Gas
Quartz+Carb	(Vol %)	>90	65 - 90	60 - 70
TOC	(Wt. %)	3 - 6	2 - 10	2 - 4

> Lower Besa River First Black Shale
> Best gas-shale reservoir evaluated in North America
> Excellent vertical and lateral reservoir continuity

Paramount and Nexen are actively drilling in this basin. Nexen have 128,000 acres of highly prospective shale gas lands in the Liard basin, with between 5 and 23 Tcf of unrisked prospective resource. Paramount evaluation, after a few wells, suggests that the Laird Basin holds an original GIP of 170–500 BCF per section with an expected recovery of 20%. Transeuro Energy is proceeding to develop the pre-existing Beaver River gas field. They continue to develop this field with the hope that the surrounding shales will feed these reservoirs. Currently they are producing 1.1 MMCFD down from previous production. Future plans include drilling 6 more wells to target 14 horizons.

The Cordova Embayment area, an area of 936,000 acres where most blocks of land were purchased in 2007, is now being drilled. British Columbia has an experimental scheme ownership where operations are kept confidential for 3 years of which Nexen, Penn West, and Canadian Natural Resources Ltd. have participated. To date, September 2014, 22 wells are producing at a combined average rate of 42 MMCFD from the Devonian Sequences in the Muskwa, Otter Park, and Evie. Nexen has drilled two vertical and three horizontals and Penn West has drilled 28 wells so far. Penn West just announced an $850 million Joint Venture with Mitsubishi to help develop their property in this area.

Recently, Encana entered into an agreement with Enbridge Inc. for the sale of its majority stake at the Cabin Gas Plant in Horn River Basin. Phase 1 of the development will have 400 MMCFD of natural gas processing capacity. This phase is to be deferred. Phase 2 will add an additional 400 MMCFD of capacity but this has been deferred as well.

Spectra Energy Corp. transportation system stretches from Fort Nelson, in northeast British Columbia and Gordondale at the British Columbia/Alberta border, to the southernmost point at the British Columbia/U.S. border at Huntington/Sumas. They have about 2800 km (1700 miles) of natural gas transmission pipeline which can transport 2.9 BCFD. TransCanada Corp. has filed an application for an Alberta pipeline extension subject to regulatory approvals, the approximate $310 million project is expected to be operational early in second quarter 2012 with commitments for contracted gas rising to approximately 540 MMCFD by 2014.

Triassic Doig and Montney Fort St. John/Dawson Creek Area

The Montney is a liquids-rich tight gas/shale gas play, now producing at more than 1.7 BCFD as of March 2013. This Montney Play Trend, of 6.6 million acres, is now one of the most active natural gas plays in North America. The primary zones are the Upper Middle and Lower Montney as well as the Doig and Doig Phosphate. The major facies include fine-grained shoreface, shelf siltstone to shale, fine-grained sandstone turbidities, and organic-rich phosphatic shale. This play varies from the traditional distal shale facies along the Alberta/British Columbia border to a tight calcareous siltstone and sandstone in Central Alberta. The current trend

for companies is to explore up dip towards the "oil window" in search of liquids-rich gas. The main Montney players in order of rig utilization are Shell Canada Ltd, Encana Corp., Progress Energy Ltd Talisman Energy Ltd, Tourmaline Oil Corp, Murphy Oil Co. Canadian Natural Resources Ltd., ARC Resources Ltd as well as, at least, another 23 operators. This report will just cover the most active players.

Shell Canada Limited keeps expanding their program in the Sunset Prairie-Groundbirch area where they have only tapped into 5% of their resource estimate of 8 Tcf for the area. Shell's Groundbirch venture currently includes five natural gas-processing plants, over 250 wells and 900+ km of pipeline which are producing over 190 million standard cubic feet of raw gas per day as of mid-2012. The heart of this tight gas formation lies in a wide layer of siltstone, sandstone, and shale some 8200–9800 ft below ground, and the producing zone is approximately 500-ft thick Groundbirch is still being explored, so the current drilling program includes a mix of single and multiple well pads. Eventually, most of the wells will be drilled on pads containing up to 26 wells, with two such pads for every 3 square miles of land.

Encana is one of the biggest players with 482 rig releases since 2005. They drilled 90 wells in 2009 with 8–10 wells per section, 62 wells in 2010, 43 in 2011, and 41 wells in 2012. The horizontal sections are up to 2400 m long with up to 17 fracs per well with some recent IP at 10 MMCFD. They believe they have an estimated 70 Tcf of GIP in their trend. Their Montney, Cutbank Ridge area was producing at the end of 2011 at 428 MMCFD and 1.1 MB of liquids/D and in 2012 it was 433 MMCFD and 1.5 MB of liquids/D with their current forecast for 2014 at 600 MMCFD. They have advanced their resource hub design to a new level with 8–12 wells per hub, 100–200 completions per hub, up to 17 stages per well, laterals of 650–100 ft, horizontals of 6500–10,000 ft, and completions every 450 ft. They have announce one well with a 8935 ft lateral, 12 frac, and an IP of 19 MMCFD. Their completion costs have been reduced by approximately 60% in the last 5 years. They have recently announced 40% sale of some of their acreage in the area to Mitsubishi Corporation.

Murphy Oil Corporation, in their Tupper Creek Area, has now reached a production level of over 197 MMCFD. They have almost 144,000 net acres in the Montney Trend and drilled 60 wells last year. Their Tupper West gas plant has the capacity of 180 MMCFD with the current production of 150 MMCFD from 59 wells. Recently, because of price, they have curtailed production and reduced spending.

ARC Resources Ltd. has now increased their daily production to 235 MMCFD and 1800 barrels per day of liquids. They have stated that they could sustain rates of up to 800 MMCFD and 17,000 BPD of liquids for a period of 10 years based on their resource portfolio. The first two of three 60 MMCFD gas plants are on stream at full capacity. Some of their wells are producing 30–200 barrels per MMCF and another recent well is producing 4.7 MMCFD from a 100 m section in the Upper Montney. The total BC Montney 2012 Q4 production was 235 MMCFD and 2600 BOPD.

This play keeps expanding both aerially and stratigraphically as operators are searching for the more liquids-rich sections (see the map below). Two other operators are partnering up with other international players where Progress Energy Resources Corp. created 50/50 partnership with the Malaysian PETRONAS and Talisman Energy Inc. partnered up with South Africa's Sasol Ltd. Other companies have

reported liquid yields per MMCF of 20 BBL (Terra Energy Inc.), 30 BBL (Canadian Spirit Resources Inc.), 35–50 BBL (Tourmaline Oil Corp.), and 60–85 BBL (Crew Energy Inc.). A very extreme gas rate of 24.5 MMCFD was announced by Painted Pony Petroleum Ltd.

The graph below shows the well production in the Montney from the Adams, 2013 report.

The graph below shows the gas deliverability from 2004 to 2015. Note the increasing Montney contribution, http://www. neb-one.gc.ca/clf-nsi/rnrgynfmtn/nrgyrprt/ntrlgs/ntrlgsdlvrb lty20132015/ntrlgsdlvrblty20132015ppndc-eng.pdf

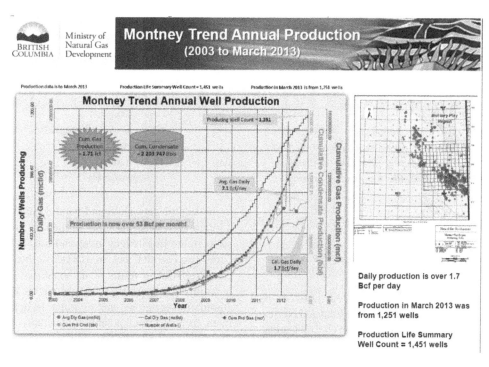

Daily production is over 1.7 Bcf per day

Production in March 2013 was from 1,251 wells

Production Life Summary Well Count = 1,451 wells

> Large proportion of the Montney play has economically benefitted by the presence of liquids rich gas.

> Focused development of the Montney wet gas trend began in 2009/2010 as natural gas commodity prices were falling.

> The current rate of development is not only an indication of the more favourable reservoir characteristics *(porosity, organic carbon content, pressures)*, but also a reflection of the higher natural gas liquid *(C2–C4)* and condensate *(C5–C12)* concentrations of the formation.

> Producers continue to push the limits of the northeastern, liquids-rich portion of the fairway.

Zones of dry gas, wet gas, wet gas+oil, and oil within the Montney Formation (derived from sales gas data). *Source Geoscience Reports 2013, BC Ministry of Natural Gas Development*

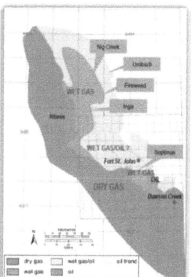

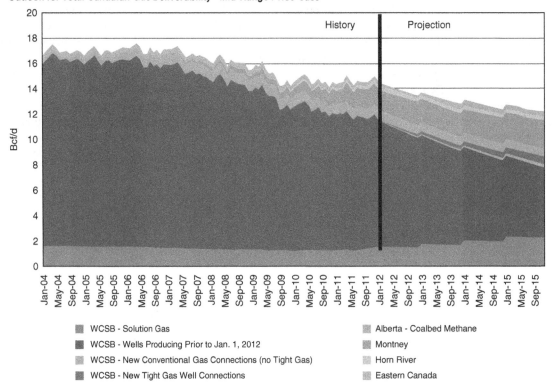

Outlook for Total Canadian Gas Deliverability - Mid-Range Price Case

The following table is from Adams (2013).

Lower Cretaceous – Gething and Buckinghorse N.E. British Columbia

Shale gas activity directed towards Cretaceous horizons is being assessed in several areas of the Fort St. John and Northern Foothills regions. The Blair Creek and Farrell Creek areas in the Northern Foothills region have seen a steady increase in the sale of petroleum and natural gas rights over the last 4 years. Lower Cretaceous sequences are the exploration focus in the Beg/Jedney areas and further south in the Blair Creek and Farrell Creek areas. Each of these areas has unique characteristics in terms of its shale gas potential. Companies currently operating in these areas are evaluating fracture stimulation programs and continue to optimize completion methods that could potentially increase well productivity. The Buckinghorse Formation is about 1000 m thick in some

Parameter	Shale Gas Play					
	Horn River Basin	Montney (B.C. only)	Barnett	Marcellus	Haynesville	Fayetteville
Basin Area (km²)	11 500	1 600 to 10 000	1 900	95 000	9 000	9 000
Depth (m)	1 800 to 3 000	1 000 to 3 500	2 000 to 2 600	2 500 to 3 000	3 000	350 to 2 300
Thickness (m)	50 to 350	50 to > 300	15 to 182	12 to 275	75	16 to 180
Porosity (%)	2 to 8	1 to 6	4 to 5	10	8 to 9	2 to 8
Total Organic Content (%)	1 to 8	1 to 7	4.5	3 to 12	0.5 to 4.0	4.0 to 9.8
Reservoir hydrocarbons	dry gas	wet gas, dry gas	wet gas, dry gas	wet gas, dry gas	dry gas	dry gas
Natural Fracturing	Yes	Yes	Yes	Yes	Yes	Yes
Pressure regime	Overpressured	Overpressured	Overpressured	Normal to overpressured	Overpressured	Normally pressured
Proximity to major consuming markets	Distant	Distant	Close	Very close	Close	Close
GIP (billion m³)	12 629 (10 466 to 14 894)	2 270 to 19 800	9 263	42 492	2 0311	1 473
GIP (Tcf)	448 (372 to 529)	80 to 700	327	1 500	717	52
Marketable (billion m³)	2 198 (1 715 to 2 714)	Uncertain	1 246	7 422	7 110	1 178
Marketable (Tcf)	78 (61 to 96)	Uncertain	44	262	251	41.6
State of development	very early	early	medium	early	medium	medium
Notes:		hybrid shale/tight-gas				

Source: U.S. data from U.S. Department of Energy Modern Shale Gas Development in the United States: A Primer

LNG Exports and Pipeline Proposals

places. Painted Pony Petroleum has 82,465 net acres of Buckinghorse potential with recompletion and testing of three wells and two more wells drilled. No production numbers announced yet as they experiment with drilling and completion techniques. They have announced that two existing wells will be fracked in 2013. Canadian Spirit is another player in the area, mostly with experimental schemes, on the Gething. No production volumes reported yet. Unconventional Gas Resources is experimenting with the Buckinghorse shale. The Asian gas market is being targeted by 11 joint venture export groups with the building of LNG terminals in Kitimat, Prince Rupert, and Grassy Point BC, 643 km north of Vancouver. These projects are summarized below:

Canadian LNG export projects in development, as of 11/2013:

Kitimat LNG (Chevron, Apache), 1.4 Bcf/D, Permits received, awaiting investment decision; **BC LNG Export Cooperative**, 0.125 Bcf/D, Permits received; **LNG Canada** (Shell, KOGAS, Mitsubishi, PetroChina), 2.0–3.2 Bcf/D, Feasibility stage, applied for some permits; **Pacific Northwest LNG** (Progress/Petronas, Japex), 2.0 Bcf/D (Merger approval granted), Completed feasibility, progressing to pre-FEED; **Nexen/Inpex**, Conducting feasibility; **Prince Rupert LNG** (BG Group), 3.0 Bcf/D, Advancing feasibility, applying for permits; **AltaGas/Idemitsu Kosan**, 0.27 Bcf/D, Conducting feasibility; **ExxonMobil/Imperial Oil (WCC LNG Ltd)**, 4.0 Bcf/D, Applied for export license; **Woodfibre LNG**, 0.3 Bcf/D, Applied for export license. Expressions of interest to proceed with a project have also been made by **Woodside Petroleum & South Korea E&S.**

Canada's Minister of Natural Resources, on March 27, 2014, announced the approval of government for four long-term liquefied natural gas (LNG) export licenses for Pacific

NorthWest LNG, Prince Rupert LNG, WCC LNG, and Woodfibre LNG.

http://bit.ly/1pyT8DD

http://www.capp.ca/getdoc.aspx?DocId=233813&DT=PDF

http://www.csur.com/misc/Oilweek_Top_%20100_%20Breakfast_%20JunE%2012.pdf

http://engage.gov.bc.ca/lnginbc/

http://www.theglobeandmail.com/report-on-business/industry-news/energy-and-resources/british-columbias-potential-lng-terminals/article8956483/

B.C Shale information link: There is a wealth of data on this website.

http://www.empr.gov.bc.ca/OG/OILANDGAS/PETROLEUMGEOLOGY/SHALEGAS/Pages/default.aspx

http://www.offshore-oil-and- gas.gov.bc.ca/OG/oilandgas/petroleumgeology/UnconventionalGas/Documents/C%20Adams.pdf

http://www.bcogc.ca/publications/reports

Adams, C. (2012): Summary of shale gas activity in northeast British Columbia 2011, BC Ministry of Energy and Mines, p. 1–19.

http://www.empr.gov.bc.ca/Mining/Geoscience/PublicationsCatalogue/OilGas/OGReports/Documents/2012/Summary%20of%20ShalE%20Gas%20Activity%20in%20NEBC%202011%20Version%20HQ.pdf

Adams, C., 2013, Summary of shale gas activity in Northeast British Columbia 2012; in Oil and Gas Reports 2013, British Columbia Ministry of Natural Gas Development, p. 1–27.

http://www.empr.gov.bc.ca/OG/oilandgas/petroleumgeology/statsactivity/Pages/default.aspx

BC MEM/NEB Report, May 2011: Ultimate Potential for Unconventional Natural Gas in Northeastern British Columbia's Horn River Basin

http://www.em.gov.bc.ca/OG/Documents/HornRiverEMA_2.pdf

http://www.aapg.org/explorer/2010/10oct/regsec1010.cfm

Geoscience BC is an industry-led, industry-focused, applied geoscience organization. Their mandate is to encourage mineral and oil and gas exploration investment in British Columbia through the collection, interpretation, and marketing of publically available. Some of their major projects have been aquifer studies.

http://www.geosciencebc.com/s/AboutUs.asp

This link below summarizes news items concerning the Horn River area.

http://hornrivernews.com/

Alberta The shales and tight rocks of the Western Canada Sedimentary Basin have been under investigation for the last number of years. The Alberta portion of this basin, Alberta Basin, has been studied thoroughly by Alberta Energy Resources Conservation Board (ERCB), Alberta Geological Survey (AGS), Geological Survey of Canada (GSC), and National Energy Board (NEB).

Alberta has extensive experience in the development of energy resources and has a strong regulatory framework already in place. Shale gas and liquids is regulated under the same legislation, rules, and policies required for conventional natural gas. The Energy Resources Conservation Board (ERCB) regulates exploration, production, processing, transmission, and distribution of natural gas within the province. Estimates of shale resources within the Western Canada Sedimentary Basin (see map below) vary from 86 to 1000 Tcf. This early estimate did not include liquid phase. While there is a huge potential in Alberta, commercial shale production is at early stages but additional new plays have suddenly begun to emerge.

In October 2011, the NEB published the "Tight Oil Developments in the Western Canada Sedimentary Basin" which included Plays highlighted are the Bakken/Exshaw Formation (Manitoba, Saskatchewan, Alberta, and British Columbia), Cardium Formation (Alberta), Viking Formation (Alberta and Saskatchewan), Lower Shaunavon Formation (Saskatchewan), Montney/Doig Formation (Alberta), Duvernay/Muskwa Formation (Alberta), Beaverhill Lake Group (Alberta), and Lower Amaranth Formation (Manitoba). The list did not include potential formations, such as the Second White Specks, Nordegg, Pekisko, and others, largely because these new developments are at very early stages. The NEB estimated that Canadian tight oil production, at March 2011, to be over 160,000 BBL/D. It is too early to estimate with any degree of confidence what the ultimate impact of exploiting tight oil plays in western Canada might be; however, there are some indications. The Alberta Energy Resources Conservation Board's latest Supply and Demand report estimates that Alberta's tight oil plays will add an additional 170,000 BBL/D to conventional light oil production by 2014. In Saskatchewan, tight oil production in the first quarter of 2011 was 90,000 BBL/D, while Manitoba, reached 25,000 BBL/D. Companies have so far identified just over 500 million barrels of proved and probable reserves in their plays of interest and not all companies active in those plays have issued formation-specific reserves. This is enough oil to support about 134,000BBL/D of production over 10 years. As well, the technologies used to develop tight oil will continue to evolve, likely increasing the amount of recoverable oil from these plays. Since 2007, the various governments have been collecting and still in the progress of collecting data on the following formations: Colorado Group-First White Speckled Shale, Puskwaskau, Wapiabi, Colorado

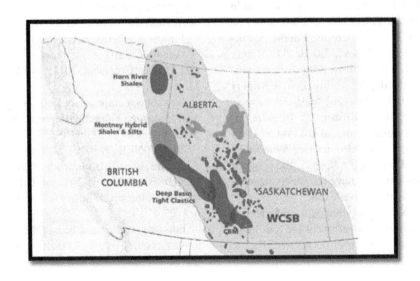

TABLE 76.1   Potential shale gas formations in northeast British Columbia.

	Formations	Description	Depth	Average thickness	Total organic carbon	Gas in place
		Prospective horizons				
Lower Cretaceous	Wilrich and Buckinghorse shales	Potential interbedded sand/siltstone	800–1200 m	100 m	2.3%	60Bef Per section
Jurassic	Nordegg and Fernie shales	Recognized source rocks	1200–2500 m	up to 30 m organic-rich section	pu to 14%	>20 Bcf per section
Triassic	Dolg, Dolg Phosphate and Montney	Montney turbidites may increase permeabilityPhosphate units have high TOC and are excellent source rocks	1200–3000	300–500 m	0.5 to >10%	10 to 100 Bcf per section
Devonian	Exshaw, Besa River, Frot Simpson Horn River and Muskwa	Exshaw and Muskwa are widely distributed organic shalesFort Simpson and Besa River are thick basinfilling shales	1800–3500 m	Huge thicknesses are common with some high TOC intervals	0.5 to >10%	10 to 100 Bcf per section
		Geologic analogue				
Mississippian	Barnett Shale (Fort Worth Basin)	Marine-shelf deposit	2000–2500 m	100 m	4.5%	140 Bcf per section

Shale, Muskiki, Second White Speckled Shale, Blackstone, Kaskapau, Fish Scales, Shaftesbury, Joli Fou, Wilrich Formation, Bantry Shale member, Fernie Formation, Fernie Shale, Pokerchip Shale, Nordegg, Rierdon, Montney, Lower Banff, Exshaw, Duvernay, and Muskwa.

In October 2012, a very comprehensive study was published by Rokosh et al. titled "Summary of Alberta's Shale- and Siltstone-Hosted Hydrocarbons". This study concluded that the shale gas resources (hydrocarbon endowment) in Alberta alone are estimated to be 3424 Tcf of natural gas,

TABLE 76.2   Summary of estimates of Alberta shale- and siltstone-hosted hydrocarbon resource endowment.

Unit	Adsorbed gas content %[a]	Natural gas (Tcf)	Natural-gas liquids (billion bbl)	Oil (billion bbl)
Duvernay P50	6.8	443	11.3	61.7
Duvernay P90-P10	5.6–8.5	353–540	7.5–16.3	44.1–82.9
Muskwa P50	6.9	419	14.8	115.1
Muskwa P90-P10	4.1–10.5	289–527	6.0–26.3	74.8–159.9
Montney P90-P10	10.8–26.0	1630–2828	11.7–54.4	78.6–220.5
Basal Banff/Exshaw P50	5.7	35	0.092	24.8
Basal Banff/Exshaw P90-P10	3.2–10.0	16–70	0.034–0.217	9.0–44.9
North Nordegg P50	33.7	246	2.1	47.9
Wilrich P90-P10	6.2–59.2	115—568	0.689–4.449	20.2–172.3
Total P50 (medium estimate resource endowment)	n/a	3424	58.6	423.6

[a]The percentage of adsorbed gas represents the portion of natural gas that is stored as adsorbed gas.

TABLE 76.3 Ultimate potential for Montney unconventional petroleum in British Columbia and Alberta.

Hydrocarbon type	In-place			Marketable		
	Low	Expected	High	Low	Expected	High
Natural Gas-billion m³ (Tcf)	90,559 (3197)	121,080 (4274)	153,103 (5405)	8952 (316)	12,719 (449)	18,257 (645)
NGLs-million m³ (million barrels)	13,884 (87,360)	20,173 (126,931)	28,096 (176,783)	1540 (9689)	2308 (14,521)	3344 (21,040)
Oil-million m³ (million barrels)	12,865 (8049)	22,484 (141,469)	36,113 (227,221)	72 (452)	179 (1125)	386 (2430)

58.6 billion barrels of NGLs, and 423.6 billion barrels of oil. They evaluated the geology, distribution, characteristics, and hydrocarbon potential of key shale and/or siltstone formations (units) in Alberta. Five units show immediate potential: the Duvernay Formation, the Muskwa Formation, the Montney Formation, the Nordegg Member, and the basal Banff and Exshaw formations (sometimes referred to as the Alberta Bakken by industry). The study also includes a preliminary assessment of the Colorado, Wilrich, Rierdon, and Bantry Shale units. These units were systematically mapped, sampled, and evaluated for their hydrocarbon potential. In total, 3385 samples were collected and evaluated for this summary report. Table 76.2 and four maps are from this report.

The resource estimates listed above provide an estimate of total hydrocarbons-in-place. Geological and reservoir engineering considerations will ultimately determine the potential recovery of this large resource.

In November, the National Energy Board (NEB) in conjunction with their provincial agencies, an ultimate potential of the Montney was released (see below): "The Ultimate Potential for Unconventional Petroleum from the Montney Formation of British Columbia and Alberta"

http://www.neb-one.gc.ca/clf- nsi/rnrgynfmtn/nrgyrprt/ntrlgs/ltmtptntlmntnyfrmtn2013/ltmtptntlmntnyfrmtn2013-eng.html

Thermal Maturity Maps of the Montney, Muskwa, Duvernay, and Nordegg from Rokosh et al. (2012).

Cretaceous Colorado Group Eastern Alberta

This play is potentially widespread but there has been limited shale gas activity and production within this interval mostly as a result of the gas price. The shale gas intervals are normally comingled so numbers are difficult to grasp for the shales. There have been small companies producing gas from this zone but they are limited and some are selling their interests. Some companies are now focusing on the liquids potential of the Second White Specs.

Lower Jurassic Nordegg (Gordondale) West Central Alberta

Anglo Canadian Oil Corp. now Tallgrass Energy Corp. has been playing the potential of the Nordegg Member, which is a source rock composed of basinal shales, silts, and carbonates. They feel that the Nordegg Member contains a huge amount of oil. They drilled a horizontal well to test this play producing limited liquids. Athabasca Oil Sands second Kaybob Nordegg horizontal well, at 04-11-63-20w5, offsetting its first Nordegg horizontal well. After a 16-stage slickwater frac and 4 days of cleanup, the 04-11 well made 335 b/d of 41° gravity oil and 500 MCFD of gas at 910 psig flowing pressure. There are others in this play but information is tight: Penn West, EOG, Apache, Surge, Nordegg, Petro-Bakken, Altima, Long Run and others (see Meloche (2011)).

Triassic Montney Shale Western Alberta

The Montney fairway extends in Alberta where this play is being picked up for both gas and liquids rich gas. Some of this Montney is classified as conventional because of facies change. Companies actively testing oil-prone Montney exploration acreage include ARC Resources Ltd. at Ante Creek and Tower, Athabasca Oil Sands at Kaybob, Celtic Exploration at Karr, CIOC at Karr and Simonette, Canadian Natural Resources Ltd. at Tower, Crew Energy Inc. at Tower, Harvest at Ante Creek, Imperial Oil at Berland, Long Run at Girouxville, RMP Energy Inc. at Grizzly and North Waskahigan, Seven Generations at Karr, and Triology Energy Corp. at Kaybob West (http://www.ogj.com/articles/print/volume-111/issue-4/general-interest/montney-duvernay-will-be-key-to-canada.html).

Devonian Duvernay/Muskwa Shales Western Alberta

The exciting new liquids play, Duvernay Shale is the stratigraphic equivalent to the Muskwa in NE BC. The Duvernay has been credited as the source rock for most of the gigantic Devonian oil and gas pools of Alberta. This zone

---

compares favorably to other North American shale plays with its position in the liquids window, organic content, porosity, thickness and over pressuring. The Duvernay is often compared to the prolific Eagle Ford of Texas because they both are shale plays that offer a full spectrum, from dry gas through liquids-rich gas to oil. According to the Energy Resources Conservation Board, the Duvernay holds an estimated 443 Tcf of gas, 11.3 billion barrels of natural

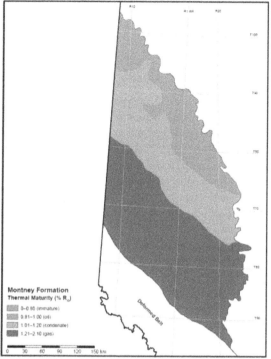

Figure 2.3.7. Thermal maturity map of the Montney Formation.

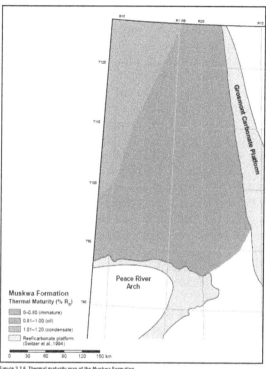

Figure 2.2.6. Thermal maturity map of the Muskwa Formation.

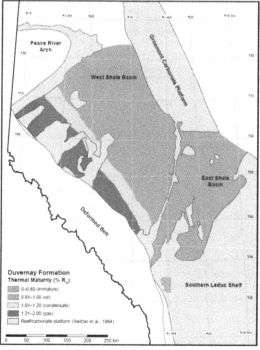

Figure 2.1.7. Thermal maturity map of the Duvernay Formation.

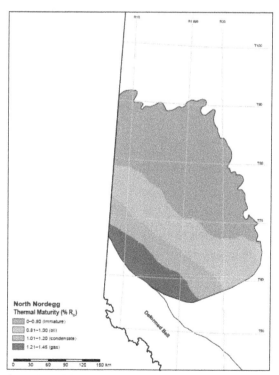

Figure 2.5.7. Thermal maturity map of the north Nordegg.

gas liquids and 61.7 billion barrels of oil. It is estimated that $4.2 has been spent on this play as of June 2012. This BMO Capital markets research report, June 2012, has a wealth of data on this play. Encana believes that this play is two times the size of the Eagle Ford Play (https://bmo.can.idrsite .com/Login%20PagE%20Document%20Library/AD Library/Presentations/BMO%20Duvernay%20DUG% 20JunE%202012.pdf; http://www.junewarren-nickles.com/ alberta-oil-and-gas-update/).

The Duvernay play is divided into the Western and Eastern Shale Basin with the West divided into three drilling districts, Kaybob, Edson, and Pembina. The companies involved in this deep and expensive play of 3100 to 3700 m are numerous, some of which are Celtic now Exxon (paid C$2.6 billion), Encana, ConocoPhillips, Husky, Athabasca, Chevron, Trilogy, Shell, Talisman, Yoho, Taqa North amongst others.

Encana has accumulated a 253,000 acre position in this play. Initially, they announced that their first two wells produced 2–5 MMCFD and 158–390 bbl per day, respectively. Their well costs are $12 million with a EUR per well at 3–6 Bcf and 350–600 MBBL over a lateral length: 3500–6500 ft and at a TVD of 8300–13,000 ft. Their yield is 159 to 320 bbl (50–60° API) per MMCF. Just recently they announced that one well had an initial production rate of 1400 B/D of condensate and 4 MMCFD of gas over its first 30 days of production and continues, after 160 days, at 350 B/D of condensate and 2 MCF/D of gas. They planned 20 wells in 2013 with

an exit rate of 1500 BD of liquids. They recently announced a joint working interest with PetroChina. Encana's Duvernay comparison is shown below.

Talisman has drilled and released information on four wells into this play and has reported condensate rates of 120–733 B/D of 50–57° API. They report that their liquids yield is 200–1000 BBL per MMCF. They spent $510 million in June increasing their footprint to 352,000 acres.

Late Devonian and Early Mississippian Alberta Bakken – Exshaw Southern Alberta

The Alberta Bakken (Exshaw) is another emerging tight oil resource play in SW Alberta to NW Montana consisting of three zones, Big Valley/Stettler Carbonates, Bakken/Exshaw dolomitic siltstones, and Banff carbonates. This play gained momentum south of the border in Montana and has recently emerged into Alberta and there is rush to get a position. In a report late last year, research firm Wood Mackenzie said the tight oil play that straddles the Alberta-Montana border could contain a recoverable 2.6 billion barrels of oil. Production of about 300–350 BOPD has been published. There are a number of companies in this play. Over 30 horizontal wells have been drilled so far but with little publication of results. Crescent Point, Shell, Murphy, Torc, Argosy, Primary Nexen, Bowood/Legacy, Rosetta, and Newfield are some of the companies involved. Crescent Point Energy has 1,000,000 acres,

Duvernay – World Class Reservoir

Encana controls one third of the top tier land in the condensate window*

Reservoir Attribute	Duvernay	Marcellus	Eagle Ford
TVD Depth (ft)	8,200 – 13,100	400 – 8,400	8,800 – 13,800
Gross Thickness (ft)	65 – 230	50 – 200	50 – 280
Porosity (%) / **Permeability** (nD)	3 – 8 / 10 – 400	10 / 20 – 55	10 – 11 / 50 – 1200
Reservoir Temperature (°F)	220 – 250	130 – 170	220 – 280
Gas Composition	Wet / Dry	Wet / Dry	Wet / Dry
Total Organic content (TOC %)	1 – 20	3.0 – 12.0	4.5
Maturation (Ro %)	0.6 – 2.9	1.4 – 3.0	0.6 – 2.2
Pressure Gradient (PSI/ft)	0.72 – 0.96	0.45 – 0.60	0.60 – 0.80
EUR (MMboe / well) (6:1)	0.7 – 1.6	0.3 – 1.0	0.5 – 1.3

*Encana estimate
Source: RBC Richardson Barr / RBC Rundle estimates, industry estimates, DOE April 2009 publication "Modern Shale Gas: A Primer"

drilled eight wells in the fourth quarter of 2012. Murphy is drilling 6–9 wells with 5 drilled to date: 3 producers, one being evaluated, and one awaiting completion. They have announced tests of 415 to 800 BOPD. Deethree Exploration said it has two drilling rigs operating on the lands of 200,000 acres, where they have tested 600–950 BBL/D of 30° API oil. They have drilled 17 horizontal wells into this zone. Torc has reported that two of their wells have yielded IP rates of 510 and 514 BOPD (see Zaitlin 2011, 2012).

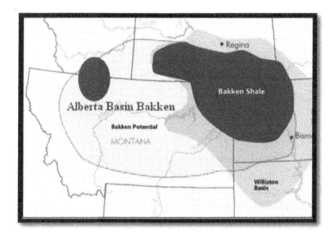

The Alberta Energy Resources Conservation Board (ERCB) just recently published a document to clarify the definition of shale for shale gas development and to identify the geological strata from which any gas production will be considered to be shale gas:http://www.ercb.ca/docs/documents/bulletins/Bulletin-2009-23.pdf

Alberta Energy Shale Gas: http://www.energy.alberta.ca/NaturalGas/944.asp

The Alberta Geological Survey (AGS) is active in publishing geological studies including a number of studies on shales.

AGS Shale Gas Section: http://www.ags.gov.ab.ca/energy/shale-gas/shale-gas.html

AGS Reports : http://www.ags.gov.ab.ca/publications/pubs.aspx?tkey=shalE%20gas

AGS Conference Papers and posters: http://www.ags.gov.ab.ca/conferences/geology-poster-ppt.html

Alberta Duvernay/Muskwa and Montney Formations Shale Analysis poster by the ERCB and Alberta Geological Survey: http://www.ags.gov.ab.ca/conferences/shale_gas_cspg_2009_poster.pdf

The Alberta Geological Survey has this link with documents on the Colorado Play: http://www.ags.gov.ab.ca/publications/pubs.aspx?tkey=colorado%20group

In 2012 the Alberta Geological Survey just published a new document (see Rokosh et al., 2012) The ERCB is the regulator for Alberta: http://www.ercb.ca/portal/server.pt

Saskatchewan

Upper Cretaceous Colorado Group – Biogenic Gas Central Saskatchewan

As in Alberta, the Colorado Group shales have been produced in Saskatchewan at low volumes for a 100 years but the recent gas price decline has kept this play minimized. In this province, the past exploration focus has been primarily on two types of biogenic shale gas potential within the Upper Cretaceous. The first type is a hybrid shale gas play along the Saskatchewan–Alberta border, where thin laminae of sand and silt lie within the shales of the Upper Colorado Group. Other intervals within the Colorado Group that were once lumped and dismissed as 'non-productive shale' are also now being re-evaluated. The second type of play currently being evaluated is the Colorado shale gas play in the eastern half of the province. These highly organic shales have been the focus of exploration in the past, prior to World War II, when gas seeps were reported near the towns of Kamsack and Hudson Bay. Several wells near Kamsack produced from the early 1930s to late 1940s with total gas production of 168 MMCF. From 2001 to September 2008, 59 new wells, licensed for gas, were drilled in the Hudson Bay and Kamsack areas. There are still no major commercial discoveries and not much news out of Saskatchewan this year as a result of the lower gas price and the economy. There are, however, around 13 wells in SW Saskatchewan that under production from the Colorado shales.

Between 2004 and 2008, more than 50 test wells were drilled for shale gas in all areas in the province, including Watrous, Moose Jaw, Strasbourg, Foam Lake, Smeaton, Shell Lake, and Big River but no commercial discoveries have been announced (http://www2.canada.com/reginaleaderpost/news/business_agriculture/story.html?id=c41a6b5b-b892-40cc-8cb4-902156681111&k=18412)

PanTerra Resource Corp. have drilled and cased 36 wells within their more than one million acres of land. They feel they have 3 Tcf of recoverable gas. They had been coring, logging, and fracture stimulating but no rates have been announced to date. Because of the low gas prices, they have put this project on hold. There has also been some activity in the Pasquia Hills in central east Saskatchewan. Pasquia Hills has a huge potential for oil shale in this area but there have been about 23 wells drilled by various operators with gas shows and some limited gas tests. Nordic announced recently that survey work has now commenced for a five-well drilling program on the Company's land in Preeceville. Nordic believes that with new drilling technology available, it will be successful in unlocking the enormous reserves of shale and natural gas. After drilling two unsuccessful wells, they will be returning in the fall for another well. It is unclear whether this play is unconventional or conventional or both with both gas and oil as their targets. Recently, Questerre announced a Pasquia Hills program. They acquired 100% interest in

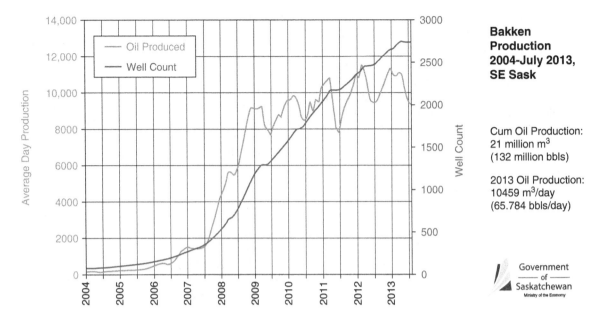

**Bakken
Production
2004-July 2013,
SE Sask**

Cum Oil Production:
21 million m³
(132 million bbls)

2013 Oil Production:
10459 m³/day
(65.784 bbls/day)

Government
of
Saskatchewan
Ministry of the Economy

39,000 acres. Situated in one of the Canada's largest oil shale deposits with plans of a 2012–2013 work program to include core holes and trenching to assess potential. They have partnered up with a United States firm that uses the EcoShale In-situ capsule process, which is an innovative approach that moves the machines to the rocks instead of moving the rocks to the machines to extract the oil. Drilled 16 wells in 2012 and analysis of core indicates recoveries between 10–20 gallons/ton with select intervals of up to 16–20 gallons/ton within a 20–35 m section. Ten more wells to be drilled beginning fall 2012 and then integrate core data and develop resource estimate by early 2013. Xtra Energy Corp. announced today that it has entered into a Letter of Intent with two corporations which own a 55% working interest in the Pasquia Hills oil shale permit SHP00008 (http://www.saturnminerals.com/i/pdf/oilshalesaskatchewan .pdf).

Upper   Devonian–Lower   Mississippian   Bakken
Saskatchewan is also reaping the benefits of the boom in horizontal and fracturing techniques drilling, especially in the Bakken. Production has risen from about 1–2000 BOPD in 2005 to about 65,784 BOPD at the end of 2013 with a cumulative production of 21 MM m³ or 132 MM BBL. The Bakken production comes from the tight siltstone and sandstone beds within the shales (Kreis and Costa 2005) so it is not really a shale oil play. The Bakken wells tend to highly productive at 200 BOPD producing a light sweet crude oil with 41° API gravity. There are many players in this zone. One of the two bigger players are Crescent Point with 704,000 net acres, 3800 drilling locations, 198 wells to be drilled in 2012 and 4.6 billion BOIP. They plan to drill 163 wells in 2013. Currently, they are producing at 42,000 BPD. Their new water flood has increased recovery factor

by 3%. PetroBakken now Lightstream is the other one with 256,200 net acres, 979 producing wells, 700 net drilling locations, and about 19,000 BPD.

Saskatchewan Government energy and resources is the regulator.

http://www.er.gov.sk.ca/Default.aspx?DN=4c585c56- 193a-485a-91fd-7c49f0104a60

Manitoba

Cretaceous Colorado Group There is the potential of shale gas in Manitoba, but no activity or production. There have been a number of publications on the shallow shale potential by Nicholas and Bamburak. (http://www.wbpc .ca/assets/File/Presentation/11_Nicolas_Manitoba.pdf   and Nicholas 2011 http://www.wbpc.ca/assets/File/2011%20Pre sentations/Tuesday/Nicolas%20WBPC%202011_ShalE& Percnt;20gas%20to%20ThreE%20Forks.pdf)

Upper Devonian–Lower Mississippian Bakken The production of oil from the comingled Bakken/Torquay, which began in the mid-1980s, continues, with about 640,740 BBL per month or 21,385 BBL/D. Cumulative historical production is 42,364,754 BBL from about 1831 producing wells. The Bakken produces more water than oil so water disposal is a continuing issue. The following graph shows production from the Bakken, Mississippian, and Triassic (Lower Amaranth).

The Manitoba oil and gas is the regulatory agency: http://www.gov.mb.ca/stem/petroleum/index.html

Manitoba Mineral Resources: http://www.manitoba.ca/ iem/mrd/index.html

Manitoba Geological Survey: http://www.manitoba.ca/ iem/mrd/geo/index.html

Annual Production

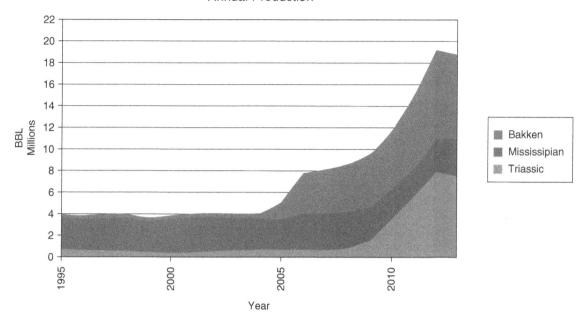

Ontario

Upper Devonian Kettle Point Shale (Antrim Shale Equivalent) Middle Devonian Marcellus Shale

Upper Ordovician Blue Mountain and Collingwood Shale (Utica Equivalent) Exploitation of these shales has been very quiet with only a few operators discussing the evaluation of these shale targets. These shales are mostly considered secondary targets but only one well has been drilled to test these zones to date. The only drilling activity is by the Ontario Geological Survey. They drilled two stratigraphic tests last year to assess the shale gas potential of the Kettle Point Formation. They have just released a request for proposals to drill two more stratigraphic test wells to test the Collingwood-Blue Mountain. No results have been published yet. In the spring of 2010, two boreholes were drilled through the Kettle Point Formation. Core samples were collected to evaluate gas concentration and other key parameters. Similar work was performed in 2011 near Mount Forest in the County of Wellington to assess the shale gas potential of the Ordovician shale succession. Furthermore, in the summer of 2012, additional rock samples were collected from previously drilled wells from Southern Ontario and were analyzed for mineralogy and Rock-Eval® 6 pyrolysis parameters. These analyses may assist in refining stratigraphic correlations across provincial and international borders. This project is referenced in Béland Otis 2012.

The Ministry of Natural Resources of Ontario is the regulator:

http://www.ogsrlibrary.com/government_ontario_petroleum.html

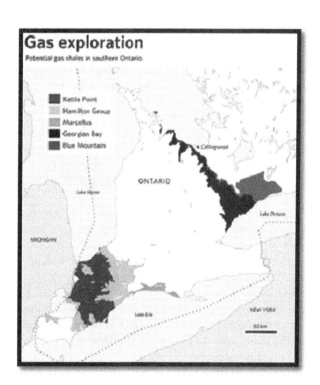

http://www.ogsrlibrary.com/downloads/Ontario_Shale_Gas_OPI_2009_Nov11.pdf

http://www.wnloilandgas.com/media/uploads/Carter_Presentation.pdf

http://www.ogsrlibrary.com/

Ontario Geological Survey:

http://www.mndm.gov.on.ca/en/mines-and-minerals/geo
science/non-renewable-energy

http://www.geologyontario.mndmf.gov.on.ca/mndmfiles/
pub/data/imaging/OFR5716/OFR5716.pdf

http://www.geologyontario.mndmf.gov.on.ca/mndmfiles/
pub/data/imaging/OFR5817/OFR5817.pdf

Quebec – St. Lawrence Lowlands

Ordovician Lorraine and Utica Shale The other potential
bright light in Canadian shale exploration in 2008 was in
Quebec, within a 300 km by 100 km fairway between Montreal and Quebec. The Upper Ordovician Utica and Lorraine
shales are the targets. On March 8, 2011, the Quebec Provincial Government effectively declared a temporary moratorium on the use of chemical fracturing during shale gas
drilling pending stricter a full environment assessment audit.
http://www.bape.gouv.qc.ca/sections/rapports/publications/
bape273.pdf in French only

As well, no new wells will be drilled without local
approval. This review conducted by an 11-person committee could take up to 30 months. The government had
previously awarded permits for 29 drilling sites, where
fracking has taken place on 18 locations. A year ago, when
Quebec Environment Minister Pierre Arcand announced the
$7-million Strategic Environmental Assessment (SEA) on
shale gas, he said it was in part to reassure the public that
shale gas development will not go ahead in Quebec unless
it is determined to be safe. But a year into that process,
opposition remains as fierce as ever. The committee's final
report is to be submitted to the government by November
2013. It is now estimated that it might be complete in 2014
(http://www.montrealgazette.com/technology/year+later+
opposition+shalE&Plus;exploration+still+fierce/6327547/
story.html#ixzz1qkV4IfFA).

Less than 24 hours after being sworn in, Quebec's
new natural resources minister has shut the door nearly
definitively on shale gas exploration and commercialization in the province (http://business.financialpost.
com/2012/09/20/quebec-hints-at-long-term-shale-gas-ban-
citing-ecological-risks/)

As of February 2013, Quebec announced a moratorium
on shale gas development in the St. Lawrence Valley until
after the above review is complete. Talisman has suspended
shale gas exploration here.

Industry has drilled or evaluated 23 wells and spent $200
million. Assuming a green light after the environment review
finishes industry is saying that it would take 3–4 years before
the production stage is reached. The CERI published Potential Economic Impacts of Developing Quebec's Shale Gas
in March 2013 (http://www.ceri.ca/images/stories/2013-03-
08_CERI_Study_132_-_Quebec_Shale.pdf).

Both Forest Oil Corporation and their partners and
Talisman and their partners have drilled to evaluate

both the Lorraine (up to 6500 ft thick) and the Utica
(300–1000 ft thick). Talisman with their partners and a
771,000 acre land position has drilled six vertical wells
with tested rates at from 300 to 900 MCFD. In 2009
and 2010, they drilled or will be drilling five horizontals, which were currently being evaluated. Talisman
has since suspended its shale gas exploration in Quebec
(http://www.theglobeandmail.com/globe-investor/talisman-
suspends-shale-gas-exploration-in-quebec/article4753334/).
Forest, after drilling two vertical wells with production
rates up to 1 MMCFD and three horizontals, is waiting
for the rock work and the analysis before proceeding
further. The horizontal rates range from 100 to 800 MCFD
with four-stage fracs. These are 10 year leases. Forest
estimated 4.1 Tcf resource potential at 20% recovery. These
black shales of 1–3% TOC are 500 ft thick within the gas
window. Canbrian, Gastem, Junex, Questerre, Molopo,
Intragaz, Petrolympic, and Altai are among the other
interest holders in this play. Questerre Energy Corporation
reported on the test results from the St. Edouard No.
1A horizontal well. The horizontal well was successfully
completed with an 8-stage fracture job to stimulate the
rock. Cleanup and flowback commenced January 29,
2010. During the test, the well flowed natural gas at an
average rate of over 6 MMCFD (http://www.questerre.com/
assets/files/PDF/101121_QEC_Presentation-Update.pdf).

Upper Ordovician Macasty Shale In addition, the Upper
Ordovician Macasty Shale (Utica Equivalent) drilled by Corridor and Petrolia on Anticosti Island in the Gulf of St.
Lawrence has seen some interest, largely as a secondary
target, with results from recent coring identifying shale oil
potential. Corridor reported the results of an independent
resource assessment of the Macasty Shale that resulted in
a best estimate of the Total Petroleum Initially-In-Place
33.9 billion barrels of oil equivalent (BBOE) for Corridor's
land holdings with the low estimate at 21.4 BBOE and the
high estimate at 53.9 BBOE. Corridor and Petrolia have
announced a new program where coring, water wells, and
data collection are expected to be completed by the end of
2012, with the final analytical results due in 2013. These
results were just announced in January 2013. Junex has a
position in Anticosti Island as well.

http://www.corridor.ca/investors/documents/March2011
TechnicalPresentationonAnticostiIsland.pdf

http://www.corridor.ca/documents/Martel_Anticosti_
NAPE_2012.pdf

http://www.corridor.ca/media/2011-press-releases/
20110713.html

http://www.petroliagaz.com/imports/pdf/en/18168.Petro
lia.FinalReport.pdf

Utica Emerges in Quebec Shale Play Extends to Canada
by Susan Eaton

http://www.aapg.org/explorer/2010/01jan/shale0110.cfm

St. Lawrence Lowlands, Quebec: Shale Gas Area (Séjourné et al 2013)

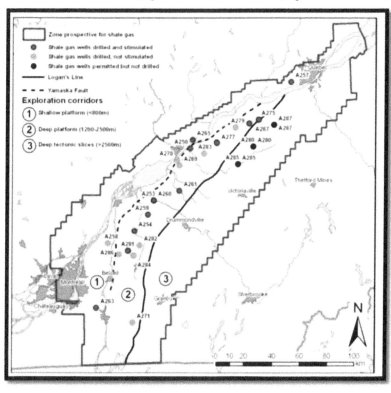

In a surprise move Quebec announced, February 13, 2014, that it would move ahead with oil exploration on Anticosti with the province pledging $115-million to finance drilling for two separate joint ventures. These shales will have to be fracked. Quebec would become an equity partner with five separate firms, controlling 50% of the licenses and rights to 60% of any commercial profit. Shale gas plays in the province's St. Lawrence Lowlands enjoy another advantage in being close to the northeast U.S. gas market. The *Association pétrolière et gazière du Québec* (Quebec Oil and Gas Association). APGQ/QOGA membership: Altai Resources Inc., Canadian Forest Oil, Canadian Quantum, Canbriam, Dessau (Associate member), Gastem, Intragaz, Junex, Molopo, Questerre, Roche (Associate member), SNC-Lavalin (Associate member), and Talisman Energy Inc.

Quebec Shale Conference 2013, 2012, 2010, and 2009

http://www.apgq-qoga.com/en/

http://www.apgq-qoga.com/en/2011/07/20/qoga-annual-conference/

http://www.apgq-qoga.com/wp-content/uploads/2012/09/Programme-APGQ-2012-HR.pdf

http://www.apgq-qoga.com/en/conference-2013-2/

Ministère des Resources naturelles et de la Faune de Québec is the regulator.

http://www.mrnf.gouv.qc.ca/english/energy/oil-gas/oil-gas-potential.jsp

St. Lawrence Lowlands, Quebec: Shale Gas Area (Séjourné et al. 2013)

New Brunswick

Lower Mississippian Fredrick Brook Shale Moncton Basin
The Lower Mississippian Fredrick Brook Shale in the Moncton Basin has been the focus of thermogenic gas exploration in this province. The Green Road G-41 well was drilled by Corridor Resources in November 2009 and tested in two zones in the Fredrick Brook, after fracing with propane. The lower black shale interval of the formation flowed at a rate of 0.43 MMCFD, whereas the upper silty/sandy shale zone of the formation tested at initial peak rates of 11.7 MMCFD with a final rate of 3.0 MMCFD. Corridor also announced the farmout of 116,018 acres this shale-potential land to Apache. Apache drilled their second well into this play and proceeded to run five slickwater stimulations per well with no gas recovery. Apache has left the project. Ten wells have been drilled into this play with seven completed and six testing gas. The rates have not been consistent. Another appraisal well has been recently spudded. Their plans are to try to develop this thick play of greater than 500 m vertically. During 2011, Corridor completed the drilling of the vertical Will DeMille O-59 shale gas appraisal well to a total depth of 3188 m measured depth. Strong gas shows were

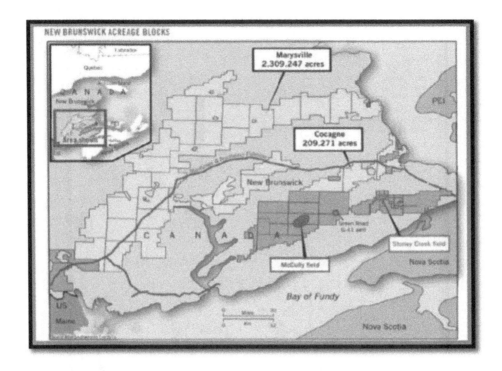

encountered within Hiram Brook sandstones and the Upper Frederick Brook shale. Based upon initial analysis of well log information, the well intersected at least eight intervals with significantly elevated gas shows that are considered frac candidates. Corridor plans to evaluate these intervals with logs and sidewall cores in order to select the intervals for future fracture stimulation. The Will DeMille O-59 well is located north of Elgin, New Brunswick. The Province of New Brunswick has recently issued recommendations for new proposed environmental requirements that allow for the exploration and development of oil and gas in New Brunswick. Corridor is working with the government and other stakeholders to ensure best practices are followed and oil and gas activities can be completed in a safe and responsible manner. Corridor plans to stimulation in 2014 at their Elgin shale gas project using the new government guidelines. Details of their plays can be found at http://www .corridor.ca/documents/CorridorOverviewMemorandum UpdateFB.pdf.

Contact Exploration and PetroWorth Resources are also re-evaluating their shale gas potential in the Fredrick Brook. On March 16, 2010, Southwestern Energy Company bid $47 million for 2.5 million acres in two areas for both conventional and unconventional resources of the Mississippian Horton Group. The company has completed airborne magnetic and gravity acquisition and is in the second phase of surface geochemical sampling and the acquisition phase of approximately 250 miles of 2D data. Interpretation of the data is underway. $10.7 million was invested in 2010 with $14.2 million investment planned for 2011 and then $14.2

million in 2012 with possible well(s). In consultation with the New Brunswick Department of Natural Resources and other key government officials, SWN Resources Canada will defer there planned 2012 exploration program until 2013 to provide additional time for public engagement and completion of the permitting process. Just recently they have applied to the government for a 1 year extension on their permits with the plans for a further 130 miles of 2D seismic with the plans of future drilling. They just finished their seismic program in December 2013.

"Frederick Brook Shale spurs Canadian exploration," by Susan Eaton *AAPG Explorer*, August 2010, p. 6–10.

http://www.aapg.org/explorer/2010/08aug/fredrick0810 .cfm

New Brunswick Natural Resources, Minerals and Petroleum is the regulator for this province.

http://www.gnb.ca/0078/minerals/index-e.aspx

http://www.gnb.ca/0078/minerals/GSB_Hydrocarbon_ Basin_Analysis-e.aspx#Objective

Shale Gas Website http://www2.gnb.ca/content/gnb/en/ corporate/promo/natural_gas_from_shale.html   Update on New Brunswick by Steven Hinds http://www .wnloilandgas.com/media/uploads/Hinds_Nfld2011.pdf

Update on Fracking

Energy and Mines Minister Craig Leonard is promising to review two new reports on the shale gas industry before the provincial government acts on any of their recommendations.

http://www2.gnb.ca/content/dam/gnb/Corporate/pdf/ ShaleGas/en/ExecSummary.pdf

In February 2013, New Brunswick released their new rules for industry by pushed ahead with dozens of new regulations governing shale gas exploration.

In bringing forward 97 new rules that will cover the oil and gas industry's practices, the government said it is striking a balance between environmental protection and the economic potential that fracking for shale gas represents.

http://www2.gnb.ca/content/dam/gnb/Corporate/pdf/ShaleGas/en/RulesforIndustry.pdf

http://www2.gnb.ca/content/gnb/en/corporate/promo/natural_gas_from_shale.html

Nova Scotia

Upper Devonian/Lower Mississippian Horton Bluff Kennetcook Basin The Upper Devonian–Lower Mississippian Horton Bluff Shale in the Kennetcook Basin has been the primary target for thermogenic shale gas exploration in the province by Triangle (Elmworth) Petroleum since May 2007. A 2D and 3D seismic program was initiated and a total of 5 vertical exploration wells have been drilled since May 2007. Various fracture treatments have been performed although none have successfully produced gas so far. On April 16, 2009, Triangle executed a 10-year production lease on its Windsor Block in Nova Scotia that covers 474,625 gross acres (270,000 net acres) with a potential of 20 Tcf recoverable. They have agreed to drill at least 7 more wells in this block before 2014. In 2009, they conducted a 30 km 2D seismic program to try to pinpoint areas with structure for future shale targets. Currently, there has been no work this year as they are looking for partners. The Government is appointing in the Spring of 2011, an internal committee of officials from the Departments of Energy and Environment to examine the environmental issues associated with hydraulic fracturing in shale gas formations and make recommendations on any additional required regulatory measures was to be finished in early 2012 but now has been extended to 2014 with another independent review pending starting in the fall of 2013 (http://www.gov.ns.ca/nse/pollutionprevention/docs/Hydraulic.Fracturing.Web.Update-2012.07.pdf; http://www.cbc.ca/news/canada/nova-scotia/story/2012/04/17/ns-fracking-decision-delay.html; http://novascotia.ca/news/release/?id=20130828001).

There are currently no applications for hydraulic fracturing of shales in Nova Scotia and none are anticipated soon until this review is complete (http://www.novascotia.ca/nse/pollutionprevention/consultation.hydraulic.fracturing.as).

This abstract is from "The Horton Bluff Formation Gas Shale Opportunity, Nova Scotia, Canada, Adam MacDonald, 2012 AAPG Search and Discovery (http://www.searchanddiscovery.com/abstracts/html/2011/annual/abstracts/Mac

Donald.html?q=%2BtitleStrip%3Acanada+%2ByearSort%3A%5B2011+TO+2012%5D).

The Horton Bluff Formation gas shales are within the Carboniferous lacustrine Horton Group of the Maritimes Basin. Gas in place estimates are 69 Tcf and leading indicators of a prospective shale gas play such as TOC at >5.5%, Maturity (Ro) of 1.6, thickness of >500 m and estimates of 100 Bcf per section across an area of >2 million acres have generated an increased interest in the Horton Bluff Formation within this frontier basin. Comparison of this shale play characteristics to many others (mineralogy, gas-filled porosity, pressure gradient, adsorbed gas) across North America ranks the Horton Bluff shale as among some of the most prospective. The Nova Scotia Department of Energy (NSDOE), working closely with industry, has recently undertaken the task of trying to understand the resource potential. The GIP or "size of the prize" is determined by the shales' gas generating potential, the mineralogy that may dictate the fracing techniques and lead into the engineering solutions that need to be achieved through the drilling and piloting phase to reach commercial producabilty.

The energy trader who co-founded Galveston LNG Inc. and later sold the Kitimat LNG scheme to Apache Canada and EOG Resources for roughly $300 million is back with a new plan to export natural gas from Canada's east coast. Alfred Sorensen said today that his new company, Pieridae Energy Canada, plans to build an export terminal at Goldboro, Nova Scotia. It is contemplated that the gas source comes from the Marcellus, New Brunswick, and offshore Nova Scotia (http://www.albertaoilmagazine.com/2012/10/pieridae-energy-proposes-east-coast-lng-facility/)

The Goldboro LNG Facility is to include a gas liquefaction plant and facilities for the storage and export of LNG, including a marine jetty for off-loading, and upon completion, is expected to ship approximately five million metric tons of LNG per year and have on-site storage capacity of 420,000 cubic meters of LNG. The Goldboro LNG Facility is to be located adjacent to the Maritimes & Northeast Pipeline, a 1400-km transmission pipeline system built to transport natural gas between Nova Scotia, Atlantic Canada, and the North eastern United States. Pieridae Energy (Canada) Ltd. ("Pieridae") is pleased to announce that the Province of Nova Scotia has issued environmental assessment (EA) approval, with conditions, for company's proposed Goldboro LNG project in March of 2014.

The Nova Scotia Department of Energy is the regulator for the province.

http://www.gov.ns.ca/energy/oil-gas/onshore/

Newfoundland

Ordovician Green Point Shale Western Newfoundland The Cambro-Ordovician Green Point Formation is the focus of exploration activity for oil-bearing shale in the western parts of the province. This Green Point interval has been studied in outcrop by the Geological Survey of Canada and is summarized in Hamblin (2006). Oil seeps have been documented along the entire coastline and some oil production from as early as the 1900s have been recorded. A well drilled in 2008 from the onshore to the near offshore by Shoal Point Energy and partners encountered about 500–1000 m of shale with siltstone stringers with high gas and oil shows throughout the formation but no testing was attempted then. The geochemistry analysis indicates that this zone is in the oil window. Further drilling of the shale oil potential in this formation was undertaken by reentry of the previous well bore, sidetracking, and testing. These plans were unsuccessful and discontinued because of severe formation damage. The company has recently locked up contiguous blocks of land of about 720,000 acres to the north stretching more than 160 km. Just recently Black Spruce Exploration made an agreement to drill up to 12 wells in this play over the entire length of this play.

These projects are delayed since the government announced a study of the future of fracking.

http://www.releases.gov.nl.ca/releases/2013/nr/1104n06.htm

The Newfoundland Department of Natural Resources is the regulator for the onshore portion of the province. http://www.nr.gov.nl.ca/mines&en/oil/

http://www.nr.gov.nl.ca/nr/energy/petroleum/onshore/onshore.html

This is the latest publication by the DNR on the Shale Oil Potential of the Anticosti Basin.

http://www.internationalpavilion.com/NAPE%202012%20Presentations/J%202.48-3.00%20Wednesday%20Newfoundland%20and%20Labrador.pdf

The Canada-Newfoundland Labrador Offshore Board is the regulator for the offshore portion.

http://www.cnlopb.nl.ca/

Northwest Territories

Devonian Canol Shale The Northwest Territory Geoscience Office commissioned Dr. Brad Hayes of Petrel Robertson Consulting Ltd. of Calgary to undertake a regional-scale study of the unconventional shale gas and shale oil potential of the southern and central Northwest Territories. His report assembles available outcrop and subsurface data to systematically assess shale gas and oil potential and is available as NWT Open File 2011-08 (see below). The work follows on an earlier unconventional natural gas scoping study for the NWT (see Hayes, 2010). Canada's Northern Oil & Gas Directorate held lease sales in 2011 and 2012 where industry has committed $628 million in work commitments on 13 exploration licenses in the central Mackenzie region. It is speculated the Canol shale play was the main target. The Canol shale formation could be

as big as the prolific Bakken light oil play. Initial estimates peg the Canol play at 2–3 billion barrels of recoverable crude in a region that has seen drilling activity for almost a century but has yet to reap substantial economic benefit because of its remote and challenging terrain. The plan is for companies such as Imperial Oil, Shell Canada and MGM Energy, ConocoPhillips, and Husky Energy to continue activity to prove up the resource and eventually produce crude for southern market (http://m.theglobeandmail.com/globe-investor/husky-prepares-an-arctic-   expedition/article4179898/?service=mobile).

The MGM in partnership with Shell, who farmed in, were the first to announce the results of drilling into this new play. Their vertical well into the Canol shale resulting in the recovery of approximately 140 barrel of fluid consisting of frack fluid, crude oil and natural gas. According to MGM, the Canol/Hare Indian shale is 30–170 m thick at a depth of 1000–2500 m. In addition, the Bluefish Shale is 15–25 m thick at a depth of 1000–2700 m. Both are highly brittle, which is a key attribute for successful fracturing. There independent reserve estimate on four exploration licenses are about 11 Billion Barrels oil in place, mean. Drilling is restricted to the months of January to March. Husky drilled two vertical exploratory wells into the oil mature Devonian-aged Canol and Hare Indian/Bluefish Shales south of the community of Norman Wells in the Central Mackenzie Valley. They plan to test them this year. They have built an all-weather road into the drill site. Meanwhile, ConocoPhillips drilled two wells on its lone license in the Canol shale this winter and is planning to drill two horizontal wells in the winter of 2013–2014 (http://www.albertaoilmagazine.com/2013/03/mgm-energy-releases-underwhelming-results-for-canol-shale-well/).

The Enbridge Norman Wells to Alberta pipeline runs through the area flowing 40,000 BOPD. The Norman Wells oil field discovered in the 1920s has been in decline for decade and the pipeline is running at 33% capacity (see Hadlari and Issler, 2012; Pyle and Gal, 2012).

http://www.ihs.com/products/oil-gas-information/source-newsletter/international/jan2012/canadian-alaskan-exploration.aspx

Geoscience Office

http://www.nwtgeoscience.ca/petroleum/

http://www.nwtgeoscience.ca/petroleum/unconventional_gas.html

Yukon

Currently, the Yukon Geological Survey is conducting studies to determine the potential of shale gas in the territory. Shale gas has not been explored for or produced in Yukon; however, future oil and gas projects will most likely consider shale gas reservoirs as potential targets. Shale is likely found in all of Yukon's oil and gas basins. Whether or not the shale formations contain natural gas in sufficient quantity to produce has yet to be determined. The Yukon Geological Survey conducted a scoping study to identify the presence of shale gas and other unconventional oil and gas resources in the Yukon. The results of this study were published in 2012 (http://ygsftp.gov.yk.ca/publications/miscellaneous/Reports/YGS_MR-7.pdf).

Northern Cross Yukon acquired 15 exploration permits in Northern Yukon. There will be 4 wells drilled in the far north for conventional targets as well as consideration of the shale potential in the Devonian. In the south, in the Laird Basin, which extends into BC, EFLO Energy and partners are planning to exploit the Devonian/Mississippian shales near the Kotaneelee conventional field. This resource has the potential of 500 to 100 Bcf of conventional gas and 7.2–13 Tcf of shale gas. A sales gas pipeline exists to Ft. Nelson.

The Yukon government has established a committee to review hydraulic fracturing before it is permitted.

http://www.legassembly.gov.yk.ca/rbhf.html

Yukon Energy, Mines and Resources http://www.geology.gov.yk.ca/ http://www.emr.gov.yk.ca/oilandgas/

Nunavut

There are 12 basins with potential and discovered hydrocarbons through to the Paleozoic. Nothing is being worked on but shale plays would exist within the many source rock intervals. It is too isolated to be commercial at present.

Canada-Nunavut Geoscience Office (http://cngo.ca/)

Other Important Canadian Websites

National Energy Board of Canada
http://www.neb-one.gc.ca/clf-nsi/rcmmn/hm-eng.html
Geological Survey of Canada
http://www.nrcan.gc.ca/earth-sciences/about/organization/organization-structure/geological-survey-of-canada/9590
Canadian Association of Oil Producers
http://www.capp.ca/Pages/default.aspx

Societies, Conferences and Courses

Canadian Society for Unconventional Gas (CSUR)
http://www.csur.com/
Annual Unconventional Resources Conference
2014 October 15 and 16 at the TELUS Convention Centre in Calgary, AB. Note that they have technical luncheons for members. http://www.csur.com/events/technical-conference
Canadian Society of Petroleum Geologists (CSPG)
Note the CSPG has technical luncheons throughout the year.
http://www.cspg.org/

GeoConvention 2014: Focus: Calgary TELUS Convention Centre, May 12–16, 2014
http://www.geoconvention.com/
CSPG Courses
http://www.cspg.org/CSPG/Education/Education_Week/CSPG/ConEd/Courses_By_Event.aspx?h key=aa56013f-7c87-4cfd-af15-8aa5c9cece60

Other Meetings

CI Energy Group's 10th Annual Shale Oil & Gas Symposium, January 28–29, 2014 in Calgary http://www.shalegassymposium.com/

2014 International LNG in B.C.: British Columbia's LNG in the Global Market May 21–23, 2014|Vancouver Convention Centre http://engage.gov.bc.ca/lnginbc/files/2014/02/LNG_BC_agenda_download1.pdf

8th Unconventional Gas Technical Forum on June 5 and 6, 2014 at the Victoria Conference Centre, Victoria, B.C.
http://www.empr.gov.bc.ca/OG/oilandgas/petroleumgeology/UnconventionalGas/Pages/default.aspx

2014 Williston Basin Conference, Bismarck, North Dakota, May 20–22, 2014
http://www.wbpc.ca/

Hart Energy's Developing Unconventionals (DUG™) Canada conference and exhibition, Feb 25–27 2012 at the TELUS Convention Centre, in Calgary, Alberta, Canada.

Hart Energy's Developing Unconventionals (DUG™) Canada conference and exhibition, Feb 24–26 2013 at the TELUS Convention Centre, in Calgary, Alberta, Canada.

Feb 24–26, 2014 at the TELUS Convention Centre in Calgary, Alberta, Canada
http://www.dugcanada.com/

Exploration, Mining and Petroleum New Brunswick 2013, Delta Hotel Fredericton, November 3–5, 2014
http://www2.gnb.ca/content/gnb/en/departments/energy/conference/Conf_home.html

9th International Symposium of West Newfoundland Oil and Gas 10-11 September 2014, Corner Brook area of West Newfoundland.
http://www.wnloilandgas.com/international-symposium/

REFERENCES

Adams, C. (2010). Shale gas activity in British Columbia – exploration and development of BC's shale gas areas: Presentation to 4th Annual Unconventional Technical Forum, Victoria, BC.

Adams, C. (2011). Overview of shale gas operator activity in the Horn River Basin: Presentation to 5th Annual Unconventional Technical Forum, Victoria, BC.

Adams, C. (2012). Summary of shale gas activity in northeast British Columbia 2011: BC Ministry of Energy and Mines, p. 1–19. http://www.empr.gov.bc.ca/Mining/Geoscience/PublicationsCatalogue/OilGas/OGReports/Documents/2012/Summary

%20of%20Shale%20Gas%20Activity%20in%20NEBC%
202011%20Web.pdf

Adams, C. (2013). Summary of shale gas activity in Northeast British Columbia 2012; in Oil and Gas Reports 2013, British Columbia Ministry of Natural Gas Development, p. 1–27. http://www.empr.gov.bc.ca/OG/oilandgas/petroleumgeology/statsactivity/Pages/default.aspx

Adams, C. (2013). The Status of Drilling and Production in BC's Shale Gas Plays September 23, 2013 http://www.empr.gov.bc.ca/OG/oilandgas/petroleumgeology/statsactivity/Documents/5th%20NE%20BC%20Nat%20Gas%20Summit.pdf

BC MEM/ NEB Report. (2011). Ultimate Potential for Unconventional Natural Gas in Northeastern British Columbia's Horn River Basin, http:// www.em.gov.bc.ca/ OG/ Documents/ Horn-RiverEMA_2.pdf

B.C. Oil and Gas Commission. (2012). Montney Formation Play Atlas NEBC http:// www.bcogc.ca/ node/ 8131/ download

Beaton, A. P., Pawlowicz, J. G., Anderson, S. D. A., Berhane, H., Rokosh, C. D. (2010). Organic petrography of the Montney Formation in Alberta: shale gas data release. Energy Resources Conservation Board, ERCB/ AGS Open File Report 2010-07, 129 p. http:// www.ags.gov.ab.ca/ publications/ OFR/ PDF/ OFR_2010_07.PDF

Beaton, A. P., Pawlowicz, J. G., Anderson, S. D. A., Berhane, H., Rokosh, C. D. (2010). Organic petrography of the Duvernay and Muskwa formations in Alberta: shale gas data release. Energy Resources Conservation Board, ERCB/ AGS Open File Report 2010-06, 147 p. http:// www.ags.gov.ab.ca/ publications/ OFR/ PDF/ OFR_2010_06.PDF

Béland Otis, C., Carter, T., Fortner, L. (2011). Preliminary Results of a Shale Gas Assessment Project in Ontario, Canada: Search and Discovery Article #50390

Béland Otis, C. (2012). Preliminary results: potential Ordovician shale gas units in Southern Ontario. Summary of Field Work and Other Activities 2012, Ontario Geological Survey, Open File Report 6280, p. 29-1 to 29-12. http://www.geologyontario.mndm.gov.on.ca/ mndmfiles/ pub/ data/ imaging/ ofr6280// OFR6280.pdf

Carter, F., Béland-Otis (2009). Ordovician and Devonian Black Shales and Shale Gas in Southwestern Ontario, Canada, http://www.ogsrlibrary.com/downloads/Ontario_Shale_Gas_OPI_2009_Nov11.pdf

CBM Solutions. (2005). Gas shale potential of Devonian strata, northeastern British Columbia. B.C. Ministry of Energy, Mines and Petroleum Resources, Resource Development and Geoscience Branch: Petroleum Geology Special Paper 2005-01,110 p.

Chalmers, G. R. L., Bustin, R. M. (2008). Lower Cretaceous gas shales in northeastern British Columbia, Part II: evaluation of regional potential gas resources. *Bull Canadian Petrol Geol*, 56:22–61.

Chalmers G. R. L, Bustin R. M. (2012). Light volatile liquid and gas shale reservoir potential of the Cretaceous Shaftesbury Formation in northeastern British Columbia, Canada, *AAPG Bull*, 96(7):1333–1367.

Churcher, P. L., Johnson, M. D., Telford, R. G., Barker, J. F. (1991). Stratigraphy and oil shale resource potential of the upper Ordovician Collingwood member, Lindsay formation, southwestern Ontario: Ontario Geological Survey, Open File Report 5817, 98 p. http:// www.geologyontario.mndmf.gov.on.ca/ mndmfiles/ pub/ data/ imaging/ OFR5817/ OFR5817.pdf

Dixon, J. (2009). The Lower Triassic Shale member of the Montney Formation in the subsurface of northeast British Columbia. Geological Survey of Canada, Open File 6274, 9 p.

Dunn, L., Schmidt, G., Hammermaster, K., Brown, M., Bernard, R., Wen, E., Befus, R., Gardiner, S. (2012). The Duvernay Formation (Devonian): Sedimentology and Reservoir Characterization of a Shale Gas/ Liquids play in Alberta, Canada: CSPG Convention, http:// www.cspg.org/ documents/ Conventions/ Archives/ Annual/ 2012/ core/ 280_GC2012_The_Duvernay_Formation.pdf

Faraj, B., Williams, H., Addison, G., McKinstry, B., Donaleshen, R., Sloan, G., Lee, J., Anderson, T., Leal, R., Anderson, C., Lafleur, C., Ahlstrom, J. (2002). Gas shale potential of selected Upper Cretaceous, Jurassic, Triassic and Devonian shale formations in the WCSB of western Canada: Implications for shale gas production: Des Plaines, IL, Gas Technology Institute, GRI-02/ 0233, compact disc, 285 p.

Faraj, B., Williams, H., Addison, G., McKinstry, B. (2004). Gas potential of selected shale formations in the western Canadian sedimentary basin. *GasTIPS*, 10(1):21–25.

Ferri, F., Hickin, A. S., Huntley, D. H. (2011a). Besa River Formation, western Liard Basin, British Columbia (NTS 094N): geochemistry and regional correlations. In: Geoscience Reports 2011, B.C. Ministry of Energy and Mines, pp. 1–18.

Ferri, F., Hickin, A. S., Huntley, D. H. (2011b). Geochemistry and shale gas potential of the Garbutt Formation, Liard Basin, British Columbia (Parts NTS 094N, O; 095B, C). In: Geoscience Reports 2011, B.C. Ministry of Energy and Mines, p. 19–36.

Fiess, K. M., Ferri, F., Fraser, T. L. Pyle, L. J., Rocheleau, J. (2013). Liard Basin Hydrocarbon Project: Shale Gas Potential of Devonian-Carboniferous Strata in the Northwest Territories, Yukon and Northeastern British Columbia, 2013 CSEG CSPG CWLS Conference http://geoconvention.org/archives/2013abstracts/263_GC2013_Liard_Basin_Hydrocarbon_Project.pdf

Fraser, T. A., Allen, T. L., Lane, L. S., Reyes, J. C. (2012) Shale gas potential of devonian shale in north Yukon: results from a diamond drillhole study in western Richardson Mountains. In: MacFarlane, K. E., Sack, P. J., eds. Yukon Exploration and Geology 2011, p. 45–74. http://www.geology.gov.yk.ca/pdf/04_fraser_etal.pdf

Hadlari, T., Issler, D. R. (2012). Natural fracturing of the Canol Formation oil shale: an unconventional spin on unconventional spin on the Norman Wells oilfield: CSPG Convention, http://www.cspg.org/documents/Conventions/Archives/Annual/2012/109_GC2012_Natural-Fracturing_of_the_Canol_Formation_Oil_Shale.pdf

Hamblin, A. P. (2006). The "shale gas" concept in Canada: a preliminary inventory of possibilities. Geological Survey of Canada, Open File 5384, 108 p.

Hannigan, P. K., Morrow, D. W., MacLean, B. C. (2011). Petroleum resource potential of the northern mainland of Canada (Mackenzie Corridor). Geological Survey of Canada, Open File 6757, 271 p.

Hayes, B. J. R. (2010). Northwest Territories Unconventional Natural Gas Scoping Study: Northwest Territories Geoscience Office, NWT Open File 2010-03, 81 p. http://gateway.nwtgeoscience.ca/advancedsearch.php?rptnums=2010-03&authors=&rpttype=&datestart=&dateend

Hayes, B. J. R. (2011). Regional Characterization of Shale Gas and Shale Oil Potential, Northwest Territories: Northwest Territories Geoscience Office, NWT Open File 2011-08, 34 p. plus maps and cross-sections, http://gateway.nwtgeoscience.ca/ advancedsearch.php?rptnums=2011-08&authors=&rpttype=&datestart=&dateend

Hayes, B. J. R., Archibald, H. B. (2012). Scoping study of unconventional oil and gas potential, Yukon. Yukon Geological Survey, Miscellaneous Report 7, 100 p. http:// ygsftp.gov.yk.ca/ publications/ miscellaneous/ Reports/ YGS_MR-7.pdf

Johnson, M. D., Telford, P. G., Macauley, G., Barker, J. F. (1989). Stratigraphy and oil shale resource potential of the Middle Devonian Marcellus Formation, southwestern Ontario. Ontario Geological Survey, Open File Report 5716, 149 p. http:// www.geologyontario.mndmf.gov.on.ca/mndmfiles/pub/data/imaging/OFR5716/OFR5716.pdf

Kreis, L. K., Costa, A. (2005). Hydrocarbon potential of the Bakken and Torquay formations, southeastern Saskatchewan. In: Summary of Investigations, 2005, Vol. 1, Saskatchewan Geological Survey, Sask. Industry Resources, Misc. Rep. 2005-4.1, CD-ROM, Paper A-10, 10 p.

Kreis, L. K., Costa, A. L., Osadetz, K. G. (2006). Hydrocarbon potential of Bakken and Torquay formations, southeastern Saskatchewan. In: Gilboy, C.F., Whittaker, S.G., eds., *Saskatchewan and Northern Plains Oil & Gas Symposium 2006*, Saskatchewan Geological Society Special Publication 19, p. 118–137. http:// www.manitoba.ca/ iem/ mrd/ geo/ willistontgi/ downloads/ kreis_et_al_bakken- torquay_paper.pdf

Kohlruss, D., Nickel, E. (2009). Facies analysis of the upper Devonian – lower Mississippian Bakken formation, southeastern Saskatchewan. In: Summary of Investigations 2009, Vol. 1, Saskatchewan Geological Survey, Sask. Ministry of Energy and Resources, Misc. Rep. 2009-4.1, Paper A-6, 11 p.

Kohlruss, D., Nickel, E., Coolican, J. (2012). Stratigraphic and Facies Architecture of the Saskatchewan Oil Sands, Williston Basin Conference. http:// www.ndoil.org/ image/ cache/ 230pm_Wed_Kohlruss_WBPC_2012.pptx

Lavoie, D., Hamblin, A. P., Thériault, R., Beaulieu, J., Kirkwood, D. (2008). The upper Ordovician Utica shales and Lorraine group flysch in southern Quebec: tectonostratigraphic setting and significance for unconventional gas. Geological Survey of Canada, Open File 5900, 56 p. http:// geopub.nrcan.gc.ca/ moreinfo_e.php?id=225728

Lavoie, D., Pinet, N., Dietrich, J., Hannigan, P., Castonguay, S., Hamblin, A. P., Giles, P. (2009). Petroleum Resource Assessment, Paleozoic successions of the St. Lawrence Platform and Appalachians of eastern Canada: Geological Survey of Canada, Open File 6174, 2009; 275 p. http:// geopub.nrcan.gc.ca/ moreinfo_e.php?id=248071

Lavoie, D. (2012). Lower Paleozoic shale gas and shale oil potential in Eastern Canada: Geological Settings and Characteristics of the Upper Ordovician Shales: Search and Discovery Article #80242

Lavoie, D., Thériault, R. (2012). Upper Ordovician shale gas and oil in Quebec: sedimentological, geochemical and thermal frameworks; oral presentation, Canadian Society of Petroleum Geologist, GeoConvention 2012: Vision, Calgary, Alberta, May 14–18, 2012.

Marcil, J.-S., Dorrins, P. K., Lavoie, J., Mechti, N., Lavoie, J.-Y. (2012). Utica and Other Ordovician Shales: Exploration History in the Quebec Sedimentary Basins, Eastern Canada: Search and Discovery Article #10451

McDonald, A. (2012) The Horton Bluff Formation Gas Shale Opportunity, Nova Scotia, Canada; AAPG Search and Discovery Article #90124 http:// www.searchanddiscovery.com/ abstracts/ html/ 2011/ annual/ abstracts/ MacDonald.html?q=%2BtitleStrip%3Acanada+%2ByearSort%3A%5B2011+TO+2012%5D

Meloche, D. (2011). How deep is shallow? re-evaluation of depositional models for the Jurassic Nordegg and rock creek members of the fernie group: CSPG Convention, Calgary, http:// www.cspg.org/ documents/ Conventions/ Archives/ Annual/ 2011/ 263-How_Deep_is_Shallow.pdf

National Energy Board (NEB). (2009). A Primer for Understanding Canadian Shale Gas

National Energy Board (NEB) (2011), Tight Oil Developments in the Western Canada Sedimentary Basin, http:// www.neb-one.gc.ca/clf-nsi/rnrgynfmtn/nrgyrprt/l/tghtdvlpmntwcsb2011/tghtdvlpmntwcsb2011-eng.pdf

National Energy Board (NEB), (2013). The Ultimate Potential for Unconventional Petroleum from the Montney Formation of British Columbia and Alberta, http:// www.neb-one.gc.ca/ clf- nsi/ rnrgynfmtn/ nrgyrprt/ ntrlgs/ ltmtptntlmntnyfrmtn2013/ ltmtptntlmntnyfrmtn2013-eng.html

Nicolas, M. P. B., Bamburak, J. D. (2009). Geochemistry and mineralogy of Cretaceous shale, Manitoba (parts of NTS 62C, F, G, H, J, K, N): preliminary results. In Report of Activities 2009, Manitoba Innovation, Energy and Mines, Manitoba Geological Survey, p. 165–174.

Nicolas, M. P. B., Bamburak, J. D. (2009). Shallow unconventional Cretaceous shale gas in southwestern Manitoba: CSPG, CSEG, CWLS Convention, Calgary.

Nicolas, M. P. B. (2011) Stratigraphy and regional geology of the Late Devonian–Early Mississippian Three Forks Group, southwestern

Manitoba (NTS 62F, parts of 62G, K); Manitoba Innovation, Energy and Mines, Manitoba Geological Survey, Geoscientific Report GR2012-3, 92 p http:// www.manitoba.ca/ iem/ mrd/ info/ libmin/ GR2012-3.pdf,

Pawlowicz, J. G., Anderson, S. D. A., Rokosh, C. D., Beaton, A. P. (2009a). Mineralogy, permeametry, mercury porosimetry and scanning electron microscope imaging of the Banff and Exshaw formations; shale gas data release. Energy Resources Conservation Board, ERCB/ AGS Open File Report 2008-13.

Pawlowicz, J. G., Anderson, S. D. A., Rokosh, C. D., Beaton, A. P. (2009b). Mineralogy, permeametry, mercury porosimetry and scanning electron microscope imaging of the Colorado Group; shale gas data release. Energy Resources Conservation Board, ERCB/ AGS Open File Report 2008-14.

Pyle, L. J., Gal, L. P. (2012). Devonian Horn River Group in Mackenzie Plain Area, Northwest Territories, http://www.empr.gov.bc.ca/OG/oilandgas/petroleumgeology/Unconventional Gas/Documents/Devonian%20Horn%20River%20Group%20in%20Mackenzie%20Plain%20Area_NWT.pdf

Rokosh, C. D., Pawlowicz, J. G., Berhane, H., Anderson, S. D. A., Beaton, A. P. (2009). What is Shale Gas? An introduction to shale gas geology in Alberta. Energy Resources Conservation Board/ Alberta Geological Survey, Open File Report 2008-08, 26 p. http://www.ags.gov.ab.ca/publications/OFR/PDF/OFR_2008_08.PDF

Rokosh, C. D., Lyster, S., Anderson, S. D. A., Beaton, A. P., Berhane, H., Brazzoni, T., Chen, D., Cheng, Y., Mack, T., Pana, C., Pawlowicz, J. G. (2012). Summary of Alberta's shale- and siltstone-hosted hydrocarbon resource potential. *Energy Resources Conservation Board*, ERCB/ AGS Open File Report 2012-06, 327 p. http:// www.ags.gov.ab.ca/ publications/ abstracts/ OFR_2012_06.html

Ross, D., Chalmers, G., Bustin, R. M. (2005). Reservoir characteristics of potential gas shales in the western Canadian sedimentary basin: In: P.Lufholm, D.Cox, eds. *2005 WTGS Fall Symposium: West Texas Geological Society*, Publication No. 05–115, p. 41.

Ross, D. J. K., Bustin, R. M. (2007). Shale gas potential of the Lower Jurassic Gordondale member, northeastern British Columbia, Canada. *Bull Canadian Petrol Geol*, 55:51–75.

Ross, D. J. K., Bustin, R. M. (2008) Characterizing the shale gas resource potential of Devonian- Mississippian strata in the western Canada sedimentary basin: application of an integrated formation evaluation. *AAPG Bull*, 92, 87–125.

Russum, D., (2011). The Montney, A world class resource with a myriad of challenges and opportunities. *CSPG CSEG CWLS Convention*. http:// www.searchanddiscovery.com/ documents/ 2011/ 80128russum/ ndx_russum.pdf

Séjourné, S., Lefebvre, R., Malet, X., Lavoie, D. (2013). Synthèse géologique et hydrogéologique du Shale d'Utica et des unités sus-jacentes (Lorraine, Queenston et dépôts meubles), Basses-Terres du Saint-Laurent, Province de Québec. Geological Survey of Canada, Dossier Public 7338, 156 p.

Stephan, S., Lefebvre, R., Lavoie, L., Malet, X. (2013). Geological and Hydrogeological Synthesis of the Utica Shale and the Overlying Strata in Southern Quebec Based on Public Data in a Context of a Moratorium on Exploration CSPG Convention 2013 http:// www.geoconvention.com/ uploads/ 2013abstracts/ 066_GC2013_Geological_and_Hydrogeological_Synthesis.pdf

Walsh, W., Adams, C., Kerr, B., Korol, J. (2006). Regional "shale gas" potential of the Triassic Doig and Montney formations, northeastern British Columbia: BC Ministry of Energy, Mines and Petroleum Resources, Resource Development and Geoscience Branch, Petroleum Geology Open File 2006-02, 19 p.

Wrolson, B. M., Bend, S. L. (2012). High Resolution Organic Facies of the Bakken Formation, Williston Basin, Saskatchewan, Canada: Search and Discovery #50645.

Yurkowski, M., Wilhelm, B. (2009). Oil Shales in Saskatchewan http://www.saturnminerals.com/i/pdf/oilshalesaskatchewan.pdf

Zaitlin, B., Berger, Z., Kennedy, J., Kehoe, S. (2011). The Alberta Bakken: A Potential New, Unconventional Tight Oil Resource Play. CSPG Convention, http://www.cspg.org/documents/Conventions/Archives/Annual/2011/137-The_Alberta_Bakken.pdf

Zaitlin, B., (2012). An Emerging Unconventional Hybrid Light Tight Oil Play within the Southern Alberta Bakken-Exshaw Petroleum System: A Comparison between the Williston and Southern Alberta Basins. Williston Basin Conference, Regina, http://wbpc.ca/+pub/document/2013_presentations/Zaitlin%20-%20An%20Emerging%20Unconventional%20Hybrid.pdf

SHALE GAS/SHALE OIL IN EUROPE

BY KEN CHEW

Summary of the 5 Months November 2013–March 2014

Europe remains relatively unexplored for shale gas and, especially, shale liquids. In total, some 110 exploration and appraisal wells with a shale gas exploration component have been drilled, including horizontal legs from vertical wells. Twenty-five of these wells were shallow gas tests drilled in Sweden, largely using mineral exploration equipment. Up to 6 wells have been drilled to target shale liquids. Significant shale gas exploration activity since November 2013 has been limited to Poland, where 6 shale gas exploration wells were spudded, and England where one coal seam gas exploration well is also investigating shale gas potential. The first shale liquids test results have been reported from Poland.

Opposition to hydraulic fracturing and shale oil and gas exploration at grassroots level in general remains strong and major protests have taken place in the United Kingdom and Romania. Public pressure has resulted in moratoria being placed on some or all aspects of shale gas exploration and production in Bulgaria, Czech Republic, France, Germany, and Netherlands, plus certain administrative regions in Spain and Switzerland. Proposed new environmental legislation has led OMV to abandon its plans for shale gas exploration in Austria. But at a European and national political level, one can detect a desire to permit and even encourage exploration. While Germany effectively introduced a moratorium on hydraulic fracturing in November 2013, the Spanish government has moved to explicitly legalize hydraulic fracturing and the European Commission clarified its position in January 2014 without issuing binding legislation.

Political institutions and major companies have noted the impact of shale gas and shale oil production on the U.S. economy and fear that European consumers are suffering from unduly high gas prices and that European companies are becoming uncompetitive compared with their North American rivals and even considering relocating businesses to North America. Politicians, especially in Eastern Europe, are expressing a desire for more energy independence and recent political upheavals in Ukraine have added to those concerns. In addition, technical experts such as the French Académie des Sciences and German Federal Institute for Geosciences and Natural Resources have challenged the views and reports of environmental authorities.

The decision by ExxonMobil to withdraw from its six licenses in Poland, considered to be the most prospective European country for shale gas, has also had the effect of downgrading expectations in some quarters. Talisman has also withdrawn from Poland and Marathon and Eni have indicated their intention to do so. Nevertheless, Poland, in terms of current and proposed drilling activity, and the United Kingdom, in terms of farm-in activity and government support for the emerging shale gas industry, seem to be the two countries that offer most prospectivity in the medium term. The perceived prospectivity of the U.K. is indicated by farm-ins to acreage with shale gas potential by Centrica, GDF Suez and French major Total, all within the past 9 months.

Shale Gas in Europe

Europe is particularly well-suited to gas resource play exploitation on account of its large market, established pipeline infrastructure, increasing demand and current dependence on gas imports. Relatively high natural gas prices add to the attraction. Shale gas exploration in Europe is in its infancy. The first exploratory well was spudded in Scotland in 2005 and since then shale-specific exploratory drilling has been limited to five countries, with most wells being drilled from 2010 onwards. As a consequence, little is known about Europe's ultimate potential.

Resources Rogner's 1996 estimate of the in-place shale gas resource of Europe (including Turkey) was 550 Tcf. More recent studies indicate significantly larger in-place resources. In its assessment of the world's shale gas resource, the U.S. Energy Information Administration (EIA) estimated the European shale gas in-place resource for 9 countries at 2314 Tcf with a combined technically recoverable resource of 563 Tcf (U.S. EIA, 2011). A revised and extended report published in June 2013 (U.S. EIA, 2013) increased the study to 12 countries and included additional areas within countries such as Poland. The GIP estimate increased to 2408 Tcf but the recoverable estimate decreased to 472 Tcf (Table 76.4).

The EIA study estimates a gross shale rock volume based on maps of areal distribution and cross-sections indicating lateral extent, unit thickness, and depth of burial. Free gas and adsorbed GIP are then estimated using available organic content and maturity data. Technically recoverable gas is then calculated by applying one of three recovery factors (favorable: 25%; average: 20%; less favorable: 15%) based on clay content, geological complexity, reservoir pressure, and gas-filled porosity. Clearly, these estimates must be treated with caution. Much of the detailed information required to make accurate assessments is simply not available in many areas and so the assessments are still relatively speculative as the following individual country reports frequently indicate.

Austria: OMV has suggested a potential recoverable shale gas resource of 15 Tcf in the Vienna Basin, Austria, from an in-place resource of 200–300 Tcf.

Denmark: The U.S. Geological Survey (2013) has estimated mean technically recoverable resources of Denmark's Alum Shale to be 6.935 Tcf, of which 0 to 4.848 Tcf (mean: 2.509 Tcf) occur onshore and 0 to 8.492 Tcf (mean: 4.426 Tcf) lie offshore. The much larger EIA recoverable shale gas estimates (2011: 23 Tcf; 2013: 32 Tcf) are for the onshore area only.

Germany: In Germany, the Federal Institute for Geosciences and Natural Resources (BGR) has estimated recoverable shale gas to be in the range 24–80 Tcf from an in-place resource of 240–800 Tcf.

Netherlands: TNO's "best estimate" for "producible gas in place" in "high potential" areas of the Netherlands is 198 Tcf from an estimated in-place resource of 3950 Tcf. This is substantially greater than either of the EIA estimates (Table 76.4).

Poland: Four estimates of Polish Shale Gas resources have been made public through end-2013. The variance between the estimates is, in large part, a result of differences in methodology, which are outlined in Table 76.5.

Polish Geological Institute – National Research Institute (PGI) The study was based on archive data from 39 wells drilled between 1950 and 1990. Cut-offs of 15 m shale thickness (minimum), 2% TOC (minimum), and thermal maturity (Ro) in the range 1.1–3.5% were chosen. In the absence of key data such as porosity, permeability, mineral composition, reservoir pressure and initial production (IP), Estimated Ultimate Recovery (EUR) and average well drainage area were based on U.S. analogues. The range of estimates therefore reflects the combination of two uncertainties: EUR per well, for which the range uses the results of the most poorly performing, the main cluster, and best performing of U.S. basins; potentially productive area conforming to the cut-off criteria, for which the greatest uncertainty lies in estimating TOC, which was largely unrecorded and which therefore relied on extrapolation from Gamma Ray and other logs. A revised PGI estimate, based on testing of at least one significant horizontal leg, was planned for 2014 but it has also been

TABLE 76.4 Shale gas initially in place and technically recoverable from selected European countries rounded to nearest trillion cubic feet (EIA, 2011, 2013).

	EIA 2011		EIA 2013	
	Gas initially in place	Technically recoverable	Gas initially in place	Technically recoverable
Bulgaria			66	17
Denmark	92	23	159	32
France	720	180	727	137
Germany	33	8	80	17
Lithuania	17	4	4	0
Netherlands	66	17	151	26
Norway	333	83	0	0
Poland	792	187	763	148
Romania			233	51
Spain			42	8
Sweden	164	41	49	10
United Kingdom	97	20	134	26
Total	2314	563	2408	472

suggested that a new summary report will await the drilling of 100 exploratory wells, which is still some way off.

U.S. Geological Survey (USGS) The study was based on archive data from 56 wells drilled before 1990. The methodology was broadly similar to that of the PGI study (as the USGS was involved in training PGI staff). There were, however, significant differences in some of the criteria adopted, as follows:

1. The PGI study includes offshore, whereas the USGS study is limited to onshore. Whereas this, and the estimation of other cut-off criteria, resulted in a potentially productive area in the range 18,540–41,140 km^2 in the PGI study, the modal value in the USGS study was 4850 km^2 with a maximum value of 20,250 km^2.

2. Based on U.S. analogues, the PGI study estimated EUR per well in the most likely cases as 0.4 Bcf/well, whereas the USGS estimates it as 0.2 Bcf/well.

3. The USGS also applies an average success factor to drilled wells which, in the most likely case, it estimates at 50%.

The combination of these differences accounts for the variance between the PGI and USGS estimates.

TABLE 76.5 Estimates of polish in-place and technically recoverable shale gas resources.

Poland shale gas		Gas initially in place (Tcf) Risked/Mean	Technically recoverable (Tcf)			Recovery Percent
Study	Basin		P95/Min	Risked/Mean	P5/Max	
EIA 2011	Baltic	514		129		25.1
EIA 2011	Lublin	222		44		19.8
EIA 2011	Podlasie	56		14		25.0
EIA 2011	**Polish Foredeep Total**	792		187		23.6
Polish Geological Institute March 2012	**Polish Foredeep Total**		1.22–2.71	12.2–27.1	30.5–67.8	
U.S. Geological Survey July 2012	**Polish Foredeep Total**		0	1.345	4.086	
EIA 2013	Baltic	532.1		105.2		19.8
EIA 2013	Lublin	45.8		9.2		20.1
EIA 2013	Podlasie	53.6		10.1		18.8
EIA 2013	**Polish Foredeep Total**	631.5		124.5		19.7
EIA 2013	Fore Sudetic	106.7		21.3		20.0
EIA 2013	**Poland Total**	738.2		145.8		19.8

U.S. Energy Information Administration (EIA)   The EIA reports (U.S. EIA, 2011, 2013) were prepared on behalf of the EIA by Advanced Resources International (ARI).

Their methodology is fundamentally different from those of the PGI and USGS. As global studies, the reports rely on geological information and reservoir properties assembled from the technical literature and data from publically available company reports and presentations. Depths and thicknesses, for example, are typically assembled from regional cross-sections. In addition, the methodology is based on making an estimation of gas initially in place and applying a recovery factor to this. The reports use one of three recovery factors: 15%, 20%, and 25%. In the case of Poland, they have generally selected a 20% recovery factor which is said to be applicable to plays with a medium clay content, moderate geologic complexity, and average reservoir pressure and properties. In the case of Poland, this may be a rather optimistic outlook given that the only North American Lower Paleozoic analogue, the Utica Shale, is considered to have superior gas shale character to the Polish Foredeep shales but has recovery factors varying between 10% and 20%.

The 2011 study was limited to onshore but still considered the potentially productive area to be 53,115 km^2, greater than the maximum combined onshore and offshore area estimated by the PGI. All gas shale characteristics were based on those of the Lower Silurian. In the 2013 study, the potentially productive area of the onshore Polish Foredeep was reduced to 36,080 km^2, which is nevertheless some 3000 km^2 greater than the maximum potentially productive onshore area estimated by the PGI. The 2013 report assessed the combined Lower Silurian, Ordovician and Upper Cambrian intervals.

Sweden: The initial (2011) EIA report provided an estimated technically recoverable resource of 41 Tcf for Sweden's Alum Shale, which Shell's three wells found to have a very limited content of natural gas which it was not possible to produce. In the 2013 report, the estimate of technically recoverable resources for the Alum Shale was reduced to 9.8 Tcf.

United Kingdom: Based on analogy with comparable shale plays in the United States, in 2010, the British Geological Survey (BGS) tentatively estimated recoverable reserves (England only) at approximately 5.3 Tcf. In June 2013, the BGS and Department of Energy and Climate Change (DECC) produced an estimate for GIP in the Bowland Shale in the Craven Basin across Central England (Andrews, 2013). Total GIP is estimated to fall within the range 822 Tcf (P90) to 2281 Tcf (P10) with the most probable estimate (P50) being 1329 Tcf. Compared with the EIA analyses, the BGS study is very much "bottom-up," based on geological parameters, volumetrics, and gas contents. A similar study is under way for the Weald Basin in southeast England and this will be followed by a study of the Carboniferous potential of central Scotland. The Midland Valley of Scotland, where Europe's first certification of recoverable shale gas resources has taken place, was considered in the initial (2011) EIA report to be non-prospective.

General: Given the potential size of the in-place resource, it is not surprising that investigations have been proposed in at least 17 countries. Company interest has ranged from super-majors, such as ExxonMobil and Shell, through majors (Chevron; ConocoPhillips; Eni; Total) and major independents (e.g., Marathon Oil; Talisman) to small niche players (e.g., Cuadrilla Resources) and coal seam gas explorers who may have some shale gas potential on their acreage (e.g., IGas Energy). Some of these companies have subsequently withdrawn from Europe (Talisman) or indicated their intention to do so (Marathon Oil; Eni). A number of companies have published resource estimates for their own acreage and these are reported in the shale gas plays section for individual plays by country (2.1 below).

Major shale gas plays in Europe   Organic-rich "bituminous" shales that form the self-contained source–reservoir petroleum systems that can be exploited for natural gas and hydrocarbon liquids occur at many stratigraphic levels. Global oceanic anoxic events (OAEs), which favor the preservation of organic matter, took place in the Late Ordovician and into the Silurian, in the Early Jurassic (Toarcian), and in the Late Cretaceous (Cenomanian–Turonian). In Europe, additional anoxic events occurred on a regional scale in the Late Cambrian, Visean (Middle Mississippian) to Namurian (Early Pennsylvanian), and Late Jurassic (Kimmeridgian-Tithonian). All of these units are under investigation in Europe, though the younger units tend to be more prospective for shale liquids than shale gas. There are three potentially major regional shale gas plays in Europe plus a number of others with more restricted distribution.

Lower Paleozoic   The oldest is a Lower Paleozoic play that occurs in northwest Europe running from eastern Denmark through southern Sweden to north and east Poland. The organic-rich shales with shale gas potential lie on the south western margin of the Baltica paleocontinent and tend to thicken towards the bounding Trans-European Suture Zone. This play was first tested in Sweden in 2009 and has since been the focus of exploratory drilling in Poland. In Denmark and Sweden, the principal target is the kerogenous Alum Shale of Middle Cambrian to Early Ordovician (Tremadoc) age.

Denmark: Natural gas was first found onshore Denmark in Nordjylland (North Jutland) in 1873 in association with water wells. The first successful well was drilled in 1905, finding gas at intervals down to 600 ft. Commercial gas production took place from the late 1930s to the early 1950s in the Frederikshavn area. The source, however, is probably shallow biogenic gas. Today, licenses have been awarded over the Fennoscandian Border Zone and Norwegian-Danish Basin onshore Denmark.

Total S.A./Nordsøfonden. Total S.A. has been awarded two licenses. In March 2012, Total S.A. applied for a third area relinquished by Schuepbach Energy in November 2011 but by September 2012 this application appears to have lapsed. Total and the Danish North Sea Fund (Nordsøfonden) commenced evaluation of the Alum Shale in the North Jutland area (Fennoscandian Border Zone) during 2012. Prior to the planned drilling of Vendsyssel-1 in Nordjylland, a well work program, including environmental studies, was submitted to Frederikshavn Municipality in October 2012. Despite reviews by the Danish Energy Agency, Environmental Protection Agency, Nature Agency, and the Administration of Frederikshavn Municipality, which raised no comments, Frederikshavn City Council decided on February 27, 2013 to request a full Environmental Impact Assessment (EIA). While disagreeing with the City Council's decision Total and Nordsøfonden decided not to contest it as this would just lead to further delays. On February 26, 2014, the Frederikshavn City Council agreed the EIA reports for publication and it entered an 8-week public consultation period. In August/September 2013, Total and Nordsøfonden undertook an airborne gravity and magnetic survey over the area of their Nordsjælland (North Zealand) license.

Sweden: Fennoscandian Border Zone

Shell. On 28th November 2009, Shell spudded the first well in a three-well test program in Sweden's Colonussänkan permit. The permit overlies the Colonus Shale Trough, Fennoscandian Border Zone (also known as the Sorgenfrei-Tornquist Zone), southern Sweden. Lövestad A3-1, Oderup C4-1, and Hedeberga B2-1 ranged in depth from 2448 to 3134 ft.

The wells encountered Alum Shale (Alunskiffer) ranging in thickness from 225 to 345 ft. The TOC ranged from 3% to 16%, averaging 7%. Porosity averaged 6.5% and permeability was approximately 40 nanodarcies. Water saturation, however, was high (80%) and gas analysis (94% methane) indicated that gas content was approximately 30 scf/ton and that the Alum Shale is undersaturated. Vitrinite reflectance measurements from 1.7% to over 2% indicate that the shale is postmature with little capacity for further gas generation. In May 2011, therefore, Shell announced that its investigations had been completed, that the rock samples from the three wells found only very limited gas traces that are not producible, and that the licenses would not be renewed when they expired at end-May 2011 (Svenska Shell 2011).

Östergötland Lower Paleozoic Basin

The Östergötland Lower Paleozoic Basin is an E-W faulted synclinal outlier within which the Alum Shale occurs at shallow depth and is thermally immature but the 45–80 ft thick shale has high TOC contents of up to 20%. It has been considered by some to be analogous to the biogenic-sourced shale gas of the Antrim Shale in the Michigan Basin. In both basins, methanogenesis may be a consequence of dilution of saline formation brines by meltwater from overlying Pleistocene glaciers. Although the Alum Shale is thermally immature in Östergötland, bitumen is present in limestone concretions and thin sandstones. The origin of these oils is thought to be either long-distance migration or more local heating during the intrusion of Permo-Carboniferous sills. Gas analysis indicates a mixture of thermogenic gas and secondary biogenic gas resulting from the breakdown of these pre-existing hydrocarbons. Gas flows are known from water wells and seeps in the area and flows of up to 40,000 cf/d have been reported from wells. Local farmers use the gas as a heating source and the Linköping commune has a mining (processing) concession, valid until 2033. Four companies own a total of 24 licenses in the Östergötland Lower Paleozoic Basin.

Aura Energy. In October 2011, Aura Energy, an Australian uranium exploration company that is investigating the uranium potential of Sweden's Alum Shale, commenced a 5-hole drilling program at its Motala shale gas project in the Östergötland Lower Paleozoic Basin, south-central Sweden, on the east shore of Lake Vättern near the town of Linköping. The 5 shallow wells were completed during Q4 2011 and gas samples were sent for analysis.

Gripen Oil & Gas. In April 2012, Gripen Gas (now Gripen Oil & Gas), the largest license holder with 17 permits, announced that it had tested biogenic gas from the Alum Shale at a depth of around 300 ft in 4 shallow wells drilled in the Ekeby permit in Östergötland. The best well, GH-2, flowed 97.5% methane and in Q3 2012 was appraised by two successful step-out wells and a further well drilled adjacent to GH-2 that cored the entire Alum Shale section. A further three appraisal wells were drilled in June 2013. All three wells flowed gas to surface from a 3″ hole. The OPC Ltd. has assigned a 2C contingent resource estimate of 51.0 Bcf raw gas to the Östergötland onshore licenses. Gripen Oil & Gas has identified three 36-hole development areas on the Ekeby permit. Development Area One (2C resource: 1.70 Bcf of 100% CH4) will be tested by extended well tests in three wells before development proceeds. Gripen Oil & Gas has also been granted the Sandön licence, in Lake Vättern, where the Alum Shale is thought to be deepest and thickest. Water depths range from 30 to 100 ft. Drilling has already taken place in the lake for mining exploration purposes and it is expected that a first shale gas exploration well will be drilled in Q4-2013 or early 2014. Gripen Oil & Gas also has five (5) concessions in the Baltic Depression on the island of Öland.

Siljan Ring Depression

AB Igrene. Further north, AB Igrene has 18 concessions with Lower Paleozoic shale potential in the Siljan Ring, where Lower Paleozoic rocks have been preserved around the margin of a depression formed by a major Late Devonian meteor impact. The concessions have been renewed until June 15th 2015. To date five percussion holes have been drilled followed by five core holes of about 1600 ft each, three of them in the Mora area on the west of the ring, which is now the focus of exploration, with the last two core holes having been drilled there in Q2 2013. Produced gas is dry, exceeding 90% methane with the remainder dominantly nitrogen. Gas occurs at depths below 1180 ft in Mora-001 and below 1000 ft in Solberga-1. Identified units with shale gas potential include the Tøyen Formation (Lower Ordovician), Fjäcka Shale (Upper Ordovician) and a Llandovery (Lower Silurian) shale (Kallholn Formation?). [*Note*: In the late 1980s, the Gravberg-1 well was drilled through a fractured granite within the impact crater to a total depth of 22,000 ft to test Thomas Gold's theory of the abiogenic origin of petroleum.]

Poland: Further to the southeast, in Poland, the most widespread Lower Paleozoic target has been Lower Silurian-age graptolitic shale, with the Upper Cambrian to Lower Ordovician and Upper Ordovician being secondary targets. The Silurian in particular thickens towards the southwest in the area of the Gdansk Depression (Baltic Depression) and the Danish-Polish Marginal Trough which defines the southwest margin of the Baltic Depression. In parts of the Trough, such as the Warsaw Trough and Lublin Trough, more than 10,000' ft of Silurian section may be present. Burial depths to base Silurian range from 3000 ft in the east and on the East European Platform Margin to 15,000 in the west, being deepest in the Warsaw Trough. Depths in the Lublin Trough are variable because of the presence of a number of separate fault blocks, but maximum depth to base Silurian is about 10,000 ft.

To date, this play has been the most sought after in Europe. Some 39 concessions have been awarded in the Baltic Depression, of which 10, largely operated by LOTOS Petrobaltic, are offshore in the Baltic Sea and 29 lie onshore in the Gdansk Depression. Six of the most easterly concessions, such as the four held by Wisent Oil & Gas, were considered to be more prospective for shale liquids than for shale gas. Four concessions have subsequently been relinquished by BNK Petroleum, Eni, and PGNiG. Another 40 concessions have been awarded in the Danish-Polish Marginal Trough and 14 on the East European Platform Margin, northeast of the Marginal Trough. Seventeen of these awards (12- Marginal Trough; 5- Platform Margin) have since been relinquished by Cuadrilla, Dart Energy, DPV Service, ExxonMobil, and Marathon.

Baltic Depression

Thirteen different companies are active in the onshore Gdansk Depression including ConocoPhillips and the Polish state company, PGNiG, plus a number of small niche players.

3Legs Resources/ConocoPhillips. The first tests of the Polish Lower Paleozoic commenced in the Gdansk Depression. Between June and October 2010, Lane Energy (a subsidiary of 3Legs Resources) drilled two vertical wells, Łebień LE-1 on its western Lębork concession and Legowo LE-1 (tested with a DFIT) on the eastern Cedry Wielkie concession. In January 2011, Netherland, Sewell & Associates estimated gross GIP in the Silurian/Ordovician section of Lane's six licenses at 170 Tcf. Following a decision to prioritize Lane's three western Baltic Basin concessions, Legowo LE-1 was plugged and abandoned.

A 3300 ft horizontal leg drilled within the Upper Ordovician Sasino Formation in a second Łebień well (LE-2H) in May 2011 was the first horizontal shale gas well drilled in Poland. After a 13-stage slickwater frac, the well flowed an unstabilized 2.2 MMscf/d on September 8, 2011 using coiled tubing and N2 lift. It was recompleted with a tubing string on September 17, and flowed from 380 up to 520 Mscf/d on N2 lift, plus frac fluid. On average, 15% of the total frac fld had been recovered by the end of the test. A second flow test in early November 2012 using a $3\frac{1}{2}''$ string flowed at rates of up to 780 Mscf/d, averaging 550 Mscf/d. A third phase of testing commenced on July 8, 2013 using a 2 3/8" tubing string. Downhole pressure gauges were installed and pressurized samples collected. Flowing commenced on July 21 at an initial rate of 470 Mscf/d and had declined to 230 Mscf/d by September 11, by which time 44% of the frac fluid had been recovered. The productive intervals in all three wells were in the Lower Silurian and Upper Ordovician.

In July 2011, Lane spudded Warblino LE-1H, in a third concession (Damnica). A vertical pilot was drilled to 10,570 ft. This was followed by a horizontal leg of 4088 ft within the top 16 ft of a presumed new Cambrian prospective interval which was then redrilled with a 1650 ft horizontal leg (12,610 ft MD) because of hole stability issues. Subsequent analysis has revealed that the horizontal leg was actually drilled within the Lower Ordovician Sluchowo Formation instead of the intended Cambrian Alum Shale (also known as the Piasnica Formation). A 7-stage gel frac test was suspended after 5 days during which flow declined from 60–90 Mscf/d to 18 Mscf/d. On retest in summer 2012, the well produced at a rate of 90 Mscf/d after 20 days of flow. Log analysis confirms that the Alum Shale (Piasnica Formation) is prospective in this area re-entering the well to drill and test a lateral within the Cambrian section is under consideration.

Lane's initial seismic and drilling program on its six Gdansk Depression concessions was funded by ConocoPhillips (see Ownership Transactions: Farm-ins) after which ConocoPhillips opted to retain an interest (and operatorship) in only the three western concessions. The companies spudded the Strzeszewo LE-1 vertical well on the Lębork concession on October 4, 2012 and drilled to a TD of 10,040 ft. A DFIT was carried out in the Cambrian interval

TABLE 76.6 **Saponis wells, Baltic depression.**

Location	Play	Age	Thickness range (ft)	SiO$_2$%	Carbonate	Clay %	Porosity %	Total organic carbon %
Saponis Concessions Baltic Depression	Graptolitic Shale	Lower Silurian	300–485	28–30	8–27	44-50	1–9.6 (avg 3.8)	0.1–4.2
	Graptolitic Shale	Ordovician	75–90	32–54	8–18	33–43	1.4–6.9 (avg 4.2)	0.05–6
	Alum Shale	Upper Cambrian	45	25	30	39	4.1–5.2 (avg 4.6)	5.0–9.2 (avg 7.2)

in January 2013 followed by a single-stage hydraulic frac in May. Cleanup using nitrogen lift commenced on August 4. The well was shut in on September 7, 2013 after recovering 22.5% of the frac fluid and flowing gas intermittently. A second DFIT, single-stage hydraulic frac and flow test took place in the primary target, the Upper Ordovician Sasino Formation, in December 2013, flowing gas at modest rates and recovering 63% of frac fluid.

Lublewo LEP-1 was spudded in the Lebork concession in September and drilled to a TD of 9593 ft. The rig then spudded Slawoszyno LEP-1 in February 2014 on the Kawia concession, drilling to a TD of 9250 ft. The wells confirmed the thickness and prospectivity of the Upper Ordovician Sasino Formation and Upper Cambrian Piasnica Formation. Based on these results and recently acquired 2D seismic, the rig was then mobilized to drill a 5250 ft horizontal sidetrack in the Sasino formation in Lublewo LEP-1. A stimulation program (cross-linked gel) of at least 20 frac stages is planned for this horizontal leg in Q3 2014.

BNK Petroleum – Saponis Investments. The drilling contractor, NAFTA Pila, which drilled the first two Lane wells spudded Wytowno S-1 (Slawno concession) in December 2010 on behalf of Saponis (BNK; RAG; Sorgenia: LNG Energy). The US$ 6 million well reached TD at 11,745 ft in mid-February 2011. The well encountered gas shows in a shallower 130 ft Lower Silurian section and over a deeper 300 ft Lower Silurian hot shale section. The well appears to have been drilled on a localized paleo-topographic high which accounts for the absence of a Cambro-Ordovician section. The strongest shows were recorded in the deeper Lower Silurian interval (124 scf/ton), while the shallower interval averaged 77 scf/ton. Wytowno S-1 was followed by a 11,780 ft well, Lebork S-1, on the Slupsk concession which encountered gas shows over a 935 ft interval from Lower Silurian to Cambrian Alum Shale. The Lower Silurian averaged 40 scf/ton while the 155 ft Cambro-Ordovician interval averaged 268 scf/ton. The TOC is also significantly higher in the Cambro-Ordovician interval.

In July 2011, Saponis spudded a third well, Starogard S-1 which had reached a TD of 11,560 ft by early September.

The well encountered a similar Lower Silurian to Cambrian section to that of Lebork S-1 with a gross thickness of some 820 ft. Gas contents (Lower Silurian: 38 scf/ton; Ordovician: 17 scf/ton) were lower than in the first two wells. Completion of the first two wells commenced in mid-September 2011 with fracking of the Cambrian interval in Lebork S-1 commencing on September 30. The fracturing of the Cambrian and Ordovician intervals did not permit an effective test to take place as insufficient proppant was injected as a result of higher than expected overpressures. The gas that did flow and was flared contained methane, ethane, and propane. The shale character in the three Saponis wells is indicated in Table 76.6.

In April 2012, GIP in the Saponis concessions was estimated to lie within the range 45.4–66.8 Tcf with a best estimate of 55.5 Tcf. Prospective recoverable resources were estimated in the range 4.5–13.2 Tcf with a best estimate of 8.0 Tcf. In December 2013, BNK Petroleum and Esrey Energy (formerly LNG Energy) acquired the Saponis interests of RAG and Sorgenia. It was also announced that Saponis will relinquish the Starogard and Slawno concessions, retaining the Slupsk concession on which Lebork-S1 was drilled.

BNK Petroleum – Indiana Investments. BNK announced that it would commence the drilling of three wells on its wholly owned blocks to the south of the Saponis Slawno and Slupsk concessions in February 2012 and on February 28 spudded Miszewo T-1 in the Trzebielino concession. The well drilled to a TD of ~17,700 ft but only muted gas shows were recorded. The well appears to have been drilled on the downthrown side of a major fault and to have encountered a different depositional environment from wells further to the northeast.

Gapowo B-1 was then spudded in May 2012 in the Bytow concession. It lies on the upthrown side of the fault and was drilled to a TD of some 14,000 ft. The well encountered a 400 ft Lower Silurian interval and 155 ft Ordovician interval, both overpressured. Core data suggest that prospective shale is 130–250 ft in thickness and has higher porosity (3.9–6.1%; 5.1% avg.), permeability and TOC (1.1 – 4.2%; 2.5% avg.) than any of the other BNK-operated wells in the

Baltic Depression. The average gas readings from these fractured overpressured shales were over 20 times greater than those encountered in Lebork S-1. Gas in place for the most prospective Lower Silurian/Ordovician interval is estimated at up to 86 Bcf/section with total GIP for the well estimated at up to 135 Bcf. Permission to drill and fracture stimulate a horizontal leg was obtained and a 5900 ft horizontal section was drilled and cased in January/February 2014. A fracture stimulation over 30% of the hole length is planned for Q4 2014.

San Leon. San Leon/Talisman commenced a two vertical well Gdansk Depression drilling program with the spudding of the Lewino-1G2 well in the Gdansk-W concession in late September 2011. Strong gas shows were encountered over an interval in excess of 3300 ft ranging from Middle Silurian to Upper Cambrian. After reaching a TD of 11,810 ft, the rig moved to the Rogity-1 location on the Braniewo concession. This well drilled to 9147 ft, encountering shows of rich gas over a 1600 ft interval from Lower Silurian to Middle Cambrian. Oil shows were also recorded in Lower Silurian shale, Ordovician limestone and shale, and Middle Cambrian sandstone. Following Talisman's announcement of its withdrawal from its Polish Operations, on May 8, 2013 San Leon reported that it had assumed 100% ownership of the Gdansk West and Braniewo concessions through its acquisition of the shares of Talisman Energy Polska. San Leon has announced a proposed pilot development program for the Gdansk West concession. The concession appears to fall within the gas/condensate window and the pilot area of 183 km^2 (70 square miles) is estimated to contain 1.3 Tcf and 40 million bbl of condensate recoverable. The entire Gdansk West concession has an estimated 12–18 Tcf shale GIP.

On July 2, 2013, United Oilfield Services (UOS) performed a hydraulic fracture through the $5\frac{1}{2}''$ liner of Lewino-1G2, pumping over 11,000 barrels of fluid and 95 tons of sand at 120 barrels/minute with a maximum pressure of 12,200 psi. The frac was conducted between 11,632 and 11,647 ft in the Upper Ordovician (Caradoc) Sasino Formation. On average, 25% of frac fluid was recovered along with a small but consistent flow of burnable gas. The UOS performed two further fracs in Q4 2013. The first was a refrac of the initial interval but using ceramic proppant. The second frac was conducted in a higher part of the Sasino Formation using a slickwater design and flowed gas throughout the duration of the cleanup period (1 week). In January 2014, after 6 weeks flow, the well was producing 45–60 Mscf/d (cond/gas: 20 bbl/MMscf). Flow is believed to have been from the upper fractured interval only (frac 3) with total Upper Ordovician potential for 200–400 Mscf/d. A 5000 ft horizontal well is now planned from the same well pad. San Leon/Hutton had planned to drill a vertical well on the South Prabuty concession in 2013 but this has not taken place.

PGNiG (Polskie Górnictwo Naftowe i Gazownictwo – state-controlled). A promising gas flow was also reported by PGNiG from a single-stage frac test of the Silurian/Ordovician at 9500 ft on its Lubocino-1 well on the Wejherowo concession, completed in March 2011. Gas quality was good with heavier hydrocarbons reported, no H2S and low N2. A second test was subsequently conducted at 9200 ft. A horizontal well (Lubocino-2H) was spudded in August 2012. After hydraulic fracturing, testing was continuing in Q2 2013. The company drilled Lubocino-3H between August and December 2013. PGNiG has also drilled a vertical well, Opalino-2, in the Wejherowo concession. The well, which was spudded in September 2012, with a targeted depth of 10,000 ft, was reported to be production testing in July 2013. Gas flows were reported from the Cambrian at around 10,000 ft. Wells Opalino-3 and 4 were drilled between November 2013 and March 2014 and PGNiG has indicated that it may start production from the Opalino area in 2015. In conjunction with a consortium of Polish companies (KGHM; PGE; Tauron Polska Energia; Enea), PGNiG also drilled Kochanowo-1 on the Wejherowo concession between April and July 2013. Between March and May, the company drilled Wysin-1, its first well on the Stara Kiszewa concession, some 20 miles southeast of Gdansk. Borcz-1 was drilled on the Kartuzy-Szemud concession in the period July–September 2013.

Eni. Eni completed an initial 3 vertical well program on its Malbork (Kamionka-1) and Elblag (Bagart-1; Stare Miasto-1) concessions. A horizontal leg was drilled on one of the wells and frac testing commenced in the second half of 2012. Eni is reported to be withdrawing from Polish shale gas exploration and did not renew two of its three licenses when they expired in December 2013 and January 2014. The third license is valid until June 2018.

Danish-Polish Marginal Trough & East European Platform Margin.

There are 37 currently valid (March 1, 2014) concessions covering the Platform Margin and Marginal Trough, the most prominent participants being Chevron, Marathon, Polish state company PGNiG, and PKN Orlen, another Polish company. Marathon has announced that it intends to dispose of its Polish assets as part of its global portfolio management program.

ExxonMobil. The first wells in the Podlasie Depression of the East European Platform Margin (Siennica-1) and Lublin Trough of the Danish-Polish Marginal Trough (Krupe-1) were drilled by ExxonMobil in Q4-2010 and Q1-2011. The wells were fracced in September/October 2011 but the wells failed to flow commercial volumes of gas. In June 2012, it was reported in the Polish press that ExxonMobil will discontinue its Polish shale gas exploration operations. The company had the option to relinquish or transfer its six concessions. Total SA, partner in two of them, announced in October 2012 that it will become operator of one of the concessions and drill a

TABLE 76.7   Silurian shale character in Lublin Trough wells.

Location	Play	Age	Thickness (ft)	SiO_2%	Carbonate %	Clay %	Effective porosity %	Total organic carbon %
PKN Orlen Lublin Trough	Lublin Shale	Lower Silurian	200	16–33	10–30	40–56	4.7 max	0.8–3.5

further well, while relinquishing the other. Of the other four concessions, two were relinquished and two transferred to PKN Orlen.

PKN Orlen. On October 24, 2011, PKN Orlen commenced its drilling program in the Lublin Trough of the Danish-Polish Marginal Trough spudding its first well, Syczn-OU1 in the Wierzbica concession. Based on the results of this vertical well, Syczn-OU2K was spudded in September 2012, with a planned horizontal leg of some 3600 ft. A 12-stage fracture operation and 4-week production test was conducted in Q2 2013 but did not yield commercial flow rates. Two other wells were completed in the Wierzbica concession in April (Streczyn-OU1) and September 2013 (Dobryniów-OU1). In mid-December 2011, PKN Orlen spudded Berejow-OU1 in the Lubartów concession, followed later that month by the Berejow-OU2K horizontal well. Both wells have now been completed and the horizontal well was frac tested in November/December 2013 using 19,000 cubic metres (5 million U.S. gallons) of water and 160 tons sand. Seven frac stages were conducted over a horizontal length of 2300 ft at a depth of 8200 ft. Between May and July 2013, Uscimów-OU1 was also drilled in the Lubartów concession. In July 2012, Orlen spudded Goździk-OU1 in the Garwolin concession, drilling to a TD of 13,830 ft.

On the East European Platform Margin PKN Orlen drilled the Stoczek-OU1 vertical well to a depth of 10,300 ft on the Wodynie-Lukow concession between November 2013 and January 2014. This was immediately followed by the Stoczek-OU1K horizontal well with a measured depth of 14,130 ft, drilled at a depth of 9500 ft in the Silurian. The planned length of the horizontal leg was 3600 ft.

Chevron. Chevron also commenced its Lublin Trough program in Q4-2011 with a well in the Grabowiec concession at Lesniowice, spudded on October 31, 2011. A second well at Andrzejow on the Frampol concession was spudded in March 2012. A third well was spudded on the Zwierzyniec concession in December 2012. Krasnik-1 was then drilled on the Krasnik concession between May and August 2013. The company has now drilled a Lower Silurian test on each of its four Lublin Trough concessions. Zwierzyniec is apparently the most promising as the first well was fracced and a second well is planned. Further development work on the Grabowiec concession, where Chevron conducted a DFIT in Grabowiec-6, has been delayed by protests by local villagers at Żurawlów.

Marathon/Nexen/Mitsui. In Q4-2011 and Q1-2012 Marathon drilled Cycow-1 (Orzechow concession) and Domanice-1 (Siedlce concession) on the East European Platform Margin. The latter well was plugged and abandoned. Drilling activity then moved to the Danish-Polish Marginal Trough where Lutocin-1 (Rypin concession) and Prabuty-1 (Kwidzyn concession) were drilled in the Pommeranian Trough, followed by Lubawskie-1 (Brodnica concession), spudded in September 2012. The final well in the initial 6-well program, SOK-Grębków-1 in the Sokolow Podlaski concession in the Podlasie Depression, East European Platform Margin, was plugged and abandoned in January 2013. Diagnostic fracture injection tests were conducted in the four non-abandoned wells. Three wells were hydraulically fractured using hybrid slickwater/gel fracs, with one well flowing gas for a week. Marathon has announced that it intends to dispose of its Polish assets as part of its global portfolio management and relinquished 7 of its 11 concessions in October/November 2013. Nexen (40%) and Mitsui (9%) have an interest in the remaining 4 concessions.

San Leon. San Leon/Talisman spudded Szymkowo-1, the final well in their 3-well drilling program, on the Szczawno concession, Danish-Polish Marginal Trough (Pomeranian Trough), in early March 2012. The well drilled to a depth of 14,930 ft and recorded wet gas shows over some 2000 ft of Lower Paleozoic shale. The strongest shows were encountered in the Lower Silurian and Ordovician over a combined thickness of some 350 ft. San Leon has reported that a 1650 ft horizontal leg was drilled in this well. Following Talisman's withdrawal from Poland, San Leon is now operator and now owns 50% of the concession, the other 50% being held by Greenpark Energy.

PGNIG. PGNiG spudded the Lubycza Królewska-1 well on the Tomaszów Lubelski concession, Lublin Trough on March 26, 2012. The well was completed in August 2012 and may be frac tested. Also in the Lublin Trough, Kościaszyn-1 (Wiszniow-Tarnoszy concession) and Wojcieszków-1 (Kock-Tarkawica concession) were drilled between September 2013 and January 2014.

Dart Energy. Dart Energy has published a "best estimate" of 9.485 Tcf shale GIP in its Milejow concession, where a seismic program was carried out in Q3-2011. Dart considered its Polish assets to be non-core and relinquished the concession in October 2013.

Poland General. An interesting feature revealed by sampling and gas shows from the three Lane Energy, three Saponis, and two San Leon/Talisman well locations in the Gdansk Depression is that thermal maturity appears to decrease in an east to northeast direction leading to an increase in the content of NGLs. The Starogard well produced hydrocarbons up to pentane and Rogity-1 produced C1–$nC8$ while the western wells in general produced only methane, ethane, and propane. This does suggest that there is the potential for significant liquids production from some concessions. Rogity-1 also discovered a 30 ft oil column in tight Middle Cambrian sandstone, confirming the decrease in thermal maturity toward the northeast of the Gdansk Depression. The earliest deposition of organic-rich shales is diachronous from northwest to southeast. Only to the northwest of Gdansk (Łeba High) are the organic-rich Upper Cambrian Alum Shale (Piasnica Formation) and Lower Ordovician Sluchowo Formation. Organic-rich shales also occur in the Łeba High region from Middle Ordovician (Llandeilian) through to Lower Silurian (Llandovery). Elsewhere, organic-rich Upper Ordovician shale occurs only to the southeast of the Łeba High in the central part of the Gdansk Depression. Organic-rich Silurian shale occurs throughout the region though once again this is diachronous, with only the Llandovery tending to be organic-rich in the northwest while further to the south and east in the eastern part of the Gdansk Depression, the Podlasie Depression and the Lublin Trough, both the Llandovery and Wenlock contain organic-rich shales. The Polish Treasury Ministry is said to be targeting commercial production from at least one pad by late 2014 or early 2015. It is assumed that the initial production of 20–30 million cf/d will come from the Lower Paleozoic play.

Lithuania: The Cambrian to Lower Silurian succession is also thought to have potential in southwest Lithuania. The Lithuanian Geological Survey has estimated in-place shale gas resources at up to 20 Tcf, with 10–15% recoverable.

Chevron. Chevron plans 2D and 3D seismic and multiple wells primarily for shale gas/oil exploration (see Farm-ins: Lithuania) on the Rietavas onshore license. On June 25, 2012 the Lithuanian Geological Survey opened two areas to tender for exploration with a submission deadline of October 31, 2012. In January 2013, it was announced that Chevron was the only applicant. Award of the application was delayed until Parliament amended laws to strengthen environmental regulations and on September 16 Chevron was announced winner of the tender but on October 8 Chevron announced that it was withdrawing from the tender. It is understood that this was because changes in the legal and financial environment covering hydrocarbon exploration had been and will continue to be made since the date Chevron was granted the award. On October 29, 2013, the Environment Ministry announced that a repeat tender will be announced, possibly in Q1 2014 if all legal and tax changes have been determined. Terms will be eased, especially in the area of experience

which was extremely stringent in the initial tender. Chevron continues to hold a 50% interest in LL Investicijos, holder of the Rietavas licence.

Romania Chevron. Chevron has acquired a concession (Block EV-2 Barlad) on the platform margin in northeast Romania where the Silurian foredeep shales that are prospective in Poland and Ukraine are also believed to occur at depths between 10,000 and 13,000 ft. Chevron has committed to 400 km of 2D seismic and 3 wells. Prior to the introduction of a shale gas drilling moratorium, the first well in the multi-well drilling program had been planned for late 2012. Following the expiry of the moratorium, Chevron moved ahead with seismic exploration and the permitting required for drilling. Having obtained all the required permits, in early October 2013 Chevron commenced preparatory work at the drill site near the village of Silistea in Vaslui County but halted operations after 5 days of protests by villagers. On October 19, 1700 locals and environmental activists held protests. The government, while accepting the right to peaceful protest, has warned against violence. President Traian Basescu has appeared on TV to stress that failure to source indigenous gas resources plays into the hands of Russia's Gazprom while Prime Minister Victor Ponta has stated that the government's desire and political decision is to have energy independence for Romania by exploiting all of the resources that the country has. Chevron currently plans the first well on the Barlad concession in Q2 2014. A second Lower Paleozoic play occurs on the composite Saxothuringian-Barrandian-Moldanubian terranes (Bohemia) that probably detached from Gondwana at around the time of the Ordovician–Silurian boundary.

Czech Republic Hutton Energy. BasGas (now Hutton Energy) has applied for acreage in the Prague and Intra-Sudetic basins of the composite Bohemian terranes. The Silurian pelagic shale is reported to be the target in both basins. The Trutnov application in the Intra-Sudetic Basin was approved on December 21, 2011 but the Trutnov award was cancelled in April 2012 and sent back to the Ministry of Environment regional department to be decided again. In September 2012, the Minister of Environment announced a moratorium on shale gas exploration in the Czech Republic until June 30, 2014.

Spain

Silurian black shale has also been identified as a potential play in Spain's Ebro Basin (by San Leon) and in central Spain.

Carboniferous The second major play is a Carboniferous basinal marine shale play that extends eastwards from western Ireland and includes the East Irish Sea/Cheshire Basin in northwest England, the Anglo-Dutch Basin, the Northwest German Basin, the Fore-Sudetic Monocline (Northeast German- Polish Basin) in southwest Poland, and the Culm Basin in eastern Czech Republic. The age of the most prospective shales appears to young westwards from the Visean (Middle Mississippian) Kulm facies of Poland, the Czech Republic and northeast Germany to the Namurian (Upper Mississippian to Lower Pennsylvanian) of northwest Germany, the Epen Formation of the Netherlands, the Bowland Shale in northwest England, the Black Metals Marine Band of the Midland Valley Scotland, and the Clare Shale in western Ireland. Visean (Middle Mississippian) shale may also be prospective in Scotland and northwest Ireland. Tests of the Namurian Black Metals Marine Band in the Midland Valley of Scotland by three wells drilled in 2005 and 2007 were the earliest investigations of shale gas potential in Europe. The Carboniferous play has since been drilled in England, Wales, and Poland.

Czech Republic   Cuadrilla Resources. Cuadrilla Resources has received preliminary notification of the award of the Mezerici license in which the target is considered to be deep marine sediments present in the Lower Carboniferous of the Culm Basin, where the Variscan foreland basin reaches its most easterly extent on the eastern flank of the Bohemian Massif. In September 2012, the Minister of Environment announced a moratorium on shale gas exploration in the Czech Republic until June 30, 2014.

Germany: The nature of German E&P reporting is such that it can be difficult to establish the activity taking place on long-held licenses. It is assumed that ExxonMobil, both directly and indirectly through the BEB ExxonMobil/Shell joint venture, will be examining the potential of Visean (Middle Mississippian) shale in eastern Germany and Namurian (Upper Mississippian to Lower Pennsylvanian) shale in the west. Some of BNK Petroleum's initial eight concessions were also targeting Carboniferous shale gas, as are Wintershall's Rhineland and Ruhr concessions and Dart Energy's Saxon I West and Saxon II concessions. Dart reports combined shale gas in-place estimates for the two concessions in the range 0.25–2.95 Tcf with a best estimate of 0.97 Tcf. As part of its restructuring, Dart is looking to joint-venture, farm out or sell its German assets. The BNK has announced that it intends to relinquish five of its eight German concessions, retaining the three concessions (Adler; Falke; Falke South) in North Rhine – Westphalia's Munsterland Basin, which appear to have primarily Carboniferous potential.

Ireland (Republic of Ireland and Northern Ireland)   Enegi Oil. In February 2011, Enegi Oil was awarded Licensing Option ON11/1 to evaluate the shale gas potential of the Namurian (Upper Mississippian–Lower Pennsylvanian) Clare Shale in western Ireland. The Clare Shale is known to have high TOC (3–8%) but also high levels of thermal maturity. The main issues are whether it is over-mature for gas and gas leakage during Late Carboniferous uplift. In September 2012, Enegi stated that vitrinite reflectance analysis indicates that the shale is of lower maturity than recorded in the literature and that it had engaged Fugro to undertake further testing of the prospectivity. The report submitted to the Irish Petroleum Affairs Division (PAD) in November 2012 indicated that within the area of seismic coverage and assuming a porosity of 7%, GIP is estimated at 3.62 Tcf. The in-place estimate for the entire option area is 13.05 Tcf and for the high-grade area it is 1.23 Tcf. Having completed the work program, Enegi Oil announced on February 21, 2013 that it had applied to the PAD for an Exploration License. The final award decision is subject to further research being conducted by the Environmental Protection Agency.

Tamboran Resources. In the Northwest Ireland Carboniferous Basin (Lough Allen Basin), which straddles the border between the Irish Republic and Northern Ireland, Tamboran Resources, and the Lough Allen Natural Gas Co. have taken out licenses on both sides of the border to evaluate the potential of the Visean (Middle Mississippian) Bundoran and Benbulben shales, both of which yielded strong gas shows in wells drilled in the mid-1980s. TOCs are lower than in Clare (<2%).

Netherlands   Cuadrilla Resources. Cuadrilla Resources has been awarded a license (Noord Brabant) on the margin of the London-Brabant High and West Netherlands Sub-basin of the Anglo-Dutch Basin. It is assumed that the Namurian (Upper Mississippian to Lower Pennsylvanian) Geverik Member of the Epen Formation shale is one of the targets in this location. Two wells, at Boxtel and Haaren, are planned. It is also possible that one of these wells may be targeting shale oil in the Lower Jurassic Aalburg and Posidonia formations in the Roer Valley Graben while another also targets tight gas in the Triassic. Cuadrilla's other Netherlands license (Noordoostpolder) in the Northwest German Basin is a Namurian gas shale play. Drilling of the first well (Boxtel) is unlikely to take place before 2015 as a result of permitting delays and the need to await the outcome of studies commissioned by the Dutch Ministry of Economic Affairs, Innovation and Agriculture on the risks associated with unconventional gas drilling and production. The first study, which was delivered on August 27, 2013, concluded that risks remained very small and is being followed by a 12- to 18-month study to identify preferred locations for shale gas exploration.

Poland

A total of 21 concessions considered to have shale gas potential have been awarded in the Fore-Sudetic Monocline in southwest Poland but four (4) have subsequently been relinquished (Lane Energy: 3; San Leon: 1). Lane Energy withdrew from the area to focus on its Baltic Basin concessions. San Leon remains the largest concession holder with nine concessions covering this play. Other operators include Petrolinvest SA (through Silurian Sp. and ECO Energy), PGNiG, PKN Orlen, and Canadian International Oil Corp. Although all 17 remaining Fore-Sudetic Monocline concessions are considered to have some shale gas prospectivity, some are also being investigated for their conventional oil and gas prospects.

PGNiG. On behalf of the Polish state company, PGNiG, Halliburton frac tested Upper Carboniferous shale in Markowola-1 in the Lublin Trough in July 2010 but the flow rates are said to have been lower than expected.

San Leon. The next test of the Carboniferous, Siciny-2, was spudded on November 10, 2011 by San Leon in the Gora concession, Fore-Sudetic Monocline. This well was located close to Siciny 1G-1, drilled in the 1970s, which had encountered a 3266 ft Carboniferous section and was still in Carboniferous at TD. Siciny-2 was drilled to a depth of 11,550 ft, encountering some 3300 ft of Carboniferous. Continuous gas shows were encountered across three prospective shale intervals and two tight sandstone intervals encountered below 9400 ft. In the first instance, testing will focus on the deeper of the tight sandstone intervals. The three highly fractured shale intervals in Siciny-2 lie between 6775 and 8560 ft with a gross thickness of 1400 ft. The TOC values range from 1.2% to 3.25% and vitrinite maturity between 1.2% and 1.5%. Porosities are in the range 1.4–8.5% and average permeability is between 80 and 100 nD. Silica content is about 45%. Further prospective shale intervals are expected beneath the deeper of the two tight sandstones in which drilling terminated. Shale GIP is estimated at up to 70 Bcf/section. San Leon's estimate 61 Tcf of net risked recoverable tight gas and shale gas across its 13 concessions in the Fore-Sudetic Monocline. In March 2013, San Leon signed a Memorandum of Understanding with Halliburton to jointly explore and develop the Carboniferous and deeper in three of these concessions (Wschowa; Gora; Rawicz). Halliburton perform 3 DFITs on the tight gas sand in Siciny-2 between May and August 2013. These will be followed by a hydraulic fracture treatment. Subject to the execution of a further binding agreement, completion of the DFIT and hydraulic fracturing phase shall give Halliburton the option to earn up to a 25% working interest in the Carboniferous and deeper sections within the Concessions by fully funding two vertical exploration wells, including full technical evaluation with core, DFIT, and vertical hydraulic fracturing if technically warranted.

United Kingdom **England**: In central England, Visean-Lower Namurian (Middle to Upper Mississippian) organic-rich deep-water marine shales of the Craven Group were deposited in a complex set of tectonically active grabens and half-grabens ("troughs") which developed as a result of

N-S tension during the onset of subduction in the Variscan Foreland Belt to the south. Laterally, on horsts and tilt-block highs, the shales grade into shelf limestones and deltaic sandstones but in the depocentres hemipelagic marine shales and interbedded mass flow deposits may be as much as 16,000 ft thick.

Rifting ceased in the late Visean but subsidence continued in the Namurian. A number of phases of Late Carboniferous uplift associated with the Variscan orogeny brought about basin inversion, and a complex set of hydrocarbon maturity conditions. The primary target has been informally named the Bowland-Hodder unit by the British Geological Survey, and comprises the Bowland Shale, Hodder Mudstone, Edale Shales, Holywell Shales, and other local unit names. The unit has been subdivided into a thick lower syn-rift unit (Lower Bowland Shale; Hodder Mudstone) and thinner upper post-rift basin infill unit (Upper Bowland Shale). The lower unit is known to reach thicknesses of up to 11,500 ft in depocentres and may be thicker. The upper unit transgressed across the highs and platforms but is considerably thicker and more organic rich within the basins. It is typically about 500 ft thick but reaches up to 2900 ft. Depths to the top of the Bowland-Hodder unit range from 0 to 16,000 ft. Depth to gas maturity (Ro > 1.1%) is estimated at about 9500 ft but will be less where there has been subsequent basin inversion and uplift. The TOC ranges from 0.2% to 8% but normally falls within the range 1–3%. For reasons of sample availability, most of these values come from the upper unit. Present-day kerogen is dominated by mixed type II-III and type III but may have been modified by maturation. The most prospective areas are considered to be the Bowland Basin, Edale Gulf, Gainsborough Trough, and Cleveland Basin.

Cuadrilla Resources. Cuadrilla Resources, through its Bowland Resources subsidiary, has interests in the onshore portion (Bowland Basin) of the East Irish Sea Basin in PEDL 165 in Lancashire, northwest England. Spudded on August 16, 2010, the company's Preese Hall-1 well targeted a Visean-Namurian (Middle to Upper Mississippian) interval with the Bowland Shale the primary target. Drilled to a depth of 9098 ft, the vertical well encountered over 4000 ft of shale between 4400 and 9004 ft. The shales contained both vertical and horizontal fractures and produced "substantial gas flows." The well encountered three prospective shale formations with a net thickness of 2411 ft: Sabden Shale of Arnsbergian (Late Mississippian) age (approximately 170 ft); Bowland Shale of Brigantian (Middle to Late Mississippian) age (1685'); Hodder Mudstone of Visean (Middle Mississippian) age (554 ft). The well was due to have a 12 frac-stage completion over an interval from 5260 to 9000 ft but after 5 fracs, fracking was suspended due to two small earthquakes in the vicinity of the well (2.3 and 1.5 Richter Local Magnitude). The company commissioned a study to determine the relationship, if any, between the fluid injection and seismicity (see Above-Ground Issues: United Kingdom). The first three fracs (perforated intervals from 8420 to 8949 ft in the Hodder Mudstone) were tested on co-mingled flow and produced satisfactory amounts of gas and frac flowback water. Fracs 4 and 5 (7810–8259 ft in the base of the Lower Bowland Shale) were being flowed in mid-August 2011. In December 2013, Cuadrilla announced that no further exploration work will take place on the Preese Hall site. Between January and August 2011, the rig drilled a second well 3 km NE of Preese Hall-1 at Grange Hill-1, where top Lower Bowland Shale was forecast at ~6500 ft, slightly shallower than in Preese Hall-1. Preliminary core analyses suggest similar gas contents to Preese Hall-1 but over a thicker series of possible pay zones, as indicated by the final TD of 10,775 ft compared with the forecast TD of 9500 ft. In July 2013, Cuadrilla announced that it intended to apply for planning permission to hydraulically fracture and test the well but subsequently announced that it does not intend to frac the well at this time but rather to use it as an observation well for seismic monitoring of two new sites. Becconsall-1, 15 km south of Preese Hall-1, spudded on August 16, 2011 and represented a substantial step-out from the locations of the first two wells. Top Lower Bowland Shale was forecast at ~8,000 ft, significantly deeper than in the previous two wells. On October 13, a vertical sidetrack, 1Z, was spudded and the well was completed on December 21, 2011. No results have been announced other than the TD of 10,500 ft. Cuadrilla plans a DFIT on this well.

On October 6, 2012, drilling commenced on a fourth well (Anna's Road-1), some 5 km southwest of the Preese Hall-1 location. Top Bowland Shale was estimated at 9000 ft and TD at 11,500 ft. On November 16, however, it was reported that the well had been abandoned at 2000 ft because of a stuck packer. Plans to respud the well in January 2013 were subsequently altered to allow the company to modify its planning application to include the vertical well, a 3000 ft horizontal leg, hydraulic fracturing and flow testing. Cuadrilla subsequently decided to abandon the Anna's Road well and restore the site to its previous condition. On February 4, 2014, Cuadrilla announced that the company intends to apply for planning permission to drill, hydraulically fracture and flow test up to four exploration wells on each of two sites, one at Roseacre Wood, Roseacre, and the other at Preston New Road, Little Plumpton. Separate applications will also be made to install two seismic arrays that will be used to monitor the hydraulic-fracturing process.

Because of the substantial thickness of the Bowland Shale in PEDL 165, shale character shows significant variation. The shale is thought to have undergone an early period of oil generation prior to Variscan (Late Carboniferous) uplift. The subsequent deposition of the Manchester Marl and anhydrite (Upper Permian) formed a regional seal. Peak maturity occurred during the Jurassic–Cretaceous and was followed by Alpine uplift. The TOC ranges from 1% to 6%, averaging 2–4%. Thermal maturity ranges from wet gas (C1–C5) at the top of the shale to dry gas, with Ro range of 0.8–2.0% and

Btu in the range 990–1450. Porosity can range from 1% to 6% but gas-filled porosity is typically 4–5%. Silica/carbonate content is high with less than 50% clay, though the younger Sabden Shale, which is generally within the oil window, has higher clay content.

Based on gas desorption and geochemical studies undertaken at the Preese Hall well and a net shale thickness of 2411 ft in that well, original GIP at the Preese Hall location was estimated at 538.6 Bcf/square mile. On September 22, 2011, Cuadrilla Resources announced a preliminary GIP estimate of 200 Tcf for its 1130 km^2 (436 square miles) PEDL 165 license in Lancashire. The uncertified estimate was based on the two wells drilled at that time by Cuadrilla plus historical data from three wells drilled between 1987 and 1990 by British Gas. At the Shale UK 2014 conference on March 4, 2014, Cuadrilla announced an increase in GIP to 330 Tcf with potential for 100 Tcf recoverable.

IGas Energy. On November 4, 2011, IGas Energy spudded a joint coal seam gas/shale gas exploration well in the Carboniferous Rossendale Basin (beneath the Permo-Triassic Cheshire Basin) on PEDL 190 south of the River Mersey opposite Liverpool. The Ince Marshes-1 well was completed on January 21, 2012 having encountered about 1000 ft of Bowland Shale in which gas indications were observed throughout. The well was still in shale at TD with total thickness estimates at ~1650 ft.

The TOC generally fell in the range 1.2–3.7%, averaging 2.73%. Previous independent analysis suggested 4.6 Tcf GIP in this area but on the basis of the well results IGas said that its potential in-place resource could be doubled to 9.2 Tcf. On June 3, 2013, IGas reported that its in-house estimate of shale gas initially in place in its northwest England licenses was 15.1 to 172.3 Tcf with a most likely in-place resource of 102 Tcf. In June 2012, IGas announced the beginning of a formal process to find a farm-in partner for its Cheshire shale gas prospect. On January 15, 2013, IGas announced a successful share placing, part of which is intended to fund a two-well shale gas appraisal program intended to "augment value ahead of any farm-out." ExxonMobil was one of the companies rumored to have been in discussions with IGas. On January 10, 2014, IGas spudded a second Rossendale Basin joint coal seam gas/shale gas exploration well at Irlam, west of Manchester. Irlam-1 has a planned TD of 10,100 ft and is forecast to penetrate a 3175 ft Bowland Shale sequence (with a further possible 300 ft cut out by faulting).

Dart Energy. Dart Energy has 11 licenses in the Cheshire and Stafford basins. Netherlands Sewell & Associates (NSAI) has a best estimate of 30.5 Tcf original GIP over six of these licenses. The Bowland Shale may also be prospective east of the Pennine High in the East Midlands, Humber, and Cleveland basins, where it is a known source rock for oil and gas. Since 2010, a complex set of acquisitions, farm-ins and ownership exchanges has taken place on licenses PEDL 139 and 140 in the Gainsborough Trough, East

Midlands Basin, reflecting the potential of this acreage. Of the original licenses, only Egdon Resources remains, with eCorp International, IGas, Dart Energy and, most recently, French major Total all acquiring interests. Total (40%) is the largest interest holder and IGas is the operator. Total will fund a work program including the drilling of a vertical exploration well. Total has also taken an option to farm into the adjacent PEDL 209 operated by Egdon. Dart Energy bought into the Gainsborough Trough acreage through its acquisition of Greenpark Energy's unconventional gas assets (see Licence Acquisitions: United Kingdom). Dart's (NSAI) best estimate of Original Gas in Place for 7 of its 13 licenses east of the Pennines is 32.4 Tcf net to Dart (47.6 Tcf gross). The above estimates of net GIP predate the GDF SUEZ farm-in. A number of wells that have been drilled in Yorkshire to explore for conventional Permian and Carboniferous carbonate and sandstone reservoirs have also drilled through and sampled the Bowland Shale.

Rathlin Energy (UK). Between April and September 2013, Rathlin Energy (UK), a wholly owned subsidiary of Canada's Connaught Oil & Gas, drilled Crawberry Hill-1 and West Newton-1 in the Humber Basin, north of Hull, East Riding of Yorkshire. The wells were conventional exploration wells designed also to appraise the Bowland Shale. Planning permission for testing the two wells has been applied for. The applications are for flow tests of the Permian carbonate and Carboniferous (Lower Namurian) sandstone plus mini fall-off tests (DFITs) in the Upper Visean/Lower Namurian (presumed Bowland Shale) at 8783 ft in Crawberry Hill-1 and 10,023 ft in West Newton-1.

Viking UK Gas. Between June and October 2013, Viking UK Gas, a wholly owned subsidiary of Third Energy, which in turn is 97% owned by a private equity arm of Barclays Bank, drilled Kirby Misperton-8 as a deep Bowland Shale appraisal well on the Kirby Misperton conventional field in the Cleveland Basin, North Yorkshire. The neighboring Kirby Misperton-1 had encountered ~2500 ft of Bowland Shale when drilled in 1985. Planning permission has also been obtained to investigate Carboniferous shale potential in southern England.

Coastal Oil & Gas. On PEDL 252 on the southern margin of the Wales – Brabant High near Woodnesborough in Kent (north of the Kent Coalfield), Coastal Oil & Gas has received planning permission for a well to take core samples of some eight Westphalian (Middle Pennsylvanian) coal seams and the Lower Limestone Shales of the Tournasian (Lower Mississippian) Avon Group. It is not known when this well will be drilled.

Wales
IGas Energy. IGas Energy has identified 1.14 Tcf of 2P contingent resources of GIP in the Bowland Shale equivalent on its acreage in North Wales.

Coastal Oil & Gas. In South Wales Coastal Oil & Gas applied for permission to drill the Llandow gas shale exploration well to a depth of 2130 ft to log and core the Namurian Millstone Grit Shale Group, the Dinantian Upper Limestone Series and Lower Limestone Series, and possible gas shale in the Ordovician, in addition to Devonian tight gas. Despite this well being drilled on the same basis as previous coal seam gas exploration wells drilled in the area by Coastal in 2007/8, the company was obliged to withdraw the application in the face of local opposition to the drilling. When resubmitted, the application was rejected by Vale of Glamorgan Council but has since been approved on appeal (see Above-Ground Issues: United Kingdom). Although the principal shale gas target in the Llandow well appears to have been the Lower Limestone Shales of the Courceyan (Lower Mississippian) Avon Group, Coastal's partner, Eden Energy has identified the Namurian as the principal target over its acreage. The most prospective unit is presumed to be the Pendleian (basal Namurian or Upper Mississippian) Aberkenfig Formation. Eden has reported a gross unrisked P90 estimate of 34.2 Tcf shale GIP in the Namurian of its seven South Wales licenses (17.1 Tcf net to Eden).

U.K. Methane. In August/September 2011, U.K. Methane (a company with similar management to Coastal Oil & Gas) spudded St. Johns-1 and Banwen-1, targeting Namurian shale. The target depths are believed to be relatively shallow, about 2000 ft in the case of St. Johns-1.

Scotland
Dart Energy. The basal Namurian (Upper Mississippian) Black Metals Marine Band in license PEDL 133 in the Midland Valley of Scotland was cored by Composite Energy in Airth-6 (2005) and Longannet-1 and Bandeath-1 (2007). These were the earliest shale gas investigations in Europe. The Black Metals Member (Limestone Coal Formation) of the Kincardine Basin occurs at depths of 1000–4000 ft. The Black Metals was 120 ft thick in Airth-6 and the core analysis results are assumed to have been promising, as BG Group subsequently farmed into the license. In June 2011, Australia's Dart Energy (formerly Composite Energy) announced the results of an independent assessment by NSAI of shale resources in PEDL 133, in the Midland Valley of Scotland. This indicates an estimated GIP of 0.8 Tcf in the Black Metals Member, and a potential resource of 0.1 Tcf. The deeper Visean (Middle Mississippian) shales of the Lawmuir and Lower Limestone formations are estimated to contain 3.6 Tcf GIP with a gross resource of 0.5 Tcf. Dart Energy owns 100% of the Namurian prospect but BG retains a 51% interest in the Visean prospect. Shale gas exploration of PEDL 133 is still at an early stage while the company focuses on the start-up of coal seam gas production from the license.

Liassic (Lower Jurassic) The third major regional play comprises Lower Jurassic bituminous shales that are being targeted in the Weald Basin (southern England), Paris Basin, the Netherlands, northern Germany, and Switzerland's Molasse Basin. In continental Europe, the principal target is the Lower Toarcian Posidonia Shale. In eastern Germany and Poland, the Lower Toarcian grades into a terrestrial facies and loses its source potential. In southern England, the principal bituminous shales (Blue Lias; Black Ven Marls; Green Ammonite Beds) are older and occur in the Lower Lias. These bituminous shales are clearly oil-prone. The principal limitation regarding their shale gas potential therefore lies in finding locations in which they have been sufficiently deeply buried to have entered the gas window. Locations where this may have occurred include the flexural foreland basin of the Swiss Molasse and the Mesozoic depocentres of the Lower Saxony Sub-basin (Northwest German Basin) and the offshore Broad Fourteens Basin and Central Graben of the Netherlands.

A number of companies are thought to be investigating Lower Jurassic shale gas potential. These include Cuadrilla Resources in *England*'s Weald Basin and Schuepbach Energy in *Switzerland*'s Molasse Basin. Whether the Liassic shales will be within the gas window in the Weald Basin remains to be seen, though it is possible that they may have generated biogenic gas at shallow depths. To date, the only known exploratory tests of the Lower Jurassic shale gas play have taken place in Germany where ExxonMobil drilled wells between 2008 and 2011.

France: Schuepbach Energy was awarded two permits in the Languedoc-Provence Basin where Total was awarded the Montélimar permit. The Lower Jurassic (Toarcian) Schistes Carton is considered to have potential in the area covered by all three permits. Both the Schuepbach and Total permits have since been cancelled (Above-Ground Issues: France) and Schuepbach is reported to have asked the French government for 1 billion Euros by way of compensation. A number of other companies have also applied for permits in Languedoc-Provence, many of them overlapping.

Germany ExxonMobil. The ExxonMobil/Shell co-venture (BEB) commenced shale gas exploratory drilling in 2008 in the Lower Saxony Basin, drilling Damme-2/2A in the Munsterland concession and Oppenwehe-1 in Minden. Schlahe-1 was drilled in 2009 in the Scholen concession. Posidonia Shale is known to have been at least one of the targets for these wells. ExxonMobil is believed to have spudded Lünne-1 (Bramschen concession, Emsland) around January 17, 2011 and reached the Posidonia Shale at about 4720 ft. In March 2011, the Lünne-1A horizontal sidetrack was drilled in the Posidonia Shale to a total length of 760 ft. (The well was planned to have a 1600 ft horizontal leg.) A frac test is planned but has not yet been applied for. The thickness of the Posidonia Shale ranged from 80 ft (Lünne-1) to 115 ft (Oppenwehe-1; Schlahe-1). ExxonMobil has also announced plans to drill and frac test a 3300 ft horizontal leg (Z14b) at

a depth of 3380 within the Posidonia Shale from well Z14 in the Bahrenborstel Upper Permian Zechstein carbonate sour gas field. The Bahrenborstel Z14b sidetrack is one of a number of future horizontal shale gas exploratory wells planned by ExxonMobil in Lower Saxony, also including Leese-Ost-1 and Ortland 26. No dates for drilling any of these wells are available at this time

BNK Petroleum. BNK Petroleum has announced that it intends to relinquish five of its eight German concessions, retaining only the three concessions (Adler; Falke; Falke South) in North Rhine – Westphalia's Munsterland Basin where the Posidonia Shale is a possible target.

San Leon. Realm Energy, a wholly owned subsidiary of San Leon has a single concession (Aschen) in the Lower Saxony Basin where the Posidonia Shale is believed to be within the gas window.

Other plays with shale gas potential Austria OMV. OMV has investigated the potential of the Upper Jurassic Mikulov Formation in the Deep Vienna Basin. The company estimates that the formation contains 200–300 Tcf of GIP of which 15 Tcf may be recoverable. The target occurs at depths greater than 14,700 ft and a temperature of 160° C. Two initial wells had been planned near Herrnbaumgarten and Poysdorf in the Mistelbach District of Lower Austria at a combined cost of EUR 130 million. But subsequent proposed changes to Austrian environmental legislation mean that the project is no longer economically viable.

Bulgaria: The Lower to Middle Jurassic of the Moesian Platform, especially the basal Stefanetz Member of the Middle Jurassic Etropole Formation, is a target in northern Bulgaria, where Direct Petroleum (TransAtlantic Petroleum) has a license and Chevron successfully applied for a license that was subsequently revoked. Chevron has indicated that the Silurian was also a target in its Novi Pazar license.

TransAtlantic Petroleum/LNG Energy. Direct Petroleum/LNG Energy spudded the 10,500 ft Goljamo Peshtene R-11 well in the A-Lovech exploration licence in late September 2011. The well (TD 10,465 ft) encountered 375' of net pay in the Etropole Formation with numerous gas shows in the C1–C3 range. TransAtlantic has estimated the gross unrisked prospective undiscovered recoverable resource at 11 Tcf (best estimate). Operations in Bulgaria are constrained by the decision of the Bulgarian Parliament in January 2012 to ban hydraulic fracturing. Permission from the Bulgarian Government has not yet been received to resume completion and testing operations on the Peshtene R- 11 well.

Croatia: Hungary's MOL and its part-owned subsidiary INA have indicated that the Miocene of the Mura and Drava sub-basins (Pannonian Basin) of eastern Croatia has shale gas potential.

France: Permo-Carboniferous basins in the Languedoc such as the Stephanian-Autunian (Upper Pennsylvanian–Lower Permian) Lodève Basin may have some potential in bituminous Autunian (Lower Permian) shale. Schuepbach Energy was awarded two permits in the Languedoc-Provence Basin, one of which also incorporated part of the Lodève Basin. The Schuepbach permits have since been cancelled (see Above-Ground Issues: France) and Schuepbach is reported to have asked the French government for 1 billion Euros by way of compensation. Realm (San Leon) has identified Stephanian-Autunian potential in the Bresse-Valence Basin, where it has submitted an application. Elixir Petroleum is exploring for shale gas (and tight gas) in the Permo-Carboniferous of the Moselle concession in the eastern Paris Basin, where in the past at least two wells have produced gas to the surface from the target interval (probably Carboniferous). In the main Paris Basin, many conflicting applications have been filed. While the main focus of these is probably Liassic shale oil, a number are presumably also targeting shale gas potential in underlying Permo-Carboniferous half-grabens.

Germany: The Upper Devonian Kellwasser shale has been touted as having potential in northern Germany, as have Wealden paper shales of Berriasian age in the Lower Saxony Sub-basin.

ExxonMobil. The ExxonMobil/Shell co-venture (BEB) commenced shale gas exploratory drilling in 2008 in the Lower Saxony Basin, drilling Damme-2 and 3 in the Munsterland concession and Oppenwehe-1 in Minden. Schlahe-1 was drilled in 2009 and Niedernwöhren-1 was spudded in the Schaumburg permit in October 2009. ExxonMobil is believed to have spudded Lünne-1 (Bramschen concession, Emsland) around January 17, 2011. The Wealden is known to have been at least one of the targets in all of these wells and Damme-3 and Niedernwöhren-1 targeted the Wealden exclusively. Damme-3 is known to have been frac tested (3 fracs). Wealden thickness ranges from 800 (Schlahe-1) to 2300 ft in the Damme wells. Realm Energy (San Leon) also sees the Wealden as a potential target on its Aschen concession.

3Legs Resources. In the Bodensee Trough, north of the Swiss-German border, Parkyn Energy, another 3Legs Resources subsidiary, took out two licenses in 2009 in which the principal prospect appears to be lacustrine shale of Permian age. The company acquired 2-year extensions to these licenses in December 2013 and then transferred its interest to Rose Petroleum in order to maximize its focus on its Baltic Basin concessions in Poland.

Hungary: The shale gas exploration situation in Hungary is unclear.

Falcon Oil & Gas. In September/October 2009, Falcon Oil & Gas/ExxonMobil/MOL tested an Upper Miocene basin-centerd gas prospect in the Makó Trough (Pannonian Basin) with only limited success, after which ExxonMobil and MOL exited the project. But Falcon has suggested that its acreage

holds a "potential fractured oil and gas play." Previously, in 2007, Falcon had tested a naturally fractured marl-rich section of the Upper Miocene Endröd Formation in Szekkutas-1. After fracture treatment at about 11,100 ft, the well flowed at an unstabilized rate of 1.577 million scf/d plus 50 to 100 ppm H2S. RPS Energy (January 2013) estimated the 2C gas resources of the Lower Endröd at 1.11 Tcf but with the qualification that there is a less than or equal to 25% chance that the contingent resources will be converted to reserves.

MOL/INA. MOL and its part-owned subsidiary INA have indicated that the Miocene of the Mura and Drava sub-basins (Pannonian Basin) of eastern Croatia has shale gas potential and it can be assumed that this extends into western Hungary.

Delcuadra. In September 2009, Austria's RAG (Rohöl-Aufsuchungs Aktiengesellschaft) acquired Toreador Hungary Ltd. Toreador had just drilled the Balotaszallas-E-1 (Ba-E-1) well in the Kiskunhalas Trough of the Pannonian Basin. Ba-E-1 encountered an over-pressured 1840 ft gross gas-bearing interval in an interbedded Karpatian (Lower Miocene) sequence of siltstone, shale and sandstone below 10,000 ft. The two lowest zones were fractured and are believed to have produced gas-condensate. At that time, the tested lithology was reported as tight sandstone (Shaoul et al., 2011). In July 2011, the Delcuadra Kft consortium (Delta Hydrocarbons 53%; RAG 25%; Cuadrilla 22%) recompleted an additional 3 zones of the Lower Miocene reservoir in Ba-E-1. At the Global Shale Gas Plays Forum in September 2011, RAG reported this as a shale gas frac and has subsequently confirmed that the completions were carried out in "a thick heterolithic sequence of shales and (very) fine clastics." Testing produced a gas flow rate of 1 million cf/d plus small amounts of condensate. Both are being sold and a long-term production test commenced in August 2011 and full gas-condensate production should commence before end-2011. Cuadrilla has the option to earn a further interest by drilling and completing a second well in the Ba-IX Mining Block. This well, Ba-E-2, was planned for the second half of 2012.

Italy   Independent Resources. A shale gas/coal seam gas combination play is being investigated by Independent Resources in the Ribolla Basin, Tuscany. Upper Miocene (Messinian) gas shale straddles a coal seam of up to 20 ft thickness over a distance of tens of kilometers along the basin axis. The play was discovered in the course of evaluating the results of the Fiume Bruna-1 & 2 coal seam gas exploration wells drilled in 2009–2010. Farm-out discussions with companies that have experience of analogous plays were undertaken but have not produced a suitable partner, in part, the company believes, due to the public opposition in Europe to unconventional gas exploration and exploitation. In-place and recoverable 2C contingent resources are estimated at 300 and 160 Bcf, respectively. New seismic is scheduled for late 2013 and any early development will probably be based on coal seam gas to avoid concerns about hydraulic fracturing.

Netherlands

The Upper Jurassic Kimmeridge Clay is sufficiently deeply buried in the Central Graben in the northern Netherlands offshore to have reached the gas window. In view of the high well cost and drilling density likely to be required, it seems unlikely that offshore shale gas development will be economic in the foreseeable future unless an existing platform and wells happen to be fortuitously located in an optimal location for shale gas development.

Romania

Chevron and Sterling Resources/TransAtlantic Petroleum have acquired a number of licenses in the Moesian Platform of the East European margin in the south of the country, along the Bulgarian border. The targets are believed to be shale of Silurian to Lower Devonian age (Tandarei Formation) and Middle Jurassic age (Bals Formation). Sterling Resources/TransAtlantic Petroleum is reprocessing existing 2D seismic to identify a drillable location and evaluate re-entering a legacy well on a Silurian prospect in Sud Craiova Block EIII-7.

Romgaz. State-controlled Romgaz says it has made an unconventional discovery, which includes shale gas in the Transylvanian Basin. It has been encountering gas in drilling since the mid-1990s. A moratorium was imposed on shale gas exploration in May 2012 but that has now been lifted. Chevron is proceeding with seismic exploration and acquiring the necessary authorizations for drilling.

Spain

Applications that are presumed to be for shale gas exploration have been submitted in the Basque-Cantabrian Basin (BNK; Realm Energy (San Leon)), Pyrenean Foothills (Cuadrilla Resources), Ebro Basin (Realm Energy (San Leon)), and the Campo de Gibraltar (Schuepbach Energy/Vancast). The focus of interest appears to be the Basque-Cantabrian Basin and the area of the Pyrenean Foothills immediately to the east. Trofagas Hidrocarburos (BNK) has been awarded three concessions in the basin, Realm (now San Leon, operating as Frontera Energy Corp.) has two awards plus two pending awards, Leni Oil & Gas has interests in four and while SHESA (owned by the Basque Energy Board, the regional government of the Basque Country) has interests in a substantial number of permits and it seems to be focusing on the Enara permit. There does, however, appear to be a divergence of opinion regarding the most prospective targets. BNK, Leni, and San Leon believe that the Jurassic is most prospective (especially the Lias) while SHESA is targeting Albian–Cenomanian shales. San Leon sees the Middle Albian–Lower Cenomanian Valmaseda Formation and Carboniferous shale as objectives in its Geminis license on the Basque Country coast.

SHESA. SHESA and its partners, HEYCO Energy and True Oil, plan to drill two vertical wells, Enara 1 and 2,

to evaluate the Albian–Cenomanian Valmaseda Formation where it estimates there are 200 Tcf in place.

BNK Petroleum. BNK has submitted five Environmental Impact Assessments on its Sedano and Urraca concessions as part of the exploratory drilling permitting process and plans to drill in Q1 2014, pending permitting.

San Leon. In the Ebro Basin, San Leon's six pending awards are primarily targeting organic-rich Paleozoic shales (Ordovician; Silurian; Carboniferous) but Eocene shale is also a target.

Switzerland   Schuepbach. In addition to the Lower Jurassic Posidonia Shale, Schuepbach has also targeted the Aalenian (Middle Jurassic) Opalinuston in the Molasse Basin. It is understood, however, that the cantonal authorities in Fribourg would not renew the Fribourg license when it expired at end-2011, over environmental concerns. Schuepbach still hopes to explore for shale gas in Canton Vaud, to the south of Fribourg.

United Kingdom

The Upper Jurassic Kimmeridge Clay has been proposed as a possible target in the Weald Basin, England, but there is considerable doubt that it will be mature for significant gas generation in this basin, although biogenic shale gas may be a possibility. Cuadrilla's interest in the Kimmeridge Clay is for shale oil rather than shale gas. If there is shale gas potential in the basin it seems more likely that it will come from older shales (Rhaetic or older). For example, Esso's 1963 Bolney 1 well is reported to have found a marine Middle Devonian interval within the gas window.

Some general gas resource play issues Most plays are "statistical" in nature. Every coal unit and shale unit is "different" and also generally displays inhomogeneity. Statistical distributions can be obtained for parameters such as estimated ultimate recovery (EUR) and peak production from analagous wells. With a large enough sample size (number of wells) the geology of the play and the best drilling and completion strategies can be understood sufficiently well to make performance of a play and its recoverable resources predictable. European exploration is still some considerable way from achieving these levels of understanding. To convert recoverable resources into reserves requires good technology, smart wells, fracture and stimulation, and real-time micro-seismic mapping.

Shale Liquids in Europe

The principal shale liquids (tight oil) target in Europe is the Liassic (Lower Jurassic) that is considered by many to be an analogue to the Bakken Formation of the Williston Basin. It is being investigated in France, Germany, The Netherlands, and Portugal. The Upper Jurassic is understood to be a target in the south of the United Kingdom and central Poland, while a liquids-rich area has also been identified in the Polish Lower Paleozoic play.

France

There are four Liassic (Lower Jurassic) targets in the Paris Basin: Schistes Carton (Toarcian); Banc de Roc (Pliensbachian); Amaltheus Shale (Pliensbachian); Sinemurian-Hettangian Shale. The Liassic section is similar to the Bakken Formation in that the bituminous shales also contain a middle calcareous member (Banc de Roc). The TOC ranges from 1% to 12%, averaging 6%, and thickness ranges from 30 to 230 ft.

Hess. Toreador Resources (now ZaZa Energy Corp.) investigated the fractured shale oil potential of the Liassic interval. Shows had previously been detected in 11 conventional exploration wells drilled from the 1950s onwards and six wells have produced oil on test. On May 10, 2010, Toreador signed an investment agreement with Hess Corp. whereby each partner would hold a 50% interest in Paris Basin unconventional oil exploration and production. In July 2012, ZaZa Energy transferred its remaining 50% interest to Hess (see Ownership Transactions: Farm-ins). Toreador/Hess had planned to drill six wells in 2011, at least two of them horizontal, but as a result of the French government study into the economic, social, and environmental impact of shale gas and shale oil drilling and the introduction of the resultant legislation, the program was suspended. Hess had scheduled a three (3) vertical well drilling program for 2013 but it is unclear whether this will test the Liassic play.

San Leon. Realm Energy (now wholly owned by San Leon), although focused on shale gas, has shale oil potential on the nine permits for which it has applied in the Paris Basin. San Leon estimates the potential of these licenses at greater than 100 million BOE. Realm had in the past indicated that the Toarcian Schistes Carton may also have shale oil potential within the area of its Blyes permit application in the Bresse-Valence Basin.

Vermilion Energy. In spring and autumn 2010, Vermilion Energy fracture tested two vertical wells in the Toarcian Schistes Carton, producing 32–38° oil from both wells. In February 2011, these wells were believed to be producing about 63 bbl/d. Vermilion had planned to drill another two vertical wells in 2011 to evaluate all four zones and to drill a horizontal well in 2012 based on 2011 results but it is understood that Vermillion has now suspended all Paris Basin shale oil evaluation activity. On September 22, 2011, Vermillion withdrew three permit applications in the Paris Basin, possibly as a consequence on the ban on hydraulic fracturing introduced on July 13, 2011 (see Above-Ground Issues: France).

Germany

Outcrop work by BNK Petroleum has identified samples of the Toarcian-age Posidonia Shale with thermal maturities ranging from below the oil window to within the gas window. It therefore seems probable that over some of BNKs acreage the Posidonia Shale will fall within the oil window and have potential for tight shale oil. The BNK has announced that it intends to relinquish five of its eight German concessions, retaining only the three concessions (Adler; Falke; Falke South) in North Rhine – Westphalia's Munsterland Basin, where the Posidonia Shale is a possible target.

Hungary

Falcon Oil & Gas has indicated that the Upper Miocene Endröd Formation in the Makó Trough (Pannonian Basin) has shale oil potential. In 2007, Magyarcsanad-1 flowed 48° API oil and gas from natural fractures in argillaceous marl and siltstone of the Upper Endröd at 13,320 ft. The initial rate of 387 bo/d and 655 Mscf/d declined to 63 bo/d and 137 Mscf/d, respectively, after 23 days. The well produced a total of 850 barrels of oil and 2 million cf gas intermittently between November 2009 and July 2012. The RPS Energy (January 2013) estimated the 2C oil resources of the Upper Endröd at 76.71 million barrels but with the qualification that there is a less than or equal to 25% chance that the contingent resources will be converted to reserves.

Lithuania *Minijos Nafta.* In mid-May 2012, local oil producer Minijos Nafta spudded a Cambrian sandstone oil exploration well, Skomantai-1, on its Gargzdai concession, which was also intended to test the Ordovician and Silurian shales for unconventional hydrocarbons. Core samples were sent abroad for analysis of natural fractures, induced fracturing potential, porosity, permeability, source rock quality, and maturity. In this location, the Lower Paleozoic probably has shale liquids potential.

Netherlands

Cuadrilla Resources, in partnership with Dutch state company EBN, is targeting multiple unconventional hydrocarbon prospects on its Noord Brabant concession. The first well to be drilled (Boxtel-1) will evaluate the shale oil potential of the Posidonia Shale (Lower Toarcian) in the Roer Valley Graben at a depth of about 11,500 ft.

Poland *Wisent Oil & Gas.* Wisent Oil & Gas has four of the most easterly concessions in the Gdansk Depression, along the border with the Russian enclave of Kaliningrad, where Lukoil has been producing conventional oil for some time. As is noted earlier, these concessions appear to be situated in a more liquids-prone part of the basin (see Lower Paleozoic: Poland). In addition to tight shale oil, Wisent expects there to be conventional prospects in Cambrian and Ordovician carbonates. Wisent drilled its first well, Rodele-1, on the Kętrzyn concession between February and March 2013 to a depth of 5075 ft. The Silurian shale was found at the depth and thickness expected. The well was fracced in September

2013. Wisent spudded Babiak-1 on the Lidzbark Warmiński concession in March. The well drilled to a true measured depth of 9160 ft, including a 1865 ft horizontal leg. The well was fracced in July 2013.

San Leon. Following Talisman's announcement of its withdrawal from its Polish Operations, on May 8, 2013, San Leon reported that it had assumed 100% ownership of the Gdansk West and Braniewo S concessions through its acquisition of the shares of Talisman Energy Polska. On July 3, San Leon then announced an agreement whereby Wisent Oil & Gas could earn a 45% interest in the Braniewo S concession by conducting a three-stage fracture on the Rogity-1 well followed by drilling and testing a multi-stage horizontal well on the concession. The first Rogity-1 frac was completed in tight oil-bearing Middle Cambrian quartzitic sandstone on August 4, 2013. The fluid produced from the 9055-9070 interval indicated that the sandstone is in the vicinity of the oil–water contact in this location. The second (8860–8925 ft in Ordovician shale and marl) and third (8630–8710 ft in Lower Silurian Llandovery shale) frac stages were then tested co-mingled with the first, oil (39°) flowing to surface. Geochemical analysis indicated that oil different from Cambrian oil was recovered and that mobile oil had therefore been recovered from the shales. Wisent will now drill a vertical well, Rogity-2, either on the crest of the Rogity Cambrian structure or to the south of Rogity-1, where the Silurian shale is forecast to have higher TOC. Wisent will also undertake 3D seismic or additional drilling to compensate for the horizontal multi-fracced well originally envisaged.

Hutton Energy. In 2011, Hutton Energy's Polish subsidiary Strzelecki Energia acquired three concessions in the Mogilno-Łódź Trough of the Northeast German-Polish Basin in central Poland. In addition to conventional traps in Jurassic and Triassic sandstone, the company considers that the concessions have unconventional oil and gas potential in Jurassic shale, most probably of Middle Jurassic (Dogger) and Late Jurassic (Kimmeridgian) age. In February 2013, Hutton engaged Challenge Energy to find a partner to progress exploration activity on the three concessions. The prospectus indicated an upside of 100 million barrels of Jurassic shale oil potential.

Portugal

On March 1, 2012, Porto Energy Corp., holder of five licenses on and offshore the Lusitanian Basin, announced that it had entered into a definitive joint venture agreement with Sorgenia International of the Netherlands and Austria's Rohöl-Aufsuchungs Aktiengesellschaft (RAG) to evaluate the unconventional resource potential of the Lower Jurassic (Liassic) basal Brenha Formation within Porto's concessions. The organic-rich Lias is the source of oil seeps along the coast and has historically been surface mined for bitumen. In September 2012, the company announced that it has received approval from the Portuguese oil and gas authority

for its development and production plan for the Company's concessions onshore Portugal. The plan covers a period of 5 years during which Porto will execute a work program focused on commercializing the Lias resource play in one or more areas within its concessions. In October 2012, Porto announced the conclusion of a 23-well stratigraphic drilling program (Phase 1) to evaluate the unconventional resource potential of the Lower Jurassic (Lias) stratigraphic interval. The preliminary results demonstrated thicknesses and presence of organic-rich intervals consistent with pre-drill estimates and cores taken within the Lias interval showed higher than expected TOC. The farm-in partners had, until December 31, 2012, to elect to proceed to Phase 2 activities as contemplated under the joint venture but on April 5, 2013 Porto announced that the partners had elected not to do so. Porto therefore expects to commence a farm-out process for its unconventional onshore Lias acreage.

United Kingdom

The Bowland Shale is an oil source rock in the East Midlands area of England and can be expected to lie within the oil mature zone at a number of locations. In the younger Namurian (Upper Mississippian–Lower Pennsylvanian) Millstone Grit Group in Lancashire, basinal shales such as the 2000 ft "Sabden Shale" (Samlesbury Fm) and the 230 ft "Caton Shale" (Silsden Fm) are believed to lie within the oil window in some locations. In southern England, the Upper Jurassic Kimmeridge Clay (thickness > 1600 ft; TOC > 20%) and Lower Oxford Clay (thickness 300 ft maximum TOC 12%) and Blue Lias, Black Ven Marls, and Green Ammonite Beds of the Lower Jurassic Lower Lias (thickness > 1600 ft; maximum TOC 12%) have shale liquids potential.

Cuadrilla Resources. Cuadrilla Resources is investigating the shale oil potential of the Upper Jurassic Kimmeridge Clay in its Bolney project on PEDL 244 in the Weald Basin, southern England, where Esso found gas shows at shallow depth in Bolney-1 (1963). In April 2010, Cuadrilla received planning permission to drill the Lower Stumble test of the Kimmeridge Clay using the well pad of Balcombe-1, drilled by Conoco in 1986 on the Bolney (Lower Stumble) anticline. Top Kimmeridge Clay is estimated to occur at a depth of around 1830 ft at this location and to lie within the relatively small sweet spot where the Kimmeridge Clay has reached oil maturity. Cuadrilla spudded Balcombe-2 on August 2, 2013, drilling to a TD of 2700 ft on September 5, despite interruptions caused by protesters. The 1700 ft Balcombe-2Z horizontal leg was then drilled within the Mid-Kimmeridge "I" Micrite at 2500 ft and the well completed on September 22, 2013. The well encountered hydrocarbons and has been suspended while Cuadrilla applies for planning permission for testing. The planning application excludes hydraulic fracturing and is being treated as a conventional well producing from natural fractures. In May 2011, AJ Lucas reported that Cuadrilla had fracced the Cowden 2 gas discovery well in the Weald

Basin. The well was drilled by Independent Energy in August 1999 on a separate license, EXL 189. The results were said to be inconclusive. It is not known if this was a test of the well's shale oil or shale gas potential since an oil discovery, Lingfield-1, was also made within the EXL 189 license area in 1999. AJ Lucas indicated that a further well would be drilled later but it is unclear whether this refers to the Lower Stumble shale oil test on PEDL 244 or a well on EXL 189.

Celtique Energie. Celtique has applied to drill the Fernhurst-1 well on PEDL 231 in the Weald Basin as a test of a conventional Kimmeridge Limestone and a Great Oolite stratigraphic trap play but also to establish the shale liquids potential of the Kimmeridge Clay and Middle and Lower Lias.

Ownership Transactions There have been a substantial number of business deals in Europe as late entrants tried to gain a foothold in promising acreage and smaller companies sought additional financing. Full company M&A activity has also occurred though most transactions have taken the form of license purchases or farm-ins.

Company Mergers, Acquisitions, and Restructuring On February 28, 2011, Dart Energy Ltd. of Australia announced that it would acquire with immediate effect the 90% of the shares in the UK's Composite Energy Ltd. that it did not already own. Although primarily a coal seam gas explorer, Composite Energy had acreage with shale gas potential in both Scotland and Poland. On takeover, Composite Energy became Dart Europe Ltd. On August 10, 2011, Toreador Resources Corp. announced a merger with ZaZa Energy LLC of Houston, TX, combining ZaZa's Eagle Ford and Eagle Ford/Woodbine ("Eaglebine") interests with Toreador's Paris Basin interests. The new company will be called ZaZa Energy Corp. On August 26, 2011, the UK's San Leon Energy plc and Canadian company Realm Energy International Corp. announced an agreement whereby San Leon would acquire all of the shares of Realm, resulting in Realm becoming an indirect subsidiary of San Leon. The acquisition was completed on November 10, 2011. On completion of the deal, San Leon acquired 3 licenses in Poland, one in Germany, and two in Spain. In addition, Realm had 10 outstanding license applications in France and eight applications in Spain. With the exception of nine applications in the Paris Basin focused on shale oil, the primary target of the Realm licenses and applications was shale gas. On January 16, 2012, Dart Energy announced the formation of Dart Energy International Shale. This wholly owned subsidiary will manage and develop the company's growing portfolio of shale gas interests. With the exception of one asset in China, these are held in Europe. On January 24, 2013, San Leon Energy plc completed the acquisition of Aurelian Oil & Gas plc. Aurelian

owned unconventional gas assets and acreage with unconventional gas potential in Poland plus other largely conventional assets in several European countries. On April 2, 2013, Dart Energy announced the cancellation of the planned IPO of Dart Energy International (Dart Energy's non-Australian operations).

On September 17, 2013, Eden Energy announced that it had entered into a conditional agreement to sell all of its U.K. coal seam methane and shale gas portfolio to Shale Energy plc. The assets comprise Eden's 50% interest in 17 licenses in England and South Wales and a 100% interest in one license in South Wales. Seven licenses in South Wales have an estimated P90 shale GIP of 17.1 Tcf net to Eden (6.35 Tcf recoverable). On condition that Shale Energy raised the capital to make the acquisition by November 2013, Eden Energy would hold a 29.9% direct interest in Shale Energy. Although Shale Energy failed to complete by the deadline, Eden has continued to extend the conditional agreement with Shale Energy Plc for the sale of its entire UK coal seam methane and shale gas portfolio for £11.5 million (approximately A$19.3 million), and at the date of this report, that contract, conditional upon Shale Energy raising £7 million, remains in existence. On November 13, 2013, Poland's PKN Orlen completed the acquisition of Canadian company TriOil Resources Ltd., a move designed to buy in experience in horizontal drilling and hydraulic fracturing.

License Acquisitions Germany
On December 23, 2011, in the course of acquiring BG's UK coal seam gas interests, Dart Energy obtained an exclusive option to acquire for no additional consideration a 100% interest in two license areas held by BG in Germany (Saxon I West and Saxon II), which are prospective for both CBM and shale gas. Dart exercised the option in May 2012.

Poland
On November 15, 2010, the UK's San Leon Energy plc announced that it had agreed to acquire Mazovia Energy Resources (a EurEnergy Resources Corp. subsidiary), holder of three concessions in the Fore Sudetic Monocline, southwest Poland. The concessions are thought to have Carboniferous shale gas potential. On December 10, 2010, Eni S.p.A. announced that it had agreed to acquire Minsk Energy Resources (a EurEnergy Resources Corp. subsidiary), holder of three concessions in the Baltic Depression. On January 21, 2013 PKN Orlen announced that it has been assigned two concessions on the East European Platform Margin, previously held by ExxonMobil.

United Kingdom
On December 28, 2011, Dart Energy announced that it had agreed to acquire all of the unconventional gas assets of Greenpark Energy Ltd., comprising 22 onshore licenses in

the UK. Seven of these licenses are considered to have shale gas potential.

Farm-ins and Interest Transfers Bulgaria

On August 29, 2011, LNG Energy Ltd. announced that it had entered into an agreement with TransAtlantic Petroleum Ltd. to earn a 50% interest in the A-Lovech exploration license in northwest Bulgaria. The LNG Energy was to provide up to US$ 7.5 million to drill, core, and test a 10,500 ft Middle Jurassic shale gas exploration well. Closure of the deal was announced on September 22, 2011.

France

On May 10, 2010, Toreador Resources Corp. (now ZaZa Energy Corp.) and Hess Corp. announced an agreement, whereby Hess would make a $15 million upfront payment and invest up to $120 million in a two-phase work program on Toreador's awarded and pending shale oil exploration permits in the Paris Basin. Phase 1 consists of an evaluation of the acreage and drilling of six wells. Depending on the results of Phase 1, Phase 2 was expected to consist of appraisal and development activities. Following Phase 2, provided contractual obligations had been met, Hess would hold a 50% share of Toreador's working interest in the covered permits. On July 26, 2012, ZaZa Energy announced that it had transferred its remaining 50% interest to Hess Corp., retaining a 5% overriding royalty interest. On September 26, 2013, however, the Minister for Ecology, sustainable Development and Energy refused to authorize the transfer to Hess Corp. on the grounds that (a) Hess's French subsidiary did not have the technical capacities required by the Mining Code and (b) the licenses could be exploited only by the use of hydraulic fracturing. On July 15, 2011, Realm Energy International Corp. (now San Leon) announced that it had entered into a farm-out agreement with ConocoPhillips covering its nine exploration license applications in the Paris Basin. The agreement provided Realm with a limited carry on exploration expenditure conditional on actual acreage acquired and required activity commitments. Realm was designated operator for the initial exploration phase with ConocoPhillips having an operatorship option thereafter. The nine licenses are considered to be primarily prospective for tight oil. It is assumed that this agreement lapsed with the acquisition of Realm by San Leon.

Germany

Following renewal of its two licenses in the Bodensee Trough, north of the Swiss-German border on December 20, 2013, Parkyn Energy, a 3Legs Resources subsidiary, transferred its interests to Rose Petroleum in exchange for a 2% royalty and contribution of €400,000 toward past costs.

Lithuania

In October 2012, it was reported that Chevron had taken a 50% interest in LL Investicijos, holder of the Rietavas oil field license, with an option to acquire the other 50%. Tethys Oil announced that Chevron had also acquired a further 6% stake in the Rietavas field from Tethys with an option to acquire a further 8.5% within 3 years. It is understood that Chevron's primary interest in the license is in shale exploration.

Poland

In August 2009, ConocoPhillips reached an agreement to farm into 3Legs Resources' six Baltic Depression concessions. ConocoPhillips is funding the initial exploration and evaluation program but 3Legs Resources remains the operator. ConocoPhillips had until March 20, 2012 to determine whether to exercise an option to take a 70% interest in the concessions. If exercised, operatorship would transfer to ConocoPhillips. On March 20, 2012, 3Legs Resources announced that ConocoPhillips would exercise its option in respect of the three western concessions. Completion of the option exercise took place on September 14, 2012, whereupon operatorship of the three western concessions passed to ConocoPhillips. It was also announced that the two companies were considering options for the three eastern Baltic Depression concessions that are situated in a more liquids-prone part of the basin (see Lower Paleozoic: Poland). In order to develop an appropriate strategy for the three eastern concessions, they were to be divested into a separate Polish legal entity, which would be a wholly owned subsidiary of 3Legs Resources. ConocoPhillips has opted not to acquire a 70% interest in the three eastern concessions. On March 1, 2010, Irish company San Leon Energy Ltd. disclosed that it had entered an agreement with Talisman Energy Inc. whereby Talisman would acquire a 60% interest in San Leon's three Baltic Depression concessions in exchange for covering 60% of the cost of a seismic program and drilling one well on each of the three concessions with an option to follow up with a further three wells. If the second three wells are not drilled, Talisman's interest will reduce to 30%. On March 6, 2013, Talisman announced that it was evaluating its options in Poland and on May 8, San Leon reported that it had reacquired 100% ownership of Talisman's Polish interests. On July 3, San Leon then announced an agreement whereby Wisent Oil & Gas could earn a 45% interest in the Braniewo concession by conducting a three-stage fracture on the Rogity-1 well followed by drilling and testing a multi-stage horizontal well on the concession. On April 26, 2011, Marathon Oil Corp. announced that Nexen Inc. will take a 40% interest in 10 of Marathon's 11 concessions in the Lower Paleozoic play, eastern Poland. On June 9, 2011, Mitsui & Co. Ltd. reported that it had agreed to acquire a 9% interest in the 10 concessions, reducing Marathon's interest to 51%. Marathon remained operator. The one concession excluded from the farm-outs was Plonsk SE in the Danish-Polish Marginal Trough. On May 13, 2011, Total SA announced an agreement with ExxonMobil to farm-in to two concessions in the Lublin Trough, Danish-Polish Marginal Trough. Total acquired a 49% interest while ExxonMobil retained a 51% interest and operatorship. The farm-in was approved in July 2011. On August 14, 2011, Hutton Energy plc (formerly BasGas Pty Ltd.) announced that through its Polish subsidiary Strzelecki Energia, it intended to take a 49% interest in four ExxonMobil concessions in the Podlasie Depression of the East European Platform margin. Exxon-Mobil would retain 51% and operatorship. Although the deal was approved subsequently by the Polish Office of Competition and Consumer Protection, the deal was never closed. On June 6, 2012, San Leon announced that it had taken a 75% interest in three concessions owned by Hutton Energy plc through its Polish subsidiary Strzelecki Energia, one in the Danish-polish Marginal Trough and two with Carboniferous prospectivity in the Fore-Sudetic Monocline. In both cases, the farm-in concessions are adjacent to concessions in which San Leon already has an interest. Total SA has indicated that it will become 100% owner and operator of the Chelm concession on the East European Platform Margin. ExxonMobil previously held 51% of the concession and was operator.

United Kingdom

On June 13, 2013, it was announced that Centrica plc has acquired a 25% interest in the Bowland exploration license PEDL 165 from operator Cuadrilla Resources (56.25%) and AJ Lucas (18.75%). Centrica acquired the interest for GBP 40 million and will pay a further GBP 60 million in exploration and appraisal costs. If Centrica elects to continue into the development phase, it will then pay a further GBP 60 million. On October 22, 2013 Dart Energy announced that GDF SUEZ will farm into 13 of its U.K. licenses in Central England and North Wales. GDF SUEZ will acquire a 25% interest with Dart retaining 75% and operatorship. The funding will support an unconventional exploration and appraisal programme of up to 4 shale gas wells and 10 coal seam gas wells. The farm-in was completed on November 28, 2013. On January 13, 2014 it was announced that French major Total SA would take a 40% interest in PEDLs 139 and 140 in the Gainsborough Trough in Lincolnshire. Total will pay $1.6 million in back costs to the other partners, Dart Energy (17.5%), Egdon Resources (14.5%), eCorp Oil & Gas (13.5%) and IGas (14.5%) and fund a fully carried work program of up to $46.5 million. The program includes the acquisition of 3D seismic, drilling and resting a vertical exploration well and, conditional on its success, a second horizontal well. The farm-in was completed on February 4, 2014, at which time IGas became operator. On January 30, 2014, Egdon Resources announced that Total SA has signed an opt-in agreement for PEDL 209, adjacent to PEDLS 139 and 140 (above) in the Gainsborough Trough. Total has the option to farm-in until December 31, 2015 and on doing so will earn a 50% interest by paying 100% of an exploration program up to £13.47 million. Three conventional prospects on the acreage are excluded from the deal.

Relinquishments

Germany   BNK Petroleum. BNK Petroleum has announced that it intends to relinquish five of its eight German concessions, retaining the three concessions in North Rhine – Westphalia's Munsterland Basin. Some of the licenses to be surrendered contained primarily conventional prospects.

Poland   By February 28, 2014, a total of 24 shale gas and 1 shale liquids concessions had been relinquished, 12 in the Danish-Polish Marginal Trough, 5 on the East European Platform Margin, 4 on the Fore-Sudetic Monocline and 3 in the Baltic Depression.

BNK Petroleum (Indiana Investments). In December 2013, BNK relinquished its Darlowo concession in the Baltic Depression.

Cuadrilla. In February 2014, the company relinquished its Lukow concession on the Danish-Polish Marginal Trough.

Dart Energy. In October 2013, Dart Energy relinquished its Milejow concession in the Lublin Trough, to focus on its U.K. operations.

DPV Service. Between April and July 2013, DPV Service relinquished the five concessions with unconventional potential that it held in the Danish-Polish Marginal Trough, retaining only a conventional license.

Eni. Eni chose not to renew two of its three concessions in the Baltic Depression when they expired in December 2013 and January 2014. The third license is valid until June 2018.

ExxonMobil. After the transfer of two concessions to PKN Orlen and surrendering its 51% interest in another to its partner, Total SA, ExxonMobil has relinquished its other three Polish concessions, one in the Danish-Polish Marginal Trough and two on the East European Platform Margin.

Lane Energy. Lane Energy (3Legs Resources) has relinquished without drilling the three concessions that it held in the Fore-Sudetic Monocline.

Marathon Oil. In October and November 2013, Marathon relinquished four concessions in the Danish-Polish Marginal Trough and three on the East European Platform Margin.

PGNiG. PGNiG relinquished the Bartozyce shale liquids concession in the Baltic Depression in December 2013.

San Leon. In April 2013, San Leon relinquished one concession in the Fore-Sudetic Monocline, which it had held through its Vabush Energy Subsidiary. It retains interests in nine other concessions in that basin.

Saponis (BNK Petroleum; Esrey Energy). In 2013, Saponis announced its intention to relinquish the Starogard and Slawno concessions, retaining the Slupsk concession. The relinquishments had not taken place by February 28, 2014.

Above-Ground Issues

Other than general fiscal, legal, and environmental regulation of hydrocarbon exploration and exploitation, there are a number of issues that specifically face most gas and liquid resource play developments. Per-well reserves and productivity can be low and benefit from an established gas compression and distribution infrastructure. To convert resources into reserves also requires large numbers of wells. Some North American resource plays employ 10-acre spacing as opposed to the 640-acre spacing typical of conventional wells. This could pose a problem in densely populated areas of Europe but horizontal wells drilled from a single pad can be used to reduce the well footprint. In British Columbia's Horn River Basin, Apache Corp.'s well design concept should recover gas from two different stratigraphic intervals over an area of 7 km^2 from a single 28-well pad. Other environmental issues such as water availability and water disposal capacity may also impact on ultimate recovery. Almost inevitably, the concerns that have been raised in the United States about potential contamination of groundwater supplies from chemicals used in hydraulic fracturing of shale gas reservoirs are being echoed in Europe. In addition, the potential of fracking to induce local seismicity has also been raised. A major public misconception appears to be that the word "unconventional" implies new, untested, and therefore risky, drilling, and completion technology. Public disquiet has manifested itself in a number of countries, most notably Bulgaria, France, Germany, Romania, and the United Kingdom. The issues have now entered the political realm, creating a further condition of uncertainty. While vested commercial interests (e.g., the coal, nuclear and renewable energy industries; importers of conventional gas; natural gas storage operators) are almost certainly a factor, populism in advance of elections is undoubtedly playing a part and environmental groups are using the controversy to advance their own agendas. Until there is public recognition that the drilling and fracturing technology that is in use has been applied for decades in hundreds of thousands of wells and that all that is "unconventional" is the mode of subsurface occurrence of the natural gas, there are likely to be deferrals and delays in the evaluation of shale gas potential in a number of countries. It remains a problem of perception. "People overestimate the dangers of what is new and underestimate those of what they're used to" (Rudolf Huber, CEO of NeXtLNG Ltd.).

The commissioning on November 8, 2011 of the first of two 1224 km (760-mile) Nord Stream gas pipelines across the Baltic Sea from Portovaya Bay in Russia to Lubmin in Germany, effectively created a divergence of interests between the western European countries served by Nord Stream (e.g., Germany; Denmark, United Kingdom, The Netherlands, Belgium, France, and Czech Republic) and those countries still dependent on Russian gas from the overland route transiting through Ukraine (e.g., Poland, Bulgaria, and Romania). Gazprom's announcement that it is considering further Nord Stream pipelines and its downbeat remarks about European shale gas exploitation suggest that it sees shale gas development in Europe as a threat to its position as largest gas supplier to the continent and is keen to divert governments away from shale gas and back towards Russia as a guaranteed supplier.

International Energy Agency: Conscious of the impact that negative publicity has on realizing the potential of unconventional gas, on May 29, 2012 the Paris-based International Energy Agency released a World Energy Outlook special report on "Golden Rules" that are needed to support a potential "Golden Age of Gas." The report provides insights into the environmental challenges linked to unconventional gas production and how best to deal with them.

European Union (EU): On February 4, 2011, the European Council announced a number of priority actions in its Conclusions on Energy (PCE 026/11). Priority 7 stated "In order to further enhance its security of supply, Europe's potential for sustainable extraction and use of conventional and unconventional (shale gas and oil shale) fossil fuel resources should be assessed." In September 2011, the EU Energy Commissioner, Guenther Oettinger of Germany, said that he hopes to put forward proposals in Spring 2012 to standardize regulations on hydraulic fracturing. This followed a report published in July for the European Parliament by six German authors entitled *Impacts of shale gas and shale oil extraction on the environment and on human health*. Herr Oettinger's announcement produced a strong reaction from the Polish Treasury Minister who stated that exploration for unconventional hydrocarbon resources is already adequately regulated and that the possibility of European Union wide regulation is not provided for in the Lisbon Treaty (Treaty on the Functioning of the European Union or TFEU). (Both the Lisbon Treaty and the Energy Treaty Charter recognize state sovereignty in the use of a county's energy resources.) On September 22, 2011, Herr Oettinger's spokeswoman, Marlene Holzner, said that the commission is studying whether the current European Union environmental laws would apply to shale gas production, but is not planning to propose any new legislation. On October 13, 2011 EU Climate Action Commissioner, Connie Hedegaard, said that she was not inclined toward a moratorium on shale gas drilling based on the information that she had heard so far. The European Commission (EC) selected a Brussels law firm, Philippe & Partners, to analyze how the relevant applicable European legal framework, including environmental law, is applied to the licensing, authorization, and operation of shale gas exploration and exploitation, using a sample of four Member States: France, Germany, Poland, and Sweden. The 104-page report was published on November 8, 2011. On January 27, 2012 Energy Commissioner Oettinger stated that "the legal study confirms that there is no immediate need for changing our EU legislation." On September 7, 2012, the EC published three new studies on unconventional fossil fuels, in particular shale gas. The studies look at the potential effects of these fuels on energy markets, the potential climate impact of shale gas production, and the potential risks shale gas developments and associated hydraulic fracturing may present to human health and the environment.

Unconventional Gas: Potential Energy Market Impacts in the European Union. The study on energy market impacts shows that unconventional gas developments in the United States have led to greater Liquefied Natural Gas supplies becoming available at global level, indirectly influencing EU gas prices.

Climate impact of potential shale gas production in the EU. The study on climate impacts shows that shale gas produced in the EU causes more GHG emissions than conventional natural gas produced in the EU, but – if well managed – less than imported gas from outside the EU, be it via pipeline or by LNG due to the impacts on emissions from long-distance gas transport.

Support to the identification of potential risks for the environment and human health arising from hydrocarbons operations involving hydraulic fracturing in Europe. The study on environmental impacts shows that extracting shale gas generally imposes a larger environmental footprint than conventional gas development. Risks of surface and ground water contamination, water resource depletion, air and noise emissions, land take, disturbance to biodiversity, and impacts related to traffic are deemed to be high in the case of cumulative projects.

In launching the EU's green paper on energy and climate aims for 2030, Energy Commissioner Günther Oettinger took a favorable position on shale gas, quoting the gas prices in the United States compared with European prices. Anne Glover, chief scientific adviser, gave a scientific green light to shale while noting that there are risks involved with all energy production, including wind and coal. But she also noted that in terms of extraction and production, there are non-scientific issues to be debated. Connie Hedegaard, Climate Commissioner, was less positive, stating that different geological conditions and environmental rules mean that shale gas exploitation in Europe will bear little comparison with the United States separately from the European Commission, the German chairman of the European Parliament's committee on the Environment, Public Health and Food Safety indicated in July 2011 that he wants a new "energy quality directive" that would introduce stringent regulations to cover fuels with what are deemed to be adverse environmental impacts – tar sands oil and shale gas among them. As might be expected, given the variety of political positions represented, the European Parliament's opinions are more diverse than those of the Commission and in some cases contradictory. These were expressed in two resolutions adopted by committee in mid-September 2012. The emergence of exploration for shale oil and shale gas in some EU countries should be backed up with "robust regulatory regimes" according to separate non-binding resolutions by the Energy Committee (on industrial aspects) and the Environment Committee (on health and environment ones). Member states should be "cautious" pending

further analysis of whether EU level regulation is adequate, according to environment MEPs. Each EU country has the right to decide for itself on whether to exploit shale gas, said the Energy Committee. Member states should have robust rules on all shale gas activities, including hydraulic fracturing of rock ("fracking"). The MEPs also advised the EU to learn from U.S. experiences, with a view to using environmentally friendly industrial processes and "best available technologies." The European Commission previously concluded that EU rules adequately cover licensing and early exploration and production of shale gas but "a thorough analysis" of EU regulation on unconventional fossil fuels is needed, given the possible expansion of their exploitation, noted Environment MEPs. In April 2013, the chair of Parliament's science and technology options assessment panel said that Europe is in the "denial phase" on shale gas. On January 22, 2014, the European Commission adopted and published a non-binding Recommendation and Communication for its Shale Gas Enabling Framework (thereby avoiding legislative proposals). Member states are invited to implement it within 6 months (thereby bypassing the European Parliament elections in May 2014) and the EC will review the Recommendation 18 months after publication.

The Recommendation invites Member States to ensure that:

- There is an integrated approach to the granting of permits;
- Risk assessments are undertaken on potential drilling sites;
- Baseline studies and subsequent monitoring are undertaken on drilling sites;
- Operators apply best practice;
- Use of chemicals in frac fluid is minimized and fluid content used is made publicly available on a per well basis;
- An EIA is carried out where required under EU Directive 2011/92/EU (i.e., when gas production exceeds 17.5 MMscf/d).

Background comment. Individual EU member states have the right to determine exploitation of energy resources and their energy mix (TFEU Article 194). Member states are also free to set more stringent environmental protection measures than required by EU legislation (TFEU Article 193). Most aspects of hydrocarbon exploration and production are covered by existing EU legislation: Hydrocarbon Directive; Water Framework Directive; Groundwater Directive; Environmental Impact Assessment; Registration, Evaluation, Authorisation and Restriction of Chemical substances (REACH); Natura 2000 (protected areas); and other regulations covering waste, noise, etc.

Austria

OMV's plans to drill one or two shale gas exploratory wells in the wine quarter of Lower Austria ran into substantial opposition. Despite seeking community support, the company's plans were resisted not only by environmental and community groups but also by politicians, including the Environment Minister and Governor of Lower Austria. On March 2, 2012, OMV announced that it would suspend drilling plans pending the completion of a comprehensive environmental and social study by the Federal Environmental Agency and TÜV Austria Group, a technical and environmental safety consultancy. In July 2012, the Minister for Environmental and Agricultural Affairs announced plans to reform the environmental impact assessment act to incorporate shale gas exploration. The cost of producing a detailed environmental inspection and assessment for any proposed well effectively makes shale gas drilling uneconomic and OMV currently has no further plans for shale gas exploration in Austria.

Bulgaria

The shale gas debate featured in the October 2011 presidential elections with the two principal opposition candidates both indicating that they opposed shale gas development. The election was won, however, by the candidate of the ruling party (Citizens for European Development of Bulgaria). Environmental organizations and opposition parties wished to impose a temporary moratorium on shale gas exploration and called for a referendum on allowing such activities. Although the Ministry of Economy, Energy and Tourism indicated that it planned a thorough assessment of the risks involved in shale gas development it appeared to be broadly supportive of shale gas exploration. On October 19, 2011, a delegation representing a number of ministries and regional governors visited Poland to learn from the Polish experience. In the face of massive public protests, on January 18, 2012 parliament placed an indefinite ban on the use of hydraulic fracturing. The previous day, prior to final execution of the license agreement, the government announced the withdrawal of the Novi Pazar permit for shale gas exploration awarded to Chevron in June 2011. Chevron continues in discussions with the government to provide assurances that hydrocarbons can be produced safely from shale. The Minister for Economy and Energy has said that he believes powerful financial interests were behind the mass protests. The pro-Russian Centre Left party played a leading role in opposing shale gas research. (Gazprom provides 98% of Bulgaria's gas.) Since the moratorium was imposed, a Movement for Energy Independence has been established and has called for cancellation of the moratorium. But informed opinion suggests that at present, the government does not believe that it is politically worthwhile to confront public opinion.

Czech Republic

The Náchod District assembly and some 50 local administrations submitted formal objections to the Ministry of Environment's award of the Trutnov permit to Basgas Energia Czech, a subsidiary of Hutton Energy. In April 2012, the Trutnov award was cancelled and sent back to the Ministry of Environment regional department to be decided again. In September 2012, the Minister of Environment announced a moratorium on shale gas exploration in the Czech Republic until June 30, 2014. Prime Minister Petr Necas has said that suitable legislation that will define the framework for prospecting and exploitation is required, and only after such legislation has been adopted could further steps be discussed.

France

In February 2011, shale gas and shale oil drilling in France was suspended by the authorities pending a progress report on the environmental consequences of shale exploitation. The ultimate outcome of this process was the passing of a law on July 13, 2011 (Law 2011-835) that prohibited the exploration for, and production of, liquid or gaseous hydrocarbons by hydraulic fracturing. Permit holders had 2 months in which to advise the administrative authorities of the techniques that they use or intend to use in their exploration activities. Failure to respond or an intention to use hydraulic fracturing would result in withdrawal of the permit. A national commission would also be established to evaluate the environmental risks associated with hydraulic fracturing and to set out the conditions under which scientific research under public control can take place. The government is to report annually to parliament on the evolution of exploration and production technology in France, Europe, and internationally and also on the results of the scientific research undertaken. In September 2011, major French E&P company Total SA announced as part of its report to the authorities that it would continue the evaluation of its Montélimar exploration license but that the work program does not envisage the use of hydraulic fracturing. Other companies were expected to adopt a similar approach.

On October 3, 2011, the ministers of Ecology, Sustainable Development, Transport & Housing and Industry, Energy & the Digital Economy announced in a joint press release that the three permits issued specifically for exploration of shale gas would be cancelled. These are the Total SA Montélimar exploration and the Schuepbach Villeneuve-de-Berg and Nant licences. Total expressed surprise as it had undertaken not to use hydraulic fracturing and was awaiting the government's notification to understand the legal basis for the cancellation. The official confirmation of the repeal of the three licences was gazetted on October 13, 2011. On November 26, the CEO of Total SA announced that the company would appeal against the revocation of the Montélimar license and on December 12, the company filed an appeal in the Paris Administrative Court in order to clarify the situation, on the grounds that the company had complied with the Act of July 13, 2011. In October 2012, Christophe de Margerie, Total CEO, stated that the Group was no longer willing to spearhead the shale gas quest in France and that it is up to politicians and government officials to decide on the future of shale gas exploration in the country.

The French Union of Petroleum Industries declared that the cancellation decisions will send a negative signal to international investors and are prejudicial to an economy, which imports 99% of its oil and 98% of its gas consumption. The CEO of French company GDF Suez said that while it was appropriate that the government evaluate technology and processes, closing the door forever to shale gas development would be "a major mistake." On October 11, 2011, the National Assembly rejected a bill submitted by the parliamentary opposition which set out to prohibit exploration for, and exploitation of, unconventional hydrocarbons irrespective of the techniques used. The proposed bill was deemed to contain several flaws and to be incompatible with the law of July 13, 2011. On March 21, 2012, a new decree established a National Commission to evaluate shale oil and gas exploration techniques and operations. The commission would be consulted on conditions for implementation of hydraulic fracturing research projects; managing risk and environmental protection during experimentation on new techniques to exploit shale oil and gas; the government's annual report to parliament set out in the Act of July 13, 2011 (above). At end September 2012, the members of the commission had not yet been appointed. Since the presidential election in May 2012 and the formation of a new government, official statements have sent a variety of messages, some of them contradictory. Most recently, on September 14, President Hollande stated that, as far as the exploration and exploitation of nonconventional hydrocarbons are concerned, hydraulic fracturing will be banned throughout his 5-year term in office and instructed the Environment Minister to reject seven applications for permits to explore for shale gas, citing potential impacts on health and the environment. (It should be noted that he did not order the rejection of a number of pending applications for permits with shale oil potential.) The president's forceful statement seems to have taken industry representatives and even some government sources by surprise, as there are major concerns in France regarding the long-term impact of such a ban on the economy and energy security.

At a parliamentary hearing in April 2013, senators, company representatives, scientists, and energy policy analysts broadly supported a resumption of exploration for shale oil and gas so that at the very least France's resource potential could be evaluated. On October 11, 2013, ruling on a challenge submitted by Schuepbach Energy, the French Constitutional Court ruled that the ban on hydraulic fracturing introduced in Law 2011-835 is not disproportionate and that the law conforms to the Constitution. Only Arnaud Montebourg, Minister of Industrial Renewal, has supported

development of shale gas, subject to the proviso that the industry finds alternative ways of bringing gas to surface by methods that do not risk the pollution and other dangers for which hydraulic fracturing has been blamed. There is, however, a groundswell of support for lifting the ban from industry. A consequence of the ban on hydraulic fracturing has been an accumulation of unprocessed applications for exploration permits. On January 31, 2014 a total of 110 permit applications were awaiting decision.

Background comment. In 2007, 78% of all French electricity production came from nuclear power. Two new European pressurized water reactors (EPRs) are due to be commissioned by 2017, so it can be assumed that the nuclear industry will not be supporting shale gas development!

Germany

Fracking was first used in conventional wells in 1955 (Schleswig-Holstein) and 1977 (Lower Saxony). Between 1977 and 2010 some 140 frac operations were conducted in Germany. The first fracking of unconventional gas wells (tight gas) occurred in the mid-1990s in the Söhlingen Field, Lower Saxony, and fracking was conducted in at least three other tight gas fields in Lower Saxony in the period 2005–2010. Despite a 55-year history of fracking, there was no public interest in the application of the technology in Germany until 2010. Unlike France, where governance is highly centralized, the German Länder (constituent states of the Federal Republic of Germany) has a high degree of autonomy. The political strength of the Grüne (Green environmental party) is at an all-time high both federally and at state level, and environmental groups have exerted considerable pressure on politicians in areas where shale gas development is proposed. In March 2011, the state Environment minister of North Rhine-Westphalia, a member of the Grüne, introduced a moratorium on shale gas exploration. To date, however, most shale gas exploration has taken place in Lower Saxony, which has not introduced such a moratorium. The Minister for Environmental Protection in the federal government announced on July 29, 2011 that an expert survey on the environmental impact of shale gas production would be ordered and that changes to the geological and mining laws are likely.

On August 4, 2011 the Federal Environment Agency published an opinion entitled *Einschätzung der Schiefergasförderung in Deutschland* (Assessment of shale gas production in Germany). The report is generally negative towards shale gas and appears to selectively quote, for example, sources such as the Tyndall Centre for Climate Change at the University of Manchester and Robert Howarth at Cornell University that are generally considered to exaggerate the impact of natural gas as a source of greenhouse gas emissions. If the planning and legislative requirements proposed in the report are implemented, they will probably have the effect of making shale gas production uneconomic in Germany. In September 2012, the German Federal Ministry for the Environment, Nature Conservation and Nuclear Safety (BMU) produced a report entitled *Environmental Impacts of Fracking in the Exploration and Production of Natural Gas from Unconventional Deposits*. Although the report does not recommend a ban on hydraulic fracturing, some of the proposed conditions are sufficiently prescriptive to cast doubt on the viability of much unconventional E&P activity. In the same month, following a risk study published by the Environment and Economic ministries of the state of North Rhine – Westphalia the state authorities banned hydraulic fracturing operations until more evidence on the risks involved is available. In December 2012, the German parliament voted to put forward a bill permitting hydraulic fracturing to resume, subject to strict controls (e.g., limited to areas where water resources will not be impacted, mandatory EIA, mandatory information on fluid treatment and flowback). In January 2013, the Federal Institute for Geosciences and Natural Resources (BGR) responded to the September 2012 Environment Ministry (BMU) report, criticizing it for its lack of geosciences expertise, its inconsistency and subjectivity, and failure to use the available broad knowledge base of existing technology. Though current legislation permits requests for hydraulic fracturing to be filed under existing water rights and mining law, none of the public authorities would grant such a request, so there was effectively a moratorium on hydraulic fracturing until the new bill was introduced. The government failed to introduce the bill prior to the September 2013 federal elections. During the subsequent coalition-building process, in November 2013, it was announced that the partners in government had agreed to place a moratorium on hydraulic fracturing until environmental and health concerns are resolved. In February 2014, spokespersons for the German gas industry pointedly indicated that the decline in German gas production can be attributed to the political reluctance to permit hydraulic fracturing, despite it having been used in German tight sandstone reservoirs for over 50 years (see above).

Background comment. The German unconventional hydrocarbons industry is not well developed and domestic companies lack the necessary technology. These companies are focused on the production, importation, and storage of conventional gas. Germany also has a substantial renewable energy industry. All of these interests would be threatened by large volumes of low-cost indigenous natural gas. It is therefore in the interests of German industry not only to make shale gas production as unprofitable as possible in Germany but to use its influence to restrict large-scale gas production elsewhere in Europe (see the direction of the German strategy in the *European Union*).

Ireland (Republic of Ireland)

The principal prospect in Ireland lies in the Northwest Ireland Carboniferous Basin (Lough Allen Basin), which straddles the border between the Irish Republic and Northern Ireland but most of the opposition has come from the Republic side. At a company information meeting in early September 2011, the operator, Tamboran Resources, offered to conduct fracking without chemical additives but this did nothing to soften the opponents of the scheme. The government subsequently asked the Environmental Protection Agency (EPA) to conduct a study on the effects of fracking. The 26-page report prepared by the University of Aberdeen, entitled *Hydraulic Fracturing or 'Fracking': A Short Summary of Current Knowledge and Potential Environmental Impacts*, will be used as the groundwork for a more comprehensive study by the EPA. The major environmental trust, An Taisce, has called for fast track regulation to clarify the currently uncertain regulatory position regarding onshore drilling. In April 2013, Energy Minister Pat Rabitte reaffirmed that no decision on permitting hydraulic fracturing will be taken until the EPA's study is concluded at the end of 2014.

Background comment. As confirmed by the Minister of State at the Department of Communications, Energy and Natural Resources on March 21, 2012, conventional fracking has already been applied by Dowell Schlumberger in the case of three Irish onshore wells: Dowra-1 Re- entry (1981), Dowra 2 (2002), and Thur Mountain 1 (2002).

Netherlands

Although the provincial authorities in Noord Brabant were opposed to Cuadrilla Resources' plans to drill two wells, in early 2011, the Dutch Minister for Economic Affairs, Agriculture and Innovation granted a license for drilling to proceed. (The Dutch state, through its wholly owned company EBN, has a 40% interest in the license.) On June 29, 2011, however, the ministry indicated that shale gas exploration in the Netherlands would not move ahead until the results of the UK's inquiry into hydraulic fracturing had been assessed. "If it appears that there are unacceptable risks, no drilling for shale gas will occur," the Minister, Maxime Verhagen, said in a letter to parliament. "Concerns regarding shale gas are understandable and I take them very seriously."

In October 2011, Cuadrilla encountered another setback when a court ruled that Boxtel town council was wrong to grant a temporary exemption from zoning for its Boxtel-1 well since that was based on activities concluding within 5 years and, if commercial production had been established, it was likely that operations would exceed this time span. Cuadrilla must now return to Boxtel council to resolve the situation and their spokesman expected a few months' delay to a well that was due to spud early in 2012. The most recent company estimate is that this well will not now spud until 2013.

The Dutch Ministry of Economic Affairs, Innovation and Agriculture commissioned an independent study of all possible risks and consequences of shale gas exploitation, including methane emissions from drilling, the presence of heavy metals in drilling mud, and the risk of induced seismicity. The study was delivered on August 27, 2013 and concluded that while total risks were slightly greater than the risks associated with conventional exploration due to the large number of wells involved, they remained very small. The government will now undertake a study to identify all likely shale gas exploration sites on Dutch soil to determine where locations with the highest likelihood of success and least environmental risk occur. This study is likely to take 12–18 months and in the meantime no shale gas exploration applications will be processed and where permits have already been granted, companies will not proceed until the study is complete. Some 60 out of 400 local authorities had declared their opposition to shale gas production.

Poland

Unlike most other countries, the major political debate in Poland has been about maximizing the benefit of shale gas exploitation to the state. In advance of the October 2011 parliamentary election, the opposition Law and Justice Party prepared draft legislation covering Polish shale gas. In the election, however, the ruling Civic Platform–Polish People's Party coalition won sufficient seats to continue in government. Draft regulations regarding a hydrocarbon extraction tax on conventional and unconventional hydrocarbon production were published on October 16, 2012 but will not come into effect until 2015. The implied tax burden is 40% of gross profits. More contentious, however, has been the licensing regime and the process of granting shale gas concessions, with six persons, three Ministry of Environment officials and three company employees, detained and released on bail on suspicion of offering or receiving bribes for the allocation of licenses. *The Economist* has noted that the existing rules were designed for a system in which a small number of state-controlled companies were operating, and not for the current exploration environment. With most of the prospective shale gas acreage now under license, however, any changes will be taking place after the horse has bolted.

It was expected that new regulations on shale gas extraction would be announced in November/December 2012 and implemented from 2013. It is thought that a simplification of environmental requirements and general reduction in red tape would form a part of the new regulatory environment. Companies would no longer require to hold a license before conducting non-drilling energy exploration operations, but energy companies looking to enter the Polish shale market will have to be pre-approved by the Polish government to buy existing licenses. A state-owned National Energy Minerals Operator (NOKE) will also be created. The NOKE will participate in shale gas projects, where it is intended that it strengthens administrative oversight of licence obligations and have right of first refusal on secondary trade in exploration licenses. The NOKE will pay its net profit to the Polish Treasury and to municipal governments, thereby involving local communities in successful shale gas development. The

NOKE's profits will also go to a planned Hydrocarbon Generations Fund, a form of sovereign wealth fund.

In mid-February 2013, the new hydrocarbon regulations were published in a draft form for a 1-month public consultation period. They involve a changed Geology and Mining Law and amendments to eight other bills. Some environmental requirements will be loosened. The final draft of the regulations was submitted to the Prime Minister's office in mid-June. Parliament was expected to approve the bill before the end of 2013, after which it will require the signature of the President.

In order to accelerate shale gas development, in March 2014, the Polish government offered six-year tax breaks to shale gas projects, avoiding the introduction of special taxes. According to Prime Minister Donald Tusk, the proposal would go through Parliament's approval as early as early April. It has also been reported that the government will drop plans to create the proposed National Energy Minerals Operator (NOKE), in a move to cut red tape and reduce regulatory hurdles.

Romania

The protests in Bulgaria (above) have been echoed in Romania. Bulgarian activists demonstrated outside the Romanian embassy in Sofia (capital of Bulgaria) and have been in contact with like-minded groups in Romania. The Barlad municipality, where Chevron planned to drill later in 2012, opposed shale gas exploration and in March 2012, some members of the parliamentary opposition filed a legislative initiative which, if passed, would ban hydraulic fracturing. The parliamentary opposition came to power in April 2012 and in May introduced a moratorium on shale gas exploration, due to run until December 2012. In June, the March proposal to ban shale gas exploration and exploitation by hydraulic fracturing and cancellation of licenses in which fracking would be used, was overwhelmingly rejected by the Romanian Senate. Comments by the Environment Minister in August 2012 that shale gas exploitation by hydraulic fracturing will not be approved unless the results of EU studies on its environmental and health implications indicate that it is acceptable, suggested that the moratorium was likely to be extended but when it expired at end-December 2012 it was not renewed.

Spain

In October 2012, the Government of Cantabria in northern Spain published a draft law which, if implemented, would prohibit the use of hydraulic fracturing as long as the doubts and uncertainties surrounding the use of the technique that exist today persist. The Cantabrian regional parliament passed the proposals into law on April 8, 2013. It should be noted that the bulk of the exploration permits in the Basque-Cantabrian basin, especially those in which shale gas exploration is proposed, lie in other regions: Basque Country (where the Autonomous Government is an active participant); Castilla y Leon; La Rioja; Navarra. On October 30, 2013, it is reported that the Spanish Government explicitly legalized hydraulic fracturing by amending a 1998 hydrocarbon exploration law to include hydraulic fracturing under permitted exploration techniques. The EIAs will not be required before conducting fracturing operations. A law streamlining environmental requirement for industrial projects, which could accelerate shale gas exploration approvals, was published on December 5, 2013. In January 2014, the Spanish government then announced that it was taking Cantabria's ban on hydraulic fracturing to the Constitutional Court, arguing that it violates national law on hydrocarbon exploitation.

Sweden

In the September 2010 parliamentary election campaign, the opposition center-left alliance comprising the Social Democrats, the Left Party, and the Green Party pledged to oppose large-scale fossil fuel production in Sweden, including Shell's planned exploitation of shale gas in southern Sweden. In the event, the ruling centre-right Alliance coalition was re-elected.

Switzerland

In Switzerland, the cantons have a substantial degree of independence and E&P is solely a cantonal responsibility. The Swiss Federation could have an indirect influence on shale gas through its responsibility for environmental legislation but there is no legislation specifically targeted at shale gas at the present time. The federal government's environmental focus is currently on carbon capture and storage (CCS). In April 2011, the cantonal authorities in Fribourg suspended all shale gas prospecting activities and refused the renewal of Schuepbach's exploration license, due to expire at end-2011. The explanation given was that the environmental impact and pollution risk accompanying drilling had not yet been clearly identified and that the canton preferred to focus on renewable energies. In the canton of Jura, the Green party has questioned the authorities on their policies regarding shale gas. In Neuchatel, the Grand Council has decided that in the event of a discovery, in principle an exploitation concession will be awarded to Celtique Energie and that shale gas is not specifically excluded from this decision. The Celtique Energie web site, however, suggests that their only unconventional prospects (shale oil and shale gas) are in the Weald Basin in southern England.

United Kingdom

On November 24, 2010, the House of Commons Energy and Climate Change Committee launched an evidence-based enquiry into the prospects for shale gas in the United Kingdom, the risks and hazards associated with shale gas, and the potential carbon footprint of large-scale shale gas extraction.

The committee visited Fort Worth and Austin, Texas, Washington, DC, and two Cuadrilla Resources drilling sites near Blackpool, Lancashire.

The voluminous report (223 pages in two volumes) that was published on May 23, 2011 produced a number of conclusions and 26 recommendations. In its summary, however, the committee stated that "on balance, we feel that there should not be a moratorium on the use of hydraulic fracturing in the exploitation of the UK's hydrocarbon resources, including unconventional resources such as shale gas" (House of Commons Energy and Climate Change Committee, 2011). Nevertheless, a number of issues have arisen in the United Kingdom and some examples are given below. On October 21, 2011, the Vale of Glamorgan Council (South Wales) rejected a planning application submitted by Coastal Oil & Gas to drill Llandow-1, a shallow (2600 ft) conventional and shale gas exploratory well situated on an industrial estate. Despite Environment Agency Wales indicating that it had "no objection to the application as submitted," the Welsh Government declining to get involved as the issues were "not of more than local importance" and the application itself stating "This application does not include fracking," the local environmental group "The Vale says No" supported by the local member of the UK Parliament put sufficient pressure on the councillors to ensure that all 17 members of the planning committee opposed the application.

Although in debate the councillors spoke of their concerns about pollution if fracking followed a positive exploration outcome, this does not represent a valid reason for rejection. The official reason given was therefore that "the applicant has submitted insufficient information to satisfy the Local Planning Authority that the quantity and quality of groundwater supplies in the vicinity of the site, would be protected." The council leader indicated subsequently that better guidelines were required from the Welsh Assembly (regional government) for test drilling and fracking.

On July 7, 2012 an appeal against the decision to reject the planning application was upheld by the Welsh Planning Inspectorate who concluded that the main issue was the potential effect on the quantity and quality of groundwater supplies. The inspector concluded that the proposal would not harm groundwater supplies and Llandow-1 can now be drilled. As was indicated earlier (Section Shale Gas in Europe: Carboniferous), fracking operations at Cuadrilla Resources' Preese Hall drilling site were halted after two small earthquakes (2.3 and 1.5 Richter Local Magnitude) were reported on April 1, and May 27, 2011. The British Geological Survey (BGS) subsequently determined that the earthquakes at depths of 12,000 and 6500 ft were within a few thousand feet of the drilling site and that the correlation between the earthquakes and their proximity to, and the timing of, hydraulic fracturing operations pointed to the earthquakes being the result of the fracking process.

On November 2, 2011, Cuadrilla Resources (well operator) presented a geomechanical report (de Payter and Baisch, 2011) on the causes of the seismicity and future mitigation procedures to the Department of Energy and Climate Change (DECC). The report concluded that the repeated seismicity resulted from direct injection of fluid into the same critically stressed fault zone and that this could be avoided in future by rapid flowback after treatment and reduction in treatment volume, accompanied by real-time seismic monitoring to initiate appropriate action when seismic magnitude exceeds predefined thresholds. The DECC sought input from the BGS and other expert sources before taking any decision on the resumption of fracking operations. A BGS spokesman did, however, indicate that earthquakes of the magnitude reported in Lancashire have been occurring for hundreds of years as a result of coal mining and generally go unnoticed. The independent report prepared for DECC agreed "that a suitable traffic light system linked to real-time monitoring of seismic activity is an essential mitigation strategy" allowing adjustments to be made to the injection volume and rate during the fracturing procedure, thereby preventing noticeable seismic activity (Green et al., 2012). On December 5, 2012, in a move widely read as encouraging shale gas exploitation, the government announced the creation of a new Office of Unconventional Gas and Oil, with the intention of focusing regulatory effort to meet the needs of future production. On December 13, it was announced that hydraulic fracturing can resume, subject to controls to mitigate the risk of seismic activity. In the March 2013 Budget, the UK Chancellor of the Exchequer stated that tax arrangement for companies involved in shale gas exploration would be "generous." Planning clarity should be available by summer 2013 and proposals would be developed to ensure that local communities benefit from shale gas projects in their area.

On December 6, 2011, the Northern Ireland Assembly passed a motion calling for a moratorium on hydraulic fracking. But no legislation exists to compel a Northern Irish Minister to act upon a moratorium and as the Minister for Enterprise, Trade and Investment has pointed out, no application has been submitted. She will, however, be in a difficult position if one is submitted. It should be noted that hydraulic fracturing has already been used in Fermanagh in 2001, in three tight gas wells.

United Kingdom – General. A more general concern on the part of United Kingdom environmentalists is that development of an extensive low-cost shale gas industry threatens the development of renewable energy within the country, by rendering the latter uneconomic. There is also the argument on the one side that gas provides the most sustainable bridge to a low-carbon future while others see that ready availability of gas will simply result in increasing use of fossil fuel-based energy. As there are divisions even within the British government on these issues we can expect that, in the UK at least, this debate is set to run for some time!

REFERENCES

Andrews, I. J. (2013). The Carboniferous Bowland Shale Gas Study: Geology and Resource Estimation. British Geological Survey for Department of Energy and Climate Change, London, UK.

Green, C. A., Styles, P., Baptie, B. J. (2012). Preese Hall shale gas fracturing: review & recommendations for induced seismic mitigation. Department of Energy and Climate Change, 2012.

House of Commons Energy and Climate Change Committee. (2011). *Shale Gas.* Fifth Report of Session 2010–12, London, The Stationery Office Limited.

Lovell, M., Davies, S., Macquaker, J. (2010). Petrophysics in unconventional gas reservoirs. Presentation at The Geology of Unconventional Gas Plays, The Geological Society, London, 5–6 October, 2010.

Polish Geological Institute – National Research Institute. (2012). Assessment of shale gas and shale oil resources of the Lower Paleozoic Baltic-Podlasie-Lublin Basin in Poland. Warsaw, March 2012.

Rogner, H.-H. (1996). *An Assessment of World Hydrocarbon Resources*, International Institute for Applied Systems Analysis, Working Paper WP-96-056, p. 44.

Shaoul, J. R., Spitzer, W., Dahan, M. W. (2011). Case study of unconventional gas well fracturing in Hungary, SPE Paper 142751.

Svenska Shell. (2011). World Wide Web Address: http://www.shell.se/home/content/swe/naturgas/.

TNO (2009). Inventory non-conventional gas. TNO Report TNO-034-UT-2009-00774/B.

U.S. Energy Information Administration. (2011). World Shale Gas Resources: An Initial Assessment of 14 Regions Outside the United States.

U.S. Energy Information Administration. (2013) Technically Recoverable Shale Oil and Shale Gas Resources: An Assessment of 137 Shale Formations in 41 Countries Outside the United States.

U.S. Geological Survey (2012). Potential for Technically Recoverable Unconventional Gas and Oil Resources in the Polish-Ukrainian Foredeep, Poland, 2012. Fact Sheet 2012–3102, July 2012.

U.S. Geological Survey. (2012). Undiscovered Gas Resources in the Alum Shale, Denmark, 2013. Fact Sheet 2013–3103, December 2013.

APPENDIX 1.

Distribution of known shale gas drilling in Europe. *Base map courtesy of IHS.*

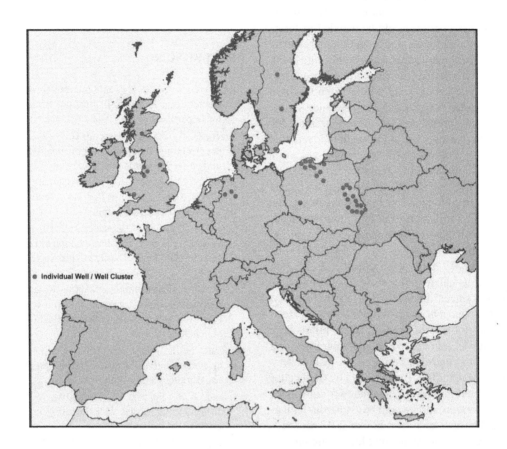

APPENDIX 2

Shale gas exploration and appraisal wells drilled in Europe

Bulgaria

Geological Province	Sub-Province	Concession	Well Name	No	Operator	Spud	Compl	TD ft	Horiz	Fracs	Target Fm	Result - Content
Moesian Platform		A-Lovech	Goljamo Peshtene	R-11	LNG Energy	27-Sep-11	End Nov-11 @TD	10,466			Etropole	numerous shows C1 - C3
Moesian Platform		A-Lovech	Deventci	R-2	TransAtlantic Petroleum	27-Oct-11					Etropole	

Germany

Geological Province	Sub-Province	Concession	Well Name	No	Operator	Spud	Compl	TD ft	Horiz	Fracs	Target Fm	Result - Content
Northwest German Basin	Lower Saxony Basin	Münsterland	Damme	2	ExxonMobil	2008	2008	10,950			Wealden, Posidonia	
Northwest German Basin	Lower Saxony Basin	Münsterland	Damme	2A	ExxonMobil	2008	2008	10,935			Posidonia Shale	
Northwest German Basin	Lower Saxony Basin	Münsterland	Damme	3	ExxonMobil	2008	2008	5,280		3	Wealden	
Northwest German Basin	Lower Saxony Basin	Minden	Oppenwehe	1	ExxonMobil	Jun-08	2008	8,730			Wealden, Posidonia	
Northwest German Basin	Lower Saxony Basin	Schaumberg	Niederwöhren	1	ExxonMobil	2009	2009	3,394			Wealden	
Northwest German Basin	Lower Saxony Basin	Scholen-Barenburg II	Schlahe	1	ExxonMobil	2009	2009	4,870			Wealden, Posidonia	
Northwest German Basin	Lower Saxony Basin	Bramschen	Lünne	1	ExxonMobil	17-Jan-11	Mar-11	5,170			Wealden, Posidonia	
Northwest German Basin	Lower Saxony Basin	Bramschen	Lünne	1A	ExxonMobil	Mar-11	2011	5,503	Y		Posidonia Shale	

Poland

Geological Province	Sub-Province	Concession	Well Name	No	Operator	Spud	Compl	TD ft	Horiz	Fracs	Target Fm	Result - Content
Baltic Depression	Gdansk Depression	Trzebielino	Miszewo	T-1	Indiana Investments (BNK)	28-Feb-12	Sep-12	17,700			Low Paleozoic	Muted gas shows
Baltic Depression	Gdansk Depression	Bytow	Gapowo	B-1	Indiana Investments (BNK)	Mid-late May 12	Jul-12	14,100			Low Paleozoic	Major gas shows
Baltic Depression	Gdansk Depression	Bytow	Gapowo	B-1A	Indiana Investments (BNK)	Mid-Jan 14	23-Feb-14		Y	7 planned	Low Paleozoic	High gas shows
Baltic Depression	Gdansk Depression	Slawno	Wytowno	S1	Saponis Investments	Dec-10	14-Feb-11	11,750			Low Paleozoic	significant gas shows: C1 - C3
Baltic Depression	Gdansk Depression	Slupsk	Lebork	S1	Saponis Investments	11-Mar-11	25-Apr-11	11,780		Y	Low Paleozoic	significant gas shows: C1 - C3
Baltic Depression	Gdansk Depression	Starogard	Starogard	S1	Saponis Investments	16-Jul-11	Sep-11	11,560			Low Paleozoic	significant gas shows: C1 - C5
Baltic Depression	Gdansk Depression	Lebork	Łebień	LE-1	Lane Energy (3Legs)	Mid Jun-10	28-Jul @ TD	10,120		1; DFIT	Low Paleozoic	gas
Baltic Depression	Gdansk Depression	Lebork	Łebień	LE-2H	Lane Energy (3Legs)	10-May-11	Jun-11		Y	13	Low Paleozoic	gas shows
Baltic Depression	Gdansk Depression	Lebork	Strzeszewo	LE-1	Lane Energy (3Legs)	04-Oct-12	Early Dec-12	10,040		DFIT; 2	Low Paleozoic	
Baltic Depression	Gdansk Depression	Lebork	Lublewo	LEP-1	Lane Energy (3Legs)	13-Dec-13	20-Jan-14	9,583	Y	20 fracs planned	Sasino Fm	
Baltic Depression	Gdansk Depression	Lebork	Lublewo	LEP-1ST1H	Lane Energy (3Legs)							
Baltic Depression	Gdansk Depression	Kawia	Slawoszyno	LEP-1	Lane Energy (3Legs)	15-Feb-14	17-Mar-14	9,250			Low Paleozoic	
Baltic Depression	Gdansk Depression	Cedry Wielkie	Legowo	LE-1	Lane Energy (3Legs)	27-Aug-10	Q4-2010	11,270		2 DFITs	Low Paleozoic	gas shows
Baltic Depression	Gdansk Depression	Damnica	Warblino	LE-1H	Lane Energy (3Legs)	17-Jul-11	Sep-11	10,570	Y	7	Low Paleozoic	gas shows
Baltic Depression	Gdansk Depression	Gdansk-W	Lewino	1G-2	Talisman Energy Polska	26-Sep-11	17-Nov-11	11,800		3	Low Paleozoic	C1+ small C2-C5
Baltic Depression	Gdansk Depression	Elblag	Bagart	1	Eni Polska	01-Dec-11	Mar-12				Low Paleozoic	
Baltic Depression	Gdansk Depression	Elblag	Stare Miasto	1	Eni Polska	Apr-12	Sep-12			Y	Low Paleozoic	
Baltic Depression	Gdansk Depression	Malbork	Karnionka	1	Eni Polska	May-12	Aug-12				Low Paleozoic	
Baltic Depression	Gdansk Depression	Wejherowo	Lubocino	1	PGNiG	Dec-10	Mar-11	11,500		2	Low Paleozoic	promising gas flow. No H2S and low N2. Heavier hydrocarbons
Baltic Depression	Gdansk Depression	Wejherowo	Lubocino	2H	PGNiG	Aug-12	Nov-12		Y	DFIT; frac	Low Paleozoic	
Baltic Depression	Gdansk Depression	Wejherowo	Lubocino	3H	PGNiG	Aug-13	Dec-13		Y		Low Paleozoic	
Baltic Depression	Gdansk Depression	Wejherowo	Opalino	2	PGNiG	Sep-12	Dec-12				Low Paleozoic	flowed gas
Baltic Depression	Gdansk Depression	Wejherowo	Opalino	3	PGNiG	Nov-13	Jan-14				Low Paleozoic	
Baltic Depression	Gdansk Depression	Wejherowo	Opalino	4	PGNiG	Jan-14					Low Paleozoic	
Baltic Depression	Gdansk Depression	Wejherowo	Tepcz	1	PGNiG	Planned					Low Paleozoic	
Baltic Depression	Gdansk Depression	Stara Kiszewa	Wysin	1	PGNiG	Mar-13	May-13	13,100			Low Paleozoic	
Baltic Depression	Gdansk Depression	Wejherowo	Kochanowo	1	PGNiG	Apr-13	Jul-13	10,650			Low Paleozoic	
Baltic Depression	Gdansk Depression	Kartuzy-Szemud	Borcz	1	PGNiG	Jul-13	Sep-13				Low Paleozoic	

APPENDIX 2 (*Continued*)

Shale gas exploration and appraisal wells drilled in Europe

Geological Province	Sub-Province	Concession	Well Name	No	Operator	Spud	Compl	TD ft	Horiz	Fracs	Target Fm	Result - Content
Poland (continued)												
Danish-Polish Marginal Trough	Lublin Trough	Grabowiec	Grabowiec	G6	Chevron Polska	31-Oct-11	Feb-12			DFIT	Low Paleozoic	
Danish-Polish Marginal Trough	Lublin Trough	Frampol	Frampol	1	Chevron Polska	Mar-12	Apr-12				Low Paleozoic	
Danish-Polish Marginal Trough	Lublin Trough	Zwierzyniec	Zwierzyniec	1	Chevron Polska	Dec-12	Mar-13			DFIT; frac	Low Paleozoic	
Danish-Polish Marginal Trough	Lublin Trough	Krasnik	Krasnik	1	Chevron Polska	May-13	Aug-13				Low Paleozoic	
Danish-Polish Marginal Trough	Lublin Trough	Chelm	Krupe	1	ExxonMobil E&P Poland	03-Dec-10	Jan-11	12,490		Y	Low Paleozoic	
Danish-Polish Marginal Trough	Lublin Trough	Pionki-Kazimierz	Markowola	1	PGNiG	Apr-10	Jun-10			1	Upp Carb	
Danish-Polish Marginal Trough	Lublin Trough	Tomaszów Lubelski	Lubycza Królewska	1	PGNiG	26-Mar-12	Aug-12				Low Paleozoic	
Danish-Polish Marginal Trough	Lublin Trough	Wiszniow-Tarnoszyn	Kościaszyn	1	PGNiG	Oct-13	Jan-14				Low Paleozoic	
Danish-Polish Marginal Trough	Lublin Trough	Kock-Tarkawica	Wojcieszków	1	PGNiG	Sep-13	Dec-13				Low Paleozoic	
Danish-Polish Marginal Trough	Lublin Trough	Wierzbica	Syczyn	OU1	Orlen Upstream	24-Oct-11	Dec-11	9,445			Low Paleozoic	
Danish-Polish Marginal Trough	Lublin Trough	Wierzbica	Syczyn	OU2K	Orlen Upstream	24-Sep-12	Early Nov-12	13,450	Y	12	Low Paleozoic	
Danish-Polish Marginal Trough	Lublin Trough	Wierzbica	Stręczyn	OU1	Orlen Upstream	22-Feb-13	Apr-13	11,480			Low Paleozoic	
Danish-Polish Marginal Trough	Lublin Trough	Wierzbica	Dobryniów	OU1	Orlen Upstream	26-Jun-13	Sep-13				Low Paleozoic	
Danish-Polish Marginal Trough	Lublin Trough	Lubartów	Berejow	OU1	Orlen Upstream	Mid Dec-11	Jan-12	8,573			Low Paleozoic	
Danish-Polish Marginal Trough	Lublin Trough	Lubartów	Berejow	OU2K	Orlen Upstream	12-Dec-12	Feb-13	12,565	Y	7	Low Paleozoic	
Danish-Polish Marginal Trough	Lublin Trough	Lubartów	Uscimów	OU1	Orlen Upstream	May-13	04-Jul-13				Low Paleozoic	
Danish-Polish Marginal Trough	Lublin Trough	Garwolin	Goździk	OU1	Orlen Upstream	16-Jul-12	Oct-12	13,830			Low Paleozoic	
Danish-Polish Marginal Trough	Pomeranian Trough	Rypin	RYP-Lutocin	1	Marathon Oil Poland	Early Apr-12	Jul-12			Y	Low Paleozoic	
Danish-Polish Marginal Trough	Pomeranian Trough	Kwidzyn	KWI-Prabuty	1	Marathon Oil Poland	18-Jul-12	Sep-12			DFIT; frac	Low Paleozoic	
Danish-Polish Marginal Trough	Pomeranian Trough	Brodnica	BRO-NM Lubawskie	1	Marathon Oil Poland	Sep-12	Dec-12			DFIT	Low Paleozoic	
Danish-Polish Marginal Trough	Pomeranian Trough	Szczawno	Szymkowo	1	Talisman Energy Polska	07-Mar-12	Jun-12	14,930	Y		Low Paleozoic	C1 - C3
East European Platform Margin	Podlasie Depression	Minsk Mazowiecki	Siennica	1	ExxonMobil E&P Poland	20-Feb-11	Apr-11			Y	Low Paleozoic	
East European Platform Margin	Podlasie Depression	Siedlce	SIE-Domanice	1	Marathon Oil Poland	Jan-12	Mar-12				Low Paleozoic	
East European Platform Margin	Podlasie Depression	Sokolow Podlaski	SOK - Grębków	1	Marathon Oil Poland	Dec-12	Jan-13				Low Paleozoic	
East European Platform Margin	Podlasie Depression	Wodynie-Lukow	Stoczek	OU1	Orlen Upstream	18-Nov-13	Jan-14	10,300			Low Paleozoic	
East European Platform Margin	Podlasie Depression	Wodynie-Lukow	Stoczek	OU1K	Orlen Upstream	Jan-14	Mid Mar-14	14,130	Y		Silurian	
East European Platform Margin	Volhyno-Podolian Monocline	Orzechow	ORZ-Cycow	1	Marathon Oil Poland	Dec-11	Jan-12			DFIT; frac	Low Carb	
Fore-Sudetic Monocline		Gora	Siciny	2	Gora Energy (San Leon)	10-Nov-11	Mid Feb-12	11,550		DFIT; 2 fracs	Low Carb	C1 - C3

APPENDIX 2 (*Continued*)

Shale gas exploration and appraisal wells drilled in Europe

Geological Province	Sub-Province	Concession	Well Name	No	Operator	Spudded	Completed	TD ft	Horiz	Fracs	Target Fm	Result - Content
Sweden												
Östergötland Lower Paleozoic		Motala area	5 wells		Aura Energy	Early Oct-11	Q4-2011				Alum Shale	
Östergötland Lower Paleozoic		Ekeby	Brunneby	BY-1	Gripen Energy	Mar-12	Mar-12	282			Alum Shale	weak flow of flammable gas
Östergötland Lower Paleozoic		Ekeby	Bobergs	BH-1	Gripen Energy	Mar-12	Mar-12	328			Alum Shale	weak flow of flammable gas
			Häradallmänings									
Östergötland Lower Paleozoic		Ekeby	Rocklunda	RL-1	Gripen Energy	Mar-12	Mar-12	282			Alum Shale	weak flow of flammable gas
Östergötland Lower Paleozoic		Ekeby	Ekebyborna	GH-2	Gripen Energy	Mar-12	Mar-12	328			Alum Shale	21 Mscfd in 2-hour flow - 97.5% CH4
Östergötland Lower Paleozoic		Ekeby	Ekebyborna	GH-1A	Gripen Energy	Sep-12	Sep-12	300			Alum Shale	strong gas flow
Östergötland Lower Paleozoic		Ekeby	Ekebyborna	GH-2A	Gripen Energy	Sep-12	Sep-12	340			Alum Shale	gas flow
Östergötland Lower Paleozoic		Ekeby	Ekebyborna	GH-3	Gripen Energy	Sep-12	Sep-12	305			Alum Shale	intermittent gas flow
Östergötland Lower Paleozoic		Ekeby	Fossala	FA-2	Gripen Energy	22-Jun-13	28-Jun-13	352			Alum Shale	
Östergötland Lower Paleozoic		Ekeby	Kullen KN-1	GH-5	Gripen Energy	18-Jun-13	24-Jun-13	366			Alum Shale	3.1 Mscfd after 30 minutes
Östergötland Lower Paleozoic		Ekeby	Uddenä	UD-1	Gripen Energy	10-Jun-13	14-Jun-13	355			Alum Shale	no details
Siljan Ring Depression			5 percussion holes		AB Igrene	2009	2009				Low Paleozoic	
Siljan Ring Depression			Mora	001	AB Igrene	2010	2010	~1,650			Low Paleozoic	
Siljan Ring Depression			Solberga	1	AB Igrene	2011	2011	~1,650			Low Paleozoic	
Siljan Ring Depression			Stumsnäs	1	AB Igrene	2011	2011	~1,650			Low Paleozoic	
Siljan Ring Depression			Mora	002	AB Igrene	Q2-2013	Q2-2013	1,670			Low Paleozoic	
Siljan Ring Depression			Mora	003	AB Igrene	Q2-2013	Q2-2013	1,450			Low Paleozoic	
Fennoscandian Border Zone	Colonus Shale Trough	Colonussänkan	Lövestad	A3-1	Shell			3,134			Alum Shale	
Fennoscandian Border Zone	Colonus Shale Trough	Colonussänkan	Oderup	C4-1	Shell			3,010			Alum Shale	
Fennoscandian Border Zone	Colonus Shale Trough	Colonussänkan	Hedeberga	B2-1	Shell	28-Nov-09		2,448			Alum Shale	
United Kingdom												
Anglo-Dutch Basin	Cleveland Basin	PL 080	Kirby Misperton	8	Viking UK Gas	06-Jun-13	04-Oct-13	~10,000			Bowland Shale	
Anglo-Dutch Basin	Humber Basin	PEDL 183	Crawberry Hill	1	Rathlin Energy (UK)	15-Apr-14	12-Aug-14	~9,000		DFIT	Bowland Shale	
Anglo-Dutch Basin	Humber Basin	PEDL 183	West Newton	1	Rathlin Energy (UK)	27-Jun-14	06-Sep-14	~10,420		DFIT	Bowland Shale	
Cheshire Basin	Rossendale Basin	PEDL 190	Ince Marshes	1	IGas	04-Nov-11	21-Jan-12	5,174			Bowland Shale	gas indications
Cheshire Basin	Rossendale Basin	PEDL 193	Irlam	1		10-Jan-14					Bowland Shale	
East Irish Sea Basin	West Bowland Basin	PEDL 165	Preese Hall	1	Cuadrilla	16-Aug-10	08-Dec-10	9,098			Bowland Shale	substantial gas flows
East Irish Sea Basin	West Bowland Basin	PEDL 165	Grange Hill	1	Cuadrilla	15-Jan-11	15-Apr-11			5	Bowland Shale	
East Irish Sea Basin	West Bowland Basin	PEDL 165	Grange Hill	1Z	Cuadrilla	15-Apr-11	Early Aug-11	10,775			Bowland Shale	
East Irish Sea Basin	West Bowland Basin	PEDL 165	Becconsall	1	Cuadrilla	23-Aug-11	13-Oct-11				Bowland Shale	
East Irish Sea Basin	West Bowland Basin	PEDL 165	Becconsall	1Z	Cuadrilla	13-Oct-11	21-Dec-11	10,500			Bowland Shale	
East Irish Sea Basin	West Bowland Basin	PEDL 165	Anna's Road	1	Cuadrilla	06-Oct-12	21-Nov-13	2,000			Bowland Shale	
Midland Valley of Scotland	Kincardine Basin	PEDL 133	Airth	6	Composite Energy	15-Oct-05	05-Dec-05	3,524			Black Metals M.B.	Junked & Abandoned
Midland Valley of Scotland	Kincardine Basin	PEDL 133	Longannet	1	Composite Energy	16-Feb-07	29-Apr-07				Black Metals M.B.	
Midland Valley of Scotland	Kincardine Basin	PEDL 133	Bandeath	1	Composite Energy	15-May-07	18-Jun-07				Black Metals M.B.	
South Wales Carboniferous		PEDL 148	Banwen	1	UK Methane	07-Sep-11	16-Sep-11				Aberkenfig?	
South Wales Carboniferous		PEDL 149	St.Johns	1	UK Methane	26-Aug-11	23-Mar-12				Aberkenfig?	

APPENDIX 3

Selected companies with potential interest in shale gas exploration in Europe, by country.

Country	Companies
• **Austria**	• OMV
• **Bulgaria**	• Chevron; TransAtlantic Petroleum / Esrey Energy
• **Croatia**	• INA-MOL
• **Czech Republic**	• Applications: Cuadrilla Resources; Hutton Energy
• **Denmark**	• Total / Danish North Sea Fund (Nordsøfonden)
• **France**	• No valid permits as a result of moratorium on hydraulic fracturing: at 28 February 2014, there were 110 outstanding permit applications (many overlapping) from multiple companies.
• **Germany**	• BNK Petroleum; Dart Energy; ExxonMobil / Shell (BEB); Rose Petroleum; San Leon; Wintershall
• **Ireland**	• Enegi Oil; Tamboran Resources
• **Italy**	• Independent Resources plc
• **Lithuania**	• Chevron (through LL Investicijos)
• **Netherlands**	• Cuadrilla Resources / EBN; Hutton Energy
• **Poland**	• 3Legs Resources; BNK Petroleum; CalEnergy Resources; Canadian International Oil Corp; Chevron Corp; ConocoPhillips; Cuadrilla Resources; Eni; Esrey Energy; EurEnergy Resources; Greenpark; Hutton Energy; LOTOS Petrobaltic; Mac Oil; Marathon; Mitsui; Nexen; PETROLINVEST; PGNiG; PKN Orlen; RAG; San Leon; Sorgenia; Total SA
• **Romania**	• Chevron; TransAtlantic Petroleum/Sterling Resources
• **Spain**	• BNK Petroleum; Leni Gas & Oil; Repsol; San Leon; Schuepbach Energy; Sociedad de Hidrocarburos de Euskadi (SHESA); Vancast Exploración
• **Sweden**	• Aura Energy; AB Igrene; Gripen Oil & Gas AB; Tekniska Verken i Linköping AB
• **Switzerland**	• Schuepbach Energy
• **United Kingdom**	• Centrica; Coastal Oil & Gas; Cuadrilla Resources; Dart Energy; eCORP; Eden Energy; Egdon Resources; GDF Suez; IGas; Tamboran Resources; Total; UK Methane

• Certain of the companies listed have indicated their intention to exit shale gas exploration in the country indicated but retained their interest at the time of compilation of the table.

CHINA SHALE GAS AND SHALE OIL

BY SHU JIANG (ENERGY & GEOSCIENCE INSTI-
TUTE,UNIVERSITY OF UTAH)

The shales spanning from Pre-Cambrain Sinian (a period
right before Cambrian) to Quaternary are widely distributed
in China. The Pre-Cambrian to Upper Paleozoic organic-
rich shales with maturity in gas window and shallow Quaternary shales have shale gas potentials and Mesozoic to
Cenozoic organic-rich shales with maturity in oil window
have shale gas potentials (Figure 76.33). In 2010, The Strategic Research Center of Oil and Gas, Ministry of Land and
Resources and China University of Geosciences at Beijing
used an analog assessment regime to announce that China
Shale Gas resource is predicted to be about 30 billion cubic
meters (BCM) (1050 Tcf). In 2011, the U.S. Energy Information Administration (EIA) assessed that China could have
1275 Tcf technically recoverable shale gas, in March 2012,
China Ministry of Land and Resources announced China had
25.08 trillion cubic meters (886 Tcf) of recoverable onshore
shale gas reserve. Recently, EIA reduced China recoverable

shale gas reserve to 1115 Tcf in June 2013 and gave a number of 32 billion barrel recoverable shale oil for China. Either
number indicates China's shale resource is comparable with
United States updated 665 Tcf recoverable shale gas and
58 billion barrels of shale oil resource. One recent breakthrough is that China is the only country outside of North
America that has reported commercially viable production
of shale gas based on the latest commercial quantity shale
gas production reported from Sichuan Basin by Sinopec and
PetroChina, although the volumes contribute less than 1%
of the total natural gas production in China. In comparison, the shale gas as a share of total natural gas production
in 2012 was 39% in the United States and 15% in Canada
(EIA, October 2013). The successful development evidence
in Sichuan Basin especially in Fuling area recently makes the
China's 2015 output target of 6.5 BCM possible, which was
thought impossible before last month. This will encourage
Beijing government and draw interests of international oil
companies.

For U.S. producing shales, they were deposited marine
depositional setting. But for hydrocarbon-related shales in
China were formed in diverse paleo-environments. The

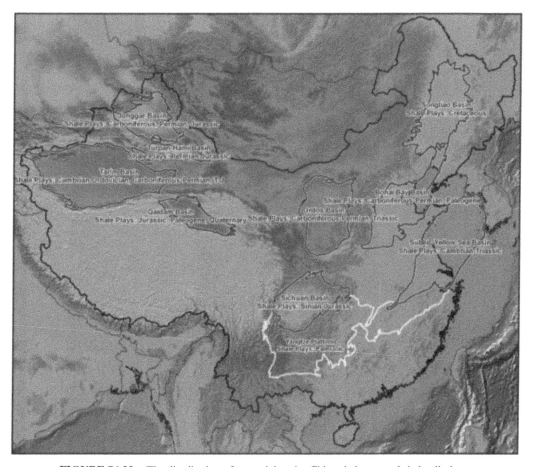

FIGURE 76.33   The distribution of potential major China shale gas and shale oil plays.

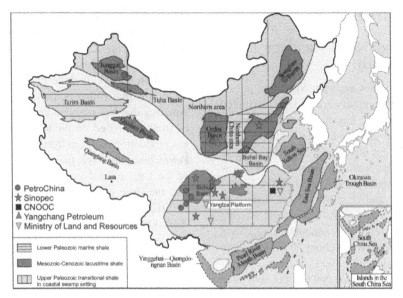

FIGURE 76.34 The shale exploration activities in China.

Pre-Cambrian to Lower Paleozoic shales distributed all over China were deposited in marine setting. The Upper Paleozoic (Carboniferous to Permian) shales mainly in North China and NW China were deposited in transitional (coastal swamp associated with coal) setting. The Meso-Cenozoic sporadically distributed shales were deposited in lacustrine setting (Figure 76.34). The typical marine shale, transitional shale, and lacustrine shale can be represented by Lower Paleozoic Sichuan Basin, Carboniferous to Permian Ordos

Basin, and Cenozoic Bohai Bay Basin, respectively (Figure 76.35). The Table 76.8 summarizes the depositional settings and distribution in time and space for the potential gas and oil shales in China.

China has investigated shale gas and shale oil for nearly 6 years. So far, 2013, 2D seismic data covering 9000 km^2 and 3D seismic data covering 800 km^2 were acquired, and 150 shale gas wells (including shallow parameter wells behind outcrop) targeting marine, lacustrine, and transitional

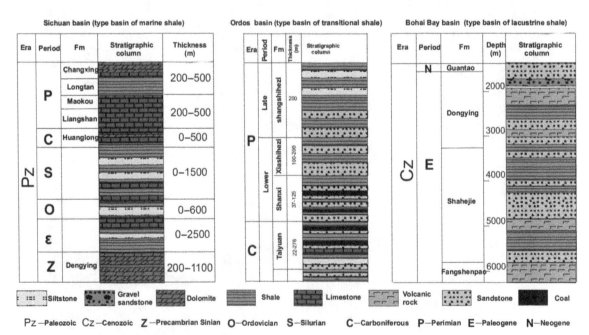

FIGURE 76.35 Three kinds of potential shales (marine, lacustrine, and transitional/coastal setting) and their type basins.

TABLE 76.8 **Depositional setting and distribution in time and space of potential China shales.**

Depositional setting	Age and formation		Distribution area
Lacustrine	Cenozoic	Neogene	Qaidam Basin
		Paleogene	Bohai Bay Basin, QaidamBasin
	Mesozoic	Cretaceous	Songliao Basin
		Jurassic	Turpan-Hami, Junggar,Tarim, Qaidam, Sichuan Basin
		Triassic	Ordos Basin, Sichuan Basin
	Paleozoic	Late Permian	Junggar, Turpan-Hami
Transitional (coastal setting associated with coal)	Paleozoic	Late Permian (Longtan Fm)	Yangtze Platform including Sichuan in Upper Yangtze
		Early Permian (Taiyuan, Shanxi Fm)	North China
		Late Carboniferous (Benxi Fm)	North China
Marine	Paleozoic	Early Silurian (Longmaxi Fm)	Yangtze Platform including Sichuan in Upper Yangtze
		Late Ordovician (Wufeng Fm)	Yangtze Platform including Sichuan in Upper Yangtze, Tarim Basin
		Early Cambrian (e.g., Qiongzhusi Fm)	Yangtze Platform, Tarim Basinasin
	Pre-Cambrian	Sinian (e.g., Doushantuo Fm)	Upper and Middle Yangtze Platform

(coastal swamp setting associated with coal) shales were drilled so far by the PetroChina, Sinopec, CNOOC, Yanchang Petroleum, other state or private companies who recently got shale blocks, Ministry of Land and Resources and foreign partners of Chinese state oil companies. The exploration activities have been mainly focused in Sichuan Basin, Yangtze Platform outside Sichuan Basin, Ordos Basin, Bohai Bay Basin, and Nanxiang Basin (Figure 76.34). Recently, the Junggar basin has also become target basin for shale oil associated with tight dolomite oil play. So far, almost half of drilled shale wells have good shale gas and shale oil show. Among 30 horizontal wells, several wells were reported very successful based on test results rate of over 100,000 cubic meters daily production. The rate from well Yang201-H2 in Luzhu, Sichuan was reported to produce at 430,000 cubic meters/day and recently a well in Fuling area in E Chongqiong was reported to produce 547,000 cubic meters/day. Sinope drilled 30 shale gas wells in Fuling pilot area and 6 wells were reported to have high rate commercial production with acreage 180,000 cubic meters/well. Sinopec claimed the 2015 output target in Fuling area will be doubled based on recently encouraging results. For lacustrine shales, PetroChina and Sinopec recently speeded up lacustrine shale oil exploration in Junggar and Sichuan Basin, for example, Sinopec drilled Shipping 2-1H horizontal wells targeting Jurassic lacustrine shale and got 33.79 tons condensate production after 5 stages fracing in 864 m lateral.

Geological investigation and exploration show that most potential shales in China had and still have high organic content and marine shales have high maturity for gas generation and lacustrine (lake) shales have low maturity for oil generation. Characteristics of high organic matter content, high maturity, high brittle minerals (Figure 76.36), and high intra-organic nano-porosity (Figure 76.37) make China marine shales same to U.S. shales and potentially producible. The drilled shale gas wells targeting marine shale in Sichuan Basin show the similar favorable shale properties, for example, high TOC, high brittleness, etc. as U.S. producing marine shales (Figure 76.38). Generally, China lacustrine shales have high clay content than marine shales (Figure 76.36), this is why many experts think it is much more difficult to frac the lacustrine shale. Since lacustrine basins contribute 90% oil production in China and they are expected to pay a more significant role in shale oil production, we need new technologies to develop the gas or oil trapped in lacustrine shales. Recently, the tight dolomite oil production from Permian source rock interval in Junggar Basin in NW China (Figure 76.39) and tight sand oil from Ordos Basin (Figure 76.40) in North China showed the potentials of lacustrine tight oil potential similar to Bakken shale oil that is mainly produced from middle Bakken dolomite equivalent tight reservoirs. But the oil production from lacustrine shale is still in early stage. In the future, the shale gas and shale oil and tight sand and tight carbonate reservoirs within the organic-rich shale could consist of hybrid reservoirs (e.g., shale oil and tight sand oil in Triassic source rock interval in Ordos Basin, Figure 76.40).

What made shale gas or shale oil work is hydraulic fracturing or fracing, but every shale in the world is different, the shale depositional settings and geologic history made each

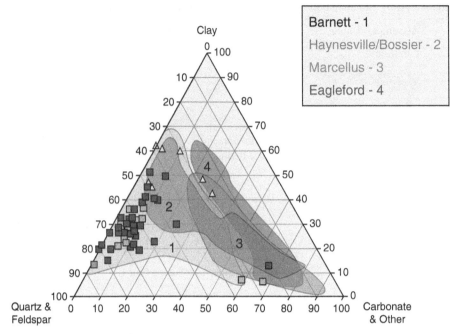

FIGURE 76.36   Ternary diagram for mineralogy of marine shale (square legend) and lacustrine shale (triangle legend) in China and its comparison with mineralogy of typical U.S shales.

shale with unique mechanical property. Shale gas and shale oil are produced from marine shales, fine-grained chalks and dolomite interbedded in source rock intervals in U.S. basins. These basins have relatively simple tectonic settings than China. Even promising marine shales in China are similar to brittle Barnett shale in United States regarding mineralogy, the complex tectonic setting, much more complex diagenetic history and harsh ground conditions make shale gas extracting in China more challenging than that in the United States. In some areas in China, the shale resources are either located in the subsurface below the rugged mountain or desert, also, the historical multi-stages of strong tectonic compression, extension in China cause shales in China have different stress fields than those in United States, for example, the maximum principal stress is horizontal in some areas in China and the maximum principal stress is vertical in United States, this is why the fracing experiences in United States may not work very well in China. We need to investigate more about the geology, geomechanics, and hydraulic fracturing design for unique China shales.

Since shale gas exploration and production is technically challenging and China basins have complex tectonic activities and different properties for shales, China has been collaborating with international oil firms and service companies to achieve the ambitious shale gas production plan. Chinese state-owned oil, coal, and power energy companies and privately owned junior companies with non-energy experience have tied up with foreign oil companies such as Shell, ExxonMobil, Chevron, ConocoPhillips, Eni, BP, Total, Statoil, Schlumberger, etc. to gain hydraulic fracturing technology in shales. Even though the recent second round bidding blocks located at the margin or outside conventional oil and gas producing basins disappointed many companies. With the speeding and recent good exploration result of Paleozoic marine shale gas exploration and Meso-Cenozoic lacustrine shale oil exploration and very exciting test result in Sichuan Basin, Shell's production-sharing contract with CNPC (parent company of PetroChina) got approved by Chinese government and Hess signed PSC with CNPC in

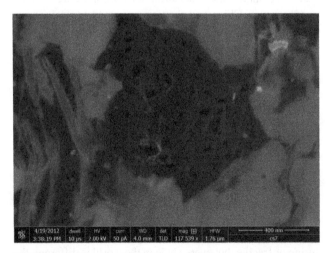

FIGURE 76.37   The SEM of ion polished sample showing intra-organic nano-pores of a marine shale Sichuan Basin, China.

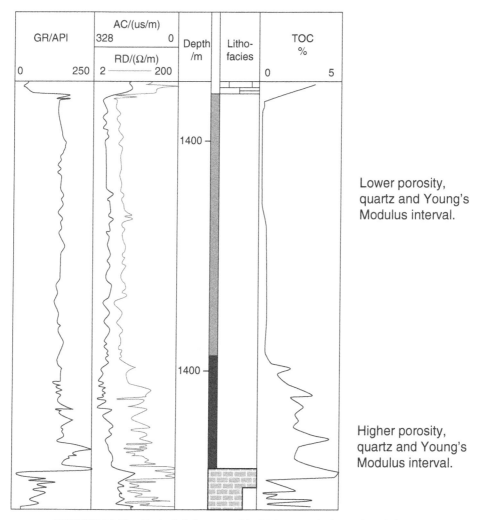

FIGURE 76.38   Typical shale gas well in Sichuan Basin for marine shale gas.

Langma shale oil block in NW China recently, which are inspiring for many companies. Shell will spend $1 billion developing China's shale resources in Fushun-Yongchuan block covering 3500 sq km in Sichuan Basin. PetroChina plans to drill 113 horizontal shale gas development wells in the next 2 years in Sichuan Basin. Sinopec has planned to drill more in SE Chongqing/E Sichuan Basin and NE Sichuan Basin due to recently commercial shale gas from both marine and lacustrine shales in Sichuan Basin. Based on these, the coming third shale gas bid round will be better than the first two.

The complex geologic setting and different geomechanics regime in China basins did challenge many international companies with successful U.S. shale experiences to frack shales in China. The trial-and-error in the in pilot shale gas areas in Changning-Weiyuan in Sichuan, Fuling in Chongqing, Yanchang block in Ordos has helped companies know better and better to frack shales in China. With limited

participation from established global service companies such as Baker Hughes and Schlumberger, Sinopec's Jianghan oilfield has improved in key areas of fracturing and logging. At one well in Sichuan Basin, Sinopec-Jianghan did 22-stage fracturing at a depth of 1500 m and the test result showed commercial flow of shale gas. So far, the horizontal drilling and hydraulic fracturing of shales have been reported to generate large stimulated reservoir volume (SRV) (Figure 76.41).

At the same time, The China National Energy Administration (NEA) issued the Shale Gas Industry Policy (Policy) in late October, 2013. The Policy recommends certain reforms to encourage more companies besides oil companies to get access to shale gas exploration and development in China. Also, the new policy gives subsidies and tax incentives to shale gas production companies.

In summary, China has huge potential for both shale gas and shale oil potential, even though the geological setting and

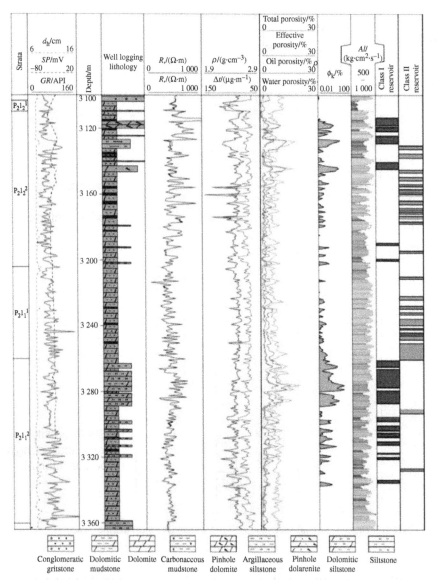

Comprehensive well logging evaluation chart of the Lucaogou Formation reservoir in the Jimsar Sag. d_h—hole diameter; contaneous potential; GR—gamma ray; R_1— resistivity of intrusion zone; R_t—formation resistivity; ρ—density; Δt—interval trans-; ϕ_N— neutron porosity; AI—acoustic impedance.

FIGURE 76.39   Tight dolomite oil from Permian source rock interval with shale oil show, Junggar Basin, NW China (L. Kuang, 2012).

geomechanics regime are more complex than U.S. producing shales for hydraulic fracturing, with the learning curve for the lacustrine shale gas in Ordos Basin and shale oil in Nanxiang Basin in central China and tight/shale oil in Santanhu and Junngar Basin in Northwest China, and recent commercial shale gas flow from marine in Sichuan Basin by PetroChina and Sinopec, technology advancement and policy support for incentives and reforms from Chinese government, China has commercially produced 200 million cubic meters in 2013 from pilot areas and the China vast shale resources are expected to be produced on a large scale.

Valuable links

Maps

- Active Shale Gas Plays, lower 48 http://www.eia.gov/oil_gas/rpd/shale_gas.pdf
- Various shale gas plays (Barnett, Fayetteville, Haynesville-Bossier, Marcellus, Woodford) and shale oil (Bakken). http://www.eia.gov/pub/oil_gas/natural_gas/analysis_publications/maps/maps.htm

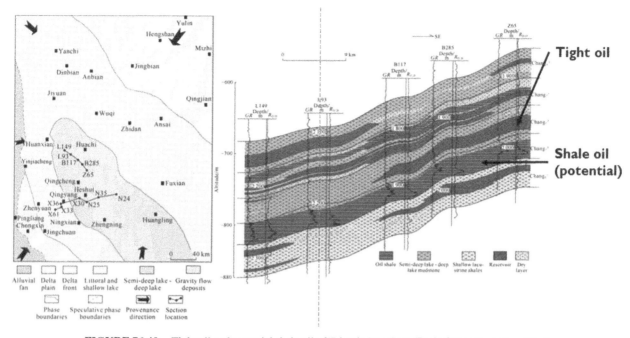

FIGURE 76.40   Tight oil and potential shale oil of Triassic Yanchang Fm in Ordos Basin (modified from YAO Jingli, 2013).

FIGURE 76.41   Horizontal drilling and hydraulic fracturing for both marine and lacustrine shales in China

Assessments

- Assessments of undiscovered oil and gas resources, onshore U.S. http://energy.cr.usgs.gov/oilgas/noga/
- Assessments of undiscovered oil and gas resources, World http://certmapper.cr.usgs.gov/rooms/we/index.jsp
- Assessment of Australian energy resources https://www.ga.gov.au/image_cache/GA17412.pdf (courtesy of Gesocience Australia)

INDEX

Note: Page locators followed by f and t indicate figure and table respectively.

Absorbers, 206–207
Absorption systems, 207
AC. *See* Alternating current
Acceleration factor, 602
  wind turbines, 20
Accident risks in hydropower development
  in different stages, 413, 414t
  influencing factors, 416–417
  management of
    business development, 417
    construction, 417
    contracting, 418, 419f
    design safety management, 417–418
    engineering and tendering, 417
    feasibility assessment, 417
    international standards and guidelines, 418
    lockout/tag-out (LO/TO), 420, 421f
    maintenance, 419
    operation, 417, 419
    permit-to-work system, 419–420
    plant's safety standards and procedures, 419
    risk assessment methods, 418
    systematic approach, 418, 419f
  mitigation
    high-voltage systems, work on, 421
    proximity of water under pressure, work in, 421–422
    public safety related to dams and waterways, 422
    transportation, 420
    underground work, 420–421, 421f
  natural hazards, 414, 415f

occupational accidents
  data collection, 415
  hydropower plant construction, 415
  natural hazards, 415
  site investigations, 415, 415f
  statistics, 415f, 416, 416f
plant-specific hazards, 416
powerhouse flooding, 414
third-party, 416
transformers explosion, 414
transportation, personnel and materials, 414, 416
tunnel work related hazards, 414
Acoustic screen model
  data analysis, 454–456, 454f, 455f
  data collection, 452–454, 453f
  depiction of, 452, 452f
  detectability, 452, 455
  detection, 452
  identification, 452
  weighting, 452
Activated reactive evaporation (ARE), 231
Actuator disc model with no rotation (ADM-NR), 44
Actuator discs, 28, 143–144
  and aerofoil data, 142–143
  blade element momentum (BEM) theory, 39–40
  numerical discretization, 40
  result, 146, 149f, 150f
  uniform thrust, 38
Actuator disc with rotation (ADM-R), 44
Actuator line model (ALM), 44
  and aerofoil data, 142–143
  result, 145, 145f, 146f

Alternative Energy and Shale Gas Encyclopedia, First Edition. Edited by Jay H. Lehr and Jack Keeley.
© 2016 John Wiley & Sons, Inc. Published 2016 by John Wiley & Sons, Inc.

WITHDRAWN